D1752278

Transition to Renewable Energy Systems

Edited by
Detlef Stolten
and Viktor Scherer

Related Titles

Ladewig, B., Jiang, S. P., Yan, Y. (eds.)

Materials for Low-Temperature Fuel Cells

2012
ISBN: 978-3-527-33042-3

Bagotsky, V. S.

Fuel Cells
Problems and Solutions

2012
ISBN: 978-1-118-08756-5

Stolten, D., Scherer, V. (eds.)

Efficient Carbon Capture for Coal Power Plants

2011
ISBN: 978-3-527-33002-7

Wieckowski, A., Norskov, J. (eds.)

Fuel Cell Science
Theory, Fundamentals, and Biocatalysis

2010
ISBN: 978-0-470-41029-5

Crabtree, R. H.

Energy Production and Storage
Inorganic Chemical Strategies for a Warming World

2010
ISBN: 978-0-470-74986-9

Kamm, B., Gruber, P. R., Kamm, M. (eds.)

Biorefineries – Industrial Processes and Products
Status Quo and Future Directions

2010
ISBN: 978-3-527-32953-3

Stolten, D. (ed.)

Hydrogen and Fuel Cells
Fundamentals, Technologies and Applications

2010
ISBN: 978-3-527-32711-9

Transition to Renewable Energy Systems

Edited by
Detlef Stolten and Viktor Scherer

WILEY-VCH
Verlag GmbH & Co. KGaA

Editors

Prof. Detlef Stolten
Juelich Research Center
Institute for Electrochemical
Process Engineering (IEK-3)
52425 Juelich
Germany

Prof. Viktor Scherer
Ruhr-University Bochum
Department of Energy Plant Technology
Universitätsstraße 150
44780 Bochum
Germany

All books published by **Wiley-VCH** are carefully produced. Nevertheless, authors, editors, and publisher do not warrant the information contained in these books, including this book, to be free of errors. Readers are advised to keep in mind that statements, data, illustrations, procedural details or other items may inadvertently be inaccurate.

Library of Congress Card No.:
applied for

British Library Cataloguing-in-Publication Data
A catalogue record for this book is available from the British Library.

Bibliographic information published by the Deutsche Nationalbibliothek
The Deutsche Nationalbibliothek lists this publication in the Deutsche Nationalbibliografie; detailed bibliographic data are available on the Internet at <http://dnb.d-nb.de>.

© 2013 Wiley-VCH Verlag GmbH & Co. KGaA, Boschstr. 12, 69469 Weinheim, Germany

All rights reserved (including those of translation into other languages). No part of this book may be reproduced in any form – by photoprinting, microfilm, or any other means – nor transmitted or translated into a machine language without written permission from the publishers. Registered names, trademarks, etc. used in this book, even when not specifically marked as such, are not to be considered unprotected by law.

Print ISBN: 978-3-527-33239-7
ePDF ISBN: 978-3-527-67390-2
ePub ISBN: 978-3-527-67389-6
Mobi ISBN: 978-3-527-67388-9
oBook ISBN: 978-3-527-67387-2

Cover Design   Formgeber, Mannheim
Typesetting   Manuela Treindl, Fürth
Printing and Binding   Betz-druck GmbH, Darmstadt

Printed in the Federal Republic of Germany
Printed on acid-free paper

Foreword

The Federal Government set out on the road to transforming the German energy system by launching its Energy Concept on 28 September 2010 and adopting the energy package on 6 June 2011. The intention is to make Germany one of the most energy-efficient economies in the world and to enter the era of renewable energy without delay. Quantitative energy and environmental targets have been set which define the basic German energy supply strategy until 2050.

Central goals are an 80–95% reduction in greenhouse gas emissions compared with 1990 figures, increasing the use of renewable energy to reach a 60% share of gross final energy consumption and 80% of gross electricity consumption, and reducing primary energy consumption by 50% relative to 2008 levels.

The *Energiewende*, as we call it, is among the most important challenges confronting Germany today – it is an enormous task for society as a whole. Urgent technological, economic, legal, and social issues need to be addressed quickly. Science and research bear a special responsibility in this process.

I very much welcome the comprehensive approach of the Third International Conference on Energy Process Engineering, which brings together international experts to discuss the potential of different technological options for a sustainable modern energy supply. This systemic perspective will help us find out whether individual technologies such as electrolysis can provide a sound basis for a new energy supply system or for closing existing infrastructure gaps.

The results of this international conference will be of great importance for further development, both in Germany and elsewhere. I would be happy to see our concept of a sustainable energy supply also gain ground in other countries.

Dr. Georg Schütte
State Secretary
Federal Ministry of Education and Research

Contents

Foreword *V*

Preface *XXIX*

List of Contributors *XXXI*

Part I   Renewable Strategies *1*

1 **South Korea's Green Energy Strategies** *3*
   Deokyu Hwang, Suhyeon Han, and Changmo Sung
1.1   Introduction *3*
1.2   Government-Driven Strategies and Policies *5*
1.3   Focused R&D Strategies *7*
1.4   Promotion of Renewable Energy Industries *9*
1.5   Present and Future of Green Energy in South Korea *10*
       References *10*

2 **Japan's Energy Policy After the 3.11 Natural and Nuclear Disasters – from the Viewpoint of the R&D of Renewable Energy and Its Current State** *13*
   Hirohisa Uchida
2.1   Introduction *13*
2.2   Energy Transition in Japan *14*
2.2.1   Economic Growth and Energy Transition *15*
2.2.2   Transition of Power Configuration *15*
2.2.3   Nuclear Power Technology *17*
2.3   Diversification of Energy Resource *17*
2.3.1   Thermal Power *18*
2.3.2   Renewable Energy Policy by Green Energy Revolution *18*
2.3.2.1   Agenda with Three NP Options *18*
2.3.2.2   Green Energy Revolution *19*
2.3.2.3   Feed-in Tariff for RE *21*

2.3.3	Renewable Energy and Hydrogen Energy	22
2.3.4	Solar–Hydrogen Stations and Fuel Cell Vehicles	22
2.3.5	Rechargeable Batteries	23
2.4	Hydrogen and Fuel Cell Technology	24
2.4.1	Stationary Use	24
2.4.2	Mobile Use	25
2.4.3	Public Acceptance	25
2.5	Conclusion	26
	References	26

3 The Impact of Renewable Energy Development on Energy and CO_2 Emissions in China 29

Xiliang Zhang, Tianyu Qi and Valerie Karplus

3.1	Introduction	29
3.2	Renewable Energy in China and Policy Context	30
3.2.1	Energy and Climate Policy Goals in China	30
3.2.2	Renewable Electricity Targets	31
3.3	Data and CGEM Model Description	31
3.3.1	Model Data	33
3.3.2	Renewable Energy Technology	33
3.4	Scenario Description	35
3.4.1	Economic Growth Assumptions	35
3.4.2	Current Policy Assumptions	37
3.4.3	Cost and Availability Assumptions for Energy Technologies	38
3.5	Results	39
3.5.1	Renewable Energy Growth Under Policy	39
3.5.2	Impact of Renewable Energy Subsidies on CO_2 Emissions Reductions	40
3.5.3	Impact of a Cost Reduction for Renewable Energy After 2020	42
3.6	Conclusion	44
	References	45

4 The Scottish Government's Electricity Generation Policy Statement 47

Colin Imrie

4.1	Introduction	47
4.2	Overview	47
4.3	Executive Summary	48
	References	65

5 Transition to Renewables as a Challenge for the Industry – the German Energiewende from an Industry Perspective 67

Carsten Rolle, Dennis Rendschmidt

5.1	Introduction	67
5.2	Targets and current status of the Energiewende	67
5.3	Industry view: opportunities and challenges	69

5.4	The way ahead	73
5.5	Conclusion	74
	References	74

6 The Decreasing Market Value of Variable Renewables: Integration Options and Deadlocks *75*

Lion Hirth and Falko Ueckerdt

6.1	The Decreasing Market Value of Variable Renewables	75
6.2	Mechanisms and Quantification	77
6.2.1	Profile Costs	78
6.2.2	Balancing Costs	83
6.2.3	Grid-Related Costs	83
6.2.4	Findings	83
6.3	Integration Options	84
6.3.1	A Taxonomy	84
6.3.2	Profile Costs	85
6.3.3	Balancing Costs	88
6.3.4	Grid-Related Costs	89
6.4	Conclusion	90
	References	90

7 Transition to a Fully Sustainable Global Energy System *93*

Yvonne Y. Deng, Kornelis Blok, Kees van der Leun, and Carsten Petersdorff

7.1	Introduction	93
7.2	Methodology	94
7.2.1	Definitions	95
7.3	Results – Demand Side	97
7.3.1	Industry	97
7.3.1.1	Industry – Future activity	97
7.3.1.2	Industry – Future Intensity	98
7.3.1.3	Industry – Future Energy Demand	99
7.3.2	Buildings	99
7.3.2.1	Buildings – Future Activity	99
7.3.2.2	Buildings – Future Intensity	101
7.3.2.3	Buildings – Future Energy Demand	102
7.3.3	Transport	103
7.3.3.1	Transport – Future Activity	103
7.3.3.2	Transport – Future Intensity	105
7.3.3.3	Transport – Future Energy Demand	107
7.3.4	Demand Sector Summary	107
7.4	Results – Supply Side	108
7.4.1	Supply Potential	108
7.4.1.1	Wind	109
7.4.1.2	Water	109
7.4.1.3	Sun	110

7.4.1.4	Earth	*110*
7.4.1.5	Bioenergy	*110*
7.4.2	Results of Balancing Demand and Supply	*111*
7.5	Discussion	*112*
7.5.1	Power Grids	*112*
7.5.2	The Need for Policy	*113*
7.5.3	Sensitivity of Results	*113*
7.6	Conclusion	*114*
	References	*115*
	Appendix	*118*

8 The Transition to Renewable Energy Systems – On the Way to a Comprehensive Transition Concept *119*
Uwe Schneidewind, Karoline Augenstein, and Hanna Scheck

8.1	Why Is There a Need for Change? – The World in the Age of the Anthropocene	*119*
8.2	A Transition to What?	*121*
8.3	Introducing the Concept of "Transformative Literacy"	*122*
8.4	Four Dimensions of Societal Transition	*123*
8.4.1	On the Structural Interlinkages of the Four Dimensions of Transitions	*124*
8.4.2	Infrastructures and Technologies – the Technological Perspective	*125*
8.4.3	Financial Capital – the Economic Perspective	*127*
8.4.4	Institutions/Policies – the Institutional Perspective	*129*
8.4.5	Cultural Change/Consumer Behavior – the Cultural Perspective	*131*
8.5	Techno-Economists, Institutionalists, and Culturalists – Three Conflicting Transformation Paradigms	*132*
	References	*135*

9 Renewable Energy Future for the Developing World *137*
Dieter Holm

9.1	Introduction	*137*
9.1.1	Aim	*137*
9.2	Descriptions and Definitions of the Developing World	*138*
9.2.1	The Developing World	*138*
9.2.2	The Developing World in Transition	*138*
9.2.3	Emerging Economies – BRICS	*140*
9.3	Can Renewable Energies Deliver?	*141*
9.4	Opportunities for the Developing World	*142*
9.4.1	Poverty Alleviation through RE Jobs	*142*
9.4.2	A New Energy Infrastructure Model	*143*
9.4.3	Great RE Potential of Developing World	*144*
9.4.4	Underdeveloped Conventional Infrastructure	*144*
9.5	Development Framework	*145*
9.5.1	National Renewable Energies Within Global Guard Rails	*145*

9.5.2	The International Context: Global Guard Rails	*145*
9.5.2.1	Socio-Economic Guard Rails	*145*
9.5.2.2	Ecological Guard Rails	*146*
9.6	Policies Accelerating Renewable Energies in Developing Countries	*148*
9.6.1	Regulations Governing Market/Electricity Grid Access and Quotas Mandating Capacity/Generation	*148*
9.6.1.1	Feed-in Tariffs	*149*
9.6.1.2	Quotas – Mandating Capacity/Generation	*149*
9.6.1.3	Applicability in the Developing World	*149*
9.6.2	Financial Incentives	*151*
9.6.2.1	Tax relief	*152*
9.6.2.2	Rebates and Payments	*152*
9.6.2.3	Low-Interest Loans and Guarantees	*152*
9.6.2.4	Addressing Subsidies and Prices of Conventional Energy	*152*
9.6.3	Industry Standards, Planning Permits, and Building Codes	*153*
9.6.4	Education, Information, and Awareness	*153*
9.6.5	Ownership, Cooperatives, and Stakeholders	*153*
9.6.6	Research, Development, and Demonstration	*154*
9.7	Priorities – Where to Start	*154*
9.7.1	Background	*154*
9.7.2	Learning from Past Mistakes	*154*
9.8	Conclusions and Recommendations	*156*
	References	*157*

10 An Innovative Concept for Large-Scale Concentrating Solar Thermal Power Plants *159*

Ulrich Hueck

10.1	Considerations for Large-Scale Deployment	*159*
10.1.1	Technologies to Produce Electricity from Solar Radiation	*160*
10.1.2	Basic Configurations of Existing CSP Plants	*160*
10.1.3	Review for Large-Scale Deployment	*161*
10.1.3.1	Robustness of Technology to Produce Electricity	*161*
10.1.3.2	Capability to Produce Electricity Day and Night	*161*
10.1.3.3	Type of Concentration of Solar Radiation	*162*
10.1.3.4	Shape of Mirrors for Concentration of Solar Radiation	*163*
10.1.3.5	Area for Solar Field	*164*
10.1.3.6	Technology to Capture Heat from Solar Radiation	*165*
10.1.3.7	Working Fluids and Heat Storage Media	*165*
10.1.3.8	Direct Steam Generation	*168*
10.1.3.9	Inlet Temperature for Power Generation	*168*
10.1.3.10	Type of Cooling System	*169*
10.1.3.11	Size of Solar Power Plants	*169*
10.1.3.12	Robustness of Other Technologies	*169*
10.1.4	Summary for Comparison of Technologies	*170*
10.2	Advanced Solar Boiler Concept for CSP Plants	*171*

10.2.1	Summary of Concept *171*	
10.2.2	Description of Concept *172*	
10.2.2.1	Direct Solar Steam Generation *172*	
10.2.2.2	Rankine Cycle for Steam Turbine *172*	
10.2.2.3	Solar Boiler for Steam Generation *174*	
10.2.2.4	Solar Steam Generation Inside Ducts *175*	
10.2.2.5	Arrangement of Heat-Transfer Sections *177*	
10.2.2.6	Utilization of Waste Heat *177*	
10.2.2.7	Thermal Storage System for Night-Time Operation *178*	
10.3	Practical Implementation of Concept *179*	
10.3.1	Technical Procedure for Implementation *179*	
10.3.2	Financial Procedure for Implementation *181*	
10.3.3	Strategic Procedure for Implementation *181*	
10.4	Conclusion *182*	
	References *182*	
11	**Status of Fuel Cell Electric Vehicle Development and Deployment: Hyundai's Fuel Cell Electric Vehicle Development as a Best Practice Example** *183*	
	Tae Won Lim	
11.1	Introduction *183*	
11.2	Development of the FCEV *183*	
11.2.1	Fuel Cell Stack Durability and Driving Ranging of FCEVs *184*	
11.2.2	Packing of FCEVs *184*	
11.2.3	Cost of FCEVs *185*	
11.3	History of HMC FCEV Development *185*	
11.4	Performance Testing of FCEVs *188*	
11.4.1	Crashworthiness and Fire Tests *188*	
11.4.2	Sub-Zero Conditions Tests *189*	
11.4.3	Durability Test *190*	
11.4.4	Hydrogen Refueling *190*	
11.5	Cost Reduction of FCEV *191*	
11.6	Demonstration and Deployment Activities of FCEVs in Europe *192*	
11.7	Roadmap of FCEV Commercialization and Conclusions *194*	
12	**Hydrogen as an Enabler for Renewable Energies** *195*	
	Detlef Stolten, Bernd Emonts, Thomas Grube, and Michael Weber	
12.1	Introduction *195*	
12.2	Status of CO_2 Emissions *196*	
12.3	Power Density as a Key Characteristic of Renewable Energies and Their Storage Media *197*	
12.4	Fluctuation of Renewable Energy Generation *199*	
12.5	Strategic Approach for the Energy Concept *200*	
12.6	Status of Electricity Generation and Potential for Expansion of Wind Turbines in Germany *200*	

12.7	Assumptions for the Renewable Scenario with a Constant Number of Wind Turbines *202*	
12.8	Procedure *205*	
12.9	Results of the Scenario *206*	
12.10	Fuel Cell Vehicles *207*	
12.11	Hydrogen Pipelines and Storage *208*	
12.12	Cost Estimate *210*	
12.13	Discussion of Results *212*	
12.14	Conclusion *213*	
	References *214*	

13 Pre-Investigation of Hydrogen Technologies at Large Scales for Electric Grid Load Balancing *217*
Fernando Gutiérrez-Martín

13.1	Introduction *217*
13.2	Electrolytic Hydrogen *218*
13.2.1	Electrolyzer Performance *219*
13.2.2	Hydrogen Production Cost Estimate by Water Electrolysis *221*
13.2.3	Simulation of Electrolytic Hydrogen Production *224*
13.3	Operation of the Electrolyzers for Electric Grid Load Balancing *226*
13.3.1	The Spanish Power System *228*
13.3.2	Integration of Hydrogen Technologies at Large Scales *230*
13.3.2.1	Hourly Average Curves *230*
13.3.2.2	Annual Curves *232*
13.4	Conclusion *236*
13.5	Appendix *238*
	References *238*

Part II Power Production *241*

14 Onshore Wind Energy *243*
Po Wen Cheng

14.1	Introduction *243*
14.2	Market Development Trends *244*
14.3	Technology Development Trends *246*
14.3.1	General Remarks About Future Wind Turbines *246*
14.3.2	Power Rating *247*
14.3.3	Number of Blades *247*
14.3.4	Rotor Materials *248*
14.3.5	Rotor Diameter *249*
14.3.6	Upwind or Downwind *250*
14.3.7	Drive train Concept *250*
14.3.8	Tower Concepts *253*
14.3.9	Wind Turbine and Wind Farm Control *254*

14.4	Environmental Impact	256
14.5	Regulatory Framework	257
14.6	Economics of Wind Energy	258
14.7	The Future Scenario of Onshore Wind Power	261
	References	262

15 Offshore Wind Power 265
David Infield

15.1	Introduction and Review of Offshore Deployment	265
15.2	Wind Turbine Technology Developments	271
15.3	Site Assessment	273
15.4	Wind Farm Design and Connection to Shore	274
15.5	Installation and Operations and Maintenance	276
15.6	Future Prospects and Research Needed to Deliver on These	278
	References	281

16 Towards Photovoltaic Technology on the Terawatt Scale: Status and Challenges 283
Bernd Rech, Sebastian S. Schmidt, and Rutger Schlatmann

16.1	Introduction	283
16.2	Working Principles and Solar Cell Fabrication	284
16.2.1	Crystalline Si Wafer-Based Solar Cells – Today's Workhorse Technology	286
16.2.2	Thin-Film PV: Challenges and Opportunities of Large-Area Coating Technologies	288
16.3	Technological Design of PV Systems	290
16.3.1	Residential Grid-Connected PV System: Roof Installation	290
16.3.2	Building-Integrated PV	292
16.3.3	Flexible Solar Cells	294
16.4	Cutting Edge Technology of Today	295
16.4.1	Efficiencies and Costs	296
16.4.2	Crystalline Silicon Wafer-Based High-Performance Solar Modules	297
16.4.3	Thin-Film Technologies	298
16.5	R&D Challenges for PV Technologies Towards the Terawatt Scale	300
16.5.1	Towards Higher Efficiencies and Lower Solar Module Costs	301
16.5.2	Crystalline Silicon Technologies	301
16.5.3	Thin-Film Technologies	302
16.5.4	Concentrating Photovoltaics (CPV)	302
16.5.5	Emerging Systems: Possible Game Changers and/or Valuable Add-Ons	303
16.5.6	Massive Integration of PV Electricity in the Future Energy Supply System	303
16.5.7	Beyond Technologies and Costs	304
16.6	Conclusion	304
	References	305

17	**Solar Thermal Power Production** *307*	
	Robert Pitz-Paal, Reiner Buck, Peter Heller, Tobias Hirsch,	
	and Wolf-Dieter Steinmann	
17.1	General Concept of the Technology *307*	
17.1.1	Introduction *307*	
17.1.2	Technology Characteristics and Options *308*	
17.1.3	Environmental Profile *311*	
17.2	Technology Overview *312*	
17.2.1	Parabolic Trough Collector systems *312*	
17.2.1.1	Parabolic Trough Collector Development *312*	
17.2.2	Linear Fresnel Collector Systems *317*	
17.2.3	Solar Tower Systems *320*	
17.2.4	Thermal Storage Systems *324*	
17.2.4.1	Basic Storage Concepts *325*	
17.2.4.2	Commercial Storage Systems *327*	
17.2.4.3	Current Research Activities *327*	
17.3	Cost Development and Perspectives [17] *328*	
17.3.1	Cost Structure and Actual Cost Figures *328*	
17.3.2	Cost Reduction Potential *331*	
17.3.2.1	Scaling Up *331*	
17.3.2.2	Volume Production *331*	
17.3.2.3	Technology Innovations *331*	
17.4	Conclusion *332*	
	References 332	
18	**Geothermal Power** *339*	
	Christopher J. Bromley and Michael A. Mongillo	
18.1	Introduction *339*	
18.2	Geothermal Power Technology *341*	
18.3	Global Geothermal Deployment: the IEA Roadmap and the IEA-GIA *342*	
18.4	Relative Advantages of Geothermal *343*	
18.5	Geothermal Reserves and Deployment Potential *344*	
18.6	Economics of Geothermal Energy *346*	
18.7	Sustainability and Environmental Management *346*	
	References 350	
19	**Catalyzing Growth: an Overview of the United Kingdom's Burgeoning Marine Energy Industry** *351*	
	David Krohn	
19.1	Development of the Industry *351*	
19.2	The Benefits of Marine Energy *352*	
19.3	Expected Levels of Deployment *354*	
19.4	Determining the Levelized Cost of Energy Trajectory *357*	
19.4.1	The Cost of Energy Trajectory *357*	

19.5	Technology Readiness	*360*
19.5.1	Tidal Device Case Study 1	*361*
19.5.2	Tidal Device Case Study 2	*362*
19.5.3	Tidal Device Case Study 3	*363*
19.5.4	Tidal Device Case Study 4	*364*
19.5.5	Tidal Device Case Study 5	*365*
19.5.6	Tidal Device Case Study 6	*366*
19.5.7	Tidal Device Case Study 7	*367*
19.5.8	Tidal Device Case Study 8	*368*
19.5.9	Tidal Device Case Study 9	*369*
19.5.10	Tidal Device Case Study 10	*370*
19.5.11	Wave Device Case Study 1	*371*
19.5.12	Wave Device Case Study 2	*372*
19.5.13	Wave Device Case Study 3	*373*
19.5.14	Wave Device Case Study 4	*374*
19.5.15	Wave Device Case Study 5	*375*
19.5.16	Wave Device Case Study 6	*376*
19.5.17	Wave Device Case Study 7	*377*
19.5.18	Wave Device Case Study 8	*378*
19.6	Conclusion	*378*
	References 379	

20 Hydropower *381*
Ånund Killingtveit

20.1	Introduction – Basic Principles	*381*
20.1.1	The Hydrological Cycle – Why Hydropower Is Renewable	*382*
20.1.2	Computing Hydropower Potential	*383*
20.1.3	Hydrology – Variability in Flow	*383*
20.2	Hydropower Resources/Potential Compared with Existing System	*385*
20.2.1	Definition of Potential	*385*
20.2.2	Global and Regional Overview	*385*
20.2.3	Barriers – Limiting Factors	*387*
20.2.4	Climate-Change Impacts	*387*
20.3	Technological Design	*388*
20.3.1	Run-of-River Hydropower	*388*
20.3.2	Storage Hydropower	*388*
20.3.3	Pumped Storage Hydropower	*389*
20.4	Cutting Edge Technology	*389*
20.4.1	Extending the Operational Regime for Turbines	*390*
20.4.2	Utilizing Low or Very Low Head	*391*
20.4.3	Fish-Friendly Power Plants	*391*
20.4.4	Tunneling and Underground Power Plants	*391*
20.5	Future Outlook	*394*
20.5.1	Cost Performance	*394*
20.5.2	Future Energy Cost from Hydropower	*396*

20.5.3	Carbon Mitigation Potential 396
20.5.4	Future Deployment 397
20.6	Systems Analysis 398
20.6.1	Integration into Broader Energy Systems 398
20.6.2	Power System Services 398
20.7	Sustainability Issues 398
20.7.1	Environmental Impacts 399
20.7.2	Lifecycle Assessment 399
20.7.3	Greenhouse Gas Emissions 399
20.7.4	Energy Payback Ratio 400
20.8	Conclusion 400
	References 401

21 The Future Role of Fossil Power Plants – Design and Implementation 403
Erland Christensen and Franz Bauer

21.1	Introduction 403
21.2	Political Targets/Regulatory Framework 403
21.3	Market Constraints – Impact of RES 406
21.4	System Requirements and Technical Challenges for the Conventional Fleet 407
21.4.1	Flexibility Requirements with Load Following and Gradients 408
21.4.2	Delivery of System Services 410
21.4.2.1	Primary Reserve/Control 411
21.4.2.2	Secondary Reserve/Control 411
21.4.2.3	Tertiary or Manual Reserve 411
21.4.2.4	"Short-Circuit Effect," Reactive Reserves, and Voltage Regulation, Inertia of the System 412
21.4.2.5	Secure Power Supply When Wind and Solar Are Not Available 412
21.4.3	District Heating 413
21.4.4	Co-combustion of Biomass 414
21.5	Technical Challenges for Generation 416
21.6	Economic Challenges 418
21.6.1	Principles Underlying the Data on CAPEX and OPEX 418
21.7	Future Generation Portfolio – RES Versus Residual Power 421

Part III Gas Production 423

22 Status on Technologies for Hydrogen Production by Water Electrolysis 425
Jürgen Mergel, Marcelo Carmo, and David Fritz

22.1	Introduction 425
22.2	Physical and Chemical Fundamentals 426
22.3	Water Electrolysis Technologies 430

22.3.1	Alkaline Electrolysis	*430*
22.3.2	PEM Electrolysis	*433*
22.3.3	High-Temperature Water Electrolysis	*436*
22.4	Need for Further Research and Development	*438*
22.4.1	Alkaline Water Electrolysis	*440*
22.4.1.1	Electrocatalysts for Alkaline Water Electrolysis	*441*
22.4.2	PEM Electrolysis	*442*
22.4.2.1	Electrocatalysts for the Hydrogen Evolution Reaction (HER)	*442*
22.4.2.2	Electrocatalysts for the Oxygen Evolution Reaction (OER)	*443*
22.4.2.3	Separator Plates and Current Collectors	*443*
22.5	Production Costs for Hydrogen	*446*
22.6	Conclusion	*446*
	References	*447*

23	**Hydrogen Production by Solar Thermal Methane Reforming**	***451***
	Christos Agrafiotis, Henrik von Storch, Martin Roeb, and Christian Sattler	
23.1	Introduction	*451*
23.2	Hydrogen Production Via Reforming of Methane Feedstocks	*453*
23.2.1	Thermochemistry and Thermodynamics of Reforming	*453*
23.2.2	Current Industrial Status	*455*
23.3	Solar-Aided Reforming	*456*
23.3.1	Coupling of Solar Energy to the Reforming Reaction: Solar Receiver/Reactor Concepts	*456*
23.3.2	Worldwide Research Activities in Solar Thermal Methane Reforming	*460*
23.3.2.1	Indirectly Heated Reactors	*461*
23.3.2.2	Directly Irradiated Reactors	*468*
23.4	Current Development Status and Future Prospects	*476*
	References	*478*

Part IV Biomass *483*

24	**Biomass – Aspects of Global Resources and Political Opportunities**	***485***
	Gustav Melin	
24.1	Our Perceptions: Are They Misleading Us?	*485*
24.2	Biomass – Just a Resource Like Other Resources – Price Gives Limitations	*485*
24.3	Global Food Production and Prices	*487*
24.3.1	Production Capacity per Hectare in Different Countries	*488*
24.4	Global Arable Land Potential	*490*
24.4.1	Global Forests Are Carbon Sinks Assimilating One-Third of Total Carbon Emissions	*491*
24.4.2	Forest Supply – the Major Part of Sweden's Energy Supply	*492*
24.5	Lower Biomass Potential If No Biomass Demand	*493*

24.6	Biomass Potential Studies *494*	
24.7	The Political Task *494*	
24.8	Political Measures, Legislation, Steering Instruments, and Incentives *495*	
24.8.1	Carbon Dioxide Tax: the Most Efficient Steering Instrument *495*	
24.8.2	Less Political Damage *496*	
24.8.3	Use Biomass *496*	
	References *497*	

25 Flexible Power Generation from Biomass – an Opportunity for a Renewable Sources-Based Energy System? *499*
Daniela Thrän, Marcus Eichhorn, Alexander Krautz, Subhashree Das, and Nora Szarka

25.1	Introduction *499*
25.2	Challenges of Power Generation from Renewables in Germany *500*
25.3	Power Generation from Biomass *507*
25.4	Demand-Driven Electricity Commission from Solid Biofuels *510*
25.5	Demand-Driven Electricity Commission from Liquid Biofuels *511*
25.6	Demand-Driven Electricity Commission from Gaseous Biofuels *512*
25.7	Potential for Flexible power Generation – Challenges and Opportunities *515*
	References *518*

26 Options for Biofuel Production – Status and Perspectives *523*
Franziska Müller-Langer, Arne Gröngröft, Stefan Majer, Sinéad O'Keeffe, and Marco Klemm

26.1	Introduction *523*
26.2	Characteristics of Biofuel Technologies *524*
26.2.1	Biodiesel *528*
26.2.2	HVO and HEFA *529*
26.2.3	Bioethanol *529*
26.2.4	Synthetic BTL *530*
26.2.5	Biomethane *532*
26.2.5.1	Upgraded Biochemically Produced Biogas *532*
26.2.5.2	Thermochemically Produced Bio-SNG (Synthetic Natural Gas) *532*
26.2.6	Other Innovative Biofuels *532*
26.2.6.1	BTL Fuels Such as Methanol and Dimethyl Ether *533*
26.2.6.2	Biohydrogen *533*
26.2.6.3	Sugars to Hydrocarbons *533*
26.2.6.4	Biobutanol *534*
26.2.6.5	Algae-Based Biofuels *534*
26.3	System Analysis on Technical Aspects *534*
26.3.1	Capacities of Biofuel Production Plants *534*
26.3.2	Overall Efficiencies of Biofuel Production Plants *535*
26.4	System Analysis on Environmental Aspects *537*

26.4.1	Differences in LCA Studies for Biofuel Options 537
26.4.2	Drivers for GHG Emissions: Biomass Production 538
26.4.3	Drivers for GHG Emissions: Biomass Conversion 540
26.4.4	Perspectives for LCA Assessments 541
26.5	System Analysis on Economic Aspects 542
26.5.2	Total Capital Investments for Biofuel Production Plants 542
26.5.3	Biofuel Production Costs 543
26.6	Conclusion and Outlook 545
26.6.1	Technical Aspects 545
26.6.2	Environmental Aspects 545
26.6.3	Economic Aspects 546
26.6.4	Future R&D needs 546
	References 547

Part V Storage 555

**27 Energy Storage Technologies –
Characteristics, Comparison, and Synergies** 557
Andreas Hauer, Josh Quinnell, and Eberhard Lävemann

27.1	Introduction 557
27.2	Energy Storage Technologies 558
27.2.1	Energy Storage Properties 558
27.2.2	Electricity Storage 559
27.2.3	Storage of Thermal Energy 561
27.2.4	Energy Storage by Chemical Conversion 564
27.2.5	Technical Comparison of Energy Storage Technologies 565
27.3	The Role of Energy Storage 567
27.3.1	Balancing Supply and Demand 568
27.3.2	Distributed Energy Storage Systems and Energy Conversion 570
27.3.2.1	Distributed Energy Storage Systems 570
27.3.2.2	In/Out Storage Versus One-Way Storage 571
27.3.2.3	Example: Power-to-Gas Versus Long-Term Hot Water Storage 571
27.4	Economic Evaluation of Energy Storage Systems 572
27.4.1	Top-Down Approach for Maximum Energy Storage Costs 572
27.4.2	Results 573
27.5	Conclusion 575
	References 576

28 Advanced Batteries for Electric Vehicles and Energy Storage Systems 579
Seung Mo Oh, Sa Heum Kim, Youngjoon Shin, Dongmin Im, and Jun Ho Song

| 28.1 | Introduction 579 |
| 28.2 | R&D Status of Secondary Batteries 581 |

28.2.1	Lithium-Ion Batteries	581
28.2.2	Redox-Flow Batteries	582
28.2.3	Sodium–Sulfur Batteries	583
28.2.4	Lithium–Sulfur Batteries	584
28.2.5	Lithium–Air Batteries	585
28.3	Secondary Batteries for Electric Vehicles	587
28.4	Secondary Batteries For Energy Storage Systems	590
28.4.1	Lithium-Ion Batteries for ESS	591
28.4.2	Redox-Flow Batteries for ESS	592
28.4.3	Sodium–Sulfur Batteries for ESS	593
28.5	Conclusion	594
	References	595

29 Pumped Storage Hydropower 597
Atle Harby, Julian Sauterleute, Magnus Korpås, Ånund Killingtveit, Eivind Solvang, and Torbjørn Nielsen

29.1	Introduction	597
29.1.1	Principle and Purpose of Pumped Storage Hydropower	597
29.1.2	Deployment of Pumped Storage Hydropower	598
29.2	Pumped Storage Technology	599
29.2.1	Operational Strategies	601
29.2.2	Future Pumped Storage Plants	602
29.3	Environmental Impacts of Pumped Storage Hydropower	602
29.4	Challenges for Research and Development	604
29.5	Case Study: Large-Scale Energy Storage and Balancing from Norwegian Hydropower	605
29.5.1	Demand for Energy Storage and Balancing Power	606
29.5.2	Technical Potential	607
29.5.3	Water Level Fluctuations in Reservoirs	609
29.5.4	Environmental Impacts	611
29.6	System Analysis of Linking Wind and Flexible Hydropower	612
29.6.1	Method	612
29.6.2	Results	613
29.7	Conclusion	616
	References	617

30 Chemical Storage of Renewable Electricity via Hydrogen – Principles and Hydrocarbon Fuels as an Example 619
Georg Schaub, Hilko Eilers, and Maria Iglesias González

30.1	Integration of Electricity in Chemical Fuel Production	619
30.2	Example: Hydrocarbon Fuels	621
30.2.1	Hydrocarbon Fuels Today	621
30.2.2	Hydrogen Demand in Hydrocarbon Fuel Upgrading/Production	622
30.2.3	Hydrogen in Petroleum Refining	623
30.2.4	Hydrogen in Synfuel Production	624

30.2.5	Example: Substitute Natural Gas (SNG) from H_2–CO_2	624
30.2.6	Example: Liquid Fuels from Biomass	625
30.2.7	Cost of Hydrogen Production	626
30.3	Conclusion	627
30.4	Nomenclature	627
	References	628

31 Geological Storage for the Transition from Natural to Hydrogen Gas 629
Jürgen Wackerl, Martin Streibel, Axel Liebscher, and Detlef Stolten

31.1	Current Situation	629
31.2	Natural Gas Storage	631
31.3	Requirements for Subsurface Storage	633
31.4	Geological Situation in Central Europe and Especially Germany	636
31.5	Types of Geological Gas Storage Sites	639
31.5.1	Pore-Space Storage Sites	639
31.5.2	Oil and Gas Fields	640
31.5.3	Aquifers	642
31.5.4	Abandoned Mining Sites	644
31.5.5	Salt Caverns	646
31.6	Comparisons with Other Locations and Further Considerations with Focus on Hydrogen Gas	652
31.7	Conclusion	653
	References	654

32 Near-Surface Bulk Storage of Hydrogen 659
Vanessa Tietze and Sebastian Luhr

32.1	Introduction	659
32.2	Storage Parameters	661
32.3	Compressed Gaseous Hydrogen Storage	662
32.3.1	Thermodynamic Fundamentals	662
32.3.2	Hydrogen Compressors	662
32.3.3	Hydrogen Pressure Vessels	663
32.4	Cryogenic Liquid Hydrogen Storage	669
32.4.1	Thermodynamic Fundamentals	669
32.4.2	Liquefaction Plants	670
32.4.3	Liquid Hydrogen Storage Tanks	671
32.5	Metal Hydrides	675
32.5.1	Characteristics of Materials	675
32.6	Cost Estimates and Economic Targets	677
32.7	Technical Assessment	679
32.8	Conclusion	684
	References	685

33	**Energy Storage Based on Electrochemical Conversion of Ammonia**	*691*
	Jürgen Fuhrmann, Marlene Hülsebrock, and Ulrike Krewer	
33.1	Introduction *691*	
33.2	Ammonia Properties and Historical Uses as an Energy Carrier *692*	
33.3	Pathways for Ammonia Conversion: Synthesis *693*	
33.3.1	Haber–Bosch Process *694*	
33.3.2	Electrochemical Synthesis *697*	
33.4	Pathways for Ammonia Conversion: Energy Recovery *698*	
33.4.1	Combustion *698*	
33.4.2	Direct Ammonia Fuel Cells *699*	
33.4.3	Energy Recovery via Hydrogen *699*	
33.5	Comparison of Pathways *700*	
33.6	Conclusions *702*	
	References *703*	

Part VI Distribution *707*

34	**Introduction to Transmission Grid Components** *709*	
	Armin Schnettler	
34.1	Introduction *709*	
34.2	Classification of Transmission System Components *710*	
34.2.1	Transmission Technologies *710*	
34.2.1.1	Overhead Lines *710*	
34.2.1.2	Underground Lines *711*	
34.2.2	Conversion Technologies *712*	
34.2.2.1	Switchgears/Substations *712*	
34.2.2.2	Power Transformers *714*	
34.2.2.3	FACTS Devices *714*	
34.2.2.4	HVDC Converters *715*	
34.2.3	System Integration of Transmission Technologies *717*	
34.3	Recent Developments of Transmission System Components *720*	
	References *721*	

35	**Introduction to the Transmission Networks** *723*	
	Göran Andersson, Thilo Krause, and Wil Kling	
35.1	Introduction *723*	
35.2	The Transmission System – Development, Role, and Technical Limitations *724*	
35.2.1	The Development Stages of the Transmission System *724*	
35.2.2	Tasks of the Transmission System *727*	
35.2.3	Technical Limitations of Power Transmission *728*	
35.3	The Transmission Grid in Europe – Current Situation and Challenges *729*	
35.3.1	Historical Evolution of the UCTE/ENTSO-E Grid *729*	
35.3.2	Transmission Challenges Driven by Electricity Trade *730*	

35.3.3	Transmission Challenges Driven by the Production Side	*731*
35.3.4	Transmission Challenges Driven by the Demand Side and Developments in the Distribution Grid	*731*
35.3.5	Conclusion	*732*
35.4	Market Options for the Facilitation of Future Bulk Power Transport	*732*
35.4.1	Cross-Border Trading and Market Coupling	*732*
35.4.2	Cross-Border Balancing	*733*
35.4.3	Technological Options for the Facilitation of Future Bulk Power Transport	*733*
35.5	Case Study	*735*
	References	*739*

36 Smart Grid: Facilitating Cost-Effective Evolution to a Low-Carbon Future *741*

Goran Strbac, Marko Aunedi, Danny Pudjianto, and Vladimir Stanojevic

36.1	Overview of the Present Electricity System Structure and Its Design and Operation Philosophy	*741*
36.2	System Integration Challenges of Low-Carbon Electricity Systems	*743*
36.3	Smart Grid: Changing the System Operation Paradigm	*744*
36.4	Quantifying the Benefits of Smart Grid Technologies in a Low-Carbon future	*746*
36.5	Integration of Demand-Side Response in System Operation and Planning	*749*
36.5.1	Control of Domestic Appliances	*750*
36.5.2	Integration of EVs	*755*
36.5.3	Smart Heat Pump Operation	*761*
36.5.4	Role and Value of Energy Storage in Smart Grid	*762*
36.6	Implementation of Smart Grid: Distributed Energy Marketplace	*768*
	References	*770*

37 Natural Gas Pipeline Systems *773*

Gerald Linke

37.1	Physical and Chemical Fundamentals	*773*
37.2	Technological Design	*776*
37.3	Cutting Edge Technology of Today	*780*
37.4	Outlook on R&D Challenges	*784*
37.5	System Analysis	*791*
	References	*794*

38 Introduction to a Future Hydrogen Infrastructure *795*

Joan Ogden

38.1	Introduction	*795*
38.2	Technical Options for Hydrogen Production, Delivery, and Use in Vehicles	*796*
38.2.1	Hydrogen Vehicles	*796*

38.2.2	Hydrogen Production Methods	797
38.2.3	Options for Producing Hydrogen with Near-Zero Emission	800
38.2.4	Hydrogen Delivery Options	800
38.2.5	Hydrogen Refueling Stations	801
38.3	Economic and Environmental Characteristics of Hydrogen Supply Pathways	802
38.3.1	Economics of Hydrogen Supply	802
38.3.2	Environmental Impacts of Hydrogen Pathways	805
38.3.2.1	Well-to-Wheels Greenhouse Gas Emissions, Air Pollution, and Energy Use	805
38.3.2.2	Resource Use and Sustainability	805
38.3.2.3	Infrastructure Compatibility	806
38.4	Strategies for Building a Hydrogen Infrastructure	806
38.4.1	Design Considerations for Hydrogen Refueling Infrastructure	806
38.4.2	Hydrogen Transition Scenario for the United States	807
38.5	Conclusion	809
	References	810

39 Power to Gas 813

Sebastian Schiebahn, Thomas Grube, Martin Robinius, Li Zhao, Alexander Otto, Bhunesh Kumar, Michael Weber, and Detlef Stolten

39.1	Introduction	813
39.2	Electrolysis	814
39.2.1	Alkaline Water Electrolysis	814
39.2.2	Proton Exchange Membrane Electrolysis	817
39.2.3	High-Temperature Water Electrolysis	818
39.2.4	Integration of Renewable Energies with Electrolyzers	819
39.3	Methanation	820
39.3.1	Catalytic Hydrogenation of CO_2 to Methane	820
39.3.2	Methanation Plants	821
39.3.3	CO_2 Sources	823
39.3.3.1	CO_2 via Carbon Capture and Storage	823
39.3.3.2	CO_2 Obtained from Biomass	824
39.3.3.3	CO_2 from Other Industrial Processes	825
39.3.3.4	CO_2 Recovery from Air	826
39.4	Gas Storage	828
39.4.1	Porous Rock Storage	829
39.4.2	Salt Cavern Storage	830
39.5	Gas Pipelines	831
39.5.1	Natural Gas Pipeline System	831
39.5.2	Hydrogen Pipeline System	833
39.6	End-Use Technologies	834
39.6.1	Stationary End Use	835
39.6.1.1	Central Conversion of Natural Gas Mixed with Hydrogen in Combustion Turbines	835

39.6.1.2 Decentralized Conversion of Natural Gas Mixed with Hydrogen in Gas Engines *835*
39.6.1.3 Conversion of Hydrogen Mixed with Natural Gas in Combustion Heating Systems *835*
39.6.2 Passenger Car Powertrains with Fuel Cells and Internal Combustion Engines *836*
39.6.2.1 Direct-Hydrogen Fuel Cell Systems *836*
39.6.2.2 Internal Combustion Engines *837*
39.7 Evaluation of Process Chain Alternatives *838*
39.8 Conclusion *841*
References *843*

Part VII Applications *849*

40 Transition from Petro-Mobility to Electro-Mobility *851*
David L. Greene, Changzheng Liu, and Sangsoo Park

40.1 Introduction *851*
40.2 Recent Progress in Electric Drive Technologies *853*
40.3 Energy Efficiency *854*
40.4 The Challenge of Energy Transition *856*
40.5 A New Environmental Paradigm: Sustainable Energy Transitions *858*
40.6 Status of Transition Plans *859*
40.7 Modeling and Analysis *862*
40.8 Conclusion *870*
References *871*

41 Nearly Zero, Net Zero, and Plus Energy Buildings – Theory, Terminology, Tools, and Examples *875*
Karsten Voss, Eike Musall, Igor Sartori and Roberto Lollini

41.1 Introduction *875*
41.2 Physical and Balance Boundaries *876*
41.3 Weighting Systems *878*
41.4 Balance Types *879*
41.5 Transient Characteristics *881*
41.6 Tools *882*
41.7 Examples and Experiences *883*
41.8 Conclusion *887*
References *888*

42 China Road Map for Building Energy Conservation *891*
Peng Chen, Yan Da, and Jiang Yi

42.1 Introduction *891*
42.2 The Upper Bound of Building Energy Use in China *892*
42.2.1 Limitation of the Total Amount of Carbon Emissions *893*

42.2.2	Limitation of the Total Amount of Available Energy in China	894
42.2.3	Limitation of the Total Amount of Building Energy Use in China	895
42.3	The Way to Realize the Targets of Building Energy Control in China	897
42.3.1	Factors Affecting Building Energy Use	897
42.3.1.1	The Total Building Floor Area	897
42.3.1.2	The Energy Use Intensity	899
42.3.2	The Energy Use of Northern Urban Heating	900
42.3.3	The Energy Use of Urban Residential Buildings (Excluding Heating in the North)	902
42.3.4	The Energy Use of Commercial and Public Buildings (Excluding Heating in the North)	904
42.3.5	The Energy Use of Rural Residential Buildings	906
42.3.6	The Target of Buildings Energy Control in China in the Future	908
42.4	Conclusions	909
	References	910
43	**Energy Savings Potentials and Technologies in the Industrial Sector: Europe as an Example**	**913**
	Tobias Bossmann, Rainer Elsland, Wolfgang Eichhammer, and Harald Bradke	
43.1	Introduction	913
43.2	Electric Drives	916
43.2.1	E-Drive System Optimization	919
43.2	Steam and Hot Water Generation	922
43.3	Other Industry Sectors	926
43.4	Overall Industry Sector	931
	References	935

Subject Index   937

Preface

Renewable energy gets increasingly important for its increasing share in the energy supply, the urgency to act on global climate change and not the least for its increasing competitiveness. Already today, renewable energies deliver substantial shares to the global final energy consumption. As of 2010 16.7% were generated by renewables, out of which 8.2% were accounted for modern renewables, comparing favorably to three times the share of nuclear energy. Worldwide over 20% of the electricity was produced by renewable energies in 2011, with 15.3% generated by hydropower and 5% by other renewables, breaking down into 2.1% wind power, biomass and 0.3% of solar electricity. Whereas hydropower and biomass for electricity were just slightly increasing in 2011, wind power increased over 30% and solar over 60%.

If available, hydropower is the most effective to use with installations from the kw-range to the biggest power plants of all kinds with 55 plants above 2 GW peaking in the 22.5 GW installed capacity at the Three Gorges Dam in China. Hydropower is also the most reliable renewable energy source that can be operated driven by consumer demand even better than most fossil power plants. Nonetheless, there are restrictions owing to ecological consequences of flooding larger areas and regulating rivers, relocating local people and topographic availability. Hydropower provides the majority of electricity in some countries, peaked by Norway with 95%, Brazil with 85%, Austria with 5 5% and Sweden with over 50%.

It is a major challenge though, to reach such high levels of other renewables for their fluctuating nature and their lower energy density that increases the necessary efforts for harnessing them. In other words, strong efforts for cost reductions are necessary to make them competitive to fossil power generation and additional measures are required to integrate them into the electric power grid.

Hence, modern renewables other than hydropower require a broader view of the energy pathway beyond electricity generation including transmission, storage and end-use if a transition to renewables – meaning the reliance on major shares of renewables – is attempted. Opportunities and complexity rise at the same time when electric transportation via batteries or fuel cells are included for propulsion of passenger cars and additionally biofuels are considered as a substitute for diesel in trucks, trains and aircraft.

This book provides a part on energy strategies as examples how a secure, safe and affordable energy supply can be organized relying on renewable energies.

It provides descriptions, data, facts and figures of the major technologies that have the potential to be Game Changers in power production considering the varying climates and topographies worldwide. It addresses biomass, gas production and storage in the same technical depth and includes chapters on power and gas distribution, including smart grids, as well as selected chapters on end-use of energy in transportation and the building sector.

These papers are based on the overview presentations of the 3rd ICEPE 2013, Transition to Renewable Energy Systems, held in Frankfurt, Germany.

DECHEMA is gratefully acknowledged for organizing the conference and supporting this book by making it part of the conference proceedings. The scientific support of the subdivision Energy Process Engineering of ProcessNet is gratefully acknowledged.

The contributions of the chapter authors are gratefully acknowledged as well as the support of Anke Wagner, Bernd Emonts and Michael Weber who helped us considerably in handling the issues associated with this book.

Not the least the great effort of the Wiley team is to be mentioned since they made it possible to have a fully copy-edited book within a time frame of twelve months from the concept to print.

We wish that this book will help professionals – be it in science, industry or politics – to complement their knowledge of technologies, and their scope of strategies to generate a transition to renewable energy systems.

Detlef Stolten
Juelich Research Center and RWTH Aachen University, Germany

Viktor Scherer
Ruhr-University Bochum, Germany

January 2013

List of Contributors

Göran Andersson
ETH Zürich
Institut für El. Energieübertragung
ETL G 26
Physikstr. 3
8092 Zürich
Switzerland

Karoline Augenstein
Wuppertal Institut für Klima, Umwelt,
Energie GmbH
Döpperberg 19
42103 Wuppertal
Germany

Kornelis Blok
ECOFYS GERMANY
Am Wassermann 35
50829 Köln
Germany

Harald Bradke
Fraunhofer-Institut für
System- und Innovationsforschung ISI
Breslauer Straße 48
76139 Karlsruhe
Germany

Christopher J. Bromley
GNS Science
Wairakei Research Centre
Private Bag 2000
Taupo 3352
New Zealand

Marcelo Carmo
Forschungszentrum Jülich GmbH
Institut für Energie- und
Klimaforschung
IEK-3: Elektrochemische
Verfahrenstechnik
52425 Jülich
Germany

Peng Chen
Tsinghua University
Department of Building Science and
Technology
Beijing 100084
P.R. China

Po Wen Cheng
University of Stuttgart
Allmandring 5b
70569 Stuttgart
Germany

List of Contributors

Erland Christensen
VGB Power Tech
Klinkestraße 27–31
45136 Essen
Germany

Yan Da
Tsinghua University
Department of Building Science and Technology
Beijing 100084
P.R. China

Yvonne Y. Deng
ECOFYS GERMANY
Am Wassermann 35
50829 Köln
Germany

Bernd Emonts
Forschungszentrum Jülich GmbH
IEK-3 Institut für En. & Klimaforschung
Wilhelm-Johnen-Str.
52428 Jülich
Germany

David Fritz
Forschungszentrum Jülich GmbH
Institut für Energie- und Klimaforschung
IEK-3: Elektrochemische Verfahrenstechnik
52425 Jülich
Germany

Jürgen Fuhrmann
Weierstrass Institute for Applied Analysis and Stochastics
Mohrenstraße 39
10117 Berlin
Germany

David L. Greene
Oak Ridge National Laboratory
Center for Transportation Analysis
2360 Cherahala Blvd
Knoxville, Tennessee 37932
USA

Thomas Grube
Forschungszentrum Jülich GmbH
IEK-3 Institut für En. & Klimaforschung
Wilhelm-Johnen-Str.
52428 Jülich
Germany

Fernando Gutiérrez-Martín
Universidad Politécnica de Madrid
Rda Valencia 3
28012 Madrid
Spain

Atle Harby
Stiftelsen SINTEF
Sem Sælands vei 11
7465 Trondheim
Norway

Andreas Hauer
ZAE Bayern
Walther-Meißner-Str. 6
85748 Garching
Germany

Lion Hirth
Strategic Analysis (FYCA)
Vattenfall GmbH
Chausseestraße 23
10115 Berlin
Germany

Dieter Holm
ISES Africa
P.O. Box 58
0216 Hartbeespoort
South Africa

Ulrich Hueck
DESERTEC Foundation
Ferdinandstr. 28-30
20095 Hamburg
Germany

Colin Imrie
Scottish Government
Energy and Climate Change
Directorate
4th floor, 5 Atlantic Quay
Broomielaw, Glasgow
Scotland
UK

David Infield
University of Strathclyde
16 Richmond Street
Glasgow G1 1XQ
Scotland
UK

Anund Killingtveit
Department of Hydraulic &
Environmental Engineering
S. P. Andersens veg 5
7491 Trondheim
Norway

Wil Kling
Eindhoven University of Technology
Department of Electrical Engineering
Den Dolech 2
5600 Eindhoven
The Netherlands

Thilo Krause
ETH Zürich
Institut für El. Energieübertragung
ETL G 26
Physikstr. 3
8092 Zürich
Switzerland

David Krohn
RenewableUK
Greencoat House
Francis Street
London SW1P 1DH
UK

Bunesh Kumar
Forschungszentrum Jülich GmbH
IEK-3 Institut für En. &
Klimaforschung
Wilhelm-Johnen-Str.
52428 Jülich
Germany

Eberhard Lävemann
ZAE Bayern
Walther-Meißner-Str. 6
85748 Garching
Germany

Tae Won Lim
Hyundai Motor Company's Fuel Cell
Vehicle Group
104, Mabuk-Dong, Giheung-Gu,
Yongin-Si,
Gyunggi-Do, 446-912
South Korea

Gerald Linke
E.ON New Build & Technology
Global Engineering
Alexander-von-Humboldt-Str. 1
45896 Gelsenkirchen
Germany

Roberto Lollini
Bergische Universität Wuppertal
Fachbereich D – Architektur
Campus – Haspel
Haspeler Str. 27
42285 Wuppertal
Germany

Gustav Melin
Svebio
Holländargatan 17
111 60 Stockholm
Sweden

Jürgen Mergel
Forschungszentrum Jülich GmbH
Institut für Energie- und
Klimaforschung
IEK-3: Elektrochemische
Verfahrenstechnik
52425 Jülich
Germany

Michael A. Mongillo
GNS Science
Wairakei Research Centre
Private Bag 2000
Taupo 3352
New Zealand

Franziska Müller-Langer
DBFZ Deutsches
Biomasseforschungszentrum
gemeinnützige GmbH
Torgauer Str. 116
04347 Leipzig
Germany

Eike Musall Musall
Bergische Universität Wuppertal
Fachbereich D – Architektur
Campus – Haspel
Haspeler Str. 27
42285 Wuppertal
Germany

Joan Ogden
University of California Davis
Institute for Transportation Studies
One Shields Avenue
Davis, CA 95616
USA

Seung Mo Oh
Seoul National University
Chemical and Biological Engineering
599 Gwanak-ro, Gwanak-gu
Seoul 151-744
Republic of Korea

Alexander Otto
Forschungszentrum Jülich GmbH
IEK-3 Institut für En. &
Klimaforschung
Wilhelm-Johnen-Str.
52428 Jülich
Germany

Carsten Petersdorff
ECOFYS GERMANY
Am Wassermann 35
50829 Köln
Germany

Robert Pitz-Paal
Deutsches Zentrum für Luft- und
Raumfahrt (DLR)
Institut für Solarforschung
Linder Höhe
51147 Köln

Josh Quinnell
ZAE Bayern
Walther-Meißner-Str. 6
85748 Garching
Germany

Bernd Rech
Helmholtz Zentrum Berlin
Keküléstrasse 5
12489 Berlin
Germany

Martin Robinius
Forschungszentrum Jülich GmbH
IEK-3 Institut für En. &
Klimaforschung
Wilhelm-Johnen-Str.
52428 Jülich
Germany

Carsten Rolle
Bundesverband der Deutschen
Industrie
11053 Berlin
Germany

Igor Sartori
Bergische Universität Wuppertal
Fachbereich D – Architektur
Campus – Haspel
Haspeler Str. 27
42285 Wuppertal
Germany

Christian Sattler
Deutsches Zentrum für Luft- und
Raumfahrt e.V.
Institut für technische
Thermodynamik – Solarforschung
Linder Höhe
51147 Köln
Germany

Georg Schaub
Engler-Bunte-Institut
Bereich Chemische Energieträger –
Brennstofftechnologie
Engler-Bunte-Ring 1 (Geb. 40.02)
76131 Karlsruhe
Germany

Hanna Scheck
Wuppertal Institut für Klima, Umwelt,
Energie GmbH
Döpperberg 19
42103 Wuppertal
Germany

Sebastian Schiebahn
Forschungszentrum Jülich GmbH
IEK-3 Institut für En. &
Klimaforschung
Wilhelm-Johnen-Str.
52428 Jülich
Germany

Uwe Schneidewind
Wuppertal Institut für Klima, Umwelt,
Energie GmbH
Döpperberg 19
42103 Wuppertal
Germany

Armin Schnettler
RWTH Aachen
Institut für Hochspannungstechnik
Schinkelstraße 2
52056 Aachen
Germany

Detlef Stolten
Forschungszentrum Jülich GmbH
IEK-3 Institut für En. &
Klimaforschung
Wilhelm-Johnen-Str.
52428 Jülich
Germany

Goran Strbac
Department of Electrical and
Electronic Engineering
Imperial College London
South Kensington Campus
London, SW7 2AZ
UK

Chungmo Sung
Korea Hwarangno
14-gil 5 Seongbuk-gu
Seoul, 136-791
Republic of Korea

List of Contributors

Daniela Thrän
Helmholtz-Zentrum für
Umweltforschung – UFZ
Permoserstr. 15
04318 Leipzig
Germany

Vanessa Tietze
Forschungszentrum Jülich GmbH
IEK-3 Institut für En. &
Klimaforschung
Wilhelm-Johnen-Str.
52428 Jülich
Germany

Hirohisa Uchida
Tokai University
Department of Nuclear Engineering
1117 Kitakaneme, Hiratuka-shi
Kanagawa 259-1292
Japan

Kees van der Leun
ECOFYS GERMANY
Am Wassermann 35
50829 Köln
Germany

Karsten Voss
Bergische Universität Wuppertal
Fachbereich D – Architektur
Campus – Haspel
Haspeler Str. 27
42285 Wuppertal
Germany

Jüergen Wackerl
Forschungszentrum Jülich GmbH
IEK-3 Institut für En. &
Klimaforschung
Wilhelm-Johnen-Str.
52428 Jülich
Germany

Michael Weber
Forschungszentrum Jülich GmbH
IEK-3 Institut für En. &
Klimaforschung
Wilhelm-Johnen-Str.
52428 Jülich
Germany

Zhang Xiliang
Institute of Energy, Environment and
Economy
Tsinghua University
Beijing 100084
China

Jiang Yi
Tsinghua University
Department of Building Science and
Technology
Beijing 100084
P.R. China

Li Zhao
Forschungszentrum Jülich GmbH
IEK-3 Institut für En. &
Klimaforschung
Wilhelm-Johnen-Str.
52428 Jülich
Germany

Part I
Renewable Strategies

1
South Korea's Green Energy Strategies

Deokyu Hwang, Suhyeon Han, and Changmo Sung

1.1
Introduction

The purpose of this chapter is to present an overview of South Korea's green energy strategies and policy goals set under the National Strategy for Green Growth: (1) government-driven strategies and policy towards green growth; (2) to narrow down the focus and concentrate on R&D for a new growth engine; and (3) to promote renewable energy industries.

The Republic of Korea is the world's fifth largest importer of oil and the third largest importer of coal [1] (see Table 1.1). Our green growth plan is to increase the share of new and renewable energy in the total energy supply from 2.7% in 2009 to 3.78% in 2013; we aim to double that share to 6.08% by 2020 and 11% by 2030 (Figure 1.1). The statistics of energy consumption from 2000 to 2010 in South Korea are presented in Table 1.2. The energy policy has focused on dealing with oil prices and supply during the post-oil shock period in the mid-1970s [2], but today's energy policy includes the plan and actions for addressing climate change and environment

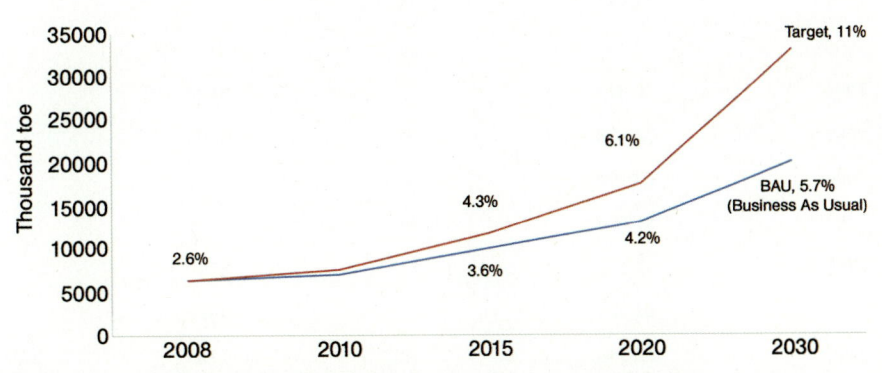

Figure 1.1 A scenario of renewable energy utilization plan from 2008 to 2030; toe, tonnes of oil equivalent. Source: MKE [3].

Transition to Renewable Energy Systems, 1st Edition. Edited by Detlef Stolten and Viktor Scherer.
© 2013 Wiley-VCH Verlag GmbH & Co. KGaA. Published 2013 by Wiley-VCH Verlag GmbH & Co. KGaA.

Table 1.1 Producers, net exporters, and net importers of crude oil, natural gas, and coal.

Oil		Gas		Coal	
Net importers	Mt	Net importers	Bcm[a]	Net importers	Mt
United States	513	Japan	116	China	177
China	235	Italy	70	Japan	175
Japan	181	Germany	68	*South Korea*	*129*
India	164	United States	55	India	101
South Korea	*119*	*South Korea*	*47*	Taiwan	66
Germany	93	Ukraine	44	Germany	41
Italy	84	Turkey	43	United Kingdom	32
France	64	France	41	Turkey	24
The Netherlands	60	United Kingdom	37	Italy	23
Singapore	57	Spain	34	Malaysia	21
Others	483	Others	279	Others	213
Total	2053	Total	834	Total	1002

a Billion cubic meters.
Source: IEA [1].

Table 1.2 Statistics of energy consumption (thousand toe) from 2000 to 2010 in South Korea.

Year	Electricity	Heat	Renewable energy
2000	20 600	1119	2130
2001	22 165	1150	2456
2002	23 947	1223	2925
2003	25 250	1300	3210
2004	26 840	1343	3928
2005	28 588	1530	3896
2006	29 990	1425	4092
2007	31 700	1438	4491
2008	33 116	1512	4747
2009	33 925	1551	4867
2010	37 338	1718	5346

Source: MKE [13].

protection and securing energy resources. The Korean government has strategically emphasized the development of 27 key national green technologies in areas such as solar and bio-energy technologies, and pursued the target through various policy measures, such as the Renewable Portfolio Standard (RPS), waste energy, and the One Million Green Homes Project.

Thus, Korea's plan is to reduce carbon emissions, improve energy security, create new economic growth engines, and improve the quality of life based on green technologies.

In August 2008, Korean President Lee announced a "low-carbon, green growth" strategy as a new vision to guide the nation's long-term development. Five months later (January 2009), the Korean government responded to the deepening recession with an economic stimulus package, equivalent to US$ 38.1 billion, of which 80% was allocated towards the more efficient use of resources such as freshwater, waste, energy-efficient buildings, renewable energies, low-carbon vehicles, and the railroad network. In July 2009, a Five-Year Plan for Green Growth was announced to serve as a mid-term plan for implementing the National Strategy for Green Growth between 2009 and 2013, with a fund totaling US$ 83.6 billion, representing 2% of Korea's GDP. It was expected to create 160 000 jobs in the green sector, providing opportunities for both skilled and unskilled labor; the forecast rate was ~35 000 additional jobs per year between 2009 and 2013 [4].

The national goals had been established through strategies and policies such as the Presidential Committee on Green Growth [4], the National Basic Energy Plan and Green Energy Industry Development Strategy [5], the Basic Act on Low Carbon Green Growth and Related Legislation [6], and National Strategy and Five-Year Implementation Plan [4]. Eventually, the goal for Korea is to move away from the traditional "brown economy" to a "green economy" model where long-term prosperity and sustainability are the key objectives.

1.2
Government-Driven Strategies and Policies

In an effort to push forward the national goals, the Presidential Committee on Green Growth (PCGG) [4] was launched to facilitate collaboration in deliberating and coordinating various green growth policies across ministries and agencies. Green growth committees were also set up under local governments. Both the central government and local governments worked out 5 year green growth plans and have invested 2% of the GDP annually. Also, the Korean government was the first in the world to lay the groundwork for the continued pursuit of green growth by enacting the Framework Act on Low Carbon, Green Growth. It paved the way for reducing greenhouse gas (GHG) emissions in a groundbreaking manner through a market system by legislating the Greenhouse Gas Emissions Trading Act, supported across various political parties. Thus the government prepared the legal and institutional groundwork and also the framework for putting green growth as the new paradigm for national progress into practice.

The National Basic Energy Plan [7] established specific measures to increase energy efficiency, decrease energy intensity, and achieve the target to increase the renewable energy portfolio to 11% by 2030. The government plans on reaching this target by implementing programs such as the Smart Grid, the Two Million Homes strategy (which aims to have two million homes run on a mix of renewable energy resources by the end of 2018) and an 11 year renewable energy portfolio standard (RPS), which will replace the Renewable Portfolio Agreement (RPA) and feed-in tariffs (FITs) currently in operation by 2012. In 2005, the Ministry of Knowledge Economy (MKE)'s predecessor, the Ministry of Commerce, Industry and Energy, established the RPA, signing an agreement with the nine largest energy suppliers to provide financial support of US$ 1.1 billion between 2006 and 2008 and administrative support for clean and renewable energy projects. The aim was to increase the use of clean and renewable energy in the industrial sector and reduce 170 000 tons of GHG FIT regulations mandate electricity utilities to buy electricity generated by clean and renewable energy at a price fixed by the government, which then compensates the utility to offset the difference in price from conventional energy supplies. It has been noted that the FIT market-based instrument has been the driver behind the increased supply of clean and renewable energy in the nation but has also been criticized as being anti-competitive and causing difficulty in forecasting electricity generation. Because of this, the government planned to replace the FIT in 2012 with the RPS that will mandate utilities to generate a specific amount of clean and renewable energy. The RPS will be operated by the MKE and will mandate utilities with generation capacity over 2000 MW to obtain certain amount of renewable energy. The amount of renewable generation mandated was 2% in 2012, increasing to 10% in 2022. Participants will be able to meet their quotas either by buying renewable energy certificates (RECs) from independent power providers, or by earning RECs through their own generation. The expected share of the individual green energy sources for the 11% for 2030 is illustrated in terms of photovoltaics (PV), wind, bioenergy, and so on in Table 1.3.

There were two approaches leading this green energy technology effort: (1) select 27 key green technologies to concentrate on while bridging the technology gap, and (2) establish an assistance program for green technology R&D to lead emerging green technology for the future. The Green Energy Industry Development Strategy focused on both early growth engine technologies, such as PV, wind, smart grids and LEDs, and next-generation growth engines, including carbon capture and storage, fuel cells, and integrated gasification and combined cycle technologies.

PCGG developed the legislative framework for green growth, called the Basic Act on Low Carbon Green Growth. In January 2010, the Korean President signed and promulgated this Act, which mandated a target for GHG emission reductions, renewable energy supply, and energy savings and security.

Table 1.3 Prediction of renewable energy demand (thousand toe) and (in parentheses) the expected share of the individual green energy sources (%).

Energy	2008	2010	2015	2020	2030	Average annual increase (%)
Solar thermal	33 (0.5)	40 (0.5)	63 (0.5)	342 (2.0)	1882 (5.7)	20.2
PV	59 (0.9)	138 (1.8)	313 (2.7)	552 (3.2)	1364 (4.1)	15.3
Wind	106 (1.7)	220 (2.9)	1084 (9.2)	2035 (11.6)	4155 (12.6)	18.1
Bioenergy	518 (8.1)	987 (13.0)	2210 (18.8)	4211 (24.0)	10357 (31.4)	14.6
Water power	946 (14.9)	972 (12.8)	1071 (9.1)	1165 (6.6)	1447 (4.4)	1.9
Geothermal	9 (0.1)	43 (0.6)	280 (2.4)	544 (3.1)	1261 (3.8)	25.5
Marine	0 (0.0)	70 (0.9)	393 (3.3)	907 (5.2)	1540 (4.7)	49.6
Waste	4688 (73.7)	5097 (67.4)	6316 (53.8)	7764 (44.3)	11021 (33.4)	4.0
Total	6360	7566	11731	17520	33027	7.8
Total primary energy supply (million toe)	247	253	270	287	300	0.9
Ratio (%)	2.58	2.98	4.33	6.08	11.0	

Source: MKE [3].

1.3
Focused R&D Strategies

For the growth of renewable energy, strategic R&D is required. The Korean government has identified renewable energy as its next engine for growth by focusing on selected R&D investments and increasing its budget (Figure 1.2 and Table 1.4). In 2011, the MKE announced the strategy of renewable energy R&D [8] by selecting five core sectors for power generation technologies: PV, wind power, bioenergy, coal, and fuel cells.

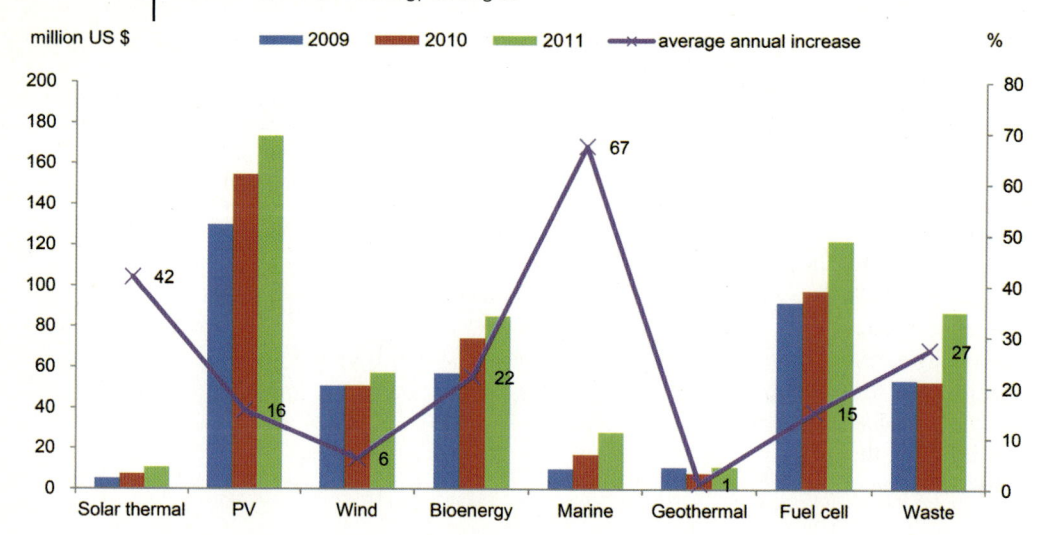

Figure 1.2 R&D budget of renewable energy in South Korea.

Table 1.4 R&D budget of renewable energy in South Korea.

Energy	Budget (million US$)			Average annual increase (%)
	2009	2010	2011	
Solar thermal	5	7	11	42
PV	130	155	173	16
Wind	51	51	57	6
Bioenergy	57	74	85	22
Marine	10	17	28	67
Geothermal	11	8	11	1
Fuel cell	92	98	122	15
Waste	54	53	87	27
Total	404	455	564	18

Source: GTC-K [9].

The commercial and technical feasibility of renewable energy requires a considerable level of R&D and field demonstration. It also requires a fully integrated approach across many interdepartmental agencies. For example, offshore wind projects [10] have been planned for both South Korea's southwest coast and the southern island of Jeju. Its target has been to generate 100 MW offshore wind capacity by 2013, and to achieve 600 MW by 2016 and 2.5 GW by 2019. This included not only an increased R&D budget, but also an intensive field demonstration project plan for global applications.

1.4
Promotion of Renewable Energy Industries

Based on a consensus among public and private stakeholders, the national strategy for renewable energy envisaged three main directions: (1) a technology roadmap, (2) dissemination and commercialization of technologies, and (3) promotion of export and revenue growth. The Technology Roadmap [3] for green energy placed periodic goals for the industrialization of technology development in three phases: phase I (2008–2010), phase II (2011–2020), and phase III (2021–2030). This roadmap linked a product from commercialization to the global market. From the 27 green technologies, several core technologies were selected to promote global market domination through renewable energy convergence strategies. In the past, the government has driven the dissemination and commercialization of technologies [3]; however, the policy was changed to promote private sector-led approaches for competitiveness. This was mainly because the government-driven policy appeared to limit the effectiveness of performance.

For the past few years, supporting strategies for export and business growth [11] have been successful. For example, major energy firms with both financial assurance and tax support mechanisms successfully built a system for corporate growth. Today, private sector participation has been promoted actively through drastic regulatory improvements. The Korean government has established the "Reregulation Support Centre" to support SMEs entering overseas markets. As a result, Korea's relative clean technology ranking has improved from eighth to fifth, being one of the world's top five fastest climbers (Figure 1.3).

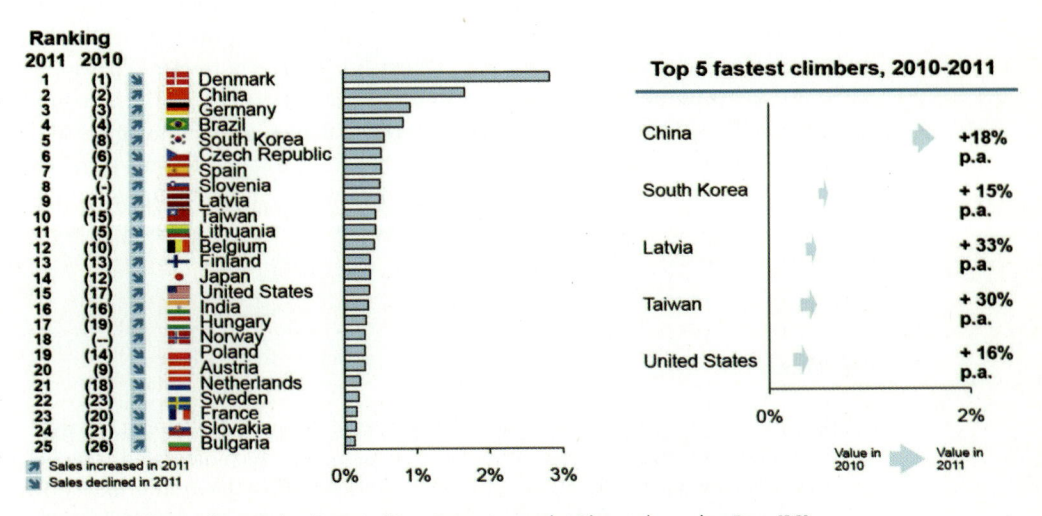

Figure 1.3 Relative clean technology ranking. Source: van der Slot and van den Berg [12].

1.5
Present and Future of Green Energy in South Korea

Although there is still much to be accomplished, South Korea has successfully pushed its green technology initiatives in the last 4 years. With continued support by the government and private sector, South Korea should expect further momentum with the changing economic landscape, as green industries are emerging as a new growth engine. As a consequence, the green industry has been growing rapidly, and the export of green products is rising sharply. The government's efforts in expanding R&D in green technology has transformed the way in which companies have invested in top-ranking green technologies, which is now attracting the attention of the global markets.

With regard to the international proliferation for green growth, South Korea will increase its green Official Development Assistance (ODA) to more than US$ 5 billion from 2013 to 2020. South Korea's green ODA will shift to the Global Green Growth Partnership. To support green growth systematically in developing countries, South Korea will work through the Global Green Growth Institute (GGGI), which was founded in June 2010. The Institute will expedite cooperation between developing and developed nations, while encouraging partnerships between the private and public sectors. In this way, developing countries will receive the necessary policy support, in addition to skills and know-how, more efficiently.

In March 2012, the Green Technology Center Korea (GTC-K) was launched to become the hub for technical cooperation needed to support green growth in the developing world. GTC-K is also responsible for training and educating international experts in relevant fields. The GGGI will be the centerpiece of the global green growth strategy, whereas GTC-K will be the technology arm of green growth in developing countries. The Green Climate Fund was created as the result of the United Nations Climate Change Conference in Durban, South Africa, in December 2011. The fund will provide financial resources for green growth strategies and technologies. With strategy, technology, and finance addressed in this green triangle, it is the hope that green growth initiatives will escalate. South Korea will continue to strive to build on this green triangle, so that this architecture can be utilized by all developing and developed countries.

References

1 International Energy Agency (2012) *Key World Energy Statistics. Statistics Book*, IEA, Paris, pp. 11–15.
2 Lee, H. J. and Won Yoon, S. W. (2010) *Renewable Energy Policy in Germany and Its Implications for Korea*, Korea Institute for Industrial Economics and Trade (KIET), Seoul, p. 13.
3 Ministry of Knowledge Economy (2008) *The 3rd Basic Renewable Energy R&D and Utilization Plan*. Government Report, Ministry of Knowledge Economy, Seoul.
4 Presidential Committee on Green Growth (2009) *National Strategy and Five-Year Implementation Plan (2009–2013)*. Government Report, Presidential Committee on Green Growth, Seoul.

5 Ministry of Knowledge Economy (2008) *Green Energy Industry Development Strategy*. Government Report, Ministry of Knowledge Economy, Seoul.
6 (2010) *Basic Act on Low Carbon Green Growth and Related Legislation*.
7 Prime Minister's Office (2008) *National Basic Energy Plan (2008–2030)*. Government report, Prime Minister's Office, Seoul.
8 Ministry of Knowledge Economy (2008) *Renewable Energy R&D Strategy*. Government Report, Ministry of Knowledge Economy, Seoul.
9 Green Technology Centre Korea (2012). *National Green Technology R&D Analysis*, Green Technology Centre Korea, Seoul.
10 Ministry of Knowledge Economy (2010) *Offshore Wind Projects*. Government Report, Ministry of Knowledge Economy, Seoul.
11 Ministry of Knowledge Economy (2012) *Renewable Energy R&D and Utilization Action Plan*. Government report, Ministry of Knowledge Economy, Seoul.
12 van der Slot, A. and van den Berg, W. (2012) *Clean Economy, Living Planet*, Roland Berger Strategy Consultants, Amsterdam, p. 6.
13 Ministry of Knowledge Economy (2011) *Yearbook of Energy Statistics*, Ministry of Knowledige Economy, Seoul.

2
Japan's Energy Policy After the 3.11 Natural and Nuclear Disasters – from the Viewpoint of the R&D of Renewable Energy and Its Current State

Hirohisa Uchida

2.1
Introduction

On 11 March 2011, Japan was hit by one of the most powerful earthquakes in recorded history, with a magnitude 9. That earthquake shifted the northeast part of the Japanese islands by 5 m to the east on average. The earthquake induced massive tsunamis, and these natural disasters killed 15 882 people, and forced 324 858 people to evacuate from their homes. A further 2668 are still missing (as of 11 March 2013). The earthquake induced tsunamis higher than 10 m, and these tsunamis attacked four reactors of the Fukushima Dai-Ichi nuclear power plant (FDNPP) of Tokyo Electric Power Co., Inc. (TEPCO). The seawater attacked emergency cooling water pumps, and swept away diesel oil tanks for the emergency power generators. In addition, the earthquake destroyed electric towers and grids supplying electricity from outside to the plant. The whole plant was in total blackout. Subsequently, the cooling water level inside the reactors decreased, and the fuel rods in reactors Nos. 1, 2, and 3 were exposed to the air, resulting in meltdown. Judging from raised reactor temperatures and deficient amounts of cooling water, the fuel rods probably melted through the reactor pressure vessels. All cooling water systems were destroyed, and TEPCO decided to pour seawater into the reactors. Later, fresh water was poured into the reactors from outside using pumps and helicopters, and circulated by additional pumps. Still nobody can enter these reactors, only robots, because of the extremely high radioactivity, and no exact investigation has been made on the accident process of each reactor. On 24 November 2012, the radioactivity inside the three reactors was around 10 000–70 000 mSv h^{-1}, and the amount of contaminated water inside each reactor was estimated to be 1.4×10^4–2.3×10^4 t. Before the attacks of the earthquake and tsunamis, the No. 4 reactor was out of operation for inspection and no fuel rods were fitted in the reactor vessel. Therefore, the radioactivity level there is as low as 0.1–0.6 mSv h^{-1}. However, the 1535 fuel rods used are maintained in a water pool inside the No. 4 reactor. The water pool was critically damaged by the earthquake, and is maintained by temporary repairs inside the reactor. Intentional venting of the

Transition to Renewable Energy Systems, 1st Edition. Edited by Detlef Stolten and Viktor Scherer.
© 2013 Wiley-VCH Verlag GmbH & Co. KGaA. Published 2013 by Wiley-VCH Verlag GmbH & Co. KGaA.

reactors was performed to reduce internal pressure. However, hydrogen gas seemed to have been produced inside the reactors by radioactive rays and/or by reactions of water vapor with the metal surface of the rods, and hydrogen explosions took place. In addition, radioactively contaminated water leaked out of the reactors. These incidents scattered huge amounts of radioactive substances into the air, forest, soil, ground water and sea. More than 150 000 people (93 864 from Fukushima Prefecture, and 58 608 from other Prefectures near Fukushima) are still evacuated from their homes (as of 9 March 2013). The exact number of evacuees is still unknown because more people left independently of governmental evacuation orders. The government estimates that the cost of the damage caused by the FDNPP accident will rise to over JPY 20×10^{12} in the next 20 years.

In Japan, electric power cannot easily be interchanged among electric power companies because of the different power frequencies, 50 and 60 Hz in eastern and western Japan, respectively. Therefore, TEPCO area scheduled (rolling) blackouts. The unification of the power frequencies among the electric companies would surely contribute to power interchange and stabilization throughout Japan.

After a public opinion poll in August 2012, 70% of the Japanese people voted against the further use of nuclear power (NP). Based on that result, the government announced a new policy, the "Green Energy Revolution," as mentioned in Section 2.3.2. This should be reflected in a new governmental basic energy plan to be published in the near future.

When the low energy self-sufficiency of 4% is taken into account, NP has been important for Japan, and dominated around 30% of the total electric power configuration until 2010 (Figure 2.1). However, the FDNPP accident clearly demonstrated, first, that the cost of NP generation becomes extremely high on including damage costs compared with other power generations, and second, that NP generation is highly risky for human and social security. An official report on the accident was published in English by the Japanese National Diet legislature body [1]. This chapter reviews Japan's historical energy transition and the governmental energy policy plans proposed, and reports on the current state of research and development (R&D) of renewable energy (RE) and hydrogen technology in Japan.

2.2
Energy Transition in Japan

About 96% of the total energy supply in Japan depends upon imports from abroad. Therefore, the diversification of energy resources has been inevitable in order to secure energy supply security. In this section, Japan's energy transition is described and discussed.

2.2.1
Economic Growth and Energy Transition

From the 1960s to the 1970s, Japan experienced rapid economic growth thanks to crude oil at a low price. However, frequent occurrences of regional conflicts in the Middle East steeply raised the oil price from US$ 3 per barrel to more than $ 30 per barrel. Because of these oil problems, Japan started to strengthen its fragile energy supply structure by introducing NP, coal and liquid natural gas (LNG) as alternatives to oil. The Ministry of International Trade and Industry (MITI) launched the Sunshine Project in 1974 with R&D on new energy technologies such as coal liquefaction, use of geothermal heat, photovoltaic (PV) energy generation, and hydrogen production and utilization. During that period of rapid economic growth, Japan suffered heavily from air pollution due to oxidant smog and water pollution due to industrial activities. People's awareness of the environment increased, and Japan focused on R&D on new and clean energy. From 1993 to 2000, the New Sunshine Project was introduced with R&D on large-scale wind power generation, fuel cells, applications of superconducting materials, ceramic gas turbines, and power storage technology. The current R&D on RE such as PV power generation, wind power generation, and geothermal energy, started in that project.

2.2.2
Transition of Power Configuration

During Japan's economic growth since the 1960s, the amount of electric power generated increased steadily: 1×10^{11} kWh (1960), 2.939×10^{11} kWh (1970), 4.850×10^{11} kWh (1980), 7.386×10^{11} kWh (1990), 9.396×10^{11} kWh (2000), 10.064×10^{11} kWh (2010) [2] (Figure 2.1).

Electricity generation has dominated with about 40% of the total primary energy supply. Electric power was supplied in the 1960s by oil/coal-fired thermal and hydroelectric power, in the 1970s by oil/coal-fired thermal power, in the 1980s–1990s by oil/coal/LNG-fired thermal power and NP, and in 2000s by oil/coal/NG-fired thermal power and NP. In 1966, the first NP plant started electricity generation in Japan. From 1985 to 2010, NP generation dominated with about 30% of the power configuration. Figure 2.1 shows a sudden decrease in NP generation in 2011, which reflects the effect of the shutdown of NP reactors. After the FDNPP accident, NP reactors were gradually shut down, and by May 2012 all 54 reactors had been totally shut down. At present, only two PWR reactors (118×10^4 kW each) in Oi, Fukui Prefecture, have been in operation since July 2012. New energy types in the configuration include RE and heat storage, biomass, and waste power. The proportion of RE was as low as 0.3–0.4% in the configuration in 2011 (Figure 2.2). When the fact that Japan's energy self-sufficiency is only 4% is taken into account, the R&D and the active use of NP so far have been well grounded from the viewpoint of deprived energy resources in Japan.

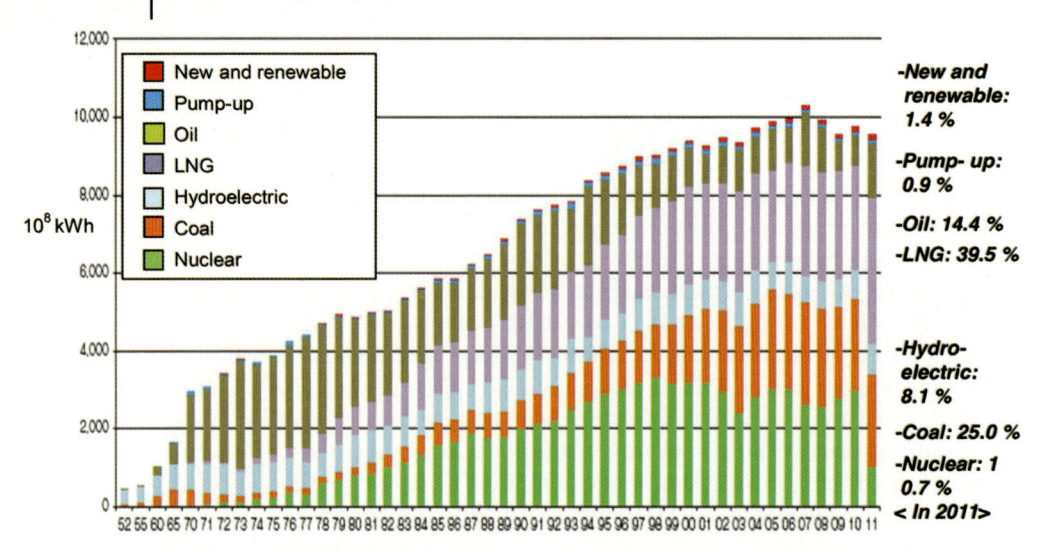

2 Japan's Energy Policy After the 3.11 Natural and Nuclear Disasters

Figure 2.1 Change in the electric power configuration in Japan from 1952 to 2011 [2]. The figures shown are based on the output record of the power companies. The contribution of private generation by enterprises and cogenerations is excluded.

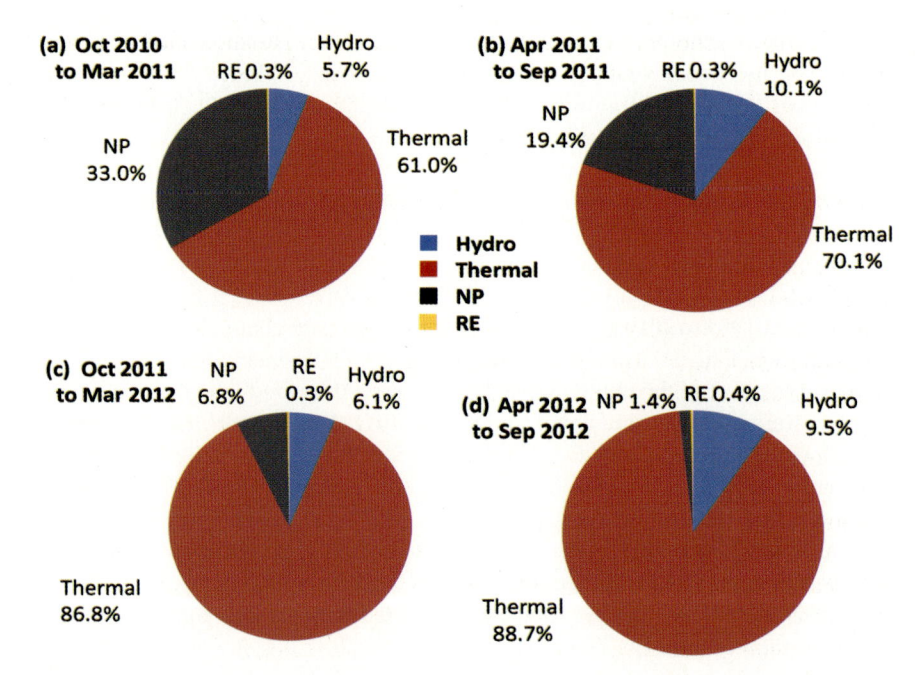

Figure 2.2 The power configurations in Japan before and after the FDNPP accident on 11 March 2011.

2.2.3
Nuclear Power Technology

NP generation has been said to be clean, without CO_2 emissions, and friendly to the environment in respect of global climate change. These catchphrases accelerated the active use of NP in Japan and other countries. However, the FDNPP accident confronted us with the terrible fact that the accident contaminated the Japanese islands extensively and violently disturbed human life and social systems. This should be seriously considered in terms of "clean NP" by other countries considering the introduction and use of NP.

Figure 2.2 shows the power configuration situation before and after the accident on 11 March 2011. The share of NP was 33% before the accident (Figure 2.2a), and then decreased to (b) 19.4%, (c) 6.8%, and (d) 1.4% in April–September 2012 as NP reactors were gradually shut down. The power generation by TEPCO was $26\,407 \times 10^4$ MWh in 2010, the highest compared with those of the eight other Japanese electric companies, namely from 673×10^4 to $13\,152 \times 10^4$ MWh. The power deficit due to the shutdown of NP has been made up mainly by coal/LNG thermal and hydroelectric power without NP since March 2012.

The reserve to production ratio of uranium is estimated to be about 100 years. However, this may be shortened as developing countries introduce more NP in the future. Therefore, NP generation is not exactly stable over a long time from the viewpoint of uranium resources.

The market of nuclear technology can be divided into two fields: (1) NP generation using fission reaction of uranium, and (2) application of radiation such as X-rays, γ-rays, and heavy ion beams to nondestructive inspection, industrial processing, agriculture, medical examination, and radiotherapy. The approximate economic size of each market in Japan is JPY 7×10^{12} per year for NP generation, and JPY 8×10^{12} per year for industrial and medical applications.

From the viewpoint of nuclear technology education, we need to foster young people who can contribute to the design of safer reactors, safe operation and management of reactors, and dismantlement of old reactors. In addition, we need people who can measure correctly the radioactivity of the environment and foods, and decontaminate radioactive substances in contaminated areas. Needless to say, R&D on robots, which work inside and dismantle broken reactors with extremely high radioactivity, is urgently needed in Japan.

2.3
Diversification of Energy Resource

International political conflicts often lead to unstable energy supplied to Japan. Therefore, diversification of energy resources has been essential. Japan may introduce energy alternatives to NP in the future. The solution may be the use of coal/NG thermal power and RE. In order to avoid risks with energy supply, the diversification of energy resources is a significant issue in Japan.

2.3.1
Thermal Power

The Japanese government declared the active introduction and use of thermal power in the realization of the "Green Energy Revolution" (see Section 2.3.2.2) where the dependence on NP is assumed will be reduced until 2030.

Fuji Electric and Siemens are constructing a gas turbine combined cycle (GTCC) thermal power station in Okinawa, Japan. The energy conversion efficiency of this GTCC is in the range 60–65%, which is higher by a factor 1.5 than that of conventional thermal power systems with efficiencies of 40%. In Japan, judging from energy density, thermal power instead of RE is expected to be the alternative to NP for the time being. In addition, the CO_2 emissions from coal/oil/NG thermal power generation have become extremely low owing carbon capture and storage (CCS) technology, which reduces CO_2 emission by 80–90% compared with conventional thermal power systems. In Japan, coal-fired thermal power plants are emitting 0.2 g SO_x kWh^{-1} and 0.2 g NO_x kWh^{-1} on average, and an advanced ultra-supercritical (USC) plant is emitting 0.02 g SO_x kWh^{-1} and 0.06 g NO_x kWh^{-1}. These emissions are 1–10% of the world average SO_x and NO_x emissions compared in 2006. In 2012, Tokyo Gas declared its intention to generate over 5×0^6 kW up to 2020 by constructing new LNG-fired thermal power plants with other enterprises in order to enter the electricity generation market.

The finding of shale gas has introduced a gas scenario in the world. However, a serious problem is the fact that Japan has been importing the most expensive NG in the world as LNG. The NG price to Japan is $ 14–15 MMBtu^{-1}, higher than Germany (NG) by a factor of 1.4 and the United States (NG) by a factor of 3.6. Increased production of shale gas may reduce the NG price in the future. For the time being, however, LNG thermal power generation as an alternative to NP may raise electricity charges in Japan. We have the choice of whether we will accept the increase in electricity charges by the active use of thermal generation in order to secure human security, or to use further NP to maintain economic and industrial activity.

2.3.2
Renewable Energy Policy by Green Energy Revolution

After the FDNPP accident, active introduction and rapid distribution of RE are strongly expected in Japan, although the present share of RE in the power configuration is as low as 0.3–0.4% (Figure 2.1 and Figure 2.2), or around 10% including hydropower. The Japanese government announced three options for NP in June 2012, and a new concept, the "Green Energy Revolution," in September 2012 for the rapid diffusion of RE. In any case, the government intends to reduce the dependence upon NP in the future.

2.3.2.1 Agenda with Three NP Options
In June 2012, the government published the agenda for the power configuration in 2030. The figures given in Table 2.1 are based on the actual power configuration

Table 2.1 The Japanese governmental agenda for the power configuration with three options for NP supply in 2030, under three NP options, 0%, 15%, and 25% (June 2012).

Power configuration	(2010)	Nuclear power option		
		0%	15%	20–25%
NP (%)	26	0	15	20–25
RE (%)	10	35	30	25–30
FE (%)	63	65	55	50

NP, nuclear power; RE, renewable energy including hydropower; FE, fossil energy such as coal, oil and LNG.

in 2010 before the FDNPP accident. Up to 2030, the government will introduce RE actively from 25 to 35% in the power configuration, independent of the NP dependence, as can be seen from Table 2.1.

2.3.2.2 Green Energy Revolution

On 14 September 2012, the government published an energy policy, the Green Energy Revolution, towards realizing a society independent of NP. In order to realize such a society in the 2030s, the following three regulations are introduced: (1) the maximum period of use of an NP reactor is strictly limited to 40 years, (2) the restart of the shutdown NP reactors is permitted only after confirmation by the Nuclear Regulation Authority (NRA), Japan, and (3) no additional new NP plant will be constructed for the future. For the realization of the Green Energy Revolution, the government set up the following three principles:

1. active energy saving;
2. rapid diffusion of RE;
3. cogeneration by the effective use of waste heat and power.

The government images for the principles are summarized in Table 2.2, Table 2.3, Table 2.4 and Table 2.5, where the data are compared with the actual data in 2010. In these tables, the government assumes the 15% NP option in Table 2.1.

Table 2.2 The Japanese governmental image for energy saving until 2030 (September 2012).

Energy	(2010)	2015	2020	2030
Power generation (10^8 kWh)	11 000	−250	−500	−1100
Power plant capacity (kl crude oil)	3.9×10^8	-1600×10^4	-3100×10^4	-7200×10^4

Table 2.3 The Japanese governmental image for the introduction of RE until 2030 (September 2012).

Energy	(2010)	2015	2020	2030
Power generation (10^8 kWh)	1 100	1 400	1 800	3 000
Power plant capacity (10^4 kW)	3 100	4 800	7 000	13 00
Power generation without hydropower (10^8 kWh)	250	500	800	1 900
Power plant capacity without hydropower (10^4 kW)	900	2 700	4 800	10 800

Table 2.4 The Japanese governmental image for the introduction of RE in 2030 under three NP options, 0%, 15%, and 25% (Ministry of Economy, Trade, and Industry, October 2012) [3].

Power	Total RE generation (10^8 kWh)			
	(2010)	NP option		
		0%	15%	25%
	1060 (9.7%)	3500 (37%)	3000 (31%)	2500 (26%)
Solar PV	38 (0.3%)	721 (7%)	666 (7%)	561 (6%)
Wind	43 (0.4%)	903 (9%)	663 (7%)	333 (3%)
Geothermal	26 (0.2%)	272 (3%)	219 (2%)	168 (2%)
Hydropower	809 (7.4%)	1200 (12%)	1095 (11%)	1095 (11%)
Biomass, etc.	144 (1.3%)	350 (4%)	328 (3%)	328 (3%)

The percentages in parentheses are the breakdown of the total RE generation.

Table 2.5 The Japanese governmental image for the introduction of cogeneration by the effective use of waste heat and power until 2030 (September 2012).

Energy	(2010)	2015	2020	2030
Power generation (10^8 kWh)	300	400	600	1500
Power plant capacity (10^4 kW)	900	1200	1500	2500

Active Energy Saving

The government will actively promote energy saving, as shown in Table 2.2. Typical actions are the active introduction of smart meters, a home energy management system (HEMS), a building energy management system (BEMS), the rapid diffusion of next-generation vehicles such as hybrid and fuel cell (FC) vehicles, and the active utilization of waste heat.

Rapid Diffusion of RE

The government vision for the introduction of RE in 2015, 2020, and 2030 is shown in Table 2.3. The breakdown for the introduction of RE in 2030 under three NP options, 0, 15, and 25%, is given in Table 2.4. In 2010, RE contributed 9.7% of the total electric power configuration. Table 2.4 shows the breakdown of each RE power source, namely PV power, wind power, geothermal power, hydropower, and biomass/waste energy, in 2030. The introduction of a feed-in tariff (FIT) for RE is promoting the rapid spread of solar PV systems, as mentioned in Section 2.3.2.3. For the further introduction of wind energy, removal of restrictions is necessary from the viewpoint of environmental assessment. Otherwise, conventional environmental restrictions would delay the spread of wind energy in Japan.

Cogeneration

Cogeneration by the effective use of waste heat is essential for energy saving. Table 2.5 shows the government's intention for the introduction of cogeneration up to 2030. As an example, the production of cold water (273–278 K) and the manufacture of a freezer (down to 243 K) have been successfully achieved by the effective use of waste heat and the reaction heats of hydrogen storage alloys by cyclic hydriding and dehydriding [4]. Such a system can save energy and CO_2 emissions by more than 80% compared with a conventional cooling freezer system.

R&D on the use of waste heat from industrial facilities, power generators, and incinerators is attracting high interest. Further investigations are advancing in the utilization of high-temperature heat from the Sun and low-temperature heat from rivers, drainage, snows and ice.

2.3.2.3 Feed-in Tariff for RE

Since the government introduced a FIT for RE in July 2012 [5], large-scale PV systems with levels of several tens of megawatts are being actively constructed throughout Japan. Recently, the price of PV systems has been rapidly decreasing: the construction cost of a PV system has fallen from JPY 10×10^8 MW^{-1} to JPY 3×10^8 MW^{-1} system. Since Japan has only small areas of available land, the diffusion of large-scale wind energy systems is slower than that of PV systems. The construction of floating types of offshore large-scale wind energy systems is being planned in Japan.

The present FIT is valid until the end of March 2013, then it is to be revised each year. Typical tariffs and durations are as follows: PV, JPY 42 kWh^{-1} for output over 10 kW (20 years) and output below 10 kW (10 years); wind, JPY 23.1 kWh^{-1} for output over 20 kW (20 years) and JPY 57.75 kWh^{-1} for output below 20 kW (20 years); small- and medium-scale hydroelectric, JPY 25.2 kWh^{-1} for output from 1000 to 3,000 kW

(20 years), JPY 30.45 kWh^{-1} for output from 200 to 1000 kWh (20 years), and JPY 35.7 kWh^{-1} for output below 200 kW (20 years); and geothermal, JPY 27.3 kWh^{-1} for output over 15 000 kW (15 years) and JPY 42.0 kWh^{-1} for output below 15 000 kW (15 years). The classification of the FIT of biomass ranges from JPY 13.65 kWh^{-1} to 40.95 kWh^{-1} for 20 years according to sources such as biogas, woody materials, and waste and recycled woods [5].

2.3.3
Renewable Energy and Hydrogen Energy

RE such as PV and wind power is clean and safe, but fluctuating and intermittent. The unstable and low-density output of RE should be stored as electric power in rechargeable batteries or as hydrogen using hydrogen storage materials [6, 7]. At present, many large-scale PV systems are being constructed in order to connect output to conventional grids to sell generated power. However, in such a case, unstable and intermittent output should be controlled by peak cut or peak shift, as is well known. By the storage of fluctuating and intermittent RE output, the stored energy can be used independent of weather conditions, and contribute to peak cut or peak shift according to demand. The storage of RE is indispensable for smart grid networks and higher utilization efficiency. The storage of RE as hydrogen has the great advantages that hydrogen can be easily and safely stored in hydrogen storage alloys at low pressures on a large scale, or can be transported as H_2 gas using pipelines over long distances. If output electricity from RE has to be transported over a long distance, the energy utilization rate becomes low due to energy loss. In principle, RE should be produced and used in each local area, so that the output energy can be utilized effectively. The combination of RE with rechargeable battery or hydrogen storage will contribute to smart grids and smart communities for efficient energy use and energy saving.

Hydrogen can be produced by electrolysis, independent of electricity sources. The produced and stored hydrogen can generate electricity using FC systems. Thus, electricity and hydrogen are compatible with each other. This high compatibility between electricity and hydrogen will create tremendously diverse ways of efficient utilization of RE.

2.3.4
Solar–Hydrogen Stations and Fuel Cell Vehicles

Honda has been testing and demonstrating solar–hydrogen stations for fuel cell vehicles (FCVs) in the United States and Japan (Figure 2.3). This is a typical practical application of the efficient use of solar and hydrogen energy.

At present, using a 10 kW PV system, 1.5 kg of hydrogen gas per day with a pressure of 35 MPa is produced inside a high-pressure electrolyzer, and can be directly supplied to the vehicle. This Honda FCV runs 100 km per kilogram of H_2, and 620 km with a full charge. If required, using an inverter, this FCV can supply AC power to external equipment with an output of 9 kW over a 7 h period. Solar–hydrogen power stations may spread rapidly as the number of FCVs is increasing.

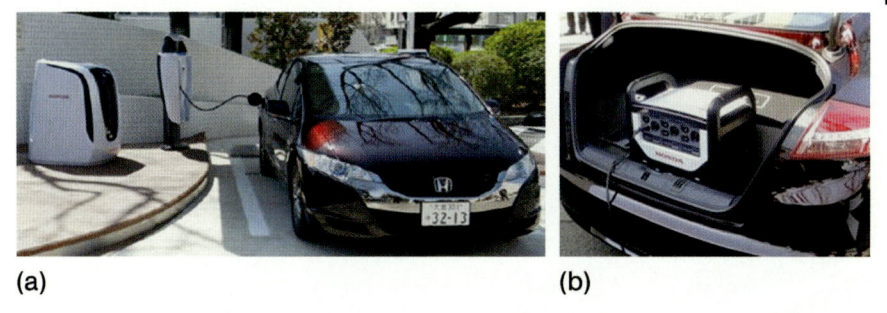

Figure 2.3 A solar–hydrogen power station (Honda) for FCVs at Saitama Prefectural Office, Japan (a). The generated power from FCVs can be used as an AC power source to external systems using an inverter (b).

2.3.5
Rechargeable Batteries

Rechargeable batteries are used for power peak cut and/or peak shift by power storage, and are essential for the efficient use of unstable RE. Typical rechargeable batteries for power storage are sodium–sulfur (NaS), redox flow (RF), lithium (Li) ion, nickel–metal hydride (Ni–MH), and lead (Pb) cells in Japan. Typical costs per battery are JPY 4×10^4 kWh^{-1} for NaS, JPY 20×10^4 kWh^{-1} for Li ion, JPY 10×10^4 kWh^{-1} for Ni-MH, and JPY 5×10^4 kWh^{-1} for Pd. No cost for the RF battery is given because large-scale RF batteries are still at the stage of demonstration testing. The RF battery using vanadium ions for positive and negative electrodes is attracting high interest because of its high and stable power storage capacity. Sumitomo Electric Industries started demonstration testing of a 5 MWh solar power storage using an RF battery system in July 2012. NGK Insulators has applied the NaS battery to megawatt-class power storage. The NaS battery is used rather in industrial areas because it is operated at 573 K. For residential use, the Li ion battery with 6 kWh is commercialized. Typical features of the Ni–MH battery, such as high chemical stability and high cyclic charge–discharge durability under high current use and load, have been confirmed by many commercialized hybrid vehicles within in and outside Japan. The high diffusivity of H atoms inside the negative electrode of hydrogen storage alloys permits high current use, and the rare earth metal- based hydrogen storage alloys used for the negative electrode are responsible for the high chemical and cyclic stabilities [8–10]. Large-scale Ni–MH batteries are used for power storage and control in mobile and residential use because of their high safety and reliability under high load. Kawasaki Heavy Industries has developed Ni–MH rechargeable batteries with high capacities for power storage and control (Figure 2.4) [11].

Many applications using the Ni–MH rechargeable battery are demonstrating prominent results in mobile use such as light rail vehicles, trains, and hybrid vehicles, and in residential use such as smoothing of fluctuating RE. Many companies are constructing and demonstrating smart communities or towns where Li ion and the Ni–MH batteries are widely used for power storage and grid control.

Figure 2.4 Applications of GIGACELL Ni–MH rechargeable batteries (Kawasaki Heavy Industries)
(a) to power storage and control for monorail trains of Tokyo Monorail and (b) to power smoothing of a wind power system in Akita Prefecture: a 1500 kW wind power system combined and controlled with a 102 kWh Ni–MH battery since 2007 by Kawasaki Heavy Industries [11].

2.4
Hydrogen and Fuel Cell Technology

FC technology is at the stage of actual application and commercialization rather than R&D. Recent advances can be seen in the mobility and residential use of FC systems [4].

2.4.1
Stationary Use

The number of ENE FARMs, a residential FC system, reached over 40 000 by 2012. The high price of over JPY 250×10^4 per system is subsidized by about JPY 50×10^4–100×10^4 by the central and local governments. Electric generation by gas using FCs seems to have attracted greater attention after the FDNPP accident. The ENE FARM is operated with city gas, LPG or NG because these gases can easily be reformed into H_2 gas inside each FC system. A typical output is 0.7 kW per system. A solid oxide fuel cell (SOFC) type was added to the conventional polymer electrolyte fuel cell (PEFC) type in October 2011. The ENE FARM with SOFC is operated at 973 K and higher efficiency: 45% electric generation and 42% heat use. A combination of PV and ENE FARM is called "double generation" (JX Nippon Oil and Energy) where electricity generation from PV power is used during the day, and electricity from FCs, ENE FARM, is used during the night. The double generation is a subject of the FIT in Japan.

2.4.2
Mobile Use

Automobile technology is changing from combustion engine drive using oil to electric motor drive using electricity. The number of hybrid or plug-in hybrid vehicles with gasoline engines and electric motors is expanding. On the other hand, automobile companies will also produce electric vehicles (EVs). However, the cruising range of small-sizes EVs is limited to 150 km. Many automobile companies are developing EVs for short-range driving and FCVs for long-range driving. An FCV can run for 500–600 km with a full capacity of H_2. At present, FCVs are fueled with 350 MPa H_2, but this is expected to rise to 700 MPa in the near future.

Japanese companies involved with automobiles, oil, gas, and relevant machines are negotiating with local and central governments to introduce regulations for 700 MPa H_2 gas handling in filling stations and on the road. In April 2013, new H_2 filling stations (700 MPa) will be constructed in the center of the cities Ebina in Kanagawa Prefecture and Kaminokura in Aichi Prefecture. JX Oil and Energy has declared that these new filling stations will supply not only gasoline to conventional vehicles, but also H_2 to FCVs and electricity produced by FCs to EVs. Such stations are called "multi-fuel stations."

In 2015, the Japan Automobile Manufacturers Association (JAMA) will start to commercialize FCVs. Improvements to the infrastructure are urgently needed for the start of FCV commercialization. The number of H_2 filling station is planned at around 100 until 2015, then 1000 for 2×10^6 FCVs until 2025, and 3000 for 7×10^6 FCVs after 2030.

The selling price of H_2 is expected to decrease from JPY 145 Nm^{-3} to JPY 80 Nm^{-3} in the future. However, the price depends on the number of FCVs coming for filling per day. For the rapid diffusion of FCVs, and the enhancement of public acceptance of hydrogen technology, 18 private and public sectors established an organization, "The Research Association of Hydrogen Supply/Utilization Technology" (HySUT), in 2009. HySUT has been active in Japan in order to realize a hydrogen mobility society.

2.4.3
Public Acceptance

In introducing hydrogen technology and realizing a hydrogen society, public acceptance of hydrogen and relevant systems is of great importance. As mentioned in Section 2.3.3 and Section 2.3.4, FCVs can supply electric power to the outside. After the tsunami and FDNPP incidents, many people were forced to move to evacuation sites such as school buildings and large halls. The Honda FCV in the Section 2.3.4 can continuously supply electric power at more than 63 kWh, corresponding to use by a family for 6 days. In August 2012, Toyota demonstrated power supply from an FCV bus of 9.8 kW for over 50 h to an evacuation hall. Japan often experiences natural disasters such as typhoons, floods, and earthquakes, hence evacuation and survival training are routine for the Japanese. Electric power supply from FCVs may contribute considerably to the enhancement of public acceptance of hydrogen technology in Japan.

2.5 Conclusion

The FDNPP accident was a wakeup call to the Japanese people to rethink energy and human security. The Japan Buddhist Federation appealed for "a lifestyle without dependence on nuclear power" on 1 December 2011. A similar appeal was made by the priests of the Eiheiji temple, the centre of Zen meditation. Previously, such official appeals from Buddhist groups have never been made in Japan, indicating that the priests were greatly shocked at the nuclear power accident with its major influence on human life. Still over 150 000 people are forced to live far away from home because of the high local radioactivity since the FDNPP accident. A severe nuclear accident may take place in Japan again, since gigantic earthquakes are strongly expected along the Pacific side of Japan.

In the election for the Lower House of the Diet of Japan on 16 December 2012, the Liberal Democratic Party (LDP) held the majority. This resulted from the fact that the 11 different political parties except the LDP had been independently insisting on stopping or decommissioning the reactors, and that they could not obtain the legally required minimum number of votes. The LDP has been insisting on the further use of NP with the nuclear fuel cycle for the promotion of the economy, and will restart the NP reactors that were shut down after the FDNPP accident.

The International Atomic Energy Agency (IAEA) and the Japanese government held the Fukushima Ministerial Conference on Nuclear Safety at Fukushima on 15–17 December 2012. They stressed the necessity for strengthening nuclear safety, including emergency preparedness and response [12]. This point is very important when the fact that an NP accident is not solely a problem for the country where it occurs, but can also raise serious problems for neighboring countries, as we learned from the Chernobyl NP accident in 1986. An NP accident in China or Korea would immediately contaminate the Japanese islands via the prevailing westerly winds.

Independent of any political change, the active introduction of RE and hydrogen/FC technology is advancing in Japan because Japanese industries are eager to expand their market in these fields. This will lead to the creation of new jobs and the enhancement of industrial and economic activity. Energy transition in Japan proceeds slowly but steadily.

References

1. Kurokawa, K. (2012) *The Fukushima Nuclear Accident Independent Investigation Commission*, National Diet of Japan, Tokyo.
2. METI (2012) *Energy White Paper 2012*, Ministry of Economy, Trade, and Industry, Tokyo, http://www.enecho.meti.go.jp/topics/hakusho/2012/gaiyou_2012.pdf (last accessed 4 January 2013).
3. Niihara, H. (2012) Transition of energy policy – the 3.11 as a turning point, Plenary Lecture from the Agency of Natural Resources and Energy, METI, in the Seminar Energy Revolution from Kanagawa Prefecture, 29 October, Yokohama, http://www.pref.kanagawa.jp/cnt/f421075/#player24 (last accessed 4 January 2013).

4 Uchida, H. (2010) Policy and action programs in Japan – hydrogen technology as eco technology, *Schriften des Forschungszentrum Jülich, Energy & Environment, Plenary Talk Section*, vol. 78 (ed. D. Stolten), Forschungszentrum Jülich, Jülich, ISBN 978-3-89336-658-3, pp. 105–115.

5 METI (2012) *Feed-in Tariff Scheme in Japan*, Ministry of Economy, Trade, and Industry, Tokyo, http://www.meti.go.jp/english/policy/energy_environment/renewable/pdf/summary201207.pdf (last accessed 4 January 2013).

6 Huang, Y., Goto, H., Sato, A., Hayashi, T., and Uchida, H. (1989) Solar energy storage by metal hydrides. *Z. Phys. Chem. N. F.*, **164**, 1391–1395.

7 Uchida, H., Haraki, T., Oishi, K., Miyamoto, T., Abe, M., and Kokaji, T. (2005) A wind and solar hybrid energy storage system using nanostructured FeTi hydrogen storage alloy, in *Proceedings of the International Hydrogen Energy Congress and Exhibition, IHEC2005*, Center for Hydrogen Energy and Technologies (ICHET), Istanbul, pp. 41–44.

8 Hoshino, H., Uchida, H., Kimura, H., Takamoto, K., Hiraoka, H., and Matsumae, Y. (2001) Preparation of a nickel–metal hydride rechargeable battery and its application to a solar vehicle. *Int. J. Hydrogen Energy*, **26**, 873–877.

9 Matsumura, Y., Sugiura, L., and Uchida, H. (1989) Metal hydride electrodes compacted with organic compounds. *Z. Physik. Chem. N. F.*, **164**, pp. 1545–1549.

10 Uchida, H., Matsumoto, T., Watanabe, S., Kobayashi, K., and Hoshino, H. (2001) A paste type negative electrode using a $MnNi_5$ based hydrogen storage alloy for a Ni–metal hydride battery. *Int. J. Hydrogen Energy*, **26**, 735–739.

11 Kawasaki Heavy Industries (2012) *GIGACELL*, http://www.khi.co.jp/english/gigacell/index.html (last accessed 4 January 2012).

12 IAEA (2012) *IEAE and Japan Host Fukushima Ministerial Conference on Nuclear Safety*, www.iaea.org/newscenter/news/2012/fukushconference.html (last accessed 4 January 2012).

3
The Impact of Renewable Energy Development on Energy and CO_2 Emissions in China

Xiliang Zhang, Tianyu Qi and Valerie Karplus

3.1
Introduction

China has adopted targets for the deployment of renewable energy through 2020. Compared with many nations, China's targets are sizable in terms of both total installed capacity and the anticipated contribution of renewable energy to total electricity generation. An important objective of renewable energy development in China is to reduce CO_2 emissions and the reliance on imported energy by decoupling rising fossil energy use from economic growth over the next several decades. This decoupling will also have a positive impact on local air and water quality – environmental pollution is estimated to cost over 4% of GDP each year [1]. Emphasis on renewable energy is also designed to promote China's competitiveness as a leading global supplier of clean, low-cost renewable energy technologies. In this chapter, we quantify the impact of China's renewable energy targets on energy use from both renewable and fossil sources and also the impact on CO_2 emissions, which are outcomes of significant interest to policymakers in China.

Targets for renewable energy deployment form part of a broader set of energy and climate policies that China's central government has defined for the period through 2020. National goals have been set through 2020 for energy and carbon intensity[1]) reduction, and also for the contribution of nonfossil sources to total primary energy. These broad goals are then supported by measures that target increases in specific types of generation – targets applied specifically to wind, solar, and biomass electricity generation are the focus of this analysis. As officials begin to consider policies for the period beyond 2020, there is a strong need to understand how such supply-side targets for renewable energy could contribute to China's broader energy and climate policy goals. In order to understand what role renewable energy

---

1) Carbon intensity is the ratio of carbon dioxide emissions per unit of output, which in this case is economic activity measured as gross domestic product (GDP).

Transition to Renewable Energy Systems, 1st Edition. Edited by Detlef Stolten and Viktor Scherer.
© 2013 Wiley-VCH Verlag GmbH & Co. KGaA. Published 2013 by Wiley-VCH Verlag GmbH & Co. KGaA.

could play in achieving China's low-carbon development, we assess the impact of renewable energy targets.

This analysis is organized as follows. First, we discuss in detail recent developments in China's energy and climate policy, the expected contribution of renewable energy and related policies, and the status of renewable energy development in China. Second, we describe the model used in this analysis, the China-in-Global Energy Model (CGEM). We include a detailed discussion of how renewable energy is represented. Third, we describe the policy scenarios and how they are implemented in the modeling framework. Fourth, we present the results, which explore the impact of China's renewable energy targets on energy use, CO_2 emissions, and consumption under alternative economic growth and technology cost assumptions. Fifth, we discuss the relationship between China's renewable energy targets and the nation's long-term energy and climate policy goals.

3.2
Renewable Energy in China and Policy Context

3.2.1
Energy and Climate Policy Goals in China

China's energy and climate policy sets forth a national carbon intensity reduction target of 17% for the Twelfth Five-Year Plan (2010–2015). This target is consistent with the nation's commitment at the Copenhagen climate talks of achieving a 40–45% CO_2 intensity reduction by 2020, relative to a 2005 baseline. The Twelfth Five-Year Plan was the first time a CO_2 intensity target was included. Previous Five-Year Plans relied on energy intensity targets without including an explicit reduction target for CO_2 intensity. Looking forward, reducing CO_2 emissions remains an important energy-related policy goal alongside energy security, air quality improvement, and balancing economic development across rural–urban and east–west dimensions.

Alongside carbon and energy intensity goals, China also has a goal to increase the contribution of nonfossil energy (including renewable sources and hydro and nuclear energy) in total primary energy use. In 2010, the target was 10% (actual nonfossil energy was 9.1% in 2010), with the target rising to 11.4% in 2015 and 15% in 2020. The nonfossil energy goal is viewed as way to reinforce the goal of carbon reduction specifically through the deployment of low-carbon energy (and especially electricity) sources. While the nonfossil energy goal focuses on expanding the contribution of technology to CO_2 emissions reduction, broad mandates for improving industrial and building energy efficiency have also been strengthened and expanded during the Eleventh and Twelfth Five-Year Plans.

3.2.2
Renewable Electricity Targets

Broad targets for energy and carbon intensity, nonfossil energy, and energy efficiency are typically implemented by directly assigning responsibility for target implementation at the sectoral, industry, or company level. Renewable energy quotas belong to the category of measures expected to support the achievement of the government's overall carbon intensity and nonfossil energy goals. China's *National Renewable Energy Law of 2006* provides for renewable energy targets at the national level, a feed-in tariff, and special measures to support target achievement, tax relief for developers, and public R&D support.

The expansion of China's renewable energy development in recent years has been substantial. China's renewable energy supply from wind, solar, and nontraditional biomass (including biomass for electricity, biogas, and biofuels) increased threefold between 2000 and 2010, from 95 Mtce (million tons of coal equivalent) to 293 Mtce. The composition of renewable energy in China in 2010 is shown in Figure 3.1.

Current renewable energy targets foresee a sixfold increase in wind power, a 62.5-fold increase in solar power, and a 5.4-fold increase in biomass electricity by 2020 relative to 2010 (for wind, some expect this deployment to occur even faster). Targets for 2015 and 2020 are discussed in Section 3.4.2.

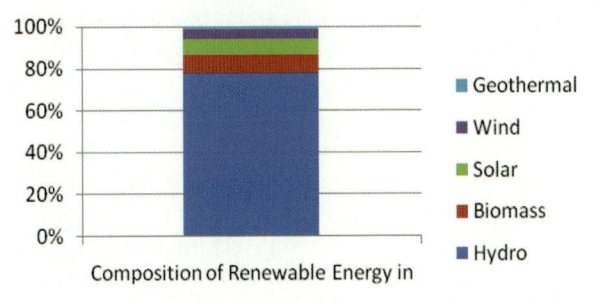

Figure 3.1 Composition of renewable energy in China in 2010 (excluding traditional biomass).

3.3
Data and CGEM Model Description

This chapter employs the China-in-Global Energy Model (CGEM) to evaluate the impact of China's renewable energy development on energy and CO_2 emissions. The CGEM is a multi-regional, multi-sector, recursive-dynamic computable general equilibrium (CGE) model of the global economy that separately represents 19 regions and 18 sectors as shown in Table 3.1.

Table 3.1 Sectors and regions in the China-in-Global Energy Model (CGEM).

Sector	Description	Region	Additional description
Crops	Crops	China	Chinese mainland
Forest	Forest	United States	
Livestock	Livestock	Canada	
Coal	Mining and agglomeration of hard coal, lignite and peat	Japan	
Oil	Extraction of petroleum	South Korea	
Gas	Extraction of natural gas	Developed Asia	Hong Kong, Taiwan, Singapore
Petroleum and coke	Refined oil and petrochemical products, coke production	European Union	Includes EU-27 plus countries of the European Free Trade Area (Switzerland, Norway, Iceland)
Electricity	Electricity production, collection, and distribution	Australia–New Zealand	Australia, New Zealand, and other territories (Antarctica, Bouvet Island, British Indian Ocean Territory, French Southern Territories)
Nonmetallic minerals products	Cement, plaster, lime, gravel, concrete	India	India
Iron and steel	Manufacture and casting of basic iron and steel	Developing South-East Asia	Indonesia, Malaysia, Philippines, Thailand, Vietnam, Cambodia, Laos, rest of South-East Asia
Nonferrous metals products	Production and casting of copper, aluminum, zinc, lead, gold, and silver	Rest of Asia	Rest of Asian countries
Chemical rubber products	Basic chemicals, other chemical products, rubber and plastics products	Mexico	Mexico
Fabricated metal products	Sheet metal products (except machinery and equipment)	Middle East	Iran, United Arab Emirates, Bahrain, Israel, Kuwait, Oman, Qatar, Saudi Arabia
Food and tobacco	Manufacture of foods and tobacco	South Africa	South Africa
Equipment	Electronic equipment, other machinery and Equipment	Rest of Africa	Rest of African countries

Table 3.1 (continued)

Sector	Description	Region	Additional description
Other industries	Other industries	Russia	
Transportation services	Water, air, and land Transport, pipeline transport	Rest of Europe	Albania, Croatia, Belarus, Ukraine, Armenia, Azerbaijan, Georgia, Turkey, Kazakhstan, Kyrgyzstan, rest of Europe
Other service	Communication, finance, public service, dwellings, and other services	Brazil Latin America	Brazil Rest of Latin American Countries

3.3.1 Model Data

The CGEM is a recursive-dynamic general equilibrium model of the world economy developed collaboratively by the Tsinghua Institute of Energy, Environment, and Economy and the MIT Joint Program on the Science and Policy of Global Change. Much of the sectoral detail in the CGEM model is focused on providing a more accurate representation of energy production and use as they may change over time or under policies that limit CO_2 emissions. CGEM is parameterized and calibrated based on the latest Global Trade Analysis Project global database (GTAP 8) and China's official datasets, and the base year of the CGEM model is 2007. The GTAP 8 dataset is includes consistent national accounts on production and consumption (input–output tables) together with bilateral trade flows for 57 sectors and 129 regions for the year 2007 [2, 3]. CGEM has replaced the GTAP 8 data with data from China's official data sources, the national input–output tables, and energy balance tables for 2007. To maintain consistency between these two datasets, we have re-balanced the revised global database with a least-squares recalibration method [4].

The model is solved recursively in 5 year intervals starting with 2010. The EPPA (Emissions Prediction and Policy Analysis) model represents production and consumption sectors as nested constant elasticity of substitution (CES) functions (or the Cobb–Douglas and Leontief special cases of the CES). The model is written in the GAMS software system and solved using the MPSGE modeling language [5].

3.3.2 Renewable Energy Technology

We represent 11 types of advanced technologies in CGEM as shown in Table 3.2. Three technologies produce perfect substitutes for conventional fossil fuels (crude oil from shale oil, refined oil from biomass, and gas from coal gasification). The remaining eight technologies are electricity generation technologies.

Table 3.2 Advanced technologies in the CGEM model.

Technology	Description
Wind	Convert intermittent wind energy into electricity
Solar	Convert intermittent solar energy into electricity
Biomass electricity	Convert biomass into electricity
IGCC	Integrated coal gasification combined cycle to produce electricity
IGCC-CCS	Integrated coal gasification combined cycle with carbon capture and storage to produce electricity
NGCC	Natural gas combined cycle to produce electricity
NGCC-CCS	Natural gas combined cycle with carbon capture and storage to produce electricity
Advanced nuclear	Nuclear power with new technology
Biofuels	Converts biomass into refined oil
Shale oil	Extracts and produce crude oil from oil shale
Coal gasification	Converts coal into gas with perfect substitute for natural gas

Electricity generated from wind, solar, and biomass is treated as an imperfect substitute for other sources of electricity due to their intermittency. The final five technologies – NGCC, NGCC with CCS, IGCC, IGCC with CCS, and advanced nuclear – all produce perfect substitutes for electricity output.

Wind, solar, and biomass electricity have similar production structures, as shown in Figure 3.2. As they produce imperfect substitutes for electricity, a fixed factor is introduced on the top level of CES layers to control the penetration of the technologies [6]. Like biofuels, biomass electricity also needs land as a resource input and competes with the agricultural sectors for this resource. Other inputs, including labor, capital, and equipment, are intermediate inputs and are similar to shale oil and biofuel.

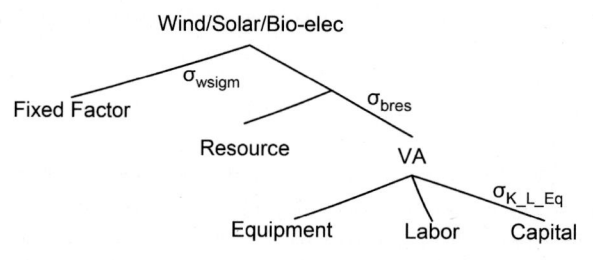

Figure 3.2 CES production structure for wind and solar power.

To specify the production cost of these new technologies, we set input shares for each technology for each region. This evaluation is based on demonstration project information or expert elicitations [7, 8]. A markup factor captures how much more expensive the new technologies are compared with traditional fossil technologies. All inputs to advanced technologies are multiplied by this markup factor. For electricity technologies and biofuels, shown later in Table 3.6, we estimate the markups for each technology based on a recent report by the Electric Power Research Institute [9] that compares the technologies on a consistent basis.

3.4
Scenario Description

We design scenarios to assess the impact of China's renewable energy policy under several economic growth assumptions. We first simulate energy use and CO_2 emissions under three economic growth assumptions in the absence of policy. These scenarios provide a basis for comparing three corresponding "Current Policy" scenarios in which existing renewable energy targets through 2020 are implemented. The goal is to understand the interaction between baseline economic growth and the requirements of current policies. As economic growth through 2050 will influence the level of energy use, which will in turn impact energy prices and the relative prices of various generation types (including the competitiveness of renewable electricity), we consider economic growth as an important source of uncertainty. The six main scenarios considered in this analysis are shown in Table 3.3.

Table 3.3 Scenario description.

Economic growth	Renewable energy policy	
	No Policy (NP)	Current Policy (CP)
High	No Policy-H	Current-H
Medium	No Policy-M	Current-M
Low	No Policy-L	Current-L

3.4.1
Economic Growth Assumptions

We design high, low, and medium economic growth trajectories that diverge after 2015, assuming that the Twelfth Five-Year Plan growth rate of 7.5% is achieved in all scenarios. After 2015, the growth rate is less certain, so we design the scenarios to include three potential trajectories. The high and low growth scenarios represent roughly 25% above and below the medium growth trajectory through 2035, and the detailed growth rates assumed in each period are shown in Table 3.4.

3 The Impact of Renewable Energy Development on Energy and CO_2 Emissions in China

Table 3.4 Annualized growth rate assumptions for the low, medium, and high GDP scenarios.

Scenario	Annualized growth rate assumption (%)[a]							
	2007–2010	2010–2015	2015–2020	2020–2025	2025–2030	2030–2035	2035–2040	2040–2045	2045–2050
Low	9.3	7.5	5.7	4.4	4.0	2.9	2.7	2.4	2.4
Medium	9.3	7.5	7.3	5.7	5.2	3.9	3.4	3.2	2.9
High	9.3	7.5	9.0	7.4	6.8	4.7	2.8	1.8	1.2

a Annualized growth rate assumptions are set for the previous 5 years, unless specified otherwise.

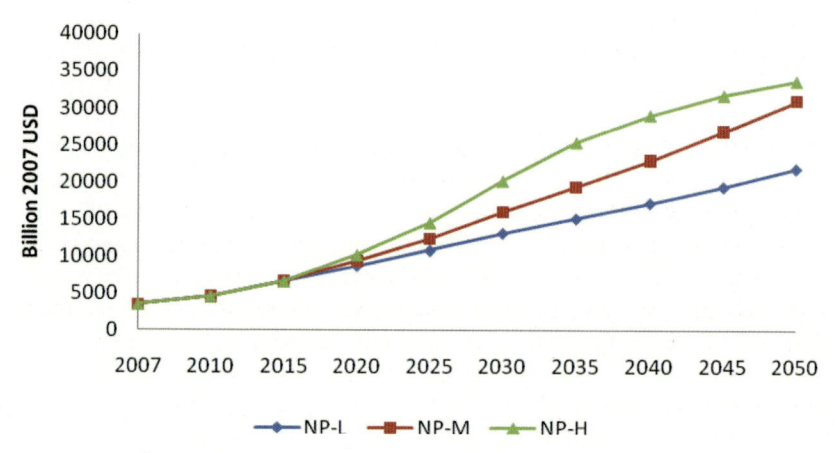

Figure 3.3 Economic growth trajectories in the high (H), medium (M), and low (L) cases.

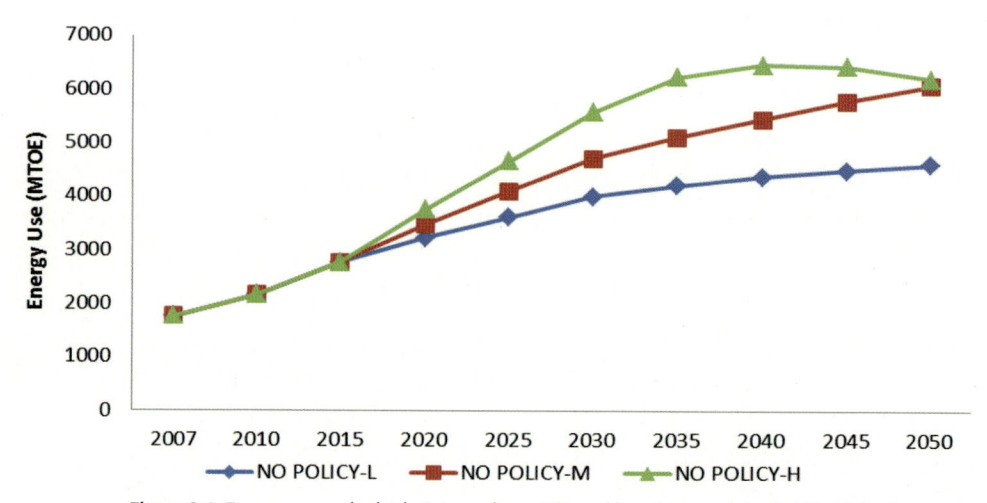

Figure 3.4 Energy use under high (H), medium (M), and low (L) growth in the "No Policy" scenario.

After 2035, we adjust the growth rate downwards, consistent with the developed state of the Chinese economy by that point. Using these growth rate assumptions produces the GDP trajectories and energy consumption patterns in the high, medium, and low cases as shown in Figure 3.3 and Figure 3.4.

3.4.2
Current Policy Assumptions

We then run the low, medium, and high growth scenarios assuming Current Policy for renewable energy through 2020 in China, which is described in Section 3.2. Current policy includes targets specified for wind, solar, and biomass generation. The policy targets are stated in terms of installed capacity with the exception of wind, which also has a target for generation. We convert capacity targets to generation targets as shown in Table 3.5. To obtain generation targets, we assume that the ratio of kilowatt hours generated to installed capacity remains constant as installed capacity is scaled up to meet the target. We use values for 2010 to compute this ratio.[2] The assumptions for installed capacity and generation from 2010 to 2020 are shown in Table 3.5. After 2020, no capacity or generation target being proposed for renewable energy in China. Hence the activity of renewable energy is based on their cost competitiveness.

To model the implementation of targets, we introduce an endogenous subsidy to the production of renewable energy from each type until the generation target is achieved. The subsidy is assumed to be financed from household income, which is consistent with the current practice of financing the feed-in tariff for renewable energy through fixed increases in the electricity price. In our modeling strategy, the generation target does not depend on the economic growth assumption. After 2020, we assume that the subsidies are phased out linearly through 2030, and that no subsidies remain in place after 2030.

Table 3.5 Published targets for installed capacity and conversion to generation target through 2020.

Type	Renewable energy target[a]					
	Installed capacity (GW)			Generation target (TWh)		
	2010	2015	2020	2010	2015	2020
Wind	31	100	200	58.9	190	390
Solar	0.8	21	50	0.95	25	59.5
Biomass	5.5	13	30	33	78	180

a   2010, actual; 2015 and 2020, authors' projections.

2) In 2010 it was widely acknowledged that a fraction of installed capacity was not yet connected to the grid, and so our assumption may have the effect of underestimating the ratio of generation to installed capacity in the future.

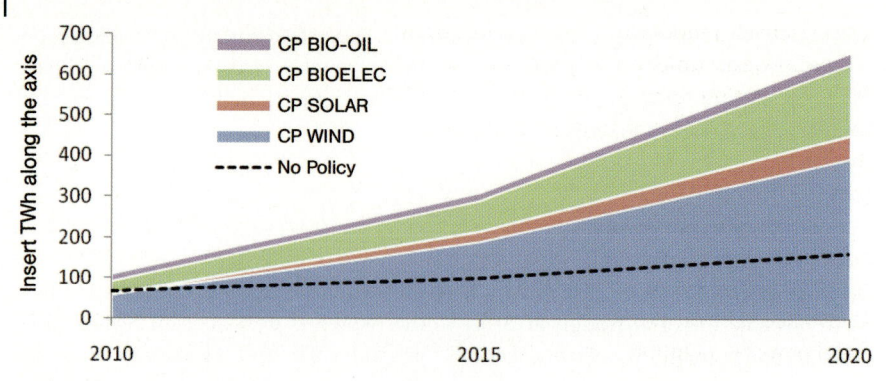

Figure 3.5 Renewable energy generation target by type and relative to total renewable generation in the No Policy scenario (dashed black line).

3.4.3
Cost and Availability Assumptions for Energy Technologies

We assume that all three renewable energy technologies are available in the base year 2007 at a higher cost relative to fossil generation sources. Each generation type has an associated cost markup as shown in Table 3.6, which captures the incremental cost relative to the levelized cost of conventional fossil fuel generation. Renewable energy can enter the market when its cost falls relative to fossil fuel electricity, which can occur either as the fossil fuel price rises (due to increased demand) or if renewable energy is subsidized. To simulate realistic rates of adoption once renewable energy become cost competitive, we included an additional resource input in the production function of each renewable electricity type that simulates limits on early adoption due to the need to repurpose production facilities, train the labor force, and other startup costs. This resource input, which is parameterized for each renewable energy type, is treated identically in all scenarios [8, 10].

Renewable energy subsidies are often justified as supporting the technology in its early stages, allowing developers to gain experience and scale up production in ways

Table 3.6 Markups expressed in percentage terms as the additional cost for each renewable electricity type relative to fossil fuel electricity.

Type	Markup (%)		
	2010–2020	2020–2050	
	All scenarios	Six main scenarios	Low-cost scenario
Wind	207	20	10
Solar	200	200	50
Biomass	60	60	30

that effectively reduce the future cost of each renewable energy type. In our six main scenarios we assume that the markup on renewable energy relative to conventional fossil generation remains constant over time. However, we also include a scenario in which the subsidized development of renewable energy results in lower costs. In this scenario, the wind markup is 10% (compared with 20%), solar 50% (compared with 200%), and biomass 30% (compared with 60%).

Both No Policy and Current Policy cases include growth assumptions for nuclear and hydro power that are currently set forth by government plans.[3] As we are interested in the impact of supporting renewable energy specifically, we do not explore alternative cost or availability assumptions for nuclear, hydro, and conventional fossil power generation.

3.5
Results

We now consider the impact of the renewable energy targets against the background of the three alternative GDP growth trajectories. We find that the level of GDP growth results in different renewable energy development trajectories. For each scenario, we consider the impact of renewable subsidies on energy use, CO_2 emissions, and economic growth. We find that although renewable energy subsidies result in an increase in renewable energy, the impact on CO_2 emissions is relatively modest. This is because displacing some fossil fuel use by renewable energy in the electricity sector puts downward pressure on fossil fuel prices, leading to increased use in other sectors. We further find that if the cost of renewable energy is successfully reduced during the subsidy period, it will compete successfully without subsidies through 2050 and supply a large share of energy in China. However, our analysis suggests that subsidies alone will not be sufficient to realize the emissions reduction potential available from renewable energy. This analysis demonstrates that it is important to consider impacts on the integrated energy–economic system when designing renewable energy policy.

3.5.1
Renewable Energy Growth Under Policy

Current policies result in significant growth in renewable energy under all three growth scenarios. In all scenarios, renewable energy growth follows the target trajectory through 2020, significantly above the level of renewable energy generation under the No Policy scenario. After 2020, the differences between the No Policy and Current Policy scenarios are less pronounced. In both scenarios, the renewable growth trajectories diverge under different growth assumptions and affect both energy demand and the relative prices of energy types. In the Current Policy case,

---

3) The government plan for the installed capacity of nuclear power is 40 GW in 2015 and 70 GW in 2020, of hydro power is 290 GW in 2015 and 420 GW in 2020.

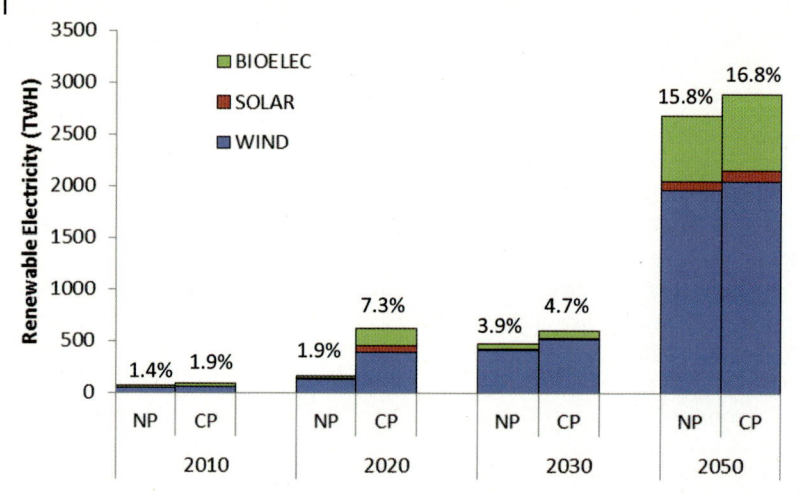

Figure 3.6 Renewable energy output and percentage of total generation in the No Policy (NP) and Current Policy (CP) scenarios under medium economic growth assumption.

as subsides are phased out between 2020 and 2030, the total generation from renewable energy begins to fall, and its contribution into the future depends on its cost competitiveness relative to other generation types.

Figure 3.6 compares the renewable electricity generation and its share of total electricity use in 2010, 2020, 2030, and 2050. The target is met in both cases through 2020. After 2020, under slower economic growth, fossil energy prices increase more slowly, and so renewable energy is less competitive relative to fossil sources. However, if large demand pressure causes energy prices to increase more rapidly in the high growth scenario, renewable energy will be more cost competitive and by 2050 may make a significant contribution to overall generation that is almost three times as large as in the low growth scenario. These results demonstrate how GDP growth can strongly influence the prospects for renewable energy through its impact on fuel demand and competition among fuels – higher growth puts more pressure on fossil fuel resources, so there is more market pressure to increase renewable energy. Although renewable energy makes a slower start without current policies, its eventual contribution by 2050 is about the same under the No Policy and Current Policy scenarios.

3.5.2
Impact of Renewable Energy Subsidies on CO_2 Emissions Reductions

Our modeling framework allows us to assess the impact that current renewable energy subsidies will have on total CO_2 emissions from China's energy system. We consider two periods, 2010–2020 and 2020–2050, and compute the total reduction achieved, focusing on the medium growth case only for simplicity.[4] We compare this

---

4) Using instead the low or high growth assumption does not change the policy results significantly.

with an "idealized" reduction that assumes that all new renewable energy generation displaces fossil fuel generation and that there is no incentive to increase the use of carbon-intensive fuels in other sectors as a result of displacing them from electricity.

We compute the CO_2 emissions reduction achieved in the medium growth case by comparing the No Policy and Current Policy scenarios for each. We find that the renewable electricity target has the effect of lowering emissions intensity by 2% in 2015 and 3.5% in 2020 compared with the No Policy scenario. From 2020 to 2050, we find an average 1.5% reduction in CO_2 emissions intensity after 2020 in the Current Policy scenario (although no targets are being imposed in this period).

In terms of the total CO_2 emissions reduction, the model predicts that cumulative CO_2 emissions will be lower by 1173 mmt (million metric tons) (1.2%) over the period 2010–2020. After 2020, we find that the impact of a target from 2010 to 2020 on future CO_2 emissions is more complex. Cumulative emissions from 2020 to 2050 are slightly higher with early renewable deployment (Current Policy scenario) relative to a No Policy scenario by 8628 mmt or 1.8%. Comparing the total cumulative reduction over the period 2010–2050, we find a net increase of 7455 mmt (1.3%) under the Current Policy scenario. We note that economic growth is slightly higher after 2020 in the Current Policy scenario, so despite a slight increase in CO_2 emissions under policy, the emissions intensity remains reduced relative to the No Policy scenario.

Another reason why CO_2 emissions reductions are not larger is sectoral leakage. For this analysis, we use a CGE model with energy system detail in order to capture how the renewable subsidy policy interacts with fuel prices, fuel demand, and the broader evolution of the energy–economic system and its associated CO_2 emissions. The total CO_2 emissions reductions measured using this model will reflect how the policy affects underlying energy prices, and how these effects are transmitted across markets through economic activity and trade linkages in China and on a global scale. The objective is to capture all of the real-world factors that will affect the impact of renewable energy on CO_2 emissions outcomes. It is instructive to compare the results of this model with a calculation that focuses on renewable energy only and assumes that renewable energy directly displaces fossil energy use and associated CO_2 emissions, which can be taken as an "ideal" upper bound on emissions reductions. Table 3.7 compares the actual simulated emissions reductions with the ideal calculation. The simulated "actual" reduction is the reduction that we expect given the interactions of the renewable target with the broader economy, including relative energy prices. The simulated reduction is sizable in 2015 and 2020 (although still smaller than ideal). After the subsidies are phased out from 2020 to 2030, we find a slight increase in total CO_2 emissions in every future period, reflecting higher economic growth and sectoral leakage. In the model, we further observe that the

Table 3.7 Reduction in CO_2 emissions due to Current Policy, relative to the No Policy scenario.

Reduction	2015	2020	2025	2030	2035	2040	2045	2050
Simulated ("actual") reduction (mmt)	150	141	−305	−542	−396	−302	−213	−76
Ideal reduction (mmt)	173	454	411	204	207	205	194	199

prices for fossil generation types remain lower under the Current Policy scenario for much of the next half century, which provides an incentive to increase their use. The result suggests that once dynamics in the broader economic–energy system are taken into account, the total CO_2 reduction predicted due to the deployment of renewable is significantly smaller than the ideal reduction.

3.5.3
Impact of a Cost Reduction for Renewable Energy After 2020

Earlier scenarios assumed that the markup for renewable energy remains constant after 2020. If we instead assume that the cost of each renewable energy type will decrease significantly after 2020, we find that renewable electricity generation increases significantly by 2050 as the cost of renewable electricity falls (as shown in Figure 3.7). This increase could be dramatic: under the Current Policy + low cost scenario, we find that renewable generation increases to 30% of the total compared with 17% under the Current Policy scenario only and 16% under the No Policy scenario.

We also consider the impact that the assumed cost reduction has on renewable generation by type and on total CO_2 emissions relative to the Current Policy case with no cost reduction. Focusing on the period 2010–2050, we find that the cumulative CO_2 reduction is significantly larger, reaching 5385 mmt or 1% relative to the No Policy scenario. As shown in Table 3.8, an average 5.8% emission intensity reduction is observed in Current Policy with cost reduction scenario, compared with 1.8% in Current Policy only scenario. The total CO_2 emissions in the No Policy, Current Policy, and Current Policy + low cost scenarios are shown in Figure 3.8. As discussed above, for the reason of GDP stimulation and "sector leakage" of the energy use, the total CO_2 reduction predicted due to the deployment of renewable is significantly smaller than the ideal reduction, in some periods even increasing CO_2 emissions, once dynamics in the broader economic–energy system are taken into account.

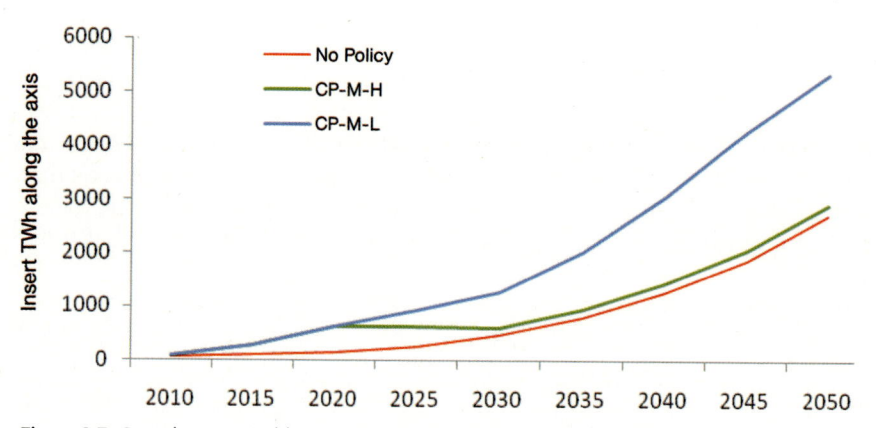

Figure 3.7 Growth in renewable energy in the No Policy, Current Policy, and Current Policy + low cost scenarios under the medium growth assumption.

However, it is important to realize that the leakage effects associated with the supply-side cost shock are also more pronounced. This result is consistent with the fact that in the Current Policy + low cost scenario we find that in 2050 the electricity price is 4% lower and the coal price is 10% lower relative to the Current Policy scenario.

Table 3.8 Impact on renewable energy generation and CO_2 emissions intensity reductions (No Policy, Current Policy, and Current Policy + low cost, broken down by type).

Scenario	Renewable electricity type	Renewable energy generation (Mtoe)[a]			
		2015	2020	2030	2050
No Policy	Wind	81	136	414	1971
	Solar	1	2	6	81
	Biomass	17	23	54	638
Current Policy	Wind	191	394	518	2052
	Solar	24	57	8	98
	Biomass	74	173	71	745
	CO_2 emission intensity reduction	2.0%	3.5%	1.8%	0.8%
Current Policy + low cost	Wind	191	394	735	2288
	Solar	24	57	334	2110
	Biomass	75	173	203	932
	CO_2 emission intensity reduction	2.0%	3.5%	5.4%	8.6%

a Mtoe: million tons of oil equivalent.

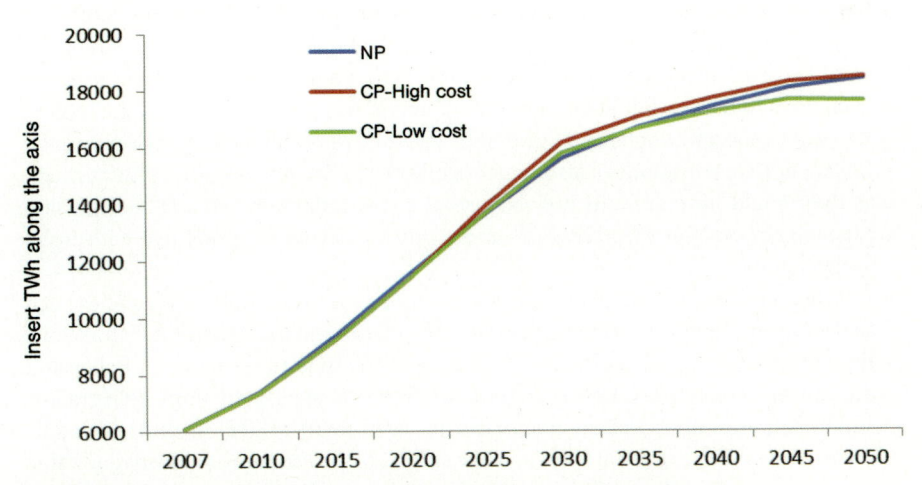

Figure 3.8 Total CO_2 emissions in the No Policy (NP), Current Policy (CP), and Current Policy + low cost scenarios under the medium growth assumption.

3.6
Conclusion

China's renewable energy policy is currently focused on increasing the installed capacity of wind, solar, and biomass electricity and increasing its contribution to total generation. When the current policy is simulated in the CGEM model, we find that the policy does have the effect of increasing the renewable electricity generation from 2010 to 2020 in both absolute (from 92 to 629 TWh) and relative terms (from 1.9% to 7.3% of total generation). Owing to the introduction of renewable energy over the period 2010–2020, the overall CO_2 emissions intensity falls by 2%.

After 2020, the impact of renewable energy largely depends on the economic growth and cost assumptions. We find that high economic growth results in a higher energy demand and prices, which create more favorable conditions for the adoption of renewable energy sources. The low economic growth assumption, by contrast, alleviates the price pressure of fossil fuels and so renewable sources are less competitive – but total energy use and CO_2 emissions are also lower overall. In this respect, renewable energy may be expected to respond automatically to price signals, delivering a low-cost substitute when fossil demand is high, but playing a less prominent role when fossil fuel demand is lower. If renewable energy is to respond in this way, it will be important to allow the prices of fossil fuels to reflect their true cost of production. In our model, we assume that energy prices are determined by the market. If we assume instead that end-user fuel or electricity prices are managed by the government (which is currently the case in China), growth in renewable energy will lower over the time period we consider.

Since subsidies carry a cost to the government (ultimately borne by the population through taxes and electricity tariffs), it is also important to explore their potential benefits. Many are optimistic that current policies may further reduce the cost of renewable energy after 2020, for instance, through materials substitution, manufacturing advances, and additional reductions in installation costs. We explore a scenario that reduces the markup for renewable generation after 2020, which we assume has occurred as a result of renewable generation expansion under the policy from 2010 to 2020. After 2020, the cost reduction has a large impact on the level of renewable energy adoption. With higher levels of renewable energy adoption, the impact of CO_2 emissions is also larger, while electricity prices do not rise as much as they would have done in the absence of a cost reduction. This is because less expensive renewable electricity can more quickly become competitive with fossil fuel generation.

When it comes to reducing CO_2 emissions, we find that supply-side policies such as the current renewable target may have a more modest impact on total emissions than many expect, due to offsetting leakage effects. In both the Current Policy and the Current Policy + low cost scenarios, we find that ideal reductions delivered by additional renewable capacity is partially (or even totally) offset in future years by increases in the use of fossil fuels in other sectors of the economy. Adding renewable generation in the electricity sector reduces the need to build more fossil-fired generation capacity, placing downward pressure on fossil fuels. The greater the contribution to generation, the greater is the downward pressure on fossil fuel prices, and the greater are the leakage effects. Policymakers would be well served to consider the impact of these offsetting effects as they design complementary or alternative policies to bring renewable energy into the generation mix. One such approach would be to include electricity and other sectors under a cap-and-trade system for CO_2 emissions, an approach that is already being piloted on a limited basis in some Chinese provinces.

Finally, we consider the contribution of the renewable electricity target to China's national carbon and nonfossil energy goals. Our model results suggest that the renewable electricity targets will make a relatively modest contribution to the Twelfth Five-Year Plan carbon intensity reduction goal of 17%, accounting for about 12% of the total reduction in 2015. We further find that the targets contribute about 11% to China's Copenhagen commitment of a 45% CO_2 intensity reduction by 2020, relative to CO_2 intensity in 2005. We point out that if the ideal reduction numbers are used instead, this reduction looks much larger. This analysis cautions against the use of sector-by-sector calculations of CO_2 reduction opportunities that ignore broader economy-wide interactions and the resulting impact. A policy approach that covers all sectors and allows substantial flexibility to reduce CO_2 at lowest cost – such as a cap-and-trade system – will prevent emissions leakage and ensure that targeted reductions in CO_2 emissions are achieved over the long term.

References

1 The World Bank and China Ministry of Environmental Protection (2007) Cost of Pollution in China: Economic Estimates of Physical Damages, World Bank, Washington, DC, http://siteresources.worldbank.org/INTEAPREGTOPENVIRONMENT/Resources/China_Cost_of_Pollution.pdf (last accessed 16 February 2013).

2 Narayanan, B., Dimaranan, B., and McGougall, R. A. (2012). *GTAP 8 Data Base Documentation, Chapter 2: Guide to the GTAP Data Base*, Center for Global Trade Analysis, Purdue University, West Lafayette, IN.

3 Narayanan, B. (2012) *GTAP 8 Data Base Documentation, Chapter 3: What's New in GTAP 8*, Center for Global Trade Analysis, Purdue University, West Lafayette, IN.

4 Rutherford, T. F. and Paltsev, S. V. (2000) *GTAPinGAMS and GTAP-EG: global datasets for economic research and illustrative models*, University of Colorado, Department of Economics Working Paper, www.gamsworld.org/mpsge/debreu/papers/gtaptext.pdf (last accessed 16 February 2013).

5 Rutherford, T. F. (2005) GTAP6inGAMS: the dataset and static model, presented at the Workshop on Applied General Equilibrium Modeling for Trade Policy Analysis in Russia and the CIS, Moscow, 1–9 December 2005, http://www.mpsge.org/gtap6/gtap6gams.pdf (last accessed 16 February 2013).

6 McFarland, J. R., Reilly, J. M., and Herzog, H. J. (2004) Representing energy technologies in top-down economic models using bottom-up information. *Energy Econ.*, **26** (4), 685–707.

7 Babiker, M. H., Reilly, J. M., Mayer, M., Eckaus, R. S., Sue Wing, I., and Hyman, R. C. (2001) *The MIT Emissions Prediction and Policy Analysis (EPPA) Model: Revisions, Sensitivities, and Comparisons of Results*, Report No. 71, MIT Joint Program on the

Science and Policy of Global Change, Massachusettets Institute of Technology, Cambridge, MA.

8 Paltsev, S., Reilly, J. M., Jacoby, H. D., Eckaus, R. S., McFarland, J., Sarofim, M., and Babiker, M. H. (2005) *The MIT Emissions Prediction and Policy Analysis (EPPA) Model: Version 4*, Report No. 125, MIT Joint Program on the Science and Policy of Global Change, Massachusetts Institute of Technology, Cambridge, MA.

9 Electric Power Research Institute (2011) *Program on Technology Innovation: Integrated Generation Technology Options*, Electric Power Research Institute, Palo Alto, CA.

10 Karplus, V. J., Sergey, P., and Reilly, J. M. (2010) Prospects for plug-in hybrid electric vehicles in the United States and Japan: a general equilibrium analysis. *Transport. Res. Part A: Policy Pract.*, 44 (8), 620–641.

4
The Scottish Government's Electricity Generation Policy Statement

Colin Imrie

4.1
Introduction

This chapter provides a composite of several documents, primarily the Electricity Generation Policy Statement [1], but also the Energy in Scotland: a Compendium of Scottish Energy Statistics and Information [2] document, and the ISLES Executive Summary [3].

4.2
Overview

Scotland can achieve its target of meeting the country's electricity needs from renewables as well as more from other sources by 2020.

Renewable generation will be backed up with thermal generation progressively fitted with carbon capture and storage – ensuring that Scotland's future electricity needs can be met without the need for new nuclear power stations.

The Electricity Generation Policy Statement (EGPS) sets out the Scottish Government's plans for renewable energy and fossil fuel thermal generation in Scotland's future energy mix. The report is based on research studies looking at future energy supply, storage, and demand.

The statement shows that low-carbon energy policies will not only benefit the environment and create jobs, but also lead to lower household bills. Low-carbon energy policies and measures could lead to an average annual household energy bill of £1285 by 2020 – whereas carrying on with "business as usual" will lead to bills of £1379.

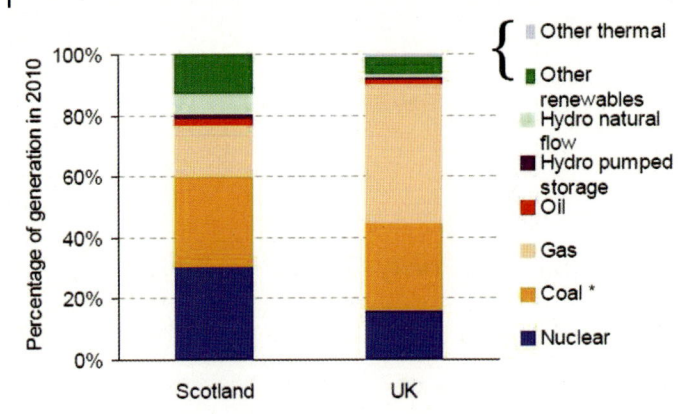

Figure 4.1 Percentage of electricity generated by fuel, 2010.
Source: DECC, Coal includes a small amount of non-renewable waste, *Energy Trends*, December 2011, www.decc.gov.uk/en/content/cms/statistics/publications/trends/trends.aspx.

The Scottish Government aims to develop an electricity generation mix built around four key principles:

- a secure source of electricity supply;
- an affordable cost to customers;
- decarbonized by 2030;
- achieves the greatest possible economic benefit and competitive advantage for Scotland.

4.3
Executive Summary

1. Electricity plays a central role in the life and lives of the nation. Its generation, and the economic and the environmental benefits which could arise from a shift from fossil fuel generation to a portfolio comprising renewable and cleaner thermal generation, are matters of considerable importance to the Scottish Government.

2. This draft *Electricity Generation Policy Statement* (EGPS) examines the way in which Scotland generates electricity, and considers the changes that will be necessary to meet the targets which the Scottish Government has established.

3. It looks at the sources from which that electricity is produced, the amount of electricity that we use to meet our own needs, and the technological and infrastructural advances and requirements that Scotland will require over the coming decade and beyond.

4. The Scottish Government's policy on electricity generation is that Scotland's generation mix should deliver:

- a secure source of electricity supply;
- at an affordable cost to consumers;
- that can be largely decarbonized by 2030;
- and that achieves the greatest possible economic benefit and competitive advantage for Scotland, including opportunities for community ownership and community benefits.

5. The draft EGPS is constructed around a number of relevant targets and related requirements:

- delivering the equivalent of at least 100% of gross electricity consumption from renewables by 2020 as part of a wider, balanced electricity mix, with thermal generation playing an important role though a minimum of 2.5 GW of thermal generation progressively fitted with carbon capture and storage (CCS);
- enabling local and community ownership of at least 500 MW of renewable energy by 2020;
- lowering final energy consumption in Scotland by 12%;
- demonstrating CCS at a commercial scale in Scotland by 2020, with full retrofit across conventional power stations thereafter by 2025–2030;
- seeking increased interconnection and transmission upgrades capable of supporting projected growth in renewable capacity.

6. Scotland's renewables potential is such that, should the relevant technologies be developed successfully, it could deliver up to £46 billion of investment and be much more than enough to meet domestic demand for electricity. The remainder could be exported to the rest of the United Kingdom and continental Europe to assist other countries in meeting their binding renewable electricity targets.

7. The draft EGPS is structured as follows:

- *Energy demand reduction* – Summary look at the Scottish Government's Energy Efficiency Action Plan (EEAP), against the backdrop of a fall in final energy consumption of **7.4%** in 2009 compared with the previous year.
- *Renewables* – The importance of renewables in the light of the Scottish Government's 100% target mentioned above, the target for at least 500 MW of renewable energy (electricity and heat) to be in local and community ownership by 2020; and in the context of the Renewables Routemap and the related heat and transport targets.
- *CCS* – The Scottish Government's policy is that renewable generation should operate alongside upgraded and more efficient thermal stations, and that there should be a particularly strong role for CCS, where Scotland has the natural advantages and resources that could enable it to become a world leader.

- *Nuclear* – The draft EGPS confirms that nuclear energy will be phased out in Scotland over time, with no new nuclear build taking place in Scotland. This does *not* preclude extending the operating life of Scotland's existing nuclear stations to help maintain security of supply over the next decade while the transition to renewables and cleaner thermal generation takes place.
- *Bioenergy* – Confirmation that biomass should be used in small heat only and CHP (combined heat and power) applications, off gas-grid, the better to contribute to meeting the Scottish Government's target of 11% of heat demand to be sourced from renewables by 2020.
- *Role of electricity storage* – Developments in this area, while financially and technologically challenging, can help address the variability of certain forms of renewable generation.
- *Transmission and distribution* – The draft EGPS reaffirms the important role that Scotland can play in developing greater onshore and offshore grid connections within and across the United Kingdom and Europe. It continues to press for a sensible regulatory regime – in particular an equitable outcome on charging – and also looks at the importance of (and need to build upon) the recently published Irish–Scottish Links on Energy Study (ISLES) and the importance of developing North Sea grid.
- *Modeling the target of the equivalent of 100% of gross electricity consumption from renewables by 2020* – Modeling commissioned by the Scottish Government confirms that this target is technically feasible. The work, summarized in Annex B of the draft EGPS, also looks at the changes to the generation mix and power flows that will be required.
- *Market factors* – The draft EGPS also reiterates the need for a sensible outcome to the current process of Electricity Market Reform (EMR) and the need for that outcome to respect the devolution settlement and help deliver Scotland's renewable and CCS potential.

8. A Strategic Environmental Assessment (SEA) of the draft EGPS (and Renewables Routemap [4]) has also been completed, balancing the objectives and targets contained within those documents with more localized effects on environmental features such as landscapes and biodiversity.

9. The Scottish Government is seeking comments on the policies set out in these documents, as well as on the possibility of implementing an Emissions Performance Standard (EPS) for power generation in Scotland, distinct from that proposed by the UK Government under its proposals for Electricity Market Reform (see Annex A of the draft EGPS).

10. Figure 4.2 highlights that there is potentially up to 30 GW of renewable capacity in various stages of project planning and development – an increase of around 600% on the level of capacity currently deployed.
We are wholly confident that our objectives for a resilient energy system, with a high proportion of renewable energy, can be delivered by the market and can

4.3 Executive Summary | 51

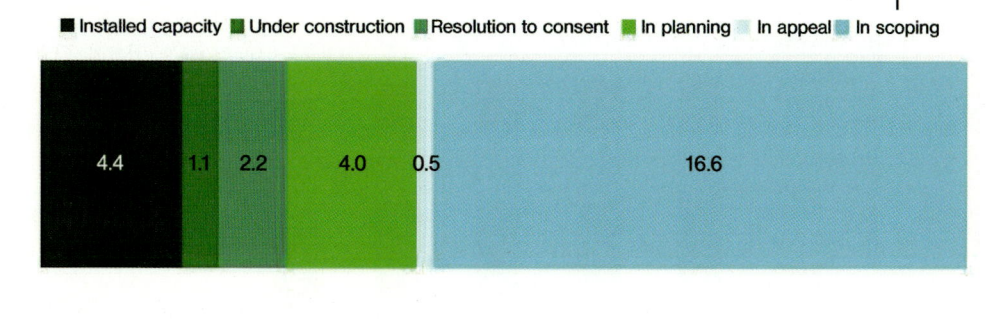

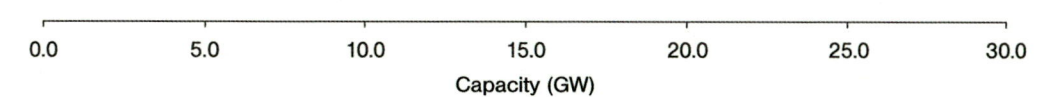

■ Installed capacity ■ Under construction ■ Resolution to consent ■ In planning ■ In appeal ■ In scoping

| 4.4 | 1.1 | 2.2 | 4.0 | 0.5 | 16.6 |

0.0   5.0   10.0   15.0   20.0   25.0   30.0
Capacity (GW)

Figure 4.2 Renewable capacity at various stages of project planning.

be achieved whatever constitutional changes may occur over the next few years. Scotland is, and will remain, a net exporter of electricity owing to renewable deployment. For example, the UK targets to produce 15% of all energy and an estimated 30% of electricity from renewable sources by 2020 will require connection to Scotland's vast energy resource and we will continue to work to connect Scotland to an ever more integrated UK and EU market. Indeed, the countries of the British Isles are working towards an All Islands electricity market, and the EU has designated the North Sea as a priority corridor for energy infrastructure that will enhance Scotland's ability to export low-carbon energy in the longer term.

Energy Demand Reduction
11. Scotland's ability to supply sufficient renewable electricity and heat to meet its targets in a cost-effective way depends critically on *reducing demand*. High demand requires more generating capacity to be built. As a consequence of uncertainties over individual behaviors, electricity demand could vary, but it is likely to rise in the long term as more electricity is used for transport and heat reasons. This means that energy efficiency measures across all three energy sectors will be crucial.
We published *Conserve and Save*, the Scottish Government's Energy Efficiency Action Plan (EEAP), in Autumn 2010 and the first progress report in October 2011. It established a target to reduce the total final energy demand in Scotland by 12% by 2020 from a 2005–2007 baseline, covering all fuels and sectors. The data for 2009 show that final energy consumption fell by 7.4% compared with 2008 and 9.6% against the target baseline. Although this is due in part to the temporary impact of the global recession on energy demand, the reduction indicates that Scotland is on track to meet the 2020 final energy reduction target.

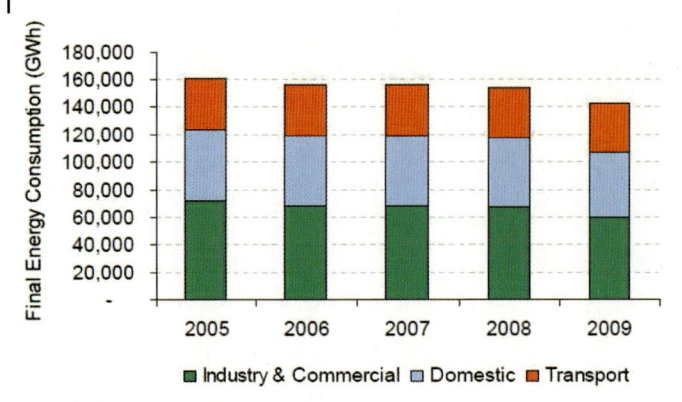

Figure 4.3 Scottish final energy consumption.
Source: DECC, *Total Final Energy Consumption at Sub-National Level*, www.decc.gov.uk/en/content/cms/statistics/energy_stats/regional/total_final/total_final.aspx.

12. The key actions relating to energy efficiency include to:

    - improve the energy efficiency of all our housing stock to meet the demands of the future;
    - establish a single energy and resource efficiency service for Scottish businesses;
    - develop a public sector that leads the way through exemplary energy performance and provides the blueprint for a low-carbon Scotland;
    - promote infrastructure improvements, for example, by developing a sustainable heat supply; and
    - ensure that people are appropriately skilled to take advantage of the opportunities.

13. Energy efficiency is at the top of our hierarchy of energy policies as the simplest and most cost-effective way to reduce emissions while seeking to maximize the productivity of our renewable resources. Energy efficiency complements our other energy-related strengths, and works across areas such as housing, business, and transport, all of which are major consumers of fuel, to help us create a more sustainable Scotland with opportunities for all to flourish.

Renewables

14. We published our *2020 Routemap for Renewable Energy* in Scotland in June 2011, a document which goes hand-in-hand with our continuing drive to reduce demand.

    Because the pace of renewables development has been so rapid in Scotland, and because we have a potential resource capable of powering Scotland several times over, the Renewables Routemap commits to a new renewable electricity target.

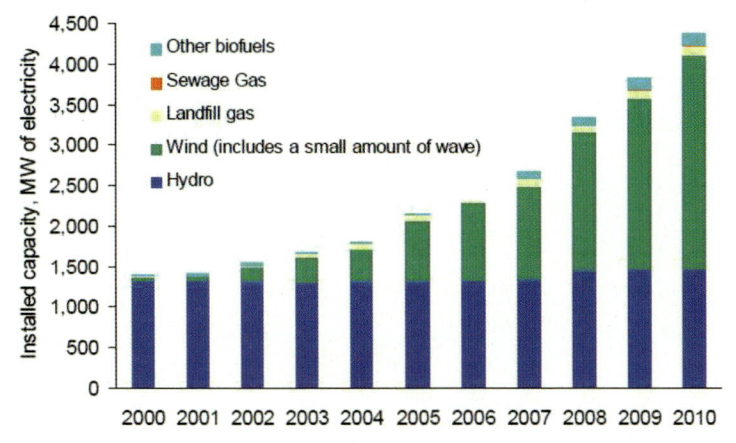

Figure 4.4 Installed capacity of renewable energy in Scotland. Source: DECC, *Energy Trends*, December 2011, www.decc.gov.uk/en/content/cms/statistics/publications/trends/trends.aspx.

15. We believe that Scotland has the capability and the opportunity to generate a level of electricity from renewables by 2020 that would be the equivalent of 100% of Scotland's gross electricity consumption. We have set out our new target to reflect this ambition. Achieving the target will require the market to deliver an estimated 14–16 GW of installed capacity. This new target does *not* mean an energy mix where Scotland will be 100% reliable on renewables generation by 2020; but it supports Scotland's plans to remain a net exporter of electricity. Owing to the intermittent nature of much renewables generation, we will need a balanced electricity mix to support security of supply requirements with efficient thermal generation continuing to play an important role.

16. Scotland has the largest offshore renewable energy resources in the EU (25% of EU offshore wind, 25% of EU tidal, and 10% of EU wave power). With 10 GW of offshore wind and 1.6 GW of wave and tidal projects (see the map in Figure 2.5) currently planned offshore renewables, Scotland has the potential to make a major contribution to the EU's overall renewables target [5]. This is why we have developed clear links with our neighboring governments in Ireland, Northern Ireland, and across the North Sea to promote the development of offshore grid connections to harness the vast renewable energy potential of the North and Irish Seas.

Thermal Generation – CCS

17. Our analysis demonstrates that while renewable energy will play the predominant role in electricity supply in Scotland by 2020, the Scottish electricity generation mix cannot currently, or in the foreseeable future, operate without baseload and balancing services provided by thermal electricity generation. The scheduled closure of existing plants and the construction of a minimum of 2.5 GW of new or replacement efficient fossil fuel electricity generation

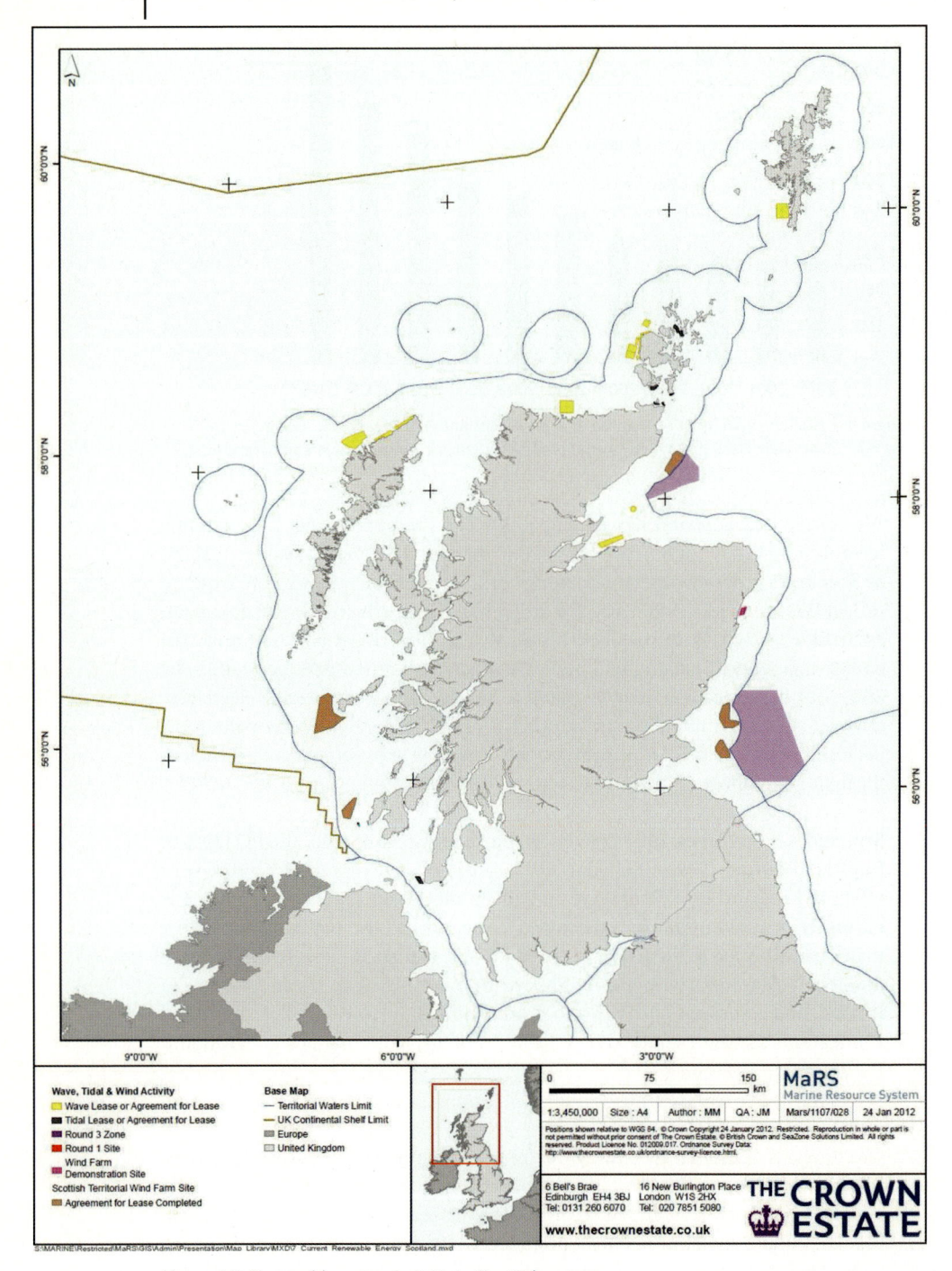

Figure 4.5 Renewable energy activity in Scottish waters.

4.3 Executive Summary

progressively fitted with CCS would satisfy security of supply concerns and, together with renewable energy, deliver large amounts of electricity exports. This generation portfolio would be consistent with our climate change targets and reporting under the net Scottish emissions account

18. The introduction of the 300 MWe CCS requirement, the UK Government's Carbon Price Floor, and its proposals for an Emissions Performance Standard mean that thermal plants will – rightly – be operating in a highly regulated and increasingly constrained market.

19. The Scottish Government has never intended to support unabated new coal plants in Scotland, as this would be wholly inconsistent with our climate change objectives. We have made it absolutely clear that any new power station in Scotland must be fitted with a minimum CCS on 300 MWe of its generation from day one of operation. CCS can potentially reduce emissions from fossil fuel power stations by up to 90% and will be a vital part of our commitment to decarbonize electricity generation by 2030.

20. As with renewables, Scotland has the opportunity to become one of the world's leaders in the development of CCS. The successful demonstration of CCS in Scotland over the next decade could create up to 5000 jobs and be worth £3.5 billion to the Scottish economy. There are well-developed proposals for a CCS demonstration at Peterhead, and Scotland's R&D capability in our universities or test sites gives us a leading position to develop projects in other markets.

21. Scotland has considerable natural advantages in CO_2 storage, alongside our world-leading research and development expertise. *Opportunities for CO_2 Storage Around Scotland* [6], published in 2009, showed that Scotland has an extremely large CO_2 storage resource:

    - Offshore saline aquifers, together with a few specific depleted hydrocarbon fields, can easily accommodate the industrial CO_2 emissions from Scotland for the next 200 years.
    - Scotland's offshore CO_2 storage capacity is the largest in the EU, comparable to that of Norway, and greater than that of The Netherlands, Denmark, and Germany combined.
    - *Progressing Scotland's CO_2 Storage Opportunities*, published in March 2011 [7], confirms the European significance of Scotland's CO_2 storage resource with more detailed evaluation of the Captain Sandstone (beneath the Moray Firth).

22. The building of any new thermal-based stations above 50 MW requires consent from Scottish Government Ministers under Section 36 of the Electricity Act 1989. We made the following announcement in November 2009:

- From 9 November 2009, any application for a new coal plant in Scotland will need to demonstrate CCS on a minimum of 300 MW (net) of capacity from their first day of their operation.
- Further new builds from 2020 will be expected to have full CCS from their first day of operation.
- A "rolling review" of the technical and economic viability of CCS will take place by 2018, looking specifically at retro-fitting CCS to existing coal plants, with the likelihood of having existing plants retro-fitted by no later than 2025.
- If CCS is not proven to be technically or financially viable, then we will consider low-carbon alternatives that would have an equivalent effect.

23. This policy relates to coal stations only. The Scottish Government's position on gas, oil, and thermal stations is that for stations over 300 MWe, applicants will have to demonstrate that any new applications demonstrate carbon capture readiness.

24. With the help and expertise of the relevant regulators, agencies, and competent authorities, including DECC offshore licensing, SNH (Scottish Natural Heritage), Crown Estate, Marine Scotland, SEPA (Scottish Environment Protection Agency), and the Health and Safety Executive, the Scottish Government has been able to identify and list the approvals required for a large-scale CCS project in Scotland. The CCS regulatory framework aims to inform future development plans, help raise public awareness of CCS legislative and regulatory obligations, encourage early developer engagement with local communities and regulators, and also enable joint working between regulators and planners to manage closely multiple consent applications being progressed simultaneously.

25. The regulatory framework has been shared with governments all over the world and been used by the UK Government, European Commission, International Energy Agency, and the Global CCS Institute to promote regulatory best practice and a useful guide for counties to develop and test their regulatory provision to enable emerging CCS projects to be managed and processed efficiently. The framework remains a live document and is subject to review and update to reflect any further legislative or regulatory changes that may come on-stream in the future.

Thermal Generation – Nuclear

26. The two large-scale nuclear power stations currently operating in Scotland make up a large proportion of the baseload electricity currently supplied to the national grid. Both these stations will continue to provide important baseload generation over the coming years as we make the transition to renewables and other low-carbon electricity-generating technologies.

27. We are determined that nuclear energy will be phased out in Scotland over time, with no new nuclear build taking place. This does *not* preclude extending the

operating life of Scotland's existing nuclear stations to help maintain security of supply over the next decade while the transition to renewables and cleaner thermal generation takes place.

Thermal Generation – Bioenergy

28. Estimates suggest that heat accounts for around 50% of the current total energy demand in Scotland. We have placed a high priority on achieving our target of 11% of heat demand to be sourced from renewables by 2020 (the current level of renewable heat is around 2.8%). Scottish Ministers are also obliged to publish a Renewable Heat Action Plan and to keep it updated through to 2020. The first update was published in December 2011 [8].

Table 4.1 Scottish renewable heat capacity and output, 2010, by technology.

Technology	2010 total capacity (MW)	2010 total output (GWh)
Biomass primary combustion	203	941
Biomass CHP	138	601
Waste treatment (energy from waste, landfill gas, and anaerobic digestion)	23	74
Solar thermal	17	9
GSHP	24	60
ASHP	5	11
WSHP	0.1	0.1
Total	411	1696

CHP, combined heat and power; GSHP, ground source heat pumps; ASHP, air source heat pumps; WSHP, water source heat pumps.
Source: *Renewable Heat in Scotland, 2010*. A Report by the Energy Saving Trust for the Scottish Government, www.scotland.gov.uk/Topics/Business-Industry/Energy/Energy-sources/19185/Heat/RHIS.

29. Our policy on biomass is set out in the National Planning Framework 2 [9], Section 36 Thermal Guidance, and in Section 36 Biomass Scoping Opinion Guidance. Essentially, because of the multiple energy uses to which biomass can be put, the limits to supply, and the competition for that supply from other nonenergy sectors, we need to encourage the most efficient and beneficial use of what is a finite resource. We would prefer to see biomass used in heat-only or CHP schemes, off gas-grid, and at a scale appropriate to make the best use of both the available heat and local supply.

There are several reasons for this:
- Evidence suggests that the use of biomass for heat-only or CHP use will be essential in order to meet Scotland's target for renewable heat.

- Use of available heat in heat-only and CHP schemes achieves 80–90% energy efficiency for the former and 50–70% for the latter, compared with 30% in electricity-only schemes. Given the limited resource, we have to ensure that it is used as efficiently as possible.
- Concentrating biomass use in areas which are off gas-grid will deliver the highest carbon savings (given that in most cases it will be displacing oil or coal), and can also make the greatest impact on alleviating fuel poverty.
- We are not categorically opposed to large-scale development. However, we believe that operators of large biomass stations will find it more difficult to use the heat generated and to source supply locally.
- Our view is that developments should be scaled appropriately so that they can make efficient use of the available heat and local supply. We believe that this will enhance security of supply, minimize carbon emissions and reduce the impact on other sectors competing for biomass material.
- There may be a significant role for imported biomass. However, the global market is an immature one and is likely to be volatile given projections of increased global demand. Its use will be dependent upon price, availability, and evidence of sustainability. As with the local resource, it should be used in plants that support maximum heat use and decentralized energy production.

Thermal Generation – Energy from Waste

30. We believe that energy generated from waste (EfW) can play a role in meeting Scotland's energy requirements. Anaerobic digestion, for instance, can help Scotland become a zero waste society, diverting food, garden, and other organic waste from landfill, reducing methane emissions, producing fertilizer or soil additives for use on local farms, reducing climate change impacts, and creating biogas, which can be used as a renewable energy source.

31. EfW combustion processes (i.e., incineration, pyrolysis, and gasification) can also contribute to both renewable energy and climate change targets, offsetting consumption of virgin fossil fuels and recovering value from resources that cannot be reused or recycled and which would otherwise be lost in landfill. Our Zero Waste Plan includes a commitment to regulate the types of waste that may be used in energy from waste combustion processes, ensuring that only materials that cannot be reused or recycled to yield greater value are used.

Thermal Generation – Waste Heat from Large Electricity and CHP Generators

32. Our Energy Efficiency Action Plan [10] highlights the opportunity for waste heat to increase energy efficiency and reduce Scotland's greenhouse gas emissions. We have commissioned research looking at the economic and technical potential for using waste heat from large-scale fossil fuel power stations in Scotland to provide heating through local district heat networks. The findings show that it is technically possible to recover significant amounts of heat from the existing large power station sites at Cockenzie, Longannet, and Peterhead, and also the proposed site at Hunterston.

33. However, the research also highlights that the main challenges to heat recovery are economic, and that there is no easy solution to make commercial investment attractive. Direct financial incentives from the public sector were shown to be an expensive and impractical route to support, whereas accelerating the connection of heat loads offers a more cost-effective route to encourage commercial deployment.

34. The report contains a number of recommendations aimed at helping to remove some of these barriers, which we will be considering carefully – including through the work of the forthcoming Expert Commission on the delivery of district heating and the Sustainable Heat Project, which we are conducting with DECC and OfGEM.

35. Thermal power stations generating electricity are ~35% efficient in converting fuel to electricity, with the remainder being discharged as waste heat. If this waste heat is captured then significant amounts of fossil fuel use can be avoided. There are ~2.4 million households in Scotland, each using on average 20 MWh of heat energy per year. There are other large-scale users of heat, such as public buildings, sports and leisure facilities, hospitals, schools, and commercial buildings. These buildings have varying heat requirements for space heating or hot water and have peak loads at different times.

36. As part of any future application, either for new or significant retrofitting for any thermal electricity generating station (gas, coal, biomass, etc.), developers will need to provide evidence that they have demonstrated how waste heat from any thermal station could be used by residential or non-domestic developments including public buildings and industry. The application will need to demonstrate that a feasibility study on the use of heat has been carried out and that discussions with local authorities have taken place to investigate the potential demand and to identify users of the heat.

Electricity Storage
37. Electricity storage could play an important and growing role in renewable electricity production, helping to address the intermittency of certain forms of renewable generation, alongside interconnection and demand-side response. The benefits of increased use of storage include:

  - allowing the best use of existing generation and in particular renewable energy resources;
  - reduced reliance on fossil fuel stations as back-up capacity;
  - helping to stabilize the transmission and distribution grid – using stored energy to avoid temporary constraints on the network and to improve power quality;
  - benefits for generators who could store electricity when prices are low and sell it when prices are high;

- potential savings in greenhouse gas emissions; and
- the potential for storage to provide "black start" capacity.

38. We conducted an Energy Storage and Management Study [11] in 2010. It did not include a scenario which exactly matched our 100% renewable electricity target, although it *did* find that, in the event of renewable generation reaching 120% of demand, there could be a role for storage from 2020 onwards, even with planned upgrades to interconnectors.

39. The study also concluded, however, that – *at least with the current existing market and regulatory framework* – storage was not economic in comparison with alternatives such as constraining generation or investment in greater interconnection.

Transmission and Distribution

40. *Delivering Scotland's Future Transmission Grid* – our vision is to connect, transport and export Scotland's full energy potential. Scotland can and must play its part in developing onshore and offshore grid connections to the rest of the United Kingdom and to European partners – to put in place the key building blocks to export energy from Scotland to national electricity grids in the United Kingdom and Europe.

41. We are working with transmission systems owners in Scotland, developers, the energy sector, local authorities, the UK energy regulator OfGEM and governments in other parts of the United Kingdom and across the EU, to deliver a strategically planned onshore and offshore electricity transmission network to connect and transport Scotland's renewable energy potential.

42. The period 2012–2020 will see significant activity to reinforce and develop the UK system (and those between Scotland and England in particular) and to connect both our onshore and offshore renewable generators. The Scottish Government is part of the Electricity Networks Steering Group (ENSG), led by DECC and OfGEM, which works closely with industry to identify, plan, and deliver the grid reinforcement necessary across the United Kingdom to meet the Scottish and UK Governments' 2020 targets.

43. The ENSG's updated Vision for 2020 [12], published in January 2012, identifies a range of grid reinforcement needed in Scotland. It reconfirms the scale of the need for reinforcement across Scotland, clarifies the costs, supports what SSE and SP have in their network development plans, emphasizes rightly that these are TSO-led plans and thus do not pre-empt any planning processes, reiterates how important these grid upgrades will be to meeting Scotland's renewables ambitions, and improves the capability on Scotland's main interconnector assets by adding around a further 3 GW of import and export capacity in central Scotland, therefore strengthening security of supply and system stability as the generation portfolio moves to a greater balance of renewable energy sources.

The Regulatory Challenge

44. We support electricity regulatory frameworks that accelerate renewable deployment, improve grid access, and remove barriers to grid connection and use. To address the unacceptable waiting times for renewable projects waiting for a grid connection, the Scottish Government worked with the UK Government to support a "connect and manage" approach to give developers more reasonable connection dates ahead of reinforcement work to the transmission system, with socialization of the constraints management costs across all grid users.

45. Since it was introduced in August 2010, 73 large generation projects, totaling ~26 GW, have advanced their expected connection dates as a result of the Connect and Manage regime. Fifty-eight of those projects are in Scotland given early connection and an average reduction in connection date of 6 years.

46. Scotland plays a key part in the UK electricity market and is a net exporter of electricity. We will continue to work for ever closer integration of electricity markets, and stronger grid connections and interconnections, at both UK and EU levels.

47. Through our work on the North and Irish Seas grid, we believe that delivering closer market integration and interconnection requires a strategic, coordinated, and collaborative approach between countries, regions, and members states. It also requires significant and sustained working with other UK and EU countries to standardize electricity markets, transmission, and energy regulation. We are therefore working closely on these issues with governments in the United Kingdom and Europe.

48. A fine example of this is the work that we have done in partnership with the Ireland and Northern Ireland Governments on a feasibility study of offshore transmission grid to exploit offshore energy off Scotland's west coast. This Irish–Scottish Links in Energy Study (ISLES) project will become a key building block in delivering a sub-sea grid in the Irish Sea, the emerging outcomes of which was published on 23 November 2011. The full technical study was published in April 2012.

The Irish–Scottish Links on Energy Study (ISLES)

49. ISLES is an EU INTERREG IVA-funded, collaborative project between the Scottish Government, the Northern Ireland Executive, and the Government of Ireland. Scotland is the lead partner. The Executive Summary from the ISLES Feasibility Study and conference presentations are available from the ISLES website: http:///www.islesproject.eu.

50. The study has assessed in detail the feasibility of an offshore interconnected transmission network and sub-sea electricity grid to support renewables generation in coastal waters off western Scotland and in the Irish Sea. The ISLES

project is an important milestone in understanding this work. It shows that such a network is technologically feasible and economically viable with a supportive regulatory framework, coordinated policy, and political will. It raises issues of EU relevance and which will require EU-wide solutions.

51. ISLES is a good forensic assessment of the opportunities and challenges around an offshore grid and will help inform the work of the DECC/OfGEM led Offshore Transmission Coordination Group, which is assessing ways of delivering offshore interconnected networks. The ISLES project highlights the importance of cooperation between industry, regulators, and governments with a shared commitment to achieving a low-carbon economy.

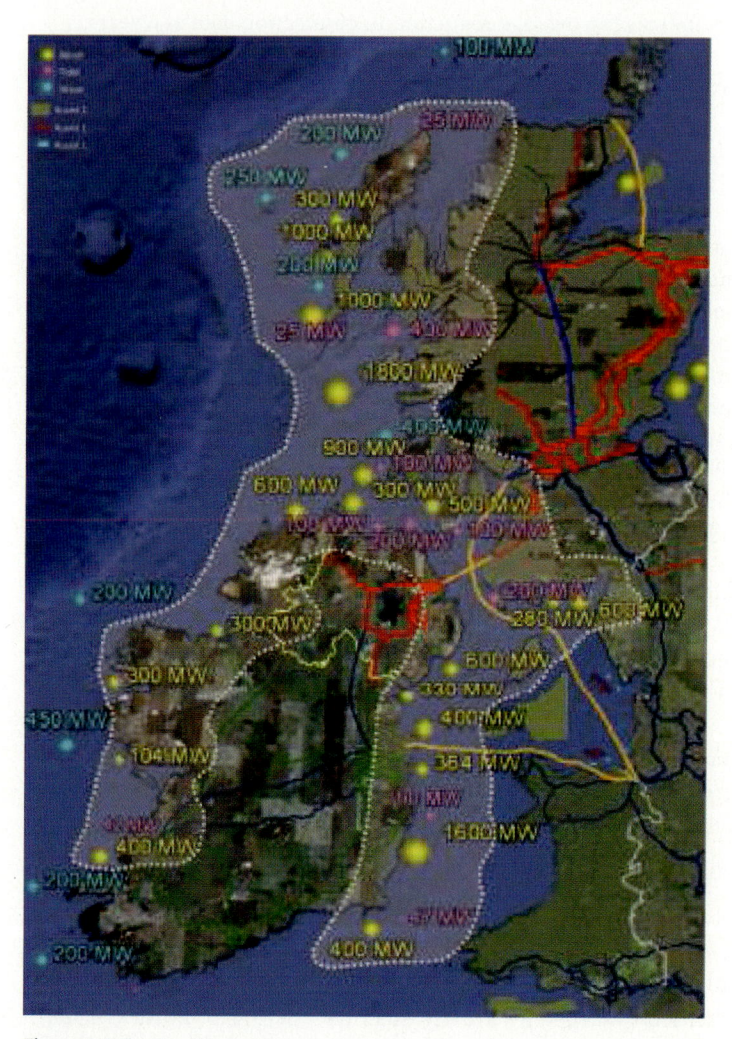

Figure 4.6 Caption to come.

Key Points
- There are no technological barriers to the development of an ISLES network.
- There is sufficient onshore network capacity in the United Kingdom for the connection of ISLES on the scale and within the timeframe envisaged by 2020.
- Two zones are proposed for offshore development: Northern ISLES (2.8 GW resource is realistic) and Southern ISLES (3.4 GW is achievable).
- There are no significant environmental constraints that cannot be adequately mitigated.
- It presents a key body of evidence to inform the debate on regulatory harmonization, which is a significant factor in permitting cross-jurisdiction projects throughout Europe.
- The economic findings are complex and are modeled assumptions of alignment of key regulated subsidies (where this is not the case at present).
- Business case projects a subsidy level of £85 MWh^{-1}, commensurate with current offshore ROCs levels and thus viable.

Resource potential of ISLES zone

52. The ISLES study also further demonstrates that offshore interconnected transmission networks will require industry, political, and policy support across jurisdictions. We are now working with UK and EU counterparts to take these discussions forward at UK/Irish level through the British–Irish Council and the EU level via participation in the EU-led North Seas Countries Offshore Grid Initiative. In these discussions, we are working to help inform EU-level work where the Commission is prioritizing energy corridors within the Northern Seas of Europe for infrastructure support.

53. ISLES offers a model that could deliver a transparent mechanism for the trading of renewable subsidies between countries and member states in exchange for counting renewable output against targets – thus contributing to domestic and EU obligations. Again, this is a key issue at EU level that is now under active consideration in the British–Irish Council – where we are working jointly with BIC partner countries grid and regulatory development.

54. In December 2010, Nine EU member states and Norway signed a Memorandum of Understanding committing to the development of a blueprint for a North Sea grid in the following areas:

- grid configuration and integration;
- market and regulatory issues; and
- planning and authorization procedures.

55. The Scottish Government is part of this work. We are working with UK and EU Governments in the working groups in each of these areas, and will continue to do so. We are working closely with the UK and Irish Governments on this, ensuring that Scotland's perspective and experience help to formulate long-term European policy in this area. This remains a priority area for us.

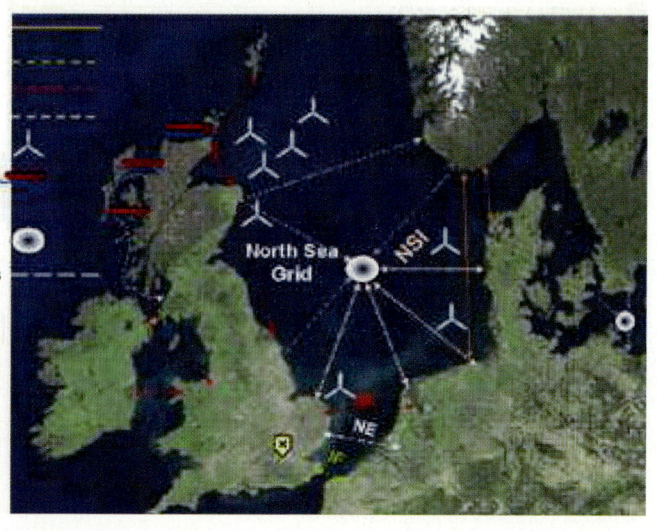

Figure 4.7 Post-2020 North Sea grid.

56. The Scottish Government has also been represented on the Adamowitsch Working Group and its follow-up body on North Sea grid connections. This is a unique forum for sharing information and learning about projects, developments, and studies across member states, helping deepen collective knowledge of offshore development, and promoting Scotland's potentially critical role in ensuring Europe-wide security of supply. It has identified significant issues to be addressed – around interconnection, standardization of regulatory and legal frameworks, financing, development, and political will.

57. A Scottish–Norwegian–Swedish consortium is now working towards connecting Scotland's renewable resource with Norway's hydro storage capacity by the construction of a 1.4 GW HVDC (high-voltage direct current) interconnector between Peterhead and Samnanger [13]. The NorthConnect cable is planned to be energized by 2021.

Delivering the Scottish Government's Objectives for Electricity Generation

58. The previous sections have explained our overarching electricity and energy policy objectives, and in particular the role we see for various technologies within the electricity mix. Scotland's potential resource and expertise mean that our 100% target is both technically achievable and desirable, but reaching that level of generation will still be extremely challenging.

59. Some of the actions needed fall within our control, such as technology and market support, planning, and consenting; but success will also be heavily dependent on regulatory processes that we will seek to influence but over which we do not currently have any direct control.

References

1. Scottish Government (2012) *Electricity Generation Policy Statement*, http://scotland.gov.uk/Topics/Business-Industry/Energy/EGPS2012/DraftEPGS2012 (last accessed 5 January 2013).
2. Scottish Government (2012) *Energy in Scotland 2012 Statistical Compendium*, http://scotland.gov.uk/Topics/Business-Industry/Energy/EGPS2012/2012StatisticalCompendium (last accessed 5 January 2013).
3. Scottish Government (2011) *ISLES Executive Summary (Draft)*, http://www.scotland.gov.uk/Topics/Business-Industry/Energy/Action/leading/iles/exec-summary-draft (last accessed 5 January 2013).
4. Scottish Government (2011) *2020 Routemap for Renewable Energy in Scotland*, http://www.scotland.gov.uk/Publications/2011/08/04110353/0 (last accessed 5 January 2013).
5. Offshore Valuation Group (2012) *The Offshore Valuation*, http://www.offshorevaluation.org/ (last accessed 5 January 2013).
6. Scottish Government (2009) *Opportunities for CO_2 Storage Around Scotland*, http://www.scotland.gov.uk/Publications/2009/04/28114540/0 (last accessed 5 January 2013).
7. Scottish Government (2011) *Progressing Scotland's CO_2 Storage Opportunities*, http://www.scotland.gov.uk/Topics/Business-Industry/Energy/Energy-sources/traditional-fuels/new-technologies/SGactionCCS/ScotlandsCO2Storage (last accessed 5 January 2013).
8. Scottish Government (2009) *Renewable Heat Action Plan for Scotland*, http://www.scotland.gov.uk/Publications/2009/11/04154534/0 (last accessed 5 January 2013).
9. Scottish Government (2009) *National Planning Framework for Scotland 2*, http://www.scotland.gov.uk/Publications/2009/07/02105627/0 (last accessed 5 January 2013).
10. Scottish Government (2010) *Conserve and Save: Energy Efficiency Action Plan*, http://www.scotland.gov.uk/Publications/2010/10/07142301/0 (last accessed 5 January 2013).
11. Scottish Government (2010) *Energy Storage and Management Study*, http://www.scotland.gov.uk/Publications/2010/10/28091356/0 (last accessed 5 January 2013).
12. Department of Energy and Climate Change (DECC) (2012) *Electricity Networks Strategy Group (ENSG)*, http://www.decc.gov.uk/en/content/cms/meeting_energy/network/ensg/ensg.aspx (last accessed 5 January 2013).
13. NorthConnect (2012) *NorthConnect project*, http://www.northconnect.no/ (last accessed 5 January 2013).

5
Transition to Renewables as a Challenge for the Industry – the German Energiewende from an Industry Perspective

Carsten Rolle, Dennis Rendschmidt

5.1
Introduction

The German government decided to change the German energy system to a more sustainable one. Based on a completely new energy concept the Fukushima disaster in March 2011 only served as an accelerator. This endeavor is threefold: the energy system in the long-term is to become sustainable from an ecological, a social and an economical point of view. This comprises both all energy generation aspects and all energy consumption aspects like electricity, housing and mobility/transport and others. Thus it requires to remodel and modify a country's entire energy system – technically and economically. To reach this goal the German government has framed a whole bunch of political targets and measures. It obviously affects all groups of energy consumers and producers in different sectors (households as well as enterprises as well as public authorities).

5.2
Targets and current status of the Energiewende

These targets comprise amongst others:

- nuclear power phase out until 2022: no nuclear power plant is going to generate electricity in Germany from 2022 onwards
- reduced greenhouse emissions by 40% in 2020 (baseline 1990)
- in 2050 80% of electricity will be generated by renewables
- reduced primary energy consumption by 80% in buildings in 2050 (baseline 2008)
- reduced total energy consumption by 40% for transportation in 2050 (baseline 2005)
- 1 million electrical vehicles on German roads in 2020

Transition to Renewable Energy Systems, 1st Edition. Edited by Detlef Stolten and Viktor Scherer.
© 2013 Wiley-VCH Verlag GmbH & Co. KGaA. Published 2013 by Wiley-VCH Verlag GmbH & Co. KGaA.

5 Transition to Renewables as a Challenge for the Industry

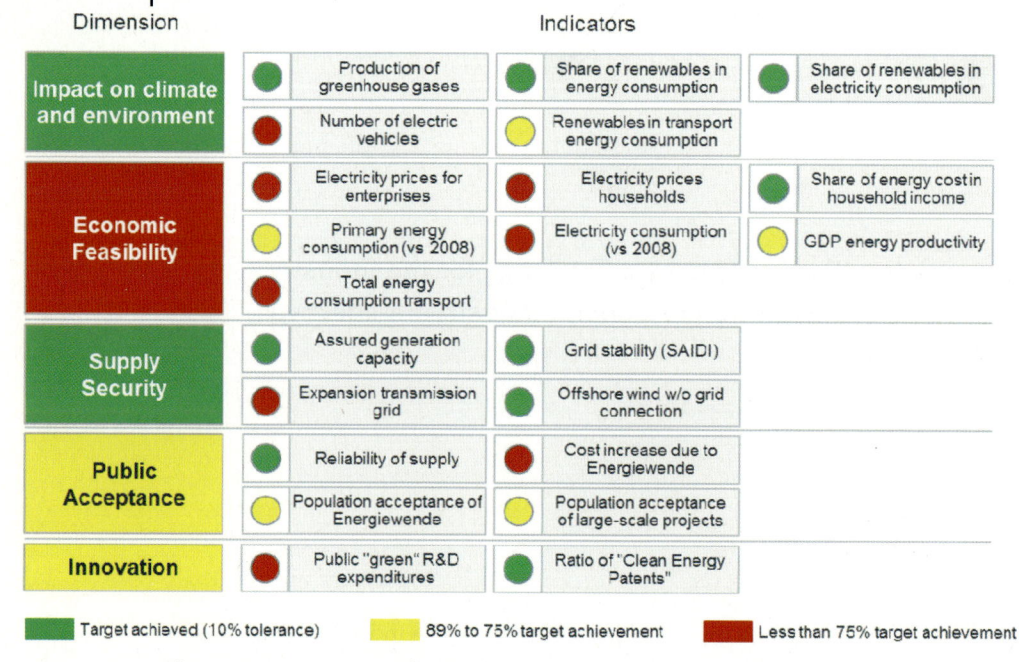

Dimension	Indicators					
Impact on climate and environment	🟢	Production of greenhouse gases	🟢	Share of renewables in energy consumption	🟢	Share of renewables in electricity consumption
	🔴	Number of electric vehicles	🟡	Renewables in transport energy consumption		
Economic Feasibility	🔴	Electricity prices for enterprises	🔴	Electricity prices households	🟢	Share of energy cost in household income
	🟡	Primary energy consumption (vs 2008)	🔴	Electricity consumption (vs 2008)	🟡	GDP energy productivity
	🔴	Total energy consumption transport				
Supply Security	🟢	Assured generation capacity	🟢	Grid stability (SAIDI)		
	🔴	Expansion transmission grid	🟢	Offshore wind w/o grid connection		
Public Acceptance	🟢	Reliability of supply	🔴	Cost increase due to Energiewende		
	🟡	Population acceptance of Energiewende	🟡	Population acceptance of large-scale projects		
Innovation	🔴	Public "green" R&D expenditures	🟢	Ratio of "Clean Energy Patents"		

🟩 Target achieved (10% tolerance)   🟨 89% to 75% target achievement   🟥 Less than 75% target achievement

Figure 5.1 Current status of German Energiewende from industry perspective [1].

Evaluating the current status of the Energiewende on its way to a successful implementation is reasonable on five dimensions; based on the three objectives of energy policy (supply security, environmental sustainability and competitiveness) we added *Public Acceptance* (as it is a crucial support factor) and *Innovation* (as it is a main driver for the industry). This evaluation has been done by taking into account different indicators (see Figure 5.1).

Thus, the status for *impact on climate and environment* is on a good way and has even more improved recently – especially because of over-achievement of installed renewable generation capacities. *Economic feasibility* has slightly improved but is still a critical issue – e.g. German households are facing the second highest electricity prices in Europe; prices for commercials are also on high levels (e.g. Germany 10–11 €cts/kwh vs. USA 5–6 €cts/kwh). *Supply security* remains stable at the moment, but decreased during the last months – Germany is internationally still top-notch regarding avoidance and resolving of blackouts and Germany's generation capacities are currently sufficient; yet regionally there has been an increased number of required interventions into grid stability. Also, the realization of grid extensions plan lags behind schedule. *Public acceptance* exhibits a complex current status: 60% of the German population support the objectives of the Energiewende, but only one quarter is willing to accept significantly higher electricity prices in the long-run. Industry-wise cost efficiency is even more important: 77% of the German companies expect increasing electricity prices, on the other hand one third is looking forward to rising revenues entailed by the Energiewende. *Innovation* status stays stable but not on a high level – public R&D expenditures have been stagnating. Still, there is emerging improvement due to future governmental research programs. The ratio of clean energy patents has been climbing. Additionally, it has become evident that

5.3
Industry view: opportunities and challenges

For German industry players the Energiewende holds both many challenges but also enormous opportunities. Particularly additional future revenue potentials are counting for credit postings. German companies may well be able to benefit from building up knowhow today and generating profits from an increasing international market in green technology tomorrow. In 2020 potential global sales volumes for German suppliers of green technology will rise up to more than € 70 billion per year in 2030 (see Figure 5.2).

Coming from approx. € 40 billion per year in 2011 this means an annual growth rate of 4%. Grid technologies in 2020 account for around € 31 billion, potentials for electricity generation sum up to € 25 billion per year (thereof about 2/7 will be from renewables). As in this calculation we assume stable German market shares in the world, additional market revenues might arise from:

- increasing global market shares of German companies,
- increasing markets,
- new intermediate products (to be developed – not yet identified).

Furthermore, significant additional potentials exist in the field of energy efficiency technologies: taking into account a growing energy efficiency for intermediate and end products we find extra gross value added of approx. € 3 to 6 billion per year in 2020. Last not least, Germany is assembling comprehensive expertise and a first mover advantage by remodelling its entire energy system towards a more decentralized and complex one.

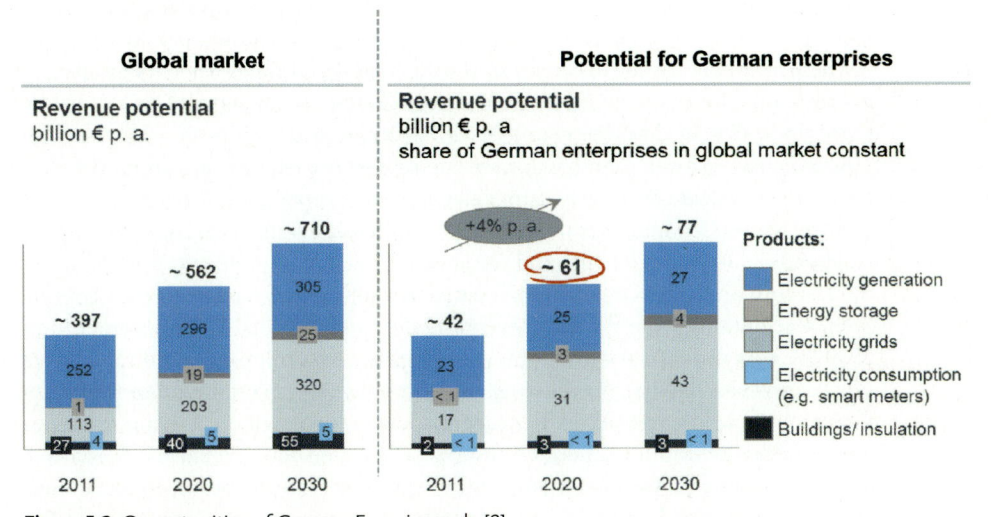

Figure 5.2 Opportunities of German Energiewende [2].

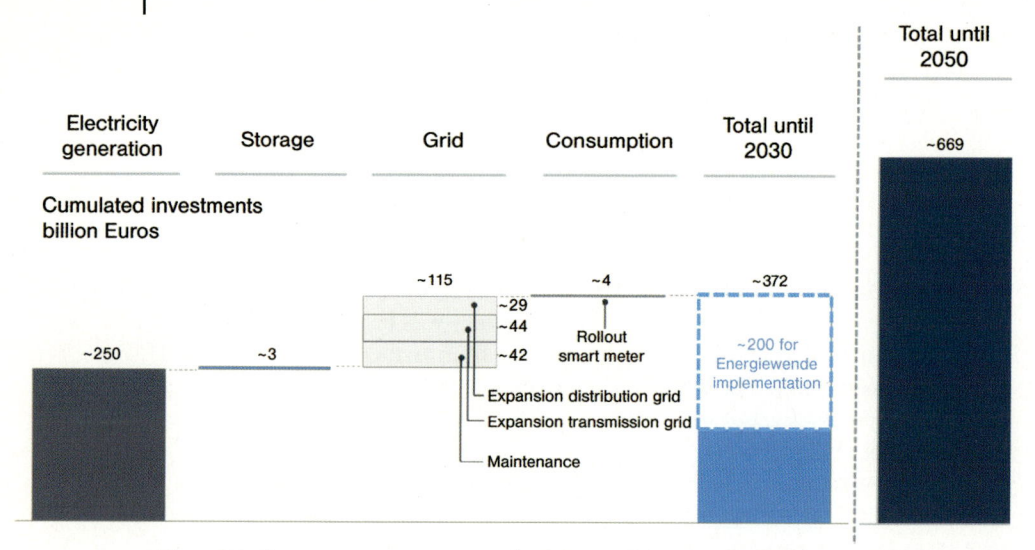

Figure 5.3 Necessary investments into the German electricity system [2].

Besides industrial effects the Energiewende also means a higher energy independence for Germany and less greenhouse gas emissions. Fuel imports will decrease by 2% per year and CO_2 emissions will be reduced almost by half until 2030.

Implementing the Energiewende also requires considerable investments in the German energy system. In order to achieve all targets a cumulated amount of more than € 350 billion have to be invested until 2030. Until 2050 this absolute value increases up to more than € 650 billion of which a large portion consists of replacement investments. Renewable energies – especially PV and wind (both onshore and offshore) – account for around € 200 billion, conventional generation for less than € 50 billion. For stability reasons and in order to integrate the fluctuating electricity generated by renewable energies the German grid will have to be expanded. This grid expansion will cost approx. € 70 billion, maintaining the existing grid another € 40 billion. Figure 5.3 gives an overview of the necessary cumulated investments.

These numbers are due to certain assumptions and may differ when using altering assumptions, for instance investment cost in the distributions networks could be saved up to 15% by developing and installing smart grids.

The displayed investments will translate into future cost for the supply of electricity. In 2030 unit cost of generating electricity and directing it through a network will be 15% to 35% higher compared to a continuation of the current fossil system – depending on prices for fuel and CO_2 (see Figure 5.4). On the other hand, under the assumption that prices for fossil fuels (especially gas and hard coal) will continue their development from the last 20 years and calculating with € 70 per ton CO_2, in 2030 the cost difference would only be marginal. Assuming fuel and CO_2-prices increasing even stronger in the long-run cost advantages could emerge from implementing the Energiewende. In order to allow for different underlying strategies when implementing the Energiewende we calculated four scenarios: Scenario 1 assumes all Energiewende-targets completely achieved (including reduced electricity consumption by about 23%). Scenario 2 also assumes all Energiewende-tar-

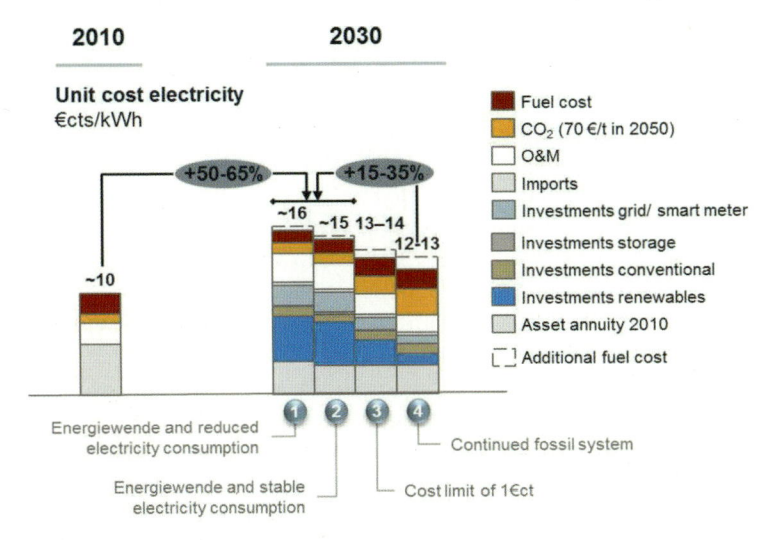

Figure 5.4 Expected development of unit cost for electricity (scenarios) [2].

gets achieved, but with a stable electricity consumption. Compared to scenario 1 we modeled a more excessive development of renewables and storage capacities. Generation capacities in scenario 4 are arranged to secure extensive CO_2-reduction limiting the increase of cost to 1 €ct per kilowatt-hour compared to a continued fossil system. Therefore we assume less PV and wind offshore capacities but expanded usage of conventional generation capacities than in scenario 2. Scenario 4 serves as a reference scenario for extrapolating the current system, but taking into account the nuclear phase-out by allowing for longer life-times of gas generation capacities. All scenarios presume CO_2-cost of 70 € per ton and a supply security regarding domestic generation capacities (imports and exports in the model have been limited to 10% of the hourly load curve in order to assure robustness of the results regarding biases in neighboring countries).

This cost analysis comprises cost of capital of existing assets (conventional power plants, renewable generation capacities and transmission and distribution grids), cost of capital of future assets, cost of net imports as well as cost of operation and maintenance, fuels and CO_2. Not being able to confirm a cost benefit in the medium-term the scenario calculations nevertheless illustrate the opportunity for hedging risk of fuel and CO_2 prices. Due to declining usages of fuel and due to the increase of other cost the portion of fuel cost decreases from 20% in 2010 to 8–11% in 2030 – a reduction of approx. 50%. The lower level of fuel usage also reduces the imports of fossil fuels by more than 30% until 2030, mainly for gas and oil.

Taking into account that so far no alternative pricing model is underway we translated this increase in cost into electricity prices. Under the assumption of the current pricing model staying in place the average electricity prices for enterprises will increase by 25% until 2020 within the four scenarios (compared to 2010). This increase is solely driven by rising tax and other allocations – whereas the spot price at the wholesale market will be sinking by approx. 25% due to the increasing portion of low marginal cost wind and PV. The assumed price increase is higher for households than for commercial customers, as wholesale prices have a higher impact on commercial prices. Electricity is a crucial input factor for manufacturing – especially for energy-intense industries (e.g. metals, chemicals). Therefore prices for electricity are a major driver of competitiveness. German commercial electricity prices compared to other countries have significantly increased over the last decade (3.6% p.a.) – in 2010 they amounted to 10.3 €cts/kWh on average and are thus ranging much higher than in the relevant economic areas worldwide (see Figure 5.5): twice as high as in the US (€cts 5.1 per kWh) and South Korea (€cts 4.2 per kWh). Compared to other European economies German commercial electricity prices are also rather high: UK (€cts 9.5 per kWh), France (€cts 6.9 per kWh) and Italy (€cts 11.5 per kWh).

Even though in many of the regions prices are likely to increase in the future Germany will not improve its relative position as prices here are expected to rise significantly as well (see discussion above). Infrastructure investments for conventional power plants renewable energy capacities and networks lead to increasing prices in many European countries during the last years, whereas industry customers in the US were able to benefit from the exploration of large shale gas reservoirs and thus decreasing prices. Furthermore, rising cost for fossil fuels, tax and the introduction of CO_2-certificates influenced European prices. These trends are likely to persist in the mid-term.

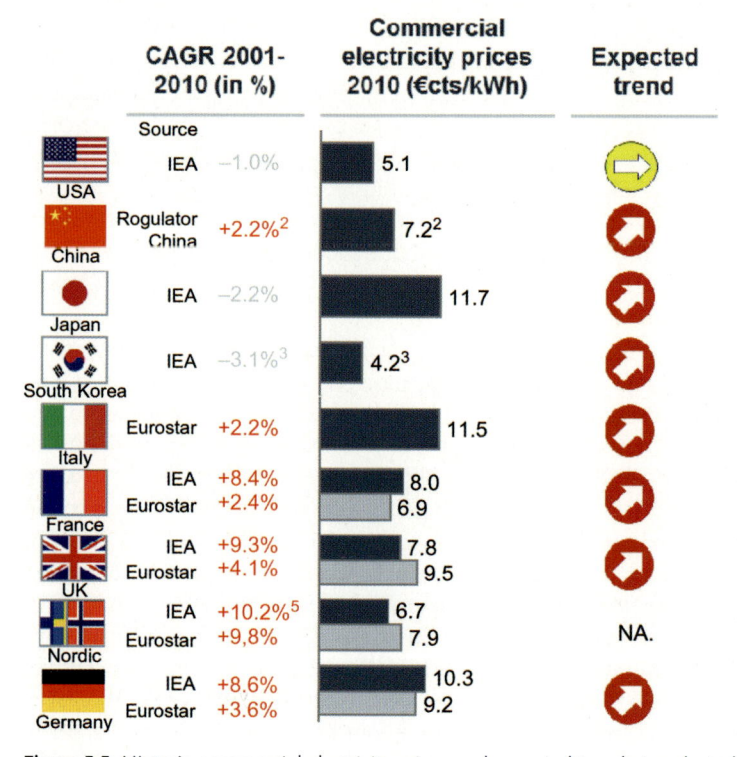

Country	Source	CAGR 2001-2010 (in %)	Commercial electricity prices 2010 (€cts/kWh)	Expected trend
USA	IEA	−1.0%	5.1	→
China	Rogulator China	+2.2%[2]	7.2[2]	↗
Japan	IEA	−2.2%	11.7	↗
South Korea	IEA	−3.1%[3]	4.2[3]	↗
Italy	Eurostar	+2.2%	11.5	↗
France	IEA / Eurostar	+8.4% / +2.4%	8.0 / 6.9	↗
UK	IEA / Eurostar	+9.3% / +4.1%	7.8 / 9.5	↗
Nordic	IEA / Eurostar	+10.2%[5] / +9.8%	6.7 / 7.9	NA.
Germany	IEA / Eurostar	+8.6% / +3.6%	10.3 / 9.2	↗

Figure 5.5 Historic commercial electricity prices and expected trends in selected regions [2].

5.4
The way ahead

Taking into consideration both challenges and opportunities the Energiewende holds for German enterprises some next steps have to be taken in order to minimize risk and exploit chances. From an industry perspective, amongst others four crucial points have to be tackled in the near future to get the Energiewende on the next level of implementation:

1. Foster and accelerate extension of electricity grid:
   In order to adjust the fluctuations of renewable energies and to integrate the increasing number of decentralized generation capacities the German electricity grid has to be stabilized by extending it. This process has been initiated but is currently behind schedule.
2. Develop and implement new market design:
   Currently the market design of the German electricity sector does not integrate output generated by renewable energies. Two separate systems exist in parallel – the renewable one and the conventional one. Still the renewable energies affect the market price restraining electricity output from conventional generation to enter the market. A new market design accounts for these aspects. It has to ensure economic viability of conventional generation capacities, integrate renewable and other flexibility options (e.g. storage, demand-side-management) and to support the cost recovery of grids.
3. Limit cost to an acceptable level:
   Besides all opportunities the Energiewende is offering it might get on the verge of coming with higher cost than expected. In order to strengthen and secure the competitive position of the German industry in the international context all effort has to be made to limit the additional cost to an acceptable level.
4. Think and act European:
   The Energiewende is not only a national issue as Germany is not an independent, closed entity – neither socially nor politically nor economically. This also holds true for the electricity market. Germany interacting with its European neighbors when importing and exporting electricity is reliant on other nations' energy strategies. Therefore the German Energiewende has to be discussed and coordinated on a supranational level.

Beside these tangible recommendations there exists another overriding field where to take action in; that is the alignment of diverging objectives. Despite substantial interfaces and correlations the different targets in the arenas of energy, climate and environmental policy are standing side by side in an unaligned manner. Moreover, on different political levels (e.g. European vs. national vs. regional) there are diverging goals and instruments in place. For affected enterprises this often implies multiple burdens culminating into electricity and emission trading cost, absorbing funds, hampering investments and thus weakening their international competitiveness. In order to clear the plurality of expressions and declarations and to avoid multiple burdens there has to be established a distinct differentiation between political overall objectives. Those have to be broken down into intermediary goals (in order to allow for fine adjustment) and finally refined to practical instruments which can

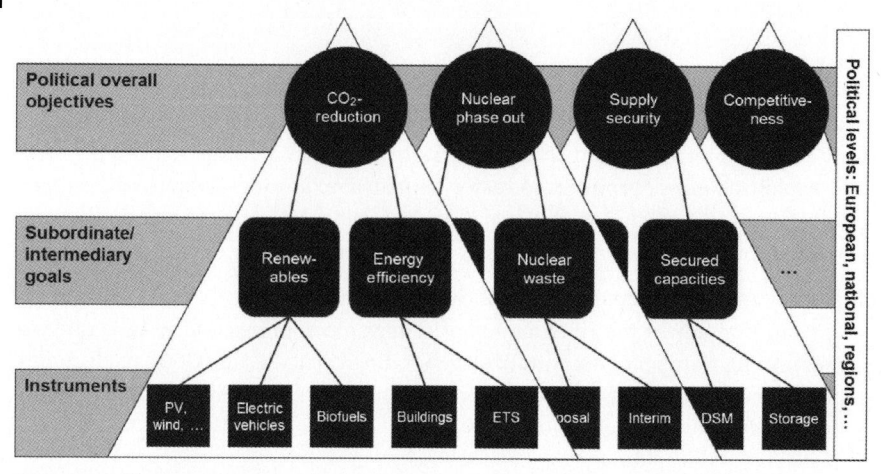

Figure 5.6 Schematic overview of diverging objectives in energy and climate policy.

be implemented in real terms. The realization of ambitious political objectives is cost-intensive. Therefore cost efficiency has to be a future criterion when setting these objectives and shaping the respective instruments. This means energy policy, climate policy and environmental policies have to be coordinated and evaluated regarding their cost related to climate protection. At the same time targets on the European level have to be aligned with national and regional levels. Figure 5.6 gives a schematic overview of exemplary objectives in energy and climate policy and on various political levels and on different target/instrument levels.

5.5
Conclusion

If arranged and implemented in the right way the German Energiewende has the chance to serve as a stimulation giving an example how to evolve a country with a highly developed industry, which has been heavily relying on fossil fuels and nuclear power, to a sustainable one. Nevertheless, to get there means overcoming a couple of challenges, technical, economic and political ones. The remodeling of Germany's energy system goes far beyond what has been expected by policy decision makers. The future development of cost has to be regarded very carefully by a rigorous monitoring process. The current market design has to be evolved in order to support the future generation landscape. Policy decision-takers need to support and foster these necessary measures also by taking a supranational view and aligning objectives and instruments.

References

1 BDI (2012) *Energiewende-Navigator 2012 des BDI*, Berlin.
2 Boston Consulting Group (2013) *Kompetenzinitiative Energie – Trendstudie 2030+*, Munich.
3 BDI (2013) *Energiewende auf Kurs bringen*, Bundesverband der Deutschen Industrie, Berlin.
4 ZEW (2012) *Indikatoren für die energiepolitische Zielerreichung*, Zentrum für Europäische Wirtschaftsforschung, Mannheim.

6
The Decreasing Market Value of Variable Renewables: Integration Options and Deadlocks[1]

Lion Hirth and Falko Ueckerdt

6.1
The Decreasing Market Value of Variable Renewables

Electricity generation from renewables has been growing rapidly in recent years, driven by technological progress, economies of scale, and deployment subsidies. Renewables are one of the major options for mitigating greenhouse gas emissions and are expected to grow significantly in importance throughout the coming decades [1–3]. As the potential of hydro power is widely exploited in many regions, and biomass growth is limited by supply constraints and sustainability concerns, much of the growth will need to come from wind and solar power. Wind and solar are variable[2] renewable energy (VRE) sources in the sense that their output is determined by weather, in contrast to "dispatchable" generators that adjust output as a reaction to economic incentives. Following Joskow [4], we define the market value of VRE as the revenue that generators can earn on markets without income from subsidies. The market value of VRE is affected by three intrinsic technological properties:

- The supply of VRE is *variable*. Owing to storage constraints and supply and demand variability, electricity is a time-heterogeneous good. Thus the value of electricity depends on *when* it is produced. In the case of VRE, weather determines *when* electricity is generated, which affects their market value.
- The output of VRE is *uncertain* until realization. Electricity trading takes place, production decisions are made, and power plants are committed the day before delivery. *Forecast errors* of VRE generation need to be balanced at short notice, which is costly. These costs reduce the market value.

---

1) The findings, interpretations, and conclusions expressed herein are those of the authors and do not necessarily reflect the views of Vattenfall or the Potsdam Institute.
2) Variable renewables have also been termed intermittent, fluctuating, or non-dispatchable.

Transition to Renewable Energy Systems, 1st Edition. Edited by Detlef Stolten and Viktor Scherer.
© 2013 Wiley-VCH Verlag GmbH & Co. KGaA. Published 2013 by Wiley-VCH Verlag GmbH & Co. KGaA.

6 The Decreasing Market Value of Variable Renewables: Integration Options and Deadlocks

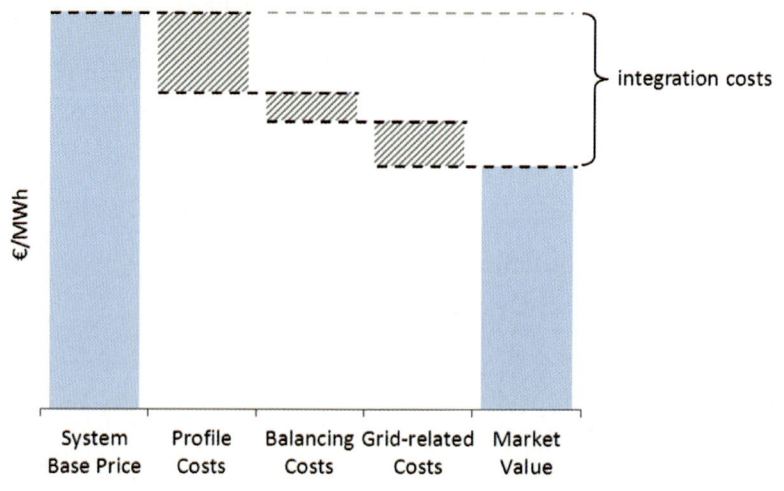

Figure 6.1 The system base price and the market value of wind power. The difference between these two can be decomposed into profile, balancing, and grid-related costs.

- The primary resource is bound to certain *locations*. Transmission constraints cause electricity to be a heterogeneous good across space. Hence the value of electricity depends on *where* it is generated. Since good wind sites are often located far from load centers, this reduces the value of wind power (but might increase the value of solar, if located close to loads).[3]

At high penetration rates, these three properties reduce the market value of VRE. We compare the market income of a VRE generator with the system base price (Figure 6.1). The system base price is the average wholesale electricity price during 1 year. The effect of variability is called "profile costs," the effect of uncertainty "balancing costs," and the effect of locations "grid-related costs". The sum of all three are "integration costs" [6].

Profile, balancing, and grid-related costs are not constant, but depend on a large number of factors and parameters. Most important, they are a function of the VRE penetration rate. The market value of wind and solar power decreases with the penetration rate. Equivalently, one can say that integration costs of wind and solar power increase with penetration (Figure 6.2).

This chapter is based on a recent journal paper [7] and two working papers [6, 8]. The techno-economic mechanisms that cause integration costs are discussed in Section 6.2. Quantifications from the literature, market data, and model results are presented. Section 6.3 extends this work by discussing integration options.

---

3) Of course, all types of generation are to some extent subject to expected and unexpected outages and are bound to certain sites, but VRE generation is much more uncertain, location-specific, and variable than thermal generation. Also, although weather conditions limit the generation of wind and solar power, they can always be downward adjusted and are in this sense partially dispatchable. The fourth typical property of VRE that is sometimes mentioned [5], low variable costs, does not impact the value of electricity.

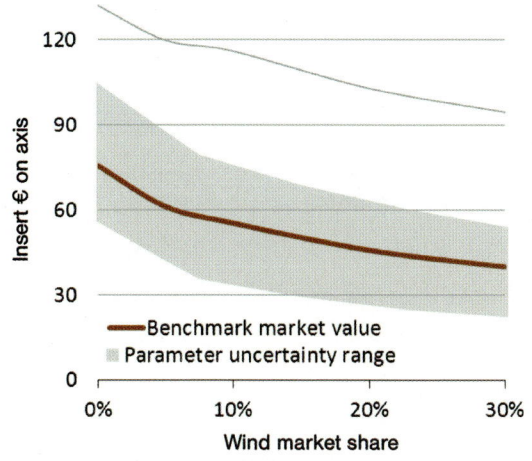

Figure 6.2 The market value of wind decreases with higher penetration. According to the European Electricity Market Model EMMA model results [7], wind power is worth €75 MWh^{-1} at low penetration, but only €40 MWh^{-1} at 30% market share. In all but one scenarios the value is between 20 and €50 MWh^{-1}.

We use the term "integration options" as an umbrella term that encompasses all measures that help to mitigate the value drop. Although the principle mechanisms that we discuss apply for all power systems, most examples are taken from the European context.

6.2 Mechanisms and Quantification

This section discusses the economic mechanisms and the underlying technological constraints that cause the market value to decrease. We complement that with quantifications from previously published studies, model results, and market data.

There are two branches of the literature on which we build. On the one hand, there is economic literature on the market value of VRE, such as [4, 9–15]. On the other hand, there is the "integration cost" literature, often found in engineering journals. Good overviews of this branch of the literature are given in [16–18]. We recently tried to translate the findings of these two schools into a common terminology [8].

For the analysis, we sometimes report the market value not in absolute terms (€ per MWh), but relative to the system base price. We call this relative price the "value factor."

Profile, balancing, and grid-related costs can be quantified from models or from market data. For example, profile costs can be estimated either from dispatch models or from observed spot prices. Figure 6.3 summarizes VRE properties, respective costs, and quantification strategies. Results of quantification exercises are discussed in the following subsections. Because of their large size, we will discuss profile costs in most detail.

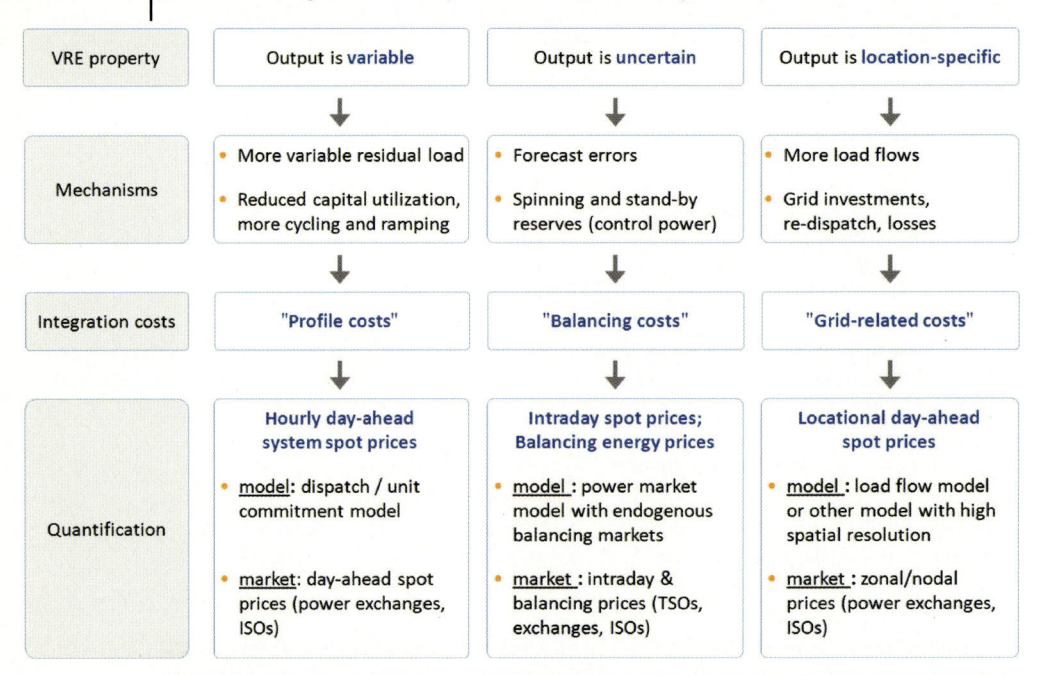

VRE property	Output is **variable**	Output is **uncertain**	Output is **location-specific**
Mechanisms	• More variable residual load • Reduced capital utilization, more cycling and ramping	• Forecast errors • Spinning and stand-by reserves (control power)	• More load flows • Grid investments, re-dispatch, losses
Integration costs	"Profile costs"	"Balancing costs"	"Grid-related costs"
Quantification	Hourly day-ahead system spot prices • <u>model</u>: dispatch / unit commitment model • <u>market</u>: day-ahead spot prices (power exchanges, ISOs)	Intraday spot prices; Balancing energy prices • <u>model</u> : power market model with endogenous balancing markets • <u>market</u> : intraday & balancing prices (TSOs, exchanges, ISOs)	Locational day-ahead spot prices • <u>model</u> : load flow model or other model with high spatial resolution • <u>market</u> : zonal/nodal prices (power exchanges, ISOs)

Figure 6.3 The three inherent properties of VRE, the corresponding costs, and possibilities for quantification from model and market data.
TSO = Transmission System Operator; ISO = Independent System Operator.

6.2.1
Profile Costs

Wind and solar power have variable costs of close to zero. Power is produced when the wind is blowing and the sun is shining – independently of the power price. In times of high wind speeds or solar radiation, VRE sources generate so much electricity that they reduce the electricity price. As a consequence, VRE "cannibalizes itself." The more VRE capacity is installed, the stronger is this effect.

Figure 6.4 displays the average price paid for electricity from wind and solar power in Germany relative to the base price (value factor) for the period 2001–12. As the wind penetration rate grew from 2% to 8%, the price of wind power fell from 1.02 to 0.89 of the base price. As the solar penetration grew from 0% to 4%, the solar value factor decreased from 1.3 to 1.05. These historical market prices confirm the model results presented in Figure 6.2.

The impact of solar power can be easily seen in the daily price structure (Figure 6.5). Whereas historically prices used to be high around noon due to high demand, now they are much reduced because of the solar in-feed during those hours. In that way, VRE sources reduce their own revenues and thus the market value decreases.

The value drop can be explained by the way in which the equilibrium price of electricity is determined. The price settles where the merit-order curve (short-term supply curve) intersects with residual demand (demand net of VRE generation).

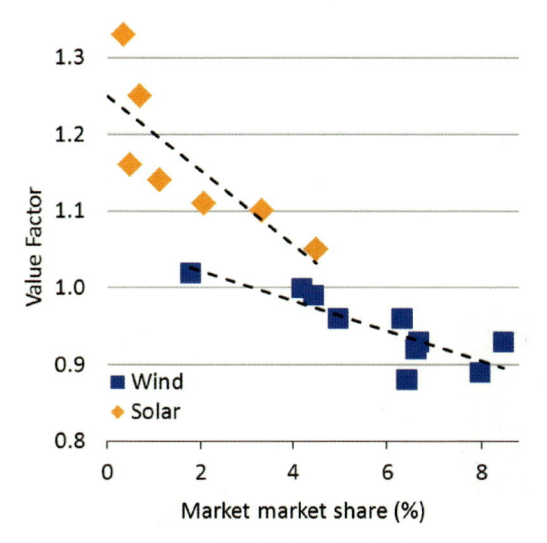

Figure 6.4 Historical wind and solar value factors in Germany. Data sources and methodology are discussed in [7].

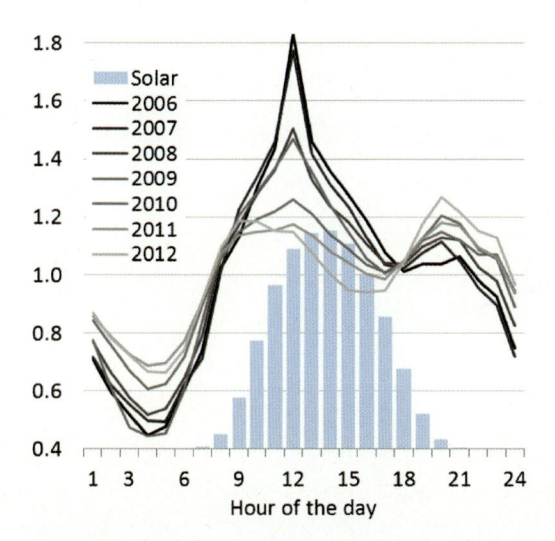

Figure 6.5 The daily price structure in Germany during summers from 2006 to 2012. The bars at the bottom display the distribution of solar generation over the day [7].

During windy and sunny hours the residual load curve is shifted to the left and the equilibrium price is reduced, which we call the "merit-order effect" (Figure 6.6). The more capacity is installed, the larger the price drop will be. This implies that the market value of VRE falls with higher penetration (Figure 6.7).[4]

4) In economic terms, the equilibrium price clears the market by equalizing demand and supply. This mechanism is a universal principle and not confined to power markets. However, because electricity is very costly to store, its price varies strongly on short time scales (minutes to hours).

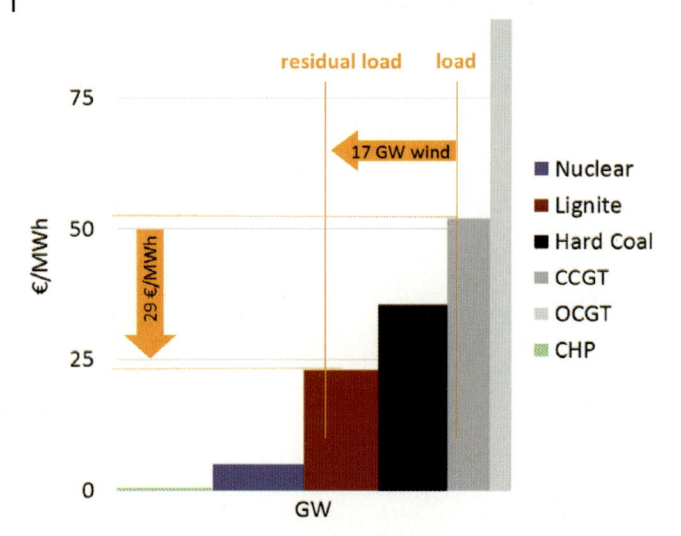

Figure 6.6 Merit-order effect during a windy hour: VRE in-feed reduces the equilibrium price (numbers are illustrative). This is the case in thermal[5] power systems. CCGT = combined-cycle gas turbine; OCGT = open-cycle gas turbine; CHP = combined heat and power.

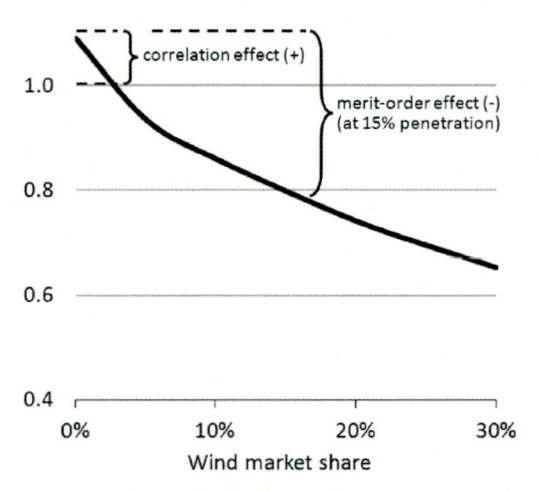

Figure 6.7 The wind value factor. The positive seasonal correlation increases the value of wind at low penetration rate. The merit-order effect reduces it at higher penetration rate.

---

5) "Thermal" (capacity-constrained) power systems are systems with predominantly thermal generators. These systems offer limited possibility to store energy. In contrast, (energy-constrained) "hydro" systems have significant amounts of hydro reservoirs that allow storage energy in the form of water.

Table 6.1 Utilization of the residual generation capacity (RES) with increasing share of VRE.

Parameter	VRE share (% of consumption)					
	No RES	10% RES	20% RES	30% RES	40% RES	50% RES
Peak residual load (GW$_{thermal}$)	80	74	73	73	72	71
Residual generation (TWh$_{residual}$)	489	440	391	342	293	244
Utilization of residual capacity [in FLH (full load hours)]	70% (6100)	68% (6000)	61% (5300)	54% (4700)	47% (4100)	39% (3500)
Average utilization effect (€ MWh$_{VRE}^{-1}$)[a]	0	10	20	24	30	39

[a] Assuming €80 MWh$_{VRE}^{-1}$, a constant average capital cost of the residual system of €200 kW a^{-1}, and a ratio of wind to solar energy of 2:1. For details, limitations, and sources, see [8].

At low penetration rates, before the merit-order effect comes into force, the value of wind and solar power is above the base price (at least in Europe). The reason is that VRE and demand are positively correlated, wind on seasonal time scales and solar on diurnal scales.

Looking at profile costs from a system cost perspective, there are two underlying techno-economic mechanisms that cause these costs to arise. Intuitively, more variability causes more ramping and cycling of thermal plants ("flexibility effect"). This is costly because of part-load efficiency losses, start-up fuel costs, and increased wear and tear.

However, there is a less obvious, but economically more important mechanism: high VRE shares reduce the average utilization of plants ("utilization effect" [14]). This is costly, because the capital embodied in these plants is costly. Table 6.1 provides illustrative calculations based on German load and in-feed data. As the market share of VRE increases from 0% to 50%, the average utilization of thermal capacity is reduced from 70% to 39%. These are also the fundamental reasons why prices fall if residual demand decreases.

Having discussed the mechanisms, we now come to quantifications of profile costs. Wind value factor estimates of around 30 published studies are summarized in Figure 6.8. At low penetration rates, wind value factors are reported to be close to unity. They are estimated to decrease to around 0.7 at 30% market share. Solar value factors are reported to decrease faster, so they reach 0.7 at 10–15% penetration rate (not shown). Figure 6.9 displays model results from the north European power system model EMMA [7]. The value factor is estimated to decrease to 0.5–0.8 at a penetration rate of 30%. The model results are consistent with the literature, and also consistent with the market data shown in Figure 6.4.

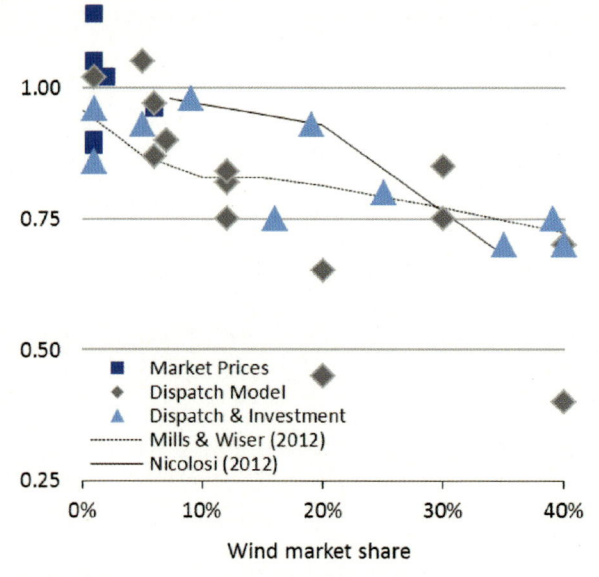

Figure 6.8 Wind value factors as reported in the literature. For a list of references, see [8].

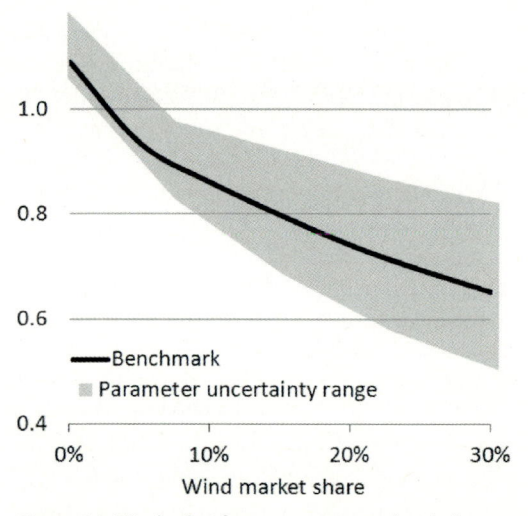

Figure 6.9 Wind value factors as estimated with the energy system model EMMA [7]. The benchmark runs are best-guess parameter assumptions.

6.2.2
Balancing Costs

Wind speeds and solar radiation are fundamentally stochastic processes, hence wind and solar predictability will always be limited. Realized VRE generation deviates from day-ahead forecasts. Balancing costs arise because balancing these forecast errors is costly. Those cost are caused by the capital costs of idle stand-by reserves, wear and tear due to cycling and ramping, and part-load efficiency losses.

Grubb [10] estimated balancing costs statistically to be around 3.6% of the value of electricity. Mills and Wiser [15] and Gowrisankaran *et al.* [19] modeled balancing costs in unit commitment models and reported them to be 3–5% of the base price. Wind integration studies [16, 17, 20, 21] reported balancing costs below 10% of the base price, sometimes below €1 MWh^{-1}.

Studies based on observed prices for balancing energy often find much higher balancing costs, for example [22]–[24]. However, market balancing costs well below 10% of the base price have been reported [8, 25].

Despite some conflicting evidence, we are confident in concluding that balancing costs are significantly smaller than profile costs at high penetration rates.

6.2.3
Grid-Related Costs

The quality of renewable energy resources varies across space. For example, windy sites with cheap land and little acceptance issues are typically located far away from load centers. This implies that adding large amounts of VRE to a power system increases load flows, which in turn increase network losses and tighten grid constraints. These are the reasons for "grid-related costs." On markets, these costs are represented as locational marginal spot prices (nodal or zonal) or as geographically differentiated grid fees.

Quantifications of grid-related costs are sometimes reported in wind integration studies [17, 26], and there are a few studies that used location prices [27, 28]. However, the results are very diverse for different power systems and methodologies. In general, grid-location costs are higher in geographically widespread power systems that can be found in the United States and in Nordic countries, but lower in continental European systems that feature a more dense transmission network. In the thermal systems of continental Europe, grid-related costs are probably significantly smaller than profile costs.

6.2.4
Findings

Three robust findings emerge from this review. First, integration costs are high. They can reduce the market value of wind and solar power to half of the system base price or less. Second, the market value decreases with penetration. Finally, profile costs are, under most conditions, higher than balancing costs, despite the latter seeming to attract much more attention. Within profile costs, the utilization effect is more important than the flexibility effect.

An important consequence of the decreasing market value is that it is unlikely that wind and solar power will become competitive if deployed on a large scale. However, there are a multitude of options that increase the market value of VRE.

6.3
Integration Options

The previous section explained why the value of wind and solar power decreases with penetration, and showed that this fall in value can be very significant. However, there are a number of options to mitigate this. We call these options collectively "integration options." In this section, we first introduce a new taxonomy of integration options, then we discuss options to tackle profile, balancing, and grid-related costs one by one.

During that discussion, it should be kept in mind that increasing the VRE market value is not an end in itself. Most (but not all) integration options are costly, and it is not clear if and to what extent these options are economically efficient. Only an integrated welfare analysis of the power system can reveal which integration options should be pursued.

Moreover, in the absence of externalities, market prices will incentivize all efficient integration options. Hence this section should *not* be read as a list of things that policy should subsidize, but rather as a starting point for further research.

6.3.1
A Taxonomy

Integration options can be classified along at least three dimensions:

1. *Which* integration challenge is addressed: variability, uncertainty, locational specificity.
2. *How* the challenge is addressed: the challenge itself is mitigated, or its economic impact reduced.
3. The *type* of integration option: technological innovation, investments, market design.

Table 6.2 summarizes this taxonomy as a matrix along the first two dimensions. Take the example of profile costs (first column). Profile costs arise because VRE variability increases the variability of residual load, thereby increasing specific (€ per MWh) capital costs and cycling of plants. Some integration options, such as increased long-distance transmission, or a different wind turbine design with more even output, reduce the variability of VRE. Other integration options, such as a shift of the thermal generation mix from capital-intensive base-load to peak-load generators, does not change VRE variability itself, but reduces the economic impact by reducing capital costs.

Table 6.2 A taxonomy of integration options.

	Profile costs	Balancing costs	Grid-related costs
Challenge	*Variability*: residual load becomes more unevenly distributed and more volatile	*Uncertainty*: forecast errors increase in absolute terms	*Locational specificity*: geographical distance between generation and consumption increases
Economic impact (cost driver)	Reduced capital utilization (utilization effect) and more ramping and cycling of plants (flexibility effect)	Reservation and activation of fast-reacting reserves, e.g., control power	Grid congestion
Mitigate the challenge	*Increase utilization of capital*: storage, transmission, DSM (demand-side management), different turbine layout, reduce must-run	*Make forecast errors smaller*: improve weather forecasts, joint load–VRE forecasts, transmission	*Reduce mismatch*: move generation closer to loads, e.g., via technology shift from offshore wind to solar photovoltaics
Reduce economic impact	*Reduced capital intensity*: shift generation mix from base to peak load	*Provide quickly responding capacity*: more flexible thermal plants, improve spot market design, change control power market design, integrate control areas	*Reduce congestion*: grid investments, introduce locational price signals on spot markets

6.3.2
Profile Costs

The important driver of profile costs is the reduced utilization of the capital stock embodied in the power system, especially in thermal plants. To reduce specific (per MWh) capital costs, on the one hand, utilization can be increased. On the other hand, the capital intensity of the system can be reduced.

By far the most important means of reducing capital intensity is a shift in the thermal generation mix from capital-intensive base-load technologies such as nuclear power, but also lignite and lignite CCS (carbon capture and storage), to less capital-intensive mid- and base-load generators such as open cycle and combined cycle gas turbines. Simple back-of-the envelope calculations indicate that at a VRE penetration rate of 50% no thermal plant will run more than 7000 full load hours (FLH), hence base-load technologies are not needed at all (Figure 6.10). Turned around, investing today in long-living base-load plants creates a barrier to high VRE deployment. This might be the most important potential deadlock for the transition to renewable energy systems.

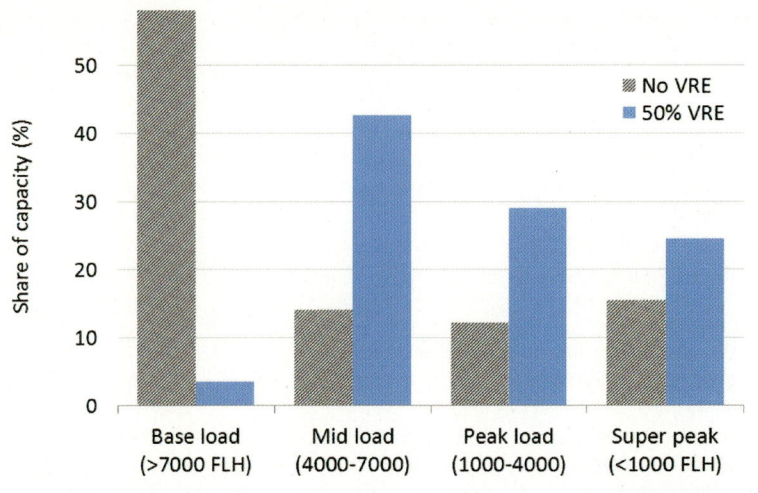

Figure 6.10 The cost-optimal distribution of thermal capacity without VRE and at a VRE share of 50% [8].

A wide range of options exist to increase the utilization of capital in thermal plants and the rest of the power system:

- electricity storage
- demand response
- market integration of different thermal power systems
- market integration of thermal and hydro power systems
- options at the electricity–heat interface
- unconventional ancillary service provision
- different wind turbine design
- a more balanced mix of variable renewables.

A discussion of technological characteristics, cost structures, or learning potentials of these options is beyond the scope of this chapter. Instead, we focus on the qualitative impact that each option has on the market value of wind and solar power.

Very intuitively, *electricity storage* and *demand response* even out fluctuations of renewables and load and increase the utilization of thermal plants. Solar power fluctuates mainly on daily time scales, such that daily storage helps in integrating solar power, as do demand response activities that shift demand for a few hours. Wind power fluctuates more irregularly over a wide range of time scales, hence it requires more long-term storage. Since long-term storage is costly, electricity storage and demand response could be a more important option for solar power than for wind power.

Integrating different thermal power systems via transmission investments and/or market design changes such as (flow-based) market coupling helps to keep up the market value, because fluctuations are smoothed over a larger geographic area [29].

However, weather systems in Europe typically have a size of 1000–1500 km, such that transmission grids have to cover fairly long distances for effective smoothing. For example, model results [7] indicate that doubling the interconnector capacity between north-western European countries (not including Nordic) would increase the value of wind by less than €1 MWh^{-1} at a penetration rate of 30%. The impact on solar power is even less, since solar generation is better correlated than wind.

Integrating thermal with hydro systems is more promising. Reservoir hydro power can offer intertemporal flexibility and can readily attenuate VRE fluctuations. In Europe, flexible hydro plants are located in the Nordic countries, the Alps, and also in France and Spain. Making existing hydro flexibility in Norway and Sweden available to the European power system could be one of the crucial options to stem the fall in value of VRE. Hydro power is an important integration option both for wind and solar power.

A failure to integrate markets in Europe could produce an important deadlock. National solutions for market rules, capacity markets, and balancing settlement and also sluggish interconnector capacity expansion would create a long-lasting barrier for VRE integration.

The *interface between heat and electricity* offers a number of flexibility options (which might also be classified as storage or demand response). A prominent example is the application of heat storage units at combined heat and power (CHP) plants, which has been pioneered in Denmark. Model results [7] indicate that the impact on both solar and wind market value could be very large in systems with significant CHP generation. Other possibilities are to include heat pumps or direct electrical heating in heating grids, or to combine heat storage units with heat pumps of micro-CHP in small-scale heating systems. However, some of these measures are contrary to the aim of increasing energy efficiency.

Ancillary services such as control power and voltage support are today usually provided by synchronized generators. During the time they provide these services, generators typically have to be dispatched ("must-run"). As an alternative, control power can also be provided by loads, variable renewables, or storage units. Voltage support can be provided by phase-shift transformers or power electronics such as the converters that are already installed in wind turbines and photovoltaic systems. It is important that grid codes and market design do not prevent the entry of these unconventional technologies into ancillary service markets.

The *design of wind turbines* has a large impact on the variability of their output. By increasing hub heights and the ratio of swept area to electrical capacity, wind turbines are able to provide more stable output. Modern wind turbines are already designed to run about 3000 FLH, whereas historically they often delivered only 2000 FLH or less.

Finally, *the renewables mix* could be adjusted to reduce overall VRE variability. Wind and solar output combined is less variable than the output of each technology separately. Future technologies that are variable, but uncorrelated or even anticorrelated with wind and solar power generation, could improve the mix further. However, potential technologies such as wave power are far away from being commercially deployed.

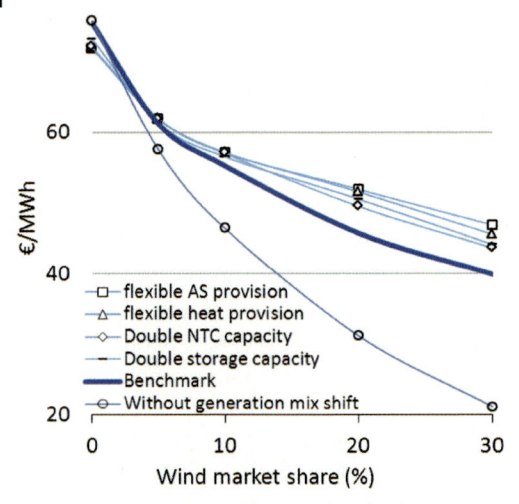

Figure 6.11 Some integration options modeled in EMMA. The benchmark value is the same as in Figure 6.2. A flexible provision of ancillary services (AS) or district heating in addition to increasing interconnector (NTC) or storage capacity increases the market value. Not allowing the capital stock to adjust dramatically reduces the value.

Some have argued that because of the decreasing VRE market value, the current energy-only wholesale markets should be transformed [30, 31]. However, the value drop is *not* the consequence of a flawed market design. It efficiently reflects the economic costs of variability. We believe that energy-only markets are well suited to integrate large amounts of VRE.

Figure 6.11 summarizes the effect of four of the discussed integration options as modeled in EMMA. All integration options increase the long-term market value of wind significantly, by €4–7 MWh^{-1} or 10–18%. However, although integration options can mitigate the value drop, they cannot prevent it.

6.3.3
Balancing Costs

There are four broad options for reducing balancing costs: reduce forecast errors, make existing flexibilities available for provision of balancing services, create new sources of flexibility, and geographical integration of control areas. Whereas the first option tackles the problem itself (forecast errors), the other three reduce its economic impact. Each option is discussed in turn.

Improved meteorological models and the use of real-time generation data can help to reduce forecast errors of wind and solar power significantly, especially for short prediction horizons ("nowcasting"). In systems where VRE generation and load are negatively correlated (solar power and cooling demand; wind power and heating demand), the joint forecast of VRE generation and load can reduce overall forecast errors. Sometimes market design changes might be necessary to set the right incentives for these technological changes. As has been emphasized [32],

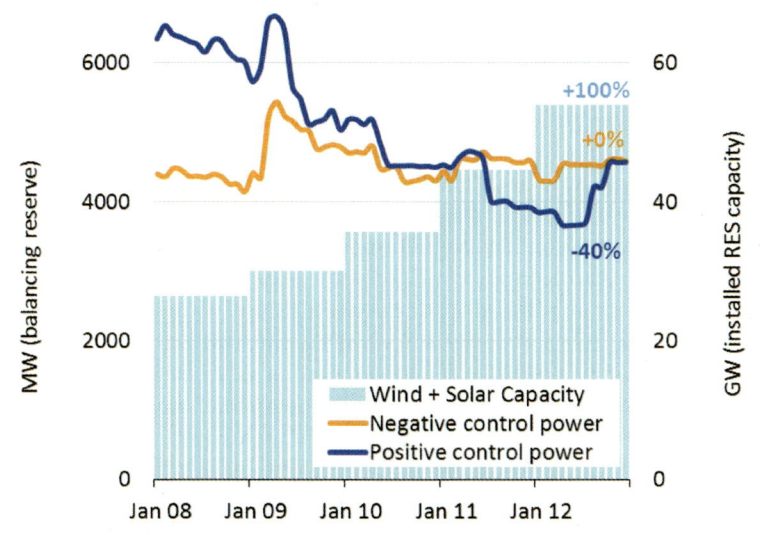

Figure 6.12 Reserved control power capacity could be reduced by 20% despite a doubling of VRE capacity in Germany. The main reason for this is a cooperation of TSOs. Source: [33].

a single price balancing settlement system with marginal pricing provides efficient incentives.

Existing flexibility resources that can provide balancing services at low costs should be activated by opening the respective markets. Liquid intra-day markets with short gate-closure times (1 h and less) and short contract durations (15 min and less) allow VRE generators to use continuously improving weather forecasts. Intra-day markets could replace day-ahead auctions as the most important spot market. Lowering entrance barriers to regulating power markets is crucial to allow small generators, loads, storage facilities, and foreign suppliers to bid into the market.

In the long run, existing sources of flexibility might not be sufficient. Thermal and hydro plants with higher ramping capabilities and lower minimum load might be needed. In addition, many of the options that were listed in Section 6.3.2 could also provide fast-responding flexibility.

Integrating a larger geographic region into one control area helps in balancing VRE and other forecast errors. Since 2009, the four German TSOs have cooperated closely, which helped to bring down the need for regulating power provision despite a strong growth of VRE capacity (Figure 6.12).

6.3.4
Grid-Related Costs

Probably there is only one sensible measure to reduce grid-related costs: investment in transmission grids. Back-of-the envelope calculations suggest that relocating generators or even loads is almost always more expensive than building transmission lines. Model results confirm this [34].

6.4
Conclusion

This chapter has discussed the decreasing market value of wind and solar power as a barrier for the transition to renewable energy systems. VRE sources feature three distinct properties, variability, uncertainty, and locational specificity. These characteristics cause the market value of electricity from wind and solar power to decrease with higher penetration. Equivalently, one can say that "integration costs increase." In many cases, the decrease in market value is so strong that it probably overcompensates learning effects and cost decreases. Hence, without changes in the energy system, wind and solar power will never become competitive at a large scale and subsidies would be needed to achieve ambitious policy targets.

We have propose a taxonomy to structure "integration options" and discussed a number of them. Integration options are measures that increase the market value of VRE at high penetration rates. However, increasing the market value is not an end in itself, and without a cost–benefit analysis we cannot say if these options make sense from a welfare perspective. Having said that, we believe that there are a number of "no regret options" that should be adopted independently of renewable deployment: transmission investment, making intra-day markets more liquid, lowering entrance barriers to control power markets, and market coupling of spot and control power markets.

On the other hand, there are a few actions that decrease the market value of variable renewables. Important deadlocks are large investments in base-load generation technologies and national solutions instead of European market integration.

References

1 IPCC (2011) *Special Report on Renewable Energy Sources and Climate Change Mitigation*, Cambridge University Press, Cambridge.
2 IEA (2012) *World Energy Outlook*, International Energy Agency, Paris.
3 GEA (2012), *Global Energy Assessment – Toward a Sustainable Future*, Cambridge University Press, Cambridge, and the International Institute for Applied Systems Analysis, Laxenburg.
4 Joskow, P. (2012) Comparing the costs of intermittent and dispatchable electricity generation technologies. *Am. Econ. Rev.*, **100** (3), 238–241.
5 Milligan, M. and Kirby, B. (2009) *Calculating Wind Integration Costs: Separating Wind Energy Value from Integration Cost Impacts*, NREL Technical Report TP-550-46275, National Renewable Energy Laboratory, Golden, CO.
6 Ueckerdt, F., Hirth, L., Luderer, G., and Edenhofer, O. (2013) *System LCOE: What Are the Costs of Variable Renewables?*, USAEE Working Paper 2200572, United States Association for Energy Economics, Cleveland, OH.
7 Hirth, L. (2013) The market value of variable renewables. *Energy Econ.*, in press, http://dx.doi.org/10.1016/j.eneco.2013.02.004; an earlier version is available as USAEE Working Paper 2110237, United States Association for Energy Economics, Cleveland, OH.
8 L. Hirth, Integration Costs and the Value of Wind Power. Thoughts on a valuation framework for variable renewable electricity sources. USAEE Working Paper 12–150, 2012.
9 Flaim, T., Considine, T. J., Witholder, R., and Edesess, M. (1981) *Economic Assessments of Intermittent, Grid-*

Connected Solar Electric Technologies, a Review of Methods, NASA STI/Recon Technical Report N, vol. 82, p. 30737.

10 Grubb, M. J. (1991) Value of variable sources on power systems, in *Generation, Transmission and Distribution, IEE Proceedings C*, 1991, vol. 138, pp. 149–165.

11 Hirst, E. and Hild, J. (2004) The value of wind energy as a function of wind capacity. *Electr. J.*, **17** (6), 11–20.

12 Lamont, A. D. (2008) Assessing the long-term system value of intermittent electric generation technologies. *Energy Econ.*, **30** (3), 1208–1231.

13 Twomey, P. and Neuhoff, K. (2010) Wind power and market power in competitive markets. *Energy Policy*, **38** (7), 3198–3210.

14 Nicolosi, M. (2012) The economics of renewable electricity market integration. an empirical and model-based analysis of regulatory frameworks and their impacts on the power market, PhD thesis, University of Cologne.

15 Mills, A. and Wiser, R. (2012) *Changes in the Economic Value of Variable Generation at High Penetration Levels: a Pilot Case Study of California*, Ernest Orlando Lawrence Berkeley National Laboratory, Berleley, CA.

16 Smith, J. C., Milligan, M. R., DeMeo, E. A., and Parsons, B. (2007) Utility wind integration and operating impact state of the art. *IEEE Trans. Power Syst.*, **22** (3), 900–908.

17 Holttinen, H., Meibom, P., Orths, A., Lange, B., O'Malley, M., Tande, J. O., Estanqueiro, A., Gomez, E., Söder, L., Strbac, G., Smith, J. C., and van Hulle, F. (2011) Impacts of large amounts of wind power on design and operation of power systems, results of IEA collaboration. *Wind Energy*, **14** (2), 179–192.

18 Milligan, M., Ela, E., Hodge, B. M., Kirby, B., Lew, D., Clark, C., DeCesaro, J., and Lynn, K. (2011) Integration of variable generation, cost-causation, and integration costs. *Electr. J.*, **24** (9), 51–63.

19 Gowrisankaran, G., Reynolds, S. S., and Samano, M. (2011) *Intermittency and the Value of Renewable Energy*, Working Paper 17086, National Bureau of Economic Research, Cambridge, MA.

20 Gross, R., Heptonstall, P., Anderson, D., Green, T., Leach, M., and Skea, J. (2006) *The Costs and Impacts of Intermittency: an Assessment of the Evidence on the Costs and Impacts of Intermittent Generation on the British Electricity Network*, UK Energy Research Centre, London.

21 DeMeo, E. A., Jordan, G. A., Kalich, C., King, J., Milligan, M. R., Murley, C., Oakleaf, B., and Schuerger, M. J. (2007) Accommodating wind's natural behavior. *Power Energy Mag. IEEE*, **5** (6), 59–67.

22 Holttinen H. (2005) Optimal electricity market for wind power. *Energy Policy*, **33** (16), 2052–2063.

23 Pinson, P., Chevallier, C., and Kariniotakis, G. N. (2007) Trading wind generation from short-term probabilistic forecasts of wind power. *IEEE Trans. Power Syst.*, **22** (3), 1148–1156.

24 Obersteiner, C., Siewierski, T., and Andersen, A. N. (2010) Drivers of imbalance cost of wind power: a comparative analysis, presented at EEM 10, 7th International Conference on the European Energy Market, Madrid.

25 Holttinen, H. and Koreneff, G. (2012) Imbalance costs of wind power for a hydro power producer in Finland. *Wind Eng.*, **36** (1), 53–68.

26 Deutsche Energie-Agentur (dena) (2010) *dena Grid Study II. Integration of Renewable Energy Sources in the German Power Supply System from 2015–2020 with an Outlook to 2025*, dena, Berlin.

27 Brown, S. J. and Rowlands, I. H. (2009) Nodal pricing in Ontario, Canada: implications for solar PV electricity. *Renew. Energy*, **34** (1), 170–178.

28 Lewis, G. M. (2010) Estimating the value of wind energy using electricity locational marginal price. *Energy Policy*, **38** (7), 3221–3231.

29 Obersteiner C. (2012) The influence of interconnection capacity on the market value of wind power. *Wiley Interdisciplin. Rev. Energy Environ.*, **1** (2), 225–232.

30 Kopp, O., Esser-Frey, A., and Engelhorn, T. (2012) Können sich erneuerbare Energien langfristig auf wettbewerblich organisierten Strommärkten finanzieren? *Z. Energiewirtsch.*, **36**, 243–255.

31 Winkler, J. and Altmann, M. (2012) Market designs for a completely renewable power sector. *Z. Energiewirtsch.*, **36** (2), 77–92.

32 Vandezande, L., Meeus, L., Belmans, R., Saguan, M., and Glachant, J.-M. (2010) Well-functioning balancing markets: a prerequisite for wind power integration. *Energy Policy*, **38** (7), 3146–3154.

33 Hirth, L. and Ziegenhagen, I. (2013) Control power and variable renewables: a glimpse at German data, presented at EEM 13, the 10th International Conference on the European Energy Market, Stockholm.

34 Göransson, L. and Johnsson, F. (2012) Cost-optimized allocation of wind power investments: a Nordic–German perspective. *Wind Energy*, in press, http://onlinelibrary.wiley.com/doi/10.1002/we.1517/abstract.

7
Transition to a Fully Sustainable Global Energy System

Yvonne Y. Deng, Kornelis Blok, Kees van der Leun, and Carsten Petersdorff

7.1
Introduction

The last 200 years have witnessed an incredible increase in energy use worldwide. In recent decades, it has become clear that the way in which this energy is supplied is unsustainable and both short- and long-term energy security are at the top of the political and societal agenda.

Scenario studies, which chart possible futures, typically show small incremental changes against a "business-as-usual" (BAU) reference. In contrast, evidence suggests that we should be able to meet our energy demand entirely from renewable sources, given their abundance: Global final energy use was ~310 EJ in 2007 (~500 EJ in primary energy terms) [1], whereas technical potentials for renewable energy sources range in the order of hundreds to thousands of exajoules per year.

In an attempt to reconcile these figures, the Ecofys energy scenario provides a comprehensive analysis, examining all energy uses worldwide, all carrier forms (e.g., electricity or fuel), and all purposes (heat in buildings or heat in industry).

The key question which guided our study was: "Is a fully sustainable global energy system possible by 2050?"

We conclude that an (almost) fully sustainable energy supply is technically and economically feasible, given ambitious, but realistic, growth rates of sustainable energy sources. The path to achieving this system deviates significantly from BAU and sometimes difficult choices must be made on the way. The work presented in this chapter is an adapted version of a larger report published in 2011, and subsequent scientific publications [2–4].

7.2 Methodology

Energy demand is the product of

- the volume of the activity requiring the energy (e.g., travel or industrial production) and
- the energy intensity per unit of activity (e.g., energy used per volume of travel):

$$E(t) = A(t) I(t) \tag{7.1}$$

where $E(t)$ is the total energy demand at a given time, $A(t)$ is the energy-requiring activity at that time, and $I(t)$ is the energy intensity of the activity at that time.

This energy scenario forecasts future global demand and supply by following this prioritization (Figure 7.1):

1. Future energy demand scenario:
   a. Future demand side activity is based on existing studies or projected from population and GDP growth.
   b. Future demand side energy intensity is forecast assuming fastest possible roll-out of the most efficient technologies.

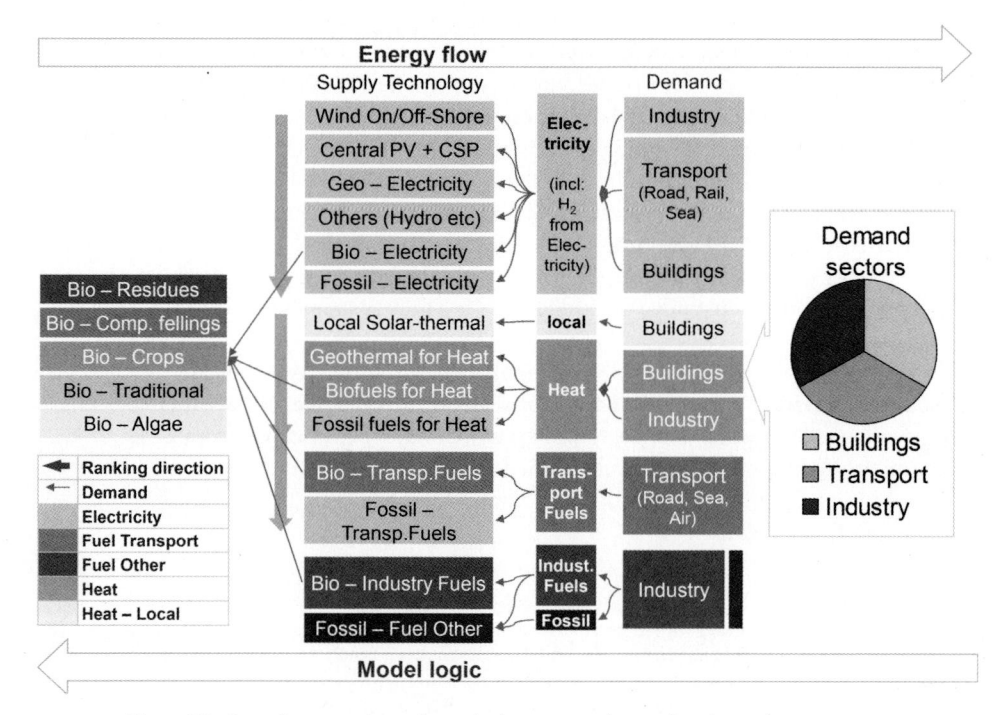

Figure 7.1 Overall approach used to calculate energy demand and supply.

c. The resulting energy demand is summed up by carrier (electricity, fuel, heat) and calibrated against International Energy Agency (IEA) energy statistics at sector-carrier level.
2. Future supply scenario:
    a. The potential for supply of energy is estimated by energy carrier.
    b. Demand and supply are balanced in each time period according to the following prioritization:
        i. Renewables from sources other than biomass (electricity and local heat) are used first, if available.
        ii. Biomass up to the sustainable potential is used second.
        iii. Conventional sources, such as fossil and nuclear, are used last.

Energy flows have been characterized by carrier type (electricity, heat and fuels), consistent with the IEA energy balances, to which this work is calibrated with 2005 as the base year [1]. Unless stated otherwise, all energy in this publication is final energy.

There is always a choice to be made of the technologies to include, depending on their stage of development. In this study, we have tried to rely solely on existing technologies or technologies for which only incremental technological development is required.

7.2.1
Definitions

We differentiate the various renewable power sources into two types (Table 7.1): the share of supply-driven power sources[1] is constrained to a ceiling to reflect the fact that a certain minimum share of balancing, or demand-driven, sources are required to ensure continuous supply (see also Section 7.5.1 Power Grids).

Table 7.1 Classification of renewable energy supply power into supply-driven and demand-driven (balancing) options.

Source	Type	Notes
Wind onshore	Supply-driven	
Wind offshore		
Tidal and wave		
Photovoltaics (PV)		
Concentrating solar power (CSP)	Demand-driven	With storage
Geothermal electricity		
Hydropower		But ensuring minimum water flows
Bioelectricity		

1) Supply-driven (also known as "inflexible") power options are those options whose generation at any given time depends on the availability of the energy source, for example, wind power, photovoltaic power, or ocean power. Demand-driven (also known as "flexible") power options are those options which can be tailored to demand, such as geothermal electricity, hydropower, CSP with storage, and electricity from biomass. Note that the classification into these two groups is somewhat arbitrary as all sources fall on a continuum of flexibility.

We have chosen to distinguish between energy demand in industry, buildings, and transport. These three sectors cover ~85% of total energy use and were studied in detail; they are congruent with the sectors in the IEA statistics, which form the basis of this work. The remaining sectors (e.g., agriculture, fishing) are included in this study, but were not examined separately.

The definition of demand side subsectors as used in this scenario is given in Table 7.2.

Table 7.2 Demand sector definitions.

Sector	Subsector (IEA)	Subsector (Scenario)	Scenario category
Industry	Iron and steel	Steel	A
	Nonferrous metals	Aluminum	A
	Nonmetallic minerals	Cement	A
	Paper, pulp, and print	Paper	A
	Chemical and petrochemical	Chemicals	B
	Food and tobacco	Food	B
	[All others]	Other	B
Buildings	Residential	Residential	n/a
	Commercial and public services	Commercial	n/a
Transport	Road	Passenger – PTWs (personal two- and three-wheelers)	n/a
	Road	Passenger – car, city	n/a
	Road	Passenger – car, non-city	n/a
	Road	Passenger – bus + coach	n/a
	Rail	Passenger – rail	n/a
	Aviation	Passenger – airplane	n/a
	Road	Freight – truck	n/a
	Rail	Freight – rail	n/a
	Aviation	Freight – airplane	n/a
	Domestic navigation + world marine bunkers	Freight – ship (includes both freight and passenger transport, but most energy use is presumed to be in the freight share)	n/a

7.3
Results – Demand Side

An understanding of our energy system begins with a detailed look at the demand for energy:

- Where is energy used?
- In what form and with what efficiency?
- Which functions does this energy deliver?
- Can this function be delivered differently?

A typical example to illustrate this approach is our energy use in buildings. A large fraction of our total energy demand, especially in cooler climates, comes from the residential built environment. The energy is used in the form of heat, often with large losses.

The desired function is a warm home, but does it need to be delivered in the current form? In fact, this function can be delivered with much less energy input if the building is insulated to reduce heat loss.

Asking these questions in all three demand sectors, industry, buildings, and transport, leads to our scenario for future energy demand.

Most activity projections are based on established forecasts of population and GDP growth [1, 4, 5].

7.3.1
Industry

7.3.1.1   Industry – Future activity
Activity levels for the "A" sectors (in tonnes produced) are linked to population growth. Activity levels for the "B" sectors are based on GDP growth.

A Sectors

For these sectors, an assumption is made on the future evolution of tonnes produced per capita. This per capita value is then multiplied by the future population to estimate total future production levels.

B Sectors

For these sectors, data on current production are difficult to find and the aggregation of many different production processes makes this a very heterogeneous sector to treat at a technological level.

We therefore followed a similar approach to that taken in many econometric models and assumed a future physical activity level linked to, but increasing less strongly than, GDP per capita.

Figure 7.2 shows the resulting activity evolution for both A and B sectors, indexed on 2005 levels. The reductions in OECD A sectors should not require a compromise of living standards, but reflects an increased re-use of materials at the consumer

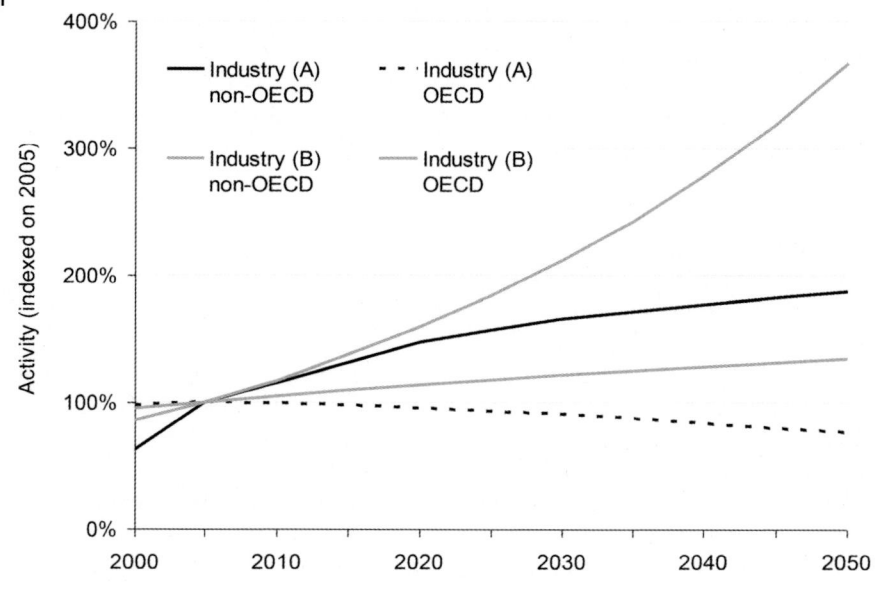

Figure 7.2 Evolution of indexed absolute activity levels in industry.

end and advances in material efficiency at the producer's end, for example, building cars with lighter frames. Note that production increasingly comes from recycling disused feedstock: stocks of energy-intensive materials have grown over the past decades. As large parts of the stock reach the end of their life, it is expected that recycling will increase as the availability of recoverable materials increases. This might result in a situation where production from primary resources will be needed only to offset losses due to dissipative use (e.g., hygienic papers, fertilizer), quality loss (e.g., paper fiber, plastics) or other losses.

7.3.1.2 Industry – Future Intensity

Future energy intensity is projected based on key marker processes. We adopt a decrease in energy intensity, measured in energy per tonne produced for A sectors, and in energy per economic value for B sectors.

A Sectors

The energy intensity evolution was examined in detail for the four A sectors and the results are shown in Figure 7.3. Although the individual technologies vary by sector, all sectors follow these common assumptions [6–9]:

- Recycling of steel, paper and aluminum and alternative input materials into the clinker process in cement production.
- Ambitious refurbishment of existing plants to meet performance benchmarks.
- Requirements for using best available technology (BAT) in all new plants.
- Continuing improvements of BAT over time.

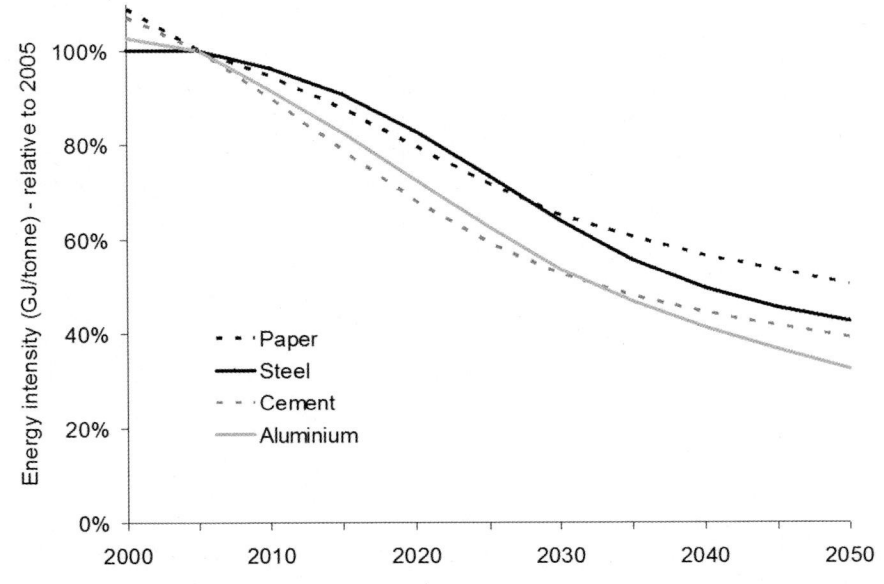

Figure 7.3 Evolution of energy intensity in industry (A sectors).

B Sectors

For the B sectors, an annual efficiency improvement of 2% was assumed, which may be obtained through improved process optimization, more efficient energy supply, improved efficiency in motor-driven systems and lighting, and also sector-specific measures.

7.3.1.3 Industry – Future Energy Demand

Figure 7.4 shows how total industrial energy demand would develop, resulting from the evolution of activity and intensity.

7.3.2 Buildings

The buildings sector provides high potential for energy savings and electrification. The sector is also marked by longevity: decisions on building design today influence building energy use for many decades [10, 11].

7.3.2.1 Buildings – Future Activity

Activity is expressed in terms of total floor area of buildings, measured in square meters. These steps are followed to establish future activity levels:

1. Total future floor area is projected based on population growth and increasing living space per capita. The future residential floor space per capita is informed by a relationship between living space and GDP per capita.

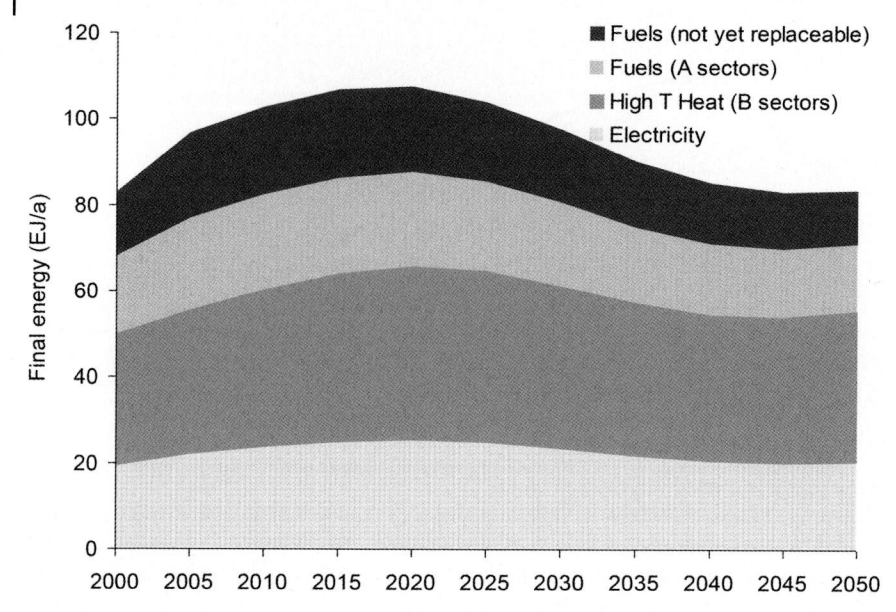

Figure 7.4 Evolution of energy demand in industry by energy carrier.

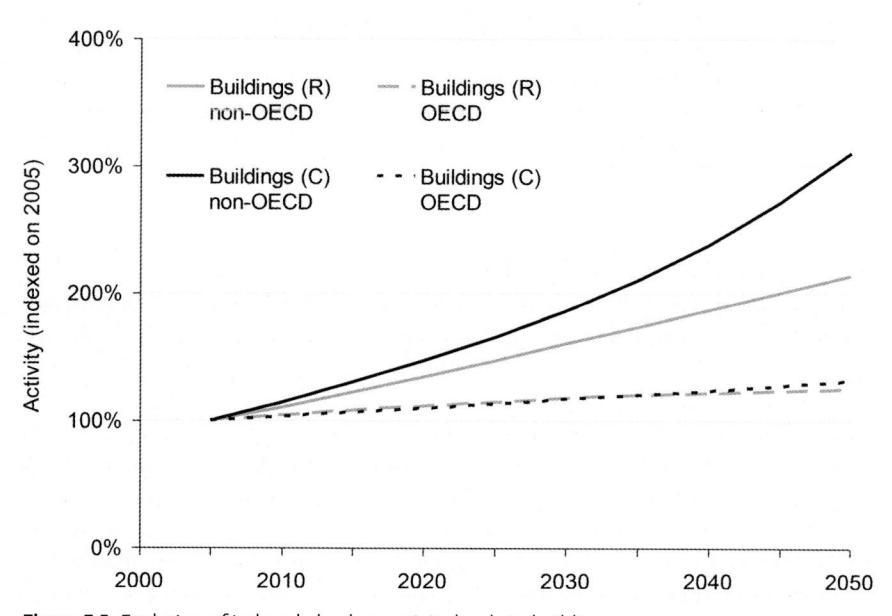

Figure 7.5 Evolution of indexed absolute activity levels in buildings.

2. Assumptions on typical, historical demolition rates are then used to divide the total area into floor area that exists today (pre-2005 stock) and floor area yet to be built (new stock).

For the residential buildings sector, this results in a decrease of existing building stock by ~10% to 2050, and an increase in newly built floor area of ~110% of the existing stock. This means that more than half the building stock in 2050 will have been built after 2005 and the majority of today's buildings will still be present.

For commercial floor space, a similar approach is followed but, rather than using population growth as a marker, this evolution is pegged to GDP growth with a decoupling factor.

The overall indexed evolution of residential and commercial floor space is shown in Figure 7.5.

7.3.2.2 Buildings – Future Intensity

The steps below are followed to project the future evolution of energy intensity, that is, the possible future heat and electricity demand per square meter of living or commercial floor space.

Existing Building Stock
1. All existing buildings will have to be retrofitted by 2050 to ambitious energy efficiency standards. This requires retrofit rates of up to 2.5% of floor area per year, which is high (compared with current practice), yet feasible [12].
2. For any given retrofit, it is assumed that, on average, 60% of the heating requirements are abated by insulating walls, roofs, and ground floors, replacing old windows with highly energy efficient windows, and by installing ventilation systems with heat recovery mechanisms [13–18].
3. One-quarter of the remaining 40% of heating and hot water need is met by local solar thermal systems, and the rest by heat pumps. This is an average global assumption; in practice, these values will differ by region due to differences in the structure of heat demand and the availability of solar energy throughout the year [19, 20].
4. Cooling is provided by local, renewable solutions where possible [21–23].
5. Increased electricity needs per unit floor area due to increased cooling demand, increased use of appliances (per unit area), and heat pump powering have been partially offset by increasing efficiency [21, 22].

New Building Stock
1. Increasingly, new buildings will be built to a "near zero energy use" standard, reaching a penetration of 100% of new buildings by 2030 (earlier in OECD regions). By "near-zero energy use" we mean buildings which have an energy use at levels comparable to the passive house standard developed in Germany. These highly energy-efficient buildings have very low heat losses through the building envelope (insulation and improved windows) and almost no losses from air exchange (use of heat recovery systems) [24].

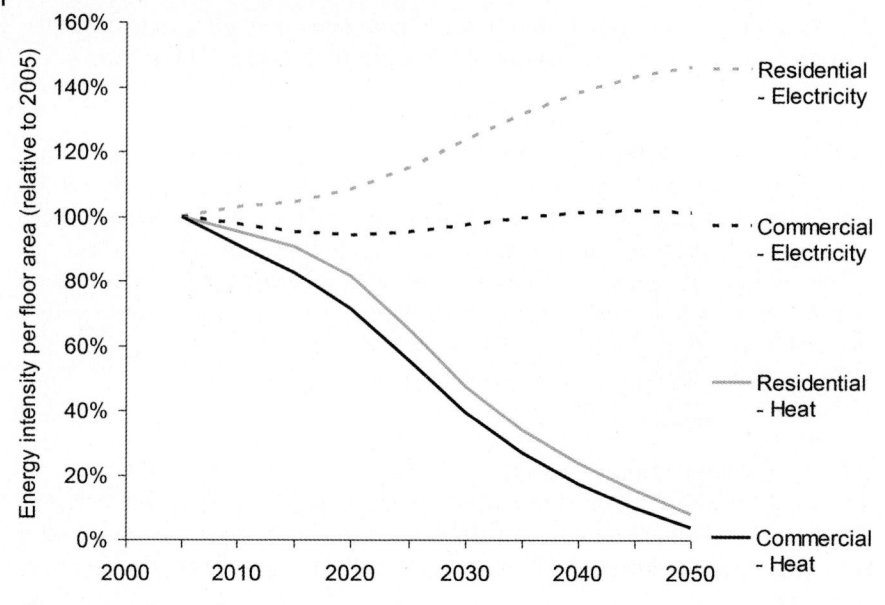

Figure 7.6 Evolution of energy intensity in buildings.

2. In these building types, most of the heat demand is met by passive solar (radiation through windows) and internal gains (from people and appliances). Any residual heat demand is assumed to be met by renewable energy systems in the form of solar thermal installations and heat pumps [11]. This building type only requires electric energy [24].
3. The near zero-energy concept is also applied to warm/hot climates, often returning to traditional building approaches. These include external shading devices and an optimal ventilation strategy [25]. There is a residual cooling demand in warm/hot climates, especially in nonresidential buildings with high internal loads from computers (offices) or lighting (retail). Increased electricity needs from increased cooling and appliances, and also the use of heat pumps, have been estimated and included in this scenario [26, 27].

The resulting overall evolution in energy intensity is shown in Figure 7.6 [28, 29]. Note that for heat pumps, Figure 7.6 does not show the heat provided, but the electricity required to drive the heat pump.

The increased electricity demand stems from heat pump operation, and also increased use of appliances, lighting, and cooling, which can only be partially offset by efficiency improvements.

7.3.2.3 Buildings – Future Energy Demand

Despite the strong increase in activity, namely floor space, the energy intensity reductions lead to a drastically reduced need for building heat in the form of fuels or heat delivered directly to buildings. At the same time, an increase in electricity

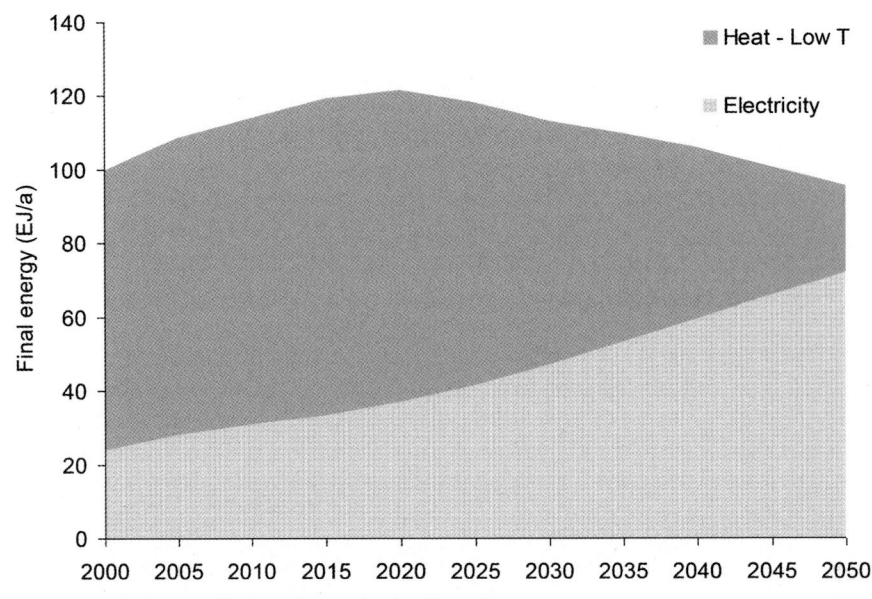

Figure 7.7 Evolution of energy demand in buildings by energy carrier.

use is expected. These results are shown in Figure 7.7. Note that as in Figure 7.6, for heat pumps this figure does not show the heat provided, but the electricity required to drive the pump.

7.3.3
Transport

7.3.3.1 Transport – Future Activity
This scenario uses an established BAU transport activity forecast for traffic volumes [5] (except shipping), with a marked increase in worldwide travel volumes, in accordance with population and GDP projections. Shipping activity projections were based on a link to overall global GDP growth.

Modal shifts are applied to this BAU forecast to arrive at final activity volumes by mode. Total transport volumes were not changed, with the following exceptions for passenger travel:

- a shift from vehicle travel to human-powered travel such as walking and cycling (short distances, therefore small volume);
- a shift from (business) aviation travel to alternatives such as videoconferencing (business travel represents the minority of passenger air travel [30]).

The results are shown in Table 7.3.

In the BAU case, transport volumes are expected to increase substantially, especially in developing economies, with a clear emphasis on individual road transport.

Table 7.3 Annual transport activity including modal shifts.

Mode	Units[a]	BAU 2000	BAU 2050	Scenario 2050
Passenger – PTWs	bn pkm	2.5	6.2	8.4
Passenger – car, city	bn pkm	5.5	14.1	7.2
Passenger – car, non-city	bn pkm	9.9	21.4	13.6
Passenger – bus + coach	bn pkm	9.2	9.1	13.0
Passenger – rail	bn pkm	2.0	6.0	16.2
Passenger – airplane	bn pkm	3.4	16.9	12.3
Freight – truck	bn tkm	7.9	27.1	18.7
Freight – rail	bn tkm	6.5	18.8	27.3
Freight – air plane	bn tkm	0.1	0.7	0.5
Total passenger	bn pkm	32.3	73.6	70.7
Total freight	bn tkm	14.5	46.5	46.5

Baseline from [5].
a bn, billion; pkm, person-kilometres; tkm, tonne-kilometres.

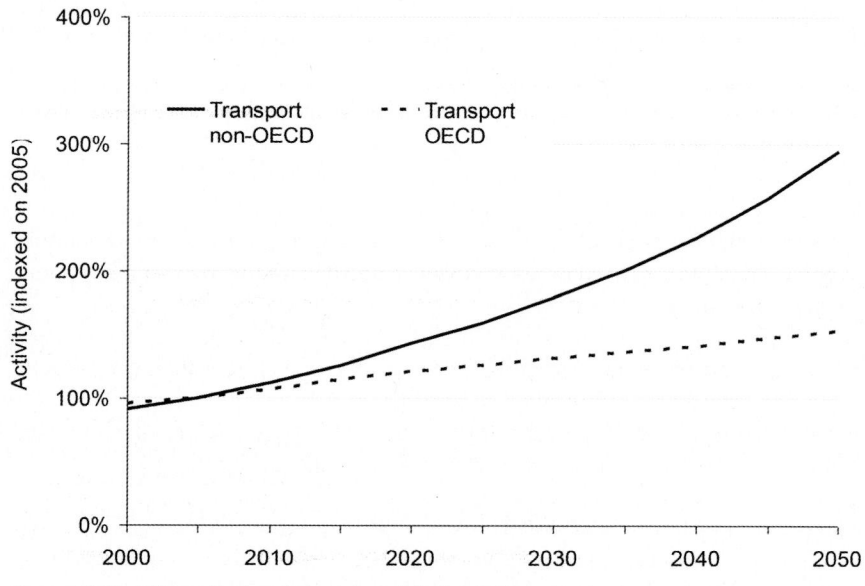

Figure 7.8 Evolution of indexed absolute activity levels in transport.

The scenario postulates substantial modal shifts away from inefficient individual road and aviation modes and towards the more efficient rail and shared road modes (see Figure 7.8).

7.3.3.2 Transport – Future Intensity

The following steps ensure that the scenario employs the most efficient transport modes, and preferably modes that are suitable for a high share of renewable energy:

1. Move to efficient technologies and modes of employment, for example, trucks with reduced drag, improved air traffic management, or reduced fuel needs in hybrid buses.
2. Electrification where possible, for example, electric cars in urban environments and electric rail systems.
3. Finally, providing the fuel from sustainable biomass, where possible.

Table 7.4 Energy intensity assumptions in transport.

Mode	Efficiency gains 2050 versus 2000 (%)	Electrification (%)	Comments
Passenger – PTWs	50	40–90	E.g., scooters
Passenger – car, city	75	90	90%, i.e., most transport done by electric vehicles or on electric portion of plug-in hybrid vehicles
Passenger – car, non-city		70	
Passenger – bus + coach	50–65	50–70	Hybrids/electric, especially in cities
Passenger – rail	30	95–100	Shift to fully electrified rail, resistance reduction, space optimization
Passenger – airplane	~50	n/a	Improvements in airframe and engine design, gains from air traffic management optimization [30]
Freight – truck	65	30	"Last-mile" delivery vans electric
Freight – rail	30	95–100	Shift to fully electrified rail
Freight – airplane	~50	n/a	See passenger airplane travel above
Freight – ship	~50	0 (but ~5 (H_2) from electricity)	Propeller and hull maintenance and upgrades, retrofits including towing kite, operational improvements including speed reduction; small share of hydrogen fuel in ships [32]

Assumptions for aviation, shipping, and road freight are based on [32–35].

Table 7.4 summarizes the most noteworthy fuel shift assumptions:

- A complete shift to plug-in hybrids and/or electric vehicles as the primary technology choice for light-duty vehicles.
- Long-distance trucks undergoing large efficiency improvements due to improved material choice, engine technology, and aerodynamics (no substantial electrification due to the prohibitive size and weight of batteries required with current technology). Only delivery vans covering "the last mile" are electrified, leading to an electric share estimate of 30% for trucks.
- A (small) share of shipping fuel gradually being replaced by hydrogen, won from renewable electricity. This has been deemed a feasible option due to the centralized refueling of ships.

Where a shift takes place from fuel to electricity, it is assumed that the electrically powered vehicle will need 1.5–2.5 times less final energy on average, since the energy is delivered to the vehicle in an already converted form [31].

The resulting overall evolution of energy intensity in the transport sector is shown in Figure 7.9. When interpreting Figure 7.9 (and Figure 7.10 later), it is important to remember that since demand is shown in final energy, the share of electricity in transport looks small, even though for many modes it delivers the vast majority of mechanical energy to the wheels. This is because the fuels still undergo conversion in the vehicle's combustion engine and therefore represent a larger energy content in these graphs.

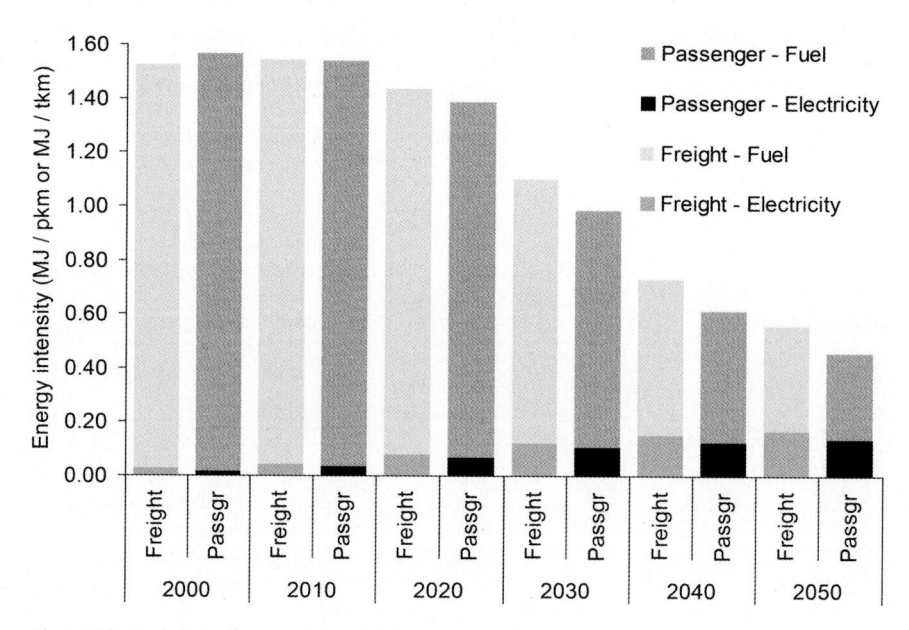

Figure 7.9 Evolution of energy intensity in transport.

7.3.3.3 Transport – Future Energy Demand

The assumptions on activity evolution with modal shift and on energy intensity evolution with fuel shift lead to the overall energy demand evolution in the transport sector shown in Figure 7.10. Of the total energy saving in comparison with BAU in 2050, more than 80% is due to efficiency and electrification, the rest to modal shifts.

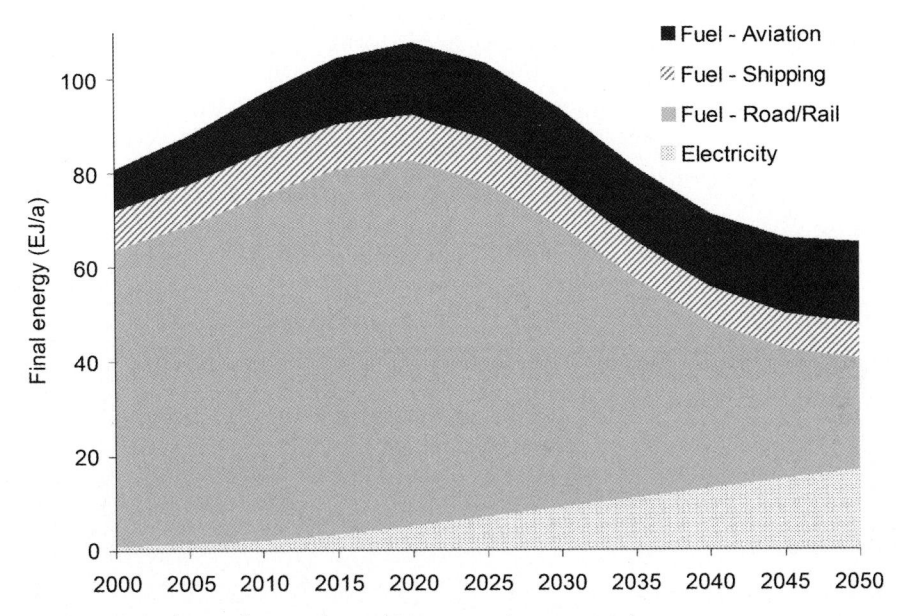

Figure 7.10 Evolution of energy demand in transport by energy carrier.

7.3.4
Demand Sector Summary

The methodology for the demand side described above leads to:

- a much reduced demand overall compared to BAU;
- a much higher electrification rate than BAU.

The resulting overall demand split is presented in Table 7.5. Note in particular that the demand for power rises steadily from just below 50 to over 120 EJ a^{-1}.

It is imperative to understand that the reduction in total energy demand in this scenario is not derived from a reduction in activity. It depends primarily on the reduction in energy intensity through aggressive roll-out of the most efficient technologies.

Table 7.5 Global energy demand in all sectors, split by energy carrier.

Energy carrier	Energy demand (EJ a^{-1})					
	2000	2010	2020	2030	2040	2050
Electricity	45.7	60.0	71.9	85.7	103.5	127.4
Heat – high T	30.8	36.7	40.5	38.1	33.1	32.9
Heat – low T	77.7	86.0	87.4	67.8	47.4	24.1
Fuel – road/rail	69.2	80.9	86.3	66.4	39.3	26.5
Fuel – shipping	8.2	9.4	9.7	8.6	7.3	7.2
Fuel – aviation	8.8	12.3	15.6	16.3	15.6	17.1
Fuel – industry	18.0	22.0	21.8	19.3	15.8	13.6
Fuel – nonreplaceable fossil	14.9	20.4	19.9	17.2	14.0	12.5
Total	273.4	327.6	353.3	319.4	276.2	261.4

7.4
Results – Supply Side

Once total demand has been established, it is matched with supply. In the following, we first describe our findings for the total potential supply available, followed by the results for the matching of this supply with the demand calculated above.

7.4.1
Supply Potential

This scenario is based on the *deployment potential*, shown in Figure 7.11 (for all sources except bioenergy). This is the potential that can be captured at any time, considering technical barriers and ambitious, yet feasible, market growth. The deployment potential does not necessarily represent the most cost-effective development, that is, it does not account for market barriers or competition with other sources.

The *realizable potential* is the fully achievable potential of the resource with a long-term development horizon beyond 2050. It is equivalent to the technical potential as defined by the *Intergovernmental Panel on Climate Change (IPCC)* [36].

In the following we discuss the potential for each of the sources shown in Figure 7.11 in turn.

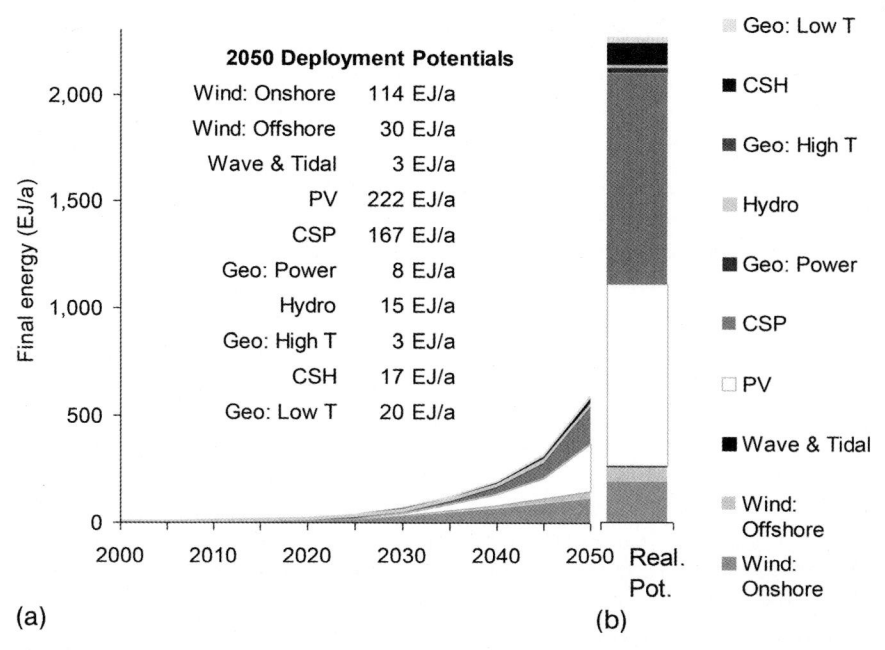

Figure 7.11 Global deployment potential (a) and realizable potential (b) of renewable energy sources (excluding bioenergy).

7.4.1.1 Wind

The scenario includes power generation from both onshore and offshore wind. The growth of onshore wind power has been remarkable in the last decade, with annual growth rates exceeding 25% in most years. Several offshore wind parks are already in operation worldwide and many more are currently in planning phases.

For offshore wind potential annual growth rates of ~30% and for onshore wind rates nearer 20% are used [37–42].

7.4.1.2 Water

Hydropower is the largest renewable power source to date, providing almost 15% of worldwide electricity; over 980 GW installed capacity existed in 2009. [1, 41]

Although hydropower can be produced sustainably, past projects have suffered from ecological and societal side effects. We have therefore restricted future growth to small hydropower and efficiency gains in existing large hydropower schemes to reflect the need for an evolution that respects existing ecosystems and human rights [39, 43].

Potentials for wave and tidal power, also called "ocean power," are less dense than other forms of power, such as wind or solar, but can be heavily concentrated, for example, on windy coastlines. There are several on-going pilot projects to harness wave energy and to design sustainable tidal systems. The scenario incorporates a wave and tidal energy potential estimated at around 5% of the potential of offshore wind. Annual growth rates in the scenario do not exceed 20% [38, 44, 45].

7.4.1.3 Sun

The largest realizable technical potential for renewable power and heat generation is from direct solar energy. This energy scenario includes four different sources of solar energy:

- solar power from photovoltaics (PV);
- concentrating solar power (CSP);
- concentrating solar high-temperature heat for industry (CSH);
- (solar thermal low-temperature heat for buildings – this is treated in Section 7.3.2).

PV is a well-established source of electric energy, with around 21 GW of capacity installed worldwide at the end of 2009 [41]. The scenario contains a PV potential characterized by continuing annual growth rates of 25–30% for the next two decades, including outputs from both building-integrated and large-area PV installations [39, 46–49].

With increasing storage times, CSP is attracting attention for its potential to provide power on demand, even after dark. Systems with up to 15 h of storage are now being trialed, such as Gemasolar in Spain. Although still in its infancy, the expectations for CSP are large and the scenario is based on growth rates of up to 20% over the time horizon of the study [50].

CSH would enable industrial installations to utilize directly the high temperatures generated by concentrated solar farms. This technology is not yet on the market and is therefore only included at a very small potential in this study, at around one-tenth of the potential of CSP.

7.4.1.4 Earth

Geothermal energy from the high temperatures found below the Earth's surface can be used directly ("direct use") to produce building heat [51]. At sufficiently high temperatures, it can also be used for power generation ("indirect use") and/or process heat.

Geothermal energy has been exploited for many years, with around 10 GW of power production capacity installed worldwide at the end of 2007. Given the lack of attention paid to this option in the past and its high potential, specifically for demand-driven power, the scenario is based on the premise that the current 5% annual growth rate could be doubled to reach the levels of other renewable power options. This potential could be captured with existing ("conventional") technology, but new approaches can widen the geographical applicability further [52–54].

7.4.1.5 Bioenergy

The scenario incorporates a significant share of sustainable bioenergy supply to meet the remaining demand after other renewable energy options have been deployed. Bioenergy requires a more elaborate approach than most other renewable energy options because:

- Bioenergy requires a more thorough analytical framework to analyze sustainability, as cultivation, processing, and use of biomass have a range of interconnected sustainability issues.
- Bioenergy encompasses energy supply for a multitude of energy carrier types using a multitude of different energy sources. Therefore, a detailed framework of different possible conversion routes is needed.

For a full treatment of the bioenergy approach in this scenario, the reader is referred to a separate publication [4].

The scenario only includes bioenergy supply that is sustainable and leads to high greenhouse gas emission savings in comparison with fossil references. The potentials used are shown in Table 7.6.

Table 7.6 Bioenergy deployment potential.

Source	Sustainable deployment potential (primary) (EJ a^{-1})	
	2000	2050
Traditional biomass	35	0
(Additional) residues and waste	< 1	101
Complementary fellings[a]	< 1	38
Energy crops	< 1	115
Algae	< 1	90

a Complementary fellings include the sustainable fraction of currently used traditional biomass.

7.4.2
Results of Balancing Demand and Supply

Demand and supply are balanced in the following order in each time period:

1. Where available, non-bioenergy renewable options are used first.
2. If demand cannot be fully satisfied, bioenergy is used up to the sustainably available potential in that year.
3. All residual demand is supplied by conventional sources, such as fossil and nuclear energy.

Following this strict prioritization of options, the overall evolution of energy supply is determined, as shown in Figure 7.12.

Stabilizing energy demand, driven by strong energy efficiency, combines with fast renewable energy supply growth, resulting in an energy system that is 95% sustainably sourced.

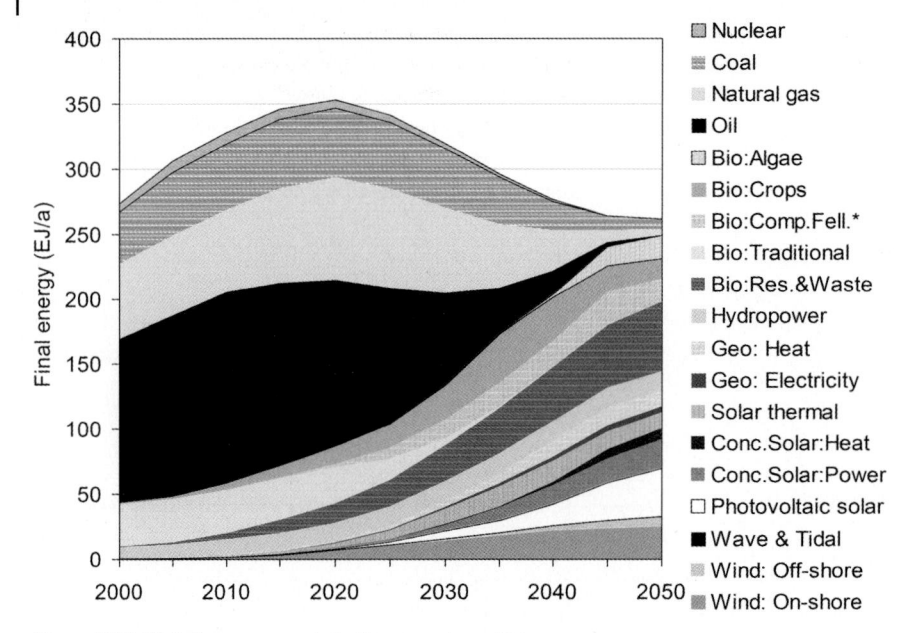

Figure 7.12 Global energy supply in the scenario, split by source.
* Complementary fellings include the sustainable share of traditional biomass use.

7.5
Discussion

The energy scenario we have presented combines the most ambitious efficiency drive on the demand side with strong growth of renewable source options on the supply side to reach a fully sustainable global energy system by 2050. Both are important: the transition cannot be achieved on the supply side alone.

7.5.1
Power Grids

The scenario constrains the solar and wind power share to 20–30% initially, gradually lifting this constraint over time. For high penetrations of renewable energy, only limited analysis is available. Based on a number of studies [55–58], we expect that the limiting share could rise to 60% by 2050 for all regions, provided that electricity systems are redesigned to offer much more flexibility than they do today.

This requires that full use is being made of all the following levers:

- grid capacity improvements to remove bottlenecks and increase transmission capacities;
- demand side management, particularly for wholesale customers, but also at individual consumer level;

- storage, in the form of pumped hydro, centralized hydrogen storage, battery and heat storage;
- conversion of excess renewable electricity to hydrogen for use as a fuel in specific applications.

7.5.2
The Need for Policy

This energy scenario presents a radical departure from our current system of energy use. It is clear that current policies would not be able to deliver on this scenario. Requirements for additional policies include objectives for both *public* and *private* institutions:

Public bodies should:
1. create the long-term framework enabling the transition;
2. invest in large infrastructure and early-stage R&D projects.

Private actors, both consumers and companies, should:
1. operate under a long-term perspective, resulting in adoption of best practices in energy efficiency;
2. channel investments into the most efficient and sustainable energy options.

Some policy needs are more pressing than others because they represent enabling factors: failing to address them will have repercussions for the likelihood of success in other sectors.

The two *key enabling factors* for this energy scenario are:

- strong energy efficiency measures coupled with electrification for remaining demand;
- the preparation of our energy grids to cope with the increasing demand for renewable, often supply-driven, electricity.

In addition, a set of policy and market rules needs to be enforced, which ensures that the sustainability of biomass use for energy is safeguarded [3]. For a more detailed discussion of the policy requirements, the reader is referred to [2].

7.5.3
Sensitivity of Results

The results presented here depend directly on the assumptions made regarding the scale and rate of change in all sectors. However, some inputs have a larger effect than others, especially with respect to the overall aim of reaching a 100% renewable energy supply.

Most critical for our analysis are the assumptions on energy efficiency improvements. If they are not attained, the decrease in global energy use will not materialize.

Another important assumption is the postulated economic growth. We have applied medium growth rates; alternative rates could result in higher or lower overall energy demand growth.

Whether a higher energy demand would result in a lower renewable energy share in 2050 depends on the availability of a suitable energy supply. We have already seen that for some energy carriers, e.g. supply-driven electricity, large contingencies exist by 2050, so renewable energy supply would not limit the renewable energy share, even in a scenario with higher demand. However, for some energy carriers the situation looks different and increased demand would reduce the overall renewable energy share attainable by 2050. The energy carriers with the least room for maneuvre are the fuel carriers, specifically when supplied from biomass. For electricity, there could be bottlenecks in some of the demand-driven power sources, but no bottlenecks are foreseen for the bulk of the supply-driven sources.

7.6
Conclusion

A fully renewable global energy system is possible: we can reach a 95% sustainably sourced energy supply by 2050 (see Figure 7.13). To achieve this goal, we need to combine aggressive energy efficiency on the demand side with accelerated renewable energy supply from all possible sources. This requires a paradigm shift towards long-term, integrated strategies and will not be met with small, incremental changes.

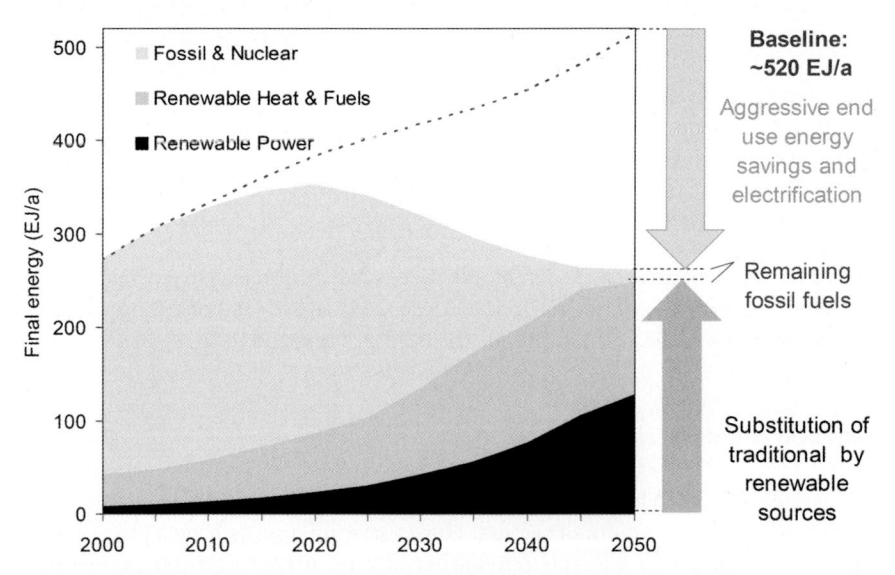

Figure 7.13 Key developments of the transition to a fully sustainable energy system.

References

1. International Energy Agency (2009) *Statistics and Balances*. World Energy Balances, OECD and non-OECD country databases, www.iea.org/stats/index.asp.
2. World Wide Fund for Nature, Ecofys, Office for Metropolitan Architecture (2011) *The Energy Report: 100% Renewable Energy by 2050*, WWF International. Gland.
3. Deng, Y. Y., Blok, K., and van der Leun, K. (2012) Transition to a fully sustainable global energy system. *Energy Strategy Rev.*, **1** (2), 109–121.
4. Cornelissen, S., Koper, M., and Deng, Y. Y. (2012) The role of bioenergy in a fully sustainable global energy system. *Biomass Bioenergy*, **41**, 21–33.
5. Population Division of the Department of Economic and Social Affairs of the United Nations Secretariat, *World Population Prospects: The 2006 Revision and World Urbanization Prospects: The 2007 Revision*, http://esa.un.org/unup, 19 December 2008.
6. International Energy Agency (2004) *Model Documentation and Reference Case Projection for WBCSD's Sustainable Mobility Project (SMP)*, IEA, Paris.
7. Martin, N., Worrell, E., Ruth, M., Price, L., Elliot, R. N., Shipley, A. M. et al. (2000) *Emerging Energy-Efficient Industrial Technologies*, LBNL-46990, Lawrence Berkeley National Laboratory, Berkeley, CA, and American Council for an Energy-Efficient Economy, Washington, DC.
8. Kim, Y. and Worrell, E. (2002) International comparison of CO_2 emission trends in the iron and steel industry. *Energy Policy*, **30**, 827–838.
9. Worrell, E., Price, L., Neelis, M., Galitsky, C., and Nan, Z. (2008) *World Best Practice Energy Intensity Values for Selected Industrial Sectors*, LBNL-62806, Lawrence Berkeley National Laboratory, Berkeley, CA.
10. Gielen, D., Bennaceur, K., Kerr, T., Tam, C., Tanaka, K., Taylor, M. et al. (2007) *Tracking Industrial Energy Efficiency and CO_2 Emissions*, IEA, Paris.
11. Levine, M., Ürge-Vorsatz, D., Blok, K., Geng, L., Harvey, D., Lang, S. et al. (2007) – Residential and commercial buildings, in *Climate Change: Mitigation. Contribution of Working Group III to the Fourth Assessment Report of the Intergovernmental Panel on Climate Change*, Cambridge University Press, Cambridge, Ch. 6.
12. Hegger, M., Liese, J., and Söffker, G. (2008) *Energy Manual: Sustainable Architecture*, Birkhäuser, Basel.
13. Friedrich, M., Becker, D., Grondey, A., Laskowski, F., Erhorn, H., Erhorn-Kluttig, H. et al. (2008) CO_2 *Buildings Report 2007*. co2online gemeinnützige GmbH, Fraunhofer-Institut für Bauphysik for Bundesministerium für Verkehr Bau and Stadtentwicklung (Federal Ministry of Transport Building and Urban Development) (in German), available at http://www.buildup.eu/publications/2088.
14. Eichhammer, W., Fleiter, T., Schlomann, B., Faberi, S., Fioretto, M., Piccioni, N. et al. (2009) *Study on the Energy Savings Potentials in EU Member States, Candidate Countries and EEA Countries*, http://ec.europa.eu/energy/efficiency/studies/doc/2009_03_15_esd_efficiency_potentials_final_report.pdf (last accessed 5 January 2013).
15. Boermans, T. and Petersdorff, C. (2007) *U-Values for Better Energy Performance of Buildings*, Eurima/Ecofys, Eurima, Brussels.
16. Petersdorff, C., Boermans, T., Stobbe, O., Joosen, S., Graus, W., Mikkers, E. et al. (2004) *Mitigation of CO_2 Emissions from the Building Stock – Beyond the EU Directive on the Energy Performance of Buildings*, Eurima/Euroace/Ecofys, Ecofys, Utrecht.
17. Petersdorff, C., Boermans, T., Joosen, S., Kolacz, I., Jakubowska, B., Scharte, M. et al. (2005) *Cost-Effective Climate Protection in the Building Stock of the New EU Member States*, Eurima/Ecofys, Ecofys, Utrecht.

18 Wesselink, B. and Deng, Y. Y. (2009) *Sectoral Emission Reduction Potentials and Economic Costs for Climate Change (SERPEC-CC)*, European Commission DG-RTD and DG-ENV, http://www.ecofys.com/en/publication/133 (last accessed 5 January 2013).

19 Schimschar, S., Blok, K., Boermans, T., and Hermelink, A. (2011) Germany's path towards nearly zero-energy buildings – enabling the greenhouse gas mitigation potential in the building stock. *Energy Policy*, **39**, 3346–3360.

20 World Business Council for Sustainable Development (2011) *Energy Efficiency in Buildings: Business Realities and Opportunities*, World Business Council for Sustainable Development, Geneva.

21 World Business Council for Sustainable Development (2009) *Energy Efficiency in Buildings: Transforming the Market*, World Business Council for Sustainable Development, Geneva.

22 Harvey, D. (2006) *A Handbook on Low-Energy Buildings and District-Energy Systems: Fundamentals, Techniques and Examples*. Earthscan, Oxford.

23 Harvey, D. (2010) *Energy and the New Reality 1 – Energy Efficiency and the Demand for Energy Services*. Earthscan, Oxford.

24 Parker, D., Sherwin, J., and Hermelink, A. (2009) NightCool: an Advanced Cooling Technology for Passivhaus, in Tagungsband der 13. Internationalen Passivhaustagung, Frankfurt, pp. 367–372.

25 Boermans, T., Hermelink, A., Schimschar, S., Grözinger J., Offermann, M., Engelund Thomsen, K. et al. (2011) *Principles for Nearly Zero-Energy Buildings. Paving the Way for Effective Implementation of Policy Requirements, for Buildings Performance Institute Europe (BPIE)*, http://dl.dropbox.com/u/4399528/BPIE/publications/HR_nZEB%20study.pdf (last accessed 5 January 2013).

26 MED-ENEC (2009) *The 2006–2009 Phase of the MED-ENEC Initiative Delivered 10 Pilot Projects for Innovative Building Concepts in the Mediterranean and the Middle East*, http://www.med-enec.com/building-projects/pilot-projects (last accessed 5 January 2013).

27 Bettgenhäuser, K., Boermans, T., Offermann, M., Krechting, A., and Becker, D. (2011) *Climate Protection by Reducing Cooling Demands in Buildings, for German Umweltbundesamt*, http://www.uba.de/uba-info-medien/3979.html (last accessed 5 January 2012).

28 Bettgenhäuser, K. and Boermans, T. (2011) *The Environmental impacts of Heating Systems in Germany; for German Federal Environment Agency (UBA)*, http://www.uba.de/uba-info-medien/4070.html (last accessed 5 January 2013).

29 International Energy Agency (2004) *Oil Crises and Climate Challenges: 30 Years of Energy Use in IEA Countries*, OECD/IEA, Paris.

30 International Energy Agency (2007) *Energy Use in the New Millennium: Trends in IEA Countries*, OECD/IEA, Paris.

31 UK Committee on Climate Change (2009) Alternatives to air travel: highspeed rail and videoconferencing, in *Meeting the UK Aviation Target: Options for Reducing Emissions to 2050*, Ch. 3, http://www.theccc.org.uk/reports/aviation-report (last accessed 5 January 2013).

32 Åhman, M. (2001) Primary energy efficiency of alternative power trains in vehicles. *Energy*, **26**, 973–989.

33 International Maritime Organization, Marine Environment Protection Committee (2009) *Prevention of Air Pollution from Ships: Second GHG Study*, IMO, London.

34 US Department of Energy (2010) *Technology Roadmap for the 21st Century Truck Program*, US Department of Energy, Washington, DC.

35 Air Transport Action Group (2010) *Beginner's Guide to Aviation Efficiency*, Air Transport Action Group, Geneva.

36 Cullen, J. M., Allwood, J. M., and Borgstein, E. H. (2011) Reducing energy demand: what are the practical limits? *Environ. Sci. Technol.*, **45**, 1711–1718.

37 Banuri, T., Barker, T., Bashmakov, I., Blok, K., Christensen, J., Davidson, O. et al. (2001) Working Group III (Mitigation), in *Third Assessment Report of the Intergovernmental Panel on Climate*

38. Global Wind Energy Council (2007) *Global Wind Report 2007*, GWEC, Brussels.
39. Ecofys (2008) *Internal Assessment*, Ecofys, Utrecht.
40. Hoogwijk, M. and Graus, W. (2008) *Global Potential of Renewable Energy Sources: a Literature Assessment*, Ecofys, Utrecht.
41. Leutz, R., Ackermann, T., Suzuki, A., Akisawa, A., and Kashiwagi, T. (2001) *Technical offshore wind energy potentials around the globe*, presented at the European Wind Energy Conference and Exhibition, Copenhagen.
42. REN 21 (Renewable Energy Network for the 21st Century) (2010) *Renewables 2010 Global Status Report*, REN 21 Secretariat, Paris.
43. WWF China, Chinese Renewable Energy Industries Association (2008) *China Wind Power Report*, WWF China, Chinese Renewable Energy Industries Association, Beijing, ISBN 978-82-90980-29-9.
44. Ayukawa, Y., Naoyuki, Y., Chen, D., Chestin, I., Kokorin, A., Denruyter, J. P. et al. (2006) *Climate Solutions*, WWF International, Gland, http://assets.panda.org/downloads/climatesolutionweb.pdf (last accessed 5 January 2013).
45. EU-OEA (European Ocean Energy Association) (2010) *European Ocean Energy Roadmap*, OEA, Brussels.
46. International Energy Agency (2010) *Implementing Agreement on Ocean Energy Systems. 2009 Annual Report*, IEA, Paris.
47. European Photovoltaic Industry Association (2008) *Solar Generation V*, EPIA, Brussels.
48. European Photovoltaic Industry Association (2009) *Global Market Outlook for Photovoltaics Until 2013*, EPIA, Brussels.
49. Hofman, Y., de Jager, D., Molenbroek, E., Schillig, F., and Voogt, M. (2002) *The Potential of Solar Electricity to Reduce CO_2 Emissions. Report Writing for IEA Greenhouse Gas R&D Programme*, Report No. PH4/14, Ecofys, Utrecht.
50. Li, J.-F. and Wang, S. C. (2007) *China Solar PV Report*, WWF China, Greenpeace, and European Photovoltaic Industry Association, China Environmental Science Press, Beijing.
51. DLR (German Aerospace Centre) (2005) *Concentrating Solar Power for the Mediterranean Region – MED-CSP*, http://www.dlr.de/tt/med-csp (last accessed 5 January 2013).
52. Lund, J. W., Freeston, D. H., and Boyd, T. L. (2005) Worldwide direct uses of geothermal energy 2005. *Geothermics*, **34**, 691–727.
53. Bertani, R. (2005) World geothermal power generation in the period 2001–2005. *Geothermics*, **34**, 651–690.
54. Bertani, R. (2007) World geothermal generation in 2007, in *Proceedings European Geothermal Congress 2007*, 30 May–1 June 2007, Unterhaching, Germany, http://www.geothermal-energy.org/pdf/IGAstandard/EGC/2007/083.pdf (last accessed 5 January 2013).
55. Forseo (2008) *The Investor's Guide to Geothermal Energy*, http://www.forseo.eu/english/publications.html (last accessed 5 January 2013).
56. Blok, K. (1984) A renewable energy system for The Netherlands, in *Proceedings of the 5th International Solar Forum* (eds F. Auer and T. Lanz), DGS Sonnenenergie, Munich.
57. Sørensen, B. (2000) *Renewable Energy*, 2nd edn., Academic Press, London.
58. European Climate Foundation (2010) *Roadmap 2050: a Practical Guide to a Prosperous, Low-Carbon Europe*, European Climate Foundation. The Hague.
59. Börner, J., Burges, K., and Nabe, C. (2010) *All Island TSO Facilitation of Renewables Studies*, http://www.ecofys.com/en/publication/21 (last accessed 5 January 2013).

Appendix

Global energy provided by source and year (final EJ a^{-1}).

Source	2000	2010	2020	2030	2040	2050
Total electricity	45.7	60.0	71.9	85.7	103.5	127.4
Wind power: on-shore	0.2	1.4	6.7	14.3	22.0	25.3
Wind power: off-shore	0.0	0.0	0.5	1.3	3.4	6.7
Wave and tidal	0.0	0.0	0.0	0.1	0.3	0.9
Photovoltaic solar	0.0	0.1	0.7	6.5	16.9	37.0
Concentrated solar: power	0.0	0.1	0.6	3.9	13.7	21.6
Hydropower	7.9	11.3	13.4	14.4	14.8	14.9
Geothermal	0.1	0.3	0.7	1.7	3.4	4.9
Biomass	0.0	0.0	0.0	0.0	1.7	16.2
Coal	18.2	21.5	14.8	10.0	5.4	0.0
Gas	8.6	14.0	25.6	28.3	20.1	0.0
Oil	4.2	3.1	2.5	1.4	0.5	0.0
Nuclear	6.5	8.2	6.5	3.8	1.2	0.0
Industry fuels and heat	63.7	79.1	82.3	74.6	63.0	59.0
Concentrated solar: Heat	0.0	0.0	0.1	0.4	2.6	8.8
Geothermal	0.0	0.1	0.2	0.6	1.6	2.9
Biomass	1.0	6.1	16.9	31.3	40.7	34.8
Fossil fuels	62.7	72.9	65.0	42.2	18.0	12.5
Building fuels and heat	77.7	86.0	87.4	67.8	47.4	24.1
Solar thermal	0.0	0.7	3.3	11.9	16.0	12.6
Geothermal	0.2	0.5	1.5	4.1	10.5	8.4
Biomass	33.4	33.2	29.2	14.2	10.2	3.1
Fossil fuels	44.1	51.6	53.5	37.6	10.6	0.0
Transport fuels	86.2	102.6	111.6	91.3	62.3	50.8
Biomass	0.7	4.8	12.9	29.7	45.7	50.8
Fossil fuels	85.5	97.8	98.8	61.7	16.6	0.0
Grand total	273.4	327.6	353.3	319.4	276.2	261.4

8
The Transition to Renewable Energy Systems – On the Way to a Comprehensive Transition Concept

Uwe Schneidewind, Karoline Augenstein, and Hanna Scheck

8.1
Why Is There a Need for Change? – The World in the Age of the Anthropocene

For thousands of years, humanity has had to adapt to its natural surroundings. Human beings had to wrest space and opportunities from an often inhospitable Nature, in order to secure their well-being. Over the past 5000 years, this has time and again led to cases of local overexploitation of Nature (e.g., forest clearance), but has not extended to a global dimension of depleting ecological systems. This changed with the beginning of the Industrial Revolution and the increased use of fossil resources during the mid-19th century. Owing to the tremendous economic success and massive population growth, humanity has itself become a major factor directly influencing global eco-systems, such as the atmosphere, the oceans, and global soil systems. Nobel Prize winner Paul Crutzen thus described the current phase in the Earth's geological history as the "Anthropocene" [1], in order to illustrate the central impact that human beings have on global ecological systems.

Over the course of the last and the current century, the human impact on global eco-systems has been increasingly approaching critical thresholds. Within international environmental research, these thresholds are described as "tipping points" [2]. They refer to the limits of ecological bearing capacities and it is shown that once these are exceeded, the future development of a particular ecological system cannot be estimated with great certainty any longer and that there is a growing threat of system functions collapsing altogether. In a seminal paper published in *Nature* in 2009, Johann Rockström, together with almost 30 renowned environmental researchers, determined which specific ecological threats are of particular global relevance, where the "planetary boundaries" for these specific threats lie, and where these have already been crossed [3]. The planetary boundaries (see the two inner circles in Figure 8.1) denote the threshold where certain tipping points are at risk of being exceeded. The analysis covered global challenges such as climate change, acidification of the oceans, loss of biodiversity, land use patterns, and freshwater systems.

Transition to Renewable Energy Systems, 1st Edition. Edited by Detlef Stolten and Viktor Scherer.
© 2013 Wiley-VCH Verlag GmbH & Co. KGaA. Published 2013 by Wiley-VCH Verlag GmbH & Co. KGaA.

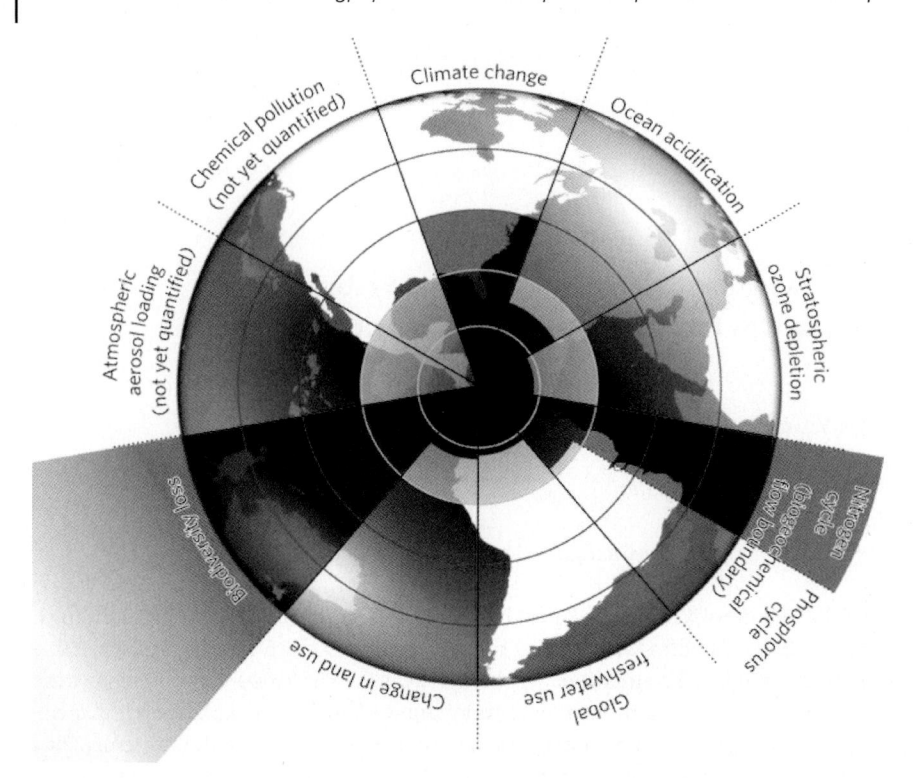

Figure 8.1 Planetary boundaries [3].

Figure 8.1 provides an overview of central results of the study by Rockström *et al.* [3]. Two findings are of major importance:

1. In more than one area, planetary boundaries have already been exceeded. Apart from climate change, this is the case with regard to the loss of biodiversity and with regard to the ecological impact of global nitrogen cycles, especially in the context of large-scale agriculture. Other areas are close to reaching global boundaries.
2. The different problem areas are tightly interlinked. This may result in problem shifting, which means that a solution for one problem can be found only at the expense of aggravating another problem. An example of this is the debate about biofuels: an increase in the production of biofuels – intended to contribute to a solution to problems of climate change – is causing massive ecological problems of deforestation in Asia and Latin America.

Taking these findings, there is growing evidence that guaranteeing prosperity for nine billion people in 2050 requires more than technological solutions: What is needed is a comprehensive transition process, including cultural and social issues in addition to economic issues of distribution, and thus addressing questions of a globally fair prosperity [4, 5].

Continuing along established development pathways (increasing prosperity at consistent land use-, resource, and carbon intensity) will not allow for such a transition. This is why the German Advisory Council on Global Change called for a "Great Transformation" in its Flagship Report 2011 [6].

It is the aim of this chapter to help gain a better understanding of how "great" or comprehensive the transformation actually needs to be and across which societal dimensions it should extend.

8.2
A Transition to What?

A "Great Transformation" needs a direction. The overall vision of "prosperity for nine billion people within planetary boundaries in 2050" has been debated with regard to differing views on its concrete implications. One widely known formula expressing this vision is the definition of "sustainable development" that was coined in the Brundtland Report [7] and then became increasingly popular in the aftermath of the United Nations Conference on Environment and Development in Rio de Janeiro in 1992: "Sustainable development is development that meets the needs of the present without compromising the ability of future generations to meet their own needs."

This vision can be specified with a view to particular challenges: the vision of 100% renewable energy supply [6], the vision of a low-resources world [8], or the vision of a degrowth society [9, 10]. All of these visions point to fundamental change processes, which extend over a number of societal domains and have not only national, but also global implications.

In the following, the focus will be on the example of the German "Energy Transition," that is, the transformation of the German energy system by 2050 with the goal of providing energy from mainly renewable energy sources and without causing problem shifts. Avoiding problem shifting in this case means, first, dealing with possible "rebound" effects. These refer to situations where an increase in efficiency leads to overall growth effects, which then hamper the initial ecological benefits of the increase in efficiency [11–13]. Second, problem shifts are also trade-offs between different ecological dimensions. A prominent case is the increased use of biofuels mentioned above, causing severe problems with regard to land use.

The German "Energy Transition" is a good example of an integrated transformation project, because it necessarily depends not only on technological changes, but just as much on social and institutional change. The political decision in 2011 for a comprehensive "Energy Transition" makes Germany an important global pioneer for such a transformation project, which will make an important contribution to the development of a "Transformative Literacy."

8.3
Introducing the Concept of "Transformative Literacy"

The term "literacy" describes the ability to read and write, and thus to understand and take part in different forms of communication. It also implies an ability to understand and make use of cultural and symbolic elements that are often an implicit aspect of language and the way in which language is used. Thus, "literacy" refers to more than the physical or technical ability to read and write – it also implies a more general cultural capacity and competence to act in a specific context. For instance, "computer literacy" describes the ability to use computers adequately for a specific purpose, which goes beyond factual knowledge about their technical features and functionalities.

Roland Scholz was the first to use the concept of literacy within the debates on environmental protection and sustainability. He coined the term "environmental literacy" [14] and defined it as "the ability to read and utilize environmental information appropriately, to anticipate rebound effects, and to adapt to changes in environmental resources and systems, and their dynamics." According to this definition, environmental literacy goes beyond understanding environmental information and includes the ability to take this information into consideration when acting – as an individual, a company, or a political entity.

The concept of "transformative literacy" departs from Scholz's basic idea. Following Scholz's definition, transformative literacy can be conceptualized as the ability to read and utilize information about societal transformation processes, and accordingly to interpret and get actively involved in these processes. Thus, it refers to the ability to grasp adequately the multi-dimensional nature of transformation processes and to incorporate this type of knowledge in concrete action in actual transformation processes.

The notion of literacy contains a specific metaphoric power when considering its implications with regard to the acquisition of language. Achieving mastery of a language is a gradual and multi-stage process: learning first words and terms of a foreign language provides a basic orientation. Actual command of the new language is achieved with a continuously expanding vocabulary and increasing knowledge of grammar and familiarity with its correct usage. Eventually, mastery of a language may reach a level where it is possible to express oneself in diverse literary forms such as novels or poems.

What are the parallels with this type of literacy in the field of language when transferring the literacy concept to questions regarding transitions to sustainable development? Transitions are changes of socio-technical systems, which include technological, social, institutional, and economic dimensions and their interrelations [15].

It is argued here that there are four discernible dimensions with regard to which we have to be "literate," in order to understand transitions and actively get involved in the relevant change processes:

- a technological dimension;
- an economic dimension;

- an institutional dimension;
- a cultural dimension.

An understanding of these four dimensions, individually and in their interrelations, together amounts to a transformative literacy. It is important to note that even within the individual dimensions, different levels of understanding coexist. This can be shown in the context of the current economic debate. Ever louder calls for a "New Economics" (http://www.neweconomics.org/) illustrate that basic concepts and theories of economic processes are being challenged and that the economic dimension itself is in a period of transition.

Sustainability-related transition processes, such as the "Energy Transition," are predominantly discussed in terms of technological change. Knowledge does exist with regard to different technological options and possible technological scenarios; however, there is a kind of "illiteracy" with regard to the concrete economic and overall societal change processes that are needed.

Thus, transformative literacy needs to include a combination of natural science-based technological, economic, institutional, and cultural analyses – and therefore requires a high degree of interdisciplinary competences. Producing knowledge in this way is in conflict with many organizational and institutional traditions in science, which explains why a transformative literacy has barely developed so far.

The aim of this chapter is to introduce the concept of transformative literacy. It will be explored how such an integrated concept or "language" provides a useful new perspective. Since this is a first exploratory approach, a comprehensive transformation "vocabulary" or "grammar" is still to be developed in further research.

8.4
Four Dimensions of Societal Transition

The need for a wide-ranging sustainability-oriented transition is articulated by the German Advisory Council on Global Change in terms of a "Great Transformation" [6], defined as a radical change in different societal domains, such as production structures, consumption patterns, and institutions. In this chapter, such complex and interlinked types of radical change are systematized analytically along four dimensions: a technological, an economic, an institutional and a cultural dimension. The study of transitions therefore requires an interdisciplinary perspective, because processes of social change relating to these dimensions are addressed differently within the respective academic disciplines. Depending on the specific perspective, overall change is explained as a predominantly technological, economic, institutional, or cultural phenomenon, respectively. In this chapter, an integrated and interdisciplinary perspective is outlined, which refers to a variety of relevant theories and approaches from different disciplines, for example established (macro-) economic approaches (for instance, endogenous growth theory [16]) and approaches from the social and cultural sciences (for instance, structuration theory [17]).

These dimensions need to be addressed in a twofold way:

1. As an interrelated set of structures within a socio-technical system: technological structures are embedded in economic structures, which in turn are influenced by institutional structures.
2. As independent starting points for fostering processes of change: cultural change, for instance in individual lifestyles, can have a direct impact on overall processes of change, thus a transition or a "Great Transformation."

In the following, the four central dimensions will be depicted and it will be shown

- how the dimensions are systematically interlinked with each another,
- what the potentials and limits are in each dimension with regard to a transition,
- what the current state with regard to a transition is in each of the dimensions, and
- what the future research and development needs are in the different dimensions.

Finally, it will be shown how the conceptual framework of a transformative literacy that has been developed here helps systematizing the different dominant strands of the transitions towards sustainability debate and provides a starting point for more profound analyses.

8.4.1
On the Structural Interlinkages of the Four Dimensions of Transitions

"Transition" in this context implies thinking about ways of actively shaping change processes. In order to understand transition processes, one has to focus on the involved actors, routinized action and behavior, and innovations (in the sense of new practices). Therefore, a first important element of transformative literacy is an understanding of the relation between "agency" and "structure." Anthony Giddens dealt with this relation in his *Theory of Structuration* [17] (for an application in the field of sustainability research, see Schneidewind [18] and Schneidewind and Scheck [19]) and developed the concept of the "duality of structure." This explains how action is always related to and influenced by structure, while at the same time structure is produced and reproduced through action. In this context, structures can be "rules" or "resources."

How can this concept of structuration contribute to a better understanding of transformation processes? For instance, mobility behavior is affected by a number of different types of structure: existing transport infrastructures, public transport timetables, types of dominant vehicles, and so on. The same is true for everyday practices in private households, for instance, with regard to heating or electricity consumption. At the same time, there is a certain margin of flexibility with regard to behavior change, for instance, in the case of cycling instead of using a car in specific situations. A change in behavior can then contribute to structural change (for instance, the introduction of measures for improving conditions for cycling in urban traffic).

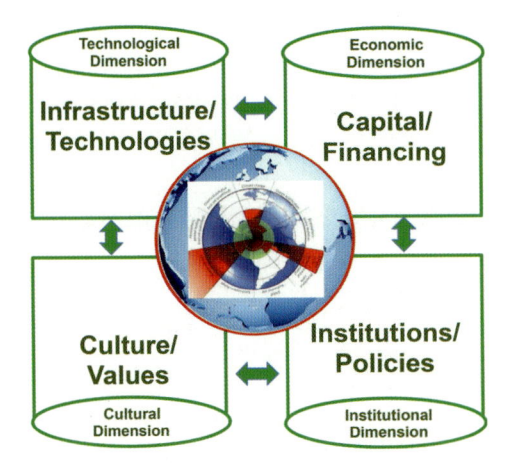

Figure 8.2 The four dimensions of transitions.

The four dimensions of a transition can in this way be understood as structures according to the conceptualization of Giddens. Infrastructures/technologies, capital, institutions, and cultural values and practices are structures on which concrete actions are based and to which behavioral patterns are related. Changes in these structures are important elements and driving forces of transition processes.

Figure 8.2 highlights that the different structural dimensions are tightly interlinked and mutually influence each other: technologies and infrastructures are influenced by economic structures (i.e. by the availability of capital), which in turn are influenced by institutional framework conditions. Institutions ultimately reflect norms and values shared by a society, thus its culture. Transformative literacy therefore requires knowledge about all four dimensions and their mutual interrelations. A closer look will be taken on each of the four dimensions in the following.

8.4.2
Infrastructures and Technologies – the Technological Perspective

Infrastructures and technologies are important driving factors of societal welfare and at the same time they determine the ecological consequences of modern economic activity. They play a major role in the creation of path dependencies and lead to a fixation of ecological trajectories for often decades ahead. Examples for this can be found in the energy system, industry infrastructures, the transport system, and building infrastructures. For instance, the infrastructures that are currently being built in the large and growing megacities in China, India, and other emerging economies determine their future energy consumption to a considerable extent. If infrastructures and technologies were to become radically more sustainable, guaranteeing prosperity at much less ecological strain would become feasible. Concepts of increasing energy and resource productivity [13] and of a green economy [20] depart from this insight.

The energy system is a typical example of a technology- and infrastructure-based system. It centers around infrastructures and technologies for energy production and distribution (e.g., power plants and energy grids) and energy consumption (e.g., industrial plants, heating installations, cars, and other means of transportation). Assuming a given future development of consumption, the question is how the energy system can be transformed, in order to provide energy based on renewable sources. This requires a change in energy production (from conventional power plants to wind, solar, and water power), the development of energy storage capacities and new forms of grid control (e.g., smart grids), and technological and infrastructure innovations on the consumption side (e.g., energy-efficient buildings, developing electric mobility), in order to adapt to a situation where energy supply completely based on renewable sources is possible.

Focusing on technological change and rebuilding infrastructures on the way to a "Great Transformation" is attractive in many ways: evoking technological progress as an important driver for economic and societal development has been a common theme established over many centuries. It is therefore also an established practice for businesses to invest in technological innovations and for politics to rely on proven instruments of funding and promoting technology development. Conceiving of the "Great Transformation" as a technological challenge, as a consequence means influencing the direction of technological innovation. There are no other implications for policy going beyond established instruments and practices.

The vision of a green economy thus originates from the alluring prospect of reconciling ecological requirements and economic development through technological innovations – without any need for more complex economic, institutional, or even cultural change.

It is tempting to rely on the potential of technological innovation that seems so promising at first glance – there are severe risks, however, that emerge only at a second glance:

- Technological solutions usually involve the danger of causing *problem shifting*. Many technologies have unintended side effects in other ecological dimensions than the one originally addressed: the debate on biofuels has demonstrated this; in the case of developing electric mobility, questions of resource use are hardly assessable at the moment, and the use of energy-saving light bulbs, containing harmful substances, resulted in problems with their disposal because of these substances.
- Even more problematic is the *rebound effect*: technological solutions are generally most successful when they are also economically efficient. The larger the economic efficiency gain, the greater is the incentive to increase consumption, which may significantly reduce the ecological benefits caused by the original increase in efficiency [21]. For instance, the Volkswagen Beetle launched in 2010 uses almost exactly as much fuel per 100 km as the original Beetle built in 1960 – even though the current model's engine is much more efficient. The efficiency gain has, however, been compensated by the increase in horsepower, equipment, size, and weight of the modern version.

- Finally, a major problem with regard to changes in infrastructures is that this is usually a very long-term undertaking. Substantial changes in infrastructures come at large costs for national economies and often give rise to protests by those who profit from the already existing infrastructures and technologies. An example of this is the strategic behavior of incumbent electric utility companies in the context of the German energy transition and their interest in preserving centralized energy system infrastructures and the related business models.

In spite of these risks, the "Great Technological Transformation" is the most profoundly understood and the most precisely calculated transformation scenario. A vast number of technological analyses at global, national, and urban levels show that an energy transition towards a system based on renewable energies by 2050 is feasible [6, 22].

However, technology-based system analyses in the end always remain static in nature. They depict (technologically) possible futures. What they cannot provide is a proposition about how to implement and actively shape this transition process. Especially over the course of the last couple of years, it has become obvious that the mere existence of technological options is not a sufficient precondition for actual change to take place. Even though the technological feasibility has been proven, the envisaged energy transition is still in its infancy.

It is therefore evident that what is needed is an integration of the technological perspective with economic, institutional, and cultural aspects, in order to gain a more comprehensive understanding of transition processes.

8.4.3
Financial Capital – the Economic Perspective

Infrastructures and new technologies are generally capital intensive. Investments are needed to build up research and development processes and infrastructures. Only where the necessary capital can be made available – in companies, by the state, or by private consumers – will change processes actually take shape.

How difficult this is can be seen in the context of the German energy transition: energy-saving renovations in the building sector could significantly contribute to an overall reduction in energy consumption. However, the actual renovation rate remains much lower than the envisaged 2–3% of the building stock per year, because in spite of the energy-saving potential, sufficient private capital cannot be mobilized.

Similarly, battery-electric cars are not selling as well as had been hoped, because their purchase price is often much higher than that of a comparable conventional internal combustion engine car.

In contrast, the development of renewable energy has been relatively successful in Germany over the last couple of years, because the Renewable Energies Act (EEG) has created incentives for investments via guaranteed feed-in remuneration. However, from a systemic point of view, it is often argued that by this transfer system and the massive promotion of relatively inefficient energy production technologies, as in the case of solar power, capital that should have been used for a more cost-efficient transformation of the energy system has been wasted. This has spawned a heated debate about reforms of solar energy subsidies in Germany. These examples show that transformation processes such as the energy transition require a good understanding of the economic dynamics involved.

Based on an aggregated economic perspective, it can be shown that a transition to a completely renewables-based energy system is possible at a reasonable effort. The Stern Review [23] showed that on a global scale, investing 1% of global GDP would suffice for avoiding the worst effects caused by climate change. The German Advisory Council on Global Change has presented an overview of scenarios and estimates that arrived at similar conclusions [6]. The same is true at an urban scale. The Wuppertal Institute has found that the vision of a CO_2-free city of Munich by 2058 would require investments at a similar scale [22].

Nevertheless, mobilization of the financial capital needed for an ecological transformation is far from being realized at present. At a global scale, massive investments are much rather directed to the exploration of new fossil energy resources, such as oil sands, shale gas, and deep-sea oil and gas. Many, also economically rational, energy efficiency measures are not implemented, because there are other investment opportunities that achieve higher returns and because there is a lack of information among companies and individual consumers. Also in emerging economies where urban infrastructures are currently being built up, the full potential of implementing high energy efficiency standards from the beginning is not realized.

Increasing pressure on return on investment leads to global allocations of capital that are often opposed to the requirements of a "Great Transformation." Investment decisions are influenced by a number of framework conditions, articulated in questions such as:

- What are the specific incentives for investments (e.g., investment security, tax incentives, low-cost loans)?
- What are the regulatory framework conditions of capital/financial markets (e.g., what types of real and financial investments are allowed/prohibited)?
- What role do capital transfers by nation states play (e.g., in the context of development policy or global climate protection policy)?

In order to deal with these questions, what is needed is "economic literacy," which improves the understanding of the differing logics of action of investors, companies, and consumers. It focuses on research in the fields of business models, competitiveness, financial markets, and generally the basic functions of modern money economies [24]. The debate about de-growth or post-growth societies [10] also needs to be considered in this context, assuming that societies that find alternatives to a constant need to grow could contribute significantly to a reduction in global ecological pressure.

This type of analysis shows how much economic dynamics are interrelated to the demands of a "Great Transformation." Against this background, it becomes obvious that even a green economy, which merely aims at a transformation based on green technologies, is still far from being realized.

Economic dynamics are nonetheless a central impetus for a "Great Transformation." Without utilizing the creative force of markets, achieving a wide-ranging transition is hardly realizable. At the same time, today's market dynamics actually contribute to the stability and continuation of established unsustainable developments. Therefore, one also needs to focus on the institutional framework conditions in which economic market dynamics are embedded.

8.4.4
Institutions/Policies – the Institutional Perspective

Institutions are rules and mechanisms for the realization of socio-political goals, facilitating social cohesion in a society as a whole. One important type of formal institutions is laws and regulations. However, due to increasing internationalization, the direct impact and range of national legislation are decreasing. In this context, it can also be observed that traditional processes of law-making and the major role played by the state in these ("government") are being eroded and instead new patterns of institutional "governance" have developed, which include a multitude of actors and structures involved in these processes [25, 26].

Especially economic processes are shaped by institutional structures. Duties and taxes, such as green taxes or regulations on emission protection and the return of products, have an impact on business decisions regarding the economic feasibility of supplying specific products and carrying out specific production processes. They also have an impact on companies' decisions about their location. In a globalized economy, these kinds of decisions are based not only on national regulation, but also increasingly on international regulatory frameworks (e.g., EU legislation, WTO agreements).

Formal institutions, such as laws, are the outcome of political processes. An "institutional literacy" therefore requires a good understanding of these increasingly complex political processes: how do specific issues manage to get on the political agenda? How and under what conditions are they then becoming part of legislation? What are effective and efficient policy instruments? How can specific socio-political goals be achieved?

Institutional governance extends over various political levels that need to be considered and included in the analysis: They range from global climate agreements, European or international emissions trading regimes, national laws on the promotion of renewable energies, regional government grants for the renovation of building structures, to local measures, such as public parking space management or sustainability-oriented urban development plans.

The institutional setting of societies reflects their overall values and orientations. Our modern institutional system has developed around three central questions [27]:

- the economic question (securing prosperity in a society);
- the social question (securing justice and a fair distribution of wealth);
- the democratic question (securing of comprehensive political participation).

Especially in the era after the Second World War, these questions were the central motivating forces for the development of institutions in Europe.

However, in the process of the increasing differentiation and globalization of modern societies, it becomes obvious that by striving for these economic, social, and democratic goals, more and more unintended ecological, social, and economic side effects emerge. These threaten to contradict the original aspirations and have led to the current ecological crisis, the financial and economic crisis, and the increasingly unequal distribution of wealth.

Against this background, the "Great Transformation" is predominantly a fundamental institutional reform project [27, 28]. It aims at building capacities within the institutional setting of modern societies for reflexivity, participation, balancing power, and inducing innovation (so-called institutional basic strategies; see Figure 8.3). To some extent, parallels can be detected between this approach and the idea of "inclusive institutions" that has been developed by Daron Acemoglu and James Robinson in their analysis of the success or failure of nations ("Why nations fail") [29].

"Institutional Literacy" for a "Great Transformation" therefore includes a number of multi-dimensional competences. It needs to be considered that in the context of modern governance structures, developing and shaping institutions no longer depend on the state as the single or most important actor [26]. New patterns in this field can be observed to range from bottom-up strategies by growing social movements (e.g., Transition Towns, initiatives for introducing local currencies) to new forms of standard setting by businesses and industrial sectors in global supply.

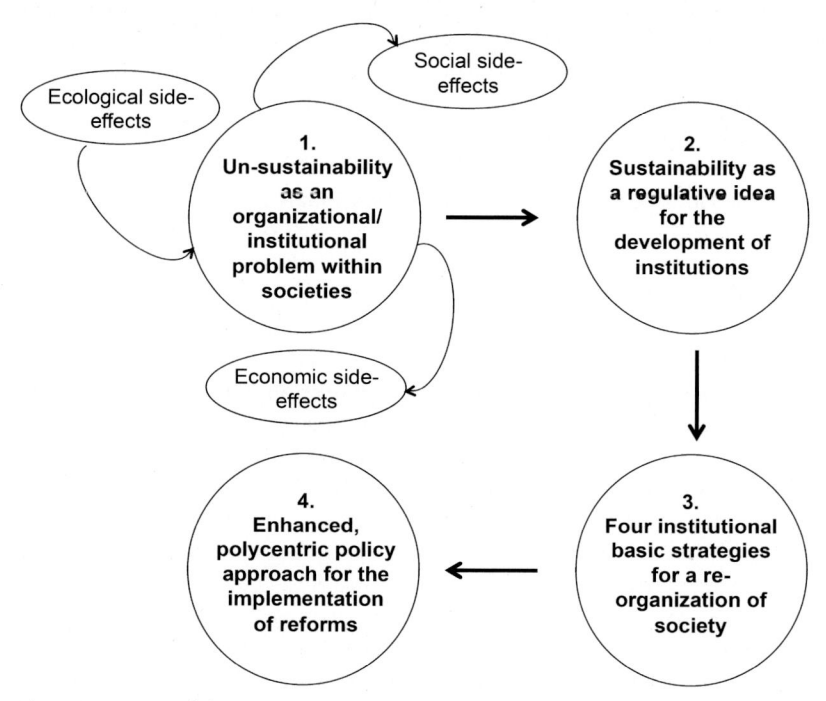

Figure 8.3 Sustainability transitions as institutional challenge [27].

Regarding these issues, comprehensive approaches have been developed in the field of governance theory and policy analysis, and knowledge has been gained on effective forms of policy mixes. Still, many questions remain unanswered, especially concerning the success conditions for sustainability policies, the need for their democratic legitimization, and ways of initiating "real-world-experiments" [30] that could enhance knowledge about improved forms of governance [31].

8.4.5
Cultural Change/Consumer Behavior – the Cultural Perspective

Ultimately, institutions reflect the norms and values of a society. Only those institutions that are compatible with the specific norms and values of a society can be enforced in the political process and will prevail over time.

"Transformative Literacy" therefore also implies an understanding of cultural settings and value orientation. In the end, all institutional, economic, and technological processes and dynamics are embedded in the culture of a society. Cultural analysis can thus be seen as the most fundamental kind of a system analysis [32]. This insight is related to the concept of humanity as a "narrative telling species," that is, a species characterized by its ability and desire to tell stories. Norms and values have a stabilizing effect on societies and individuals, and they can develop a strong motivational impetus – an important example is the driving force of values of freedom or equality for societal change over the last 300 years.

Changes in cultural values can have a direct impact on sustainability-oriented change in economic processes. For instance, sustainable consumption patterns become increasingly dominant in certain sectors and have significant effects on market development (e.g., organic food), and in some smaller niches new and alternative lifestyles begin to establish (Transition Towns, renewable energy based cities and regions).

Modern social psychology emphasizes that individual behavior usually does not result from knowledge, but rather it is the other way around. This implies that a "Great Transformation" depends on opening up new possibilities for acting differently, in order to break routines and old habits and facilitating behavior that is compatible with new norms and values. Generally, this requires institutional and political support and framing.

Changes in cultural values and lifestyles are in principle highly compatible with democratic forms of change. Global value research [33] has shown that the importance of values such as "autonomy" and also the willingness to assume societal responsibility are increasing on an international scale. Especially in developed democratic societies, the increasing significance of post-material values can be observed. This can be an important basis for a broad acceptance of wide-ranging political measures fostering transition processes in a global perspective. An example of this is the broad societal support for the energy transition in Germany. Civil-society organizations are important drivers in this transition, which amounts to changes not only in technology, institutions, and markets, but also in cultural values. The overall success of the German energy transition also depends on substantial energy savings. The required level of energy savings will not be realized by implementing energy efficiency measures alone (for instance, also due to the occurrence of rebound effects). Achieving a sustainable level of energy consumption requires a reduction in energy use in absolute numbers, which – apart from increasing overall efficiency – depends on changes in consumption and use patterns, and thus changes in social practices and culturally shaped habits.

"Cultural Literacy" as an element of "Transformative Literacy" implies the need to develop a better understanding of the common cultural norms and values on which modern societies are based. These are rooted in historical, religious, and often also region-specific backgrounds. What is also needed is a better understanding of cultural change processes, which are often catalyzed by specific societal events or broader developments (e.g., technological innovations, such as the diffusion of the Internet or the growing importance of post-material values). Finally, it is essential to grasp the role played by different actors in shaping cultural developments. For instance, it can be shown that global businesses with their brands become culture-shaping actors [34].

8.5
Techno-Economists, Institutionalists, and Culturalists – Three Conflicting Transformation Paradigms

In 2012, Niko Paech introduced and discussed a "history of dogmas in the discourse on sustainability" [35]. He distinguished three transformation dogmas that currently shape the discourse on sustainability and illustrated them with the debate about economic growth (Figure 8.4). This characterization can also be applied to the discourse on the German energy transition:

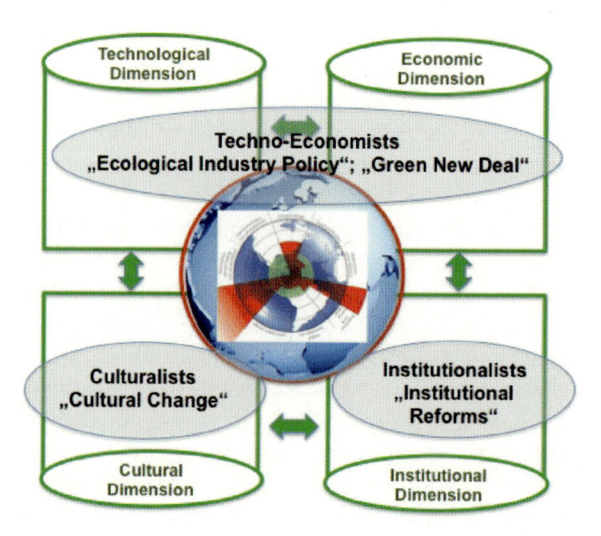

Figure 8.4 Three transformation paradigms. Adapted from Paech [35].

1. The first dogma identified by Paech is the perspective on "green growth." This is currently the dominant strand in the debate, according to which an absolute decoupling of economic growth and environmental impact is possible by way of technological efficiency increases. This dogma creates the vision of a green growth, that is, a technological revolution towards massive ecological efficiency increases, which provide a solution for both ecological challenges and economic development and growth. This perspective of a green industrial revolution is dominant in national environmental policies and also in debates at the Rio+20 Conference in June 2012. Transferring this to the four dimensions of transformation introduced above, this dogma is equivalent to a techno-economic transformation perspective. This would mean that a sustainability transition is possible within established cultural and institutional settings by combining technological innovation potential and economic development perspectives. Consequently, techno-economists deal with the energy transition as a major economic investment program. Their rationale for implementing the energy transition is primarily economic: major cost savings can be gained in the future (the Renewable Energy Research Association calculates cost savings of €570 billion for Germany by 2050 [36]), roughly 300 000 jobs have been created in the renewable energy sector [37], and there is a future industrial and economic development perspective in this field.
Apart from "green growth," Paech distinguishes two further dogmas that are both more critical with regard to the growth paradigm – these are the dogma of "institutional change" and the dogma of "substantial change."

2. The dogma of institutional change is represented by those who are skeptical with regard to a merely techno-economic transformation. It is argued that so far an absolute decoupling of economic growth and environmental pressure has not been achieved in relevant ecological dimensions. It is explained that this is due to the efficiency increases being offset by rebound and growth effects that are inherent to the system. A transition is therefore only possible when there is institutional change in the economic system: Depending on the specific institutionalist tradition, it is argued that institutional reform should address the internalization of external effects [13, 38], the question of property rights [39], or the organization of modern money economies [24, 40]. Transferring this to the transformation dimensions introduced in this chapter, this dogma is equivalent to an institutionalist transformation perspective. Such an approach could, for instance, include the four institutional basic strategies of sustainability-oriented policy mentioned above. In the context of the German energy transition, institutionalists focus on different types of institutional levers: these include broader emissions trading regimes, ecological tax reforms aiming at the internalization of external effects caused by fossil electricity generation [13], and the further development of the Renewable Energy Act, as well as a number of supplementing institutional reforms, such as the development of a capacity market for reserve power plants (especially in the field of natural gas).

3. Finally, "substantial change" refers to a dogma most critical with regard to the growth paradigm. It is argued that a sustainability transition is only possible when there is a comprehensive cultural change throughout society as a whole. Important elements of this dogma are concepts of "sufficiency," "subsistence," industrial deconstruction, and deglobalization. This implies a fundamental breach with today's capital-based and globally organized economic system and a redirection towards local resilience. According to this dogma, an absolute technology-based decoupling is impossible and institutional change is only feasible in a situation where alternative values and lifestyles have developed bottom-up. In contrast to concepts of "top-down" institutional change, this perspective focuses on cultural change that develops from the bottom, as for instance in the Transition Town movement and similar initiatives. Transferring this to the transformation dimensions introduced in this chapter, this dogma is equivalent to a culturalist transformation perspective. In the context of the German energy transition, culturalists focus on questions of regionalization and energy self-sufficiency. Various projects exist that show how the energy transition can be understood as a process of cultural change: "100% renewable energy regions" are finding ways of regaining regional autonomy by becoming energy self-sufficient; similarly, energy cooperatives have been founded and initiatives have been built that repurchase local grids.

In reality, transition processes relating to all of the three dogmas can be identified: there are different techno-economic, institutional, and cultural transformations taking place at the same time. The conflict between these three dogmas or paradigms relates more to the question of which one of the different dimensions provides the central impetus for a "Great Transformation." It is nonetheless clear that an actual transition will in the end involve cultural, institutional, and techno-economic changes.

This illustrates the importance and role of a "Transformative Literacy." If we understand it as "the ability to read and utilize information about societal transformation processes, to accordingly interpret and get actively involved in these processes," it increases societal reflexivity when observing and actively contributing to transition processes. It also has the potential to open up new perspectives, facilitating a broad debate on shaping and steering transitions – going beyond a dogmatic conflict. The overall aim is to develop theories, utilize empirical observations, and also to engage actively in social and institutional experiments, in order to understand better the role of the different dimensions of transitions and to provide actors that are involved in transition processes with an orientation. The German energy transition is a promising field of application and experimentation for improving our understanding of comprehensive transition processes. The energy transition involves technological, economic, institutional, and cultural change taking place simultaneously and this will provide valuable lessons for the future development of a "Transformative Literacy" towards a more sustainable development.

References

1. Crutzen, P. J. and Stoermer, E. F. (2000) The 'Anthropocene.' *Global Change Newsl.*, **41** (1), 17–18.
2. Lenton, T. M., Held, H., Kriegler, E., Hall, J. W., Lucht, W., Rahmstorf, S., and Schellnhuber, H. J. (2008) Tipping elements in the Earth's climate system. *Proc. Natl. Acad. Sci. U.S.A.*, **105** (6), 1786–1793.
3. Rockström, J. et al. (2009) A safe operating space for humanity. *Nature*, **461**, 472–475.
4. Wuppertal Institute (2006) *Fair Future – Begrenzte Ressourcen und Globale Gerechtigkeit*, Beck, Munich.
5. Sachs, W. and Santarius, T. (2007) *Fair Future: Resource Conflicts, Security and Global Justice*, Zed Books, London.
6. Wissenschaftlicher Beirat Globaler Umweltveränderungen (WBGU) (2011) *Welt im Wandel. Gesellschaftsvertrag für eine Große Transformation*, WBGU, Berlin.
7. United Nations (1987) *Our Common Future. Report of the World Commission on Environment and Development*, United Nations, New York.
8. Bringezu, S. and Bleischwitz, R. (2009) *Sustainable Resource Management. Global Trends, Visions and Policies*, Greenleaf Publishing, Sheffield.
9. Jackson, T. (2009) *Prosperity Without Growth. Economics for a Finite Planet*, Earthscan, London.
10. Seidl, I. and Zahrnt, S. (2010) *Postwachstumsgesellschaft. Konzepte für die Zukunft*, Ökologie und Wirtschaftsforschung, vol. 87, Metropolis, Marburg.
11. Santarius, T. (2012) *Der Rebound-Effekt. Über die unerwünschten Folgen der erwünschten Energieeffizienz*, Impulse zur Wachstumswende No. 5, Wuppertal Institute, Wuppertal.
12. Enquete-Commission (2012) *Zwischenbericht der Projektgruppe 3 "Entkopplung" der Enquete-Kommission "Wachstum, Wohlstand, Lebensqualität"*, Drucksache PG 3/33, Enquete-Kommission, Berlin.
13. von Weizsäcker, E. U., Hargroves, K., and Smith, M. (2010) *Faktor Fünf: die Formel für nachhaltiges Wachstum*, Droemer, Munich.
14. Scholz, R. W. (2011) *Environmental Literacy in Science and Society. From Knowledge to Decisions*, Cambridge University Press, New York.
15. Grin, J., Rotmans, J., and Schot, J. (eds) (2010) *Transitions to Sustainable Development. New Directions in the Study of Long Term Transformative Change*, Routledge, New York.
16. Aghion, P., and Howitt, P. (1997) *Endogenous Growth Theory*, MIT Press, Cambridge, MA.
17. Giddens, A. (1984) *The Constitution of Society. Outline of the Theory of Structuration*, Polity Press, Cambridge.
18. Schneidewind, U. (1998) *Die Unternehmung als strukturpolitischer Akteur*, Metropolis, Marburg.
19. Schneidewind, U. and Scheck, H. (2012) *Die Stadt als "Reallabor" für Systeminnovationen*, Institut für Sozialinnovation, "Soziale Innovation und Nachhaltigkeit," VS Verlag, Wiesbaden.
20. United Nations Environment Programme (2011) *Towards a Green Economy: Pathways to Sustainable Development and Poverty Eradication*, UNEP, Nairobi, www.unep.org/greeneconomy (last accessed 6 January 2013).
21. Alcott, B. (2005) Jevons' paradox. *Ecol. Econ.*, **54** (1), 9–21.
22. Siemens AG, Wuppertal Institute for Climate, Environment, Energy (2009): *Sustainable Urban Infrastructure: Ausgabe München; Wege in eine CO_2-freie Zukunft*, Siemens, Munich.
23. Stern, N. (2007) *The Economics of Climate Change: the Stern Review*, Cambridge University Press, Cambridge.
24. Binswanger, H.-C. (2009) *Die Wachstumsspirale: Geld, Energie und Imagination in der Dynamik des Marktprozesses*, Metropolis, Marburg.
25. Voss, J. P., Bauknecht, D., and Kemp, R. (eds) (2006) *Reflexive Governance for Sustainable Development*, Edward Elgar Publishing, Cheltenham.

26 Kooiman, J. (2003) *Governing as Governance*, Sage Publications, London.

27 Schneidewind, U., Feindt, P. H., Meister, H. P., Minsch, J., Schulz, T., and Tscheulin, J. (1997) Institutionelle Reformen für eine Politik der Nachhaltigkeit: vom Was zum Wie in der Nachhaltigkeitsdebatte. *GAIA*, **6** (3), 182–196.

28 Minsch, J., Feindt, H. P., Meister, P., and Schneidewind, U. (1998) *Institutionelle Reformen für eine Politik der Nachhaltigkeit*, Springer, Berlin.

29 Acemoglu, D., Robinson, J. (2012) *Why Nations Fail: the Origins of Power, Prosperity, and Poverty*, Crown Publishing, New York.

30 Gross, M., Hoffmann-Riem, H., and Krohn, W. (2005) *Realexperimente. Ökologische Gestaltungsprozesse in der Wissensgesellschaft*, Transcript, Bielefeld.

31 Biermann, F. et al. (2009) *Earth System Governance: People, Places and the Planet. Science and Implementation Plan of the Earth System Governance Project*, Earth System Governance Report 1, IHDP Report 20, IHDP, Bonn.

32 Moebius, S. (2009) *Kultur. Einführung in die Kultursoziologie*, Transcript, Bielefeld,.

33 Inglehart, R. and Welzel, C. (2005) *Modernization, Cultural Change and Democracy. The Human Development Sequence*, Cambridge University Press, Cambridge.

34 FUGO – Forschungsgruppe Unternehmen und gesellschaftliche Organisation, Universität Oldenburg (2004), *Perspektiven einer kulturwissenschaftlichen Theorie der Unternehmung*, Metropolis, Marburg.

35 Paech, N. (2012) Dogmenhistorie des Nachhaltigkeitsdiskurses, presented at Akademie Loccum, 13 December 2012.

36 FVEE (Renewable Energy Research Association) (2012) *Forschung senkt Kosten der Energiewende*, FVEE Stellungnahme, http://www.fvee.de/fileadmin/publikationen/Politische_Papiere_FVEE/12.10.EE_Kosten/2012_10_10_FVEE_stellungnahme_kosten.pdf (last accessed 6 January 2013).

37 BMU (Federal Ministry for the Environment, Nature Conservation and Nuclear Safety) (2011) *Erneuerbar beschäftigt! Kurz- und langfristige Wirkungen des Ausbaus erneuerbarer Energien auf den deutschen Arbeitsmarkt*, 2nd edn, BMU, Berlin.

38 Scherhorn, G. (2012) Nachhaltigkeit und Eigentum, *Gegenblende. Das gewerkschaftliche Debattenmagazin*, http://www.gegenblende.de/++co++5f6759f2-af1e–11e1–49bd–52540066f352 (last accessed 6 January 2013).

39 Helfrich, S. and Heinrich-Böll-Stiftung (eds) (2012) *Commons: für eine neue Politik jenseits von Markt und Staat*, Transcript, Bielefeld.

40 North, P. (2009) *The Transition Guide to Money: Creating Alternative Systems for Your Community*, Green, Totnes.

9
Renewable Energy Future for the Developing World

Dieter Holm

9.1
Introduction

9.1.1
Aim

This chapter presents the unique window of opportunity of establishing the use of Renewable Energies (REs) in the Developing World in a practical and cost-effective manner. The approach is based both on theoretical studies for the ISES White Paper *Renewable Energy Future for the Developing World* [1] and also – and this is important – on a lifetime's experience as a Renewable Energy consultant in this part of the world. For obvious reasons, the material presented is not necessarily found in the reports of development agencies. Some philosophers argue that we can learn from our mistakes – provided that we survived them.

The chapter starts with working definitions of the terms Developing World and Transition in the Developing World. This is followed by the question of whether REs can deliver the quantities needed, and the special opportunities that are waiting to be realized. Then a framework is presented with which developments should be implemented. This is followed by a section on proven policies and measures with special reference to their applicability in the Developing World. In conclusion, there is a section on priorities for action, and how to restructure them.

9.2
Descriptions and Definitions of the Developing World

9.2.1
The Developing World

One definition of the Developing World would be: "The rest of the world, except Australia, Canada, the European Union (EU), Israel, Japan, New Zealand, the Southern African Customs and Trade Union and the United States of America." The International Monetary Fund [2] lists 157 nations, but forgot Cuba and North Korea. The Developing World consists of "Developing Nations" or "Developing Economies" which are undergoing industrialization, experiencing economic growth, and receiving foreign investments. "Emerging Economies," "Developing World," and "Third World" are mostly taken as synonyms, even if they might be considered either euphemistic or abusive language.

Geographically, the Developing Countries are concentrated in Latin America, Africa, and Southern Asia. About three-fifths of the world's population live in Developing Countries. Typically, their economies are still heavily dependent on agriculture, often at the subsistence level, with some mining. Beneficiation through secondary industries is rarely found in the initial stages. Tourism plays an important role.

Infrastructure is often elementary, with a scarcity of skills in engineering, technology and the professions. Politics, human cultural aspects, and the arts of language, music, crafts, and religion are often more appreciated and nurtured. There is a more relaxed attitude towards time, planning, and the future.

Scientific surveys are infrequent and discontinuous. Unsurprisingly, statistics and data for countries in the Developing World are problematic. On the scale of Transparency International's *Corruption Perceptions Index* [3], they tend to occupy the bottom ranks, and similarly with the per capita GDP, literacy, life expectancy, labor productivity, and others – inasmuch as such statistics are reliable. Generally, the fertility rate, violent crime rates, and political instability are higher compared with industrialized countries.

The primary energy source for many millions in the Developing World is firewood and biomass, often used unsustainably. The transition to sustainable energies and the simultaneous elimination of material poverty pose a huge challenge to Developing Countries – and the industrialized world.

9.2.2
The Developing World in Transition

Industrialization in the West was accomplished on the material basis of fossil energies. If the Developing World were to follow the example set by some industrialized nations, the global impact would be unthinkable. Hence the transition to a sustainable energy world must be rapid and orderly. This requires both national policies and international cooperation.

9.2 Descriptions and Definitions of the Developing World | 139

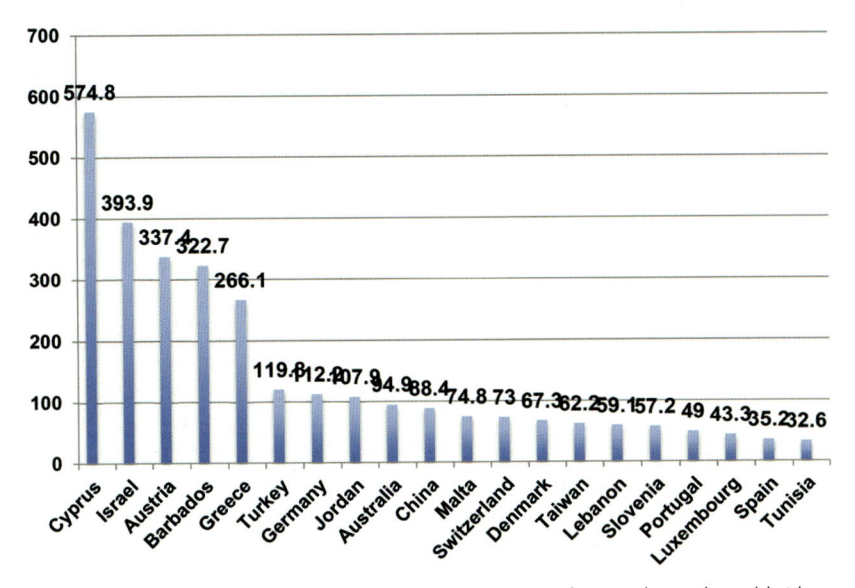

Figure 9.1 Solar thermal capacity per capita (Wth/cap) [4]. China ranks tenth worldwide.

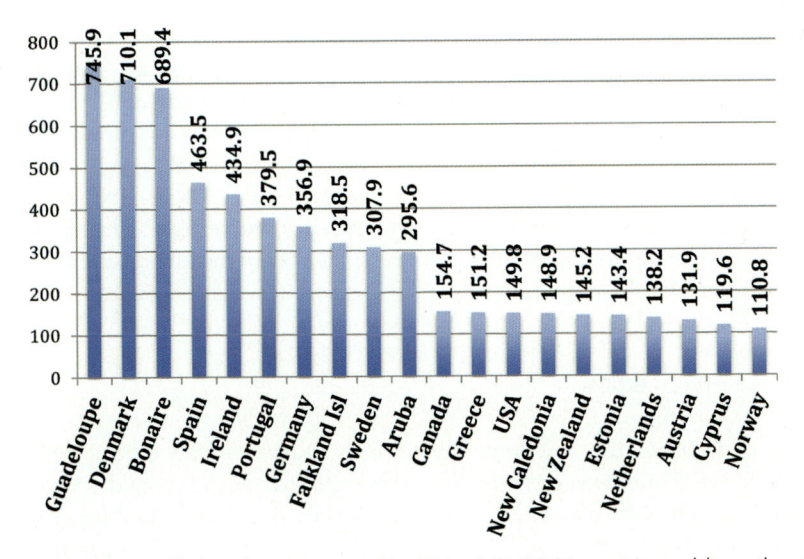

Figure 9.2 Installed wind capacity per capita (W/cap) [5]. BRICS countries rank lower than twentieth.

Many RE technologies have been proven for feasibility in the world markets, and since suitable policies have been tried and tested, the near-term risks of adoption are lower than those of procrastination. The laggards lack awareness and political will. Neither China nor other BRICS countries are prominent in per capita RE installations (see Figures 9.1 and 9.2).

REs are expected to be in the mainstream worldwide by 2030 or earlier.

The tide is turning inexorably towards renewable energies. To the Developing Nations, this transition offers unique opportunities:

- Significant population and/or business growth occurs in the Developing World, but the infrastructure is underdeveloped. Instead of investing in the technologies of the past, Developing Nations can leapfrog to most modern renewable energy technologies (RETs). The use of cellular telephones illustrates how the old technology of expensive and vulnerable landlines has been bypassed by a new technology. Likewise, the concept of large centralized coal- or gas-fired power stations with large transmission and distribution networks is making way for on-site distributed generation serving a mini-grid.
- Most Developing Countries have great renewable energy resources, notably solar radiation.
- REs and other renewable energy resources are more equally distributed than fossils fuels. This means that by transitioning to renewable energy, Developing Nations are less exposed to imported energy costs and energy wars.
- Since Developing Nations generally do little R&D, they will benefit from mature, low-maintenance technologies without having contributed to the R&D costs.

The transition to renewable energies has been retarded by the inertia of established systems and by artificial government-endorsed market distortions. Nonrenewable sources seem to be cheap because they do not account for the real social, environmental, or military costs. Accounting for externalities would approximately double fossil energy prices. Furthermore, governments routinely give massive direct and indirect subsidies through protecting monopolies, granting financial backups, and ignoring the cost to future generations.

A combination of energy conservation, energy efficiency, and renewable energies presents a much more environmentally, socially, and economically sustainable energy path to the Developing World where job creation is a high priority.

9.2.3
Emerging Economies – BRICS

Brazil, Russia, India, China, and South Africa (BRICS) have quickly moved from the category of international nonentities to market significance. Their rapid growth is sometimes viewed with alarm. In 2012, their Standard and Poor rankings [6] were BBB, BBB, BBB–, AA–, and BBB+, respectively, while their Transparency International *Corruption Perceptions Index* rankings of governance are 73, 143, 95, 75, and 64 respectively, out of 182 ranked. BRICS account for one-fifth of global trade and half of current growth. Brazil, Russia, and South Africa focus on export of raw materials, while India and China concentrate on manufacturing and services. During 2001–2010, China's agricultural contribution to GDP halved, while services grew from 33 to 43.6%. The recent GDP growth rate was exceptional, albeit it from a relatively low basis. BRICS countries have fossil-based industries, except Brazil with hydropower. Pollution levels reflect this. All BRICS counties except Russia have a broad-based population pyramid, and youth unemployment is therefore a great risk. The official unemployment rate varies between 4% for China to 24.8% for South Africa. *Youth unemployment and the associated political risks are a common threat.*

9.3
Can Renewable Energies Deliver?

Normally one differentiates between the technical and economic potential of REs. Technologies are not static and their economic competitiveness also is continually changing. RE technologies are becoming cheaper as a result of development and economies of scale, while fossil technologies are on a course of diminishing efficiency returns. Long-term predictions about demand and supply have been known to be far off-beam.

A graphic representation by Perez and Perez [7] dramatically illustrates the RE resources and the recent global demand (Figure 9.3). It should be borne in mind that the scale is *volumes* of spheres, not areas. Also, the four fossils spheres on the right represent total reserves, whereas the nine on the left are annual figures. The illustration leaves no doubt that the availability of REs does not constitute a limitation.

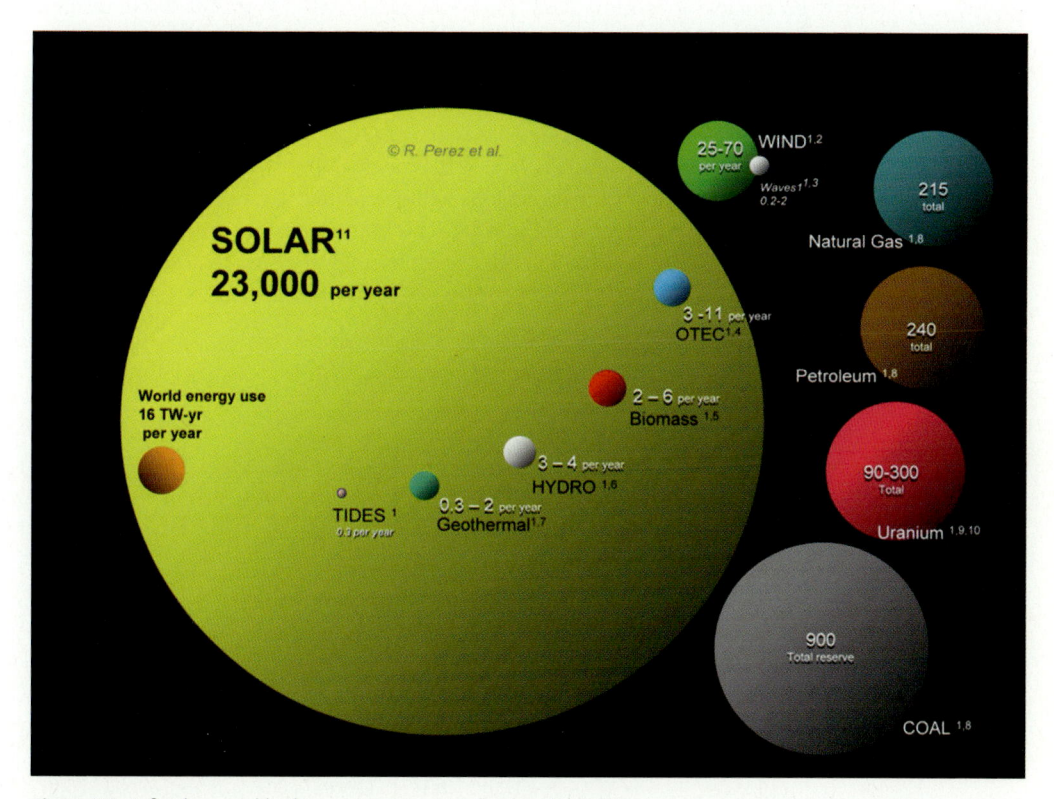

Figure 9.3 A fundamental look at energy reserves for the planet [7].

9.4
Opportunities for the Developing World

The Developing World has at least four differentiating opportunities in the arena of energy. These are Poverty Alleviation through RE Jobs, a New Energy Infrastructure Model, the Great RE Potential in the Developing World, and the Underdeveloped Conventional Infrastructure.

9.4.1
Poverty Alleviation through RE Jobs

Reducing poverty and unemployment are key priorities in the Developing World. Renewable energies offer more new work opportunities than other energies. This fact still has to be realised by many Developing Nations.

For example, the use of solar water heaters (SWHs) creates low capital and low risk work opportunities, with the greater job creation opportunities being situated in the business of selling, installing and servicing SWHs. The main barriers to higher market penetration in the Developing World are lacking awareness of policy makers.

In Germany with less favourable resources, renewable energy electricity jobs surpassed nuclear jobs in to 2002. They will create 250 000 to 350 000 renewable energy jobs by 2050.

Table 9.1 shows that electricity generated by Renewable Energy generally creates many times more direct work opportunities in the Developing World than fossil or nuclear energies. For example, solar thermal (10.4 jobs/GWh) creates 35 times more than current coal (0.3 jobs/GWh). When nuclear energy and gas are compared with PV, then 620 times more direct jobs will be created. This excludes significant numbers of indirect and induced jobs [8]. It is also important where the jobs are created. With most Renewable Energies it is possibly to generate power where the people are and want to be.

Table 9.1 Summary of direct work opportunities from conventional and renewable sources [8].

	Conventional Energy Technology	Total		Renewable Energy Technology	Total	
		/MW	/GWh		/MW	/GWh
Jobs directly created from generating electricity	Coal (current)	1.7	0.3	Solar thermal	5.9	10.4
	Coal (future)	3.0	0.7	Solar Panels	35.4	62.0
	Nuclear	0.5	0.1	Wind	4.8	12.6
	Pebble Bed Modular Reactors	1.3	0.2	Biomass	1.0	5.6
	Gas	1.2	0.1	Landfills	6.0	23.0

It might be expected that more jobs associated with REs would make those energies more expensive than fossil energies. In fact, there is a great risk to socio political stability as a result of very high unemployment rates (more than 30%) in many parts of the Developing World. Especially the younger generation is restive because unachievable promises have been made to them. This leads to relatively low salaries, unless governments artificially increase salaries, thereby destroying work opportunities. Also labour tends to be less productive compared to the Industrialised World. Finally, externality costs of fossil energies are routinely ignored. However, ignored costs are not avoided costs.

The Developing World is characterised by large expanses of land with dispersed settlement patterns, originally intended for agriculture. With limited financial resources, the energy supply and distribution infrastructure did generally not keep up with the demand. Recent rapid population growths, combined with urbanisation and higher demands of electricity put additional stress on the systems. Very often, little maintenance is being done. Today the energy infrastructure is in need of upgrading. Instead of providing dated, inefficient and polluting energies and their infrastructure, there is now a unique opportunity to provide modern energy service technologies, combined with modern information technologies.

9.4.2
A New Energy Infrastructure Model

In the past, many Developing Nations had centralist systems with government utility monopolies forced to deliver power at non-sustainable prices for the sake of political expediency. Brownouts and blackouts at great national costs resulted.

Developing Nations will soon find that renewable energy powered Distributed Generation (DG) and Combined Heat and Power (CHP) Plants create local jobs, are environmentally more benign, and are not dependant on the weaknesses of a centralised generation system.

Typical energy conversion losses of conventional systems from mined coal to end-use are significant: With incandescent lighting the useful percentage may be less than 6%. All the rest is pollution in the form of ash, gases and reject heat.

Distributed generation plants can be built in smaller increments, closely following the demand profile. By contrast, this is impossible with conventional power plants, which come in big chunks, tying down big chunks of capital, long before it is actually needed ("stranded assets" in poor countries).

DG in rural areas enables secondary industries, thereby assisting beneficiation, adding value to local products and creating local jobs. This, in turn, stems the tide of rural depopulation and urban squatters – serious issues with Developing Nations.

Non-polluting DG systems generate heat and power at the point of energy consumption, and are much more efficient than the old centralised fossil-fired power plants. They also reduce line losses. Pakistan's line losses are between 35 to 42% [9]. In Sub-Saharan Africa "the hidden costs of distribution losses are usually more than 0.5 percent of GDP, and may be as much as 1.2 percent of GDP in some countries" [10]. In addition there are "non-technical losses" (euphemism for electricity theft)

... "the highest ... in Nigeria where the utility is capturing only 25 percent of the revenues owed" [10].

By contrast, distributed renewable energy (co)generation driven by the private sector and cooperatives contributes towards sustainable development and the democratisation of power, in both senses of the word.

9.4.3
Great RE Potential of Developing World

Two thirds of the global hydropower potential is found in the Developing World. 95% of the world's best daily winter sunshine above 6,5 kWh/m^2 is found in Africa. 59% falls on Southern Africa, and 24% on South Africa. Because of its high altitude and weaker solar radiation, it is the winter season that is decisive for solar water heating, solar space heating, solar thermal power and photovoltaics [11].

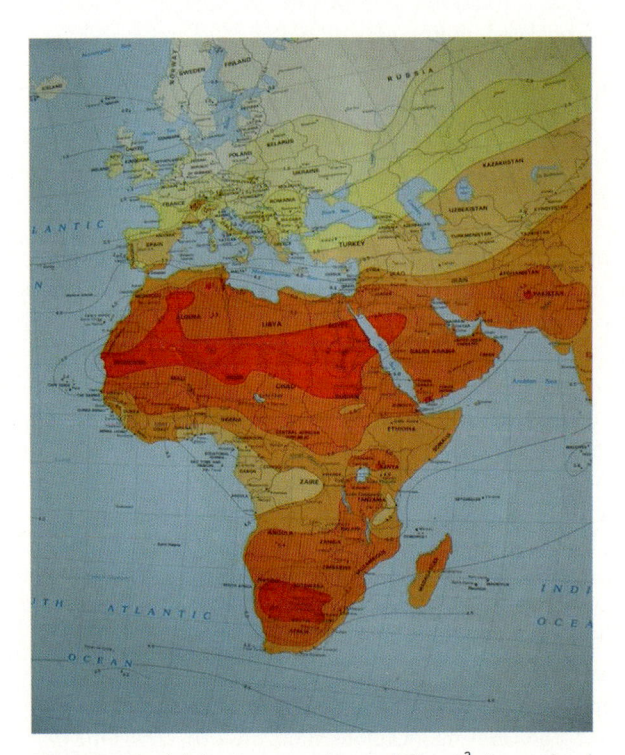

Figure 9.4 Total daily winter sunshine in kWh m^{-2}. Europe has less sunshine [9].

9.4.4
Underdeveloped Conventional Infrastructure

One of the reasons why the Developing World is called "developing" is because it has not invested much in the conventional fossil-based or nuclear energy infrastructure.

Therefore there is less resistance to change and less opposition from vested interests. Without having contributed to the R&D investment in Renewable Energy Technologies, the Developing World enjoys the unique opportunity to leapfrog to the most modern clean renewable technologies, thereby avoiding the need to go down the dirty fossil fuel path.

9.5 Development Framework

9.5.1 National Renewable Energies Within Global Guard Rails

The great energy transition requires a change of heart. Responsible governments consider our common future – and have a strong hand in shaping it. They set inspiring long-term stretch targets that attract commitments of entrepreneurs and resources and money by industry and academia. They also encourage educational institutions and bureaucracy to adapt. While the availability and composition of local RE resources differ, and the demands in time and space are changing, it is important to respect the boundaries of what is desirable rather than of what is possible.

9.5.2 The International Context: Global Guard Rails

The German Advisory Council on Global Change (WBGU) produced a comprehensive report, *World in Transition – Towards Sustainable Energy Systems* [10], introducing the innovative concept of "guard rails" bounding the paths towards global energy sustainability. Guard rails are those levels of damage which should not be crossed. There are six economic and five ecological guard rails (not goals).

9.5.2.1 Socio-Economic Guard Rails

Access to Advanced Energy for All
Everyone should have access to advanced energy. This includes access to electricity, and substituting health-endangering biomass use by advanced fuels.

Meeting the Individual Minimum Requirement for Advanced Energy
The following final energy quantities are considered to be the minimum for elementary individual needs: by 2020 at the latest, everyone should have at least 550 kWh final energy per person and year, by 2050 at least 700 kWh, and by 2100 at least 1000 kWh.

Limiting the Proportion of Income Expended for Energy
Poor households should not need to spend more than one-tenth of their income on meeting elementary energy requirements.

Minimum Microeconomic Development
To meet the macroeconomic minimum per capita energy requirement (for energy services utilized indirectly), all countries should be able to develop a per capita GDP of at least about US$ 3000, in 1999 values.

Keeping Risks Within a Normal Range
A sustainable energy system needs to build upon technologies whose operation remains within the "normal range" of environmental risk. Nuclear energy does not meet this requirement, particularly because of its intolerable accident risks and unresolved waste management, but also because of the risks of proliferation and terrorism.

Preventing Disease Caused by Energy Use
Indoor air pollution from the burning of biomass and air pollution in towns and cities resulting from fossil energy sources cause severe damage worldwide. The overall health impact caused by this should, in all WHO regions, not exceed 0.5% of the total health impact in each region (measured in DALYs = disability-adjusted life-years).

9.5.2.2 Ecological Guard Rails

Climate Protection
A rate of temperature change exceeding 0.2 K per decade and a mean global temperature rise of more than 2 K compared with pre-industrial levels are intolerable parameters of global climate change.

Sustainable Land Use
Around 10–20% of the global land surface should be reserved for nature conservation. Not more than 3% should be used for bioenergy crops or terrestrial CO_2 sequestration. As a fundamental matter of principle, natural ecosystems should not be converted to bioenergy cultivation. Where conflicts arise between different types of land use, food security must have priority.

Protection of Rivers and Their Catchment Areas
In the same vein as terrestrial areas, about 10–20% of riverine ecosystems, including their catchment areas, should be reserved for nature conservation. This is one reason why hydroelectricity – after necessary framework conditions have been met (investment in research, institutions, capacity building, etc.) – can only be expanded to a limited extent.

Protection of Marine Ecosystems
The use of the oceans to sequester carbon is not tolerable, because the ecological damage can be major, and knowledge about the biological consequences is too fragmentary.

Prevention of Atmospheric Air Pollution
Critical levels of air pollution are not tolerable. As a preliminary quantitative guard rail, it could be determined that pollution levels should nowhere be higher than they are today in the EU, even though the situation there is not yet satisfactory for all types of pollutants. A final guard rail would need to be defined and implemented by national environmental standards and multilateral environmental agreements.

A test run demonstrates that turning energy systems towards sustainability is technically and economically feasible.

A target of 10/20/50% of renewable primary energy by 2010/2020/2030 has also been corroborated by the European Renewable Energy Council (EREC), stating that renewable energy could supply 50% of the global energy by 2040 [11]. The EREC also states that more initiative should be taken at nation state level, rather than waiting for international agreements. Others consider this to be a timid target.

The key findings of the WBGU study are as follows:

- The transition will only work with intensified capital and technology transfer from industrialized to Developing Countries. Market maturity of RE and energy efficiency (EE) need to be accelerated in the industrialized countries, for instance through redirecting R&D resources and demonstration and implementation strategies. This will reduce the entrance barriers to all, especially the Developing Nations.
- Worldwide cooperation and convergence of living standards are likely to facilitate rapid technology development and dissemination.
- Binding CO_2 reduction commitments are a prerequisite.
- Further greenhouse gas (GHG) reduction policies by other sectors (e.g., NO_x and NH_4 from agriculture) are required.
- 450 ppm of CO_2 may not be sufficient for climate stabilization, and should not be taken as a safe stabilization level.
- An alternative reduction path by fossil and nuclear energy entails substantially higher risks and environmental impacts, and is more costly, mainly because of CO_2 sequestration costs.
- The next 10–20 years offer a window of opportunity in a system with a long time lag. Transitioning at a later stage will cost disproportionately much more.
- The currently most cost-effective technologies such as wind and biomass have to be used to the maximum in the short and medium term.
- Efficient use of fossil fuels is part of the transition, in particular the efficient use of natural gas.
- A certain amount of carbon sequestration in geological caverns will be necessary during this century.

A roadmap with goals and policy options for the transformation highlights the following:

- Eradicating energy poverty and establishing minimum global supply.
- This requires a new World Bank policy to integrate energy in poverty reduction strategies as well as strengthening regional development banks.

- Promoting socio-economic development.
- Combining regulatory and private sector initiatives.
- Protecting natural life-supporting systems: Reduced global CO_2 emissions by at least 30% from 1990 levels by 2050. For established industrialized nations, this entails a reduction of 80%, while developing and newly industrialized countries' emissions should rise by no more than 30%.
- Improved energy productivity (GDP to energy input ratio) of initially 1.4% annually is required, followed by 1.6%, to reach triple the current productivity by 2050 from 1990 levels. This requires international standards for fossil-fuelled power stations, and 20% RE based electricity in the EU by 2012; mandatory labeling; phased-out nonrenewable energy subsidies, and primary energy targets for buildings.
- Expanding RE substantially from the current 12.7% to 20% by 2020.
- Phasing out nuclear energy by 2050, with stricter monitoring of all sites.

9.6
Policies Accelerating Renewable Energies in Developing Countries

Fossil and nuclear fuel prices are not the result of free market mechanisms, nor do they reflect their true costs. Barriers to REs are their relatively higher capital cost, lack of economies of scale and access to affordable credit, selective punitive grid connection costs, lack of standards, and lack of training and awareness. In Developing Countries, the barrier of perceived investor risk is even higher due to political, regulatory, and market risks. Developing Nations can profit from the substantial body of knowledge accumulated by the world leaders in renewable energy.

There are six categories of relevant *policy mechanisms*:

1. regulations governing market or grid access
2. financial interventions and incentives
3. industry standards, planning permits, and building regulations (codes)
4. education and information dissemination
5. public ownership and stakeholder involvement
6. awareness, R&D.

These will now be considered in more detail.

9.6.1
Regulations Governing Market/Electricity Grid Access and Quotas Mandating Capacity/Generation

Preferential access for REs to the grid crucial. There are two general types of regulatory policies: one mandates the feed-in price and the other mandates quotas.

9.6.1.1 Feed-in Tariffs

With the feed-in law, electricity grid operators have to accept all electricity generated by RE, and pay fixed tariffs. These are differentiated according to technology, size, and location. This prevents only the currently cheapest technology from being promoted. Finally, it also encourages equitable access to all, ranging from the poor single-parent household to the multi megawatt offshore wind farm developer.

Prices are fixed over typically 20 years, and are adjusted biannually to new entrants, reflecting the price-learning curve. This attracts long-term investors and also encourages participants to join early. Utilities also qualify. The cost is distributed over all national end users. About 90 industrial and developing nations use the feed-in law, and by far the best RE market successes have been achieved with the feed-in system.

9.6.1.2 Quotas – Mandating Capacity/Generation

This is the reverse of the feed-in system. Instead of the government fixing the price, it fixes the target, and trusts that the market will determine the price. The government may force producers, distributors, or end consumers to accept a minimum share (quota) to come from REs. Quotas can be applied to any RE such as biofuels, solar thermal energy, or electricity. With obligation/certificate/Renewable Portfolio Standards (RPS) systems, the government will only see what happened at the end of a period, when it is too late. With the *tendering system*, the government sets targets in addition to a maximum electricity price. The abandoned Non-Fossil Fuel Obligation (NFFO) of the UK was such a system. Normally, tenders are awarded from the lowest group up until the quota is filled. The difference between the market reference and the winning tender is subsidized by government. RPS and tender systems are of shorter duration than feed-in systems.

9.6.1.3 Applicability in the Developing World

Much of the debate between feed-in tariff proponents and tendering or quota system adherents appears to be motivated more by ideology than fact. Developing Nations cannot always afford the luxury of ideological debates. The question is rather what fits and what works in the real world.

Renewable Energy Capacity and Generation

From a government perspective, it appears that prices are determined with feed-in systems, while energy output is felt to be uncertain. Conversely, quotas are determined with quota systems while prices are seen to be uncertain. For Developing Nations, steady energy prices are more important than precisely achieving predetermined quotas by a predetermined date. In fact, countries with feed-in systems have regularly outperformed.

Governments are not the only players in energy. The private sector is indispensable in the Developing World. These people remain in business because they understand how to assess risks. Feed-in systems are less risky than quota systems. The Developing World tends to have poor risk ratings.

Innovation, Domestic Industries, and Benefits Accrued

Theoretically, feed-in systems could discourage innovation and competitiveness. In reality, once companies have achieved a certain level of income, they start to invest in R&D to enhance their competitive edge and increase profits. This proceeds at no cost to the government, that is, the taxpayer.

Under quota systems, the surplus – if any – tends to accrue to the end-user, with the producer not having sufficient margin to invest in the uncertain future inherent in the quota and tender systems.

Even worse, overseas companies that have grown strong on feed-in systems at their home base compete successfully in foreign countries with quota systems.

The transaction costs and stop-and-go scenarios of quota systems discourage the establishment of national industries and limit the growth of national jobs.

Geographic and Ownership Equity

Under quota systems, the cheapest projects dominate, gravitating to the geographic areas where the cheapest renewable energy sources and renewable energy technologies are available. The RPS also favor large, capital-intensive companies which can manipulate the market in order to eliminate smaller local entrepreneurs. These are serious issues in Developing Countries with weak and nascent industries.

The feed-in system does not have these disadvantages. The fact that feed-in laws lower the market entry barriers to small producers, while at the same time welcoming large investors, is of immense interest to Developing Countries wishing to attract foreign investors while fostering their smaller domestic industries.

Pricing laws also enhance the participation of local farmers and communities. This increases local ownership and buy-in. The World Wind Energy Association strongly endorses local ownership in the form of community power projects.

Technology and Diversity of Supply

Because quota systems focus on the cheapest technology, there is little or no diversity of energy supply; for example, an over-exposure to hydropower bears great risks, as has repeatedly been illustrated in a number of cases with African hydropower schemes. This exposure is increasing with the effects of climate change in the Developing World.

Costs, Prices, and Competition

In theory, one would expect a lack of competition and higher energy prices with the feed-in system. However, in reality, the economies of scale and the better predictability of the market led developers to invest in R&D, enhancing competitiveness and cost reductions. In addition, the declining tariffs of the feed-in law ensure lower electricity prices. Quota systems tend to reduce participation to a limited number of players, which form cartels and abuse market power. Quota-based systems are not inherently cheaper, nor are feed-in systems inherently more costly [12].

Financial Security

Under the feed-in system, the long-term certainty resulting from guaranteed prices (typically 20 years) causes companies to invest in technology R&D, train staff, and maintain resources and services with a longer term perspective. This in turn makes it more attractive to financiers.

By contrast, quota systems harbor political and procedural uncertainties. The stop-and-go RE policies of many countries are disruptive to industry and unnerve potential investors. This is of great concern in Developing Countries where the local industries are underdeveloped and often cannot compete with established global players in a capital-intensive environment. The fact that government officials in Developing Countries are often challenged by tender procedures exposes local bidders and developers to additional uncertainty.

Green Certificates have lost value with the fall of the carbon market.

Ease of Implementation

Feed-in laws are easier to implement and are highly transparent. For obvious reasons, this is important with Developing Nations. Under the quota system, the requirements are much more demanding. Developing Nations typically do not have the data, expertise, resources, and time required. If the quota target is too low, then local economies of scale will not be attained, jobs will be lost, and the costs to the national economy are consequential. If the target is too high, prices will be pushed up dramatically. This supports the argument that quota/certificate systems, by their nature, are more complex, difficult to administer, and open to manipulation – and that such problems could even more pronounced in Developing Countries. In South Africa, an RE bidding process was repeatedly delayed because of a lack of government capacity.

In summary, bidding processes are bureaucratic, cause high transaction costs, and are time consuming for both developers and public authorities. This makes them inappropriate for Developing Nations.

Feed-in systems fix prices of new entrants into the market. This means that new entrants have certainty about the price over the contract duration. Should a government find that the price was too high/low, it can easily adjust the price to new entrants. With the quota system, it is not as easy to tamper with targets and timetables because lead times of several years are required.

Feed-in laws are the proven way to advance REs in the Developing World.

9.6.2
Financial Incentives

Financial incentives are one way in which governments can address the energy market failures, thereby attempting to level the playing field. They may be tax credits (relief), rebates, investment, or production support. The nuclear industry survives on these.

9.6.2.1 Tax relief
- *Investment and production tax credits (PTCs)* are designed to encourage investment in renewable energy technologies, and can cover either the total installed costs or the plant costs only. Reduced income is only of interest to high-income groups – hardly a problem in the Developing World. In the United States (1980s) and India (1990s), investment tax breaks helped to jump-start the wind industry, but also led to sub-standard design. The tax cycle – and not the energy demand – determines investments. PTCs only worked in countries with additional incentives.
- *Other forms of tax relief*, such as environmental (carbon) taxes are a more impact-related incentive, as is accelerated depreciation. Import duties can be managed on REs. Carbon taxes should be introduced according to a long-term and well-publicized plan, allowing industry to make long-term adjustments. Tax revenues should be ring-fenced to be used on energy transition initiatives.

9.6.2.2 Rebates and Payments
If the rebate or subsidy on installed domestic SWH is changed unpredictably, it will create instability in the market. The administrative effort is considerable.

9.6.2.3 Low-Interest Loans and Guaranties
In the Developing World, many more poor people could have access to REs if they had access to reasonable loans. These are feasible if the monthly installments are comparable to the expenditure on candles, paraffin (kerosene), and appliances. Without such finance, only 2–5% of the population in the Dominican Republic, India, Indonesia, and South Africa could have access to modern energy, whereas it would be 50% with suitable loans, a 10-fold increase. Such schemes tend to be country and culture specific.

9.6.2.4 Addressing Subsidies and Prices of Conventional Energy
During the mid-1990s, US$ 250–300 billion in subsidies were paid each year to the fossil fuel and nuclear industries of the world. Surprisingly, about 80–90% of these global subsidies to the fossil fuels and nuclear industries are paid out by the Developing World [12] to keep the energy price well below the true costs of production and delivery.

Even small subsidies for petroleum products in Developing Countries send out the wrong signals and direct nations down unsustainable energy paths, eventually trapping the poor.

Most of the diesel transported to remote regions is spent on transport [13]. US$ 50–60 billion is projected to be spent on subsidized power projects in the Developing World by 2030.

Even if all subsidies on fossils were to be stopped forthwith, the inertia of the government subsidies in existing infrastructure is still biased towards nuclear and fossil fuels.

Mostly it would be better policy to channel resources towards energy efficiency, energy conservation, and REs instead of trying to find new money streams to subsidize established sunset technologies.

Governments are large energy consumers through their energy-inefficient buildings, vehicles, transport systems, and infrastructure. They should lead by example.

9.6.3
Industry Standards, Planning Permits, and Building Codes

Energy efficiency and REs are furthered by technology standards and certification, siting and permit standards, grid connection standards, and building regulations (codes).

Technology standards foster fair competition and build investor confidence. New technologies demand new standards. They also facilitate export and import, such as the EU Solar Keymark or the ISO standards. Industry standards can also be an intentional barrier to entry for importers. *Siting standards* and environmental impact assessments can delay the process of establishing renewable energy technologies. Proactive siting regulations are the answer. *Grid-connection standards* should not be onerous. *Building regulations (codes)* should promote energy efficiency and REs, over the cradle-to-cradle lifecycle. Building laws demanding SWH can be significant and carry no costs to the fiscus. That is how Israel achieved a high market penetration of SWH. *Energy-efficient appliance* and *building ratings* support efficiency and awareness. Daylighting and efficient artificial lights also save air-conditioning costs. Buildings have longer lifetimes than most power stations; they should be energy producers and consumers (prosumers).

9.6.4
Education, Information, and Awareness

The above measures need end-user awareness. Educational institutions have the task of enlightening the new generation. For example, the Indian Solar Finance Capacity Building Initiative enlightens Indian bank officials about solar technologies, encouraging investment. IRENA, the International Solar Energy Society, and nongovernmental organizations contribute to knowledge dissemination.

9.6.5
Ownership, Cooperatives, and Stakeholders

Many Developing Nations have a tradition of communal ownership. This can be applied to renewable energies in the Developing World. In capitalist or free-market countries such as Denmark and Germany, cooperatives play an important role as owners and developers. Local farmers pool resources and obtain an additional harvest from renewable energies. This greatly enhances local buy-in and support. Public buy-in engenders public pride and avoids obstruction or vandalism. It also supports government RE policies when these are periodically attacked by vested interests.

9.6.6
Research, Development, and Demonstration

The Developing World has a backlog in RE R&D. The most pressing are *nontechnological aspects (sociological, cultural, political)*: perceptions of awareness, desirability, status, accessibility and affordability dominate over *techno-economic aspects (efficiency, economy, innovation)*. Here adaptation of existing RETs to harsh climatic conditions and low/no maintenance are priorities. Demonstration projects are essential in the in the Industrialized World, and even more so in the Developing World.

9.7
Priorities – Where to Start

9.7.1
Background

In spite of their attraction, democracies have the problem that decisions are very difficult to reach, and often end in compromises that make all stakeholders equally unhappy. Enthusiasm with implementation then becomes luke-warm.

If the leadership of a Developing Nation came to power through the barrel of a gun, then they are less likely to have the experience, qualifications, or inclination required for long-term energy planning. For fear of committing mistakes in full view of the public, they would rather procrastinate.

In the face of limited time, resources, and competing demands by various vociferous groups, it is not always easy to come to reasoned decisions.

9.7.2
Learning from Past Mistakes

On the subject of learning from previous mistakes, the best qualified to comment would be governments, international aid agencies, and their consultants. It appears that here they are not exactly prolific authors.

Third-world governments – like all governments – wish to be re-elected and would be inclined to support populist projects that impress the greatest number of their subjects – which is invariably the lowest income bracket. They also like to insist that foreign development monies be channeled exclusively through government officials' hands. In addition, they are tempted to micro-manage, in spite of limited capacities.

International development agencies, on the other hand, have their own agendas, which may include poverty alleviation, gender issues, creating markets for their donor countries, and job creation for their home consultants.

Consultants wish to do a good job in order to be reappointed. They also have to make a living.

The result of rather complex administrative procedures, a percentage of the finance does eventually reach a percentage of the target group like a drop of water on a hot stone. The following examples in South Africa illustrate the point. An international donor agency might decide that the provision of solar cookers to the rural poor is a good idea. The logic is that this would reduce over-harvesting of firewood and consequent desertification. It would free women and girls from the chore of collecting heavy loads of firewood over increasingly long distances, thereby enabling them to receive education. This would enhance birth control and also reduce indoor air pollution caused by burning firewood. Solar stoves would be manufactured locally to reduce unemployment and crime that is believed to be related to unemployment. Very convincing.

The outcome of a multi-year well-funded project was that the uptake by the target group was very limited and the idea of local production had to be abandoned. Apart from cultural and technical issues, the rural poor did not see solar stoves being used by the rich, and understandably regarded the technology to be second-class.

Similarly, the large multi-year solar home project was funded by international donors. The target group was the rural poor, with the expectation that this would provide an entry to the modern world and enable the target group to use solar lighting for study and some light essential services. A 52 W PV panel with regulator and battery was installed per home. Because of the remoteness, low density, and rural infrastructure, it was difficult to communicate with the end-users, and not easy to provide maintenance.

Before installation, the population had unrealistically been promised "electricity for all" by the national power utility, which happened to have a surplus capacity at that stage. The remote rural poor expected to receive free grid electricity. Hence they looked at "solar home systems" with a wary eye.

In practice, many of the installed regulators were bridged to tap the last bit of power from the battery that went flat. There is no record of any functional solar systems left or of the whereabouts of the defunct lead-acid batteries, which may have been discarded into the environment.

The money flow controlled by government was unilaterally stopped long before completion of the program. Apart from technical problems, the poor did see either the government or the well-heeled urban population using PV. The technology was in danger of being stigmatized as the poor man's technology.

Likewise, a domestic solar water heating initiative was focused on the low-income sector. Low-cost, low-pressure SWH systems were installed in a mass rollout, randomly facing north, south, east or west. The workmanship was not consistent, with predictable outcome.

A similar but smaller program was launched with biogas digesters. The results were disappointing. It remains to be seen whether the public will continue buying SWHs when subsidies are stopped.

In all cases, interventions were focused on areas where the need was seen to be most dire. With the best of intentions, and considerable investments of time, effort, and money, the impact was negligible, if not negative, in relation to the input, By now, the lesson should be clear – albeit counter-intuitive: *It is a strategic mistake to introduce a novel technology in remote, poor rural areas. Rather start with the urban, arrived and grid-connected population sector. This sector represents the aspirational role model.*

By choosing this priority, it allows the local industry to find its feet, overcome teething problems, and rapidly build service capabilities using easy and low-cost transport and communication channels. Prices drop much faster as a result of the learning curve. Thereby the new technologies become affordable to the poor. Sustainable jobs are created. Above all, RETs become associated with class, and attain a desirability status, especially if popular sport or entertainment stars are seen to be using them.

It should be noted that this strategy of initially prioritizing the better-off urban areas also leads to a higher level of awareness, desire, availability, and affordability. The deployment will reach the rural poor much sooner and at a lower social, economic, and political cost.

This is not to say that energy efficiency initiatives should not be run simultaneously.

9.8
Conclusions and Recommendations

Developing Nations wish to occupy their rightful place in the concert of nations. People of the Developing World require *energy services*. Electricity by itself does not provide new income sources. Electricity follows, rather than leads, economic development [14].

The Developing World cannot follow the dirty energy path even if it wanted to. There are simply not enough fossil resources, nor can the world absorb the environmental impact. This insight, combined with the fact that the energy infrastructure in Developing Countries is at present underdeveloped, enables technology leapfrogging.

Given sufficient stakeholder awareness and political support, the following priority policy recommendations for Developing Nations can be made:

1. Establish long-term, consistent and transparent, renewable energy targets and a grid-feeder law.
   1.1 Produce a Renewable Energy and Energy Efficiency White Paper.
   1.2 Publicize and workshop the draft White Paper, obtain buy-in of stakeholders.
2. Institute supportive finance mechanisms.
3. Establish and enforce standards.
4. Support research, development and demonstration of REs.
5. Encourage stakeholder/public ownership, participation.

Producing ambitious RE Development Plans is good, implementing them is even better. As Developing Nations emerge into economic prominence, they can be expected to accept greater global environmental accountability as the EU does. Nations can improve their own well-being by being more energy efficient and less reliant on fossils. This means enhancing one's own benefit while contributing to the common good.

In Europe, the driver of the Renewables Revolution was the environmentalism of young voters protesting against their governments' connivance with nuclear interests. In the Developing World, the overwhelming young generation of today is frustrated with unemployment. It is to be hoped that their governments will read the writing on the wall and use REs for job creation. Proven governance policies and measures are available that empower small- and medium-sized entrepreneurs in the private sector and attract investment.

Many independent scientists have confirmed that the transition to RE is necessary, urgent, and techno-economically feasible. Times of transition are always turbulent times. In such times, a natural reaction is to panic and cling to the habitual, procrastinating in the fear of making the wrong decision. The decision to procrastinate is also a decision – most often the wrong one.

As the world transitions to the new era, it will not wait for the Developing World to catch up. One chooses whether one wants to be a loser or winner in the dawning solar age.

References

1 Holm, D. (2005) *Renewable Energy Future for the Developing World.* ISES White Paper, ISES, Freiburg.
2 International Monetary Fund (2012) *IMF Economics List. World Economic Outlook.* IMF, Washington, DC, p. 179.
3 Transparency International (2011) *Corruption Perceptions Index 2011*, ISBN 978-3-943497-18-2, Transparency International, Berlin, www.transparency.org (last accessed 6 January 2013).
4 Weiss, W. and Mauthner, F. (2012) *Solar Heat Worldwide: Markets and Contribution to Energy Supply 2010*, IEA Solar Heating and Cooling Programme May 2010, p. 12.
5 World Wind Energy Association (2012) *World Wind Energy Report 2011*, World Wind Energy Association, Bonn, p. 8.
6 Rogers, S., Sedghi, A., and Burn-Murdoch, J. (2012) *Credit Ratings: How Fitch, Moody's and S&P Rate Each Country*, http://www.guardian.co.uk/news/datablog/2010/apr/30/credit-ratings-country-fitch-moodys-standard (last accessed 6 January 2013).
7 Perez, R. and Perez, M. (2009) A fundamental look at energy reserves for the planet. *IEA SHC Solar Update*, **50**, 2–3.
8 Holm, D., Banks, D., Schäffler, J., Worthington, R., and Afrane-Okese, Y. (2008) *Renewable Energy Briefing Paper – Potential of Renewable Energy to Contribute to National Electricity Emergency Response and Sustainable Development*, Sustainable Energy and Climate Change Project, Earthlife Africa, Johannesburg.
9 Solarex Inc. (1992) *World Design Insolation*, Solarex, Frederick, MD.
10 German Advisory Council on Global Change (WBGU) (2004) *World in Transition – Towards Sustainable Energy Systems*, ISBN 1-85383-802-0, Earthscan, London.
11 European Renewable Energy Council (2004) *Renewable Energy Scenario to 2040 – Half of the Global Energy Supply from Renewables in 2040*, http://www.rethinking 2050.eu/fileadmin/documents/REThinking2050_full_version_final.pdf (last accessed 2 March 2013).
12 Sawin, J. L. (2004) National Policy Instruments – Policy lessons for the advancement and diffusion of renewable energy technologies around the world. Thematic background paper presented at Renewables 2004. Secretariat of the International Conference for Renewable Energy, Bonn, www.renewables2004.de (last accessed 6 January 2013).
13 Perlin, J. (1999) Electrifying the unelectrified. *Solar Today*, November/December.
14 Schramm, G. (1998) Electrification Programmes and the Role of International Development Banks. In Holm, D. and Berger, W. *ISES Utility Initiative for Africa – Selected Proceedings of the Initial Implementation Conference.* ISES, Johannesburg.

10
An Innovative Concept for Large-Scale Concentrating Solar Thermal Power Plants

Ulrich Hueck

10.1
Considerations for Large-Scale Deployment

In some countries, such as the United States and Spain, the deployment of concentrating solar power (CSP) had temporarily turned into a booming power plant market. The installed and announced capacity of CSP plants was rapidly increasing worldwide.[1]

After a time of initial subsidizing, significant cost reductions for electricity generated in CSP plants are required in order to make large-scale deployment independent of such public subsidies.

If cost reduction is crucial, then the related technology is likely to undergo significant improvements and changes. Within the scope of such developments, a technology that is dominant today may dramatically change or even disappear by tomorrow.[2]

This section summarizes existing technologies and basic configurations for solar power plants. The review refers in particular to the proposition of large-scale deployment[3] of numerous units in deserts.[4]

Section 10.1.4 provides a summary for comparison of different technologies.[5] This leads to the advanced solar boiler concept for CSP plants, which is described in Section 10.2.

---

1) See http://www.nrel.gov/csp/solarpaces and http://en.wikipedia.org/wiki/List_of_solar_thermal_power_stations for lists of CSP plants.
2) Some historical examples of once dominant technologies, which were superseded and then disappeared, are galleys, horse-drawn carriages, steam trains, muzzle loaders, mechanical cash boxes, and punched card computers.
3) China provided an example of large-scale deployment of power generation: It increased the installed capacity of its coal-fired power plants by more than 90 GW in 2006 alone (www.pewclimate.org/global-warming-basics/coalfacts.cfm). With 174 new plants of the 500 MW class, the pace of installation was ~250 MW per day. A similar pace of installation is desirable for renewables.
4) In this chapter, the term "deserts" refers to dry regions with high levels of insolation, sparse vegetation and large open spaces, which do not have intensive living creature activity.
5) The urgent reader may skip via Section 10.1.3.3 to Section 10.1.4.

Transition to Renewable Energy Systems, 1st Edition. Edited by Detlef Stolten and Viktor Scherer.
© 2013 Wiley-VCH Verlag GmbH & Co. KGaA. Published 2013 by Wiley-VCH Verlag GmbH & Co. KGaA.

10.1.1
Technologies to Produce Electricity from Solar Radiation

Several different basic technologies can produce electricity from solar radiation, such as:

- concentrating solar power with steam turbines
- concentrating solar power with gas turbine configurations
- photovoltaic
- parabolic mirrors with Stirling engines
- solar updraft towers with wind turbines.

This evaluation mainly refers to basic features of concentrating solar thermal power plants equipped with steam turbines. Photovoltaic is only considered in Section 10.1.3.2 for comparison of the technologies used to produce electricity day and night.

Concentrating solar power with gas turbine configurations is evaluated in Section 10.1.3.7 on the basis of working fluids and in Section 10.1.3.12 regarding robustness of technologies. Parabolic mirrors with Stirling engines and solar updraft towers are only assessed in Section 10.1.3.12.

10.1.2
Basic Configurations of Existing CSP Plants

The following features differentiate basic configurations of existing CSP plants [1]:[6]

- robustness of technology
- capability to produce electricity day and night
- type of concentration of solar radiation
- shape of mirrors for concentration of solar radiation
- area for solar field
- technology to capture heat from solar radiation
- working fluids and heat storage media
- inlet temperature for power generation
- type of cooling system
- size of solar power plants.

Direct steam generation from solar radiation is also investigated in Section 10.1.3.8. Further relevant features may exist or arise in the future.

For the configuration of CSP plants, numerous combinations derived from the above features are possible and have already been implemented in the field. However, so far it appears unclear which combination of which features is likely to best suit the large-scale deployment of solar power plants in deserts.

---

6) For numerous details on solar power plants, see [1].

10.1.3
Review for Large-Scale Deployment

This section provides a review of the basic differentiating features for the configuration of solar power plants as outlined above. For each feature, the focus is on its main purpose and key stumbling blocks, considering large-scale deployment in deserts.

The criteria for review are as follows:

- physical feasibility
- robust technical practicability
- capability for large-scale deployment in deserts
- efficiency of power generation
- risk of environmental impact
- installation, operating and maintenance costs
- availability of resources
- Basic requirements of the energy market.

The selection of review criteria is based on the assumption that their positive fulfillment will lead to a plant configuration that should best suit large-scale deployment in deserts.

A summary of the presumably most advantageous plant features appears in Section 10.1.4. In conclusion of the comparison, Section 10.2 describes an innovative plant configuration that has the potential to fulfill all criteria best.

10.1.3.1 Robustness of Technology to Produce Electricity

For good reason, steam turbines with generators are used in most existing CSP plants. In closed systems, water/steam is a suitable medium to drive turbines. Steam turbine–generators with high power output are in operation in numerous fossil-fueled and nuclear plants.[7] In principle, large-scale deployment of steam turbines in solar thermal power plants is also possible. This review therefore primarily refers to CSP plants equipped with steam turbines.

The robustness of other solar technologies for power generation is assessed in Section 10.1.3.12.

10.1.3.2 Capability to Produce Electricity Day and Night

The installation of large solar power plants without the prospect of night-time operation requires power balancing from other plants. These plants are likely to run on non-renewable resources, which would be counterproductive. Therefore, the ability to run day and night is decisive for the large-scale deployment of solar power plants.

---

7) Large steam turbine–generators in fossil-fueled plants produce power output above 500 MW. In large nuclear units, steam turbine–generators produce power output above 1000 MW.

In CSP plants, the solar energy must therefore be stored during the day for operation at night. The thermal storage of heat in material and its retrieval are basically inexhaustible processes, although not trivial and by no means fully explored for large-scale deployment. However, efficient operation of CSP plants is in principle possible day and night with thermal heat storage in material.[8]

The storage of directly produced electricity from photovoltaic energy typically requires double conversion: from electricity to another form of energy and back to electricity. These can be chemical energy (e.g., in batteries or hydrogen), potential energy (e.g., as pumped hydro storage), or others. However, energy is lost in theory and in practice during both conversions.

For large-scale deployment of solar power, energy storage should not continuously exhaust resources, such as batteries, and should rather utilize inexhaustible processes with minimum energy losses, such as thermal storage of heat. This is the main reason for concentrating solar power retaining an edge over photovoltaic for large-scale deployment in deserts.

From this point of view, thermal storage of heat is the preferred technology for large-scale night-time production of electricity from solar power in deserts.

10.1.3.3 Type of Concentration of Solar Radiation

The type of concentration of solar radiation influences the amount of heat transmitted across the irradiated surface. The type of concentration also leads to specific advantages and disadvantages.

For point concentration, the amount of heat per receiver surface is comparably large. The temperature difference between the surface for solar radiation and the inlet temperature of the heat transfer medium is also large. Efficient point concentration therefore requires rather sophisticated solutions in terms of materials, heat-transfer capabilities, cooling technology, cleanliness, and protection against overheating. However, with point concentration, high temperatures are possible, which is beneficial for plant efficiency.

For linear concentration, the amount of heat transmitted per receiver surface is comparably lower. Consequently, the total surface for the solar heat transfer must be larger. But large surfaces easily result in higher heat losses. If high temperatures from point concentration were applied to linear concentration, heat losses are then likely to increase significantly if long uncovered lines were placed in the field.

In conclusion, an optimized configuration for the concentration of solar radiation operates at high temperatures, uses simple technology, and exhibits minimum heat losses.

For such a design, total surfaces for the heat transfer should be larger than with point concentration but must not become too large in order to avoid heat losses.

---

8) Section 10.2.2.7 provides an example of a thermal storage system for night-time operation.

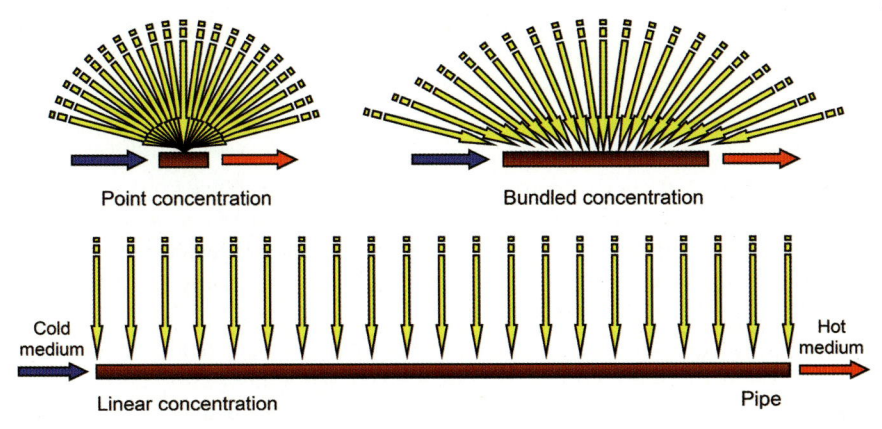

Figure 10.1 Different types of solar radiation concentration.

An optimized configuration for such bundling of solar heat is called "bundled concentration."[9]

Figure 10.1 depicts the basic design of point concentration, linear concentration, and bundled concentration.

10.1.3.4 Shape of Mirrors for Concentration of Solar Radiation

Flat mirrors, called heliostats, Fresnel-shaped configurations with flat mirrors, parabolic trough-shaped mirrors and parabolic mirrors can be used to concentrate solar power for subsequent power generation.

Established combinations for different ways of concentrating solar radiation and differently shaped mirrors are depicted in Figure 10.2: parabolic-trough mirrors and linear Fresnel mirror configurations refer to linear concentration. Point concentration applies to flat mirrors for central receiver technology and parabolic mirrors, which can be equipped with Stirling engines.

Although the cost difference might not be very large, the production and testing of flat mirrors are cheaper than for curved mirrors.

Flat mirrors can track the Sun on two axes. The reflected solar beams can thus always reach the receiver surfaces at the same optimized angle. In contrast, parabolic trough mirrors and Fresnel mirrors can track the Sun only on one axis. Therefore, in winter and in the morning and evening, the angle between the reflected solar beam and the receiver surface can be less effective.

---

[9] A simple example explains the concept: Consider heating water for a hot shower by burning natural gas. With point concentration, you have to heat the cold water inside a very short section of the water pipe. For that, a welding torch might be required and you have to hold it close to the pipe's surface. To avoid damaging the pipe, you will already need special material in that section of piping. Alternatively, if you use linear concentration, a large quantity of small gas flames would be placed underneath long sections of pipework. The water also gets hot in this way but before reaching the shower a considerable amount of heat will be lost since the ambient air cools the pipe. At best, you can burn the gas in a closed boiler underneath a bundle of water pipes. In this way there is hardly any risk of overheating these simple pipes and loss of heat from the water to the ambient air is very small.

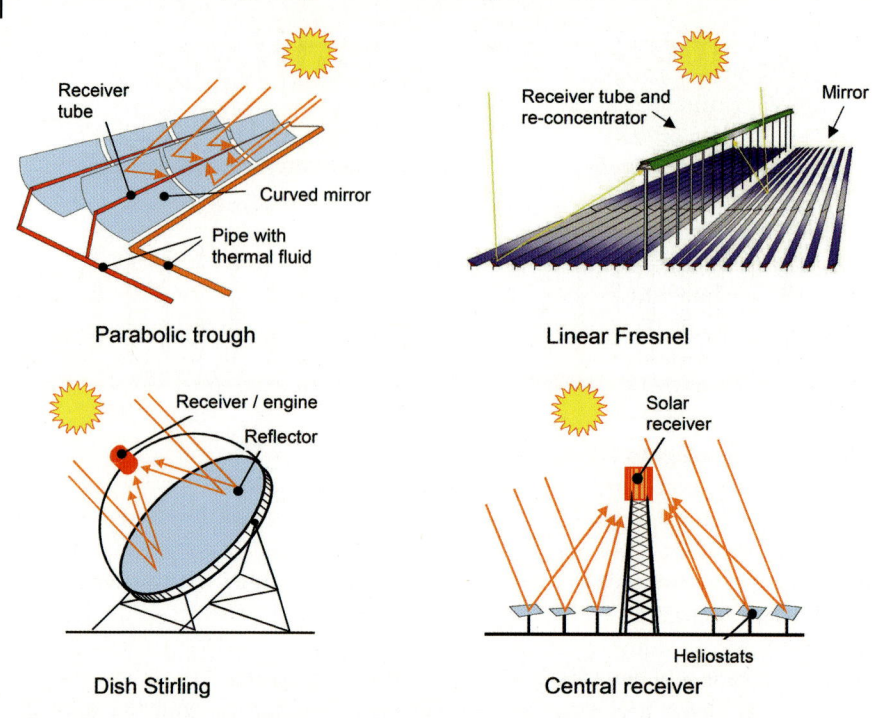

Figure 10.2 Established configurations for concentrating solar thermal power plants.

The parabolic trough shape is a purely geometric feature for concentrating solar power. Fast computing capabilities for exact focusing and repeated alignment calibration of unlimited numbers of flat mirrors are likely to supersede the geometric parabolic trough design, as electronics often simplify the design of mechanical systems.[10]

In conclusion, an optimized plant configuration should comprise flat mirrors in the solar field.

10.1.3.5 Area for Solar Field

At the current stage of deployment of solar power plants, the slope of the area for a solar field may not be a pressing issue. So far, enough flat spaces are available. However, with increasing deployment of solar power plants, large flat areas that meet all other requirements for installation may become rare. The flexibility for selection of area will sooner or later gain importance.

Therefore, over the long term, technologies that can be deployed regardless of the slope of the area are likely to supersede technologies that require large flat spaces.

Point concentration or bundled concentration of solar radiation is almost independent of the slope of the area. Mirrors can be placed on any skew hillside provided

---

10) Some examples of articles of daily use where electronics simplify the design of the mechanical system are watches, weighing machines, sewing machines, and cameras.

that the reflected solar beams reach the central location and the mirrors can be easily reached for cleaning.

In contrast, solar power plants with horizontal receiver tubes in large solar fields require flat areas of land. This constraint is due to the heat-transfer fluid that is pumped through the receiver tubes. With different temperatures throughout the field, caused, for example, by the misalignment of mirrors, missing heating at connecting pipes between mirrors, or clouds, the density and viscosity of the heat transfer fluid change. Keeping the flow constant and stable in flat areas is therefore already an engineering challenge in itself. The varying influence of gravity with uphill and downhill flow makes such systems significantly more unstable, and flow control becomes more costly or even impossible. Consequently, an optimized solar plant configuration should not be bound to flat areas of land.

10.1.3.6 Technology to Capture Heat from Solar Radiation

Receiver tubes, metal surfaces, and porous materials are predominantly used to capture heat from solar radiation.

At receiver tubes, the solar beam typically passes through an outer glass and hits a steel tube on the inside. Vacuum between that steel tube and the outer glass suppresses losses through heat conduction and heat convection. Thermal radiation passes through vacuum and is therefore not easily insulated this way. That effect is only significant at high temperatures.

In contrast, solar heat transfer across open metal surfaces suffers reasonable losses to the surrounding air by heat conduction and heat convection. Thermal radiation losses also take place.

Another approach to point concentration uses air for heat transfer. At the central receiver the air may pass through porous materials, such as ceramics, which make heat transfer very effective. However, in particular with ambient air and even with closed cycles of air, the holes in such porous material can clog. Additional sophisticated cleaning technology can therefore make this fragile technology more costly. Heating air without passing it through porous materials is difficult.

The design of solar receivers has to be cost-effective and robust and must minimize solar heat loss to the environment. This includes capturing losses through heat convection, heat conduction, and thermal radiation in addition to the maximum utilization of such waste heat.

10.1.3.7 Working Fluids and Heat Storage Media

Air, water/steam, thermal oil, and molten salts are established working fluids for the transportation of solar heat.

Water/steam drives a steam turbine in the traditional Rankine cycle. Hot air may also run a turbine similar to the gas turbine configuration.

Solar heat storage is possible with thermal oil, molten salts, which can serve as a phase change material (PCM), and solids such as concrete or ceramics. Effective storage of solar heat does not work with air or water/steam since large vessels would have to withstand high pressure from these media.

Table 10.1 Working fluids and heat storage media for solar thermal power.

Available media[a]	Working fluid for solar heat	Driving of turbine/engine	Storage of solar heat
Air	Possible	Possible	Not feasible
Water/steam	Possible	Possible	Not feasible
Thermal oil	Possible	Not possible	Possible
Molten salt	Possible	Not possible	Moved/fixed
Phase-change material	Not possible	Not possible	Fixed
Solids	Not possible	Not possible	Fixed[b]

a Media such as CO_2 or N_2 are excluded as working fluids for solar heat or for driving turbines owing to their debatable suitability and unclear costs for such applications. Thermochemical heat storage is excluded since large-scale deployment should avoid the complex use of special materials such as magnesium hydride (MgH_2), manganese dioxide (MnO_2), antimony pentoxide (Sb_2O_5), or barium peroxide (BaO_2).

b Sand for storage of solar heat is excluded owing to tremendous technological difficulties regarding effective charging and discharging of heat.

In principle, solar heat storage and release can take place in two different configurations: By moving the storage medium as working fluid through heat exchangers or by keeping storage media at fixed locations, through which the working fluid passes.

All of the options are summarized in Table 10.1.

Several options from Table 10.1 for working fluids and driving of turbines are excluded as follows:

- If air is used in open cycles and/or air has to pass through porous media for heat exchange, then technology for particle separation is required in deserts. This makes the use of air as working fluid for solar heat more costly.
- Use of thermal oil is restricted by temperature with adverse impact on plant efficiency.
- Thermal oil is rather expensive and requires replacement after a certain period in use.
- In the event of leaks, thermal oil can harm the environment.
- A molten salt as heat transfer fluid requires auxiliary heating during commissioning and in particular during longer periods of plant shutdown to keep the material in its molten state in tanks and also in pumps, valves, and pipework, if not drained properly. Hardening of salt may endanger the equipment.
- Thermal oil and molten salts are rather inert media, which may slow plant start-up.

In conclusion, water/steam in a closed cycle seems to be the best suited working fluid for solar heat and for driving turbines in CSP plants that are located in deserts. Using the same medium for both purposes makes complementing heat exchangers superfluous.

Some of the above options for solar heat storage are excluded by the following considerations:

- The heat storage capacity of thermal oil is comparably low.
- Auxiliary energy supply is required during long shutdowns to keep a molten salt in its liquid state. However, a reliable thermal storage medium should not depend on external energy supply to be available during long shutdowns in remote desert regions.
- Pumping a molten salt through heat exchangers requires auxiliary energy for constantly moving huge masses of material.

The preferred option for thermal storage of solar heat in deserts therefore seems to be at fixed locations, that is, the heat storage material remains unmoved and is not circulated. Each storage module must then be a heat exchanger. This increases the initial investment, but reduces the auxiliary energy consumption and improves reliability of solar thermal power plants in deserts.

Molten salts and solid materials such as powdery salt or concrete are apparently suitable for thermal storage of solar heat at high temperatures, for example, for superheating of steam at night.

Salts and other suitable substances appear beneficial as phase-change material for thermal storage of solar heat. Phase-change materials are typically compounds that melt and solidify at certain temperatures.[11] The phase change provides a high thermal storage capacity per unit volume and heat recovery takes place at almost constant temperature. Phase change is an inexhaustible process, theoretically without any loss of energy. Based on the physical principle, phase-change material therefore has good prospects for fulfilling the desired function of storing large amounts of solar heat for night-time operation. However, predictions of high costs for the numerous heat exchangers that would be needed at fixed locations are currently the major stumbling block for commercial use of storage systems with phase-change material.

Still, the best combination of options in Table 10.1 appears to be a closed water/steam Rankine cycle with direct solar steam generation and with thermal storage of solar heat at fixed locations in phase-change material, molten solids, and solids. Since all of the above-mentioned technical drawbacks are avoided, such a configuration has numerous advantages, including the following:

- Maintaining the cleanliness of water/steam in a closed cycle is comparably easy.
- The temperature restrictions for water/steam and for pressurized steel pipework are similar.
- Leaks in containments of working fluid and storage media cause no harm to the environment.
- The working fluid and storage media are cheap and abundantly available.[12]

---

11) See http://freespace.virgin.net/m.eckert for references on phase-change materials.
12) If water/steam is used in a closed cycle, then only small volumes are lost during operation.

- No particular auxiliary heating is required for the working fluid and storage media.[13]
- After shutdown, all systems reach a stable state with minimum auxiliary energy supply.
- Plant start-up is very fast.
- Load changes allow for immediate redirection of steam to the thermal storage system.
- The same feedwater pump provides the pressure for driving the steam turbine and for charging or discharging the thermal storage system.
- After conversion of solar radiation to heat, the only relevant heat exchange on the way to the turbine takes place between water/steam and heat storage materials for operation at night.[14]

10.1.3.8 Direct Steam Generation

As concluded above, direct steam generation from solar heat is advantageous.

Direct steam generation in horizontal tubes generates two-phase flow for fairly long distances, which is not beneficial to plant efficiency. Direct steam generation should therefore preferably take place in vertical tube arrangements. This provides a sharp phase transition.

In long horizontal configurations, steam hammering can also harm the mechanical integrity of pipework. Furthermore, steam leaks at outdoor locations can put nearby plant personnel in deadly peril. Direct solar steam generation should therefore be limited to compact pipework arrangements at safe locations.

10.1.3.9 Inlet Temperature for Power Generation

Based on Carnot's theorem, high inlet temperatures in thermal power machines are advantageous for the efficiency of thermal power plants. Therefore, no basic plant design feature should constrain operation at high temperatures.

An example of such a restriction is the transportation of solar heat through pipes in large fields of mirrors. An increase in temperature escalates heat losses by thermal radiation, increasing exponentially to the fourth power of the temperature. Therefore, at a certain threshold, a further increase in temperature in the field has an inverse impact on plant efficiency. Such counterproductive dependency should be avoided for large-scale deployment of solar power.

For an optimized solar plant configuration, an increase in operating temperature should not significantly increase heat losses. Instead, inevitable heat losses should be captured to a great extent and subsequently utilized within the thermal cycle for power generation. In practice, after the conversion of solar radiation to heat, it should be possible to insulate all downstream pipework and components that transport or contain solar heat at any elevated temperature.

---

[13] For start-up, steam turbines require sealing steam, which is typically provided by an auxiliary boiler.
[14] Cooling of condensate also requires exchange of heat in the condenser. This heat exchange is common for all steam turbine plant configurations.

Theoretically, an optimized plant configuration does not have to be constrained by any of its subsystems in the development and deployment of cost-effective technology for operating at increasingly high temperatures.

10.1.3.10 Type of Cooling System

Condensers with cooling water and air-cooled condensers are used for the operation of steam turbines. Cooling with water is more effective but water is rare in deserts. Dry cooling provides an alternative in dry regions[15] but lowers the overall plant efficiency during daytime operation. The choice of cooling system therefore mainly depends on the availability of cooling water.

10.1.3.11 Size of Solar Power Plants

For large-scale deployment in deserts, the underlying technology of solar power plants should permit the construction of large units.

Some factors can restrict the size of solar thermal power plants: long-distance transportation of solar heat with a heat transfer fluid leads to increasing heat losses, and it is also difficult to provide stable heat-transfer fluid flow in very large piping systems.

Alternatively, mirrors can divert the solar radiation to a central location. However, with increasing distances, correct alignment of such heliostats becomes more and more difficult.

Of the two technologies described above, the one that provides a simple and cost-effective solution to increase plant sizes will be the superior technology. Less accurate focusing of distant heliostats to larger central surfaces might be relatively easy to address. Then the distance of remote mirrors is only restricted by radiation losses through the low-lying atmosphere.

In conclusion, for large solar thermal power plants, reflecting solar radiation with heliostats to a central location appears advantageous over long-distance pumping of heat transfer fluid.

10.1.3.12 Robustness of Other Technologies

Complementing Section 10.1.3.1, the robustness of other solar technologies is assessed in this section, considering large-scale deployment in deserts.

Solar power might be utilized with configurations similar to gas turbines. The solar heat is transmitted to pressurized air and replaces the heat from burning fuel gas in the gas turbine's combustion chamber. With a less complicated setup, hot air alone may heat a water/steam cycle for running a steam turbine.

Owing to the low heat storage capacity of air and its low thermal conductivity, sophisticated technology is required for solar heat transfer. The disadvantages of using porous materials for such an application were outlined in Section 10.1.3.6.

---

15) The largest coal-fired plant in the world, Kendal power station in South Africa, comprises six units each rated at 686 MW and uses an indirect dry-cooling system, which means that it uses significantly less water in its cooling processes than the conventional wet-cooled power stations (www.eskom.co.za).

Stirling engines can be placed at the focal point of parabolic mirrors. Point concentration of solar radiation then provides very high temperatures to run the Stirling cycle with air or gas. High efficiency is already possible but the size of each single unit is rather limited.

Cost savings in operation and maintenance require simple technology for mass products, such as mirrors. Complex engines for power generation should preferably be installed in smaller quantities of larger size.[16] It may therefore not be cost-effective to base the large-scale deployment of solar power in remote desert areas on thousands of small Stirling engine systems, each one also to be equipped with a suitable storage facility for night-time supply of electricity.

Solar updraft towers utilize the reduced density of ambient air that is heated by solar radiation on the ground. The stack effect drives wind turbines at the inlet of a tall central tower. For this technology, a large sealing structure is required as a "collector" above the ground, under which hot ambient air is directed to the central stack. Leaks in such a huge, fragile, horizontal structure are likely in harsh weather conditions with high wind speed and diminish the function of the entire plant. This makes the solar updraft tower technology an unlikely candidate for large-scale deployment.

Solar updraft towers with wind turbines also require flat areas of land, which is disadvantageous as discussed in Section 10.1.3.5, and the concentration of solar energy without mirrors is rather low.

In conclusion, large-scale utilization of solar radiation for production of electricity should be based on robust technology with reasonable unit size. Debris in the working fluid, harsh weather conditions or other typical environmental impacts should not place the operation of an entire solar power plant at risk.

10.1.4
Summary for Comparison of Technologies

Summarizing the review in Section 10.1.3, a suitable plant configuration for large-scale deployment of solar power in deserts should integrate the following features:

- Robustness of technology: concentrating solar power with steam turbine and generator.
- Capability to produce electricity day and night: thermal storage of heat for night-time operation.
- Type of concentration of solar radiation: bundled concentration.[17]
- Shape of mirrors for concentration of solar radiation: flat mirrors.
- Area for solar field: plant configuration must not be restricted to flat areas.
- Technology to capture heat from solar radiation: maximum utilization of waste heat.

---

[16] Consider, for example, the quantity of specialized service personnel required for implementing product changes or upgrading components.
[17] See Section 10.1.3.3 for the definition of the term "bundled concentration" in the context of converting solar radiation to heat.

- Working fluid for concentrating solar power: water/steam.
- Storage media for solar heat: phase-change material and solids at fixed locations.
- Direct steam generation: compact vertical arrangements of tubes at safe locations.
- Inlet temperature for power generation: increase not constrained in principle by any subsystem.
- Type of cooling system: water or air depending on the availability of cooling water.
- Size of solar power plant: reflection of solar radiation from distant positions to central surfaces.

10.2
Advanced Solar Boiler Concept for CSP Plants

This section describes the integration of basic physical and technical features for an optimized configuration of CSP plants. The resulting advanced solar boiler concept for CSP plants is consistent with the conclusions summarized in Section 10.1.4. The configuration therefore seems to best fulfill the review criteria detailed in Section 10.1.3:

- physical feasibility
- robust technical practicability
- capability for large-scale deployment in deserts
- efficiency of power generation
- risk of environmental impact
- installation, operating and maintenance costs
- availability of resources
- basic requirements of the energy market.

None of the individually optimized plant design features for the advanced solar boiler impair any other aspect of its performance. Consequently, the integration of these optimized individual contributions leads to a CSP plant configuration with outstanding performance in terms of high efficiency, low costs – except for the thermal storage system – and robust operation.

10.2.1
Summary of Concept

The advanced solar boiler concept can be summarized as follows: A solar field with flat heliostats directs the solar radiation to several segments in a central tower [2].[18] These separate segments provide preheating, evaporation, superheating, and optional reheating of water/steam. The superheated steam drives a steam turbine–generator for power generation or is used for other industrial processes.

---

18) For a description of concept and efficiency calculation regarding a basic solar boiler design, see [2].

The following complementing features make this concept feasible:

- Receiver tubes in vertical arrangements convert the solar radiation to heat.
- An insulated structure surrounds the receiver tubes for heat loss reduction.
- High temperatures are handled at the bottom of tower; low temperatures prevail at the top.
- Waste heat from the tower can be recovered for the preheating of condensate and feedwater.
- Ambient air can provide controlled convective heat recovery.
- Hybrid operation with conventional boilers is an optional possibility.
- Thermal storage in phase-change material and solids allows for night-time operation.
- Thermal storage tanks are charged by condensation and discharged by generation of steam.

This innovative configuration for an advanced solar boiler with bundled concentration of solar radiation bridges the prevailing technical gap between point concentration in solar tower CSP plants and linear concentration in CSP plants with parabolic trough mirrors or Fresnel-shaped mirror configurations. The configuration already lowers the costs of solar-based daytime electricity by omitting several components and power consumers found in many existing CSP plant configurations [1].

10.2.2
Description of Concept

The different aspects of the advanced solar boiler concept are described in the subsequent sections. The summary of all aspects integrates the new design for CSP plants.

10.2.2.1 Direct Solar Steam Generation
According to Section 10.1.3.7, water/steam is apparently the best option regarding a working fluid for the transportation of solar heat. A pump[19] provides the pressure for the liquid feedwater in a closed water–steam cycle. Solar radiation heats the feedwater for direct steam generation. The expansion of that steam drives a turbine to power a generator for the production of electricity. Cooling in the condenser at almost vacuum liquefies the saturated steam from the turbine. Downstream from the condenser, the water/steam cycle starts again.

10.2.2.2 Rankine Cycle for Steam Turbine
The Rankine cycle for steam turbines consists of several steps, as shown in the temperature–entropy diagram in Figure 10.3. Figure 10.4 depicts the corresponding configuration of steam plant components.

---

[19] In reality, parallel condensate pumps and parallel feedwater pumps are connected in series to increase the condensate's pressure in at least two steps.

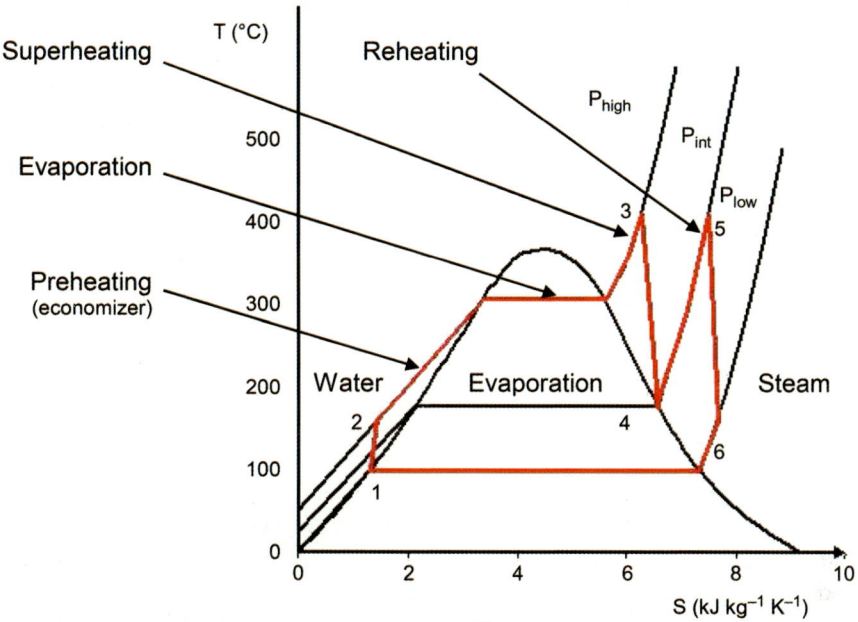

10.2 Advanced Solar Boiler Concept for CSP Plants | 173

Figure 10.3 Temperature–entropy diagram for steam.[20]

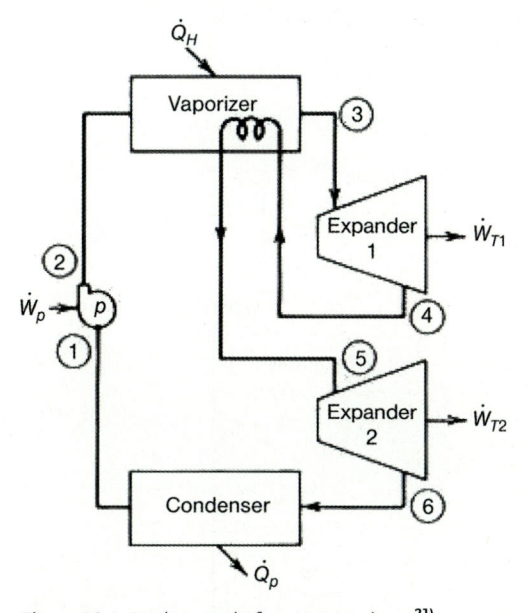

Figure 10.4 Rankine cycle for steam turbine.[21]

---

20) Graphic retrieved from http://commons.wikimedia.org/wiki/File:Rankine_cycle_with_reheat.jpg.
21) Graphic retrieved from www.powerfromthesun.net/Book/chapter12/chapter12.html.

The feedwater pump P in Figure 10.4 provides the pressure increase between points 1 and 2. The preheating and evaporation of water and superheating of steam take place between points 2 and 3. A high-pressure turbine expands the steam from point 3 to point 4. Optional steam reheating takes place between points 4 and 5, which increases the plant efficiency. An intermediate- and/or low-pressure turbine expands the reheated steam between points 5 and 6. Condensing of saturated steam through cooling of condensate is carried out between points 6 and 1 in the condenser.

10.2.2.3 Solar Boiler for Steam Generation

The design of the solar boiler suits the different phases for the generation of superheated steam as shown in Figure 10.5. In the conversion of solar radiation to heat, preheating, evaporation and superheating take place at different segments of the boiler. By variation of size and heat supply, each segment is adapted to the desired temperatures, heat flux and mass flow. A conventional drum-type phase separator (3) ensures that only the steam phase of the working fluid reaches the segment for superheating.

For direct steam generation with solar heat, such a breakdown of the heat transfer into several segments is advantageous. In particular, the thermal load on the surfaces of the segments is lower than the thermal load on a central receiver in a tower with point concentration. At the solar boiler, very high temperatures are limited to the small receiver segment for steam superheating. Separation of evaporation and superheating ensures high steam quality. However, sophisticated control of disturbances through clouds is required.

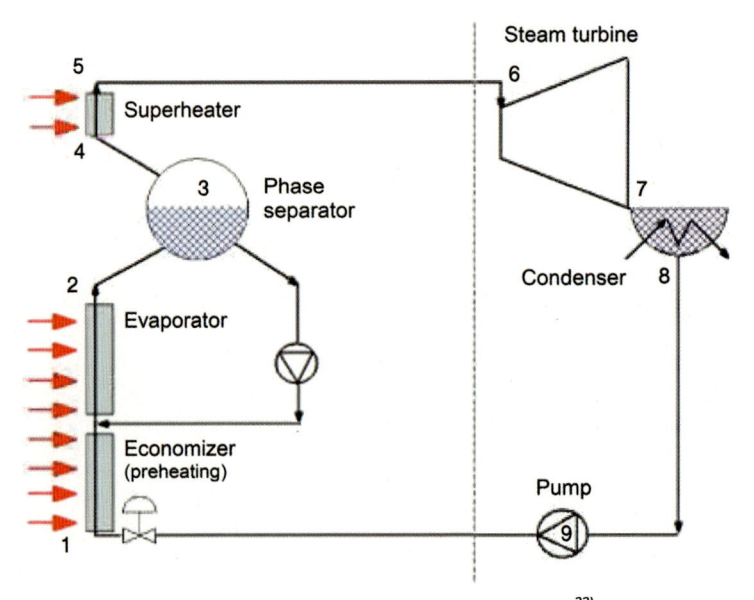

Figure 10.5 Flow diagram of solar boiler for steam generation.[22]

22) Template for graphic retrieved from [2] and adjusted.

Figure 10.5 is only a flow diagram, which does not anticipate the spatial positions of the different receiver segments for solar heat transfer. Optional steam reheating is excluded from the flow diagram for the sake of readability.

10.2.2.4 Solar Steam Generation Inside Ducts

This section describes a robust design for direct solar steam generation with minimum thermal losses.

Arrangements of receiver tubes as shown in Figure 10.6 provide the transfer of solar heat to the plant's working fluid. The receiver tubes are aligned vertically, which means that Figure 10.6 is a top view of the segment. Fluid flows upwards in the tubes. For preheating, the vertical orientation of the receiver tubes provides inherent support of the flow because warm water is lighter than cold water.[23] Steam generation in vertical tubes is advantageous due to the avoidance of obstructing two-phase flow.

Plain steel pipes can be used, similar to the tubing of conventional steam boilers. Use of the latest high-tech vacuum receiver tubes might also be possible, provided that their operation at high temperatures is admissible.

With increasing temperature, the receiver tubes will emit more fractions of the heat radiation. Thermal losses through radiation increase exponentially to the fourth power of the temperature. Therefore, an insulated duct encases the receiver tubes

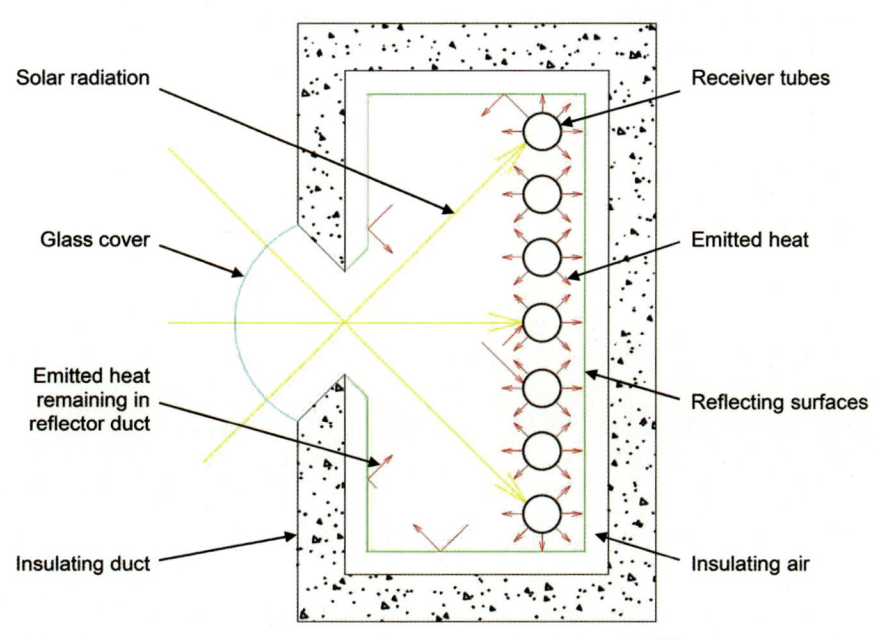

Figure 10.6 Top view of sketch for solar steam generation inside of duct.[24]

23) Warm water is lighter than cold water only beyond 4 °C.
24) Graphics and patents: Dirk Besier, Wiesbaden, Germany, dirk.besier@arcor.de; www.solarsteamer.de; www.hanssauerstiftung.de/neu/index.php?option=com_content&view=article&id=79&Itemid=108.

to minimize heat losses. Surfaces inside the duct are either equipped with mirrors or colored white. The inner surfaces will then reflect most of the emitted heat back to the receiver tubes.

Insulating air cools the reflecting inner surface. With forced convection, this waste heat can be used for the preheating of condensate and/or feedwater.

The solar radiation from Sun-tracking mirrors has to pass through the duct inlet. A small inlet is advantageous for encasement at high temperatures. The inlet size varies for the different segments of the tower: nearby mirrors can track a small inlet, whereas remote mirrors focus with lower accuracy on a wide inlet size.[25]

In contrast, at a solar tower with point concentration, all mirrors have to focus the sunlight on to a small central receiver. Therefore, compared with point concentration, the advanced solar boiler allows a larger solar field and correspondingly a higher power output from the CSP plant. The distance of remote mirrors is only restricted by radiation losses through the low-lying atmosphere.

Additional tubes with condensate and/or feedwater can provide cooling of inlet edges and inner surfaces of the duct. This feature is not depicted in Figure 10.6. Furthermore, glass may cover the duct inlet. This closure leads to better utilization of emitted heat inside the advanced solar boiler.

The shape and dimensions of the duct in Figure 10.6 and also the quantity and diameter of the receiver tubes serve solely for explanation of the physical principle. Figure 10.7 illustrates a rotund duct with several inlets and corresponding tube arrangements inside. Such a design or other arrangements may suit a solar tower that is surrounded by Sun-tracking mirrors.[26]

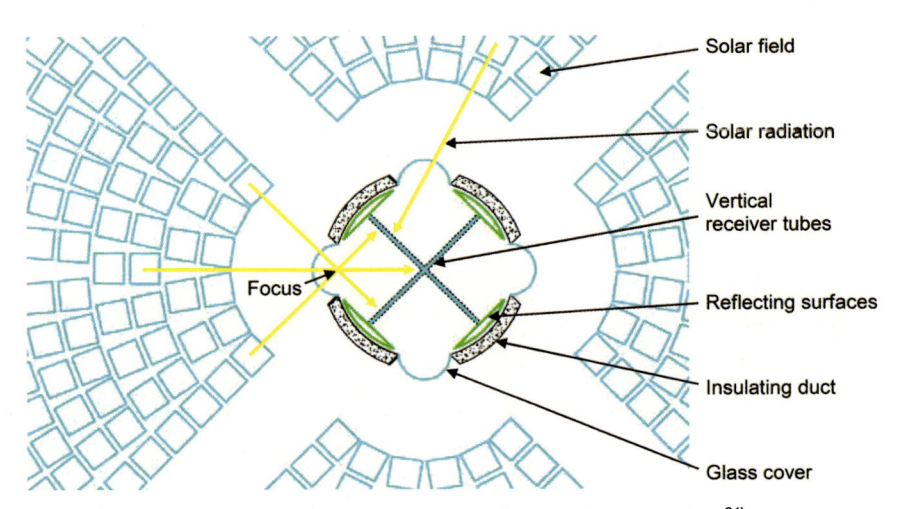

Figure 10.7 Sketch of rotund solar boiler segment with encased receiver tubes.[24]

---

[25] Section 10.2.2.5 describes the resulting spatial arrangement of heat transfer segments with their relation to nearby and remote mirrors.

[26] The described multistage cavity receiver is not restricted to water/steam. It could also operate in a solar tower plant configuration where a molten salt is used as heat transfer fluid.

Accidental steam leaks at receiver tubes will occur at encased locations. Plant personnel are therefore not at risk during such incidents.

10.2.2.5 Arrangement of Heat-Transfer Sections

Figure 10.8 shows the arrangement of boiler segments for the different steps of solar steam generation as described in Section 10.2.2.2 and Section 10.2.2.3. High-temperature segments are situated at the bottom of the solar boiler and low-temperature segments at the top. This spatial configuration coincides with the accuracy of diverting solar radiation from different distances to a central location:

- Small duct inlets for high temperatures are possible only with the precise tracking of nearby heliostats. These mirrors provide superheating and optional reheating.
- Lower temperatures for preheating and evaporation coincide with less accurate focusing of sunlight from remote mirrors to wider inlets at upper sections of the tower.

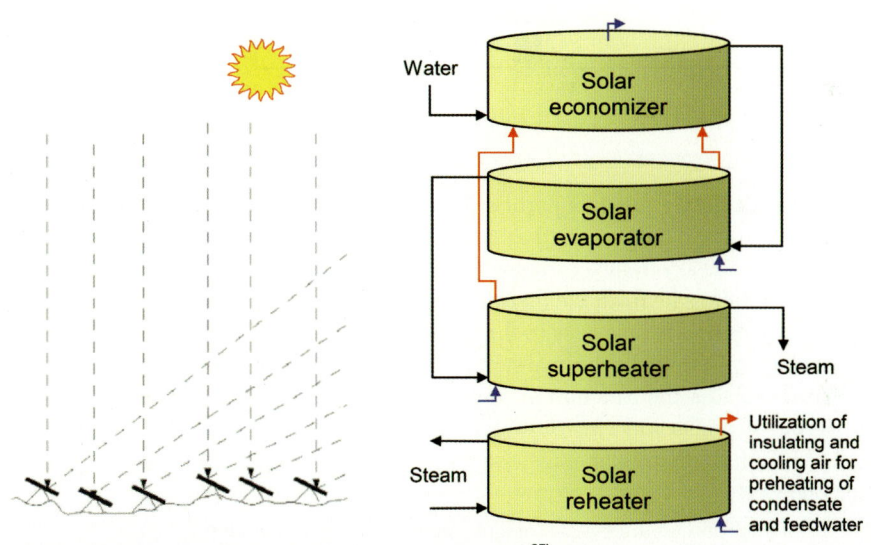

Figure 10.8 Spatial configuration of solar boiler segments.[27]

10.2.2.6 Utilization of Waste Heat

It is possible to utilize the waste heat from different segments of the tower by means of vertical convective heat transfer. Waste heat that is transported by the insulating air can be used, for example, for preheating of condensate and feedwater.

At the advanced solar boiler, the convective transfer of waste heat can mostly take the natural upward direction as shown in Figure 10.8. Ambient air is therefore used for this purpose. The air enters the segments at their bottom and could leave the boiler through a small stack at the top.

---

27) Left side of graphic with mirrors retrieved from [2].

10.2.2.7 Thermal Storage System for Night-Time Operation

For night-time operation, the advanced solar boiler is equipped with a thermal storage system, which consists of multiple heat storage tanks. Figure 10.9 depicts integration into the power plant's water/steam cycle. For the sake of simplicity, pumps, valves, and optional reheating are not shown in the abridged system diagram.

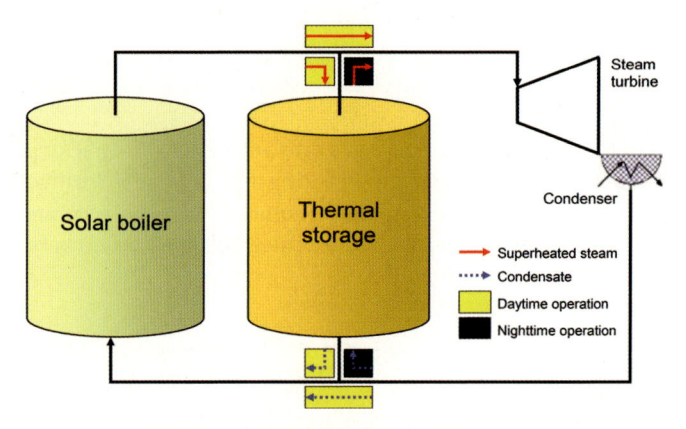

Figure 10.9 Integration of thermal storage system for night-time operation.

During daytime, a certain portion of the superheated steam drives the turbine. The other portion charges the thermal storage system. Steam turbine load changes or clouds in the sky induce corresponding changes of the steam flow to the storage system. Condensate from the thermal storage system and pressurized condensate from the condenser flow back to the solar boiler.[28]

During night-time the thermal storage system is discharged for steam generation. Then it replaces the function of the solar boiler by preheating, evaporation and superheating of steam. The feedwater pump provides the required steam pressure.

Details of the thermal storage system are shown in Figure 10.10. Its function is as follows [3]:

> The proposed three-part storage system consists of a concrete preheater unit, a phase change material (PCM) evaporator unit,[29] and a concrete superheater unit:[30]
>
> During discharging feedwater enters the preheater (...) and is heated up close to the boiling curve (...), then the water is circulated through the PCM storage, where part of the water evaporates (...). The steam is separated from the water in the

---

28) Solar boiler steam exchanges heat by condensing in the thermal storage system. During charging, the outlet of the thermal storage system can feed to the pressure side of the condensate pumps, which is the suction side of the downstream feedwater pumps. Footnote 19 refers to this.
29) Sodium nitrate ($NaNO_3$) as PCM melts at 306 °C, which leads to steam for charging at ~107 bar, ~320 °C and water for discharging at ~81 bar, ~295 °C.
30) The utilization of high-temperature concrete as sensible heat storage material up to 500 °C seems feasible but costly. Molten salt or powdery salt could be cheaper and serve the same purpose.

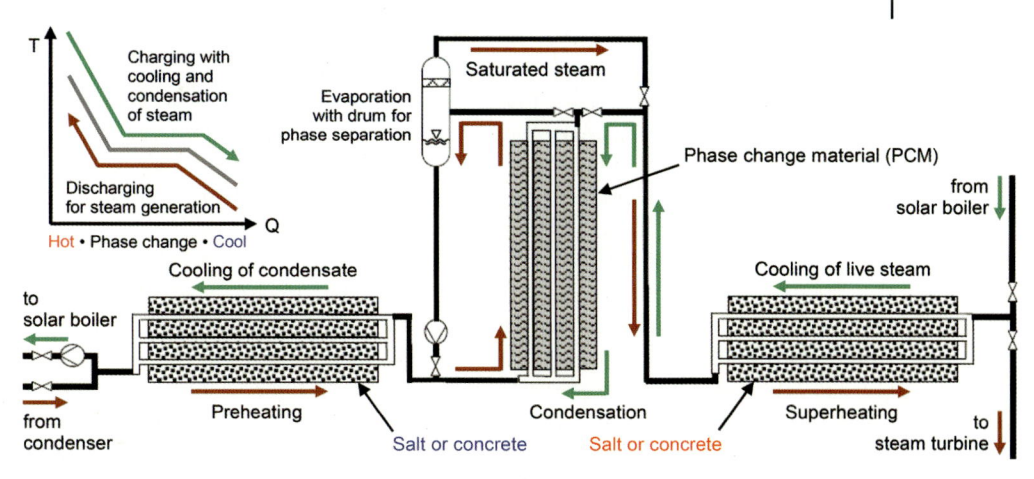

Figure 10.10 Three-part thermal energy storage system for direct steam generation [3].

steam drum and is superheated in the concrete unit (...), while the remaining water is recirculated through the PCM storage. (...) During discharge, the flow direction is from bottom to top. (...)

In charging mode, steam at a temperature slightly above saturation properties (...) is routed into the PCM module where it condenses. The flow direction during charging is from top to bottom so that the condensate is removed by gravity. A condensate drain ensures that the medium leaves the module only in liquid form.

The complete thermal storage system for the advanced solar boiler consists of multiple three-part storage subsystems, each equipped with individual tanks of reasonable size. Charging and discharging can take place by individually directing steam or feedwater to these different subsystems.

10.3
Practical Implementation of Concept

This section outlines the practical implementation of the advanced solar boiler concept for large-scale deployment in deserts. Technical, financial, and strategic aspects are taken into account.

10.3.1
Technical Procedure for Implementation

In summary, the concept for the advanced solar boiler is new and not yet tested, although each of its subsystems is based on well-known physical principles and/or established technologies.

The process of technical implementation has to start with a feasibility study, review of design features, determination of the expected efficiency,[31] plant size considerations, cost estimates, computerized simulations, scalable engineering layout, and detailed design. These are complex and challenging tasks.

Subsequent practical prototyping is required. It should start on a small scale.[32] Fast completion provides the essential reference for similar subsequent projects.

A small prototype of the innovative thermal storage system with phase change material is already being tested at a fossil-fueled steam plant in Spain [3].

The advanced solar boiler provides water/steam conditions that come close to conventional boilers in fossil-fueled power stations. Testing in hybrid operation with an existing fossil plant is therefore an option as shown in Figure 10.11.[33]

Step size is of major importance for bringing an innovative technology to market. The size of subsequent advanced solar boiler plants must therefore only increase in incremental steps.[34] Learning from prototypes is the best guarantee for the smooth roll-out of an innovative technology to large-scale deployment.

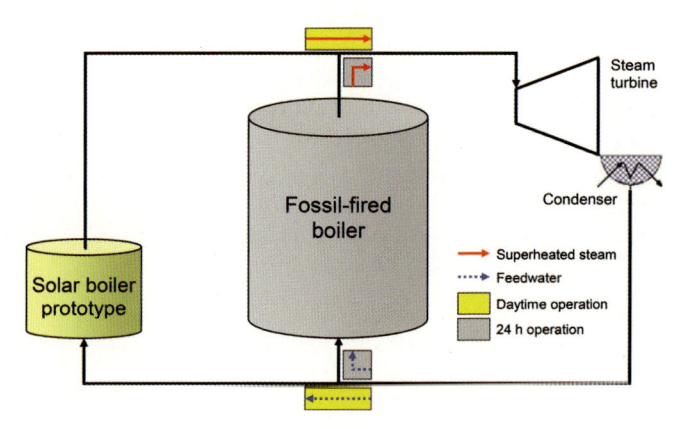

Figure 10.11 Small prototype of a solar boiler at a fossil-fired steam plant.

31) Preliminary estimates for the advanced solar boiler indicate an overall efficiency of the conversion from direct solar irradiation energy to electricity of about 25%.
32) The initial prototype is usually far more costly than subsequent models.
33) For hybrid operation, the initial investment will only refer to the solar field, the solar boiler itself, and some connecting pipework for water/steam. The conventional plant is already available, including pumps, steam turbine and condenser, cooling system, generator. and transformer for connection to the grid.
34) The determination of too large step sizes for bringing prototypes to the market was the repeated pivotal management mistake that resulted in the failure of several major investments by Germany's energy sector in the past: The GROWIAN wind turbine was planned for 3 MW while other national units where about 10 times smaller at that time. The size of the high-temperature gas-cooled reactor prototype at Jülich was 13 MW while the next unit at Hamm-Uentrop was already designed for 308 MW. The fast breeder reactor prototype at Karlsruhe had 20 MW while 327 MW was envisioned for the follow-up unit at Kalkar. The prototype for a nuclear fuel reprocessing plant at Karlsruhe had a capacity of 35 t a^{-1} while the unit at Wackersdorf was already announced for 350 t a^{-1}. No engineer can build an 830 m-high tower based solely on the experience of testing a tower with a height between 35 and 83 m. The same holds for the scaling up of energy facilities.

10.3.2
Financial Procedure for Implementation

To date, the costs of electricity generated in solar thermal power plants have been higher than the direct costs of electricity from fossil-fueled or nuclear units. Subsidies and long-term venture capital are therefore required for further introduction of renewable solar power to the market.

Remuneration for feed-in of electricity from renewables can stimulate national markets. However, application of this financial instrument is difficult across geographic and economic regions. For example, a guaranteed feed-in tariff from central Europe for solar power from deserts in northern Africa would only work if the required high-voltage power lines are also in place. If the power lines are not completed on time, for example due to protracted approval processes, then the financing model for related solar power plants could be at risk.[35]

Another aspect can also fulfill the expectations of public sponsors and private investors: the prospects for significant cost reduction and improvement of efficiency through the introduction of an innovative technology. Then the investment for research and development not only supports implementation and testing of the prototype, it also stimulates the market by forcing all competitors to provide better technology at lower prices. The arrangement of seed capital for an advanced solar boiler prototype could therefore include applications for public research money and involvement of investors, who want to promote decisive cost reduction for CSP plants.

Even for small prototypes, saving of fossil fuel in hybrid operation as shown in Figure 10.11 is likely to provide a fast return on investment.

Other methods for financing prototypes of the advanced solar boiler are possible.

10.3.3
Strategic Procedure for Implementation

The large-scale deployment of solar power plants in deserts is part of the DESERTEC concept for Europe, the Middle East and North Africa (EUMENA).[36] In close cooperation with numerous partners, the industrial initiative Dii GmbH, founded in Germany, is preparing the implementation of that concept.[37]

High costs of electricity from solar thermal power plants are a stumbling block for implementation of the DESERTEC concept. Therefore, determined cost reduction of solar thermal power plants should not be left alone to market developments. An innovation strategy with decisive action for cost reduction could excite market forces and result in faster implementation of the DESERTEC concept.

The concept of the advanced solar boiler allows such an innovation strategy: As a practical research and development project, it could help to bundle the know-how and

---

[35] A 2×700 MW, 400 kV AC power line designed for 450 kV DC already exists between Spain and Morocco.
[36] DESERTEC Foundation: www.desertec.org.
[37] For more information on the private industry joint venture Dii GmbH, see www.dii-eumena.com.

resources of numerous partners in industry, the public sector and science for testing and implementing with incremental steps an innovative solar thermal power plant design that appears suitable for cost-effective large-scale deployment in deserts.[4]

10.4 Conclusion

The described configuration of the advanced solar boiler comes close to the established and well-proven design of heat-recovery steam generators in fossil-fueled steam plants. The concept integrates the latest results of research work on a basic solar boiler concept explored by scientists at the Universidad Politécnica de Madrid [2], a thermal heat storage system designed and tested by the German Aerospace Center (DLR) in Stuttgart [3], and developments on efficient solar steam generation supported by the Hans-Sauer-Stiftung in Deisenhofen near Munich.[24]

Each of the physical principles of the advanced solar boiler is already known and well understood. The plant design is robust and mostly cost-effective. It refers to known technologies, but offers some innovative rearrangements compared with existing applications. The advanced solar boiler provides high efficiency for production of electricity and allows large-scale deployment in deserts.[4] The concept also fulfils requirements of the energy market, such as the capability for night-time operation. Furthermore, the concept does not involve any risks to the environment and its large-scale deployment is based solely on resources with abundant availability.

Practical detailed development, implementation and testing of the advanced solar boiler concept will support the fast and cost-effective large-scale deployment of solar power supply from deserts as planned with the DESERTEC concept.[36]

References

1 Romero-Alvarez, M. and Zarza, E. (2007) Concentrating solar thermal power, in *Handbook of Energy Efficiency and Renewable Energy* (eds F. Kreith and D. Y. Goswami), CRC Press, Boca Raton, FL, Ch. 21.

2 Muñoz, J., Abánades, A., and Martínez-Val, J. M. (2009) A conceptual design of solar boiler, *Solar Energy*, **83**, 1713–1722.

3 Laing, D., Bahl, C., Bauer, T., Lehmann, D., and Steinmann, W.-D. (2009) German Aerospace Center (DLR), Thermal energy storage for direct steam generation, presented at the 15th International SolarPACES Symposium 15–18 September, Berlin.

11
Status of Fuel Cell Electric Vehicle Development and Deployment: Hyundai's Fuel Cell Electric Vehicle Development as a Best Practice Example

Tae Won Lim

11.1
Introduction

The world energy consumption has dramatically increased owing to the steady increase in the global population, higher standards of living, and demands for enhanced industrial production and transportation for decades. Fossil fuel represents the highest proportion of world energy in the current energy market. As there is a limited amount of fossil fuel in the world, it is essential to find alternative energy resources. For this reason, it is necessary to make an energy paradigm shift from fossil fuel to alternative and sustainable energy resources. Among many alternative and sustainable energy resources, hydrogen is one of the most promising for the next generation. A polymer electrolyte membrane fuel cell (PEMFC) is a promising candidate for generating electricity for vehicle powertrains owing to its high efficiency and power density. A hydrogen fuel cell electric vehicle (FCEV) that adopts a PEMFC for the powertrain has attracted much attention for next-generation automotive applications for reducing carbon emissions and is suitable for zero-emission mobility. Hyundai Motor Company (HMC) has been developing FCEVs since the end of the 1990s. For an FCEV to be commercially viable, however, several barriers, such as cost, durability, and hydrogen refueling infrastructure, must first be overcome. This chapter addresses the recent advances and progress in overcoming these challenging barriers and the future roadmap for the commercialization of FCEVs.

11.2
Development of the FCEV

In 1966, General Motor developed a first FCEV, the Chevrolet Electronvan, with a driving range of 120 miles and a top speed of 70 mph. After that initial development, the project did not proceed well owing to the prohibitive costs. However, the FCEV has subsequently attracted new attention for automobile OEMs (original equipment

Transition to Renewable Energy Systems, 1st Edition. Edited by Detlef Stolten and Viktor Scherer.
© 2013 Wiley-VCH Verlag GmbH & Co. KGaA. Published 2013 by Wiley-VCH Verlag GmbH & Co. KGaA.

manufacturers) owing to environmental problems and fossil fuel depletion from the 1990s. Many automobile OEMs have developed several different types of FCEVs since then, and not only have the technologies for FCEVs improved, but also cost reductions of FCEVs have been achieved.

11.2.1
Fuel Cell Stack Durability and Driving Ranging of FCEVs

The US Department of Energy (DOE) carried out an FCEV demonstration program with several automobile OEMs from 2004. The National Renewable Energy Laboratory (NREL) collected data on and analyzed the performance of FCEVs within the DOE demonstration program; four automobile OEMs participated. The NREL divided FCEVs into two groups, Gen 1 and Gen 2, based on manufacturing time. Based on the NREL data analysis, the average fuel cell stack durabilities to 10% voltage degradation of Gen 1 and 2 were 821 and 1062 h, respectively. The durability of current fuel cell stacks is expected to be much greater, thanks to materials development in the chemical industry. The driving ranges of Gen 1 and Gen 2 were 103–190 and 196–254 miles, respectively. These driving ranges were reasonably satisfactory with respect to the target driving range in 2009 of 250 miles. Since then, the driving ranges of FCEVs have been much improved, and the current driving ranges of four FCEVs from different automobile OEMs are given in Table 11.1.

Table 11.1 Driving ranges of FCEVs from different OEMs.

Model	Driving range (km) (NEDC)
Honda FCX Clarity	460
Daimler B-Class F-Cell	380
Toyota FCHV-adv	650
Hyundai ix35 FCEV	594

11.2.2
Packing of FCEVs

In the early stage of FCEV development, the FCEV was a kind of modified vehicle based on a conventional vehicle such as with the powertrain system being exchanged from an internal combustion (IC) engine to a fuel cell stack. For this reason, the FCEV was developed focusing only on a new powertrain system and was not well designed for enjoying a drive. As the technical performance of FCEV improved, especially the fuel cell stack, automobile OEMs started to develop FCEVs for commercial use. A smaller size of the fuel cell stack and improved packing technology such as modulation and optimization of the fuel cell system led to the development of FCEVs for practical use. Currently, OEMs are targeting the development of a fuel cell system that is comparable in size to a conventional IC engine. Downsizing of the fuel cell system makes it possible to manufacture FCEVs on a mass production line similarly to conventional vehicles.

11.2.3
Cost of FCEVs

The cost of FCEVs is still high compared with conventional vehicles because of the high technical investment and limited production numbers. The costs of specific FCEV components such as the fuel cell stack and hydrogen storage system are comparatively high as they are in early stage of development. Additionally, conventional components such as the pressure sensor, water pump, and so on are also high because they are developed only for FCEVs and produced in small numbers. The cost of these components is ~5–10 times higher than the corresponding components in conventional vehicles. Generally, mass production processes will lead to cost reductions of common components in the manufacturing industries. Hence for cost reductions in FCEVs there is a high potential for decreasing the cost of conventional components if there is a large enough production volume in the FCEV market. However, the production volume of FCEVs is comparatively small because of the limited number of hydrogen refueling stations worldwide. To increase the number of hydrogen refueling stations and reduce the cost of FCEVs, strong governmental support is essential in order to enter a new technical era: a fossil fuel-free and hydrogen-based society.

11.3
History of HMC FCEV Development

HMC started the development of FCEVs at the end of the 1990s. Figure 11.1 shows the history of HMC FCEV development. The first FCEV to be developed was the Santa Fe with a 75 kW fuel cell stack installed in 2000. Since then, HMC has developed several different kinds of FCEVs, including buses. HMC developed an FCEV with an 80 kW in-house stack based on the Tucson in 2006. After developing an in-house stack, HMC accelerated the development of FCEVs. Based on its subsequent technical advances, HMC will introduce the world's first small series production FCEV from the end of 2012. Figure 11.2 shows an overview and the package layout of the ix35 FCEV. The specifications of this FCEV, which will be produced with a small series production process, are given in Table 11.2.

The ix35 FCEV adopts a 700 bar hydrogen storage system for a long driving range: 594 km with a single charge of hydrogen. The fuel efficiency is 0.95 kg H_2 per 100 km, equivalent to 28 km l^{-1} for a gasoline engine vehicle. Its maximum speed is 160 km h^{-1} and its acceleration from 0 to 100 km h^{-1} is 12.5 s. The ix35 FCEV employs several advanced technologies to improve the competitiveness for commercialization. A combination of a metallic bipolar plates, AC induction motor and Li ion battery not only improved it technically but also led the a cost reduction.

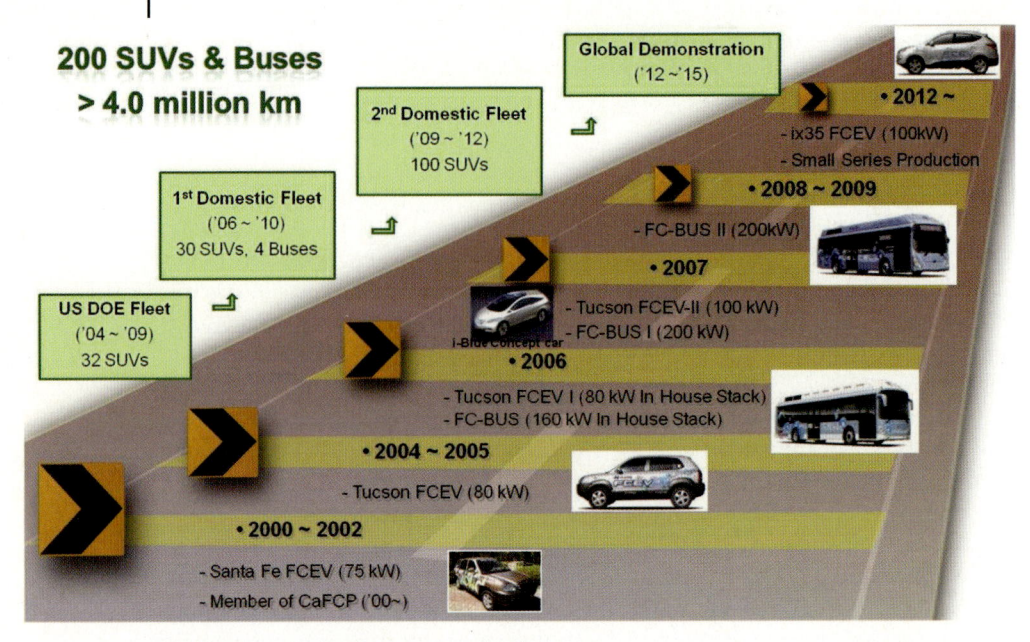

Figure 11.1 History of HMC FCEV development.

Table 11.2 Specification of ix35 FCEV.

Item	Specification
Fuel cell power	100 kW
Battery, Li ion	24 kW
Motor system	AC induction
Hydrogen tank	700 bar
Fuel efficiency	0.95 kg H_2 per 100 km (28 km l^{-1})
Driving range	594 km
Acceleration (0 $\rightarrow$ 100 km h^{-1})	12.5 s
Maximum speed	160 km h^{-1}

In addition, integration and modulation of the fuel cell system imparts more efficiency to the FCEV production system. A balance-of-plant (BOP) system is also essential for optimizing the FCEV operation. The BOP system of the ix35 FCEV mainly consists of a thermal management system and air and fuel processing systems and typically operates in connection with a hydrogen fuel storage system as shown in Figure 11.3.

11.3 History of HMC FCEV Development | 187

Figure 11.2 Overview and package layout of the ix35 FCEV.

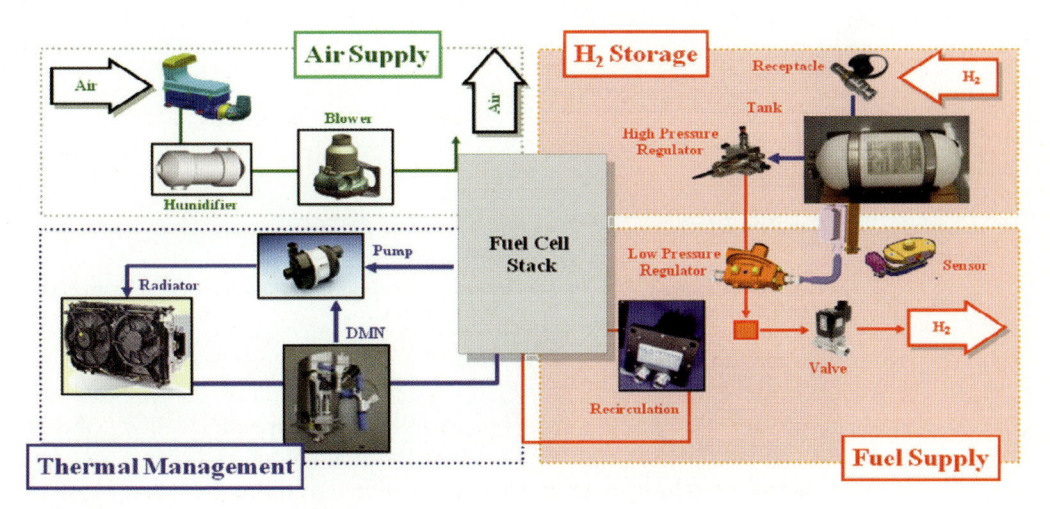

Figure 11.3 BOP system of the ix35 FCEV.

11.4
Performance Testing of FCEVs

11.4.1
Crashworthiness and Fire Tests

The public perception of the dangers and safety concerns about FCEVs is still widespread because FCEVs are mainly operated using hydrogen. To change public awareness of FCEVs, it is necessary to verify their safety convincingly. Therefore, HMC has performed a variety of crashworthiness tests and fire tests of the hydrogen storage tank system. Figure 11.4 shows the crashworthiness testing of the ix35 FCEV: HMC performed the front, rear, and side crash tests to verify its safety. In the front crashworthiness test, HMC especially tested the electrical insulation of the stack, and it was verified that the ix35 FCEV is electrically well insulated. Additionally, the stack voltage dropped drastically after the crash, as shown in Figure 11.4. The stack voltage drop is one of the electrical safety regulations for electric vehicles, including FCEVs, and the drastic voltage drop of the ix35 FCEV well satisfied the regulation. In the rear crashworthiness test, HMC tested the safety of the hydrogen storage tank system such as for leakage of hydrogen. The rear crashworthiness test verified that there was no hydrogen leakage after the crash, which means no chance of a hydrogen explosion. This testing proved the robust safety of the hydrogen storage tank system of the ix35 FCEV.

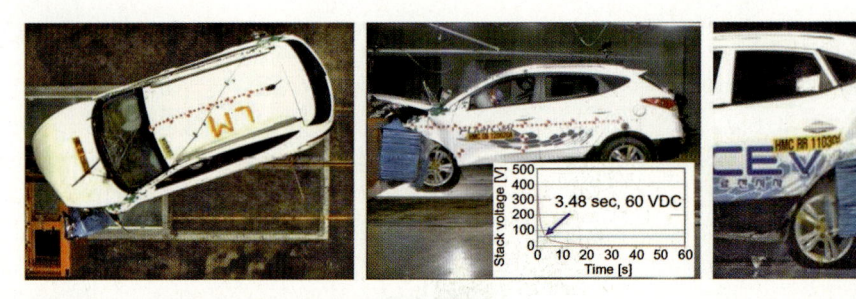

Figure 11.4 Crashworthiness testing of the ix35 FCEV.

	Gasoline vehicle	FCEV	
Condition		Fire initiated from the ashtray	
Result	Fuel tank exploded after 40 minute.	PRD activated after 22 minutes. (Type III tank)	PRD activated after 13 minutes. (Type IV tank)

Figure 11.5 Fire testing of the ix35 FCEV.

HMC also performed fire and explosion testing of the hydrogen storage tank of the ix35 FCEV, as shown in Figure 11.5. An artificial fire was set under the vehicle to check the safety of the hydrogen storage tank system. The conventional fuel tank of a corresponding gasoline engine vehicle exploded within 40 min, whereas the FCEV was protected by activating the pressure relief device in 22 min (type III hydrogen tank) or 13 min (type IV hydrogen tank). The results indicated that the fire safety of typical hydrogen storage tanks for FCEVs is at least comparable to that of conventional gasoline engine vehicles.

11.4.2
Sub-Zero Conditions Tests

The cold start-up capability and the durability of FCEVs under sub-zero conditions have been one of major concerns over the years since the fuel cell generates water by the oxygen reduction reaction at cathode. The cold start-up capability of the ix35 FCEV is in the range –20 to 40 °C. To verify the cold start-up capability, HMC has performed not only indoor laboratory-scale but also field tests under sub-zero conditions. As shown in Figure 11.6, FCEVs were placed on the road at Mt. Taebaek, which is one of coldest places in Korea. The tested FCEVs were successfully started at –19 °C, verifying the excellent cold start-up capability of the FCEV. HMC performed cold start-up tests on the ix35 FCEV in Sweden in 2011, and it started successfully at –20 °C. In addition, the fuel cell stack was tested robustness down to –41.5 °C in Sweden. Although cold start-up tests at lower temperatures were not performed, the robustness of the ix35 FCEV fuel cell stack was verified.

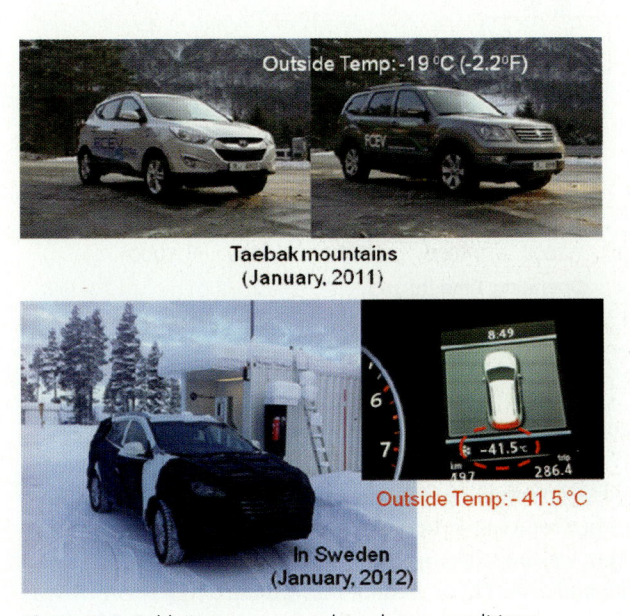

Figure 11.6 Cold start-up tests under sub-zero conditions.

11.4.3
Durability Test

Durability is an important factor for the commercialization of FCEVs, and HMC has performed both laboratory-scale bench tests and real road vehicle tests. A clear understanding of the correlation between bench test results and vehicle test results is of paramount importance to estimate the durability of FCEVs efficiently. This could facilitate the validation of a proposed degradation mechanism and the development of novel materials and components for FCEV commercialization. HMC has developed bench test methodology to simulate the real status of an FCEV on the road. The durability correlation between the bench and the vehicle test results shows good agreement, as shown in Figure 11.7. HMC has deployed two ix35 FCEVs in Denmark since early 2011. In Figure 11.7, the performances of the two vehicles are shown in yellow symbols. Two organizations are currently operating ix35 FCEVs on a daily basis. Comparison of the results for the bench test vehicle and the daily basis driving test are in good agreement, as shown in Figure 11.7. From these durability test results, it is anticipated that the ix35 FCEV has more than a 5 year durability guarantee if we assume 500 h for 1 year of driving.

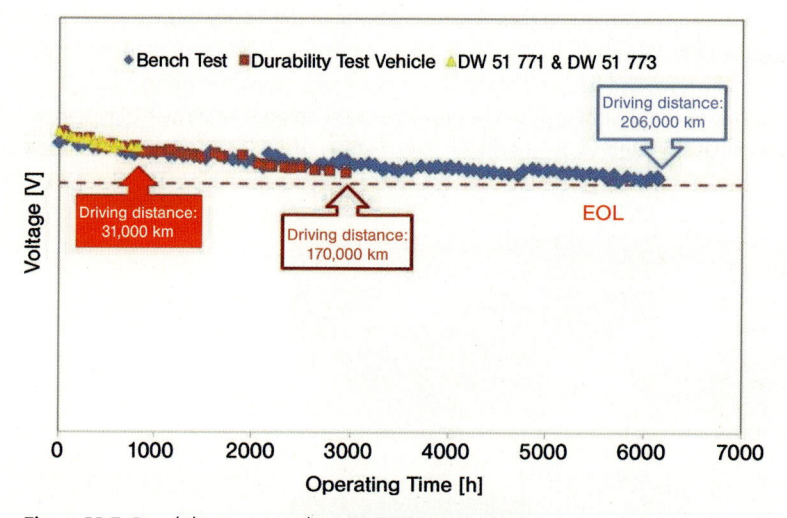

Figure 11.7 Durability tests on the ix35 FCEV.

11.4.4
Hydrogen Refueling

Currently constructed hydrogen refueling stations, especially in Germany, required IR communication tools when refueling hydrogen. To satisfy this requirement, an IR communication component was installed to communicate between a vehicle and a hydrogen refueling station and collect hydrogen refueling data for the ix35 FCEV, as shown in Figure 11.8. A short hydrogen refueling time is one of advantages of FCEVs.

11.5 Cost Reduction of FCEV | 191

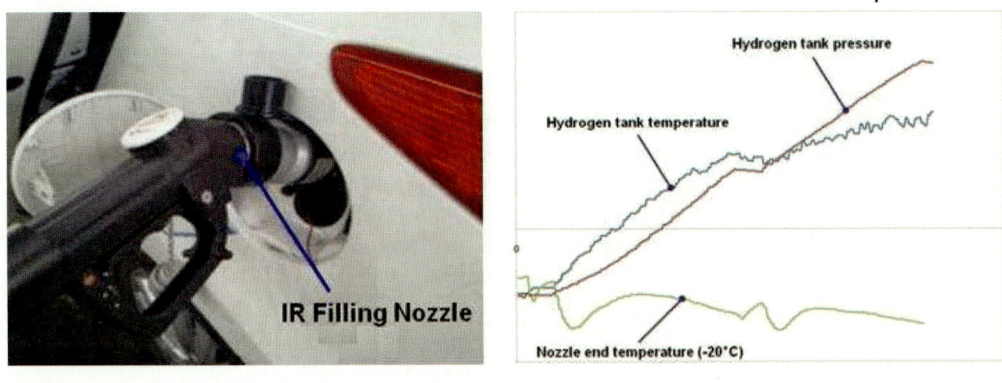

Figure 11.8 IR filling nozzle and hydrogen refueling time.

The IR communication and precooling system of the hydrogen tank leads to a short hydrogen refueling time. As shown in Figure 11.8, the hydrogen refueling time of ix35 FCEV is ~3–4 min, which is competitive with the conventional IC engine vehicle.

11.5
Cost Reduction of FCEV

Recently, the cost of FCEVs has fallen to approximately one-fifth of the level in 2005, as shown in Figure 11.9. The cost reduction is mainly due to both fuel cell system

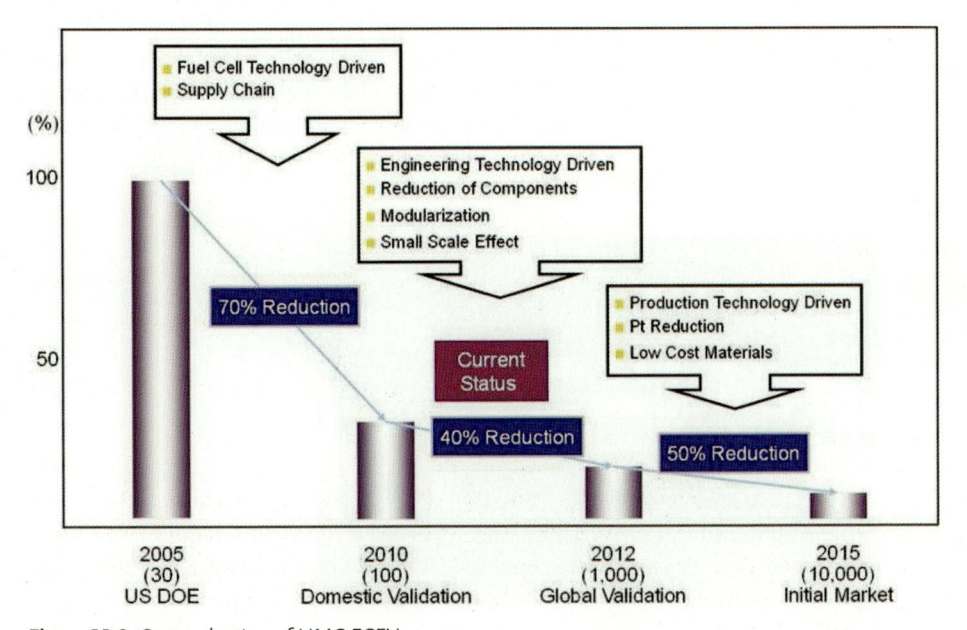

Figure 11.9 Cost reduction of HMC FCEV.

optimization and the development of cost-effective components, including stamped metallic bipolar plates, advanced membrane-electrode assembly (MEA) and gas diffusion layer (GDL), injection-molded end plates, simplified BOP components, and so on. However, the cost of FCEVs is still too high for ordinary customers because of some technical challenges and the limited market size in the world. The cost of FCEV should therefore be lowered by at least a further 50% by 2015, by benefiting from technical innovation of key components (i.e., MEA, GDL, gasket, bipolar plate, power electronics, hydrogen tank, etc.) and economies of scale.

11.6
Demonstration and Deployment Activities of FCEVs in Europe

HMC participated in the US DOE fleet program with 32 FCEVs in the California and Michigan areas from 2004 to 2009 (Figure 11.10). Through this fleet program, HMC achieved a driving distance of 835 212 km (518 765 miles) and an average fuel economy of 16.3 km l^{-1}. HMC also participated in the Korean domestic fleet program with 30 FCEVs and 4 FCs from 2006 to 2010. The total driving distance of the vehicles was 1 297 799 km (806 086 miles) and the average fuel economy reached 19.2 km l^{-1}.

Since early 2011, HMC has participated in European hydrogen activities such as H2Mobility, and is a member of both H2Mobility and UK H2Mobility. As a member of H2Mobility and UK H2Mobility, HMC has devoted efforts to anticipate the forthcoming market demand for hydrogen refueling stations and FCEVs in Europe. HMC has also provided opinion on governmental policy for hydrogen development in Europe through H2Mobility. HMC's activities as a member of H2Mobility help to establish a roadmap of the hydrogen transport network and the development of hydrogen technology in Europe. With abundant and deeply understood experience of hydrogen transport technologies, including hydrogen infrastructure, HMC has a strong willingness to support and collaborate to promote the development of a hydrogen transport network in Europe.

Figure 11.10 DOE fleet program.

Figure 11.11 EU road tour in Cardiff, UK.

HMC will have a small series production of ix35 FCEVs from the end of 2012 to 2015. During this period, HMC will produce 1000 units of ix35 FCEVs. The European market is a main market for the HMC ix35 FCEV. HMC will participate in an EU-funded demonstration and deployment program and EU tenders for FCEV deployments. HMC has been participating in an EU-funded demonstration and deployment program called H2moves Scandinavia since 2011. HMC has provided two ix35 FCEVs in Denmark and Norway. Under the H2moves Scandinavia program, HMC participated in an EU Road Tour (Figure 11.10) with other OEMs, Daimler, Toyota, and Honda, from Hamburg to Copenhagen which visited five countries and 10 cities covering a total distance of ~4000 km. During this EU Road Tour, HMC participated in several VIP events for promoting a hydrogen society and public events for enhancing public awareness of hydrogen and FCEVs. This program will be terminated at the end of 2013. Through the program, HMC has been involved in EU governmental policies on dissemination of hydrogen refueling stations and FCEVs and enhancement of public awareness of a hydrogen society by collaborating with other automotive OEMs and hydrogen production companies.

HMC also participated in the EU Parliament Test Drive Program (Figure 11.12), during which a Member of EU Parliament test drives an ix35 FCEV during alternate weekends. These test drives are intended to lead to enhanced hydrogen awareness of European policy makers and an appropriate European energy policy. HMC has also participated in Clean Energy Partnership (CEP) as a full member from 2012. Currently, two ix35 FCEVs are operated in Germany under the CEP program to promote a hydrogen society.

HMC also has a strong willingness to participate in a future EU-funded FCEV deployment program with other automotive OEMs and hydrogen production companies for promoting a hydrogen society in Europe. HMC has been chosen as an FCEV provider in the cities of Copenhagen, Skane, and Skedsmo from 2013 with official vehicles for the city council. In addition to these regions, several EU municipalities have proposed deploying FCEVs in their region. London has a strong willingness to deploy FCEVs as official vehicles such as police cars. Aberdeen also

Figure 11.12 EU Parliament test drive.

has a plan to deploy ix35 FCEVs as official vehicles of the city council. These deployment activities lead to dissemination and opening up of the market for FCEVs in Europe. They also enhance public awareness of a hydrogen society in Europe. HMC will actively participate in all possible FCEV deployment activities in Europe.

11.7
Roadmap of FCEV Commercialization and Conclusions

Recently, fleets of FCEVs from the major automotive OEMs (OEMs) have been introduced in the United States, in cities throughout Europe, and in Japan. In South Korea, HMC has been operating over 200 FCEVs worldwide for the past 12 years, with the aid of the Korean government's funding program. So far, the total driving distance of HMC FCEVs has exceeded 4 million km. Owing to all the efforts exerted by the major automotive OEMs to develop FCEVs, the fuel cell market is expected to introduce FCEVs in thousands per OEM after 2015 and tens of thousands of units between 2018 and 2020. HMC is now in the middle of the pre-commercial production stage and will continue to pursue the commercialization of FCEV with a next-generation system that will provide more cost-effective and highly durable vehicles for its customers.

12
Hydrogen as an Enabler for Renewable Energies

Detlef Stolten, Bernd Emonts, Thomas Grube, and Michael Weber

12.1
Introduction

Energy technology is currently undergoing a considerable transformation worldwide. The factors driving this change forward are climate change, supply security, industrial competitiveness, and local emissions. Although these driving forces are recognized throughout the world, their priority varies from country to country. In the aftermath of the nuclear disaster in Fukushima, caused by a natural catastrophe, some countries decided to discontinue their nuclear power programs. In Germany, these events led to a broad political consensus among political parties against the further use of nuclear power. In Japan, discontinuation was decided upon, a decision the newly elected government is going to reverse. At the same time, greenhouse gas (GHG) emissions are to be reduced even further. Compared with 1990 levels, GHG emissions are to be cut by 40% by 2020, 55% by 2030, 70% by 2040 and 80–95% by 2050 [1]. The topics of renewable energies, electric mobility, efficient power plants, and combined heat and power generation are widely acknowledged as the grand challenges in this area. However, in view of the GHG reduction targets outlined above, only two of these four grand challenges remain: renewable energies and electromobility based on renewable energies. Neither combined heat and power generation nor highly efficient centralized fossil-fired power plants bear the potential to meet the above-mentioned targets for 2040 or 2050. Cogeneration might serve as a bridging technology, however. The new technologies to be applied need to be "game changers." This means that they must have the potential to deliver quantitatively and at competitive cost to the future energy mix. Moreover, these technologies need to be available in time. As a lower limit for market penetration, a 10 year time frame can be considered and approximately 10 years for industrial development will be needed to make a product out of a demonstrator leaving industrial or institutional research. Hence the seemingly far-away target of 2050 for CO_2 reduction comes down to a relatively close target of 2030 for research. Considering that this is important when selecting the new technologies, basically there is little

Transition to Renewable Energy Systems, 1st Edition. Edited by Detlef Stolten and Viktor Scherer.
© 2013 Wiley-VCH Verlag GmbH & Co. KGaA. Published 2013 by Wiley-VCH Verlag GmbH & Co. KGaA.

room for completely new developments, yet improvements in size, performance, and cost of existing technologies will be paramount and require intense research. This is not to say that completely new second-generation technologies are out of range for implementation after 2050, but whatever is not ready for development by 2030 will be very unlikely to contribute to the above-mentioned goal of 2050. Wind turbines, solar plants, and fuel cell vehicles are three examples. Beyond these game changers, technologies representing "missing links" are to be considered, in other words, technologies connecting game changers with each other and/or the system.

Danish policy serves as a cornerstone in reducing CO_2 emissions. It aims to generate electricity and heat fossil-free by 2035 [2]. This is particularly noteworthy because Denmark has no nuclear power and must therefore rely almost entirely on renewable energies. During a transition period, the intermittent power input from renewable energy sources can be efficiently and cost-effectively balanced by natural gas power plants. The latter can be converted to carbon-free operation at a later stage using hydrogen.

12.2
Status of CO_2 Emissions

In Germany, a total of 912 million tonnes of CO_2 equivalent were emitted in 2009. Table 12.1 shows the CO_2 emission sources and their contribution to total emissions. Electricity generation accounts for the largest share with 30%. The energy sector is responsible for a total of 37%; 11% of emissions originate in the residential sector and 23% in industry, trade, and commerce. About 11% of CO_2 emissions are caused

Table 12.1 CO_2 emission sources and their share of total emissions in 2009 [3].

CO_2 emission source	Share of total emissions (%)
Energy sector (thereof electricity)	37 (30)
Transportation:	17
Passenger cars	11
Goods/other transport	6
Residential	11
Industry (thereof electricity)	19 (2)
Trade and commerce	4
Agriculture	8
Other	4
Total	100

by passenger cars and 6% by heavy-duty trucks and rail, ship, and air transportation [3]. CO_2 emissions in Germany have already been reduced by 26% between 1990 and 2009. However, in order to reduce emissions by at least 80%, large sectors must become CO_2 neutral. Examples include the entire electricity sector and passenger car transportation. By implementing the concept proposed here, they can be transformed in a comparatively simple and cost-effective manner. Furthermore, there is also potential for reducing CO_2 emissions in the residential sector, and also in trade and commerce. As heavy-duty trucks, rail, ship, and air transportation rely on middle distillates, namely different diesel qualities, marine gas oil, kerosene, and even bunker oil, biofuels look more promising for these applications than hydrogen for higher energy density, which is about 10 times that of gaseous hydrogen at 700 bar.

Direct CO_2 emissions account for the largest share of total emissions with ~800 million tonnes. However, it should be noted that other substances also have a significant influence despite their very low quantities as they have a much higher equivalent factor. The equivalent factor indicates how much greater the impact on the climate would be for the same amount of a substance in relation to CO_2. CO_2 is given a reference value of 1. As these values also depend on time, their impact can change over the time span considered. Table 12.2 shows the global warming potentials (GWPs) for different GHGs as a function of time. In the literature, GWPs – sometimes referred to as CO_2 equivalent factors – are usually used in relation to the 100 year time horizon.

Table 12.2 Global warming potentials (GWPs) of greenhouse gases [4].

Greenhouse gas	GWP		
	20 years	100 years	500 years
CO_2	1	1	1
CH_4	72	25	7.6
N_2O	289	298	153
Hydrofluorocarbons (HFCs)	437–12 000	124–14 800	38–12 200
Perfluorocarbons (PFCs)	5200–8 630	7390–17 700	9500–21 200
SF_6	16 300	22 800	32 600

12.3
Power Density as a Key Characteristic of Renewable Energies and Their Storage Media

As the primary argument for choosing renewable technologies, their power density rather than their potential average power provided by Nature in the form of primary

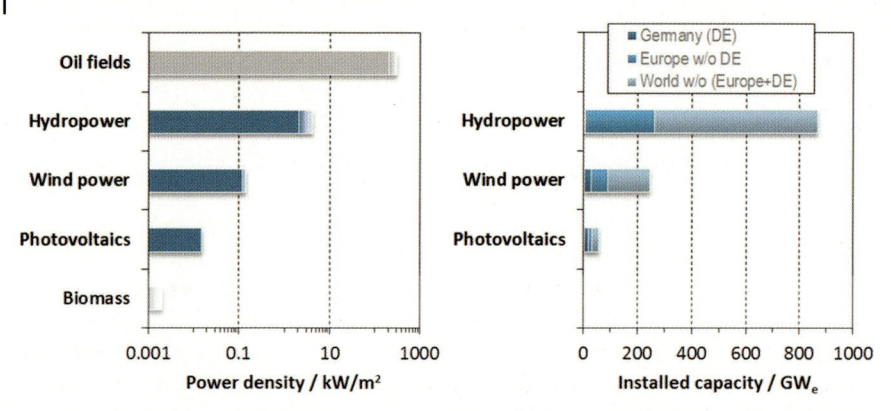

Figure 12.1 Comparison of the power densities and installed capacities of renewable energies.

energy is suggested. The average annual power density represents a measure of the effort required to concentrate energy and convert it into electricity. Here, the active area of the technical device is taken as the reference point, for example, the cell area in the case of photovoltaics (PV). Whereas hydropower has a power density in the region of a few kilowatts per square meter, wind power has an average power density of ~150 W m^{-2} and PV provides a mere 15 W m^{-2}. Hence there is about an order of magnitude difference between these technologies, indicating economic advantages of the technology with higher power densities. The levels of installed capacities in Figure 12.1 support this approach.

Hydropower in Germany has been expanded almost to full capacity, whereas wind power and PV still have huge potential for expansion. According to conservative estimates, the potential associated with wind power is at least 189 GW [5, 6]. This includes both onshore and offshore wind turbines. We have chosen wind power for the following scenario owing to its higher average annual power density. Other renewable primary energies retain their current status in the scenario. These assumptions were made to keep the scenario as simple as possible.

Similar considerations hold for the selection of the preferred storage medium. Today, lithium-ion batteries have a storage density of around 5 MJ l^{-1} when the electrode materials of common LiCoO$_2$ systems are considered. In order to achieve longer lifetimes of the batteries in vehicles, the state of charge is kept between 90 and 80% as an upper level and between 10 and 20% as a lower level. Hence about 70% depth of discharge is achieved today. Thereby the technical storage density drops to ~1.0 MJ l^{-1} or ~0.5 MJ kg^{-1}. Hydrogen in a fuel tank at 700 bar has a volume-specific storage density of ~3 MJ l^{-1} or ~6 MJ kg^{-1}, both including the tank, and a physical storage density in the liquid state of 8.46 MJ l^{-1}. These are two suitable energy storage options for electric mobility. Gasoline, in contrast, has an excellent storage density of 32 MJ l^{-1} as a pure fuel with a negligible tank volume and around 30 MJ kg^{-1} including the tank. We chose hydrogen as the storage medium because the storage density of hydrogen is 4–6 times greater than that of batteries. An overview of the storage densities of different materials is given Table 12.3.

Table 12.3 Energy density of gasoline and ethanol compared with hydrogen and batteries.

Product	Physical storage density		Technical storage density	
	MJ l^{-1}	MJ kg^{-1}	MJ l^{-1}	MJ kg^{-1}
Gasoline	32 [7]	43 [7]	30	35[a]
Ethanol	21 [7]	27 [7]	19	22[a]
Hydrogen (700 bar)	5	120	3 [8]	6 [8]
Li ion batteries	5[b]	1.5[b]	1.0	0.36 [9]–0.5

a   Technical storage densities of gasoline and ethanol are estimated for a tank storage with a capacity of 60 l and a mass of 10 kg.
b   Theoretical storage density of the electrode system graphite–LiCoO$_2$ for lithium ion batteries.

12.4
Fluctuation of Renewable Energy Generation

Figure 12.2 shows an example of the strong fluctuation of wind power over time. There are three basic options to adapt the fluctuating input to the original determined by the customers. First, the wind power can be used correctly in the grid when provided as needed. Further demand in times of low wind power production can be compensated efficiently and comparably cheaply by natural gas power plants. Since the fluctuations of renewable energies such as wind and solar in particular are very high, it is assumed that electricity produced in excess should be stored. Batteries and pumped hydro storage can provide only limited capacities. Hence electrolysis with subsequent hydrogen storage is suggested as a means of mass storage of excess electricity.

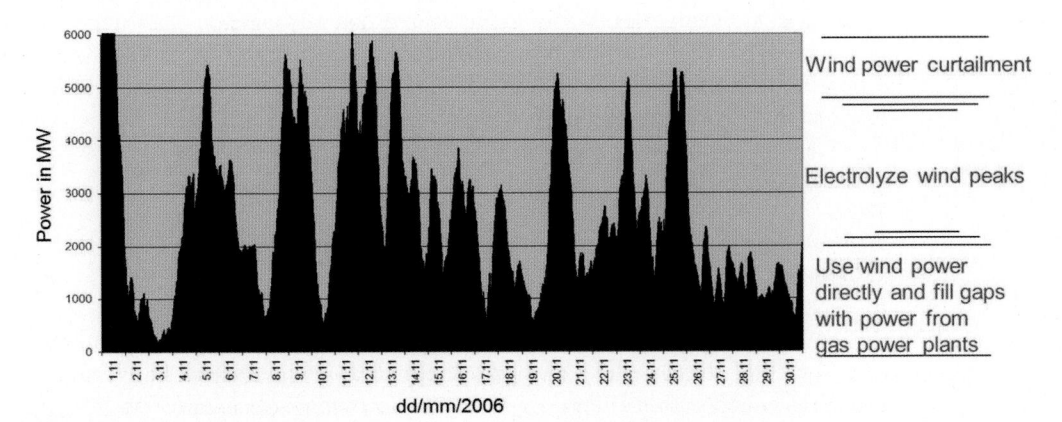

Figure 12.2 Example for the fluctuation of wind input based on real wind input data into the EoN grid in 2006 (quarter hourly resolution).

In order to keep the investment for electrolyzers in check, a third regime of power curtailment is suggested. Since very high power is produced over only a short period in the year, the greatest savings and electrolyzer investment can be expected at a relatively low curtailment level. Therefore, the three regimes will be applied in the following scenario.

12.5
Strategic Approach for the Energy Concept

The facts compiled above provide some strategic guidance for a renewable energy concept:

- Power production, road traffic, and residential heating are the sectors in which CO_2 reductions can most easily be attained.
- The short timeline requires a focus on large technologies, mentioned above as game changers, and missing links, technologies needed to connect power production and end use, such as electrolysis.
- Only electoral mobility can deliver a CO_2 reduction of 80% in passenger transportation.
- Wind power is most attractive as a renewable energy source owing to its higher power density compared with PV. Hydropower provides an even higher power density, and the German landscape form does not provide the resources for a further substantial increase.
- Fluctuating renewable energy input needs mass storage.
- Natural gas power plants can compensate renewable power production gaps.
- Cost efficiency of the new energy system is paramount and will only be achieved if all energy sectors are taken into consideration.

The system to be investigated for further reducing the CO_2 output will consist of wind power as the main renewable energy source, natural gas power plants for compensation of power production gaps, and hydrogen for storage of the excess energy. The use of hydrogen in the gas grid and for fuel cell vehicles will be investigated for its effect on CO_2 emissions and cost.

12.6
Status of Electricity Generation and Potential for Expansion of Wind Turbines in Germany

Currently, electricity generation is primarily based on fossil and nuclear energy. Almost 80% of electricity is produced using nuclear energy, lignite, hard coal, or gas. Wind energy accounts for only ~6%, with a strong increase in that share over the last year, however. The share of primary energies in electricity generation is shown for 2010 in Figure 12.3.

12.6 Status of Electricity Generation and Potential for Expansion of Wind Turbines in Germany

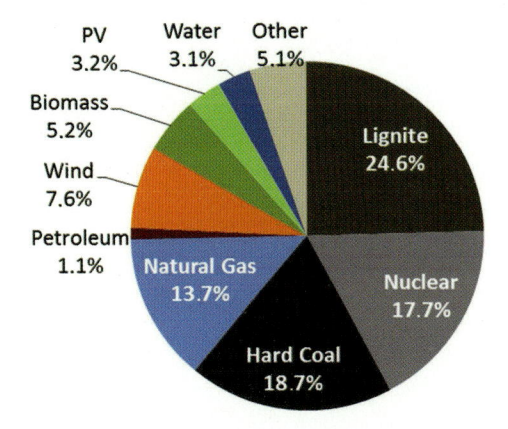

Figure 12.3 Share of primary energies in the generation of 628 TWh of electricity in Germany in 2010.

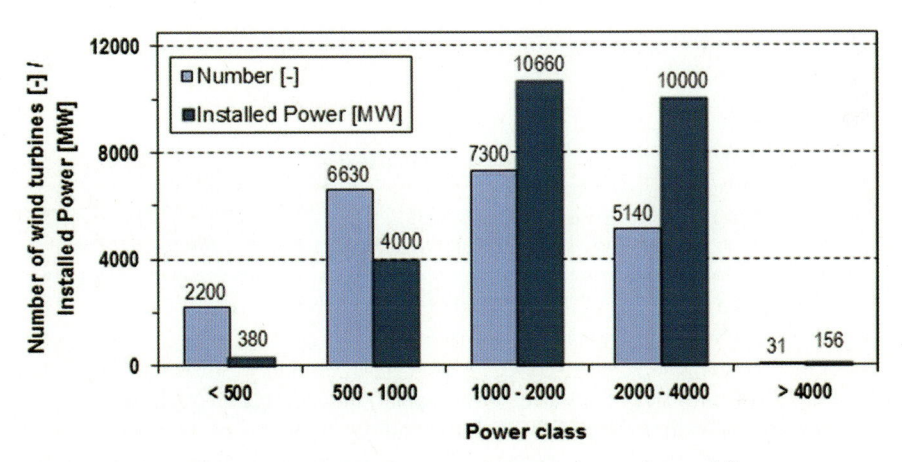

Figure 12.4 Distribution of installed wind turbines in Germany according to different power classes in 2010. Source: data from [11].

The distribution of installed wind turbines in Germany is shown for 2010 in Figure 12.4. This results in a weighted average capacity of 1.23 MW per wind turbine[1].

The following section discusses the assumptions of the concept outlined here in more detail. Wind power is the cornerstone of this proposed electricity supply. Nuclear and coal-fired power plants are completely dispensed with. Gas power plants compensate for no-wind periods and excess electricity is converted into hydrogen via electrolysis. At present, wind power accounts for ~8% of electricity generation [10]. Substituting almost 80% of today's electricity production is extremely challenging and necessitates a massive improvement in wind energy in the form of expansion and repowering.

---

1) Installed capacity/number of turbines.

Table 12.4 Predicted installed wind capacity in Germany.

Year	Source/prognosis	Total capacity (GW)	Onshore (GW)	Offshore (GW)
2009	[12]	26	26	0.06
2015	[13]	48	35	13
	[14]	36	26	10
	Constant expansion (2 GW a^{-1})	38		
2020	[15]	38	28	10
	[14]	48	28	20
	[16]	47	37	10
	Constant expansion (2 GW a^{-1})	48		
2030	[16]	62	37	25
	Constant expansion (2 GW a^{-1})	68		
Potential	[5]	80	45	35
	[17]	198+	198	No data
	[18][a]	79–108+	79–108	No data
2050	Assumptions of the concept	239	169	70

a Assumption: Relationship between potential and installed capacity for Baden-Württemberg and Lower Saxony in 2002 can be projected onto the whole of Germany.

Table 12.4 shows the predicted installed capacity of onshore and offshore facilities. The potential is predicted very differently depending on the source. The assumptions of our concept are therefore optimistic data that are in line with literature data.

In the literature, we found only one study that dealt with the potential of wind power beyond 2030, conducted by the Fraunhofer Institute for Wind Energy and Energy System Technology (IWES). Commissioned by the German Wind Energy Association (BWE), detailed geographic studies were performed and it was determined that at least 2% of land in Germany could be used to produce electricity, taking into account all restrictions such as nature reserves and proximity to built-up areas. On this land, turbines with a capacity of 3 MW would provide a total of 198 GW. Offshore turbines were not investigated.

12.7
Assumptions for the Renewable Scenario with a Constant Number of Wind Turbines

On the basis of the argumentation and data outlined, a scenario was designed with deliberately simple parameters. The following assumptions were made:

12.7 Assumptions for the Renewable Scenario with a Constant Number of Wind Turbines

- The number of onshore wind turbines remains constant at the 2011 level, namely ~22 500. The average capacity per turbine is increased from 1.23 to 7.5 MW and the capacity utilization from almost 1400 to 2000 full-load hours[2]. The national mean for 3 MW turbines is already slightly higher than this last value [6].
- Offshore wind energy is expanded to 70 GW [19] and full-load hours are assumed to be 4000 h a^{-1}.
- PV is taken into account with an installed capacity of 24.8 GW at the end of 2011.
- Electric power feed-in from onshore and offshore wind turbines and PV installations is considered as time-dependent values and is balanced with the time-dependent vertical grid load of the German electric transmission grid.
- The contribution of other renewable energies remains constant at the 2010 level[3]. They therefore play no role in causing variations or balancing them.
- Variations in wind and solar energy are balanced as required by gas power plants. Other fossil energy carriers are no longer used. Combined cycle turbine power plants will operate for more than 700 h a^{-1}. Their partial load behavior is accounted for with an efficiency deduction of 15% compared with the efficiency at nominal power.
- Excess electricity is used to produce hydrogen using electrolysis, pumped through hydrogen pipelines to filling stations, and used in fuel cell vehicles. An efficiency of 70% (in relation to heating value) and a minimum operating time of 1000 h a^{-1} are assumed for electrolysis. When peaks that would lead to a shorter operating time were assumed to be curtailed. Salt caverns provide seasonal storage. It is estimated that a vehicle will consume 1 kg of hydrogen per 100 km. Mean driving distance is taken as 11 400 km a^{-1} for each vehicle.
- In the residential heating sector, the amount of natural gas consumed is cut by half compared with 2010. This could be used to produce electricity as required.

Figure 12.5 shows two "snapshots" from the scenario calculation. The vertical grid load, in other words the electricity required that is transferred from the power plants to consumers via the grid, is compared with the amount of power fed in. Any surplus is converted into hydrogen by electrolysis and deficits are balanced by gas power plants. It can be seen that there is a larger surplus in the winter months than in the summer months, which can be explained by the strong seasonal character of wind.

The scenario was deliberately designed to be simple, with few components. It aims to use an integrated concept to determine the feasibility of a renewable energy supply. The contribution of other energy storage systems and the transnational exchange of electricity is considered to be comparatively low and is therefore neglected here as a first approach. The key feature involves relating the stationary sectors to the transportation sector. The two are linked by hydrogen. The large quantities involved mean that there are two basic options for hydrogen utilization: reconversion to electricity for the grid and its use as a fuel. The efficiency of reconversion with or without feed-in into the natural gas grid will correspond to no more than generating electricity with

---

2) As the hub height and size increase, so do the capacity and full-load hours.
3) Input work (TWh a^{-1})/8760 (h a^{-1}).

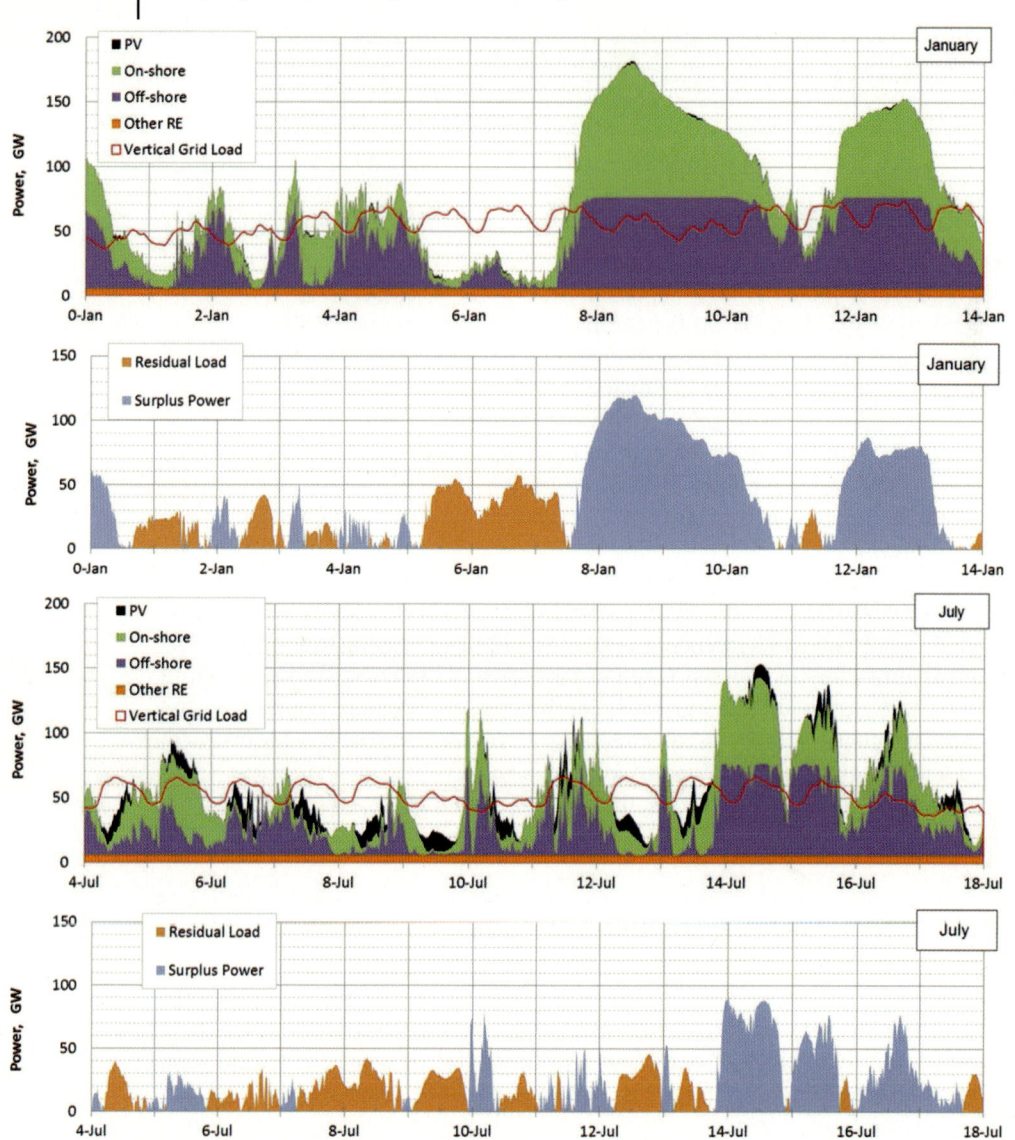

Figure 12.5 Comparison of calculated energy feed-in, vertical grid load 2010, and the resulting excess/deficit.

natural gas. If hydrogen is used for the methanation of CO_2 beforehand, the efficiency decreases. When used as a fuel in fuel cell vehicles, hydrogen reduces the energy consumed (tank-to-wheel) by around 50% compared with vehicles run on gasoline (Figure 12.6). At the same time, the CO_2 emissions that are avoided in relation to the lower heating value are 25% higher for the oil-based fuels than for natural gas, which means that the use of hydrogen in the road transportation sector prevents in total 2.5 times more CO_2 than its reconversion into electric power for the grid.

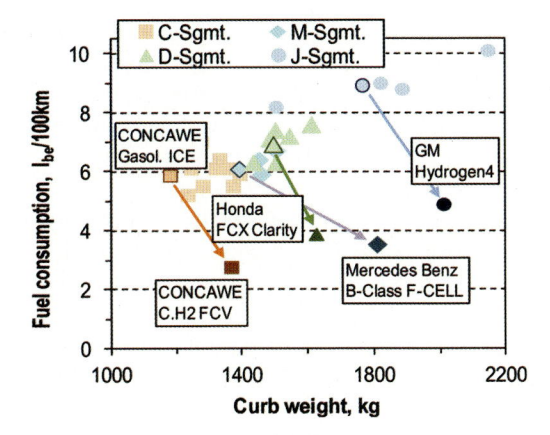

Figure 12.6 Energy required by fuel cell vehicle prototypes and present-day gasoline vehicles.

Methanation is currently being discussed as an energy storage option as it would allow existing technologies and methods of transport to be used. However, it should be noted that methanation would simply shift CO_2 emissions from coal-fired power plants to more flexible gas power plants and would not facilitate CO_2 mitigation at the levels required. Furthermore, it would also involve considerable technical effort. It is therefore more attractive both ecologically and economically to use hydrogen to replace gasoline and diesel in the long run in the transportation sector.

The change-over from combustion engines to fuel cells has impacts on the fuel consumption and weight of vehicles in particular. Figure 12.6 compares the fuel consumption and the curb weight in different vehicle segments as of today. C represents medium cars, D large cars, M vans and multipurpose cars, and J encompasses sport utility cars and off-road vehicles. In all segments, the fuel consumption decreases while curb weight increases. Overall, the consumption is approximately halved while the weight increases by around 10–20%.

12.8
Procedure

The calculations are based on quarter-hour data provided by transmission grid operators for wind and PV feed-in and also for vertical grid load. The wind profiles for 2010 are scaled according to the specifications described earlier; annual full-load hours are increased using an amplification function that mainly affects the lower power range. The PV profiles from 2010 predicted by 50 Hz are taken for those periods for which data have not yet been published by the individual grid operators, and all profiles are scaled to ensure that the exact measured annual sum is obtained for each grid operator. They are then linearly scaled as a function of date in order to account for expansion during the year before finally being scaled to the installed capacity as per the scenario. An efficiency of 70%$_{LHV}$ is assumed for electrolyzers and

1000 annual full-load hours are taken as the minimum. For gas power plants, current nominal load efficiency averaged over various manufacturers – 58.5% (combined cycle) and 36.5% (gas turbines) – is reduced by 15% due to dynamic operation. This means that power plants are generally assumed to operate at 85% of their nominal load efficiency. The share of each power plant is discussed in the following.

Using the current status of existing technologies as a basis, a specific fuel consumption of 3.3 l of diesel equivalent or 1 kg of hydrogen per 100 km was assumed for fuel cell vehicles [20]. An annual driving distance of 14 900 km was assumed. In comparison, the average annual value today is around 11 400 km. Expected reductions in fuel consumption would increase the number of vehicles that could be fueled by the quantity of hydrogen determined by the scenario. However, as the fleet of vehicles increases over time, the annual mileage decreases slightly. The GermanHy study, for example, predicts a total of 52.1 million passenger cars in 2050 [21], and the Statistisches Bundesamt [22] indicates 41.7 million in 2010. According to Shell passenger car scenarios, the annual distance averaged over all drive types will decrease from around 12 200 km in 2012 to around 11 900 km in 2030 [23]. Another source [21] refers to an annual distance of 11 400 km for passenger cars in 2012. To a first approximation, it is assumed that the mentioned effects will cancel each other out. The same assumption applies to limiting the number of wind turbines to the current level and increasing the mean wind turbine capacity to 7.5 MW, which may not be achieved everywhere. The grid losses on a transmission level are not part of the vertical grid load, and to date they have not been detailed separately in official statistics. If gas power plants are built in close proximity to consumers, these grid losses will only become highly relevant if more renewables are fed into the grid. The losses can therefore be balanced to a large extent by expanding renewables slightly, as they amount to only a few percentage points.

12.9
Results of the Scenario

This energy system can cover the grid load totaling 488 TWh (2010) and supply the transportation sector with 5.4 million tonnes of hydrogen. The grid load will be covered as follows: 75% wind and PV, 10% other renewables, and 15% natural gas (Figure 12.7). Hydrogen will supply the majority of vehicles in the road transportation sector. Based on the assumptions outlined in the GermanHy study for 2050 [21], these vehicles can be broken down into 28 million passenger cars, 2 million light commercial vehicles, and 47 000 buses, with respective shares of 68, 62, and 55% of the relevant German sectors in 2011. The natural gas volume used today is sufficient to cover the residual load.

Substituting oil-based fuels in the transportation sector with the given quantities of hydrogen reduced the total CO_2 emissions in 2009 by almost 9%. In the electricity sector, emissions were reduced by 27%. The proportion of CO_2 emissions produced by generating electricity for public supply decreased to 5.4% of the remaining total.

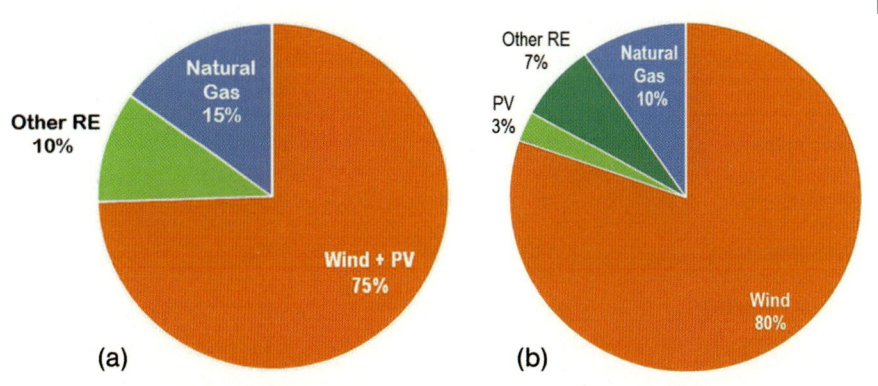

Figure 12.7 Share of energy carriers in electricity generation according to the scenario.
(a) Vertical grid load (488 TWh$_{el}$)
(b) total generation, including electricity for electrolysis (745 TWh$_{el}$).

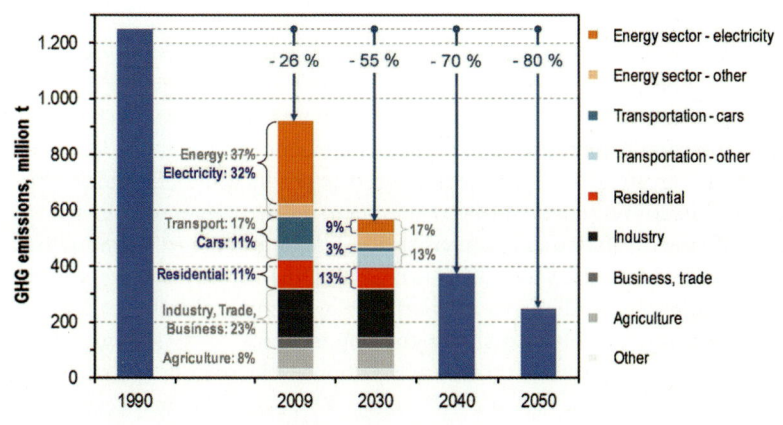

Figure 12.8 Contributions to reducing CO_2 emissions.

Taking account of the total reduction of 26.5% achieved between 1990 and 2009, 697 million tonnes of CO_2 or 55% could therefore be saved compared with 1990 (Figure 12.8). Emissions would then amount to 567 million tonnes of CO_2 equivalent. The emission targets for 2030 can therefore be met with the proposed measures. Further reductions are technologically possible but their feasibility and economic impact must be investigated first.

12.10
Fuel Cell Vehicles

Annual passenger car emissions of more than 100 million tonnes of CO_2 are to be reduced to a minimum in future by the use of fuel cells and hydrogen. Ideally, no more GHGs will be emitted by passenger cars. However, in order to make this a reality, technical requirements and customer needs must both be considered.

Table 12.5 Technical data for Mercedes-Benz B-Class with an electric motor and fuel cell [27].

Criterion	Units	Value
Rated output	kW (hp)	100 (136)
Rated torque	Nm	290
Maximum speed	km h^{-1}	170
Fuel consumption	l$_{diesel\ equivalent}$ per 100 km	3.3
CO$_2$ emissions	g km^{-1}	0
Range	km	385
Battery capacity	kWh; kW	1.4; 35
Cold-start capability	°C	−25

Although it is also conceivable that the mobility patterns of people will change in the future, if a system involves severe limitations it will not prevail in the long run. It is therefore in the interests of car manufacturers to produce fuel cell vehicles that are comparable to conventional vehicles. This includes the range, refueling time, dynamics, cold-start capability, and purchase price. The state of the art is exemplified in Table 12.5 by the Mercedes-Benz B-Class F-CELL. Other manufacturers, such as Honda, General Motors, and Toyota are also developing fuel cell vehicles that are technically comparable. Table 12.5 shows that the technical specifications of the vehicles are already comparable to those of conventional vehicles. There is still room for improvement in terms of cost and service life. The cost of the fuel cell system is to decrease in the long term from \$ 49 kW^{-1} to \$ 30 kW^{-1} and the service life is to be increased from 2500 h to 5000 h [25]. This corresponds to an engine life of around 200 000–250 000 km. The market introduction of these fuel cell vehicles is planned for 2014 [26].

12.11
Hydrogen Pipelines and Storage

In addition to cars and hydrogen production facilities, a transmission and distribution infrastructure is also required to supply hydrogen in a safe, energy-efficient, and cost-effective manner. Hydrogen has a very high gravimetric energy density (see Table 12.3) but a very low density, and must therefore be elevated to high pressure to ensure a sufficient range. Transporting it in pipelines similarly to natural gas is therefore an ideal solution. It is too expensive to transport large quantities by truck, and liquefaction requires almost one-third of the energy stored in the hydrogen, making it just as expensive.

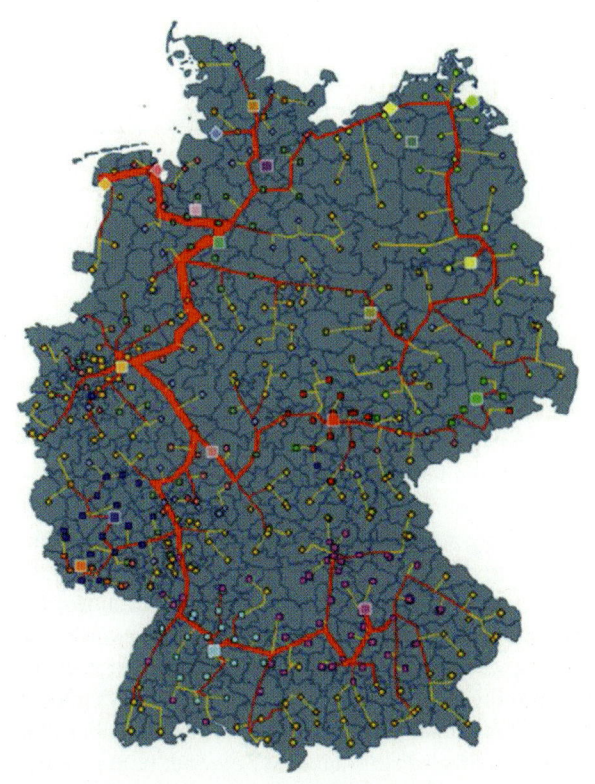

Figure 12.9 Hydrogen pipeline system for Germany [28].

A pipeline system (see Figure 12.9) for nationwide supply in Germany would necessitate some 12 000 km in the transmission grid and 31 000–47 000 km in the distribution grid [28]. A clear separation is made as this is easier to implement technically and offers several advantages in terms of grid expansion [29]. It is assumed that 9800 existing gas stations will make the changeover to hydrogen or at least expand to include it. The total investment volume for the transmission and distribution pipeline systems will be in the region of €6–7 billion and €13–19 billion, respectively, including compressors [28].

In the scenario, 5.4 million tonnes of hydrogen are produced annually. The installed electrolyzer capacity is 84 GW. If constant consumption is assumed, this results in a storage capacity of 800 000 t, which corresponds to a storage volume of 9 billion m^3 and a chemically bound energy of 27 TWh$_{LHV}$. Approximately 90 TWh$_{LHV}$ are required for a 60 day reserve. Pumped-storage hydroelectricity in Germany currently accounts for a total storage capacity of 0.04 TWh$_{el}$. If natural gas storage is also taken into account, the number of salt domes must be doubled. Other types of storage such as aquifers or porous reservoirs are excluded because they are not as impervious and they contaminate the hydrogen to a greater extent. In old oil and gas reservoirs, for example, hydrogen sulfide is formed, which is highly corrosive and must therefore be removed before hydrogen can be transported.

12.12
Cost Estimate

Implementing a hydrogen infrastructure involves several components. The costs are detailed in Table 12.6. The biggest factor is producing hydrogen using electrolysis. Filling stations account for ~€20 billion and the generation of peak electricity ~€24 billion. Between €19 billion and €25 billion will be required for the grid. As a comparison, the German natural gas grid incurred investment costs of €37 billion between 1995 and 2010 [30]. Another €5–15 billion will be required for salt dome storage, depending on whether just seasonal average or a 60 day reserve is assumed. An extra in price for fuel cell vehicles at the level of a diesel hybrid car is assumed, which might be reduced over time [25]. The investment costs for wind turbines were not considered since future feed-in tariffs are already established and a feed-in tariff of €0.06 kWh^{-1} was assumed in the study.

In addition to the use of hydrogen as an energy carrier for fuel cell vehicles, other applications are also conceivable. For example, hydrogen could be fed directly into the gas grid or as methane after methanation. Both options are technologically feasible. As explained earlier, the reduction in CO_2 emissions would be much smaller, making these less attractive ecologically. A monetary assessment shows that replacing gasoline and diesel with hydrogen rather than with natural gas would also make economic sense. If hydrogen were to be used in the transportation sector, fuel consumption would be halved and there would be a difference of roughly €0.63 l^{-1} gasoline between the direct cost of hydrogen (€0.77 l^{-1} gasoline) and the allowable fuel cost (€1.40 l^{-1} gasoline). This would also allow taxes to be levied. In the natural gas grid or in the case of methanation, in contrast, hydrogen would count as an additional cost driver. There would be a markup of €0.41–0.50 l^{-1} gasoline (Table 12.7).

The price of natural gas is too low for a positive difference under the given conditions. The reason for this is the low market value of an energy carrier that is primarily used for heating. In other words, electrolytic hydrogen produced from more expensive renewable energy should not be used to replace comparably cheap natural gas. Instead of using such methane for heating, other options, such as improved insulations, are ecologically and economically more attractive.

Even though subsidies for the market introduction of new technologies in the energy sector are necessary, there must be a clear perspective for every technology to be economically viable after mass market penetration. This is not the case for hydrogen feed-in to the gas grid or even worse for methanation with subsequent feed-in to the gas grid. Table 12.8 shows the subsidy trap for these two technologies, amounting to about €8 billion and €13 billion per annum, respectively. On the other hand, using hydrogen in the transportation sector is a viable option since oil as a much more valuable fuel than natural gas is substituted; cf. Table 12.8. Based on the assumption that 5.4 million tonnes of hydrogen will be needed per year in 2050 to supply 40 million passenger cars with fuel.

Direct feed-in requires annual subsidies of €8.27 billion compared with current costs of €14.1 billion per year [32], posing an additional burden comparable to today's feed-in tariff subsidies.

Table 12.6 Components and costs for hydrogen supply.

Component	Assumptions	Cost (€ billion)
Water electrolyzers	84 GW @ €500 kW^{-1}	42
Pipeline system	43 000–59 000 km	19–25
Dome storage	Seasonal compensation 60 day reserve	5 15
Filling stations (9800)	New stations: €2 million per filling station Retrofitting: €1 million per filling station	20
Peak electricity generation systems (gas turbines, combined cycle)	Total 42 GW (both systems)	24
Total cost of infrastructure		110–126
Fuel cell vehicles	28 million vehicles @ €5000 per vehicle	140
Total costs		250–266

Table 12.7 Comparison of revenues for hydrogen in different areas of application.

Assumptions	Calculation	Cost differential
Direct energy costs for hydrogen production with wind power		
Electricity costs: €0.06 kWh$_{el}^{-1}$ Electrolysis efficiency: 70% 1 l gasoline = 9 kWh [31]	€0.06 kWh$_{el}^{-1}$ / 0.7 = €0.086 kWh^{-1} H$_2$ ~€0.77 l^{-1} gasoline	–
Revenue for fuel in the road transportation sector		
Production costs: €0.70 l^{-1} gasoline Efficiency improvement of a factor of two (fuel cells/comb. engine)	€0.70 l^{-1} gasoline × 2 = €1.40 l^{-1} gasoline *Surtax of ~100% possible*	€0.63 l^{-1} gasoline
Revenue for direct feed-in into the natural gas grid		
Purchase price €0.04 kWh^{-1}	€0.04 kWh × 9 kWh l^{-1} gasoline = €0.36 l^{-1} gasoline *In addition, tax of €0.18 l^{-1} gasoline must be levied*	–€0.41 l^{-1} gasoline
Revenue for replacing natural gas with methanation		
Incentive: €0.04 kWh^{-1} Efficiency: 75%	€0.36 l^{-1} gasoline × 0.75 = €0.27 l^{-1} gasoline	–€0.50 l^{-1} gasoline

Table 12.8 Economic comparison of different possible uses of hydrogen.

Amount of hydrogen	Method	Assumptions	Economic impact
5.4 million t a^{-1}	Direct feed-in into natural gas grid	Costs: €0.086 kWh^{-1} Avails: €0.04 kWh^{-1}	−€8.27 billion
	Methanation and natural gas grid	Costs: €8.6 kWh^{-1}/0.75 = €11.47 kWh^{-1} Avails: €0.04 kWh^{-1}	−€13.4 billion
	Use in transportation sector	Costs: €8.6 kWh^{-1} Benefit: 40 million passenger cars; 12 000 km a^{-1}; 6 l per 100 km; € 100 per barrel	€2.5 billion

In addition, further costs apply for a pipeline upgrade in case of higher concentrations of hydrogen. A study of transmission pipeline operators concluded that a share of only 10% of hydrogen in the natural gas grid would demand modifications costing €3.6 billion. It is assumed that the natural gas grid will no longer be able to accommodate hydrogen from as early as 2022 onwards owing to material problems [33]. The option of direct feed-in might therefore not even be a long-term solution.

Methanation removes all material problems by converting H_2 with CO_2 into CH_4. However, this process involves conversion losses. The thermodynamic efficiency is 83% and the technical efficiency, depending on the size and complexity of the plant, is ~70–75%. In the literature, an efficiency of ~51–65% can be found for the energy chain for the production of methane from electricity [34].

Finally, hydrogen can be used as fuel in the transportation sector, and thus reduce oil imports; 97% of the oil in Germany is imported [35]. According to our assumptions in the study in 2050, 40 million passenger cars will consume ~5.4 million tonnes of hydrogen every year. If we assume that these vehicles are supplied with fossil fuels, travel 12 000 km in a year and consume 6 l of fuel per 100 km, then imports worth €18 billion per year would be necessary given an oil price of €100 per barrel. Against this, the costs of hydrogen production are €15.5 billion. This represents savings of €2.5 billion for the national economy. Furthermore, these measures would also reduce if not avoid geostrategic dependencies.

12.13
Discussion of Results

The proposed scenario meets the target set by the German federal government to reduce CO_2 emissions by 55% compared with 1990 levels. Only measures affecting the power production, the transportation sector, and less natural gas consumption in households for heating were considered. The reduction scenario meets the 2030 timeline. Because of the major changes in the energy sector and the introduction of two new technologies, namely electrolysis and fuel cells, it is not likely that the timeline for the 55% scenario of 2030 could be kept. Hence the study is to be considered a building block for an 80% CO_2 reduction scenario with a timeline of 2050.

Important elements include the massive expansion of wind energy and the storage of excess energy in the form of hydrogen, which can then be transferred to the transportation sector and used in high-efficiency fuel cell drives. Such a use of hydrogen has the greatest potential for reducing CO_2 emissions compared with all options of energy recirculation in the grid, and it also yields the highest avails among mass-market applications.

In addition to 22 GW in the form of existing gas power plants, 42 GW of new capacity are also required. Beyond 700 full-load hours, combined cycle power plants will be used that produce 68 TWh. Simple gas turbines, in contrast, produce only 5.5 TWh. Furthermore, the set cap of 1000 annual full-load hours necessitates an installed electrolyzer capacity of 84 GW, 9 billion Nm^3 gas storage capacity, 9000 filling stations, and 43 000–59 000 km of pipelines if all filling stations are to be connected by pipeline, with the higher value including a detour factor of 1.5 for local distribution grids.

12.14
Conclusion

Comparing annual variations in grid load and electricity supply after a considerable expansion of wind power onshore and offshore shows the following:

- The demand for electricity in Germany can be relatively easily covered when nuclear power, coal, and mineral oil are no longer used without increasing natural gas imports.
- For economic reasons, the different sectors of power production, transportation, and residential heating are to be considered together.
- Hydrogen is the most suitable large-scale storage option, because
  – other options such as pumped storage and batteries do not have sufficient potential capacities and
  – methanation is not feasible economically.
- Renewable energies then account for ~90% of electricity generation, where 34% of the electricity generated is excess energy for electrolysis.
- Excess wind power must be put to good economic use, since the amount is too large to be wasted. In this scenario, 257 TWh of 745 TWh turned out to be excess energy, translating into 34% off all of the electricity produced.
- This excess electricity generation is sufficient to supply 28 million fuel cell vehicles with hydrogen. Based on typical consumption data from 2010, the use of hydrogen in the transportation sector will cut CO_2 emissions by 81 million tonnes, representing a 6.5% decrease in total emissions compared with 1990.

- In the electricity sector, a reduction of 20% is achieved compared with 1990, and in the domestic heating sector 2.2%. Combined with the 26.5% already achieved in 2009, this results in a reduction of 55%.
- Investments are manageable. The expected costs for the complete hydrogen infrastructure, that is, including pipelines, electrolyzers, storage systems, and filling stations, are around €100 billion. Around €37 billion were spent maintaining, servicing, and installing new pipelines in the natural gas grid between 1995 and 2010.
- The study was made with static assumptions and is to be considered as a first approach. It does not include dynamic effects and spatial resolution. Only investment costs are considered. Follow-up studies will also assess electricity generation costs.
- Further measures for reducing emissions beyond the 55% scenario may include:
  – the use of biofuels in applications where fuel cells and batteries are unsuitable
  – energy-saving measures
  – a combination of different measures, such as smart grids and heat pumps
  – an increased focus on wind energy instead of solar energy.

References

1 German Federal Government (2011) *Framework Paper. Resolutions of 6 June 2011*, http://www.bmu.de/energiewende/downloads/doc/47467.php (last accessed 16 January 2013).

2 Danish Government (2011) *Danish Government's Program*, http://www.stm.dk/publikationer/Et_Danmark_der_staar_sammen_11/Regeringsgrundlag_okt_2011.pdf (last accessed 16 January 2011).

3 UBA (2011) *National Trend Tables for the German Atmospheric Emission Reporting – 1990–2009 (Final version: 2011.01.17)*, Umweltbundesamt (Federal Environmental Agency), Dessau-Rosslau.

4 Intergovernmental Panel on Climate Change (2007) *4th Assessment Report, Climate Change 2007: Working Group I: The Physical Science Basis, Technical Summary*, IPCC, Geneva, pp. 33–34.

5 BMU (2009) *Renewable Energy Sources in Figures – National and International Development*, Bundesministerium für Umwelt, Naturschutz und Reaktorsicherheit (Federal Ministry for the Environment, Nature Conservation and Nuclear Safety), Berlin.

6 BWE (2011) *Studie zum Potenzial der Windenergienutzung an Land – Kurzfassung*, Bundesverband Windenergie, Berlin.

7 JEC – Joint Research Centre–EUCAR–CONCAWE Collaboration (2011) *Well-to-Wheels Analysis of Future Automotive Fuels and Powertrains in the European Context – Tank-to-Wheels Report*, European Commission, Joint Research Centre, Brussels.

8 von Helmolt, R. (2009) Brennstoffzellen- oder Batteriefahrzeug? Ähnliche Antriebe, unterschiedliche Infrastruktur, presented at f-cell 2009, Stuttgart.

9 Jossen, A. and Weydanz, W. (2006) *Moderne Akkumulatoren richtig einsetzen*. Reichardt Verlag, Untermeitingen, ISBN 3-939359-11-4.

10 German Wind Energy Association (BWE) (2012) *Statistics*, http://www.windenergie.de/en/infocenter/statistics (last accessed 27 July 2012).

11 Fraunhofer Institute for Wind Energy and Energy System Technology (IWES) (2010) *Windmonitor*, http://windmonitor.iwes.fraunhofer.de/windwebdad/www_reisi_page_new.show_page?page_nr=48&lang=en (last accessed 16 January 2013).

12. Neddermann, B. (2009) *Status der Windenergienutzung in Deutschland – Stand 31.12.2009*, DEWI GmbH, http://www.dewi.de/dewi/fileadmin/pdf/publications/Statistics%20Pressemitteilungen/31.12.09/Anhang_Folien_2009.pdf (last accessed 5 November 2010).

13. EEG/KWK-G (2009) *Information Platform for German Transmission System Operators, 2009: EEG Medium-Term Forecast: Developments from 2000 to 2015*, http://www.eeg-kwk.net/cps/rde/xbcr/eeg_kwk/2009-05-11_EEG-Mittelfristprognose-bis-2015_mitUeNB-Logos_Fussnote-korr.pdf; (last accessed 5 November 2010).

14. Deutsche Energie-Agentur GmbH (dena) (2005) *Energiewirtschaftliche Planung für die Netzintegration von Windenergie in Deutschland an Land und Offshore bis zum Jahr 2020 (Dena Grid Study)*, http://www.offshore-wind.de/page/fileadmin/offshore/documents/dena_Netzstudie/dena-Netzstudie_I_Haupttext.pdf (last accessed 4 November 2010).

15. Nitsch, J. (2008) *Leitstudie 2008 – Weiterentwicklung der "Ausbaustrategie Erneuerbare Energien" vor dem Hintergrund der aktuellen Klimaschutzziele Deutschlands und Europas*, Federal Ministry for the Environment, Nature Conservation and Nuclear Safety (BMU), Berlin.

16. Hundt, M., Barth, M., Sun, N., Wissel, S., and Voss, A. (2009) *Verträglichkeit von erneuerbaren Energien und Kernenergie im Erzeugungsportfolio*, Study commissioned by E.ON Energie AG, Munich, Institute for Energy Economics, and the Rational Use of Energy (IER), University of Stuttgart, Stuttgart.

17. Fraunhofer Institute for Wind Energy and Energy System Technology (IWES) (2011) *Studie zum Potenzial der Windenergienutzung an Land*, IWES, Kassel.

18. Krewitt, W. and Nitsch, J. (2003) The potential for electricity generation from on-shore wind energy under the constraints of nature conservation: a case study for two regions in Germany. *Renewable Energy*, 28, 1645–1655.

19. BMU (2011) *Renewable Energy Sources in Figures – National and International Development*, Bundesministerium für Umwelt, Naturschutz und Reaktorsicherheit (Federal Ministry for the Environment, Nature Conservation and Nuclear Safety), Beerlin.

20. Daimler (2012) *Specifications for Mercedes-Benz B-Class F-CELL*, http://www.daimler.com/dccom/0-5-1228969-49-1401156-1-0-0-1401206-0-0-135-0-0-0-0-0-0-0-0.html (last accessed 3 February 2012).

21. BMVBS (2009) *GermanHy – Woher kommt der Wasserstoff in Deutschland bis 2050?* A study on behalf of the Federal Ministry of Transport, Building and Urban Development (Bundesministerium für Verkehr, Bau und Stadtentwicklung), Berlin.

22. DESTATIS Statistisches Bundesamt (2012) *Number of passenger cars in Germany*, https://www.destatis.de/DE/ZahlenFakten/Wirtschaftsbereiche/TransportVerkehr/Unternehmen InfrastrukturFahrzeugbestand/Tabellen/Fahrzeugbestand.html (last accessed on 3 February 2012).

23. Shell Deutschland (2012) *Shell Passenger Car Scenarios up to 2030*, http://s07.static-shell.com/content/dam/shell/static/deu/downloads/publications-2009shellmobilityscenariossummaryde.pdf (last accessed 3 February 2012).

24. DESTATIS. The annual distance is taken from reference 21.

25. Papageorgopoulos, D. (2012) Fuel cells – session introduction, presented at the 2012 Annual Merit Review and Peer Evaluation Meeting, Arlington, VA, 15 May.

26. Zetsche, D. (2011) Interview, *Frankfurter Allgemeine Zeitung*, 10 September.

27. Daimler (2012) Downloads, http://www.daimler.com/Projects/c2c/channel/documents/1837055_Daimler_Elektrifizierung_des_Antriebs_2012_de.pdf, http://media.daimler.com/dcmedia/ (last accessed 16 January 2013).

28. Baufume, S., Grube, T., Grüber, F., Hake, J.-F., Krieg, D., Linßen, J., Stolten, D., and Weber, M. (2012) *GIS-based Analysis of Hydrogen Pipeline Infrastructure for Different Supply and*

Demand Options, presented at the 12th Symposium on Energy Innovation, Graz, 15–17 March.

29 Krieg, D. (2012) *Konzept und Kosten eines Pipelinesystems zur Versorgung des deutschen Strassenverkehrs mit Wasserstoff*, PhD thesis, RWTH Aachen University, Jülich.

30 BDEW (2012) *Energy Data – Gas Networks in Germany*, Bundesverband der Energie- und Wasserwirtschaft (German Association of Energy and Water Industries), Berlin.

31 JEC – Joint Research Centre–EUCAR–CONCAWE Collaboration (2011) *Well-to-Wheels Analysis of Future Automotive Fuels and Powertrains, WtT-Appendix 1, Version 3c*, European Commission, Joint Research Centre, Brussels.

32 BDEW (2011/2012) *Erneuerbare Energien und das EEG: Zahlen, Fakten, Grafiken*, Bundesverband der Energie- und Wasserwirtschaft (German Association of Energy and Water Industries), Berlin, 15 December 2011, revised edition 23 January 2012.

33 Netzentwicklungsplan Gas (2012) *Entwurf der deutschen Fernleitungsnetzbetreiber, Netzentwicklungsplan Gas 2012 (Study of Transmission Grid Operators)*, http://www.netzentwicklungsplan-gas.de/files/netzentwicklungsplan_gas_2012.pdf (last accessed 16 January 2013).

34 Fraunhofer Institute for Wind Energy and Energy System Technology (IWES) (2011) *Energiewirtschaftliche und ökologische Bewertung eines Windgas-Angebots*, IWES, Kassel.

35 Landesamt für Bergbau, Energie und Geologie des Landes Niedersachsen (LBEG) (2011) *Erdöl und Erdgas in der Bundesrepublik Deutschland 2010*, LBEG, Hannover.

13
Pre-Investigation of Hydrogen Technologies at Large Scales for Electric Grid Load Balancing

Fernando Gutiérrez-Martín

13.1
Introduction

Current energy systems are not sustainable owing to resource limitations, environmental impacts, and management aspects such as the low use of technologies or transporting electricity. Therefore, efficient utilization of existing infrastructures should first be considered to address the increasing power demands and external costs; for example, electric grids are designed to meet the highest expected demand, which occurs only for short intervals while for the remaining time, in particular during off-peak hours at night, the power plants are underutilized and can deliver substantial amounts of energy to other sectors, with great advantages for the reliability of the system at the same time [1].

On the other hand, the potential of renewable energies is higher than the global demands in many countries and it is usually well distributed throughout the territory. However, the benefits of these energy resources, such as solar or wind, are overshadowed by their intermittent nature, the incompatibility with base load technologies, or the competition with ordinary thermal utilities, which can result in low utilization factors (e.g., currently the power capacity in Spain exceeds 100 GW, while the peak consumption reaches about 45 GW) [2].

This makes the management of the grid a challenging task, which requires electricity storage or load-leveling options to improve the efficiency based on more holistic energy approaches. In this context, the production of hydrogen by electrolysis using the power grid mix has been considered a promising option in several countries that have sufficient energy surpluses as an alternative to other operational procedures or exporting the electricity [3–5].

In fact, the "hydrogen economy" has been defined as a future economy in which hydrogen is adopted for mobile applications and electric grid load balancing: it acts as an storable "energy carrier" that can be can be used as a "zero emission" fuel for other sectors, such as transport, and improve the energy quality by smoothing the

Transition to Renewable Energy Systems, 1st Edition. Edited by Detlef Stolten and Viktor Scherer.
© 2013 Wiley-VCH Verlag GmbH & Co. KGaA. Published 2013 by Wiley-VCH Verlag GmbH & Co. KGaA.

power output, providing reserve service, as well as avoiding plant curtailments or costly grid upgrading when electricity production is high.

The need and potential for integrating energy storage–conversion in power systems with high wind penetration is widely recognized within electric utilities; electrolyzers, being both flexible loads and conversion devices, can increase the flexibility of the system and be an important measure to allow the integration of additional renewable energy [6].

This study is based on electricity-powered hydrogen production, which is a proven technology, thus providing a realistic analysis of the transition to a hydrogen economy. It requires a holistic approach to the generating systems, the demand profiles, and their variations in space–time, together with an evaluation of the electrolysis technologies that are applicable for managing the grid loads at large scales; the strategy is to control the power inputs, taking into account the efficiency and dynamics of electrolyzers, to demonstrate the benefits that could be achieved by using surplus electricity at a low price, as well as the leveling effects on the energy balances of the plants and their economic, technological, and environmental consequences as a whole.

Barriers for implementation can arise from political willingness, economic priorities, or technical limitations to put into practice a task of this breadth, but whether this option is more feasible as an alternative to existing procedures, when sufficient power surplus or grid constraints exist, is the prime question that we try to answer in this chapter.

13.2
Electrolytic Hydrogen

There is wide agreement that water can be considered the most interesting feedstock to obtain sustainable hydrogen; the main reason is that renewable energy sources can then be easily integrated in the process. Electrolysis represents the most important path to obtain hydrogen from water; it is a mature technology [7] based on dissociation of water molecules by applying direct current electrical energy. The hydrogen obtained has high purity, which is a great advantage since it is suitable for being directly used in low-temperature fuel cells (see Figure 13.1a).

The technology and sizes of commercially available electrolyzers vary greatly. In this chapter, the focus is on technology that can be useful to the power grid; consequently, the review covers alkaline water electrolyzers (AWEs) because they are the only type available for large-scale hydrogen production. Advanced alkaline electrolyzers are at present sufficiently developed to start the production of hydrogen at significant rates; however, the massive production required by the hydrogen economy will need electrolysis units with capacities higher than the existing ones.

Each type of electrolyzer has its own advantages and drawbacks: bipolar cells have a more complex design due to the combination in series, but their compact configuration allows them to reduce the volume and external connections and achieve greater current densities and gas pressures; in addition, the electrolyte channels offer many paths that make the flow of internal (parasitic) currents easier. Most manufacturers have developed their electrolyzers from bipolar modules since they are considered more suitable than monopolar modules [8].

Nevertheless, one of the greatest drawbacks to using hydrogen for electric grid load balancing is that electrolyzers, fuel cells, and engines using current technology have relatively low efficiencies, so the total system has low round-trip efficiency. However, allowing for coproduction of hydrogen and power, utilities could improve their production, storage, and use, and optimize the system based on economic and reliability factors; therefore, what we are trying to ascertain is whether systems that are optimized for hydrogen production and power generation are competitive, depending on the prices at which electricity and hydrogen are bought and/or sold.

13.2.1
Electrolyzer Performance

Several factors influence the performance of the electrolytic processes; these usually involve tradeoffs between the operating voltages, the rates of hydrogen production, and capital costs.

The analysis show the importance of maximizing the "capacity factors" of hydrogen systems and minimizing the "electricity costs" from the power systems:

- The capacity factor has a significant effect due to the investment costs, even if scaled at larger sizes; this factor will have a lesser effect in the future because of the lower capital cost (i.e., it costs more to have the present systems idle than future systems).
- Electricity prices have larger effects in the present systems because they are less efficient; finally, increasing the efficiency reduces the overall cost of hydrogen more at larger installations (in relative percentage terms).

Although electrolyzers have been used for a long time, their future applications will probably often have to deal with variable energy sources to generate clean hydrogen and contribute to the operation of the electric grid. In this respect, there is still a long way to go in some respects.

In recent years, significant advances have been achieved regarding alkaline electrolysis, mainly in two directions: on the one hand, the electrolyzer efficiency has been improved with the aim of reducing the operating costs associated with electricity consumption; on the other, the operating current densities have been increased in order to reduce the investment costs (it should be noted that for large units, these are almost proportional to the cell's surface area).

Several studies on AWEs have been reported, although there is still a paucity of publications related to the performance of electrolyzers; hence more data are required to model the electrolyzers based on their physical behavior, to provide a good basis for the semi-empirical parameters for calculating the current–voltage curve, thermal model, and so on [9].

To evaluate water electrolysis in this study, a concise model was selected to describe the current–voltage characteristics of an electrolytic cell by means of incorporating thermodynamic, kinetic, and electrical resistance effects; these are expressed quantitatively with three main parameters: the thermodynamic parameter, which is the water dissociation potential, the kinetic parameter, which reflects the overall electrochemical effect of both electrodes, and the ohmic parameter, which reflects the total resistance of the cell.

The model assumes that power consumed in the water electrolysis process is proportional to the square of the potential difference between the cell and the water dissociation potentials, also considering the cell resistance, in such a form that the relationship between the voltage and current density can be expressed as

$$V = \frac{J + 2K(Jr + E_0) + (J^2 + 4KE_0 J)^{\frac{1}{2}}}{2K} \tag{13.1}$$

where the value of E_0 decides the starting point of the electrolysis curve which represents the water dissociation potential, the value of the resistance r is relevant to the slope of the curve, and the value of the kinetic parameter K influences both the slope and the curve shape.

Using Equation 13.1, different electrolyzers with various operating conditions can be compared with each other, avoiding the difficulties inherent in Butler–Volmer and Tafel equations, while the results are found to agree well with experimental data and previously published work [10]. Figure 13.1b shows the curve for a typical AWE, where a nearly lineal relationship at elevated current densities can be observed.

On the other hand, the Faraday efficiency represents the ratio of hydrogen production and its theoretical value due to the parasitic losses, which are important at

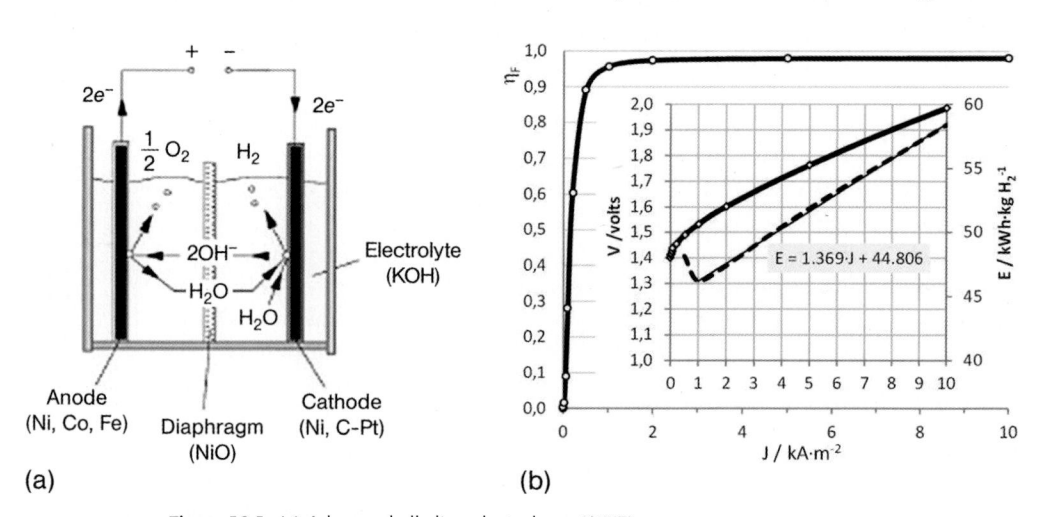

Figure 13.1 (a) Advanced alkaline electrolyzer (AWE); (b) current–voltage and energy efficiency curves with $E_0 = 1.40$ V, $K = 120$ mΩ^{-1} m^{-2}, $r = 0.020$ mΩ m^2, $f_1 = 0.98$ and $f_2 = 0.025$ kA^2 m^{-4} (at 80 °C), and $\eta_{BOP} = 0.93$.

13.2 Electrolytic Hydrogen

elevated temperatures and low current densities [11, 12]; this phenomenon can be described by the following equation and is illustrated in Figure 13.1:

$$\eta_F = \frac{f_1 J^2}{f_2 + J^2} \qquad (13.2)$$

Therefore, it should be noted that electrolyzers must be operated above a minimal intensity, in such a form that the current efficiency is practically not affected; this is especially important for AWEs that are designed to manage fluctuating input currents and can only operate down to about 10% of their rated power (idling current).

13.2.2
Hydrogen Production Cost Estimate by Water Electrolysis

The direct capital cost of the plant is one of the three significant parameters in calculating the total costs from electrolysis, the others being the electrolyzer efficiency and the cost of electricity. Process plants generally have a nonlinear relationship between the cost of the plant and its production capacity ($C = W^n$), holding up to a maximum value of W that reflects the practical size of the limiting process unit. For electrolyzers, the limiting unit is the cell stack; the area of electrodes is limited by manufacturing and fluid dynamics, and the number of cells is limited by tolerances in manufacturing and the need to avoid excessive voltages across the stack.

The largest commercial electrolyzer stacks today have a capacity on the order of 1 ton of H_2 per day; for greater capacities, several parallel cells stacks can be placed in one electrolyzer and/or several electrolyzers can be installed, with some sharing of utilities such as power electronics, controls, and other balance-of-plant components; sources in industry have confirmed a power law relationship with an exponent n between 0.6 and 0.7 for a wide range of capacities in today's market (up to about 1 ton d^{-1}) and costs increase nearly linearly beyond this point (suggesting that maximum unit size can be a target in itself for large installations) [13–15].

Looking ahead at markets for the current technologies requires methods for estimating cost reductions as the number of units increases by orders of magnitude. The consensus seems to be that a developed market will see unit costs coming down by a factor of two or more: a fair number to be used as the purchased capital cost is between €200 and €400 per kilowatt of capacity, especially for large electrolyzers which operate at increased current densities:

$$C_{EL}(\text{€}) = a\, Q_0 \left(\text{kg}_{H_2}\ \text{h}^{-1}\right)^b \times J_0 \left(\text{kA m}^{-2}\right)^{-c} \qquad (13.3)$$

Alkaline electrolyzer improvements typically target reduced capital costs by increasing pressures, reducing complexity, using new materials, increasing current densities, or by a combination of such methods; the current densities can be increased from the typical conventional values of 2 kA m^{-2} to ~10 kA m^{-2} by using new membranes and reducing the gaps between electrodes.

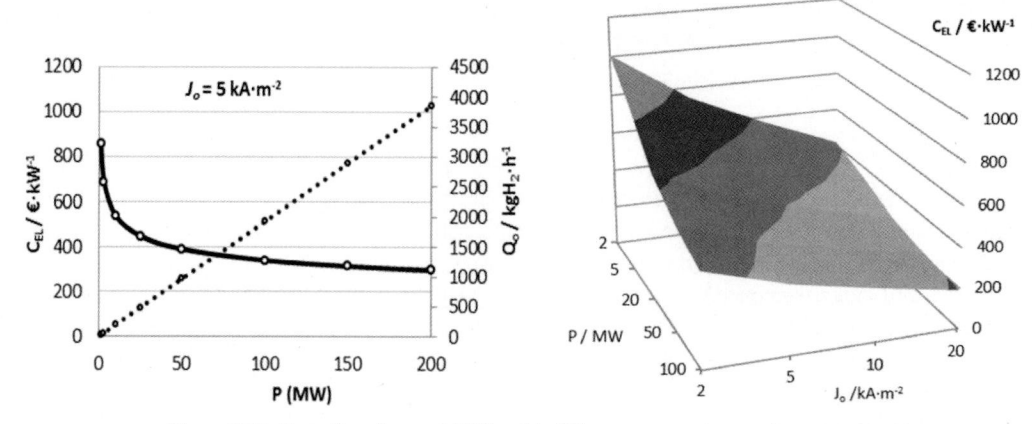

Figure 13.2 Costs for advanced AWEs with different power sizes and current densities ($a = 1.20 \times 10^5$, $b = 0.80$, $c = 0.25$).

Balance-of-plant costs (BOP)– dominated by items such as transformers, rectifiers, and controllers – comprise a significant proportion of the total installed costs; they also include water purification, the hydrogen dryer, and a gas purifier if needed (the percentage of BOP varies considerably between 34% and 86% of the total cost excluding storage and dispensing).

Efficiency improvements are more limited, as there are thermodynamic principles that cap the top efficiency of a system; once this near-ideal efficiency is reached, cost reductions could only come from other areas, although this should be balanced with efficiency improvement for installation in larger production facilities where electricity dominates the cost of hydrogen.

The energy yield of the electrolyzer ($\text{kWh kg}_{H_2}^{-1}$) can be obtained directly from the cell voltage characteristics and current efficiency:

$$E = \frac{26.8 \text{ V}}{\eta_F \, \eta_{BOP}} \tag{13.4}$$

Based on information provided by suppliers for their state-of-the art technologies, electrolyzers are now capable of producing hydrogen using less than 50 kWh kg^{-1}, representing an efficiency greater than 67% based on the heating value of hydrogen (LHV: 33.55 kWh kg^{-1}); this refers to the complete operation, including power electronics and other BOP components (the energy use outside the electrolyzer represents an additional 5–10%). The stack efficiency is higher, with cell yields as high as 74% LHV that reflect that the development works to lower the energy consumption of electrolyzers: reduced membrane thickness and compact designs reduce the ohmic losses, and better catalysts and improved hydrodynamics at electrode surfaces reduce the overvoltages on both anodes and cathodes. It is not likely, however, that cell technologies will achieve significantly greater yields, as they are at a point of diminishing returns on efficiency, and the scope for further improvement is likely to be more focused on capital cost reduction, reliability/durability and scale-up to greater capacities; all the technologies evaluated for this study operate at temperatures below 90 °C, using only electric energy to drive the process.

For electrolysis to be operated competitively for hydrogen production, it must be run in areas having low-priced electricity for the industrial sector, in addition to off-peak and renewable power; one additional approach to reduce electricity cost is the use of interruptible power. Sensitivity analysis indicates that electricity price is the most important variable, followed by energy use and then purchased capital cost; replacement and plant staffing can also have a measurable impact on the hydrogen cost. The importance of electricity price is a key reason why utilities need to be involved if the future hydrogen economy includes electrolysis.

It can be discussed whether the cost of the electrolysis plant should include some capital cost for the power lines and other electrical servicing equipment. It could be costly to run lines from the electricity to electrolysis plants unless they are closely located; these capital costs might be absorbed by the utility companies but the cost would be reflected in the electricity price.

Then, hydrogen production costs (C_H, € kg^{-1}) can be estimated using the expressions below, which include the capital, operation, and maintenance costs ($C_{CC} + C_{O\&M}$, € a^{-1}) and the inputs of electrolysis, namely electricity and water ($C_E + C_W$, € a^{-1}) (other costs such as electrolytes and resins are not significant compared with the former); Q_H is the production of hydrogen (kg a^{-1}).

$$C_H = \frac{C_{CC} + C_{O\&M} + C_E + C_W}{Q_H} \tag{13.5}$$

The capital costs are the purchased cost of electrolyzers (Equation 13.3) and BOP, by using a factor F based on the annual discount rate and the number of years to recover the investment capital; the costs concerning the plant operation and maintenance, which also include the workforce, may be calculated as a rate of the investment capital (OM):

$$C_{CC} + C_{O\&M} = (F + OM) C_{EL+BOP} \tag{13.6}$$

The annual costs for electricity and water can be calculated from the power of the electrolytic plants (P), their availability in hours per year (u), and the prices of electricity and water (C_{Ai}):

$$C_E + C_W = 8760\, P\, u\, C_{AE} + 9\, Q_H\, C_{AW} \tag{13.7}$$

Finally, the annual production of hydrogen is the power input divided by the energy yield of the electrolyzer stack (E, kWh kg$_{H_2}^{-1}$), which depends of the current density, the current–voltage characteristics, and the Faraday efficiency (as discussed in Section 13.2.1); the additional use of energy outside the electrolyzer is the BOP efficiency (η_{BOP}):

$$Q_H = \frac{8760\, P\, u}{E} \tag{13.8}$$

13.2.3
Simulation of Electrolytic Hydrogen Production

We have elaborated a worksheet with all the equations to model the electrolyzer performance and hydrogen production costs, as described above. The main inputs are the installed power, the utilization ratio and nominal current density, the parameters for calculating the efficiency, and the factors related to the investment capital, the annual depreciation, the O&M rates, and the prices of energy and water. Compression and storage can be included in the BOP terms.

Table 13.1 shows the output for an installation with a high capacity, running the electrolyzers at a moderately high current density and part utilization periods, and using low-priced electricity; the results are a large electrolysis plant with high production, a diminished energy efficiency, and reduced costs of hydrogen due to the equipment and energy (the proportions are shown).

Table 13.1 Worksheet for calculating the electrolyzer performance and hydrogen production costs.

Data inputs	Outputs
Power capacity P = 50 MW, u = 0.50 J_0 = 5 kA m^{-2}	V_0 = 1.763 V, η_F = 0.979 C_H = €4.00 kg$_{H_2}^{-1}$
Electrolyzer performance and costs: see Figure 13.1 and Figure 13.2	E = 51.9 kWh kg$_{H_2}^{-1}$ E_f(LHV), η = 64.7%
BOP cost: 60% of electrolyzer costs	C_{EL+BOP} = €31.3 million
Depreciation d = 10%, n = 20 years, O&M rate = 6%	Q_H = 4220 t$_{H_2}$ a^{-1}
Electricity and water, C_{AE} = €0.05 kWh^{-1}, C_{AW} = €10 m^{-3}	

Table 13.2 Simulation of hydrogen production costs by the composite design (k = 4, n_0 = 1, α^2 = 2) [16].

Factors and levels		Z_{j0}	ΔZ_j	Regression coefficients of codified variables, $f(X_j)$			b_0: 4.614
Z_1	P (MW)	50.0	30.0	b_1: −0.309 b_{11}: −0.271	b_{12}: 0.192	b_{13}: 0.049	b_{14}: 0.002
Z_2	u	0.50	0.30	b_2: −1.678 b_{22}: 1.309	b_{23}: 0.216	b_{24}: 0.001	
Z_3	J_0 (kA m^{-2})	5.00	3.00	b_3: −0.143 b_{33}: −0.236	b_{34}: 0.128		
Z_4	C_{AE} (€ kWh^{-1})	0.05	0.03	b_4: 1.551 b_{44}: −0.384			

13.2 Electrolytic Hydrogen

In addition, we simulated the cost of hydrogen by an orthogonal second-order composite design with four parameters (Table 13.2): the regression coefficients show very significant effects of the utilization ratios and electricity prices, with relevant interactions of P/u, J_0/u, and J_0/C_{AE}, whereas P/J_0 are much less correlated and the relations are negligible for P/C_{AE} or u/C_{AE}.

Figure 13.3 shows the production costs for AWEs with advanced operation and the economic factors used in the model, which allow the effects of the key parameters, their interactions, and the curvature of the response surfaces to be analyzed graphically.

We conclude that a reduction in the production cost of hydrogen by electrolysis to below €4 kg^{-1} can be reached by using low-priced electricity during surplus periods which are long enough, with higher power capacities as the utilization ratio decreases, and increasing the current density as the energy prices, the utilization ratios, or the power of installations become lower.

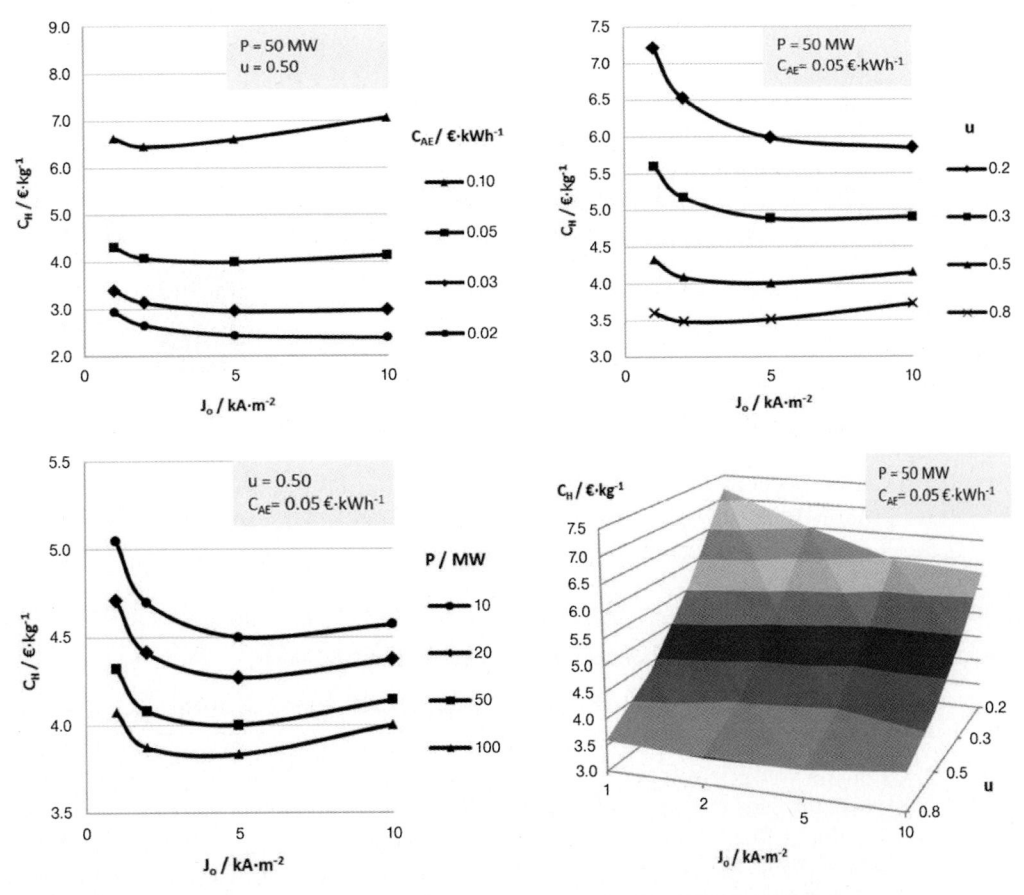

Figure 13.3 Costs of hydrogen showing the favorable effects of electricity prices and utilization ratios, the moderate effect of the plant capacity, and the optimal values for the current densities.

Summing up, high efficiency is beneficial, but economically optimized production is usually more important; the more the cell voltages are increased above E_0, the higher are the current densities and in turn the production rates.

In new advanced alkaline electrolyzers, the operational cell voltage has been reduced and the current density increased compared with the more conventional electrolyzers. Reducing the cell voltage reduces the unit cost of energy and thereby the operating costs, while increasing the current density reduces the investment costs. However, there is a conflict of interest because the ohmic resistance in the electrolyte increases with increasing current due to gas bubbling; increased current densities also lead to higher overpotential at the anodes and cathodes.

Three basic improvements are implemented in the design of advanced alkaline electrolyzers: (1) new cell configurations to reduce the surface-specific resistance despite increased current (e.g., zero-gap cells and low-resistance diaphragms), (2) higher temperatures (up to 160 °C) to increase the conductivity of the electrolyte, and (3) new catalysts to reduce the overpotentials (e.g. mixed-metal coating containing cobalt oxide at the anode and Raney nickel at the cathode).

If the efficiency is assumed to be 100%, the cost reduction is limited; this is not an argument against research on electrolyzers, but a strong indication that there is no reason to wait for more efficient electrolyzers to start business development.

The energetic efficiency on converting electricity to hydrogen is reasonably high on modern plants (80–90% based on the higher heating value, HHV). The lifetime is as long as 20 years, with a major service check every 6 years. However, the relatively high costs of equipment are a barrier for the use of low-priced electricity for hydrogen production; therefore, there will be an increasing use of electrolysis if cheap units are brought to the market, which is characterized by few plants sold at high prices for industrial use, contrary to what we will see in the future with a large number of plants for energy use.

In the past, electrolyzers with capacities up to 20 000 $Nm^3\ h^{-1}$ (100 MW electrical absorption) have been built. In the future, if the demands for large quantities of clean hydrogen increase, larger electrolyzer units could be built and in this way the costs would be strongly reduced. Developing electrolyzers with capacities in excess of say 1000 $Nm^3\ h^{-1}$ is just an engineering matter and not a feasibility question (a reasonable size could be 5000 $Nm^3\ h^{-1}$).

13.3
Operation of the Electrolyzers for Electric Grid Load Balancing

An interesting system aspect of electrolytic processes for the production of hydrogen is the possibility of power management in electric grids. Like all electrochemical energy converters, they can respond to load changes almost instantaneously; highly dynamic electrolyzers could therefore be used for hydrogen production in the case of fluctuating production from power plants; they can be operated, in combination with a hydrogen storage tank, for load management in the same way as variable electricity consumers (leading to higher utilization of power systems, since electrolyzers can be used to balance the grid).

The integration of electrolysis plants would thus add significantly to the flexibility of the power systems. Increasing renewable energy capacities tend to give larger fluctuations in electricity prices and increase problems of power overflow when they cover the complete loads. A more flexible demand will mitigate these problems and give more stable prices; thus, the integration of electrolyzers (or other flexible load, storage, and energy conversion systems) would increase the value of the power generation both as a flexible load responding to market signals, but also as a provider of ancillary services (active and reactive power control).

Large electrolyzers in the range of hundreds of megawatts will have to connect at the transmission level at very high voltages, whereas medium-scale electrolyzers (50 MW or less) may connect to the distribution system at lower voltages. This can mitigate transfer capacity problems in the system, which occur, for example, in periods with high wind generation; in this way, significant costs of transmission line upgrades are avoided. In addition, electrolyzers may offer possibilities for improving security by, for example, including them in remedial action schemes as significant loads that can be disconnected [17].

The grid electricity (AC) has to be rectified before it can be used to power the electrolyzer (DC), and the voltage has to be adapted by a transformer to the level required by the electrolyzer. If it is required that the plant can be regulated continuously from zero to full power, a device for this purpose is also necessary.

In this case, the choice of the equipment has an influence on the power control capabilities of the system. Large electrolyzers currently in operation are not designed for fast control, although they permit the voltage to be adjusted in the interval 80–100% so as to change the current from 20% to 100%; small alkaline electrolyzers are available which offer 20 s start-up time and better performance in terms of control range and regulation times (a dynamic range of 5–100% of rated capacity has been achieved and they will have response times from 5 to 100% in just a few seconds). The electrolyzer's loads are controlled by the current, which is proportional to the load. If feed water and cooling are available, 20% overload should not be a problem, but at lower efficiency. IHT (Industrie Haute Technologie) produces atmospheric and pressurized (32 bar) electrolyzer units from 3 to 760 $Nm^3\ h^{-1}$ [18]; the latter are the largest units on the market and they have the highest discharge pressure; the diameter is 1.6 m and 139 electrodes are stacked into one unit (weighing ~15 t); the largest electrolyzer with a consumption of 3.5 MW consists of four units (two half stacks in parallel), the operating voltage is 540 V, and each unit consumes 3240 A; the power density of the electrolyzer may be calculated as the maximum power consumption divided by the volume (with a length of 10 m and zero-gap structure this is: 174 $kW\ m^{-3}$); the gas produced is led to the back of the electrodes and therefore does not diminish the conductivity between the electrodes. Dynamic operation of the pressurized units, which are of a type that may be of interest for large installations, is possible: it takes few minutes for a heated depressurized electrolyzer to reach the operating pressure, after which it may in principle be controlled freely; however, the load range of the electrolyzer is stated to be 25–100%, and at lower loads problems may arise in connection with the purity of the gas. At a current density of 2 $kA\ m^{-2}$ and a cell voltage of 1.9 V, the plant consumes 4.61 $kWh\ Nm^{-3}\ H_2$ at full load; at part load, the efficiency is increased and the consumption for each Nm^3 is reduced (this means that it is better to reduce the loads on all units instead of shutting down part of the electrolyzers). The discontinuous operation may also induce some additional degradation and higher maintenance of the electrolyzers [19].

The cost of hydrogen production is calculated on the basis of the electricity prices (varying) and electrolyzer depreciation, that is, the cost for hydrogen varies with the use of the equipment (e.g., part time production at lowest energy prices).

Electricity is traded at hour-to-hour prices on the spot market and typically varies from €0 to 0.15 kWh^{-1}. Since the average price increases by the number of hours of operation per year and the depreciation decreases according to the operating hours, there will be an optimum number of hours per year; these hours are most likely to be during the night, because the lower consumption will cause energy prices to decrease. However, the curves tend to be flat at operating times up to 100%, so if it serves a purpose to increase production, the additional cost is only marginal; this might well be desired when the demand for fuel is growing in the transport sector.

Hence, if the economy converts to hydrogen from clean energy sources, the electric power industry is in a unique position to increase the efficiency and reduce carbon and pollution levels. If electrolysis is used to produce the hydrogen, utilities could multiply the amounts of electricity that they provide, and many could be provided without adding capacity; also, there are more than sufficient solar and wind resources to produce all the hydrogen needed, so utilities can provide a clean, carbon-free fuel. However, this requires a holistic approach to the current and future power systems, which must properly involve the hydrogen technologies, as depicted in following sections.

13.3.1
The Spanish Power System

For the study we used the data provided by the Spanish operator REE (Red Eléctrica de España), relative to the energy demands and structure of power generation, with a disaggregation of hours. The load curves show a deep valley (0–7 h) and daily peaks at 12 h and 21 h; the curves also vary over the year, showing relevant decreases in electricity demand during feast days and spring–autumn seasons. The generation curves exceed the hourly demands of electricity, with some surpluses (less than 5%) which are used for hydro storage or exported to neighboring countries.

The electrical system installed capacity was 106 GW in 2011 and it was running at an overall capacity factor of only 30.9%. Conventional utilities represent 66% of the total power capacity, with a coverage ratio of 1.59 over the peak hourly demand; moreover, the load duration curve shows that within the 300 h of greatest annual consumption the power demand exceeds 5 GW. This is added to the geographical asymmetries of generation and demand throughout the country, which means costly grid reinforcements for transporting the electricity.

As electricity cannot be stored in large amounts (at least with current technology), we have to maintain the grid stability by fitting the instantaneous power generation and the expected load demands, at the same time that we have to deal with growing needs of infrastructures and the integration of non-manageable resources, such as solar and wind power [20, 21].

After introducing the special generation rules in 1997, the contribution of renewable energies to the power capacity in Spain has risen considerably, leading the country to a prime position in the world, although it is necessary to go further according to the energy targets for 2020. At the same time, a large number of natural gas-powered utilities were built in this period, up to the point where the coverage ratio with conventional generation only is nowadays near 60% over the maximum of annual demands (as mentioned above). If we consider the utilization factors of each technology, which depend on the availability of the installations to produce the energy when it is required, the maintenance periods and shut-downs per year, or the meteorological conditions in the case of renewables, the great overcapacity running in the country leads to a short utilization of the installations, except the nuclear facilities, whereas the penetration of renewables and the decrease in of consumption due to the "financial crisis" have accentuated this situation. Table 13.3 shows the factors for all the technologies, where if we take into account that some installations have priority of evacuation (special energies) we can foresee that they are conditioning the others, especially the investments in thermal utilities (if used only as rolling power); on the other hand, because of the intermittent nature of some renewables, the CO_2 reduction would be less than anticipated due to cycling of the fossil fuel plants that make up the balance of the grid [22].

The increasing penetration of wind, especially, and also photovoltaic (PV) energy, brings with it problems related to the integration of such highly variable resources; when conventional base load and rolling power utilities are included in the analysis, this provides a completely new perspective for management of the whole power system. Thus, depending on the control strategies for the generating plants and the electrolyzers, the capacity factors can achieve high values.

Table 13.3 Distribution of power capacities and electricity generation per technology (Spain, 2011).

Primary sources	Power capacity (MW)	Utilization factor (%)	Generation (GWh)	Energy mix (%)	Demand (GWh)
Coal	12 210	43.4	46 427	16.1	–
Fuel + gas	5 425	15.8	7 491	2.6	–
Gas combined cycle (GCC)	27 123	23.2	55 074	19.1	–
Nuclear	7 777	84.7	57 670	20.0	–
Hydro + pumping	17 538	18.0	27 650	9.6	–
Wind	20 881	23.0	42 060	14.6	–
Other (special regime)	15 340	38.2	51 383	17.9	–
Total (net)	10 6295	30.9	279 711	100.0	270 361

13.3.2
Integration of Hydrogen Technologies at Large Scales

In this section, we analyze the management strategies for surplus grid power using electrolytic hydrogen, by means of two approaches: the hourly average curve representative of one year and the annual curves that represent the real power generation and demands during the year. Hence several production scenarios using base-load technologies and intermittent resources are superimposed on the power demands in each case, to create different electricity balances that can be converted to hydrogen by "dynamic electrolyzer operations," as described above. The analyses include the primary energy balances and design parameters of the installations throughout the country, together with estimates of the global economy and feasibility of the project (e.g., capital and electricity costs, fuels and utility savings, emissions).

13.3.2.1 Hourly Average Curves

Figure 13.4 shows various power generation scenarios superimposed on the load-average curve for the year 2009 in Spain. For each scenario we can estimate the hydrogen producible in the "valleys" and also the hydrogen consumable in the "peak" hours, using the electrochemical conversion systems with electrolyzers and hydrogen; looking at the maximum surplus power value, we calculate the number and size of the electrolytic plants ($P_{0i} = 50$ MW, $J_0 = 10$ kA m^{-2}); then, with the parameters of the electrolyzer (Figure 13.1), running continuously in the range 10–120%, we approximate the daily production of hydrogen by.

$$Q_p\left(t_{H_2}\ d^{-1}\right) = \sum \frac{P_t}{E_t} = \sum \frac{u_t\ P_0}{k_0 + k_1\ J_t} \tag{13.9}$$

where we regulate the current density of the electrolyzers according to the utilization factor in each period: $J_t = u_t\ J_0$; $u_t = P_t/P_0$ ($0.10 \le u_t \le 1.20$) (see the Appendix).

The hydrogen consumable by fuel-cell generators is estimated from the power imbalances in the peak hours, using the typical electricity efficiency ($\eta = 0.60$):

$$Q_c\left(t_{H_2}\ d^{-1}\right) = \frac{P_c\left(\text{MWh}\ d^{-1}\right)}{\eta_{\text{LHV}}\left(\text{MWh}\ t_{H_2}^{-1}\right)} \tag{13.10}$$

Finally, the hydrogen available for other uses, for example, vehicle fuel, stationary energy, or industrial processes, is obtained as

$$Q_a = Q_p - Q_c \tag{13.11}$$

Detailed profiles of electricity and hydrogen balances for the scenario 2 are shown in Figure 13.5, while the tables in Figure 13.4 summarize the main results for all the generation scenarios, revealing that the strategies will depend on the limiting factors in each case, for example, the size and utilization ratios of the electrolyzer plants, the hydrogen reserve capacities to regenerate peak electricity and its availability for other uses, and the economic outcomes of the whole systems.

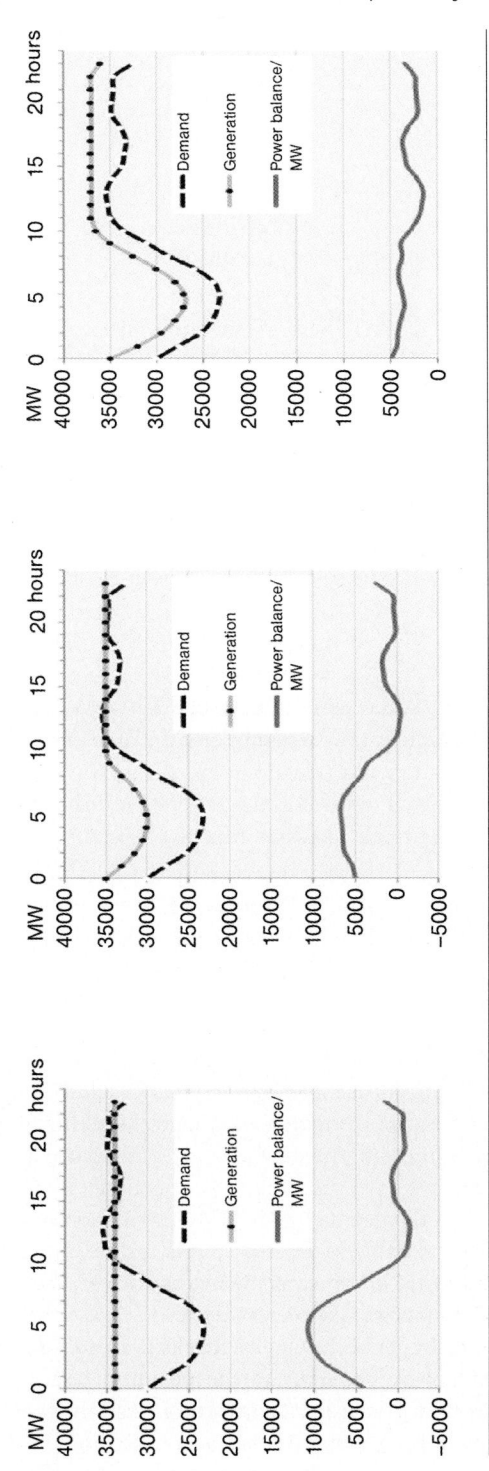

13.3 Operation of the Electrolyzers for Electric Grid Load Balancing

Scenario 1		Scenario 2		Scenario 3	
Electrolyzer installations	179	Electrolyzer installations	112	Electrolyzer installations	83
Utilization ratio, u	0.42	Utilization ratio, u	0.52	Utilization ratio, u	0.78
Production cost, C_H (€ kg^{-1})	4.08	Production cost, C_H (€ kg^{-1})	3.87	Production cost, C_H (€ kg^{-1})	3.51
Peak capacity, P_c (MW)	2335	Peak capacity, P_c (MW)	1002	Peak capacity, P_c (MW)	0
Surplus, Q_a (t_{H_2} d^{-1})	778	Surplus, Q_a (t_{H_2} d^{-1})	1062	Surplus, Q_a (t_{H_2} d^{-1})	1382
Sales − costs (M€ a^{-1})	−818	Sales − costs (M€ a^{-1})	−189	Sales − costs (M€ a^{-1})	247

Figure 13.4 Simulation of power generation scenarios with the resulting energy balances, the number, utilization, and production cost of electrolysis, the peak-power sizes, the surplus hydrogen, and the economic differences with electricity prices at €0.05 kWh^{-1} (off-peak) and €0.10 (peak), the hydrogen sold at €4 kg^{-1} and +€900 kW^{-1} for the purchased cost of the peak capacities.

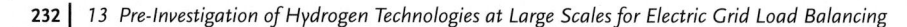

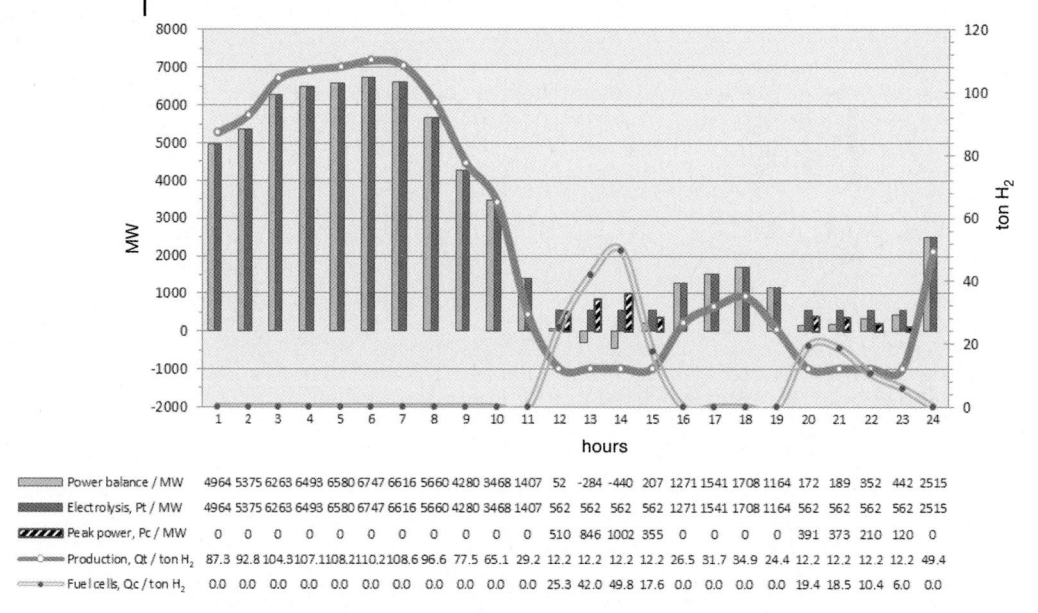

		1	2	3	4	5	6	7	8	9	10	11	12	13	14	15	16	17	18	19	20	21	22	23	24
Power balance / MW		4964	5375	6263	6493	6580	6747	6616	5660	4280	3468	1407	52	−284	−440	207	1271	1541	1708	1164	172	189	352	442	2515
Electrolysis, Pt / MW		4964	5375	6263	6493	6580	6747	6616	5660	4280	3468	1407	562	562	562	562	1271	1541	1708	1164	562	562	562	562	2515
Peak power, Pc / MW		0	0	0	0	0	0	0	0	0	0	0	510	846	1002	355	0	0	0	0	391	373	210	120	0
Production, Qt / ton H_2		87.3	92.8	104.3	107.1	108.2	110.2	108.6	96.6	77.5	65.1	29.2	12.2	12.2	12.2	12.2	26.5	31.7	34.9	24.4	12.2	12.2	12.2	12.2	49.4
Fuel cells, Qc / ton H_2		0.0	0.0	0.0	0.0	0.0	0.0	0.0	0.0	0.0	0.0	0.0	25.3	42.0	49.8	17.6	0.0	0.0	0.0	0.0	19.4	18.5	10.4	6.0	0.0

Figure 13.5 Balances of electricity and hydrogen for the scenario 2 using electrolyzers and fuel cells.

This analysis illustrates how variations of the generation scenarios serve the purpose of obtaining a balance of the efficiency, the economy, and the ease of operations; the capacity of electrolytic installations, the electricity consumption, and the fuel cells determine the costs of the system, whereas they generate returns by selling hydrogen and electricity, and there are also savings in conventional fuels and power utilities.

Nevertheless, such a preliminary approach deserves to be completed with the whole analyses of the load curves and hydrogen operations, which consider the real generation and demand annually, and also the utilization ratios of technologies or other factors such as the storage.

13.3.2.2 Annual Curves

Figure 13.6a shows the hourly profiles, fixing a base-load power which is added to the production of nonmanageable resources and compared with the real electricity demands during a year, where one can appreciate the daily, weekly, and seasonal variability of power uses. In this scenario, we have simulated a base generation of 26 500 MWh each hour of the year, which represents 92.4% of the annual power consumption, while the production from variable sources accounts for 30.7% of consumption; the total surplus ratio is 123.1% and means the net balance of electricity theoretically available for other uses (58042 GWh a^{-1}).

If this energy is used in relation to the hydrogen technologies for electric grid load balancing, the model has to include the following elements: (1) the maximum power surplus (21.537 GW), which determines the capacity and utilization of electrolyzers (we use a limit value of 15 GW); (2) the hydrogen production, taking into account the

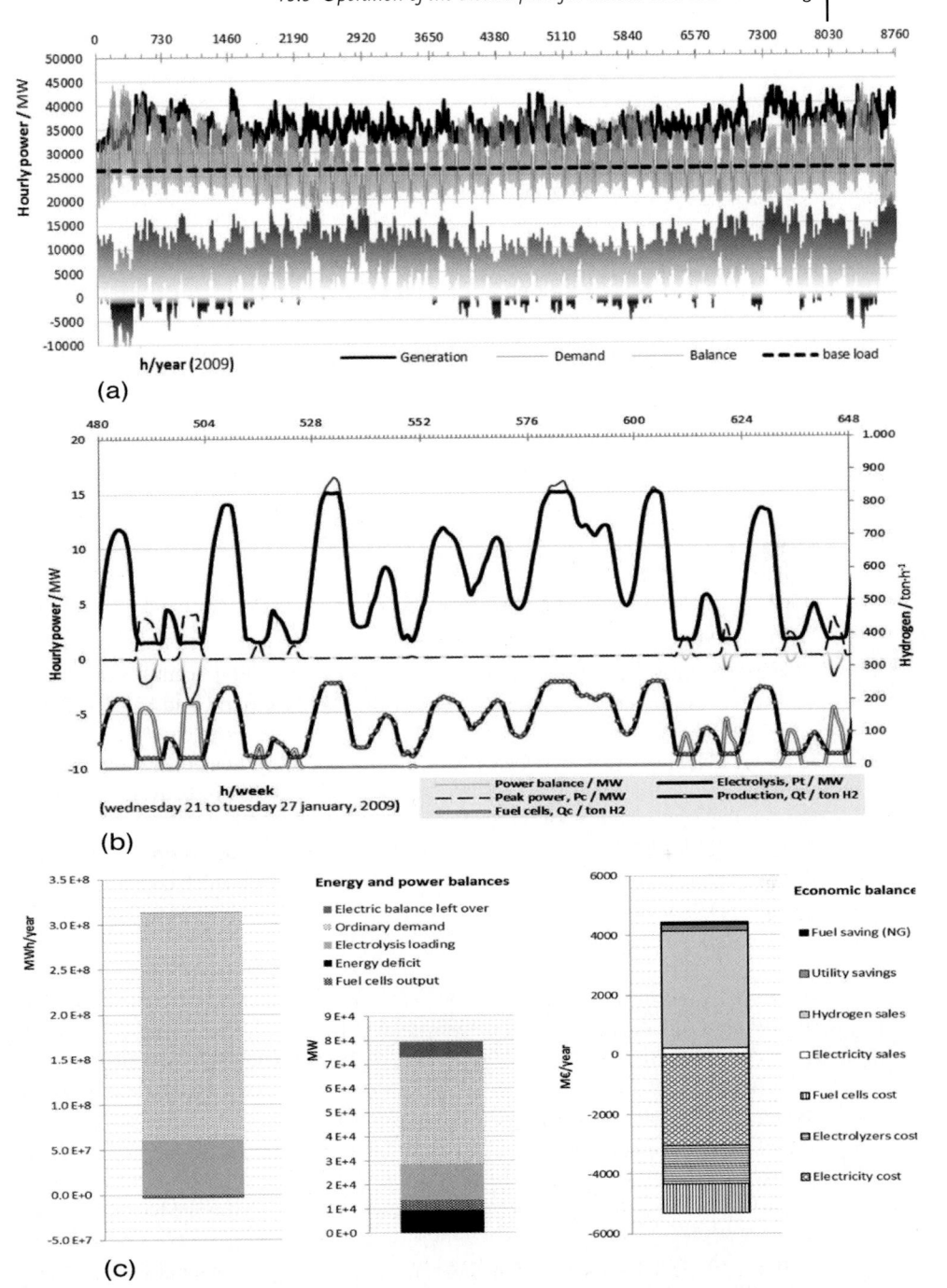

Figure 13.6 (a) Power generation scenario and load curves with disaggregation of hours (Spain, 2009); (b) weekly balance of electricity and hydrogen using the electrolyzer and fuel-cell system; (c) cumulative energy balances, power capacities, and global economic results (annually).

power inputs and the efficiency and dynamic range of electrolyzers (1146.9 kt a^{-1}; $u = 47\%$); (3) the deficits of electricity originated by the power imbalances and electrolyzer's operation, which determine the peak generation with fuel cells or other reserve capacities (13.436 GW) and the hydrogen consumed in these devices (14.7%). Figure 13.6b displays the balances of electricity and hydrogen at each hour of the year, showing the operation of the electrolyzers, which is proportional to the loads within the dynamic range, and the fuel cells, which are activated from time to time to supply the power shortfalls (7.0%). Figure 13.6c summarizes the cumulative energy balances, the maximum power capacities, and the global economic results for this "base-case" scenario (M€ – 908 a^{-1}).

Finally, we elaborated a worksheet in order to simulate distinct scenarios, taking into account all the variables of the model, namely the patterns of generation and demand in different periods, the electrolysis parameters, including the maximum power and current density to the electrolyzers, and the economic factors such as the costs of electrolyzers, ancillaries, and fuel cells, the prices of electricity and hydrogen, and the savings in conventional fuels and utilities which are avoided. This serves the purpose of parametric analyses to study the sensitivity of the size, operation, and economy of the electricity and hydrogen processes to these factors, as detailed in Table 13.4.

The economic balances are dominated by electricity costs and hydrogen sales in such a form that, allowing for coproduction of hydrogen and power, utilities could improve their production, storage, and use, and optimize the system based on economic or reliability factors; thus, what we are trying to ascertain is whether power systems that are optimized for electricity generation and hydrogen production are competitive, depending on the prices at which electricity and hydrogen are bought and/or sold. The analyses show that sufficient hydrogen market prices and low surplus electricity values render the installation of large electrolyzers cost-effective, compensating for the drop in capacity factors and hence the ratios of hydrogen produced to capital costs. Also, the use of hydrogen for peak-power regeneration is only favorable when these units are less expensive and can be used more efficiently at higher utilization ratios; nevertheless, this must also be balanced with the extra benefits of selling electricity in highly fluctuating spot markets, the avoided costs in conventional utilities and their associated environmental impacts, and the improved operation and management of the whole power systems.

In this time-step model, a hierarchy can be assumed for handling the surplus or deficit of each form of energy to minimize installation sizes and conversion losses; therefore, whenever possible, energy is used in a form as close as possible to that in which it is created; for example, power is used as electricity as it is better to save fuel in this way than to make hydrogen for electricity with low round-trip efficiencies. This is simulated in the model by entering the maximum capacities for the electrolyzers and fuel cells; when the sum of uncontrolled supplies exceeds the demand, the power surplus is exported for other uses, whereas in cases of power shortfalls we can import the electricity, use dispatchable generators, or shed loads. For electrolytic hydrogen, there are also the industrial, transport, and other fuel uses to substitute fossil fuels [23].

Table 13.4 Values used in the scenarios and results obtained by simulation of the different factors.

j	z				
	Values in the base scenario (Z_{j0})	Range of values and sensitivity to the cost balance			
		(ΔZ_j)	M€ a^{-1} (−)	M€ a^{-1} (+)	S_j^b
Load demands and generation from RE[a]	Year: 2009 Loads/RE: 251.4/77.3 TWh	Year: 2010 Year: 2011	−927 −654		−2.1% 28.0%
Base load generation profiles	Annual, fixed: 26.5 GWh h^{-1}, i.e. 75.0% of the generation energy balance: +18.8%	±1.5 GWh h^{-1} Seasonal Weekly Daily	−1274 −891 −681 −685	−572	6.83 0.17 2.21 2.17
Electrolyzer and fuel-cell parameters	P_0 = 50 MW	±25	−1096	−809	0.32
	E_0 = 1.4 V	±0.1	−674	−1115	−3.40
	K = 120 mΩ^{-1} m^{-2}	±10	−933	−887	0.30
	r = 0.020 mΩ m^2	±0.005	−855	−960	−0.23
	f_1 = 0.98	±0.01	−945	−872	3.94
	f_2 = 0.025 kA^2 m^{-4}	±0.005	−905	−913	−0.02
	η_{BOP} = 0.93	±0.03	−1024	−793	3.94
	J_0 = 10 kA m^{-2}	±5	−901	−1034	−0.15
	u_t ≥ 0.10	±0.05	−706	−1138	−0.48
	u_t ≤ 1.00	±0.20	−923	−908	0.04
	P_t ≤ P_{limit} (15 000 MW)	±5000	−665	−1286	−1.03
	η_c = 0.60	±0.10	−1043	−812	0.76
	$P_{c,t}$ ≤ $P_{c\,limit}$ (4000 MW)	±2000	−379	−1307	−1.02
Cost factors	Electrolyzer: a = €1.2×10^5,	±6×10^4	−271	−1546	−1.40
	b = 0.80,	±0.10	−282	−2138	−8.18
	c = 0.25	±0.05	−1064	−770	0.81
	BOP = 60%	±26	−701	−1116	−0.53
	Fuel cells[c]: €1400 kW^{-1}	±600	−483	−1334	−1.09
	GCC utilities: €500 kW^{-1}	±100	−953	−864	0.25
	Annual discount rate: 10%	±5	−479	−1396	−1.01
	Depreciation: 20 years	±5	−1070	−824	0.54
	O&M annual rate: 6%	±3	−562	−1254	−0.76
	Electricity prices (€ kWh^{-1}):	±0.025	625	−2442	−3.38
	valley 0.050, peak 0.100	±0,050	−1017	−800	0.24
	Hydrogen price: €4.0 kg^{-1}	±2.0	−2865	1049	4.31
	Natural gas price: €0.034 kWh^{-1}	±0.017	−945	−871	0.08

a RE includes all the energy sources which constitute the "special regime" in the Spanish power system.
b The parametric sensitivity on the cost balances is obtained as:
S_j = {[M€ a^{-1} (+) − M€ a^{-1} (−)] / 908} / (2ΔZ_j/Z_{j0}).
c Electrocatalysts represent a large proportion of fuel-cell costs. Researchers have claimed a new deposition method, to be used at low-cost industrial scale, which permits high utilization of platinum (10 kW g^{-1}); this could help to foster the technology for electric grid and vehicles, running for more than 1000 h [24]

Anyhow, the simulations show the benefits that can be achieved by using low-priced power, and also their leveling effects on the electricity balances of the generating plants.

As summarized in Table 13.4, the most sensitive factors are the base loads, the thermodynamic potential for electrolysis, the Faraday and BOP efficiencies, the limiting power to electrolyzers and fuel cells, the costs of the electrochemical cells, and the annual rates, but particularly the electricity and hydrogen prices that are the only values capable of resulting in positive economic balances within the ranges considered. The base load profiles can provide some advantages if we discriminate the peak and valley periods, especially during the weekend or night hours. Finally, the variation of the loads in different years shows consistent effects on the responses.

13.4 Conclusion

This report is the result of a preinvestigation of hydrogen technologies in the power systems, which considers two main aspects:

- state-of-the-art for electrolysis technology at large scales with respect to performance and costs, as well as future developments for the electrolyzer techniques;
- the potential for introduction of electrolysis in the utility system in connection with an extended production of renewable electricity.

It is a prospective study, where the scope was to define scenarios of hydrogen for electric grid load balancing, which are based in the efficiencies, the economy, and ease of operations. The proposed model includes a holistic approach to the power generation and loads profiles, together with an evaluation of the electrolysis technologies that are applicable at large scales; the strategy is to control the power inputs, taking into account the dynamics of electrolyzers, to show the benefits of using surplus electricity at low prices, and also the leveling effects on the energy balances of the plants.

The electricity systems in many countries hold an excess of capacity that shows the failures of the current liberalized markets to allocate resources efficiently, and also that the form of fixing of some tariffs in the electric pool causes an excess of retribution of conventional power utilities. What will be required is a long-term policy oriented to the security and sustainability of energy supply, which at the same time facilitates innovation in the sector; for example, if there is no need for new thermal plants, we can use the investments to promote energy efficiency, renewables, and storage options, with neat externalities and competitive advantages especially in Spain.

As discussed, hydrogen production by water electrolysis using the power grid mix is a promising option, depending of the prices of electricity in different periods, when sufficient electrolyzer efficiency and cost reductions are achieved. The analysis shows the effects of all relevant factors, including the variability of the loads, renew-

Table 13.5 Summary of main results.

	State-of-the-art (full loading)	Target scenario base (variable)
Electrolyzer efficiency	67% LHV at 2 kA m^{-2}	57.5% LHV at 10 kA m^{-2}
Sizes and investment costs	3.5 MW, €900 kW^{-1}	50 MW, €200–400 kW^{-1}
Production cost of hydrogen	€6.5 kg^{-1} at €0.10 kWh^{-1} ($u = 1.00$)	€4.0 kg^{-1} at €0.054 kWh^{-1} ($u = 0.47$)
Price of electricity when production of hydrogen becomes viable		€3.0 kg^{-1} at €0.035 kWh^{-1} ($u = 0.47$)
Peak power value when utilization of hydrogen becomes viable		€3.0 kg^{-1}, (η_c LHV)$^{-1}$ = €0.14.9 kWh^{-1}
Estimated natural gas costs for electricity generation		€0.034 kWh^{-1}

ables, and base power generation, the parameters related to the sizes, operation, and energy use of electrolyzers and fuel cells, the costs of the installations, and the prices of the energies. Hence a feasible economic result can be anticipated for the different scenarios provided that the electrolyzers are built by orders of magnitude from current capacities and they are utilized at optimal current densities within the dynamic ranges, being also dependent on the energy balances to arrive at the best utilization of each form of energy (i.e., the surplus and shortfalls of electricity and hydrogen).

The main results that provide important quantitative numbers at a glance are summarized in Table 13.5.

The results achieved using the annual scenarios show that the most sensitive factors are the prices of surplus electricity and hydrogen, the costs of electrolyzers depending of their sizes, the cell voltage and current efficiency, and the base power generation patterns associated with the fixing of a limit capacity to electrolyzers and fuel cells for improving their utilization costs. Depending on the preferred targets in the future, different strategies can be implemented that favor the reduction of fossil-fuel utilities, increase storage capacity and hydrogen production, and so on.

Following this preliminary analysis, the main difficulties could be to attain the synchronism of all parts of the systems and the great investments in such a large enterprise without precedent. It is interesting to go deeper into the distinct scenarios, which can reach an adequate balance of the design variables depending on the predominant criteria in each case, for example, minimization of the capital costs, higher electricity consumption, greater fuel demands (H$_2$), or both.

Finally, there are some limitations of the study that should be taken into account for further research and development; namely the whole components of installations – such as the electrical controls and downstream hydrogen processes – should be considered, and also important is the estimation of the economic parameters, which must always bring up-to-date the clear profit values and externalities, the effects on employment and social welfare, the geographical depicture of the scenarios to represent the regional deployment of the subsystems better, and so on.

As a concluding remark, with a large fraction of renewable sources in the energy system, water electrolysis is unavoidable even though the technology is not perfect. Efficiencies, short-term costs and the advantages and drawbacks of the hydrogen technologies must be considered. However, a more fundamental question could be: what is the alternative, if business as usual is not an option?

13.5
Appendix

The exact calculation of hydrogen production depends on the hourly available power and current density, which are not strictly proportional to the power use, leading to a second-order equation that also contains the dependent term $\eta_F (J_t)$:

$$Q_p = \frac{P_t}{E_t} = \frac{u_t \, P_0}{k_0 + k_1 \, J_t}$$

$$J_t = \frac{I_t}{A} = \frac{26.8 \, V_0 \, J_0 \, Q_p}{\eta_F \, \eta_{BOP} \, P_0}$$

$$Q_p = \frac{-k_0 + \left[k_0^2 + \dfrac{4 \, k_1 \times 26.8 \, V_0 \, J_0}{(\eta_F \, \eta_{BOP} \, P_0) \, u_t \, P_0} \right]^{\frac{1}{2}}}{2 \left(\dfrac{k_1 \times 26.8 \, V_0 \, J_0}{\eta_F \, \eta_{BOP} \, P_0} \right)}$$

However, using a model in which current density is approached such as $J_t = u_t J_0$ leads to similar results with differences in hydrogen production of less than +1%; therefore, we use this simplified method which still has sufficient accuracy and avoids cumbersome or iterative calculations.

References

1 Hajimiragha, A. H., Cañizares, C. A., Fowler, M. W., Moazeni, S., Elkamel, A., and Wong, S. (2011) Sustainable convergence of electricity and transport sectors in the context of a hydrogen economy. *Int. J. Hydrogen Energy*, **36** (11), 6357–6375.

2 REE (2012). *The Spanish Power System. Annual Report*, http://www.ree.es/ingles/sistema_electrico/informeSEE.asp (last accessed 10 September 2012).

3 Floch, P., Gabriel, S., Mansilla, C., and Werkoff, F. (2007) On the production of hydrogen via alkaline electrolysis during off-peak periods. *Int. J. Hydrogen Energy*, **32** (18), 4641–4647.

4 Gutiérrez, F. and Atanes, E. (2010) Power management using electrolytic

hydrogen. *SciTopics*, 1 October, http://www.scitopics.com/Power_management_using_electrolytic_hydrogen.html (last accessed 27 June 2012).

5 Gutiérrez-Martín, F. and Guerrero-Hernández, I. (2012) Balancing the grid loads by large scale integration of hydrogen technologies: the case of the Spanish power system. *Int. J. Hydrogen Energy*, **37** (2), 1151–1161.

6 Kroposki, B., Levene, J., Harrison, K., Sen, P. K., and Novachek, F. (2006) *Electrolysis: Information and Opportunities for Electric Power Utilities*, Technical Report NREL/TP-581-40605, National Renewable Energy Laboratory, Golden, CO, http://www.nrel.gov/docs/fy06osti/40605.pdf (last accessed 3 April 2012).

7 Kellersohn, T. (2001) *Ullmann's Encyclopedia of Industrial Chemistry*, 6th edn, Wiley-VCH Verlag GmbH, Weinheim.

8 Ursúa, A., Gandía, L. M., and Sanchis, P. (2011) Hydrogen production from water electrolysis: current status and future trends. *Proc. IEEE*, **100** (2), 410–426.

9 Mazloomi, S. K. and Sulaiman, N. (2012) Influencing factors of water electrolysis electrical efficiency. *Renew. Sustain. Energy Rev.*, **16** (6), 4257–4263.

10 Shen, M., Bennett, N., and Ding, Y. (2011) A concise model for evaluating water electrolysis. *Int. J. Hydrogen Energy*, **36** (22),14335–14341.

11 Ulleberg, Ø. (2003) Modeling of advanced alkaline electrolyzers. A system simulation approach. *Int. J. Hydrogen Energy*, **28** (1), 21–33.

12 Khater, H. A., Abdelraouf, A. A., and Beshr, M. H. (2011) Optimum alkaline electrolyzer-proton exchange membrane fuel cell coupling in a residential solar stand-alone power system. *ISRN Renewable Energy*, article ID 953434, doi:10.5402/2011/953434.

13 Harrison, K., Martin, G., Ramsden, T., and Saur, G. (2008) *Wind-to-Hydrogen Project: Electrolyzer Capital Cost Study*, Technical Report NREL/TP-550-44103, National Renewable Energy Laboratory, Golden, CO.

14 NREL (2009) *Current State-of-the-Art Hydrogen Production Cost Estimate Using Water Electrolysis*, Independent Review NREL/BK-6A1-46676, National Renewable Energy Laboratory, Golden, CO.

15 Da Silva, E. P., Marin Neto, A. J., Ferreira, P. F. P., Camargo, J. C., Apolinário, F. R., and Pinto, C. S. (2005) Analysis of hydrogen production from combined photovoltaics, wind energy and secondary hydroelectricity supply in Brazil. *Solar Energy*, **78** (5), 670–677.

16 Akhnazarova, S., and Kafarov, V. (1982) *Experiment Optimization in Chemistry and Chemical Engineering*, Mir, Moscow.

17 Department of Chemistry, Technical University of Denmark (KI/DTU), Fuel Cells and Solid State Chemistry Department, Risø National Laboratory, Technical University of Denmark, and DONG Energy (2008) *Pre-investigation of Water Electrolysis*. Energinet 2006-1-6287, http://www.risoe.dk/rispubl/NEI/NEI-DK-5057.pdf (last accessed 10 March 2012).

18 IHT Industrie Haute Technologie (2012) *Clean Hydrogen Solutions*, http://www.iht.ch/technologie/electrolysis/industry/clean.html (last accessed 3 July 2012).

19 Mansilla, C., Dautremont, S., Shoai Tehrani, B., Cotin, G., Avril, S., and Burkhalter, E. (2011) Reducing the hydrogen production cost by operating alkaline electrolysis as a discontinuous process in the French market context. *Int. J. Hydrogen Energy*, **36** (11), 6407–6413.

20 RED (2010) *Power Demand Tracking in Real Time*, http://www.ree.es/ingles/operacion/curvas_demanda.asp (last accessed 8 January 2012).

21 Elygrid (2011) *Improvements to Integrate High Pressure Alkaline Electrolyzers for Electricity/H_2 Production from Renewable Energies to Balance the Grid*, http://www.elygrid.com/ (last accessed 18 July 2012).

22 Gutiérrez-Martín, F., et al., *Effects of wind intermittency on reduction of CO_2 emissions: The case of the Spanish power system*, Energy (2013), http://dx.doi.org/10.1016/j.energy.2013.01.057 (in press).

23 Barton, J., and Gammon, R. (2010) The production of hydrogen fuel from renewable sources and its role in grid operations. *J. Power Sources*, **195** (24), 8222–8235.

24 Madri+d (2012) Una Pila Española Supera la Meta de Potencia Marcada por Estados Unidos, www.madrimasd.org/informacionidi/noticias/noticia.asp?id=53229 (last accessed 1 May 2012).

Part II
Power Production

14
Onshore Wind Energy

Po Wen Cheng

14.1
Introduction

Wind energy is regarded as the one of the most cost-effective renewable energy sources, if not *the* most cost-effective (excluding hydropower). The use of wind energy for electricity generation started in the nineteenth century with experimental turbines in Denmark and Scotland. The oil crisis in 1973, the energy crisis in 1979, and the incident at the US Three Mile Island nuclear power plant in 1979 renewed the interest of governments in alternative energy sources, among others wind energy. Several megawatt prototypes were built in North America and Europe, but none of them were sold commercially because the technology was not ripe advanced for such large wind turbines and the subsequent decline in oil prices made electricity generation with wind turbines uneconomical.

With the nuclear accident at Chernobyl in 1986, the mounting evidence of climate change through the use of fossil fuels and the urge towards greater energy independence of the OECD countries at the end of the 1980s, clean energy sources has regained focus and wind energy has since then enjoyed a relatively long period of continuous expansion and technical improvement. The continuous development of the technology starting from the robust concept, the so-called Danish concept, (stall-regulated wind turbine where the output power is limited by the aerodynamic stall phenomenon, fixed rotational speed, three blades, upwind orientation of the rotor where the undisturbed wind field reaches the rotor first, unlike the downwind concept where the wind field reaches the tower first), towards the current concept of three blades, upwind orientation, variable speed, and pitch control has contributed to the steady fall decrease in the cost of energy from above $ 150 MWh^{-1} to the current level of around $ 70 MWh^{-1}, depending on the mean wind speed at the site [1].

However, the development of wind power is concentrated in a handful of countries. Approximately 75% of the world installed wind power is concentrated in just five countries, China, the United States, Germany, Spain, and India. The top 10 countries with the most installed wind power represent 86% of the world installed

Transition to Renewable Energy Systems, 1st Edition. Edited by Detlef Stolten and Viktor Scherer.
© 2013 Wiley-VCH Verlag GmbH & Co. KGaA. Published 2013 by Wiley-VCH Verlag GmbH & Co. KGaA.

wind capacity [2]. This is a direct consequence of the policy instruments that these countries have put in place to encourage the development of wind energy. By the end of 2010, the electricity generated by wind turbines amounted to 1.6% of total electricity production worldwide (including electricity generated from fossil fuels) and 8.3% of the electricity production from renewable energy (including hydropower) [3].

The wind energy sector has enjoyed steady and double-digit growth for more than a decade. With the economic crisis from 2009, a rapid expansion of production capacity, particularly in China, and the uncertainties in the policies of major wind power countries such as the United States and Spain, the growth of the wind energy sector in 2013 will be less than in previous years [4]. The decline will be especially pronounced in the United States due to the uncertainty of the Production Tax Credit (PTC) that was due to expire at the end of 2012. However, this gap in growth is being mitigated by higher demand from emerging countries such as Brazil, Turkey, and South Africa. Over the long term, it is expected that the growth of onshore wind energy will be sustained but the figure will not be of the same magnitude as the growth rate between 2000 and 2010 [5].

This chapter concentrates on the development of onshore wind energy; the role of offshore wind energy is treated in Chapter 13. The subjects treated here include the current state-of-the-art, future technological developments, the environmental aspects of wind energy utilization, the economics of onshore wind energy, system integration of onshore wind energy within the electricity grid, and a short discussion on the policy instruments that allow further increases in onshore wind power. It should be noted that it is not possible to discuss here all the details related to the specific subjects mentioned above, owing to constraints on length, and a deeper understanding can be obtained from the references cited.

14.2
Market Development Trends

Wind energy has enjoyed almost 20 years of double-digit growth owing to the changes in the policies and renewable energy targets set by governments, especially those in OECD countries. According to the International Energy Agency (IEA) [3], the contribution of wind energy to the total primary energy supply (TPES) is still marginal, less than 1%. However, one needs to keep in mind that there are different conventions for calculating the primary energy balance and that significantly affects the representation of the wind energy and other renewable energy sources as a percentage of the TPES.

Since wind power is mostly used in electrical power generation, it makes more sense to look at the contribution of wind energy to the total installed power generation capacity and its contribution to the total electricity supply. The world installed capacity of wind power increased from around 24 GW in 2001 to about 240 GW at the end of 2011 [2] (Figure 14.1). About 2% of the installed wind generation capacity belongs to offshore wind power. Offshore wind power will grow significantly in northern Europe, especially in the United Kingdom and Germany. Outside northern

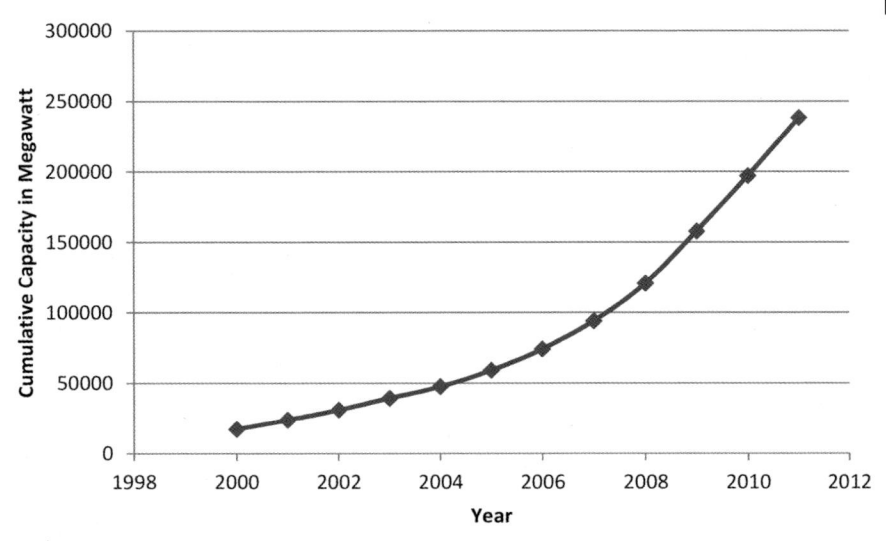

Figure 14.1 Cumulative installed capacity of wind power. Data from BWE (Bundesverband WindEnergie).

Europe, the role of offshore wind energy will remain small for the next 5–10 years. The generation cost of offshore wind energy represents the greatest obstacle for the expansion of offshore wind energy, as the installation cost per installed kilowatt is two to three times higher than for onshore wind energy and the levelized cost of energy (LCOE) (i.e. cost per kilowatt hour of energy produced taking into account the overnight capital cost and also other running costs such as operation and maintenance costs) is about double the LCOE onshore. More information can be found in Chapter 13.

Wind power represented about 4% of the total installed capacity of power generation at the end of 2010. According to an estimate from BTM Consult [5], wind energy will produce ~2.26% of the world's electricity in 2012. It is projected that wind-generated electricity could meet 8% of the world electricity demand by 2021.

The contribution of wind energy to the total power generation mix differs strongly from region to region. In the EU [6], the wind generation capacity represents ~10% of the power generation capacity and the wind-generated electricity contributes, in a normal wind year, ~6.3% of the total electricity generation in the EU. This is a much higher figure than the world average, mainly due to the effective policy instruments put in place that promote the installation of renewable energies. However, even within the EU there are major differences in terms of the wind electricity penetration in the electricity grid. Wind electricity in Denmark has the highest contribution to the total electricity consumption, ~26%, followed by Spain and Portugal, ~15%, Ireland, ~12%, and Germany, ~10%.

The future growth of installed wind capacity will take place in Asia, where it is expected that the region will overtake Europe in terms of installed wind generation capacity within the next 2 years [2, 4]. Within Asia, the growth is centered on China and India, which together represented 50% of the global market in 2011. The bottleneck for future growth will be the grid infrastructure for the transportation of the electricity generated by wind in remote areas to the areas of consumption. China still has a significant amount of installed wind capacity that has not been connected to the grid owing to the transmission bottleneck. For this reason, the current policy is moving away from large wind farms in remote areas with excellent wind resources towards southern and coastal areas with lower wind speeds but close to the centers of electricity consumption with adequate transmission capacity.

The development of wind energy in the United States depends heavily on the tax incentive, mainly on the so-called Production Tax Credit (PTC). This was due to expire by the end of 2012 but it seems likely that the PTC will be extended now that the uncertainty over the next administration is over. However, it is unlikely that 2013 will achieve a similar level of activity to that in 2012 as many projects were put on hold due to regulatory uncertainty. However, even if the PTC is extended and there are other existing policy instruments in place, such as the Renewable Portfolio Standard (RPS), the wind energy market will still face significant challenges in the United States owing to the low electricity prices. One of the main factors for the downward pressure on electricity prices in the United States is the nonconventional gas. It is projected that the United States will benefit from the exploitation of nonconventional gas and oil exploration, and that it will become a net exporter by 2021 [7]. The cost pressure from this type of natural gas can reduce the financial incentive to install new wind farms unless carbon emission pricing is taken fully into account such that cleaner energy sources become attractive.

Outside North America, Europe, and Asia, there will be a few pockets of growth from the emerging countries, notably Australia, Brazil, Chile, New Zealand, and South Africa [4].

14.3
Technology Development Trends

14.3.1
General Remarks About Future Wind Turbines

The functional principle of wind energy utilization is fairly straightforward and has been used for many centuries to carry out mechanical work. Currently, most wind turbines are designed to produce electricity, hence the main goal is to have a high rotational speed and low mechanical torque. In very simple terms, the torque is created by the aerodynamic forces resulting from the pressure differences around the blade profile. In this way, the air flow is slowed and the kinetic energy of the air flow has been extracted. The aerodynamic torque drives the main shaft and the generator that convert mechanical energy in electrical energy.

Currently there are many different wind turbine manufacturers that produce a variety of wind turbine concepts. One can use the following turbine characteristics to illustrate the evolution of the concepts and future technology trends.

14.3.2
Power Rating

The wind turbine rating has been increasing continuously from a few hundred kilowatts to the currently highest power rating of 7.5 MW. One needs to be aware that power rating alone is not an indicator of how productive or efficient the wind turbine is and it does not say anything about the LCOE. In general for high wind speed sites, that is, sites with a mean wind speed above 9 m s^{-1} at hub height, a high power rating is usually preferred because it is associated with a high energy output with relatively small rotor diameter. However, with the trend towards low wind speed sites, a high power rating is not necessarily the most cost-effective concept. The main reason is that the total amount of energy may well be higher than for a wind turbine with a lower power rating, and the extra cost associated with the large generator and oversized electrical components may not justify the amount of time that the wind turbine is producing the rated power. Furthermore, one needs to consider the power rating together with the rotor size (i.e., specific power expressed as megawatts per unit rotor swept area) and the resulting capacity factor (which is a measure of the equivalent full load hours) in order to find an optimum configuration with the lowest cost of energy [8].

This means that unlike offshore wind energy, the growth of the power rating is likely to be limited. The current development trend of low wind speed wind turbines with relatively large rotor diameter paired with a smaller generator agrees with the reported cost optimization trend [8]. In practical terms, it means that the power rating of onshore wind turbines will likely remain in the range 3–5 MW with relatively large rotors of 120–150 m during the next 5 years unless advances are made in transportation and logistics that can overcome the difficuty of installing and transporting very large rotor blades to difficult sites, for example, forests and complex terrains. This is not to say that there will be no wind turbines with 5 MW or higher power ratings installed onshore because the economics of the wind park must always be considered on a site-specific basis. Nevertheless, the installation of very large wind turbines (larger than 5 MW) will remain a very small proportion of the total installed onshore capacity.

14.3.3
Number of Blades

Currently, most wind turbines have three blades, and a small number have two blades. The main reason is that the dynamics of two-blade wind turbines are more challenging than those of three-blade wind turbines. This is even more true for large wind turbines where the dynamics and instability become a more important issue. It is to be expected that for onshore wind turbines, the three-blade option will still be the main concept used by th wind turbine manufacturers because of issues such as noise and visual impact.

The sound power level generated by the rotating blade is directly proportional to the fifth power of the tip speed and two-blade wind turbines have an inherently higher tip speed and therefore a higher sound power level. Since noise is an important criterion for the siting of onshore wind farms, two-blade wind turbines automatically carry this disadvantage. For offshore wind farms this may not be an issue and therefore it is likely that there will be a market for offshore wind turbines with a two-blades rotor if the technical challenges faced in the early development of two-blade wind turbines are solved. Another subjective aspect that needs to be taken into account is the general perception that two-blade wind turbines appear to rotate less smoothly because of the higher rotational speed. Furthermore, environmental protectionists prefer three-bladed wind turbines with a lower rotational speed since bird kill rates have been decreasing owing to the lower speed of modern wind turbines compared with the higher speed types in the 1990s [9].

14.3.4
Rotor Materials

Currently the rotors are mostly made of composite materials, with glass fiber as reinforcement. With the growth of the rotor diameter to increase the energy capture area of the wind turbine, the load on the rotor also increases. It is assumed that the load increase is proportional to the cubic power of the rotor diameter. However, data from the scaling of existing rotor blades show that the load increase is below the theoretical estimate, closer to an exponent of 2.5 [10]. Nevertheless, this poses a significant challenge to the structural design of the blade. One of the main limiting factors for the blade design is the maximum blade tip deflection. This is especially true for upwind wind turbines, where the deflection of the blade has to be limited in order to prevent it from striking the tower in extreme load events. On the other hand, in order to reduce the edgewise cyclic loads, mainly caused by gravity, it is important to limit the mass of the blade. These reasons led to a trend towards more flexible blade design with a large blade tip deflection. To limit the maximum blade tip deflection, higher structural stiffness is needed.

Carbon fiber is an ideal candidate for the role of blade stiffening material owing to its high tensile strength compared with glass fiber. However, the cost of carbon fiber has been the major obstacle to its application in rotor blades. Some manufacturers (vertically integrated) have being using carbon fiber to keep the blade mass low, which in turn reduces the loads for other components, such as the tower, the main shaft, and the bedplate. Other manufacturers have avoided the use of carbon fiber and pursued design optimization through better modeling and reduction of the uncertainties regarding the materials properties. With the increased popularity of carbon fiber in the automobile and aerospace industries, the upward price pressure on carbon fiber is likely to be sustained for some time, and for this reason some manufacturers have returned to the glass fiber only design. Currently the world's longest rotor blade of 75 m with a mass of ~25 t is made completely of fiber glass [11]. The use of carbon also brings another technical challenge to wind turbine designer, namely that it is an electrical conductor and the blade lightning protection system needs to be redesigned to take this into account when carbon fiber is used for the blades.

The future trend for blade materials is that glass fiber will remain the most popular and economically attractive option for the medium term. Manufacturers will carefully evaluate the merit of carbon fiber, and only in those cases where the stiffness becomes a critical issue and all the other design options have been exhausted will carbon fiber be considered. Carbon fiber will most likely be used only for highly loaded structural elements such as the spar cap.

14.3.5
Rotor Diameter

The size of the rotor is going through a new growth wave. The main driver is the so-called low wind speed turbines. The available sites with very good wind resources and with easy access to the grid are increasingly rare in countries with high wind energy penetration, such as Germany. Most of the wind turbine manufacturers are using a relatively larger rotor within an existing product line of similar power rating to increase the capacity factor significantly.

The capacity factor of a wind turbine is the ratio between the total amount of energy produced at a specific site in one year and the amount of energy that the turbine is capable of producing if operating at the rated power (the name plate power) all the time throughout the year. Since the wind resource is variable, it is not possible that the turbine can operate at the rated power all the time, therefore the capacity factor is always less than one. Typical capacity factors for onshore sites vary from 25 to 40%. The historical trend shows that the capacity factor has been on the rise; for US wind projects, the increase in the capacity factor is clearly observable in the last 10 years, from the upper 20% to the upper 30% for a relatively good wind regime [12].

This added value comes with a reasonable cost increase related to the use of a larger rotor, while most of the wind turbine components remain almost the same as for the existing product platform. This combination of a large rotor with medium-sized generator leads to a better LCOE for low wind speed sites [8]. This strategy can be observed from the evolution of the GE wind turbine development (Figure 14.2).

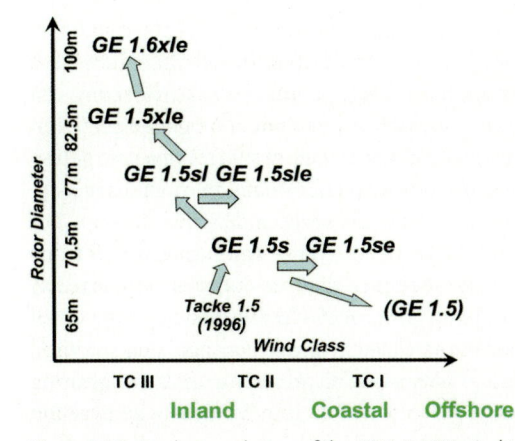

Figure 14.2 Product evolution of the GE 1.5 MW wind turbine family.

It can be observed that the rotor diameter has increased by more than 40% in 15 years whereas the generator size has hardly changed increased.

However, the increase in the rotor diameter will be limited by other constraints, such as material limitation of glass fiber because a further increase in the rotor diameter may require the use of carbon fiber which can diminish the benefits of the higher energy capture. Other constraints such as transportation and logistics can also limit the further growth of the rotor diameter. In order for the blades to be transported on the road, the dimensions of the blade need to meet the constriction of bridges, turning radius of the roads, and so on. For the blade to increase further in length, an inexpensive modular blade concept has to be developed in order to allow the blade to be transported in pieces. This is not an easy engineering task, as the joints between the blade section need to withstand a high number of fatigue load cycles and the extreme loads caused by extreme wind conditions or faults.

14.3.6
Upwind or Downwind

Downwind wind turbines are those where the air flow first hits the tower structure before hitting the rotor blades, whereas for upwind wind turbines the air flow encounters the rotor blades first without additional disturbance of the tower structure. Again the development tendency is clear: the majority of the wind turbines are upwind and this is likely to remain so since the disadvantages of the downwind turbine, such as higher fatigue loads for the blade and the turbine components outweigh most of the time the advantages (e.g., the clearance between the rotor blade and the tower is always guaranteed).

14.3.7
Drive train Concept

Here the main characteristic is the rotational speed of the generator.

1. *Concept with a gearbox:* The geared concept, as the names suggests, requires a gearbox to raise the rotational speed of the main shaft to the rotational speed of the generator. There are a few variations of this concept. The two most common types used these days are the asynchronous doubly fed induction generator (DIG) and the synchronous generator with permanent magnet (PM). These are generators that have many industrial applications with relatively high rotational speed, which require a gearbox with a higher gear ratio with usually three stages. For very large wind turbines, the gearbox can be very large and heavy, which leads to a high tower top mass with the consequence that the tower and foundation have to be sized accordingly to take into account the high loads.
Few manufacturers have developed the so-called hybrid or compact gear solution. The main idea here is the balance between the generator and the gearbox mass. Instead of using a three-stage gearbox with high gear ratio to achieve the 1200–1500 rpm (rotations per minute) needed for the conventional generator,

it uses a single- or two-stage gearbox to produce 400–600 rpm. In this concept, the gearbox is lighter while the increase in the generator mass still makes the complete drive train mass less than that in the traditional gearbox-driven type. This concept generally uses a synchronous generator with Ps. A disadvantage of using a synchronous generator with PM is the material cost of the PM, which increased 10-fold between July 2009 and July 2011 [13].

The advantage of a synchronous generator with full power conversion is the higher efficiency at partial load with higher grid integration capability. However, advances in the power control of asynchronous doubly fed generators means that the cost increase of the PM has eroded the advantages of the synchronous generator with PM. For this reason, GE Wind announced recently that it will return to the DIG concept 10 years after it first introduced the synchronous generator with PM for its 2.5 MW wind turbine, citing the cost advantage as the main reason [14].

2. *Direct drive:* The generator speed of the direct drive concept is the same as the main shaft rotational speed, which means that there is no need for a gearbox. This is a huge simplification of the mechanical drive system, which in theory increases the overall system reliability of the wind turbine. Until recently, there are only a handful of wind turbine manufacturers that offer direct drive wind turbines, Emerson being the longest established since about 20 years. Worldwide the direct drive wind turbine constitutes about 21.2% of the total installed capacity [5]. The main disadvantage of the direct drive wind turbine is the higher cost that is related to the large ring generator. In order to accommodate the large number of pole pairs, the size of the radial flux generator is very large, hence the typical shape of the Enercon wind turbine with a large generator directly behind the rotor. The simplicity of the concept comes with a disadvantage, namely the relatively higher tower top mass compared with the tower top mass of a turbine with gearbox of a similar power rating. The tolerance on the generator air gap is fairly small, which means that for a large ring generator the allowable mechanical deformation of the generator has to be limited by using a very stiff structural design. This in turn adds a significant amount of passive structural mass to the generator design.

The success of Enercon wind turbines in the onshore wind energy market is also connected with the unique combination of turbine sale coupled with a long-term service agreement. For investors with little appetite for risk, this combination significantly reduces the uncertainty of the cost for every kilowatt hour produced by the turbine. This is especially attractive for small investors who do not have the technical capability to carry out maintenance of the wind turbines. In the last few years, about a dozen new direct drive wind turbines have been introduced to the market; however, most of them are mainly due to the different cost structure of offshore wind energy, where the operation and maintenance costs amount to one-third of the energy production cost [15]. With the implicit assumption that direct drive wind turbines are inherently more reliable as a system than geared wind turbines, this represents a potential cost reduction opportunity.

This assumption is subject to dispute from manufacturers that have opted for geared offshore wind turbines, claiming that the complexity of the electrical system of the direct drive wind turbine does not necessarily offer higher system reliability as research into reliability has shown that electrical components are the most vulnerable part of the wind turbine in term of reliability [16].

3 stages
+ proven technology
+ supply chain
+ good availability
− maintenance effort

1-2 stages

Geared

IG
+ grid events
+ easy grid frequency adjustment
+ raw material supply
+ long track record
− converter costs
− efficiency

Siemens (old)

+ compact design
− maintenance effort

DFIG
+ converter costs
+ power quality
+ long track record
+ raw material supply
− grid events

Vestas (old), Sinovel, GE, Suzlon, Repower, GUP…

Vestas V164, Areva, WinWind …

PMSG
+ grid events
+ easy grid frequency adjustment
− use of rare earth material
− shorter track record
− converter costs
− higher cogging

EESG
+ proven technology
+ raw material supply situation
+ simple converter design possible
+ easy grid frequency adjustment
− small air gap required
− electrical excitation unit required
− air-gap torque control (with simple converter)

Kenersys

− magnets limit the allowable rotor temperature

Vestas

Enercon, MTorres

Goldwind/Vensys, Siemens DD (new), XEMC…

− currently high production costs
− high tower head mass

+ high integration level possible
− rare earth costs
− not yet proved for mass production

+ low number of mechanical parts
− currently deeper knowledge for turbine integration necessary

Direct Drive

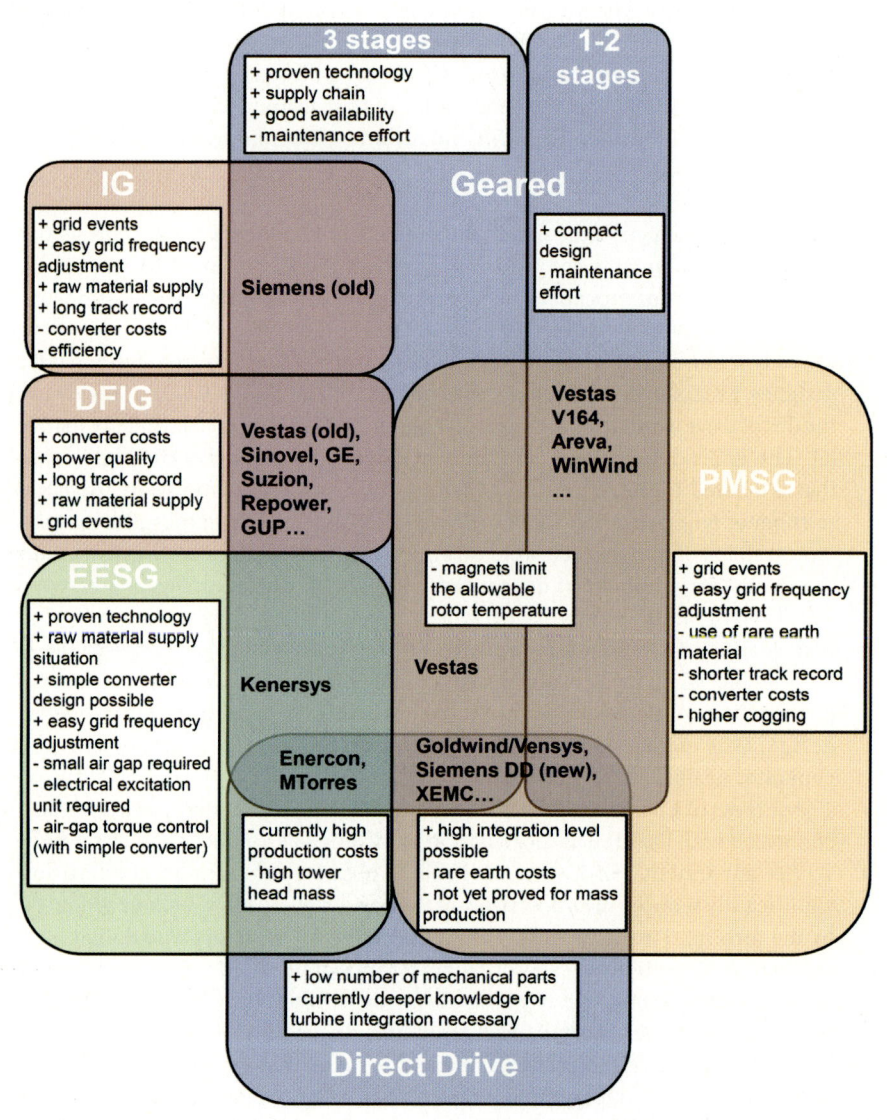

Figure 14.3 Advantages (+) and disadvantages (−) of the different drivetrain concepts. Reproduced from [17] with permission from J. Wenske, Fraunhofer IWES.

Nevertheless, it is generally agreed that direct drive wind turbines that use PM generators will suffer in the short to medium term from the impact of high magnet prices (neodymium). Here it is worth mentioning that Enercon is one of the few exceptions that kept to its original concept of using electromagnets which was not impacted by the price rise of PMs, a strategic decision that proved to benefit from significant cost advantages. It is unlikely that the price of PMs will decrease in the next few years owing to the high demand from other application areas such as electrical mobility, even with additional new supplies of rare earth materials from Australia and the United Sates.

For onshore wind energy, the direct drive concept will remain a viable concept, as Enercon has shown. However, the majority of the wind turbines installed onshore will be different variations of geared wind turbines, and this will remains so unless new concepts such as transverse flux generators with low manufacturing costs, different generator architecture, and so on, can break the weight and cost barriers of the traditional direct drive wind turbines.

Figure 14.3 presents a graphical summary of the main drive concepts with their advantages and disadvantages.

14.3.8
Tower Concepts

The traditional tower concept for the wind turbine is the tubular steel tower. It is easy to manufacture and has a relatively small footprint compared with the lattice type of tower. For most people it is also perceived as the visually most pleasant option for wind turbine towers. Two factors have driven new developments in tower design in recent years. One is the continuous search for better wind resources in low wind speed areas, which leads to the demand for taller towers in order to capture the higher wind resources at higher altitudes. The other is the continual growth of the wind turbine size that also requires higher hub heights. The steady increases in the price of steel and the demand for high structural stiffness mean that the tubular steel tower is reaching its limit. To develop low wind speed sites it is necessary to reach a greater hub height, typically between 100 and 150 m, in order to capture the better wind conditions at that height. For a tubular steel tower above 100 m it is difficult to keep the first natural frequency of the wind turbine above the rotational frequency of the turbine without substantial increases in the amount of steel materials, consequently increasing the weight and the cost. The other reason that limits the use of steel tubular towers with large hub heights is that the diameter of the tower section should be within the 4.3 m upper limit set by the transportation constraints. Therefore, tubular steel towers with hub heights > 100 m are not really an economic option.

Hybrid tower concepts use concrete tower sections that are fabricated in modules and transported to the site for assembly. To facilitate the connection with the nacelle and the yaw bearing, a transition piece is installed between the concrete sections of the tower with the steel sections. The steel section is usually bolted to the specially prepared transition piece. This type of hybrid tower offers the advantage of a cost-efficient way to raise the hub height, and together with the increase in rotor diameter, this makes low wind speed sites onshore economically more attractive for the deployment of wind energy [18].

A lattice tower is another way to raise the hub height of the wind turbine without a substantial mass increase. More importantly, the footprint of the tower base can be much larger than for the steel tubular tower since it will be assembled onsite and is not subject to the constraints of road transport. This increase in footprint also increases the moment of inertia and effectively reduces the stress of the members of the lattice tower. However, in terms of visual impact and public acceptance, the lattice tower still lags behind the more slender concrete and hybrid towers.

It is likely that more of the taller turbines with hub heights > 100 m will be installed in the near future owing to the wind energy development at lower wind speed sites, and more of the innovative tower concepts will be introduced to the market that can further reduce the cost of the taller towers.

14.3.9
Wind Turbine and Wind Farm Control

The dynamic behavior of the wind turbine and the need to operate in an uncontrolled environment means that the turbine must possess a robust and stable control system that can operate under very different wind conditions without sacrificing the power performance. The control system exercises a very significant influence on the wind turbine system loads and the power performance, hence it has to be carefully designed to make sure that the loads are minimized and the power output is maximized.

With the growth of the wind turbine size, the load on the components also increases; more specifically, the load generally increases faster than the increase in rotor diameter. This poses a problem for the scale-up of the wind turbine, where the ratio between energy outputs per kilogram mass becomes less attractive for larger wind turbines if the load increase is disproportionate. Therefore, the development of advanced control system has attracted much more attention with the increase in wind turbine size because of the potential that it offers for load reduction.

Currently, the main actuator of the wind turbine is the pitch system, which changes the blade angle of attack in order to keep the power output constant (i.e., maximum power) at wind speeds above the rated level. Most turbines operate with a so-called collective pitch system, where the pitch angle is the same for all the blades. Because of the wind shear, the mean wind speeds are not constant over the height; therefore, the blades experience asymmetric loads in each rotation. In order to reduce this load imbalance, individual pitch control has been introduced, where each of the blades is allowed to operate with a slightly different pitch angle, with the aim of minimizing the load imbalance. Individual pitch control has been adopted by many wind turbine manufacturers for wind turbines with large rotors, mainly to keep the load level as low as possible, and hence to keep the increase in tower top mass within a tolerable limit.

Recently, new control mechanisms have been introduced, notably the active rotor with aerodynamic control. The principle behind all the active rotors is the same, namely using electromechanical devices to influence the aerodynamic properties of the blade [19]. To be more precise, the lift and drag coefficients of the rotor are no longer fixed but can be shifted up and down by actuating these electromechanical devices. By adapting lift and drag forces, one can change the loads and/or the power output. The real potential of the active rotors is not so much in increase in the power performance but a substantial decrease in the loads, therefore making the design of very large rotor blades possible. The electromechanical devices could be flaps, slats, micro tabs, active suction, plasma actuators, and so on. The main challenges for the active rotor blades are the cost and reliability aspects. The cost should be lower than the potential benefits of deploying such active rotors, which include mass reduction or power capture increase due to the rotor size increase without a load increase. The aspect of reliability should not be neglected because adding more sensors and actuators to the system will automatically decrease the system reliability with everything else being equal.

Another approach to reduce the load is to predict the load ahead of time. This can be done, for example, by using a LIDAR (light detection and ranging) system that can measure the line of sight of wind speed at several hundred meters to several kilometers ahead of the wind turbine. By providing the wind turbine control with the wind information in advance, the control system can predict the dynamic behavior of the wind turbine and adjust the control system accordingly. This can be a wind gust that could have caused a larger load increase if no wind preview information had been provided to the wind turbine. Experiments with research wind turbines using LIDAR-assisted control showed that it is possible to reduce the fatigue loads of the wind turbine considerably together with a collective pitch system [20]. However, the results need to be verified also for larger wind turbines of megawatt size. Another challenge that the LIDAR-assisted control system faces is the cost of the LIDAR system. Currently, the deployment of a LIDAR system for load reduction purposes is economically unjustifiable. The cost of the LIDAR system will have to decrease significantly in order to make it a viable option for the wind turbine manufacturers.

So far, only the control of a single wind turbine has been discussed. With the increased significance of wind power in the power generation system, the wind park is no longer a simple collection of wind turbines but a wind power plant. The control of such a wind power plant need to fulfill the requirements set by the grid operators, mainly to guarantee the system stability and robustness. Modern wind farms have shown that the grid requirements, such as fault ride through, fluctuations in frequency and voltage, and power curtailment, can be met with the existing technology. One of the potential areas that can bring substantial benefits to wind farm operators is optimization of the wind farm with respect to the wake losses [21].

The wind speed behind a wind turbine is reduced compared with the undisturbed ambient wind speed, since kinetic energy has been extracted by the wind turbine. This creates a velocity deficit in the wind profile that in turn reduces the available wind power for the wind turbine in the second row. In the worst case, where the wind turbine in the second row is directly in the wake of the first turbine, it can mean up to 40% less power compared with the first turbine [22]. There are several ways to increase the power of the wind turbine in the second row, for example, by reducing the thrust, and hence the power extraction of the first wind turbine, or by giving a slight yaw misalignment to the first wind turbine so that the wake is deflected. The main challenge in this field of research is that the wake behind the wind turbine is a highly complex and dynamic phenomenon. Without a reliable model to describe the evolution and interaction of the wakes behind the wind turbine, it will be very difficult to devise an optimization strategy to increase the wind farm output. For this reason, most of the research effort is currently focused on the modeling of the wake and predicting the impact of the wake on the wind turbine. It is expected that in the long term this can yield an optimization strategy that could increase the overall wind park output by 10–15%.

14.4
Environmental Impact

The use of wind energy carries some consequences for the environment. The most notable and often mentioned environmental impacts are noise emission, danger to the bird population, loss of habitat, and ice throws, and a more subjective impact is the change in the landscape.

In terms of noise, modern turbines of the megawatt class produce much less noise than the turbines in the 1990s due to the advances in the aerodynamic and aeroacoustic aspects that improve the power capture and reduce the aerodynamic noise [23]. Aerodynamically efficient rotor blades are also those that produce the least noise. At the source, that is, the wind turbine, the sound power level is in the order of 105 dB(A), which is comparable to that of a lawnmower. For inhabitants close to the wind park, the most important measure is the sound power level perceived at the reception point. For a megawatt-sized wind turbine 300 m from the reception point, the sound power level is reduced to ~40 dB(A), which is comparable to that of a refrigerator [24].

For mixed residential areas, the noise level is usually regulated and should not exceed 55 dB(A) during the day and 40 dB(A) during the night. The sound power emission from the wind turbine increases as a function of the wind speed. However, above a wind speed of 12 m s^{-1}, the background noise from the wind is so dominant that the noise contribution from the wind turbine is insignificant compared with the overall background noise. If the noise generated from the wind turbine operation exceeds the level prescribed for the reception point, the modern wind turbine can be programmed to operate at a reduce power output with lower rpm during the night so that the maximum noise limit at the reception points is observed.

With the respect to bird impact, it is widely known that overall impact of wind turbines on bird fatality is several orders of magnitude lower than the fatality caused by other human factors, such as buildings, windows, transmission lines, vehicles, and domestic cats. There is no single siting rule for the wind farm with respect to the potential impact on the bird or bat population. Therefore, it is important to assess the site during the planning stage to estimate the potential impact of the wind farm on the bird and bat population. In general, wind turbines should be sited away from the migration routes of migrating birds, and avoiding large population densities of raptors and raptor prey. The use of modern large wind turbines has shown that the bird impact probability decreases with increase in the size of the wind turbine, and slower rotational speeds and an increased distance between the wind turbines are thought to have shown positive effects on bird fatality [25].

For wind turbines in cold climates, the formation of ice can be a hazard for the environment if it is located close to roads or human dwellings. Currently, wind turbines are operated in many parts of the world where icing occurs regularly. Modern wind turbines are equipped with special devices to cope with the effect of icing, including special steels to deal with the brittleness of the materials at low temperatures, specialized anemometers that prevent them from freezing (to prevent the anemometers from giving a wrong wind speed measurement), and additional heating for electrical equipment and the gearbox. Most wind turbines operated in cold climates are also equipped with ice sensors that detect the formation of the ice on the turbine. This is necessary for turbines sited close to roads where ice throw can pose a danger to traffic. When ice is detected, the turbine is shut down until a visual inspection has confirmed that the blade is ice free and can be operated again. However, ice detection on the rotor blades is still far from perfect since most of the ice detectors are installed on the nacelle and not on the blades themselves where the ice formation is most critical [26].

Currently, there are commercially available technologies that mitigate the effect of icing on the operation of the wind turbine, such as blade coatings that prevent ice formation on the blade and deicing devices installed on the blade to remove the ice mass through heating. The economics of such deicing precautions depend on the number of hours per year that the turbine has to be stopped due to icing and the associated loss of energy production.

14.5
Regulatory Framework

It is undeniable that without regulatory support, renewable energies, including wind energy, would have not expanded at the speed that has been seen in the past 20 years. While many see direct subsidy or tax credits as market distortions, it should be noted that the energy market and energy production have always been heavily regulated and subsidized regardless of the energy form, fossil, nuclear, or renewable. The true cost of the energy consumption cannot be determined based on the direct observable cost, since many of the goods that are impacted by the consumption of fossil fuels do not possess a market value, such as clean air, reduced health risk, and the contribution to mitigation of climate change. The difficulty that has been observed with the emission trading scheme in Europe shows how complex it is to put a market value even on something measurable such as the CO_2 emissions. Therefore, one should view the policy instruments in a broader socio-economic perspective as instruments to steer the energy production and consumption pattern to a desirable state that maximizes the benefits for the whole of society.

There are many policy instruments that have been used for promoting wind energy production. The most successful are without doubt the feed-in tariff pioneered by Germany where the installation of wind power has surged since 1990.

Feed-in tariff systems are characterized by a fixed guaranteed price paid to producers that feed in electricity to the grid gained from renewable energy sources [27]. This fixed price has been determined independently to account for the true generation costs, therefore giving the investor the certainty that if the plant is operated efficiently, a positive return can be expected. Furthermore, the wind energy producers have been granted prioritized access to the grid. Through tariff regression and price reviews every few years, the cost reduction effect due to technological advances and changing market conditions can be taken into account.

Whereas the feed-in tariff systems are price driven, a different policy instrument of quota obligations can be used to stimulate wind energy production [28]. The producers, sellers, or end users of electricity are obliged to cover a certain percentage of their electricity portfolio with renewable electricity. The non-fulfillment of the quota is sanctioned with fines. Those who cannot meet the quota are required to purchase renewable energy certificates from producers of renewable electricity at the market certificate price. The income from such energy producers is composed of two parts: the market price for the electricity produced and the price from trading with the certificates. The expected profits from both incomes are variable and the estimated long-term risk eventually determines the attraction of such renewable energy investments.

In the United States, a Production Tax Credit (PTC) has been implemented, which is a direct tax incentive of a fixed amount (US\$ 0.022 kWh^{-1}) given to wind energy generators for a predetermined period (10 years). The nature of the PTC as policy instrument has produced mixed results: on the one hand, it has encouraged significant increases in wind power installations so that the United States has become the second biggest market of wind power; on the other, the uncertainty with the extension of the PTC every time it expires has created a highly volatile market that discourages long-term planning of wind energy project developments.

14.6
Economics of Wind Energy

Energy is probably one of the few industries that have relied on constant government support in terms of direct or indirect subsidies, regardless of the source of the energy [29]. Nevertheless, one of the most frequently raised objections is that wind energy is not economic and requires substantial subsidies. Therefore, it is important to take a closer look at the cost of the wind energy. In this case, the cost of wind energy is defined as the cost to produce 1 kWh of electricity at a given site. Grid integration and balancing costs are not included as these will be discussed separately.

The cost of wind energy has decreased significantly due to performance improvements and equipment cost reductions. On the other hand, the cost of wind energy depends strongly on the site conditions, mainly the wind conditions, which will determine how much wind energy can be produced at the site. With the latest estimate from IEA Wind Task 26 [1], it is expected that the LCOE (levelized cost of energy) for wind onshore will be in the range $ 70–90 MWh^{-1}, with mean wind speeds ranging from 7.5 to 6 m s^{-1} at 50 m height. This low cost is partially due to the fact that currently it is a buyers' market, with overcapacity on the manufacturing side and the current equipment cost is significantly lower than that between 2005 and 2009.

For the long-term development of the LCOE, several scenarios have been analyzed and the results show a range of cost reductions between 0 and 40% [1, 4]. The main assumption behind the 0% cost reduction is that the upward cost pressure on the equipment will be maintained and the performance and technical improvements are used to neutralize the upward trend of the equipment cost that was observed in the period between 2004 and 2009 where the demand experienced a strong growth. A more reasonable range of cost reduction is the 20–30%, while 40% can be deemed overly optimistic.

For wind turbines, the largest part of the cost, roughly 75% [30], is related to the capital investment, as the wind turbine must be purchased, installed, and connected to the grid. This is in contrast to the thermal power plant, where 40–70% of the cost is related to the fuel and operation and maintenance (O&M) costs. The O&M cost of a wind turbine is estimated to be around €0.01–0.015 kWh^{-1} depending on the turbine type and the reliability track record.

When comparing the cost of wind electricity with that of electricity generated from thermal power, mostly the comparison is unfavorable for wind and for renewable energy in general. One of the reasons is how the electricity cost is calculated. Normally the electricity cost is calculated without taking into account the historical fuel risk, that is, fluctuations of the fuel price over the lifetime of the power plant. This represents a static view of the generation cost at the moment of the cost calculation. This was not an issue in the past when most of the electricity was generated from fossil fuels as they were all subject to fuel price fluctuations. However, with renewable energy entering the energy mix, this calculation method gives a skewed view of the actual cost of electricity generation. Essentially, almost risk-free electricity generation, onshore wind energy in this case, was not given any credit for offering long-term price stability over the rapidly changing gas or oil prices. In a study by the economist Awerbuch [31], it was shown that if the price of the fuel risk has been correctly discounted in the cost calculation, then the overall picture changes. Because the cost of electricity from wind turbines is more predictable in the long term than that of electricity from a gas or coal power plant, it should be discounted at a different rate than for the coal or gas power plant owing to the financial risk associated the fluctuating fuel prices. As consequence, the risk-adjusted cost of wind electricity (Figure 14.4) is lower than that from coal- or gas-fired power plants. Figure 14.4 shows the different power generation costs according to different calculation methods: the standard IEA approach, the fuel risk-adjusted method, and a third

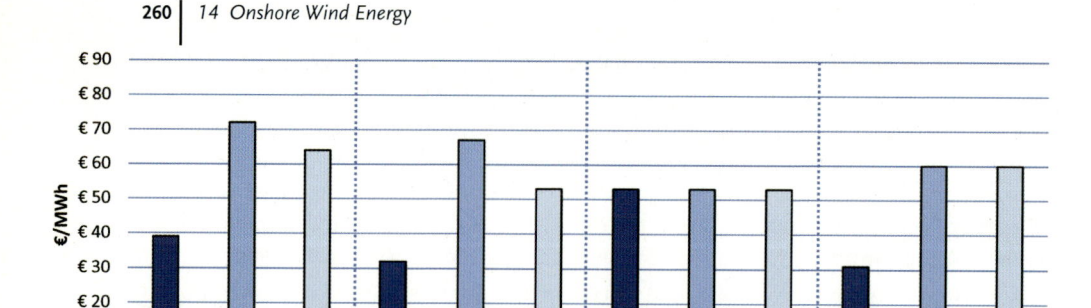

Figure 14.4 Risk-adjusted cost of energy. Data from [23].

scenario with a no-cost long-term fuel contract. The fuel risk-adjusted approach is consistent with the pricing of financial products where the risks of different products are taken into account by applying adjusted discount rates.

So far, only the electricity generation cost at the source (that is, at the wind farm) has been discussed. However, the cost of the electricity for the consumer needs to include also the other cost components, namely the need for balancing power and the upgrading of the grid infrastructure. The balancing power is needed in order to compensate for the variable wind power to meet fluctuating electricity demands. The grid infrastructure needs to be expanded because possibly the electricity is generated in one area and consumed in another area, which requires an expansion of the grid capacity to transport the electricity. However, it should be noted that an overall improvement of the grid infrastructure will benefit the transmission system and the power generation plants as a whole regardless of the source of the power generation. Therefore, this cost should not be attributed solely to wind power.

The balancing cost of wind energy depends on the characteristics of the reserve balancing power; in a grid with large hydropower as balancing reserve the cost of balancing power can be very low, less than €0.005 MWh^{-1} in Norway with 20% wind power penetration, whereas for the United Kingdom the cost of balancing power can rise to €4 MWh^{-1} owing to the lack of cheap balancing power such as hydropower. However, this value can be considered to be moderate as it amounts to less than 10% of the wholesale price of 1 MWh of wind energy [30]. In many European countries, the cost of balancing power is expected to be closer to the order of €2 MWh^{-1} of wind energy, with an assumed 20% wind power penetration.

The cost for grid infrastructure upgrading to accommodate more wind power will depend on the existing grid topology, capacity, and the transportation distance and the amount of wind power that needs to be transported. For 30% wind power in the total power mix, the cost for grid infrastructure upgrading is estimated to be in the order of €0.01–5 MWh of wind energy. This is similar to the cost of balancing power mentioned earlier. The current challenge with major grid upgrading to accommodate the offshore wind power, for example, is that many of the national grids are now controlled by several private operators as part of the market liberalization process. The cost of the grid upgrading is a significant burden on the local grid operator where the offshore wind power is being fed to the grid while the benefits of a grid upgrade can be felt system wide and not only by the specific grid operator. Therefore, grid upgrading is a task that needs to be undertaken/endorsed at the national level and preferably at the European level in order to divide the risk and the cost more evenly among all the beneficiaries of the grid upgrade.

14.7
The Future Scenario of Onshore Wind Power

How will onshore energy develop in the next 20 years and what role can it play in the transition to renewable energy? It is clear that onshore wind energy is currently the most affordable renewable energy source. It is available almost everywhere and can be harvested in a reliable way. Currently, the installed capacity of wind power is very concentrated in a handful of countries. These countries have implemented policies with financial incentives that encourage the development of wind energy on a large scale. The number of countries that will have a significant amount of wind energy will increase over time as the subsidies on fossil fuels become less and less sustainable and the pressure to reduce greenhouse gas emissions increases. It has been shown that a 30% wind power penetration is technically viable and can be implemented with reasonable investment in grid infrastructure [32].

Furthermore, the worldwide increase in the production capacity of wind turbines means that the cost of energy from wind will remain low and competitive. The challenge for countries with relatively high wind power penetration will be to find more wind resources, most likely in lower wind speed regions where deployment of wind energy has not been economically viable so far. Fortunately, the wind turbine manufacturers have recognized this market shift and have introduced a number of wind turbines specifically made for low wind speed areas. These high-capacity wind turbines will be able to capture more energy at lower wind speeds and therefore reduce the cost of energy.

Technology developments will contribute continuously to the reduction in the cost of energy, by finding the optimal ratio between the rating of the wind turbine and the capacity factor. Advances in rotor design and innovative materials will enable larger and lighter rotor blades to be developed. Innovative concepts and manufacturing technology for major components/subsystems of the wind turbine, drivetrain, tower, and control that can contribute further cost reductions of wind turbines and improvements in wind park control to reduce wake losses could add up to a 15% increase in power output.

The availability of affordable capital for development of wind energy projects will be crucial for the continuous growth of onshore wind power, especially for developing countries. The simple reason is that onshore wind power projects are capital-intensive investments in the initial stages with low running costs once the wind power plant is producing power. Access to affordable capital has become more difficult since the financial crisis in 2008. This effect has been mitigated as many governments have made additional capital available for infrastructure projects during the last few years and at the same time the level of financial incentives was maintained. However, with the lingering crisis, tepid recovery, and mounting government debts, the wind energy market is under pressure as financial incentives have been cut due to budget constraints (e.g., Spain) or policy uncertainties (e.g., United States). However, some emerging countries that were less affected by the financial crisis and still possess sufficient financial leeway to stimulate the development of wind energy (e.g., China, Brazil, South Africa) will see still substantial increases in wind power installations in the next few years.

Without a long-term and stable policy framework with clear rules, the risks of investing in wind energy would be deemed too high for many private investors and therefore it is necessary that clear long-term targets for climate goals and renewable energy are set by governments and international bodies while the financial incentives for investing in wind power need to be kept in place. It is important that these financial incentives are measured objectively to avoid being perceived as too generous, and at the same time the incentives should be reviewed and adjusted on a periodic basis to account for the cost reductions due to technological advances and changing market conditions. The growth of onshore wind energy will continue regardless of the difficulties mentioned earlier; the real uncertainty is how fast it will be able to grow to meet the prediction of 25% wind electricity by 2030 [4].

References

1 International Energy Agency (2012) *IEA Wind Task 26. The Past and Future Cost of Wind Energy*, Technical Report NREL/TP-6A20-53510, IEA, Paris.
2 Global Wind Energy Council (2012) *Global Wind Report. Annual Market Update 2011*, Global Wind Energy Council, Brussels.
3 International Energy Agency (2012) *Renewable Information 2012*, IEA, Paris.
4 Global Wind Energy Council (2012) *Global Wind Energy Outlook 2012*, Global Wind Energy Council, Brussels.
5 BTM Consult (2012) *International Wind Energy Development: World Market Update 2011: Forecast 2012–2016*, BTM Consult, Copenhagen.
6 European Wind Energy Association (2012) *Wind in Power – 2011 European Statistics*, European Wind Energy Association, Brussels.
7 International Energy Agency (2012) *World Energy Outlook 2012*, IEA, Paris.
8 Molly, J. P. (2011) Rated power of wind turbines: what is best? *DEWI Mag.*, **38**, 49–57.
9 National Wind Coordinating Collaborative (2010) *Wind Turbine Interactions with Birds, Bats, and Their Habitats: a Summary of Research Results and Priority Questions. Birds and Bat Factsheet*, National Wind Coordinating Collaborative, Washington, DC.
10 Griffin, D. A. (2004) *Blade System Design Studies. Volume II: Preliminary Blade Designs and Recommended Test Matrix*, SAND2004-0073, Sandia National Laboratories, Albuquerque, NM.
11 Siemens (2012) *World's Longest Turbine Blade*, http://www.siemens.com/press/en/presspicture/?press=/en/presspicture/pictures-photonews/2012/pn201204.php (last accessed 11 November 2012).
12 Wiser, R. and Bolinger, M. (2011) *2010 Wind Technologies Market Report*,

DOE/GO-102011-3322, US Department of Energy Office of Energy Efficiency and Renewable Energy, Washington, DC.

13 de Vries, E. (2012) *The Evolution of Wind Turbine Drive Systems*, http://www.windpowermonthly.com/news/rss/1129015/evolution-wind-turbine-drive-systems/(last accessed 11 November 2012).

14 Quilter, J. (2012) *After 10 Years GE Goes Back to DFIGs*, http://www.windpowermonthly.com/go/worldwide/news/1153928/10-years-GE-goes-back-DFIGs/(last accessed 11 November 2012).

15 Kaiser, M. J. and Snyder, B. F. (2012) *Offshore Wind Energy Cost Modeling*, Springer, Berlin.

16 Spinato, F., Tavner, P. J., van Bussel, G. J. W., and Koutoulakos, E. (2009) Reliability of wind turbine subassemblies, *IET Proc. Renew. Power Gener.*, 3 (4), 387–401.

17 Wenske, J. (2011) *Wind Energy Report 2011 – Special Report Direct Drives and Drive-Train Concepts Development Trends*, Fraunhofer IWES, Kassel.

18 O'Brian, H. (2012) *Towers – the Next Area for Innovation?*, http://www.windpowermonthly.com/news/indepth/1124461/Towers---next-area-innovation/(last accessed 11 November 2012).

19 Barlas, T. (2010) *Knowledge Base Report for UpWind WP 1B3: Smart Rotor Blades and Rotor Control for Wind Turbines – State of the Art*, Project Report Upwind, EU Contract Number 019945(SE S6), European Commission, Brussels.

20 Schlipf, D., Fleming, P., Haizmann, F., Scholbrock, A. K., Hofsäss, M., Wright, A., and Cheng, P. W. (2012) Field testing of feedforward collective pitch control on the CART2 using a nacelle-based lidar scanner, presented at Science of Making Torque from Wind, Oldenburg, Germany, October 2012.

21 Barthelmie, R. J., Politis, E., Prospathopoulos, J., et al. (2008) Power losses due to wakes in large wind farms, presented at WREC 2008, Glasgow.

22 Barthelmie, R., Frandsen, S., Jensen, L., Mechali, M., and Perstrup, C. (2005) Verification of an efficiency model for very large wind turbine clusters, presaented at the Copenhagen Offshore Wind Conference 2005, 26–28 October 2005.

23 Larsson, C. and Ohlund, O. (2011) Measurements of sound from wind turbines, presented at the 4th International Meeting Wind Turbine Noise, 11–14 April 2011, Rome.

24 GE Reports (2010) *How Loud Is a Wind Turbine?*, http://www.gereports.com/how-loud-is-a-wind-turbine/ (last accessed 11 November 2012).

25 National Wind Coordinating Collaborative (2010) *Wind Turbine Interactions with Birds, Bats, and Their Habitats: a Summary of Research Results and Priority Questions. Birds and Bat Factsheet*, National Wind Coordinating Collaborative, Washington, DC.

26 Laakso, T., Baring-Gould, I., Durstewitz, M., Horbaty, R., Lacroix, A., Peltola, E., Ronsten, G., Tallhaug, L., and Wallenius, T. (2010) *State-of-the-Art of Wind Energy in Cold Climates*, VTT Working Paper 152, VTT, Espoo.

27 Couture, T., Cory, K., Kreycik, C., and Williams, E. (2010) *A Policymaker's Guide to Feed-in Tariff Policy Design*, Technical Report NREL/TP-6A2-44849, US Department of Energy, National Renewable Energy Laboratory, Golden, CO, http://www.nrel.gov/docs/fy10osti/44849.pdf (last accessed 25 January 2013).

28 Fouquet D. (2007) *Prices for Renewable Energies in Europe: Feed-in Tariffs Versus Quota Systems – a Comparison*, Report 2006/2007, European Renewable Energy Federation, Brussels, http://www.eref-europe.org/attachments/article/49/EREF_price_report_2007.pdf (last accessed 25 January 2013).

29 European Environment Agency (2004) *Energy Subsidies in the European Union: a Brief Overview*, Technical Report No 1/2004, European Environment Agency, Copenhagen.

30 European Wind Energy Association (2009) *Economics of Wind Energy*, European Wind Energy Association, Brussels.

31 Awerbuch, S. (2003) Determining the real cost – why renewable power is more cost competitive than previously believed. *Renew. Energy World*, **6** (2), 53–61.

32 NREL (2011) *Eastern Wind Integration and Transmission Study*, Subcontract Report NREL/SR-5500-47086, National Renewable Energy Laboratory, Golden, CO.

15
Offshore Wind Power

David Infield

15.1
Introduction and Review of Offshore Deployment

Offshore wind is the latest, and possibly the most technically challenging, phase of wind power exploitation. The move from onshore to offshore is driven by a combination of technical and policy imperatives, not least the desire to avoid controversial planning applications for large wind farms onshore. There are clear technical attractions, including the generally stronger and more persistent wind speeds, lower turbulence levels, and the relative ease of handling the very large blades of multi-megawatt turbines that can be problematic to transport by road. In addition, there is the further attraction, for the many nations with coastlines, of proximity to major electricity load centers. It is well known that a disproportionate number of people live in large coastal cities, with London, New York and Shanghai being archetypal examples. It is often difficult for a combination of reasons related to wind resource and planning to locate onshore wind farms near such population centers, but offshore wind farms are often feasible, as for example the London Array wind farm located off the Kent coast in the United Kingdom. Early offshore wind farms understandably were located in modest water depths relatively close to shore. Vindeby, the world's first offshore wind farm, installed in 1991, is sited less than 2 km from the Danish coast at Lolland in the Great Belt (Storebælt), which is the largest strait between the Baltic Sea and the Kattegat, and in ~1–2 m of water (Figure 15.1). The 11 Bonus 450 kW turbines are still operational. They stand on gravity base foundations and have a hub height of 35 m. Since then, the Danes have constructed larger and more ambitious offshore wind farms, the first truly large one being the 160 MW Horns Rev 1 wind farm completed in 2002.

Both Belgium and The Netherlands with accessible offshore sites have significant capacity operational now and are building more. Belgium's first wind farm, completed at Bligh Bank in 2010, is known as Belwind, and comprises 55 3 MW Vestas V90 turbines. The Netherlands' first offshore farm was built at Egmond aan Zee in 2006 and also features 3 MW Vestas V90 turbines, in this case 36 of them.

Transition to Renewable Energy Systems, 1st Edition. Edited by Detlef Stolten and Viktor Scherer.
© 2013 Wiley-VCH Verlag GmbH & Co. KGaA. Published 2013 by Wiley-VCH Verlag GmbH & Co. KGaA.

Figure 15.1 Vindeby – the world's first offshore wind farm. Photograph copyright Bonus Energy A/S.

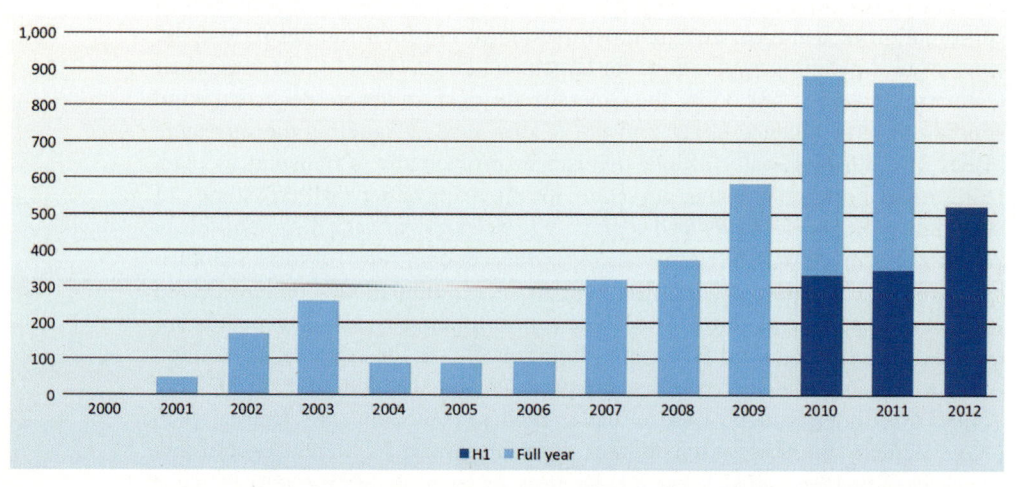

Figure 15.2 Growth of European offshore installed capacity in megawatts up to mid-2012. H1 indicates the first 6 months of the year, and Full year indicates up to 2009, and after that the remaining 6 months of the year. Source: EWEA [1].

The second Dutch offshore development, in 2008, is the Princess Amalia wind farm comprising 60 Vestas V80 2 MW turbines.

The growth of European offshore wind capacity, which accounts for most of the capacity worldwide, is illustrated in Figure 15.2.

Germany has more ambitious plans, currently centered around the Alpha Ventus site, also known as Borkum West. This is a highly challenging site in the North

Sea, 45 km from the nearest land (the island of Borkum). The first phase of the development was officially opened in April 2010. It consists of 12 turbines, of which six are 5 MW Areva Multibrid (M5000) turbines and the other six 5M models from REpower. The turbines stand in 30 m of water and are not visible from land. The REpower turbines are installed on jacket foundations (OWEC Jacket Quattropods) and the Areva turbines are installed on tripod-style foundations. Foundations will be discussed in more detail later but the professional view now is that tripod structures are too expensive to fabricate relative to the alternatives. It has been reported that the Areva Multibrid turbines have suffered from technical problems, in particular overheating, believed to be of the gearbox.

The United Kingdom started on wind relatively late but now leads the world in offshore wind deployment, with nearly 2 GW of capacity operational by late 2012. The process started just 12 years ago with the installation of two 2 MW turbines just off the north-east coast at Blyth. These were quickly followed by the so-called Round One sites that were generally within about 12 km of the shore and in water depths between 10 and 25 m. The first Round One wind farms were supported by direct Government grant in addition to the electricity trading incentive scheme[1]. These sites are clearly visible from shore and there are potential conflicts with other uses of the locations, such as fishing and shipping. These issues, combined with the better wind resource further from the shore, is driving the move to more remote sites, and these also open up the possibility of much larger wind farms. Dogger Bank, for example, is a licensed UK Round Three site that can accommodate a total wind capacity up to 13 GW spread over nearly 9000 km^2. However, there are severe engineering challenges associated with such developments, not least the repair of turbines remote from the shore and where average sea and wind conditions at these sites conspire to make access and heavy lifting difficult, and at times impossible.

According to the recent IPCC SRREN report on renewables, Chapter 7 on wind [2], estimates of the global technical potential for offshore wind energy range from 15 to 130 EJ a^{-1} (4000–37 000 TWh a^{-1}) considering only relatively shallower and near-shore applications. Of course, greater technical potential is available if deeper water sites are considered. It is impossible to provide definitive estimates of the offshore resource potential since in the end the economics will depend on technological developments without which possible sites cannot be feasibly exploited. New technologies, for example, in the field of turbine design, foundations, and connection to shore, would all be required to exploit far offshore sites with water depths over 40 m, and these developments would have to deliver electricity that is competitive with other forms of generation. There is already talk of floating wind turbines, and at least two prototypes are currently under test, the best documented being a conventional 2 MW Siemens turbine on a spar-type buoy located off the Norwegian coast. We will return to the prospects of such technology developments in Section 15.6.

---

1) In exchange for direct grant funding, these first farms had to supply operational performance data. This data are examined later in this chapter. It is regrettable that subsequent offshore wind farms were not required to disclose such data since this perpetuates a lack of public understanding as to how these investments are performing.

Figure 15.3 367.2 MW Walney offshore wind farm completed 2012. Copyright ecoGizmo.

As already mentioned, the United Kingdom is the world leader in terms of installed offshore capacity with 1858 MW online in 2012 (see, e.g., Figure 15.3), and also has 2359 MW currently under construction, with more than 42 000 MW in the pipeline. Other European countries have significant capacity and are reviewed above.

Asia is increasingly interested in offshore wind. South Korea, for example, has plans to build a 100 MW wind farm by 2014, a 400 MW project by 2016, and a 2000 MW development by 2019, totaling 2500 MW. China has plans to build its first genuinely offshore wind farm (as opposed to inter-tidal), and Japan, since the Fukushima nuclear disaster, has high hopes for offshore wind in the context of the Japanese public's demand for a non-nuclear future. By 2015, Japan plans to demonstrate its first offshore turbines: one 2 MW and two 6 MW machines to be located 16 km from the coast. If these are successful, the plan is to construct a 1 GW offshore array by 2020. It has been estimated that the Japanese offshore technical potential resource exceeds 1500 GW, which is more than eight times the present Japanese installed electricity generation capacity. Therefore, like the United Kingdom and many other countries with a coastline, lack of electricity will not be a problem in the long term. This, however, is not to underestimate the challenge of integrating this power into the various national, albeit increasingly interconnected, electricity supply systems. Such considerations are beyond the scope of this chapter. Chapter 8 of the IPPC SRREN report [3] provides a useful summary of the integration issues.

The total installed global offshore capacity is now (late 2012) approaching 5 GW. There is much under construction and much more in the pipeline, making the offshore wind sector a magnet for investors worldwide and also for companies with hardware or skills to contribute. There are, however, some surprising omissions from the list of countries with offshore wind assets in place. The most notable is the United States, where, despite a number of attempts to gain approval, offshore wind development has so far failed to persuade the sceptics and detractors.

It is hoped that this situation may change with the re-election of President Obama and an apparent re-invigoration of American plans for sustainable energy.

Although onshore wind is widely agreed to generate electricity on windy sites at costs comparable to conventional generation[2] (and indeed is sometimes claimed to be cheaper), offshore wind at this stage of development is considerably more expensive. Figures vary with the details of the site and the local wind resource but are often regarded at present as being roughly double onshore generation costs at the point of connection to the power system. Table 15.1 includes actual data provided on the UK Round 1 sites alongside data and estimates for other sites, and Figure 15.4 shows how the capital costs for the UK sites are distributed across the key elements identified. It should be borne in mind that such costs and their distribution are highly dependent on local site conditions, but they do provide a rare glimpse of actual costs.

Since offshore wind-generated electricity is currently so much more costly than that from onshore wind or conventional generation, the United Kingdom, with its major policy commitment to offshore wind, must find a way to reduce generation costs substantially. In support of this goal, in 2011 a group of stakeholders from industry, government, and the Crown Estate was formed into the Offshore Wind Cost Reduction Task Force (CRTF). Their recent report [5] is optimistic about the

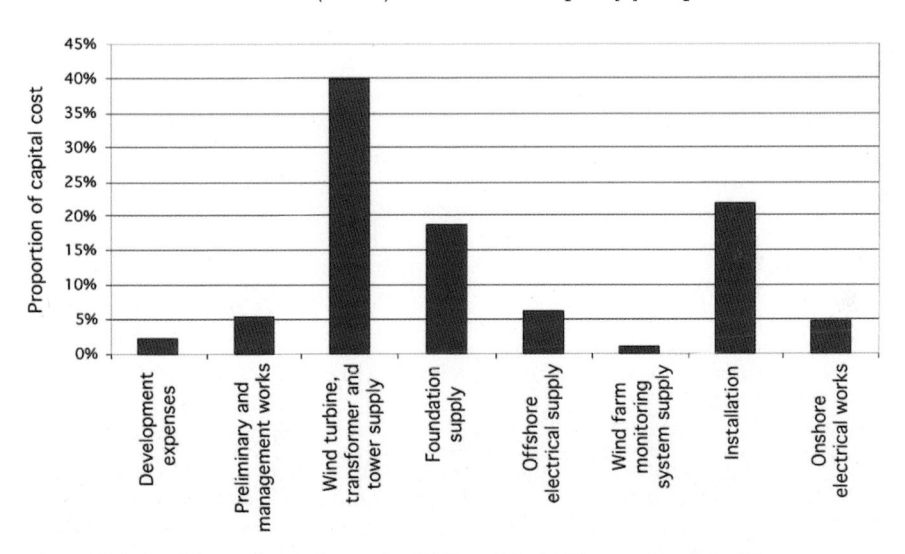

Figure 15.4 Breakdown of capital costs for UK Round 1 wind farms, taken from [4].

2) Fossil fuel costs for conventional power generation have been rather unstable in recent years, driven mainly by violent changes in the cost of traded gas. This makes comparison with wind generation problematic, as does the intrinsic subsidy of fossil fuels through the failure to date to properly internalize the external environmental costs of the associated CO_2 emissions. Nuclear costs are opaque to say the least, and although some proponents claim nuclear is cheaper than wind, this is impossible to prove (or disprove). A recent report in the Financial Times (4 December 2012) that the cost of EDF's flagship reactor under construction at Flamanville in Normandy has doubled from the €3.3 billion forecast to €8.5 billion confirms that real care must be taken when talking about the cost of nuclear power.

15 Offshore Wind Power

Table 15.1 Offshore capital costs (real and also estimated).

Development/estimate	Capital cost (£/kW)	O&M* (£/kW)	O&M* (p/kWh)	Life (years)	Cost of capital (%)	Load factor (%)	Levelised cost (£/MWh)
Future Offshore (DTI 2002)	1000		1.2	20	10	35	51[a]
Energy Review (DTI 2006d)	1500	46		20	10	33	79[a,c]
Danish Wind Industry (see footnotes)	1100		0.7	20	7.5	47[0]	33[b]
Horns Rev (DK)	1310[1]		0.7[2]	20	7.5	45[3]	40[a]
Nysted (DK)	1190[1]		0.7[2]	20	7.5	37[3]	42[a]
North Hoyle (UK)	1350[4]	35[5]		20	10	37[4]	60[a]
Scroby Sands (UK)	1250[6]	25[6]		20	10	34[6]	58[a]

Notes:

Exchange rates: £1 GBP = 10.9766 DKK

Discount rate:
- a 10% nominal (DTI 2003) and (DTI 2006d)
- b 5% real, assumed to be 7.5% nominal (as quoted in Danish Wind Industry 2003 and by assumption for Horns Rev and Nysted)
- c All costs in this table calculated 'overnight' – for simplicity neglecting interest during construction.

DTI 2006 published levelised costs include interests during construction, and on this basis their central estimate of costs is £83/MWh.

Technical data:
- 0 Approximation implied by data published by Danish Wind Industry – see http://www.windpower.org/en/tour/econ/offshore.htm
- 1 From (Garrad Hassan 2003)
- 2 From Danish Wind Industry (see above)
- 3 Operational data. Published by Wind Stats Newsletters (Vols. 18–20, 2005–2007 – see http://www.windstats.com), quoted figure averaged from the following quarterly data: Winter 2005 (0.57 Horns Rev, 0.5 Nysted), Spring 2005 (0.40, 0.33), Summer 2005 (0.30, 0.27), Autumn 2005 (0.54, 0.4), Winter 2006 (0.45, 0.35), Summer 2006 (0.27, 0.23), Autumn 2006 (0.58, 0.54)
- 4 npower 2006 report to DTI – http://www.dti.gov.uk/files/file32843.pdf
- 5 Long run estimate from npower's 2nd report to DTI http://www.dti.gov.uk/files/file32844.pdf.
- 6 Scroby Sands report to DTI, 2005: http://www.dti.gov.uk/files/file34791.pdf

* O&M costs may be annualized, capitalized or expressed per unit. We have used two conventions here following the relevant studies. In principle each convention can be converted to the other.

Table 15.2 Overview of offshore wind costs.

Sub-area	Descriptions	% COE
Turbines	• Current turbines are less than or equal to 5–6 MW, with 3 blades on a horizontal axis, and most designs use gearboxes to drive the generator • Turbines are installed in arrays to create large wind farms	28%
Foundation	• < 30 m foundations are usually simple steel tubes e.g. monopoles, although larger turbines may use more sophisticated foundations • Foundations suited for 30–60 m water depth are more sophisticated than monopoles and are fixed to the sea floor e.g. concrete gravity bases, tripods or jackets • Floating platforms could potentially be used in 60–100 m depths – for example, tension leg platforms; various spar buoy concepts are being developed outside of the UK for depths $\gg$ 100 m	19%
Collection & Transmission	• Currently high voltage AC (HVAC) cables are used to link turbines to an offshore substation, with power clean-up at each turbine • HVAC cables are also used to transmit power to the onshore substation as current wind farms are relatively close to shore within (60–80 km)	13%
Installation	• Currently oil & gas vessels that jack-up from the seabed to install the foundation and turbine. Dynamic positioning (DP2) vessels have also been used to a certain extent, but this is not yet the norm	22%
O&M	• Current access technologies involve helicopter transfers and direct boat access from shore which works best in calm seas • Limited remote condition monitoring	18%

potential for cost reduction, concluding that, "based on the evidence gathered ... the CRTF concludes offshore wind can reach £100/MWh by 2020.". To achieve this cost reduction, significant progress is required on a number of fronts, not least reduced operation and maintenance costs (O&M). This report conveniently summarizes the breakdown of total offshore wind costs and comments on them based on interviews with experts conducted by BVG Associates for a 2008 Carbon Trust report [6] (Table 15.2).

15.2
Wind Turbine Technology Developments

Wind turbines have developed remarkably rapidly during the last 25 years or so, commercial machines having typically increased in rated capacity over this period from 100 kW to around 3 MW. The latter is probably the largest cost-effective size for onshore sites, although this does vary according to site constraints and cost of access to land, being smaller typically in China than Europe.

Table 15.3 Selected planned offshore prototypes showing wind technology evolution.

Manufacturer	Model	Date of full-scale prototype test[a]	Rated power (MW)	Generator type	Gearbox stages[b]
Areva	M5000	2004	5.0	PM	2
Repower	5M	2004	5.0	DFIG	3
Bard	Bard 5.0	2008	5.0	DFIG	3
Repower	6M	2009	6.2	DFIG	3
Bard	Bard 6.5	2011	6.5	PM	3
Siemens	SWT 6-120	2011	6.0	PM	0
Sinovel	SL 6000	2011	6.0	DFIG	3
GE Energy	4.1-133	2011	4.1	PM	0
Guodian UP	UP-6000	2012	6.0	DFIG	3
Alstom	Haliade 150	2013	6.0	PM	0
Gamesa	G11X-5.0	2013	5.0	PM	2
Vestas	V164/7.0	2013	7.0	PM	2
Nordex	N150/6000	2013	6.0	PM	0
Gamesa	G14X	2014	7.0	PM	2
Sway	Sway	2015	10.0	–	0
AMSC	Sea Titan	2015	10.0	PM	0

a Note that forward dates may change.
b 0 indicates direct drive.
Adapted from work by Ander Madriaga and colleagues at the University of the Basque Country, Spain.

The economics of offshore operations, as outlined in Section 15.1, are driving another round of scaling up of the technology, well beyond the size of the largest onshore machines. This scaling up to around 6 MW today with anticipated developments taking turbine ratings to perhaps 15 MW over the next decade or so is also driving changes in design. Whereas a mature 2 MW turbine typically uses a doubly fed induction generator (DFIG) to achieve variable-speed operation and includes a gearbox to take the generator speed to around 1500 rpm, larger turbines tend to use fully rated converters. Present designs include direct-drive arrangements (with no gearbox) with large multi-pole, usually permanent magnet (PM), synchronous generators, and conventional drivetrains with nominally 1500 rpm generators (either induction or conventional wound-rotor synchronous machines, or hybrids with PM generators and low ratio gearboxes). It has been claimed by Romax, the drivetrain modelers and analysts, that the latter option is the most cost-effective for turbines over 7 MW, but in truth, until there is more operational experience, this is a hard judgment to make. Table 15.3 summarizes the characteristics of a number of large turbines under development around the world, specifically for offshore application (indeed, these turbines would be too large to be cost-effective onshore where different economies of scale apply). It can be seen that there is a clear trend in time towards larger turbines, and that there would appear also to be a convergence towards the use of PM electrical machines. Whether this tendency persists will depend on the costs of the rare earth magnets required, and these have been rising steeply recently.

15.3
Site Assessment

Most wind site analysis onshore draws on simplifications regarding the atmospheric boundary layer. Generally the atmosphere is regarded as neutrally stable, and for high wind speeds onshore this is a reasonable assumption as mechanical mixing tends to dominate over convection in determining the characteristics of the fluid motion and turbulence. The atmosphere follows the adiabatic lapse rate (i.e., the gas laws apply with rising height) and the wind profile can be described by a simple logarithmic law with no stability terms. Offshore these simplifications cannot be relied upon as the atmospheric process is more complex and in particular is strongly affected by the difference in temperature between the sea surface and the air. Either stable or unstable conditions may persist for longer times and distances and with higher wind speeds, complicating significantly the physics and also changing the height of the boundary layer.

These complex and interconnected factors also affect wake development in ways that are still not fully understood. In general, turbulence levels seem to affect the rate of mixing between a wind turbine's wake and the undisturbed air around it, thus diluting the velocity deficit in the wake and dissipating its vortex structure. The generally lower turbulence levels offshore mean that wind turbine rotor wakes persist for much greater distances than onshore and as a result wake effects are much greater than had been anticipated by the offshore wind farm designers, and as a result higher packing densities (i.e., inter-turbine separation distances that are too small) have been used than would have been economically optimal. Reported yields from some major offshore wind farms are thus significantly lower than planned, undermining confidence in the present design tools used by the industry. Of course, it is not difficult to incorporate this experience in the design and layout of the next generation of offshore wind farms, guided by these operational data, and in the fullness of time improved offshore meteorologically informed wake models and planning tools will be developed.

An improved understanding of the offshore meteorology is also required at the stage of site assessment. Wasp, as originally developed for onshore wind site appraisal, has been extended to try to compensate for offshore atmospheric stability, but despite this no offshore wind farm can be developed without measurements made on-site. A full-height offshore mast is expensive but necessary at this time. Research is under way, however, to explore measurements made from sea level, for example, using LIDAR (a light detection and ranging technique based on the use of lasers and the Doppler shift of light reflected off moving particles in the flow), and these may well become trusted enough in the years to come to replace met masts. Even when the data have been collected, this needs to be extrapolated to estimate the long-term resource at the site, just as for onshore. This is usually done by looking backwards through the use of met data. The challenge offshore is a lack of credible offshore data, and using Measure–Correlate–Predict (MCP), the standard hind casting methodology, can be problematic if onshore met data are used as the reference. This is because of the lack of similarity in meteorological conditions between land and sea, and also due to greater distances of potential sites from the land, simply because the correlation breaks down with distance. Both of these factors reduce the temporal correlation of wind speed values at the two sites that is essential to effective MCP.

15.4
Wind Farm Design and Connection to Shore

The inter-machine spacing aspect of offshore wind farm design has already been mentioned. Clearly, larger spacing requires more cable to interconnect the turbines, so there is a design trade-off here. Various different designs of offshore collector arrangements are possible, the main issue being whether ring-type connections are used in place of more conventional radial configurations, with the differences being primarily in terms of reliability in the face of sub-sea cable faults. Similarly, there are choices regarding the connection to shore, with multiple cables allowing a degree of robustness in the event of cable faults. As offshore wind farm size has grown, so has the connection voltage. The first offshore wind farms were connected to shore at 33 kV or similar. Today's wind farms are typically connected at 125 kW with 33 kV collector systems within the wind farm. An offshore substation, usually supported on a jacket structure, is used to raise the wind farm collector circuit to the voltage of the shore connection. This arrangement or a similar one is fine until the distance to shore exceeds about 80 km. At this stage, AC sub-sea cabling becomes problematic and the use of high-voltage DC (HVDC) is anticipated. Point-to-point HVDC technology is readily available but of course costly. Because of this, and because some parts of the sea, especially around the United Kingdom, are expected to see a number of wind farms, it makes sense to consider whether interconnected HVDC links offshore might not be more cost-effective. Illustrations of such meshed HVDC systems are often included in articles in technical magazines, but it is salutary to note that this technology is not yet commercially available and there remain

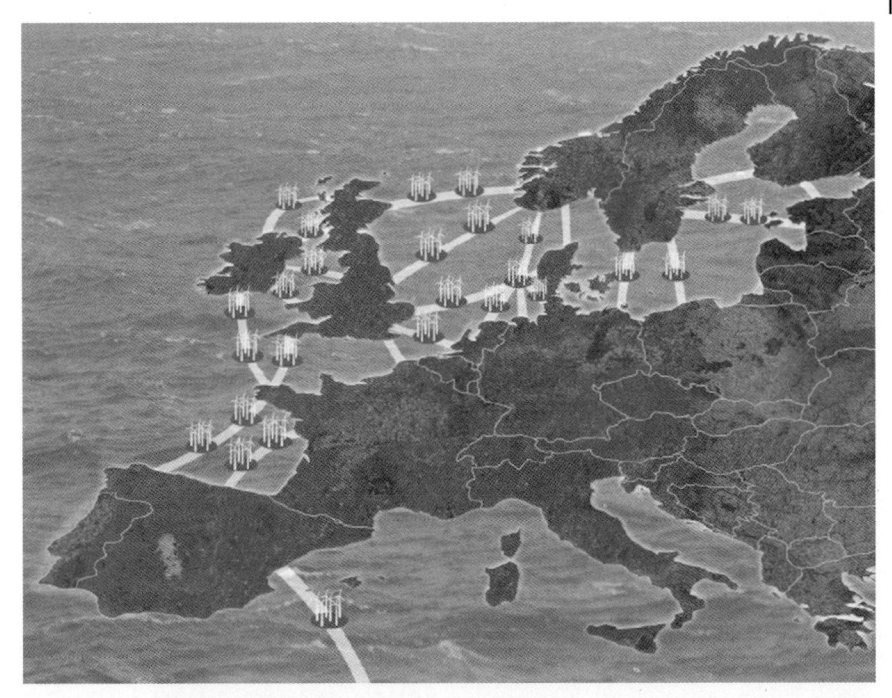

Figure 15.5 European DC supergrid concept.
Source: taken from [7] and reproduced with permission of Airtricity.

serious concerns as to how such HVDC networks would be protected in the event of cable faults. An example of a meshed HVDC network in the North Sea is shown in Figure 15.5 [4]. The ideas are not completely fanciful and sub-sea HVDC will be used increasingly in UK waters over the next few years, first to provide so-called HVDC bootstraps to the UK's existing north to south AC transmission system to ease the flow of onshore wind power from Scotland to load centers in the south of England, and second to supplement the existing HVDC connection to France with HVDC links to The Netherlands and Norway.

Foundation choice and detailed design can significantly affect the cost of offshore wind since the cost accounts typically for around 20% of the cost of wind power (Table 15.2). The most suitable foundation type will depend primarily on the water depth, but is also affected by the sub-sea terrain and soil. The range of possible foundation types is illustrated in Figure 15.6; the most common to date is the monopile by a large margin. Nevertheless, this approach has not been without technical challenges, a particular problem being the reliability of the grouting between the sub-sea pile and the transition piece that links it to the turbine tower, and allows for any inaccuracy in the verticality of the piling to be compensated for. However, for the largest turbines (5 MW and above) in water deeper than 35 m, monopiles (or at least today's monopile technology) cannot be used and tripods or jackets need to be considered. Recent analysis suggests that jackets will be more cost-effective owing to the relative ease of fabrication using standard welding technology.

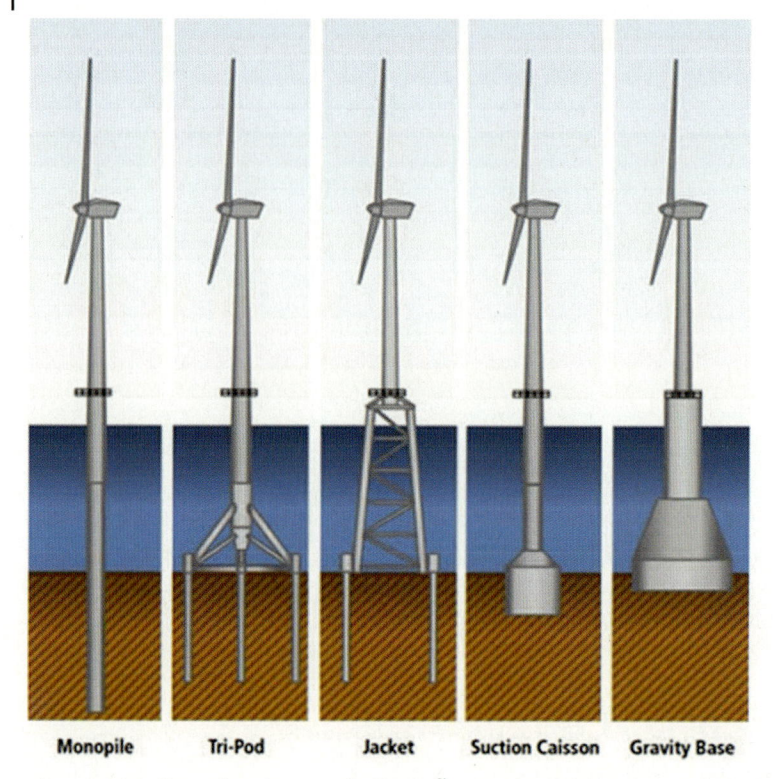

Figure 15.6 Different foundation technologies[3] currently available. Source: taken from [3].

Monopile • Tri-Pod • Jacket • Suction Caisson • Gravity Base

15.5
Installation and Operations and Maintenance

Because the cost of O&M is much more significant offshore than onshore, it is important to consider how this may be reduced. The O&M figure of 18% in Table 15.2 reflects current offshore wind farms; UK Round 3 sites, like Germany's Alpha Ventus, are further from the land and also experience more challenging sea conditions, with mean annual significant wave heights of over 2 m, in contrast to the relatively benign near-shore or sheltered sites with significant wave heights more typically around 1 m. The key issue impacting on O&M and lowering turbine availability is access. For more remote sites, it may well prove impossible to access turbines for months at a time, and even if access can be managed, say by use of helicopters, major repairs cannot be completed in high seas and winds owing to the need for heavy lifting. Figure 15.7, taken from some recently published research [8],

---

3) All terms are self-explanatory except perhaps the suction caisson where sand and loose material are sucked into a steel or concrete structure both to embed the foundation in the sub-sea material and also to add mass to the structure.

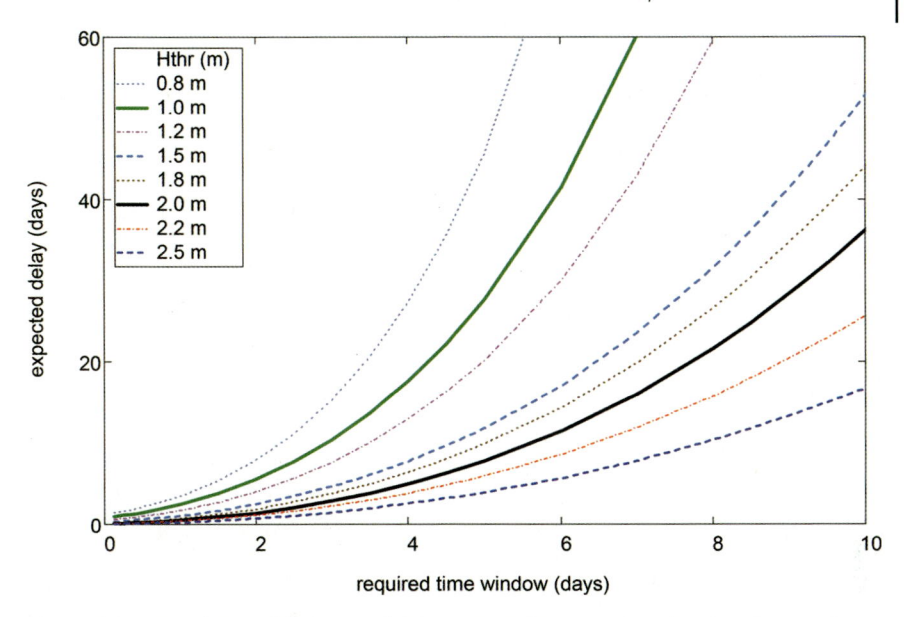

Figure 15.7 Dependence of the expected delay time on the time required and on the vessel operational limits.

shows how the expected delay time to repair caused by waiting for a suitable weather window depends critically on both the time taken to effect the repair and also the significant wave height that the repair vessel can cope with, H. This example was for Barrow, one of the Round 1 wind farms located near shore and subject only to modest seas. It is apparent, however, even for this benign site, how tens of wind turbine operational days can be lost waiting to undertake the more major repairs such as generator or gearbox replacement that in themselves take only days to complete. Round 3 sites will be far worse, and months of operation may be lost waiting for repair, resulting in low turbine availability. Onshore availability of 97% or higher is often these days taken for granted, but it would be foolish to assume that similar figures can be achieved offshore. High onshore availability depends in part on a rapid response to faults, but as can be seen from the preceding discussion, this will not always be possible. It is therefore essential that steps be taken to improve offshore availability through a combination of better, more reliable design, design for quick and easy maintenance and repair (sometimes at odds with low-cost integrated design solutions), and improved access and repair vessels able to operate in more extreme wind and wave conditions. Most analysts also expect a move from reactive to preventative maintenance, the latter ideally based on accurate and robust condition monitoring systems.

15.6
Future Prospects and Research Needed to Deliver on These

Since one vision of future offshore connection technology has already been presented, this last section will focus on the next generation of offshore turbines.

Future offshore wind turbines may well depart from the onshore upwind Danish concept design (in which the turbine rotor is maintained upwind of the tower, usually by a yaw drive system and controller). They are just as likely to be downwind (with the rotor operated downwind of the tower) as upwind. They may have three blades like most contemporary turbines, but alternatively they may have just two blades to reduce weight and to ease installation, the higher aerodynamic noise emissions from such rotors being unlikely to be an issue offshore. They may use conventional drivetrains or hybrid drivetrains or be direct drive, or even possibly utilize multiple very high-speed PM machines or hydraulic transmissions as under investigation by Mitsubishi. These future turbines may well be larger and more flexible[4], and thus lighter, than anything seen to date, and they may even comprise multiple rotors. Vertical axis turbine may become competitive owing to their ability to scale up, even to, say, 50 MW. Nothing is certain here and it will inevitably take time for offshore turbine designs to converge, as they have done onshore.

One area that perhaps has not received enough attention is the use of power electronics, or rather its excessive use, in offshore wind systems. Large modern turbines operate at a variable speed for a number of good reasons, drive load reduction and improved energy capture being perhaps the most important. As already mentioned, the very large offshore wind farms are likely to be connected to shore using HVDC. Simply combining contemporary turbines with HVDC would result in conversion from AC from the PM machine to DC to AC for connection to the offshore AC collector system, then to DC for HVDC transmission to shore, and then finally to AC for connection to the onshore power system. This is far from optimal and, when recognition of the relatively poor reliability of power electronic conversion is considered, patently stupid. Over 20 years ago, Infield and Leithead proposed a system using stall regulated turbines with low-cost, robust, squirrel cage induction machines connected to a floating frequency local offshore network. This concept has the further advantage of removing the complexity of pitch regulation and its associated limited reliability. As might be expected, there are minor reductions in wind turbine performance, but recent studies conducted at Strathclyde University indicate that these losses are minor provided that the turbines are connected in groups of 10 or fewer.

The wind chapter of the IPCC SRREN report [2] concluded that, "The cost of offshore wind energy exceeds that of onshore wind energy due, in part, to higher O&M costs as well as more expensive installation and support structures." Component-level technological advances can contribute to lower offshore wind energy costs, and some of those advances may even be driven by the unique aspects of offshore

---

4) Mechanical flexibility is a well-proven engineering approach to strength and thus cost reduction, but it requires much improved models, especially where unsteady aerodynamics are involved.

wind energy applications, such as distance from people and transportation to sites not directly restricted by road or other land-based infrastructure limits. New turbine concepts, as reviewed above, may make an impact but this is very hard to judge. However,, as summarized in Table 15.2, it is most likely that the major reductions in cost will be achieved by improved installation techniques and maintenance strategies, together with improved support structure design and manufacturing.

The most predictable change will be the move to floating support structures to give access to a much greater wind resource. The first floating turbine (Figure 15.8), a 2 MW Siemens turbine, was installed in 2009 by Statoil, the Norwegian national oil company, just off the Norwegian coast in a water depth of 210 m and using a spar buoy support structure. For coastlines such as Norway's that fall away steeply, there is little alternative to floating turbines offshore, hence their interest in developing the technology. However, in other locations, floating turbines can, for example off parts of the UK coastline, allow turbines to be located closer to land than the shallow water sites available in the region, thereby reducing the costs of cable connection to shore. Such savings, however, will be required to offset the considerably higher present costs of floating support structures and the associated tethering. This first floating demonstration machine is conservative in the sense that very little movement of the turbine is permitted; future lower cost structures may well allow greater displacements in order to reduce structural loads on the support structure, but at the expense of potentially higher unsteady aerodynamic loads. Such an approach (Figure 15.9) has been utilized in the world's second demonstration ~5 km off the

Figure 15.8 The world's first full-scale floating wind turbine, Hywind, being assembled in the Åmøy Fjord near Stavanger, Norway, in 2009, before deployment in the North Sea. Image downloaded from Wikipedia, 2012.

Figure 15.9 The world's second full-scale floating wind turbine, the 2 MW WindFloat. Image downloaded from Wikipedia, 2012.

coast of Agucadoura, Portugal, that has the further distinction of the being the first to be installed without the use of a heavy lift vessel, highlighting a further advantage of this concept.

The hidden challenge with such developments is the creation of a design process that is reliable. This in turn requires much improved aerodynamic models that can cope with both flexible rotors and large rotor displacements caused by both wind loads and the motion of the sea, even to the extent that the rotor may intercept its own wake, causing blade vortex interaction. Truly interactive hydrodynamic and aerodynamic models will also be required. This should keep researchers busy for many years, but more seriously will require major resources and thorough experimental validation, which are never cheap. Still, compared with the size of the planned offshore investments that are measured these days in billions of dollars, they should be easily affordable. Whether such essential underpinning research is supported will depend on the research funding agencies, which sadly are not known for their foresight.

References

1. European Wind Energy Association (2012) The European Offshore Wind Industry – Key Trends and Statistics, 1st half 2012, EWEA, Brussels.
2. Wiser, R., Yang, Z., Hand, M., Hohmeyer, O., Infield, D., Jensen, P. H., Nikolaev, V., O'Malley, M., Sinden, G., and Zervos, A. (2011) Wind energy, in *IPCC Special Report on Renewable Energy Sources and Climate Change Mitigation* (eds O. Edenhofer, R. Pichs-Madruga, Y. Sokona, K. Seyboth, P. Matschoss, S. Kadner, T. Zwickel, P. Eickemeier, G. Hansen, S. Schlomer, and C. von Stechow), Cambridge University Press, Cambridge, pp. 535–607.
3. Sims, R., Mercado, P., Krewitt, W., Bhuyan, G., Flynn, D., Holttinen, H., Jannuzzi, G., Khennas, S., Liu, Y., O'Malley, M., Nilsson, L. J., Ogden, J., Ogimoto, K., Outhred, H., Ulleberg, O., and van Hulle, F. (2011) Integration of renewable energy into present and future energy systems, in *IPCC Special Report on Renewable Energy Sources and Climate Change Mitigation* (eds O. Edenhofer, R. Pichs-Madruga, Y. Sokona, K. Seyboth, P. Matschoss, S. Kadner, T. Zwickel, P. Eickemeier, G. Hansen, S. Schlomer, and C. von Stechow), Cambridge University Press, Cambridge, pp. 609–705.
4. Gross, R., Heptonstall, P., and Blyth, W. (2007) *Investment in Electricity Generation: the Role of Costs, Incentives and Risks*, UK Energy Research Centre, London.
5. UK Department of Energy and Climate Change (2012) *Offshore Wind Cost Reduction Task Force Report*, DECC, London.
6. Carbon Trust (2008) *Offshore Wind: Big Challenge, Big Opportunity*, Carbon Trust, London.
7. Freris, L. and Infield, D. (2008) *Renewable Energy in Power Systems*, John Wiley & Sons, Ltd, Chichester, 2008.
8. Feuchtwang, J. and Infield, D. (2012) Offshore wind turbine maintenance access: a closed-form probabilistic method for calculating delays caused by sea-state. *Wind Energy*, in press; Article published online: 1 August, 2012, DOI: 10.1002/we.1539.

16
Towards Photovoltaic Technology on the Terawatt Scale: Status and Challenges

Bernd Rech, Sebastian S. Schmidt, and Rutger Schlatmann

16.1
Introduction

Photovoltaics (PV) (direct conversion of solar radiation into electricity) has developed into a mature technology during the past few decades. The Sun delivers almost 10 000 times more energy per year to the Earth's surface than the global population consumes during the same period. Considering this vast amount of energy, PV has almost unlimited potential to contribute to the transformation to a fully renewable energy system. The growth rate of the PV industry and related global PV installations was more than 40% per year during the last decade, and thus a significant new industry has been created. However, since 2011, oversupply and overcapacity have had a serious impact on many solar companies and the industry is now in a consolidation phase. On the other hand, the resulting dramatic price decrease of solar modules will open up new markets.

Before going into the outline of this chapter and into the technical section, it is useful to highlight a few facts:

- The unit Wp (watt-peak) characterizes the output of a solar cell under so-called standard measurement conditions (see Section 16.2). This typically reflects illumination at noon on a clear and sunny summer's day. As a rule of thumb, 1 Wp of PV power produces about 1 kWh of electricity per year in central Europe and around 2 kWh in the sunbelt of the Earth. Accordingly, full operating hours are 1000 and 2000 h per year, respectively.
- 29 GWp of PV modules were installed in 2011 [1]. The total cumulative worldwide installation of PV modules amounted to 70 GWp in 2011 [1].
- PV systems produced and installed using today's state-of-the-art technology produce much more energy than is required for their production. Depending on production technology and the region where modules are installed, energy payback times are between 0.5 and 5 years [2], while solar modules are designed for a typical lifetime of 25 years.

Transition to Renewable Energy Systems, 1st Edition. Edited by Detlef Stolten and Viktor Scherer.
© 2013 Wiley-VCH Verlag GmbH & Co. KGaA. Published 2013 by Wiley-VCH Verlag GmbH & Co. KGaA.

The Section 16.2 briefly describes basic solar cell operation principles and discusses physical limits of energy conversion efficiency. The production sequence of today's mainstream technologies will be given with a focus on the "classic" silicon wafer-based technology, which currently represents almost 85% of the world market. The Section 16.3 provides up-to-date information on the technical design and cost structure of PV systems. A solar system installed in Berlin in 2012 will serve as a "best practice example" because the German PV market is currently the largest and most developed worldwide. A simple scenario based on market and technology data will be given to show opportunities for large-scale use in different areas and applications. Section 16.4 looks ahead and provides a description of cutting-edge technologies which may take over market segments in the years to come. Specific examples selected by the authors reveal efficiency potential or new application fields that are currently underdeveloped. Finally, in Section 16.5, strategies and requirements for next-generation PV technology are discussed, addressing very high conversion efficiencies at low cost and the challenge of grid integration and solar energy storage.

16.2
Working Principles and Solar Cell Fabrication

Solar cells directly convert light into electricity. The thermodynamic efficiency limit of this conversion process is above 80%, determined by the surface temperature of our Sun of ~5800 K. The world record efficiency realized so far under concentrated sunlight with a multi-junction solar cell is 43% [3, 4]. An introduction to the physics, operating principles, and fundamental limits of solar cells can be found in many textbooks on semiconductor devices and dedicated books on solar cells (see, e.g., the book by Würfel and Würfel [5]).

Figure 16.1a sketches the basic working principle of a solar cell based on a p–n semiconductor junction. Light can be absorbed in the solar cell and generate electron–hole pairs if the photon energy is larger than the bandgap of the semiconductor. However, the carriers lose their excess energy (photon energy minus bandgap energy) in a very fast process called thermalization. Further loss processes are reflection losses at the front side of the solar cells and so-called recombination losses, when carriers recombine before they reach the contacts. In a famous paper, Shockley and Queisser [6] derived an efficiency limit for a single-junction solar cell of somewhat above 30% under standard solar irradiation and about 40% under highly concentrated sunlight. The theoretical limit for multi-junction cells with an infinite (or very large) number of component cells, each adapted to a specific part of the solar spectrum, is higher than 80% [5] because virtually no excess photon energy is lost. The world record efficiencies of different solar cell technologies are updated and published on a regular basis [3]. The laboratory benchmark for crystalline silicon technology – the PV mainstream today – is 25% in conversion efficiency [3, 7].

In general, solar cells are characterized by their current–voltage (j–V) characteristic measured under illumination. Such a j–V curve is plotted in Figure 16.1b.

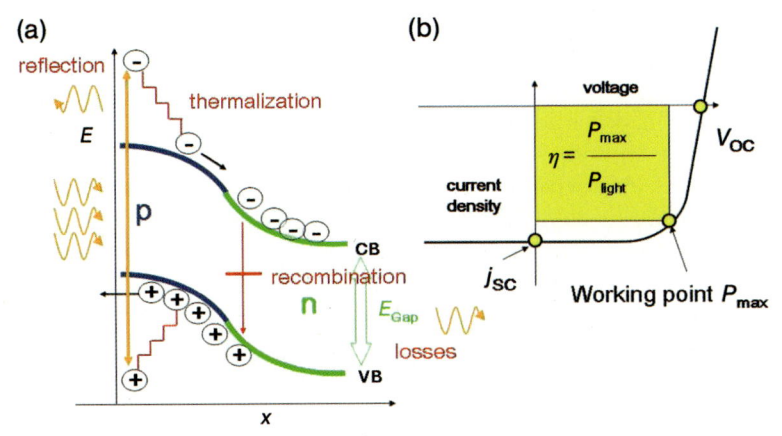

Figure 16.1 (a) Energy-band diagram of a p–n homo-junction solar cell. Reflection and incomplete absorption of the sunlight, together with thermalization and recombination of free charge carriers in the conduction band (CB) and valence band (VB), lead to losses during the conversion of solar to electrical energy.
(b) Current density j as a function of the applied voltage V of a solar cell. The solar cell conversion efficiency is given by the ratio of the maximum power output of the solar cell P_{max} to the power of the insolation P_{light}. The open-circuit voltage V_{oc} and the short-circuit current density j_{sc} are parameters that may be analyzed to obtain information about the working principle of a particular solar cell.

Key parameters usually used to characterize solar cells and modules can be extracted from this curve (see Figure 16.1 caption). Standard Test Conditions (STC) are 1000 W m^{-2} solar irradiance and 25 °C PV cell or module temperature. The power output of a solar module derived under these conditions is the so-called "peak power" Wp already mentioned. To calculate the energy produced by 1 Wp of installed PV capacity, many more parameters have to be considered, including the real operating temperature, solar spectrum, and sunshine hours per year. Solar modules that face full sunlight typically operate at temperatures of 50 °C, reducing their efficiency by 5–15% (relative) compared with STC. The most significant contribution to the energy yield per year is the average number of sunshine hours. This number is typically 1000 h under moderate and above 2000 h under very sunny climate conditions. The energy yield provided by a solar energy system can only be predicted with any certainty by using exact meteorological data and a detailed knowledge of how solar modules will operate under these climate conditions and their specific method of installation. Software solutions are commercially available that can be used assuming climatic averages for specific application regions. However, exact techniques for the prediction of energy yield remain a highly active research and development topic owing to the need for very precise calculations for large investments, new PV technologies, harsh climates or installation conditions, and very long operating times.

16.2.1
Crystalline Si Wafer-Based Solar Cells – Today's Workhorse Technology

The world-wide PV market is dominated by solar modules consisting of individual crystalline silicon (c-Si) wafer-based solar cells. Even though more efficient or cheaper technologies (in cost per square meter) are known, the price–performance ratio of c-Si-based solar modules has proven itself to be highly competitive. c-Si PV technology has benefited strongly from the wide use of silicon in micro-electronics. Most process steps in c-Si PV are relatively simple, well known, and readily available from equipment manufacturers. In addition, the process sequence has proven robust and easily expandable using new process developments in research laboratories. Usually, additional product features become available with only limited additional investment cost, since complete reinvestment in the entire production line is unnecessary. Thus, evolutionary development, especially in processing wafers to solar cells, has been pushed much further than was expected only a decade ago, and is expected to last for perhaps another decade.

The process route for c-Si technology can be divided into a few elemental steps as depicted in Figure 16.2: first, raw quartz material is refined to metallurgical grade Si (98–99% purity), which is then highly purified to obtain electronic grade Si (more than 99.9999% pure). From this purified Si, mono- or polycrystalline wafers are formed, which are subsequently processed into solar cells. Finally, these cells are electrically connected in series and encapsulated into a (usually glass-based) solar module. Such modules are typically around 1.0–1.6 m^2 in size, and produce some 120–280 W, at various voltages.

The process for refining quartz sand into metallurgical grade silicon was developed in the late 1900s, and industrialized by, among others, Rathenau, who also founded the company AEG. In the basic process, quartzite (SiO_2) is reduced by charcoal in an arc furnace [8]. Only a few percent of the reaction product, metallurgical grade silicon (metal Si), is used for PV or electronic (integrated circuits) purposes, whereas

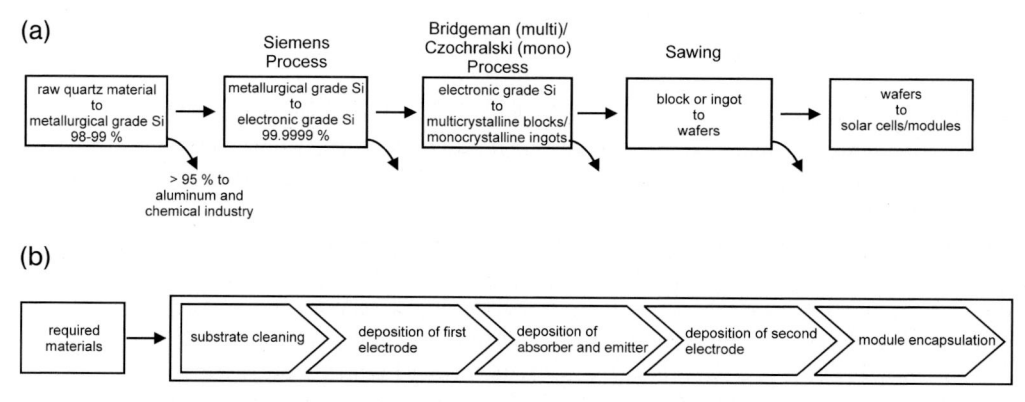

Figure 16.2 (a) Process route of c-Si technology and (b) process route of thin-film technologies. Whereas the different steps in (a) are typically performed by different manufacturers, the steps of the route in (b) are subsequently performed in one factory.

the aluminum and chemical (silicones) industries divide up the remaining world production roughly equally. Since the refining process is very energy intensive, production tends to be located in places where suitable quartz can be found and energy costs are low (China, United States, Norway, Brazil).

To qualify as PV or electronic stock material, the metallurgical Si must be highly purified, reducing impurities to ppm (oxygen, carbon) or even ppb (metal impurities, phosphorus) levels. To this end, the metallurgical Si is converted to gaseous chlorosilane compounds, which are subsequently fractionally distilled to very high purity, and finally reduced at high temperature in an H_2 atmosphere to ultrapure, fine-grained solid Si ("poly-Si"). This complete purification process is known as the Siemens process [8]. It is still the industry standard, although many alternatives have been investigated and tested to improve on its material and energy efficiency.

In the next process step, the refined poly-Si grains are made molten and from the liquid phase, either multicrystalline blocks or monocrystalline ingots are grown, which are intentionally lightly doped with donor or acceptor atoms (resulting in n-type or p-type material, respectively). The standard for multicrystalline blocks is the Bridgeman process [8], in which the liquid silicon is slowly cooled within a very large Si_3N_4-coated quartz crucible ($90 \times 90 \times 30$ cm). The final crystal quality process is determined by the poly-Si material, impurities from the crucible walls and surroundings, and structural properties in the crystal, determined by the cooling process. For monocrystalline Si, in the standard Czochralski process [8], a defect-free seed crystal is used to pull a growing ingot very slowly out of the melt. In this process, control of the temperature and surrounding atmosphere is of the utmost importance. In both processes, some additional purification is achieved by the fact that under quasi-thermodynamic equilibrium, impurities tend to remain in the melt rather than incorporated in the growing crystal.

From the block or ingot, individual wafers of 160–200 μm thickness are cut, usually losing as much Si in process as is left in the wafer. Slurry-based wire-sawing processes are the standard today, but alternatives are under investigation, notably diamond-wire sawing. In all varieties, great emphasis is placed on minimizing kerf loss and recycling of the valuable Si material.

To make a cell from the wafer, the first step is to etch the sawing damage from the wafers outer surfaces. In the next step (or in some cases simultaneously), the wafer top surface is etched to achieve a textured surface, which serves to enhance light transmission into the wafer by reducing its reflection. Then, a thin emitter region is created by doping with atoms of opposite polarity to those in the bulk of the wafer. Subsequently, an anti-reflection (AR) film is deposited on top of the (textured) emitter surface, to reduce reflection losses further. On top of the AR coating, a fine grid of metal fingers, based on metal particles (mostly Ag), is screen printed as a front contact. On the back side of the wafer, a metal film is screen printed as a back contact electrode, often consisting of Al and Ag. The Al in this mixture also reacts with the Si (in that case p-doped) in the wafer to form a highly p-doped region, called the back surface field (BSF), which allows for selective extraction of the holes. Finally, a rapid thermal step ("firing") is performed to establish proper electrical contact at both front and back surfaces.

All individual cells are tested and sorted for subsequent processing into solar modules. As all cells deliver about 0.5 V and a few ampères under illumination, usable electrical output for the solar module can be created by soldering the (40–60) individual cells into a series-connected string. To guarantee an outdoor lifetime of 20–30 years, such strings are mechanically, electrically, and chemically protected by encapsulation between two glass plates or a glass plate and a protective film.

16.2.2
Thin-Film PV: Challenges and Opportunities of Large-Area Coating Technologies

Thin-film PV technologies provide the potential for significant cost reductions and, accordingly, huge investments were made worldwide into a multiplicity of new production plants. Thin-film solar cells have the shortest energy payback time and lowest materials usage among the present PV technologies. Thin-film solar cells provide new options for PV applications, and unique esthetics for building integration or roll-to-roll production of light-weight and flexible products. The current market shares of the different PV technologies are shown in Figure 16.3.

Many PV materials, in contrast to c-Si, are direct semiconductors and have very high absorption coefficients in the relevant part of the electromagnetic spectrum (UV–VIS–NIR) where most of the solar energy is to be captured. These materials usually have a charge carrier lifetime and mobility properties that allow charge separation towards the electrodes, that is, over a distance of a few micrometers at most. To be processed into working solar cell devices, lateral current transport (over several millimeters to centimeters) must be assisted by a metal back contact and a transparent front contact, usually a highly doped transparent conductive oxide such as FTO (fluorine-doped tin oxide, SnO_2:F), AZO (aluminum-doped zinc oxide, ZnO:Al), or ITO (indium tin oxide, In_2O_3:SnO_2). A less common alternative for thin-film solar cells is to use a very fine grid of metal fingers as front contact (as is usual for wafer-based crystalline cells).

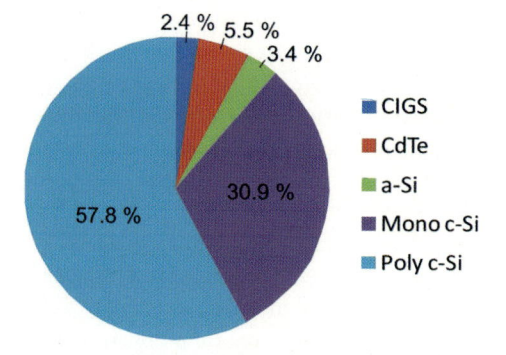

Figure 16.3 Relative contributions of different PV technologies to the global production in 2011 [9]. The three thin-film technologies Cu(In,Ga)(S,Se)$_2$ (CIGS), CdTe, and a-Si, share about 11.3% of the PV market.

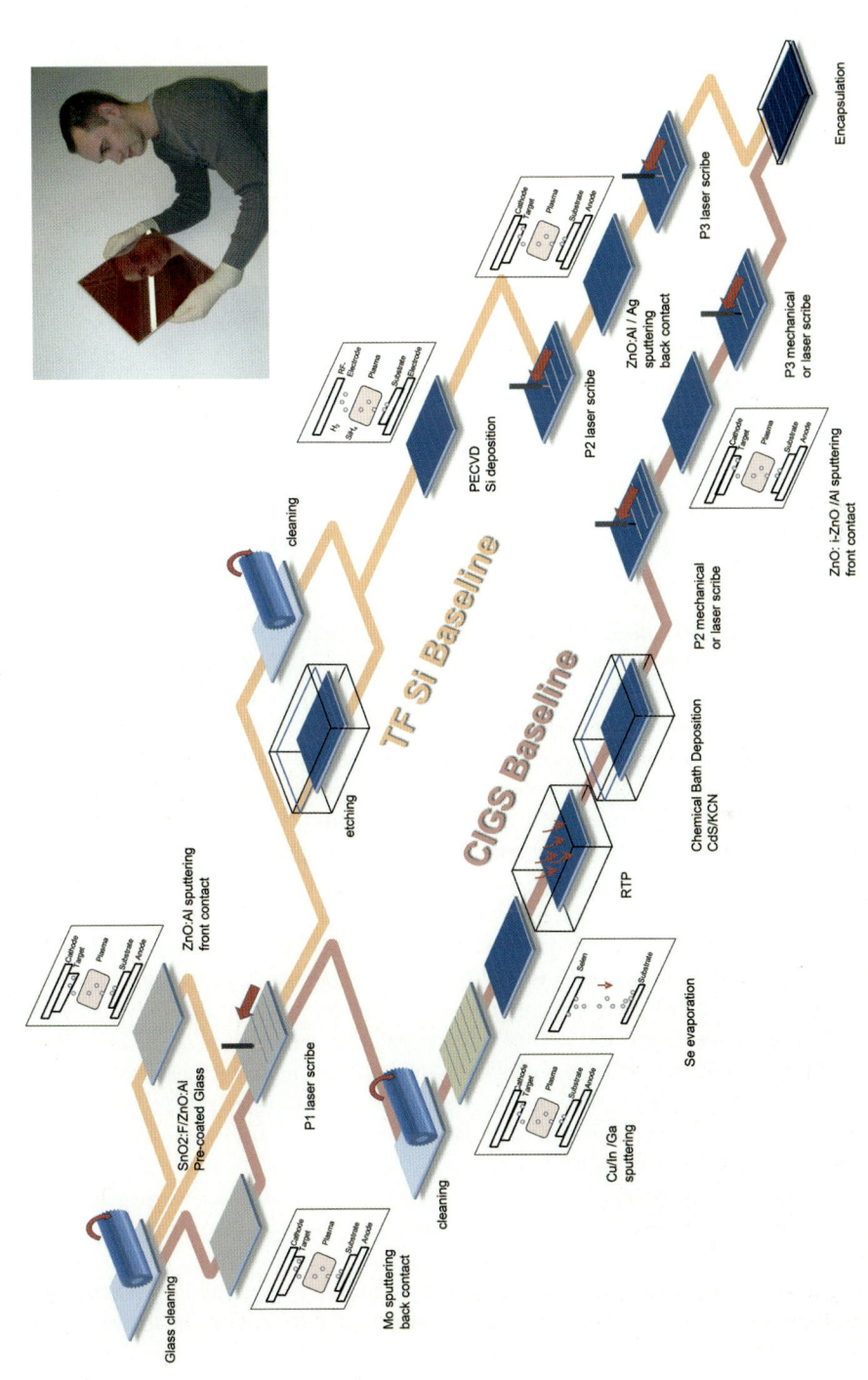

Figure 16.4 Flow chart for thin-film PV module processing at PVcomB. Note that both thin-film technologies, CIGS [Cu(In,Ga)(S,Se)$_2$] and TF Si, share substrate cleaning, transparent conductive oxide (TCO) processing, and laser scribing, whereas the semiconductor fabrication for thin-film Si and CIGS relies on dedicated processes.

Hence the main technological steps in a thin-film solar cell production process are substrate cleaning, first electrode deposition, absorber and emitter deposition, second electrode deposition (each deposition is typically followed by a cell structuring step), and finally module encapsulation (Figure 16.2b). The processing principle for two different technologies is nicely demonstrated by the flow chart of the R&D baseline of PVcomB at the Helmholtz-Zentrum Berlin (Figure 16.4). When the cell is deposited bottom up on the substrate, the first electrode is the metal and the second, sun-facing, electrode is the transparent oxide. If the cell is deposited downwards on the sun-facing substrate (in which case the substrate is sometimes referred to as "superstrate"), the first electrode must be the transparent oxide and the second electrode the metal back electrode. As thin-film deposition processes are generally compatible with large-area substrates, thin-film solar modules (i.e., a number of solar cells electrically connected in series) are usually processed at once, in contrast to wafer-based solar modules, where cell production is separated in time and space from module production.

16.3
Technological Design of PV Systems

16.3.1
Residential Grid-Connected PV System: Roof Installation

Some prerequisites, challenges, and financial aspects, such as investment and energy generation costs, of current PV systems can be illustrated with the help of an example residential PV system installed in Germany in October 2012. Germany serves as the lead market with strong competition along the value chain from module supply towards a fully installed and operating system.

Figure 16.5 sketches the key components of this grid-connected PV system. The modules are roof mounted (facing south) based on monocrystalline silicon wafers and have a black appearance for esthetic reasons. The electricity is either used in-house or fed into the grid. In both cases, the DC electricity produced by the solar modules is converted to 50 Hz AC electricity by use of an inverter. It should be noted that the inverter and grid connection technology requires only very little space. Key parameters and cost of the components are listed in Table 16.1. Note that factory prices of solar modules in October 2012 ranged between US$ 0.4 and 0.9 Wp^{-1} [10], depending on quantity, supplier, and module performance. Note that the lowest prices are from the spot market and these modules are sold at prices significantly below production costs.

Operating costs of about €250 (€3.2 m^{-2}) per year and financing costs (e.g. interest, assumed to be 5% per year) of the PV system need to be considered in addition to the nonrecurring costs listed in Table 16.1. Assuming a depreciation rate of 5% over the duration of 20 years (reflecting the minimum lifetime of the PV system), this leads to yearly capital costs (interest and depreciation) of 10% of the total investment. Note that the operating costs represent only a minimal addition to those yearly costs.

16.3 Technological Design of PV Systems

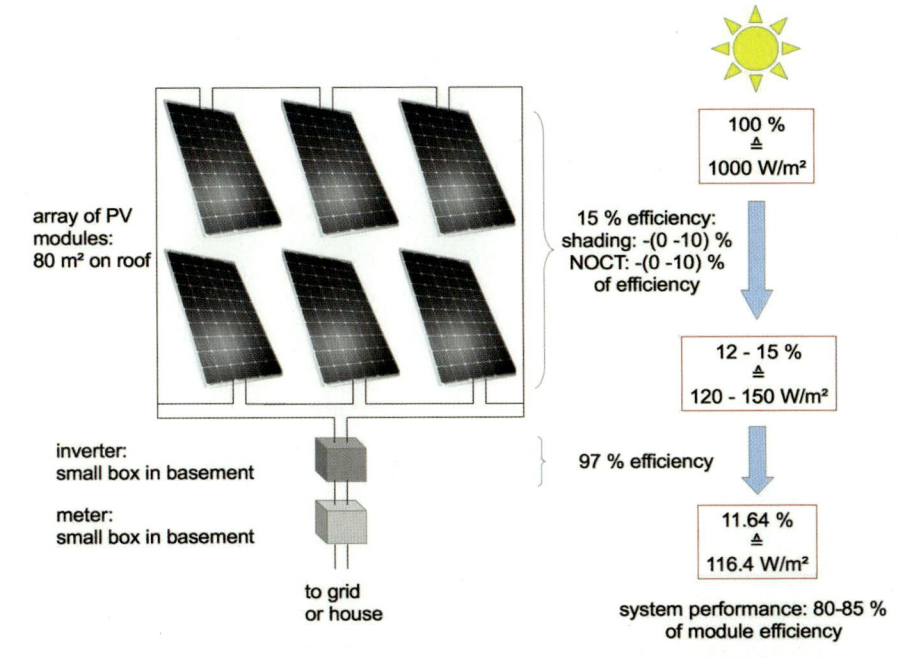

Figure 16.5 Basic layout of a residential roof-mounted and grid-connected PV system (see explanations in the text). NOCT means nominal operation cell temperature, reflecting in this example that a solar module operated under 800 W m^{-2} light intensity will heat up to about 50 °C. The efficiency under NOCT conditions is 13.5% compared with 15% under STC conditions (relative loss of 10%) – a typical value for c-Si modules.

Table 16.1 Investment costs of a residential PV system installed in Berlin in 2012 and resulting energy generation costs assuming capital costs of 10% (5% interest and 5% depreciation based on 20 years lifetime) per year.

Component	Costs		
	€	€ Wp^{-1}	€ kWh^{-1}
Modules	11500	1.00	0.10
Inverter	2500	0.22	0.02
Installation	7500	0.65	0.06
Total	21500	1.87	0.18

The energy generation costs of the PV system, that is, costs per kilowatt hour, are then given by the ratio of the total costs during the 20 years to the total amount of energy produced by the system.

The expected performance was calculated using the software PV*SOL Expert 5.0 [11] (Table 16.2). According to the calculation, the 11.5 kWp system is expected to deliver 10 MWh of electricity. The resulting value of 869.2 kWh per installed kWp is

Table 16.2 Expected performance of the residential PV system calculated by use of the software PV*SOL Expert 5.0 [11][a]. The calculation is based on the climate data of Berlin 1981–2000.

Parameter	Value
PV power	11.52 kWp
PV area	79 m^2
Insolation on PV area	85 730 kWh
PV-generated energy	10 020 kWh
Shading losses	3%
System performance	11.7%
Performance ratio	79.9%
Annual yield	869.2 kWh kWp^{-1}

a The specific energy generation costs were calculated to ~€0.18 kWh^{-1}, almost exactly reflecting the feed-in tariff for residential solar systems installed in Germany in October 2012. In comparison with the simple approximation in Table 16.1, interest rates of only 3% were considered, but maintenance, service, and a small yearly degradation of the system are included.

about 10% lower than achievable under optimum conditions owing to three almost equally contributing reasons:

1. The inclination of the roof is 15° and slightly below optimal.
2. Shading from neighboring buildings.
3. Limiting output power to 70% of the peak power according to the German feed-in law [12] for systems not taking part in grid management.

Whereas reasons 1 and 2 have to be typically considered for residential systems, reason 3 may serve as an example for a specific regulation related to the integration of PV in an existing grid infrastructure.

Note that the cost breakdown for solar electricity generated in residential systems is very similar to that for large rooftop systems and the cost structure, (i) modules, (ii) inverters, (iii) grid connection, (iv) installation, and (v) maintenance and service, has to be considered in a similar way for very large-scale solar power plants. In the latter case, usable land area must be optimized as part of the cost calculation instead of being a boundary condition, as is the case for a rooftop system. In large systems, labor and investment costs are reduced due to the economy of scale, which leads to different shares between the cost segments and results in lower costs for electricity generation in large-scale applications.

16.3.2
Building-Integrated PV

A second example is a building-integrated PV (BIPV) system as realized by Glaswerke Arnold and Masdar PV, modifying 5.7 m^2 large PV modules based on thin-film

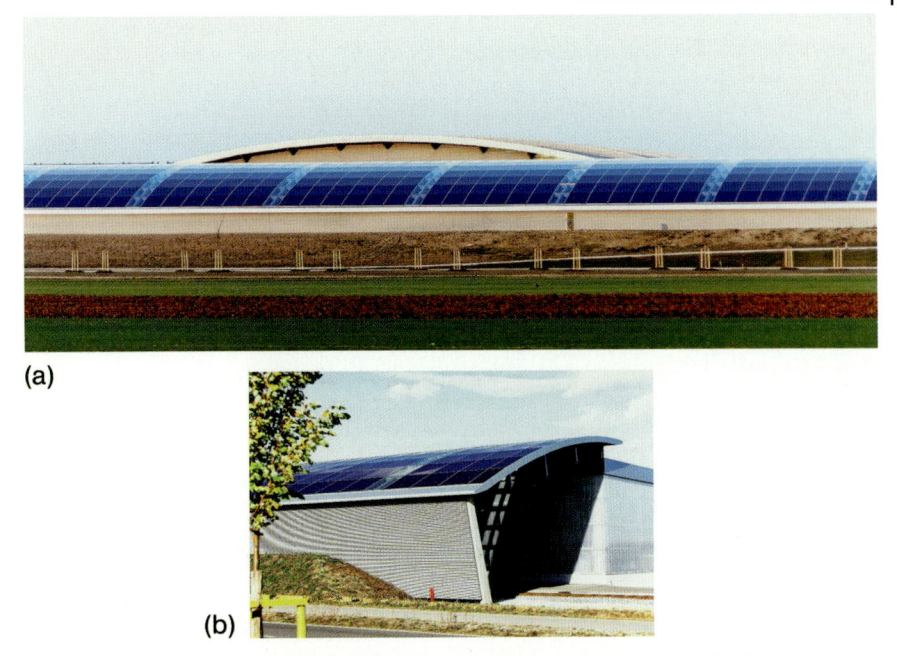

Figure 16.6 Train station in Ingolstadt, Germany: (a) full view and (b) detailed side view. Courtesy of Masdar PV GmbH.

silicon in such a way that they are appropriate for walkable overhead glassing construction (Figure 16.6) [13]. The entire glass roof has an area of 11 600 m² with the 1728 PV modules covering an area of 9900 m² and representing an installed nominal power of roughly 760 kWp. This PV installation provides at least two features in addition to the electricity generation:

- The glass-based modules are part of the roofing construction.
- They are an esthetic eye catcher.

Usually, as in the example given, a BIPV construction represents a unique architectural project. In that case, the esthetic function plays a leading role and electricity generation is of secondary concern and not the driving force. This is even more the case for different types of semi-transparent modules which are created by the combination of large-area thin-film coating and laser-based interconnection technologies. Small parts of the module are partially removed by laser ablation allowing for nicely designed patterns and see-through optics with different degrees of transparency. Figure 16.7 shows an example of the "see-through" effect. Such modules can be applied in overhead roof or facade constructions. Obviously, the deliberately created optical transmission decreases the amount of electricity generated by the solar modules. The cost calculation of such PV systems is more complicated, and does not allow for direct comparison to the residential or solar power plant systems. Usually, the electricity generation cost for BIPV is higher.

Figure 16.7 Semi-transparent solar module. Courtesy of Masdar PV GmbH.

Although long recognized as a PV market with an enormous growth potential, no systematic optimization approach for BIPV has been established. As a result, BIPV still remains a relatively small part of the PV market. There are two important reasons for this. Solar modules are completely independently produced items in more or less optimized manufacturing process, whereas for modules used in BIPV, customized solutions have to be created for almost every individual BIPV project, for reasons of size or specifics of the electrical connections. Customization generally increases the cost of the electricity generated. Second, a solar module, even with its mandatory IEC or UL certification, which is sufficient for residential or solar power plant applications, is not automatically a permitted construction element. In the example above, the BIPV modules in the railway station needed additional glass reinforcement (two 10 mm heat-strengthened glass panels) to qualify for overhead application. In addition, the standard encapsulation material for these solar modules, EVA (ethyl vinyl acetate), must be replaced by PVB (polyvinylbutyral), since that is an allowed building material known from its applications in safety glass. Furthermore, construction materials usually have to comply with national, sometimes even regional or local building codes, making standardization very difficult. To exploit the potential of BIPV fully, solar modules should be optimized as part of the value chain for construction materials, instead of as isolated electronic devices, and building codes need to be standardized.

16.3.3
Flexible Solar Cells

Thin-film PV can be produced on flexible substrates as well as on glass (see Section 16.2.2). As such, the resulting modules are equivalent to glass-based modules, and can compete in the same markets. There are three potential advantages connected to using flexible substrates, which allow for a roll-to-roll production process of solar modules:

1. Roll-to-roll processes can achieve lower production costs. Production equipment with the equivalent throughput can have a smaller footprint, since flexible substrates have significantly lower thermal mass, and heating and cooling can be achieved over very small distances.
2. Flexible modules can, in principle, be installed more easily on roof tops – even on curved surfaces. Ideally, they could be integrated in building or roofing elements directly. This potentially reduces installation costs to a very large extent, since only electrical connections have to be made on the building site; the mechanical construction for the solar modules is taken care of "automatically" by installation of the construction element. Again, this potential benefit has not been fully exploited. Individual producers had to develop their own market channels for this purpose.
3. A further potential benefit of flexible modules lies in their lower weight (1–2 kg m^{-2} instead of 12–20 kg m^{-2} for glass-based modules), and also in smaller storage and transportation volumes. In addition to a growing number of mobile power applications, the weight difference also allows access to an enormous area of large industrial roofing structures. Very often, these roofs cannot support larger weights, but could certainly carry the additional load of flexible modules.

Thus, flexible solar modules might allow lower production costs when more standardized production processes can find their way into the market. In addition, they have a number of very interesting potential markets, most notably BIPV, when they can be fully integrated into the industrial production of standardized building or roofing elements. Still, only a few companies have managed to survive the present market situation, although many have worked very hard on a market launch of flexible products.

16.4
Cutting Edge Technology of Today

PV is still a very new energy technology and its maturity within the energy system, that is, contributing more than a small fraction to the total energy supply, is still to be established on a global scale. Perhaps Germany is one exception to this, since the transition from margin to mainstream is currently taking place. The lead market today is grid-connected PV systems (roof-mounted PV systems for residential houses, and much larger commercial buildings or large solar farms). This section highlights the importance of improved conversion efficiencies for very low electricity generation costs on the basis of a simple cost model, gives examples of c-Si modules already approaching 20% efficiency (at still increased costs), and discusses leading thin-film technologies.

16.4.1
Efficiencies and Costs

Although it is beyond the scope of this chapter and the ability of the authors to provide detailed insight into the exact cost structure and future cost perspectives of PV, we developed a very simplified cost model for the energy generation costs, EGC (costs per kWh), partly based on the 2012 residential system described in Section 16.3.1. The model nicely emphasizes the demands on current and future (cutting edge) technologies and illuminates future challenges. EGC are defined as the ratio of yearly costs to the yearly amount of energy provided by the system:

$$EGC = \frac{\text{costs}/a}{\text{energy}/a} = \frac{0.1\, MC + BoS/a}{\eta\, ISR},$$

where η is the conversion efficiency of the PV modules and ISR the incident solar radiation per year (determined by solar constant, hours of sunshine and area of PV system). The costs of a fully installed PV system can be divided into the costs of the PV Module MC and the remaining costs for the Balance of Systems (BoS). The latter comprises investment costs for the inverter and the installation of the PV system, in addition to recurring operating costs. For this model, operating costs of about €250 (€3.2 m^{-2}) per year and costs arising from the financing of the PV-system may be considered in addition to the nonrecurring costs listed in Table 16.1. Similarly to the residential PV system described in Section 16.3.1, we assume an interest rate and a depreciation rate of 5% per year each over 20 years of operation leading to yearly expenses of 10% of the total investment costs resulting in the factor 0.1 in the equation.

Under these simple assumptions, electricity generation costs increase linearly with increase in the solar module price per square meter for a certain efficiency and insolation (Figure 16.8). During recent decades, the price of solar modules has decreased dramatically with increasing production capacity. Reflected by the PV learning curve [14], the price of PV modules has been driven down by 20% for each doubling of the cumulative PV module production over the past three decades. The horizontal dotted lines in Figure 16.8 indicate rough PV electricity generation costs based on the module prices in 1980, 1990, 2000, and 2010 taken from the learning curve.

Today's lower limit of module costs is given by the zero-coating case, that is, if no costs arise due to the fabrication of the solar cells. In this case, the minimum costs are given by the substrate and encapsulation costs, usually two glass panels, which can be estimated to be around €20 m^{-2}. The model nicely highlights the fact that a higher module efficiency is the crucial factor in reducing the cost of PV electricity. However, note that doubling the efficiency, for example from 10 to 20% at 1000 h under standard conditions in Germany, is equivalent to a doubling of the (equivalent) hours under standard conditions and keeping the efficiency constant. The latter could indeed be achieved by building the PV system on a rooftop, for example in northern Africa or southern United States.

It should be noted that these scenario curves may serve as a rule of thumb. Certainly, higher thermal loads can lead to higher module temperatures, which will reduce the performance ratio and may lead to shorter inverter lifetimes.

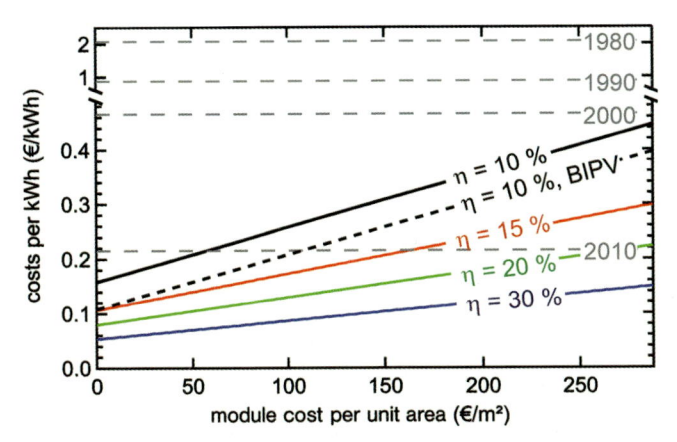

Figure 16.8 Energy generation costs as a function of module costs per unit area for different conversion efficiencies in a region with 1000 sunshine hours. The gray dashed horizontal lines is calculated using the learning curve of module prices from 1980 to 2010. The black dashed line corresponds to the 10% efficiency curve under the assumption that €50 m^{-2} can be saved by a BIPV installation. Note that an efficiency of 20% and 1000 sunshine hours are equivalent to a 10% system in a region with 2000 sunshine hours.

Additional constraints may be given by the local conditions such as wind load, soiling, sand blasting, snow load, solar array orientation, or shading. In our simple model, the energy generation costs arising from the BoS cost are given by the intersection of the scenario curves with the y-axis. Even when assuming minimal costs for the module (MC), including glass substrates and junction box, solar cell efficiencies of 7.5% (such as historically delivered by some thin-film technologies) are not high enough to compensate for the BoS costs to reach the kilowatt hours cost level of 2010 (gray dotted horizontal line). A detailed investigation of the status of BoS costs and an outlook towards optimization was recently published by Ringbeck and Sutterlüti [15]. They discussed the main applications, that is, ground-mounted systems, commercial roof top systems, and residential roof top systems. Moreover, different locations in Europe, Asia, Africa, and America were compared.

As discussed in Sections 16.3.2 and 16.3.3, BIPV applications, flexible products, mobile applications, and remote applications often have completely different cost structures which cannot be discussed in a simple way. In Figure 16.8, the BIPV case is illustrated by the dashed line for 10% efficiency by assuming that the modules reduce costs of roofing material (as in the case presented in Section 16.3.2.) by, for example, €50 m^{-2}.

16.4.2
Crystalline Silicon Wafer-Based High-Performance Solar Modules

Already today, the most efficient commercially available silicon-based solar modules approach 20% conversion efficiencies. Pioneers of these product families are the companies Sunpower and Panasonic, relying on only back-contacted cells [interdigitated back contact (IBC)] or so-called Si hetero-junction cells, respectively.

Table 16.3 High-efficiency crystalline Si solar modules compared with a reference monocrystalline Si solar module[a].

Technology	Cell open-circuit voltage (mV)	Cell efficiency (%)	Module efficiency (%)
"Standard" mono-c-Si	620–630	18	15
Sunpower	680	22.9	20.4
Panasonic	720–730	21.6	19.0

a All values are taken from the data sheets of the producers (Sunpower E20 series, Panasonic HIT N240-SE10). The mono-c-Si reference cell/module data are taken from the reference system described in Section 16.3.

In both cases, high-quality Si wafers with excellent electronic properties serve as the base material. In the IBC concept, both contacts are on the back side of the solar cell. Hence it follows that no metallic grid structure is required on the front side of the solar cell. In the case of the hetero-junction concept, hydrogenated amorphous silicon layers form the emitter and back surface field. Contacting is realized by the combination of a TCO film and a grid structure on the front side of the solar cells. The cell production is almost completely based on technologies developed for a-Si:H solar cell processing. The astonishing feature of a-Si:H/c-Si hetero-junction cells is the high open-circuit voltage, reaching 745 mV in the laboratory [16] and more than 700 mV in the product. Table 16.3 summarizes some key data to compare the latter "cutting edge" technologies with the mono c-Si "state-of-the-art" standard technology applied in the reference system (see Section 16.3). Not surprisingly, research activities combining the advantages of IBC and the hetero-junction concepts are ongoing [17].

Higher production costs and selling prices currently counterbalance the efficiency advantage of the latter high-efficiency modules. According to the SEMI International Roadmap for Photovoltaics (ITRPV) [14], improvements in cell technology and manufacturing are expected within the next few years leading to efficiencies of more than 17.5% and 21% for c-Si modules in standard and back-contact configurations, respectively, with significantly reduced production costs.

16.4.3
Thin-Film Technologies

Over the past few decades, industrial development has mainly been in the field of amorphous and microcrystalline silicon [18] (a-Si/µc-Si) and also compound semiconductors [19] such as CIGS [$Cu(In,Ga)(S,Se)_2$, chalcopyrite compounds] and CdTe [20]. Further very active fields of research and development, predominantly still in the academic world, have been organic photovoltaics (OPV) [21] and dye-sensitized solar cells (DSSCs) [22]. The latter are usually based on TiO_2 with an organic dye as absorber material. OPV and DSSCs are also thin-film technologies, but will not be covered further here.

The three thin-film solar cell absorber materials under large-scale industrial development (all in the gigawatts range) are based on CdTe, thin-film Si (mostly a-Si/µc-Si), and CIGS. All three technologies rely on vacuum processing for most of the crucial device processes.

CdTe is a semiconductor material having an energy gap close to the optimum value for single-junction solar cells (about 1.44 eV). In addition, the phase diagram of CdTe allows for congruent sublimation, that is, the material can be evaporated and deposited with unchanged composition. Hence the process, usually at substrate temperatures above 500 °C, displays some intrinsic stability that none of the competing technologies possesses. Still, to make a fully functional solar cell device based on CdTe absorber layers, including top TCO (usually SnO_2 based), the incorporation of a CdS buffer layer, and relatively complex back-contact stack, has proven to be a far more complex matter to control. The record laboratory efficiency was recently improved from the long-standing NREL value of 16.7% to 17.3% by the United States-based company First Solar [23], which produced pre-commercial modules with a record efficiency of 14.4%.

Thin-film Si attracted industrial attention from the late 1970s onwards. In its various forms, both amorphous and nano- or microcrystalline, the material is easily processed at relatively low temperatures (around 200 °C). Amorphous Si itself has a large bandgap of around 1.7 eV, and hence only absorbs in the spectral region up to wavelengths in the red. By using it in combination with microcrystalline Si with its crystalline bandgap of 1.1 eV, a "micromorph" tandem device can be created that utilizes the solar spectrum into the infrared region. This material combination, or even triple-junction devices, are the present state-of-the-art of thin-film Si solar cells, and will be the trend for the future, owing to a strong market pressure to increase module efficiency above 10%. In 2012, LG Electronics, using a triple-junction device, announced a new world record efficiency of 13.4% for thin-film Si solar cells [23]. Pre-commercial modules produced by Oerlikon Solar [24] (now part of TEL – Tokyo Electron Ltd) reached efficiency values up to 10.7%.

Production technology for thin-film Si was pushed by parallel developments in the flat panel display (FPD) industry and the advent of turn-key equipment suppliers such as Oerlikon, Applied Materials, Ulvac, and Ju Sung. These suppliers have lowered the entry barrier for many newcomers in the past few years, although some companies chose to continue to rely on in-house specified equipment. A total production capacity of well over 1 GW has been installed.

The compound semiconductor family CIGS was first researched in the 1970s, with industrial-scale activities taking off from the 1980s. The bandgap of the material can be varied between 1.1 and more than 2 eV, depending on whether S or Se is used and the Ga content. Furthermore, for good devices, additional Na is needed from the substrate glass (or deliberately added). Even though many interesting scientific questions regarding the material remain unsolved, CIGS solar cells have reached conversion efficiency values above 20% [25], thus moving into the league of crystalline Si.

Solar cells based on CIGS absorbers are produced by a multitude of deposition methods, which can be broadly classified in two groups: one using the co-evaporation of Cu, In, Ga, and Se from individual sources at temperatures above 500 °C, and the other using sequential processing routes. The steps used in sequential processing routes include, first, depositing the metal precursor layer stack (usually by sputtering, but alternatively by nonvacuum methods such as electroplating), optionally followed by the deposition of a layer of Se, and subsequently, thermally assisted semiconductor formation in a controlled atmosphere containing Se and/or S. The co-evaporation approach intrinsically offers better compositional control throughout the thickness of the layer, whereas sequential processing allows for better lateral homogeneity and greater long-term process stability and control. The two methods reach comparable module efficiencies, with the present record value of 14.6% produced by the equipment company Manz AG, based on a co-evaporation process [26].

16.5
R&D Challenges for PV Technologies Towards the Terawatt Scale

Solar energy can be used by PV systems in almost any location. However, sunlight has a low energy density of 1 kW m^{-2} in full sunlight and about 100–300 W on average, depending on the geographic location; therefore, large areas have to be covered by solar cells. Residential applications, BIPV, applications in harsh environments, and mobile applications require dedicated research starting with basic solar cell technology, through module design and system integration. In addition, fabrication technologies have to be developed that allow mass production of all elements mentioned above.

Regarding the target for PV technologies to provide power on the terawatt scale, next-generation PV devices must provide conversion efficiencies as high as possible and, in the ideal case, provide added value, for example, by being part of the building. Necessary prerequisites are the use of high-quality and abundant materials and processing technologies suitable for mass production. Several R&D roadmaps for PV exist. An industry-driven group of researchers used a methodology developed for the semiconductor industry to draft a roadmap for short- to mid-term developments for c-Si wafer-based technology (up to 2020). Coordinated by the PV section of SEMI, the outlined developments ranged from "implemented in industry at present" to "industrial solution not yet known." A broader roadmap was developed by a group of leading scientists from academia and industry covering R&D goals up to 2030. It is the second edition of "A Strategic Research Agenda for Solar Energy Technology" (SRA) [27]. It covers the full value chain from cell and module technologies via PV components and systems towards the socio-economic aspects and enabling research activities. The SRA was published in 2011 (relying on data from 2010) and gives quantitative goals for solar module costs and related electricity generation costs. It should be noted that the dramatic increase in production capacity, and the resulting decrease in solar module price in 2011 and 2012, have already led to electricity generation costs equal to those set as targets for 2015–2020 in the SRA and other recently published roadmaps.

16.5.1
Towards Higher Efficiencies and Lower Solar Module Costs

Enhanced conversion efficiencies have the highest impact on cost reduction. Consequently, each PV technology follows efficiency goals while trying to maintain or even reduce the production costs related to the module area. Figure 16.9 illustrates efficiency goals according to the SRA [27] for the three classes of solar modules commercialized today, and selected R&D topics for these technologies are briefly listed in the following.

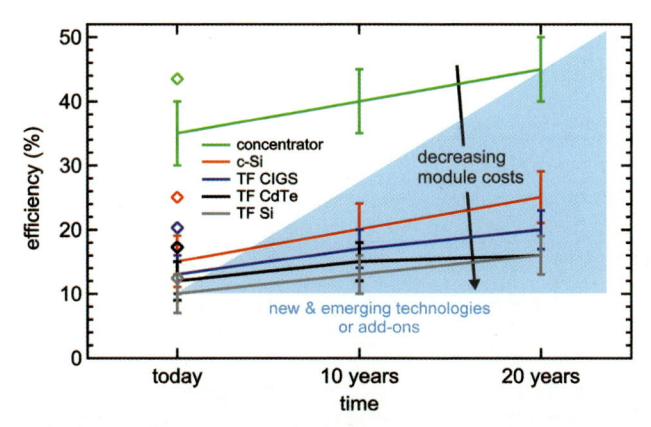

Figure 16.9 Efficiency targets for the three classes of commercialized solar module technologies according to the SRA [26]. The error bars reflect different materials and cell concepts within each class. For example, in the c-Si case these are monocrystalline silicon wafers or multicrystalline Si wafers, double-sided or back-contacted cells; in the TF case, thin-film Si (lower values), CdTe, or CIGS. The open diamonds represent the world records reported for laboratory solar cells for each class. Concentrating PV will be introduced in Section 16.5.4.

16.5.2
Crystalline Silicon Technologies

The key R&D targets (Table 16.4) for the classical Si wafer technology adapted from the SRA for the next decade and for 2030 are to reduce the costs by cheaper silicon based material and additionally by thinner wafers: Furthermore, module lifetimes beyond 35 years are envisaged.

Table 16.4 Quantitative milestones for crystalline Si PV.

Component	In 10 years	In 20 years
Materials	< 3 g W^{-1} Thickness < 100 µm	< 2 g W^{-1} Thickness < 50 µm
Cells and modules	Efficiency > 20%	Efficiency > 25%
Modules	Lifetime > 35 years	

16.5.3
Thin-Film Technologies

For Si thin-film technologies, the amount of high-quality Si used is significantly lower and the advantages of integrated module manufacturing and large-area coating technologies provide unique advantages. However, today's module efficiencies are in the range 7–10%. Hence it follows that the key issue for this technology is an enhancement of the conversion efficiency above 14 and 16% in 10 and 20 years, respectively.

Flexible products can be realized by almost any thin-film technology and provide, as mentioned before, many new application fields. However, intense R&D is required in order to exploit the principal advantages of thin-film technologies discussed in Section 16.3.3. Table 16.5 shows the quantitative milestones for the compound semiconductor technologies CdTe and CIGS. One of the main R&D goals is the reduction of scarce elements, such as In, Ga, and Te. In the case of CIGS, the replacement of In and Ga by Zn and Sn has become an important research topic in the past few years [28], and efficiencies of 11.1% [29] could already be demonstrated.

Table 16.5 Quantitative milestones for compound semiconductor (CIGS, CdTe)-based thin-film PV according to SRA [27].

Component	In 10 years	In 20 years
Materials and processing	Reduction of scarce or expensive elements (In, Ga, Te); High-throughput equipment	Nonvacuum processing
Cells and modules	Efficiency > 16%	Efficiency > 20%
Modules	Lifetime > 35 years Flexible modules	

16.5.4
Concentrating Photovoltaics (CPV)

CPV systems rely on direct solar radiation which is focused by cost-effective optical elements on highly efficient PV devices. Commercially available III/V semiconductor-based multi-junction solar cells already approach 40% efficiency; the world record for such a device is 43% [3]. In recent years, the cumulated capacity installed for CPV systems using high-efficiency multi-junction solar cells has reached 100 MWp and system efficiencies approaching 30% have been reported [30].

R&D topics for the emerging CPV technology relate to three components, efficiency improvement for the concentrator cell, realization of cost-effective and highly efficient optical systems (optical efficiencies above 90%) for concentrating sunlight, and adapted and effective technologies for module assembly.

Within the next 10–20 years, module conversion efficiencies above 40% are feasible and, if the thermal energy is also used, even significantly higher system efficiencies are possible [30].

16.5.5
Emerging Systems: Possible Game Changers and/or Valuable Add-Ons

The NREL chart of cell efficiencies [23] carefully monitors the development of solar cell conversion efficiencies for the different technology lines and is regularly updated. Solar cells and modules based on organic semiconductors have shown a steep gradient and many records have been reported in the past years. Module efficiencies above 10% seem feasible within the next few years and the first products may enter the market. Promising features of such "plastic" solar cells are the easy realization of flexible products and the application of cost-effective printing technologies [22]. However, solution-based processing and printing technologies are not limited to organic semiconductors and today's record efficiencies for solution-based PV have been realized by CIGS compound semiconductors. A recent review on solution-based PV appeared recently [31].

Progress in PV cells will not only rely on the device structure but may also benefit from add-on technologies. Instead of using the solar spectrum in an optimal way by multi-junction solar cells, up- and down-conversion of photons are another path to boost conversion efficiencies towards the physical limits. An example of a recent approach was described by Cheng *et al.* [32]. A review on so-called third-generation PV was published by Green [33].

Many groups worldwide are working to develop cost-effective photo-catalytic systems that directly convert sunlight into chemical fuels such as hydrogen and methane. This challenge is often called the "holy grail" of solar energy research, because this approach would help to solve the storage problems encountered with the fluctuating supply of solar energy (see, e.g., van de Krol and Grätzel [34]).

16.5.6
Massive Integration of PV Electricity in the Future Energy Supply System

The broad portfolio of PV applications requires research and development at different levels of integration comprising, for example, small and independent isolated applications, integration in residential and commercial buildings, and large ground-based systems. These applications will rely fully or partly on dedicated storage technologies, management and control systems during electricity generation and usage chain, and new BIPV elements that are not yet available.

In almost all cases, low costs, long lifetimes (exceeding 20 years), and extremely high reliability are mandatory. An esthetic appeal will be increasingly important considering the wider penetration of PV. More details can be found elsewhere [27].

One key R&D topic is innovative storage solutions with small capacities (1–10 kWh), for example, for residential houses, and large long-term capacities (> 1 MWh). Here we refer to the dedicated sections in this book about energy storage.

16.5.7
Beyond Technologies and Costs

Within the energy supply system today, PV is still a relatively new contributor. In Germany, PV is already highly visible when traveling through the country. Moreover, PV has an impact on electricity generation, the employment sector, and also the socio-economic balance when considering the cost of electricity generation. Other countries such as Japan and China have started or restarted strong initiatives for enhanced utilization of PV. Considering the transformation of the energy supply on a global scale, there is a strong need for socio-economic and enabling research. PV technology has been dramatically improved during the past few decades and there is still plenty of room for technological innovation. However, to make solar energy a pillar of the global energy supply requires both a bottom-up approach utilizing solar radiation in a large variety of applications, and a top-down approach to integrate PV in regional, national and even international grids.

16.6
Conclusion

During recent decades, PV has emerged from a few specialty applications to a fully developed mature technology, on the brink of being a very significant supplier of energy worldwide. Cost reductions have been significant and a huge global industry has been developed. The use of PV on a terawatt scale is one of the most promising possibilities for mitigating global warming and for providing renewable energy in an economically sensible and sustainable manner. However, PV still plays a significant role in only a few countries or regions. Especially in those areas of the world where sunlight is abundantly and predictably available, PV has contributed only marginally to the countries' energy supply.

Key elements to support and enhance a massive deployment of PV on the global scale are as follows:

- R&D for efficiency improvements, cost reductions, and lifetime enhancement along the value chain. The learning curve for wafer-based silicon PV is still not at its limit and new technologies allowing it to serve similar but also complementary market segments are emerging. It should be noted that there is no known limitation hindering solar cells from providing very high efficiencies at very low cost. New materials and device concepts may either boost existing technologies or open up new alternatives. Prerequisites for very large-scale production are carefully performed lifecycle analysis along the value chain mentioned earlier considering the availability of resources.
- The fluctuation of solar radiation requires solutions for grid integration and small- and large-scale storage. The interplay between the demand side and other renewable energy carriers provides both additional challenges for system design and also synergies arising from, for example, the peak demand for electricity

during daytime or higher wind speeds during cloudy weather conditions favoring wind energy production. Finally, manifold socio-economic factors have to be considered for PV implementation within the future energy system.

Solar energy is ready to make a substantial contribution to the global energy supply within the next few decades and bears the potential to become the major energy source of the future.

Acknowledgments

The authors thank Masdar PV for providing practical examples and information in the field of BIPV. Thanks are due to Daniel Lück of Solapeak for his contributions regarding residential PV installations. The authors are grateful to David Starr for critical manuscript revisions. This work was supported by the Federal Ministry of Education and Research (BMBF) and the State Government of Berlin (SENBWF) in the framework of the program Spitzenforschung und Innovation in den Neuen Ländern (grant No. 03IS2151).

References

1 EPIA (2012) *EPIA: European Photovoltaic Industry Association*, http://www.epia.org (last accessed 26 January 2013).
2 Raugei, M., Fullana-i-Palmer, P., and Fthenakis, V. (2012) The energy return on energy investment (EROI) of photovoltaics: methodology and comparisons with fossil fuel life cycles. *Energy Policy*, **45**, 576–582.
3 Green, M. A., Emery, K., Hishikawa, Y., Warta, W., and Dunlop, E. D. (2012) Solar cell efficiency tables (version 40). *Prog. Photovolt. Res. Appl.*, **20**, 606–614.
4 Solar Junction (2012) *Creating Legacy Through Innovation*, http://www.sj-solar.com (last accessed 26 January 2013).
5 Würfel, P. and Würfel, U. (2009) *Physics of Solar Cells*, Wiley-VCH Verlag GmbH, Weinheim.
6 Shockley, W. and Queisser, H. J. (1961) Detailed balance limit of efficiency of p–n junction solar cells. *J. Appl. Phys.*, **32**, 510–519.
7 Zhao, J., Wang, A., Green, M. A., and Ferrazza, F. (1998) 19.8% efficient "honeycomb" textured multicrystalline and 24.4% monocrystalline silicon solar cells. *Appl. Phys. Lett.*, **73**, 1991–1993.
8 Luque, A. and Hegedus, S. (2010) *Handbook of Photovoltaic Science and Engineering*, John Wiley & Sons, Ltd., Chichester.
9 Photon Europe GmbH (2012) *Photon Int.*, (3), 142.
10 Photon Europe GmbH (2012) *Photon Profi*, (11), 80.
11 Valentin Software (2012) *PV*SOL Expert 5.0*, www.valentin.de (last accessed 26 January 2013).
12 German Feed-In Law (2012) *Bundesgesetzblatt*, 23 August 2012, I (38) 1754.
13 Jung, H., Helmke, C., Döllmeier, M., and Kirchner, J. (2012) The largest PV modules of the world for the world's largest over-head glassing PV system, in *27th European Photovoltaic Solar Energy Conference and Exhibition*, pp. 3859–3861.
14 Fischer, M., Metz, A., and Raithel, S. (2012) Semi international technology roadmap for photovoltaics (ITRPV) – challenges in c-Si technology for

suppliers and manufacturers, in *27th European Photovoltaic Solar Energy Conference and Exhibition*, pp. 527–532.

15 Ringbeck, S. and Sutterlüti, J. (2012) BoS costs: status and optimization to reach industrial grid parity, in *27th European Photovoltaic Solar Energy Conference and Exhibition*, pp. 2961–2975.

16 Kinoshita, T. Fujishima, D., Yano, A., et al. (2011) The approaches for high efficiency HIT solar cells with very thin silicon wafer over 23%, in *Proceedings of the 26th EU PVSEC, Hamburg*, pp. 871–874.

17 Mingirulli, N., Haschke, J., Gogolin, R., Ferre, R., Schulze, T. F., Dusterhoft, J., Harder, N. P., Korte, L., Brendel, R., and Rech, B. (2011) Efficient interdigitated back-contacted silicon heterojunction solar cells. *Phys. Status Solidi Rapid Res. Lett.*, **5** (4), 159–161.

18 Shah, A. (2012) Thin-film silicon solar cells, in *Practical Handbook of Photovoltaics: Fundamentals and Applications*, 2nd edn (eds A. McEvoy, T. Markvart, and L. Castañer), Elsevier Academic Press, Waltham, MA, Chapter IC-1, pp. 209–281.

19 Scheer, R. and Schock, H.-W. (2011) *Chalcogenide Photovoltaics: Physics, Technologies, and Thin Film Devices*, Wiley-VCH Verlag GmbH, Weinheim.

20 Gessert, T. A. (2012) Cadmium telluride photovoltaic thin film – CdTe, in *Comprehensive Renewable Energy*, vol. 1 (ed. W. van Sark), Elsevier, Amsterdam, Chapter 1.22, pp. 423–438.

21 Brabec, C., Dyakonov, V., Parisi, J., and Sariciftci, N. S. (eds) (2003) *Organic Photovoltaics: Concepts and Realization*, Springer, Berlin.

22 Grätzel, M. (2012) Status and progress in dye sensitized solar cells, in *27th European Photovoltaic Solar Energy Conference and Exhibition*, pp. 2158–2162.

23 NREL (2012) *Best Reseach-Cell Efficiencies*, National Renewable Energy Laboratory, Golden, CO, http://www.nrel.gov/ncpv/images/efficiency_chart.jpg (last accessed 26 January 2013).

24 Kluth, O., et al. (2011) presented at the 26th EU PVSEC, Valencia.

25 Jackson, P., Hariskos, D., Lotter, E., Paetel, S., Wuerz, R., Menner, R., Wischmann, W., and Powalla, M. (2011) New world record efficiency for $Cu(In,Ga)Se_2$ thin-film solar cells beyond 20%. *Prog. Photovolt: Res. Appl.*, **19**, 894–897.

26 Manz AG (2012) *Manz*, http://www.manz.com (last accessed 26 January 2013).

27 Science, Technology and Applications Group of the EU Photovoltaic Technology Platform (2011) *Strategic Research Agenda*, EU Photovoltaic Technology Platform.

28 Schubert, B.-A., Marsen, B., Cinque, S., Unold, T., Klenk, R., Schorr, S., and Schock, H.-W. (2011) Cu_2ZnSnS_4 thin film solar cells by fast coevaporation. *Prog. Photovolt. Res. Appl.*, **19**, 93–96.

29 Todorov, T. K., Tang, J., Bag, S., Gunawan, O., Gokmen, T., Zhu, Y., and Mitzi, D. B. (2013) Beyond 11% efficiency: characteristics of state-of-the-art $Cu_2ZnSn(S,Se)_4$ solar cells. *Adv. Energy Mater.*, **3** (1), 34–38.

30 Wiesenfarth, M., Helmers, H., Philipps, S. P., Steiner, M., and Bett, A. W. (2012) Advanced concepts in concentrating photovoltaics (CPV), in *27th European Photovoltaic Solar Energy Conference and Exhibition*, pp. 11–15.

31 Graetzel, M., Janssen, R. A. J., Mitzi, D. B., and Sargent, E. H. (2012) Materials interface engineering for solution-processed photovoltaics. *Nature*, **488**, 304–312.

32 Cheng, Y. Y., Fückel, B., MacQueen, R. W., Khoury, T., Clady, G. C. D. R., Schulze, T. F., Ekins-Daukes, N. J., Crossley, M. J., Stannowski, B., Lips, K., and Schmidt, T. W. (2012) Improving the light-harvesting of amorphous silicon solar cells with photochemical upconversion. *Energy Environ. Sci.*, 5, 6953–6959.

33 Green, M. A. (2007) *Third Generation Photovoltaics: Advanced Solar Energy Conversion*, Springer, Berlin.

34 van de Krol, R. and Grätzel, M. (2011) *Photoelectrochemical Hydrogen Production*, Springer, Berlin.

17
Solar Thermal Power Production

Robert Pitz-Paal, Reiner Buck, Peter Heller, Tobias Hirsch, and Wolf-Dieter Steinmann

17.1
General Concept of the Technology

17.1.1
Introduction

Concentrating solar thermal power (CSP) systems use high-temperature heat from concentrating solar collectors to generate power in a conventional power cycle instead of – or in addition to – burning fossil fuel. Only direct radiation can be concentrated in optical systems. In order to achieve significant concentration factors, sun tracking is required during the day, involving a certain amount of maintenance. Therefore, the concept is most suitable for centralized power production, where maintenance can be performed efficiently, and in areas with high direct solar radiation levels. The concentration of sunlight is achieved by mirrors directing the sunlight on to a heat exchanger (receiver/absorber) where the absorbed energy is transferred to a heat-transfer fluid (HTF). Owing to their high reflectivity, low cost, and excellent outdoor durability, in practice glass mirrors have become more widely accepted than lenses.

A variety of different CSP concepts exist in which the HTF is either used directly in the power cycle (steam/gas) or circulated in an intermediate secondary cycle (e.g., as thermal oil or molten salt), in which case an additional heat transfer to the power cycle is required. In the late 1960s and early 1970s, when it became clear that fossil fuel resources are limited and their unequal distribution leads to strong dependencies, systematic research work was started on this technology concept in a number of industrialized countries. Today's concepts are based on the experiences gained with a variety of prototype and research installations that were mainly erected in the 1970s and 1980s and enjoyed early commercial success in the United States [1]. A break of around 20 years due to unfavorable market conditions ensued before a new commercial dawn in Spain after the turn of the century [2] that was driven by the debate on climate change. The US plants continued to operate reliably in the

Transition to Renewable Energy Systems, 1st Edition. Edited by Detlef Stolten and Viktor Scherer.
© 2013 Wiley-VCH Verlag GmbH & Co. KGaA. Published 2013 by Wiley-VCH Verlag GmbH & Co. KGaA.

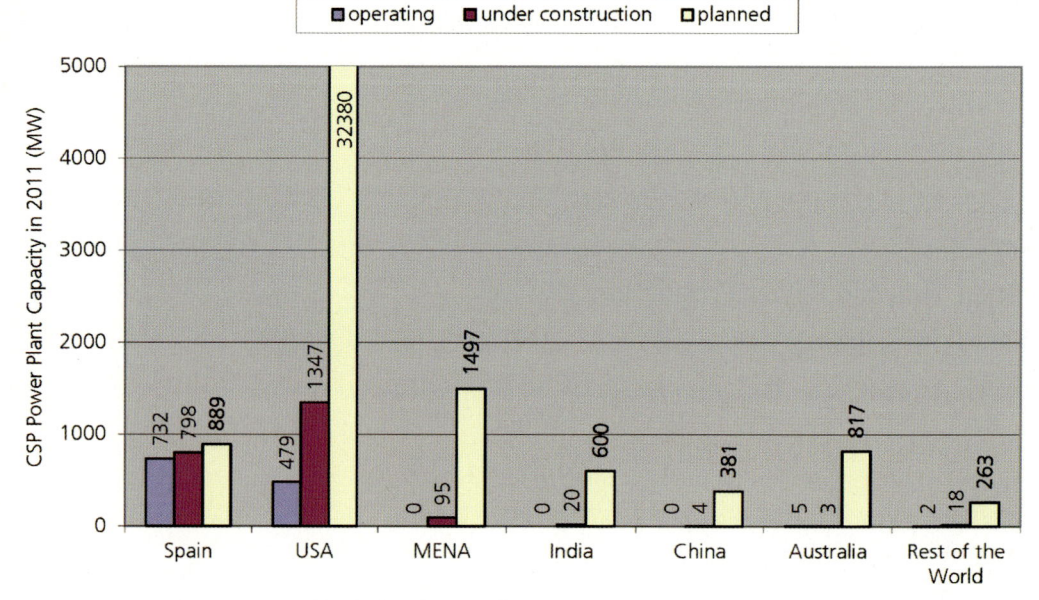

Figure 17.1 Worldwide distribution of CSP plants which are operational, under construction and planned (MENA = Middle East and North Africa). Status mid-2011. Data from [17].

interim period. Rapid technological innovation followed, leading to developments in CSP components, and the development and commercialization of new CSP technologies [3–8]. A significant initiative, cutting across technical, commercial and political fields, has been the Desertec concept [9, 10], in which it is proposed that the developed and developing world work together to harness the solar potential of the world's deserts for mutual benefit.

By mid-2011, as illustrated in Figure 17.1, there was 1.3 GW of CSP operational worldwide, 2.3 GW under construction, and 31.7 GW planned (data derived from [11–15]). Spain has been the leading exponent of CSP in Europe to date.

17.1.2
Technology Characteristics and Options

CSP systems can be distinguished by the arrangement of their concentrator mirrors: line focusing systems such as parabolic troughs or linear Fresnel systems (Figure 17.2a) only require single axis tracking in order to concentrate the solar radiation on to an absorber tube. Concentration factors of up to 100 can be achieved in practice. Point focusing systems such as parabolic dish concentrators or central receiver systems (Figure 17.2b) – using a large number of individually tracking heliostats to concentrate the solar radiation on to a receiver located on the top of a central tower – can achieve concentration factors of several thousand at the expense of two-axis tracking.

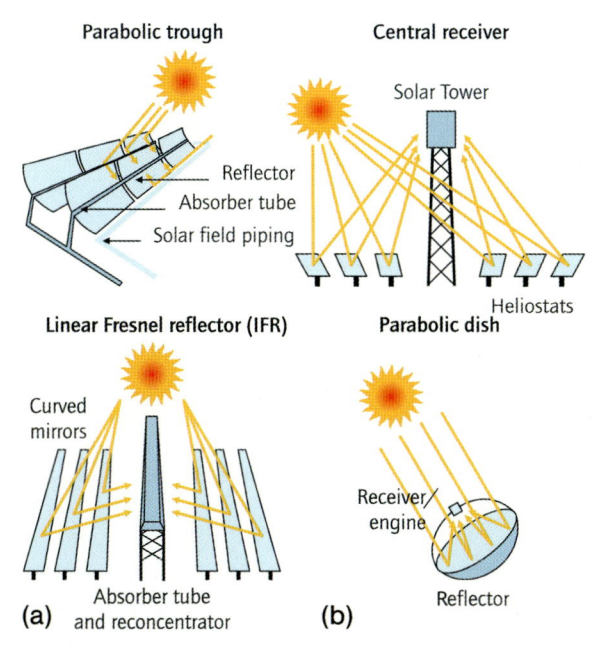

Figure 17.2 Technologies for concentrating solar radiation.
(a) Parabolic and linear Fresnel troughs; (b) central receiver system and parabolic dish.
Source: Greenpeace – awaiting permission.

According to the principles of thermodynamics, power cycles convert heat to mechanical energy more efficiently the higher the temperature. However, the collector efficiency decreases with higher absorber temperature due to higher heat losses. Consequently, for any given concentration factor there is an optimum operating temperature at which the highest efficiency of conversion from solar energy to work is achieved. With increasing concentration, higher optimum efficiencies are achievable. Figure 17.3 illustrates this characteristic assuming an ideal solar concentrator combined with a perfect (Carnot) power cycle. If the spectral absorption characteristics of the absorber are perfectly tailored to maximize absorption in the solar spectrum but avoid thermal radiation losses in the infrared part of the spectrum (selective absorber), additional efficiency gains can be expected, in particular at lower concentration factors.

In practice, the optimum operating temperatures will be lower than these theoretical figures, because power cycles with Carnot performance and ideal absorbers do not exist. Furthermore, the impact of frequent operation under part-load conditions throughout the year on the efficiency of the system has to be considered.

Similarly to domestic hot water systems, CSP systems have the important advantage of the possibility of including thermal energy storage systems (e.g., tanks with molten salt), allowing the operation of the plant to continue during cloud transients or after sunset. Thereby, a predictable power supply to the electricity grid can be achieved. In contrast to other renewable systems with electric storage, where the

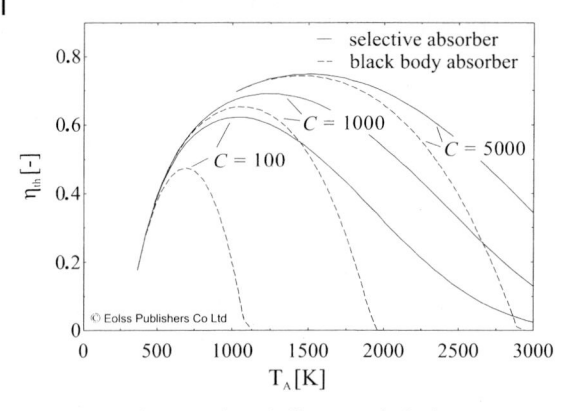

Figure 17.3 Theoretical total efficiency of a high-temperature solar concentrating system for the generation of mechanical work as a function of the upper receiver temperature for different concentration ratios and an ideal selective or a black body characteristic of the absorber [1].

inclusion of storage capacity always leads to higher investment and higher electricity prices, CSP systems with storage are potentially cheaper than CSP systems without storage. This becomes clear on comparing a solar power plant without storage of, for example, 100 MW$_{el}$ capacity that is operated for ~2000 equivalent full-load hours per year at a typical site to a system with half the capacity (50 MW$_{el}$) but the same size solar field and a suitable thermal energy storage. In this case, the smaller power block is used for 4000 equivalent full load hours so that both systems can produce the same amount of electricity per year. Assuming low storage costs, the investment in the latter system could potentially be lower than the solar-only design. In addition, the power could be sold more flexibly at times of high revenue rates.

Today, there are no power cycles specifically developed for high-temperature solar concentrating systems, but conventional fossil fuel-driven power-generation systems are adapted to solar applications. The most relevant ones are steam turbine cycles, gas turbine cycles. and Stirling engines. Currently, steam cycles are the most common choice in commercial CSP projects. They are suited to power levels beyond 10 MW and temperatures of up to 600 °C and can be coupled to parabolic trough, linear Fresnel, and central receiver systems. Stirling engines are used for small power levels (up to a few tens of kW$_{el}$) typical for dish concentrators. Gas turbines offer the potential to exploit higher temperatures than steam cycles (up to 1200 °C), covering a wide range of capacities from a few hundred kW$_{el}$ to a few tens of MW$_{el}$. At high power levels they may be combined with steam cycles to give highly efficient combined cycle systems promising to produce the same power output with 25% less solar collector area. So far, solar gas turbines have been used only in experimental facilities.

Table 17.1 summarizes some of the technical parameters of the different concentrating solar power concepts. Parabolic troughs, linear Fresnel systems, and power towers can be coupled to steam cycles of 10–200 MW electric capacity.

Table 17.1 Performance data for various CSP technologies.

Type	Capacity (MW$_{el}$)	Concentration	Peak system efficiency (%)	Annual system efficiency (%)	Thermal cycle efficiency (%)	Land use (m^2 y MWh^{-1})
Trough	10–200	70–100	21	10–16	35–42 ST	6–11
Fresnel	10–200	25–100	20	9–13	30–42 ST	4–9
Power tower	10–200	300–1000	23	8–23	30–45 ST	8–20
Dish-Stirling	0.01–0.4	1000–3000	29	16–28	30–40	8–12

ST = Steam Turbine.

17.1.3
Environmental Profile

CSP plants require large amounts of direct sunlight and hence are best constructed in arid or semiarid regions, globally known as the Sun Belt. However, CSP plants are often designed to use water for cooling at the back end of the thermal cycle, typically in a wet cooling tower. That requires roughly the same as agricultural irrigation of an area corresponding to that occupied by the CSP plant in a semiarid climate. Less than 2% of the amount of water for cooling is also used for cleaning the mirrors to maintain their high reflectivity,

Water use can be decreased by cooling with air instead, but this lowers the efficiency of the system. Switching from wet to dry cooling in a 100 MW parabolic trough CSP plant can decrease the water requirement from 3.6 to 0.25 l kWh^{-1} [16]. Using dry instead of wet cooling increases investment costs and lowers efficiency, adding 3–7.5% to the levelized electricity cost. For areas with high irradiation and available land close to the sea, such as the north coast of Egypt, using salt water for cooling could be an attractive option. It also opens up the possibility of integrating desalination with the CSP plants. Finally, there are some CSP plant designs that have inherently low fresh water requirements, such as gas turbine towers and parabolic dishes with Stirling engines.

The land use of CSP system with values between 4 and 20 m^2 (MWh y)$^{-1}$ depending on solar resource, size of the power plant, and technology option is relatively small compared with other renewables such as photovoltaic (PV) [up to 50 m^2 (MWh y)$^{-1}$], open lignite mining [60 m^2 (MWh y)$^{-1}$], or biomass [550 m^2 (MWh y)$^{-1}$] (all data from [17]). Using desert land for solar plants could in many ways be seen as better than, for instance, agricultural land for biomass energy. The areas available globally for CSP development far exceed current needs. Nevertheless, arid regions do have environmental value, and contain some biotopes or species that are threatened. Massive establishment of solar plants in an area may affect regional animal or plant populations by cutting dispersion routes and partially isolating populations from each other. This is hardly unique for CSP plants, but calls for some caution.

CSP plants are more material intensive than conventional fossil-fired plants. The main materials used are commonplace commodities such as steel, glass, and concrete whose recycling rates are high: typically over 95% is achievable for glass, steel, and other metals. Materials that cannot be recycled are mostly inert and can be used as filling materials (e.g., in road building) or can be land-filled safely. Few toxic substances are used in CSP plants: the synthetic organic HTFs used in parabolic troughs, such as a mixture of biphenyl and diphenyl ether, are the most significant. They can potentially catch fire, can contaminate soils, can create other environmental problems, and have to be treated as hazardous waste. One aim of current research activities is to replace the toxic HTF with water or molten salts.

Greenhouse gas emissions for CSP plants are estimated to be in the range 15–20 g CO_2-equivalent kWh^{-1}, similar to wind power, below values for silicon PV, and much lower than CO_2 emissions from fossil-fired plants, which range from 400 to 1000 g kWh^{-1}.

17.2
Technology Overview

17.2.1
Parabolic Trough Collector systems

17.2.1.1 Parabolic Trough Collector Development

The first experiences with parabolic trough collectors date back as far as 1870, then already driving a small engine [18]. Several installations were erected in the following decades, such as in 1912–1913 by Frank Shuman in Maadi (Meadi), Egypt. Under the influence of the oil crisis, a variety of collectors were developed in the 1980s with nonevacuated absorber tubes and aperture widths up to about 3 m by, for example, Acurex Solar, SunTec Systems, Solar Kinetics, Solel, and M. A. N. They were applied in process heat systems mainly in the United States and Europe. From these early developments, only the process heat collector PT-1 made by IST has survived and is nowadays marketed by Abengoa Solar.

The LS-1, LS-2 and LS-3 collector family developed by LUZ simultaneously with the implementation of the first large-scale solar thermal power plants (SEGS 1–SEGS 9) in California features a modular design based on steel structures with parabolic preshaped, silvered glass mirrors and improved efficiency by implementation of evacuated tube receivers. The series of commercial projects for the first time justified the investment in series production facilities for key components such as the curved mirrors and the evacuated absorber tube, establishing their dimensions as a *de facto* standard for subsequent developments.

Consequently, the EuroTrough, designed by a European consortium in the late 1990s, was based on the LS-3 concentrator geometry with a focal length (the shortest distance from the mirror to the focal line) of 1.71 m and an aperture (the projection of the concentrator area in the direction of the optical axis) width of 5.77 m, but offered advantages in stiffness and costs [19]. Increasing competition in the emerging market initiated by the attractive feed-in tariff in Spain led to the development of individual variants by the different companies of the former consortium. A different structural approach was taken by SENER, but still maintained the basic LS3 concentrator geometry [20].

Typically, these collectors are assembled from several interconnected concentrator modules (SCEs), mounted on a series of aligned pylons. The center pylon is equipped with a hydraulic drive system to allow tracking of the total collector assembly (SCA). Positioning in the direction of the Sun is controlled by means of astronomical calculations, often supplemented by a Sun position sensor. The absorber tubes mounted in the focal line move with the collector and are connected to the stationary field piping via flexible hoses or rotating joint arrangements. Recent developments show a continuing trend towards larger aperture sizes such as for the Heliotrough, Senertrough 2 and Ultimate Trough (Table 17.2). These constructions have a thick glass/silvered reflector and a steel torque tube or box in common. Solargenix and Gossamer developed a lightweight aluminum spaceframe structure that is superior in terms of shipping, handling during manufacturing, field installation, and corrosion resistance [18]. Their next generation large-aperture trough aims at an aperture width of 10 m and absorber tube diameter of 90 mm [21].

Another approach to reduce costs is the utilization of other reflector materials: the Skytrough is 6 m wide and uses reflectors made from silvered polymer film laminated to aluminum sheets mounted on a space frame [22]. Thin glass on a glass-fiber/foam sandwich was used for the SL4600 construction. Both approaches take advantage of the mechanical strength of the mirrors to reduce the amount of additional steel structure required.

Table 17.2 Characteristics of different parabolic trough collectors [19, 23–26].

Property	LS-1	LS-2	LS-3	Euro-Trough	Helio-trough	Senertrough 1	Senertrough 2	Ultimate Trough
Start of development	1984	1985	1989	1998	2005	2005	2006	2009
Aperture width (m)	2550	5000	5.77	5.77	6.78	5.77	6.87	7.51
Length per module/SCE (m)	6.3	8	12	12	19	12.27	13.23	24
SCA length (m)	50.2	47.1	99	147.8	191		158.8	242.2
Focal length (m)	0.68	1.40	1.71	1.71	1.71	1.71	2	
Torsion force carried by	Torque tube	Torque tube	V-truss framework	Torque box	Torque tube	Torque tube	Torque tube	Torque box

Optical Quality

Optical quality is one of the key performance parameters of parabolic trough collectors. It is preferably determined by a field measurement after installation of the complete collector system. Thus, influences between different components such as structure, fixing elements, and mirrors can be accounted for. With measurement systems such as photogrammetry, deflectometry, photo/video scanning, and laser systems, the optical performance of the reflecting surface may be determined. Of most interest is the slope error of the trough in the direction perpendicular to the focal line, whereas the impact on the performance parallel to the direction of the focal line is a factor of 10 lower. Prototype segments of EuroTrough-type concentrators have a root mean square (rms) value of the slope errors of ~2.7 mrad [27]. Entire modern collector modules from series production reach values of 2.5 mrad rms slope error [28]. In individually assembled modules for R&D purposes, rms values of even 1.9 mrad have been demonstrated [29]. A quality parameter that can replace the previously used rms-values of slope error is the standard focus deviation [30]. This quantifies the average deviation of the reflected beams from the ideal design focal line and is as small as 8.1 mm for test collector modules assembled under laboratory conditions [29].

Receiver Development

The receivers of parabolic troughs are composed of a stainless-steel absorber tube and a transparent glass envelope with a vacuum in between. Flexible bellows allow for thermal expansion of both components. Historically, the dimensions of the receiver used in the designs of LS2 and LS3 were still maintained in newer designs such as EuroTrough. Recently, the new generation collector designs are scaling up the geometric dimensions and demanding larger tube diameters of 80–90 mm instead of the traditional 70 mm. The vacuum is maintained over the plant's lifetime due to a completely sealed construction due to welding. Loss of vacuum by hydrogen permeation from the HTF through the metal walls is avoided by adding a getter material that absorbs the hydrogen molecules. The transparent glass envelope has an antireflective coating to permit transmission of 96% of the solar concentrated radiation [31]. Owing to the high absorber temperatures of 300–500 °C, selective absorber coatings are used by all manufacturers to avoid excessive heat losses. The absorber coatings have improved significantly from the LUZ technology in the 1990s with absorptance of 96% and emittance of 17% at 350 °C [32]. Today, several manufacturers provide receivers with similar absorptance but at much lower emittance, such as 9% [33]. New developments are under way to improve selective properties further [33]. However, it is important to note that both values must be taken into account for optimized collector yield. New tube designs have lowered the specific heat loss to 147 W m^{-1} at 350 °C [34].

Systems

Most of the commercial parabolic trough power plants today are using a diphenyl oxide–biphenyl eutectic mixture as HTF. The maximum cycle temperature is limited to values below 400 °C in order to avoid decomposition of the HTF. These plants use a steam cycle design similar to that of the SEGS VI plant [35]: a single reheat Rankine cycle with 100–105 bar and 377–383 °C live steam conditions. They are equipped with six preheating stages: three low-pressure preheaters, the deaerator, and two high-pressure preheaters.

Design cycle efficiency of these plants is slightly above 38% when a wet cooling tower is used. Dry cooling with an air-cooled condenser leads to lower cycle efficiency since the turbine exhaust pressure increases. The design efficiency of dry cooled parabolic trough power plants is about 1–2% [37] lower compared with a wet cooled plant, actual values depending on the ambient temperature at the site and the dimensions of the air-cooled condenser.

The annual efficiency of parabolic trough power plants is currently about 15% (electricity generated over incident energy) [38]. Larger power blocks would increase this efficiency (16% annual efficiency for a 110 MW dry cooled plant [38] with 39.5% design cycle efficiency is reported).

Plants with thermal storage are typically using two-tank molten salt systems with a near eutectic mixture of sodium nitrate (60%) and potassium nitrate (40%) as storage medium. Hot HTF from the solar field is used to charge the storage during time periods when the solar field delivers excess heat. Storage discharging is also done via HTF using the same heat exchangers as for charging but with reverse flow of the HTF and salt. This arrangement leads to live and reheat steam temperatures during storage discharge operation which are about 10 K lower compared with direct utilization of the solar heat. The cycle efficiency in the storage discharge mode is slightly lower since the lower HTF temperature causes a lower live steam temperature in the storage discharge mode.

Integrated solar combined cycle (ISCCS) plants are combined cycle power plants using additional solar heat generated by parabolic troughs in the bottom cycle. Three ISCCS plants are in operation, at Kuraymat, Egypt, Hassi R'Mel, Algeria, and Ain Beni Mathar, Morocco. Two further plants, Archimede in Italy and Martin Next Generation Solar in Florida, can also be counted in this category. Table 17.3 gives an overview of their overall and solar design output.

Table 17.3 ISCC power plants.

ISCCS plant	Plant design output (MW)	Design output from solar (MW)	HTF
Kuraymat, Egypt	140	20	Diphenyl oxide–biphenyl
Hassi R'Mel, Algeria	150	25	Diphenyl oxide–biphenyl
Ain Beni Mathar, Morocco	470	20	Diphenyl oxide–biphenyl
Archimede, Italy	130	5	Molten salt
FPL Martin next generation solar, Florida	3780	75	Diphenyl oxide–biphenyl

The Kuraymat, Hassi R'Mel, and Ain Beni Mathar plants produce saturated steam in the solar steam generator, which is fed into the heat recovery steam generator (HRSG) for superheating [39, 40]. The electricity generated from solar energy at the design point is less than 17% in all cases and the annual solar fractions of these plants are even lower since the solar field delivers less than the design heat input for a large portion of the year [41].

Direct Steam Generation
The limitation of the upper process temperature to about 400 °C can be overcome by changing the HTF. An obvious alternative is water/steam, which is used in the steam cycle any way. The technology is known as direct solar steam generation (DSG) and was successfully demonstrated in the European research project DISS [42]. In a DSG collector field, water is preheated, evaporated, and superheated. Steam parameters reached in demonstration loops are 110 bar and 500 °C [43]. Aside from the process temperature, the benefits of DSG compared with oil are based on savings in the heat exchanger, in reduced pumping effort, and the uncritical handing of the medium. Owing to the two-phase flow conditions in the evaporator section, the layout of the solar field differs from the traditional oil field. The solar field consists of a preheating and evaporation subfield being operated in a forced recirculation mode by means of a steam drum. From the steam drum, saturated steam enters the superheating subfield and finally a collecting header that transports the steam to the turbine. Research projects today have the aim of designing a field in once-through mode without a steam drum [44]. Thermal storage systems for DSG plants are composed of a phase change material storage part and a sensible storage part [45]. These systems have proven their functionality but have to undergo further cost reductions to be competitive. In addition to the linear Fresnel plants with DSG in operation today, one commercial 5 MW plant with parabolic trough collector technology was connected to the electricity grid of Thailand in 2012 [46]. Apart from electricity production, DSG systems are very interesting for high-temperature process heat applications such as enhanced oil recovery.

Molten Salt Applications
Salts already have a long history in solar tower power plants, with the French Themis Power Plant (1983) [47] and the US Molten Salt Electric Experiment (1983) [48] and Solar Two plant (1996) [49]. In recent years, investigations of liquid salts in parabolic trough systems have also been initiated. Liquid salts promise to overcome drawbacks of the classic biphenyl–diphenyl ether heat-transfer medium. Salts are chosen on the basis of their suitable thermophysical properties, namely high boiling/decomposition point, low vapor pressure, high specific heat capacity, high thermal conductivity, and high density at low pressures [50]. A higher process temperature leads to a significant increase in the thermodynamic conversion cycle efficiency. The high capability to store energy in small volumes has two advantages: solar field parasitics can be significantly lowered due to very low flow velocity of salt and also the amount of storage mass is drastically lowered. Furthermore, typical salt mixtures are significantly cheaper than synthetic oils. Heat can therefore be

absorbed by the salt in the irradiated receiver tubes and be stored directly in large, flat-bottomed storage tanks. Solar field and power block are fully decoupled; such a system is beneficial for satisfying the demands of full dispatchable power plants.

The current main candidates are nitrate salts. A mixture of sodium nitrate (60 wt%) and potassium (40 wt%) nitrate is used in the 5 MW-Archimede plant [51]. The upper allowed temperature is > 550 °C [52]; first crystallization of the non-eutectic melt occurs at 238 °C. Adding further substances to the salt allows lower crystallization and freezing temperatures. Very promising future nitrate salts are mixtures if calcium, sodium, and potassium nitrate and of lithium, sodium, and potassium nitrate; the eutectic mixtures achieve solidification temperatures of 133 and 120 °C, respectively [113], still at high maximum temperatures of > 480 and 550 °C, respectively [53]. The overall band of currently investigated salt candidates varies broadly [54, 55].

Nevertheless, there are also concerns about molten salt-based parabolic trough plants. Owing to the high solidification temperature, the process setup needs to be modified [56]; the solar plant is fully equipped with impedance and trace heating systems in order to ensure non-freezing of the salt. Salt melts at high temperatures are regarded as highly corrosive, but the latest research and process experience with demonstration plants show that this can be handled with the right choice of stainless steel [57]. Both the main issues and their handling lead to higher investment costs. In current research projects [51, 58], it will be demonstrated that the increase in total plant efficiency and lower storage and HTF costs will overcome the higher investment in solar field and piping and therefore lead to a decrease in electricity costs of solar thermal power plants.

17.2.2
Linear Fresnel Collector Systems

The idea for the linear Fresnel collector system goes back to the French physicist Augustin Jean Fresnel (1788–1828). For applications with short focal length, he developed an optical lens composed of a number of small lenses that are designed in such a way that they all have the same focal point. A linear Fresnel concentrator uses the same concept to bundle incoming sunlight into a focal line by a number of parallel mirror facets (Figure 17.4).

Breaking up the geometrically constant relationship between focal line, Sun, and reflector surface leads to changes in the optical path. Whereas a parabolic trough concentrator is always optically ideal despite manufacturing tolerances, the linear Fresnel concentrator suffers from inherent optical nonidealities. The shape of the reflectors can be designed for only one optical design point (i.e., position of the Sun relative to the collector). For all other Sun positions, the reflected rays are not concentrated in a common focal line but in a larger area. This so-called astigmatism typical of linear Fresnel systems explains the need for a secondary reflector arranged over the receiver tube. The secondary reflector redirects a large amount of beams that otherwise would not hit the receiver tube. Since the optical paths are highly dependent on the position of the Sun, annual calculations are required to find the best shapes of both the primary and secondary reflectors. The astigmatism is the

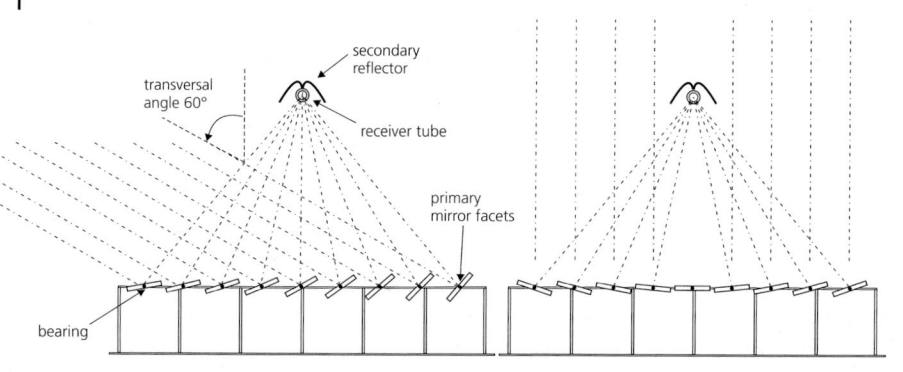

Figure 17.4 Linear Fresnel optical path for two different Sun positions.

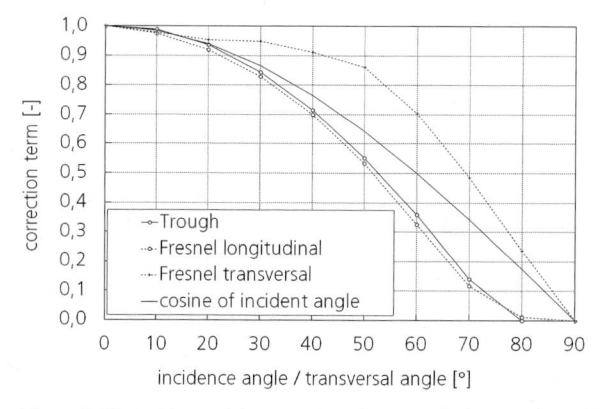

Figure 17.5 Incident angle correction for a parabolic trough and linear Fresnel system.

main reason why peak optical efficiencies of linear Fresnel systems (~67%) are lower than those for comparable parabolic troughs (~78%).

Whereas the impact of the nonideal optics is moderate under the design conditions, the optical performance decreases significantly for low Sun heights. Figure 17.5 shows the incident angle-dependent correction term for the optical efficiency for a typical parabolic trough and linear Fresnel collectors. The impact in the incident angle direction is very similar for both types of collector. The main difference comes from the correction term in the transversal direction, which is 1 for the parabolic trough due to the tracking but < 1 for the linear Fresnel. The overall efficiency is the product of both effects, which illustrates the poorer optical performance of the linear Fresnel collector the more the Run comes from the side.

The higher optical efficiency of the parabolic trough is based on the stiff, rotating collector structure. Whereas the whole concentrator structure including receiver pipe has to be tracked towards the Sun with a parabolic trough system, only the small mirror facets have to be rotated in the Fresnel case. The supporting structures, bearings, and drives have to withstand lower loads and are therefore lean in construction. Wind loads on the mirrors are very small since the facets with small aperture width are arranged in one horizontal plane near the ground. Since the optical concentrator movement of a linear Fresnel system is independent of the receiver tube, long rows with a straight receiver tube can easily be realized. In parabolic troughs, the single collectors have to be connected via flexible connections to allow individual tracking of the units. Especially for applications with high operating pressure (direct steam generation) or external heating requirements (molten salt), this systematic difference is beneficial for linear Fresnel systems.

The linear Fresnel concept was first demonstrated by Francia in 1968 [59] and revived by collector developments in Belgium [60] and Australia [61]. Demonstration collectors have been erected in recent years by Ausra [62], Novatec Solar [63], AREVA Solar [64], Solar Power Group [65], and Solar Euromed [66]. In parallel with these developments of large-scale collectors for solar thermal power plants, special collectors for process heat applications are being developed. Following the same physical principle, the collector systems differ in the construction details of the primary reflector and the receiver. Some manufacturers make use of an individual drive for each mirror segment, which allows the mirror to be turned into a safety position with the mirrors facing downwards (e.g., AREVA Solar, SPG). The alternative concept is to couple the segments of one half of the collector by connecting rods, which allows the number of drives to be reduced (e.g., Novatec Solar).

Reduction of thermal losses of the receiver is traditionally realized by means of a glass plate arranged below the receiver that avoids forced convection in the secondary reflector box (Figure 17.6). The secondary mirrors are thermally insulated on the back side. With this construction, similar or even lower aperture specific heat losses can be realized than for parabolic trough systems. High-temperature applications require stable selective surfaces on the receiver tubes, which can hardly be achieved under atmospheric conditions. Therefore, new designs foresee vacuum-type receiver systems known from parabolic trough collector systems. While most receiver designs are based on a single receiver tube, the AREVA system uses a bundle of smaller receiver tubes arranged horizontally in the aperture plane.

Linear Fresnel collector systems are traditionally used for direct steam generation in the collector field. Realized plants are operated with saturated steam at about 55 bar (Novatec Solar PE-2 with 30 MW [63], Liddell steam augmentation system with 9.3 MW$_{th}$ [67]) or superheated steam at moderate temperatures (AREVA Solar's Kimberlina plant at 7.8 MWth [64]). The linear Fresnel plants are characterized by long, nonstop collector rows with a length of 400–1000 m. Since flexible connections between the collectors as in parabolic trough technology are not required, the linear

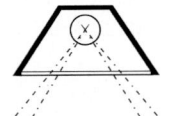

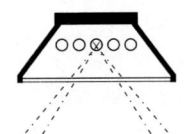

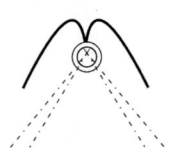

Figure 17.6 Linear Fresnel receiver concepts.

Fresnel systems represent an interesting option for high-temperature/high-pressure applications. First test facilities demonstrated the feasibility to reach steam temperatures of ~500 °C [68]. In the near future, commercial systems with high steam parameters are expected to be realized. The benefits of the linear Fresnel system are put into perspective if the overall performance is taken into consideration, since the typical annual output measured in terms of delivered electric energy divided by the aperture area is about 30% lower for a Fresnel system [60, 69–71]. With the characteristic mid-day peak, power plants without thermal storage suffer from either large thermal dumping or high part-load operating hours of the turbine. Thermal storage is an effective way to shift the high yield at noon to later hours. This, in combination with the possibility of reaching high temperatures, put the focus on linear Fresnel collector systems for molten salt applications [72–75]. Nevertheless, no molten salt demonstrations have yet been realized with linear Fresnel systems.

17.2.3
Solar Tower Systems

Solar tower systems use a large number of reflectors (so-called heliostats) to concentrate direct solar radiation to a focal point, in most cases on top of a tower. In order to reflect the incident sunlight to the focal point, the heliostats are tracked in two dimensions, that is, around two axes of rotation. The receiver is located at the focal point. The receiver absorbs the concentrated solar radiation and heats a heat transfer medium, such as a liquid or a gas. Figure 17.7 shows a scheme of a solar tower system with heliostat field, receiver, tower, power block and optional storage system.

Typical average concentration levels on the receiver are in the range 500–1000 kW m^{-2}, which are significantly higher than those achievable in linear focus systems such as parabolic trough and linear Fresnel systems. The higher concentration level allows the receiver to be operated at higher temperatures while maintaining good thermal efficiencies.

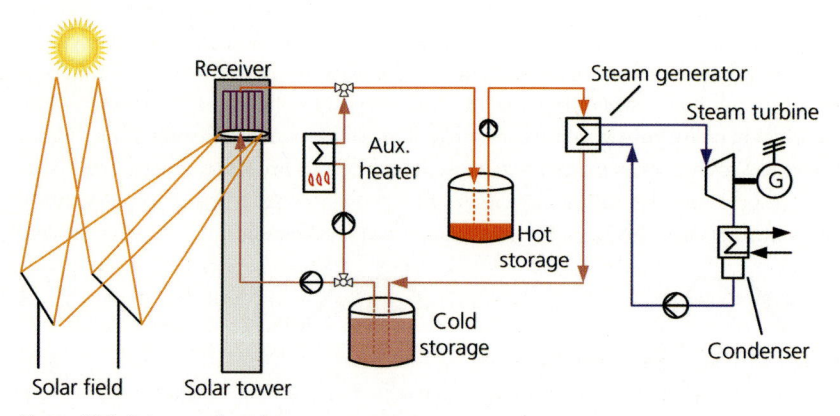

Figure 17.7 Scheme of a solar tower system.

The heliostat field consists of numerous flat or slightly curved reflectors (heliostats) that are tracked in two dimensions in order to direct the reflected radiation towards the receiver on top of the tower. Current heliostat technology differs mainly in the area of the reflecting surface. The smallest heliostats are built from single glass mirrors with an area of 1.1 m^2, each tracked independently [76]. The largest heliostats in commercial application have a total area of 121 m^2 [77] and are built of several smaller facets, mounted on a back structure. The orientation of the rotation axes may also differ between heliostat types. Special heliostat designs include an additional large curved reflector ("tower reflector") near the aim point of the heliostats to redirect the radiation to create a focal spot near ground level. This concept, called "beam-down," allows the receiver to be built close to ground level, but suffers from additional optical losses.

The field configuration, that is, the shape of the heliostat field, depends mainly on the site latitude and the power level. With higher power levels, surround fields with heliostats all around the tower are usually more economical. The higher the site latitude of a plant (i.e., the further away from the equator), the more the heliostat field tends to be shifted towards north (in the northern hemisphere) or south (in the southern hemisphere). As an example, mid-sized solar tower systems in Spain are built in a north field configuration. The type of receiver (external or cavity) also influences the optimal field configuration.

A variety of solar tower concepts exist or are under development. These concepts differ mainly in the type of heat transfer medium, which dominates the layout and selection of all other components except for the heliostat field.

Liquid or gaseous HTFs are state-of-the-art in solar tower technology. Water/steam, molten salt, and air are typical fluids. In this case, the receivers are usually built from a large number of metallic tubes, often grouped together in panels that are interconnected in parallel or serial mode. Two main configurations exist: external receivers and cavity receivers. In external receivers, the tubes are arranged on the outside of a surface, for example, a cylinder when used in a surround field configuration. Typical solar heat fluxes in this type of receiver are up to 1 MW m^{-2}. In cavity receivers, the tubes are installed at the inner walls of a larger cavity with a smaller aperture. Whereas the solar heat flux in the aperture is in the same range as for an external receiver, the flux is distributed inside the cavity over a large area with the absorber tubes, thus reducing the solar flux on the absorber tubes.

For air receivers, volumetric receivers are another option. Volumetric absorbers are structures with high open porosity, for example, ceramic matrix structures or foams. They allow the concentrated solar radiation to penetrate into the volume of the absorber, and the absorbed power is then transferred to an air flow passing through the open structure. The receivers can be operated at ambient pressure, or in pressurized mode by closing the receiver aperture with a quartz window.

State-of-the-Art in Solar Tower Systems
Solar tower (or power tower) technology is in an early phase of market introduction. The first commercial plants are in operation, and new large plants are under con-

struction. The most relevant solar tower plants that are in operation are described briefly in the following sections.

The plant PS10 [77] was the first commercial solar tower system and was commissioned in 2007 by Abengoa Solar. The plant is located near Seville, Spain, and is rated at 11 MW$_{el}$. The concentrated solar radiation heats a metallic tube receiver where saturated steam of 250 °C and 40 bar is generated to drive a turbine for power generation. A pressurized water/steam tank is used to provide some storage capacity for about 30 min. In 2009, the larger plant PS20 [79] with 20 MW$_{el}$ was commissioned, based on the same technology.

The solar tower plant Gemasolar was erected by Torresol Energy near Ecija, Spain and has a power output of 19.9 MW$_{el}$ [78, 80]. The plant uses molten salt as heat-transfer and storage medium (Figure 17.7). During solar operation, the molten salt is pumped from the cold storage tank (at about 290 °C) to the receiver, where it is heated to about 565 °C. The hot salt is then piped to the hot storage tank. For power generation, the hot salt from the tank is pumped to the steam generator, where superheated steam at 540 °C is generated to produce power. The cooled salt (290 °C) is then pumped back to the cold storage tank. This plant concept allows decoupling of solar energy collection and electricity production. The heliostat field consists of 2650 heliostats, each with a reflective area of 115 m^2. The concrete tower has a height of 140 m. With its 15 h storage system, the plant can operate in summer for 24 h, and the annual capacity factor reaches 75%.

The Sierra SunTower demonstration system [76] built by eSolar is installed near Lancaster, CA, USA. It is a modular system with two identical tower units. In the receivers, superheated steam is generated and piped to a common steam turbine for power production. The steam cycle has a power level of 5 MW$_{el}$. Larger power levels can be achieved by combining multiple modular tower units, all with identical layout of the heliostat field, tower, and receiver [81]. Mainly the power cycle must be adapted to the changed power level. Since the heliostat subfields are relatively small, high solar collection efficiencies can be achieved.

The solar tower Jülich [82] is a pre-commercial solar tower demonstration plant in Germany with a power level of 1.5 MW$_{el}$. Air at near-ambient pressure is used as heat-transfer medium. The air is heated with solar energy in an open volumetric receiver to about 700 °C. The hot air can be used directly in an HRSG to produce power in the steam cycle. Alternatively, in charge mode, hot air is directed to the regenerator-type storage system. In discharge mode, the flow direction in the regenerator is reversed to heat the entering air, which is then used in the HRSG for power generation. The heliostat field contains over 2000 heliostats with a total mirror area of about 18 000 m^2.

The above-mentioned solar tower plants all have significantly below the power level that is considered most economical. New solar tower plants that are currently under construction have higher power levels:

- Ivanpah (Brightsource): this plant is under construction near Ivanpah, CA, USA and will represent with its total power output of 390 MW the largest CSP facility worldwide [112]. Plant commissioning is planned to be in 2013. The plant consists of three independent solar tower units, each with its own power block. The receivers are used to generate superheated steam to drive a steam cycle for power production. In total, 170 000 heliostats will be installed, each of about 16 m².

- Crescent Dunes (SolarReserve): this solar tower plant is under construction near Tonopah, NV, USA and has a power level of 110 MW [83]. Molten salt is used as HTF and storage medium. The storage is designed for about 4500 full-load operating hours of the power block (capacity factor > 50%). Commissioning of the plant is planned to be in early 2014.

A number of other solar tower plants are in different phases of project development. Abengoa has announced the construction of a 50 MW solar tower plant in South Africa. SolarReserve is developing two additional projects in Spain (50 MW) and California (150 MW). Brightsource plans to erect another plant in California with a total power output of 500 MW. Several other tower projects are under way.

Current Solar Tower Development Trends

The most relevant issues for solar towers are increase in annual energy yield and cost reduction for investment, project development, and operation. Thermodynamic cycle efficiency, financing conditions, and heliostat field costs were determined to have the largest impacts on cost reduction [84]. Especially for financing conditions, it is important to prove technological maturity in order to lower the risk surcharge. Several approaches to cost reduction for heliostats exist [85], with highly automated production of heliostats being an important factor. Also, the high solar concentration enables the upper process temperature to be increased and consequently the solar-to-electric conversion efficiency to be improved.

Thermodynamic cycle efficiency can be increased by using higher process temperatures in the power cycle. Existing solar tower steam cycles have live steam conditions that are well below those common in state-of-the-art fossil-fired steam cycles. Several proposals exist for the improvement of live steam conditions, associated with an increase in the receiver temperature for steam receivers or molten salt receivers (up to 650 °C, to feed supercritical steam cycles) [86, 87]. Increasing the receiver temperature is challenging mainly with respect to material issues for the heat transfer medium and also for structural materials. Degradation and corrosion effects limit the selection of materials and/or require the use of expensive special materials.

Application of alternative power cycles is another option to increase thermodynamic cycle efficiency. Solar gas turbine systems can achieve high conversion efficiencies, either in combined cycle configuration (with bottoming cycle) or as recuperated Brayton cycles [88]. However, these cycles require receiver temperatures of up to 1000 °C to reach high solar shares. Receiver operation at such high temperatures results in a decrease in receiver efficiency, offsetting part of the advantage of increased cycle efficiency. On the other hand, gas turbine systems offer the advantage of significantly reduced cooling requirements. A first prototype of a solar–hybrid gas turbine system with a power level of 4.5 MW was built near Seville, Spain by Abengoa within the EC-cofunded project SOLUGAS, and has been in operation since

2012 [89]. Other projects involve the modification of commercial gas turbines to the specific requirements of solar–hybrid operation. New approaches are investigating the use of innovative supercritical CO_2 cycles that show promising efficiencies with moderate receiver temperatures [90].

Several new receiver concepts are proposed for maintaining high efficiencies at the envisaged higher receiver temperatures. Innovative receiver concepts aim at the reduction of the radiative losses and the convective losses (free convection, wind). This can be achieved by improved receiver geometries, selective absorber characteristics, and more efficient conversion of solar radiation to heat, for example, through direct absorption in the heat-transfer medium. Several concepts with direct absorption in molten salt or small particles are under consideration [91–93].

Multi-tower concepts are under discussion as a further possibility for cost reduction. The division of the heliostat field into several standardized subfields, each with its own tower, offers several advantages. Smaller heliostat fields result in higher solar collection efficiencies, and the shorter average distance between heliostats and receiver reduces atmospheric attenuation losses, especially important in regions with high aerosol content in the air. Some additional benefit can be achieved by redirecting selected heliostats at certain times from one receiver to another, mainly when cosine losses of the heliostat are significantly different between the two heliostat orientations. The heat-transfer medium exiting the various receivers can be transported to a central power block, with the storage being installed either at each tower or at the central power block. By modifying the number of standardized subfields, the rated power of the plant can easily be scaled to the desired level, without changing the design of the subfields with receiver and tower. However, the higher efficiency must be traded against additional costs of piping and other components [94].

17.2.4
Thermal Storage Systems

The development of storage systems has accompanied CSP technology almost from the start. While CSP systems already exhibit a short-time storage capacity due to the thermal mass of the components and the working fluid, today's commercial storage units provide energy to operate the power block of a CSP plant for several hours. The integration of storage units offers various advantages:

- The generation of electricity can follow the demand structure.
- The extent of fossil-fired, low-efficiency backup systems required for compensation of fluctuating renewables can be reduced.
- Improved efficiency by avoiding transients resulting from the passing of clouds.
- Improved efficiency by reducing part-load operation of power block components.
- Reduced start-up period by using energy from the storage unit for preheating.

Figure 17.8 shows the general concept for storage integration into CSP-plants: the storage unit is operated in parallel with the power block during the charging period using surplus heat from the solar concentrator system. During discharge, the heat provided by the storage unit is used to operate the power block.

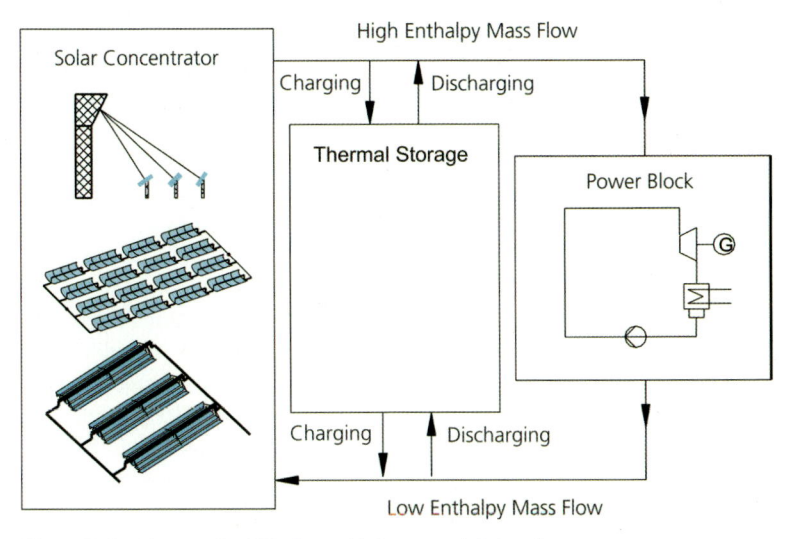

Figure 17.8 Scheme of a CSP plant with integrated thermal storage.

The adaptation of the storage system to the characteristics of both the solar concentrator and the power block is crucial. Owing to the large variety of concentrator types, working fluids, and thermodynamic cycles considered for application in CSP plants, different storage concepts have been developed to meet the specific combination of requirements for a given configuration [95, 96]. Preferably, the comparison of storage systems should be based on the effect on the "levelized electricity cost" (LEC). The characterization of the storage system requires the definition of the following key data:

- thermal capacity
- thermal power
- operating temperature
- required mechanical power
- thermal losses.

A more detailed description includes also information about the response time, footprint, effects of part-load operation and partial loading. Capital cost estimations and operating costs complete the characterization of a storage concept.

17.2.4.1 Basic Storage Concepts

The variety of storage types can be classified using different criteria. Depending on the power/capacity ratio, storage systems are distinguished into buffer storage intended for compensation of short transients and storage systems used for continuous operation over several hours. This classification can be combined with an additional system based on the physical process applied for energy storage. Here, three basic physical mechanisms can be identified, as described in the following.

Sensible Heat Storage

In sensible heat storage, a single-phase medium (either liquid or solid) are used for heat storage. The capacity of the storage system depends on the mass of the storage medium, the specific heat capacity, and the temperature variation during the charging/discharging process. If nonpressurized, cost-effective HTFs are used in the solar absorbers, direct storage of the hot HTF is possible.

Pressurized or expensive HTFs require the transfer of heat to cost-effective storage media. The selection of the storage material is crucial for the storage system (Table 17.4).

Sensible heat storage is preferred for single-phase working fluids such as thermal oil, molten salt, and air. This preference results from the efficiency considerations requiring the minimization of temperature differences between storage material and HTF. Since enthalpy variations of the HTF result in a change in temperature, the storage material must also undergo a change of temperature to fulfill this requirement.

Table 17.4 Solid and liquid materials for sensible heat storage.

Material	Density ($kg\ m^{-3}$)	Specific heat capacity ($J\ kg^{-1}\ m^{-3}$)	Thermal conductivity ($W\ mK^{-1}$)	Volume specific storage density, DT = 100 K ($kWh\ m^{-3}$)
Concrete	2200	720	1.5	44
Stone	2500	800	1.0–3.0	55
Mineral oil (< 320 °C)	800	2.4	3.5×10^{-4}	53
Nitrate salt (220–570 °C)	1950	1.5	3.4×10^{-4}	81

DT = Temperature difference used as basis for calculation of energy storage density (uniform boundary conditions)

Latent Heat Storage

Owing to efficiency aspects, isothermal processes such as evaporation and condensation require storage concepts operated at constant temperature during charging and discharging [97]. One option to fulfill this requirement is the application of storage media that undergo a phase change in the relevant temperature range. The selection of the storage material defines the operating temperature. For solar absorbers with direct steam generation, this operating temperature should be near to the saturation temperature of the steam. Table 17.5 shows data for three nitrate salts that have been used in latent heat storage systems.

The thermal conductivity of the materials considered for latent heat storage is low, so the development of concepts ensuring a sufficient heat transfer rate between steam and storage material is essential for the successful implementation of storage units based on latent heat energy storage.

Table 17.5 Solid and liquid materials for sensible heat storage.

PCM	Melting temperature (°C)	Density (kg m^{-3})	Thermal conductivity (W mK^{-1})	Heat of fusion (kJ kg^{-1})	Volume specific latent heat (kWh m^{-3})
KNO$_3$–NaNO$_2$–NaNO$_3$ (eutectic)	142	2000	0.5	60	33
KNO$_3$–NaNO$_3$ (eutectic)	222	2000	0.5	100	55
NaNO$_3$	306	1900	0.5	175	96

Chemical Energy Storage

The enthalpy change of reversible chemical reactions can also be used for energy storage [98, 99]. Compared with sensible heat storage and latent heat storage, chemical energy storage shows the potential for higher volumetric energy densities. Examples are the dehydration of salt hydrates (CaCl$_2 \cdot$6H$_2$O), decomposition of metal hydroxides [Ca(OH)$_2$], and the reduction of metal oxides (MnO$_2$). Chemical energy storage systems are currently in an earlier stage of development.

17.2.4.2 Commercial Storage Systems

Today, all storage systems used in commercial power plants are based on sensible heat energy storage. Parabolic trough power plants use indirect storage systems; the energy collected in the solar field is transferred to a mixture of sodium and potassium, which is cycled between 295 and 385 °C. The storage-based operation of a parabolic trough power plant over a period of 7.5 h requires a molten salt mass of 28 000 t [100]. The same molten salt is applied in the Gemasolar central receiver plant; the cyclic temperature difference is 275 °C and a storage mass of 7900 t is used to generate 19.9 MW over a period of 15 h [80].

The PS-10 and PS-20 central receiver plants produce saturated steam which is partly fed into steam accumulators. Here, the steam increases the temperature of pressurized water during charging [101]. The stored energy is used to produce saturated steam during discharge, allowing an operation at 50% capacity over a period of 50 min.

The Jülich central receiver power plant uses air at ambient pressure as heat-transfer medium in the absorber system. The hot air flows through a packed-bed storage unit at a maximum temperature of 680 °C. During discharge, the power cycle can provide 1.5 MW$_{el}$ over a period of 1.5 h [102].

17.2.4.3 Current Research Activities

The focus of current research activities is on cost reductions. Alternative molten salt mixtures with an extended operating temperature range are assessed [103]. The application of solid storage media is attractive owing to reduced material costs; new concepts such as the CellFlux approach are being developed to allow efficient heat transfer between HTFs and packed-bed storage materials [104]. After the feasibility of latent heat energy storage had been proven by a 700 kW module operated with steam at 100 bar [105], further research activities are aimed at the cost optimization of this approach, which uses integrated extended surface heat exchangers. For this concept, the size of the heat exchanger increases directly with the storage capacity. Innovative latent heat storage concepts try to decouple the size of the heat exchanger from the storage capacity by transporting the storage material along the heat exchanger. A further option to reduce costs is the sequential combination of sensible energy storage with latent heat storage; the cyclic specific enthalpy variation of the storage material is higher and less material is required.

17.3
Cost Development and Perspectives [17]

17.3.1
Cost Structure and Actual Cost Figures

The structure of a commercial CSP project is very similar to those of other large power plant projects and typically involves several players. An "Engineering, Procurement and Construction" (EPC) contractor and its suppliers provide and warrant the technology to the owner, who finances it through equity investors, banks, and eventually public grants. The owner gains revenues from the electricity off-taker (typically the electricity system operator) based on long-term power purchase agreements needed to pay off the debt and operating costs, and to generate a profit. An operation and maintenance company provides services to the owner to operate the plant. This approach results in a complex contractual arrangement to distribute and manage the overall project risk, as the overall project cost of several hundred million euros typically cannot be backed by a single entity. The perception and distribution of risks, in addition to local and regional factors, strongly affect the cost. There is therefore no single figure for the costs of electricity from CSP or, for similar reasons, for other generating technologies with which it needs to be compared. One approach that is often used to compare costs of electricity generation is to calculate the LEC relates average annual capital and operating costs of the plant to the annual electricity production. Recognizing the limitations of the approach, particularly when comparing fossil-fired and renewable technologies where it does not capture differences in value to the customer, it nonetheless gives a useful "first cut" view of comparative costs. For comparisons between fossil-fired plants and CSP with storage and/or supplementary firing, its limitations are less significant as the technologies offer similar services. Recent studies [106, 107] give levelized costs of electricity from CSP of €0.15–0.22 kWh^{-1} (US $ 0.20–0.29 kWh^{-1}) in 2010 monetary values, depending on technology, size, and solar resource.

In order to present an illustrative comparison of CSP electricity costs with other options, cost estimates for different technologies have been made taking data from a single source [107, 108], and a simplified equation was used to evaluate the LEC. The results are summarized in Table 17.6.

Table 17.6 Illustrative costs of generating technologies in 2010 [17] (currency conversion 2010 \$/€ = 0.755).

Technology	LEC (€ kWh$_e^{-1}$)	Capacity (MW)	EPC cost (€ kW$_e^{-1}$)	Cap factor	Fuel costs (€ kWh$_e^{-1}$)	O&M$_{fix}$ (€ kW^{-1} y^{-1})	O&M$_{var}$ (€ kWh$_e^{-1}$)
CSP: 100 MW no storage (Arizona)	0.179	100	3542	0.28	0	48	0
Pulverized coal: 650 MW base-load	0.069	650	2391	0.90	0.029	27	0.3
Pulverized coal: 650 MW mid-load	0.09	650	2391	0.57	0.029	27	0.3
Gas combined cycle mid-load	0.061	540	738	0.40	0.032	11	0.3
Wind onshore: 100 MW	0.085	100	1841	0.30	0	21	0
Wind offshore: 400 MW	0.153	400	4511	0.40	0	40	0
Photovoltaic: 150 MW (Arizona)	0.212	150	3590	0.22	0	13	0

This analysis assumed that the renewable energy systems (wind, PV, and CSP) are positioned to have a favorable solar or wind resource and financing conditions. For CSP, a Direct Normal Insolation (DNI) in Phoenix, AZ, USA (2500 kWh m^{-2} per annum) is considered. The solar resource in southern Europe is typically about 20% lower, whereas some sites in North Africa have a 5% higher resource potential. The impact on the cost is almost linear: per 100 kWh m^{-2} a of additional solar input, the LEC drops by ~4.5%.

The analysis presented in Table 17.6 gives a cost figure for CSP electricity within the range given in the studies mentioned [106, 107]. It also permits a comparison of the CSP generating cost with other conventional and renewable options under similar boundary conditions.

CSP costs in 2010 were about twice those of onshore wind farms, and slightly higher than estimates for offshore wind energy. Since then, costs for CSP and in particular for PV have decreased significantly. For large-scale projects in North Africa, electricity costs of less than €0.11 kWh^{-1} for PV and less than €0.15 kWh^{-1} for CSP have been reported.

Still, the implementation of CSP systems depends on market incentives established by governments. However, changes in fuel prices, higher CO_2 penalties, and in particular further cost reductions of CSP are expected to change this situation over time, as discussed in the following sections.

It is not just the cost of CSP generation which determines its economic competitiveness, but also its value, which has three components:

- the value of the kilowatt hours of electrical energy generated by the plant, which will vary over time in a competitive electricity market, reflecting the availability and cost of electricity from other sources;
- the contribution that the CSP plant makes to ensuring that generating capacity is available to meet peak electricity system demand; and
- the "services" provided by the plant in helping the electricity transmission system operator to balance supply and demand in the short term (typically, on timescales of seconds and minutes).

Potentially, incorporation of thermal storage in a CSP plant can be beneficial to all three components of value. In relation to the first component, system simulations [109] indicate that as the solar share rises in an energy system, there is increasing value in shifting generation to the evenings when the Sun is not shining, and hence an incentive to install CSP plants with thermal storage. The availability of such plants in the system means that higher penetrations of solar power can be achieved overall, and is an important consideration, beyond just generating costs, in determining the optimum mix of CSP and PV technology.

With regard to the second component of value, the provision of generating capacity to meet peak electricity system demand, CSP with storage can contribute to meeting peak system loads and can provide back-up capacity to cover variable renewable sources. Incorporation of supplementary firing will further increase the capability of the CSP plant to provide capacity at the system peak, although the efficiency of fossil fuel use for such supplementary firing is likely to be significantly lower than if it is used in a combined cycle power plant. The value of providing capacity to meet the system peak demand will depend on the system, so its quantification needs to be informed by system models.

Turning to the third component, the value of thermal energy storage in enabling the CSP plant to deliver grid services, CSP with storage can provide spinning reserves, being able to ramp up power if operating at part load in less than 30 min by drawing on the stored heat (the rate of ramping is limited by the thermal inertia of the equipment). Ramping down is quicker, on timescales of around 15 min by diverting heat to storage. This is used in Spain to deliver, on demand, 30% power ramps in less than 1 h, enabling the plant to be considered dispatchable by the Spanish grid operator REE.

Incorporation of thermal storage in a CSP plant results in significant added value compared with PV plants or CSP without storage, and means that the plants can match most of the dispatchability characteristics of a mid-load fossil-fired plant.

17.3.2
Cost Reduction Potential

Three main drivers for cost reduction are scaling up, volume production, and technology innovations. As an example, one of the first comprehensive studies of the potential for cost reduction of CSP was undertaken in the framework of the European ECOSTAR project [6]. The study proposed the potential relative reduction of the LEC of trough plants of up 60%. Half of this potential can be exploited by technical innovations, the other half by scaling-up and mass production effects. Further details of the cost breakdowns and the other cost can be found elsewhere [6, 106, 110]. If this potential is exploited, the cost for dispatchable solar power from CSP plants will decrease significantly below €0.09 kWh^{-1} and can be considered competitive in many commercial markets. The time to achieve this cost reduction is strongly coupled to the deployment rate of the technology. A total installed capacity between 10 and 100 GW is estimated to sufficient to achieve this target between 2020 and 2030 [17].

17.3.2.1 Scaling Up
CSP technology favors large power plant configurations [108] because:

- Procurement of large amounts of solar field components can lead to discounts.
- Engineering, planning, and project development costs are essentially independent of the scale of the plant.
- Operation and maintenance costs reduce with increase in plant size.
- Large power blocks have higher efficiency than small ones and cost less per kilowatt.

The impact of scaling up on CSP electricity costs is still under discussion. The Kearney report [106] indicates a 24% reduction of capital expenditure for an increase of trough plant size from 50 to 500 MW, and Lipman [110] estimated a 30% reduction in LEC for an increase in turbine power from 50 to 250 MW. Finally, the Sargent and Lundy study [111] points to a 14% cost reduction for a 400 MW power block.

17.3.2.2 Volume Production
For parabolic trough technology, the Sargent and Lundy study [111] estimated a cost reduction of 17% due to volume production effects when installing 600 MW per year. Expected cost decreases in the range 5–40%, depending on components, were predicted by Kearney [106].

17.3.2.3 Technology Innovations
According to the study provided by A. T. Kearney [106], technology innovations will

- increase power generation efficiency, mainly through increasing operating temperature;
- reduce solar field costs by minimizing component costs and optimizing optical design; and
- reduce operational consumption of water and parasitic power.

Expected cost reductions and plant efficiency improvements associated with technology innovations were reported by Kearney [106].

17.4
Conclusion

In solar thermal power systems, concentrating solar collectors provide high temperature to a power cycle of engine to generate mechanical energy that is than converted to electricity using a generator. It benefits from a mature power plant technology that has been optimized during almost a century using heat based on fossil or nuclear resources. Today, several options to concentrate the direct sunlight are followed. In linear concentrators such as the parabolic trough or linear Fresnel system, a line focus is achieved by one-axis tracking of one-dimensional curved reflector segments. In point focusing systems such as power tower systems, a point focus is generated by two-dimensional curved reflector segments by two-axis tracking. All options can be combined with a thermal energy storage to provide energy on demand. All offer a significant cost reduction potential that will allow them to achieve the level of competitiveness in the next 10–15 years. The different technologies all bear advantages and drawbacks that at present do not allow a clear preference for one particular technology.

Acknowledgments

The author would like to thank the members of staff of the DLR Institute of Solar Research for their support in the preparation of this chapter.

References

1 Pitz-Paal, R. (2007) High temperature solar concentrators in solar energy conversion and photoenergy systems, in *Encyclopedia of Life Support Systems (EOLSS)* (eds J. Blanco-Galvez and S. Malato) (developed under the auspices of UNESCO), EOLSS Publishers, http:www.eolss.net (last accessed 3 January 2013).

2 Aringhoff, R., Geyer, M., Herrmann, U., Kistner, R., Nava, P., and Osuna, R. (2002) AndaSol – 50 MW solar plants with 9 hour storage for southern Spain, presented at the 11th International SolarPACES Symposium, September 4–6, Zurich.

3 Kearney, D. and Price, H. (2005) Recent advances in parabolic trough solar power plant technology, in *Advances in Solar Energy*, vol. 16 (ed. D. Y. Goswami), Earthscan, London, pp. 155–232.

4 Pitz-Paal, R. (2008), Concentrating solar power, in *Future Energy: Improved, Sistainable and Clean Options for Our Planet* (ed. T. M. Letcher), Elsevier, Oxford, pp. 171–192.

5 Müller-Steinhagen, H. and Trieb, F. (2004) Concentrating solar power. *Ingenia Q. R. Acad. Eng.*, (19), 43–50.

6 Pitz-Paal, R., Dersch, J., Milow, B., Tellez, F., Ferriere, A., Langnickel, U., Steinfeld, A., Karni, J., Zarza, E.,

and Popel, O. (2005) Concentrating solar power plants – how to achieve competitiveness. *VGB PowerTech*, (8) 46–45.

7 Romero, M., Buck, R., and Pacheco, J. E. (2002) An update on solar central receiver systems, projects, and technologies. *J. Sol. Energy Eng.*, **124** (2), 98–108.

8 Mancini, T., Heller, P., Butler, B., Osborn, B., Schiel, W., Goldberg, V., Buck, R., Diver, R., Andraka, C., and Moreno, J. (2003) Dish-Stirling systems: an overview of development and status. *J. Sol. Energy Eng.*, 125 (2), 135–151.

9 German Aerospace Centre (DLR) (2007) *Trans-Mediterranean Interconnection for Concentrating Solar Power (TRANS-CSP)*, http://elib.dlr.de/52603/ (last accessed 12 April 2013).

10 German Aerospace Centre (DLR) (2005) *Concentrating Solar Power for the Mediterranean Region*, http://elib.dlr.de/3042/ (last accessed 12 April 2013).

11 California Energy Comission (2010) *Large Solar Energy Projects*, http://www.energy.ca.gov/siting/solar/index.html (last accessed 30 November 2012).

12 Greentechmedia (2011) *US CSP Project Tracker*, http://www.greentechmedia.com/images/wysiwyg/research-blogs/USCSPProjectTracker.pdf (last accessed 30 November 2012).

13 Protermosolar (2011) *Mapa de la Industria solar Termoélectirca en España (Map of the Solar Thermal Power Industry in Spain)*, http://www.protermosolar.com/mapa.html (last accessed 29 November 2012).

14 US Bureau of Land Management (2011) *Pending Arizona BLM solar Projects*, http://www.blm.gov/az/st/en/prog/energy/solar/pend-solar.html (last accessed 30 November 2012).

15 CSP Today (2011) *CSP Today World Map*, http://csptoday.com/global-tracker/content5.php (subscription required).

16 Burkhardt, J. J., Heath, G. A., and Turchi, C. S. (2011) Life cycle assessment of a parabolic trough concentrating solar power plant and the impacts of key design alternatives. *Environ. Sci. Technol.*, **45** (6), 2457–2464.

17 European Academies Science Advisory Council (2011) *Concentrating Solar Power: Its Contribution to a Sustainable Energy Future*. EASAC Policy Report 16, EASAC, Halle (Saale).

18 Fernández-García, A., Zarza, E., Valenzuela, L., and Pérez, M. (2010) Parabolic-trough solar collectors and their applications. *Renew. Sustain. Energy Rev.*, 14 (7), 1695–1721.

19 Lüpfert, E., Geyer, M., Schiel, W., Antonio, E., Osuna, R., Zarza, E., and Nava, P. (2001) EuroTrough design issues and prototype testing at PSA, presented at the ASME International Solar Energy Conference – Forum 2001, Solar Energy: the Power to Choose, 21–25 April 2001, Washington, DC.

20 Relloso, S., Calvo, R., Cácamo, S., and Olábarri, B. (2011) Senertrough-1 collector: commercial operation experience, continuous loop monitoring at Extresol 1 plant and technology deployment, presented at the 17th International SolarPACES Symposium, 20–23 September, Granada, Spain.

21 Gossamer Innovations (2012) *Solar Power Products*, http://www.gossamersf.com/solar-power-products.htm (last accessed 28 November 2012).

22 Brost, R., Gray, A., Burkholder, F., Wendelin, T., and White, D. (2009) Skytrough optical evaluations using Vshot measurement, presented at the 15th International SolarPACES Symposium, 15–18 September, Berlin.

23 Riffelmann, K. J., Kötter, J., Nava, P., Meuser, F., Weinrebe, G., Schiel, W., Kuhlmann, G., Wohlfahrt, A., Nady, A., and Dracker, R. (2009) Helitrough – a new collector generation for parabolic trough power plants, presented at the 15th International SolarPACES Symposium, 15–18 September, Berlin.

24 Castañeda, N., Váquez, J., and Castañeda, D. (2011) First commercial application of Senertrough. High performance at reduced cost, presented at the 17th International SolarPACES Symposium, 20–23 September, Granada, Spain.

25 Fernández, S. and Acuñas, A. (2012) New optimized solar collector SENERtrough-2, presented at the 18th International SolarPACES Symposium, 11–14 September, Marrakech.

26 Price, H., Lupfert, E., Kearney, D., Zarza, E., Cohen, G., Gee, R., and Mahoney, R. (2002) Advances in parabolic trough solar power technology. *J. Sol. Energy Eng.*, **124** (2), 109–125.

27 Ulmer, S., Heinz, B., Pottler, K., and Lüpfert, E. (2009) Slope error measurements of parabolic troughs using the reflected image of the absorber tube. *J. Sol. Energy Eng.*, **131** (1), 011014.

28 Prahl, C., Stnicki, B., Hilgert, C., Ulmer, S., and Röger, M. (2011) Airborne shape measurement of parabolic trough collector fields, presented at the 17th International SolarPACES Symposium, 20–23 September, Granada, Spain.

29 Ulmer, S., Weber, H., Koch, M., Schramm, H., Pflüger, H., Climent, P., and Yildiz, H. (2012) High-resolution measurement system for parabolic trough concentrator modules in series production, presented at the 18th International SolarPACES Symposium, 11–14 September, Marrakech.

30 Lüpfert, E. and Ulmer, S. (2009) Solar trough mirror shape specifications, presented at the 15th International SolarPACES Symposium, 15–18 September, Berlin.

31 Mateu, E., Sanchez, M., Perez, D., García De Jalón, A., Forcada, S., Salina, I., and Hera, C. (2011) Optical charactetrization test bench for parabolic trough receivers, presented at the 17th International SolarPACES Symposium, 20–23 September, Granada, Spain.

32 Lanxner, M. and Elgat, Z. (1991) Solar selective absorber coating for high service temperatures, produced by plasma sputtering, *Proc. SPIE*, **1272**, 240–249.

33 Kuckelkorn, T., Benz, N., Dreyer, S., Schulte-Fischedick, J., and Moellenhoff, M. (2009) Advances in receiver technology for parabolic trough collectors – a step forward towards higher efficiency and longer lifetime, presented at the 15th International SolarPACES Symposium, 15–18 September, Berlin.

34 Burkholder, F. and Kutscher, C. (2009) *Heat Loss Testing of Schott's 2008 PTR70 Parabolic Trough Receiver*, NREL/TP-550-45633, NREL, Golden, CO.

35 Lippke, F. (1995) *Simulation of the Part Load Behaviour of a 30 MW SEGS Plant*, SAND95-1293, Sandia National Laboratories, Albuquerque, NM.

36 NREL (2011) Concentrating Solar Power Projects, http://www.nrel.gov/csp/solarpaces/parabolic_trough.cfm (last accessed 29 November 2012).

37 Turchi, C. S., Wagner, M. J., and Kutscher, C. F. (2010) *Water Use in Parabolic Trough Power Plants: Summary Results from WorleyParsons' Analyses*, NREL/TP-5500-49468, NREL, Golden, CO.

38 Chamberlain, K. (2012) CSP parabolic trough Report 2013: cost, performance and key trends, *CSP Today*, http://social.csptoday.com/tracker/reports (for purchase only).

39 Brakmann, G., Mohammad, F. A., Dolejsi, M., and Wiemann, M. (2009) Construction of the ISCC Kuraymat, presented at the 15th International SolarPACES Symposium, 15–18 September, Berlin.

40 Brakmann, G., Badaoui, N., Dolejsi, M., and Klingler, R. (2010) Construction of ISCC Ain Béni Mathar in Morocco, presented at the 16th International SolarPACES Symposium, 21–24 September, Perpignan.

41 Dersch, J., Geyer, M., Herrmann, U., Jones, S. A., Kelly, B., Kistner, R., Ortmanns, W., Pitz-Paal, R., and Price, H. (2004) Trough integration into power plants – a study on the performance and economy of integrated solar combined cycle systems. *Energy*, **29** (5–6), 947–959.

42 Zarza, E., Valenzuela, L., León, J., Hennecke, K., Eck, M., Weyers, H. D., and Eickhoff, M. (2004) Direct steam generation in parabolic troughs: final results and conclusions of the DISS project. *Energy*, **29** (5–6), 635–644.

43 Eck, M., Eickhoff, M., Feldhoff, J., Fontela, P., Gathmann, N.,

Meyer-Grünefeldt, M., Hillebrand, S., and Schulte-Fischedick, J. (2011) Direct steam generation in parabolic troughs at 500 °C – first results of the REAL DISS project, presented at the 17th International SolarPACES Symposium, 20–23 September, Granada, Spain.

44 Feldhoff, J., Eickhoff, M., Karthikeyan, R., Krüger, J., León-Alonso, J., Meyer-Grünefeldt, M., Müller, M., and Valenzuela-Gutierrez, L. (2012) Concept comparison and test facility design for the analysis of direct steam generation in once-trough mode, presented at the 18th International SolarPACES Symposium, 11–14 September, Marrakech.

45 Laing, D., Bahl, C., Bauer, T., Lehmann, D., and Steinmann, W.-D., Thermal energy storage for direct steam generation. *Sol. Energy*, **85** (4), 627–633.

46 Krüger, D., Krüger, J., Panadian, Y., O'connell, B., Feldhoff, J., Karthikeyan, R., Hempel, S., Muniasamy, K., Hirsch, T., and Eickhoff, M. (2012) Experiences with direct steam generation at the Knachanaburi solar thermal power plant, presented at the 18th International SolarPACES Symposium, 11–14 September, Marrakech.

47 Drouot, L. P. and Hillairet, M. J. (1984) The Themis program and the 2500-kW Themis solar power station at Targasonne. *J. Sol. Energy Eng.*, **106** (1), 83–89.

48 Martin, M. (1985) *Molten Salt Electric Experiment (MSEE)*, SAND85-8175, Sandia National Laboratories, Albuquerque, NM.

49 Bradshaw, R. W., Dawson, D. B., De La Rosa, W., Gilbert, R., Goods, S. H., Hale, M. J., Jacobs, P., Jones, S. A., Kolb, G. J., Pacheco, J. E., Prairie, M. R., Reilly, H. E., Showalter, S. K., and Vant-Hull, L. L. (2002) *Final Test and Evaluation Results from the Solar Two Project*, OSTI ID: 793226, SAND2002-0120, Sandia National Laboratories, Albuquerque, NM.

50 Siegel, N. P. and Bradshaw, R. W. (2011) Thermophysical property measurement of nitrate salt heat transfer fluids, presented at ASME 2011, 5th International Conference on Energy Sustainability, Washington, DC.

51 Falchetta, M. and Mazzei, D. (2009) Design of Archimede 5 MW molten salt parabolic trough solar plant, presented at the 15th International SolarPACES Symposium, 15–18 September, Berlin.

52 Nissen, D. A. and Meeker, D. E. (1983) Nitrate/nitrite chemistry in sodium nitrate–potassium nitrate melts. *Inorg. Chem.*, **22** (5), 716–721.

53 Bradshaw, R. W. and Tyner, C. E. (1988) *Chemical Engineering Factors Affecting Solar Central Receiver Applications of Ternary Molten Salts*, SAND88-8686, Sandia National Laboratories, Albuquerque, NM.

54 Bauer, T., Pfleger, N., Laing, D., Steinmann, W.-D., Eck, M., and Kaesche, S. (2013) High temperature molten salts for solar power application, in *Molten Salts: Fundamentals and Application* (eds F. Lantelme and H. Groult), Elsevier, Amsterdam, in press.

55 Bradshaw, R. W., Cordaro, J. G., and Siegel, N. P. (2009) Molten nitrite salt development for thermal enregy storage in parabolic trough solar power systems, presented at the ASME International Conference on Energy Sustainability, San Francisco.

56 Kearney, D., Kelly, B., Herrmann, U., Cable, R., Pacheco, J., Mahoney, R., Price, H., Blake, D., Nava, P., and Potrovitza, N. (2004) Engineering aspects of a molten salt heat transfer fluid in a trough solar field. *Energy*, **29** (5–6). 861–870.

57 Bradshaw, R. W. and Goods, S. H. (2000) *Corrosion Resistance of Nickel-Base Alloys in Molten alkali Nitrates*, SAND2000-8240, Sandia National Laboratories, Albuquerque, NM.

58 Müller-Elvers, C., Wittmann, M., and Saur, M. (2012) Design and construction of molten salt parabolic trough HPS Project in Évora, Portugal, p[resented at the 18th International SolarPACES Symposium, 11–14 September, Marrakech.

59 Francia, G. (1968) Pilot plants of solar steam generating stations. *Sol. Energy*, **12** (1), 51–64.

60 Häberle, A., Zahler, C., Lerchenmüller, H., Mertins, M., Wittwer, C., Trieb, F., and Dersch, J. (2002) The Solarmundo line focussing Fresnel collector. Optical and thermal performance and cost calculations, presented at the 11th International SolarPACES Symposium, 4–6 September, Zurich.

61 Mills, D. R. and Morrison, G. L. (2000) Compact linear Fresnel reflector solar thermal powerplants. *Sol. Energy*, **68** (3), 263–283.

62 Mills, D. R., Morrison, G., Pye, J., and Peter, L. L. (2006) Multi-tower line focus fresnel array project. *J. Sol. Energy Eng.*, **128** (1), 3.

63 Selig, M. and Mertins, M. (2010) From saturated to superheated direct solar steam generation – technical challenges and economical benefits, presented at the 16th International SolarPACES Symposium, 21–24 September, Perpignan.

64 Conlon, W. M. (2011) Direct steam from cLFR solar steam generators, presented at the 18th International SolarPACES Symposium, 20–23 September, Granada, Spain.

65 Eck, M., Bernhard, R., De Lalaing, K., Kistner, R., Eickhoff, M., Feldhoff, J., Heimsath, A., Hülsey, H. and Morin, G. (2009) Linear Fresnel collector demonstration at the PSA – operation and investigation, presented at the 15th International SolarPACES Symposium, 15–18 September, Berlin.

66 Itskhokine, D., Lecuillier, P., Benmarraze, S., Guillier, L. and Rabut, Q. (2012) Augustin Fresnel project – design, construction and testing of a linear Fresnel pilot plant in the Pyrenees, presented at the 18th International SolarPACES Symposium, 11–14 September, Marrakech.

67 Paul, C., Teichrew, O., and Ternedde, A. (2012) Operation experience of the integration of a solar boiler based on Fresnel collector technology into a coal fired power station, presented at the 18th International SolarPACES Symposium, 11–14 September, Marrakech.

68 Morin, G., Kirchberger, J., Lemmertz, N., and Mertins, M. (2012) Operational results and simulation of a superheating Fresnel collector, presented at the 18th International SolarPACES Symposium, 11–14 September, Marrakech.

69 Dersch, J., Eck, M., and Häberle, A. (2009) Comparison of linear Fresnel and parabolic trough collector systems – system analysis to determine break-even costs of linear Fresnel collectors, presented at the 15th International SolarPACES Symposium, 15–18 September, Berlin.

70 Giostri, A., Brinotti, M., Silva, P., Macchi, E., and Manzolini, G. (2011) Comparison of two linear collectors in solar thermal plats: parabolic trough vs. Fresnel., presented at the ASME International Conference on Energy Sustainability, Washington, DC.

71 Gharbi, N. E., Derbal, H., Bouaichaoui, S., and Said, N. (2011) A comparative study between parabolic trough collector and linear Fresnel reflector technologies. *Energy Procedia*, **6**, 565–572.

72 Narula, M. and Gleckman, P. (2012) Central receivers vs. linear Fresnel: a comparison of direct molten salt CSP plants, presented at the 18th International SolarPACES Symposium, 11–14 September, Marrakech.

73 Schenk, H., Hirsch, T., Feldhoff, J., and Wittmann (2012) Energetic comparison of linear Fresnel and parabolic trough collector systems, presented at the ASME International Conference on Energy Sustainability, San Diego, CA.

74 Bachelier, C., Morin, G., Paul, C., Selig, M., and Mertins, M. (2012) Integration of molten salt systems into Fresnel collector based CSP plants, presented at the 18th International SolarPACES Symposium, 11–14 September, Marrakech.

75 Grena, R. and Tarquini, P. (2011) Solar linear Fresnel collector using molten nitrates as heat transfer fluid. *Energy*, **36** (2), 1048–1056.

76 Meduri, P. K., Hannemann, C. R., and Pacheco, J. E. (2010) Performance characterization and operation of eSolar's Sierra Suntower power tower plant, presented at the 16th

International SolarPACES Symposium, September 21–24, Perpignan, France.

77 Osuna, R., Fernández-Quero, V., and Sánchez, M. (2008) Plataforma solar Sanlúcar La Mayor: the largest European solar power site, presented at the 14th International SolarPACES Symposium, 4–7 March, Las Vegas, NV.

78 Arias, S. A. and Burgaleta, J.I (2011) A real CSP experience – GEMASOLAR, the first tower thermosolar commercial plant with molten salt storage, presented at the CSP Today Conference, Seville.

79 http://www.abengoasolar.com (last accessed 30 November 2012).

80 Garcia, E. and Calvo, R. (2012) One year operation experience of Gemasolar plant, presented at the 18th International SolarPACES Symposium, 11–14 September, Marrakech.

81 Rogers, D., Slack, M., and Cassity, B. (2012) Addressing the challenges associated with eSolar's unique approach to central receiver power plants, presented at the 18th International SolarPACES Symposium, 11–14 September, Marrakech.

82 Pomp, S., Schwarzbözl, P., Koll, G., Göhring, F., Hartz, F., Schmitz, M., and Hoffschmidt, B. (2010) The solar tower Jülich – first operational experiences and test results, presented at the 16th International SolarPACES Symposium, September 21–24, Perpignan, France.

83 http://www.solarreserve.com/what-we-do/csp-projects/crescent-dunes/ (last accessed 30 November 2012).

84 Finch, N. S. and Ho, K. (2011) Stochastic modeling of power towers and evaluation of technical improvement opportunities, presented at the 17th International SolarPACES Symposium, 20–23 September, Granada, Spain.

85 Kolb, G. J., Jones, S. A., Donnelly, M., Gorman, D., Thomas, R., Davenport, R., and Lumia, R. (2007) *Heliostat Cost Reduction Study*, SAND2007-3293, Sandia National Laboratory, Alberqueque, NM.

86 Kolb, G. J. (2011) *An Evaluation of Possible Next Generation High Temperature Molten Salt Power Towers*, SAND2011-9320, Sandia National Laboratories, Alberqueque, NM.

87 Raade, J. W., Elkin, B., and Vaughn, J. (2012) Novel 700 °C molten salt for solar thermal power generation with supercritical steam turbines, presented at the 18th SolarPACES Symposium, 11–14 September, Marrakech, Morroco.

88 Heide, S., Felsmann, C., Gampe, U., Stefano, G., Buck, R., Freimark, M., Langnickel, U., Boje, S., and Gericke, B. (2012) Parameterization of high solar gas turbine systems, presented at the ASME Turbo Expo, 11–15 June, Copenhagen.

89 Korzynietz, R., Quero, M., and Uhlig, R. (2012) SOLUGAS – future solar hybrid technology, presented at the 18th SolarPACES Symposium, 11–14 September, Marrakech.

90 Turchi, C. S., Ma, Z., Neises, T., and Wagner, M. (2012) Thermodynamic study of advanced supercritical carbon dioxide power cycles for high performance concentrating solar power systems, presented at the ASME 6th International Conference on Energy Sustainability, 23–26 July, San Diego, CA.

91 Singer, C., Buck, R., Pitz-Paal, R., and Müller-Steinhagen, H. (2012) Economic chances and technical risks of the Internal Direct Absorption Receiver (IDAR), presented at the ASME 6th International Conference on Energy Sustainability, 23–26 July, San Diego, CA.

92 Wu, W., Gobereit, B., Singer, C., Amsbeck, L., and Pitz-Paal, R. (2011) Direct absorption receivers for high temperatures, presented at the 17th SolarPACES Symposium, 20–23 September, Granada, Spain.

93 Kitzmiller, K., Miller, F., and Frederickson, L. (2012) Design, construction and preliminary testing of a lab scale small particel solar receiver, ipresented at the 18th SolarPACES Symposium, 11–14 September, Marrakech.

94 Augsburger, G. and Favrat, D. (2012) From single to multi tower solar thermal power plants: investigation of the thermo economic optimum transition size, presented at the 18th SolarPACES Symposium, 11–14 September, Marrakech.

95 Gil, A., Medrano, M., Martorell, I., Lázaro, A., Dolado, P., Zalba, B., and Cabeza, L. F. (2010) State of the art on high temperature thermal energy storage for power generation. Part 1 – Concepts, materials and modellization. *Renew. Sustain. Energy Rev.*, **14** (1), 31–55.

96 Medrano, M., Gil, A., Martorell, I., Potau, X., and Cabeza, L. F. (2010) State of the art on high-temperature thermal energy storage for power generation. Part 2 – Case studies. *Renew. Sustain. Energy Rev.*, **14** (1), 56–72.

97 Steinmann, W.-D. and Tamme, R. (2008) Latent heat storage for solar steam systems. *J. Sol. Energy Eng.*, **130** (1), 011004.

98 Ervin, G. (1977) Solar heat storage using chemical reactions. *J. Solid State Chem.*, **22** (1), 51–61.

99 Williams, O. M. and Carden, P. O. (1978) Screening reversible reactions for thermochemical energy transfer. *Sol. Energy*, **22** (2), 191–193.

100 Relloso, S. and Delgado, E. (2009) Experience with molten salt thermal storage in a commercial parabolic trough plant. Andasol-1 comissioning and operation, presented at the 15th SolarPACES Symposium, 15–18 September, Berlin.

101 Goldstern, W. (1970) *Steam Storage Installations*, Pergamon Press, Oxford.

102 Zunft, S., Hänel, M., Krüger, M., Dreißigacker, V., Göhring, F., and Wahl, E. (2011) Jülich solar power tower-experimental evaluation of the storage subsystem and performance calculation. *J. Sol. Energy Eng.*, **133** (3), 031019.

103 Bauer, T., Braun, M., Eck, M., Pfleger, N., and Laing, D. (2012) Development of salt formulations with low melting temperatures, presented at the 18th SolarPACES Symposium, 11–14 September, Marrakech.

104 Steinmann, W.-D., Laing, D., and Odenthal, C. (2011) Development of the CellFlux storage concept for sensible heat, presented at the 17th SolarPACES Symposium, 20–23 September, Granada, Spain.

105 Laing, D., Eck, M., Hempel, S., Johnson, M., Steinmann, W.-D., Meyer-Grünefeldt, M., and Eickhoff, M. (2012) High temperature PCM storage for DSG solar thermal power plants tested in various operating modes of water/steam flow, presented at the 18th SolarPACES Symposium, 11–14 September, Marrakech.

106 Kearney, A. T. (2010) *Solar Thermal Electricity 2025*, ESTELA, Brussels.

107 International Energy Agency (2010) *Technology Roadmap: Concentrating Solar Power*, IEA, Paris.

108 US Energy Information Administration (2010) Updated CapitalCost Estimates for Electricity Generation Plants, US Department of Energy, Washington, DC.

109 Kost, C. and Schlegl, T. (2010) *Stromgestehungskosten Erneuerbare Energien. Renewable Energy Policy Innovation*, Fraunhofer ISE, Freiburg.

110 Lipman, E. (2010) Reducing CSP plant LOEC with precise engineering synergies and R&D. Plenary Lecture, presented at the 16th SolarPACES Symposum: September 21–24, Perpignan.

111 Sargent and Lundy LLC Consulting Group (2003) *Assessment of Parabolic Trough and Power Tower Solar Technology Cost and Performance Forecasts*, National Renewable Energy Laboratory (NREL), Golden, CO.

112 http://ivanpahsolar.com (last accessed 18 April 2013).

113 Bradshaw, R. W. and Siegel, N. P. (2008) *Molten Nitrate Salt Development for Thermal Energy Storage in Parabolic Trough Solar Power Systems*. Energy Sustainability ES2008, Jacksonville, Florida, USA.

18
Geothermal Power

Christopher J. Bromley and Michael A. Mongillo

18.1
Introduction

Future deployment projections of geothermal energy resources have been published by the International Energy Agency (IEA) in a "Roadmap" document [1]. In volcanic and plate boundary settings, high-temperature geothermal energy is often the least expensive renewable energy option, especially in terms of long-run marginal cost (LRMC). For example, in New Zealand, the LRMC for new geothermal projects is US$ 50–67 MWh^{-1} (NZ$ 60–80) for the next 10 TWh, or 1.2 GW of installed capacity (Figure 18.1). The result is a projected replacement of coal (and gas) by geothermal (and wind) generation, reaching a target 90% renewable by 2020 (Figure 18.2).

Geothermal energy also has the added benefits of being a source of base-load power, in addition to efficiently providing on-demand heating and cooling for buildings, or for industrial and agricultural direct process heat applications. In order to assist decision-makers with future policy development and investment decisions, the IEA-GIA (Geothermal Implementing Agreement) has assisted in generating geothermal deployment projections and technology "roadmaps" through multi-party collaboration. This includes participation in the geothermal chapter of the Intergovernmental Panel on Climate Change (IPCC) special report on renewable energy [2]. The outcome of the deployment projections is location specific, but by 2050 geothermal energy could potentially contribute up to 30% of demand in volcanic countries (such as New Zealand, Philippines, and Japan) and 2–4% of electricity and heat demand for nonvolcanic countries through the development of hot sedimentary aquifer (HSA) and enhanced geothermal system (EGS) technologies (such as in Australia, Korea, China, India, and most of Europe).

In some countries, rapid geothermal deployment over the past 5–10 years is already displacing coal-fired power generation (Figure 18.2). With a concerted collaboration effort, experienced countries can help out those that are relatively inexperienced in geothermal development, especially in East Africa, South America (Chile, Peru, etc.) and the South Pacific (e.g., Papua New Guinea), and by assisting nations (such

Transition to Renewable Energy Systems, 1st Edition. Edited by Detlef Stolten and Viktor Scherer.
© 2013 Wiley-VCH Verlag GmbH & Co. KGaA. Published 2013 by Wiley-VCH Verlag GmbH & Co. KGaA.

340 | 18 Geothermal Power

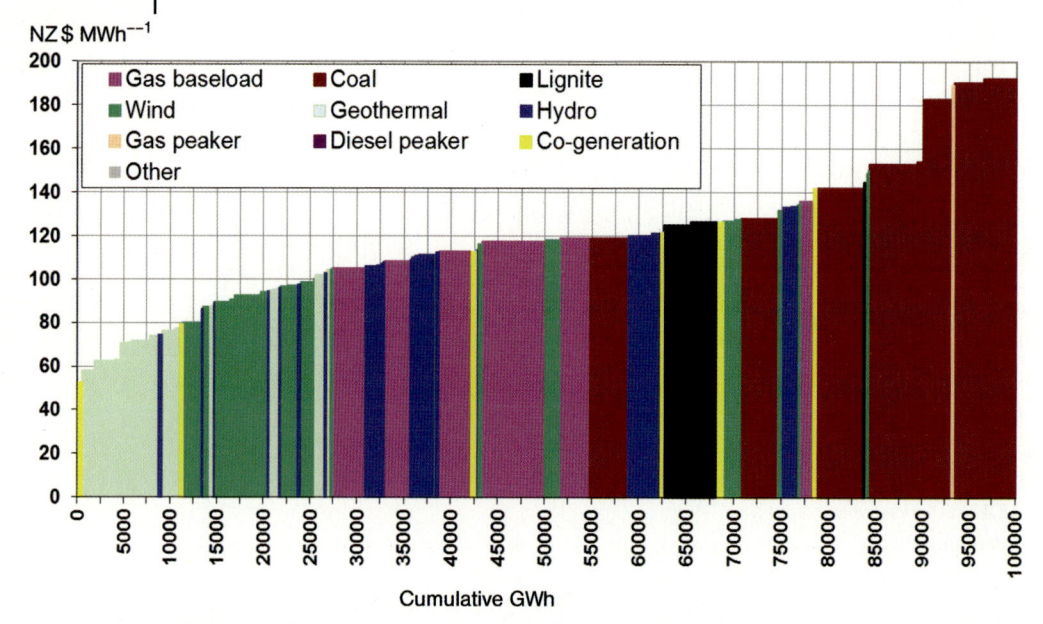

Figure 18.1 Electricity generation – LRMC for new New Zealand projects in lowest cost order. Assumes: 1.5% growth per annum (+7 TWh per 10 years); NZ$25 t^{-1} carbon tax; 8% discount rate. Geothermal predominates for ~20 years new base-load capacity. LRMC includes the costs of interest, make-up drilling, operations and maintenance. Source: www.med.govt.nz.

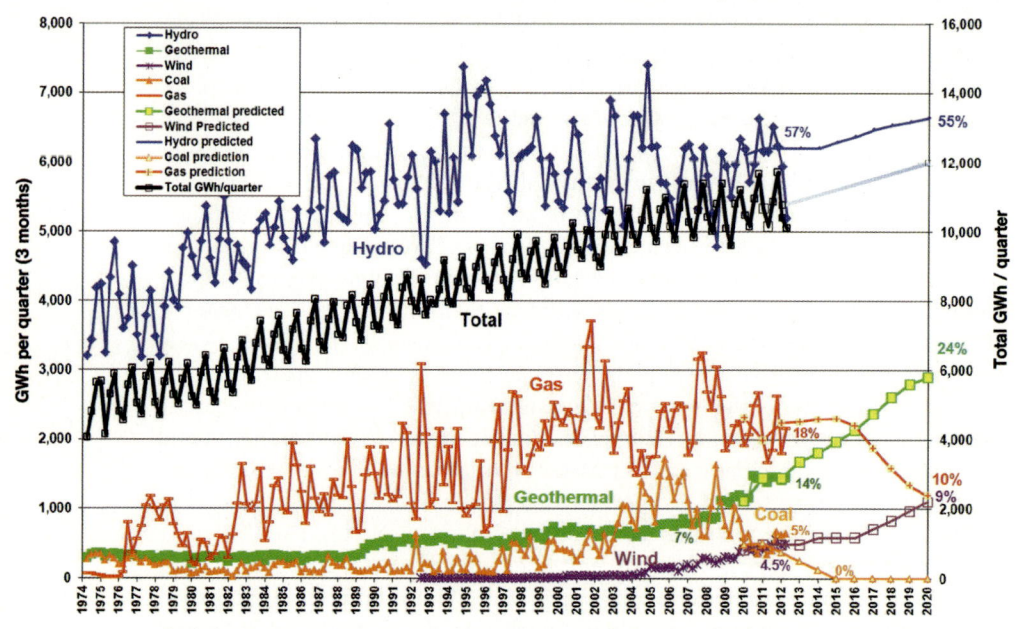

Figure 18.2 New Zealand historical and projected electricity generation trends per quarter from 1974. Between 2006 and 2012, geothermal and wind doubled, allowing coal to reduce from 15% to 5%. Source: www.med.govt.nz.

as Indonesia, China, and the Philippines) to reach their huge geothermal energy development potential [3]. The combined effect of this regional effort will be to displace significant global CO_2 emissions from fossil fuel energy sources, in both electricity and heating and cooling markets [4].

18.2
Geothermal Power Technology

The dominant types of geothermal power plant installed around the world today use technology adapted for three types of resources: (a) direct steam (from vapor-dominated systems such as at The Geysers in the United States or Larderello in Italy), (b) "flashed" steam (separated from two-phase steam–water mixtures, such as at Wairakei in New Zealand), or (c) binary-cycle heat exchangers. All these plants use the heat energy contained in water and steam discharged or pumped from geothermal wells, converting thermal and kinetic energy into electrical energy. Power plants that use high-pressure "dry" or "flashed" steam to drive turbines are the most common. They can be "back-pressure" units (exhausting steam at atmospheric pressure) or the more-efficient "condensing" units that exhaust steam into a vacuum. The condensing units require a cooling water circuit, typically consisting of forced-draft or natural-draft cooling towers where the steam condensate cools on contact with air. Evaporation in cooling towers results in net mass losses (production-injection) to the atmosphere of about 50% (±20%, depending on climate). Fluid from surrounding aquifers generally provides some recharge, which helps sustain pressures and maintain fluid flow.

Organic Rankine cycle (closed-loop binary) power plants employ secondary working fluids for geothermal power generation. These power plants do not use geothermal fluids directly in turbines. Thermal energy contained in water and/or steam produced from geothermal wells is transferred to a secondary working fluid using a heat exchanger. Organic compounds with lower boiling points than water (such as propane that boils at about 28 °C) are often used as working fluids. Alternative binary working fluids include aqueous ammonia, pure water, or refrigerants. The heat energy in the geothermal fluid boils the working fluid, changing it from a pressurized liquid to a pressurized gas within the closed loop, which can then be expanded in a turbine to spin the generator. The exhausted working fluid is cooled (typically by fan-assisted air cooling), condensed back into a liquid, pressurized, and then recycled into the heat exchanger to complete the cycle. The principal benefits of a binary plant are its higher efficiency for power generation when using lower temperature fluid, and its capability to reinject the total mass extracted (no net mass loss), thereby avoiding potential adverse environmental effects.

Combined heat and power (CHP) plants and hybrid power plants use a variety of technologies to utilize available heat energy from discharging or pumped wells in the most cost-effective and efficient manner possible. Recent advances in process engineering technology have also enabled geothermal fluids to be used to create high-quality ("boiler-grade") steam for food processing (e.g., drying milk powder) or other horticultural, aqua-cultural, or industrial processes (e.g., paper manufacture).

18.3
Global Geothermal Deployment: the IEA Roadmap and the IEA-GIA

The IEA has clearly identified the urgent need to accelerate the development of advanced energy technologies to deal with the global challenges of providing sufficient clean energy, mitigating climate change, and sustainable development [1]. In June 2008, the G8 countries acknowledged this challenge by requesting the IEA to develop a series of energy technology roadmaps. Roadmaps enable governments, industry, and the financial sector to identify practical steps that they can use to develop policy and make investment decisions that will contribute to achieving the mitigation of climate goal. The overall aim is to demonstrate the critical role of energy technologies in achieving the goal of halving energy-related CO_2 emissions by 2050, to keep the global temperature increase to < 2 °C and help mitigate global climate change effects.

The potential contribution that global geothermal deployment might make to the provision of power and direct heat, and hence to the mitigation of climate change, was highlighted recently by several international organizations, including the IEA-GIA and the International Geothermal Association (IGA). The current emphasis is on accelerating the development of naturally permeable hydrothermal resources (although EGS is considered important for the future), since there was already a long experience in their utilization for heat and power, using technologies (especially for power) considered conventional and mature. They were identified as renewable resources that could be used sustainably and produce very low CO_2 emissions.

Consequently, the IEA Technology Roadmap: Geothermal Heat and Power was prepared and published in 2011 [1]. This report provides a clear growth pathway. It shows the potential of geothermal energy to help mitigate global climate change effects via CO_2 avoidance. Deployment projections are presented and approaches and tasks are described regarding: R&D, financing mechanisms, legal and regulatory frameworks, engaging the public, and international collaboration. The major elements are Setting Goals, Identifying Gaps and Barriers, Establishing Action Items, and Specifying Priorities and Timelines.

The key findings of the IEA Geothermal Roadmap are that geothermal energy can provide low-carbon, future base-load power and heat from a mixture of renewable geothermal resources, including: hydrothermal systems, deep aquifer systems with moderate temperatures, and hot rock resources that are stimulated for energy extraction. By 2050, power generation could be 1400 TWh a^{-1}, or about 3.5% of the global total, thus avoiding 800 Mt of CO_2 emissions; geothermal heat use could contribute ~1600 TWh$_{th}$ a^{-1} (5.8 EJ a^{-1}) (excluding ground-source heat pumps) or 3.9% of projected global total; with more than 50% of the projected increase coming from hot rock resources developed as EGS. However, incentives are required to attain these goals, and key actions for the next decade are as follows:

1. Increase investor confidence and accelerate growth of geothermal heat and power by setting medium-term targets for mature and nearly mature technologies, and long-term targets for the advanced technologies. In the period to 2030, relatively

attractive economics will drive accelerated deployment of conventional, high-temperature hydrothermal resources, but in limited locations (i.e., plate boundaries, hot spots). Development of deep-aquifer, medium-temperature hydrothermal resources will also expand owing to their wider distribution/availability and increasing interest in their use for heat and power. A longer term target, by 2050, is required to deploy resources requiring advanced technologies, including hot rock (EGS), geo-pressure, offshore hydrothermal, supercritical temperature/pressure fluids, magmatic resources, and co-produced hot water from oil and gas wells.
2. Introduce differentiated economic incentive schemes (e.g., feed-in tariffs, renewable portfolio standards) that phase out as full competitiveness is reached.
3. Through collaborative organizations such as the IEA-GIA, develop and provide publicly available geothermal databases, protocols (induced seismicity, EGS), best practices (e.g., drilling handbook), and tools for resource assessment, to help spread expertise and accelerate development.
4. Initiate more simplified, timely, and effective procedures for issuing permits/consents for development, for example, introducing simplified procedures consisting of fewer steps or even a "one-stop shop" approach.
5. Provide sustained and significantly larger R&D resources for planning and developing at least 50 more EGS pilot plants in the next decade.
6. Expand and disseminate EGS technology information to increase production and improve resource sustainability and management of environmental issues.
7. Expand efforts of bi- and multi-lateral aid organizations to address economic and noneconomic barriers in developing countries, to develop their most attractive hydrothermal resources more rapidly.

18.4
Relative Advantages of Geothermal

Geothermal energy use has a number of comparative advantages in most competitive energy markets, particularly where renewable energy has a priority:

- Power plants operated using geothermal fluids have low to zero CO_2 emissions, and (thanks to directional drilling) require relatively modest land footprints (at the surface). The average direct emissions from hydrothermal-flash and direct-steam power plants amount to about 120 g CO_2 kWh_e^{-1} [5] (these emissions are of natural geological origin rather than from combustion of fossil fuels). Closed-loop binary-cycle plants and direct-use applications, with total reinjection, yield less than 1 g CO_2 kWh_e^{-1} in direct emissions. Over its full lifecycle (including the manufacture and transport of materials and equipment), CO_2-equivalent emissions range from 23 to 80 g kWh_e^{-1} for binary plants [6] (comparable to wind and hydro). This means that geothermal resources are environmentally advantageous and the net energy supplied more than offsets the environmental impacts of human, energy, and material inputs.

- Geothermal electric power plants have relatively high capacity factors (defined here as the actual, annual, net, generated power (in megawatt hours), divided by the combined name-plate generation capacity of the turbines (in megawatts), times 8760 h per year); the world-wide average for power generation is about 75% [7], but this is adversely affected by power dispatch issues and the availability status of aging turbines, whereas modern geothermal power plants exhibit capacity factors > 90%. This makes geothermal energy well suited for base-load operation (24/7), and fully dispatchable energy use. Capacity factors can be improved with smart grids, make-up drilling, and effective turbine and pipeline maintenance. Energy utilization efficiency is improved by employing combined heat and power systems or geothermal heat absorptive and vapor compression cooling technology.
- Properly managed geothermal reservoir systems are sustainable for very long-term operations, exceeding the foreseeable design life of associated surface plant and equipment.
- Displacement of fossil energy supplies with geothermal energy can also be expected to play a key role in climate change mitigation strategies.

18.5
Geothermal Reserves and Deployment Potential

Total thermal energy contained within the Earth is on the order of 12.6×10^{12} EJ and that of the crust on the order of 5.4×10^{9} EJ to depths of up to 50 km [8]. The main sources of this energy are due to the heat flow from the Earth's core and mantle, and that generated by the continuous decay of radioactive isotopes in the crust itself. Heat is transferred from the interior towards the surface, mostly by conduction, at an average of 0.065 W m^{-2} on continents and 0.101 W m^{-2} through the ocean floor. The result is a global terrestrial heat flow rate of around 1400 EJ a^{-1}. The terrestrial heat flow under continents, which cover ~30% of the Earth's surface, has been estimated at 315 EJ a^{-1} [9]. Under continents, the stored thermal energy, within depths reachable with current drilling technology, has also been estimated. Based on these estimates, the theoretically available resource is huge and clearly not a limiting factor for global geothermal deployment. Global theoretical potential estimates for the terrestrial areas in the upper 10 km were 45 EJ a^{-1} (12 500 TWh) for power and 1040 EJ a^{-1} (289 000 TWh) for direct use [2]. Technical and economic constraints will, however, be the main factors that determine the future deployment rate. Projections of future geothermal electricity production were undertaken for the IPCC report [2]. The results up to 2100 are summarized in Table 18.1. Between 2020 and 2050, deployment is projected to accelerate rapidly (approximately doubling every 10 years) as technologies are improved, particularly for EGS and HSA. For the period between 2050 and 2100, the scenario assumes steady growth in long-term deployment at about 4% per annum, in response to research-derived improvements in cost and efficiency of deep drilling performance. Beyond 2100, it is anticipated that new technologies will allow exploitation of offshore (mid-ocean ridge) and high-enthalpy magmatic energy resources. These resources should help sustain

Table 18.1 Actual (from 1995 to 2010) and expected (from 2015 to 2100) growth in the use of geothermal energy for electricity generation.

Year	Installed capacity actual or mean forecast (GWe)	Electricity production actual or mean forecast (TWh a^{-1})	Capacity factor (%)
1995	6.8	38	64
2000	8.0	49	71
2005	8.9	57	73
2010	10.7	67	75
2015	18	122	77
2020	26	182	80
2030	51	380	85
2040	90	700	88
2050	160	1300	90
2100	Up to 1100	Up to 9000	90+

growth rates, while allowing for retirements of some older projects, enabling their temperature recovery through natural recharge. In Table 18.1, capacity factors (as defined above) increase with time as older plants are retired and grid connections mature, allowing new and more efficient power plants to operate in full base-load mode, closer to their design capacity factor.

Because of the steady flow of heat to the Earth's surface, geothermal energy is considered to be a renewable resource. Geothermal projects are typically operated at production rates that cause local declines in fluid pressure and/or in temperature over the economic lifetime of the installed facilities. As thermal energy is extracted from the exploited reservoir, it temporarily creates locally cooler regions. These cooler and lower pressure zones in the reservoir lead to gradients that result in continuous recharge by conduction from hotter rock, and convection and advection of fluid from surrounding regions. Detailed modeling studies have shown that resource exploitation can be economically feasible, and still be renewable on a reasonable timescale, when recovery periods are included. This aspect of resource utilization is addressed in Section 18.7.

18.6
Economics of Geothermal Energy

The key driver to increased geothermal deployment is cost and risk reduction. Both of these factors are location specific. In some countries, incentives such as feed-in tariffs or subsidies are not needed because the economics already favor geothermal development as a renewable option. This is usually a consequence of previous indirect subsidies by the government in the form of advance exploration drilling and scientific research that typically reduced the exploration risk for commercial investors and power or heat-supply utility companies. In other countries, where the risk is borne entirely by the developer, what is often needed to spur growth is some financial incentives, guaranteed and favorable power purchase agreements, or some form of exploration drilling, risk-reduction insurance.

Typical all-up capital costs for medium-sized geothermal projects are currently in the range US$ 2.5–4 million MWe^{-1} (installed) for most conventional high-enthalpy geothermal systems. But this can easily double for smaller EGS, HAS, and lower temperature systems. For these systems, ~50% of capital costs can be the deep drilling costs. For conventional systems, it might be as low as 20%. Where the resource temperature is below about 180 °C, the need for high-temperature down-hole pumps (because the fluid will not boil and self-discharge) increases both operating and capital costs. Where fracture permeability is naturally poor, the need for reservoir stimulation increases capital costs and raises the risk of failure to establish viable, long-term, production fluid flow rates. Chemical issues with non-benign reservoir fluids can also raise the capital and operating costs for treatment or mitigation (e.g., calcite scale, silica scale, H_2S, or acid inhibition or treatment).

The interest costs associated with the relatively high capital costs of geothermal projects are often the largest component of the running costs in an LRMC assessment. Accordingly, power plants and fluid-gathering systems that are designed and built to last for 40–50 years, rather than the conventional 25–30 years, generally have lower LRMCs.

18.7
Sustainability and Environmental Management

Strategies to sustain heat and power generation output, by successfully managing geothermal fields over the longer term (i.e., 100 years and beyond), are considered to be crucial to the future of geothermal energy as a renewable source of energy. These strategies may involve some form of "cyclic utilization" or "rotational heat grazing" to allow thermal recovery of depleted reserves. This recovery originates from surrounding or deeper recharge zones. To achieve continuous geothermal generation across a region, it will be necessary to identify reserve sources of generation for use during the recovery periods. This requires advance exploration and identification of additional geothermal systems.

Adaptive management of geothermal production, by adjusting locations and rates of fluid extraction and injection, is also needed to optimize sustainable utilization of each system. This requires flexibility. Most geothermal developments that have been operating for more than 20 years have undergone changes in production–injection strategy in response to monitoring of effects on the resource. Planners and regulators need to accommodate this. Also, developers should maintain some surplus production and injection capacity throughout development. A range of future options are needed in order for adaptive and flexible management to work. Effective monitoring of reservoir conditions (temperature, pressure, and phase) is also an essential component of this strategy. Monitoring tools include borehole measurements, microgravity, seismicity, tracer studies, and pressure interference testing. The monitoring results are used to calibrate reservoir simulation models which are used for planning.

Reservoir management strategies should also emphasize the achievement of environmental balance through avoiding, remedying, or mitigating adverse effects, and encouraging beneficial effects. This will achieve greater community support. In this regard, there are some successful international experiences in developing resources using environmentally sensitive strategies. These help to promote geothermal utilization and provide excellent role models for future developers. Global awareness of these successes can be improved through information sharing and international collaboration.

The IEA-GIA provides an opportunity for international collaboration on various aspects of resource development, including environmental and sustainable management issues (www.iea-gia.org). A series of workshops, discussion documents, and papers have been produced on topical issues. These include protecting and enhancing natural thermal features, minimizing adverse effects from disposal of geothermal fluids and gases, providing policy advice, dealing with induced seismicity and subsidence, and improving sustainability of geothermal production, particularly through adaptive injection.

Within Annex I (Environmental Impacts of Geothermal Energy Development), IEA-GIA participants undertake tasks to identify environmental effects of geothermal development and devise and adopt methods to avoid or minimize their impact. These tasks include working (a) to investigate the impacts on natural features, (b) to study the problems associated with discharge and reinjection of geothermal fluids, (c) to examine methods of impact mitigation and produce environmental guidelines, and (d) to develop sustainable utilization strategies. Annex X (Induced Seismicity) investigates seismic risk from EGS fluid injection. This work has identified practical environmental strategies including (1) improved discharges from surface thermal features by targeted fluid injection, (2) creation of enhanced thermal habitats, and (3) treatment or injection of toxic chemicals or gases. Good monitoring regimes include establishment of pre-production baseline information to allow effects from such strategies to be properly identified.

Sustainable geothermal energy production is achieved by properly managing fluid production and injection rates and locations. Total energy yields achieved using low extraction rates over long duration cycles can be similar to those achieved with high extraction rates for short duration cycles. Balanced fluid/heat production that does not exceed the recharge (natural and induced) can be considered indefinitely sustainable. If extraction rates exceed the rate of recharge, reservoir depletion will occur, but following termination of production, geothermal resources will undergo asymptotic recovery towards their pre-production pressure and temperature states. Practical replenishment (~95% recovery) will occur on timescales of the same order as the lifetime of the geothermal production cycle (typically ~50–300 years). This is illustrated in Figure 18.3, and has been tested using reservoir simulations. A useful equation for estimating the thermal recovery time [10] is

$$T(\text{rec}) = (PR - 1) T(\text{ext}) \tag{18.1}$$

where T(rec) is the recovery time needed for a geothermal system to be restored to its pre-development condition, T(ext) is the time period of geothermal fluid extraction, and PR is the ratio of heat extraction rate to natural heat discharge rate. For the special case shown in Figure 18.3, $PR = 2$ and T(rec) = T(ext). In a practical situation, because the recovery process is asymptotic, a new cycle of energy extraction can begin well before full thermal recovery is achieved.

The optimum level of long-term sustainable production depends on the utilization technology (in particular the reinjection strategy) and also on the geothermal resource characteristics.

Examples of successful sustainable geothermal developments, where reservoir performance has stabilized during production, include both higher enthalpy systems (e.g., Matsukawa in Japan, Wairakei and Kawerau in New Zealand, and Larderello in Italy) and lower enthalpy systems (Laugarnes in Iceland and the Paris Basin in

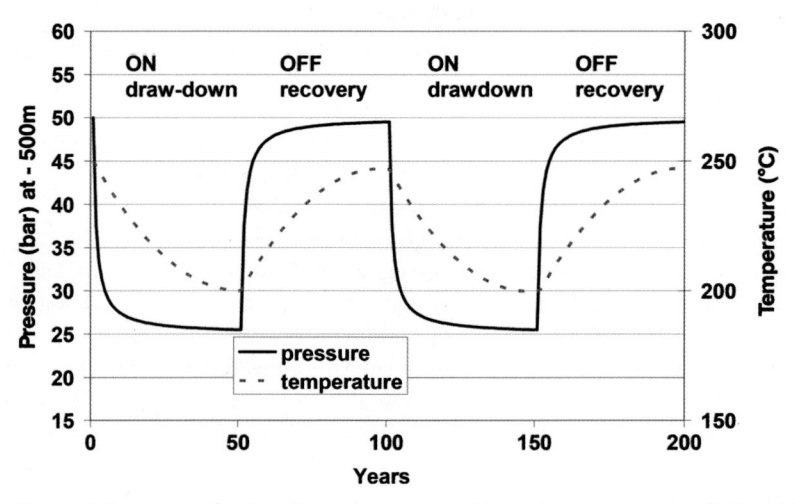

Figure 18.3 An example of the "cyclic" or "rotational heat grazing" concept. In this case, the long-term production balances the long-term energy and mass recharge. If the recharge rate is more limited, the recovery period will be longer. Temperature recovery is slower than pressure recovery.

France). Dynamic recovery factors determine the long-term response of the system to energy extraction; they change with time. Recovery is influenced by an enhanced recharge driven by the strong pressure and temperature gradients initially created by the fluid and heat extraction. Because of this dynamic recovery process, rotational utilization of geothermal resources is a viable long-term strategy, and an economic and sustainable alternative to the strategy of simply limiting extraction to maintain continuous steady-state reservoir conditions. Utilization duration can be tailored to meet demand cycles (daily or seasonal), or can be extended to periods of the order of 100 years. A term that appropriately describes this process is heat "grazing."

In high-enthalpy, liquid-dominated systems (e.g., Wairakei in New Zealand and Cerro Prieto in Mexico), a consequence of drawing down reservoir pressure, and increasing hot deep recharge, is that fluids may boil and two-phase zones may form. Conversely, resaturation of two-phase conditions may occur in reinjection sectors. This can affect the relative upflows of hot liquid and steam to the surface. Changes may include a decline in mineralized hot springs and an increase in steam-heated thermal features above production sectors, while the converse may occur above injection sectors. For reservoirs that are steam dominated (e.g., The Geysers in the United States and Larderello in Italy), a decline in reservoir pressure may reduce the natural upflow of steam to surface features. Such changes can have both adverse and beneficial effects on established users of the surface thermal features (such as hot spring resorts) and the associated thermal ecosystems.

Recent strategies of geothermal environmental management have placed an emphasis on achieving balance through avoiding adverse effects and promoting beneficial effects. A key objective of the strategies is to devise practical mitigation schemes. Production and reinjection schemes can now be planned with built-in flexibility in order to allow reaction to induced adverse effects, such as reductions in natural spring discharges or increasing subsidence, without compromising the efficient utilization of the resource. Net changes in gas emissions, from natural vents and power-plants, can be addressed in terms of effects, and technology applied for their avoidance, removal, or injection if necessary. Some examples of environmental benefits deriving from adaptive production/injection strategies include hot stream and thermal feature creation using waste hot water, increased steam-heated ground from liquid pressure drawdown, and increased hot spring discharge from shallow reinjection. Indirect environmental benefits have included wetland creation in subsidence areas and enhanced thermal ecological habitats where thermal features have increased. The key to achieving a successful balance is to adopt an adaptive resource management strategy.

To summarize, the optimum level of long-term sustainable production depends on the utilization technology and also on the geothermal resource characteristics. It can be achieved by properly managing fluid production and injection rates and locations. Upon shutdown, practical replenishment will occur on timescales of the same order as the lifetime of the production cycle. The recovery factors that determine the long-term response of these systems to energy extraction are dynamic. Recovery is influenced by an enhanced recharge driven by the strong pressure and temperature gradients created by the fluid and heat extraction. Optimized sustainable reservoir management involves countering the adverse effects of premature temperature decline with appropriate and flexible production and injection strategies. Such strategies need to be adjusted at times, in order to achieve the correct balance. Flexibility in locating and utilizing future injection wells, both inside and outside the hydrological edges of a geothermal system, is a key means of achieving a successful outcome. Optimized strategies of geothermal environmental management achieve a balance through avoiding or mitigating adverse effects and promoting beneficial effects.

References

1 IEA, (2011) *Technology Roadmap: Geothermal Heat and Power*, International Energy Agency, Paris, www.iea.org (last accessed 8 January 2013).
2 Goldstein, B., Hiriart, G., Bertani, R., Bromley, C., Gutiérrez-Negrín, L., Huenges, E., Muraoka, H., Ragnarsson, A., Tester, J., and Zui, V. (2011) Geothermal energy, In *IPCC Special Report on Renewable Energy Sources and Climate Change Mitigation (SRREN)*, Cambridge University Press, Cambridge.
3 Bromley, C. J., Mongillo, M. A., Goldstein, B., Hiriart, G., Bertani, R., Huenges, E., Muraoka, H., Ragnarsson, A., Tester, J., and Zui, V. (2010) IPCC Renewable Energy Report: the potential contribution of geothermal energy to climate change mitigation, in *Proceedings of the World Geothermal Congress 2010*, Bali, 25–29 April 2010.
4 Rybach, L. (2010) The future of geothermal energy and its challenges, in *Proceedings of the World Geothermal Congress 2010*, Bali, 25–29 April 2010.
5 Bloomfield, K. K., Moore, J. N., and Neilson, R. N. (2003) Geothermal energy reduces greenhouse gases. *Geothermal Resour. Counc. Bull.*, **32** (2), 77–79.
6 Frick, S., Schröder, G., and Kaltschmitt, M. (2010). Life cycle analysis of geothermal binary power plants using enhanced low temperature reservoirs. *Energy*, **35** (5), 2281–2294.
7 Bertani, R. (2010) World update on geothermal electric power generation 2005–2009, in *Proceedings of the World Geothermal Congress 2010*, Bali, 25–29 April 2010.
8 Dickson, M. H., and Fanelli, M. (eds) (2003) *Geothermal Energy: Utilization and Technology*, Renewable Energy Series, UNESCO, Paris, ISBN 978-92-3-103915-7.
9 Stefansson, V. (2005) World geothermal assessment, in *Proceedings of the World Geothermal Congress 2005*, Antalya, Turkey, 24–29 April 2005, ISBN 9759833204.
10 O'Sullivan, M., and Mannington, W. (2005) Renewability of the Wairakei–Tauhara geothermal resource, in *Proceedings of the World Geothermal Congress 2005*, Antalya, Turkey, 24–29 April 2005, ISBN 9759833204.

19
Catalyzing Growth: an Overview of the United Kingdom's Burgeoning Marine Energy Industry

David Krohn

19.1
Development of the Industry

The marine energy industry has continued to move towards commercial viability as an increasing number of devices move through the demonstration phase. There are now development plans for the first arrays and several world leading projects in planning.

The last year or so has seen an immense amount of activity:

- There are now 11 large-scale prototype devices deployed around the United Kingdom – more than the rest of the world combined.
- Seabed leases for over 1.8 GW of power production have been awarded.
- Major industrials such as Siemens, Andritz, Voith, Alstom, and DCNS have significantly increased their involvement in device development.
- Utility companies Scottish Power, EON, RWE, SSE, EDF, and GDF Suez are developing marine energy projects in both UK and more recently French waters. This represents ~£350 million of private finance investment excluding the significantly larger investment made by device developers in the technology.
- The European Marine Energy Centre is the world's leading marine energy test center; it has sold out all of its available testing berths, and been sought out for advice by aspiring marine energy test centers globally.
- Capital grant funding schemes have been introduced in support of the world's first arrays from UK and Scottish Governments in the form of the Marine Energy Array Demonstrator and Marine Renewables Commercialization Fund.
- NaREC has constructed a 3 MW testing rig to improve the United Kingdom's accelerated lifecycle testing capability

The expansion of the industry is thanks in no small part to the package of support that the UK and Scottish Governments have provided to industry, particularly the announcement in summer 2012 that the current level of support under the Renewables Obligation (RO) will be enhanced to five Renewable Obligation Certificates (ROCs) per megawatt hour [3]. The ROC regime adds revenue to low-carbon projects by charging carbon-intensive generation a ROC (value ~£42) and transferring it to renewable or clean generation.

This support has catalyzed a large amount of activity in the industry and enhanced the attraction of first array projects to a range of investors. The result is a thriving and diverse tidal stream energy industry in the United Kingdom with the potential to export expertise and hardware globally. This is already being demonstrated by recent projects under consideration in France which will use technology developed and tested in the United Kingdom.

In order for the immense potential of the industry to be realized, it is imperative that the mechanisms underpinning the Electricity Market Reform (EMR) are shaped with the tidal stream energy industry in mind, in addition to other forms of low-carbon generation. The creation and implementation of a market system that will provide certainty, durability, and confidence to investors, developers, and the supply chain will serve to consolidate the United Kingdom's position as the global leader, with key players capable of exporting technology and expertise. The tidal stream energy industry, in combination with the wave industry, has been forecast to be worth £6.1 billion to the UK economy by 2035, creating nearly 20 000 jobs [1]. It is essential that the EMR facilitates the growth required to make this a reality and help the United Kingdom to capture its rightful share of this global market.

19.2
The Benefits of Marine Energy

The marine energy industry holds the potential for vast socio-economic, environmental, and efficiency benefits, which has fostered a high level of political good will towards it. This situation is helped by the fact that the benefits of marine energy align closely with the stated objectives of UK energy policy, including decarbonizing of the energy mix, ensuring security of supply, and keeping energy affordable. Marine energy can contribute significantly to these objectives:

- *Decarbonization* – Renewable energy provides the greatest opportunity to decarbonize the United Kingdom's energy mix. By harnessing the forces of the ocean surrounding the island, the United Kingdom can reduce its dependence on imported fossil fuels and ensure that it leads the world in carbon reduction and halts anthropogenic climate change.
- *Security of supply* – a diversified renewables portfolio is the best way of ensuring security of supply. The predictable and consistent nature of marine energy moderates the output profile of renewables as a whole and therefore complements the intermittency of other some other forms of renewables. Using the natural resources within the United Kingdom's territorial seas ensures that dependence on imported energy sources is reduced.

Table 19.1 Total cost saving per year from diversifying the United Kingdom's energy mix by including higher proportions of marine energy. All data in £million. Source: [2].

Costs	Wind : marine (%)			
	100:0	75:25	60:40	40:60
Reduced backup capacity	0	205	201	159
Reduced costs of reserve capacity	0	108	130	137
Reduced costs of fuel and CO_2 emissions	0	211	298	337
Reduction in extra renewable capacity required to replace spill	0	192	236	234
Total savings	0	717	865	867

- *Affordability* – While the capital costs of marine energy projects are currently relatively high, investment in marine energy provides value for money for a number of reasons. Most prominently, tidal energy's moderating effect on the output profile of renewables results in reduced requirements for dispatchable balancing plant (Table 19.1).

In addition, we believe that wave energy holds the following benefits for the United Kingdom:

- *Diversification of supply system* – The key impact of diversifying the mix of renewable technologies is in reducing the variability of the hourly aggregate output levels (Figure 19.1). Wave energy provides a means of moderating the variable output of wind generation and reducing the risk of long periods of low renewables output (in essence because it will regularly be out-of-phase with wind and tidal generation).
- *Reduced volume and cost of reserve and balancing capacity* – The production profile of wave energy has a limited correlation with that of wind energy and is much less variable and more predictable. The increased predictability of wave energy, resulting in a reduction in hourly changes in output, will have a significant impact on the need for reserve capacity, as outlined in Table 19.1.

In addition to the above benefits, marine energy provides a bright spot in an otherwise bleak economic outlook as it is set to provide numerous economic and social benefits, including catalyzing an industry based in the United Kingdom, creating jobs and inward investment as well as ensuring that the United Kingdom captures a share of the potentially huge global export market. Importantly, the sector builds on the strengths of the United Kingdom's value chain and will deliver economic value to sectors such as ship building, marine operations, and mechanical, electrical and maritime engineering.

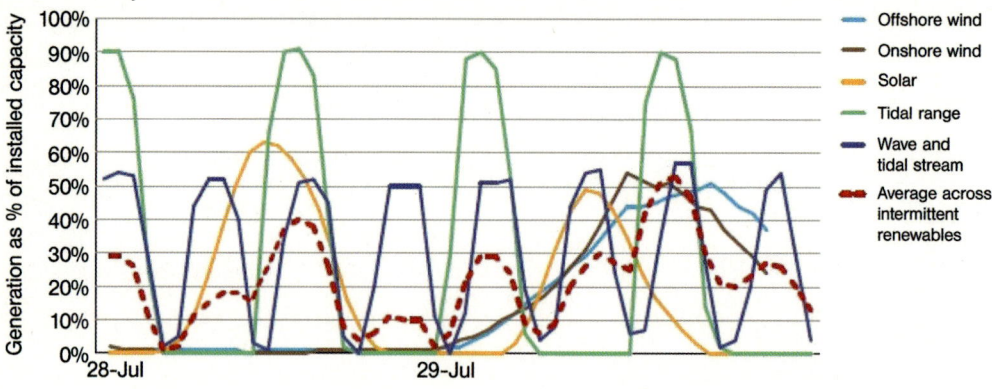

Figure 19.1 Variability of renewable generation technologies (over two illustrative days for 2030 mix). Based on observed patterns, 28–29 July 2006, scaled up to 2030 levels. The chart shows the generation that would be produced by the different renewable technologies (as a percentage of installed capacity) in the Pöyry Very High scenario over a 2 day period. Source: CCC analysis based on modeling by Pöyry.

In more detail, the effects of generating a domestic UK market are as follows:

- *First mover advantage and the creation of an industry:* The United Kingdom has worked hard to develop a world-leading industry and support is required to ensure that it maintains the global lead. As the United Kingdom learned from the wind industry, surrender of the first mover advantage can result in other countries moving ahead and capturing the benefits of a new industry.
- *Generation of inward investment:* As we have seen from the large amount of activity in the mergers and acquisitions market recently, tidal stream energy can generate significant inward investment.
- *Development of an export market:* The global marine energy resource is huge and the potential to exploit global resources provides a large opportunity to British companies. Failure to develop the local market will lead to an inability to capitalize on those markets further afield.
- *Generation of British jobs:* British expertise in marine engineering and operations means that the United Kingdom is well placed to take advantage of the employment opportunities that will be created. This is especially pertinent if the industry can be developed in British waters and home-grown skills can be nurtured.

19.3
Expected Levels of Deployment

The existing literature on the expected levels of deployment for marine energy is inconsistent and the Electricity Market Reform Expert Group (EMREG) tasked its uniquely qualified membership with assessing growth across the industry. Seeking to form an independent and authoritve deployment pipeline, EMREG used the list of

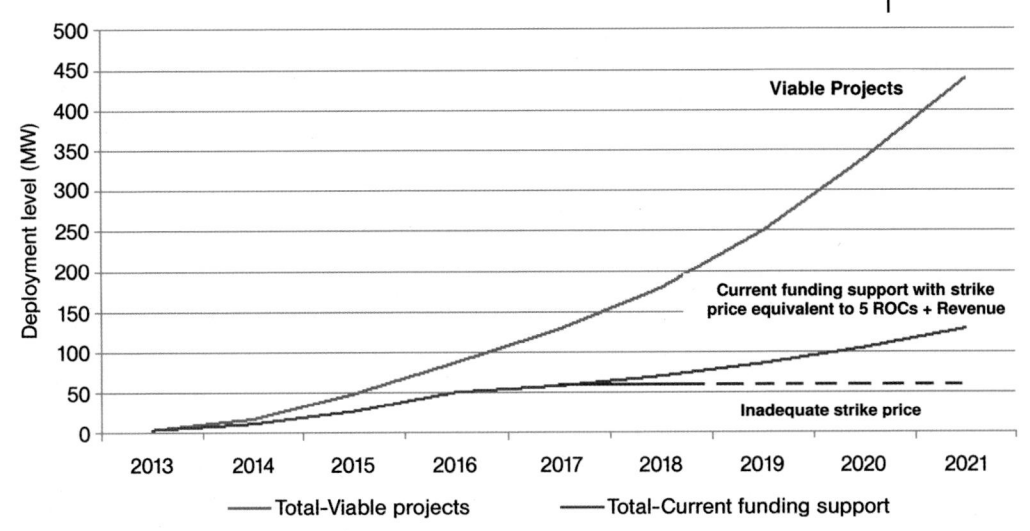

Figure 19.2 Potential scenarios for UK marine energy deployment – cumulative (post-2017 growth predicated on an appropriate level of revenue support). Current funding support scenario is based on current grant programs such as Marine Energy Array Demonstrator (MEAD), Marine Renewables Commercialization Fund (MRCF), NER300, and FP7.

wave and tidal leases from The Crown Estate as a starting point, before developing a pipeline of projects that are in a position to gain grid connection and consent approval in the relevant time frame. These projects are noted as "viable projects" in Figure 19.2.

It is noted that financing is the primary constraint to the development of the industry. The current government capital support stream will facilitate the development of a small number of projects as identified in the scenario "Current funding support" in Figure 19.2. This is based on the projects that will benefit from MEAD and MRCF support. It also accounts for various European demonstration funding streams (e.g., NER300 and FP7). To date, this is the best available forecast of anticipated deployment.

Crucially, the current investment environment is beset with uncertainty around the level of future revenue support that the industry can expect. There is a risk of a funding hiatus if the revenue support is not attractive enough to encourage project financers to take on the risk of the initial precommercial projects (plan for deployment 2014–2017). Any reduction in the support offered through the EMR would risk halting installation and jeopardizing the United Kingdom's global lead.

If an appropriate level of revenue support could be offered to marine energy, it is thought that the "actual" deployment could deliver a figure approaching full realization of the "Viable projects" scenario. Increased deployment will inevitably lead to increased innovation and also decreased vulnerability to external factors.

With this package of support, the EMREG is of the opinion that the industry can deliver on its massive potential and enable the United Kingdom to maintain its world-leading position.

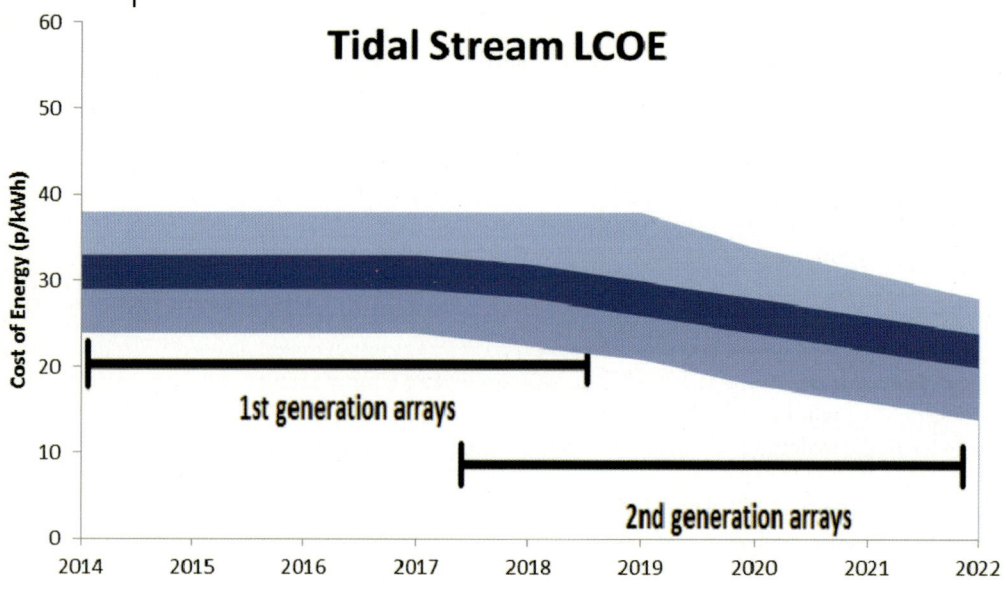

Figure 19.3 The cost trajectory for tidal stream energy.

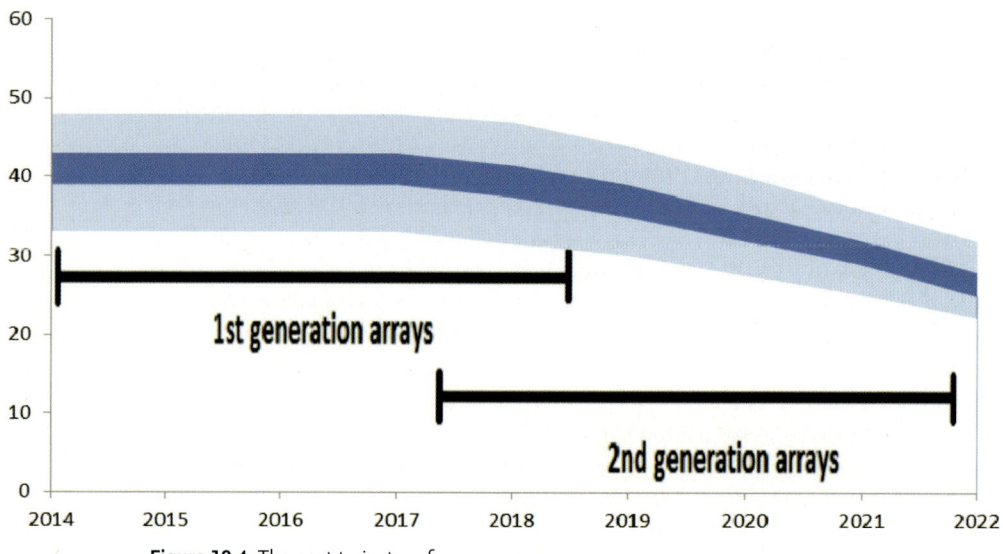

Figure 19.4 The cost trajectory for wave energy.

19.4
Determining the Levelized Cost of Energy Trajectory

The true levelized cost of energy (LCoE) of marine energy is yet to be confirmed on a large scale, owing to the lack of full-scale arrays. However, the submission of data in application for the MEAD and MRCF has crystallized the current position on LCOE and the trajectory up to 2020.

While the timing of the end of the RO regime is not ideally timed for the marine energy industry, the review period for the initial phase of the EMR period fits with the development cycle of the marine energy industry. The first generation of arrays, for which we have clear data for the cost of energy, are planned for the period 2014–2018. Following the commissioning of the first generation of arrays up to 2018, we will have a clearer picture of the cost of second-generation arrays and will be able to determine the revenue support the industry requires to continue flourishing.

Using the MEAD and MRCF data, we have been able to determine the cost of the first arrays (Figure 19.3 and Figure 19.4).

19.4.1
The Cost of Energy Trajectory

The data we have gathered represent the most current thinking on LCoE and it broadly agrees with data in established publications such as the Carbon Trust's *Accelerating Marine Energy* document [4], the Energy Technology Institute's *Marine Energy Technology Roadmap*, and the Department of Energy and Climate Change (DECC)'s own *Technology Innovation Needs Assessment* (TINA) [5], which presents the graphs shown in Figure 19.5.

While we have identified a number of issues with the trajectory outlined in the TINA document [5], we support its application of a learning rate associated with deployment. Learning rate theory is a well-demonstrated phenomenon whereby an industry will reduce costs at a fairly constant rate dependent on the experience it gains.

Analysis by the Carbon Trust puts the learning rate of wave and tidal energy at 15% (costs reduce by 15% for every doubling of an individual developer's installed capacity), but learning rates can be variable. It is therefore important that the correct mechanisms and support are put in place to ensure that the United Kingdom maintains the highest possible learning rate.

The primary motivation for providing significant and consistent revenue support is to catalyze cost reduction by building the experience of the industry and enabling learning by doing, economies of scale, and improved productivity. It is essential that the revenue support regime rewards accelerated cost reduction and provides a sufficient margin to allow continued R&D and innovation.

Most of the leading technology developers have developed their own detailed cost reduction plans. The best plans combine both high-level aspirations for cost reductions with detailed proposals for specific engineering work to reduce costs.

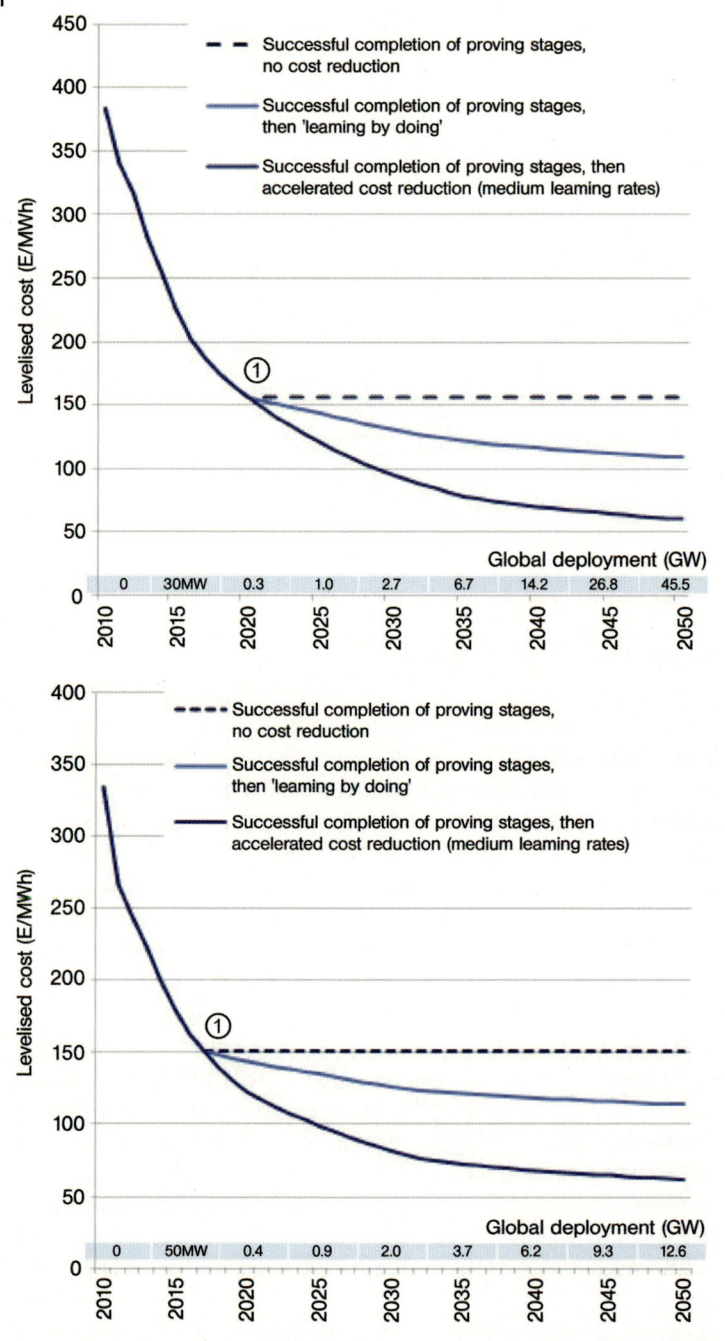

Figure 19.5 Forecast cost reduction. Potential impact of innovation on levelized costs (medium global deployment). Point 1: "proof of value" point based on the time at which a critical mass of devices has been deployed and reached a potential subsidy level of approximately 2–3 ROCs (depending on electricity prices). Source: [5].

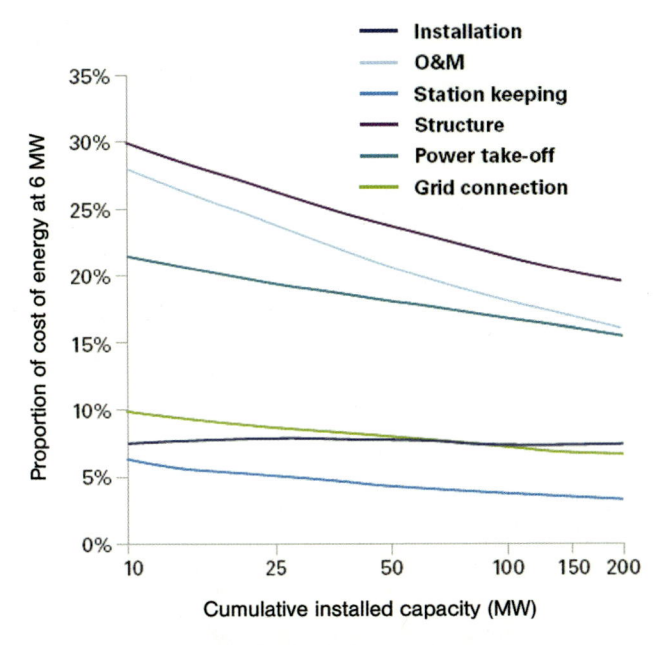

Figure 19.6 Target areas for cost reduction. Source: [4].

We are mindful of the areas that will yield the greatest reductions and feel that the Carbon Trust's *Accelerating Marine Energy* [4] highlights some of the key categories where costs can be saved (Figure 19.6).

The leading developers are pursuing an aggressive cost reduction program to ensure that costs can fall to a level that will encourage investment and catalyze development. The industry has identified the following actions they are taking to drive down their cost of energy:

- *Device power up-rating* – Following the wind industry pathway, Wave and tidal energy devices will increase in size, improving output.
- *Multiple rotors/devices* – Foundations make up a significant proportion of marine energy projects and structure innovation can reduce project costs significantly.
- *Reliability and maintainability* – Experience leads to improved availability schedules and maintenance strategies.
- *Production* – Improved scale and the establishment of specialist production facilities will reduce manufacturing costs.
- *Supply chain* – Improved deployment will enable the supply chain to scale up and permit innovation.
- *Installation* – Lessons and scale from offshore wind and also specific innovations in device installation methods and equipment will reduce installation costs.
- *Offshore grid* – Grid remains a barrier to the development of the industry but improved access, innovation, and scale can aid project development.

- *Device access* – Innovative deployment and recovery systems, offshore marine operations innovation, and scale all contribute to reduced costs.
- *Offshore wind* – Exploiting synergies in manufacturing, supply chain, grid, and O&M logistics.
- *Higher capacity sites* – Higher capacity sites become accessible with experience on more benign sites, thus improving output.
- *Reducing cost of capital* – Discount rates applied to projects decrease as investors become more accustomed to marine energy projects and the technology matures, reducing novelty and technological risk.
- *New types of investors* – As the risk profile of marine energy changes, so does the type of investor. Debt funds and equity will become more involved once the first proving arrays have been commissioned.
- *Reduced insurance costs* – As deployments increase, international standards are applied and confidence is established.

While the variables within the control of technology developers are outlined above and in the TINA document [5], there are several factors outside their control. Disruptive innovations and commodity prices will have a substantial effect on the cost trajectory and should not be discounted.

It is important to remember that an expanding industry leads to learning whereas a stagnant or contracting industry has the effect of leading to "forgetting." Examples of this are closing factories, loss of key personnel, and even the sale of intellectual property. Smaller companies and industries are particularly exposed to this and it is vital that the revenue support is set at a level that will protect innovative companies and maintain the momentum of the industry.

In order for cost reduction to take an accelerated trajectory, the industry needs certainty over the level of support that it will receive. This will allow continuous learning, investment in long-term R&D programs and efforts to alleviate policy barriers, in addition to providing each individual player in the industry an opportunity to learn. A long-term goal also allows innovation investors such as the Energy Technologies Institute, the Carbon Trust, and the Technology Strategy Board to develop programs to underpin cost reductions for the marine energy industry.

19.5
Technology Readiness

The wave and tidal energy industries are currently passing through the development phases and remain immature industries. However, significant progress has been made. Devices have moved through the technology readiness levels to the point where we have a mix of mature and innovative designs competing for project finance and the various grant programs.

A series of case studies are outlined in the following.

19.5.1
Tidal Device Case Study 1

AR1000

Manufacturer	Atlantis Resources Corp.
Type of device	Horizontal-axis turbine
Status	Full-scale prototype, 2011
Location	Currently at Narec
Rating	1 MW

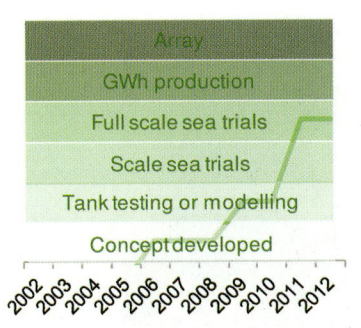

Device description
The AR1000 is a three-bladed fixed-pitch horizontal-axis turbine with active yaw mechanism and a direct-drive permanent magnet generator with variable-speed drive. Power is exported via medium voltage (3.8 kV) cables to an onshore substation. Higher efficiency blades than on the previous device have been introduced along with marine fouling treatment and structural improvements. The AR1000 has a rated for flow velocity of 2.65 m s^{-1}.

The Atlantis Resources approach is to reduce the complexity of the turbine, thus decreasing the number of potential failure points. A proprietary male and female "stab" connection arrangement allows for fast retrieval of the nacelle without having to recover the foundation structure.

Status
Following deployment of the twin rotor AK-1000 in 2010, the AR-1000 was successfully deployed and underwent sea operations during summer 2011. The device has been retrieved and is undergoing testing at Narec prior to deployment to the European Marine Energy Centre (EMEC) again to continue generating later in 2012.

19.5.2
Tidal Device Case Study 2

HS1000

Manufacturer	Andritz Hydro Hammerfest
Type of device	Horizontal-axis turbine
Status	Full-scale prototype
Location	Fall of Warness, EMEC, Orkney
Rating	1 MW

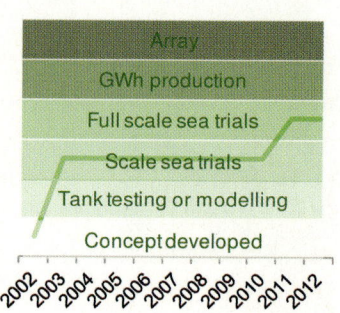

Device description

Horizontal-axis, pitch-regulated, three-bladed turbine, installed in-line with the flow. A nacelle houses the gearbox, asynchronous generator, and control systems, and subsea cables export energy back to shore. There is an onshore frequency converter. The device is seabed mounted using a gravity base foundation. The turbine is heavily instrumented to serve as a platform for future R&D activities. This will help drive improvements in reliability and efficiency and allow reductions in the cost of energy over time.

Status

The HS1000 is a full-scale precommercial device following on from the HS300, which was successfully installed, tested, and operated as a scale prototype. It is currently installed at EMEC, with measurements for product certification scheduled for 2012. This is in preparation for the 10 MW Islay array (lease granted and EIA submitted), which will begin construction activities in 2013 with ScottishPower Renewables and the 95 MW Duncansby Head array.

19.5.3
Tidal Device Case Study 3

SeaGen

Manufacturer	Marine Current Turbines
Type of device	Horizontal-axis turbine
Status	Full-scale prototype, 2008
Location	Strangford Lough
Rating	1.2 MW

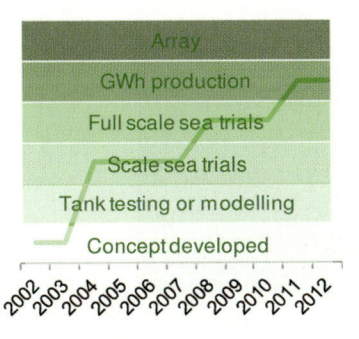

Device description

The SeaGen device comprises twin horizontal-axis rotors 16 m in diameter, each driving a gearbox and generator. The generator output is rectified, inverted, and exported to the distribution grid (in the case of the Strangford Lough project) via a step-up transformer. Twin turbines are independently operable. The rotors have full span pitch control such that they can generate on flood and ebb tides. There is no yaw system.

The structure is surface piercing so that the drive trains can be raised clear of the water for easier maintenance access.

Status

The SeaGen design is based on experience gained from the first UK sea-tested tidal energy converter, SeaFlow. SeaFlow was a single-rotor 300 kW experimental turbine that was installed 3 km north east of Lynmouth on the North Devon Coast in May 2003 and decommissioned in October 2009. The design of a 2 MW version suitable for array deployment is now complete.

Since installation in 2008, the SeaGen device has undergone modifications that most recently included replacement of one of the drive trains with a second-generation unit. The SeaGen design will be used for the proposed Skerries and Kyle Rhea deployments. A non-surface-piecing variant is under development that could form the basis of future arrays.

19.5.4
Tidal Device Case Study 4

Deep Green

Manufacturer	Minesto
Type of device	Other
Status	Scale prototype, 2011
Location	Tank test
Rating	0.5 MW (full-scale)

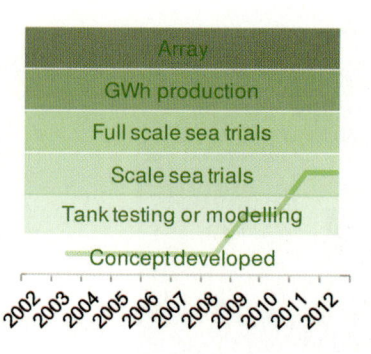

Device description

The wing is attached to the turbine and gearless-generator arrangement, with a rudder and servo system at the rear of the device. The device is tethered to the seabed in an arrangement that accommodates the power cables and communication system and is connected to a conventional mooring rig. The wing generates lift, which moves the device at speeds up to 10 times that of the water flow, and the rudder is used to steer the device in a figure-of-eight path. As this happens, the rotor is turned as the device passes through the water, generating electricity. It is designed to be economical at sites with deep water and low velocities. Deep Green is recoverable for servicing and maintenance and has a mass of less than 7 t per 0.5 MW unit).

Status

Minesto is continuing to model the device and test scale prototypes to develop the technology, prove reliability by getting hours in the water, and develop confidence in and acceptance of the concept from project developers. A one-tenth scale device was installed in Strangford Lough in 2012 as a precursor to a one-third scale device and then a full-scale device that is rated at 0.5 MW. New capital was recently injected by the company owners and a £0.5 million grant was awarded by the Swedish Energy Agency for R&D operations. Funds have also been made available through the Carbon Trust as part of the Applied Research program.

19.5.5
Tidal Device Case Study 5

Open-Centre Turbine

Manufacturer	OpenHydro
Type of device	Horizontal-axis turbine
Status	Full-scale prototype, 2011
Location	Fall of Warness, EMEC, Orkney
Rating	0.25 MW (UK device)

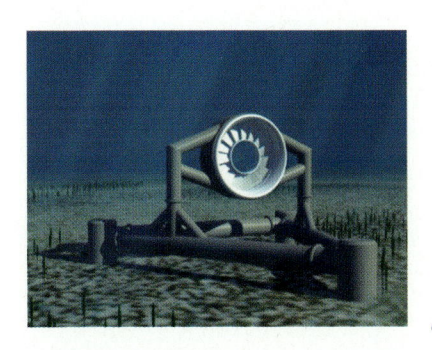

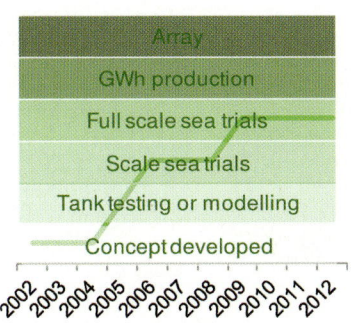

Device description

The Open-Centre tidal stream turbine is a seabed-mounted device that consists of a rotor, duct, stator, and generator. Water passes through the duct, utilizing the Venturi effect, to the slow-moving rotor with a hole to enable marine life to pass through. This is the only moving part. The permanent magnet generator is housed in the duct around the rotor.

The device has been designed to be scalable. There is no gearbox and there are no seals, which reduces the number of total components.

Status

In 2006, an Open-Centre turbine was installed on a seabed-mounted test rig at EMEC and in 2008 the installation of a second seabed-mounted turbine took place adjacent to the first using the OpenHydro Installer vessel. OpenHydro have also deployed a 1 MW (10 m diameter) device in Canada and installation of a 16 m device in Brittany in 2012 is planned.

19.5.6
Tidal Device Case Study 6

Pulse-Stream 100

Manufacturer	Pulse Tidal
Type of device	Oscillating hydrofoil
Status	Scale prototype, 2009
Location	River Humber estuary
Rating	0.1 MW

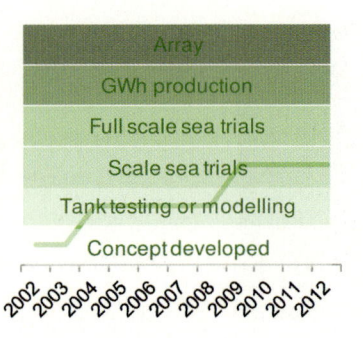

Device description

The Pulse-Stream 100 is an oscillating hydrofoil tidal stream device. It extracts power from tidal currents using horizontal blades that move up and down. This movement drives a gearbox and generator through a crankshaft. This approach enables the blade length (and therefore power generation capacity) to be comparatively large in a given water depth; the Pulse-Stream 100 can produce 1.2 MW in 18 m of water, with potential to scale up to 5 MW in 35 m of water. Two Pulse-Stream 100 machines can be installed on each foundation, offering up to 10 MW installed capacity per foundation in a water depth of 35 m.

The turbine rises out of the water to facilitate observations, learning, and maintenance.

Status

The Pulse-Stream 100 device is a scale prototype that was deployed in the Humber in 2009 and operated through to 2012. This proved the technology concept and enabled Pulse Tidal to learn from installing their device in the marine environment. The commercial device will have a buoyant base, which allows it to be commissioned onshore and then floated to site. On-site, it will be connected to a preinstalled foundation and flooded, leaving it fully submerged in operation. For maintenance, the hull can be deballasted, bringing the machine back to the surface, where it forms a stable maintenance platform. The preliminary design for this 1.2 MW device has been completed, with detailed design under way. The next step is construction and installation, ahead of performance testing and deployment into arrays.

19.5.7
Tidal Device Case Study 7

DeltaStream

Manufacturer	Tidal Energy Ltd
Type of device	Horizontal-axis turbine
Status	Full-scale prototype, 2012
Location	Ramsey Sound
Rating	1.2 MW

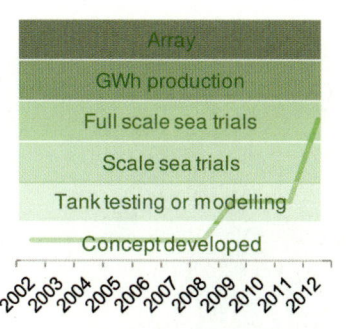

Device description

The DeltaStream device is a 1.2 MW unit which sits on the seabed without the need for a positive anchoring system. It generates electricity from three separate horizontal-axis turbines mounted on a common frame. This avoids piling or significant seabed preparation.

The use of three turbines on a single 350 t, 36 m wide triangular frame produces a low center of gravity, enabling the device to satisfy its structural stability requirements, including the avoidance of overturning and sliding. Each turbine is an upflow three-blade design with hydraulic yawing system to turn to face the oncoming tide and allow generation on both the ebb and flood tides.

Status

Following a period of design development including modeling and tank testing at Cranfield University, a full-scale device is planned for deployment in Ramsay Sound, Pembrokeshire in December 2012. Licenses for the site were granted by DECC and the Welsh Government in March 2011. A significant proportion of contracts have been awarded for the manufacture of the device and project construction.

Following a successful 12 month technical and environmental test of the device, TEL will develop a precommercial array at a site yet to be confirmed.

19.5.8
Tidal Device Case Study 8

Deep-Gen IV

Manufacturer	Tidal Generation Ltd
Type of device	Horizontal-axis turbine
Status	Scale prototype, 2010
Location	Fall of Warness, EMEC, Orkney
Rating	1 MW

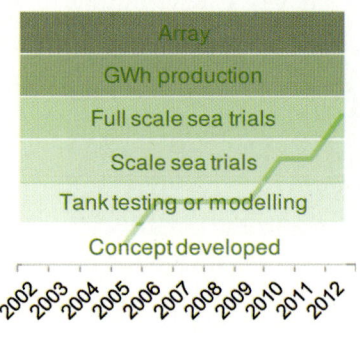

Device description

Deep-Gen IV is a three-bladed upstream tidal turbine that extracts energy during both flood and ebb tides using an active yaw system. The nacelle is a buoyant design to allow towing to the site and it is attached to a lightweight tripod foundation that is pinned to the seabed. A mechanical clamp facilitates yawing powered by a rear-mounted thruster. Once locked to face the flow, the turbine cuts in at 1 m s^{-1} and reaches rated power at 2.7 m s^{-1}. For higher flow speeds, the blade pitch and generator torque are regulated to maintain the turbine's rated power. A gearbox is used to drive an induction generator, and a frequency converter, step-up transformer, and wet mate link complete the generating system. The turbine will output grid compatible 6.6 kV three-phase power.

Status

Deep-Gen IV is a full-scale 1 MW device that will form the basis of the Tidal Generation commercial product offering. It is currently in final assembly at the Rolls-Royce Dunfermline facility in Edinburgh, and is scheduled to be deployed at EMEC in June 2012. Tidal Generation is collaborating with developers, including Meygen, with the intention of supplying devices to large array projects within the next 5 years, working with further industrial partners to improve the value proposition and develop the products that customers require.

19.5.9
Tidal Device Case Study 9

Voith HyTide 1000-13

Manufacturer	Voith Hydro
Type of device	Horizontal-axis turbine
Status	Scale prototype, 2010
Location	Fall of Warness, EMEC, Orkney
Rating	1 MW

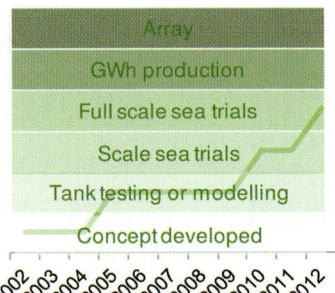

Device description
The Voith HyTide device is a seabed-mounted horizontal-axis tidal stream turbine. The three symmetrical blades capture energy from the tidal stream on the ebb and flood flow without pitch and yaw requirements. Electricity is generated using a direct-drive, permanent magnet assembly. The device is lubricated with seawater.

Ideal site characteristics in which to install the device and monopole foundation are a minimum peak current speed of 3 m s^{-1} and water depth of at least 30 m.

Status
A first turbine of this type, a 1 : 3 scale 110 kW device, is currently being installed by Voith Hydro Ocean Current Technologies in South Korean waters. There are plans to deploy a farm of ~100 MW offshore in South Korea near Jindo.

Voith Hydro and RWE Innogy completed preparatory work at EMEC in summer 2011 with the turbine installation due in autumn 2012 for a 3 year trial operation.

19.5.10
Tidal Device Case Study 10

SR250

Manufacturer	Scotrenewables Tidal Power
Type of device	Horizontal-axis turbine
Status:	Scale prototype, 2009
Location	Fall of Warness, EMEC, Orkney
Rating	0.25 MW (full-scale 2 MW)

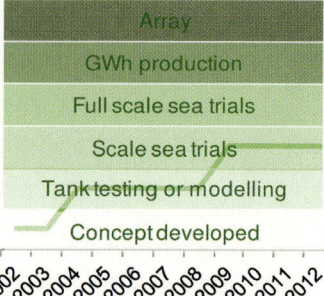

Device description

The Scotrenewables Tidal Turbine (SRTT) is a floating tidal stream turbine. The main structure of the SRTT is a cylindrical tube to which dual horizontal-axis rotors are attached via retractable legs. Two counter-rotating rotors extract the kinetic energy of the tidal flow, which is converted to electricity though the power take-off system. The system has two configurations: operational with the rotors down to generate power and transport, and survival mode with rotors retracted whereby loads are reduced in heavy seas. The system is designed to be installed, operated, and maintained using a multi-cat workboat. No specialist vessels are required at any stage in the product lifecycle. The SRTT is also suited to deployment in river flow.

Status

The SR250 was launched in Orkney in March 2011, and is currently undergoing a 24 month testing program at the EMEC Fall of Warness tidal test site. Longer term grid-connected deployments are planned for spring/summer 2012. A baseline design for the next-generation 2 MW SRTT has been completed and the company is looking to place contracts during 2012 for manufacture.

19.5.11
Wave Device Case Study 1

Oyster 800

Manufacturer	Aquamarine Power
Type of device	Oscillating wave surge converter
Status	Full-scale prototype, 2011
Location	Billia Croo, EMEC, Orkney
Rating	0.8 MW

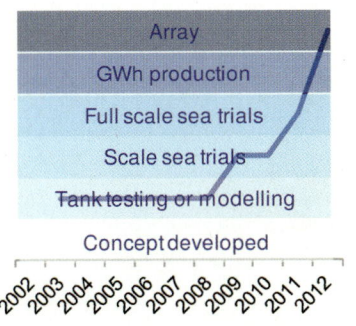

Device description

Oyster is a nearshore wave-powered pump which pushes high-pressure water to drive an onshore hydroelectric turbine. The mechanical offshore device is connected to the seabed in ~10 m water depth, typically within 1 km from the shore facility. The pump is a buoyant hinged flap that is almost entirely underwater and pitches back and forth due to wave motion. This drives two hydraulic pistons that pump high-pressure water through a subsea pipeline to shore to drive the hydroelectric turbine.

Key design features of the Oyster are that there are no offshore electronics, electricity generation equipment is located onshore using hydroelectric equipment (existing technology), and the submerged nature of the device reduces visual impact and wave loading in extreme waves.

Status

The first full-scale 0.315 MW Oyster was installed and connected to the grid at EMEC in 2009. The device withstood two winters and delivered over 6000 offshore operating hours. The full-scale next-generation Oyster 800 device was installed in 2011.

19.5.12
Wave Device Case Study 2

WaveRoller

Manufacturer	AW-Energy
Type of device	Oscillating wave surge converter
Status	Scale prototype, 2012
Location	Portugal
Rating	0.8 MW (per flap)

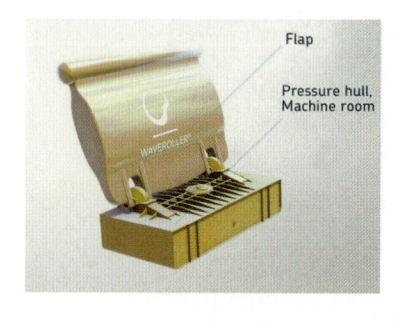

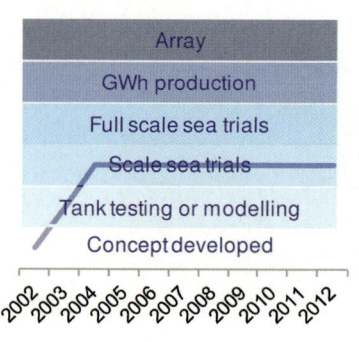

Device description

The WaveRoller device is a flap anchored to the seabed at its base. The back and forth movement of the wave surge moves the flap, transferring the kinetic energy to piston pumps, which feed into an onshore generator system. The nominal capacity of a single commercial-scale flap ranges between 0.5 and 1 MW, depending on the wave resources available on the site. The plant construction is modular; therefore, it offers a high level of scalability.

The WaveRoller is designed such that large waves pass above the device, thus reducing extreme loading. Other design considerations are accessibility for maintenance, simple installation through using a float and submerge method, and minimal visual impact on the deployment site.

Status

The first prototypes were designed as early as 1999. Open sea trials started in 2004 and included deployment in the Gulf of Finland, at EMEC, and at AW-Energy's own testing site Peniche, Portugal. The latest demonstration farm consisting of three 100 kW WaveRoller units is to be deployed also in Peniche, Portugal, during the second quarter of 2012. The company has already secured all the necessary permits and holds a license for 1 MW grid connection.

19.5.13
Wave Device Case Study 3

AWS-III

Manufacturer	AWS Ocean Energy (AWS)
Type of device	Other
Status	Scale prototype, 2010
Location	Billia Croo, EMEC, Orkney (2014)
Rating	2.5 MW

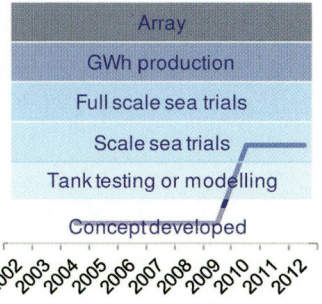

Device description
The AWS-III consists of a number of interconnected flexible wave energy absorber cells mounted on a common floating structure. The flexible membrane absorbers convert wave energy to compress air within each cell as a wave is incident on the cell. The compressed air is used to drive a turbine connected to a generator.

The device is slack moored in water depths of around 100 m using conventional mooring spreads, which reduces wave loadings on the device. The sealed air system ensures that there are no moving parts, increasing survivability, and the ancillary systems are protected within a hull that can be accessed for practical maintenance and inspection.

Status
AWS is backed by Alstom, who recently signed a joint venture agreement with SSE Renewables for the development of a 200 MW wave energy farm at Costa Head, north-west of the Orkney mainland. It is intended that this project uses the AWS-III wave energy converter technology.

AWS is conducting a program to deliver and qualify the AWS-III technology for the first phase of Costa Head, which is expected to be deployed in 2016. Accordingly, AWS expect to deploy a full-scale 2.5 MW prototype AWS-III at EMEC in 2014 and are currently constructing full-scale components including a full wave absorber cell for testing during 2012. This work is supported by the Scottish Enterprise WATERS grant scheme.

19.5.14
Wave Device Case Study 4

BOLT 2

Manufacturer	Fred. Olsen Renewables
Type of device	Point absorber
Status	Scale prototype, 2009
Location	Cornwall, UK
Rating	0.25 MW

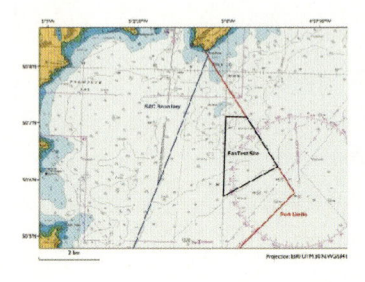

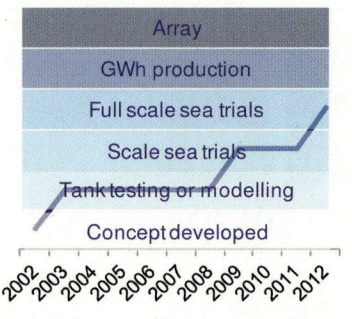

Device description

The BOLT 2 device is a near-shore, moored, floating point absorber. The unit is manufactured in composites and steel. Each unit may be autonomous or remotely operated.

Status

Fred. Olsen has tested a scale device in Norwegian waters since 2008. A prototype is planned to be launched at FaBTest in Cornwall during 2012. BOLT 2 received funding through the TSB to manufacture and deploy the prototype, working with the private sector and academia.

This test will be expanded with array testing and trial grid connection at a suitable location, potentially at WaveHub.

19.5.15
Wave Device Case Study 5

PowerBuoy

Manufacturer	Ocean Power Technologies (OPT)
Type of device	Point absorber
Status	Scale prototype, 2011
Location	Invergordon, Scotland
Rating	0.15 MW

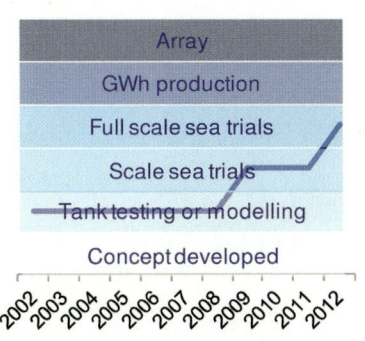

Device description
The PowerBuoy is constructed as two main hull elements, the spar and the float. The spar is designed to remain as stationary as possible whereas the float responds actively and dynamically to wave forces. The difference in motion between the two hulls is captured as mechanical energy and converted via onboard generators. The intent of the PowerBuoy control system is to tune the response of the system to maximize power capture from each wave. Conventional mooring systems are used to tether each PowerBuoy.

Status
Ocean Power Technologies has been active in the United States and Spain and has projects for both England and Scotland. It is committed to Wave Hub, Hayle, with a proposed array and successfully deployed the PB150 PowerBuoy off Invergordon in April 2011. During this project, tests were focused on issues such as examining the response of the structure and mooring system in the offshore environment and the power production, using real-time data and wave data from a nearby resource buoy.

19.5.16
Wave Device Case Study 6

P2 Pelamis

Manufacturer	Pelamis Wave Power
Type of device	Attenuator
Status	Full-scale commercial, 2010
Location	Billia Croo, EMEC, Orkney
Rating	0.75 MW

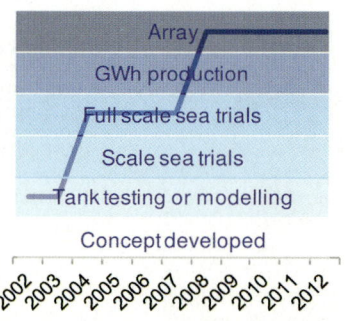

Device description

The P2 Pelamis is a semi-submerged, floating device that faces in the direction of the waves. Five tube sections are joined by universal joints that allow flexing in two directions with identical, independently operating hydraulic power take-off systems in each joint. As waves pass down the length of the machine, the tube sections move relative to each other and hydraulic cylinders in the joints resist the wave-induced motion, pumping fluid into high-pressure accumulators. This allows generation to be smooth and continuous. The resistance can be controlled to optimize the device for use in varying wave climates. Standard subsea cable and equipment are used to connect the device or array of devices to shore. A Pelamis is anchored to the seabed by a slack, chain catenary system with a yaw restraint line.

Status

The first Pelamis machine (P1) was installed in 2004 and 2011 saw the manufacture of the sixth full-scale machine and second utility customer (E.ON and ScottishPower Renewables). Vattenfall has also committed as a third. The E.ON P2 Pelamis is operating at EMEC with the ScottishPower Renewables P2 Pelamis soon to be installed alongside.

19.5.17
Wave Device Case Study 7

Limpet

Manufacturer	Voith Hydro Wavegen
Type of device	Oscillating water column
Status	Full-scale, 2000
Location	Islay
Rating	0.5 MW

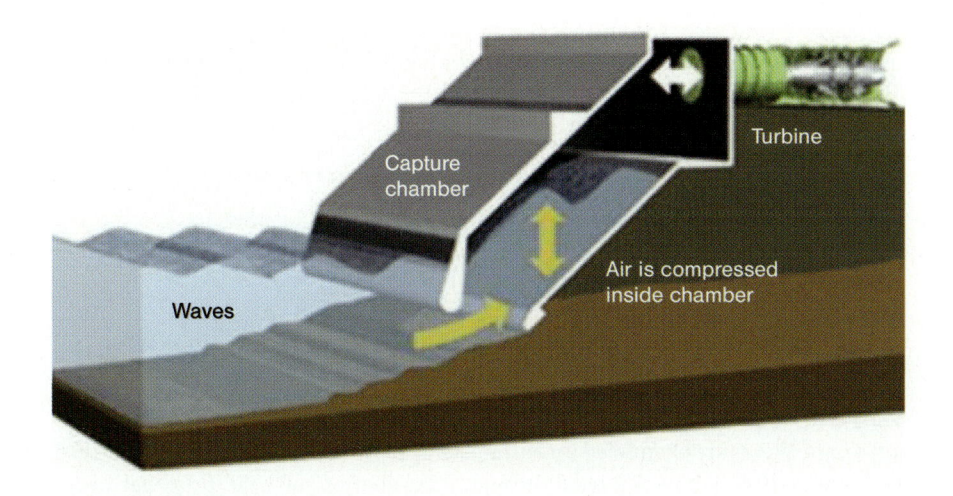

Device description
The Limpet device is a shore-mounted oscillating water column that has been tuned to capture energy from annual average wave intensities of between 15 and 25 kW m^{-1}. It comprises an air chamber with an opening below the water. The wave action moves the water level up and down the air chamber, compressing the air. The air flow is captured by two counter-rotating Wells turbines, each which has a 250 kW generator, on each surge and ebb of the wave.

Its shoreline location avoids offshore construction, installation, and service requirements.

Status
The Limpet 500 was installed in 2000 and is connected to the grid. This device is also used as a developmental facility, to demonstrate new systems and improve power output, such as testing a variable-pitch turbine. A near-shore application of the technology is under development.

19.5.18
Wave Device Case Study 8

Wave Dragon

Manufacturer	Wave Dragon
Type of device	Overtopping
Status	Scale prototype, 2003
Location	Denmark
Rating	1.5 MW

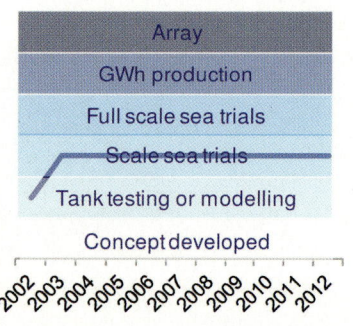

Device description

The Wave Dragon is a floating overtopping wave energy converter with wave-reflecting wings. Waves are channeled up an adjustable ramp to a reservoir, creating a head differential. Water is then passed through a number of low-head propeller hydro turbines to generate electricity. This means that there are few moving parts, aimed at reducing maintenance requirements. The device is slack-moored to the seabed.

Status

A 1:3-scale 1.5 MW device was installed offshore at Nissum Bredning in Denmark from 2003 to 2010 to model how the device behaves in the North Sea climate. Wave Dragon intends to install a precommercial demonstrator off the coast of Milford Haven, Wales, and has submitted an Environmental Impact Assessment and application for consent.

19.6
Conclusion

The marine energy industry has built up a considerable amount of momentum over the last few years and has now reached the point where it needs to scale up and commercialize. Although the challenges are immense, the successes to date show that they can be overcome. With the right support from government, the industry can continue to grow at an accelerated rate and revolutionize the way in which the world generates power.

References

1. RenewableUK (2010) *Wave and Tidal Channelling the Energy – Oct 2010*, http://www.renewableuk.com/en/publications/index.cfm/Wave-and-Tidal-Channelling-the-Energy (last accessed 2 February 2013).
2. Redpoint Energy (2009) *The Benefits of Marine Technologies Within a Diversified Renewables Mix. A Report for the British Wind Energy Association by Redpoint Energy Limited*, Redpoint Energy, London.
3. Department of Energy and Climate Change (2011) *Consultation on Proposals for the Levels of Banded Support Under the Renewables Obligation for the Period 2013–17 and the Renewables Obligation Order 2012*, p. 43, http://www.decc.gov.uk/assets/decc/11/consultation/ro-banding/3235-consultation-ro-banding.pdf (last accessed 2 February 2013).
4. Carbon Trust (2011) *Accelerating Marine Energy*, http://www.carbontrust.com/media/5675/ctc797.pdf (last accessed 2 February 2013).
5. Low Carbon Innovation Co-ordination Group (2012) *Technology Innovation Needs Assessments: Marine*, http://www.lowcarboninnovation.co.uk/working_together/technology_focus_areas/marine/ (last accessed 2 February 2013).

20
Hydropower

Ånund Killingtveit

20.1
Introduction – Basic Principles

Hydropower is generated by converting potential energy in water, as it moves from a higher to a lower elevation, into mechanical and electrical energy. The theoretical output of electrical power depends on three main factors, flow (Q), head (H) and efficiency (η):

$$P(\text{W}) = \rho\, g\, Q\, H\, \eta \tag{20.1}$$

where P is the electrical power output (J s^{-1} = W), ρ is the density of water (1000 kg m^{-3}), g is the acceleration due to gravity (9.81 m s^{-2}), Q is the water flow per unit time (m^3 s^{-1}), H is the elevation drop, usually called head (m), and η is the efficiency in the conversion process (per unit). Assuming values for density and acceleration due to gravity as given above, the equation is often simplified to give the output directly in kilowatts:

$$P(\text{kW}) = 9.81\, Q\, H\, \eta \tag{20.2}$$

The amount of energy produced depends on the duration of the flow. Assuming the duration Δt is given in hours, the amount of energy produced can be calculated as

$$E(\text{kWh}) = P\, \Delta t \tag{20.3}$$

Another useful equation gives the energy output in kilowatt hours per cubic meter of water as a function of H and η, often called the energy equivalent of water (EEKV):

$$\text{EEKV}(\text{kWh m}^{-3}) = \frac{P \cdot 1\,\text{h}}{Q \cdot 3600\,\text{s}} = \frac{9.81\, H\, \eta}{3600} \tag{20.4}$$

Transition to Renewable Energy Systems, 1st Edition. Edited by Detlef Stolten and Viktor Scherer.
© 2013 Wiley-VCH Verlag GmbH & Co. KGaA. Published 2013 by Wiley-VCH Verlag GmbH & Co. KGaA.

In addition to kilowatt hours (kWh), some other commonly used units of electrical energy are megawatt hours (MWh), gigawatt hours (GWh) and terawatt hours (TWh):

$$1 \text{ MWh} = 1000 \text{ kWh} \quad (10^6 \text{ Wh})$$
$$1 \text{ GWh} = 1000 \text{ MWh} \quad (10^9 \text{ Wh})$$
$$1 \text{ TWh} = 1000 \text{ GWh} \quad (10^{12} \text{ Wh})$$

The power output P (Eq. 20.2) in a power plant is limited to P_{max} (nameplate capacity) at the maximum flow capacity Q_{max}. If the inflow becomes larger than this, water will be lost (spilling) unless there is a reservoir for storage. In most cases, a hydropower plant will be designed with capacity considerably below the maximum inflow, and consequently spill some water. At periods where the flow is much less than the capacity, the power plant may have to run at reduced efficiency or even have to stop and bypass water to avoid damage to the turbines.

The *capacity factor* of a power plant is the ratio of the actual output of a power plant over a period of time (typically 1 year) and its potential output if it had operated at full nameplate capacity the entire time (year). To calculate the capacity factor, we take the total amount of energy the plant produced during a year and divide it by the amount of energy the plant would have produced at full capacity. Typical capacity factors for hydropower plants are in the range 0.3–0.6 [1].

For comparison with other renewable and thermal energy sources, exajoules (EJ) are often used: 1 EJ (10^{18} J) = 277.78 TWh, 1 TWh = 0.0036 EJ. The global hydropower production in 2009 was 3551 TWh or 12.78 EJ [1].

20.1.1
The Hydrological Cycle – Why Hydropower Is Renewable

Hydropower is generated from water moving in the hydrological cycle (water cycle), which is driven by solar energy. Close to 50% of all solar energy reaching the surface of the Earth is used for evaporating water and is converted into latent energy in water vapor. Air currents bring some of the water vapor in over land, where it condenses into clouds and precipitation. Precipitation on land generates runoff as water moves back to the sea under the influence of gravity. This runoff is the basis for hydropower generation and, since the water cycle is driven by solar energy, it will continue as long as the Sun shines. It has been estimated that the total annual amount of water in the cycle equals 505 000 km^3, of which 107 000 km^3 falls on land generating a surface runoff of 28 000 km^3. It is the potential energy in this water that is the theoretical basis for the potential energy that can be utilized as hydropower.

20.1.2
Computing Hydropower Potential

The potential assessment process is different from those used for other renewables (wind, solar, etc.), which usually begin with a theoretical potential that is then reduced by applying constraints to give a technical potential and further reduced to an economic potential [2].

In order to compute the potential for hydropower in an area (catchment, region, country, etc.), the usual procedure is to identify feasible sites with a suitable combination of flow (Q) and head (H) where a hydropower plant can be located. The potential at each site is calculated using Eqs. 20.2 and 20.3, considering both the total annual volume of water and its variability in time. Results are given as potential annual energy generation in GWh per year. Hydropower potential within the area is then assessed by adding up potential production from all possible sites, omitting sites with environmental restrictions, high cost, or social constraints. It has been argued that this procedure has led to underestimations of hydropower potential, both because many of the investigations were done a long time ago when limits for acceptable cost were much lower than today, and because of improved technology that makes previously infeasible projects to become feasible. On the other hand, increasing awareness of social and environmental impacts may render previously feasible projects infeasible.

20.1.3
Hydrology – Variability in Flow

The hydrological regime in most rivers shows considerable variability in time, both in the short range (hourly and daily), seasonally (summer and winter), and from year to year (dry and wet periods). Some examples are given in Figure 20.1. The variable inflow creates variable generation, which could lead to problems in meeting demand. The traditional solution to minimize this problem is to build reservoirs and store water during periods of surplus to be used in periods of low flow. Reservoirs are very useful, not only for hydropower but also for irrigation, flood control, water supply, and transport, but they are also often controversial owing to negative environmental and social impacts, and are often the reason for resistance to hydropower development [3]. In some countries in Europe, the resistance has nearly stopped the construction of new dams, and most hydropower development is done as unregulated ROR power plants, often as small plants that are considered more "environmental friendly." However, in areas where hydropower is going to supply a major share of the power, there is no substitute for reservoirs in order to balance supply and demand.

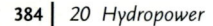

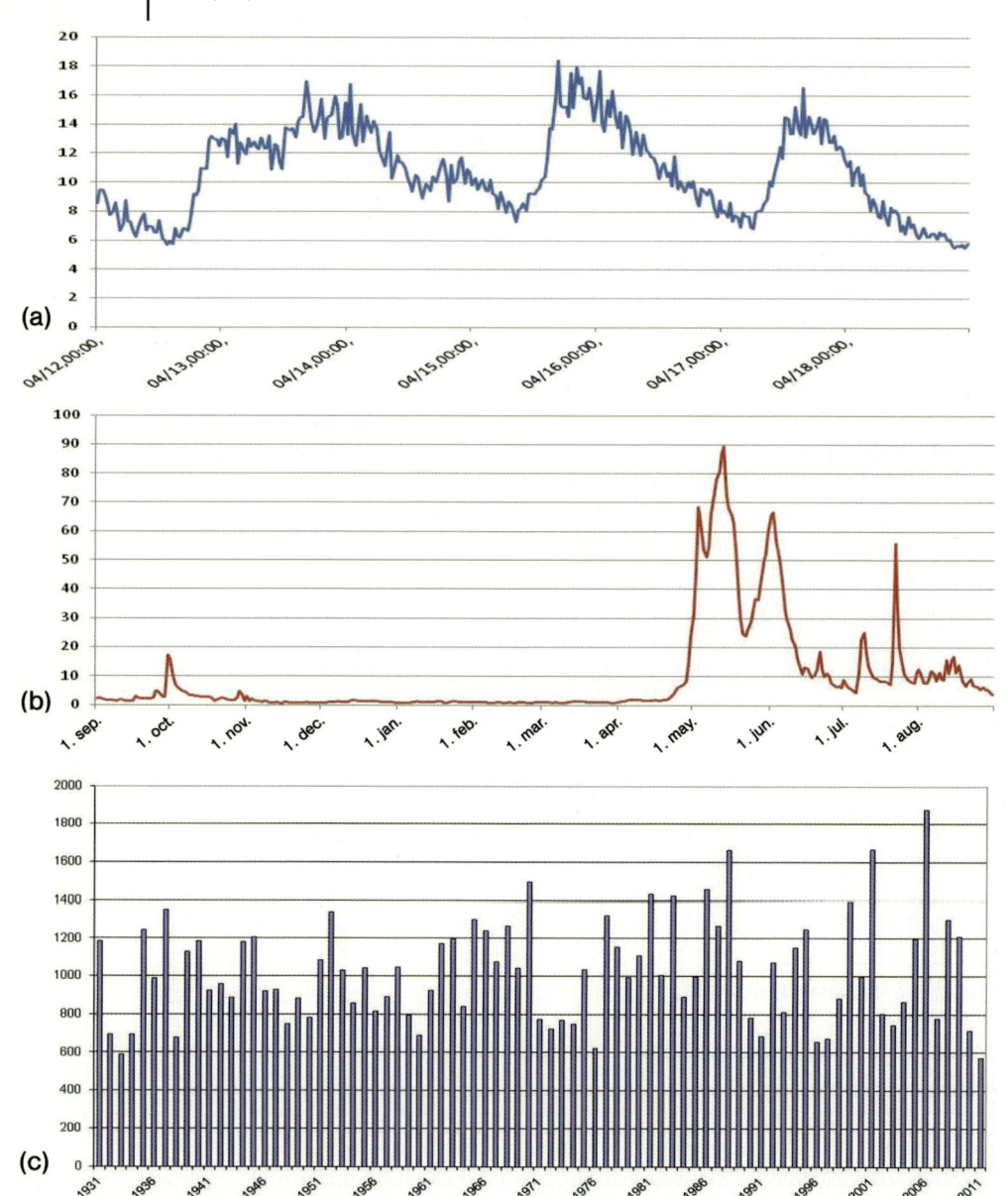

Figure 20.1 Hydrological variability in hydropower system on short, medium, and long range. (a) Flow in a small river during 7 days (Ingdalselva, Norway); (b) annual flow with typical seasonal pattern (Austbygdåi, Norway); (c) annual flow during a 70 year period (Rhine at Rheinfelden).

20.2
Hydropower Resources/Potential Compared with Existing System

20.2.1
Definition of Potential

An interesting discussion and several different definitions of potential can be found in [2]. The global annual potential in exajoules per year is given for five different definitions of potential in Table 20.1. Other sources [1, 4] give a global technical hydropower potential of 52.47 EJ per year or 14 576 TWh per year. This is the figure most commonly quoted, and even if it is not always clearly stated, it probably gives the potential at a cost < US$ 0.20 kWh^{-1}.

Table 20.1 Different estimates of global hydropower potential [2].

Estimation method	Potential (EJ)	Comments
Energy in the water cycle	504 000	40% of solar radiation at Earth's surface
Theoretical potential in runoff	200	Total mass of runoff $\times g \times H$
Technical potential	140–145	Technical potential at known sites
Technical potential at < $ 0.20 kWh^{-1}	57.4	Portion of technical potential with cost low enough to justify a site assessment
Economic potential at < $ 0.08 kWh^{-1}	29.8	Potential at sites with cost that competes with large thermal power plants

20.2.2
Global and Regional Overview

The most recent summary regarding existing hydropower and technical potential can be found in two reports [1, 2]. Both rely on statistics from the International Journal of Hydropower and Dams *World Atlas* [4]. Data in this Atlas are given for each country, based on studies of hydropower potential for identified sites, with limitations as discussed in 20.1.2.

Table 20.2 gives the technical potential as annual generation (TWh per year) for six main world regions. For each region also the recent (2009) generation and percentage of undeveloped hydropower is given. The undeveloped part varies from less than 50% in Europe to 92% in Africa. The largest undeveloped potential is found in Asia (80%, 6167 TWh) and Latin America (74%, 2248 TWh). Even in Europe, 47% or 478 TWh is still undeveloped. The least developed region is Africa with 92% or 1076 TWh that can be developed. The largest undeveloped potential in Africa is found in Congo, Ethiopia, Angola, Cameroon, and Madagascar.

Table 20.2 Regional hydropower technical potential [1].

Region	Technical potential, annual generation (TWh per year)	Generation in 2009 (TWh per year)	Undeveloped potential (%)
North America	1659	628	61
Latin America	2856	732	74
Europe	1021	542	47
Africa	1174	98	92
Asia	7681	1514	80
Australasia/Oceania	185	37	80
World	14576	3551	75

Table 20.3 Hydropower potential for 9 countries in Europe + Turkey [4].

Country[a]	Technically feasible potential (TWh per year)	Economically feasible potential (TWh per year)	Generation in 2008 (TWh per year)
Russia	1670	852	180
Norway	n/a	206	123
Turkey	216	140	48
France	n/a	98	69
Sweden	130	90	68
Austria	56	53	35
Italy	60	50	46
Iceland	64	40	12
Spain	61	37	23
Switzerland	41	n/s	38
Europe total	2885	1772	771

a The countries listed in this table differ from those included in Europe in Table 20.2.

Table 20.3 gives more detailed data for Europe; here the "economic feasible potential" is also included. Data are given for the 10 countries with the highest economically feasible potential. The largest undeveloped potential in Europe is found in Russia (672 TWh), Turkey (92 TWh) and Norway (83 TWh). However, many other countries in Europe also have economically feasible undeveloped hydropower resources, adding up to more than 150 TWh [4].

20.2.3
Barriers – Limiting Factors

The technically feasible potential includes many projects that are not yet economically feasible. This may change in the future, either because of increasing energy price or because of technological innovations that may bring the cost down. Both could make such projects economically feasible in the future.

However, even economically feasible projects may not be developed owing to other barriers and limiting factors. Most important are probably environmental and social concerns. As an example, Norway can be considered. Of the unused potential of 83 TWh, about 45 TWh has already been protected as national parks, nature conservation area, and so on, and only 38 TWh are actually available for new hydropower development. Similar restrictions are found in many countries, especially in Europe and North America.

If the projects require construction of dams and relocation of people, the public resistance usually increases and may lead to delays and cost increases, and may even stop the project completely. Omitting the reservoir may be an option, but often a reservoir will also have a multi-purpose function (e.g., flood protection, water supply, irrigation), so the decision is difficult.

Finally, climate change may lead to reduced hydropower potential in some regions, and this may limit the feasibility of new projects.

20.2.4
Climate-Change Impacts

The resource potential described in Table 20.1, Table 20.2, and Table 20.3 is based on historical hydrological conditions and does not include the effects of climate change. With a changing climate, the hydropower resource potential could also change due to:

- changes in river flow volume
- changes in river flow variability/seasonality
- changes in extreme flow events (floods, droughts)
- changes in sediment loads which could affect reservoir storage capacity.

Many studies were reviewed and summarized in [1]. Most of the studies were focused on the effect of change in river flow volume, and few have studied and quantified the effects of the other three changes listed.

In a recent study [5], the regional and global changes in hydropower generation for the existing hydropower system were investigated, based on a global assessment of changes in river flow by 2050. A summary of the main findings is given in Table 20.4. The results show that both increases and decreases can occur in regions with increasing or decreasing precipitation, but that the global effect will be small, and probably slightly positive. Regions with decreasing water resources and decreasing hydropower generation can be found around the Mediterranean,

Table 20.4 Changes in power generation due to climate change up to 2050 [1, 5].

Region	Energy generation capacity: system in 2005 (TWh per year)	Computed change by 2050 (TWh per year)
Africa	90	0.0
Asia	996	2.7
Europe	517	−0.8
North America	655	0.3
South America	661	0.3
Oceania	40	0.0
Total	2931	2.5

in southern Africa, in Australia, and in central and western America. Regions with increasing water resources and increasing hydropower generation can be found in northern Europe, in most of Russia, in East Asia, in East Africa, in Canada, and in parts of South America.

20.3
Technological Design

Hydropower plants can broadly be classified into three main types: ROR, storage hydropower and pumped storage plants. A fourth type, called in-stream or hydrokinetic, is under development but still not widely used, and will not be described further here.

20.3.1
Run-of-River Hydropower

ROR hydropower plants mainly generate electricity from the available flow in a river. Some short-term storage may be included, allowing for some adaptation to the consumption, but the generation profile will mainly be determined by the inflow profile. Most small hydropower plants are of ROR type, with no storage. The power station may be located in the intake dam, or further away, depending on topography. Transport of water from the intake to the power station (headrace) may be by a canal, a pipe, or a tunnel for large projects.

20.3.2
Storage Hydropower

Hydropower projects with a reservoir can store water for later use, typically by saving water during the high-flow season (spring, rainy season) and releasing water during the low-flow season (winter, dry season). The reservoir gives higher flexibility and allows the hydropower plant to adapt better to the demand profile, both in the short term (hours, days) and seasonally. Construction of a reservoir requires a dam and inundation of land area upstream of the dam. An alternative, widely used in Scandinavia, is to use an existing natural lake and create the reservoir by lowering the lake (lake tapping). Location of the power plant can, as for ROR plants, be either integrated with the dam, or further away, connected to the reservoir by a headrace (canal, pipe, or tunnel).

20.3.3
Pumped Storage Hydropower

A pumped storage hydropower plant consists of a reversible power plant and two reservoirs, connected by a pipe or a tunnel. The main purpose is to store energy by pumping water up into the upper reservoir during low-demand periods and generate (peaking) power by releasing the water back to the turbine during high-demand periods. Although some energy is lost in the process, typically 15–25%, the plant can still contribute very significantly by providing benefits to the grid in the form of peaking power, frequency stabilization, and load balancing. Pumped storage hydropower plants are also increasingly important for balancing other renewables, in particular wind and solar power plants. Pumped storage hydropower is presented in greater detail in Chapter 28.

20.4
Cutting Edge Technology

Although hydropower is a mature technology, there is still a need for and room for technology improvements. Since hydropower is so closely linked to water management, new requirements and policy changes can have a major impact on hydropower projects, and lead to changes in, for example, flow restrictions and reservoir operation rules and require new methods for operation and optimization. One such example is the European Water Framework Directive [6], which will have a profound effect on hydropower. With its long lifetime, typically up to 80 years or more [1], it will usually be necessary to upgrade machinery and control equipment several times during its lifetime, giving the possibility of bringing in new technology and increasing efficiency.

Cutting edge technology not only involves better "hardware" such as turbines and generators, it can also come in the form of improved planning and operation tools, and of course in improved dams, tunnels, and other civil structures. Hydropower generation can be increased by optimizing different aspects of plant operation, both for individual units and by better coordination, for example in a cascade with many interlinked reservoirs and power plants. Improved hydrological forecasts combined with advanced optimization models are likely to improve operation and at the same time consider new environmental constraints and multi-purpose use of the water. Improved or new optimization methods and software will also need to be developed for better coordination with other renewables, where hydropower could have a key role in the balancing and grid stabilization as the percentage of highly variable generation in wind and solar plants is increasing.

A few interesting areas where new technology is under development can be mentioned. This selection is mostly based on the chapter "Prospects for technology improvement and innovation" in [1] and on the *Hydropower Development* book series [7].

20.4.1
Extending the Operational Regime for Turbines

Hydropower turbines, running at fixed speed, have traditionally been optimized for operation at a "best point" defined by speed, head, and discharge. Operating outside this best point (different head and/or discharge) will usually lead to a considerable reduction in efficiency, and less energy output. Large hydropower turbines are now close to the theoretical limit for efficiency in a small region near the "best point," but may be improved for more flexible operation outside this region. This is illustrated by Figure 20.2, which shows typical efficiency curves for different types of hydro turbines [7, vol. 12]. The application of variable-speed technology offers advantages in the form of greater flexibility in handling such situations, with higher efficiency and also less damage from silt erosion in turbines as a result.

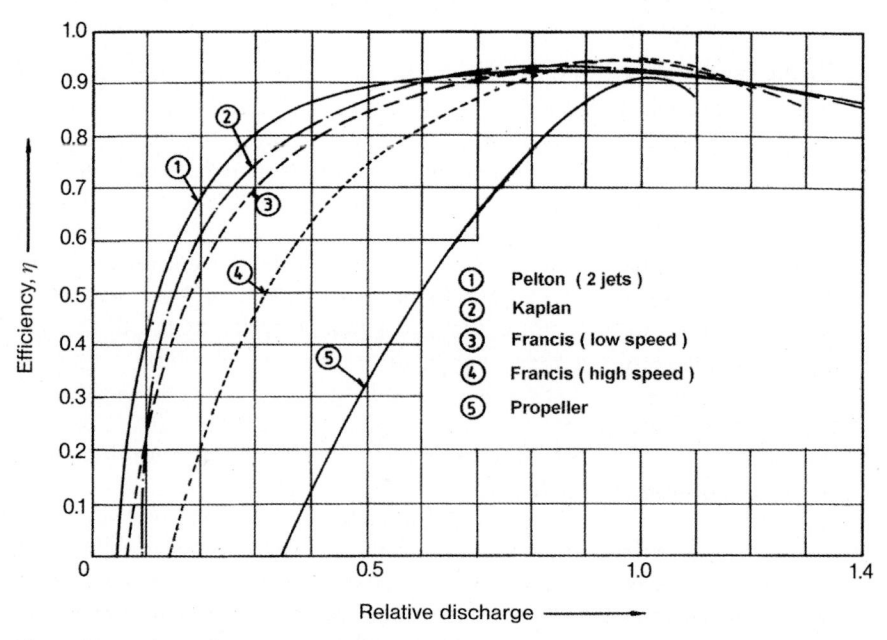

Figure 20.2 Turbine efficiency versus relative discharge for five different turbine types [7, vol. 12].

20.4.2
Utilizing Low or Very Low Head

Most existing hydropower projects were developed in times of lower energy prices, where projects with low (< 15 m) and very low (< 5 m) head were not feasible. Most such low-head projects were also excluded from hydropower potential mapping in many countries, for the same reason. Therefore, existing data on hydropower potential for low-head sites are probably not complete. As an example [1], in Canada a market potential of 5000 MW of low-head hydropower plants has recently been identified. In many countries, there is a large potential for producing electricity from irrigation dams, provided that low-head turbines can be developed at a low price. With increasing energy demand and price, and improved technology for such low-head projects, one can see growing interest in technology development. At extremely low head, hydrokinetic technology may open up new possibilities, since it can extract energy directly from moving water, in the same way as wind turbines extract energy from moving air.

20.4.3
Fish-Friendly Power Plants

Dams and hydropower plants may create obstacles for fish migration and reduce access to important rearing and spawning sites, and reduce or eradicate fish populations. Safe passage for fish is therefore a major concern in environmental analysis of hydropower projects. There are two main categories, helping both upstream and downstream migration:

1. *Downstream migration:* The main objective is to prevent fish from entering into intakes and turbines. New technology for such prevention systems is under development based on screens, acoustic cannons, light (strobe and laser light), and flow direction [8]. If it is not possible to prevent fish from entering the turbines, new type of turbines ("fish-friendly turbines") may minimize the risk of injury or death to passing fish. Some turbine manufacturers claim that the turbines may have good efficiency and still allow 90–100% of fish to pass safely.
2. *Upstream migration:* Fish ladders can be used for upstream migration, combined with flow release and adaptation in an operational regime to minimize problems when migration actually occurs. Other common devices include bypass channels and fish elevators.

20.4.4
Tunneling and Underground Power Plants

Tunnels and rock caverns are important construction elements in most large-scale hydropower projects: headrace/tailrace tunnels, access tunnels, powerhouses, surge shafts, power cables, and ventilation. In Norway, nearly all large-scale hydropower plants have been built underground since 1960 [7, vol. 14]. An example showing the typical layout of an underground powerhouse complex is given in Figure 20.3.

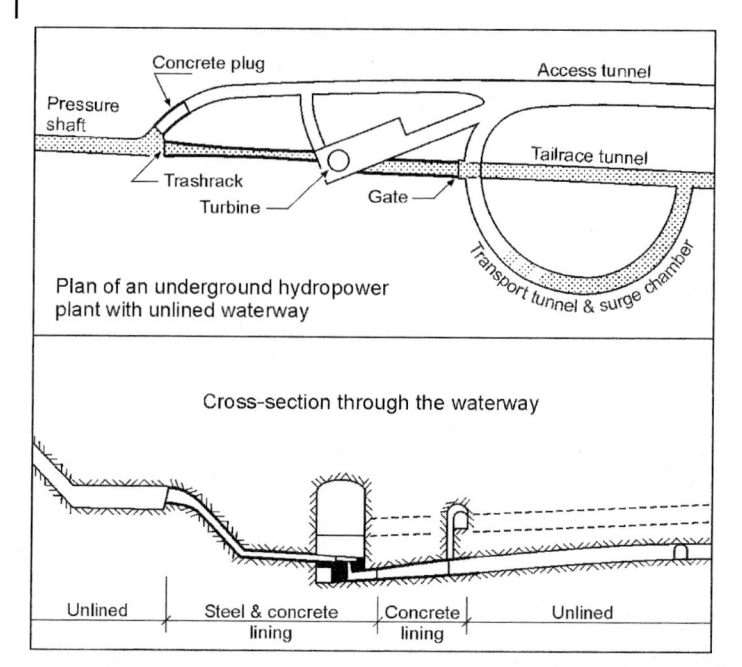

Figure 20.3 A typical underground powerhouse complex with tunnel system [7, vol. 14].

Tunneling technology has evolved from traditional drill and blast technology to full-face tunnel boring machines (TBMs), increasing the speed of tunneling and lowering costs. Still, there is a need for technology development, for example, for micro-tunnels that can be used to replace pipes and penstocks for small hydropower plants (< 1 m diameter), avoiding completely overground work and disturbances to Nature. Such technology is under development in Norway, utilizing technology from oil drilling. Another potentially important type of micro-tunneling could be used for transmission power cables, avoiding the construction of power lines through environmentally sensitive areas.

In 2002, there were 500 underground hydropower plants around the world, about 40% of them in Norway [7, vol. 14]. The use of underground stations combined with headrace and tailrace tunnels gives high flexibility in the location of power plants and construction of complex but efficient systems that are superior to traditional systems with surface powerhouses. Figure 20.4 show how the layout of high-head hydropower plants has changed gradually, mainly due to advances in tunneling technology. An example of one such system in Norway, the Ulla-Førre complex, is shown in Figure 20.5. Here, tunnels and underground power stations are combined with reservoirs and intakes in order to capture and store water from a wide area with many small watersheds that would have been very difficult to develop individually. Most of the water is collected 600 m a.s.l. (above sea level) and pumped up to a very large reservoir (Blåsjø) at 1000 m a.s.l., where it was possible to build dams and create a reservoir. Blåsjø is the largest reservoir in Norway and can store nearly 8 TWh of energy.

20.4 Cutting Edge Technology

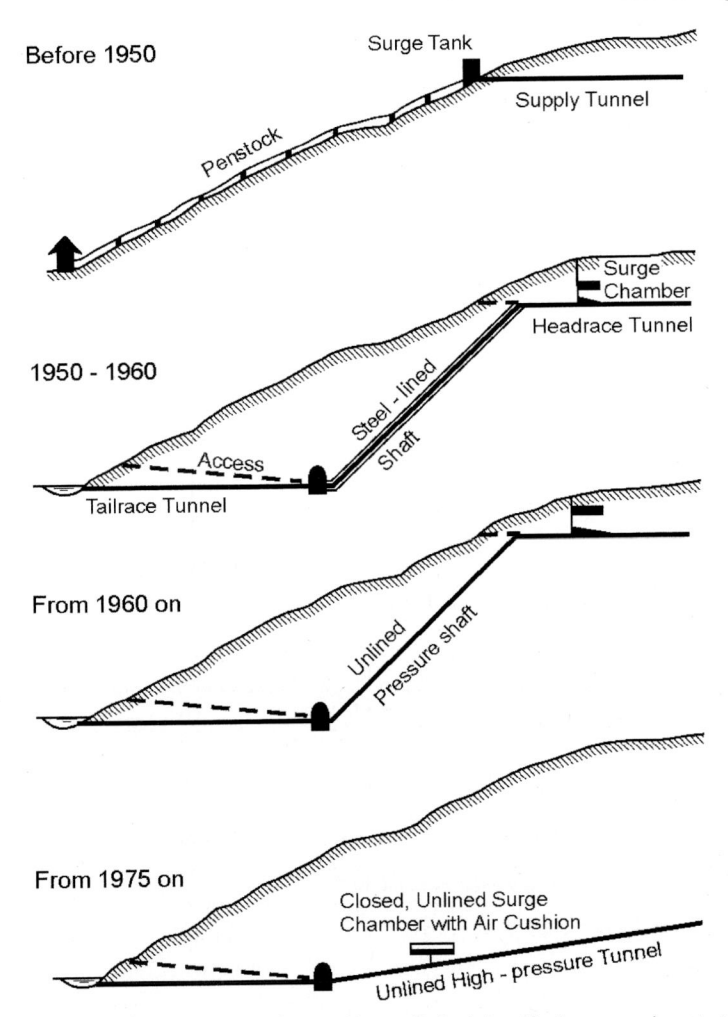

Before 1950

Surge Tank
Supply Tunnel
Penstock

1950 - 1960

Surge Chamber
Headrace Tunnel
Steel - lined Shaft
Access
Tailrace Tunnel

From 1960 on

Unlined Pressure shaft

From 1975 on

Closed, Unlined Surge Chamber with Air Cushion
Unlined High - pressure Tunnel

Figure 20.4 Development of general layout for high-head hydropower plants in Norway [7, vol. 14].

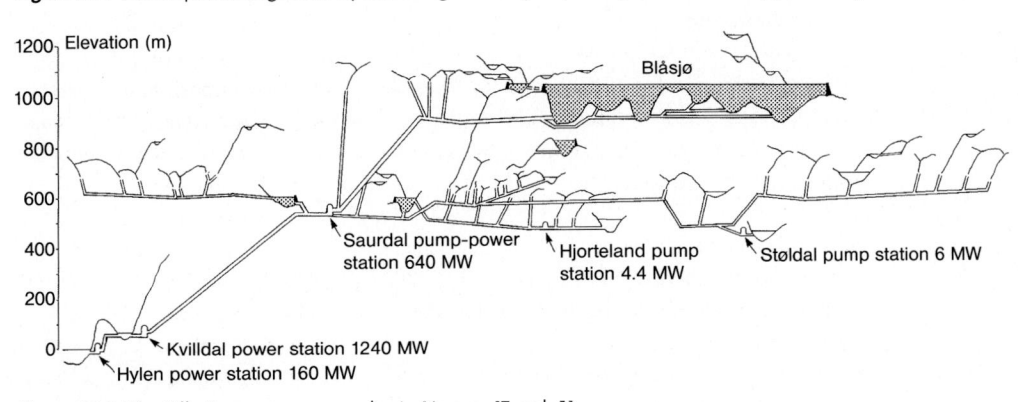

Elevation (m)

Blåsjø
Saurdal pump-power station 640 MW
Hjorteland pump station 4.4 MW
Støldal pump station 6 MW
Kvilldal power station 1240 MW
Hylen power station 160 MW

Figure 20.5 The Ulla-Førre power complex in Norway [7, vol. 1].

20.5
Future Outlook

Hydropower is already very competitive, and the only renewable technology that can compete with thermal power plants in terms of cost. In the future, there are two trends that could lead to either increasing or decreasing cost compared with other technologies: improvements in technology will probably bring costs down further, but increasing environmental concerns and restrictions could lead to increasing costs. Also, since the best project sites have already been developed, the cost of developing remaining sites will gradually increase, as illustrated by Figure 20.6.

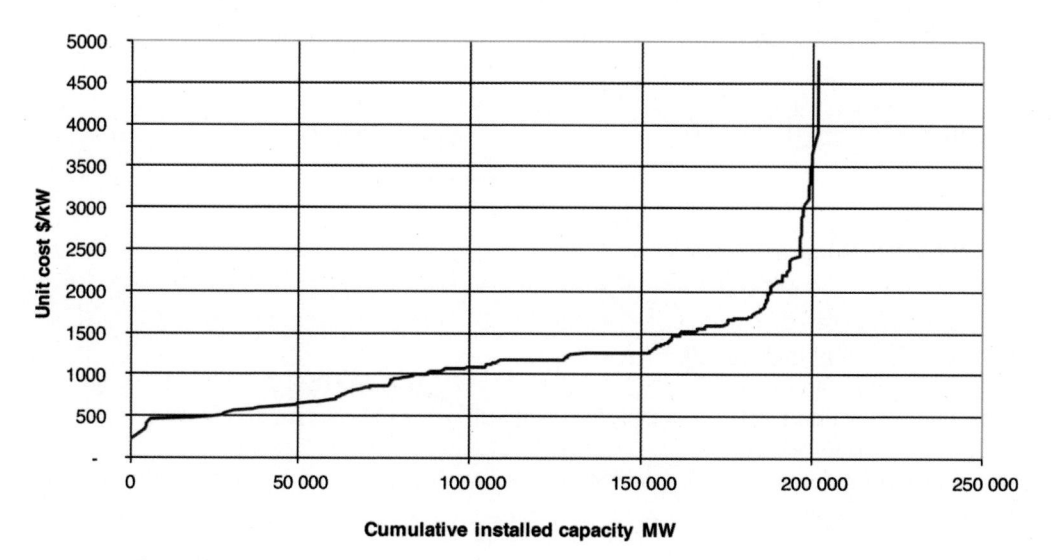

Figure 20.6 Unit cost distribution (supply–cost curve, 2005 US$) for 250 projects in Asia and Europe [9].

20.5.1
Cost Performance

Cost performance for hydropower can best be given as levelized cost of energy (LCOE). The LCOE usually includes all costs for production of energy, but not transmission and distribution costs. Important cost components for LCOE of energy projects in general include:

- planning costs
- construction (investment) costs
- operation and maintenance (O&M) costs
- fuel costs
- refurbishing costs
- decommissioning costs.

LCOE is defined as the constant unit cost (per kilowatt hour) of a payment stream that has the same present value as the total cost of building and operating a generating plant over its lifetime. It can be computed as the total cost of the project over its lifetime divided by the total energy output over the same period.

For hydropower, construction costs are by far the most important component in the cost calculation. It is common to combine planning and construction costs as investment costs, and to combine operation, maintenance, and refurbishing as O&M costs. Decommissioning costs are usually not considered, since it is rarely seen that hydropower plants are taken out of operation. In most cases, operating hydropower stations will be refurbished and kept in operation indefinitely. There are no fuel costs, so it is possible to perform an LCOE computation for hydropower with the following simplified equation:

$$\text{LCOE} = \frac{\sum_{t=1}^{n} \frac{I_t + \text{OEM}_t}{(1+r)^t}}{\sum_{t=1}^{n} \frac{E_t}{(1+r)^t}} \tag{20.5}$$

where n = lifetime of project in years, t = year number, I_t = investment costs during year t, OEM_t = O&M costs during year t, E_t = electrical energy generation during year t, which can be computed as installed capacity (P_{max}) × capacity factor (CF), and r = discount rate.

Lifetime (n) and discount rate (r) are two very important parameters that have a large impact on the LCOE for a project. A typical lifetime for large hydropower plants is 40–80 years [1], and for small plants it could be lower, 40 years or less. Hydropower plants are usually refurbished at regular intervals, and actually not taken out of operation at the end of their lifetime. There are many examples of hydropower plants that have been in operation for 100 years or more, with regular upgrading of electrical and mechanical systems, but no major upgrade of the more expensive civil structures (dams, tunnels, etc.).

To illustrate the combined influence of lifetime and discount rate, an example is given in Table 20.5. It shows that long lifetime, where hydropower is performing better than other renewables, is of little importance if the discount rate is high.

Table 20.5 LCOE for a hydropower plant: example with varying lifetime and discount rate (investment cost $ 1000 kWh^{-1}; O&M costs 2.5%; capacity factor 50%).

Lifetime (years)	LCOE ($ kWh^{-1}) for discount rate		
	3%	7%	10%
20	0.021	0.027	0.033
40	0.016	0.023	0.029
60	0.014	0.022	0.029
80	0.013	0.022	0.029

It also illustrates that the combination of low discount rate and long lifetime lowers the LCOE to nearly one-third compared with high discount rate and short lifetime.

20.5.2
Future Energy Cost from Hydropower

Many studies have tried to make estimates of future energy costs from hydropower. Most of these studies were reviewed and summarized in [1]. Since hydropower projects are very site specific, the investment costs could vary considerably from site to site, from < $ 500 to up to more than $ 5000 kW^{-1}. Most projects, however, fall in a typical range from $ 1000 to 3000 kW^{-1}. The capacity factor typically varies from 0.3 to 0.6, resulting in the typical LCOE for future hydropower given in Table 20.6.

Table 20.6 Typical cost range for future hydropower according to [1]: LCOE range for some combinations of capacity factor and investment costs.

Discount rate (%)	Capacity factor	Lifetime (years)	Investment costs ($ kW^{-1})	LCOE range ($ kWh^{-1})
3	0.3–0.6	40–80	1000–3000	0.011–0.078
7	0.3–0.6	40–80	1000–3000	0.018–0.11
10	0.3–0.6	40–80	1000–3000	0.024–0.15
Typical (7)	0.45	80	1500	0.036

An example of unit cost distribution (supply–cost curve) is shown in Figure 20.6. Here, the cumulative volume of hydropower that can be developed is plotted as a function of investment costs. Data are from a study that included 250 projects with over 200 000 MW world-wide [9]. The figure show that about 80% of these projects (160 000 MW) could be developed at a cost of $ 1500 kW^{-1}.

A typical project, based on discount rate of 7%, a capacity factor of 45%, a lifetime of 80 years and investment costs of $ 1500 kW^{-1}, results in an LCOE of $ 0.036 kWh^{-1} for future hydropower development.

20.5.3
Carbon Mitigation Potential

Hydropower, like other renewable energy sources, offers considerable potential for reduction of carbon emissions [greenhouse gas (GHG) emissions] from energy generation by replacing fossil energy generation. Coal-fired power plants typically emit 1000 g of CO_2 for each kWh produced, oil-fired plants typically 800 g kWh^{-1}, and natural gas-fired plants around 500 g kWh^{-1}. Hydropower and most other renewables have emissions very close to zero, compared with fossil fuel-based technologies.

There have been some observations of high GHG emissions from some tropical reservoirs, raising doubts about the real GHG emission for hydropower [1, Ch. 5.6], but it is still not clear if this represents a net emission or only a GHG emission that would occur anyway, for example, from a wetland. For most hydropower plants, it is safe to assume that they will contribute to reducing GHG emissions by between 500 and 1000 g of CO_2 per kWh produced, probably closest to the high estimate, since it is believed that mainly coal-fired plants that will be replaced. This means that 1 TWh of hydropower will reduce GHG emissions by 10^6 t of CO_2.

Hydropower therefore offers very high potential for carbon emission mitigation in both the near and far term. Already today, 16% of the world electricity production is supplied by hydropower, reducing global carbon emissions very significantly. In some of the highest estimates for further deployment, up to 5000 TWh of new hydropower could be put into operation before 2050. If this power is electricity from coal-fired plants, the total reduction of GHG emissions could amount to 5000 million tons of CO_2 each year.

20.5.4
Future Deployment

On a global basis, the potential for further hydropower development is very significant. The resource potential is large, in the order of 5000 TWh, and in the near and medium range resource depletion is not likely to restrict hydropower development. Hydropower is also very cost competitive, with a typical cost of $ 0.03–0.05 kWh^{-1}, although with some increasing cost in the long run. The combination of high potential and low cost could make it possible to increase hydropower generation from the recent level of ~3500 TWh per year (2009) by a factor of between 2 and 3. Some of the highest estimates from the industry reach 8700 TWh by 2050 [4], an increase of more than 5000 TWh compared to 2009. A possible scenario for future deployment is illustrated in Figure 20.7.

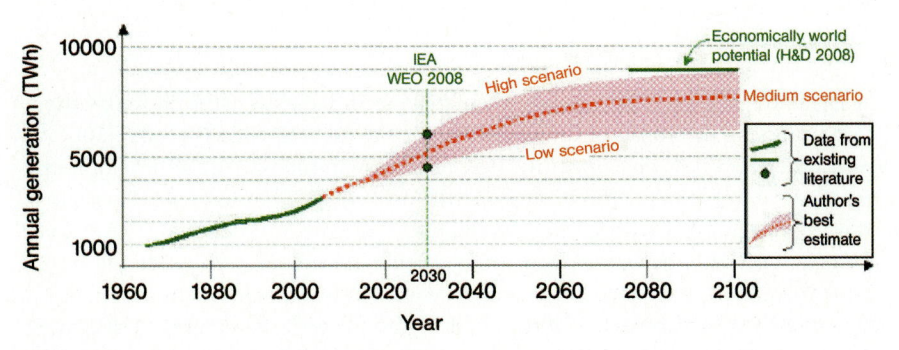

Figure 20.7 A scenario for future hydropower deployment.
Annual hydropower generation (historical data and forecast) in the world (Source EDF).

20.6
Systems Analysis

20.6.1
Integration into Broader Energy Systems

Hydropower can be successfully integrated into the power grid or operate as stand-alone plants. ROR hydropower depends on river flow, and has limited ability to follow demand, although some plants have a limited reservoir (pond) that can be used to adjust generation to fit daily load profiles. Storage hydro offers much greater flexibility and can offer significant flexibility for system operation. Hydropower systems with large reservoirs have capacity credit comparable to that of large thermal plants, and can also vary operation very rapidly, due to the fast response of water turbines. Hydropower reservoirs can store large quantities of energy and be used to balance and support other renewables.

20.6.2
Power System Services

In addition to providing energy and capacity to meet demand, hydropower offers several other advantages, such as the ability to start generation very quickly and with low start-up costs, support frequency regulation, perform a "black start," and also offer a wide range of operation where power can be generated with good efficiency. Overall, with its good capability for load following and balancing, peaking capacity, and services for supporting power quality, hydropower can play a significant role in the future electricity system. Pumped storage hydropower has a unique capability to store large quantities of energy, and thereby to support other renewables such as wind and solar power plants.

20.7
Sustainability Issues

"Sustainable development is development that meets the needs of the present without compromising the ability of future generations to meet their own needs" [10].

In order to assess sustainability, it is necessary to include social development, economic development, and environmental protection. The sustainability issue is very important for all hydropower development, and the International Hydropower Association (IHA) has developed three "sustainability tools" to guide planning, implementation, and operation of hydropower: IHA Sustainability Guidelines, IHA Sustainability Assessment Protocol, and Hydropower Sustainability Assessment Forum. The tools can be downloaded from the Internet [11].

20.7.1
Environmental Impacts

Like other types of energy generation technologies, hydropower may have negative environmental and social impacts. Hydropower projects often have impacts on the flow regime in rivers and water levels regimes in reservoirs, and may therefore have negative consequences for ecosystems and threaten biodiversity. ROR projects usually change the flow regime only marginally, whereas storage hydro projects typically lead to larger changes in the annual flow regime, for example, by decreasing summer flows and increasing winter flows. Regarding social impacts, the most severe impact occurs where relocation of people in reservoir areas is needed. Since hydropower projects are always designed and optimized for specific sites with their geographical and social characteristics, it is difficult to make general statements about impacts. However, even if impacts cannot be avoided, it is usually possible to mitigate some of the negative effects. In the planning stage, many negative impacts can be reduced or completely avoided by careful selection of good alternatives. It is very important to include environmental and social expertise in the planning process, aiming at producing "renewable energy respecting Nature" [12]. An overview of social and environmental effects of hydropower development can be found in [7, vol. 3].

20.7.2
Lifecycle Assessment

Hydropower projects have very long lifetimes of up to 80 years or more. In order to evaluate total costs, benefits, environmental effects, energy efficiency, and impact on global GHG emissions, it is important to consider the whole lifecycle for a project: investigation, planning, construction, operation, maintenance, refurbishing, and eventually decommissioning. As described in 20.5.1, the LCOE method is based on a lifecycle assessment of economic performance. In addition, two other important performance parameters can be mentioned: GHG emissions and energy payback ratio (EPR).

20.7.3
Greenhouse Gas Emissions

The majority of lifecycle GHG emission estimates for hydropower cluster between about 4 and 14 g CO_2 eq. kWh^{-1}, but under certain scenarios there is the potential for much larger quantities of GHG emissions. This can be compared with 1000 g CO_2 eq. kWh^{-1} for coal power plants, 800 for oil power plants, and 500 for gas power plants. Since 2008, UNESCO and IHA have been hosting an international research project, with the aim of establishing a robust methodology to estimate accurately the net effect on GHG emissions caused by the creation of a reservoir, and to identify gaps in knowledge. The project published GHG Measurement Guidelines for Freshwater Reservoirs in 2010 [13] to allow standardized measurements and calculations worldwide, and aims at delivering a database of results and characteristics

20.7.4
Energy Payback Ratio

The EPR is defined as the ratio of total energy produced during a system's normal lifespan to the energy required to build, maintain, and fuel it. A high ratio indicates good environmental performance. If, for example, a system has an EPR between 1 and 1.5, it consumes nearly as much energy as it generates, so it should never be developed. Hydropower has the highest EPR of all electricity generation technologies, with values of 170–267 for ROR plants and 205–280 for storage plants. This can be compared with EPRs of 1.6–7 for fossil fuels, 18–34 for large wind turbines, and 14–16 for nuclear power plants [14]. Similar results were found in a recent study in Norway [15]; the average EPR for hydropower was ~200.

20.8
Conclusion

Hydropower is a renewable energy source where power is derived from the energy of water moving from higher to lower elevations. It is a proven, mature, predictable, and price-competitive technology. Hydropower has among the best conversion efficiencies of all known energy sources (about 90% efficiency, water to wire). It requires relatively high initial investment, but has a long lifespan with very low O&M costs. The existing hydropower system has an annual generation capacity of 3500 TWh per year, and contributes 16% of the annual electricity generation worldwide. There is still a large potential for further development, as the total technical potential has been estimated to be nearly 15 000 TWh. Of this, 8000–9000 TWh can be classified as economical potential. In Europe, close to 50% of technical potential has been developed, in Asia 25%, and in Africa only 8%. Significant potential also exists to use existing infrastructure that currently lacks generating units (e.g., existing barrages, weirs, dams, canal fall structures, and water supply schemes) by adding new hydropower facilities. Only 25% of the existing 45 000 large dams are used for hydropower, and the other 75% are used exclusively for other purposes (e.g., irrigation, flood control, navigation and urban water supply schemes). Hydropower offers significant potential for carbon emissions reductions, since GHG emissions are generally very low, typically less than 1% of that from coal power plants. Hydropower is cost competitive, with LCOE typically in the range $ 0.03–0.05 kWh^{-1}, which is comparable to the cost of energy from thermal power plants. Hydropower has very good efficiency, typically 90% from water to wire, and an EPR of 200–300, the highest of all types of renewables. Hydropower can provide both energy and water management services and also help to support other variable renewable energy sources such as wind and solar, by providing storage and load balancing services. Environmental and social impacts will need to be carefully managed.

References

1 IPPC (2012) Renewable Energy Sources and Climate Change Mitigation: Special Report of the Intergovernmental Panel on Climate Change, Cambridge University Press, Cambridge.

2 GEA (2012) *Global Energy Assessment – Towards a Sustainable Future*, Cambridge University Press, Cambridge, and the International Institute for Applied Systems Analysis, Laxenburg.

3 WCD (2000) *Dams and Development: a New Framework for Decision-Making*, World Commission on Dams.

4 IJHD (2010) *World Atlas and Industry Guide*, International Journal of Hydropower and Dams, Wallington, Surrey.

5 Hamududu, B. and Killingtveit, Å. (2012) Assessing of climate change impacts on global hydropower. *Energies*, 5 (2), 305–322.

6 European Commission (2000) *Directive 2000/60/EC of the European Parliament and of the Council of 23 October 2000 Establishing a Framework for Community Action in the Field of Water Policy (Water Framework Directive)*, European Commission, Brussels.

7 NTNU (1993–2005) *Hydropower Development*: book series consisting of 17 volumes published between 1993 and 2005, Norwegian University of Science and Technology (NTNU), Department of Hydraulic and Environmental Engineering, Trondheim; the following four volumes were used in this chapter: vol. 1, *Hydropower Development in Norway*; vol. 3, *Environmental Effects*; vol. 12, *Mechanical Equipment*; vol. 14, *Underground Powerhouses and High Pressure Tunnels*.

8 Fjeldstad, H. P. (2012) Atlantic salmon migration past barriers, PhD thesis, Norwegian University of Science and Technology (NTNU), Department of Hydraulic and Environmental Engineering, Trondheim.

9 Lako, P., Eder, H., de Noord, M., and Reisinger, H. (2003). *Hydropower Development with a Focus on Asia and Western Europe: Overview in the Framework of VLEEM 2*, ECN-C-03027, Energy Research Centre of The Netherlands, Petten.

10 WCED (1987) *Our Common Future. Report from the United Nations World Commission on Environment and Development (WCED)*, Oxford University Press, Oxford.

11 International Hydropower Association (2004) *Sustainability Guidelines*, http://www.hydropower.org/sustainable_hydropower/sustainability_guidelines.html (last accessed 6 February 2013).

12 CEDREN (2012) *Centre for Environmental Design of Renewable Energy Home Page*, www.cedren.no (last accessed 6 February 2013).

13 UNESCO/IHA (2010) *UNESCO/IHA Greenhouse Gas (GHG) Research Project. From Measurement to Modelling: Interfaces Between the Field Data and the Predictive Modelling Tools, Workshop 3–4 August 2010, Oak Ridge, TN, USA*, http://unesdoc.unesco.org/images/0019/001909/190918e.pdf (last accessed 6 February 2013).

14 Gagnon, L (2008) Civilisation and energy payback. *Energy Policy*, 36, 3317–3322.

15 Raadal, H. L., Modahl, I. S., and Bakken, T. H. (2012) *Energy Indicators for Electricity Production – Comparing Technologies and the Nature of the Indicators Energy Payback Ratio (EPR), Net Energy Ratio (NER) and Cumulative Energy Demand (CED)*, Ostfold Research, Kråkerøy.

21
The Future Role of Fossil Power Plants – Design and Implementation

Erland Christensen and Franz Bauer

21.1
Introduction

The rapid growth of renewables-based power generation technologies raises questions about the future role of fossil power plants. In Europe and worldwide, fossil power plants will continue to play a major role in the coming decades, ensuring a stable and affordable supply of electricity.

In countries that have opted for a more rapid revolution of the electricity supply system with more renewables, fossil power plants still have an important role to play regarding load following, backup at times of calm wind or without solar radiation, frequency and voltage control, co-combustion of biomass, and heat delivery for district heating systems.

These are all technically challenging tasks, but the paramount challenge is economic because the surplus of renewables-based power has changed the market conditions significantly.

Although the fossil fleet is aging and its services are needed, in many countries it is no longer possible to create a business case for new fossil-fired power plants and even existing plants cannot generate sufficient revenues. This also endangers security of supply and consequently the change of the entire electricity supply system towards more renewables.

21.2
Political Targets/Regulatory Framework

The 20/20/20 targets set by the European Union, namely

- 20% cut in emissions of greenhouse gases by 2020, compared with 1990 levels
- 20% increase in the share of renewables in the energy mix
- 20% cut in energy consumption

Transition to Renewable Energy Systems, 1st Edition. Edited by Detlef Stolten and Viktor Scherer.
© 2013 Wiley-VCH Verlag GmbH & Co. KGaA. Published 2013 by Wiley-VCH Verlag GmbH & Co. KGaA.

21 The Future Role of Fossil Power Plants – Design and Implementation

Expected growth in electricity generation in billion (10^9) kWh in the EU

+32 %

Legend: Hydro power, wind, biomass, solar; Nuclear; Coal; Gas; Oil

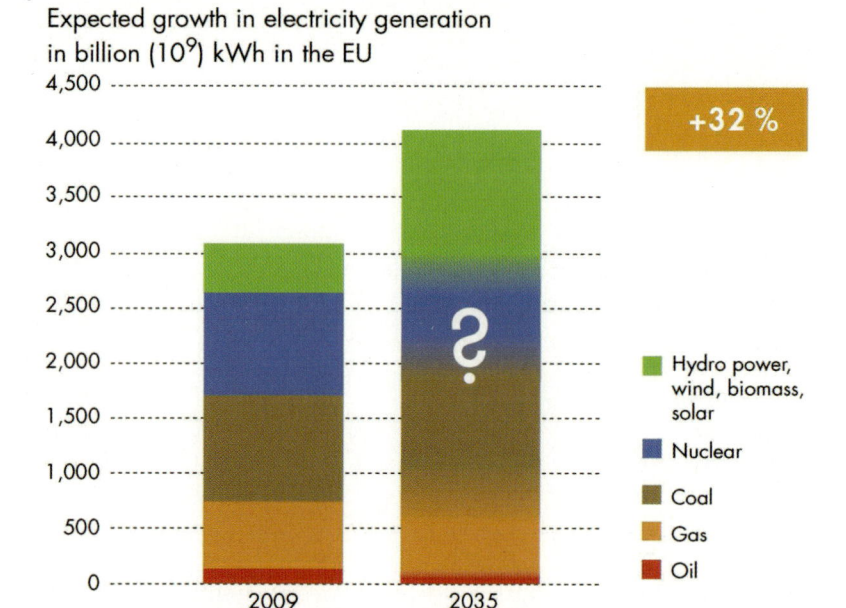

Figure 21.1 Expected development of the European electricity production and production portfolio 2008 to 2030. Source: VGB.

The share of renewables in gross final energy consumption in EU-27

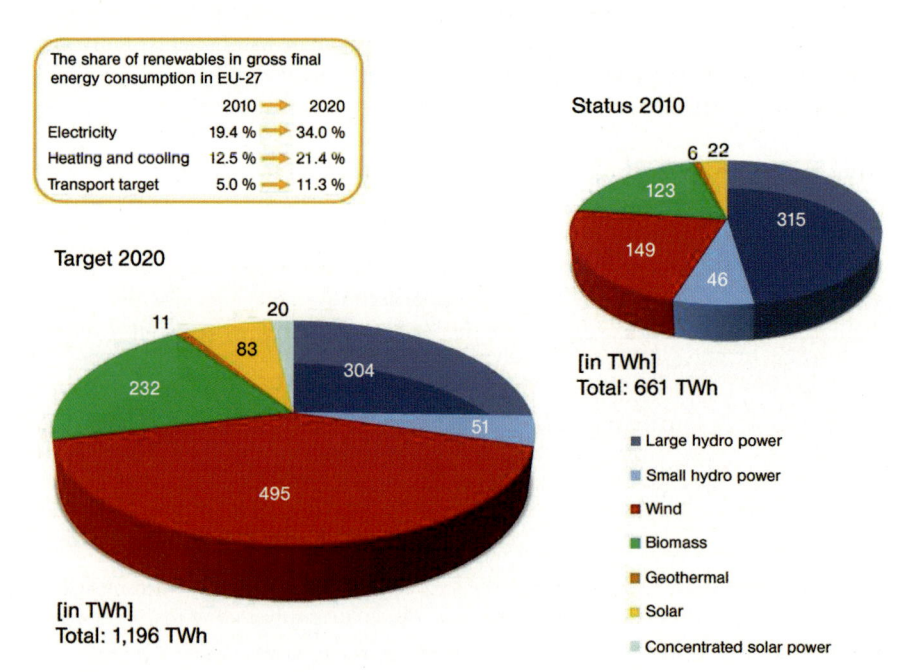

	2010	2020
Electricity	19.4 %	34.0 %
Heating and cooling	12.5 %	21.4 %
Transport target	5.0 %	11.3 %

Target 2020 [in TWh] Total: 1,196 TWh
- 304, 51, 495, 232, 83, 11, 20

Status 2010 [in TWh] Total: 661 TWh
- 315, 46, 149, 123, 6, 22

Legend: Large hydro power; Small hydro power; Wind; Biomass; Geothermal; Solar; Concentrated solar power

Figure 21.2 Renewable energy sources production 2010 and outlook for 2020. Source: VGB.

are well known. However, but even when meeting the above ambitious targets, European Union (EU) electricity consumption is still expected to grow over the coming decades, reaching about 4100 billion kWh in 2035.

In 2010, 661 TWh or about 19.4% of the total European power production of about 3400 TWh was produced by renewables with hydro power, wind, and biomass being the dominant sources. For 2020, renewables are expected to produce almost 1200 TWh or more than 30% of the expected total production of about 3800 TWh (see Figure 21.1 and Figure 21.2). Wind, hydro, biomass, and solar PV will remain the main sources.

There is still a substantial potential of 290 TWh of hydro-based electricity generation, mainly in Central Europe and Scandinavia, but it is difficult to gain public acceptance for new hydro power projects and only a few new projects are expected in the 2020 time frame.

Solar photovoltaics (PV) is expected to grow by a factor of four and wind is forecast to grow by a factor of over three. To achieve the envisaged targets, in theory six wind power plants with an installed capacity of 5 MW each will have to be erected every day to meet the 2020 targets (estimated average full load hours, onshore/offshore: < 2000 h a^{-1} / > 3000 h a^{-1}).

The production from wind and solar PV is distributed into the grid. Renewables-based electricity generation is highly dependent on season and local weather conditions and can fluctuate significantly within hours. Distributed generation also means that RES (renewable energy sources) generation is often located far away from key consumer areas. It is also very challenging to forecast production and adaptation of residual production to meet demand in due time.

Figure 21.3 is based on calculations for Germany, but according to the assumed development in renewables, it is expected that the image and figures will generally apply to most European countries within the next decade.

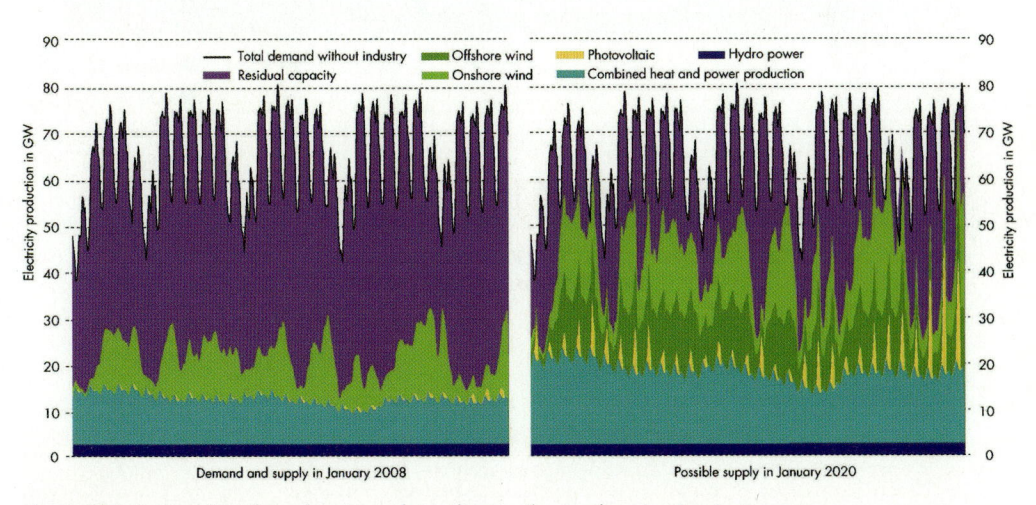

Figure 21.3 Perspective of supply pattern depending on the weather situation in Germany. Source: VGB.

21 The Future Role of Fossil Power Plants – Design and Implementation

Two main conclusions can be drawn from Figure 21.3:

- The residual load (which has to be covered by nuclear and fossil generation capacities) after hydro must run industrial plants and combined heat and power, solar PV, and onshore/offshore wind are expected to be reduced significantly.
- The gradients of load change for the residual load changes significantly. It can reach 20 000 MW h^{-1} for Germany alone compared with about 11 000 MW h^{-1} today, that is a substantial increase in flexibility is required.

A major technical challenge has to be met.

21.3
Market Constraints – Impact of RES

Over the last decade, the major part of the European electricity market was liberalized. Electricity prices are being set on an hourly basis on stock exchanges based on marginal production cost.

The most economical plant – provided that there is no congestion in the transmission system – will produce the electricity needed independent of ownership and location. The term is "merit order". Figure 21.4 shows a typical snapshot with "must-run" RES generation.

Renewables produced by wind and PV have variable production costs close to zero and priority dispatch to the market. The consequence is that the curve of the merit order is pushed to the right. As the consumption is unchanged, the market price for the whole fleet is reduced.

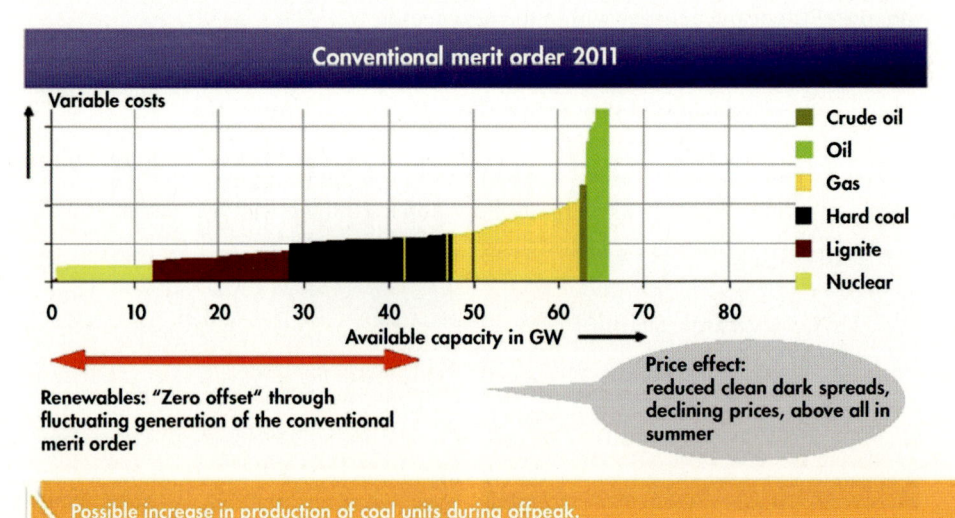

Figure 21.4 Merit order – impact of "must-run" RES. Source: RWE.

The fast-growing share of renewables driven by subsidies, supported "must-run" CHP production, wind, and solar PV are changing the market drastically:

- A large share of production/consumption is not traded on the electricity market any longer, which leads to reduced liquidity.
- Renewables have their main production period in winter (wind) or in the daytime (solar PV), that is, at times when electricity prices have traditionally been high.
- The "must-run" or priority dispatch to the right results in lower prices for electricity on the stock exchange due to zero marginal cost of RES.

All these effects result in significantly reduced production volumes and high pressure to cope with the constraints of given costs for the fossil and nuclear fleet.

Considered from the viewpoint of consumers or society, reduced electricity prices may be desirable, but the net effect after costs for:

- *Financial support for renewables:* The cost of building and operating renewables cannot be covered by current market prices for electricity, and without financial support it will not be possible to meet the 20/20/20 targets.
- *Expansion and rebuilding of the distribution and transmission network to meet the distributed new production:* The current transmission and distribution grids are designed to lead electricity from central points of production to distributed points of consumption. In the future, generation will have to be both central and distributed: central to allow 100% backup when there is little renewable production, for example, during calm wind periods of up to several weeks, and distributed with respect to the character of renewables. The major challenge for the grid will be enabling transmission of electricity from areas of surplus generation to areas of demand.
- *Upholding the necessary backup for the renewables.*

It is unlikely that consumers will see reduced electricity prices.

In this context, the role of fossil fuels is to help stabilize prices at an acceptable level for consumers and society. However, as shown in the following, operators of fossil plants currently have to meet a significant economic challenge when operating existing plants and planning expansions of the power plant portfolio.

21.4
System Requirements and Technical Challenges for the Conventional Fleet

The impact on the supply system is of high relevance for the function of the future, mainly RES-based, generation portfolio. System stability is the major issue.

The system "demand – transmission – generation" needs permanent control, which means permanent intervention of generation, that is,. by the system operator in order to keep the system stable and running.

The European transmission system covers continental Europe, including links to the United Kingdom, Ukraine, and Turkey, corresponding to the former UCTE network. The topology of the European high-voltage transmission system in connection with the RES support scheme reveals that flexibility is required in order to ensure the necessary balancing control for the grid.

In this context, functioning of the grid and its control requirements have to be considered. The impacts caused by increasing renewables-based electricity feed with its principal limitations due to the laws of physics have to be identified to describe the real constraints for the existing grid including the envisaged development.

The system can handle a high percentage of RES provided that residual back-up power is flexible in its "governing behavior," namely provision of inertia as slow governor on the one hand and as fast governor on the other.

This chapter is intended to deal with the implications for the conventional power plant portfolio; therefore, it is focused on load following and gradients, delivery of system services, and co-combustion of biomass as to reduce CO_2 emissions.

21.4.1
Flexibility Requirements with Load Following and Gradients

From the engineering point of view, hydro power is by far the cheapest, fastest, and best technology to deliver load following and load gradients to fine tune the difference between generation and consumption and to manage the grid. However, we do not have sufficient hydro power to meet the needs or there is insufficient transmission capacity available. All other technologies will also have to participate in delivering load gradients.

It is shown in Figure 21.5 that the European power plant fleet is aging. The average age of more than 30 years of the current backbone suppliers, nuclear power and hard coal- and lignite-fired power plants, gives rise to concern.

The lead time for new hard coal- and lignite-fired plants is 7–9 years. Owing to a lack of incentives for new investments, hardly any new hard coal- or lignite-fired plants are being planned; therefore, the majority of these fossil fuel plants will have to be operated for much longer than their original design lifetime of 30–40 years.

For a number of reasons, it is worrying to have to rely on old plants. They are not of the latest state-of-the-art for maximum efficiency and minimum environmental impacts and with new operating regimes with rapid load changes and many part load hours we are increasing the risk of accelerated plant aging.

Comprehensive investigations are under way to identify appropriate retrofit measures to improve plant flexibility. However, there is also a risk that many plants may reach the end of their life simultaneously and that we will face a supply gap in 5–10 years. Nuclear plants are designed for longer lifetimes, but will also need mid-life refurbishments to reach these longer time spans.

The load-following capacity of new power plants is very high with up to 32 MW min^{-1} for new fossil-fired power plants – both combined cycle plants and modern hard coal- and lignite-fired plants.

21.4 System Requirements and Technical Challenges for the Conventional Fleet

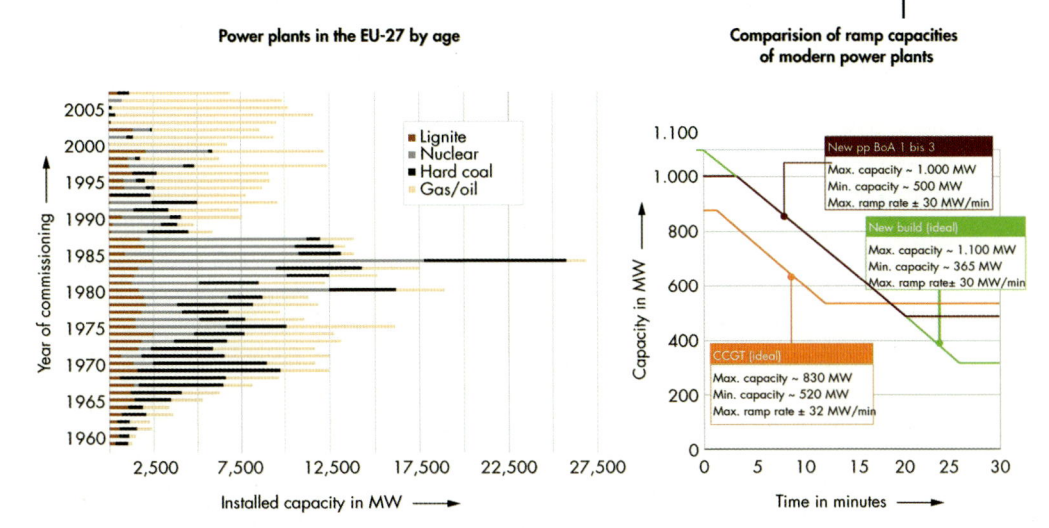

Figure 21.5 Aging fleet and flexibility potential of modern power plants.
pp = power plant; BoA = Brown coal power plants with optimized operational parameters.
Source: RWE.

However, there are challenges:

- As will be shown later, the existing conventional fleet in some countries is highly endangered by the future lower volume of generation and prices. However, we can neither as a society nor as individual plant owners afford to write off the major parts of the existing power plant fleet before the end of their life as stranded assets.
- The existing power plant portfolio is designed for load following, but by far not at the gradients and frequency of gradients that are foreseeable now.
- Building storage facilities with balancing possibilities.

An investigation in Germany shows that the frequency of load changes will be roughly 2000 times per year with ramps of about 10 GW within 15 min in 2030. Under the assumption that this German demand for balancing is to be delivered by new fossil-fired plants alone, we would need more than 20 new plants of 800 MW each just in Germany (10 000 MW/15 min/32 MW min^{-1} = 20.8 plants in the 800 MW range)! This, however, is not realistic under current market conditions.

Therefore, we will need to have new and existing fossil plants together with renewables jointly deliver the power and gradients needed – including the role of storage facilities.

With a rising share of renewables in the supply system, politicians need to reflect on adaptation of the "must-run" conditions and feed-in tariffs for renewables. The current support system in euros per kilowatt hour produced leads to economic losses for the generators of renewable power whenever the production deviates from the technically maximum possible power production.

As the share of renewables rises, we need a support system that encourages renewables to participate in delivering system services and also load following. Technically, renewables could deliver some of the services needed.

Delivering the gradients involves technical challenges for existing and new plants:

- Delivering the requested load ramp on time without:
  - over- or undershooting the requested load
  - causing design-violating temperature or pressure peaks that can damage the plant
  - causing instabilities in the plant that can lead to unwanted stops.
- Higher wear and tear due to much more frequent and faster load ramps. Almost all components in the plants will be affected.
- Staying within specifications for environmental permits.
- Maintaining the gypsum quality within specifications for sale.
- Maintaining the fly ash quality within specifications for sale to the concrete and cement industry.
- Integrating storage facilities (pump storage and new technologies such as compressed air storage or batteries that are currently not commercially available).

For new plants, the designers focus on:

- fast-reacting control loops
- frequency-controlled drives for pumps and ventilators
- advanced materials and thin-walled components in the pressure systems to reduce time for temperature equalization.

Changing the load-following capabilities for existing plants is much more difficult and expensive and must be based on an in-depth analysis of the individual plant and a business case for the costs incurred.

However, for both new and existing plants, we are entering new land. From a situation where plants changed load on a daily basis but were also operated for many hours at constant load, the future regime for many plants will be almost constant ramping with inherent technical challenges.

21.4.2
Delivery of System Services

To ensure a stable grid with stable voltage and stable 50 Hz frequency, electricity generation and consumption must be in balance at any time. System services are tools needed by the transmission system operator (TSO) to ensure this balance. Owing to their character, it is impossible or too expensive for wind and solar power to deliver system services, and hydro power, nuclear, and fossil-/biomass-/waste-fired plants will play a major role in delivering such services in the future.

The system services (Figure 21.6) are outlined in the following.

21.4 System Requirements and Technical Challenges for the Conventional Fleet | 411

Acceleration power
of inertias
of conventional
power plants

Primary control
(load-frequency control)
in rotating machines
± 3,000 MW
ENTSO-E
± 610 MW GER

Secondary control
in rotating machines
± 2,230 MW GER

Tertiary control
minute reserve and
"Intra day"
(not) rotating
+ 1,800 MW GER
− 2,500 MW GER

Tertiary control
"Day ahead planning"

Milliseconds — Seconds — 15 minutes — Hours

Time frame

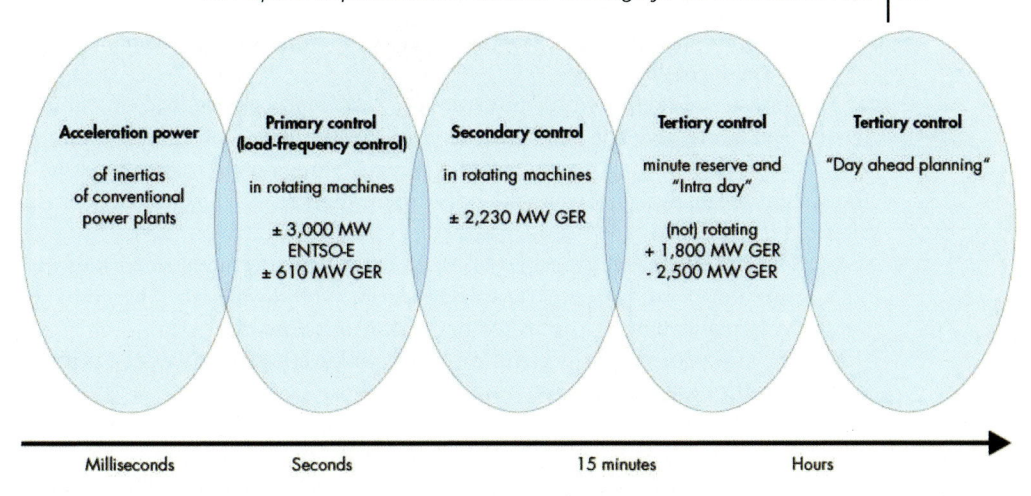

Figure 21.6 System services required for grid control. Source: University of Rostock. ENTSO-E = European Network of Transmission System Operators for Electricity.

21.4.2.1 Primary Reserve/Control

When the frequency deviates from 50 Hz, activation of the primary reserve ensures that the balance between production and consumption is restored and frequency is stabilized close to but deviating from 50 Hz.

Primary reserve is delivered at a frequency deviation of up to ±200 mHz from the 50 Hz and is delivered by production units or large consumers via automatic regulation.

The total current need for primary reserve in the entire European network of transmission system operators for electricity (ENTSO-E) region Continental Europe is ±3000 MW.

21.4.2.2 Secondary Reserve/Control

The secondary reserve is the reserve that brings the frequency back to 50 Hz after being stabilized by the primary reserve.

Secondary reserve is to relieve primary reserve for renewed activation and bring the imbalance on the connections to neighboring TSOs back to the planned exchange of power. It is activated automatically and is delivered by production/consumption units on a signal from the grid operator. As it must be delivered within 15 min, secondary reserve is primarily delivered from units that are already supplying power to the grid.

21.4.2.3 Tertiary or Manual Reserve

Manual reserves are reserves that can be activated by the TSO by asking producers or consumers to deviate from their planned hourly production/consumption. Manual reserves must be delivered within 15 min.

Manual reserves are to relieve primary and secondary reserves after activation and to ensure the system balance at trips or production limitations on production plants and connections to other TSOs.

21.4.2.4 "Short-Circuit Effect," Reactive Reserves, and Voltage Regulation, Inertia of the System

The response time required for intervention, that is, to balance load or stabilize frequency, is decisive for the controllability of the grid. The higher share of RES-based power in the grid reduces the response time due to reduced inertia of the system normally provided by the heavy rotating masses of steam turbines and alternators.

Short-circuit effect, reactive reserves, and voltage regulation are needed to ensure stable and safe operation of the transmission grid. Typically these services must be delivered by plants directly connected to the transmission grid.

The above services are important for grid stability, but cannot work as primary income for the entire fossil fleet.

21.4.2.5 Secure Power Supply When Wind and Solar Are Not Available

Figure 21.7 shows the load and wind production in a typical week in Germany. It can be seen that the wind production on Tuesday/Wednesday for about 30 h is about 10 GW higher than on Monday and that the production falls back to 5–10 GW at the end of the week. The variation of 10 GW or a total production of 300 GWh over 30 h could be compensated by the rest of the production fleet.

According to the EURELECTRIC Hydro Report, the theoretical storage capacity in German hydro power plants amounts to only 50 GWh! Also, these capacities can only be used provided that they are empty and ready to start loading when surplus

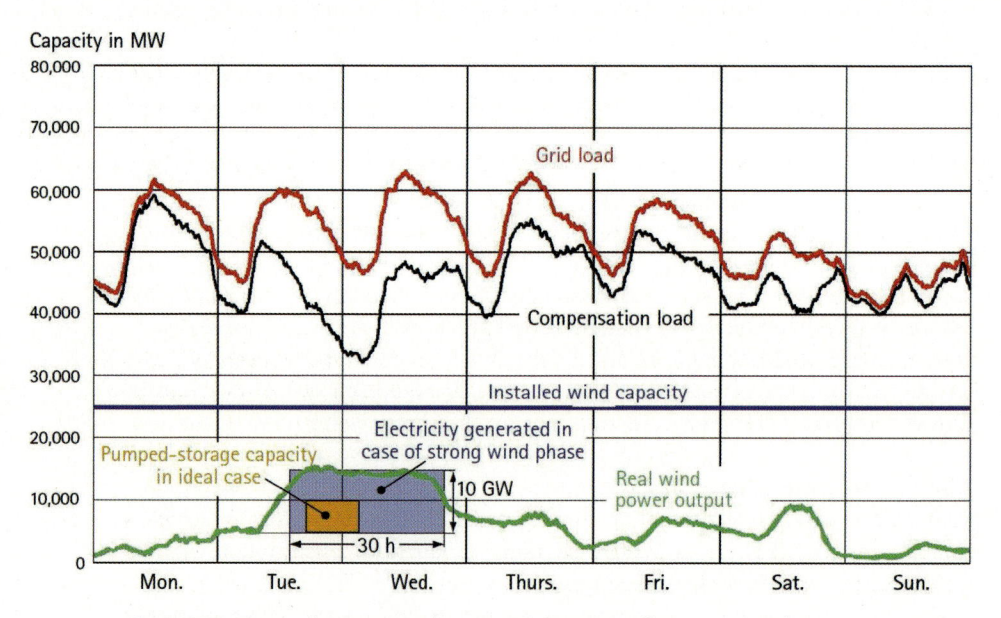

Figure 21.7 The need to store wind power to compensate for missing production as an example taken from Germany. Example of a 1 week load profile in the German high-voltage grid in 2008 and flexibility of conventional power plants. Source: RWE.

wind arises. Since existing hydro storage capacity is built and operated to fulfill other needs (arbitration on the power market or to deliver system services for the TSO), it would be pure coincidence if the storage functions were to be available at that particular time when needed.

With a rising share of wind and solar PV in the system, we will see significant load gradients that will have to be coped with by the remaining fleet within the system or by import/export from/to other regions.

Hydro power plants in Norway and in the Alpine region are often mentioned as a possible solution, which in fact is true, but:

- We will need to strengthen the transmission system significantly, the "above 10 GW" applied to the 2008 fleet of renewables in Germany only. The figure will grow with rising share of intermittent production.
- The balancing power needed to fill up the storage facilities has to be evaluated.
- Current hydro power plants will have to be adapted to the new operating regime:
  - Run-of-river hydro power plants can only be used for energy storage to a limited extent.
  - Hydro power with dam storage can, of course, withhold production and thereby function as storage as long as there is storage space available in the dams.
  - Current plants are designed to fulfill domestic needs for power and might therefore not be able to redeliver the stored energy to continental Europe when needed there.
  - A large number of new pumping, storage, and hydro power plants will have to be built to handle the intermittent production in Central Europe. Construction of transmission lines and plants will need time and a financial budget and will also cause environmental issues in the countries affected.

The only economically feasible solution is a combination of hydro power plants and local fossil-/biomass-fired central/decentral power plants in combination with an extended transmission and distribution network in continental Europe.

21.4.3
District Heating

District heating produced in CHP (combined heat and power) plants is very common in many cities in northern and eastern Europe. The fuels utilized in CHP plants range from municipal waste, coal, and gas to different biomasses.

It is possible to store heat as hot water in insulated tanks to bridge a few days of supply, but heat cannot be transported over large distances.

The first seasonal storages for heat are being developed. Mainly the Scandinavian countries are aiming at storing solar heat in summer, which is then to be utilized in the winter season. The major effect is the flattening of the load curve.

The need for district heating in areas currently supplied by district heating will be reduced over the coming decades as a result of the Energy Efficiency Directive that requires improved insulation and more energy-efficient new buildings.

However, there is still significant potential for replacing fossil-fired individual heating and district heating without electricity production with CHP-based district heating, which in many countries can also make a great contribution to the reduction of CO_2 emissions.

It is possible to integrate solar (thermal) and wind (via electric heating of water) power into district heating systems, but for the years to come the backbone of supply will remain biomass and fossil fuels for reasons of supply security.

In its latest *Energy Roadmap 2050*, the European Commission predicts moderate growth of district heating of about 1% a^{-1}. Other studies, for example the *Heat Road Map 2050* by the University of Aalborg and University of Halmstadt for Euroheat and Power in 2011, optimistically indicate growth potentials for district heating in the EU of up to a factor of two by 2030 and a factor of three by 2050.

CHP technology has been well proven over many years in many countries. The main challenge will be related to winning new areas of supply and integration of geothermal and solar heat in addition to introducing of new fuels such as. biomass.

In many European countries, fossil-fired power plants will have potential in connection with CHP production.

21.4.4
Co-combustion of Biomass

Biomass is a general term used for a wide variety of products with varying physical and chemical characteristics (Figure 21.8), such as:

- agricultural residues such as straw
- wood (pure or recovered, with or without roots, bark, leaves, and branches)
- biomass from coastal regions (contains more Cl)
- refined or not (e.g., "raw" wood chips, white or black pellets).

The use of biomass in power generation creates a variety of new issues for operators:

- Sustainability:
  - Deforestation (it will often take 50–100 years for a forest to grow to maturity; if CO_2 emission is a problem now, how can combustion of 50-year-old trees be CO_2 neutral?).
  - Destruction of rainforest?
  - Need for and recovery of fertilizer (global resources of phosphates are limited).
- Import of insects and plant diseases.
- Competition with food production.

Typically, biomass that would otherwise be left to rot in the fields is to be used:

- Residues from agricultural production (e.g., straw).
- Already "dead" wood:
  - roots and branches from industrial wood production

21.4 System Requirements and Technical Challenges for the Conventional Fleet | 415

- from, e.g., rubber plantations in Africa that will otherwise be burnt in the field due to the need to renew plantations
- from, e.g., North America where trees in vast areas have been destroyed by insect attack.

Co-firing of biomass also leads to technical challenges, as shown in Figure 21.9.

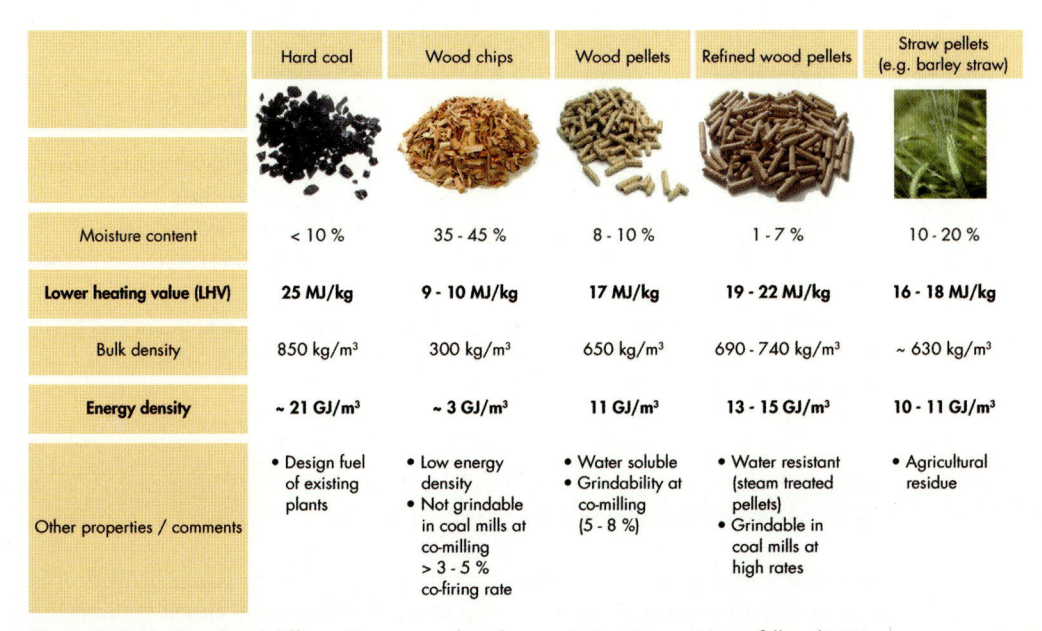

	Hard coal	Wood chips	Wood pellets	Refined wood pellets	Straw pellets (e.g. barley straw)
Moisture content	< 10 %	35 - 45 %	8 - 10 %	1 - 7 %	10 - 20 %
Lower heating value (LHV)	25 MJ/kg	9 - 10 MJ/kg	17 MJ/kg	19 - 22 MJ/kg	16 - 18 MJ/kg
Bulk density	850 kg/m³	300 kg/m³	650 kg/m³	690 - 740 kg/m³	~ 630 kg/m³
Energy density	~ 21 GJ/m³	~ 3 GJ/m³	11 GJ/m³	13 - 15 GJ/m³	10 - 11 GJ/m³
Other properties / comments	• Design fuel of existing plants	• Low energy density • Not grindable in coal mills at co-milling > 3 - 5 % co-firing rate	• Water soluble • Grindability at co-milling (5 - 8 %)	• Water resistant (steam treated pellets) • Grindable in coal mills at high rates	• Agricultural residue

Figure 21.8 Hard coal and different biomasses: key characteristics. Source: Vattenfall and VGB.

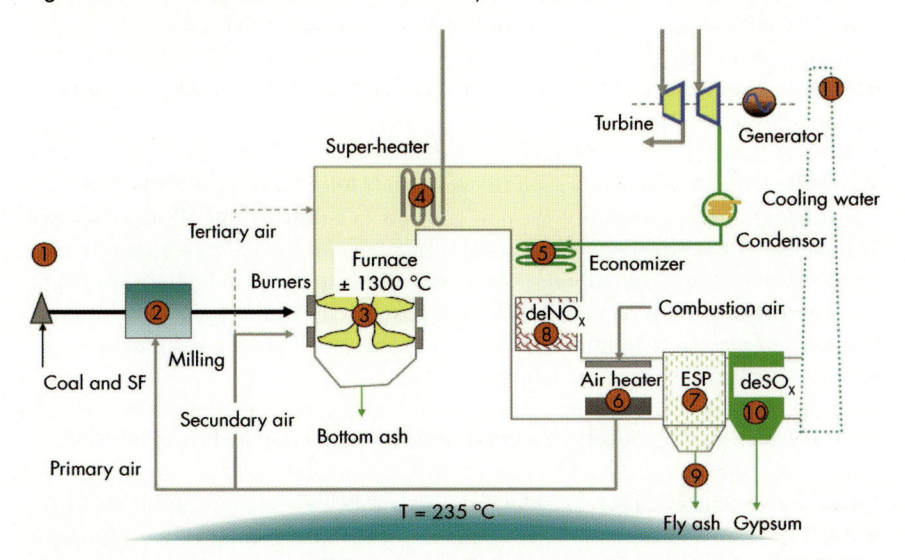

Figure 21.9 Critical areas in a coal-fired plant co-firing biomass. Source: Laborelec/GDF SUEZ.

Co-combustion of biomass and pulverized coal involves the following critical issues and challenges:

1. Fuel storage — Health (CO, fungi), spontaneous combustion
2. Milling — Mechanical problems, fire, explosion
3. Furnace — Slagging and corrosion (CO corrosion)
4. Superheater — Fouling, high-temperature corrosion (mainly by KCl)
5. Economizer — $CaSO_4$ deposits
6. High-dust de-NO_x — Poisoning of catalyst (K, P, As, Ca)
7. Air heater — Blockage, low-temperature corrosion (HCl)
8. Electrostatic precipitator (ESP) — Reduced efficiency due to lower ash resistivity
9. By-products — Continued utilization of fly ash
10. Desulfurization — Waste water and gypsum quality
11. Emissions via stack — Dust, NO_x, SO_x

Co-combustion of biomass is another future role for the existing coal-fired plant portfolio. Depending on the boiler design and biomass, the co-combustion ratio can amount to 10–25%.

21.5
Technical Challenges for Generation

It is shown in Figure 21.5 and Figure 21.10 that the existing power plant fleet is aging and a large number of potential projects are under way aimed at renewing the existing fleet and expanding it by natural gas and renewables.

As of September 2012, about 69 000 MW of new production capacity is under construction and a further 85 000 MW is in the permitting and 86 000 MW in the planning stage. It is worrying that compared with 2011, projects with a capacity of 31 000 MW were stopped.

Whether all projects will be realized is dependent, of course, on business case, permission, public acceptance, and the political frame.

Achieving public acceptance for new projects is a major issue all over Europe. The nature of the project does not seem to matter. Almost all large infrastructure projects meet strong opposition. In the power industry, we have seen significant resistance against many new projects over the last decade: hydro power, conventional, and nuclear plants. Wind turbines (both onshore and offshore) and transmission lines for gas and electricity are being opposed.

Since the delaying of projects or withdrawal of permits can completely destroy the business case, it is essential for the industry to achieve early and sustainable acceptance for projects.

Public acceptance is a prerequisite for successfully meeting the 20/20/20 targets and the much more demanding next phase of the "turnaround" of the European energy sector towards more renewables.

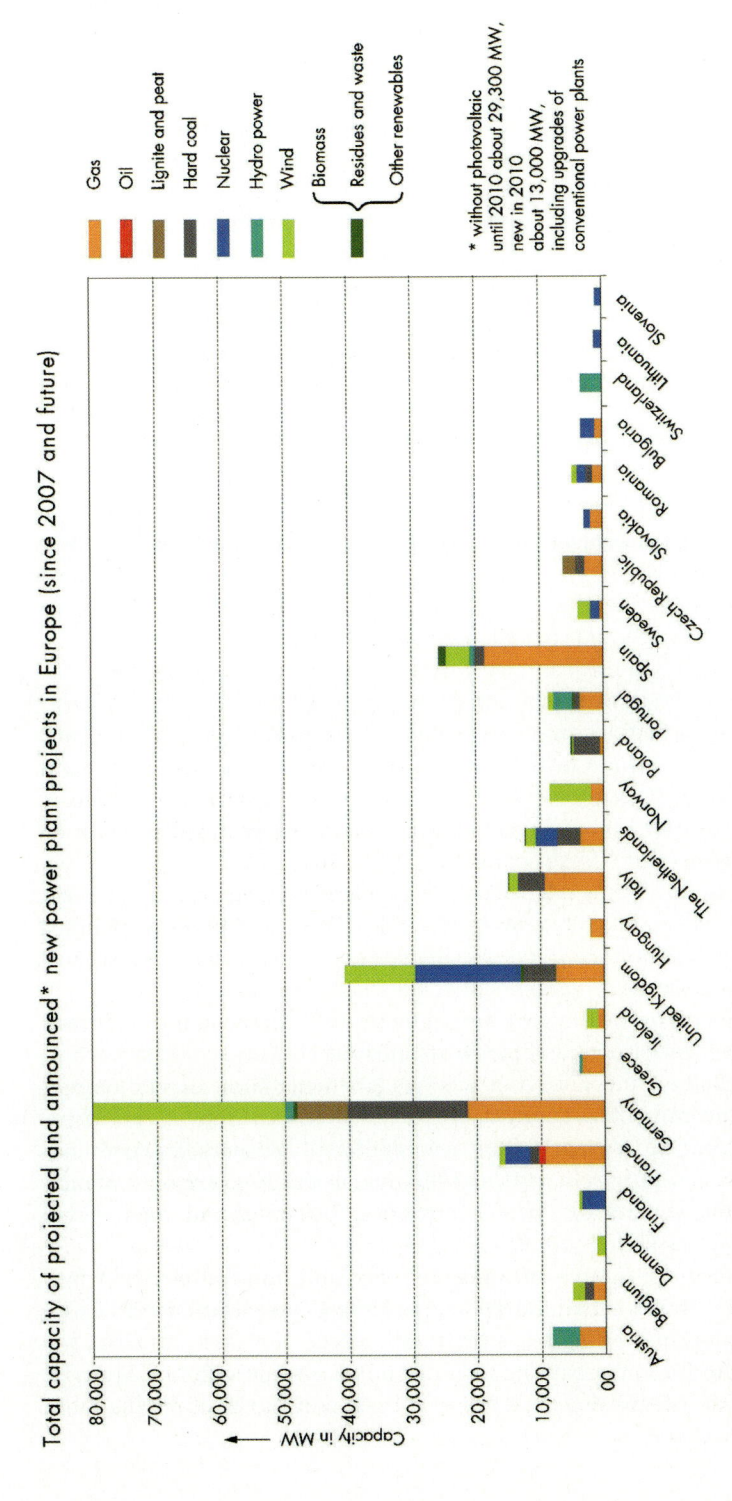

Figure 21.10 European power plant projects since 2007. Source: VGB.

21.6
Economic Challenges

As outlined earlier, the existing power plant portfolio is to be renewed, and more flexible power plants and plants working as "backup" for intermittent production are needed. Many plants are being planned, but in 2011–12 alone a total of 31 000 MW of projects have been abandoned for the lack of a business case.

VGB and EURELECTRIC are currently updating the report *Levelised Cost of Electricity* based on an extensive enquiry and consultation with generators and suppliers. The aim is to produce a comprehensive and robust set of "order of magnitude" data. The survey reflects the generators' position based on the past few years of experience with investment. It covers the entire generation portfolio: renewables (RES) (onshore/offshore wind, PV, and solar-thermal), hydro power (with pumped storage and run-of-river), and thermal power plants (coal, lignite, gas, and nuclear).

21.6.1
Principles Underlying the Data on CAPEX and OPEX

Assumptions on investment costs (CAPEX) and operating costs (OPEX) have been derived from real projects and corresponding operational data.

CAPEX is commonly defined as money spent to acquire or refurbish systems and system components such as machinery. It can be further broken down into costs for pre-development, engineering construction, all systems "within the fence," and other infrastructure. The only distinction made is to assume "grayfield" conditions for fossil and nuclear plant sites.

The expenditures for new pumped storage projects are highly dependent on geological conditions and infrastructure.

Refurbishment and lifetime extension costs have been singled out in the analysis, which are presented in a dedicated row in Table 21.1. This facilitates understanding of the increased costs of nuclear plants with lifetime extension and increased security standards, but also for the refurbishment of older hydro plants.

OPEX is commonly defined as the expenditure related to the operation and maintenance of power plants; it includes spare parts, auxiliary cost, insurance, labor costs, grid fees, taxes, and maintenance contracts. We have noted these expenditures as a percentage of investment cost per year.

The given figures are "average" values across different locations and conditions that alleviate specific impacts. In particular, the annual operating hours are strongly determined by the demand and market conditions; their impact on the levelized cost of electricity is essential.

It is obvious that the figures on investment per unit installed electric power ($€\,kW_{el}^{-1}$) scatter within a certain range. The scattering range starts in 2011 with "reliable" values that only reflect site-specific differences. However, this range will increase over time due to the growing weight of prognosis uncertainties. Thus, for 2011, variations of $\pm€100\,kW_{el}^{-1}$ were possible; beyond 2030, the scattering range rises to about $\pm€200\,kW_{el}^{-1}$.

The future development of CAPEX figures depends – beyond other factors – on technical progress. The manufacturing, physical/chemical, or market effects determine the technical progress. It is evident that the CAPEX figures are the prerequisite for the OPEX figures. Therefore, the learning curve parameters affecting the cost components of any system able to produce electricity are the following:

- physical effects
- chemical effects
- simplification of the system
- process technology
- material consumption
- quality of material, for example, rare earth, platinum
- engineering "basic"
- engineering for specific installation
- technical maturity
- standardization of components and concept
- manufacturing process → mass production Y/N
- large-scale deployment
- licensing procedure
- site impact → geology, meteorology, and so on
- installation and commissioning on-site
- lifetime impact

Figure 21.11 shows the levelized costs of electricity (LCOE) split according to investment costs, O&M costs, fuel costs, and CO_2 costs. The investment costs were calculated based on the equivalent annual cost method at a 10% discount rate. CO_2 costs were calculated based on a certificate price of €30 t^{-1} CO_2. A difference of €10 t^{-1} CO_2 changes the LCOE by 6% for gas, 11–13% for hard coal, and 13–17% for lignite.

The basic assumptions for the levelized cost of electricity in Table 21.1 are presented in the following columns: technology, lifetime, operation hours, CAPEX (balance of plant, mean regional impact), efficiency (net efficiency in terms of heat value), refurbishment (major upgrade in replacing turbine blades, pulverizer or I&C systems) and the scattering range from 2011 to 2050 as well as annual OPEX (personnel, insurance, taxes, auxiliaries, etc.) in% of invest costs.

With current and foreseeable electricity prices at the 400 kV level amounting to 50 to 60 €/MWh in a number of European countries (and in Scandinavia rather 40 to 50 €/MWh), it is obvious that it is difficult to create a business case for new projects or even the refurbishment of existing older plants.

In markets with high RES share and low electricity prices, it has been observed in recent years that completely new gas-fired combined cycle plants were mothballed after commissioning as the market prices could not cover operating costs. It is expected that in such markets the average operating hours for many fossil-fired plants will decrease from 5000–6000 h a^{-1} to 2000–3000 h a^{-1}. If no action is taken, these plants will no longer be economic to operate and they are – although the system still needs them for stability and back-up – at risk of being decommissioned for ever.

Table 21.1 Basic assumptions for levelized cost of electricity as shown in Figure 21.11. Electricity generating costs, EURELECTRIC/VGB; status 6 December 2011.

Conventional		Fossil Fuels								Nuclear	Hydro	
Type		Gas open cycle	Gas CCGT	Hard coal 600	Lignite 600	Hard coal 700[a]	Lignite 700[a]	HC 700 + CCS[a]	HC 600 + Bio cofiring	Nuclear EPR 1600[a]	Pumped storage[b]	
Lifetime	years	25	25	35	35	35	35	35	30	40	60	
Oper. Hours	hours/year	6,000	6,000	7,500	7,500	7,500	7,500	7,500	7,500	7,900	2,500	
CAPEX	EUR/kW	650	800	1,300	1,400	2,100	2,100	3,000	1,390	3,000	1,100	2,400
Invest costs	$/MWh	17.5	21.6	26.5	28.4	42.7	42.7	60.9	28.8	62.0	65.0	141.5
Invest costs	EUR/MWh	12.5	15.4	18.9	20.3	30.5	30.5	43.5	20.6	44.3	46.4	101.1
O&M costs	% of invest	3.0	2.5	2.0	2.0	2.0	2.0	2.0	2.0	2	1.0	
O&M costs	$/MWh	5.5	5.7	6.4	6.7	9.4	9.4	12.7	6.7	12.9	9.5	16.8
O&M costs	EUR/MWh	3.9	4.1	4.6	4.8	6.7	6.7	9.1	4.8	9.20	6.8	12.0
Efficiency	%	45	60	45	43	50	50	40	45	37	80	
Fuel type		gas	gas	hard coal	lignite	hard coal	lignite	hard coal	hc + biomass	nuclear		
Fuel calor. Val.	MJ/kg	40	40	27	10	27	10	27	25.6	27		
Fuel C-content	weight-%	75	75	82	30,5	82	30.5	82	73.8	0		
Fuel consumpt.	kg/MWh	200	150	296	837	267	720	333	313			
Fuel price	EUR/t	293	293	70	10	70	10	70	73			
Fuel costs	$/MWh	82.0	61.5	29.0	11.7	26.1	10.1	32.7	31.9	13.9		
Fuel costs	EUR/MWh	58.6	44.0	20.7	8.4	18.7	7.2	23.3	22.8	9.9		
CO_2 emmiss.	t/MWh	0.55	0.41	0.89	0.94	0.80	0.81	0.00	0.85	0	0	
CO_2 costs	EUR/MWh	16.5	12.4	26.7	28.1	24.1	24.2	0.0	25.4	0	0	
Fuel & CO_2	$/MWh	105.1	78.9	66.5	51.0	59.8	43.9	32.7	67.5	13.9		
Fuel & CO_2	EUR/MWh	75.1	56.3	47.5	36.5	42.7	31.4	23.3	48.2	9.9		
LCOE	$/MWh	128.1	106.2	99.4	86.2	111.9	96.0	106.3	103.0	88.8	74.5	158.3
LCOE	EUR/MWh	91.5	75.8	71.0	61.6	79.9	68.6	75.9	73.6	63.4	53.2	113.1
Exchange rate	$/EUR	1.40										
CO_2 costs	$/t	42										
CO_2 costs	€/t	30.0										
30€/tCO_2		91.5	75.8	71.0	61.6	79.9	68.6	75.9	73.6	63,4	85,4	113,1
20€/tCO_2		86	71.7	62.1	52.5	71.9	75.9	75.9	65.1	63,4	85,4	113,1
Difference	€/MWh	5.5	4.1	8.9	9.4	8.0	0.0	0.0	8.5	0.0	0.0	0.0
Difference	%	6.2%	5.6%	13.4%	16.5%	10.6%	0.0%	0.0%	12.2%	0.0%	0.0%	0.0%

a > 2015.
b Without pumping costs.
Source: EURELECTRIC/VGB.

21.7 Future Generation Portfolio – RES Versus Residual Power

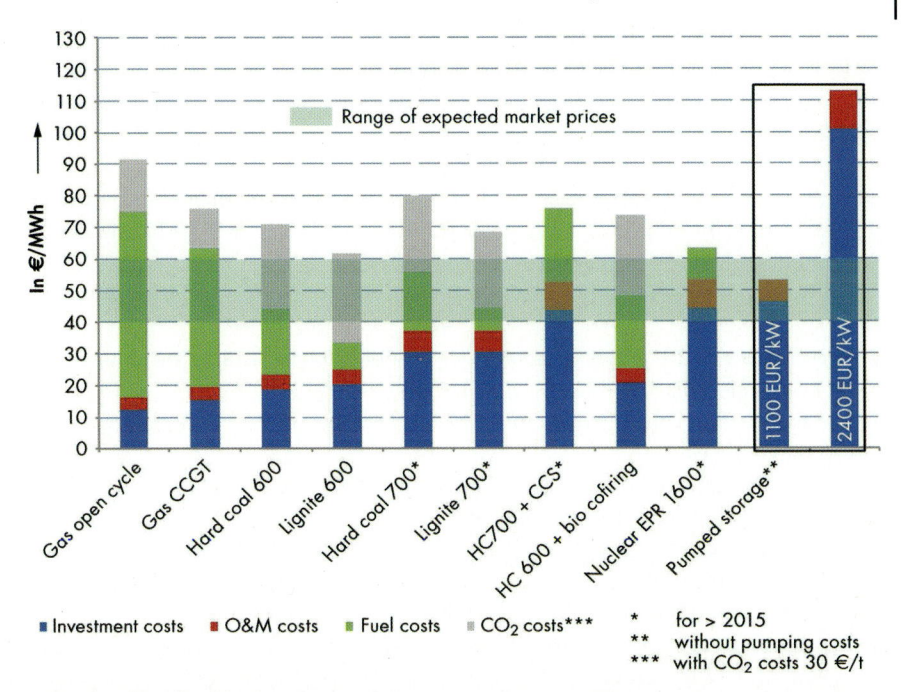

Figure 21.11 Levelized cost of electricity for a European power plant. The assumptions for the calculations are shown in Table 21.1. Source: EURELECTRIC/VGB.

This development could, if neglected, lead to tremendous problems for the entire transition away from the current fossil-based society and needs to be addressed appropriately.

21.7
Future Generation Portfolio – RES Versus Residual Power

Seen from a European and worldwide perspective, fossil-fired power plants will remain the foundation of secure electricity supply in the coming decades.

The global population is increasing by 80 million people per year. Consequently, the number of people has doubled between 1960 and today, that is, within roughly five decades. At present, approximately one-quarter of the global population of nearly 7.0 billion people does not yet have access to electricity. Electricity consumption will grow faster than any other form of energy consumption. The increase might be decelerated in the short term due to the worldwide financial and economic crisis, but in the medium term the mentioned factors will again dominate the development.

It is expected that today's electricity consumption of 17 200 billion kWh will increase by roughly 84% to 31 700 billion kWh worldwide by 2035. In 2009, about one-fifth of the electricity generated globally – roughly 3100 billion kWh – was provided in the EU. A 32% rise in demand is expected by 2035 within the EU.

Experts estimate that fossil fuels will continue to cover most of the extra demand. Fossil fuels will still account for about 70% of electricity generated worldwide in 2035. About half of the electricity generated in the EU will come from fossil fuels by that time.

Renewable energy sources will play a growing role in the global primary energy consumption structure. Likewise, nuclear power will – despite the political nuclear phase-out in some countries – maintain an important position in global electricity generation and will even grow in some countries.

Conventional power plants will also have a significant role in order to:

- meet the 20/20/20 targets by co-combustion of biomass and by combined production of electricity and heat
- guarantee stable and reliable supply of electricity to society and consumers at affordable prices
- be a substantial part of the European supply of district heating
- provide together with nuclear and hydro:
  - residual load
  - stability to the grid by compensating intermittent production from renewables
  - security of supply when sun and wind are not available
  - system services: black start, primary/secondary/tertiary control and voltage regulation of the grid.

It is among the key issues for politicians, suppliers, and operators to develop jointly technology and the electricity market – comprising all generation options according to needs – in order to guarantee supply security at affordable consumer prices and with minimum environmental and health impacts.

Part III
Gas Production

22
Status on Technologies for Hydrogen Production by Water Electrolysis

Jürgen Mergel, Marcelo Carmo, and David Fritz

22.1
Introduction

Energy technology is currently undergoing a major transformation worldwide. The generally accepted factors driving this transition are climate change, supply security, industrial competitiveness, and local emissions. With its 2010 energy concept, the German Federal Government has set course for an environmentally friendly, reliable, and affordable energy supply, in which renewable energies will account for the greatest share in the energy mix of the future. The aim is to reduce climate gas emissions compared with 1990 levels by 40% by 2020, 55% by 2030, 70% by 2040, and 80–95% by 2050 [1]. These ambitious goals can only be achieved by a highly efficient energy supply based mainly on renewable energy carriers. This will lead to a rapid rise in intermittent energy from wind and solar sources and to an ever greater decoupling between the time of electricity generation and its consumption. New challenges will thus arise with respect to considerably more flexible regulation of the electric grid, of transport, and, above all, of the reliable storage of large quantities of energy. Apart from power plants with compressed air, electrical, or thermal energy storage, great significance will be attached to storage in the form of renewable gases (power-to-gas) such as hydrogen or methane. In this case, the hydrogen will be produced by electrolysis from surplus electricity generated from renewable sources. As can be seen in Figure 22.1, potential markets for hydrogen are transport, direct reconversion, methanation, and feeding the gas into the natural gas grid, and also use as a raw material in industrial processes.

If grid-bound wind or solar hydrogen systems are to be included, different demands are made on the electrolyzer. In particular, the intermittent supply of electricity from renewable energy sources represents a special challenge for the process engineering of the electrolysis techniques.

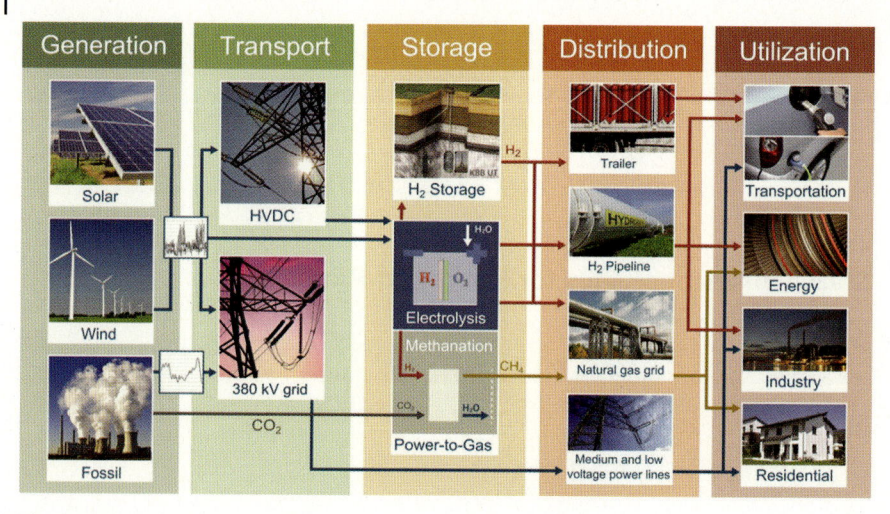

Figure 22.1 Hydrogen as storage medium for renewable energy.

22.2
Physical and Chemical Fundamentals

The production of hydrogen and oxygen from water by water electrolysis is a well-established technological process worldwide, dating back more than 100 years. However, at present only about 4% of hydrogen requirements worldwide are covered by electrolysis [2]. This is due to the higher cost of generating electrolytic hydrogen in comparison with hydrogen obtained from fossil primary energy sources such as natural gas (77%) and coal (18%) [2]. In order to decompose 1 mol of water electrolytically into its constituents of hydrogen and oxygen:

$$H_2O_{(l)} + \Delta H_R \rightarrow H_{2(g)} + \tfrac{1}{2} O_{2(g)} \tag{22.1}$$

under standard conditions (298.15 K and 1 bar), a reaction enthalpy (equivalent to the formation energy of liquid water) of $\Delta H_R = 285.9$ kJ mol^{-1} is required. In accordance with the second law of thermodynamics:

$$\Delta H_R = \Delta G_R + T\, \Delta S_R \tag{22.2}$$

part of the energy can be supplied as thermal energy. This is the maximum amount of energy corresponding to the product of the thermodynamic temperature T and the reaction enthalpy ΔS_R. The free reaction enthalpy ΔG_R corresponds to the minimum fraction of ΔH_R which has to be made available in the form of electricity. The reversible cell voltage $V_{rev}^°$ for electrolytic water decomposition can thus be calculated from the free reaction enthalpy under standard conditions (298.15 K and 1 bar) $\Delta G_R^° = 237$ kJ mol^{-1} in accordance with

$$V_{\text{rev}}^{\circ} = \frac{\Delta G_{\text{rev}}^{\circ}}{nF} = \frac{237 \text{ kJ mol}^{-1}}{2 \times 96485 \text{ C mol}^{-1}} = 1.23 \text{ V} \qquad (22.3)$$

where n is the number of electrons and F is the Faraday constant for 1 mol of hydrogen produced. However, this assumes that the fraction for $T\Delta S_R$ is integrated in the electrolysis process in the form of heat. V_{rev}° is therefore also termed the lower heating value (LHV), which corresponds to an energy content for gaseous hydrogen of 3.0 kWh Nm^{-3}. If the thermal energy is introduced in the form of electrical energy, which is usually the case with industrial electrolyzers, then the expression thermoneutral voltage is used V_{th}°, which under normal conditions results for liquid water from the reaction enthalpy in accordance with

$$V_{\text{th}}^{\circ} = \frac{\Delta H_R^{\circ}}{nF} = \frac{286 \text{ kJ mol}^{-1}}{2 \times 96485 \text{ C mol}^{-1}} = 1.48 \text{ V} \qquad (22.4)$$

which corresponds to an energy content for gaseous hydrogen of 3.54 kWh Nm^{-3}. At this voltage, the electrical energy is equal to the total reaction enthalpy of the decomposition of water. The decomposition of water by electrolysis consists of two partial reactions separated by an ion-conducting electrolyte. The three relevant processes of water electrolysis are distinguished by the electrolyte used; these are shown in Figure 22.2 together with their partial reactions for the hydrogen evolution reaction (HER) and the oxygen evolution reaction (OER), the typical temperature ranges and the ions for the related charge transport:

- alkaline electrolysis with a liquid alkaline electrolyte;
- acidic PEM electrolysis with a proton-conducting polymer electrolyte membrane;
- high-temperature electrolysis with a solid oxide electrolyte.

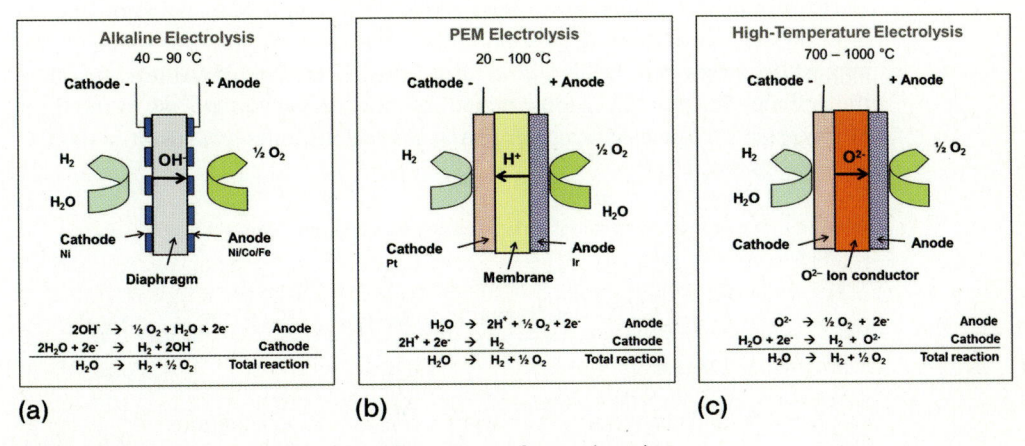

Figure 22.2 Operating principles of the different types of water electrolysis.

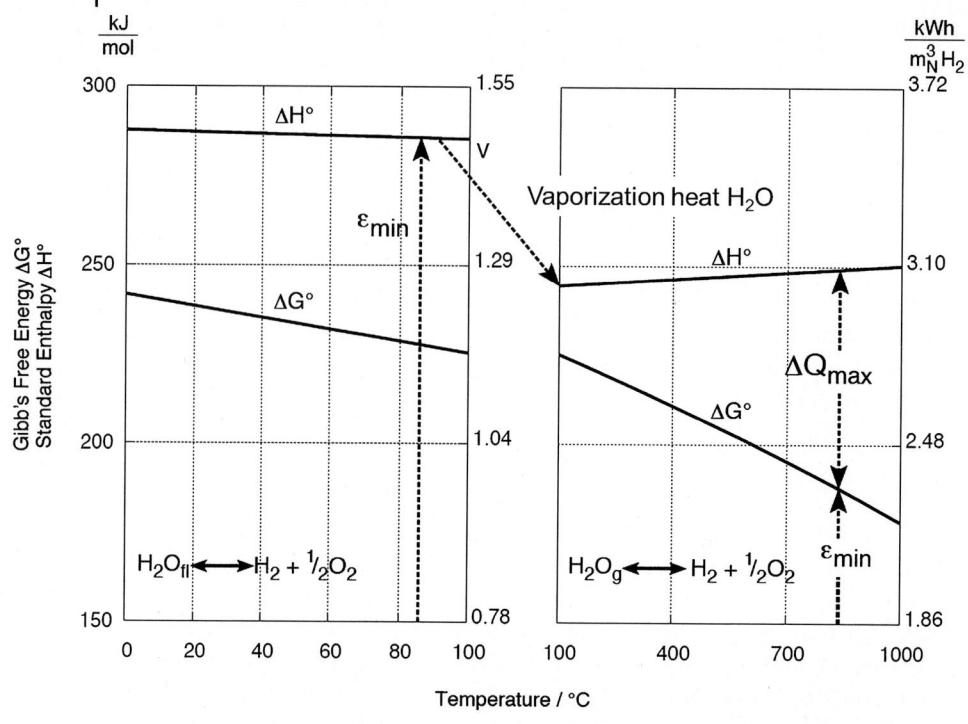

Figure 22.3 Specific energy consumption of water electrolysis as a function of temperature. Reproduced from [3] with permission from FVEE ForschungsVerbund Erneuerbare Energien.

In alkaline electrolysis the water is usually fed in on the cathode side and in PEM electrolysis on the anode side. In the case of high-temperature electrolysis, the required steam is supplied at the cathode. Figure 22.3 shows the reaction enthalpy ΔH_R as a function of temperature and of the free reaction enthalpy ΔG_R at normal pressure. The figure shows that for electrolysis processes which split the steam electrolytically at a temperatures above 700 °C, as for example in high-temperature electrolysis, according to the relation $\Delta G_R = \Delta H_R - T\Delta S_R$, the cell voltage to be applied decreases perceptibly owing to the positive entropy of the reaction while the enthalpy fraction $T\Delta S_R$ must be fed into the electrolysis process in the form of process heat. The electrical energy ΔG_R for the electrolysis of steam at 1000 °C amounts to just 0.91 V.

Table 22.1 Thermodynamic data for water electrolysis at different temperatures and normal pressure.

	ΔH_R (kJ mol^{-1})	$V_{th}^°$ (V)	ΔG_R (kJ mol^{-1})	$V_{rev}^°$ (V)
Liquid water at 298.15 K	285.9	1.48	237.2	1.23
Steam at 373.15 K	242.6	1.26	225.1	1.17
Steam at 1273.15 K	249.4	1.29	177.1	0.92

Table 22.1 summarizes the values for ΔH_R and ΔG_R with the respective voltages for V_{th}° and V_{rev}° at different temperatures.

The cell voltages achievable in practice for water electrolyzers are, however, much higher than the theoretical reversible cell voltage. This is due, first, to the overpotentials at the electrodes which are caused by irreversible processes at the electrodes and are therefore also known as activation overpotentials. Second, the resistance polarization must be overcome, which is caused by the ohmic resistance of the cell (electrolytes, separator, and electrodes). For a current density i, the real cell voltage V_{cell} is composed of the sum of the reversible cell voltage V_{rev}, the ohmic voltage drop iR and the overpotentials of the anode η_{anode} and cathode $\eta_{cathode}$:

$$V_{cell} = V_{rev} + |\eta_{anode}| + |\eta_{cathode}| + iR \tag{22.5}$$

where η_{anode} is the anode polarization (also known as the oxygen overpotential), that is, the fraction of the cell overpotential that is accounted for by the anode, $\eta_{cathode}$ is the cathode polarization, that is, the overpotential at the cathode (hydrogen overpotential), and R is the area-specific resistance of the cell. Figure 22.4 shows a typical current–voltage characteristic for alkaline electrolysis and its division into polarization curves.

An important technical evaluation criterion for electrolysis processes is the efficiency, that is, the cost:benefit ratio for an industrial electrolyzer. Since commercial products are currently available only for alkaline and PEM electrolysis where water is supplied in a liquid form, it is appropriate to use the higher heating value or the thermoneutral voltage $U_{th}^\circ = 1.48\,\text{V}$ to determine the efficiency. The efficiency relative to the higher heating value (HHV) of hydrogen (3.54 kWh Nm^{-3}) thus indicates how efficiently the electrolyzer can be operated as a technical apparatus.

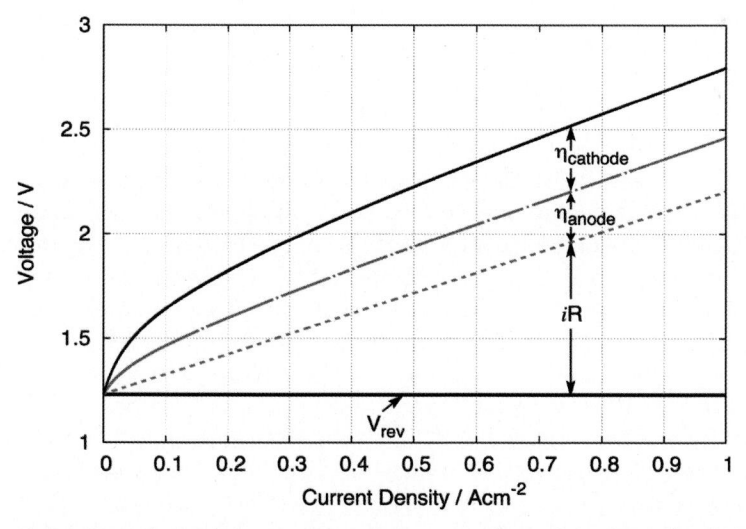

Figure 22.4 Schematic representation of a current–voltage characteristic for alkaline electrolysis and the division of the different voltage losses during operation. Modified from [4].

$$\eta_{HHV} = \frac{\text{HHV of the H}_2 \text{ produced}}{\text{electric power required}} = \frac{\dot{V}_{H_2} \times \text{HHV}}{P_{el}} \qquad (22.6)$$

If the hydrogen produced in the electrolyzer is used energetically in a subsequent application, for example by conversion into electrical energy in a fuel cell, only the lower heating value of the hydrogen is used. It is then more appropriate to relate the electrolyzer efficiency to the reversible voltage $V_{rev}^{\circ} = 1.23$ V or to the lower heating value (LHV) of hydrogen (3.00 kWh Nm^{-3}):

$$\eta_{LHV} = \frac{\text{LHV of the H}_2 \text{ produced}}{\text{electric power required}} = \frac{\dot{V}_{H_2} \times \text{LHV}}{P_{el}} \qquad (22.7)$$

In order to avoid a discussion about choosing the LHV or the HHV for calculating the efficiency, industrial facilities frequently report only the specific electric energy consumption in kWh per Nm^3 of hydrogen generated when comparing electrolyzers.

22.3
Water Electrolysis Technologies

At present, commercial electrolyzers are only available in the areas of alkaline electrolysis and PEM electrolysis. In the case of alkaline electrolysis, they have been produced commercially for several decades in a variety of power classes up to 750 Nm^3 h^{-1}, whereas product development in PEM electrolysis has only been under way for about 20 years, meaning that there are few commercial products (< 65 Nm^3 h^{-1}) on the market. High-temperature electrolysis is not currently being pursued by industry, so no commercial products are yet available.

22.3.1
Alkaline Electrolysis

Alkaline electrolyzers generally operate with an aqueous KOH solution, typically with a concentration of 20–40%. The operating temperature is usually about 80 °C and current densities are in the range 0.2–0.4 A cm^{-2}. However, although water electrolysis was introduced more than 100 years ago, only a few thousand facilities have been constructed to date. As a result of these relatively modest activities, the state-of-the-art for large-scale electrolyzers has only been marginally improved in the past 40 years [5]. Large electrolyzer facilities with a capacity of up to 30 000 Nm^3 h^{-1} of hydrogen for ammonia synthesis or fertilizer production (e.g., at Aswan, Egypt) were only constructed in the past century where cheap electricity from hydropower was locally available. These large-scale facilities mainly used pressureless bipolar electrolyzers manufactured by Norsk Hydro, BBC/DEMAG, and DeNora. The capacity of these electrolyzers was about 200 Nm^3 h^{-1} of hydrogen. Pressurized electrolyzers capable of providing hydrogen and oxygen at 30 bar were only manufactured by Lurgi [6]. The Lurgi pressurized electrolyzer shown in Figure 22.5a

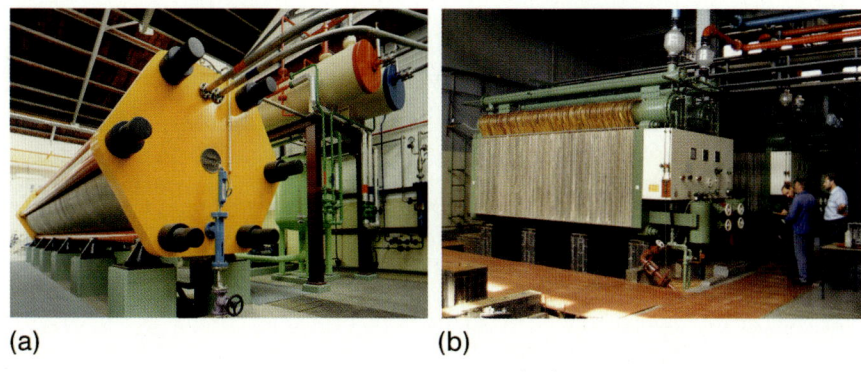

Figure 22.5 (a) Lurgi pressurized electrolyzer for 740 Nm³ h⁻¹ of hydrogen and
(b) a Bamag atmospheric electrolyzer with a capacity of 300 Nm³ h⁻¹ of hydrogen.
Source: ELT Elektrolyse Technik GmbH.

produces hydrogen at 740 Nm³ h⁻¹, which corresponds to an electrical output of ~3.6 MW. The cell stack consists of up to 560 cells with a diameter of 1.60 m and, depending on the number of cells, can be up to 10 m in length. In contrast to the Lurgi electrolyzer, the Bamag electrolyzer operated at atmospheric pressure (Figure 22.5b) uses rectangular electrodes with an active area of ~3 m² and usually has 100 cells with a capacity of ~330 Nm³ h⁻¹ of hydrogen.

In the 1980s and 1990s, large research projects were set up as a reaction to the second oil crisis, with the aim of taking innovative approaches to increase the power density of alkaline electrolysis. This was to be achieved by higher current densities, lower cell voltages and higher operating temperatures in order to reduce capital expenditure and operating costs for electrolysis facilities. The goals of this advanced alkaline water electrolysis were therefore a modification of the cell configuration to minimize the ohmic voltage drop, the development of new low-cost electrocatalysts to reduce the sum of the anodic and cathodic overpotentials while simultaneously increasing the current densities, and raising the process temperature, which can also contribute to reducing the overpotentials and the ohmic voltage drop [7]. In cooperation with Forschungszentrum Jülich, for example, Lurgi demonstrated that by using active electrodes based on Raney nickel and NiO for the diaphragm it was possible to reduce the individual voltage from 1.92 to 1.6 V at a constant current density of 0.2 A cm⁻² or to 1.72 V at twice the current density (0.4 A cm⁻²) by means of a zero-gap cell configuration. As part of a government-funded project, this technology led to the construction of a 32 bar pressurized electrolyzer with an output of 1 MW [8]. In other German national projects (HySolar [9], SWB [10], PHOEBUS [11]), different alkaline water electrolyzers were developed, constructed, and tested. This know-how is, of course, still available, but since then there have been no new innovative approaches in alkaline water electrolysis. According to the NOW (National Organization for Hydrogen and Fuel Cell Technology, NOW GmbH) study by the Fraunhofer Institute for Solar Energy Systems (ISE-Freiburg), although the voltage efficiency of the stack is 62–82%, relative to the thermoneutral voltage of 1.48 V (HHV), for commercial electrolyzers the current densities only

Table 22.2 Overview of leading manufacturers of alkaline electrolyzers.

Manufacturer	Series/operating pressure	H_2 rate (Nm^3 h^{-1})	Energy consumption (kWh Nm^{-3} H_2)	Partial load range (%)
Hydrogenics	HYSTAT/10–25 bar	10–75, max. 18/stack	5.2–5.4 (system)	40–100 5–120 (optional)
H2 Logic	x.00/atmospheric xx.00/4–12 bar	0.66–1.33 32–43	5.4 (system) 4.9–5.0 (system)	No details
NEL Hydrogen	Atmospheric Pressurized, 15 bar	10–500 60	4.1–4.35 (stack) No details	20–100 10–100
Sagim	BP 100/10 bar BP–MP/10 bar MP8/8 bar	0.1–1 1–5 0.5	5 5 5	No details
Teledyne Energy Systems	TITAN HM/10 bar TITAN EC/10 bar	2.8–11.2 28–56	No details	No details
Wasserelektrolyse Hydrotechnik	EV 01–EV 150 atmospheric	1–250	No details	15–100

range from 0.2 to 0.4 A cm^{-2} [5]. The cost of alkaline electrolyzers in the megawatt class is in the region of ~€ 1000 kW^{-1} [5, 12]. These are electrolyzers operating at atmospheric pressure or pressurized electrolyzers operated at 30 bar. Lifetimes of up to 90 000 h were specified for the stack, that is, typically alkaline electrolyzers need to be completely overhauled every 7–12 years, replacing the electrodes and diaphragms [5]. At present, alkaline electrolyzers are supplied commercially by only a few firms. Table 22.2 gives an overview of leading manufacturers.

As can be seen, most manufacturers currently specify a lower partial load range of ~20–40% for alkaline electrolyzers. This lower partial load range does not seem to be beneficial, in particular, for the flexible use of alkaline water electrolysis on the basis of renewable resources since a non-negligible fraction of the capacity for electrolysis cannot be used for hydrogen production. This can be attributed to the use of diaphragms for alkaline water electrolysis since particularly in the partial load range the diffusion of hydrogen through the diaphragm leads to safety-related shutdowns at ~2 vol.% H_2 in O_2 [13]. Today, microporous diaphragms are almost exclusively used as separators in alkaline electrolysis. Such diaphragms used to consist of inorganic serpentine asbestos [chrysotile or white asbestos: $Mg_3Si_2O_5(OH)_4$] or stabilized asbestos. Other inorganic separators include potassium titanate, polyantimonic acids, and nickel oxide on a nickel structure. In addition, efforts have been made to use organic polymers as porous separators, such as polysulfones, polyphenylene sulfides, and polytetrafluoroethylene [14]. Owing to the heterogeneous structure of inorganic diaphragms, O_2–H_2 mixing may occur, especially at high pressure, thus leading to product contamination and an increased system risk.

22.3.2
PEM Electrolysis

PEM electrolysis with proton-conducting membranes (see Figure 22.2b) has been under development for just 20 years, and in that time only a few commercial products have become available (< 65 Nm^3 h^{-1}) for industrial niche applications (e.g., for the local production of high-purity hydrogen for semiconductor manufacturing, electric generator cooling, and the glass industry). In contrast to alkaline water electrolysis, PEM uses platinum group metals for the electrodes, which are usually applied directly to the membrane. In PEM electrolysis, the oxygen evolution reaction has an intrinsic higher overpotential than alkaline electrolysis. The shift in mechanism between acidic and alkaline environments has been attributed to the kinetic facility of oxidizing hydroxide ions instead of water molecules (see Figure 22.6). PEM electrolysis requires the use of a select number of noble elements in order to sustain the acidic pH and high overpotential and only a few elements can be used.

It is well known that the activity for the oxygen evolution reaction (OER) follows the order OER = Ir ≈ Ru > Pd > Rh > Pt > Au > Nb [15]. For the HER, the catalytic activity follows the order Pd > Pt > Rh > Ir > Re > Os > Ru > Ni [15]. Since the pioneering study by General Electric on PEM electrolysis, electrocatalysts with high metal loadings have been based on iridium for the OER and platinum for the HER. To date, the loading for anode catalysts is based on an iridium range between 2.0 and 6.0 mg cm^{-2}. For the cathode, the loading ranges from 0.5 to 2.0 mg cm^{-2} [16]. Consequently, the use of large quantities of noble materials will influence the capital costs of PEM electrolyzers. Over the years, researchers have concentrated their efforts on surveying electrocatalysts in order to mitigate the irreversibility and slowness of the OER. They also aimed to reduce the metal loading and improve the durability of the catalysts against corrosion and/or dissolution. The instability is especially severe for ruthenium-based catalysts with Ru corroding at an appreciable rate with oxygen evolution [17, 18]. It is generally accepted that ruthenium

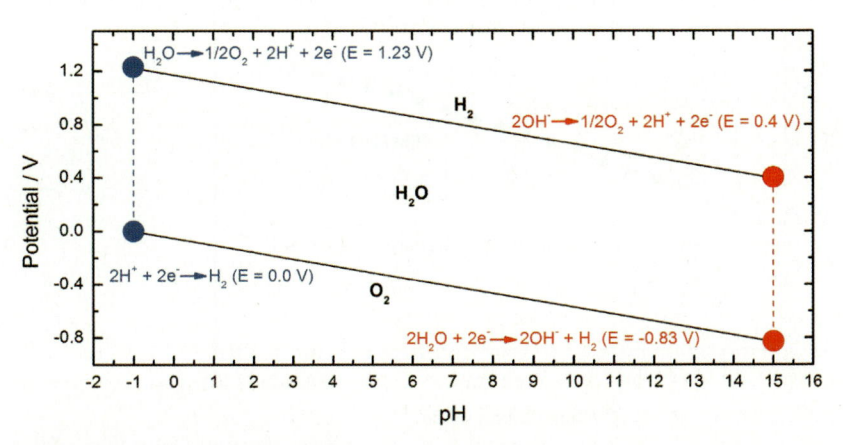

Figure 22.6 Variation of the potential for the different half-reactions in water electrolysis in relation to the pH of the electrolyte.

corrodes by forming RuO_4 in acidic electrolytes, leaching out of the catalyst layer [17, 18]. Durability is also related to the chemical and mechanical properties of the catalyst-coated membrane (CCM) in PEM electrolysis. The method of preparing the catalyst ink/paste and the method of coating the catalyst on the Nafion membrane directly affect the performance and durability of the CCM. The catalyst layer on the membrane undergoes chemical and mechanical stresses such as the harsh corrosive environment, gas evolution through the catalysts layer, cell pressure, and expansion of the membrane. At the same time, CCMs and current collectors have to provide excellent electrical and proton conductivities and also good mass transport characteristics. The feeding water must also be of high purity (< 0.1 µS). Owing to the excellent catalytic activity of platinum for the HER, any other metals coming from the feeding water and deposited on the platinum surface will increase the overpotential, thus reducing performance and durability [19]. This is especially the case if piping susceptible to corrosion is used, or if the deionized water is contaminated by impurities. Foreign ions in the feeding water can also affect the Nafion membrane and ionomer. They will face a reduction in proton conductivity due to deactivation of the sulfonic groups by the foreign ions [19].

In modern commercial systems, about 6 mg cm^{-2} of iridium or ruthenium is used on the anode and about 2 mg cm^{-2} of platinum on the cathode [20]. Under the given operating conditions, these PEM electrolyzers are operated at voltages of ~2 V, at current densities of up to about 2 A cm^{-2} and maximum operating pressures of ~30 bar [21]. Although this corresponds to the same voltage efficiency of ~67–82% (relative to the HHV) as alkaline water electrolysis, in comparison the current densities are much higher (0.6–2.0 A cm^{-2}). A comparison of the typical current–voltage characteristics for alkaline and PEM electrolysis with the typical operating ranges is shown in Figure 22.7. In contrast to alkaline electrolysis,

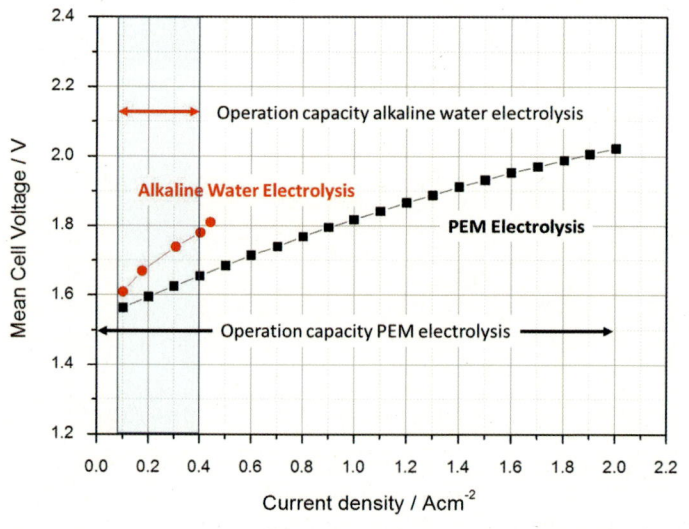

Figure 22.7 Typical current–voltage characteristics and operating ranges for alkaline and PEM electrolysis.

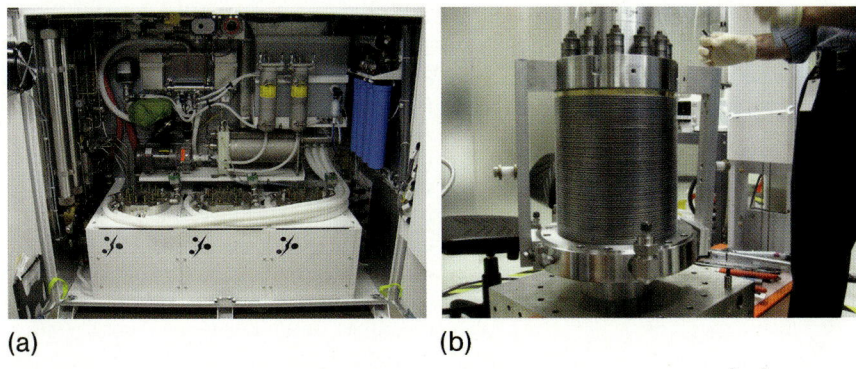

Figure 22.8 (a) Proton OnSite PEM electrolyzer of the HOGEN C type for 30 Nm^3 h^{-1} hydrogen and (b) the corresponding stack with an active single-electrode area of 213 cm^2.
Source: Proton OnSite, Wallingford, CT, USA.

long-term stabilities of only < 20 000 h are given for the lifetime of PEM electrolysis stacks. Nevertheless, Proton OnSite recently achieved a lifetime of more than 50 000 h for stacks as used, for example, in PEM electrolyzers of the HOGEN C series (Figure 22.8) [22].

As can be seen from Table 22.3, in comparison with alkaline electrolysis, PEM electrolysis permits a much larger partial load range, which is particularly beneficial for operation with renewable energy sources. On the cell and stack level, the partial load can be reduced to 0%, but the limit in industrial plants is estimated to be ~5% of the nominal capacity due to the power consumption of the peripheral components [5].

Table 22.3 Overview of leading manufacturers/developers of PEM electrolyzers.

Manufacturer	Series/operating pressure	H$_2$ rate (Nm^3 h^{-1})	Energy consumption (kWh Nm^{-3} H$_2$)	Partial load range (%)
Giner Electrochemical Systems	High pressure/85 bar 30 kW generator/25 bar	3.7 5.6	5.4 (system) 5.4 (system)	No details
H-TEC Systems	EL30/30 bar	0.3–40	5.0–5.5	0–100
Hydrogenics	HyLYZER/25 bar HyLYZER/10 bar	1–2 30	4.9 (stack) 6.7 (system) 4.8 (stack)	0–100 0–150
ITM Power	HPac, HCore, HBox, HFuel 15 bar	0.6–35	4.8–5.0 (system)	No details
Proton OnSite	HOGEN S/14 bar HOGEN H/15–30 bar HOGEN C/30 bar	0.25–1.0 2–6 10–30	6.7 6.8–7.3 5.8–6.2	0–100 0–100 0–100
Siemens	100 kW (300 kW peak) 50 bar	~20–50	No details	0–300

Table 22.4 Comparison of alkaline and PEM water electrolysis.

Alkaline water electrolysis	PEM electrolysis
Advantages • Well-established technology • No noble metal catalysts • High long-term stability • Relatively low costs • Modules up to 760 Nm^3 h^{-1} (3.4 MW)	*Advantages* • Higher power density • Higher efficiency • Simple system configuration • Good partial load toleration • Ability to accommodate extreme overloads (determining system size) • Extremely rapid system response for grid stabilization • Compact stack design permits high-pressure operation
Challenges • Increasing the current densities • Expanding the partial load range • System size and complexity (footprint) • Reduction of gas purification requirements • Overall amount of materials required (at present stacks weigh in the order of several tonnes)	*Challenges* • Increasing the long-term stability • Scale-up of stack and peripherals in the megawatt range • Cost reduction by reducing or replacing noble metal catalysts and cost-intensive components (current collectors/separator plates)

Both electrolysis techniques currently available on the market have advantages and disadvantages representing challenges for implementation of the respective process (Table 22.4):

- In contrast to PEM electrolyzers, alkaline water electrolyzers do not use platinum-group metals for the catalyst; however, their current densities are lower by a factor of five.
- The use of polymer membranes means that the gas qualities and the partial load toleration of PEM electrolyzers is more suitable in comparison to alkaline electrolyzers for intermittent operation with strongly fluctuating outputs.

22.3.3
High-Temperature Water Electrolysis

The high-temperature electrolysis of steam was developed in Germany by Dornier and Lurgi (HOT ELLY) between 1975 and 1987 [23–27]. Apart from the fast kinetics, high-temperature electrolysis has benefits from the thermodynamic perspective. The heat of vaporization of the water reduces the thermodynamic cell voltage at the transition from liquid water to steam (Figure 22.3). The total energy requirements ΔH_R increase slightly with increase in temperature but the total electricity requirement ΔG_R decreases significantly since an increasing proportion of the energy

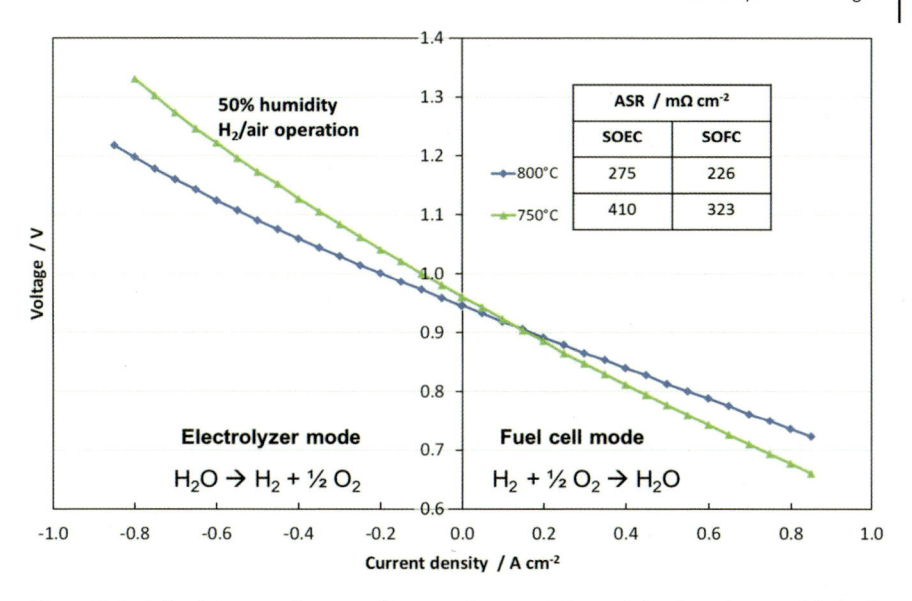

Figure 22.9 Cell voltage as a function of current density (U–I curve) for electrolysis and fuel-cell operation at different temperatures.

demand can be fed in by high-temperature heat, ΔQ_{max}, thus reducing the input of electrical energy ε_{min}. HOT ELLY used an electrolyte-supported tubular concept with yttria-stabilized zirconia (YSZ) as the electrolyte for the solid oxide electrolysis cell (SOEC). Voltages of less than 1.07 V with current densities of 0.3 A cm^{-2} were obtained in endurance tests on single cells. Reversible operation with H_2–H_2O and CO–CO_2 was also demonstrated for the first time with a 10-tube SOEC stack [27]. Dornier nevertheless terminated this development in 1990. The developments and great progress made in the field of high-temperature solid oxide fuel cells (SOFCs) in the past few years have rekindled interest in high-temperature electrolysis (solid oxide electrolysis, SOE) since almost all solid oxide cells (SOCs) are in principle reversible cells and can be used either as SOECs or SOFCs depending on the mode of operation. This trend is reflected in various projects in the United States, Europe, and Asia. However, development is still at the stage of basic research [28]. This means that published data on SOEC performance have normally been obtained exclusively in laboratory cells and stacks. Figure 22.9 shows typical U–I curves for electrolyzer and fuel cell operation at different operating temperatures. SOEC development has profited from SOFC know-how, but further work is still required, especially with respect to the optimization of electrode materials and improvement in long-term stability. Apart from materials research, process engineering studies are urgently needed on the provision of heating energy for water vaporization and preheating.

Apart from steam electrolysis, interest has recently increased in the co-electrolysis of steam and CO_2, since according to the following overall reaction:

$$\underbrace{H_2O + CO_2 \rightarrow H_2 + CO}_{\text{cathode side}} + \underbrace{O_2}_{\text{anode side}} \qquad (22.8)$$

it represents an interesting alternative for synthesis gas in order to produce synthetic fuels according to the Fischer–Tropsch process [29]. In addition to the two purely electrochemical reactions (reduction of steam and CO_2), during co-electrolysis the reversible water-gas shift reaction (WGS) also proceeds:

$$CO + H_2O \rightleftharpoons CO_2 + H_2 \qquad (22.9)$$

In contrast to the electrolysis reactions, the reversible water-gas shift reaction (Eq. 22.9) is an equilibrium reaction which is accelerated in the presence of an Ni catalyst. At high temperature, the equilibrium is on the CO side. This water-gas shift reaction thus always contributes to CO production. Figure 22.10 shows two different ways of producing synthetic fuels by Fischer–Tropsch synthesis using two different methods of high-temperature electrolysis.

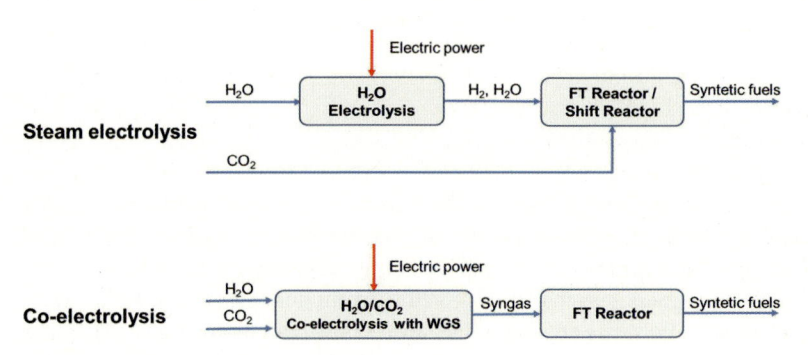

Figure 22.10 Methods of producing Fischer–Tropsch liquids from H_2O and CO_2 by steam electrolysis and co-electrolysis. Modified from [30].

22.4
Need for Further Research and Development

As a chemical energy carrier, hydrogen makes it possible to store energy from renewable sources such as wind energy and thus to decouple supply and demand. In particular, special demands are made on electrolyzers if they are used to provide an operating reserve and are thus operated at strongly fluctuating power inputs and with frequent interruptions due to low input. Normally, very rapid load changes are not problematic for electrolyzers on the stack level since the electrochemical processes react practically instantaneously to load changes. However, the time constants of the downstream system components, such as the electrolyte circuit, pressure regulators, gas separators, and heat exchangers, are considerably greater and it is therefore necessary to optimize the dynamic behavior of these components so that the load changes do not present any problems in the entire power range. This has been suc-

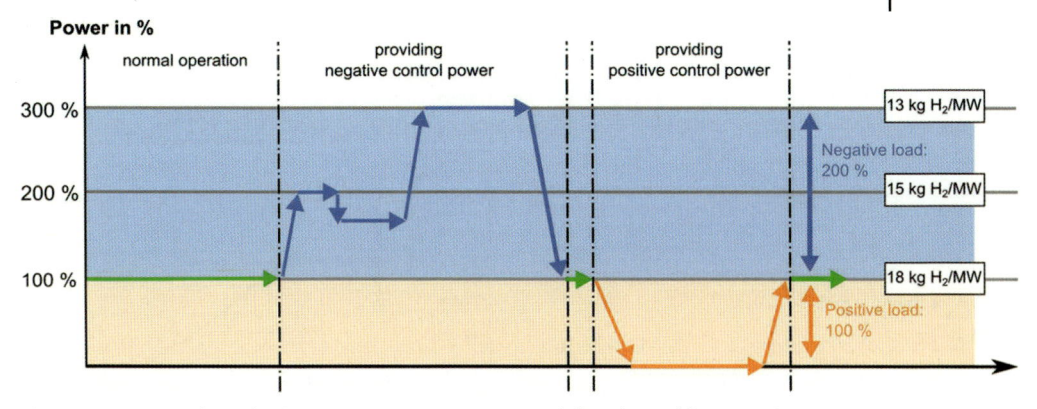

Figure 22.11 PEM electrolyzer as an operating reserve to stabilize the grid for intermittent renewable energy sources. Reproduced from [33] with permission from Siemens.

cessfully demonstrated up to a power range of ~350 kW in various projects where alkaline electrolyzers were directly coupled to photovoltaic (PV) systems [10, 11, 31]. A special problem is the partial load behavior, in particular in alkaline electrolyzers, due to increasing gas contamination. As already mentioned in Section 22.3.1, the lower partial load range of alkaline electrolyzers is only 20–40% of nominal load because contaminating gases, in particular H_2 in O_2, which can very quickly reach a critical concentration of 2 vol.%, requiring a system shutdown for safety reasons. Using special microporous NiO diaphragms, for example, can reduce this partial load range to < 10% [32]. This problem does not arise with PEM electrolysis since, in comparison with alkaline electrolysis, solid polymer membranes are employed, which allows for a greater partial load range (5–100%) and is able to accommodate extreme overloads. This is very beneficial for coupling an electrolyzer as an operating reserve to stabilize a grid fed by intermittent renewable energy sources. Hydrogenics was one of the first companies to implement utility-scale grid stabilization successfully using electrolyzer technology. Siemens are developing a PEM electrolysis system that can be operated at up to 300% overload and can therefore be used for both positive and negative operating reserve (see Figure 22.11).

Although efficiency is reduced in the overload range, the investment costs per kilowatt of installed capacity are lower (Figure 22.12a). In a study on optimizing the technology and reducing the costs of electrolyzers performed by Forschungszentrum Jülich, the impact of electricity costs, operating costs, control, and investment costs on the cost of producing hydrogen was investigated for different electrolysis technologies. Depending on different scenarios, an installed electrolysis capacity of 64–84 GW was forecast for Germany in 2030 [34]. For optimized electrolyzers in the megawatt range for 2030, alkaline electrolysis was calculated to have roughly the same investment costs of $\sim 585/kW_{installed\ capacity}^{-1}$ as PEM electrolysis with comparable efficiencies in the nominal load range, but with the option of operating PEM electrolysis with threefold overload [35]. In other words, the costs per kilowatt of installed capacity drop by a factor of three, or for an assumed electrolysis capacity of 84 GW in 2030, thus only about 28 MW nominal power of PEM electrolysis would

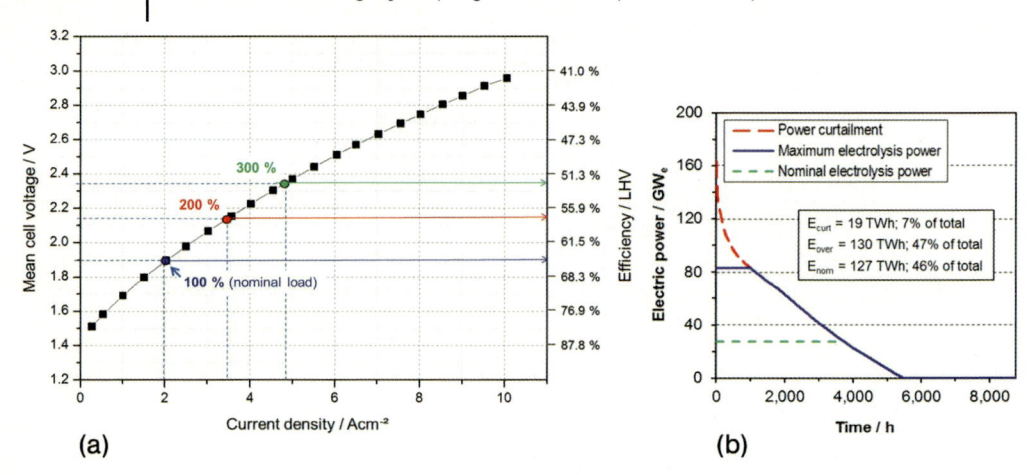

Figure 22.12 (a) Impact of overload on efficiency for a single PEM electrolyzer and (b) installed electrolysis capacity for a given energy scenario in 2030 for Germany.

have to be installed (see Figure 22.12b). With respect to the costs, however, this requires the catalyst loading to be reduced by a factor of 10 while maintaining the same efficiency for PEM electrolysis.

If water electrolysis technology is to be widely and sustainably used on the mass market for the storage of renewable energies after 2020, further steps must be taken to solve outstanding technical issues, such as improving the low power densities and the inadequate stability and reducing the high costs associated with the technologies currently in use.

22.4.1
Alkaline Water Electrolysis

In the short and medium term, relatively mature alkaline water electrolysis technology will play a part in satisfying the increasing demand for electrolytic hydrogen (Figure 22.13). To this end, there is a need for further development and optimization to expand the partial load range and increase the power density. Improved catalysts with low overpotentials and new diaphragm or membrane materials with low specific resistances must be developed in order to increase the power densities. Such materials will also need to display high mechanical stability and lifetime in intermittent operation. A good comprehensive survey of the need for research and development is given in the NOW study on the state-of-the-art and development potential of water electrolysis for producing hydrogen from renewable energy sources [5].

In addition to more materials research, on the systems level in particular, the stack size must be increased to over 1000 $Nm^3\,h^{-1}$ (H_2) by using modified electrodes and diaphragms, and the ability to handle overloads must be improved so that the plant size and the nominal load are no longer tailored to the maximum electrolysis performance, which would considerably reduce the investment costs for electrolysis.

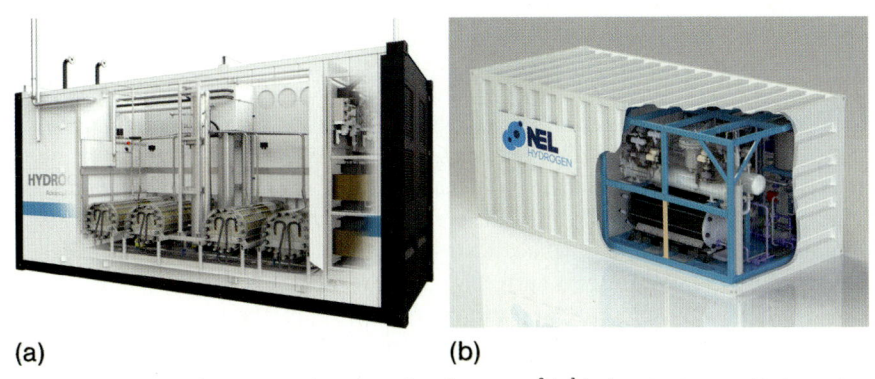

Figure 22.13 (a) Hydrogenics 10 bar electrolyze for 60 Nm^3 h^{-1} hydrogen. Source: Hydrogenics. (b) An NEL Hydrogen pressurized electrolyzer with a stack capacity of 60 Nm^3 h^{-1} hydrogen. Source: NEL Hydrogen AS.

22.4.1.1 Electrocatalysts for Alkaline Water Electrolysis

Electrocatalysts for alkaline electrolysis have attracted considerable attention since the first developments [36]. It is a well-developed technology even at the commercial level [37]. However, in recent decades, with the rise of nanotechnology, a new research trend can be observed in order to develop electrocatalysts, electrodes, and diaphragms further. Studies have also been concentrated on better understanding and improving ionic transport, gas evolution (bubble formation), and operation at higher pressure [36]. Traditionally, the most widely used electrode material for alkaline electrolysis is nickel [38, 39], owing to its stability and favorable activity in an alkaline regime. However, the main problem is the deactivation of the catalyst over time, with the formation of nickel hydride at the surface due to the high concentration of hydrogen in the cathode side. Recently, research [40–45] has been focused on stabilizing the nickel catalyst against deactivation [36] using multicatalyst systems with tuned chemical and physical properties. These multi-component catalysts modify the electronic configuration of the catalyst and facilitate the charge transfer of the chemical reaction. The reduction in the overpotential can generally be attributed to the spillover theory introduced by Bockris *et al.* [46]. In recent years, with support from new nanotechnology developments, new nanostructured electrocatalyst concepts have emerged [40–42, 44, 45, 47–64]. The main objectives are to increase the surface area and, more importantly, to provide unique catalytic properties for both hydrogen and oxygen evolution reactions. Challenges remain essentially to increase the throughput in the synthesis of these advanced materials for large-scale electrode construction. The overpotential is a function not only of temperature but also of current density [36, 37]. The liquid electrolyte and diaphragm have a direct influence on the current densities, because the hydroxyl transport has an intrinsic inertia, being the source of ohmic resistances. Understanding these resistances and developing new materials to be used as diaphragms and electrolytes could promote new opportunities to enhance the efficiency of alkaline water electrolysis.

22.4.2
PEM Electrolysis

In the long term, PEM electrolysis may become of greater significance owing to its advantages compared with alkaline electrolysis (Table 22.4). However, the electrodes and cell area must be scaled up to more than 1000 cm^2 in order to obtain systems for larger applications (> 1 MW). The still considerable investment costs (> € 2000 kW^{-1} [5]) for PEM electrolyzers must be drastically cut. As already shown in Section 1.2, commercial PEM electrolyzers currently require about 6 mg cm^{-2} of iridium on the anode and about 1–2 mg cm^{-2} of platinum on the cathode as catalysts. Owing to the high price and the limited reserves of platinum, iridium, and ruthenium, they can be used as catalyst materials to only a limited extent in the future [65]. This applies in particular to iridium, which in 2010 alone increased in price from US$ 425 to $ 780 per ounce (oz) (28.35 g), since demand increased from 81 000 oz in 2009 to 334 000 oz (9.5 t) in 2010 [66] with an annual production of just 6 t [67]. This was caused, on the one hand, by a greater demand for iridium melting pots in the electronics industry and, on the other, by China's refitting of chlor-alkali electrolyzers. If in 2030 PEM electrolysis was installed with a capacity of 30 GW, then at a catalyst loading of at present 6 mg cm^{-2} this would represent a demand for 45 t of iridium. This would mean that demand would increase by a factor of five in comparison with 2010. Activities in catalyst development for PEM electrolysis are therefore concentrated on reducing or completely replacing noble metals in the catalyst-coated membranes (CCMs) while retaining comparable performance.

22.4.2.1 Electrocatalysts for the Hydrogen Evolution Reaction (HER)

In PEM electrolysis, platinum is recognized to be the best catalyst for the HER. Commercial units and R&D studies have commonly used platinum black or platinum nanoparticles supported on carbon black (Pt/C) from different manufacturers (ETEK-BASF, Tanaka, and Johnson Matthey) as their standard catalysts for the cathode side. However, despite the lower platinum loadings compared with the loading values used on the anode side, the cathode loading still represents a component for considerable cost reductions for the total system, especially if degradation or carbon corrosion occurs. Nowadays, R&D studies are aimed at finding alternative catalysts to reduce the platinum loading, improve catalyst utilization (homogeneity, particle size), and potentially replace the noble metal, creating the so-called platinum-free catalysts. In this sense, core–shell catalyst structures [68] appear to be a very promising option. In a core–shell catalyst configuration, a non-noble element (such as copper or cobalt) formed in a variety of shapes (such as nanoparticles, nanowires, or nanotubes) is covered with a few atomic layers of a noble element (such as platinum). The well-organized/designed structure will provide higher electrochemical activities (electronic effects), higher catalyst utilization, and lower loadings of the noble metal catalyst. Carbon nanotubes are also an alternative to support core–shell catalysts. They provide higher electrical conductivities and higher corrosion resistance compared with conventional carbon black materials. Attempts to replace the noble catalyst completely have also been made. MoS_2 [69], WO_3 [70], and glyoximes [71] are among other options. These catalysts can be supported or formed on different nanostructures and compositions. Unfortunately, these materials still present significantly lower current densities. The redox potentials must be shifted to higher values compared with conventional platinum cathodes.

22.4.2.2 Electrocatalysts for the Oxygen Evolution Reaction (OER)

As previously discussed, iridium (IrO_2) is generally recognized as the state-of-the-art catalyst for the OER in PEM electrolysis [72, 73]. Ruthenium is more active than iridium but problems related to instability (corrosion) limit its use. Aiming to improve efficiency and stability and reduce costs, research groups today are searching for alternative catalysts for the OER in order to overcome the drawbacks. The approaches are concentrated on mixing IrO_2 with a cheaper oxide "diluent" forming a solid solution with less expensive and more durable materials that could be easily manufactured. To date, SnO_2 is the best choice to support IrO_2 catalysts. This support will provide higher catalyst utilization, improve the electrochemical surface area of the IrO_2 catalyst, and improve the durability against dissolution and/or corrosion. Other oxides can be used as a support for IrO_2 and/or RuO_2 catalysts, such as TiO_2, SnO_2, Ta_2O_5, Nb_2O_5, and Sb_2O_5. However, studies suggest that these "non-noble" oxides apparently do not contribute actively to the OER. In fact, SnO_2, for example, would suppress the adsorption of hydroxyl species that are directly involved in the OER. Oxide materials basically contribute by diluting the noble metal content, distributing the nanoparticles and stabilizing the noble metal particles against corrosion. These oxides will in general also negatively affect the electron conductivity of the catalysts due to the semiconductor characteristics of these oxides. Hence this drawback is normally addressed by applying a sufficient amount of IrO_2 that will homogeneously cover the SnO_2 surface, thus improving electrical conductivity [74]. However, doped supports with enhanced electron conductivity could change this picture, such as 20% RuO_2 supported on Sb doped with SnO_2 nanoparticles (ATO) [75]. The enhancement is attributed to the reduction of catalyst particle agglomeration and increased electron conductivity of the RuO_2 by the ATO support. Improvements to overcome the slowness of the OER, increase durability, and reduce costs could also be pursued by using frontier methods to fabricate advanced catalyst structures for anodes in PEM electrolysis. Core–shell structures, nano-designed catalysts, and multi-alloy compositions could play an important role in developing the next generation of catalysts for PEM electrolysis.

22.4.2.3 Separator Plates and Current Collectors

Another challenge for PEM electrolysis is the very small production runs and specific requirements leading to relatively high costs for the separator plates and current collectors typically made of titanium. These materials, due to, hydrogen embrittlement and formation of oxide layers, leading to increased contact resistance, have to be coated with additional protective layers that are also expensive. These components are responsible for 48% of the stack cost (see Figure 22.14 [76]) and a linear contributor (iR) to the required cell voltage. This is particularly important when operating at higher current densities where the internal ohmic resistance and mass transport become the dominating sources of irreversibility. It is the combination of the factors ohmic resistance, mass transport, and cost that create a difficult set of targets for R&D of the current collectors and separator plates for a PEM electrolyzer given the long list of additional constraints. The high cost of the current collectors and separator plates stems from the material, processing costs, and limited scales of economy. Currently the PEM electrolyzer employs separator plates and current collectors with some form of titanium, graphite, or a coated stainless steel. Neither titanium nor steel offers the benefit of a low cost material, and both are also plagued by a variety of operational drawbacks. It is because of these drawbacks, and the ever-important reduction of costs, that the R&D of the current collectors and separator plates face their greatest challenges. The separator plates must provide structure to the cell, insulation between the reactant gases, and a conductive path for the heat and electrons. Titanium offers excellent strength, low initial resistivity, high initial thermal conductivity, and low permeability. However, especially on the oxygen (anode) side, titanium corrodes and develops a passive oxide layer that greatly increases the contact resistance and thermal conductivity. Therefore, the performance of the electrolyzer can diminish as the stack ages [77]. To help protect the titanium separator plates, precious metal coatings and alloys have been investigated. They drastically reduce the corrosion rate but negate any cost savings over graphite as the platinum and gold coatings add an extra processing step and also an expensive coating material to the already expensive titanium base.

Graphite has frequently been used in the PEM fuel-cell field owing to its high conductivity, but its low mechanical strength, high corrosion rate, difficulty of manufacture, and very high cost cause problems for PEM electrolysis. The low strength requires increased thickness, which in turn increases the ohmic resistance, while the high corrosion rate leads to poor contact with the current collectors over time, ultimately increasing the ohmic resistance even further [77]. The high costs have led to the exploration of less expensive base metals. In most cases these base metal replacements are coated to protect against the harsh environment within a PEM electrolyzer. Stainless steel is one of these inexpensive alternatives, but it is not without drawbacks. The stainless-steel components corrode very quickly in the

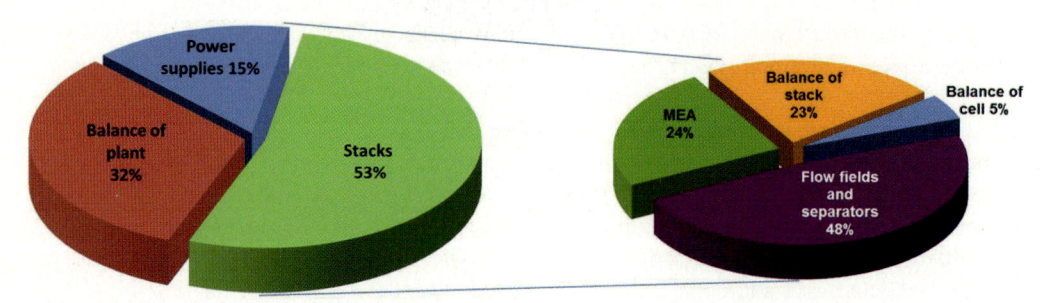

Figure 22.14 Capital cost breakdown for a Proton Onsite 13 kg day^{-1} electrolyzer. Reproduced from [76] with permission from The Electrochemical Society.

corrosive acidic environment, thus requiring a coating to maintain a reasonable lifespan. The addition of a coating typically increases the ohmic resistance and in many cases minor imperfections in the coatings can expose the base metal and thus fail to prevent corrosion and only delay its onset. Coatings can greatly improve the life of the components, but creating and applying a coating that meets the demands of this environment is not an easy task. The desired coating must exhibit low electrical resistivity, high corrosion resistance, high thermal conductivity, good durability, good adhesion with the substrate, and very low imperfection density. Having a low imperfection density is vitally important as once the base metal is exposed, corrosion will cause the coating to flake off, increasing the ohmic resistance and exposing even more of the base metal to the corrosive environment. The separator plate must be designed along with the current collectors for optimum performance. Many electrolyzers, especially those used in early development and testing, were taken directly from PEM fuel cells. Simply borrowing the design poses a problem as the fuel cell was designed to transport gas-phase reactants to the catalyst sites while evacuating the multi-phased products. In a low-temperature PEM electrolyzer, the reactants are in the liquid phase and the products are in the gas phase, altering the flow regimes. The separator plate design also depends greatly on the size of the system. Systems with relatively small cell areas (≤ 25 cm^2) typically employ designs without flow fields and rely on only the current collectors to distribute the water to the catalyst sites. The problems faced by the current collectors are similar to those of the separator plates. A good current collector must provide a rigid structure to support the membrane, a conductive path for heat and electrons, corrosion resistance in a harsh environment, and a porosity that will distribute the water to the reaction sites evenly while evacuating the oxygen or hydrogen from the reaction sites.

Similarly to the separator plates, the materials currently in use for the current collectors include a range of titanium alloys and carbon papers. Reducing corrosion in the current collector is more challenging than the separator plate owing to the difficulties in coating the complex morphology of this component. To address this issue, either new coating methods for porous media must be developed or more expensive materials must be used. A bimodal pore distribution current collector was developed and used in smaller sized electrolyzers to help reduce costs and improve performance [78, 79]. The morphology is designed with a porosity gradient in the through-plane direction so that the separator plate requires no flow fields. This reduces manufacturing costs and ohmic resistance and increases the durability of a coating on the separator plate, while still providing an even supply of water to the catalyst sites. However, as the cell area increases the bimodal current collector cannot supply the water evenly, thus limiting the effectiveness of a cell's full active area and overall capacity. Balancing the mass transport and ohmic losses in the current collector is a delicate act. For every cell configuration, geometry, and operating condition, a morphology exists that will optimize performance. When designing a current collector for optimum performance, the size, shape, and operating parameters (current density, temperature, and pressure) must all be considered before the morphology can be chosen.

22.5
Production Costs for Hydrogen

The costs of producing hydrogen depend on the energy or electricity costs, the plant size (decentralized or centralized electrolysis), and the associated investment costs for electrolysis, plant utilization, and electrical efficiency of the electrolysis. According to a 2009 study by the National Renewable Energy Laboratory (NREL), for decentralized and centralized electrolyzers electricity costs account for 79% and 76%, respectively, of the hydrogen production costs and thus represent the major factors for hydrogen costs, followed by efficiency and investment costs [80]. A centralized 100 MW electrolysis plant has hydrogen production costs of $ 2.70–3.50 kg^{-1} H$_2$ with investment costs of $ 460 kW^{-1}, an efficiency of 67% (LHV), electricity costs of $ 0.045 kWh^{-1} and 100% system utilization. It is assumed that the investment costs for a decentralized 3 MW electrolyzer plant are in the order of $ 380 kW$_{\text{installed power}}^{-1}$ and for a centralized facility of 100 MW in the order of $ 460 kW$_{\text{installed power}}^{-1}$. The investment costs for hydrogen production estimated in Section 22.4 for electrolyzers in the megawatt class amount to € 2.85 kg^{-1} H$_2$ for alkaline water electrolysis, whereas for PEM electrolysis which can be operated with threefold overload the costs are only € 1.97–2.35 kg^{-1} H$_2$ depending on the scenario. A study on the expansion of renewable energies in Germany for 2050 forecast electrolysis costs of € 3.05–3.90 kg^{-1} H$_2$ at an electrolyzer utilization of ~5000 h and an electricity price of € 0.04–0.06 kWh^{-1}, assuming € 600 kW$_{\text{installed power}}^{-1}$ at a system efficiency of 77% (LHV) [81]. Roughly the same conclusions were reached in a study by ISE-Freiburg on the current status and development potential of water electrolysis for the production of hydrogen from renewable energy sources [5], which was performed on behalf of NOW. In this way, depending on the system utilization (35–98%) and electricity costs (€ 0.030–0.050 kWh^{-1}), hydrogen production costs of € 3.17–3.85 kg^{-1} H$_2$ are estimated for a 1 MW PEM electrolyzer (efficiency 73% LHV) with investment costs of € 1200 kWh^{-1}.

22.6
Conclusion

Energy technology is currently undergoing a radical transformation worldwide. The generally accepted factors driving this transition are climate change, security of supply, industrial competitiveness, and local emissions. The expansion of energy generation capacity powered by renewable energy sources leads to new challenges with respect to the storage of large quantities of energy since the expansion of energy generated from renewable resources results in a rapid rise in intermittent energy from wind and solar plants in the electricity supply system. Apart from electrical and thermal storage, considerable emphasis is given to chemical storage in the form of the power-to-gas process, that is, hydrogen or synthetic methane that is produced by water electrolysis (sometimes including methanation). In comparison with directly feeding hydrogen into the natural gas grid or feeding in methane after methanation, the use of hydrogen for transportation purposes in high-efficiency fuel-cell powertrains promises the greatest savings in CO_2. However, if water electrolysis is to be widely and sustainably used on the mass market for hydrogen production from surplus power generated from renewable sources from 2020 onwards, further research is necessary on materials, such as alternative catalysts and membranes, and further steps are necessary to solve open technical issues. These include inadequate power density, insufficient stability and partial load capability, and excessive costs of the electrolyzers currently employed.

If these goals are achieved, production costs after 2030 in the order of € 2–4 kg^{-1} H_2 are realistic for hydrogen generated from renewable energy sources and water electrolysis.

References

1 BMWi and BMU (2010), Energiekonzept für eine umweltschonende, zuverlässige und bezahlbare Energieversorgung, BMWi and BMU, Berlin.
2 Wöhrle, D. (1991) Wasserstoff als Energieträger – eine Replik. *Nachr. Chem. Tech. Lab.*, **39** (11), 1256–1266.
3 Schnurnberger, W., Wittstadt, U., and Janssen, H. (2004) Wasserspaltung mit Strom und Wärme, in *Themenheft 2004: Wasserstoff und Brennstoffzellen – Energieforschung im Verbund 2004*, Forschungsverbund Sonnenenergie, Berlin.
4 Sandstede, G. (1989) Moderne Elektrolyseverfahren für die Wasserstoff-Technologie. *Chem. Ing. Tech.*, **61** (5), 349–361.
5 Smolinka, T., Günther, M., and Garche, J. (2011) *NOW-Studie: Stand und Entwicklungspotenzial der Wasserlektrolyse zur Herstellung von Wasserstoff aus regenarativen Energien, 2011*, NOW, Berlin.
6 Sandstede, G. (1989) Moderne Elektrolyseverfahren für die Wasserstoff-Technologie. *Chem. Ing. Tech.*, **61** (5), 349–361.
7 Winter, C.-J. and Nitsch, J. (eds) (1989) *Wasserstoff als Energieträger: Technik, Systeme, Wirtschaft*, 2nd edn, Springer, Berlin.
8 Streicher, R. and Oppermann, M. (1994) Results of an R&D program for an advanced pressure electrolyzer (1989–1994), in *Hydrogen Energy Progress X, International Conference on Hydrogen Energy Progress, 1994*, Florida Solar Energy Center, Miami.
9 Hug, W. et al. (1990) Highly efficient advanced alkaline water electrolyzer for solar operation. *Adv. Hydrogen Energy*, **8**, 681–690.
10 Szyszka, A. (1999) Schritte zu einer (Solar-) Wasserstoff-Energiewirtschaft, *13 erfolgreiche Jahre Solar-Wasserstoff-Demonstrationsprojekt der SWB in Neunburg vorm Wald, Oberpfalz*.
11 Barthels, H. et al. (1998) PHOEBUS-Julich: an autonomous energy supply system comprising photovoltaics, electrolytic hydrogen, fuel cell. *Int. J. Hydrogen Energy*, **23**, 295–301.
12 Jensen, J. O., Bandur, V., and Bjerrum, N. J. (2008) *Pre-Investigation of Water Electrolysis*, Technical University of Denmark. Lyngby, p. 196.
13 Janssen, H. et al. (2004) Safety-related studies on hydrogen production in high-pressure electrolysers. *Int. J. Hydrogen Energy*, **29** (7), 759–770.
14 VITO (2012) *Vision on Technology*, http://www.vito.be/VITO/EN/HomepageAdmin/Home/Nieuws/Persbericht/persbericht_56.htm (last accessed 1 March 2012).
15 Miles, M. H. and Thomason, M. A. (1976) Periodic variations of overvoltages for water electrolysis in acid solutions from cyclic voltammetric studies. *J. Electrochem. Soc.*, **123** (10), 1459–1461.
16 Carmo, M., Mergel, J., Stolten, D. (2012) A review on the recent progress in electrocatalysis of polymer exchange membrane water electrolysis, presented

at the 19th WHEC – World Hydrogen Energy Conference 2012, Toronto.

17 Lewerenz, H. J., Stucki, S., and Kotz, R. (1983) Oxygen evolution and corrosion – XPS investigation on Ru and RuO_2 electrodes. *Surf. Sci.*, **126** (1–3), 463–468.

18 Kotz, R., Lewerenz, H. J., and Stucki, S. (1983) XPS studies of oxygen evolution on Ru and RuO_2 anodes. *J. Electrochem. Soc.*, **130** (4), 825–829.

19 Stucki, S. et al. (1998) PEM water electrolysers: evidence for membrane failure in 100 kW demonstration plants. *J. Appl. Electrochem.*, **28** (10), 1041–1049.

20 Sheridan, E. et al. (2010) The development of a supported Iridium catalyst for oxygen evolution in PEM electrolysers, presented at the 61st Annual Meeting of the International Society of Electrochemistry, Nice.

21 Smolinka, T., Rau, S., and Hebling, C. (2010) Polymer electrolyte membrane (PEM) water electrolysis, in *Hydrogen and Fuel Cells* (ed. D. Stolten), Wiley-VCH Verlag GmbH, Weinheim, pp. 271–289.

22 Ayers, K. E., Dalton, L. T., and Anderson, E. B. (2012) Efficient generation of high energy density fuel from water. *ECS Trans.*, **41** (33), 27–38.

23 Dönitz, W. and Erdle, E. (1985) High-temperature electrolysis of water vapor – status of development and perspectives for application. *Int. J. Hydrogen Energy*, **10** (5), 291–295.

24 Dönitz, W. and Streicher, R. (1980) Hochtemperatur-Elektrolyse von Wasserdampf – Entwicklungsstand einer neuen Technologie zur Wasserstoff-Erzeugung. *Chem. Ing. Tech.*, **52** (5), 436–438.

25 Isenberg, A. O. (1981) Energy conversion via solid oxide electrolyte electrochemical cells at high temperatures. *Solid State Ionics*, **3–4**, 431–437.

26 Dönitz, W. et al. (1988) Electrochemical high temperature technology for hydrogen production or direct electricity generation. *International Journal of Hydrogen Energy*, **13** (5), 283–287.

27 Erdle, E. et al. (1992) Reversibility and polarization behaviour of high temperature solid oxide electrochemical cells. *Int. J. Hydrogen Energy*, **17** (10), 817–819.

28 Laguna-Bercero, M. A. (2012) Recent advances in high temperature electrolysis using solid oxide fuel cells: a review. *J. Power Sources*, **203**, 4–16.

29 Stoots, C. M. et al. (2009) Syngas production via high-temperature coelectrolysis of steam and carbon dioxide. *J. Fuel Cell Sci. Technol.*, **6** (1), 011014.

30 Stoots, C. M. (2010) Production of synthesis gas by high-temperature electrolysis of H_2O and CO_2, presented at *Sustainable Fuels from CO_2, H_2O, and Carbon-Free Energy*, Columbia University, New York.

31 Abaoud, H. and Steeb, H. (1998) The German–Saudi HYSOLAR program. *Int. J. Hydrogen Energy*, **23** (6), 445–449.

32 Mergel, J. and Barthels, H. (1995) Auslegung, Bau und Inbetriebnahme eines 26 kW-Wasserelktrolyseurs fortgeschrittener Technik für den Solarbetrieb, in *9. Internationales Sonnenforum: Tagungsbericht 2, 28.6.–1.7.1994*, DGS Sonnenenergie, Munich, pp. 931–1832.

33 Waidhas, M. and Käppner, R. (2011) Die Wasserstofferzeugung über Elektrolyse, presented at Workshop "Regelenergie", Darmstadt.

34 Stolten, D., Grube, T., and Weber, M. (2012) Wasserstoff: das Speichermedium für erneuerbare Energie – Eine strategische Betrachtung zur Erreichung der energiepolitischen Vorgaben der Deutschen Bundesregierung., presented at 12. Symposium Energieinnovation, Graz.

35 Mergel, J. et al. (2012) Wasserelektrolyse und regenerative Gase als Schlüsselfaktoren für die Energiesystemtransformation, presented at FVEE – Jahrestagung 2012: Zusammenarbeit von Forschung und Wirtschaft für Erneuerbare und Energieeffizienz, Berlin.

36 Zeng, K. and Zhang, D. (2010) Recent progress in alkaline water electrolysis for hydrogen production and applications. *Prog. Energy Combust. Sci.*, **36**, 307–326; Corrigendum, *Prog. Energy Combust. Sci.*, 2011, **37** (5), 631.

37 Leroy, R. L. (1983) Industrial water electrolysis – present and future. *Int. J. Hydrogen Energy*, **8** (6), 401–417.

38 Balej, J. et al. (1992) Preparation and properties of Raney-nickel electrodes on Ni–Zn base for H_2 and O_2 evolution from alkaline solutions. 2. Leaching (activation) of the Ni–Zn electrodeposits in concentrated KOH solutions and H_2 and O_2 overvoltage on activated Ni–Zn raney electrodes. *J. Appl. Electrochem.*, **22** (8), 711–716.

39 Divisek, J., Mergel, J., and Schmitz, H. (1990) Advanced water electrolysis and catalyst stability under discontinuous operation. *Int. J. Hydrogen Energy*, **15** (2), 105–114.

40 Sahin, E. A. et al. (2012) Investigation of the hydrogen evolution on Ni deposited titanium oxide nanotubes. *Int. J. Hydrogen Energy*, **37** (16), 11625–11631.

41 Ahn, S. H. et al. (2012) Electrodeposited Ni dendrites with high activity and durability for hydrogen evolution reaction in alkaline water electrolysis. *J. Mater. Chem.*, **22** (30), 15153–15159.

42 Zhang, W.-G. et al. (2011) Electrochemical Preparation of a Ni–W–P nanowire array and its photoelectrocatalytic activity for the hydrogen evolution reaction. *Acta Phys.-Chim. Sin.*, **27** (4), 900–904.

43 Tasic, G. S. et al. (2011) Characterization of the Ni–Mo catalyst formed *in situ* during hydrogen generation from alkaline water electrolysis. *Int. J. Hydrogen Energy*, **36** (18), 11588–11595.

44 Kaninski, M. P. M. et al. (2011) A study on the Co–W activated Ni electrodes for the hydrogen production from alkaline water electrolysis – energy saving. *Int. J. Hydrogen Energy*, **36** (9), 5227–5235.

45 Della Gaspera, E. et al. (2011) Structural evolution and hydrogen sulfide sensing properties of $NiTiO_3$–TiO_2 sol–gel thin films containing Au nanoparticles. *Mater. Sci. Eng. B Adv. Funct. Solid-State Mater.*, **176** (9), 716–722.

46 Bockris, J. O'M., Conway, B. M., Yeager, E. B., and White, R. E., (1981) *Comprehensive Treatise of Electrochemistry*, Plenum Press, New York.

47 Cross, M. et al. (2012) RuO_2 nanorod coated cathode for the electrolysis of water. *Int. J. Hydrogen Energy*, **37** (3), 2166–2172.

48 Zheng, H. and Mathe, M. (2011) Hydrogen evolution reaction on single crystal WO_3/C nanoparticles supported on carbon in acid and alkaline solution. *Int. J. Hydrogen Energy*, **36** (3), 1960–1964.

49 Wu, X. and Scott, K. (2011) $Cu_{(x)}Co_{(3-x)}O_{(4)}$ ($0 \leq x \leq 1$) nanoparticles for oxygen evolution in high performance alkaline exchange membrane water electrolysers. *J. Mater. Chem.*, **21** (33), 12344–12351.

50 Shui, J.-I., Chen, C., and Li, J. C. M. (2011) Evolution of nanoporous Pt–Fe alloy Nanowires by dealloying and their catalytic property for oxygen reduction reaction. *Adv. Funct. Mater.*, **21** (17), 3357–3362.

51 Lu, B. et al. (2011) Oxygen evolution reaction on Ni-substituted Co_3O_4 nanowire array electrodes. *Int. J. Hydrogen Energy*, **36** (1), 72–78.

52 Li, Y. et al. (2011) MoS_2 nanoparticles grown on graphene: an advanced catalyst for the hydrogen evolution reaction. *J. Am. Chem. Soc.*, **133** (19), 7296–7299.

53 Zhou, H. et al. (2010) Assembly of core–shell structures for photocatalytic hydrogen evolution from aqueous methanol. *Chem. Mater.*, **22** (11), 3362–3368.

54 Yamada, Y., et al. (2010) Cu/Co_3O_4 nanoparticles as catalysts for hydrogen evolution from ammonia borane by hydrolysis. *J. Phys. Chem. C*, **114** (39), 16456–16462.

55 Wanjala, B. N., et al. (2010) Nanoscale alloying, phase-segregation, and core–shell evolution of gold–platinum nanoparticles and their electrocatalytic effect on oxygen reduction reaction. *Chem. Mater.*, **22** (14), 4282–4294.

56 Wang, X., et al. (2010) Stable photocatalytic hydrogen evolution from water over ZnO–CdS core–shell nanorods. *Int. J. Hydrogen Energy*, **35** (15), 8199–8205.

57 Solmaz, R., Doner, A., and Kardas, G. (2010) Preparation, characterization and application of alkaline leached CuNiZn ternary coatings for long-term electrolysis in alkaline solution. *Int. J. Hydrogen Energy*, **35** (19), 10045–10049.

58 Raoof, J. B. et al. (2010) Fabrication of bimetallic Cu/Pt nanoparticles modified glassy carbon electrode and its catalytic activity toward hydrogen evolution reaction. *Int. J. Hydrogen Energy*, **35** (9), 3937–3944.

59 Phuruangrat, A. et al. (2010) Synthesis of hexagonal WO_3 nanowires by microwave-assisted hydrothermal method and their electrocatalytic activities for hydrogen evolution reaction. *J. Mater. Chem.*, **20** (9), 1683–1690.

60 Merga, G. et al. (2010) "Naked" gold nanoparticles: synthesis, characterization, catalytic hydrogen evolution, and SERS. *J. Phys. Chem. C*, **114** (35), 14811–14818.

61 Li, Y., Hasin, P. and Wu, Y. (2010) $Ni_xCo_{3-x}O_4$ nanowire arrays for electrocatalytic oxygen evolution. *Adv. Mater.*, **22** (17), 1926–1929.

62 Ham, D. J. et al. (2010) Hydrothermal synthesis of monoclinic WO_3 nanoplates and nanorods used as an electrocatalyst for hydrogen evolution reactions from water. *Chem. Eng. J.*, **165** (1), 365–369.

63 Dubey, P. K. et al. (2010) Hydrogen generation by water electrolysis using carbon nanotube anode. *Int. J. Hydrogen Energy*, **35** (9), 3945–3950.

64 Allam, N. K., Alamgir, F., and El-Sayed, M. A. (2010) Enhanced photoassisted water electrolysis using vertically oriented anodically fabricated Ti–Nb–Zr–O mixed oxide nanotube arrays. *ACS Nano*, **4** (10), 5819–5826.

65 Trasatti, S. (1984) Electrocatalysis in the anodic evolution of oxygen and chlorine. *Electrochim. Acta*, **29** (11), 1503–1512.

66 *Platinum 2011*, Johnson Matthey, Royston.

67 Lembke, J. and Höfinghof, T. (2011) Rohstoffe: Rhodiumpulver für den heimischen Tresor, http://m.faz.net/aktuell/finanzen/aktien/rohstoffe-rhodiumpulver-fuer-den-heimischen-tresor-14242.html (last accessed 11 June 2011).

68 Strasser, P. (2009) Dealloyed core–shell fuel cell electrocatalysts. *Rev. Chem. Eng.*, **25** (4), 255–295.

69 Li, Y. G. et al. (2011) MoS_2 nanoparticles grown on graphene: an advanced catalyst for the hydrogen evolution reaction. *J. Am. Chem. Soc.*, **133** (19), 7296–7299.

70 Rajeswari, J. et al. (2007) Facile hydrogen evolution reaction on WO_3 nanorods. *Nanoscale Res. Lett.*, **2** (10), 496–503.

71 Pantani, O. et al. (2007) Electroactivity of cobalt and nickel glyoximes with regard to the electro-reduction of protons into molecular hydrogen in acidic media. *Electrochem. Commun.*, **9** (1), 54–58.

72 Ursua, A., Gandia, L. M., and Sanchis, P. (2012) Hydrogen production from water electrolysis: current status and future trends. *Proc. IEEE*, **100** (2), 410–426; Corrections, *Proc. IEEE*, **100** (3), 811.

73 Park, S. et al. (2012) Oxygen electrocatalysts for water electrolyzers and reversible fuel cells: status and perspective. *Energy Environ. Sci.*, **5**, 9331–9344.

74 Polonsky, J. et al. (2012) Tantalum carbide as a novel support material for anode electrocatalysts in polymer electrolyte membrane water electrolysers. *Int. J. Hydrogen Energy*, **37** (3), 2173–2181.

75 Wu, X. and Scott, K. (2011) RuO_2 supported on Sb-doped SnO_2 nanoparticles for polymer electrolyte membrane water electrolysers. *Int. J. Hydrogen Energy*, **36** (10), 5806–5810.

76 Ayers, K. E. et al. (2010) Research advances towards low cost, high efficiency PEM electrolysis. *ECS Trans.*, **33** (1), 3–15.

77 Jung, H.-Y. et al. (2009) Performance of gold-coated titanium bipolar plates in unitized regenerative fuel cell operation. *J. Power Sources*, **194**, 972–975.

78 Ojong, E. T. et al. (2012) A new highly efficient PEM electrolysis stack without flow channels, operating at high pressure, presented at the World Hydrogen Energy Conference, Toronto.

79 Grigoriev, S. A. et al. (2009) Optimization of porous current collectors for PEM water electrolysers. *Int. J. Hydrogen Energy*, **34** (11), 4968–4973.

80 NREL (2009), *Current (2009) State-of-the-Art Hydrogen Production Cost Estimate Using Water Electrolysis*, National Renewable Energy Laboratory, Golden, CO.

81 BMU (2012), *Studie Langfristszenarien und Strategien für den Ausbau der erneuerbaren Energien in Deutschland bei Berücksichtigung der Entwicklung in Europa und global*, BMU, Berlin.

23
Hydrogen Production by Solar Thermal Methane Reforming

Christos Agrafiotis, Henrik von Storch, Martin Roeb, and Christian Sattler

23.1
Introduction

Hydrogen (H_2) has a long tradition as an energy carrier and as an important feedstock in the chemical industry and refineries. Hydrogen can be produced from a variety of feedstocks, including fossil fuels such as natural gas, oil, and coal and also renewables such as biomass and water with energy input from sunlight, wind, hydropower, and nuclear energy [1]. Virtually all hydrogen produced today is sourced from fossil fuels, the principal method employed being the catalytic reforming of methane (CH_4), the principal component of natural gas and other gaseous fuels such as landfill and coal seam gas. Two different reactions can be distinguished in the methane reforming process: steam methane reforming (SMR) and CO_2 (or dry) methane reforming, represented by the following equations, respectively:

$$CH_4 + H_2O \rightleftharpoons 3\,H_2 + CO \qquad \Delta H^o_{298\,K} = +206 \text{ kJ mol}^{-1} \qquad (23.1)$$

$$CH_4 + CO_2 \rightleftharpoons 2\,H_2 + 2\,CO \qquad \Delta H^o_{298\,K} = +247 \text{ kJ mol}^{-1} \qquad (23.2)$$

Both of these reactions are highly endothermic, hence the heating value of the product is greater than that of the educts and both reactions are favored by high temperatures (industrial reforming processes are carried out between 800 and 1000 °C [1]). The required energy is supplied by combustion of additional natural gas. The reaction gas product mixture is known as synthesis gas (syngas). Syngas is a gas mixture that contains varying amounts of CO and H_2. Its exothermic conversion to fuel and other products has been utilized commercially since at least the first quarter of the last century, for example, via Fischer–Tropsch technology [2, 3].

Although reforming is likely to remain the technology of choice for some time, hydrogen is ultimately seen to be the clean fuel of the future and will need to be produced entirely from renewable energy. In this perspective, the harnessing of the huge energy potential of solar radiation and its effective conversion to hydrogen are

Transition to Renewable Energy Systems, 1st Edition. Edited by Detlef Stolten and Viktor Scherer.
© 2013 Wiley-VCH Verlag GmbH & Co. KGaA. Published 2013 by Wiley-VCH Verlag GmbH & Co. KGaA.

a subject of primary technological interest. There are basically three pathways for producing hydrogen with the aid of solar energy [4]: electrochemical, photochemical, and thermochemical. The last approach is based on the use of concentrated solar radiation as the energy source for performing high-temperature reactions that produce hydrogen from the transformation of various fossil and nonfossil fuels via different routes such as water splitting (to produce hydrogen and oxygen) [5, 6], natural gas steam reforming (to produce syngas) [7–9], natural gas cracking (to produce hydrogen and carbon) [10–13], and gasification of solid carbonaceous materials such as coal or biomass (to produce syngas) [14–16]. All of these routes involve in some step endothermic reactions that can make use of concentrated solar radiation as their energy source of high-temperature process heat.

Even though, obviously, the ideal raw material for hydrogen production is water, owing to its abundance and low value and the absence of CO_2 emissions during its dissociation to hydrogen and oxygen, at least for a transition period, hydrogen supply at a competitive cost can only be achieved from hydrocarbons – essentially natural gas using well-known commercial processes such as steam reforming where methane and steam are converted to syngas. As an intermediate step, considerable effort is being spent on developing a hybrid hydrogen technology in which concentrated solar thermal energy is used to provide the heat for the high-temperature endothermic steam–methane reforming reaction. In doing so, solar energy is embodied thermochemically in the product hydrogen. This overcomes many of the limitations of solar energy, enabling it to be stored at ambient conditions, transported from the point of collection to where it is required, and used independently of daylight hours. Such a transitional technology is considered by many to be an essential stepping stone from current practice to a truly renewable-based hydrogen economy [17].

Methane is the most abundant hydrocarbon and it also has the highest hydrogen content. Water is also rich in hydrogen and therefore the steam–methane reforming reaction is the preferred reaction for hydrogen production. Both the steam reforming reaction in Eq. 23.1 and the CO_2 reforming reaction in Eq. 23.2 are highly endothermic, and therefore offer an opportunity to embed and thus store solar energy. If solar energy is used to provide the heats of reaction for reactions 23.1 and 23.2, then the product gas will theoretically contain up to 26% and 31% of solar energy embodied in chemical form, respectively (low heating value basis, assuming water stays as a vapor). In presence of water, the reforming reaction is followed by the water-gas shift (WGS) reaction according to Eq. 23.3 (see Section 23.2.1). Because the WGS reaction is exothermic, the conversion of CO to hydrogen and CO_2 actually reduces the amount of energy that can be stored in the products, to 21% in both cases (since in both cases 4 mol of H_2 are produced per mole of methane) [17].

This storage of solar energy can be further "exploited" via two process options. In the so-called "open-loop" systems, a hydrocarbon feedstock is upgraded in energy content with solar energy to produce directly on-site H_2/CO "solarized" fuel syngas for subsequent combustion in a conventional gas turbine (GT) or a combined cycle (CC) power plant. These concepts are more thermally efficient than simply using the solar energy to produce steam because they harvest the solar energy in chemical form rather than sensible heat; thus the solar energy share in the product fuel can be converted to electricity at significantly higher efficiencies in large combined gas turbine cycle plants (at 45–50% thermal efficiency) rather than just using it in the less efficient steam turbine (ST) cycle (at 30–35% thermal efficiency) [18, 19].

In the so-called "closed-loop" systems, a high-quality hydrocarbon feedstock such as methane is converted to syngas via solar reforming; the syngas is then stored or transported off-site prior to conversion back to CH_4 in a methanation reactor that recovers the solar energy as heat for industrial processes or power generation (between, for example, high-insolation solar collection sites and major industrial centers).

23.2
Hydrogen Production Via Reforming of Methane Feedstocks

23.2.1
Thermochemistry and Thermodynamics of Reforming

Both reactions 23.1 and 23.2, owing to their increase in number of moles, are favored by low pressures. As can be seen in Figure 23.1, the methane conversion for a steam reforming reaction reaches 100% around 850 °C at 1 bar, whereas at 7.5 bar the temperature has to be increased to 1200 °C to reach the same conversion. However, commercial reforming plants are operated at much higher pressure levels (usually above 25 atm) [1]. This is due to process optimization: syngas is merely an intermediate product and is further processed (e.g., to hydrogen for ammonia production or methanol). These consequent processes require high pressures. Heat transfer and mass flow requirements also favor high pressures for the reaction. Therefore, operation of reforming reactors at the pressure required for the consequent process seems to be a reasonable choice. In the presence of water, the reforming reaction is followed by the WGS reaction according to Eq. 23.3. This reaction occurs primarily at lower temperatures (250–400 °C) [20].

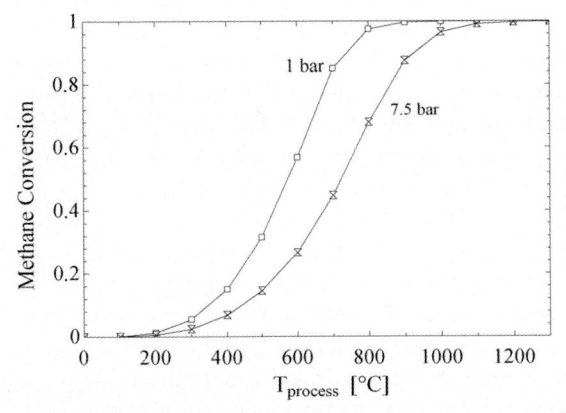

Figure 23.1 Conversion of methane in steam reforming as a function of temperature for different pressure levels.

$$CO + H_2O \rightleftharpoons H_2 + CO_2 \qquad \Delta H^\circ_{298\,K} = -41 \text{ kJ mol}^{-1} \qquad (23.3)$$

The overall reaction of steam reforming followed by the WGS is given by

$$CH_4 + 2\,H_2O \rightleftharpoons 4\,H_2 + CO_2 \qquad \Delta H^\circ_{298\,K} = +165 \text{ kJ mol}^{-1} \qquad (23.4)$$

As can be seen from reactions 23.1 and 23.2, the H_2/CO ratio in the product differs significantly: 3 and 1, respectively. Therefore, in order to provide a high hydrogen yield, steam reforming followed by the WGS reaction is most suitable, as an H_2/CO ratio of up to 4 can be achieved. However, the required properties of syngas depend on its subsequent utilization [21], since for further processing to liquid fuels, for instance, a molar ratio of H_2/CO of > 3 is unsuitable [22, 23].

Both reactions can generally be catalyzed by group VIII metals. Current state-of-the-art catalytic systems for natural gas steam reforming are based either on highly expensive precious metals such as Ru and Rh or on systems of significantly lower cost based on Ni metal. In commercial applications, Ni-based catalysts supported on mixed oxides of Ca–Al [24] or Mg–Al [25] with hexaaluminate and spinel structures (e.g., $CaAl_6O_{10}$ or $MgAl_2O_4$, respectively) have proved to be most suitable, owing to low cost and high catalytic activity. Other catalysts are neglected because of to high costs (noble metals) or technical issues (Fe and Co).

The two main issues remaining in reformer design utilizing Ni-based catalysts are sulfur compounds in the feed and carbon formation at the catalytically active site. Whereas the former can usually be eliminated by means of a hydrogenator followed by a zinc oxide bed, the latter constitutes a more complicated issue. Carbon formation in reforming reactions occurs mainly due to methane decomposition (pyrolysis) or disproportionation of carbon dioxide according to Eqs. 23.5 and 23.6. respectively:

$$CH_4 \rightarrow C + 2\,H_2 \qquad (23.5)$$

$$2\,CO \rightarrow C + CO_2 \qquad (23.6)$$

Both reactions are catalyzed by metals, hence the risk of carbon formation is high in the presence of Ni-based catalysts. Methane decomposition occurs above 650 °C in the absence of an oxygen source (i.e., steam or air). Therefore, in commercial steam reforming plants most commonly the introduction of excess steam into the feed is used to reduce the risk of carbon formation. Usually the steam to carbon ratio (S/C) is set between 2 and 5 (resulting in an atomic H/C ratio of 8–15). Therefore, the share of sensible heat in the process is largely increased and the chemical efficiency of the process is decreased. As can be seen from Eq. 23.2, considering that the aim of the process is complete conversion of methane, the maximum H/C ratio in dry reforming is 2, decreasing with an over-stoichiometric CO_2 content in the feed. Therefore, in dry reforming processes noble metals are more commonly used as catalysts, because reforming can be operated at significantly lower H/C ratios without carbon formation. Another possibility for the inhibition of carbon formation is the addition of promoters to the Ni-based catalyst (e.g., alkaline earth metal oxides) [1, 3, 5, 26–28].

23.2.2
Current Industrial Status

The commercial feedstock of choice for syngas production is natural gas and the most widely used process is steam reforming, as discussed briefly earlier. Steam reforming is usually conducted inside tubes packed with nickel catalyst. Schematics of the operation of typical industrial reformers are shown in Figure 23.2 [29]. The outer diameter of the tubes ranges typically from 100 to 150 mm and the length from 10 to 13 m. The tubes are made of high alloys (e.g., Cr25%–Ni20%) and are heated by radiation and convection from burning natural gas or refinery waste fuel gas. Typical inlet temperatures to the catalyst bed are 450–650 °C and product gas leaves the reformer at 800–950 °C depending on the application. Tubular reformers are designed with a variety of tube and burner arrangements. The heater design – either from the top or from the sides (Figure 23.2a) – and heat supply have to be such that on the one hand the tube wall temperature is high enough for the reforming reaction but on the other the tube surface is not overheated. A typical side-fired reformer (Figure 23.2b) has over 350 burners. In top-fired reformers (Figure 23.2c) the tubes are spaced to allow the down flames to fire between them. The radiant gases leave the box horizontally at the bottom and are used to generate the process steam.

In commercial applications of methane reforming, the activity of the catalyst is usually not the limiting factor. However, heat transfer and diffusion limitations decrease the effectiveness factor of the catalyst to levels even less than 10% because of transport restrictions. The optimum to achieve maximum activity with minimum pressure drop is a catalyst bed of pellets with large external diameter and high void fraction, as achieved with rings or cylinders with several holes. Other solutions may be based on the use of catalysts consisting of ceramic foams or monoliths.

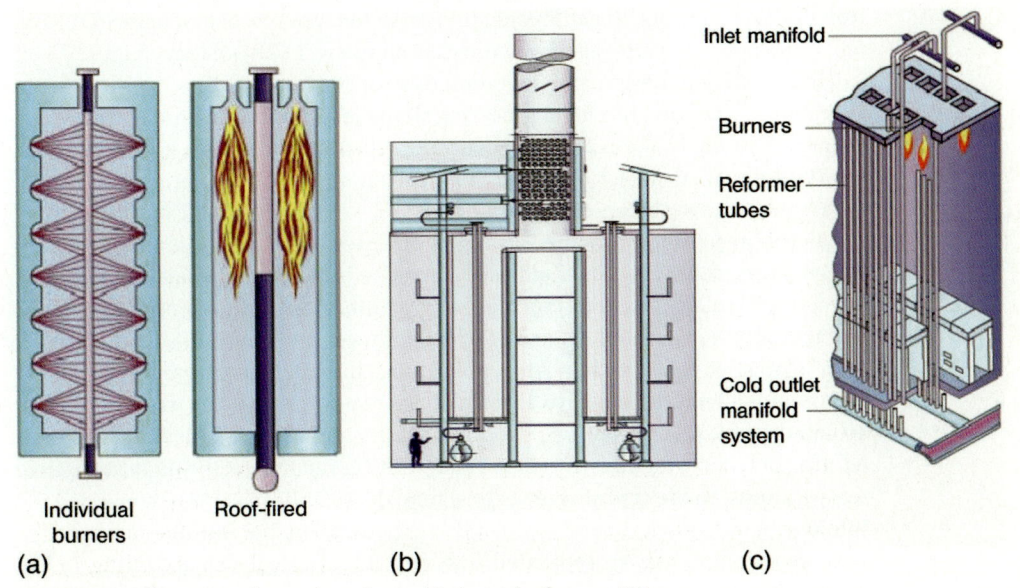

Figure 23.2 Schematics of operation of typical industrial reformers [29].

From intrinsic kinetics, it can be shown that a space velocity of 10^4 h^{-1} leads to close-to-equilibrium composition in the product gas [1–3].

The steam reforming process as practiced today faces a number of constraints. First, thermodynamics demands high exit temperatures to achieve high conversions of methane. In contrast, the catalysts are potentially active even at temperatures below 400 °C. Consequently, there have been efforts to circumvent the constraints by the use of a hydrogen-selective membrane installed in the catalyst bed [30] that continuously removes hydrogen from the reactant stream, thereby driving the equilibrium towards higher conversion at lower temperature. A low-temperature membrane reformer would be able to use low-temperature heat, a fact crucial to the coupling of reforming with solar energy, as discussed later. However, the hydrogen produced according to this concept is at low pressure and must be compressed to the usual delivery pressure of 20 bar. This limitation renders the process uneconomic except when very low electricity prices prevail, or when hydrogen is used as a feedstock for a fuel cell or as a low-pressure fuel.

23.3
Solar-Aided Reforming

23.3.1
Coupling of Solar Energy to the Reforming Reaction: Solar Receiver/Reactor Concepts

Large-scale concentration of solar energy is mainly accomplished at pilot and commercial Concentrated Solar Power (CSP) plants with four kinds of optical configuration systems using movable reflectors (mirrors) that track the sun, namely: parabolic trough (PT) collectors, linear Fresnel (LF) reflector systems, dish–engine (DE) systems, and power towers – also known as central receiver (CR) systems (Figure 23.3).

For the efficient design and operation of solar receiver/reactors, concepts from "traditional" chemical reactor engineering should be combined with ways to achieve efficient heating of the reactor via concentrated solar irradiation. First, just as in the "traditional" nonsolar chemical engineering, the catalytic reactor type can be separated into two broad categories depending on whether the catalyst particles are distributed randomly or are "arranged" in space at the reactor level: the former category includes packed and fluidized catalytic beds; the latter includes the so-called "structured" catalytic systems such as honeycomb, foam, and membrane catalytic reactors, all three of them being free of randomness at the reactor's level [31].

In solar-aided catalytic chemistry applications, the reactor concepts above have to be combined with the effective absorbance of concentrated solar irradiation to be achieved via the solar receiver. The receiver's task is to "trap" (absorb) the concentrated solar radiation and transfer it to the heat-transfer medium at the highest possible temperatures. In the case of receiver/reactors, the solar heat has to be converted not into mechanical work as in the case of power generation, but into chemical energy.

Solar receivers can be separated into two broad categories according to the mechanism of transferring heat to the heat-transfer fluid: either directly or indirectly.

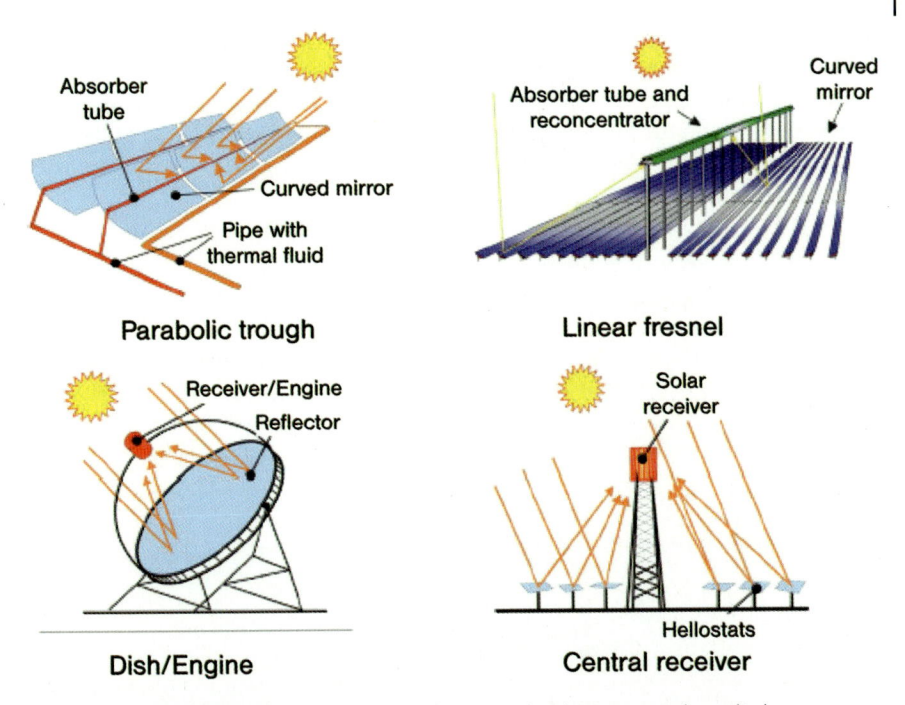

Figure 23.3 Schematics the four solar concentrating technologies currently applied at commercial CSP plants (CR) [91].

The common characteristic of the indirectly irradiated receivers (IIRs) is that the heat transfer to the working fluid does not take place on the surface which is exposed to incoming solar radiation. Instead, there is an intermediate opaque wall, which is heated by the irradiated sunlight on one side and transfers the heat to a working fluid on the other side [32]. The simplest examples of such receivers are conventional tubular receivers that consist of absorbing surfaces exposed to the concentrated solar irradiation. The heat-transfer fluid (e.g., a gas or a molten salt) is moving in a direction vertical to that of the incident solar radiation (Figure 23.4a, left); heat transfer to the thermal fluid takes place through the receiver opaque walls by conduction.

In the case of receiver/reactors, two further options are possible. In the first option the receiver and the reactor are decoupled. The latter is not located within the solar absorbing module: the enthalpy of a solar receiver-heated heat-transfer fluid is transferred via insulated tubes to a reactor where endothermic chemical reactions take place. The operating principle of this concept is shown schematically in Figure 23.4b, left. The same decoupled configuration can be used when an other-than-solar high-temperature source is available, for example, a chemical reactor can be heated via hot circulated helium coming from nuclear reactors. Alternatively, the solar-heated tubular receivers can contain the catalyst material in the form of either packed beds or structured assemblies – again in this case, the catalyst is heated via conduction through the tube walls. The operating principle of this concept is shown schematically in Figure 23.4b, middle. Finally, another version of indirectly heated receivers is the so-called heat pipe receivers where concentrated sunlight is employed for the evaporation of a liquid substance (e.g., liquid sodium), which in turn condenses on the tubes containing the heat-transfer fluid, liberating the heat of vaporization, which in this case is transferred isothermally to the heat-transfer fluid through the walls [33].

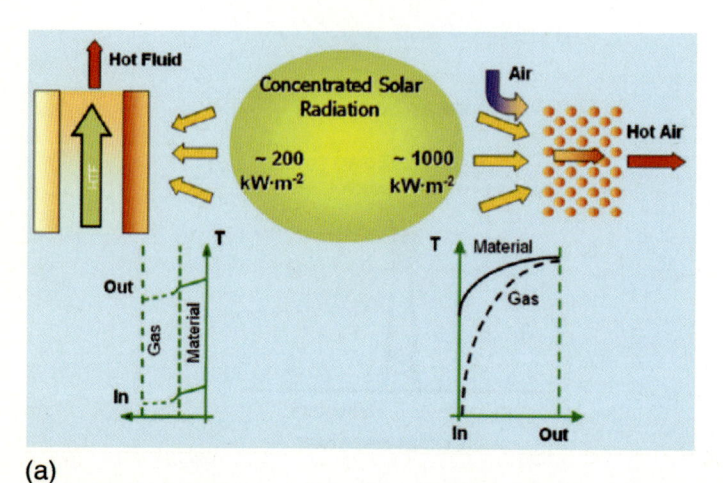

Figure 23.4 (a) Operating principles of indirectly heated tubular (left) and directly heated volumetric (right) solar receivers [37]. (b) Comparison of operating principles of indirectly irradiated receiver decoupled from the reformer (left), indirectly irradiated integrated tubular reactor/receiver (middle) and directly irradiated reactor/receiver (right). (c) Monolithic ceramic structures proposed and explored as volumetric receivers: honeycomb (left), foam (middle) [35] and pin-finned "porcupine" (right) [32].

However, the most direct and therefore potentially the most efficient method would involve direct heating of the heat-transfer fluid by the concentrated beam, thus eliminating the wall as the light absorber and heat conductor. Directly irradiated receivers (DIRs) make use of fluid streams or solid particles directly exposed to the concentrated solar radiation. A key element of all DIRs is the absorber: the component that absorbs concentrated sunlight and transports its energy to a working fluid flowing within and over it. DIRs are also called "volumetric" receivers since they enable the concentrated solar radiation to penetrate and be absorbed within the entire volume of the absorber. In different designs, the absorber is either a stationary matrix (grid, wire mesh, foam, honeycomb, etc.) or moving (usually solid) particles.

DIRs with stationary absorbers are relatively simple and the most common of the volumetric receiver family. Here, the absorbing matrix must be able to absorb highly concentrated radiation, while providing sufficient heat convection to the working gas flow. It is also required to sustain thermal stresses created by large temperature gradients in addition to thermal shock caused by rapid heating–cooling cycles. Such receivers, first proposed in 1982 [34], are compact heat exchangers comprising a pack of high-porosity material structures capable of absorbing the concentrated solar radiation and transferring it to a gaseous heat-transfer medium [35, 36]. For instance, gas (air) can be driven through the absorber, flowing past the absorbing structures parallel to the direction of the incoming radiation (Figure 23.4a, right) [37] and be heated by convection. Hence the volumetric receiver concept entails the solar irradiation and the heat extraction taking place on the same surface simultaneously (Figure 23.4b, right). This peculiarity leads to the minimization of radiation and convection losses, to the maximization of radiation absorption, and to enhanced operational flexibility and economics.

Such structures can be steel or ceramic wire meshes, ceramic foams (Figure 23.4c, left), or multi-channeled honeycomb structures (Figure 23.4c, middle). In fact, such structures have been extensively tested in CSP facilities [36, 38, 39]. Volumetric receivers based on ceramic honeycombs developed at the German Aerospace Center (DLR) have currently reached the level of commercial exploitation in the 1.5 MWe Solar Tower Jülich (STJ) thermal power plant. It is capable of heating ambient air to temperatures of ~700 °C that is further used in traditional energy cycles supplying power to the local community [40]. Researchers at the Weizmann Institute of Science (WIS) in Rehovot, Israel [32], have introduced and explored an alternative design of a structured volumetric solar absorber, nicknamed "porcupine" (Figure 23.4c, right). It is an array of pin-fins, constructed with elongated heat-transfer elements made of ceramic tubes or rods (i.e., the porcupine "quills"), implanted in a base plate The cross-flow pattern introduces turbulent mixing and enhances the rate of convective heat transfer from the absorber matrix to the fluid.

Often the operating conditions demand that the absorber be physically separated from the ambient, for example, when the flow is pressurized or the working fluid is not air. In these cases, the receiver must be equipped with a transparent window, which allows concentrated light to enter the receiver, while separating the working gas and the ambient air. Such pressurized volumetric receivers have been developed and tested at the WIS [41].

In a direct analogy with "conventional" catalytic applications, it becomes obvious that all three structured porous volumetric solar absorber modules shown in Figure 23.4c can be coated with proper functional materials capable of performing/catalyzing a variety of high-temperature chemical reactions – among them reforming – and thus be "transformed" and adapted to operate as solar chemical receiver/reactors where chemical reactions can take place in an efficient and elegant manner with the aid of the functional materials immobilized on their porous walls [42]. Such devices comprise an integrated solar receiver–chemical reactor, in which a catalyst-coated porous matrix volumetrically absorbs concentrated solar (radiant) energy directly at the catalyst sites, promoting heterogeneous reactions with gases flowing through the matrix (absorber) – hence they are frequently referred to as directly irradiated volumetric receiver/reactors (DIVRRs). In this way, absorbed radiation is converted from thermal to chemical form, thus storing solar energy in the chemical bonds of the reaction products rather than as thermal energy in a working fluid. On the other hand, direct heating in the case of receiver/reactors necessitates the use of a transparent window isolating the reactant gas streams from the ambient air and providing for reactor operation under nonatmospheric pressures if needed.

Each receiver/reactor type has its advantages and disadvantages; it is therefore not surprising that essentially all receiver/reactor concepts described earlier have been used for solar-driven reforming to a greater or lesser extent, reaching different scale-up levels. In fact, reforming of natural gas was the first process for solar conversion of hydrocarbons tested on an engineering scale of some hundreds of kilowatts of solar power input.

23.3.2
Worldwide Research Activities in Solar Thermal Methane Reforming

The concept of solar-driven reforming was first described in 1980 by T. A. Chubb of the US Naval Research Laboratory, who proposed the CO_2–CH_4 reforming–methanation cycle as a mechanism for converting and transporting solar energy via solar receivers [43], and operated a solar tubular reformer with a ruthenium catalyst at the White Sands solar furnace [44]. Interest in this technology was revived in the late 1980s with the initial research on solar-driven reforming focusing on the concept of a closed loop for storage and transport of solar energy, in analogy with that with high-temperature energy being supplied by a nuclear reactor. Parallel work took place during the same period at the Institute of Catalysis, Novosibirsk, in the former Soviet Union [45–47].

At the same time, the development of solar-based CO_2–CH_4 reforming technology was studied on the technical scale in Israel and Germany as part of the International Energy Agency Solar Power and Chemical Energy Systems (SolarPACES) R&D program. Between 1988 and 1992, in an attempt to develop more economical, compact receivers for methane reforming, experiments on a laboratory scale were started at Sandia National Laboratories (SNL), Albuquerque, NM, USA [48], at DLR, Stuttgart, Germany, and at the WIS, with both indirectly heated tubular reactors and the first windowed receiver/reactors where the catalyst was heated directly by a concentrated solar beam [49–51]. The technical characteristics and results were summarized and compared by Kirillov [52].

A significant amount of work has been conducted over the last 20 years on the development and scale-up of the technology of solar reforming of methane and other hydrocarbons at DLR, the WIS, and Sandia National Laboratory, in several cases within joint projects. Relevant research on solar reforming concepts is currently being performed all over the world, principally by laboratories and research institutes that possess pilot- to large-scale CSP facilities, such as the Plataforma Solar de Almería, Spain, operated by Centro de Investigaciones Energéticas, Medioambientales y Tecnológicas (PSA-CIEMAT), the Swiss Federal Institute of Technology (ETH) and the Paul Scherrer Institute (PSI), Switzerland, and the Ente per le Nuove Tecnologie, l'Energia e l'Ambiente (ENEA), Italy, in Europe, by the Department of Energy's National Renewable Energy Laboratory (NREL), Sandia National Laboratory, and the University of Colorado at Boulder, CO, United States, the Commonwealth Scientific and Industrial Research Organization (CSIRO), Australia, Niigata University and the Tokyo Institute of Technology, Japan, and Inha University, Korea, among others. The research activities can be divided into general reformer concepts and concepts for the improvement of the catalyst system (i.e., the catalyst and the structure to which it is applied). Regarding the catalysts in solar reforming, there are two major topics that require further development. First, the price of noble metal catalysts that are suitable for dry reforming of methane is too high for industrial application. Therefore, a nickel-based catalyst capable of performing dry reforming of methane without deactivation due to carbon deposition has to be developed [53]. The second obstacle relates to directly irradiated receiver/reactors. The catalytically activated absorber has to fulfill not only the requirements regarding the chemical reaction but also those relating to a solar absorber (i.e., high thermal shock resistance, low reflectivity and emissivity).

Overall, many technologies are tested in parallel; new concepts are first tested at the laboratory or solar-simulation scale, whereas other, more mature concepts are implemented through large bi- or multi-lateral research projects involving partners possessing large-scale solar facilities [54].

23.3.2.1 Indirectly Heated Reactors

Heat-Transfer Fluid: Air

This concept was first tested within the Advanced Steam Reforming of Methane in Heat Exchange (ASTERIX) project, a joint Spanish–German project carried out by CIEMAT and DLR in the late 1980s and early 1990s [55]. Work performed by the WIS between 1993 and 1998 has been reported [48, 56].

To address scalability issues, WIS researchers transferred the concept to the so-called "beam down" receiver (Figure 23.5) [57]; with the aid of an optical feature in the form of a 75 m^2 reflector shaped as a hyperboloid section attached to the tower at about 45 m above ground level (Figure 23.5a), about 1 MW of concentrated sunlight can be reflected down on to a ground target. These optics allow a multi-megawatt tubular reformer to be built on the ground in a way resembling the construction of a conventional, commercial reformer (Figure 23.5a) with roof burners.

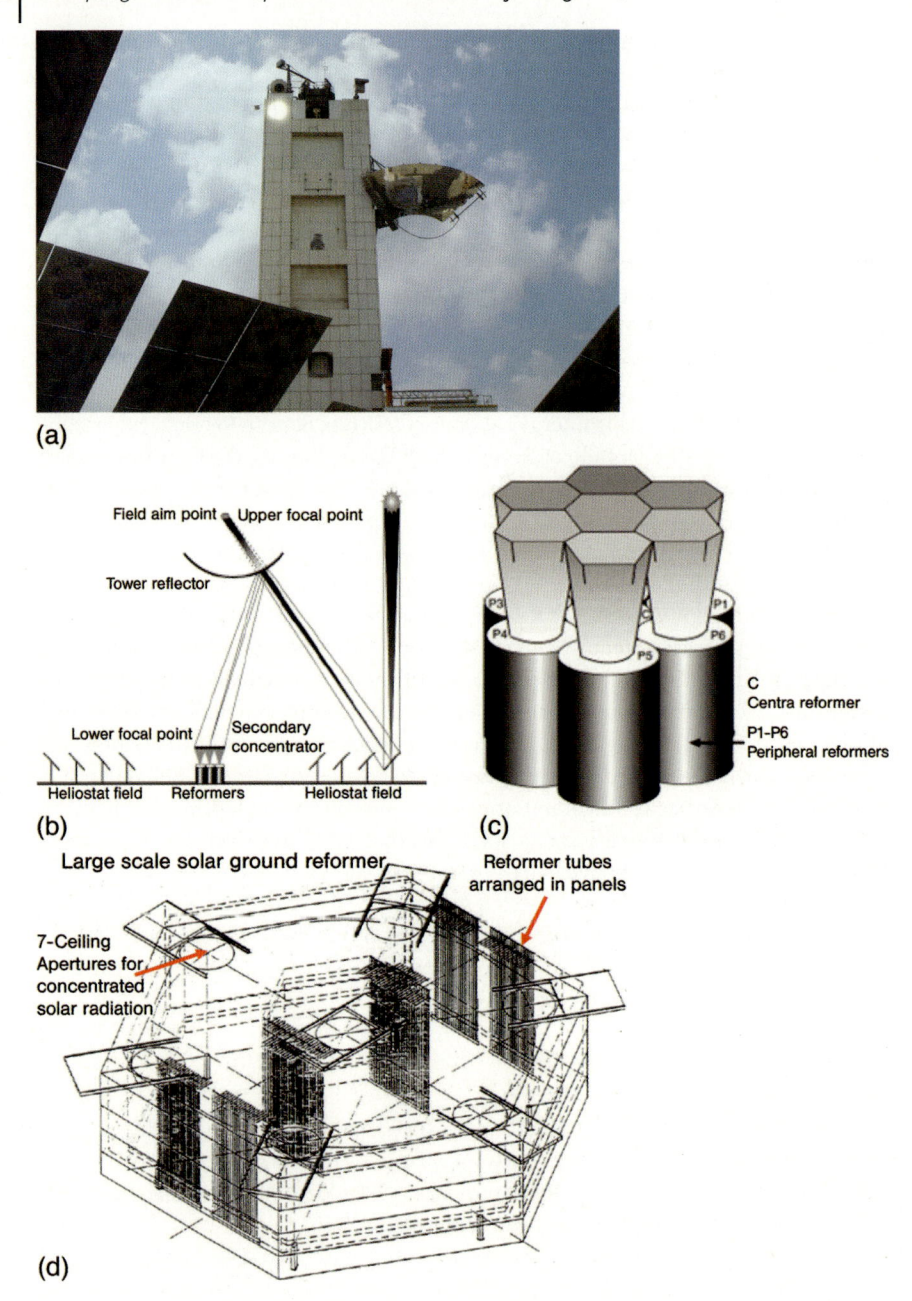

Figure 23.5 WIS's beam-down reformer technology: (a) beam down facility at WIS's solar tower at Rehovot, Israel; (b) schematic of operating principle; (c) schematic of reformers' cascade; (d) detailed schematic of construction on the ground [29, 57].

In Australia, the solar reforming of methane is particularly attractive in view of the country's enormous areas of favorable insolation and its very large reserves of natural gas and coal bed methane that are co-located in many regions. Work in Australia on solar methane reforming has been conducted by CSIRO since the early 1990s, aimed at catalyst and reactor development for conducting the CO_2-reforming reaction for application on both open- and closed-loop solar thermochemical heat pipes. CSIRO began work on solar reforming in 1999, with its 25 kW single-coil reformer (SCORE); a schematic of its operating principle is shown in Figure 23.6a. The catalyst is packed between the inner and outer tubes; the inner tube is purely for counter-current heating of the feed water stream. Production of solar-enriched fuels and hydrogen via steam reforming of natural gas (25 kW LHV) at temperatures and pressures up to 850 °C and 20 bar was demonstrated on this single-coiled reformer coupled to a 107 m^2 dish concentrator between 1998 and 2001. For the demonstration facility (2002), the reformer was designed for up to 1000 h of operation using a high-temperature stainless steel, whereas a commercial nickel-based steam reforming catalyst was used to catalyze the reforming reactions.

Between 1998 and 2001, CSIRO successfully completed a AU$7.5 million project to demonstrate a solar thermal–fossil energy hybrid concept for solar-enriched fuels and electricity, including solar steam reforming of natural gas to generate synthesis gas suitable for use as a fuel, further conversion of this gas to H_2 and CO_2 followed by recovery of CO_2 in a concentrated form, and further purification of the gas by methanation to produce fuel cell-grade hydrogen (< 10 ppm of CO). In 2004, CSIRO built a single-tower heliostat field of 500 kW_{th} capacity at its Newcastle site of the National Solar Energy Centre (Figure 23.6c) with the objective of demonstrating this solar reforming process on a larger scale. The reason for changing to a tower was the improved economics of the solar concentrator afforded by economies of scale. In 2006, CSIRO demonstrated the solar steam reforming process operating on this solar tower over a commercial catalyst at reaction temperatures of 700–800 °C and pressures of 5–10 bar. Subsequently, in June 2009, CSIRO designed and installed a much larger, hexagonal-shaped, dual-coil reformer (DCORE) capable of processing 200 kW_{th} (LHV) natural gas at 10 bar and 850 °C (Figure 6b and d). The length across the flats of the hexagonal shape coil is about 2 m, making the outer receiver dimensions about 2.5 m. Stable controlled reactor temperatures and consistent hydrogen production were reported. CSIRO has taken a different approach to most overseas investigators by focusing on developing and demonstrating solar-driven reformers that can achieve high methane conversions at lower temperatures (550–700 °C). This has a number of significant advantages, including a reduction in heat losses due to radiation and convection so that the overall utilization of solar energy is maximized. In this respect, CSIRO developed its own catalysts and claimed higher activity at lower operating temperatures – as low as 700 °C – compared with commercial reforming catalysts in three different reforming operation modes (low steam, mixed, and CO_2 reforming) [58–60].

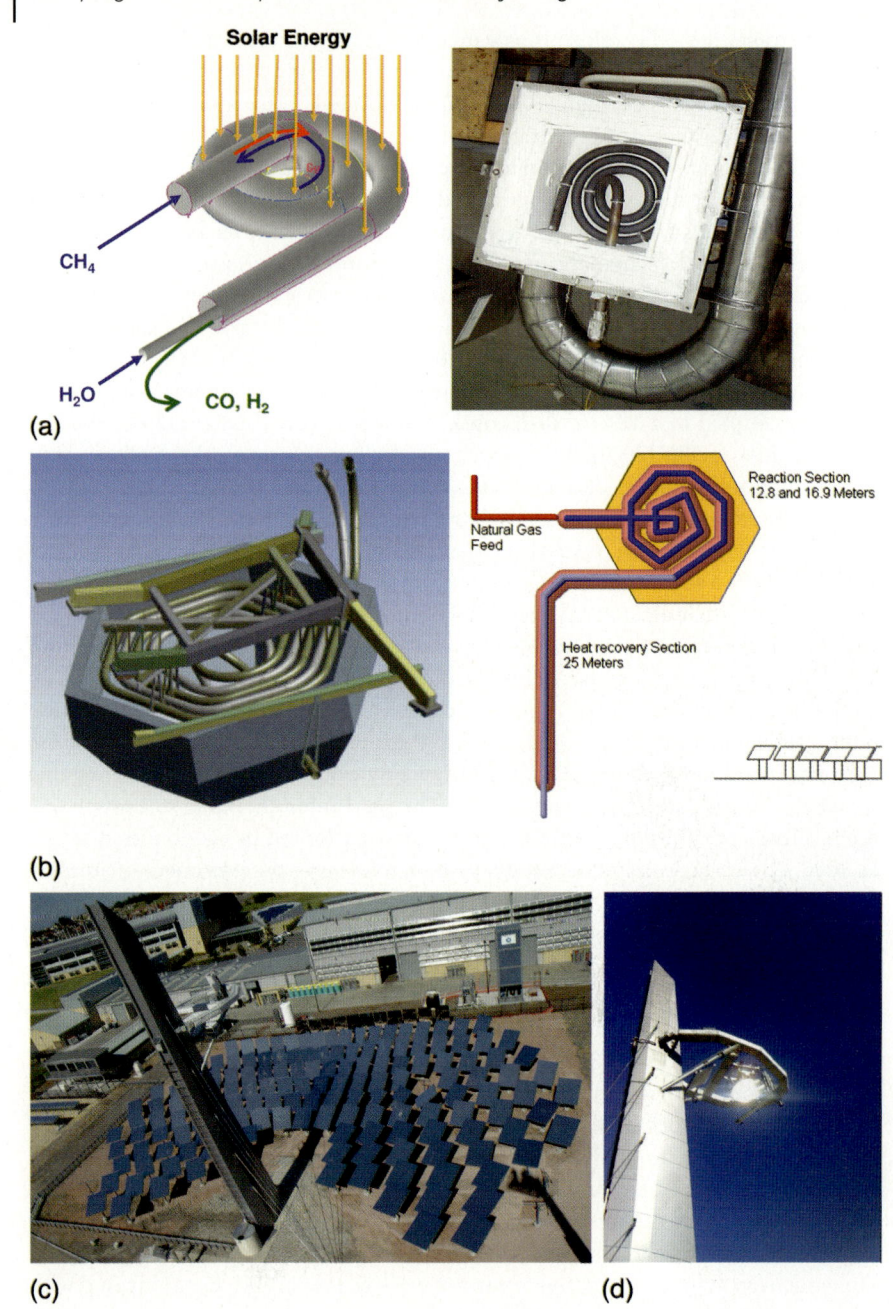

Figure 23.6 CSIRO's reformer technology development: (a) the single-coil solar reformer and its operating principle; (b) the double-coil reformer and its operating principle; (c) the single-tower, 500 kW, heliostat field at the Newcastle site, Australia; (d) the reformer on the tower during solar operation [17].

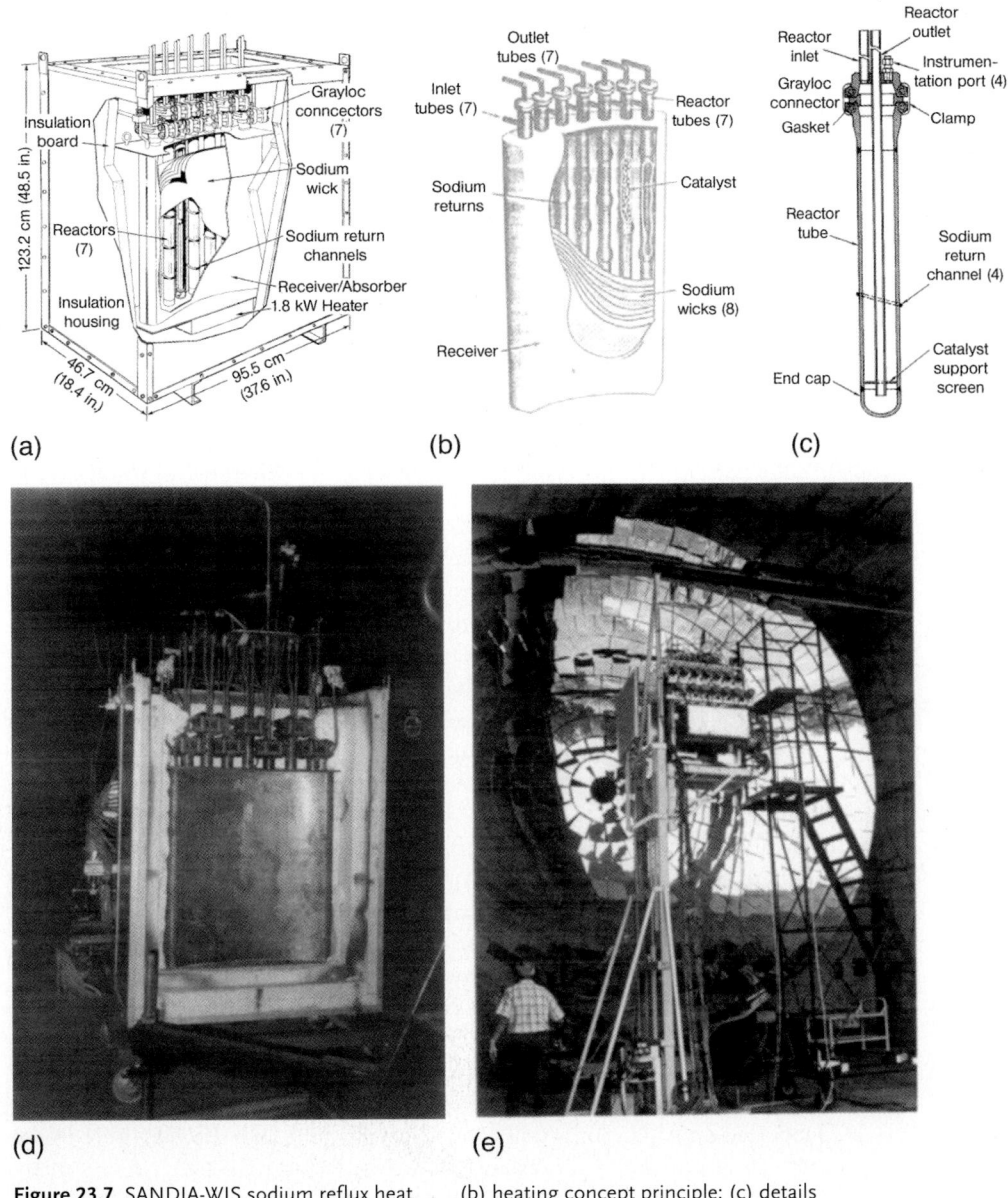

Figure 23.7 SANDIA-WIS sodium reflux heat pipe solar receiver/reformer, built and tested at WIS's solar furnace (1983–1984) for the CO_2 reforming of methane: (a) schematic diagram of the reactor; (b) heating concept principle; (c) details of a single reactor tube; (d) photograph of the receiver without the front panel; (e) the receiver installed in the facility at WIS's solar tower [62].

Heat-Transfer Fluid: Sodium Vapor

In this concept, liquid sodium contained in an evacuated chamber evaporates under the effect of concentrated sunlight impinging on one surface of the containment. The sodium vapor condenses on the reactor tubes in the chamber and liberates the heat of vaporization. Passive techniques (channels, wicks, gravity, etc.) return the liquid sodium to the absorber. Advantages claimed are the excellent heat-transfer characteristics of evaporating and condensing sodium that result in uniform temperatures throughout the chamber, thus ensuring isothermal operation of the reactor tubes, with the rate of sodium condensation determined by the local energy requirements of the reaction. The proof-of-concept of such a reactor was first demonstrated for steam reforming under simulated solar conditions using infrared lamp heating [33, 61] and subsequently under "real" solar irradiation within a Sandia Laboratories–WIS joint research project in which a 20 kW sodium reflux solar reformer was built and tested at WIS's solar furnace (1983–1984) for the CO_2 reforming of methane [62] (Figure 23.7). Recently, sodium as a heat-transfer fluid has returned to the focus of a number of research groups. New results will probably be published in the near future.

Heat-Transfer Fluid: Molten Salts

Molten salts such as $NaNO_3$–KNO_3 mixtures have been employed in the recently inaugurated Gemasolar solar tower plant as combined heat-transfer fluid–heat storage media operating at 565 °C (given that they decompose at temperatures higher than ~600 °C) [63]. Their attributes have led several research groups to propose the utilization of thermal storage with a molten salt in order to provide constant-rate solar heat supply also for an energy-demanding industrial chemical process such as steam reforming.

The group at Niigata University, Japan, has proposed the concept of dry reforming of methane employing K_2CO_3–Na_2CO_3 molten salt, which has a eutectic melting temperature of 710 °C, as the carrier material for the catalyst powder. In their initial configuration [64], the catalyst powder and the two salt components were mixed and placed in a stainless=steel cylindrical reactor (30 mm diameter and 300 mm length) heated via an infrared furnace at 950 °C. To enhance their contact with the catalyst, the preheated CH_4 and CO_2 reactant gases were "bubbled" through the molten salt containing the catalyst powder through an aperture at the bottom of the reactor. In-house-prepared non-noble metal catalysts such as Ni, Fe, Cu, and W, supported on Al_2O_3, for activity and selectivity [65], and also FeO were tested with the aim of producing H_2 from CH_4 via redox reactions [66]. Among them, Ni/Al_2O_3 was the most active and selective catalyst. Under specific conditions, about 70% of methane was converted. The H_2/CO ratio in the product gas was ~1. In subsequent studies [67, 68], the "salt bath" was replaced with Ni-loaded porous alumina or zirconia spheres impregnated with the salt material (NaCl or Na_2CO_3) to improve the heat storage/release characteristics. Currently in cooperation with Inha University, Korea [69]. this concept has culminated in a double-walled tubular receiver/reactor called MoSTAR (molten salt tubular absorber/reformer) in which the inner tube is filled with catalyst balls and the space between the inner tube and the outer absorber tube is filled with the composition of a phase-change material (carbonate) and a ceramic material (MgO) to increase the composition's thermal conductivity. Two different-sized reactor tubes were constructed and tested for dry reforming of methane during the cooling or heat-discharge mode of the reactor tube using an electric furnace, successfully sustaining a methane conversion above 90% with a feed gas mixture of CH_4–CO_2 (1:3) at a residence time of 0.36 s, 1 atm, and a reaction temperature of 920 °C. The next step of the project will be to demonstrate the performance of such "double-walled" tubular absorbers/reformers with molten salt thermal storage reactors on-Sun with a 5 kW_{th} dish-type solar concentrator.

Instead of employing the molten salt inside the reformer/reactor, the group at ENEA, Italy, has put forward the idea of heating a tubular reformer externally by a molten salt, similarly to classical industrial reformers heated by hot combustion gases. This corresponds to the ASTERIX concept described earlier for decoupling the solar receiver from the reformer but using molten salt instead of air for heating. However, this means that the particular reaction has to be performed at temperatures lower than 565 °C. Membrane reactors can be used to remove hydrogen selectively from the product stream to enhance the conversion and allow the reaction to be performed at lower temperatures [20]. ENEA continues to work in this direction and has provided an analysis of such a plant coupled to a parabolic trough solar collector, a two-tank molten salt storage system, and either a single tubular steam methane reforming (SMR) reactor with partial recirculation of the product mixture after hydrogen removal in a downstream permeator or a cascade of such SMR reactors [70]. A Pd-based thin-layer membrane supported on a ceramic or metallic material operating at 500 °C was sized so as to remove 85% of the inlet hydrogen with 100% selectivity. In subsequent studies [71], they adopted a heat exchanger-shaped (shell-and-tube) reactor, optimized its dimensions, and assessed its performance, concluding that a reactor 3.5 m long and with a diameter of 2 in is the most efficient in terms of methane conversion (14.8%) and catalyst efficiency. However, owing to the low methane conversion, this concept does not yield syngas but enriched methane (EM), that is, methane with 17% hydrogen. Engineering and experimental activities aimed at the development of a prototype apparatus are now in progress at ENEA.

Heat-Transfer "Fluid": Solid Particles
A somewhat different approach to solar-driven CO_2 reforming is being investigated at NREL and the University of Colorado at Boulder, CO. In this approach, what is heated by concentrated solar radiation is a "target" graphite tube that can reach ultra-high temperatures (around 1900 °C) under a nonoxidative atmosphere. The same reactor configuration principle can be used either for the solar thermal dissociation of methane to hydrogen and carbon black or for noncatalytic CO_2 methane reforming to produce syngas [72]. For the latter case, a triple-tube, solar-powered, fluid-wall aerosol flow reactor design has been used in the High Flux Solar Furnace (HFSF) facility at NREL to produce syngas with a lower H_2/CO ratio, as required for Fischer–Tropsch synthesis [73]. The furnace delivered a concentrated beam of sunlight ~0.1 m in diameter to the reactor, reflected via a secondary concentrator.

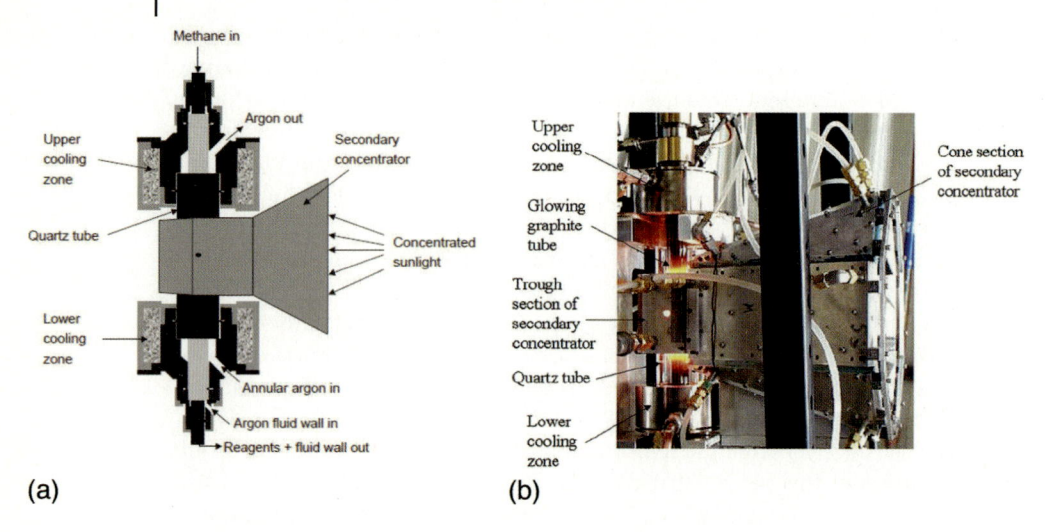

Figure 23.8 University of Colorado's solar-thermal aerosol reactor: (a) schematic and (b) photograph of the actual reactor heated to ~1200 °C, showing secondary concentrator, cooling zones, quartz tube, and solid graphite tube [11].

A schematic and a photograph of the reactor are shown in Figure 23.8a [11]. The reactor was composed of three concentric vertical tubes: an innermost tube made of porous graphite, a central tube made of solid graphite, and an outer tube made of quartz. The sunlight through the quartz tube heated the center solid graphite tube, which then radiated to the porous graphite tube. Argon was fed into the annular region between the two graphite tubes and the gas was forced through the porous tube wall. It served to protect the inner tube wall from carbon particle deposition. In this case, carbon particles are produced *in situ* as a by-product; it is claimed, however, that these are not detrimental to the process but, on the contrary, are preferably flowed with reactant gas through the reactor tube as radiation absorbers to facilitate heating and reaction – justifying the terminology "aerosol" reactor. The methane and carbon dioxide were fed into the top of the porous graphite tube, and the reaction products and the fluid-wall gas exited the bottom of the reactor. Operating with residence times of the order of 10 ms and temperatures of ~1700 °C, methane and CO_2 conversions of 70% and 65%, respectively, were obtained in the absence of any added catalysts. CH_4/CO_2 ratios of 1, 1.5, and 2 were used and it was found that ratios > 1 were needed with this system to prevent CO_2 from attacking the graphite wall of the reactor.

23.3.2.2 Directly Irradiated Reactors

Structured Reactors Based on Ceramic Honeycombs
Researchers at the WIS deposited Rh on an alumina honeycomb and irradiated it in their solar furnace through a sapphire window to catalyze the CO_2 methane

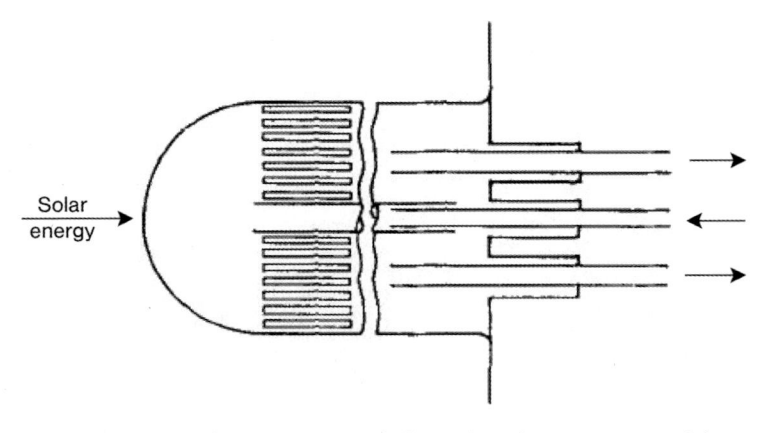

Figure 23.9 Sketch of WIS's transparent bell-jar solar reforming reactor with honeycomb catalyst [50].

reforming reaction [51]. Owing to failure issues, the sapphire window was later replaced with a fused-silica bell-jar reactor that was tested in the same facility in the configuration shown in Figure 23.9 [50]. The catalysts were two samples (12 and 25 cm thick) of cordierite honeycomb with 4 mm square holes and 0.5 mm wall thickness, first coated with a high surface area alumina wash-coat, followed by deposition of Rh. When a temperature of ~800 °C was reached on the surface of the catalyst, methane was added, keeping the CO_2/CH_4 ratio at ~1.3. Reported methane conversions reached 67%. However, scale-up of methane reformers based on honeycombs has never been reported.

Structured Reactors Based on Ceramic Foams
Reactors based on ceramic foams were the first structured reactors to be tested for solar-aided methane reforming and today are the most developed ones tested at a level of a few hundred kilowatts of solar input. The first examples of such solar reactors can be traced back to 1990 when solar reforming of methane with CO_2 in a directly irradiated volumetric receiver/reactor was first demonstrated in the "CAtalytically Enhanced Solar Absorption Receiver" (CAESAR) experiment conducted by Sandia National Laboratories (SNL) in the United States and DLR in Germany [74] between 1987 and 1990 (Figure 23.10).

A second-generation solar chemical receiver/reactor was designed for a power input of up to 300 kW and built in 1993–1994 for operation as part of a closed thermochemical storage and transportation loop within a joint project between the WIS in Israel and DLR in Germany [18, 75]. The main reactor improvements with respect to CAESAR were the installation of a higher power unit and a parabolic quartz window, behind which the receiver foam pieces were arranged in a domed cavity configuration (Figure 23.11b). For solar testing, the assembled receiver was installed on top of the tower of the WIS Solar Test Facility Unit. The receiver is now on display at DLR's facilities in Cologne, Germany (Figure 23.11c and d).

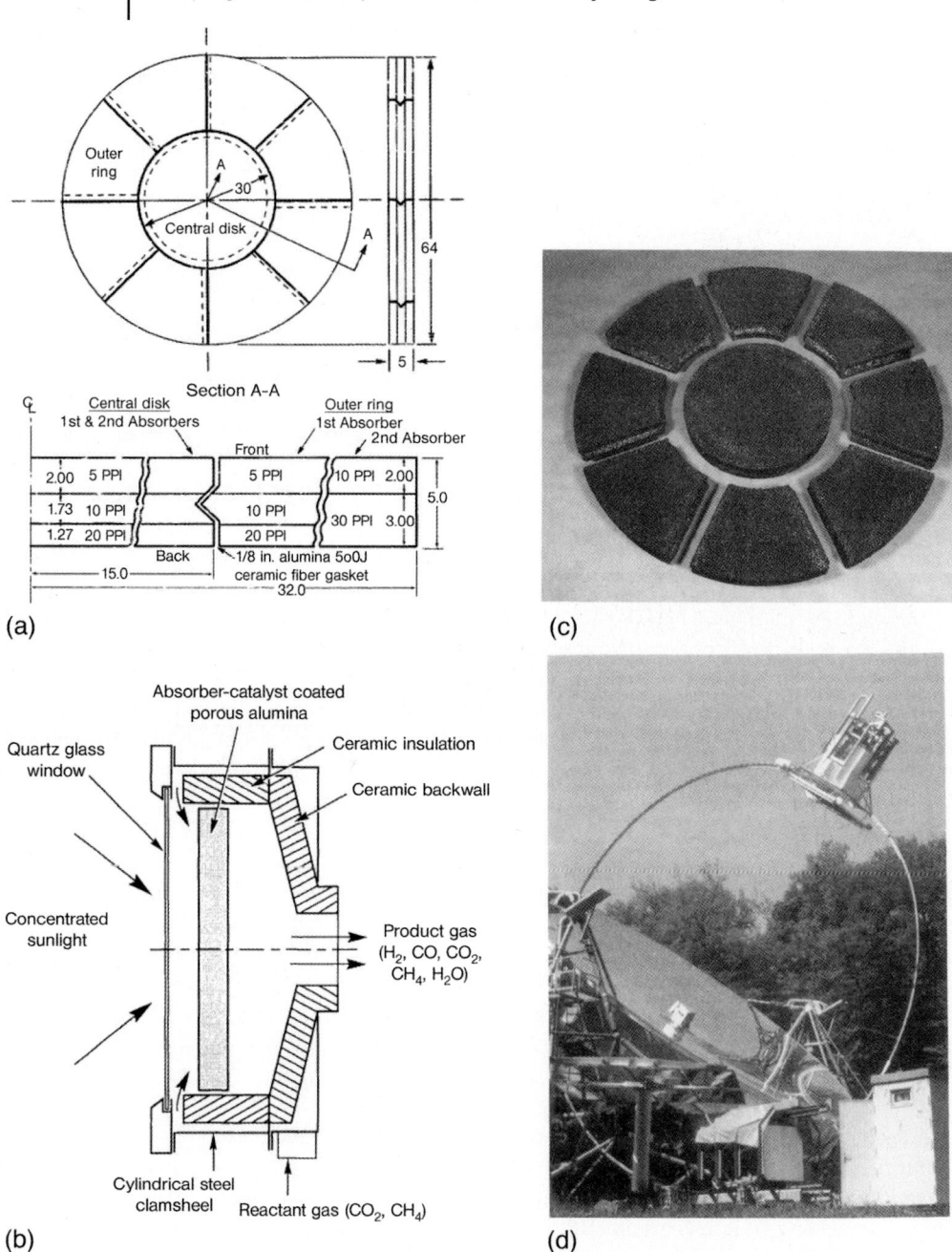

Figure 23.10 (a) Sketch of structure of CAESAR multi-layer foam absorber;
(b) schematic of receiver/reactor operation; (c) photograph of first CAESAR receiver;
(d) actual reactor on parabolic dish test facility [74].

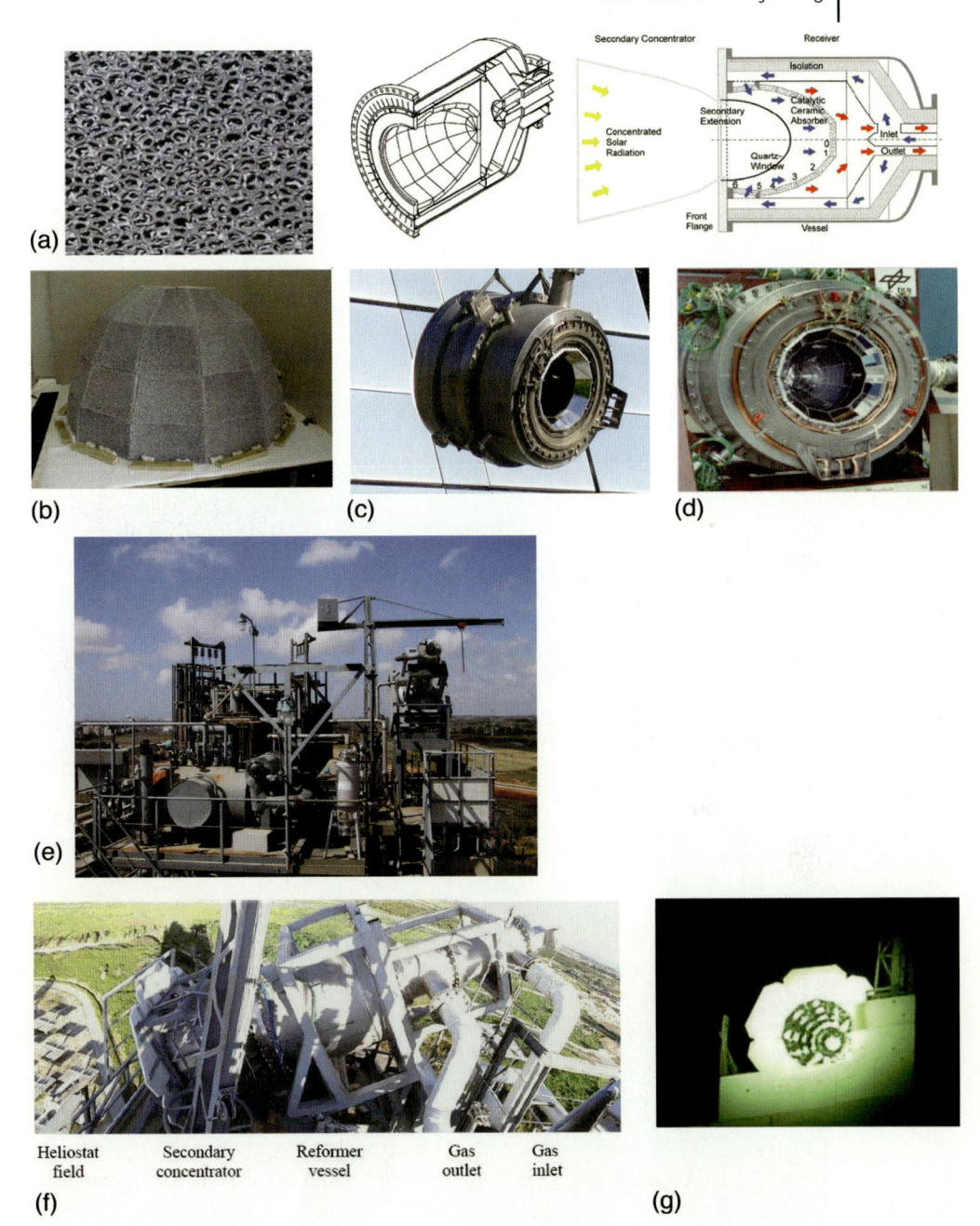

Figure 23.11 The SCR and SOLASYS ceramic foam-based directly irradiated (DIVVRR) solar steam methane reformer:
(a) sketch and operating principle;
(b) assembled dome of the solar receiver,
(c), (d) reactor photographs of the SCR;
(e) SOLASYS reactor installation on top of the solar tower of the WIS; (f) close-up photograph of the reformer installed on top of the tower; (g) reformer in operation [17].

These ceramic foam-based solar reactors were pursued further within the Project SOLASYS (partners DLR, WIS, and Ormat Pty Ltd, an Israeli company with expertise in gas turbines and related technologies), where the technical feasibility of solar reforming with liquid petroleum gas (LPG) as feedstock and combustion of the product gases (syngas mixture) in a gas turbine to generate electricity at the 300 kW$_e$ scale was successfully demonstrated. During the test period on the WIS solar tower (Figure 11f and g), the solar reformer was operated in the power range 100–220 kW, producing syngas at 8.5 bar and 760 °C with conversion rates close to chemical equilibrium.

An advanced and more compact and cost-effective volumetric receiver/reformer has been developed within the successor project SOLREF with the rationale of operating at a higher power level (400 kW$_e$) and also at higher pressure and temperature, such as 950 °C and 15 bar, which would result in a higher conversion rate of methane to hydrogen and higher efficiency (Figure 23.12a). Among the innovative modifications implemented was a new catalytic system capable of operating at

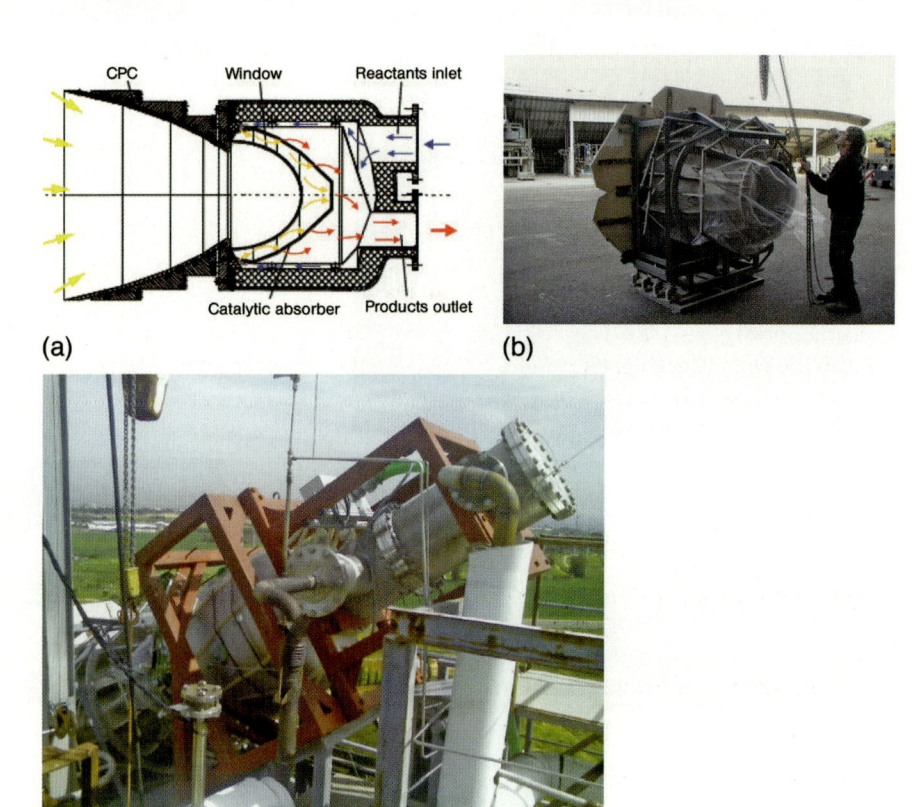

Figure 23.12 The SOLREF DIVVRR: (a) sketch of operating principle; (b) assembled solar steam reformer/reactor ready for transportation; (c) actual reactor installed on top of the solar tower of the WIS [29].

higher temperatures, thus allowing a broad range of feed compositions – biogas, landfill gas, and contaminated natural gas (CH_4 with a high content of CO_2) – to be processed and avoiding carbon deposition in the reformer/reactor. An advanced solar reformer was developed, tested, and validated under real solar conditions at the WIS (Figure 23.12b). A test campaign was carried out, demonstrating the feasibility of the SOLREF technology. Relevant publications claim that the cost of hydrogen produced by solar steam reforming of methane is very close to that of conventional steam reforming with the break-even point being at a natural gas price of €0.35 m^{-3} [9, 76].

The group at Niigata University also studied CO_2 reforming on the laboratory scale using a new type of catalytically activated "metallic foam" absorber, directly irradiated by a solar-simulated xenon lamp light and subjected to solar flux levels in the range 180–250 kW m^{-2} [77]. The absorber consisted of an Ni–Cr–Al alloy, containing Ru/Al_2O_3 as the catalyst, and was found to have a superior thermal performance in terms of absorbing solar energy and converting it into chemical energy via the reforming reaction, at relatively low solar fluxes compared with conventional ceramic foam absorbers. In a later study on the same system [78], they examined the kinetics of this reaction in the temperature range 650–900 °C and analyzed the kinetic data by four different types of kinetic model versus experimental reforming rates. In a series of parallel publications [79, 80], the same group tested in the same xenon lamp rig a series of Ni-based catalysts in combination with Ru and supported on magnesia- and alumina-based carriers (Ni–Mg–O, Ni/Al_2O_3, Ru/Al_2O_3 and Ru/Ni–Mg–O), trying to establish robust catalytic systems free of expensive metals such as Ru and Rh, first in powder form and subsequently coated on Al_2O_3 and SiC foams, and reported for the latter methane conversion reaching 80% at a reforming temperature of 950 °C. The current status of this development is that the system of choice for subsequent scale-up is an Ni/MgO–Al_2O_3 catalyst coated on SiC foams [80].

Researchers at Inha University in Korea are also investigating structured reactors based on metallic foams for CO_2 methane reforming. A double-layer metal foam absorber/reactor was developed in which the front part is not catalytically active but realizes the absorbing of solar radiation and heat transfer to the gaseous reactants fluid. The rear part is an Ni metal foam coated with Ni/Al_2O_3 catalyst. The absorber was tested in several campaigns with respect to solar CO_2 reforming of methane on a parabolic dish capable of providing 5 kW_{th} solar power (INHA DISH1, Figure 23.13) [54].The temperature at the center of the irradiated surface of the absorber ranged from 900 to 1000 °C, the total power input of the incident solar energy into the absorbers was 3.25 kW under steady-state operating conditions, and the maximum CH_4 conversion reached 60% [81]. Compared with "conventional" catalytically activated ceramic foam absorbers, direct irradiation of the metallic foam absorber was claimed to exhibit superior reaction performance at relatively low insolation or at low temperatures, and better thermal resistance preventing cracking by mechanical stress or thermal shock. In addition, the double-layer absorber helped to maintain higher temperature profiles in the reactor during the heating mode that alleviated deactivation of catalytically activated metal foam and, consequently, sustained high methane conversion for much longer during the heating and cooling mode of the reactor.

Figure 23.13 The metallic foam-based 5 kW_{th} absorber/reactor at Inha University, Korea, used for CO_2 reforming of methane on the solar dish system (INHA-DISH1) [54].

Structured Reactors Based on Fins

WIS researchers have designed a reformer based on the directly irradiated annular pressurized receiver (DIAPR) [41] and the "porcupine" concept, for operation at high temperatures and pressures. A schematic of the reformer design and operating principle is shown in Figure 23.14a [82] and a three-dimensional representation of its geometry in Figure 23.14b [83]. Catalytic elements for use in this reformer based on Ru/Al_2O_3 promoted with Mn oxides to avoid carbon deposition were developed, characterized, and supported on alumina pins by washcoating; long-term testing confirmed their chemical and thermal stability even after calcination at 1100 °C for 500 h under an argon flow [82, 84]. With input from computational models that identified initial design inefficiencies [83] and subsequent corrective actions, a prototype "porcupine" reformer using a conical quartz window was built (Figures 14b–e) [29, 85] and tested with respect to CO_2 methane reforming on the solar tower at WIS in 2010. The CO_2/CH_4 ratio was ~1 : 1.2. Eight test runs were performed, focusing on the influence of gas pressure and flow rate on the CH_4 conversion. The gas pressure was varied between 4 and 9 atm and the total inlet flow rate between 100 and 235 slpm (standard liters per minute). The maximum absorber temperature was kept below 1200 °C. The conversion of CH_4 reached 85%. Further work aimed at improving the total efficiency of the system is in progress.

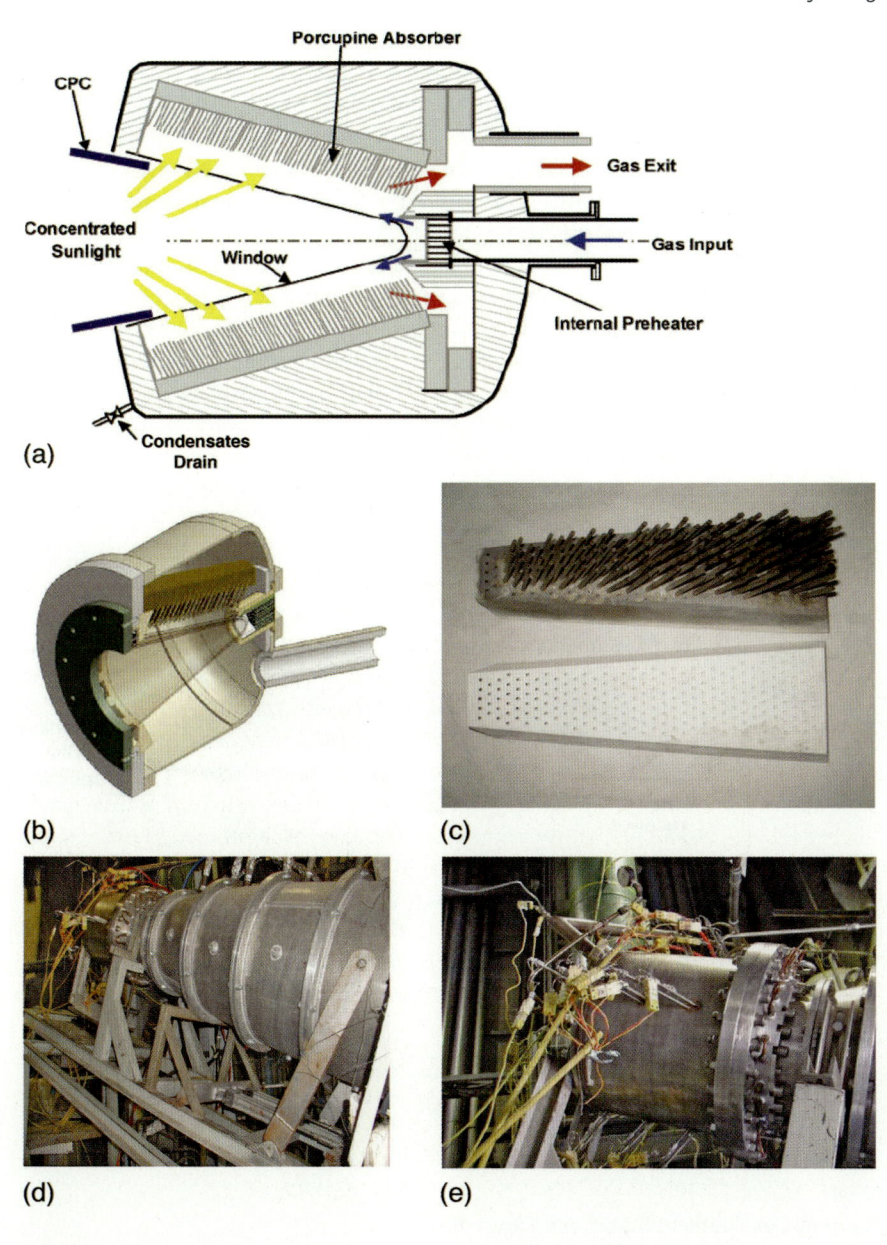

Figure 23.14 WIS "porcupine" directly irradiated solar reformer: (a) schematic of the reformer design and operating principle; (b) three-dimensional representation of the reactor geometry; (c) actual absorber sections made of alumina base, before (bottom) and after (top) inserting alumina tubes coated with the catalyst into it; (d), (e) actual 30 kW reformer as built [85].

Non-Structured Reactors Based on Solid Particles

Dry methane reforming with CO_2 in a directly irradiated particle receiver seeded with carbon black particles with a C/CO_2 molar ratio of ~0.5:100 and with CO_2/CH_4 inlet ratios varying from 1:1 to 1:6 was carried out at WIS [86]. The receiver uses a moving radiation absorber, that is, particles entrained in the reforming gas mixture that have two functions: to absorb the solar radiation and transfer it to the gas and to act as a reaction surface for the reforming reactions (Figure 23.15). Dry solar reforming was conducted without any catalyst at a temperature between 950 and 1450 °C at the WIS solar tower, using primary and secondary concentrators to reach a solar flux of up to 3 MW m^{-2} at the receiver aperture. The exit gas temperatures were 1100–1450 °C. Depending on the temperature, a proportion of the carbon particles reacted with CO_2 to form CO, that is, the particles in the receiver were gradually partially or totally consumed. The reforming experiments indicated that the carbon black particles entrained in the gas flow augmented the reaction of methane cracking.

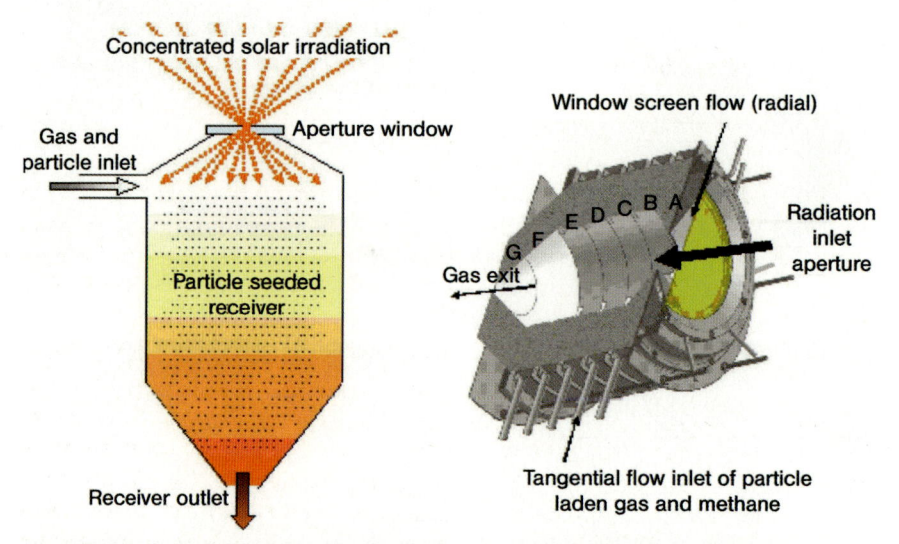

Figure 23.15 The WIS directly irradiated particle solar receiver: schematic of the operating principle of such a receiver and of the solar receiver actually used in the experiments [86].

23.4
Current Development Status and Future Prospects

Although the long-term incentives for solar thermochemistry may be apparent, there is a need to search for near-term applications to establish priorities for R&D activities and to attract support from funding agencies and industry. In this perspective, for the subsequent scale-up of current solar methane reforming technologies to demonstration level, several choices have to be made with respect to certain issues that will impact the ability of the technology to enter the marketplace.

With respect to the CSP technology to be employed, solar dish collectors focus the solar energy (typically 10–400 kW$_{th}$) to a receiver mounted at the focus of the dish. Even though the first solar-coupling step of many of the reforming studies described has been the testing of reactors positioned at the focal point of a solar dish receiver, the fact that the receiver/reactor therefore has to move with the dish as the latter tracks the Sun complicates the scale-up of the technology. Solar dish collectors are inherently modular and therefore well suited for pilot-scale experiments and distributed applications such as the destruction of toxic wastes. Therefore, it is highly probable that large-scale solar thermal reforming will require the economies of scale offered by heliostat fields with central tower receivers. These fields can comfortably generate solar thermal fluxes with megawatts capacity, although the deployment of such fields to date has been limited to electricity generation.

The choice of directly irradiated volumetric receiver versus cavity receiver for solar-driven catalytic reactions will be a key decision in technology development. Volumetric absorbers appear to offer better thermal and performance properties than conventional tubular absorbers, and with the convenience being based primarily on their intrinsic mechanism of light absorption, which is three-dimensional in nature. Directly irradiated reactors provide efficient radiation heat transfer directly to the reaction site where the energy is needed, by-passing the limitations imposed by indirect heat transport via heat exchangers. Since solar energy is absorbed directly by the catalyst, temperatures are highest at the reaction sites and the chemical reactions are likely to be kinetically limited rather than heat transfer limited as in conventional tubular reactors. A further advantage is that the intensity of the impinging solar radiation can be five times higher than in a tubular receiver (Figure 23.4a) [9]. The concurrent flow of solar radiation and chemical reactants reduces absorber temperatures and re-radiation losses.

The disadvantages or limitations of the DIVRR center largely on the need to have a transparent quartz window as the aperture to allow the ingress of concentrated solar radiation into the receiver while at the same time providing a gas seal for the reacting gases and products that in most cases are under pressure. This window must at all times be kept cool and free from contact with any gases or solids within the receiver that could damage or destroy it by either increasing its opacity or reacting with it. Methods such as protecting the window with a flow of inert auxiliary gas or by ensuring that the flow pattern of solids inside the receiver is such that they cannot contact the window have been only partially successful to date and more work needs to be done on this crucial aspect before the quartz window can be considered to have the reliability needed for a commercial large-scale receiver.

Indirectly irradiated reactors eliminate the need for a window at the expense of having less efficient heat transfer – by conduction – through the walls of an opaque absorber. Hence the disadvantages are linked to the limitations imposed by the materials of the absorber with regard to maximum operating temperature, inertness to the chemical reaction, thermal conductivity, radiative absorbance, resistance to thermal shocks, and suitability for transient operation. Therefore, "technically simpler" concepts such as the tubular indirectly heated receivers employed in the

ASTERIX project might be more attractive for large-scale implementation and demonstration of the technology.

Another set of issues relates to the choice of steam or CO_2 for reforming. There are advantages and disadvantages for each option, with a clear choice only for certain open-cycle applications. For example, if methanol is the desired end-product, the amount of steam or CO_2 used would give an optimal CO/H_2 ratio in the syngas. If H_2 is the desired product, steam reforming is the choice. A process configuration considered more recently is the mixed reforming of methane, where it reacts with a mixture of water and CO_2. This process is especially advantageous for methane sources with a high CO_2 content such as biogas (45–70 mol% CH_4 and 30–45 mol% CO_2). These feedstocks often contain close to the required amount of CO_2 for the reforming reaction and only a small amount of steam needs to be added, which also reduces carbon formation. Furthermore, the resulting H_2/CO ratio in the product can be adjusted (by the amount of steam added), according to the requirements of further processing. For the utilization of biogas, this constitutes an attractive possibility for upgrading the heating value. In that way, combustion difficulties with biogas at low engine loads can be reduced [22, 87–90].

In conclusion, significant progress has been made through the research efforts described above and the solar-driven steam reforming of methane has been demonstrated at the pilot scale. The technologies for both DIVRR and tubular reactors are both ready for small commercial-scale installations, and it is anticipated that plants in the 1–5MW$_{th}$ range will be constructed in Australia and Europe within the next couple of years. The transitional nature of this technology – coupling renewable solar thermal energy and conventional fossil fuels to make either a synthesis gas or hydrogen – is seen as a key first step on the path to sustainable hydrogen and energy production.

References

1 Liu, K., Song, C., and Subramani, V. (2010) *Hydrogen and Syngas Production and Purification Technologies*, Wiley-AIChE, John Wiley & Sons, Inc., Hoboken, NJ.

2 Rostrup-Nielsen, J. R. (1984) Catalytic steam reforming, in *Catalysis: Science and Technology*, vol. 5 (ed. J. R. Anderson), Springer, Berlin, pp. 1–117.

3 Rostrup-Nielsen, J. R., Sehested, J., and Nørskov, J. K. (2002) Hydrogen and synthesis gas by steam and CO_2 reforming. *Adv. Catal.*, **47**, 65–139.

4 Steinfeld, A. (2005) Solar thermochemical production of hydrogen – a review. *Solar Energy*, **78** (5), 603–615.

5 Kodama, T. (2003) High-temperature solar chemistry for converting solar heat to chemical fuels. *Prog. Energy Combust. Sci.*, **29** (6), 567–597.

6 Tamaura, Y., et al. (1995) Production of solar hydrogen by a novel, 2-step, water-splitting thermochemical cycle. *Energy*, **20** (4), 325–330.

7 Anikeev, V. I., et al. (1968) Catalytic thermochemical reactor/receiver for solar reforming of natural gas: design and performance. *Solar Energy*, **63** (2), 97–104.

8 Hirsch, D., Epstein, M., and Steinfeld, A. (2001) The solar thermal decarbonization of natural gas. *Int. J. Hydrogen Energy*, **26** (10), 1023–1033.

9 Möller, S., Kaucic, D., and Sattler, C. (2006) Hydrogen production by solar reforming of natural gas: a comparison

study of two possible process configurations. *J. Solar Energy Eng.*, **128** (1), 16–23.

10 von Zedtwitz, P. et al. (2006) Hydrogen production via the solar thermal decarbonization of fossil fuels. *Solar Energy*, **80** (10), 1333–1337.

11 Dahl, J. K., et al. (2004) Solar-thermal dissociation of methane in a fluid-wall aerosol flow reactor. *Int. J. Hydrogen Energy*, **29** (7), 725–736.

12 Abanades, S. and Flamant, G. (2005) Production of hydrogen by thermal methane splitting in a nozzle-type laboratory-scale solar reactor. *Int. J. Hydrogen Energy*, **30** (8), 843–853.

13 Abanades, S. and Flamant, G. (2006) Solar hydrogen production from the thermal splitting of methane in a high temperature solar chemical reactor. *Solar Energy*, **80** (10), 1321–1332.

14 Flechsenhar, M. and Sasse, C. (1995) Solar gasification of biomass using oil shale and coal as candidate materials. *Energy*, **20** (8), 803–810.

15 Trommer, D., et al. (2005) Hydrogen production by steam-gasification of petroleum coke using concentrated solar power – I. Thermodynamic and kinetic analyses. *Int. J. Hydrogen Energy*, **30** (6), 605–618.

16 Piatkowski, N., et al. (2011) Solar-driven gasification of carbonaceous feedstock – a review. *Energy Environ. Sci.*, **4** (1), 73–82.

17 Stolten, D. (ed.) (2010) *Hydrogen and Fuel Cells: Fundamentals, Technologies and Applications*, Wiley-VCH Verlag GmbH, Weinheim.

18 Tamme, R., et al. (2001) Solar upgrading of fuels for generation of electricity. *J. Solar Energy Eng.*, **123** (2), 160–163.

19 Edwards, J. H., et al. (1996) The use of solar-based CO_2/CH_4 reforming for reducing greenhouse gas emissions during the generation of electricity and process heat. *Energy Convers. Manage.*, **37** (6–8), 1339–1344.

20 Giaconia, A., et al. (2008) Solar steam reforming of natural gas for hydrogen production using molten salt heat carriers. *AIChE J.*, **54** (7), 1932–1944.

21 Spath, P. L. and Dayton, D. C. (2003) *Preliminary Screening – Technical and Economic Assessment of Synthesis Gas to Fuels and Chemicals with Emphasis on the Potential for Biomass-Derived Syngas*, National Renewable Energy Laboratory (NREL), Golden, CO, p. 160.

22 Sun, Y., et al. (2011) Thermodynamic analysis of mixed and dry reforming of methane for solar thermal applications. *J. Nat. Gas Chem.*, **20** (6), 568–576.

23 Tomishige, K., Chen, Y., and Fujimoto, K. (1999) Studies on Carbon deposition in CO_2 reforming of CH_4 over nickel–magnesia solid solution catalysts. *J. Catal.*, **181** (1), 91–103.

24 Lemonidou, A., Goula, M., and Vasalos, I. (1998) Carbon dioxide reforming of methane over 5 wt.% nickel calcium aluminate catalysts – effect of preparation method. *Catal. Today*, **46** (2), 175–183.

25 Gadalla, A. M. and Bower, B. (1988) The role of catalyst support on the activity of nickel for reforming methane with CO_2. *Chem. Eng. Sci.*, **43** (11), 3049–3062.

26 Lemonidou, A. A. and Vasalos, I. A. (2002) Carbon dioxide reforming of methane over 5 wt.% $Ni/CaO-Al_2O_3$ catalyst. *Appl. Catal. A: Gen.*, **228** (1–2), 227–235.

27 Berman, A., Karni, R. K., and Epstein, M. (2005) Kinetics of steam reforming of methane on Ru/Al_2O_3 catalyst promoted with Mn oxides. *Appl. Catal. A: Gen.*, **282** (1–2), 73–83.

28 Sehested, J. (2006) Four challenges for nickel steam-reforming catalysts. *Catal. Today*, **111** (1), 103–110.

29 Epstein, M. (2011) solar Thermal Reforming of Methane, presented at the SFERA Winter School, Zürich.

30 Kikuchi, E. (1995) Palladium/ceramic membranes for selective hydrogen permeation and their application to membrane reactor. *Catal. Today*, **25** (3–4), 333–337.

31 Cybulski, A. and Moulijn, J. A. (2005) *Structured Catalysts and Reactors*, 2nd edn., Chemical Industries, vol. 110, CRC Press, Boca Raton, FL.

32 Karni, J., et al. (1998) The "Porcupine": a novel high-flux absorber for volumetric solar receivers. *J. Solar Energy Eng. Trans. ASME*, **120**, 85–95.

33 Richardson, J., Paripatyadar, S., and Shen, J. (1988) Dynamics of a sodium heat pipe reforming reactor. *AIChE J.*, **34** (5), 743–752.

34 Olalde, G. and Peube, J. (1982) Etude expérimentale d'un récepteur solaire en nid d'abeilles pour le chauffage solaire des gaz à haute température. *Rev. Phys. Appl.*, **17** (9), 563–568.

35 Fend, T., et al. (2004) Porous materials as open volumetric solar receivers: experimental determination of thermophysical and heat transfer properties. *Energy*, **29** (5–6), 823–833.

36 Fend, T., et al. (2004) Two novel high-porosity materials as volumetric receivers for concentrated solar radiation. *Solar Energy Mater. Solar Cells*, **84** (1–4), 291–304.

37 Ávila-Marín, A. L. (2011) Volumetric receivers in solar thermal power plants with central receiver system technology: a review. *Solar Energy*, **85** (5), 891–910.

38 Chavez, J. M. and Chaza, C. (1991) Testing of a porous ceramic absorber for a volumetric air receiver. *Solar Energy Mater.*, **24** (1–4), 172–181.

39 Agrafiotis, C. C., et al. (2007) Evaluation of porous silicon carbide monolithic honeycombs as volumetric receivers/collectors of concentrated solar radiation. *Solar Energy Mater. Solar Cells*, **91** (6), 474–488.

40 Hennecke, K., et al. (2009) The Solar Power Tower Jülich – a solar thermal power plant for test and demonstration of air receiver technology, in *Proceedings of ISES World Congress 2007*, vols. I–V, Springer, Berlin.

41 Karni, J., et al. (1997) The DIAPR: a high-pressure, high-temperature solar receiver. *J. Solar Energy Eng.*, **119** (1), 74–78.

42 Agrafiotis, C. C., et al. (2007) Hydrogen production in solar reactors. *Catal. Today*, **127** (1–4), 265–277.

43 Chubb, T. A. (1980) Characteristics of CO_2–CH_4 reforming–methanation cycle relevant to the Solchem thermochemical power system. *Solar Energy*, **24** (4), 341–345.

44 McCrary, J. H., et al. (1982) An experimental study of the CO_2–CH_4 reforming–methanation cycle as a mechanism for converting and transporting solar energy. *Solar Energy*, **29** (2), 141–151.

45 Anikeev, V. I. and Kirillov V. A. (1991) Basic design principles and some methods of investigation of catalytic reactors–receivers of solar radiation. *Solar Energy Mater.*, **24** (1–4), 633–646.

46 Anikeev, V., et al. (1993) Chemical heat regeneration in power plants. *Int. J. Energy Res.*, **17** (4), 233–242.

47 Anikeev, V., et al., Theoretical and experimental studies of solar catalytic power plants based on reversible reactions with participation of methane and synthesis gas. *Int. J. Hydrogen Energy*, 1990. 15(4), 275–286.

48 Spiewak, I., Tyner, C. E., and Langnickel, U. (1993) *Applications of Solar Reforming Technology*, Sandia National Laboratories, Albuquerque, NM.

49 Hogan, R., et al. (1990) A direct absorber reactor/receiver for solar thermal applications. *Chem. Eng. Sci.*, **45** (8), 2751–2758.

50 Levy, M., et al. (1992) Methane reforming by direct solar irradiation of the catalyst. *Energy*, **17** (8), 749–756.

51 Levy, M., Rosin, H., and Levitan, R. (1989) Chemical reactions in a solar furnace by direct solar irradiation of the catalyst. *J. Solar Energy Eng.*, **111** (1), 96–97.

52 Kirillov, V. A. (1999) Catalyst application in solar thermochemistry. *Solar Energy*, **66** (2), 143–149.

53 Kumar, P., Sun, Y., and Idem, R. O. (2008) Comparative study of Ni-based mixed oxide catalyst for carbon dioxide reforming of methane. *Energy Fuels*, **22** (6), 3575–3582.

54 Meier, A. (2010) *SolarPACES Annual Report – Task II Solar Chemistry Research*, SolarPACES, International Energy Agency, Paris.

55 Böhmer, M., Langnickel, U., and Sanchez, M. (1991) Solar steam reforming of methane. *Solar Energy Mater.*, **24** (1), 441–448.

56 Epstein, M., et al. (1996) Solar experiments with a tubular reformer, presented at the 8th International Symposium on Solar Thermal Concentrating Technologies, Cologne.

57 Segal, A. and Epstein, M. (2003) Solar ground reformer. *Solar Energy*, **75** (6), 479–490.

58 McNaughton, R. (2012) Solar steam reforming using a closed cycle gaseous heat transfer loop, presented at the 18th International SolarPACES Symposium, 11–14 September, Marrakech.

59 McNaughton, R. and Stein, W. (2009) Improving efficiency of power generation from solar thermal natural gas reforming. presented at the 15th International SolarPACES Symposium, 15–18 September, Berlin.

60 Hinkley, J. T., Edwards, J. H., Stein, W. H., and Sattler, C. (2009) Hydrogen production by the solar-driven steam reformign of methane, in Encyclopedia of Electrochemical Power Sources (eds. J. Garche *et al.*), Elsevier, Amsterdam, pp. 300–312.

61 Paripatyadar, S. A. and Richardson, J. T. (1998) Cyclic performance of a sodium heat pipe, solar reformer. *Solar Energy*, **41** (5), 475–485.

62 Diver, R. B., *et al.* (1992) Solar test of an integrated sodium reflux heat pipe receiver/reactor for thermochemical energy transport. *Solar Energy*, **48** (1), 21–30.

63 Burgaleta, J. I., Arias, S., and Ramirez, D. (2011) GEMASOLAR, the first tower thermosolar commercial plant with molten salt storage, presented at the 17th International SolarPACES Symposium, 20–23 September, Granada.

64 Kodama, T., *et al.* (2001) CO_2 reforming of methane in a molten carbonate salt bath for use in solar thermochemical processes. *Energy Fuels*, **15** (1), 60–65.

65 Shimizu, T., *et al.* (2001) Thermochemical methane reforming using WO_3 as an oxidant below 1173 K by a solar furnace simulator. *Solar Energy*, **71** (5), 315–324.

66 Gokon, N., *et al.* (2002) Methane reforming with CO_2 in molten salt using FeO catalyst. *Solar Energy*, **72** (3), 243–250.

67 Kodama, T., *et al.* (2004) Ni/ceramic/molten-salt composite catalyst with high-temperature thermal storage for use in solar reforming processes. *Energy*, **29** (5), 895–903.

68 Hatamachi, T., Kodama, T., and Isobe, Y. (2005) Carbonate composite catalyst with high-temperature thermal storage for use in solar tubular reformers. *J. Solar Energy Eng.*, **127** (3), 396–400.

69 Kodama, T., *et al.* (2009) Molten-salt tubular absorber/reformer (MoSTAR) project: the thermal storage media of Na_2CO_3–MgO composite materials. *J. Solar Energy Eng.*, **131** (4), 041013.

70 De Falco, M., *et al.* (2009) Enriched methane production using solar energy: an assessment of plant performance. *Int. J. Hydrogen Energy*, **34** (1), 98–109.

71 De Falco, M. and Piemonte, V. (2011) Solar enriched methane production by steam reforming process: reactor design. *Int. J. Hydrogen Energy*, **36** (13), 7759–7762.

72 Dahl, J. K., *et al.* (2001) Solar-thermal processing of methane to produce hydrogen and syngas. *Energy Fuels*, **15** (5), 1227–1232.

73 Dahl, J. K., *et al.* (2004) Dry reforming of methane using a solar-thermal aerosol flow reactor. *Ind. Eng. Chem. Res.*, **43** (18), 5489–5495.

74 Buck, R., Muir, J. F., and Hogan, R. E. (1991) Carbon dioxide reforming of methane in a solar volumetric receiver/reactor: the CAESAR project. *Solar Energy Mater.*, **24** (1–4), 449–463.

75 Wörner, A. and Tamme, R. (1998) CO_2 reforming of methane in a solar driven volumetric receiver–reactor. *Catal. Today*, **46** (2), 165–174.

76 Pregger, T., *et al.* (2009) Prospects of solar thermal hydrogen production processes. *Int. J. Hydrogen Energy*, **34** (10), 4256–4267.

77 Kodama, T., Kiyama, A., and Shimizu, K. I. (2003) Catalytically activated metal foam absorber for light-to-chemical energy conversion via solar reforming of methane. *Energy Fuels*, **17** (1), 13–17.

78 Gokon, N., *et al.* (2011) Kinetics of methane reforming over Ru/γ-Al_2O_3-catalyzed metallic foam at 650–900 °C for solar receiver–absorbers. *Int. J. Hydrogen Energy*, **36** (1), 203–215.

79 Kiyama, A., Moriyama, T., and Mizuno, O. (2004) Solar methane reforming using a new type of

catalytically-activated metallic foam absorber. *J. Solar Energy Eng.*, **126**, 808.

80 Gokon, N., *et al.* (2010) Ni/MgO–Al_2O_3 and Ni–Mg–O catalyzed SiC foam absorbers for high temperature solar reforming of methane. *Int. J. Hydrogen Energy*, **35** (14), 7441–7453.

81 Lee, J., *et al.* (2011) Solar CO_2-reforming of methane using a double layer absorber, presented at the 17th International SolarPACES Symposium, 20–23 September, Granada.

82 Berman, A., Karni, R. K., and Epstein, M. (2006) A new catalyst system for high-temperature solar reforming of methane. *Energy Fuels*, **20** (2), 455–462.

83 Ben-Zvi, R. and Karni, J. (2007) Simulation of a volumetric solar reformer. *J. Solar Energy Eng.*, **129** (2), 197–204.

84 Berman, A., Karn, R. K., and Epstein, M. (2007) Steam reforming of methane on a Ru/Al_2O_3 catalyst promoted with Mn oxides for solar hydrogen production. *Green Chem.*, **9** (6), 626–631.

85 Rubin, R. and Karni J. (2011) Carbon dioxide reforming of methane in directly irradiated solar reactor with Porcupine absorber. *J. Solar Energy Eng.*, **133** (2), 021008.

86 Klein, H. H., Karni, J., and Rubin, R. (2009) Dry methane reforming without a metal catalyst in a directly irradiated solar particle reactor. *J. Solar Energy Eng.*, **131** (2), 021001.

87 Jun, H. J., *et al.* (2001) Kinetics modeling for the mixed reforming of methane over Ni–CeO_2/$MgAl_2O_4$ catalyst. *J. Nat. Gas Chem.*, **20** (1), 9–17.

88 Effendi, A., *et al.* (2002) Steam reforming of a clean model biogas over Ni/Al_2O_3 in fluidized-and fixed-bed reactors. *Catal. Today*, **77** (3), 181–189.

89 Lau, C., Tsolakis, A., and Wyszynski, M. (2011) Biogas upgrade to syn-gas (H_2–CO) via dry and oxidative reforming. *Int. J. Hydrogen Energy*, **36** (1), 397–404.

90 Rasi, S., Veijanen, A., and Rintala, J. (2007) Trace compounds of biogas from different biogas production plants. *Energy*, **32** (8), 1375–1380.

91 Romero, M. and Steinfeld, A. (2012) Concentrating solar thermal power and thermochemical fuels. *Energy Environ. Sci.*, **5** (11), 9234–9245.

Part IV
Biomass

24
Biomass – Aspects of Global Resources and Political Opportunities

Gustav Melin

24.1
Our Perceptions: Are They Misleading Us?

We are influenced by the perceptions we gain in life. One perception that we have been fed with over the decades through TV and newspapers is that when we see or read reports of countries and people suffering from famine, we tend to believe that we are not able to produce enough food throughout the globe. We have also heard about the global population explosion ever since we were young, and this makes it even more difficult to believe that there will be enough food in the future. Now, however, more and more people are learning that the global population is not increasing as quickly any longer, and the UNFPA (United Nations Population Fund) predicts that the human population will flatten out below 11 billion people [1]. Since we have been totally dependent on fossil fuels for a rich life, we have difficulty in understanding that a life without fossil fuels can be possible and even good. This chapter has the aim of providing information showing a different possible direction for human life on Earth than the commonly held view.

24.2
Biomass – Just a Resource Like Other Resources – Price Gives Limitations

It seems that at most bioenergy seminars and conferences in recent years at least one of the speakers has claimed that biomass resources are limited. Such statements are extremely common, and they are often delivered by speakers who are experts in areas other than agriculture or forestry, which are the two disciplines where experts are educated to obtain knowledge on the subject of biomass resources. The statement in itself is, of course, true. There are limited resources of biomass, but then there are also limited resources of oil, natural gas, coal, iron, gold, wind, solar, land, water, and all other resources on the planet.

Transition to Renewable Energy Systems, 1st Edition. Edited by Detlef Stolten and Viktor Scherer.
© 2013 Wiley-VCH Verlag GmbH & Co. KGaA. Published 2013 by Wiley-VCH Verlag GmbH & Co. KGaA.

Let us consider a resource such as gold: the amount of it is limited, but we do not consume gold. It can appear in minerals, maybe not found and enriched. It can be in jewelry and become lost or placed in a bank safe. The demand for and use of gold depend on the interest of people to invest in it and own it. If people find gold attractive, they will buy it if the price is affordable. If we had a surplus of gold, it would lose interest as jewelry, and we may instead have found other uses for gold, just as we use copper, iron, and other metals in batteries and so on. When it comes to gold and similar materials, we do not talk about limited amounts, we talk about price. Gold and other materials can be expensive, maybe too expensive to use (for a given purpose).

If we discuss energy, everyone agrees that fossil fuel resources are limited. This, however, seldom appears to be a problem since fossil fuels have been relatively cheap. They are also storable and we can use them when energy is needed. However, when using fossil fuels, we consume them and they cannot be replaced. Renewables such as solar, wind, and wave energy are unlimited but momentary. When it is dark we receive no solar energy and if there is no wind there is no wind power. Biomass is the most interesting renewable source when it comes to producing it and storing it for later use. In this sense, biomass for energy is perfect to replace fossil energy in all its current uses. Many people would like to rule out biomass as an energy source because the amount of biomass is limited. As an agronomist, the author's opinion is that sustainable biomass is available in sufficient amounts today, and the potential for increased use is tremendous. This can easily be seen in the diagrams in the following section. The limits are set not by volume but by price, as for any product on the global market. If the demand and use of biomass for energy increase, the price on biomass will also increase and then farmers will invest in production to produce more biomass and prices will fall. An increased biomass price will lead to investment and increased production, but increased price also leads to actions and changes in behavior. In a market economy, price and cost are the mechanism to catalyze the change of actions by both producers and consumers, or by both farmers and customers. If, for example, grain prices increase, farmers start to use straw instead of grain to feed animals. They may also improve crop management by irrigating, fertilize, or use better weed control to increase the yield. If farmers believe in better long-term prices, they invest in better machinery, improved seeds, or drainage systems.

24.3
Global Food Production and Prices

Since biomass volumes are not really limited but are rather a function of demand and supply, a major global discussion concerns how the demand for biomass for energy influences food prices, and if poor people find it difficult to afford food. The United Nations (UN) and the Food and Agricultural Organization (FAO) have for many years applied the strategy of trying to keep prices down to enable poor people to buy food. This may appear to be a good strategy, but a negative consequence of keeping food prices low is that it makes farming unprofitable, and unprofitable food production leads to poorer farmers with an inability to invest in improved food production. Another consequence is lack of strength to deal with crop failure, something that will happen from time to time in every country.

In a historical perspective Figure 24.1 "Wheat Price Adjusted" shows that current recent grain prices are not high, if we consider grain prices over the last 100 years. The price now registered is only one fourth of the price 100 years ago. at CBOT wheat (Chicago board of trade) in February 2013 was at the same level as in July 1995, which is half the wheat price of 1980, but the doubled price since the year 2000 which was all time low for global wheat price.

Figure 24.2 shows that global grain production has been growing steadily since the 1960s, but since grain demand has grown less than production capacity, the use of arable land peaked in 1980 and since then has decreased. The European disincentives through the Common Agricultural Policy (CAP), to pay farmers to set aside land in order to reduce the surplus production of agricultural products, was at its peak during the late 1980s and the 1990s. About 11.2 million hectares were in set aside in Europe in 2012 and several million hectares are in other "treatments" to reduce food production.

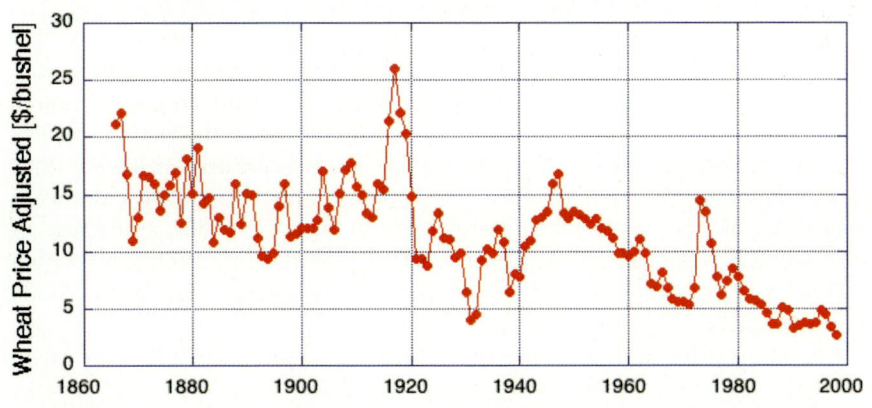

Figure 24.1 Trend in global wheat prices 1865–2000, here exampled by US wheat price adjusted for inflation, have decreased continuously over the years in real terms. The spike in 1918 was due to demand for wheat during World War I, the drop in 1933 was due to the Hawley–Smoot Tariff Act, the peak in 1944 was due to World War II, and the peak in 1974 was due to large imports by the Soviet Union. The overall downward trend since 1940 is due to more efficient farming practices and new strains of wheat. Note that these prices shown are approximate. Source: *Historical Statistics of the United States* [2].

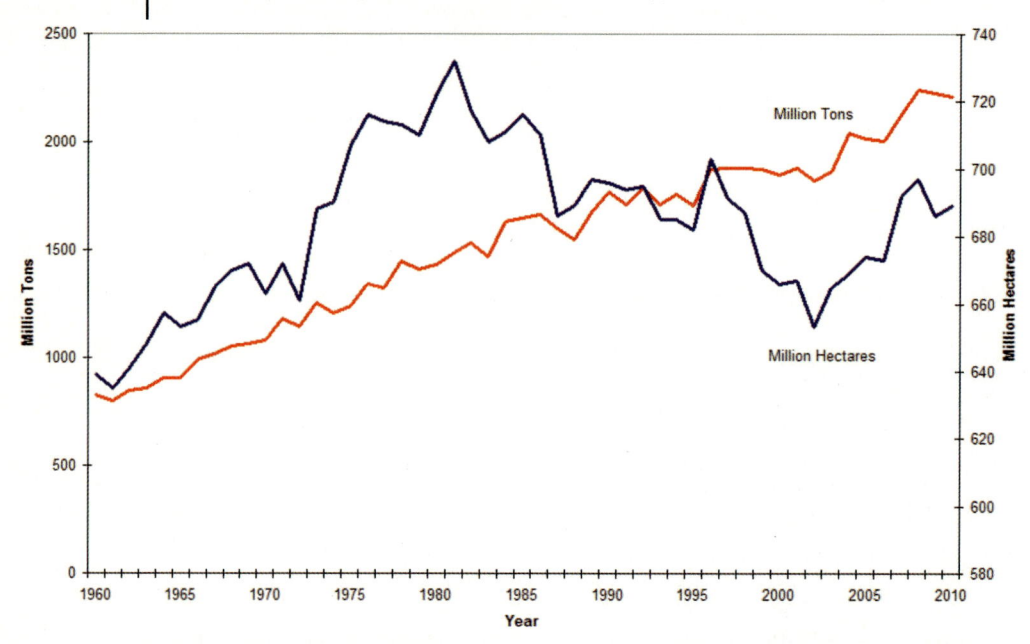

Figure 24.2 Graph of world grain (cereal) production from 1960 to 2010 measured in million hectares cultivated and million tons harvested. World grain production has had stable growth and met world demand. Since production per hectare increased, less arable land has been needed, resulting in a potential to produce even more food or energy. Source: the graph is according to Wikimedia and ultimately from the US Department of Agriculture, electronic database [3].

24.3.1
Production Capacity per Hectare in Different Countries

The background to the decreasing prices and continuously increasing yields can be clearly explained by viewing the Swedish example in Figure 24.3, which shows the development of wheat production per hectare in Sweden from 1901 to 2005. The production increased by almost 200% during the period from 1940 until 1990. This was the period when fertilizer, weed control, improved cultivars and plant material, and modern machinery were implemented into Swedish agriculture on a broad scale. Note that since 1985, the production of wheat per hectare in Sweden has not increased but remained at a stable level, when we have had a global food surplus situation.

The same production development that took place in Sweden has also taken place in the rest of Western Europe and in the United States. The yield increase per hectare in Western Europe and the United States is shown in Figure 24.4, which also shows that increased production per hectare has recently started in Brazil, going from a production level at 1.4 t ha^{-1} in the 1960s to 3.8 t ha^{-1} in 2010. The production capacity is expected to continue to grow together with the overall economic development in Brazil. Figure 24.4 also shows that it is not yet possible

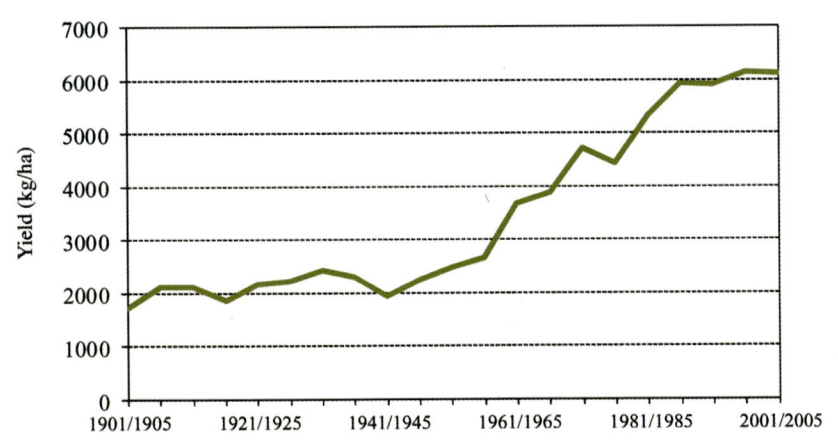

Figure 24.3 Yield of wheat in Sweden in kilograms per hectare, 1901–2005. Source: The Swedish Board of Agriculture [4].

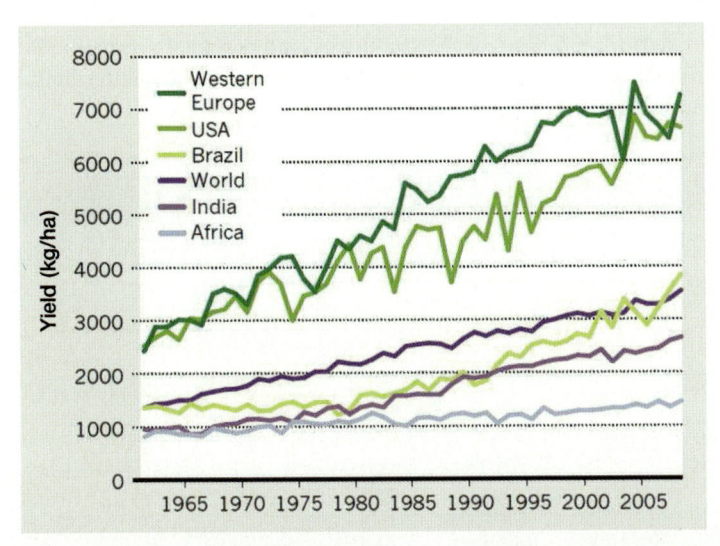

Figure 24.4 Production capacity per hectare differs widely depending on the level of development in different regions. The development in Brazil is increasing rapidly at present together with the overall positive economy in South America; the production levels in Africa are still low but will increase in the coming years. Source: [5].

to see the same development in Africa. The production of cereal in Africa was still at 1.4 t ha^{-1} a^{-1} in 2010. This is despite the fact that the growth potential per hectare in Africa under optimal conditions is much higher than the production potential per hectare in Europe.

When we discuss the potential of biomass production per hectare, it is important to remember that yields of 4000 or 6000 kg ha^{-1} of wheat, which represent bad or half good wheat production per hectare, are actually too low figures. If cereals are used, one would also harvest the straw which, is usually the same weight as grain per hectare. Since biomass for energy is only the number of tonnes produced per hectare, crops other than cereals will be preferred. Most of the available arable land is in the southern hemisphere and here other crops such as sugarcane and eucalyptus are more interesting. The current world average production of sugarcane is 71 t ha^{-1} on 23.8 million ha globally [6]. Peru is the country with the highest average yield of 124 t ha^{-1} a^{-1}. The production capacity of eucalyptus is similar to that of sugarcane.

24.4
Global Arable Land Potential

Of the world's total land surface area of 13.5 billion ha, about 8.3 billion ha are currently under grassland or forest, and 1.6 billion ha under cropland. An additional 2 billion ha are considered potentially suitable for rainfed crop production, as shown in Figure 24.5. After excluding forest land, protected areas, and land needed to meet increased demand for food crops and livestock, the FAO report on the state of food and agriculture in 2008 [7] estimates of the amount of land potentially available for expanded crop production lies between 250 and 800 million ha, most of it in tropical Latin America and in Africa.

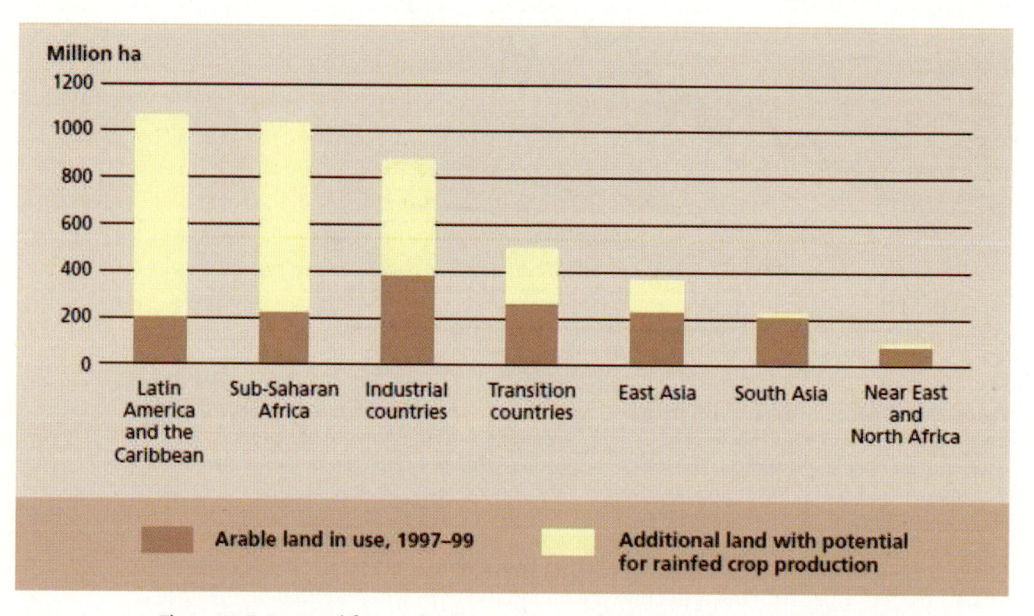

Figure 24.5 Potential for cropland expansion. Arable land in use is around 1.6 billion ha. An additional 2 billion ha are considered potentially suitable for rainfed crop production. Source: FAO [7].

In recent years, it has become obvious that not only rainfed cropland can be made available for sustainable biomass food and bioenergy production. If biomass prices were to become stable over a longer period, investment in irrigation from desalted seawater could also become profitable utilizing renewable technology. In such a scenario, additional land will be available in current desert areas.

Having in mind that the world population is forecast by UNFPA to stabilize at a level below 11 billion people, the capacity for food production globally clearly has the potential easily to exceed the required demand for food.

24.4.1
Global Forests Are Carbon Sinks Assimilating One-Third of Total Carbon Emissions

The global forest situation will not be discussed in this chapter. It is well known that in some countries there are serious problems with deforestation, which have to be addressed so that deforestation can be stopped. It is important to save rainforests and other types of forest containing rare species and unique biotopes. From environmental associations we often hear that deforestation contributed 20% of the global carbon emissions. This was true before 2000, but since then deforestation has decreased and fossil emissions have increased (Figure 24.6), and in 2010 deforestation represented 9% of global carbon emissions. At the same time, global forest growth assimilated almost one-third of all carbon emissions (Figure 24.7).

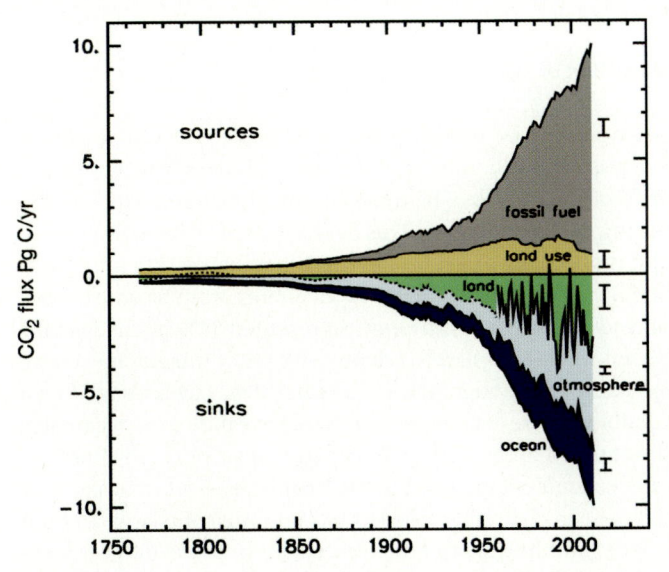

Figure 24.6 Carbon emissions from deforestation were 9% of global emissions in 2010 and are decreasing. Fossil fuel emissions continue to increase. Source: [8].

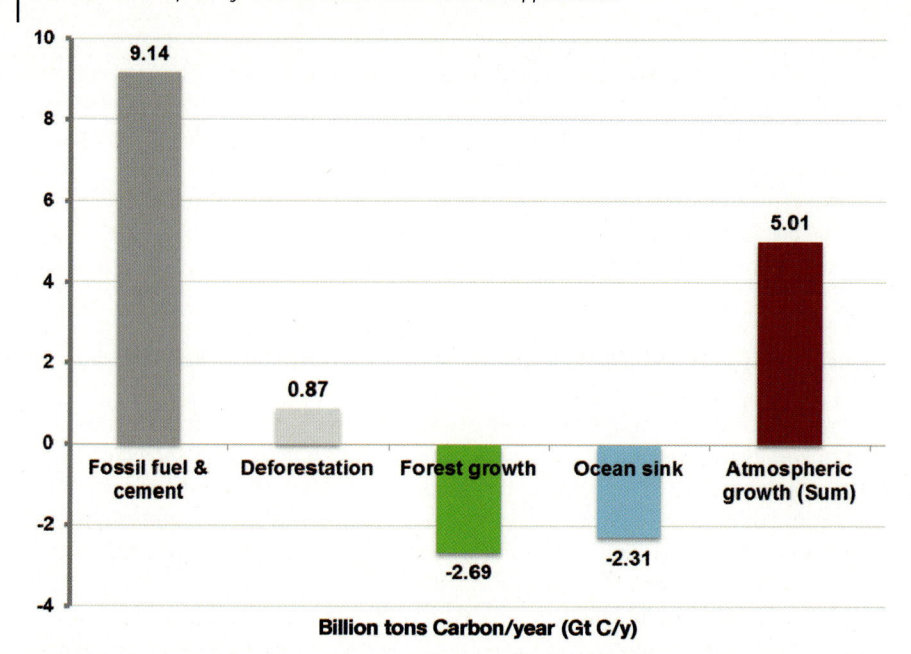

Figure 24.7 Global forest growth assimilates carbon corresponding to three times carbon emissions from deforestation. This shows that global forests are increasing today.

24.4.2
Forest Supply – the Major Part of Sweden's Energy Supply

The Swedish forest situation can be used as a typical example [9, 10]. The annual forest growth in Sweden is 120 million m^3 and the annual harvest for timber and pulp production is 75% of the growth. The total volume of standing stock in the forests has doubled during the last 100 years. This has happened despite the fact that we are using more and more of the forest for energy. When harvesting pulp wood and timber, half of the harvested amount ends up as energy from sawdust, bark, branches and tops, and so on. In 2011, bioenergy represented 32% of the Swedish energy supply. In Sweden we have to start to ask ourselves how much forest do we want, and when is it enough? Is it enough now? In 5 years? If we have enough forest now, we may almost double the use of bioenergy and still have the same sustainable forest volume. The amount of energy in the current annual surplus growth of the forest is larger than the current energy need of the total transport sector per year in Sweden. Obviously one can claim that Sweden has an unusually large amount of forests, and this is true, but the growth rate in Swedish forests is lower than in most other countries. Most countries have forests, but they are very seldom used to the same extent as they are in Sweden. The potential to use forests better globally is extremely high; many countries harvest less than 50% of their annual forest growth.

24.5
Lower Biomass Potential If No Biomass Demand

The biomass potential mainly concerns production capacity, but biomass potential itself is not any more interesting than the global grain production discussed earlier. Why produce more grain than there is demand from the market? We can understand from the price curve for wheat in the United States that there has never been such a demand for grain that prices can increase more than in the short term, and long-term prices are falling. Few long-term investments to increase production are profitable. Long-term grain volumes have always been in surplus; the difficulty is to be able to obtain enough food in the years when we have crop failure.

A similar situation is valid for global biomass potential and markets. There has been too little demand for biomass for energy owing to the availability of cheap fossil fuels. Therefore, it is difficult to say what role biomass for energy can play. The potential for biomass from agriculture and forests is certainly much higher than most people believe. The amount of land available to increase production is also tremendous. However, demand must develop in order to make people invest in producing biomass for energy. So far the price fn wood has been too low to bear any investment; actually, there is no mature market for round wood globally. More than 90% of the global industrial round wood use is locally sourced. The lack of a

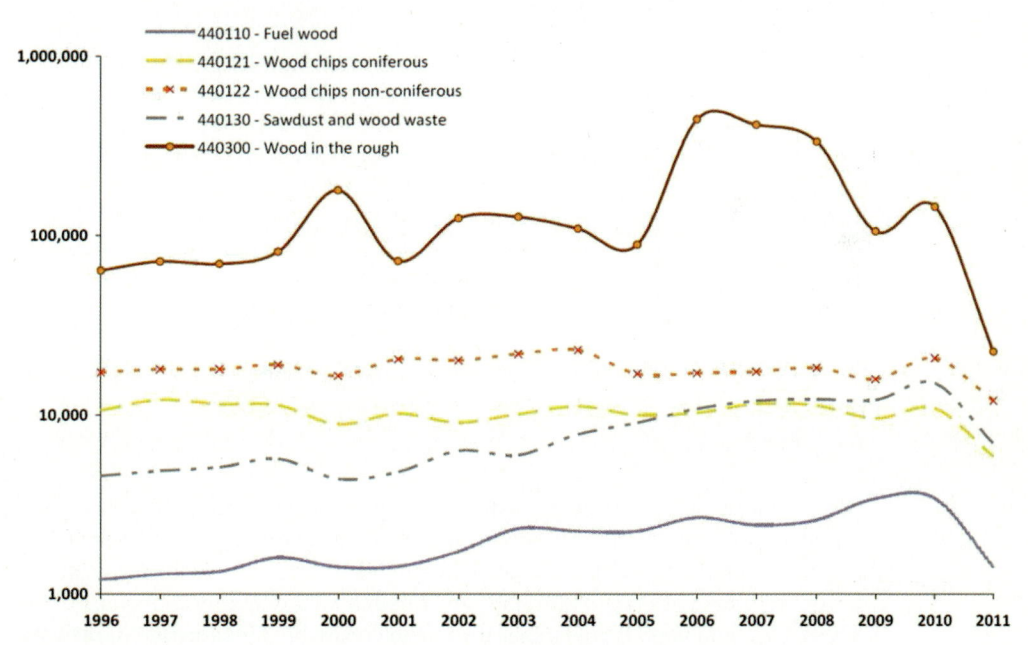

Figure 24.8 Fuel wood and waste wood trade streams in kilotonnes based on import data from Comtrade (note the logarithmic scale). The global fuel wood market is difficult to invest in, since volumes can vary from 500 to 30 million tons in just a few years. The global wood market is heavily influenced by hurricanes and storms that occasionally overflow the market with products. Source: Comtrade [11].

market and hence a lack of possibility of investing in forest plantations for such a global market becomes clear on studying the global trade streams of fuel wood in Figure 24.8.

The global forest industry is more than 90% dependent on locally sourced raw material. Obviously it will be possible to produce wood and forest material for energy or other purposes also in areas far from current forest industry plants. The key is not the biomass potential, but how to create the demand for biomass or any other competitive renewable energy source without damaging side effects.

24.6
Biomass Potential Studies

Many biomass potential studies are available; the most ambitious report is probably that from the Swedish University of Agricultural Sciences [12] in which 4911 bioenergy records were collated. The following is an excerpt the executive summary: "There are a number of scenarios predicting the future potential of biomass. There have also been many studies performed in recent decades to estimate the future demand and supply of bioenergy. However, if we compare an upper limit of the total global bioenergy production potential in 2050 of 1135 EJ, that can come available as energy supply without affecting the supply of food crops, with the highest scenarios on the global primary energy demand in 2050 of 1041 EJ, we see that the world's bioenergy potential is large enough to meet global energy demand in 2050. Unfortunately, this information is not part of the public consciousness. Supplying the public with important information about bioenergy can equip them to then put pressure on politicians to create a framework for increasing the speed with which bioenergy solutions are implemented." In 2012, the International Energy Agency (IEA) predicted a more moderate figure in their Technology Roadmap [13]. Here in a model and literature review the IEA estimated the technical potential of bioenergy to be 500 EJ in 2050, which can be compared with the current global use of energy of 490 EJ. The current global use of bioenergy is about 60 EJ.

24.7
The Political Task

The overall political mission is to strive for a good society and better health, and to protect rain forests and rare species through legislation but also human rights, proprietary rights, rights to education, equality under the law, and the possibility of borrowing money at a fair interest rate, and finally a welfare system taking care of the elderly, sick, and poor. If politicians are able to create this fundamental framework of a society, the UN will in very few years reach its millennium goals.

The overall mission for politicians is to economize with the common resources to be able to deliver all these aspects of the good society. We believe in sustainable renewable energy because it can have little negative environmental impact and few health issues. Since there is always a lack of resources in a society, politicians need to find the best way to make people take decisions that give the best benefit per unit cost. The instruments that politicians have in their tool box are legislation, fees, and taxes.

24.8
Political Measures, Legislation, Steering Instruments, and Incentives

The most damaging actions to a society are prohibited by legislation. In the environmental sector, this can be exemplified by, for example, emissions of mercury or the use of poisons such as DDT. When it comes to behavior that is not as damaging or lethal to society but is still unwanted, politicians can decide upon fees. Speeding tickets, if we drive too fast, are an example of a steering instrument to make people behave differently. In the environmental sector, we find it efficient and clever to use the "polluter pays principle" (PPP) to make companies and people act and behave in a way that reduces costs or improves the overall health situation in society. The PPP means that any company or institution that emits an emission that damages Nature in any way, such as polluting air or water, will have to pay a fee for the damage that is caused. A good example is that sulfur emissions cause acidification, and to curb the emissions an offending company may have to pay a sulfur emission fee. Nitrogen emissions of NO_x cause enhanced fertilization and acidification, and therefore may NO_x fees may be applied. Further, CO_2 emissions cause climate change and it has proven efficient to use carbon dioxide taxation to reduce CO_2 emissions.

24.8.1
Carbon Dioxide Tax: the Most Efficient Steering Instrument

In Sweden, the carbon dioxide tax has been shown to be very useful and it is one of the most important reasons why bioenergy has surpassed oil as the largest energy source, and in 2011 reached 32% of the total energy use. The carbon tax makes unwanted carbon emissions more expensive. It does not give priority to a specific renewable energy source, but gives all solutions that reduce carbon emissions better competitiveness. Another benefit is that it also improves the government's budget, since it is a governmental income and not a cost like many types of renewable support mechanisms.

Carbon tax is often criticized for creating energy poverty and making poor people even poorer, but research has shown [14] that it is upper- and middle-class people who consume the most energy that also have to pay the major part of the tax. Poor people go by bus or travel less than wealthier people and therefore pay less tax. Also, if the government introduces a carbon tax and receives the income from the tax, the state is free to use part of the increased revenue to support poor people who are having difficulty in coping with the higher cost of energy. In Sweden, the carbon dioxide tax was introduced in 1991 and the government has used the income to reduce income tax on three different occasions. Sweden now has more than 50% renewable energy, of which a major part is bioenergy, and there is now the possibility that together with wind, solar, hydro, geothermal and other types of renewable energy a society can be created giving a good life totally supplied by renewables.

24.8.2
Less Political Damage

Over the years, there have been several political proposals and decisions intended to produce good things but ending up damaging the development of renewable energy. This is unfortunately rather common. A Swedish example is when the government in a September some years ago decided to support the replacement with heat pumps and pellet boilers of oil and direct electricity heating in small houses from 1 January the following year. All households immediately stopped buying heat pumps and pellet boilers, waiting for the support measures to come into effect at the beginning of the next year. The heat pump and pellet producers had to lay off workers for a couple of months before every household that had thought of an energy investment decided to invest in January, to receive the subsidy. Since the replacement of oil and electricity was profitable already without support, the planned 5 year subsidy ended in 11 months and then new customers waited for new support that never materialized. Several pellet companies went bankrupt. To compound this failure, too many installations were done in haste, and there were not enough trained staff to do proper installations, so technical problems and a bad reputation for these new technologies occurred as a result of the poor political decision.

The political task is therefore not to decide how to act, or which technology to choose, but to create the framework where people and companies can develop their daily lives and businesses and step-by-step improve technology and behavior. We prefer politicians to use legislation and steering instruments to punish damaging emissions and bad behavior but not to interfere with current investments and market prices so that companies cannot predict market development and prices.

Political market interference makes companies insecure and postpone decisions, or make companies move to other countries for their investments. The current (January 2013) situation in the Emission Trading Scheme is an example of bad policy that prolongs a situation of uncertainty, making it difficult to take decisions on what route to take forward. Nothing is as damaging for business and company development.

24.8.3
Use Biomass

The political framework to develop a better society in which renewable energy has an important role should be based on general incentives and long-term values such as free market conditions and a level playing field between companies and sectors. If this is set up and based on sustainable environmental values, then biomass will have a much more important role than today for food, energy, and industrial use as a raw material for different products.

References

1. UNFPA, United Nations Population Fund (2011), *Globala Befolkningstrender*, http://europe.unfpa.org/webdav/site/europe/shared/images/homepage/UNFPA%20Dact%20sheet%20SE%20web.pdf (last accessed 30 January 2013)
2. Olmstead, A. L. and Rhode, P. W. (2006) Wheat, spring wheat, and winter wheat – acreage, production, price, and stocks: 1866–1999, in *Historical Statistics of the United States, Earliest Times to the Present: Millennial Edition*, (eds S. B. Carter, S. S. Gartner, M. R. Haines, A. L. Olmstead, R. Sutch, and Gavin Wright), Cambridge University Press, New York.
3. Wikimedia (2012) Created using data from Earth Policy, *World on the Edge*, http://www.earth-policy.org/datacenter/pdf/book_wote_crops.pdf (last accessed 30 January 2013); US Department of Agriculture, Foreign Agricultural Service (2012), *Production, Supply and Distribution Online*, www.fas.usda.gov/psdonline (last accessed 30 January 2013).
4. Jordbruksverket (2011) Yield wheat per hectare in Sweden, in *Jordbruket i Siffror Åren 1866–2007*, Swedish Board of Agriculture, Jönköping, ISBN 91-88264-36-X.
5. Lynd, L. R. and Woods, J. (2011) Perspective: a new hope for Africa. *Nature*, **474**, S20–S21.
6. FAOSTAT (2011) Sugarcane production, in *Crop Production*, Food and Agriculture Organization of the United Nations, Rome.
7. FAO (2008) *The State of Food and Agriculture 2008. Biofuels: Prospects, Risks and Opportunities*, Food and Agriculture Organization of the United Nations, Rome.
8. Global Carbon Project (2011) Updated from C. Le Quéré *et al.* (2009) *Nat. Geosci.*, **2**, 831–836; Canadell, J. G. *et al.* (2007), *Proc. Natl. Acad. Sci. U.S.A.*, **104**, 18866–18870.
9. UN Comtrade (2012) *United Nations Commodity Trade Statistics Database*, http://comtrade.un.org (last accessed 30 January 2013).
10. Skogsdata (2012) *The Swedish National Forest inventory*, Swedish University of Agricultural Science.
11. Andersson, K. (2012) *Bioenergy – The Swedish Experience. How Bioenergy Became the Largest Energy Source in Sweden*, Swedish Bioenergy Association, Stockholm, http://www.svebio.se/english/publikationer/bioenergy-swedish-experience (last accessed 30 January 2013).
12. Ladanai, S. and Vinterbäck, J. (2009) *Global Potential of Sustainable Biomass for Energy*, Report 013, Swedish University of Agricultural Sciences, Uppsala.
13. IEA (2012) *IEA Technology Roadmap, Bioenergy for Heat and Power 2012*, International Energy Agency, Paris.
14. Sterner, T. (ed.) (2011) *Fuel Taxes and the Poor. The Distributional Effects of Gasoline Taxation and Their Implications for Climate Policy*, RFF Press, Abingdon.

25
Flexible Power Generation from Biomass – an Opportunity for a Renewable Sources-Based Energy System?

Daniela Thrän, Marcus Eichhorn, Alexander Krautz, Subhashree Das, and Nora Szarka

25.1
Introduction

Increasing the use of renewable energy sources (RES) is part of the collective objectives of international climate policies, and also the national goals of energy security and technical innovation in many countries. Supported by strong instruments for market introduction, renewables are increasingly forming an essential part of the energy supply system [1].

Bioenergy is a form of solar energy, biochemically bonded by plants via photosynthesis. A variety of biomass resources (e.g., wood, agricultural products, and residues) forms the basis for various conversion technologies for the production of electricity, heat/cooling, and fuels. The limiting factor for the energetic use of biomass is the sustainable available potential, since bioenergy competes with food production and material use of biomass. There are several studies on the biomass potential for energetic purposes [2].

Despite differences and uncertainties in the methodological approach and in the inclusion/exclusion of drivers in the modeling of future availability [3], a global technical bioenergy potential in the range 160–270 EJ a^{-1} in 2050 (including sustainability criteria) has been derived [4]. Around 10% of the global primary energy demand is currently met by biomass and waste [5]. Biomass contributed 6% to the total inland primary energy consumption in Europe in 2010 (Denmark 14%, Germany 6.3%, Latvia 27%, Austria 25%) [6].

The power generation from renewable energy grew across the world by ~17.7% in 2011. In 2011, renewables accounted for 3.9% of global power generation, with the highest share in Europe and Eurasia (7.1%) [7]. Investments in renewable energy reached a record high of US$ 211 billion in 2010, especially in developing and transition economies. Decreasing production, installation, and maintenance costs and a simultaneous increase in deployment experience have enabled renewables to compete against fossil fuels, especially in view of the negative externalities (e.g., pollution, health impacts, and climate change) of the latter [8].

Transition to Renewable Energy Systems, 1st Edition. Edited by Detlef Stolten and Viktor Scherer.
© 2013 Wiley-VCH Verlag GmbH & Co. KGaA. Published 2013 by Wiley-VCH Verlag GmbH & Co. KGaA.

A growing share of wind power and photovoltaics (PV) within the energy system has possible side effects stemming from their intermittent character, thereby causing concerns for the system operators [9–15]. On the one hand there could be periods of low or no wind combined with strong cloud cover leading to a low feed-in into the net and a high residual load, while on the other hand a clear day with exceptionally high wind speeds can generate electricity much above the demand levels, resulting in "excess energy." In order to maintain a stable electricity system, network operators are forced to balance residual load and excess energy.

A further increase in the share of RES will therefore require the system to be highly adjustable to balance weather induced residual load and excess energy. Two options that can balance such a system are the extensive use of the range of electricity storage systems and flexible electricity generation from renewable sources.

This chapter focuses on the second option, emphasizing the concept of flexible electricity generation from biomass. It deals with the subjects of flexible generation of bioenergy, which is currently a topic of intensive debate, discussion, and future research formulation in Germany. Legislative support for the subject that currently exists only in Germany (see EEG 2012 [16]) acted as the first impulse for bioenergy plants to streamline efforts towards changing to higher flexibility. Other countries dealing with flexible generation use technologies and solutions other than bioenergy (e.g., pumped-storage hydroelectricity in the United States, China, and Japan). Section 25.2 presents an analysis of the electricity market in view of the concepts enabling/supporting the participation of biomass-driven plants in the regulating energy market and gives a detailed overview of the pitfalls of an electricity system with a high share of RES. Section 25.3 outlines the technical options and the actual development of power generation from biomass and clarifies the new incentives of the 2012 Renewable Energy Sources Act (Erneuerbare-Energien-Gesetz, EEG) [16]. Section 19.4, Section 19.5, and Section 19.6 give detailed explanations of the technical potential of the different biomass conversion technologies (based on solid, liquid, and gaseous fuels) and their potential role in flexible electricity generation. Section 25.7 discusses the overall potential for biomass-based flexible electricity generation in Germany and concludes with future challenges and open research questions with respect to system optimization, costs, and heat use.

25.2
Challenges of Power Generation from Renewables in Germany

The development of the renewable energy sector in Germany started to gain momentum in 1991 when the electricity feed-in law came into force, which has been renamed the– Renewable Energy Sources Act (EEG) in 2000 and has since undergone several amendments [16, 17]. The EEG aims to enforce technological development to lead renewable energies into the electricity market and integrate them into the energy system. Since 2001 it has been embedded in the European energy policy to promote electricity production from renewable energy sources on the domestic electricity market [6]. Currently 20.3% of the electricity consumption in Germany is produced

25.2 Challenges of Power Generation from Renewables in Germany | 501

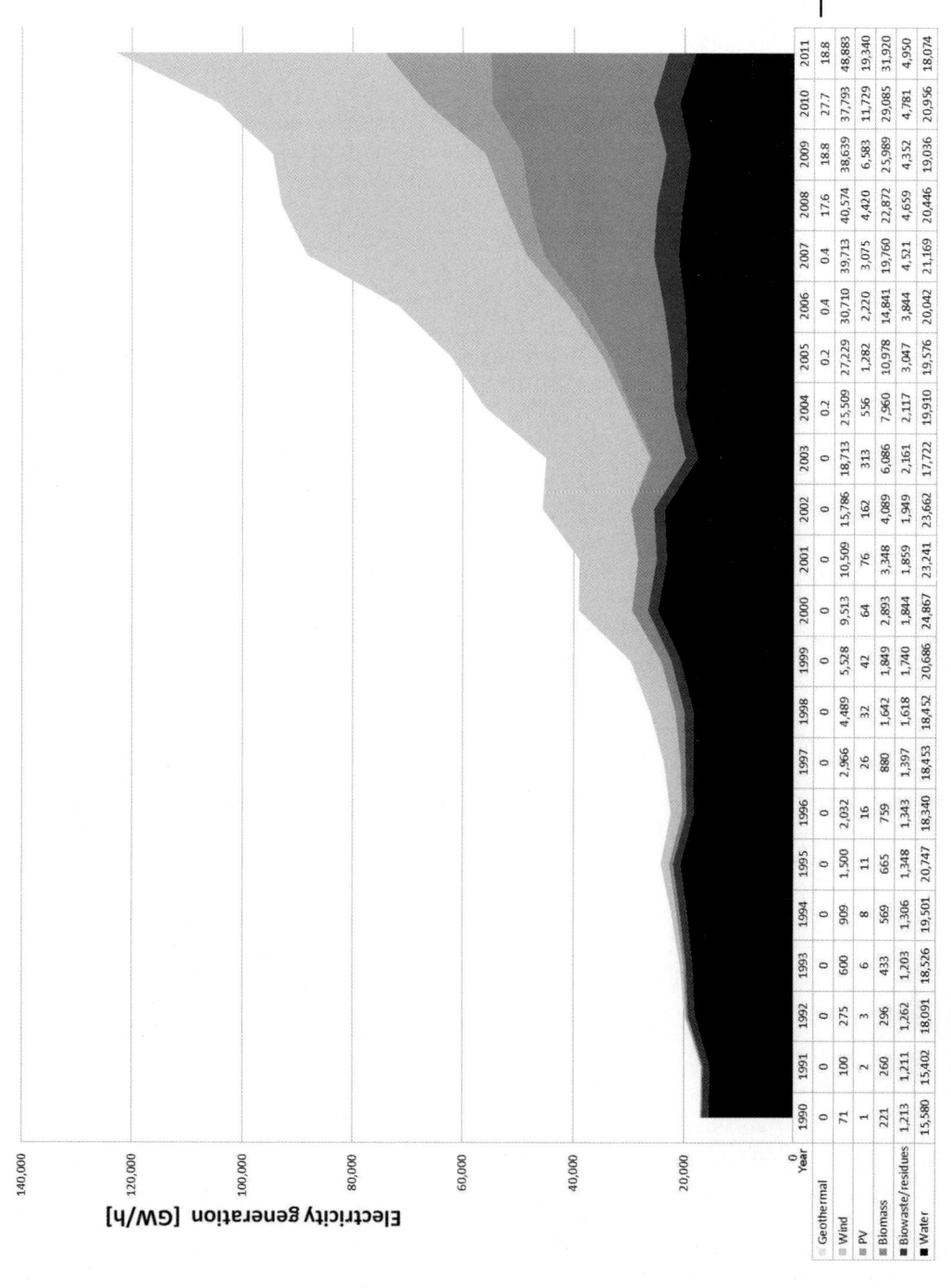

Year	1990	1991	1992	1993	1994	1995	1996	1997	1998	1999	2000	2001	2002	2003	2004	2005	2006	2007	2008	2009	2010	2011
Geothermal	0	0	0	0	0	0	0	0	0	0	0	0	0	0	0.2	0.2	0.4	0.4	17.6	18.8	27.7	18.8
Wind	71	100	275	600	909	1,500	2,032	2,966	4,489	5,528	9,513	10,509	15,786	18,713	25,509	27,229	30,710	39,713	40,574	38,639	37,793	48,883
PV	1	2	3	6	8	11	16	26	32	42	64	76	162	313	556	1,282	2,220	3,075	4,420	6,583	11,729	19,340
Biomass	221	260	296	433	569	665	759	880	1,642	1,849	2,893	3,348	4,089	6,086	7,960	10,978	14,841	19,760	22,872	25,989	29,085	31,920
Biowaste/residues	1,213	1,211	1,262	1,203	1,306	1,348	1,343	1,397	1,618	1,740	1,844	1,859	1,949	2,161	2,117	3,047	3,844	4,521	4,659	4,352	4,781	4,950
Water	15,580	15,402	18,091	18,526	19,501	20,747	18,340	18,453	18,452	20,686	24,867	23,241	23,662	17,722	19,910	19,576	20,042	21,169	20,446	19,036	20,956	18,074

Figure 25.1 Development of renewable energy electricity generation between 1990 and 2011 in Germany [17].

from renewable energy that is mainly supported by the EEG [18]. Figure 25.1 gives an overview of the increase in electricity generation from different renewable sources. The high share of renewable electricity is barely manageable by ramping up or down large fossil fuel base load power plants or cut-off of wind turbines.

At the federal level, Germany created the National Renewable Energy Action Plan (nREAP), which has targeted the production of 35% of the total electricity from renewables by 2020. In future, a further extension of RES is expected. Based on scenario calculation at the national scale, several studies [19–23] have shown an increase in power generation from renewables. Results indicate that the contribution of renewables towards electricity generation may reach between 80 and 100% by 2050. Wind and PV emerge as the dominant RES under all scenarios. In addition to the balancing of electricity supply and demand, particularly the distribution of electricity will be a challenge in a renewable energy-based power system.

In Europe, the electricity transmission networks are linked together by border crossing points. For the further development of communication and internal electricity trading between the European countries, following EU Regulation (EC) No. 714/2009, the European Network of Transmission System Operators for Electricity (ENTSO-E) was founded [24]. The ENTSO-E is supposed to ensure the security of the European electricity supply and meet the needs of the liberalized EU Internal Energy Market and facilitate market integration [25].

The EU goals are implemented in the German Energy Act (Energiewirtschaftsgesetz, EnWG) [23]. In Germany, there are four Transmission System Operators (TSOs) responsible for the security of electricity supply. The four control areas are operated by the unbundled[1] TSOs Tennet, 50Hertz, Amprion and Transnet BW (Figure 25.2).

The German power market is, according to the requirements of the EU, a liberalized market. By the introduction of market coupling into European power markets, an active trading has developed which is limited by the capacity of interconnectors at the border crossing points. Owing to sufficient interconnectors at border crossing points, Germany and Austria have one market area with one price.

The trade of electricity in the Germany/Austria market area can be organized in a bilateral OTC (over-the-counter) fashion between buyers and sellers, for forward contracts at the derivative market of European Energy Exchange (EEX) in Leipzig, or for short-term contracts at the spot market of EPEX Spot in Paris.

To bring generation and consumption together, every Balance Responsible Party (Bilanzkreisverantwortlicher)[2] has to submit the indicated day-ahead production schedules by 14:30 h to his TSO [27]. Following this, the grid capacities are calculated by the TSO. In case of network congestion, the TSO implements measures (e.g., re-dispatch) to reduce the congestion. To balance expected divergences at the time

---

1) Unbundling means a separation of the management and accounting of power production from transmission and distribution activities.
2) The Balance Responsible Party is responsible for the management of the balancing group and acts as an interface between electricity grid users and transmission system operators, and has economic responsibility for differences between feed-in and withdrawals of a balancing group (ENBW; www.enbw.com).

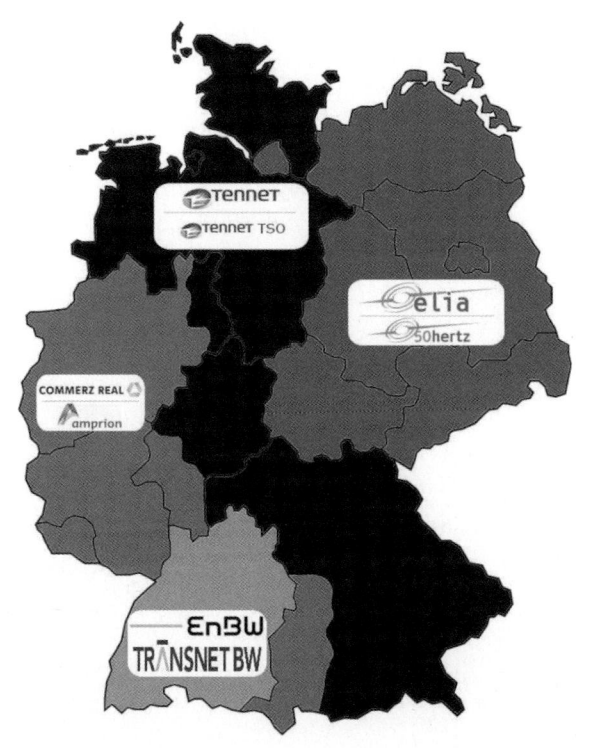

Figure 25.2 German control zones and responsible TSOs [26].

of delivery, the Balance Responsible Party has the possibility to buy or sell electricity on the intra-day market or OTC.

The TSOs in Germany have the responsibility to manage the security of the electricity system in real time, by ensuring system services (Figure 25.2). One service is the use of control power for stabilizing a permanent frequency (50 Hz) in the electricity grid. Control power is always necessary if there are net deviations between the actual customer consumption and actual power generation [28].

The different types of control power are the primary control power, secondary control power, and minute reserve, which differ principally in the particular time and the duration for which they have to be available. Primary control must be available after 30 s, secondary control after 5 min, and minute reserve after 15 min. In a prequalification process, the TSOs check if power plants achieve the minimum requirements to participate in the different control power markets. If the plants are able to fulfill the prequalification requirements, they can offer their capacities by auctions for the different control powers. Especially owing to forecast errors of the intermittent electricity generation of wind and PV, control power is needed to balance the system. In addition to the short-term system balancing through system services, such as control power, the general electricity demand–supply balance is a further challenge. Power consumption patterns and weather-induced power generation patterns show high variance.

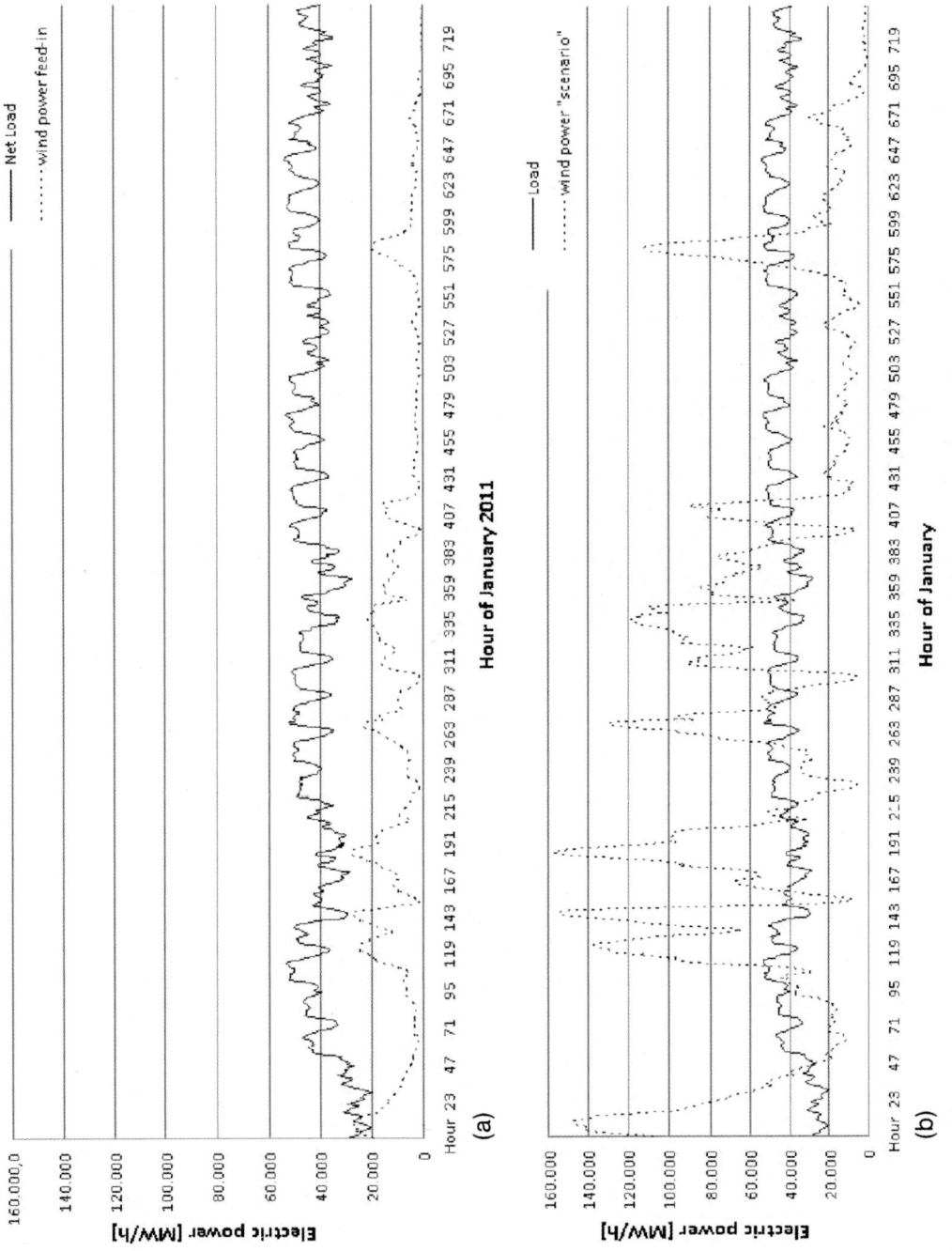

Figure 25.3 Net load and wind power generation for the area of the 50Hertz TSO: (a) hourly distribution of net load and wind power feed-in and (b) hourly distribution of net load and wind power for the feed-in scenario. Source: adapted from [27].

Figure 25.3a shows the hourly distribution of net load and wind power feed-in for the area of the 50Hertz TSO, and illustrates the relationship between net load and wind power generation observed in January 2011. On the very left side, a situation occurs where wind power exceeds the demand, but all other days show residual loads of different magnitudes. The total contribution of wind power to the total net load for January 2011 is about 18%.

Figure 25.3b displays a scenario where the total wind power generation meets the total demand based on the observed feed-in distributions of January 2011. This scenario illustrates that the sequences of excess energy alternate with sequences of residual load with considerable magnitudes, even when the RES fulfills the demand from an annual viewpoint. To balance the system will be the greatest challenge in the coming years. In addition to the translation of the excess energy into other market areas, two options are possible, storing electricity in times of excess energy and re-feed-in in times of residual load, or options for flexible generation to avoid excess energy as well as residual load. Currently, storage technologies, besides pumped storage hydro power stations, are too expensive and/or under development. Hence an interesting question is what supply technologies are available for flexible generation.

Flexibilization is possible for the existing fossil fuel-fired power plants and also for renewable power generation. Table 25.1 gives an overview of some flexibilization criteria of existing power plants. Power generation based on gaseous fuels provides a comparatively wider range of load flexibility.

Table 25.1 Flexibilization capabilities of different power plants.

Operation requirements	Definition	Steam power plants (hard coal)	Gas and steam plants	Nuclear power plants
Load changes in load-following operation	Average load gradient	3–6% min^{-1} in the range 40–100% load	4–9% min^{-1} in the range 40–100% load	10% min^{-1} from 80 to 100% 5% min^{-1} from 50 to 100% 2% min^{-1} from 20 to 100
Frequency regulation	Primary/ secondary	> 60% min^{-1} 40–100% load	180% min^{-1} 50–100% load	60% min^{-1} 60–100% load
Minimal load	% of nominal capacity	20–25% during recirculation mode 35–40% during continuous mode	30–50% single block unit 15–25% multiple block unit	20–30%
Load decrease to own needs/ island mode	Sudden, strong load reduction, without failure	Yes, in bypass operation mode	Yes, in gas turbine mode	Yes, in bypass operation mode
Plant efficiency	100% load 50% load	45–47% 42–44%	58 → 60%	36–38% EPR^a 33–35% EPR

a   Evolutionary Power Reactor
Source: adapted from [29].

Renewable sources also have the potential for flexible power production, especially biomass which is biochemically stored solar energy and therefore available if needed, independent of, for example, weather condition.

With regard to the increasing capacities for power generation from renewable energy sources, the German Renewable Energy Sources Act (EEG) [16] contains incentives for market integration and flexible electricity generation from January 2012. Hence for a wide range of technical concepts, the plant operator can switch from a fixed EEG feed-in tariff to a "market premium," which promotes operators to undertake direct selling of their share of power generation from wind, biomass, water power, or PV to the market. With this tariff model, most of the power is expected to be paid by electricity market participants such as electricity suppliers. The difference between the fixed EEG feed-in tariff and the average monthly market price will be compensated directly by the EEG. This compensation when compared with a management premium (which is paid additionally for the management and trading of the plants) results in the market premium. Like the costs of the fixed feed-in tariff, the cost of the market premium will be passed to the green electricity levy (Ökostrom-Umlage). If the operator or the trader of the operator sells the power for a higher price than the average, they can generate a higher benefit. Further, a trader can combine different plants in a pool for prequalification of control reserve, for example, for a negative minute reserve, so that the operator can generate additional income with the promotion of system services [17].

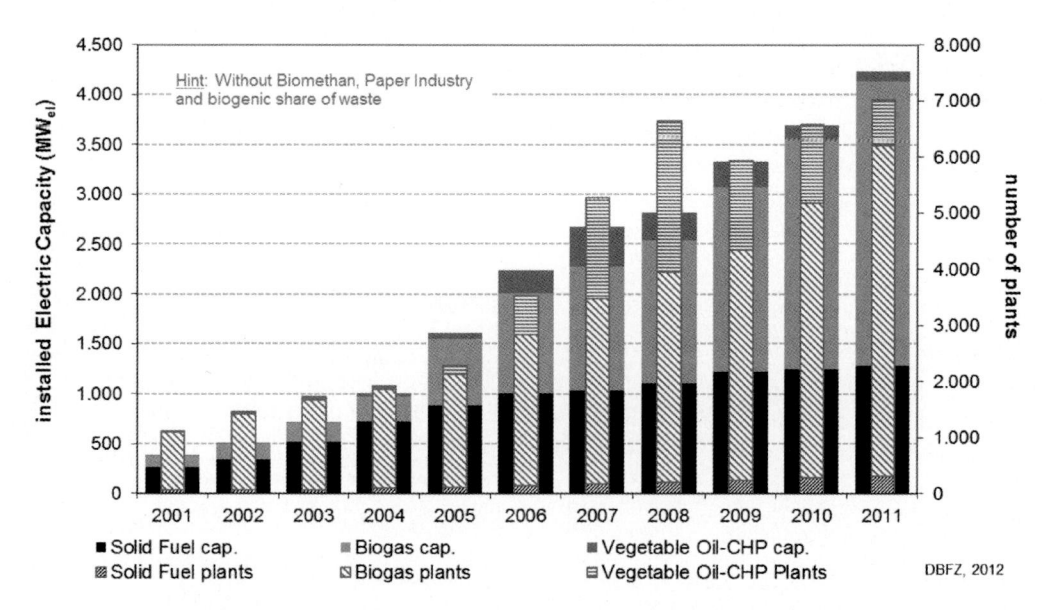

Figure 25.4 Development of installed capacities for power generation from biomass in Germany. Source: adapted from [30].

Power generation from biomass has been established under the German EEG for 15 years. Figure 25.4 gives an overview of the installed capacities up to 2011, which generated 29 TWh of power, combined with 23 TWh of heat in 2011 [30]. There has been a rapid increase in the number of installations for the provision and conversion of biogas during the last 3 years.

25.3
Power Generation from Biomass

Power generation from biomass can be realized with different conversion concepts across the range of plant sizes. In general, two main concepts can be distinguished:

- co-firing of biomass in existing lignite power plants or biogas/biomethane in existing natural gas power plants
- combined heat and power provision only by biomass-fired plants, using solid, liquid or gaseous biofuels.

Co-firing has made tremendous progress over the last decade especially in Europe and North America, and a small number of co-fired plants also exist in Asia and Australia. The concept can be utilized with a wide range of installed capacities (50–700 MW_e). The three major concepts of co-firing are (a) direct co-firing, (b) indirect co-firing, and (c) the use of a separate biomass boiler. Direct co-firing is the most commonly applied process, in which biomass and coal are burned together in the same furnace. About 82% of European plants and 95% of North American plants use direct co-firing. The primary fuel is generally coal, lignite, pulverized coal, or biomass and the co-fired fuel can be biomass, straw, sludge, peat, switchgrass, wood waste, or agricultural residues. In indirect co-firing, a biomass gasifier is used to generate clean fuel gas by converting raw biomass; the gas can then be burnt in the same furnace as coal. Indirect co-firing, although more expensive than direct co-firing, can allow the use of a wider range of biomass fuels. Table 25.2 gives an overview of the number and distribution of co-fired plants.

The second method of biomass conversion is the concept of combined heat and power (CHP) plants. Production of heat and electricity through the deployment of CHP plants is an established and well-understood technology that is over a century old [32]. The modern forms of utilizing CHP systems involve a range of thermochemical (combustion, gasification, and pyrolysis) or biochemical (fermentation) technologies. The range of technologies that can be used in CHP plants includes steam cycles, organic Rankine cycles (ORCs), gas engines, gas turbines, Stirling engines, and fuel cells [33]. For CHP systems, the overall efficiency includes the provision of electricity and the use of the co-generated heat. With the deployment of new technologies under stable frame conditions, an increase in the degree of utilization of plants can be shown for Germany as an example (Figure 25.5).

Table 25.2 Overview of co-firing plants in Europe and North America.

Country	No. of plants	Average output (MWe)	Primary fuel	Co-fired fuel
United Kingdom	21	1698	Pulverized coal, bituminous coal, coal	Wood, oil, animal feed, crop husk and pulp, grass
Denmark	5	173	Pulverized coal, wood chips	Straw, natural gas, wood chips
Finland	78	69	Biomass, peat, coal	Biomass, sludge, peat, REF[a], HFO[b], LFO[c]
Germany	3	177	Pulverized coal, lignite	Straw, wood, sewage sludge
Sweden	15	143	Coal, pulverized coal	Wood, peat, bark, wood waste, oil, rubber waste
Canada	7	362	Lignite, Pulverized coal, blended coal	Wood pellets, dry distillers grain, agricultural residues
United States	40	200	Pulverized coal	Wood, shredded rubber, wood waste, sander dust, wood chips, oil, waste paper sludge, anthracite, paper pellets, etc.

a REF: Recovered Fuel
b HFO: Heavy Fuel Oil
c LFO: Light Fuel Oil
Source: adapted from [66].

Kaltschmitt [34] provides a comprehensive overview of the major technological pathways for biomass conversion, as shown in Figure 25.6. A general problem is the low bulk density (40–200 kg m^{-3}) [35, 36] and inhomogeneous structure, which often make handling, transportation, and storage of biomass difficult [37]. In order to overcome these problems, different types of densification and homogenization technologies such as bailing, briquetting, extrusion and pelletization are used [38]. Briquetting (using a piston or screw press) and pelletization (pellet mill) are the two standard technologies using high pressure for the production of high-density solid energy carriers [37]. Torrefaction and HTC (hydro-thermal carbonization) are thermochemical conversion processes that decrease the mass while the initial energy content is only slightly reduced, resulting in a higher overall energy density [37, 39–41]. However, the optimal combination of these pretreatment concepts (torrefaction and pelletization) is currently under study.

25.3 Power Generation from Biomass | 509

- ⋯ Biogas Igniting Beam Engine (Electricity and Heat Production) [1]
- ▪ Biogas Gas Engine (Electricity and Heat Production) [1]
- ▪ Organic Rankine Cycle Turbine (ORC) (Electricity and Heat Production) [2]
- ▪ Steam Turbine (Electricity and Heat Production) [2]
- ▪ Wood Gasifier (Electricity and Heat Production) [2]
- ▪ Wood Burning Stove (Heat Production) [3]

Figure 25.5 Overall energy efficiency against different technologies. Source: adapted from [31].

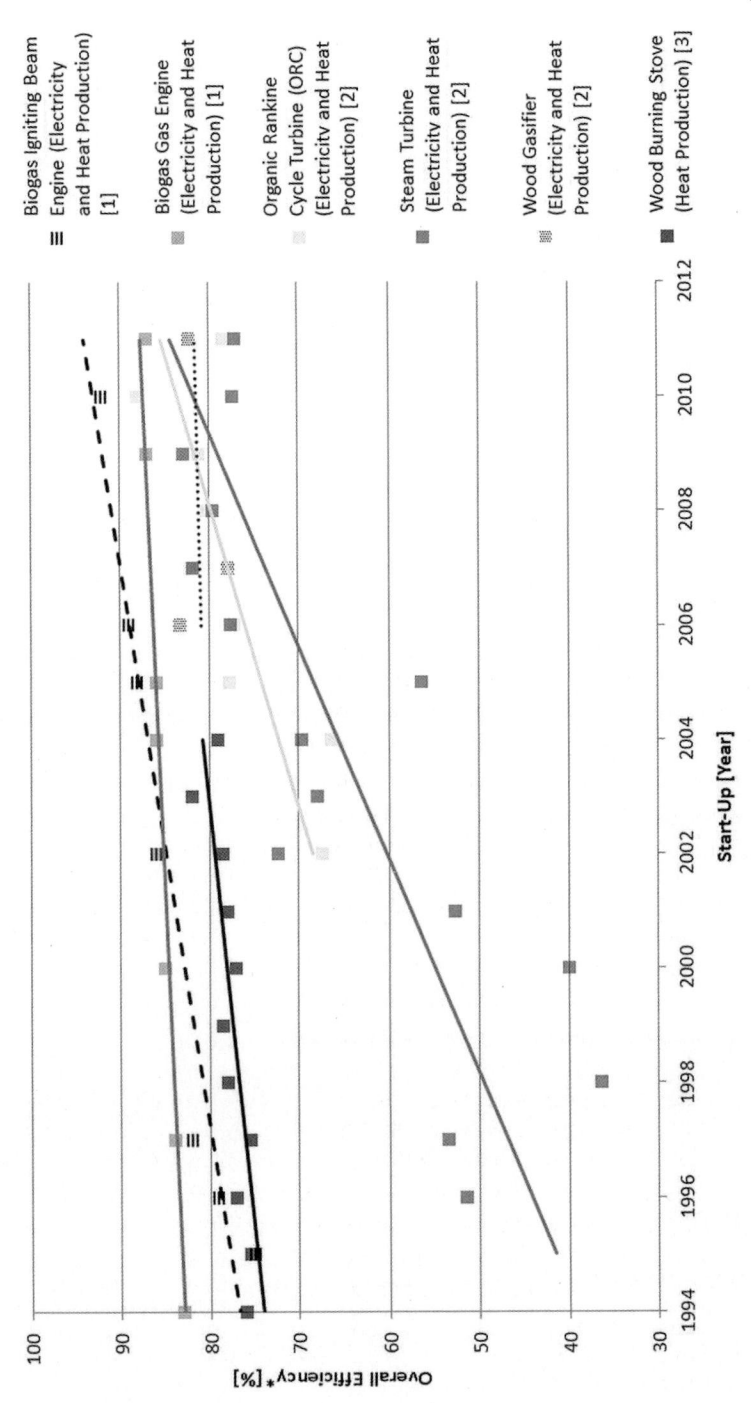

*Average value of overall efficiency for biomass CHP plants (ORC- and steam-turbine) and wood gasifier for start-up year for increased shared of hea production, especially since 2004; average value of overall-efficiency for biogas igniting beam engine and biogas gas engine from data resource of two selected producers

[1] Source: KTBL, Berlin, February 2011
[2] Source: DBFZ-Database for Biomass CHP Plants
[3] Source: http://www.bmu.de/files/pdfs/allgemein/application/pdf/bimschv1_begruendung.pdf, Version - 20.03.2008

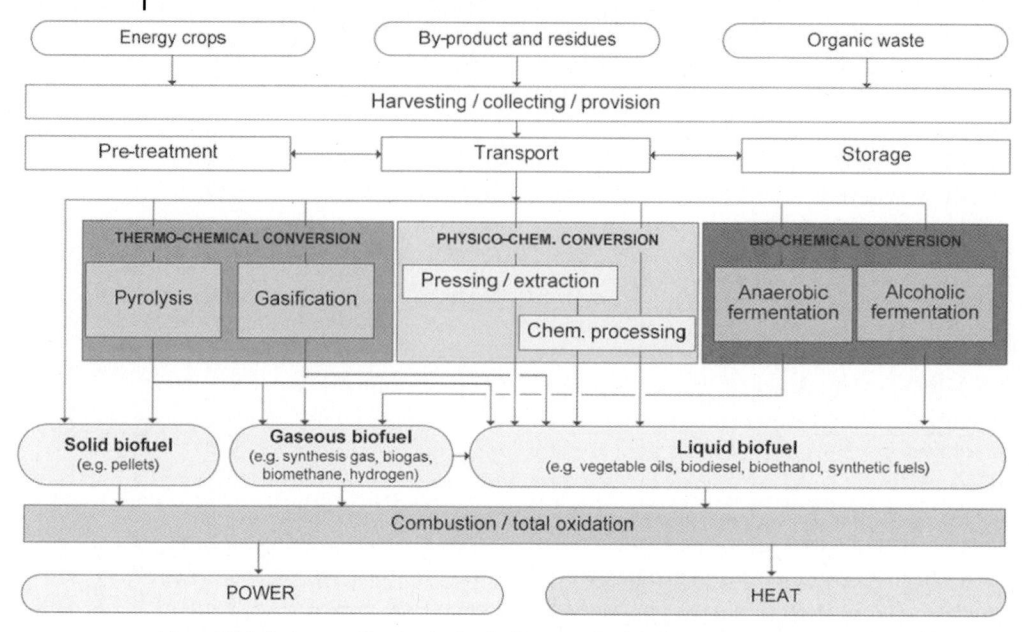

Figure 25.6 Pathways of technologies for biomass feedstock conversion. Source: adapted from [34].

The ability to store biomass and derived energy carriers is an almost unique advantage compared with other fluctuating renewable sources. The different conversion pathways can contribute to flexible power generation, depending principally on the fuel, on technology parameters (e.g., partial load efficiency), and on the available options for storability. Biomass in the form of liquid, solid, or gaseous energy carriers can be converted in ways that can fulfill the requirements for flexible generation [42]. The different technical options are described further in the following sections.

25.4
Demand-Driven Electricity Commission from Solid Biofuels

Germany has witnessed a steady rise in CHP plants, both in total number (fivefold) and installed capacity (10-fold) in the period 2000–2010 [43]. Those installations are mainly based on steam cycles and some ORC systems (Table 25.3).

Table 25.3 Installations for electricity production from solid biofuels in Germany, 2011.

No. of plants	Installed capacities	Technology options	Typical raw material	Electricity provision/ heat recovery
260	1260 MW$_{el}$	Steam turbine/ ORC turbine	Waste wood, natural wood	8.8 TWh electricity/ 14 TWh heat

Source: based on [30].

To make existing energy technologies using solid biomass more flexible, the entire pathway from solid biomass to the final products – usually heat and power – should be examined. Solid biofuels have advantageous storage characteristics due to very low energy losses and comparably high energy density, depending principally on the water content. Due to the storage, they can be made available for a flexible generation when there is a demand. Further options are the adaptation, enhancement, and changes in the technical system [42].

The combination of flexibilization and efficiency in biomass conversion has generated interest in the development of new gasification technology platforms in recent years [44]. Gasification is a versatile thermochemical conversion producing a mixture of CH_4, CO, and H_2 in various proportions depending on operational characteristics [45]. A typical product is producer gas, which can be burnt to produce heat/steam or used in gas turbines for electricity generation [46]. Further methanation of producer gas into synthetic natural gas (SNG) is also a useful pathway as storage of SNG is possible within the natural gas grid [47], thereby increasing flexibility (see also Section 25.6). Pyrolysis is the conversion of biomass to liquid (termed bio-oil, bio-crude, synoil), solid, and gaseous fractions in the absence of air, which is normally not used for CHP generation because of quality and emission problems. The bio-oil produced can be used in engines and turbines [48]; see also Section 25.5).

With the possible conversion pathway for increased flexibility through gasification, the resulting producer gas can then be converted to SNG through methanation and used in gas and steam turbines (so-called GuD). To participate in the control reserve market, for example, several technical criteria have to be fulfilled (as described in Section 25.2), including the minimum size of the plant(s) to be available within a given time frame. Single small and micro plants do not fulfill the size requirements; in this case, the so-called pool concept provides an option, in which several plants are connected to each other. It is important, however, to prove in each case the level of efficiency lost due to the emergency cooling, because of a smaller heat intake. A further challenge for micro CHP is to minimize emissions from start and stop, which can be supported by optimization of the start and stop cycles through control systems. Hence electricity generation from solid biofuels can maintain reliability, flexibility, and efficiency of the system depending on the conversion technology and storage deployment.

25.5
Demand-Driven Electricity Commission from Liquid Biofuels

Liquid fuels can be obtained from plant oil transesterification, fermentation of sucrose/starch (from sugarcane, sugarbeet, sweet sorghum, etc.), from pyrolysis/gasification of lignocellulose, and from the synthesis of dedicated biofuels from producer gas.

Power production from biomass-derived pyrolysis liquids has also been under development for a few years, liquid fuels having been produced at the laboratory scale.

Table 25.4 Installations for electricity production from liquid biofuels in Germany, 2011.

No. of plants	Installed capacities	Technology options	Typical raw material	Electricity provision/ heat recovery
550	100 MW$_{el}$	Diesel engine, heat driven	Certified plant oil/ biodiesel	0.6 TWh electricity/ 1.3 TWh heat

Source: based on [30].

Application of fast pyrolysis maximizes the liquid product yield from solid biomass (65–75 wt%) and both heat and electricity can be generated in power generation systems (PGSs) using technologies such as diesel engines, gas turbines, and co-firing. The main advantages of fast pyrolysis liquids are very low cost, possibility of decoupling solid biofuel handling from utilization, ease of storage and transportation, high energy density, and feasibility of intermittent operation [49].

The use of liquid fuels in CHP plants is an established technology in Germany, usually fed with plant oil and used based on heat demand (Table 25.4). Biogenic liquid fuels currently account for 5.4% of the biomass-based electricity generation in Germany and has shown a decreasing trend in recent years, mainly due to high oil prices [43] and the uncertain availability of certified vegetable oil produced by sustainable practices (required for remuneration via the EEG [16]). Although the transport sector is the main market for liquid biorenewables, the use of biofuels for electricity generation has been recognized. The efficiency ranges from liquid biofuel to electricity are high (η_{el} = 31–42% and η_{th} = 42–59%) and the time required for starting and stopping is short.

Liquid fuels are considered sources of flexible energy owing to their high energy density (calorific value for vegetable oil 37.6 MJ kg^{-1}, biodiesel 37.1 MJ kg^{-1}, and bioethanol 26.7 MJ kg^{-1}) [50] and ease of availability on demand. The use of small storage systems (such as are currently used in the transport sector) can be extended for use in the electricity and heating sectors. Moreover, liquid biofuels can be used in a wide range of power generation (from a few kilowatts to several megawatts) [51]. The comparatively small plant capacities and the decreasing trend of plant numbers reduce the potential for a reasonable contribution to the energy system. Despite the positive characteristics, the number and installed capacity of such plants have been decreasing drastically over the past 5 years in Germany, principally due to the high raw material price (vegetable oil).

25.6
Demand-Driven Electricity Commission from Gaseous Biofuels

The most significant gaseous biofuels for electricity generation in terms of installed capacity are biogas and biomethane. Other sources include landfill gas, SNG, and biohydrogen.

Table 25.5 Installations for electricity production from biogas in Germany, 2011.

No. of plants	Installed capacities	Technology options	Typical raw material	Electricity provision/ heat recovery
7200	2850 MW$_{el}$	Fermentation units and gas engines (83 plants with biogas upgrading to biomethane)	Manure, silage from energy crops (maize, grain, grass), wet biogenic waste fractions	19.5 TWh electricity/ 7.8 TWh heat

Source: based on [30].

Biogas (mainly $CH_4 + CO_2$) is produced by the anaerobic digestion of organic material in biodigesters/biogas plants and SNG is a second-generation fuel produced by the gasification/catalytic methanation of lignocellulosic biomass [42, 52, 53]. Biogas is primarily used in Germany for electricity production through CHP) installations. The most commonly utilized pathway is direct feed-in of electricity from biogas plants into the electricity grid; however, upcoming technologies include feed-in after upgrade of biogas to biomethane, use of the gas distribution system and centralized conversion of biomethane to electricity, and highly efficient local concepts using micro-grids for the distribution of biogas [43]. Table 25.5 shows the contribution of biogas in electricity production. Additionally, at the end of 2011 an upgrading capacity to biomethane of about 52 000 Nm^3 h^{-1} was installed [30].

Given the tremendous rise in the number of biogas plants in Germany in recent years, flexibilization opportunities are under development. The ability to shift electricity production from off-peak hours to peak hours can also significantly improve the profit from CHP plants [54], but this depends on the differences in peak- and off-peak prices, which were very low [55].

The optional pathways for increasing the flexibility of the operation of a CHP system include (i) gas and thermal storage and (ii) the development of new biological and technical solutions (Figure 25.7). Either concept can be applied at different stages of the biogas production chain after due legal and regulatory compliance. Szarka et al. [42] identified various stages at which the optimization for flexible energy production can be introduced: (i) feed management and storage of intermediates, (ii) additional storage capacity, and (iii) upgrade to biomethane.

Feed management encompasses management of the type and amount of substrate fed into the plant. Various authors have tested different mixtures of substrates and co-substrates with the goal of maximizing efficiency or environmental benefits. Schievano et al. [56] tested the substitution of energy crops with different combinations of agro-industrial by-products and residues, animal manures, and organic wastes in Italian biogas production systems. It was concluded that in terms of biogas productivity, feed mixtures that include a higher fraction of predigested swine manure perform better than energy crops. The authors emphasized that organic wastes can be used as an ideal substitute for energy crops, thus reducing a strong dependence on energy crop availability. A remarkable increase in gas production

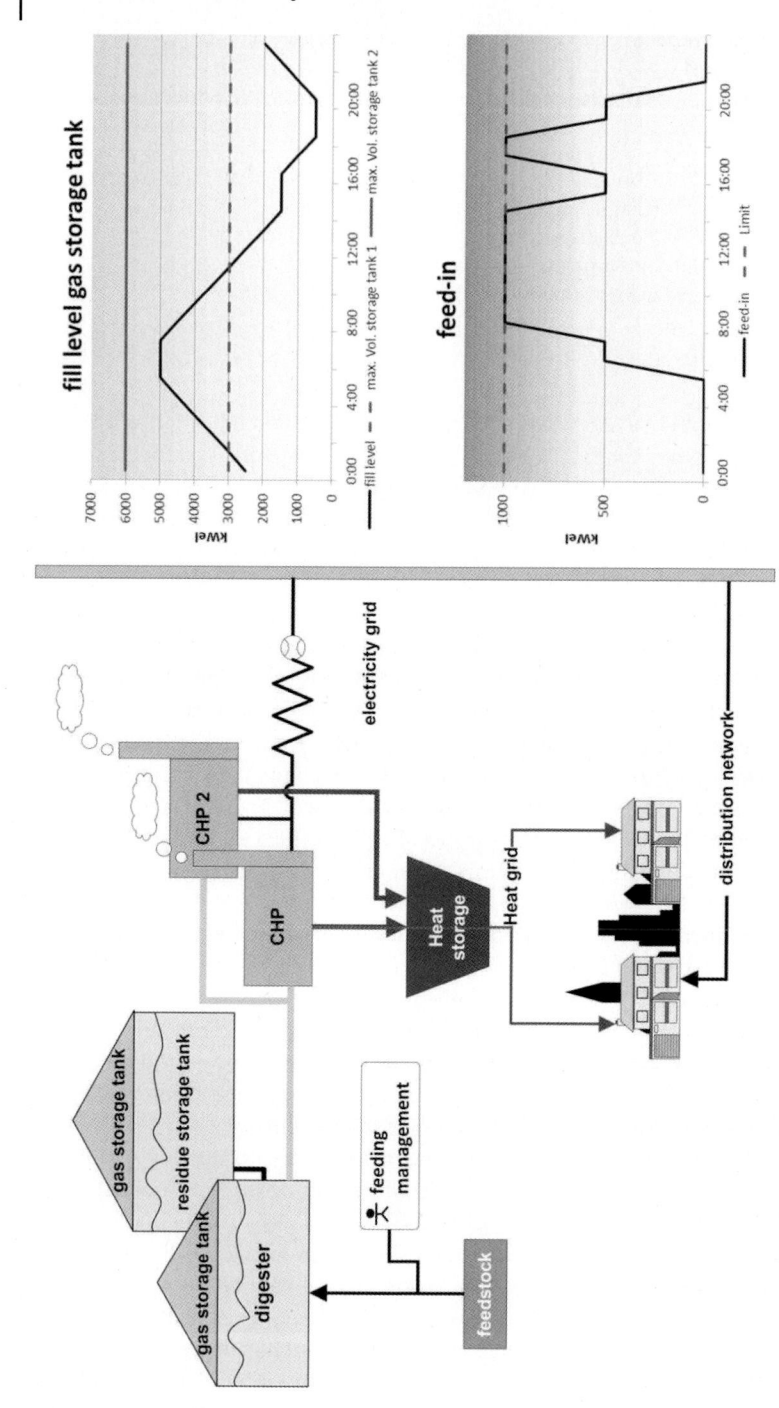

Figure 25.7 Flexible electricity generation of a biogas plant by double installed CHP and gas storage tank capacities compared with a standard plant.

by using an admixture of industrial organic waste and manure was also reported in Denmark [57]. Szarka *et al.* [42] evaluated the pilot-scale experimental results from the German Biomass Research Center (DBFZ) of using a combination of maize and sugar beet silage as substrates with different proportions and amounts. The aim of the experiment was to adjust the biogas production rate to the expected electricity price curve in a time period. The results showed a relatively high correlation. In addition to management of feed, flexibilization can be introduced by storing intermediate products (such as amino acids, sugars, and fatty acids) of biogas production for subsequent use in a methanation fermenter when demand peaks [42].

Demand-based electricity and/or heat generation can also be achieved by integrating CHP plants with adequate storage systems for biogas and heat. This essentially translates into decoupling of production and demand curves. Storage of biogas via gas storage and control units (ICT) for biogas plants can allow the storage of biogas that can subsequently be utilized for conversion to electricity when required [42]. Integrated heat storage systems from CHP plants can be achieved in the form of thermal storage tanks [54, 58] – these systems are also necessary for demand-driven CHP systems with solid and liquid biofuels.

Upgrading of biogas into biomethane for direct injection into the natural gas grid allows flexibility in terms of usage when demand arises [42]. Production of biomethane from biogas includes a cleaning process to remove trace components and an upgrading process to optimize calorific values. Upgrading is performed to meet standards for injection into the natural gas grid [59]. In Germany, there is strong legislative support (technology bonus of the EEG) for biomethane production; however, a major fraction of the biomethane is used in CHPs [60]. A further advantage of injecting biomethane into the natural gas grid is the capacity to supply the gas to densely populated areas which may be far away from production sites. Also, any increments in biogas production at remote sites are not influenced by possibilities of excess heat loss. Flexibilization can be enhanced by the choice of operation of biomethane CHP units, namely units connected to heat storage and operated preferentially at peak load, units only suitable for peak load supply, or small-scale units in smart grids [42].

Therefore, for both existing plants and new installations a wide range of options for a flexible power generation exist. Nevertheless, the most efficient mode of operation of a biogas plant in the feed-in tariff is continuous operation over the course of the entire year [61]. Which incentives will be developed by the new flexibility premium remains to be seen.

25.7
Potential for Flexible power Generation – Challenges and Opportunities

For the different concepts for power generation from biomass, increased flexibility can be achieved within the existing capacities and for new plants. To estimate the potential for flexible power generation from biomass, it is necessary to distinguish between short-term demand to ensure net stability and longer-term demand due

Table 25.6 Selected advantages and disadvantages of different fuels and technologies based on fuels for flexibilization.

Fuel type	Advantages	Disadvantages
Liquid	• Available and established technology • High energy density (calorific value) • Fast availability • High conversion efficiency	• High and unsecure substrate (vegetable oil) price • Conflicts: food versus fuel, indirect land use change • No State (EEG) support • Not economically viable
Solid	• Good storage and densification options • High energy density and low energy loss • Different combinations and pathways possible • Basic technology well established and available (e.g., in households)	• High costs • Flexible technologies need to be demonstrated • Emissions should be minimized
Gaseous	• Basic technology is available • Legislation (EEG) support • Flexible concepts already available on the market • Various technical options	• Cost optimization is required

to low availability of power from PV and wind installations or vice versa. Second, it is important to recognize the role of storage of intermediates because of high power generation from fluctuating renewable resources. Temporal attributes of storage options, that is, storage for hours/days/months, also needs to be considered. Power from biomass can contribute to the flexibilization for the short-term demand provided that there is well-adapted management for the short-term reserve.

The different bioenergy carriers are characterized by certain advantages and disadvantages (Table 25.6). For gaseous biofuels the technical potential seems to be comparatively high, but depending on the specific frame conditions other concepts can also be favorable.

Nevertheless, the potential is limited. For example, in Germany, a domestic technical biomass potential for energy purposes ranges from 1300 to 1800 PJ a^{-1} is expected for 2020 [62]. Compared with the overall German primary energy demand of 11 374 PJ [63], around about 10–15% can be covered by biomass. The current use for energy provision from biomass is in the range 900 PJ a^{-1} and provides about 730 PJ a^{-1} of bioenergy, including heat and power and also liquid biofuels (calculation based on [64]). Hence the current electricity production from biomass might increase by not more than 50–100% in future years. Figure 25.8 gives an overview of the actual resource mix for gross power generation in Germany.

Currently, the provision of the gross electricity output from biomass in short-term flexible formats can cover less than 20% of the gross demand (although it translates into cumulating the power production to 50% of the day by doubling flexible capacities).

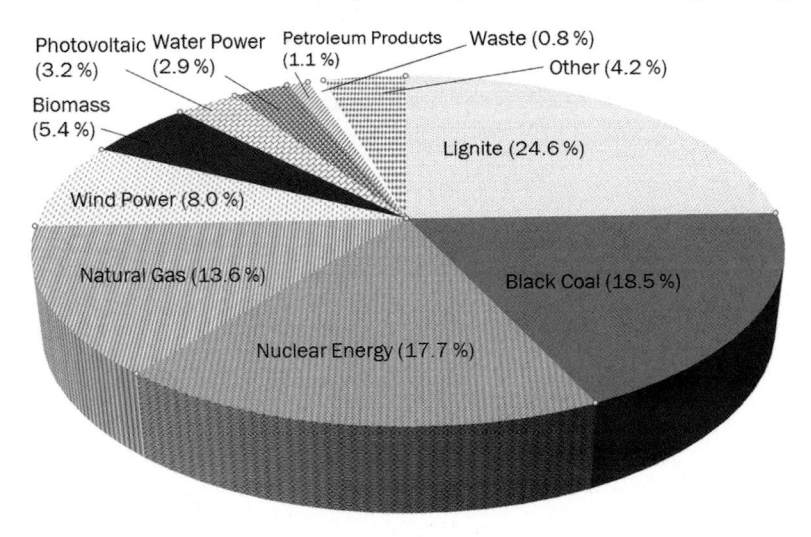

Figure 25.8 Gross power production in Germany 2011 (own compilation; data source [65]).

Pie chart data:
- Photovoltaic (3.2 %)
- Water Power (2.9 %)
- Petroleum Products (1.1 %)
- Waste (0.8 %)
- Other (4.2 %)
- Biomass (5.4 %)
- Lignite (24.6 %)
- Wind Power (8.0 %)
- Black Coal (18.5 %)
- Natural Gas (13.6 %)
- Nuclear Energy (17.7 %)

Hence the challenge is to develop highly efficient and well-integrated systems. This means:

- **System optimization:** Currently the general options for flexible power production from biomass have been described and the different compounds are available. The question of which kind of market can be supported best remains unanswered. This requires additional research on (i) regional demands of reserve capacities and other dependencies, (ii) cost effects of different levels of flexibilization, (iii) scenarios for the future bottlenecks in power production and the potential to resolve these by other renewable energies (this might be relevant for the so-called "winter peak"), and (iv) environmental effects of those modified concepts (i.e., additional efforts for transport, additional effects of greenhouse gas reduction).
- **Cost distribution:** Flexible power production from biomass is linked to additional provision costs, because the dimensions of storage units and the CHP system need to be increased. The open question is how to transfer the additional costs from the power provider to participants of the energy system, who profit from the availability of flexible energy, for example, net operators and the providers of wind/solar-based power. It remains unclear if the German Renewable Energy Sources Act (EEG) is the right arena for these cost shifts.
- **Heat storage and utilization:** The basis of flexible power production from biomass is the ~8000 CHP plants in Germany. Heat is a by-product of the conversion from biomass to energy and can increase the environmental effect, if additional fossil fuels can be substituted. Flexible power production should not reduce the efforts for efficient use of the heat. Therefore, within the plant, concepts for additional heat storage capacities have to be considered and their effects have to be assessed.

To conclude, a clearer role of power from biomass in the energy policy can foster the introduction of flexible power in the short term and may additionally support the technical development of biomass provision and conversion.

References

1 Frankl, P. (2012) Renewables – policy and market design challenges, presented at the IEA Bioenergy Conference, 13–15 November 2012, Vienna.

2 Umweltbundesamt (UBA) (2012) *Managing Biomass Sustainability – Respect the Ecological Limits of Land Use!*, Press Release 039/2012, http://www.umweltbundesamt.de/uba-info-presse-e/2012/pe12–039_managing_biomass_sustainably_respect_the_ecological_limits_of_land_use.htm (last accessed December 2012).

3 Zeller, V., Thrän, D., Zeymer, M., Bürzel, B., Adler, P., Ponitka, J., Postel, J., Müller-Lang, F., Rönsch, S., Gröngröft, A., Kirsten, C., Weller, N., Schenker, M., Wedwitschka, H. Wagner, B., Deumelandt, P., Reinicke, F., Vetter, A., Weiser, Ch., Henneberg, K., and Wiegmann, K. (2012) *Basisinformationen für eine nachhaltige Nutzung von landwirtschaftlichen Reststoffen zur Bioenergiebereitstellung*, DBFZ Report No. 13, Deutsches Biomasseforschungszentrum, Leipzig.

4 Haber, H., Beringer, T., Bhattacharya, S. C., Erb, K.-H., Hoogwijk, M. (2010) The global technical potential of bio-energy in 2050 considering sustainability constraints. *Curr. Opin. Environ. Sustain.*, **2** (5–6), 394–403.

5 Schubert, R., Schellnhuber, H. J., Buchmann, N., Epiney, A., Greisshammer, R., Kulessa, M., Messner, D., Rahmstorf, S., and Schmid, J.; German Advisory Council on Global Change (WBGU) (2008) *Future Bioenergy and Sustainable Land Use*, Earthscan, London.

6 Eurostat (2010) European Commission, Renewable Energy Statistics Available at http://epp.eurostat.ec.europa.eu/statistics_explained/index.php/Renewable_energy_statistics#Further_Eurostat_information (last accessed December 2012).

7 British Petroleum (2012) *Statistical Review of World Energy*, BP, London.

8 United Nations Environment Programme (UNEP) (2011) *Renewable Energy – Investing in Energy and Resource Efficiency*, UNEP, Nairobi.

9 Lund, H. and Clark, W.W (2002) Management of fluctuations in wind power and CHP comparing two possible Danish strategies. *Energy*, **27** (5), 471–483.

10 Lund, H. (2003) Excess electricity diagrams and the integration of renewable energy. *Int. J. Sustain. Energy*, **23** (4), 149–156.

11 Lund, H. and Münster, E. (2003) Management of surplus electricity production from a fluctuating renewable-energy source. *Appl. Energy*, **76** (1–3), 65–74.

12 Lund, H. (2005) Large-scale integration of wind power into different energy systems. *Energy*, **30** (13), 2402–2412.

13 Lund, H. (2006). Large-scale integration of optimal combinations of PV, wind and wave power into the electricity supply. *Renew. Energy*, **31** (4), 503–515.

14 Lund, H. and Munster, E. (2006) Integrated energy systems and local energy markets. *Energy Policy*, **34** (10), 1152–1160.

15 Lund, H. and Mathiesen, B. V. (2009) Energy system analysis of 100% renewable energy systems – the case of Denmark in years 2030 and 2050. *Energy*, **34**, 514–531.

16 EEG (2012) *Renewable Energy Sources Act*, http://www.erneuerbare-energien.de/files/pdfs/allgemein/application/pdf/eeg_2012_bf.pdf (last accessed December 2012).

17 Stromeinspeisungsgesetz (1991) *Electricity Feed-in Law*, http://www.dev.de/php/deutsch/gesetze/700_stromeinspeisegesetz.php (last accessed December 2012).

18 Bundesministerium für Umwelt und Reaktorsicherheit (BMU) (2012) *Erneuerbare Energien in Zahlen. Nationale und Internationale Entwicklung*, BMU, Berlin.

19 Umweltbundesamt (UBA) (2010) *Energieziel 2050: 100% Strom aus erneuerbaren Quellen*, Umweltbundesamt Dessau-Rosslau, http://www.umweltdaten.de/publikationen/fpdf-l/3997.pdf (last accessed December 2012).

20 Bundesministerium für Wirtschaft und Technologie (BMWI) (2010) *Energienszenarien für eine Energiekonzept der Bundesregierung. Studie im Auftrag des Bundesministerium für Wirtschaft und Technologie durchgeführt von Prognos AG/EWI/GWS*, http://www.bmwi.de/BMWi/Navigation/Service/publikationen,did=356294.html (last accessed December 2012).

21 Bundesministerium für Umwelt und Reaktorsicherheit (BMU)/Deutsches Zentrum für Luft und Raumfahrt (DLR) (2012) *Langfristszenarien und Strategien für den Ausbau der erneuerbaren Energien in Deutschland bei Berücksichtigung der Entwicklung in Europa und global*, http://www.fvee.de/fileadmin/publikationen/Politische_Papiere_anderer/12.03.29.BMU_Leitstudie2011/BMU_Leitstudie2011.pdf (last accessed December 2012),

22 WWF (2009) *Modell Deutschland – Klimaschutz bis 2050. Vom Ziel her denken*, World Wide Fund for Nature, http://www.oeko.de/oekodoc/948/2009-054-de.pdf (last accessed December 2012).

23 Greenpeace (2007) *Klimaschutz: Plan B – Nationales Energiekonzept bis 2050*, http://www.greenpeace.de/fileadmin/gpd/user_upload/themen/klima/Plan_B_2050_lang.pdf (last accessed December 2012).

24 EU Parliament (2009) *Regulation (EC) No. 714/2009 of the European Parliament and of the Council of 13 July 2009 on Conditions for Access to the Network for Cross-Border Exchanges in Electricity and Repealing Regulation (EC) No. 1228/2003*, http://eurlex.europa.eu/LexUriServ/LexUriServ.do?uri=OJ:L:2009:211:0015:0035:EN:PDF (last accessed December 2012).

25 ENTSO-E (2012) *European Network of Transmission System Operators for Electricity Homepage*, https://www.entsoe.eu/ (last accessed Decemeber 2012).

26 Wikipedia (2012) *Regelleistung (Energie)*, http://upload.wikimedia.org/wikipedia/commons/1/17/Regelzonen_deutscher_%C3%9Cbertragungsnetzbetreiber_neu.png (last accessed December 2012).

27 50Hertz (2012) *Transmission System Operator 50Hertz Homepage*, http://www.50hertz.com/en/109.htm (last accessed December 2012).

28 Amprion (2012) *Balancing Power Settlement with Balancing Group Managers*, http://www.amprion.net/en/control-area-balance (last accessed December 2012).

29 Balling, L., Schmid, E., and Tomschi, U. (2011) *Flexiblen Kraftwerken gehört die Zukunft. Energy 2.0 Kompendium*, http://www.energy20.net/pi/index.php?StoryID=317&articleID=179195 (last accessed December 2012).

30 Witt, J., Thrän, D., Rensberg, N., Hennig, C., Naumann, K., Billig, E., Sauter, P., Daniel-Gromke, J., Krautz, A., Wiser, C., Reinhold, G., and Graf, T. (2012) *Monitoring zur Wirkung des Erneubare-Energien-Gesetz (EEG) auf die Entwicklung der Stromerzeugung aus Biomasse*, Report No. 12, Deutsches Biomasseforschungzentrum (DBFZ), Leipzig.

31 Thrän, D. (2012) Bioenergieforschung aktuell & 2020 – Was kann Bioenergie leisten? Beitrag des Förderprogramms zur Energiewende Präsentation at the Conference Energetische Biomassenutzung, 5 November 2012, Berlin.

32 Rosillo-Calle, F. (2007) Overview of biomass energy, in *The Biomass Assessment Handbook – Bioenergy for a Sustainable Environment* (eds F. Rosillo-Calle, P. de Groot, S. L. Helmstock, and J. Woods), Earthscan, London.

33 Ortwein, A., Szarka N., and Büchner, D. (2012) Technical and systems assessment of innovative flexible micro-CHP concepts for solid biofuels, in *Proceedings of the 20th European Biomass Conference and Exhibition*, Milan, pp. 1350–1353.

34 Kaltschmitt, M. (2011) Biomass for energy in Germany: status, perspectives and lessons learned. *J. Sustain. Energy Environ.*, Special Issue, 1–10.

35 Adapa, P., Tabil, L., and Schoenau, G. (2009) Compaction characteristics of barley, canola, oat and wheat straw. *Biosyst. Eng.*, **104**, 335–344.

36 Robbins, W. C. (1982) Density of wood chips. *J. Forest.*, **80**, 567.

37 Stelte, W., Sanadi, A. R., Shang, L., Holm, J.K, Ahrenfeldt, J., and Henriksen, U. B. (2012) Biomass pelletization review. *BioResources*, **7** (3), 4451–4490.

38 Tumuluru, J. S., Wright, C. T., Kenney, K. L., and Hess, J. R. (2010) A technical review on biomass processing: densification, preprocessing, modeling, and optimization, presented at the ASABE Annual International Meeting.

39 Stelte, W., Clemons, C., Holm, J. K., Sanadi, A. R., Ahrenfeldt, J., Shang, L., and Henriksen, U. B. (2011) Pelletizing properties of torrefied spruce. *Biomass Bioenergy*, **35**, 4690–4698.

40 Kim, Y.-H., Lee, S.-M., Lee, H.-W., and Lee, J.-W. (2012) Physical and chemical characteristics of products from the torrefaction of yellow poplar (*Liriodendron tulipifera*). *Bioresource Technol.*, **116**, 120–125.

41 Mumme, J., Eckervogt, L., Pielert, J., Diakité, M., Rupp, F., and Kern, J. (2011) Hydrothermal carbonization of anaerobically digested maize silage. *Bioresource Technol.*, **102**, 9255–9260.

42 Szarka, N., Scholwin, F., Jacobi, F., Eichhorn, M., Ortwein, A., and Thrän, D. (2013) A novel role of bioenergy: a flexile, demand-oriented power supply. *Energy J.*, Article in Press.

43 Fritsche, U. R., Hennenberg, K., Hünecke, K., Thrän, D., Witt, J., Hennig, C., and Rensberg, N. (2009) *IEA Bioenergy Task 40: Country Report Germany*, Öko-Institut and German Biomass Research Centre, Darmstadt/Leipzig.

44 Ahrenfeldt, J., Thomsen, T. P., Henriksen, U., and Clausen, L. R. (2013) Biomass gasification cogeneration – a review of state of the art technology and near future perspectives. *Appl. Thermal Eng.*, **50**, 1407–4017.

45 McKendry, P. (2002) Energy production from biomass. Part 3. Gasification technologies. *Bioresource Technol.*, **83**, 55–63.

46 Pollex, A., Ortwein, A., and Kaltschmitt, M. (2011) Thermo-chemical conversion of solid biofuels. *Biomass Convers. Biorefin.*, **2**, 21–39.

47 Rönsch, S., and Ortwein, A., (2011) Methanisierung von Synthesegasen – Grundlagen und Verfahrensentwicklungen. *Chem.-Ing.-Tech.*, **83** (8), 1200–1208.

48 McKendry, P. (2002) Energy production from biomass. Part 2. Conversion technologies. *Bioresource Technol.*, **83**, 47–54.

49 Chiaramonti, D., Oasmaa, A., and Solantausta, Y. (2007) Power generation using fast pyrolysis liquids from biomass. *Renew. Sustain. Energy Rev.*, **11**, 1056–1086.

50 Agency for Renewable Resources (FNR) (2012) *Basisdaten Bioenergie Deutschland. Festbrennstoffe, Biokraftstoffe, Biogas*, Bestell-Nr. 469, FNR, Gützow-Prüzen.

51 Majer, S. (2011) Electricity generation from liquid biofuels, presented at Bioenergy Symposium, Buenos Aires.

52 Demirbas, A. (2009) Biorenewable liquid fuels, in *Biofuels: Securing the Planet's Future Energy Needs*, Springer, London, pp. 103–230.

53 Steubing, B., Zah, R., and Ludwig, C. (2011) Life cycle assessment of SNG from wood for heating, electricity, and transportation. *Biomass Bioenergy*, **35**, 2950–2960.

54 Streckienė, G., Martinaitis, V., Andersen, A. N., and Katz, J. (2009) Feasibility of CHP plants with thermal stores in the German spot market. *Appl. Energy*, **86** (11), 2308–2316.

55 EEX (2012) *European Energy Exchange Homepage*, http://www.eex.com (last accessed December 2012).

56 Schievano, A., D'Imporzano, G., and Adani, F. (2009) Substituting energy crops with organic wastes and agro-industrial residues for biogas production. *J. Environ. Manage.*, **90**, 2537–2541.

57 Mæng, H., Lund, H., and Hvelplund, F. (1999) Biogas plants in Denmark: technological and economic developments. *Appl. Energy*, **64**, 195–206.

58 Haeseldonckx, D., Peeters, L., Helsen, L., and D'haeseleer, W. (2007) The impact of thermal storage on the operational behaviour of residential CHP facilities and the overall CO_2 emissions. *Renew. Sustain. Energy Rev.*, **11**, 1227–1243.

59 Ryckebosch, E., Drouillon, M., and Vervaeren, H. (2011) Techniques for transformation of biogas to biomethane. *Biomass Bioenergy*, **35**, 1633–1645.

60 Weiland, P. (2009) Status of biogas upgrading in Germany, presented at the IEA Task 37 Workshop on Biogas Upgrading, Tulln.

61 Thrän, D., Scholwin, F., Witt, D., Krautz, A., Bienert, K., Hennig, C., Rensberg, N., Stinner, W., Schaubach, K., Gawor, M., Trommler, M., Grope, J., Daniel-Gromke, J., Richarz, V., Naumann, K., Viehmann, C., Majer, S., Schwenker, A., Wirkner, R., and Lenz, V. (2011) *Vorbereitung und Begleitung der Erstellung des Erfahrungsberichtes 2011 gemäß § 65 EEG*, http://www.bmu.de/files/pdfs/allgemein/application/pdf/eeg_eb_2011_biomasse_bf.pdf (last accessed December 2012).

62 Stecher, K. and Adler, P. (2012) *Eigene Berechnungen*, Deutsches Biomasseforschungszentrum (DBFZ), Leipzig.

63 Bundesministerium für Umwelt und Reaktorsicherheit (BMU) (2010) *Das Energiekonzept der Bundesregierung 2010 und die Energiewende 2011*, http://www.bmu.de/files/pdfs/allgemein/application/pdf/energiekonzept_bundesregierung.pdf (last accessed December 2012).

64 http://www.erneuerbareenergien.de/ (last accessed December 2012).

65 AG Energiebilanzen (AGEB) (2012) *Zusatzinformationen*, http://www.ag-energiebilanzen.de/viewpage.php?idpage=65 (last accessed December 2012).

66 International Energy Agency (2009) IEA Bioenergy Agreement Task 32, *Cofiring*, http://www.ieabcc.nl/database/cofiring.php (last accessed December 2012).

26
Options for Biofuel Production – Status and Perspectives

Franziska Müller-Langer, Arne Gröngröft, Stefan Majer, Sinéad O'Keeffe, and Marco Klemm

26.1
Introduction

The global demand for energy, especially transport fuels, will continue to increase significantly in the future, from a current demand of 93 EJ a^{-1} (2009) to an estimated 116 EJ a^{-1} by 2050 [1]. In addition to other options for meeting the increasing demand, such as improved efficiency, traffic reduction and relocation, and electro-mobility, biofuels are advocated as one of the best means to compensate for the prospected additional consumption in the years to come. However, in the short to medium term, there are a few sectors (e.g., freight transport and aviation) that are not able to apply alternatives based on nonbiogenic renewable energies (such as electro-mobility). Moreover, the implementation of biofuels is not free of concerns. The most important drivers for using biofuels are mitigation of greenhouse gases (GHGs) and security of supply, and the necessity for diversification of fuel sources to buffer against the instabilities of fossil fuel prices. With careful strategies and appropriate regulations, some biofuels could enhance energy security, aid towards achieving GHG mitigation, and give countries the chance to diversify agriculture production, raise rural incomes, and enhance access to commercial energy, especially in rural communities [2].

In the last few years, many countries have moved from voluntarily to obligatory legislation such as the EU mandatory target of 10% renewable energy in the transport sector by 2020 (which is expected to be mainly covered by biofuels) (EU RED 2009), a 7% biofuel target for 2022 in the United States and the Chinese 15% biofuel target by 2020 [3]. With respect to specific fuel quality requirements for the transport sector, the International Energy Agency (IEA) expects in its blue map scenario an increase in high-quality diesel fuels and biomethane in addition to a shift from corn-based to lignocellulosic bioethanol by 2050 (Figure 26.1). The total biofuel demand is about 27% of the total transport fuel demand in 2050, [1]. The expected development fits well with the estimated maximum technical biofuel potential for 2020 and the total

Transition to Renewable Energy Systems, 1st Edition. Edited by Detlef Stolten and Viktor Scherer.
© 2013 Wiley-VCH Verlag GmbH & Co. KGaA. Published 2013 by Wiley-VCH Verlag GmbH & Co. KGaA.

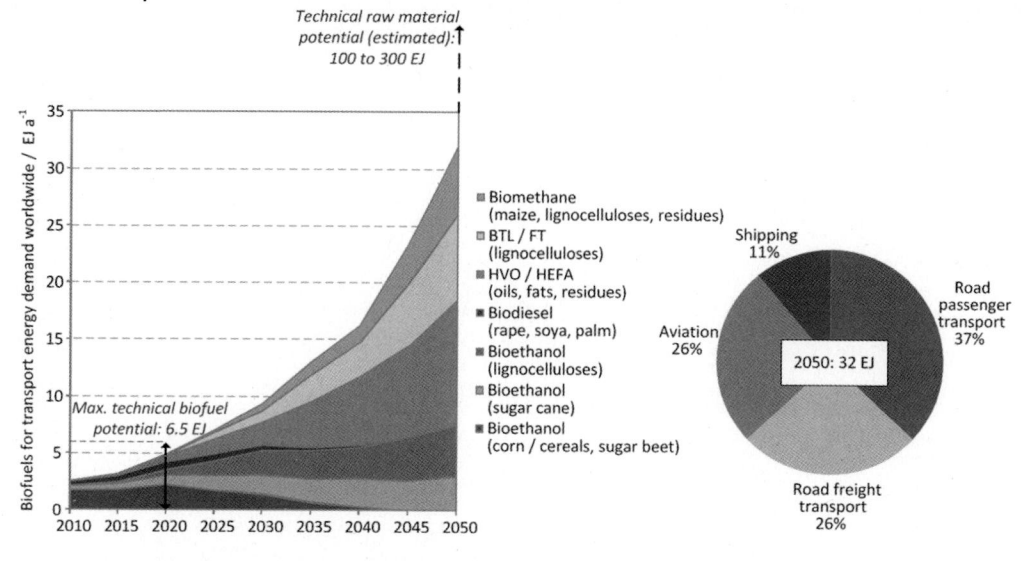

Figure 26.1 Biofuels for transport energy demand worldwide and their use in transport modes. Source: DBFZ, adapted from [4, 5].

technical raw material potential in 2050 [4]. It is expected that by 2050 the majority of biofuels will still be used for road transport, followed by aviation and shipping.

Against this background, this chapter deals with a selection of biofuel options, which are briefly characterized with regard to their production technologies and analyzed regarding certain technical, economic, and environmental aspects. The focus is on the most important drivers for efficiency, GHG emissions, and costs.

26.2
Characteristics of Biofuel Technologies

There are various options for producing alternative transportation fuels from biomass. Depending on the conversion of biomass, in principle there are three main pathways that can be considered: (i) the thermochemical pathway, (ii) the physicochemical conversion pathway, and (iii) the biochemical conversion pathway [6]. For thermochemical and biochemical conversion, future lignocellulosic options might be promising; lignocellulosic biomass includes nonfood crop varieties such as woody and herbaceous residues and energy crops. These types of conversion systems are associated with higher capacities and lower costs and are envisaged to be one of the more sustainable future technology options. An overview of the most important biofuel options under international discussion is provided in Figure 26.2.

Following the blue map scenario of the IEA (Figure 26.1), a selection of current and future biofuels are considered in more detail in the following; a summary is given in Table 26.1. These biofuels are the most important options under inter-

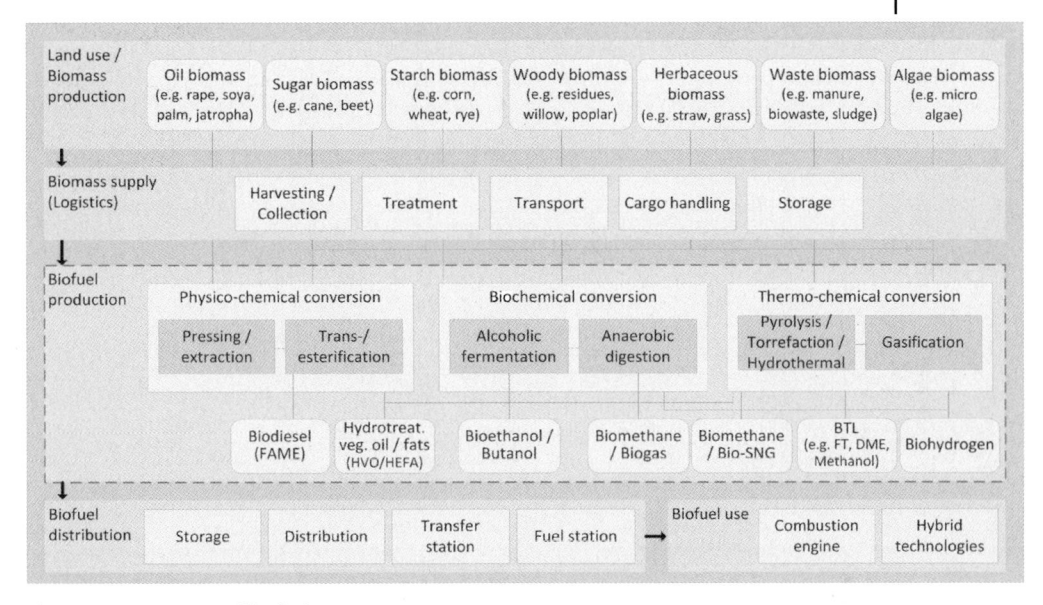

Figure 26.2 Overview of biofuel options.

national discussion and as so-called drop-in fuels (i.e., fuels that can be applied like their established fossil counterpart) they can use the existing infrastructure, distribution, and final use in different transport modes). Other biofuel options are also briefly mentioned.

So-called conventional biofuels (biodiesel and bioethanol) are currently available on the global market in considerable amounts, and well established technologies are applied for their production. Only specific parts of the crop are used as raw material, such as corn kernels or grains for their starch content, sugar cane or beet for their sugar content, and the oil in oilseeds. Residues from the production of these crops (e.g., straw) are used within agricultural cycles (e.g., as fertilizer or livestock bedding). Moreover, by-products from biofuel production are used as fodder [e.g., extraction meal, vinasse, distiller's dried grains with solubles (DDGS)] and in the chemical industry (e.g., glycerine, fertilizer) (Table 26.1).

In contrast to conventional fuel production, future biofuels can be produced via bio- and thermochemical conversion routes (i) from the whole crop (higher biofuel yield per hectare of land) and (ii) from a diverse range of raw materials, including biowaste streams that are rich in lignin and cellulose, such as straw, grass, or wood. It is expected that low-cost residues and waste sources will be the preferred raw materials in the coming years, followed by cellulosic perennial energy crops (e.g., willow, poplar, eucalyptus) [11].

Typically, future-generation biofuel production plants and the surrounding infrastructure are comparably more complex and therefore are more capital intensive (i.e., high capital risk). Technical measures show that future-generation concepts are possible in principle but have to be confirmed within demonstration plants.

Table 26.1 Technical characteristics of selected biofuel options [7–9].

Biofuel option	Typical raw materials	Main typical conversion steps	Typical by-products[a]	State of development[b]	Installed capacity/production worldwide/focus region (all 2011)[c]	R&D demand
Biodiesel	Vegetable and animal oils and fats (e.g., rape, soya, palm, jatropha, grease, algae oils)	Vegetable oil production (mechanical or solvent extraction), refining, trans-/esterification, biodiesel treatment	Press extraction, glycerine, salt	Commercial, TRL 9	50 Mt a^{-1}/17 Mt a^{-1}/US, LA soya; EU rape; SA palm	Process optimization regarding low-quality oils and fats, catalysts, and treatment technologies, methanol substitution through bioethanol
Hydrotreated vegetable oils or hydrotreated esters and fatty acids (HVO/HEFA)	Cf. biodiesel	Vegetable oil production (mechanical or solvent extraction), refining, hydrotreating, distillation	(Press extraction), propane, gasoline fractions	Commercial, TRL 9	~2 Mt a^{-1}/unknown/EU, SA palm, grease, animal fats	Raw material diversification (e.g., algae, pyrolysis or hydrothermal oil), co-refining in mineral oil refinery), process optimization regarding catalysts, hydrogen demand
Bioethanol	Sugar (beet, cane), starch (corn, wheat, rye)	Treatment, sugar extraction or hydrolysis/saccharification, C$_6$ fermentation, distillation, final dehydration	Sugar: bagasse/vinasse Starch: gluten, stillage for DDGS (distiller's dried grains with solubles), fertilizer, biogas/biomethane	Commercial, TRL 9	~90 Mt a^{-1}/70 Mt a^{-1}/US corn; BR sugar cane; EU wheat, sugar beet	Process optimization regarding process integration, e.g., upgrading by-products and stillage (e.g., recycling, biogas/biomethane, nutrient recovery)
	Lignocelluloses (straw, bagasse, wood)	Pretreatment (thermal, acid, etc.), hydrolysis, saccharification, C$_6$/C$_5$ fermentation, distillation, final dehydration	Lignin-based by-products, pentoses, stillage products such as fertilizer, biogas/biomethane	Demonstration plants, TRL 7	~0.051 Mt a^{-1}/unknown, often only test campaigns/US, EU straw; BR bagasse	Upscaling and demonstration of overall process concepts, further development for lignin, pentoses, enzyme use, and efficiency improvement

Table 26.1 (continued)

Biofuel option	Typical raw materials	Main typical conversion steps	Typical by-products[a]	State of development[b]	Installed capacity/ production worldwide/ focus region (all 2011)[c]	R&D demand
Biomethane/biogas	Sugar and starch, organic residues (e.g., biowaste, manure, stillage)	Silaging, hydrolysis (optional), anaerobic digestion, gas treatment and upgrading	Digestate, electricity	Commercial, TRL 9	~0.5 Mt a^{-1} (EU)/ unknown/EU, DE different	Lignocelluloses as co-substrate, process optimization (methane yield, enzymes, gas treatment)
Biomethane/synthetic natural gas (SNG)	Lignocelluloses (diverse, focus wood, straw)	Mechanical and thermal treatment (e.g., drying), gasification, gas treatment, synthesis (methanation), gas upgrading	Electricity and heat	Demonstration plants, TRL 7	~0.036 Mt a^{-1} (EU)/ unknown often only test campaigns/ EU wood	Upscaling and demonstration of overall process concepts, adaptation of syngas treatment to gasifier properties, efficiency increase, adapted technology for decentralized plants
Synthetic biomass-to-liquids (BTL)	Lignocelluloses (diverse, focus wood, straw, residues such as black liquor)	Mechanical and thermal treatment (e.g., drying, pyrolysis, hydrothermal), gasification, gas treatment, synthesis (e.g., Fischer–Tropsch), hydrocracking, distillation, isomerization	Waxes, naphtha, electricity and heat	Pilot plants, TRL 6	~0.033 Mt a^{-1}/ unknown/EU wood, straw	Upscaling and demonstration of overall process concepts, adaptation of syngas treatment to gasifier properties, efficiency increase and downscaling synthesis and final fuel treatment

a Usually depending on process design.
b According to technology readiness level (TRL) of the European Commission, which outlines in detail the different research and deployment steps: 1, basic principles observed; 2, technology concept formulated; 3, experimental proof of concept; 4, technology validation in laboratory; 5, technology validation in relevant environment; 6, demonstration in relevant environment; 7, demonstration in operational environment; 8, system completed and qualified; 9, successful mission operations [10].
c AT, Austria; BR, Brazil; EU, European Union; LA, Latin America; SA, Southeast Asia

In spite of the large differences between the different concepts, it should also be pointed out that none of the concepts can be referred to as a "proven technology," which can be bought "off-the-shelf." Some of these concepts show promising maturity, justifying the development of a first industrial demonstration project, together with (industrial) monitoring. For all future-generation biofuels, scale-up strategies require the integration of the different process steps along the supply chain (biomass to transportation fuel), in order to demonstrate their reliability and their effective process performance, which are important for securing financing. The focus, therefore, must be to achieve industrial reliability and technical performance through energy integration, in order to achieve high efficiency/yields and economic viability. Therefore, constructing HVO/HEFA (hydrogenated vegetable oils/hydrogenated esters and fatty acids) and BTL (biomass to liquid) plants in the vicinity of mineral oil refineries would be an option, in order to use their infrastructure (e.g., with regard to hydrogen supply required for fuel upgrading). Moreover, existing biofuel concepts (e.g., for bioethanol) show the potential to be part of a so called "bottom-up approach" (i.e., upgrading of existing biofuel concepts) for biorefineries that can function as a multiproduct provider (e.g., biofuels, bulk chemicals, energy).

26.2.1
Biodiesel

Biodiesel production based on physicochemical conversion typically consists of the production of vegetable oils and fats via mechanical and/or solvent extraction (often as a combined process), followed by (trans-)esterification and purification to biodiesel. Today, raw materials are mainly vegetable oils (produced from oil-containing crops) and animal fats and grease (only a minor share) (Table 26.1). The oil crops have different specific total oil contents in a range from 17% for soy beans (here soy oil as a by-product of soy meal production) to about 38% for rape seeds [12].

There are three basic steps in biodiesel production from oils/fats: (trans-)esterification, alcohol ester processing, and glycerine purification. For stepwise transesterification, refined oil or fat is mixed with an alcohol (usually methanol) and a catalyst (usually methylates are preferred to hydroxides) to produce a fatty acid methyl ester (FAME) and a glycerine phase.

Once separated from the glycerine, the alcohol ester is washed to remove any soap formed during the reaction and the residual free glycerol and alcohol. The alcohol ester is then dried to remove all water. To refine the glycerine further, it is neutralized with an acid to form salts and stored normally as crude glycerine. According to the applied catalyst, surplus catalyst and soaps can be separated to produce, for example, potassium sulfate that can be applied as a fertilizer. After the aqueous glycerine phase has been neutralized, the glycerine is processed further by evaporation or distillation and adsorption to crude or pharmaceutical grade, 60–88 wt% or up to 99.5 wt%, respectively. The excess methanol can be removed at different stages of the reaction – before or after the phase separation; in both cases, the methanol is recovered and reused. The type and quality of the raw material is the decisive factor determining the technical design of a plant and energy flows [6, 13, 14].

Biodiesel can be applied as blended and straight fuel. However, there are some challenges in using biodiesel in modern motor concepts fulfilling high emission standards (e.g., EURO 6).

26.2.2
HVO and HEFA

Hydroprocessing can be applied to fats and oils from plants or animal origin, in order to produce hydrotreated esters and fatty acids (HEFA), which are another biofuel option. When plant oils are used, the product is often referred to as hydrotreated vegetable oils (HVO). Through a catalytic reaction with hydrogen, the fatty acids in the triglycerides are cracked and saturated, converting the fatty acids into linear alkanes. The glycerides are converted to propane, which is usually burned for process energy provision. Water, CO_2, NH_3, and H_2S are also formed in minor amounts and have to be removed. In order to meet fuel standards, the resulting linear alkanes need to be isomerized. Through adjustments in the process, it is possible to shift the molecular weight distribution of the product [15], allowing the preferred production of either a diesel or a kerosene fraction. The separation of the fractions is achieved by distillation. The efficiency of the hydroprocessing technology is estimated to be ~1.2 t of plant oil per tonne of biofuel [16]. Currently, hydrotreatment of esters and fatty acids is the only commercially available process to produce kerosene that meets aviation standards. On comparing biodiesel and HVO/HEFA fuel regarding their specific characteristics, it was concluded that certain applications fit better to a particular fuel: the cold flow characteristics of HVO/HEFA permit their use for aviation purposes and as a neat fuel it can be a premium diesel product with a high cetane number. However, if blending with fossil diesel were applied, this premium would be lost. An open field for further research is the assessment of synergies by blending of HVO/HEFA and biodiesel [17].

26.2.3
Bioethanol

The process of bioethanol production relies on the metabolism of yeasts (usually *Saccharomyces cerevisiae*) that are capable of turning sugars into ethanol. Substrates for ethanol fermentation are therefore plant-derived sugar solutions. The different characteristics of the raw material result in distinctive technologies, showing different states of development.

The main sugar-containing feedstock is sugar cane, followed by sugar beet and sweet sorghum. The processing of sugar cane or beet to ethanol is usually integrated into sugar mills. First, sugar juice is obtained by pressing and extraction. The integration can then be realized in different steps: fresh juice, thick juice, or molasses can be processed. This leads to greater flexibility in shifting sugar-containing streams from the sugar to the ethanol production [18–20]. Another widely used feedstock is plants containing large amounts of starch, a polymer of glucose monomers. In most cases, corn, wheat, triticale, rye, and cassava are the starch-containing plants processed to alcohol. These are first ground and mixed with water to form a mash, and through the addition of amylases, the links between the glucose monomers can be cleaved. Both sugar and starch processing are commercialized, widely used, and well-understood processes [21, 22].

Alternative feedstock includes lignocellulosic materials in which the cellulose and – potentially – the hemicellulose fraction can be used. This includes herbaceous and woody biomass such as straw, bagasse, and other residues. The difficulty in processing lignocellulose to bioethanol is that cellulose and hemicellulose, together with lignin, form a recalcitrant composite material, making it difficult to cleave the macromolecules in cellulose into glucose monomers. Hemicellulose, on the other hand, is a polymer containing mainly pentoses. Processes applied usually comprise a pretreatment step at temperatures of 150–210 °C [23–26] and, in some cases, an additional acidic, alkaline, or solvent agent is also applied. The pretreatment is usually followed by an enzymatic treatment in order to degrade the cellulose specifically to glucose monomers for fermentation. An overview of pretreatment technologies was presented recently by Talebnia et al. [27]. Some of these technologies have achieved demonstration scale, and some have commercial plants announced or under construction.

Fermentation of the sugars, regardless of their origin, is usually done using yeasts. Under idealized conditions, 0.51 g of ethanol per gram of hexose is produced [28]; however, only about 95% of this value can actually be achieved [29]. The final concentration of ethanol in the broth is between 10 and 14 vol.% after fermentation. The quest to achieve similar fermentation kinetics, yields, and concentrations by fermentation of pentoses is a very active field of research [30]. The subsequent separation is achieved by distilling through 2–5 sequential distillation columns. In order to obtain anhydrous ethanol, the grade required for blending with gasoline, molecular sieves are applied.

In general, the thermochemical production of bioethanol from lignocelluloses via gasification and alcohol synthesis is also possible, but compared with the fermentation routes it is not at the focus of international R&D.

The final treatment of bioethanol is usually blending with gasoline for use in automobiles (common blending ratios are from 5 to 100 vol.% bioethanol). Blending with diesel has also been tested.

26.2.4
Synthetic BTL

The production of synthetic fuels [biomass to liquids (BTL)] is characterized by three main steps after appropriate biomass pretreatment: (i) gasification of lignocellulosic biomass to produce a raw gas, (ii) gas treatment to give synthesis gas (syngas), (iii) catalytic synthesis to give synthetic biofuels [e.g., Fischer–Tropsch (FT) fuels, methanol, dimethyl ether], and (iv) final product treatment. In the following, synthesis of FT fuels will be briefly introduced as it is the most important processing technology in an international context.

Despite a long history of development and a broad complexity of system configurations, no market breakthrough has been realized so far for the provision of synthetic fuels via biomass gasification [31].

In terms of biomass pretreatment for gasification, mechanical–thermal biomass treatments are already well established (e.g., chipping and drying of solid biofuels). Processes to generate intermediate products, which are easier to transport and use in gasification technologies (e.g., pyrolysis, torrefaction, and hydrothermal carbonization) are at the pilot/demonstration stage [32–35].

Despite the scale of a gasifier, no commercial gasification system has been introduced for medium- to large-scale use of biomass. Among other criteria, chemical characteristics and physical and mechanical properties of the utilized biomass are of importance. However, all reactors for biomass gasification (e.g., fluidized bed and entrained flow reactors) are still at the R&D stage. Depending on fuel synthesis – where reactors are available – specific qualities of syngas, with constant compositions and in large amounts, have to be achieved, primarily with regard to the gas purity and the H_2:CO ratio. Because so far no gasification system meets these requirements, appropriate gas cleaning and conditioning systems have to be applied. During gasification, in addition to the main components (CH_4, H_2, CO, and CO_2), various impurities are also generated such as tars, coarse and fine particles, sulfur compounds, alkalis, halogen- and nitrogen-containing compounds, and heavy metals. For raw gas cleaning, either low-temperature wet gas cleaning or hot gas cleaning can be applied. The effectiveness of wet gas cleaning (e.g., cyclone and filter, scrubbing based on chemical or physical absorption) has been well proven for large-scale coal gasification systems. In contrast, not all elements of hot gas cleaning (e.g., tar cracking, granular beds and filters, physical adsorption or chemical absorption, ZnO bed, physical absorption) are yet at a mature technology stage. Nevertheless, hot gas cleaning offers benefits for the overall energy and environmental performance, with regard to the avoidance of contaminated sewage. For gas conditioning, available system components can be applied: hydrocarbons in the product gas can be converted by means of an additional steam or autothermal reforming step, resulting in a higher H_2:CO ratio. To achieve the required quality for fuel synthesis, the water gas shift CO conversion is conducted as the final step of syngas production [36, 37].

FT synthesis is a catalyzed polymerization process at low temperature, whereby syngas is liquefied into hydrocarbon chains of different lengths. The FT raw products consist of a wide range of light hydrocarbons, such as naphtha, the main product FT diesel, and waxes that need to be separated in a first step. Naphtha is a gasoline fraction of minor value, a resource for the petrochemical industry. It can be upgraded to motor-applicable gasoline by isomerization. Wax upgrading is done by hydrocracking; in this catalytic oil refinery process, long-chain hydrocarbons are split into the desired diesel and middle distillates in the presence of hydrogen [38, 39]. FT fuels are of similar quality to HVO/HEFA and can be applied as a premium fuel.

26.2.5
Biomethane

For the production of the gaseous multi-fuel biomethane as a blend with or substitute for natural gas, two pathways are possible, as follows [40].

26.2.5.1 Upgraded Biochemically Produced Biogas

Regarding the biochemical conversion path, usually wet biomass or biomass with a low dry matter content is converted anaerobically to biogas (containing ~55 vol.% CH_4) in a digester. Subsequently, the raw biomethane (biogas) is cleaned to remove CO_2 and H_2S by absorptive or adsorptive methods (e.g., water or amine scrubber, pressure swing adsorption), dried, compressed, and injected into the natural gas grid. Successful commercial application has been demonstrated widely in Europe, especially in Germany.

26.2.5.2 Thermochemically Produced Bio-SNG (Synthetic Natural Gas)

In the thermochemical conversion path, a gas containing CO_2, CO, H_2O, H_2, and CH_4 is generated by gasifying solid biofuels. A subsequent gas cleaning to remove tars and sulfur-, nitrogen- and chlorine-containing compounds avoids catalyst poisoning in the subsequent methanation (synthesis). Depending on the gas composition and the methanation reactor used, the gas has to be conditioned (to increase the H_2 content in a shift reactor) before methanation. Therefore, the R&D challenges for gas production and treatment to give syngas are similar to those for BTL fuels. Experiences are existing for methane generation from synthesis gas out of the biomass gasification containing a mixture of CO and H_2. At temperatures between 200 and 700 °C and pressures up to 100 bar in combination with the assistance of a catalyst, a strongly exothermic reaction takes place, resulting in the formation of primarily methane. The desired formation of methane is enhanced at low temperatures and high pressures. Finally, the raw biomethane leaving the methanation (containing ~40 vol.% CH_4) is cleaned again to remove unwanted CO_2 (with, e.g., amine or physical scrubbing), dried, compressed, and injected into the natural gas grid. The first demonstration plant worldwide producing biomethane through thermochemical means from solid biomass started operation at the end of 2008 in Güssing, Austria, and further plants are planned, for example, in Sweden.

26.2.6
Other Innovative Biofuels

In addition to from the above technology options, many others are under development for converting biomass into different liquid or gaseous fuels. Among these, the following are regularly mentioned in the literature and each has attracted active research interest. The status of their technical development is summarized in Table 26.2.

Table 26.2 Status of the development of other innovative biofuel technologies expressed as technology readiness level (TRL) [10].

Biofuel production technology	TRL
BTL	
Methanol	9
Dimethyl ether	6
Biohydrogen	
Thermochemically	5
Biochemically	3
Sugar to hydrocarbons	4
Biobutanol	5–9
Algae-based biofuels	4

26.2.6.1 BTL Fuels Such as Methanol and Dimethyl Ether

Methanol and dimethyl ether (DME) synthesis has a similar process chain to that described above for synthetic fuels, including biomass pretreatment, gasification, gas treatment, synthesis, and final treatment. Moreover, the so-called methanol/DME-to-gasoline routes are under investigation [33, 41]. DME is currently produced in pilot plants and methanol can also be produced from glycerine or biogas, based on reforming [42, 43].

26.2.6.2 Biohydrogen

Hydrogen can be produced thermochemically by process steps such as biomass pretreatment, gasification and gas cleaning, similar to those described for synthetic fuels. However, gasification and gas cleaning have to be adjusted to generate a hydrogen-rich gas [44]. Moreover, glycerine reforming is also at the pilot stage, and like natural gas reforming, biomethane (biogas) can also be used. Another option is production via microorganisms: (i) biomass fermentation using heterotrophic or photoheterotrophic bacteria or (ii) direct or indirect water splitting by green algae or cyanobacteria [45–48]. Despite intensive research activity, biochemically produced hydrogen is still at the laboratory stage [49].

26.2.6.3 Sugars to Hydrocarbons

Some microorganisms have shown the capability to ferment sugars directly to hydrocarbons, such as olefins, alkanes, and farnesanes [50]. Little has been reported about the performance of the adapted organisms used for fermentation, so it is difficult to assess the competitiveness of these processes.

26.2.6.4 Biobutanol

The fermentation of sugars to butanol is meant to result in a fuel with superior characteristics to ethanol [51, 52]. Four structural isomers can be distinguished for butanol; of these, 1-butanol production by ABE (acetone–butanol–ethanol) fermentation was the second largest biotechnological process ever run, (exceeded in volume only by ethanol fermentation). The peak of fermentative butanol production was in the 1930's. Petrochemical processes have since replaced this pathway in industry. Today several research and development activities can be seen worldwide to make the fermentative butanol production competitive. Research to improve the classical ABE fermentation is focused mainly on the following topics: (i) butanol as the only fermentation product, (ii) improving the low butanol tolerance of the production organism, and (iii) the use of alternative, non-food substrates [53].

26.2.6.5 Algae-Based Biofuels

Applying known processes to microalgae without special adaptation is not suitable since their properties are different from those of land-based biomass. Basically, all conversion routes can also be applied with algae. Reducing total production costs (i.e., basically the algae production costs), integration of cultivation with biofuel production and efficient downstream processes are among the main issues to be resolved [54–56].

26.3
System Analysis on Technical Aspects

For the selected biofuel options, a system analysis on technical parameters was carried out, focusing on biofuel production plants. In order to compare the different biofuel options, two parameters were assessed: (i) typical plant capacities and (ii) overall energetic efficiency.

26.3.1
Capacities of Biofuel Production Plants

Usually the plant capacities of biofuel production facilities are much lower than those of, for example, typical mineral oil refinery capacities (about 6 800 MW to more than 20 000 MW crude oil) [57]. Typical ranges of present and expected biofuel capacities per production plant are summarized in Figure 26.3. Whereas for conventional biofuel options such as biodiesel and bioethanol based on sugar and starch biomasses, no significant change is expected; for other options the ranges differ owing to their stage of technology development. For options such as bioethanol, SNG and BTL, the present range represents currently installed demonstration plants, and the expected stage is based on technological requirements due to economy of scale or concepts under discussion for commercial plants.

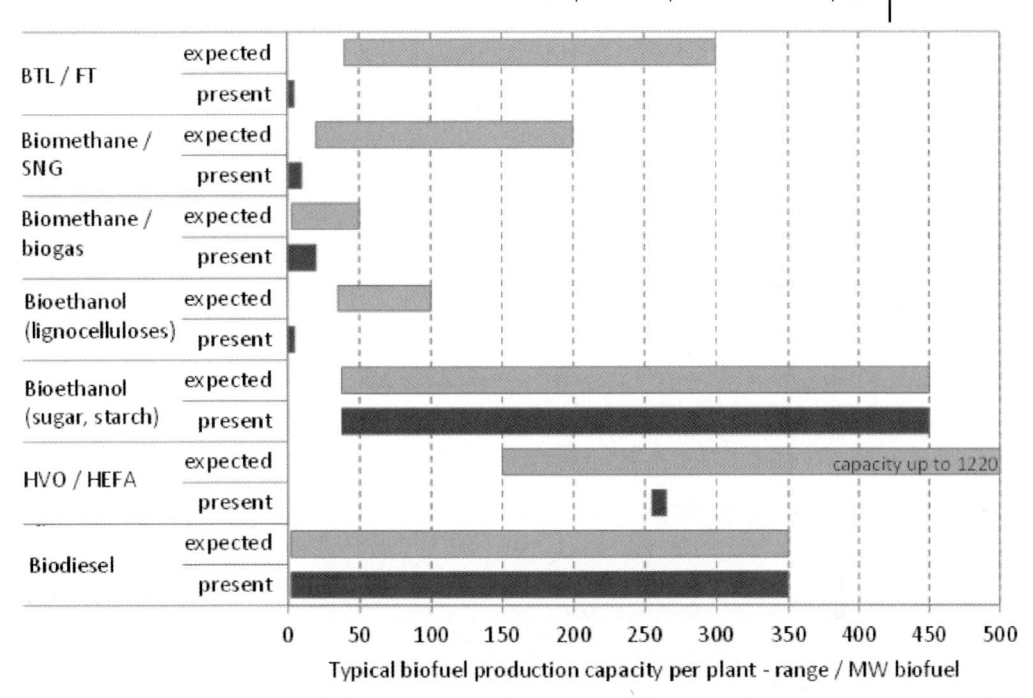

Figure 26.3 Expected and present biofuel capacities per plant.

26.3.2
Overall Efficiencies of Biofuel Production Plants

Biomass and the land utilized to produce it are limited resources, hence biomass needs to be converted into products efficiently and in a manner that is sustainable. Therefore, for a selection of biofuel options (here biodiesel, HVO/HEFA, bioethanol, BTL, and biomethane), different studies and publications were screened regarding efficiency issues. Only certain input and output streams were taken into account, making a validated comparison difficult. This is due to the different methodical approaches to the overall energetic efficiency in scientific publications and industrial practice.

In order to compare different biofuel options regarding their overall energetic efficiencies, an adequate system balance (Figure 26.4) and also a calculation approach (Eq. 26.1) are needed.

$$\eta_{\text{en,overall}} = \frac{\dot{m}_{\text{MP}} H_{\text{MP}} + \sum \dot{m}_{\text{BP}} H_{\text{BP}} + P_{\text{ne}} + Q_{\text{ne}}}{\dot{m}_{\text{RM}} H_{\text{RM}} + \sum \dot{m}_{\text{Aux}} H_{\text{Aux}} + P_{\text{ext}} + Q_{\text{ext}}} \quad (26.1)$$

where $\eta_{\text{en,overall}}$ is the overall energetic efficiency, $\dot{m}_i$ the mass flow of the main product (MP, biofuel), by-products (BP), raw material (RM, biomass free plant gate), and auxiliaries (Aux, including energy sources), H_i the heating value of the main product (MP, biofuel), by-products (BP), raw material (RM, biomass free plant gate), and auxiliaries (Aux, including energy sources), P_{ne} the net production (surplus) electricity, Q_{ne} the net production (surplus) process heat (heating value), P_{ext} the demand process electricity external supply, and Q_{ext} the demand process heat external supply (heating value).

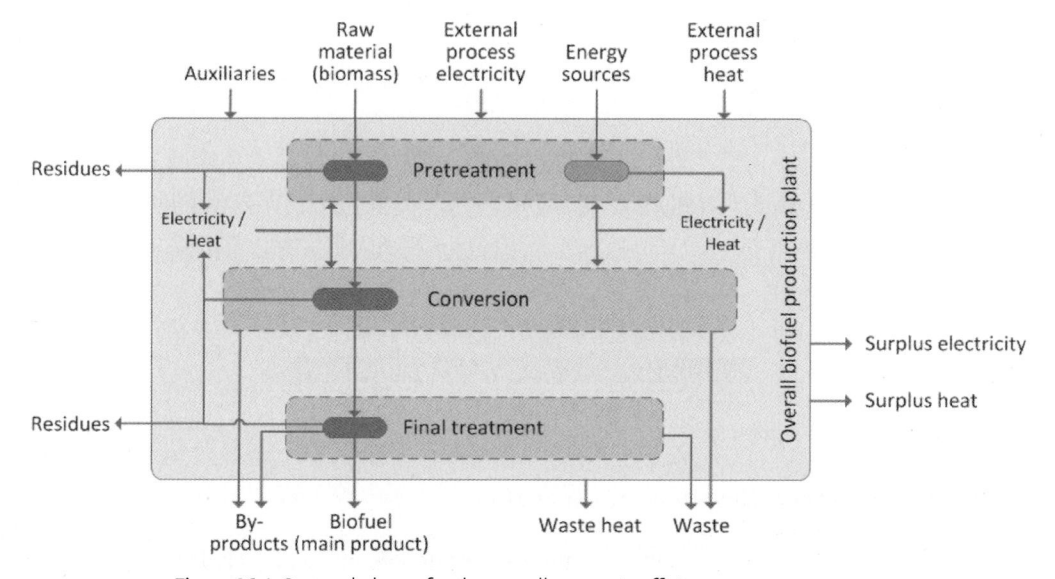

Figure 26.4 System balance for the overall energetic efficiency.

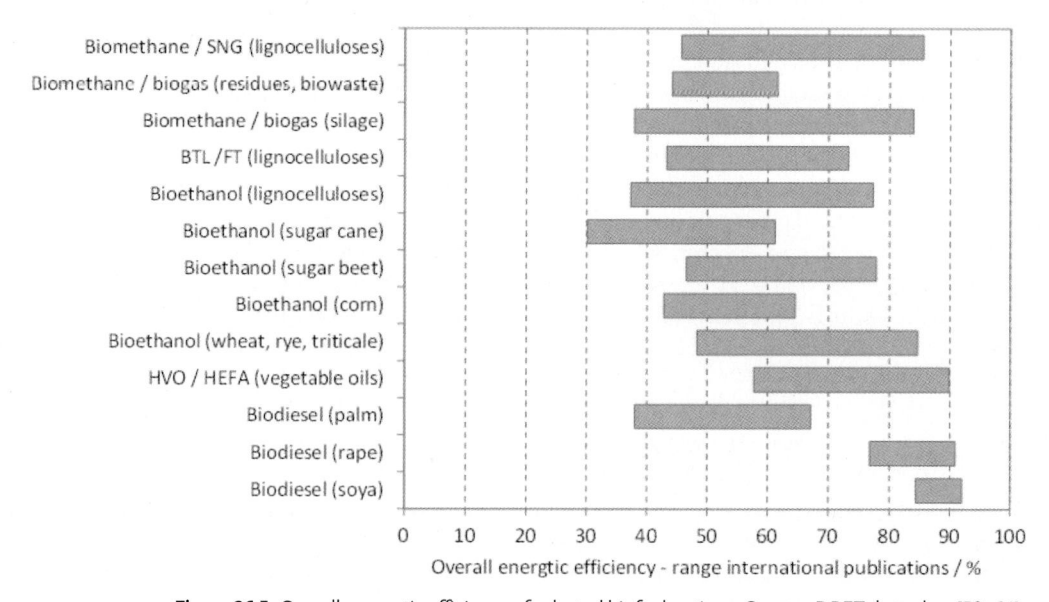

Figure 26.5 Overall energetic efficiency of selected biofuel options. Source: DBFZ, based on [58–64].

Figure 26.5 shows a comparison of minimum and maximum overall energetic efficiencies for the selected conversion technologies, the considered biofuels, and the raw materials used. The discussed system balance and also mass and energy streams (Figure 26.4) were taken into account in a comprehensive manner, based on published and our own data. The given range of efficiency is dependent on different plant designs. Accordingly, there is no general preference for an individual biofuel option, and the overall concept behind it is decisive.

In general, the most important driver of the overall efficiency is the conversion ratio of raw material to the main product biofuel. Especially for conventional biodiesel and bioethanol, by-products are also relevant. In addition, for bioethanol, the process heat demand for distillation and stillage treatment is also an influencing driver.

However, it has to be assumed that the values shown in Figure 26.5 represent production plant designs of new plants or partly theoretical expectations (e.g., for BTL and biomethane via SNG) and therefore real values of plants in operations are comparatively lower.

26.4
System Analysis on Environmental Aspects

Biofuels are advocated as alternatives to fossil transport fuels and to reduce GHG emissions in the transport sector [65–67]. However, the sustainability of large-scale biofuel use is the subject of ongoing and intense debate [68, 69], and included in this debate is the level of uncertainty associated with the methods applied to assess the lifecycle of a biofuel. Currently lifecycle analysis (LCA) is regarded as one of the best methods for assessing the GHG emissions from biofuel systems [70–73], allowing estimated GHG emissions associated with the biofuel chain to be compared with the estimated emissions associated with a fossil fuel chain [74]. LCA is a structured, internationally standardized method (see ISO 14040 and 14044, 2006) that outlines the process for evaluating the environmental performance throughout the sequence of activities executed in creating a product or performing a service.

Most LCA studies investigating the potential environmental impacts of biofuel systems cover the whole supply chain of biofuels from extraction of raw materials ("from cradle"), through to the utilization of the product and the disposal of all intermediate or waste products ("to grave").

26.4.1
Differences in LCA Studies for Biofuel Options

LCA studies on biofuels are carried out under different assumptions (e.g., system boundaries, cut-off criteria, allocation of by-products), making it very difficult to compare the results from different LCA studies [71]. For example, results for GHG emission reductions for biodiesel produced from palm oil can range between 71%

[58] to 36% (EU RED, 2009/28/EC) and 66% [75] to 33% [76] when rapeseed is used as feedstock. HVO/HEFA, as outlined above, is a promising biofuel for the aviation sector; it was also found to have similar variations in GHG emission reductions, ranging between 65% (EU RED, 2009/28/EC) to 43% [76] when the process is based on palm oil and 64% [77] to 47% [58] for rapeseed-based processes. Even though it is difficult to compare results from LCA studies, many different studies have indicated that bioethanol pathways are associated with lower GHG emissions per megajoule than biodiesel pathways [78]. For bioethanol production, available LCA studies indicate GHG emission reductions between 81% [58] to 45% when wheat is used as feedstock and 51% [79] to 11% [76] when corn is used.

26.4.2
Drivers for GHG Emissions: Biomass Production

Many LCA studies have indicated that biomass production and biomass conversion have a significant impact on the overall GHG mitigation potential of a biofuel [76, 80]. Therefore, the aim of this section is to provide a short overview of the main drivers of GHG emissions associated with biofuel production, and also to emphasize the importance of sustainability assessments for future biofuels and the associated level of uncertainty.

Current generation biofuels are produced mainly from annual crops (rapeseed, soya, palm, sugar beet, wheat, rye), whereas future-generation biofuels are largely derived from perennial herbaceous and woody plants (miscanthus, switchgrass, poplar, willow, soft/hard woods, waste wood from forestry, wood processing industries) [91]. The cropping for annual biomass systems is management intensive, requiring more inputs (e.g., fertilizers, pesticides), greater soil/land disturbances (tilling, frequency of cultivation), and less accumulation of organic matter than perennial herbaceous (lignocellulosic) and woody species [92, 93]. Hence biomass cropping systems vary extensively in relation to management, growing season, physiology, yield, feedstock conversion efficiencies, fertilizer demand, and systems losses, and all of these factors affect the magnitude of the components contributing to net GHG flux [94]. Different LCA studies for biofuel production and use have identified emissions associated with land use change and fertilizer production and application as the most crucial parameters during biomass production (Table 26.3). The issue of land use change (LUC), both direct (dLUC) and indirect (iLUC), due to increasing production area for biofuels and bioenergy, has dominated policy discussions recently, because of its significance as a source of global emissions and potential climate effects. However, the magnitude of LUC remains an important unknown in biofuel sustainability [95]. To control and possibly avoid negative effects from dLUC, the European Commission has promoted the initiation of sustainability certification schemes for biofuels used within the EU [96]. However, agricultural production systems include food/fodder and bioenergy crops, hence the introduction of sustainability criteria for bioenergy can lead to leakage (indirect) effects or iLUC. Studies including iLUC effects for different biofuel demand scenarios indicate that they can significantly reduce, or even wipe out, GHG savings from biofuels [81, 97].

Table 26.3 Overview of drivers of GHG in biomass production systems, relevant aspects, and associated uncertainties in accounting for these drivers within the LCA method.

Biomass (raw material)	Drivers of GHG emissions	Relevant aspects	Uncertainties related to drivers[a]
Oil Sugar Starch Lignocellulosic[b]	dLUC/iLUC[c]	Change in carbon stocks [78, 81]	Carbon inventory Lack of primary data
	Biomass management practices for increased yields [13, 76]	Nitrogen (N) fertilizer use and N losses[d] [82–87]	Amount of N_2O releases[e] associated with parameters mentioned
	Cultivation and transport [58]	Fuel consumption Soil compaction [88]	Parameters influencing fuel consumption Lack of primary data/site specificity for soil compaction and GHG emissions[f]
Micro algae oils	Nutrient and energy consumption [55, 89]	Upstream emissions from nutrient and energy carrier production	Lack of primary data

a Uncertainties in relation to data/data sources available on the various drivers and relevant aspects.
b Lignocellulose refers to both woody biomasses and herbaceous biomass (including waste residues such as straw and bagasse).
c dLUC refers to direct land use change; iLUC refers to indirect land use change. A dLUC occurs when areas not used for agricultural purposes (e.g., forest areas, grasslands) are converted to produce biomass. An iLUC can occur when existing agricultural and nonagricultural areas are converted to other crops/land uses to meet demands for increasing demands for bioenergy and agricultural products [90].
d N fertilizer use refers to type of fertilizer used, e.g., calcium ammonium nitrate or urea.
e The amount of N_2O emitted from biomass production depends on a number of parameters such as type of fertilizer, application technique and time, crop rotation systems, climate, and soil types.
f Variability in fuel consumption due to soil conditions at harvesting, machinery used, field structure, distance to intermediate storage or bioenergy plant, etc.

However, the calculation of emissions from iLUC is associated with a high level of uncertainty [78]. Furthermore, the models and assumptions used to quantify emissions from iLUC are still under development and are still heavily debated.

In addition to emissions from LUC, N-fertilizer application is one the major sources of GHGs in the production of biomass and can have serious implications for the GHG balances of biofuel production systems However, they are often difficult to quantify, in relation to both fertilizer production and application [71, 83, 86, 87, 94]. The most important GHG emissions from N-fertilizer application are N_2O and NH_3. The overall amount of these emissions depends on a number of factors, including type of fertilizer, application method, crop rotation, climate, and soil type. Erisman *et al.* calculated the contribution of different emissions to the total GHG emissions due to different fertilizer applications. They found that when using rapeseed produced at very high fertilizer rates (400 kg ha^{-1}) there was a net reduction of 10% GHG emissions compared with fossil diesel and at lower fertilizer input the reduction can be up to 50% [83]. Similarly, Smeets *et al.* studied several reference land uses for bioenergy (crops) production to assess N_2O emissions on the overall GHG balance [86]. In comparison with gasoline, bioethanol produced from sugar beet resulted in a GHG change from −58% to 17% and from sugar cane −103% to −62%. Bioethanol from corn resulted in a GHG change from −38% to 11% and from wheat −107% to 53%. In comparison with a fossil diesel fuel, biodiesel produced from palm oil resulted in a GHG change from −75% to −39%, for rapeseed from −80% to 72%, and for soybean from −111% to 44%. The differences were due to the different reference land uses considered and both avoided emissions when by-products were assumed to be replaced by other products.

26.4.3
Drivers for GHG Emissions: Biomass Conversion

Whereas emissions from biomass production are mostly driven by LUC and fertilizer use, the main drivers for emissions from biomass conversion are energy consumption, the use of auxiliary materials, and the overall conversion efficiency. Table 26.4 provides an overview of the most important drivers for GHG during biomass conversion.

GHG emissions from biomass conversion to biofuels are driven by the use of auxiliary materials (e.g., process chemicals), process heat [(from both the production of the energy carrier used for heat supply (e.g., natural gas), and from the heat production itself (e.g., burning of the natural gas)], in addition to the power (e.g., electricity from the public grid) required for processing biomass. To incorporate emissions from the production of the auxiliary materials and energy carriers used in the LCA of a biofuel, emission factors from an inventory database such as Ecoinvent, National Renewable Energy Laboratory (NREL), or Biograce are often used. However, these values can be generic and there can be much uncertainty in relation to this. The indirect emissions associated with upstream processes such as fossil fuel production and various auxiliaries (e.g., mineral and organic acids, enzymes, catalysts) are also full of complexities and uncertainties.

Emissions from waste treatment are usually not an important factor in the GHG balance of liquid biofuels; however, biofuels from palm oil are an exception. Emissions associated with the waste streams from palm oil production (e.g., waste water storage in open ponds, dumping or burning of empty fruit bunches) can have a significant impact on the overall result for palm oil-based biofuels [98]. However, emissions associated with these processes can be reduced through technical improvements to the waste treatment (e.g., gas-tight cover of the storage ponds) and the development of strategies for additional use of the solid waste streams (e.g., use of the empty fruit bunches as mulch or fertilizer, or as raw material for a biorefinery).

In addition to the uncertainties related to upstream emissions, it is important to note that LCAs are mostly case specific. They reflect the actual data and specific process designs of the facility investigated. This is important when comparing and assessing current biofuel options (e.g., rapeseed biodiesel) with advanced or future technologies (e.g., SNG, BTL) for which no actual data are available.

Table 26.4 Overview of drivers of GHG in biofuel conversion systems, relevant aspects, and associated uncertainties in accounting for these drivers within the LCA method.

Conversion system	Drivers of GHG emissions	Relevant aspects	Uncertainties related to drivers
Biodiesel, HVO/HEFA, bioethanol, BTL, biomethane	Energy consumption during conversion	Upstream emissions from fossil and renewable energy chains[c]	Uncertainties related to the emission factors for energy production[d]
	Auxiliary materials[a] for the conversion process	Upstream emissions due to the production of required chemicals/catalysts[c]	
	Direct emissions from conversion processes	For example, high emissions (especially CH_4) from dumping of empty fruit bunches (EFBs) and waste water treatment during palm oil production (usually stored in open ponds) [98], emissions from methane leakage and slip (e.g., from biogas upgrading) [99, 100]	Lack of primary data to identify the extent and level of emissions and leaks from developing bioenergy technologies
	Overall conversion efficiency[b]	The overall efficiency of the biomass used has an impact on the upstream emissions from biomass production per MJ of biofuel	Uncertainties related to data availability for the assessment of advanced biofuel technologies [13]

a Auxiliary materials are input materials that are necessary for the conversion process (e.g., process chemicals).
b Cf. Section 26.3.
c LCAs usually also include the so-called upstream emissions of the energy or materials used in the processes of the biofuel value chain. The term upstream emissions refer to the emissions associated with the production and provision of the energy carriers or materials used (e.g., emissions from the production of electricity provided via the public grid and used in the biomass conversion process). In some cases there are alternatives that might help to reduce emissions (e.g., methanol for biodiesel produced from biomethane instead of natural gas).
d Uncertainties associated with the upstream emissions from the production of the energy used for conversion processes (e.g., electricity from the public grid) refer to the many different processing scales and technologies involved.

26.4.4
Perspectives for LCA Assessments

Currently, the potential GHG mitigation from biofuel use is one of the main drivers for their promotion; however, other environmental and socio-economic aspects need to be considered in the overall debate on liquid and gaseous biofuels and should be an important precondition for their acceptance.

Biofuel systems are often intrinsically linked to intense agricultural production systems. Therefore, it is important to recognize that the issues for agriculture are comparable to those for the biofuel industry and the focus should not only be on climate change mitigation, but should also include other impacts relevant for agriculture, such as eutrophication and acidification.

Like agricultural and food production chains, biofuel chains are also complex systems; however, biofuels are under continuous scrutiny to prove their environmental benefits and sustainability. Therefore, in order to enhance the understanding of the potential environmental impacts associated with increasing biofuel use, tools such as LCA need to be developed further [69, 72, 74, 101] to include more complex aspects such as regional and spatial impacts [69, 102–107], biodiversity [108–110], and socio-economics impacts [111]. However, the greatest challenge for the future application of LCA is to integrate these complexities in a meaningful manner, to permit a more complete assessment for sustainability.

26.5
System Analysis on Economic Aspects

In addition to technical parameters and environmental impacts, a system analysis of biofuels production also requires consideration of economic criteria. Typically, analyzing economics is intended to evaluate different cost alternatives in order to identify relative advantages, to compare different options with regard to emissions, and to determine important influencing factors. For that, local conditions need to be taken into account. In the following, total capital investments and biofuel production costs are considered.

26.5.2
Total Capital Investments for Biofuel Production Plants

Regarding both financial risks and biofuel production costs, total capital investments (TCI) are of crucial importance. Depending on the state of technology development (cf. Table 26.1), the calculation of TCI can be based on different approaches (e.g., rough, study, or approval estimations, all with different accuracies and therefore financial uncertainties). Whereas for commercial plants approval estimations can often be applied (accuracy of ±5–15%), for plants at the pilot or demonstration stage TCI is based on study estimations (accuracy of ±20–30%). According to this, plant equipment costs are usually determined by up- or downscaling (typical scale factor of about 0.6–0.7) of the known TCI for similar technology devices [112]. Additionally, device-specific installation factors (for biofuels usually about 1.0–1.7) are taken into account [113].

The TCI figures given in Table 26.5 were calculated using published data from the last few years. The spread of published data and the influence of different plant designs, and also regional frame conditions, are reflected in the wide range of the figures. However, the tendency for biomethane and biofuels based on lignocelluloses

Table 26.5 Overview of TCI for selected biofuel options (calculations based on [6, 38, 44, 58, 113–120]).

Biofuel option	Plant capacity[a] (MW biofuel)	TCI[b] (10^6 €)	Specific TCI (€ kW^{-1} biofuel)
Biodiesel[c]	4–190	1.4–66	65–350
HVO/HEFA[c]	150–1 030	> 100	390–500
Bioethanol (starch, sugar)	7–220	16–300	1 360–2 290
Bioethanol (lignocelluloses)	15–185	30–325	1 800–2 800
BTL/FT	130–220	430–1 000	2 300–3 775
Biomethane via biogas	5–30	7.5–50	1 500–3 000
Biomethane via SNG	20–170	30–170	1 000–2 100

a Here typical capacities of commercial or expected plant capacity.
b For new plants, without land costs and surrounding infrastructure for green field installations.
c Without oil mill.

is towards increasing TCI values in comparison with conventional fuels, also due to often more complex technologies and plant designs.

Considering the effects of economy of scale, specific TCI values decrease with increasing plant size. However, in the engineering and construction industries, there is a continuous cost increase, which cannot be reflected at all. The price development of chemical plants and machinery also relates to biofuel production plants. It is commonly indexed by means of the so-called Kölbel–Schulze methodology or the Chemical Engineering Plant Cost Index (CEPCI) [121]. According to Kölbel–Schulze price index, the TCI increased by about 17% in the period 2005–2012 [122].

The approach of experience curves (i.e., models based on dynamic learning and static scale effects) is also applied for biofuel production plants. According to this, some estimates can be made using the long-term experience with similar technologies (e.g., power generation). Cost reductions of technologies (i.e., TCI reduction based on cumulative installed capacity and development stage) within a certain time period (e.g., from the present to 2020 or 2030) can be adapted to biofuel plants with a positive impact on biofuel production costs. Experience has shown that technologies display different progress factors with regard to their development stages (i.e., R&D, commercialization, and further improvement) [123, 124]. An example is bioethanol production in Brazil with a progress factor of 0.8 within the time frame from 1978 to 1995 [125].

26.5.3
Biofuel Production Costs

Especially for analyzing biofuel production costs, dynamic partial models (e.g., based on annuity) are favorably applied since the accuracy is higher than for static partial models due to periodic accounting. In order to calculate biofuel production costs effectively, different cost parameters under regional frame conditions and appropriate time horizons have to be taken into account: (i) capital expenditures (CAPEX; including TCI, equity and leverage, interest rates, lifetime, maintenance), (ii) variable operational expenditures (OPEX; raw material, auxiliaries, residues, annual full load), (iii) fixed OPEX (personnel, servicing, operation, insurance), and (iv) revenues (e.g., for by-products). Often market prices for raw materials and by-products correlate with each other (e.g., oil seeds and press extraction, starch raw materials and DDGS; cf. Table 26.1).

Usually, sensitivity analyses are carried out for the determination and optimization of influencing cost parameters, in order to obtain information about the relative change in total biofuel production costs and thus determine uncertainties. They show that in addition to the annual full-load hours of the plant, variable OPEX (especially raw materials) and CAPEX are of great importance. It is expected that production costs will increase moderately in the future owing to rising energy prices. The latter will also affect biomass prices and in general the broad implementation of biofuel strategies.

Certain plant designs and overall concepts, and also different methodical approaches with different regional frame conditions, time horizons, and cost parameters, make a comprehensive comparison of publications difficult. For a rough overview based on a publication survey, a range of available production costs for the different biofuel options, and also the price levels for crude oil, are presented in Figure 26.6. For future-generation biofuels based on lignocellulosics, the range is especially wide, which is primarily caused by their stage of development and thus many different underlying assumptions.

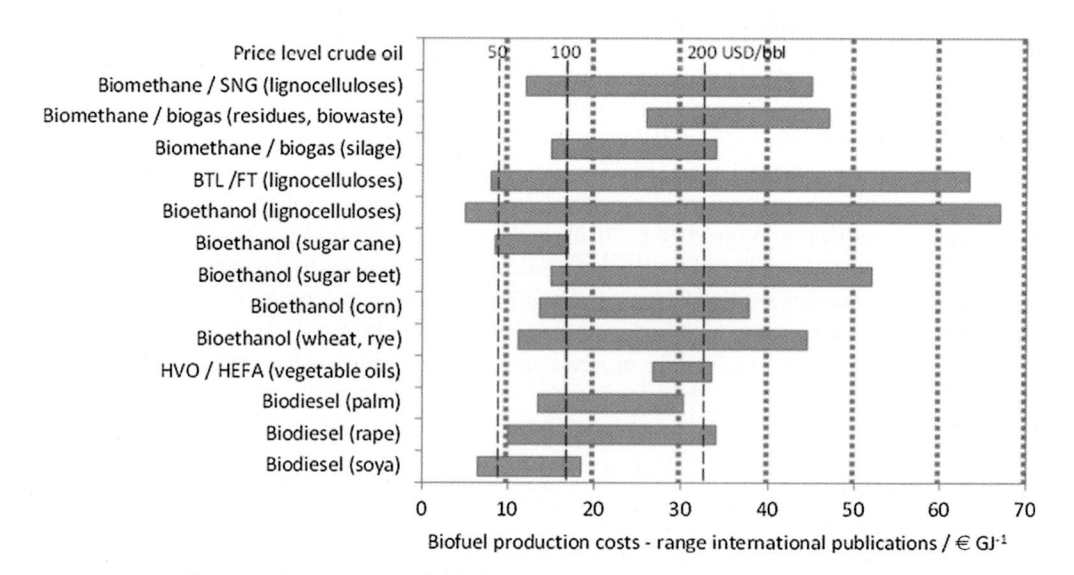

Figure 26.6 Comparison of biofuel production costs. Source: DBFZ, based on [113, 119, 126–137].

26.6
Conclusion and Outlook

This chapter is intended to provide an overview of biofuel options that are relevant for current and future transport options and thus can meet freight and person mobility requirements. According to international expert discussions summarized in the IEA roadmap scenarios for biofuels until 2050, the chapter concentrates on options such as biodiesel, HVO/HEFA, bioethanol, BTL and biomethane. Additionally, other options were considered regarding their general principles. Selected biofuel options were analyzed regarding the following aspects.

26.6.1
Technical Aspects

Current biofuels, mainly covered by bioethanol, biodiesel, and HVO/HEFA, will be the most important biofuels until 2020. It is certain that future biofuel options such as biomethane, bioethanol, and synthetic fuels based on lignocelluloses will enter the market in the EU or United States first, before becoming appropriate options for developing countries. Each biofuel option shows different benefits and drawbacks, and each of them needs to be considered with regard to available raw materials, surrounding available infrastructure, and market demand. Currently, research is focused on future-generation biofuels, with medium- and long-term perspectives that also use residues and biowaste, minimizing problems associated with current-generation biofuels. The typical biofuel plant capacity is up to 450 MW biofuel output (current bioethanol plants) or even up about 1 200 MW for refinery-similar HVO/HEFA. Lignocellulosic-based biofuel plants are intended to be smaller (up to 300 MW). The overall energetic efficiency depends greatly on the plant design concept and is in the range from 30% to about 90%.

26.6.2
Environmental Aspects

The current debate on the sustainability of liquid biofuels and the introduction of sustainability criteria within the framework of the EU RED have led to an additional incentive for producers to optimize the GHG balance of their biofuels. Although some of the main drivers for GHG emissions are usually outside the control of the biomass or biofuel producer (e.g., iLUC), others can be identified and subject to environmental improvements to allow for optimization within the biofuel value chain. Emissions from biofuel production are predominantly driven by the processes of biomass production and conversion. Emissions from biomass production are associated with fertilizer use and application, whereas emissions from biomass conversion are mostly driven by the consumption of process energy and chemicals. Hence emissions from these processes can be reduced by optimizing the type or source of auxiliary materials used in the various production steps. Furthermore, incentives for the optimization of the overall GHG mitigation potential of biofuels can lead to adjustments to the process design itself. The further development of this process is reflected in the current development of biorefineries.

26.6.3
Economic Aspects

High total capital investment associated with large plant capacities increase the risk of investment. Thus, for economic viability, the key criteria include ideal locations with appropriate infrastructure, a secure market for the product, and guaranteed long-term continuous raw material supply. In addition to these, raw material production costs and their dependence on energy prices and climatic conditions are the factors that can have a major impact on biofuel production costs. These costs will play a reduced role if residues and waste are used as raw materials. However, the supply and handling of large amounts to up-scale facilities is expected to increase the complexity of the logistics and may offset this cost advantage. Furthermore, comparably complex conversion procedures increase the operating costs and subsequently total production costs. For all these reasons, it is not expected that biofuels will be cost competitive in the foreseeable future. However, ongoing R&D is expected to reduce the conversion costs and allow production on a large commercial scale that would allow benefit from economies of scale.

26.6.4
Future R&D needs

With regard to future R&D needs, a number of challenges have to be managed in order to move towards a more sustainable transport sector; these require a combination of measures, such as the development of; innovative mobility concepts, improved and innovative alternative fuels, new vehicle technologies, and new infrastructure networks. The target and at the same time also the future challenge are to maximize the biomass-to-products ratio. For that reason, the approach of biorefineries is nowadays attracting ever increasing attention. Following the example of conventional refineries, biorefineries use biomass as raw material and produce an array of marketable products (e.g., biofuels, bulk chemicals, feed and food, energy) by integrating processes and processing different product streams with almost zero waste. In addition to the appropriate frame conditions, supporting instruments are also required to permit the targeting and planning security for large investments in pilot- and demonstration-scale units. Lessons learned during these procedures are of great importance for the realization of sustainable running of plants and further investments along the whole supply chain.

References

1 IEA (2011) *World Energy Outlook 2011*, International Energy Agency, Paris.
2 Zarrilli, S. and Burnett, J. (2008) *Making Certification Work for Sustainable Development: the Case of Biofuels*, United Nations, New York.
3 Timilsina, G. R. and Ashish, S. (2010) *Biofuel: Markets, Targets and Impacts*, Policy Research Working Paper Series 5513, World Bank, Washington, DC
4 Thrän, D., Bunzel, K., Seyfert, U., Zeller, V., Buchhorn, M., Müller, K., Matzdorf, B., Gaasch, N., Klöckner, K., Möller, I., Starick, A., Brandes, J., Günther, K., Thum, M., Zeddies, J., Schönleber, N., Gamer, W., Schweinle, J., and Weimar, H. (2011) *Global and Regional Spatial Distribution of Biomass Potentials – Status Quo and Options for Specification*, Deutsches Biomasseforschungszentrum (DBFZ), Leipzig.
5 IEA (2011) *Technology Roadmaps – Biofuels for Transport*, International Energy Agency, Paris.
6 Kaltschmitt, M. (ed.) (2009) *Energie aus Biomasse*, Springer, Berlin.
7 Bacovsky, D., Dallos, M., and Wörgetter, M. (2010) *Status of 2nd Generation Biofuels Demonstration Facilities in June 2010: a Report to IEA Bioenergy Task 39*, IEA, Paris.
8 Lorne, D., and Chabrelie, M.-F. (2010) *New Biofuel Production Technologies: Overview of These Expanding Sectors and the Challenges Facing Them*, IFP Energies Nouvelles, Rueil-Malmaison.
9 Naumann, K., Oehmichen, K., Zeymer, M., Müller-Langer, F., Scheftelowitz, M., Adler, P., Meisel, K., and Seiffert, M. (2012) *Monitoring Biokraftstoffsektor*, Deutsches Biomasseforschungszentrum (DBFZ), Leipzig.
10 European Commission (2011) *Key Enabling Technologies*, European Commission, Brussels.
11 Worldwatch Institute (2007) *Biofuels for Transport: Global Potential and Implications for Sustainable Energy and Agriculture*, Earthscan, London.
12 Thrän, D., Probst, O., Weber, M., and Müller-Langer, F. (2006) *Potenciales y Viabilidad del Uso de Bioetanol y Biodiesel para el Transporte en México*, SENER, México.
13 Majer, S., Mueller-Langer, F., Zeller, V., and Kaltschmitt, M. (2009) Implications of biodiesel production and utilization on global climate – a literature review. *Eur. J. Lipid Sci. Technol.*, **111**, 747–762.
14 Mittelbach, M., and Remschmidt, C. (2004) *Biodiesel: the Comprehensive Handbook*, Martin Mittelbach, Graz.
15 Schaub, G. (2009) Hydroprocessing of vegetable oils (HVO), presented at the Biofuel Conference "Biofuel Research – a Transatlantic Dialogue," Berlin.
16 Marker, T., Petri, J., Kalnes, T., McCall, M., Mackowiak, D., Jerosky, B., Reagan, B., Nemeth, L., Krawczyk, M., Czernik, S., Elliot, D., and Shonnard, D. (2005) *Opportunities for Biorenewables in Oil Refineries*, US Department of Energy, Des Plaines, IL.
17 Knothe, G. (2010) Biodiesel and renewable diesel: a comparison. *Prog. Energy Combust. Sci.*, **36**, 364–373.
18 Dodić, S., Popov, S., Dodić, J., Ranković, J., Zavargo, Z., and Jevtić-Mučibabić, R. (2009) Bioethanol production from thick juice as intermediate of sugar beet processing. *Biomass Bioenergy*, **33**, 822–827.
19 Keil, M., Kunz, M., and Veselka, M. (2009) Europäisches Bioethanol aus Getreide und Zuckerrüben – eine ökologische und ökonomische Analyse, 3. Teil. *Zuckerindustrie*, **134**, 114–130.
20 Krajnc, D., and Glavič, P. (2009) Assessment of different strategies for the co-production of bioethanol and beet sugar. *Chem. Eng. Res. Des.*, **87**, 1217–1231.
21 Arifeen, N., Wang, R., Kookos, I. K., Webb, C., and Koutinas, A. A. (2007) Process design and optimization of novel wheat-based continuous bioethanol production system. *Biotechnol. Prog.*, **23**, 1394–1403.
22 Mortimer, N. D., Elsayed, M. A., and Horne, R. E. (2004) *Energy and Greenhouse Gas Emissions for Bioethanol from Wheat Grain and Sugar Beet. Final Report to British Sugar plc.* Report No. 23/1, Resources Research

Unit, School of Environment and Development, Sheffield Hallam University, Sheffield.

23 Brethauer, S. and Wyman, C. E. (2010) Review: continuous hydrolysis and fermentation for cellulosic ethanol production. *Bioresource Technol.*, **101**, 4862–4874.

24 Hendriks, A. T. W. M. and Zeeman, G. (2009) Pretreatments to enhance the digestibility of lignocellulosic biomass. *Bioresource Technol.*, **100**, 10–18.

25 Mosier, N., Wyman, C., Dale, B., Elander, R., Lee, Y. Y., Holtzapple, M., and Ladisch, M. (2005) Features of promising technologies for pretreatment of lignocellulosic biomass. *Bioresource Technol.*, **96**, 673–686.

26 Sun, Y. and Cheng, J. (2002) Hydrolysis of lignocellulosic materials for ethanol production: a review. *Bioresource Technol.*, **83**, 1–11.

27 Talebnia, F., Karakashev, D., and Angelidaki, I. (2010) Production of bioethanol from wheat straw: an overview on pretreatment, hydrolysis and fermentation. *Bioresource Technol.*, **101**, 4744–4753.

28 Busche, R. M., Scott, C. D., Davison, B. H., and Lynd, L. R. (1991) *The Ultimate Ethanol: Technoeconomic Evaluation of Ethanol Manufacture, Comparing Yeast vs. Zymomonas Bacterium Fermentations*, US Department of Energy, Washington, DC.

29 Roehr, M. (ed.) (2010) *The Biotechnology of Ethanol*, Wiley-VCH Verlag GmbH, Weinheim.

30 Gírio, F. M., Fonseca, C., Carvalheiro, F., Duarte, L. C., Marques, S., and Bogel-Lukasik, R. (2010) Hemicelluloses for fuel ethanol: a review. *Bioresource Technol.*, **101**, 4775–4800.

31 Vogel, A., Müller-Langer, F., and Kaltschmitt, M. (2008) Analysis and evaluation of technical and economic potentials of BtL-fuels. *Chem. Eng. Technol.*, **31**, 755–764.

32 Bergmann, P. C. A., Kiel, J. H. A. (2005) Torrefaction for biomass upgrading, in *Proceedings of the 14th European Biomass Conference, Paris*, pp. 17–21.

33 Dahmen, N., Henrich, E., Dinjus, E., and Weirich, F. (2012) The Bioliq® bioslurry gasification process for the production of biosynfuels, organic chemicals, and energy. *Energy Sustain. Soc.*, **2**, 3.

34 Funke, A. and Ziegler, F. (2010) Hydrothermal carbonization of biomass: a summary and discussion of chemical mechanisms for process engineering. *Biofuels Bioproducts Biorefining*, **4**, 160–177.

35 Srokol, Z., Bouche, A. G., Van Estrik, A., Strik, R. C. J., Maschmeyer, T., and Peters, J. A. (2004) Hydrothermal upgrading of biomass to biofuel; studies on some monosaccharide model compounds. *Carbohydr. Res.*, **339**, 1717–1726.

36 Müller-Langer, F., Vogel, A., Kaltschmitt, M., and Thrän, D. (2007) Analysis and evaluation of the second generation of transportation biofuels, presented at the 15th European Biomass Conference and Exhibition, Berlin.

37 Vogel, A., Thrän, D., Muth, J., Beiermann, D., Zuberbühler, U., Hervouet, V., Busch, O., and Biollaz, S. (2008) *RENEW – Renewable Fuels for Advanced Powertrains, Scientific Report WP5.4. Technical Assessment – Comparative Assessment of Different Production Processes*, Deutsches Biomasseforschungszentrum (DBFZ), Leipzig.

38 Ekbom, T., Lindblom, M., Berglin, N., and Ahlvik, P. (2003) *Technical and Commercial Feasibility Study of Black Liquor Gasification with Methanol/DME Production as Motor Fuels for Automotive Uses – BLGMF*, Altener II, European Commission, Brussels.

39 Tijmensen, M. J. A., Faaij, A. P. C., Hamelinck, C. N., and Van Hardeveld, M. R. M. (2002) Exploration of the possibilities for production of Fischer–Tropsch liquids and power via biomass gasification. *Biomass Bioenergy*, **23**, 129–152.

40 Müller-Langer, F. (2009) Biomethane options for mobile use, presented at the 7th International Colloquium on Fuels – Mineral Oil Based and Alternative Fuels, Stuttgart/Ostfildern.

41 Phillips, S. D., Tarud, J. K., Biddy, M. J., and Dutta, A. (2011) *Gasoline from Wood via Integrated Gasification, Synthesis, and Methanol-to-Gasoline Technologies*,

National Renewable Energy Laboratory (NREL), Golden, CO.

42 Baghdjian, V. (2010) The promise of biomethanol. European Petrochemical Outlook, 26 I Horizon, Autumn 2010, http://www.biomcn.eu/images/stories/downloads/101001_Platts_Horizon_-_The_promise_of_bio-methanol.pdf.

43 Van Bennekom, J. G., Vos, J., Venderbosch, R. H., Torres, M. A. P., Kirilov, V. A., Heeres, H. J., Knez, Z., Bork, M., and Penninger, J. M. L. (2009) Supermethanol: reforming of crude glycerine in supercritical water to produce methanol for re-use in biodiesel plants, in *Proceedings of the 17th European Biomass Conference and Exhibition, Hamburg*, pp. 899–902.

44 Müller-Langer, F., Majer, S., and Perimenis, A. (2012) Biofuels – a technical, economic and environmental comparison, in *Encyclopedia of Sustainability Science and Technology* (ed. R. A. Meyers), Springer, New York.

45 Claassen, P. A. M. and De Vrije, T. (2006) Non-thermal production of pure hydrogen from biomass: HYVOLUTION. *Int. J. Hydrogen Energy*, **31**, 1416–1423.

46 Das, D., and Veziroğlu, T. N. (2001) Hydrogen production by biological processes: a survey of literature. *Int. J. Hydrogen Energy*, **26**, 13–28.

47 Keskin, T., Abo-Hashesh, M., and Hallenbeck, P. C. (2011) Photofermentative hydrogen production from wastes. *Bioresource Technol.*, **102**, 8557–8568.

48 Modigell, M., Schumacher, M., and Claassen, P. A. M. (2007) Hyvolution – Entwicklung eines zweistufigen Bioprozesses zur Produktion von Wasserstoff aus Biomasse. *Chem. Ing. Tech.*, **79**, 637–641.

49 Rechtenbach, D. (2009) Fermentative Erzeugung von Biowasserstoff aus biogenen Roh-und Reststoffen, Dissertation, Technischen Universität Hamburg-Harburg.

50 Westfall, P. J. and Gardner, T. S. (2011) Industrial fermentation of renewable diesel fuels. *Curr. Opin. Biotechnol.*, **22**, 344–350.

51 Qureshi, N. and Ezeji, T. C. (2008) Butanol, "superior biofuel" production from agricultural residues (renewable biomass): recent progress in technology. *Biofuels Bioproducts Biorefining*, **2**, 319–330.

52 Wu, M., Wang, M., Liu, J., and Huo, H. (2007) *Life-Cycle Assessment of Corn-Based Butanol as a Potential Transportation Fuel*, Argonne National Laboratory, Argonne, IL.

53 Hahn, H. D., Dämbkes, G., Rupprich, N., and Bahl, H. (2005) Butanols. *Ullmann's Encyclopedia of Industrial Chemistry*, Wiley-VCH Verlag GmbH, Weinheim.

54 Brennan, L. and Owende, P. (2010) Biofuels from microalgae – a review of technologies for production, processing, and extractions of biofuels and co-products. *Renew. Sustain. Energy Rev.*, **14**, 557–577.

55 Kröger, M. and Müller-Langer, F. (2012) Review on possible algal-biofuel production processes. *Biofuels*, **3**, 333–349.

56 Lam, M. K. and Lee, K. T. (2012) Microalgae biofuels: a critical review of issues, problems and the way forward. *Biotechnol. Adv.*, **30**, 673–690.

57 Mineralölwirtschaftsverband (2012) *Raffinerien in Deutschland*, http://www.mwv.de/index.php/ueberuns/raffinerien (last accessed 20 January 2013).

58 Edwards, R., Larivé, J.-F., Mahieu, V., and Rouveirolles, P. (2007) *Well-to-Wheels Analysis of Future Automotive Fuels and powertrains in the European Context – Well-to-Tank Report and Appendices*, EUCAR and CONCAWE, European Commission Joint Research Centre, Brussels.

59 Jungbluth, N., Chudacoff, M., Dauriat, A., Dinkel, F., Doka, G., Faist Emmenegger, M., Gnansounou, E., Kljun, N., Spielmann, M., Stettler, C., and Sutter, J. (2007) *Life Cycle Inventories of Bioenergy*, Swiss Centre for Life Cycle Inventories, Dübendorf.

60 Müller-Langer, F., Perimenis, A., Brauer, S., Thrän, D., and Kaltschmitt, M. (2008) *Expertise zur technischen und ökonomischen Bewertung von Bioenergie-Konversionspfaden*, Wissenschaftliche Beirat der Bundesregierung Globale Umweltveränderungen (WBGU), Berlin.

61 Müller-Langer, F., Junold, M., Schröder, G., Thrän, D., and Vogel, A. (2007) *Analyse und Evaluierung von Anlagen und Techniken zur Produktion von Biokraftstoffen*, Institut für Energetik und Umwelt, Leipzig.

62 Nikander, S. (2008) Greenhouse gas and energy intensity of product chain: case transport biofuel, MSc thesis, Helsinki University of Technology.

63 Rettenmaier, N., Reinhardt, G., Gärtner, S., and Von Falkenstein, E. (2008) *Greenhouse Gas Balances for VERBIO Ethanol as Per the German Biomass Sustainability Ordinance (BioNachV)*, Institut für Energie- und Umweltforschung (IFEU), Heidelberg.

64 Rettenmaier, N., Reinhardt, G., Münch, J., and Gärtner, S. (2007) *Datenprojekt Nachwachsende Rohstoffe*, Netzwerk Lebenszyklusdaten, Forschungszentrum Karlsruhe, Karlsruhe.

65 European Commission (2011) *Technology Map: a European Strategic Energy Technology Plan (SET-Plan) Technology Descriptions*, Joint Research Centre Institute for Energy and Transport, Luxemburg.

66 Gnansounou, E., Dauriat, A. (2011) Chapter 2 – Life-Cycle Assessment of Biofuels, in P. Ashok, L. Christian, C. R. Steven, D. Claude-Gilles and E. Gnansounou (eds.), Biofuels alternative feedstock and conversion processes. Academic Press, Amsterdam.

67 IEA (2010) *Sustainable Production of Second Generation Biofuels. Potential and Perspectives in Major Economies and Developing Countries*, IEA, Paris.

68 Dale, B. E. (2007) Thinking clearly about biofuels: ending the irrelevant 'et energy' debate and developing better performance metrics for alternative fuels. *Biofuels Bioproducts Biorefining*, **1**, 14–17.

69 McKone, T. E., Nazaroff, W. W., Berck, P., Auffhammer, M., Lipman, T., Torn, M. S., Masanet, E., Lobscheid, A., Santero, N., Mishra, U., Barrett, A., Bomberg, M., Fingerman, K., Scown, C., Strogen, B., and Horvath, A. (2011) Grand Challenges for Life-Cycle Assessment of Biofuels. *Environ. Sci. Technol.*, **45**, 1751–1756.

70 Bare, J. C. (2010) Life cycle impact assessment research developments and needs. *Clean Technol. Environ. Policy*, **12**, 341–351.

71 Cherubini, F. (2010) GHG balances of bioenergy systems – overview of key steps in the production chain and methodological concerns. *Renew. Energy*, **36**, 1565–1573.

72 Reap, J., Roman, F., Duncan, S., Bras, B. (2008) A survey of unresolved problems in life cycle assessment. *Int. J. Life Cycle Assess.*, **13**, 290–300.

73 Rebitzer, G., Ekvall, T., Frischknecht, R., Hunkeler, D., Norris, G., Rydberg, T., Schmidt, W.-P., Suh, S., Weidema, B. P., and Pennington, D. W. (2004) Life cycle assessment: Part 1: Framework, goal and scope definition, inventory analysis, and applications. *Environ. Int.*, **30**, 701–720.

74 Guinée, J., Heijungs, R., and Voet, E. (2009) A greenhouse gas indicator for bioenergy: some theoretical issues with practical implications. *Int. J. Life Cycle Assess.*, **14**, 328–339.

75 Majer, S., and Oehmichen, K. (2010) *Approaches for Optimising the Greenhouse Gas Balance of Biodiesel Produced from Rapeseed*, Deutsches BiomasseForschungsZentrum (UFOP), Leipzig.

76 Zah, R., Böni, H., Gauch, M., Hischier, R., Lehmann, M., and Wäger, P. (2007) *Ökobilanz von Energieprodukten: Ökologische Bewertung von Biotreibstoffen*, EMPA, im Auftrag des Bundesamtes für Energie, des Bundesamtes für Umwelt und des Bundesamtes für Landwirtschaft, Bern.

77 Reinhardt, G., Gärtner, S., Helms, H., and Rettenmaier, N. (2006) *An Assessment of Energy and Greenhouse Gases of NExBTL*, Institut für Energie- und Umweltforschung (IFEU), Heidelberg.

78 Laborde, D. (2011) *Assessing the Land Use Change Consequences of European Biofuel Policies*, ATLASS Consortium.

79 Fritsche, U. and Wiegmann, K. (2008) *Treibhausgasbilanzen und kumulierter Primärenergieverbrauch von Bioenergie-Konversionspfaden unter Berücksichtigung möglicher Landnutzungsänderungen externe Expertise für das WBGU-Hauptgutachten "Welt im Wandel:*

Zukunftsfähige Bioenergie und nachhaltige Landnutzung," revidierte Endfassung, Öko Institut, Freiburg.

80. Hennig, C. and Gawor, M. (2012) Bioenergy production and use: comparative analysis of the economic and environmental effects. *Energy Convers. Manage.*, **63**, 130–137.

81. Searchinger, T., Heimlich, R., Houghton, R. A., Dong, F., Elobeid, A., Fabiosa, J., Tokgoz, S., Hayes, D., and Yu, T.-H. (2008) Use of U. S. croplands for biofuels increases greenhouse gases through emissions from land-use change. *Science*, **319**, 1238–1240.

82. Crutzen, P. J., Mosier, A. R., Smith, K. A., and Winiwarter, W. (2007) N_2O release from agro-biofuel production negates global warming reduction by replacing fossil fuels. *Atmos. Chem. Phys. Discuss.*, **7**, 11191–11205.

83. Erisman, J., Grinsven, H., Leip, A., Mosier, A., and Bleeker, A. (2010) Nitrogen and biofuels; an overview of the current state of knowledge. *Nutrient Cycling Agroecosyst.*, **86**, 211–223.

84. Lisboa, C. C., Butterbach-Bahl, K., Mauder, M., and Kiese, R. (2011) Bioethanol production from sugarcane and emissions of greenhouse gases – known and unknowns. *GCB Bioenergy*, **3**, 277–292.

85. Liu, X. J., Mosier, A. R., Halvorson, A. D., Reule, C. A., and Zhang, F. S. (2007) Dinitrogen and N_2O emissions in arable soils: effect of tillage, N source and soil moisture. *Soil Biol. Biochem.*, **39**, 2362–2370.

86. Smeets, E. M. W., Bouwman, L. F., Stehfest, E., Van Vuuren, D. P., and Posthuma, A. (2009) Contribution of N_2O to the greenhouse gas balance of first-generation biofuels. *Global Change Biol.*, **15**, 1–23.

87. Smith, P., Martino, D., Cai, Z., Gwary, D., Janzen, H., Kumar, P., McCarl, B., Ogle, S., O'Mara, F., Rice, C., Scholes, B., and Sirotenko, O. (2007) Agriculture, in *Climate Change 2007: Mitigation of Climate Change. Contribution of Working Group III to the Fourth Assessment Report of the Intergovernmental Panel on Climate Change* (eds. B. Metz, O. R. Davidson, P. R. Bosch, R. Dave, and L. A. Meyer), Cambridge University Press, Cambridge, pp. 497–540.

88. Mosquera, J., Hol, J. M. G., Rappoldt, C., and Dolfing, J. (2007) *Precise Soil Management as a Tool to Reduce CH_4 and N_2O Emissions from Agricultural Soils*, Report 28, Animal Sciences Group, Lelystad.

89. Kröger, M. and Müller-Langer, F. (2011) Impact of heterotrophic and mixotrophic growth of microalgae on the production of future biofuels. *Biofuels*, **2**, 145–151.

90. Dehue, B., Cornelissen, S., and Peters, D. (2011) *Indirect Effects of Biofuel Production Overview Prepared for GBEP*, Ecofys, London.

91. Zegada-Lizarazu, W., and Monti, A. (2011) Energy crops in rotation. A review. *Biomass Bioenergy*, **35**, 12–25.

92. Fernando, A. L., Duarte, M. P., Almeida, J., Boléo, S., and Mendes, B. (2010) Environmental impact assessment of energy crops cultivation in Europe. *Biofuels Bioproducts Biorefining*, **4**, 594–604.

93. Kort, J., Collins, M., and Ditsch, D. (1998) A review of soil erosion potential associated with biomass crops. *Biomass Bioenergy*, **14**, 351–359.

94. Adler, P. R., Del Grosso, S. J., and Parton, W. J. (2007) Lifecycle assessment of net greenhouse-gas flux for bioenergy cropping systems. *Ecol. Appl.*, **17**, 675–691.

95. Tsao, C. C., Campbell, J. E., Mena-Carrasco, M., Spak, S. N., Carmichael, G. R., and Chen, Y. (2012) Biofuels that cause land-use change may have much larger non-GHG air quality emissions than fossil fuels. *Environ. Sci. Technol.*, **46**, 10835–10841.

96. Whittaker, C., McManus, M. C., and Hammond, G. P. (2011) Greenhouse gas reporting for biofuels: a comparison between the RED, RTFO and PAS2050 methodologies. *Energy Policy*, **39**, 5950–5960.

97. Fargione, J., Hill, J., Tilman, D., Polasky, S., and Hawthorne, P. (2008) Land clearing and the biofuel carbon debt. *Science*, **319**, 1235–1238.

98. Stichnothe, H. and Schuchardt, F. (2010) Comparison of different treatment options for palm oil production waste on

a life cycle basis. *Int. J. Life Cycle Assess.*, **15**, 907–915.

99 Liebetrau, J., Clemens, J., Cuhls, C., Hafermann, C., Friehe, J., Weiland, P., and Daniel-Gromke, J. (2010) Methane emissions from biogas-producing facilities within the agricultural sector. *Eng. Life. Sci.*, **10**, 595–599.

100 Majer, S., Gawor, M., Thrän, D., Bunzel, K., and Daniel-Gromke, J. (2011) *Optimierung der marktnahen Förderung von Biogas/Biomethan unter Berücksichtigung der Umwelt- und Klimabilanz, Wirtschaftlichkeit und Verfügbarkeit*, Biogasrat, Berlin.

101 Cherubini, F. and Strømman, A. H. (2011) Life cycle assessment of bioenergy systems: state of the art and future challenges. *Bioresource Technol.*, **102**, 437–451.

102 Heijungs, R. (2012) Spatial differentiation, GIS-based regionalization, hyperregionalization, and the boundaries of LCA, in *Environment and Energy* (ed. G. Ioppolo), FrancoAngeli, Milan, pp. 165–176.

103 Kim, S. and Dale, B. (2009) Regional variations in greenhouse gas emissions of biobased products in the United States – corn-based ethanol and soybean oil. *Int. J. Life Cycle Assess.*, **14**, 540–546.

104 Kim, S. and Dale, B. E. (2005) Life cycle assessment of various cropping systems utilized for producing biofuels: bioethanol and biodiesel. *Biomass Bioenergy*, **29**, 426–439.

105 Mutel, C. L. and Hellweg, S. (2009) Regionalized life cycle assessment: computational methodology and application to inventory databases. *Environ. Sci. Technol.*, **43**, 5797–5803.

106 Potting, J. and Hauschild, M. (2005) *Background for Spatial Differentitaion in LCA Impact Assessment – the EDIP 2003 Methodology*, Danish Ministry of the Environment, Copenhagen.

107 Tessum, C. W., Marshall, J. D., and Hill, J. D. (2012) A spatially and temporally explicit life cycle inventory of air pollutants from gasoline and ethanol in the United States. *Environ. Sci. Technol.*, **46**, 11408–11417.

108 Engel, J., Huth, A., and Frank, K. (2012) Bioenergy production and Skylark (*Alauda arvensis*) population abundance – a modelling approach for the analysis of land-use change impacts and conservation options. *GCB Bioenergy*, **4**, 713–727.

109 Taubert, F., Frank, K., and Huth, A. (2012) A review of grassland models in the biofuel context. *Ecol. Modell.*, **246**, 84–93.

110 Tilman, D., Hill, J., and Lehman, C. (2006) Carbon-negative biofuels from low-input high-diversity grassland biomass. *Science*, **314**, 1598–1600.

111 Guinée, J. B., Heijungs, R., Huppes, G., Zamagni, A., Masoni, P., Buonamici, R., Ekvall, T., and Rydberg, T. (2011) Life cycle assessment: past, present, and future. *Environ. Sci. Technol.*, **45**, 90–96.

112 Geldermann, J. (2006) *Mehrzielentscheidungen in der industriellen Produktion*. Kit Scientific Publishing, Karlsruhe.

113 Vogel, A., Brauer, S., Müller-Langer, F., and Thrän, D. (2008) *RENEW – Renewable Fuels for Advanced Powertrains – Deliverable D 5.3.7 – Conversion Costs Calculation*, Deutsches Biomasseforschungszentrum (DBFZ), Leipzig.

114 Brauer, S., Vogel, A., and Müller-Langer, F. (2008) *Cost and Life-Cycle Analysis of Biofuels*, Deutsches BiomasseForschungsZentrum (UFOP), Leipzig.

115 Ekbom, T., Berglin, N., and Lögdberg, S. (2005) *Black Liquor Gasification with Motor Fuel Production – BLGMF II*, Nykomb Synergetics, Stockholm.

116 Erturk, M. (2011) Economic analysis of unconventional liquid fuel sources. *Renew. Sustain. Energy Rev.*, **15**, 2766–2771.

117 Hamelinck, C. N. (2004) *Outlook for Advanced Biofuels*, University of Utrecht, Utrecht.

118 Hofmann, F., Plättner, A., Lulies, S., Scholwin, F., Urban, W., and Burmeister, F. (2005) *Evaluierung der Möglichkeiten zur Einspeisung von Biogas in das Erdgasnetz*, Forschungsvorhaben im Auftrag der Fachagentur für Nachwachsende Rohstoffe.

119 Schmitz, N. (2003) *Bioethanol in Deutschland*, Fachagentur Nachwachsende Rohstoffe (FNR), Münster.

120 Zinoviev, S., Müller-Langer, F., Das, P., Bertero, N., Fornasiero, P., Kaltschmitt, M., Centi, G., and Miertus, S. (2010) Next-generation biofuels: survey of emerging technologies and sustainability issues. *ChemSusChem*, **3**, 1106–1133.

121 VCI (2008) *Chemiewirtschaft in Zahlen 2008*, Verband der Chemischen Industrie, Frankfurt am Main.

122 Chemie Technik (2012) *Chemie Technik Exklusiv: Preisindex für Chemieanlagen*, http://www.chemietechnik.de/texte/anzeigen/118055 (last accessed 20 mJanuary 2013).

123 Franzke, S. (2001) Technologieorientierte Kompetenzanalyse produzierender Unternehmen, Dissertation, Universität Hannover, Fachbereich Maschinenbau.

124 Leible, L., Kälber, S., Kappler, G., Lange, S., Nieke, E., Proplesch, P., Wintzer, D., and Fürniss, B. (2007) *Kraftstoff, Strom und Wärme aus Stroh und Waldrestholz: eine systemanalytische Untersuchung*, Forschungszentrum Karlsruhe, Karlsruhe.

125 Wene, C.-O. (2000) *Experience Curves for Energy Technology Policy*, IEA, Paris.

126 Doornbosch, R. and Steenblik, R. (2008) Biofuels: is the cure worse than the disease? *Rev. Virtual REDESMA*, **2**, 63–100.

127 Hamelinck, C. N., van Hooijdonk, G., and Faaij, A. P. (2005) Ethanol from lignocellulosic biomass: techno-economic performance in short-, middle- and long-term. *Biomass Bioenergy*, **28**, 384–410.

128 Kavalov, B. and Peteves, S. D. (2005) *Status and Perspectives of Biomass-to-Liquid Fuels in the European Union*, European Commission, Directorate General Joint Research Centre, Institute for Energy, Petten.

129 Neste Oil (2008) *Neste Oil Financial Statements for 2007*, Neste Oil, Espoo.

130 Ong, H. C., Mahlia, T. M. I., Masjuki, H. H., and Honnery, D. (2012) Life cycle cost and sensitivity analysis of palm biodiesel production. *Fuel*, **98**, 131–139.

131 Quirin, M., Gärtner, S. O., Pehnt, M., and Reinhardt, G. A. (2004) CO_2-*neutrale Wege zukünftiger Mobilität durch Biokraftstoffe: eine Bestandsaufnahme*, Institut für Energie- und Umweltforschung (IFEU), Heidelberg.

132 Schmitz, N., Henke, J., and Klepper, G. (2009) *Biokraftstoffe: eine vergleichende Analyse*, Fachagentur Nachwachsende Rohstoffe, Gülzow-Prüzen.

133 Silalertruksa, T., Bonnet, S., and Gheewala, S. H. (2012) Lifecycle costing and externalities of palm oil biodiesel in Thailand. *J. Cleaner Prod.*, **28**, 225–232.

134 Sims, R. E. H., Mabee, W., Saddler, J. N., and Taylor, M. (2010) An overview of second generation biofuel technologies. *Bioresource Technol.*, **101**, 1570–1580.

135 Van Thuijl, E., Roos, C. J., and Beurskens, L. W. M. (2003) *An Overview of Biofuel Technologies, Markets and Policies in Europe*, Energy Research Centre of the Netherlands (ECN), Petten.

136 Vogel, G. H. (2008) Change in raw material base in the chemical industry. *Chem. Eng. Technol.*, **31**, 730–735.

137 Wakker, A., Egging, R., Thuijl, E., Tilburg, X., Deurwaarder, E. P., Lange, T. J., Berndes, G., and Hansson, J. (2005) *Biofuel and Bioenergy Implementation Scenarios. Final Report of VIEWLS WP5, Modelling Studies*, Energy Research Centre of the Netherlands (ECN), Petten.

**Part V
Storage**

27
Energy Storage Technologies – Characteristics, Comparison, and Synergies

Andreas Hauer, Josh Quinnell, and Eberhard Lävemann

27.1
Introduction

The goal of energy storage is to match the energy supply with the energy demand when they are displaced in space or time. Energy is stored and later delivered where or when it is needed. Energy storage enables otherwise wasted energy streams to be used, energy efficiency to be improved, and fluctuating renewable energy inputs to be managed. Each of these benefits will be increasingly important and necessary in future energy systems.

Higher energy efficiencies from primary energy will be the largest contribution to the aimed for CO_2 reductions in the global energy future [1]. These high efficiencies can be obtained without slowing industrial activity and without the loss of human comfort in the building sector. All energy-related transformation processes (e.g., in power plants or vehicles) are coupled to losses resulting in waste heat due to thermodynamic irreversibilities inherent to energy transformation. Storage technologies offer high potential for using this waste heat to obtain the higher overall efficiencies necessary for future CO_2 reductions.

There are many opportunities to recover and store waste heat to improve energy efficiency. Energy-intensive industries such as the production of steel, aluminum, glass, and cement are a large component of global emissions and represent a major opportunity for increased efficiency via waste heat recovery [2]. In these industries, large amounts of high-temperature energy can be stored by thermal energy storage systems, which can then deliver process heat back to industry or supply heat input to district heating grids for space heating and domestic hot water. Combined heat and power (CHP) is another example where thermal storage increases efficiency [3]. Distributed CHP units can be operated efficiently to follow the consumer electricity demand while their waste heat is stored and later recovered to meet space heating and hot water demand. Storage increases the flexibility of waste heat recovery by decoupling energy supply and demand. Both industrial waste heat storage and CHP heat storage increase the total energy efficiency of the

Transition to Renewable Energy Systems, 1st Edition. Edited by Detlef Stolten and Viktor Scherer.
© 2013 Wiley-VCH Verlag GmbH & Co. KGaA. Published 2013 by Wiley-VCH Verlag GmbH & Co. KGaA.

energy transformation and, thus, allow increased energy consumption for constant primary energy production.

The integration of storage with renewable energy systems allows the continuous availability of intermittent and unpredictable energy resources. Renewable electricity from wind turbines and solar photovoltaics (PV) can be stored in large central storages such as pumped hydro plants (PHES), compressed air energy storages (CAES), or distributed electrochemical storage (lead acid or lithium ion batteries). Surplus electricity could also be converted into fuels such as hydrogen or syngas from which other fuels can be synthesized. High-temperature thermal energy storage allows concentrating solar thermal power plants (CSP) to run for extended hours or even allow solar plants to run at night. These technologies will become increasingly important and necessary to accommodate the growing fraction of renewable energy.

In this chapter, the basic energy storage mechanisms are introduced and the opportunities created by the application of energy storage are discussed. Possibilities are presented for incorporating energy storage technologies of different energy forms and at different scales into the existing and future energy systems. An economic evaluation determines the maximum acceptable costs under a variety of economic boundary conditions and establishes the types of technologies that are appropriate for specific applications.

27.2
Energy Storage Technologies

There exist a wide range of possibilities to store the different forms of energy, including mechanical, thermal, electrical, and chemical energy. The form of energy storage depends on the final consumer demand for energy, whether it be electricity, heat and cold, or a transportation fuel. Thus, the utility of a particular form of energy storage depends on the demands of the end user. Electricity can in theory provide nearly 100% usable energy, whereas the end use for thermal energy depends on the particular application. However, electricity is typically not stored directly, but instead stored using electrochemical (e.g., batteries), mechanical energy (e.g., pumped hydro), or chemical energy (e.g., hydrogen) and the useful electricity from these storage methods depends strongly on the efficiency of the necessary energy transformations.

27.2.1
Energy Storage Properties

An energy storage system can be described by the following technical properties [4]:

- *Capacity:* The capacity defines the energy stored in the system and depends on the storage process, the storage medium, and the size of the system. An intrinsic property of capacity is the energy density, the energy per unit mass or volume.

- *Power:* The power defines how fast the energy stored in the system can be discharged (and charged). An intrinsic property of power is the power density, the power per unit mass or volume.
- *Efficiency:* The efficiency is the ratio between the energy provided to the user and the energy needed to charge the storage system. It accounts for the energy loss during the storage period and the charging/discharging operations. In complex storage systems, multiple energy transformations may each reduce the total efficiency.
- *Storage period:* The storage period defines how long (i.e., hours, days, weeks, or months for seasonal storage) the energy is stored. The storage time is indefinite for some storage mechanisms, whereas others slowly degrade (e.g., thermal losses from sensible energy storage or self-discharge losses from battery storage).
- *Charge/discharge time:* The charge/discharge time defines the duration over which the storage system is charged and discharged.
- *Cost:* The storage costs refers to either the cost per unit capacity ($ kWh^{-1}) or power ($ kW^{-1}) and it depends on the capital costs, operating costs, auxiliary equipment or systems, and the lifetime of the storage, which can be expressed in the number of cycles.

Capacity, power, and discharge time are not independent variables in general. In batteries and thermal energy storage, high-capacity and high-power systems have to be designed differently. For example, in lead acid batteries, the discharge power directly affects the capacity. A high-power thermal energy storage system may need heat transfer enhancement, for example, additional fins in the heat exchanger, which reduces the amount of active storage material and thereby the storage capacity for a fixed volume.

27.2.2
Electricity Storage

In principle, electricity can be stored as mechanical, electrochemical, or electrical energy directly. Mechanical electricity storages are by far the largest systems concerning their absolute capacity and power. Furthermore, they are currently the economically most viable solution for centralized electricity storage.

- *Pumped hydro energy storage (PHES)* systems store electricity mechanically in the gravitational potential energy of a large volume of water. Excess electricity is used to pump water to a higher elevation during charging. During discharging, water flows down elevation, through a turbine and generator pair, to convert the mechanical energy into electrical energy. The energy capacity depends on the elevation change and the volume of water. Power is determined by the size of turbines. The specific energy density is comparably low (about 0.27 kWh m^{-3} for a 100 m difference in height). The storage efficiency is in the range 0.63–0.85 [5], but more recent projects are realizing higher efficiencies due to variable-speed pumps and turbines. PHES can be used for the balancing of the demand (load leveling) and also for the utilization of surplus energy. PHES systems are in general large (hundreds to thousands of megawatts) and are used as central storage devices coupled to the power generation and the high-voltage grid. From the economics point of view, PHES is the benchmark for electricity storage today, but costs are very site specific and placement is largely dictated by geology [6].

- *Compressed air energy storage (CAES)* stores energy as a highly compressed gas, typically air. In times of low demand, excess electricity is used to compress air in underground caverns or large engineered tanks. CAES is discharged by expanding the compressed air through a gas turbine. The energy density of CAES depends on the storage pressure: at 20 and 80 bar, the storage density is 2 MWh m^{-3} and 7.5 kWh m^{-3}, respectively [7]. The efficiency of CAES depends on the efficiency of the compression and expansion processes. Traditionally, before entering the turbine, the air has to be heated using natural gas. This conventional type of CEAS results in low storage efficiencies of around 55% [8] due to the combustion of natural gas. Ongoing R&D activities are aiming at so called adiabatic CAES, whereby heat generated during the compression process is stored separately and reused during the discharging. In this case, no natural gas is needed and the storage efficiency can reach more than 70% [8]. CAES utilizing natural geology may be the most economical option, but these costs are very site specific. CAES using engineered vessels will in general be more expensive and have smaller capacity. [7].

The integration of renewable electricity from wind turbines or solar PV might be supported by electrochemical energy storage technologies in the future, particularly in the form of a network of smaller distributed storages. The need for electrochemical storage systems such as accumulator or redox-flow batteries and hybrid systems of batteries and supercapacitors with high power density and long lifetime will increase in the future due to the rising proportion of distributed renewable electricity generation. This will lead to an increasing demand for stationary storage systems. At the same time, the development of high-power and high-capacity storage systems for transportation will become more important and the possibility exists for these opportunities to converge.

The future challenges are to develop consumer-friendly cost structures and application-oriented system solutions in the field of electrochemical energy storage. In the case of energy storage for the integration of renewable energy, the development of large battery systems for surplus electricity as a service for the energy providers and their customers will become increasingly more relevant. In addition to storage capacity, such systems will be used for the stabilization of voltage and frequency within the distribution grid.

Figure 27.1 shows the range of power and the storage timescales for a number of electricity storage technologies [9]. The main distinction is the discharge time or storage period. Energy storage systems which can contribute to the increase of the overall energy efficiency and help position renewable sources within the electrical grid should be able to store electricity for at least 1 h and should have a minimum power of around 100 kW. The interesting technologies from this point of view are lead acid and redox-flow batteries, sodium–sulfur (NaS) cells, CAES, and PHES.

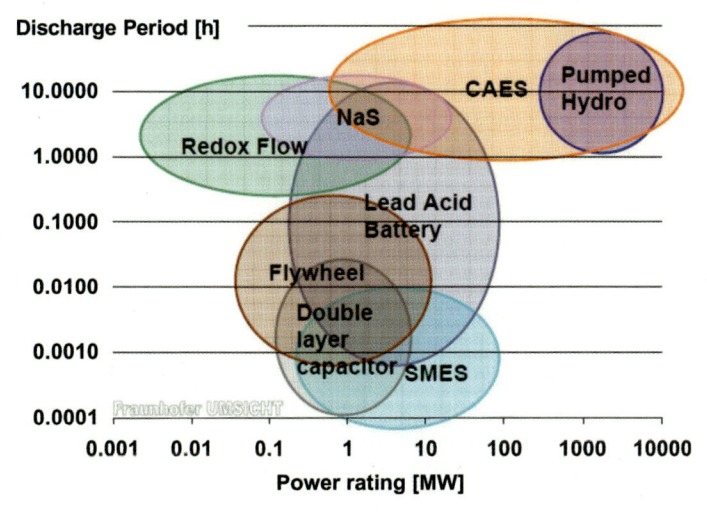

Figure 27.1 Electricity storage technologies and their discharge periods and power ratings [9].

Other technologies with shorter storage periods are better suited for power quality measures.

27.2.3
Storage of Thermal Energy

Thermal energy (heat and cold) can be stored as sensible heat in heat storage media, as latent heat associated with phase change of materials, or as thermochemical energy associated with chemical reactions (i.e., thermochemical storage) at operating temperatures from −40 to above 400 °C.

- *Sensible thermal energy storage* uses the heat capacity of a material to store thermal energy at a higher or lower temperature with respect to the temperature needed for an energy demand. Most sensible energy storage occurs at low temperatures of −10 to 100 °C, and hence is most applicable to heating and cooling in buildings and domestic hot water. Water is the most common storage medium owing to its low cost and high specific heat, which make it energy dense compared with other materials such as earth or rock. For instance the energy density of water is 90 kWh m^{-3} for a temperature change of 80 °C. Small hot water tanks are used as buffer storages for daily or weekly storage in solar-thermal installations for domestic hot water (DHW) or heating systems. Large hot water storage systems are used for long-term storage of solar heat in the building sector. Heat and cold can also be stored underground using water within aquifers, caverns, or engineered tanks, using rock or sand beds, or simply the local earth. In this case, the underground energy is discharged during winter by a heat pump to temperature level of about 5 °C. In summer, the colder ground can be used for air conditioning, effectively creating a seasonal storage in the earth.

- *Latent thermal energy storage* uses the liquid-to-solid phase change of a material in addition to the temperature change (sensible) to store thermal energy. The materials used are called phase change materials (PCM). PCM storage systems are able to reach high storage densities at small temperature changes due to the large phase change enthalpy. For example, sodium acetate trihydrate has an energy density of 120 kWh m^{-3} at 58 °C. Water (or ice storage) is a common PCM for cooling applications and has an energy density 100 kWh m^{-3} available at 0 °C. The key to PCM is that the melting temperature and enthalpy of the material match the application. This match is especially important for cold storage applications owing to the small temperature differences in these applications.

  Other applications of PCMs include micro-encapsulated PCMs within the building structure to add high-density thermal mass to make the structure less sensitive to temperature changes. In this application, melting temperatures between 20 and 25 °C are optimal. PCM slurries are another application where PCM particles are incorporated into a fluid, which allows the molten and solid phases to be easily transported. Ice slurries are an example of PCMs often used in cooling applications.

  One common limitation of PCMs is poor heat transfer due to low thermal conductivity. Power can be increased using extended surfaces or enhancing thermal conductivity, but this comes at the expense of reducing the storage capacity by displacing the PCM.

- *Thermochemical thermal energy storage (TCS)* uses a chemical reaction to store thermal energy. In most cases, these reactions involve the physical or chemical sorption of water to and from a host material. Water is separated from the host material by adding heat, which charges the storage. By bringing the two components together, the heat of reaction is released and the storage is discharged, providing heat. During charging and discharging, these systems are coupled to the ambient conditions, which makes them more complex than sensible and PCM thermal energy storage systems. TCS can reach very high storage capacities (120–250 kWh m^{-3}) and have longer storage periods than PCM and sensible energy storage because they are not subject to thermal loss. Furthermore, they are able to adapt the temperature levels at charging and discharging to the application requirements over a certain range.

  The most investigated reactions are the adsorption of water on solids (e.g., zeolites and silica gels) and absorption of water into liquids (e.g., aqueous calcium chloride and lithium chloride). An active area of research is the synthesis of new materials with sorption properties that can be tailored to specific applications. Thermochemical processes are well suited to industrial drying applications, dehumidification, and potentially other HVAC (heating, ventilation, and air conditioning) processes due to low vapor pressure, low desorption temperatures, and high cycling ability. Similarly to PCM materials, power and efficiency are typically limited by slow heat and mass transfer. These limitations can be overcome at the expense of energy density by including extended surfaces to facilitate heat and mass transfer.

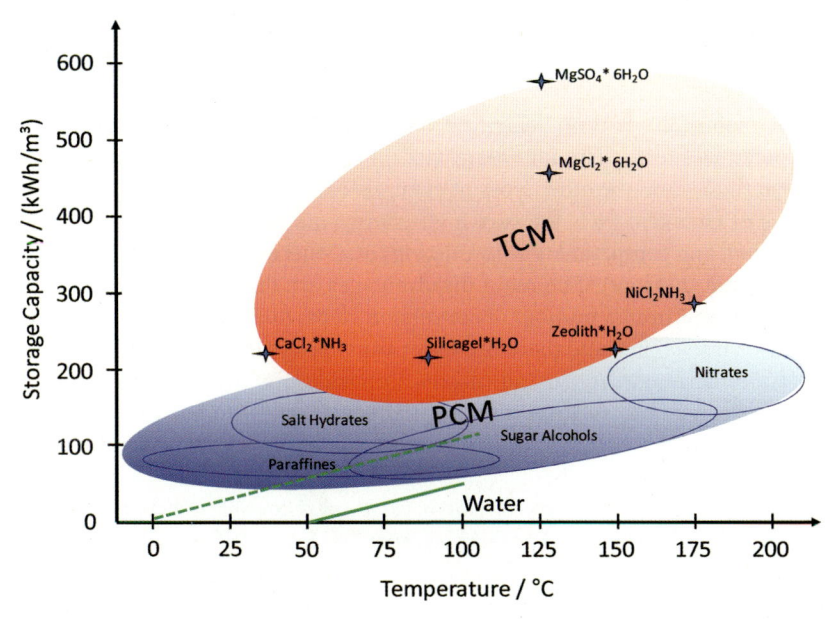

Figure 27.2 Thermal energy storage technologies and their storage capacity versus temperature [4].

In Figure 27.2, the energy storage capacity is plotted as a function of the temperature range for the three thermal energy storage technologies [4]. The trend across all thermal storage mechanisms is that higher operating temperatures are required for larger energy storage densities. The dotted green line shows the theoretical storage capacity for the sensible energy storage of water between 0 and 100 °C. The slope of this line corresponds to specific heat of water. In practice, energy density is reduced because applications restrict the temperature range. Temperature differences across heat exchangers also reduce the energy storage density. A practical energy density for water is reduced to that shown by the solid green line.

The blue and red areas show the possible materials (or material pairs) for PCM and TCS energy storage, respectively. PCMs provide in general higher capacities than sensible storage technologies. Thermochemical processes can achieve even higher values compared with PCMs, but generally require high operating (charging) temperatures.

The values given in Figure 27.2 tend to show energy storage density under the most optimistic assumptions, which is an incomplete picture of the relationship between energy density and operating temperature. Energy storage capacity is not a property of the storage material; it is an attribute of the system. Hence it depends on the system operating conditions, transport within the system, and, in the case of TCS, the ambient conditions in which the system operates.

27.2.4
Energy Storage by Chemical Conversion

The conversion of thermal or electrical energy into an energy-rich carrier fuel is another potential mechanism for energy storage. The attributes of chemical fuels distinguish them from the other types of energy storage discussed previously. The advantages of chemical fuels are very high energy and power density, indefinite storage period, and compatibility with existing fossil-fuel infrastructure for energy distribution and transport. The main disadvantages are the comparably low efficiencies that result from a number of conversion steps. Additionally, the oxidation of the fuels is irreversible and therefore they cannot be cycled. The energy carriers produced are usually hydrogen and carbon monoxide (the pair forming syngas). Through subsequent processing, syngas can be converted into higher hydrocarbons, such as methane.

- *Hydrogen energy storage* is the conversion of excess or renewable thermal or electrical energy into hydrogen gas. For example, surplus electricity from wind turbines or solar PV can run electrolysis equipment to convert water into hydrogen and oxygen. The hydrogen can be stored and be converted into electricity again by a fuel cell, combusted in a turbine, or it can be used to synthesize hydrocarbon fuels. The key point with hydrogen storage is that the source of the energy used to create the charging should be ecologically sound (e.g., renewable). Hydrogen as a fuel has a very high energy density per mass (34 kWh kg^{-1}), but a comparably low energy density per volume (2.7 kWh kg^{-1} at 1 bar), requiring very high compression to approach energy per unit volume ratios of other fuels. At large scales, the possibilities of buffering electricity peaks from offshore wind turbines by electrolysis, hydrogen storage in underground caverns, and electricity production by gas turbines are currently under investigation.
- *"Renewable methane" as energy storage* follows hydrogen, except that the hydrogen is further combined with CO_2, potentially from fossil-fuel waste streams, to convert renewable energy into methane (natural gas). As a hydrocarbon fuel, methane has a comparably high energy density per unit mass (16 kWh kg^{-1}), but requires compression (or liquefaction) to achieve high energy density per unit volume. In Germany, a 100% renewable energy scenario for 2050 was derived in which the conversion of renewable electricity into methane plays an important energy storage role [10]. The main advantage is the versatility of the fuel. Renewable methane would allow an energy supply for all sectors to be provided by a reservoir fuel, which could be a very effective complement to renewable energy systems by providing an alternative during long periods of low renewable resources.

Figure 27.3 shows how the "power-to-gas" concept works [11]. This innovative concept is based on the methanization (Sabatier process) of hydrogen by using CO_2 from fossil-fuel or renewable sources. Hydrogen comes from renewable resources and it is combined with CO_2 from fossil power plants using a carbon capture and

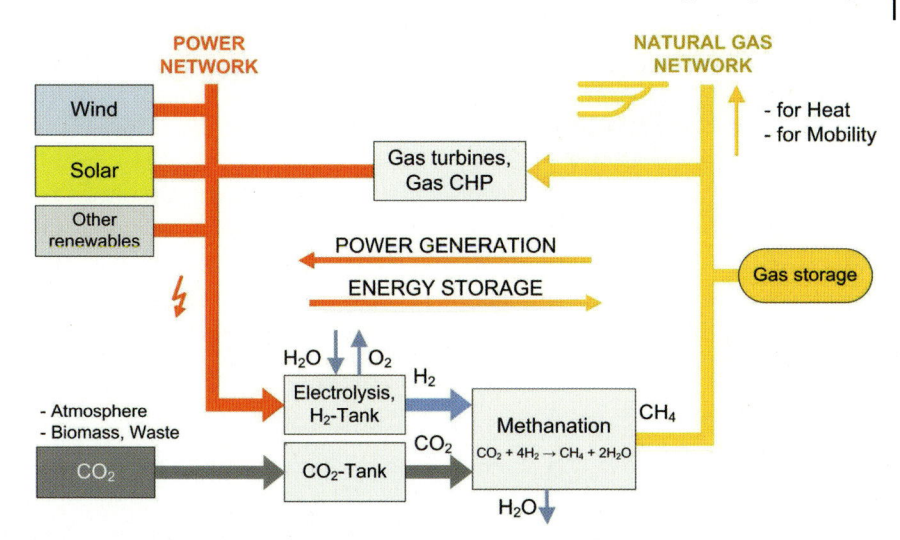

Figure 27.3 Schematic view of methane production from renewable electricity [11].

storage (CCS) process or from renewable biogas plants. The fuel is stored in the gas network and combines the electricity grid and the gas infrastructure into an intelligent and bidirectional energy system. While the German electricity grid today has a storage capacity of about 0.04 TWh, which corresponds to a storage period of 1 h, the storage capacity of the present gas network is in the region of 200 TWh, which could provide energy for months and provide an effective buffer against long periods of low renewable resources.

27.2.5
Technical Comparison of Energy Storage Technologies

The main technical parameters for several established and emerging energy storage technologies are given in Table 27.1 [5–7, 12, 13]. These parameters include typical capacity, power, and discharge times, the energy densities (by mass and volume), expected efficiencies, the lifetime, and some different cost metrics. The given data are not an inclusive list of values, but rather an example of typical values from the literature. Estimates for some parameters are not available and others are determined under different assumptions, making strict comparisons difficult. Nevertheless, differences in these parameters do indicate some major distinctions between different storage technologies.

The largest distinction among storage technologies is the form of stored energy. The main forms are electricity via mechanical or electrochemical storage, thermal energy, or chemical fuels. Thermal energy storage systems, including sensible, PCM, and TCS storage, are the least versatile energy form. Although limited to heating and cooling, there is still a very large opportunity; thermal energy loads make up over half the energy consumption in both Europe and North America.

Table 27.1 Technical and economic parameters for various energy storage technologies [5–7, 12, 13].

Storage technology	Storage mechanism	Power (MW)	Capacity (MWh)	Storage period	Density (kWh t^{-1})	Density (kWh m^{-3})	Efficiency (%)	Lifetime (No. of cycles)	Cost ($ kW^{-1})	Cost ($ kWh^{-1})	Cost ($ kWh^{-1} delivered)
Lithium ion (Li Ion)	Electrochemical	<1.7	<22	Day–month	84–160	190–375	0.89–0.98	2960–5440	1230–3770	620–2760	0.17–1.02
Sodium–sulfur (NaS) battery	Electrochemical	1–60	7–450	Day	99–150	156–255	0.75–0.86	1620–4500	260–2560	210–920	0.09–0.55
Lead acid battery	Electrochemical	0.1–30	<30	Day–month	22–34	25–65	0.65–0.85	160–1060	350–850	130–1100	0.21–1.02
Redox/flow battery	Electrochemical	<7	<10	Day–month	18–28	21–34	0.72–0.85	1510–2780	650–2730	120–1600	0.05–0.88
Compressed air energy storage (CAES)	Mechanical	2–300	14–2050	Day	—	2–7 at 20–80 bar	0.4–0.75	8620–17 100	15–2050	30–100	0.02–0.35
Pumped hydro energy storage (PHES)	Mechanical	450–2500	8000–190 000	Day–month	0.27 at 100 m	0.27 at 100 m	0.63–0.85	12 800–33 000	540–2790	40–160	0.001–0.18
Hydrogen	Chemical	Varies	Varies	Indefinite	34 000	2.7–160 at 1–700 bar	0.22–0.50	1	384–1408	—	0.25–0.64
Methane	Chemical	Varies	Varies	Indefinite	16 000	10 at 1 bar	0.24–0.42	1	—	—	0.16–0.44
Sensible storage – water	Thermal	<10	<100	Hour–year	10–50	<60	0.5–0.9	5000	—	0.1–13	0.0001
Phase change materials (PCMs)	Thermal	<10	<10	Hour–week	50–150	<120	0.75–0.9	5000	—	13–65	0.013–0.06
Thermochemical storage (TCS)	Thermal	<1	<10	Hour–week	120–250	120–250	0.8–1	3500	—	10–130	0.01–0.05

Although the remaining storage forms can provide electricity, they are generally differentiated by scale, both in size and storage period. Mechanical energy storage systems such as PHES and CAES are really only effective for large centralized storage and are highly dependent on local geology. However, they offer the largest power and capacity of all storage mechanisms with power outputs ranging between 2 MW and 2.5 GW and storage capacities ranging between 2 MWh and 190 GWh. Chemical storage is similar in that fuels require large, centralized infrastructure for efficient production. The distinction is that, once created, fuels can be easily distributed and discharged in a variety of ways and at a variety of scales, thus blurring the lines between distributed and centralized storage. The lower power and capacity of electrochemical storage are better suited to smaller scales, such as DES and transportation.

The power and capacity of energy storage are related to the physical size via the energy density. In terms of energy density, the major distinction is the very high densities of chemical fuels (16–34 MWh t^{-1}) compared with the relatively low energy densities of all other technologies. CAES and PHES typically have among the lowest energy densities (0.27–7 kWh m^{-3}). Meanwhile, at smaller DES scales, the differences in energy density between lithium ion (84–160 kWh t^{-1}) and lead acid (22–34 kWh t^{-1}) batteries, or thermochemical storage (120–250 kWh t^{-1}) and sensible energy storage (10–50 kWh t^{-1}), are likely to have a substantial impact on the volume and mass of the storage, easily becoming the critical factor for many applications.

Storage generally degrades the overall efficiency of the energy system during charging and discharging. For many technologies, the efficiency depends on the storage period due to losses. Technologies such as batteries and sensible energy storage have self-discharge or thermal losses, which effectively reduce the efficiency over time. Mechanical and chemical storage systems tend to have efficiencies that are somewhat independent of storage period due to extremely low losses.

A complete comparison among given storage technologies is not only predicated on the technical parameters, but must also consider the type of stored energy, the scale and scope of the storage, and economic factors. Technologies are much more likely to be chosen based on these measures as well as the type of energy demand and the location of the storage. Hence application-specific requirements are likely to limit the available technologies and subsequently determine the range of technical parameters that are feasible in a specific application.

27.3
The Role of Energy Storage

Energy storage systems save energy and keep it for consumption at another time or place. To a certain extent, energy storage decouples supply and demand, adding flexibility to the energy system. Within an energy system, storage serves as either a temporary demand or temporary production. Therefore, energy storage can be seen as direct competition to other energy supplies such as fossil fuels. In fact, in some ways fossil fuels are the ideal energy storage medium. Fossil fuels can be stored for long periods, energy can be recovered at high efficiency, and they provide extremely high energy and power densities (energy or power per unit mass or volume). For example, coal and oil have energy densities in the range 6.70–12.8 kWh kg^{-1}. The major drawback of fossil fuels is that there is a limited supply because charging timescales are many times longer than the present rate of discharge. Compared with fossil fuels, alternative energy storage technologies must have similar technical properties or compete in terms of other economic or ecological advantages.

An example that highlights the competition between energy storage systems and fossil fuels is the transportation sector. Energy storage enables the consumer to move independently from the energy source. Fossil fuels are well suited to this application owing to their high energy and power density. Electrochemical storage (i.e., batteries) does not yet compete on these technical merits, yet has compelling ecological advantages. Renewably produced chemical fuels such as hydrogen or methane may have similar benefits. Vehicles may also offer technical and environmental advantages as a network of highly distributed energy storage systems. Hence the evaluation of storage technologies based on the performance metrics established by fossil fuels is insufficient and the ecological advantages such as locally reduced emissions and compatibility with renewable electricity infrastructure must be evaluated alongside the technical details.

The cost of energy storage is an additional cost for energy systems. Stored energy will compete with newly generated energy and hence the price for stored energy should not exceed the price for newly generated energy. However, energy storage provides the opportunity to leverage the differences in the temporal fluctuations in cost of energy (e.g., peak–trough pricing) and pricing differences between alternative forms of energy (e.g., electricity and natural gas). In today's energy system and perhaps more so in the future, the price for energy will vary strongly with demand. Energy storage is economically attractive when low-cost, low-demand energy is stored for discharge during high-cost, high-demand periods.

27.3.1
Balancing Supply and Demand

An energy storage system can be operated whenever there is a difference in power between supply and demand. In the absence of storage, consumers are forced to take the energy as it comes or the energy provider must deliver the energy as it is needed. Figure 27.4 shows two cases in which either the power supply or the power demand is constant [4].

In Figure 27.4a, the energy is provided by a power plant running on fossil fuels (blue line) and the fluctuating demand is covered by a storage system such that the power plant is unaffected and runs constantly in its optimal operation mode. Since the storage covers the demand peaks in this scenario, higher prices for the discharged energy can be achieved compared with lower prices during charging at off-peak times. Another example of this case is a bio-gas plant running for constant electricity production, while the heat produced is charged to a thermal energy storage. The thermal energy storage may then be used to meet the seasonal space heating

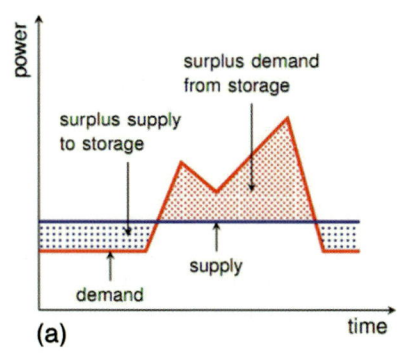

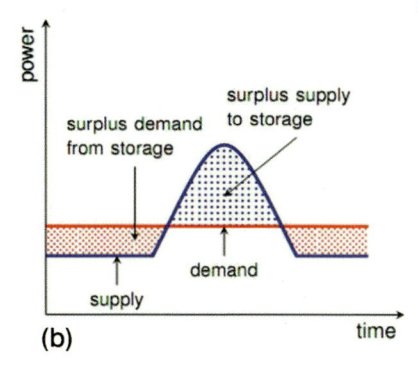

Figure 27.4 Energy storage is used for (a) supply management and (b) demand management.

demand in a district system. In this case, the energy storage can couple energy grids (electrical and thermal) operating at very different timescales.

Figure 27.4b shows a variation in energy supply such as that expected from renewable energy sources such as wind, PV, solar-thermal, or waste heat streams. In this case, energy storage provides constant base load to meet the constant power demand. Energy storage helps to integrate renewable energies into the existing energy system.

The obligation of a renewable energy storage system can be described whether it is located on the supply or the demand side. If the storage is located on the supply side, it is mainly designed for the uptake of energy. A typical application is the integration of peak output of centralized renewable energy facilities or the utilization of industrial waste heat. The focus of such systems is the storage of energy that would otherwise remain unused. In another supply-side application, storage is collocated with renewable energy generation to smooth the output. These storage systems may be sized to eliminate the intermittency of renewable energy systems and provide a lower, reliable base load power. Supply-side storages are centralized systems and in general they are large (in capacity and power) since they are directly coupled to energy sources such as fossil-fuel or renewable energy power plants.

Demand-side storage systems are installed to follow the demand profile. The demand and the supply profile are decoupled. This application can also be described as "demand-side management." Storage systems at the demand side are smaller and more distributed storage units close to the consumer. In many cases, these storage systems are comparably small units adapted to the consumer's demand. An exception is the distributed renewable energy supply (e.g., by local PV), where distributed energy storage systems also have to be implemented.

Present and future energy systems will show a time variation in both the energy supply and demand at many different frequencies and scales due to variations in renewable energy production and consumer energy demand. The characteristics of the energy demand are decoupled from the specific storage technology except for operational characteristics such as power and capacity. However, there is work to be done to identify these demand characteristics and match them to specific storage technologies. Hence there will be ample opportunity to integrate many different types of energy storage to accommodate variations in supply and demand and maintain a stable and predictable energy system.

27.3.2
Distributed Energy Storage Systems and Energy Conversion

Current discussions on energy storage are focused on large, centralized energy storage technologies such as PHES and CAES, and the conversion of surplus energy into renewable fuels. The potential of distributed energy storage technologies is mostly unexplored, but offers substantial diversity and versatility in a renewable energy system.

27.3.2.1 Distributed Energy Storage Systems
Distributed energy storage systems (DES) can be defined by their location within an energy system. Figure 27.5 shows the locations where DES can be located within an energy system [14]. They are generally located near the consumer side. For example, in an electricity grid, this location is on the low-voltage side of the distribution grid. In case of a district heating network, it is located at the level of heat substations. Owing to the growing impact of distributed renewable energy generation by PV, solar-thermal, and biomass (via CHP), DES provide one of the only options for balancing production and demand at this level. For example, only DES can effectively manage electricity peaks by distributed renewable electricity production. In this case, the distribution grid is often incapable of transporting the excess energy from the low-voltage side of the grid. It has to be stored locally, where it is generated.

Another possibility would be to build a "virtual central storage" by combining a large number of DES. Provided that the grid is able to transport the energy to the DES locations, DES can contribute to the storage demand given by the fluctuations of larger, centralized wind turbine and solar PV installations. DES allow not only the integration of different storage technologies, but also different forms of stored energy into this single virtual storage system. Incorporating a variety of energy forms

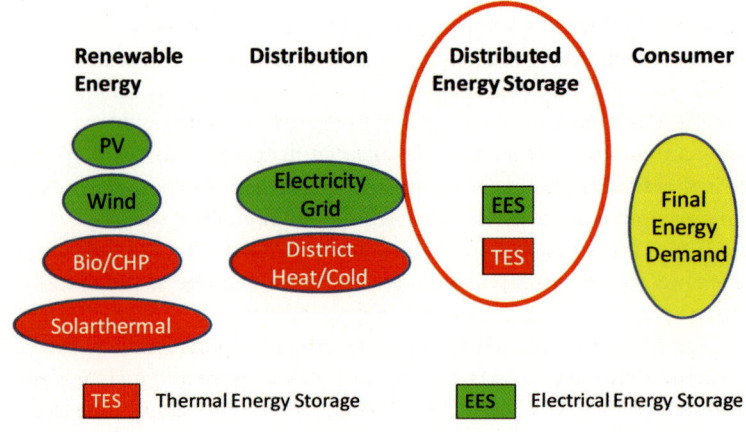

Figure 27.5 Definition of "distributed energy storage" within the energy system [14].

and technologies into a DES network will yield a more stable and robust energy system. A major advantage compared with centralized storage is that DES can be implemented incrementally, requiring initially less investment capital than large central storages. One of the main drawbacks of this concept is the communication between the distribution network and the individual DES (e.g., by "smart grids") forming the storage. These communication and coordination requirements complicate the task of including DES into effective national or even global energy plans compared with central storage technologies.

27.3.2.2 In/Out Storage Versus One-Way Storage

In general, energy storage technologies charge and discharge the same form of energy: a battery is charged by electricity and it will provide electricity during discharging. The same holds for a hot water buffer storage which is charged and discharged with sensible heat. However, fuels demonstrate a possible "one-way" storage process. For example, electricity is "charged" and methane is discharged. An energy transformation is inherent to the storage process as electricity is converted into a renewable fuel. The same is possible with thermal processes: renewable electricity can used to heat up a hot water tank. The storage is then provided by the thermal energy storage device. However, in this case the energy discharge is less flexible because it is limited to thermal processes. This idea can be expanded to consider a variety of conversions between different storage mechanisms, energy types, and storage locations. Integrating different energy conversion processes and storage mechanisms increases the number of possibilities by which primary energy is converted and delivered to the final energy needed by the consumer and hence it increases the robustness of the energy system.

In addition, there may be economic advantages when integrating different energy storage mechanisms or mixing energy networks. The cost of energy storage varies for different technologies and forms of energy storage. Thermal storage technologies tend to be less expensive than electrochemical storage. In Germany, about 60% of the final energy demand is on heating and cooling. Hence it may be both technically and economically appropriate to store thermal energy compared with electrochemical energy. This additional degree of freedom by choosing the mixing energy networks using storage could lead to large-scale economic benefits.

27.3.2.3 Example: Power-to-Gas Versus Long-Term Hot Water Storage

In this example, two storage technologies suitable for long-term-storage are compared. In both cases excess electricity is available. In one case, this electricity can be used to produce methane at a net efficiency of 60%. The methane can be used for the transportation sector, for heating, and for domestic hot water, or it can be converted into electricity again with a final efficiency of around 35% [15].

In the second case, the electricity is used to drive a heat pump with a thermal coefficient of performance (COP) of 3. At this COP, the heat pump provides 300% usable heat at temperatures below 90 °C. This heat can be stored in a large water tank with a seasonal efficiency of 75%. Hence the net effect of this surplus electricity is a storage system that is able to deliver heat at an overall efficiency of 225% [15].

Comparing both storage cases, it is clear that the performance depends strongly on the application. For heating and domestic hot water, the thermal storage system is much more efficient. Nearly seven times the energy is available for heating compared with electricity from methane. However, the main limitation is that thermal energy is strictly limited in application, whereas electricity or fuel may be used in virtually any application.

27.4
Economic Evaluation of Energy Storage Systems

One barrier to the development of renewable energy storage technologies is cost uncertainty. The typical method to discuss energy storage economics is the bottom-up approach where material and manufacturing costs, operating costs, R&D costs, and site preparation costs are lumped together for an overall estimate. This approach yields a large range of cost estimates for each given technology, reflecting the uncertainty and variation in each of these components. For instance, capital cost estimates for compressed air energy storage (CAES) vary by over two orders of magnitude ($ 15–2050 kW^{-1}) [5]. Large cost variations exist even for established storage technologies such as lead acid batteries ($ 190–1100 kWh^{-1}) and pumped hydroelectric storage (PHES) ($ 40–160 kWh^{-1}) [5]. These cost estimates make comparisons among technologies difficult if not impossible.

27.4.1
Top-Down Approach for Maximum Energy Storage Costs

An alternative, top-down approach considers the acceptable cost of energy storage by comparing it with the marginal cost of energy production[Dr Rainer Tamme, German Aerospace Center (DLR), personal communication]. The analysis assumes that the cost of energy storage should not exceed the cost of additional energy generation. Therefore, it is a direct comparison between the cost of stored energy and that energy which is newly generated. The main advantage of this approach is that it does not require detailed information about the storage technology or implementation. Instead, the maximum acceptable cost for a specific storage application is determined from the discount rate assigned to the storage capital, the storage period (or cycling time), and the storage lifetime.

The maximum acceptable cost is calculated from the discount rate of storage capital, the payback period of the investment, the frequency at which the energy storage is charged and discharged, and the cost of energy. Using the discount rate for storage capital and the storage lifetime, one calculates the present value annuity factor (PVANF) to determine the present value of the energy storage capital.

For example, in industry it is typical to seek capital improvements with rates of return greater than 25% and short payback periods of < 4 years. This requires annuity factors on the order of PVANF = 0.25–0.3. In the building sector, longer paybacks are acceptable, interest rates are more modest, and annuity factors are lower, PVANF = 0.07–0.08. One might also suggest a class of user that can tolerate extremely long recovery periods (20+ years) because the energy investment is pursued for political or ecological reasons. In these cases, annuity factors are PVANF < 0.06.

The maximum acceptable storage cost is simply the product of the number of storage cycles during the payback period and the cost of energy divided by the annuity factor. This storage cost assumes a fully realized storage capacity (i.e., storage capacity is the actual discharged energy). The maximum acceptable cost of storage is proportional to the cost of energy generation and the number of storage cycles. The analysis neglects operating costs and changes in the cost of energy production over the time period. Nevertheless, it illustrates the most important feature in energy storage economics, the relationship between acceptable storage costs, the frequency of storage cycling, and the cost of energy.

27.4.2
Results

In Figure 27.6, the storage costs ($ kWh^{-1}) are shown as a function of the length of the storage cycle (days) on a log–log plot for a few different combinations of energy costs and discount rates. The maximum acceptable storage costs decrease with increasing storage period. Energy storage with a storage cycle of 100 days must be 100 times cheaper than a storage with a 1 day cycle, which imposes a major restriction on the type of technologies that should be investigated for diurnal storage versus annual storage. For example, industrial users with energy costs between $ 0.025 and $ 0.06 kWh^{-1} can accept storage costs up to $ 0.31 and $ 0.88 kWh^{-1} for seasonal storage (100 days) and $ 31–88 kWh^{-1} for diurnal storage. The building sector with typically lower annuity factors (0.07–0.1) and a higher cost of energy (0.08–0.13 kWh^{-1}) can afford nearly 10 times the storage costs for any storage period.

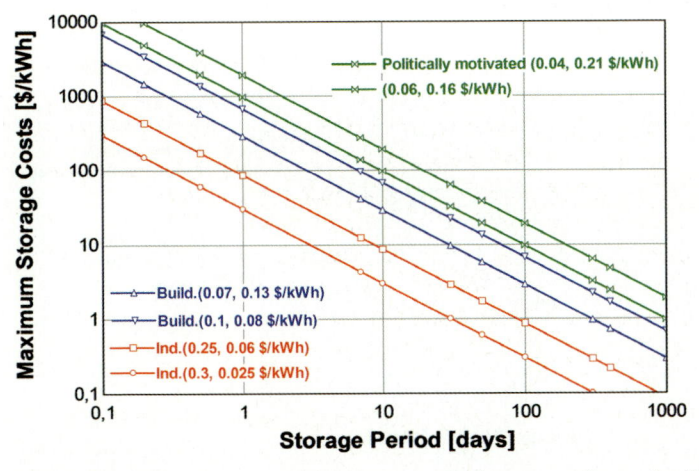

Figure 27.6 Maximum acceptable storage costs for three types of users for different energy costs and annuity factors.

For very low annuity factors (0.04–0.06) and high costs of energy ($ 0.16–0.21 kWh^{-1}), politically motivated users can accept storage costs that are three times higher than in the building sector and 30 times higher than for industrial users. The large variations in maximum acceptable storage costs based on these simple factors suggest that the specific storage application imposes an upper limit on storage costs for a capital investment.

Although published bottom-up-approach cost estimates vary widely, it is useful to compare these costs with the maximum acceptable storage costs. This comparison is plotted for several technologies in Figure 27.7 using the storage costs in Table 27.1. The frequency of storage cycling restricts the choice of storage technology based on existing cost estimates. For example, sensible storage with costs of $ 0.1–13 kWh^{-1} are currently the only possibility for economical seasonal storage (one storage cycle per year). With minor cost reductions (from $ 10 kWh^{-1}), thermochemical storage may be economical for the building sector and politically motivated implementations. ESA cost estimates for PHES and CAES storage show that it is economical for the industrial sector for storage periods up to 1 day, whereas for the building sector, PHES and CAES are economical for storage periods up to 2 weeks. The storage costs for electrochemical storage exceed the maximum allowed storage capital in the industrial sector. However, diurnal storage using lead acid batteries or possibly lithium ion batteries is economical for the building and enthusiast sectors.

Power-to-gas technologies, such as hydrogen or methane, are unique in the sense that they cannot be cycled. Hydrogen via electrolysis is estimated to cost $ 2 kg^{-1} ($ 0.05 kWh^{-1}) using surplus electricity [16], but costs may fall to $ 0.88 kg^{-1} ($ 0.022 kWh^{-1}) with emerging production technologies [17]. In this analysis, the lack of storage cycling implies an infinite storage time (hydrogen is plotted as a thin box at 1000 days), which requires very low storage costs. Hence hydrogen is not economical in the given scenarios. However, in the case of hydrogen production from electrolysis, the majority of the costs are due to the cost of the electricity.

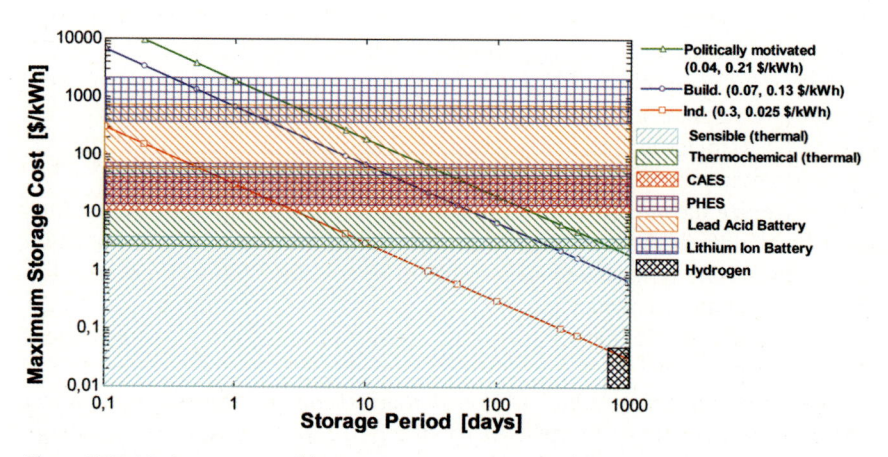

Figure 27.7 Maximum acceptable storage costs compared with estimated costs for several storage technologies.

The capital costs for the production infrastructure may be as low as $ 0.01 kWh^{-1} [16], although storage costs must also be considered. Therefore, regardless of the storage period, there may be a significant opportunity for economical hydrogen production if excess electricity from renewables (energy which would otherwise be wasted) is used for production.

This top-down analysis reveals some often overlooked points in the discussion of energy storage economics. First, there exist scenarios under which most storage technologies are economical. In these cases, it makes sense to evaluate storage technologies based on other merits such as performance parameters. Second, at present seasonal storage is only economical using thermal storage; other technologies require approximately an order of magnitude or more reduction in costs from their lowest estimates. Last, for a fixed storage period, the maximum acceptable storage costs depend on the user due to variances in payback period, discount rate, and cost of energy. As the cost to produce new energy increases, as has historically been the case, higher energy storage costs become acceptable for all users.

27.5
Conclusion

Energy storage is an increasingly important component of energy production and distribution systems because it allows decoupling between the energy supply and the energy demand. This flexibility can be used to recover waste energy, increase energy efficiency, integrate renewable energy systems, and leverage fluctuations in the price of energy.

The important parameters of energy storage are power, capacity, storage period, and cost. The many combinations of these parameters ensure that there is a storage technology for nearly every application. Furthermore, opportunities for energy storage increase when storage is integrated with the transformation between different types of energy. Excess electricity can be used to charge thermal storage for its later use in heating or cooling processes. Excess thermal energy and electricity can be used to charge chemical fuels, which in turn can be used at a different place and time in the transportation sector or converted back to electricity and thermal energy. Thus storage allows the coordination of the electrical grid, the gas network, the transportation sector, and thermal energy demands.

To maximize the benefit of energy storage, the appropriate storage technology must be identified for a specific application from both technical and economic perspectives. The type of end user, relevant energy production costs, and the storage period and frequency establish the maximum acceptable costs of storage for each application. In applications of seasonal storage for space heating, where stored energy is cycled approximately once annually, only sensible energy storage is currently economically viable. In general, thermal storage is the only economical storage mechanism for storage periods exceeding 10 days. Large, centralized storage systems, such as PHES and possibly CAES, may be economical in a variety of systems if they can be sized to cycle frequently (< 1 day). Distributed energy storages with lead acid batteries and potentially other emerging battery technologies are economically viable for applications with longer acceptable payback periods and medium storage periods (1–10 days). For transportation and other demanding applications, fuels such as hydrogen or methane are so technically superior and versatile that the comparably low efficiency and higher costs become acceptable. Traditional comparisons of storage technologies emphasize the technical metrics, but it is important to recognize that the economic boundary conditions and the final energy demand will equally guide the choice of a particular technology. Bearing in mind the economics and final energy demand of each application, a variety of energy storage implementations will contribute to reduced primary energy consumption, lowered CO_2 emissions, and mitigation of renewable energy production fluctuations, and offer a varied and robust infrastructure to achieve high-stability future energy supply.

References

1 IAEA (2008) *Energy Technology Perspectives 2008 – Scenarios & Strategies to 2050*, International Energy Agency, Paris.
2 Villar, A., Arribas, J., and Parrondo, J. (2012) Waste-to-energy technologies in continuous process industries. *Clean Technol. Environ. Policy*, **14** (1), 29–39.
3 Blarke, M. B. and Lund, H. (2008) The effectiveness of storage and relocation options in renewable energy systems. *Renew. Energy*, **33** (7), 1499–1507.
4 Hauer, A., Hiebler, S., and Reuss, M. (2013) *Wärmespeicher*, 5th edn, Fraunhofer Irb, Stuttgart.
5 ESA (2009) Electricity Storage Association, http://www.electricitystorage.org/ (last accessed October 2012).
6 Deane, J., Ó Gallachóir, B., and McKeogh, E. (2010) Techno-economic review of existing and new pumped hydro energy storage plant, *Renew. Sustain. Energy Rev.*, **14**, 1293–1302.
7 Arizona Research Institute for Solar Energy (AzRISE) (2010) *Study of Compressed Air Energy Storage with Grid and Photovoltaic Energy Generation*, University of Arizona.
8 Energietechnische Gesellschaft im VDE (ETG) (2009) *Energiespeicher in Stromversorgungssystemen mit hohem Anteil erneuerbarer Energieträger. Bedeutung, Stand der Technik, Handlungsbedarf*, VDE-ETG, Berlin.
9 Doetsch, C. (2012) Electric energy storage: future energy storage demand. Presented at Innostock 2012 – 12th International Conference on Energy Storage, Lleida, 16–18 May.
10 Stadermann, D. (ed.) (2010) *Energiekonzept 2050 – Eine Vision für ein nachhaltiges Energiekonzept auf Basis von Energieeffizienz und 100% erneuerbaren Energien, Foschungsverbund Eneuerbare Energien*, FVEE, Berlin.
11 Sterner, M. (2009) Bioenergy and renewable power methane in integrated 100% renewable energy systems. Dissertation, Kassel University Press, Kassel.
12 A. Hauer (2012) *Technology Policy Brief E17 – Thermal Energy Storage*, IEA-ETSAP and IRENA, www.etsap.org, www.irena.org (last accessed 8 January 2013).
13 Hauer, A., Specht, M., and Sterner, M. (2012) Energiespeicher als Schlüsselelemente im künftigen Energiesystem, *Energietechnol. Aktuell*, Ausgabe May.
14 Hauer, A. (2012) Integration of renewable energies by distributed energy storage systems. Presented at ECES Workshop, Paris.
15 Hauer, A. (2012) Presented at IEA Committee on Energy Research and Technology Workshop on Energy Storage Issues and Opportunities, International Low-Carbon Energy Technology Platform, Paris, 15 February.

16 Levene, J. I., Mann, M. K., Margolis, R., and Milbrandt, A. (2007) An analysis of hydrogen production from renewable electricity sources. *Solar Energy*, **81** (6), 773–780.

17 O'Donnell, L. and Maine, E. (2012) Techno-Economic analysis of hydrogen production using FBMR technology. Presented at PICMET '12: Technology Management for Emerging Technologies, San Jose, CA 28 July–1 August.

28
Advanced Batteries for Electric Vehicles and Energy Storage Systems

Seung Mo Oh, Sa Heum Kim, Youngjoon Shin, Dongmin Im, and Jun Ho Song

28.1
Introduction

Recently, the ever-increasing world energy consumption, coupled with the issue in global warming, has brought increasing awareness of the need for cleaner, more fuel-efficient electric vehicles (EVs) and energy storage systems (ESSs) that can store the electricity generated by renewable sources. EVs have already come into the market. The driving distance of these eco-friendly vehicles is double that of conventional vehicles, ensuring that energy consumption and air pollution can be mitigated in real terms [1]. Electricity generation from renewable sources such as wind, solar, and tidal energy has progressively increased worldwide. For instance, wind energy generation amounted to 240 GW in 2011, which accounts for 60% of total renewable energy generation except for hydroelectricity.

So far, however, the number of EVs actually running on the roads is limited, and so is the number of energy storage devices for renewable energy generation. It is due to the fact that the performance of current rechargeable batteries falls well short of requirements with respect to price, energy density, power density, safety, charging time, and cycle life. Hence incremental evolutionary improvements in materials, cell fabrication, and new battery systems are desperately needed. In the near future, greatly improved rechargeable batteries will very likely allow the widespread introduction of electric vehicles and energy storage systems.

In this chapter, the working principles and recent R&D status of secondary batteries, namely lithium-ion cells, redox-flow cells, sodium–sulfur cells, lithium–sulfur cells, and lithium–air cell are reviewed. In addition, recent progress in EV and ESS applications of secondary batteries is discussed. Figure 28.1 illustrates the current and future applications of secondary batteries.

Transition to Renewable Energy Systems, 1st Edition. Edited by Detlef Stolten and Viktor Scherer.
© 2013 Wiley-VCH Verlag GmbH & Co. KGaA. Published 2013 by Wiley-VCH Verlag GmbH & Co. KGaA.

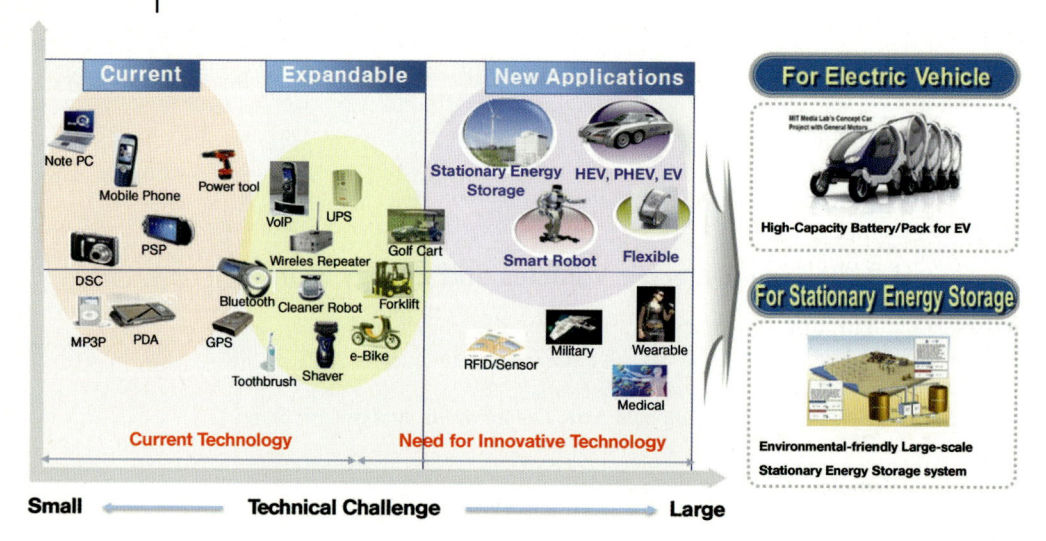

Figure 28.1 Mega-trend of applications for secondary batteries.

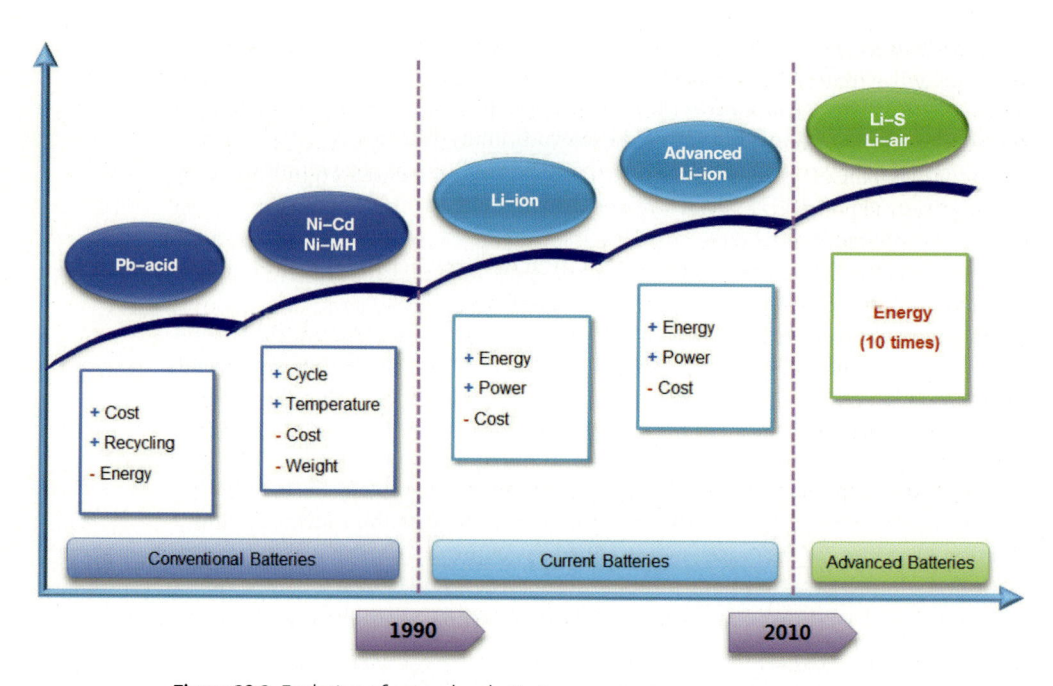

Figure 28.2 Evolution of secondary batteries.

28.2
R&D Status of Secondary Batteries

Secondary (rechargeable) battery technology has progressively developed from lead–acid, nickel–cadmium and nickel–metal hydride to lithium-ion batteries (LIBs), as shown in Figure 28.2. Secondary batteries working in aqueous electrolytes such as lead–acid, nickel–cadmium, and nickel–metal hydride batteries give limited output voltages owing to the electrochemical instability of water. LIBs are now the front-runner, largely satisfying the requirements for energy density, power density, and cycle life. Lithium–sulfur and lithium–air batteries are now being intensively developed to realize high energy-density cells, and redox-flow and sodium–sulfur cells for large-scale energy storage devices.

28.2.1
Lithium-Ion Batteries

Before the commercialization of LIBs, rechargeable lithium batteries based on a lithium metal negative electrode were developed, but were subsequently abandoned owing to safety problems: dendrite formation caused by repeated lithium metal dissolution/deposition led to internal short, thermal runaway and explosion. A search was made for active materials with good reversibility for Li intercalation/deintercalation and low charge/discharge voltage to replace Li metal. Carbonaceous materials were found to meet these requirements, and a secondary cell based on a negative carbon electrode and a $LiCoO_2$ positive electrode was commercialized by Sony in 1991. Both the negative and positive electrodes are layered compounds, into/from which Li^+ ions can be intercalated/deintercalated. This type of cell was denoted lithium-ion battery, as Li exists as the Li^+ ion rather than the metal in the carbon negative electrode.

Current LIBs consist of four key materials (Table 28.1): lithium-containing transition metal oxides ($LiCoO_2$, $LiMn_2O_4$, etc.) as the positive electrode, carbonaceous materials (graphite, hard carbon, soft carbon, etc.) as the negative electrode, nonaqueous electrolyte for lithium-ion migration, and a separator that prevents the movement of electrons between the two electrodes, but allows lithium-ion migration.

Table 28.1 Cell components and their characteristics in lithium-ion batteries.

Component	Material	Characteristics
Positive electrode	Transition metal oxide ($LiCoO_2$, $LiNiCoMnO_2$, $LiMn_2O_4$, $LiFePO_4$)	Oxidation/reduction High voltage
Negative electrode	Carbon materials (artificial graphite, surface-treated natural graphite)	Oxidation/reduction Low voltage
Electrolyte	Nonaqueous solvent + Li salt (ethylenecarbonate/diethyl carbonate + $LiPF_6$)	Lithium ion conductor
Separator	Porous polymer membrane (polypropylene, polyethylene)	Electrical insulator

Three different shapes of LIBs have been commercially developed: cylindrical, prismatic, and pouch type. For cylindrical cells, a jelly roll comprising a negative electrode plate, separator film, and positive electrode plate is inserted into a cylindrical metal case. Rectangular-shaped metal cases are used for prismatic cells. The electrodes/separator assembly is wrapped by a pouch in the pouch-type cell, which is a thin polymer-coated metal foil.

The superior performance of LIBs has made them the main power source for mobile IT devices. They also offer attractive performance advantages for EV and ESS applications. Typical LIBs have an average cell voltage of 3.6 V, which is three times larger than that for nickel–metal hydride cells. The high output power allows LIBs to compete with double-layer capacitors. They can be deep cycled with a round-trip energy efficiency of more than 90%. They show a long shelf-life and are maintenance free.

For EV applications, improvements in energy density and power density for the extension of driving distance and power output in addition to price reduction are now highly sought. For the energy storage systems, price reduction and life extension are, among others, the prime concerns for cell manufacturers [2–5].

28.2.2
Redox-Flow Batteries

Redox-flow batteries consist of a catholyte (positive redox couple in electrolyte) and an anolyte (negative redox couple in electrolyte) (Figure 28.3). The cell potential is given by the potential difference between the two redox couples. The other components are the separator membrane and a bipolar plate that supports the electrode and offers a stable current flow.

A redox-flow battery was first developed by the Lewis Research Center (LRC) at the National Aeronautics and Space Administration (NASA) in 1973, and employed Fe, Cr, and Ti redox couples [6]. A variety of redox couples that show high potential difference were tested thereafter. Of these, vanadium redox couples have been employed in cell systems with stationary energy storage capacities from tens of kilowatts to megawatts.

The most beneficial feature of redox-flow batteries is the easy control of capacity and power output, which is possible because the cell stack and electrolyte tank are separated, unlike in the other secondary batteries. The capacity of redox-flow batteries is determined by the size of the electrolyte tank and concentration of redox couples. The output power is determined by the number of stacks connected in series. In addition, as the cell stack and electrolyte tank are easily separated, there are few restrictions on the installation site. Moreover, the charging status of individual cells in a stack is identical since the electrolyte containing redox couples is supplied from the same tanks, so that cell balancing is not required.

The most critical drawback for this large-scale type of secondary battery is the requirement for periodic maintenance for ancillary components such as membrane, bearings, and pumps. The lower energy density than that for LIBs is another intrinsic drawback. The operating voltage is low when an aqueous electrolyte is used.

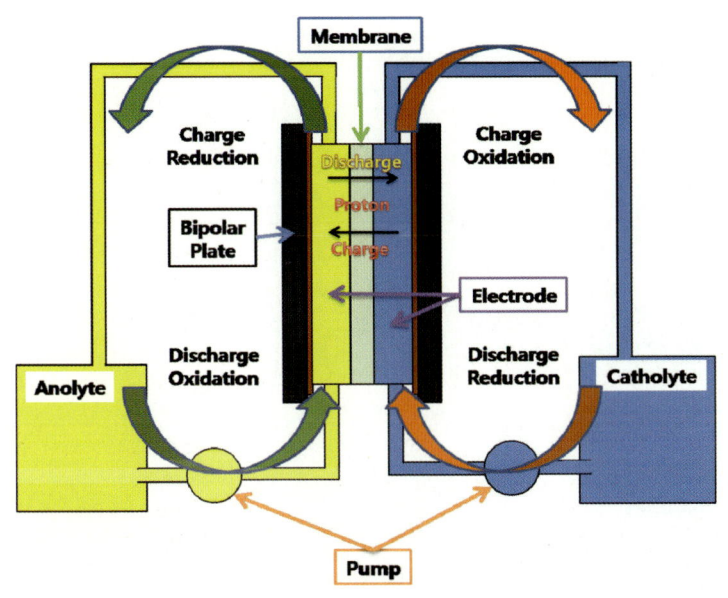

Figure 28.3 Schematic diagram of a redox-flow battery.

The capacity is also limited because the solubility of redox couples in water is limited. Efforts are now being made to explore new redox couples having a higher potential difference and high solubility in nonaqueous solvents.

28.2.3
Sodium–Sulfur Batteries

Since Ford reported the working principle of the sodium–sulfur battery in the 1970s (Figure 28.4), several companies have tried to develop this high-temperature cell for EV application. Owing to intrinsic problems such as high-temperature operation, reliability of the electrolyte tube and sealing, R&D efforts for transportation applications have diminished. Instead, several companies in Japan succeeded in the commercialization of the sodium–sulfur battery for stationary energy storage systems.

The sodium–sulfur battery consists of metallic sodium (Na) as the negative electrode, sulfur (S) as the positive electrode, and β″-alumina as the Na$^+$ ion-conducting solid electrolyte [7]. Among these, the performance of β″-alumina as an electrolyte and separator is most critical for the overall performances of this type of cell [8].

The sodium–sulfur battery has a high specific energy (theoretical value 760 Wh kg^{-1} and current practical value > 100 Wh kg^{-1}). The cells show a high charge–discharge efficiency of 87% and the battery pack system has 75% overall efficiency including inverter and heater for high-temperature (300–350 °C) operation. It has also been demonstrated that self-discharge is marginal because there is no irregular charge–discharge behavior. Moreover, sodium and sulfur are abundant in Nature and inexpensive. Owing to these features, sodium–sulfur cells are considered to be a promising energy storage device for renewable generation.

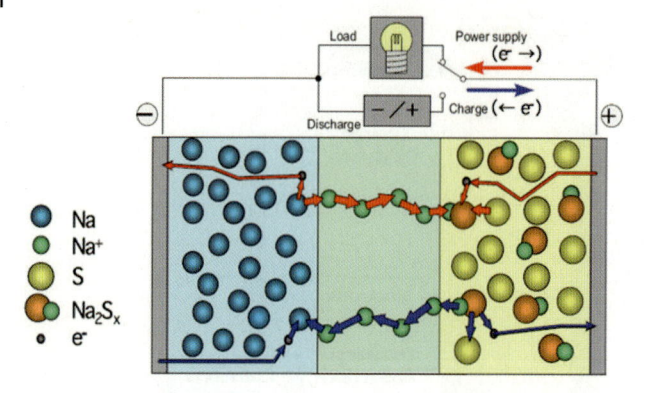

Figure 28.4 Working principle of the sodium–sulfur battery.

28.2.4
Lithium–Sulfur Batteries

Lithium–sulfur batteries have gained recognition as the next-generation battery system, largely on the basis of their high energy density, non-toxicity, and potential cost advantages stemming from the natural abundance of sulfur resources. Extensive R&D efforts are being made all over the world.

Lithium–sulfur batteries are composed of three major components: a sulfur positive electrode, Li metal negative electrode, and a nonaqueous electrolyte. Sulfur has a high theoretical specific capacity of 1672 mA h g^{-1} and a high theoretical specific energy of about 2600 Wh kg^{-1} based on the redox reaction $S_8 + 16Li \rightleftharpoons 8\, Li_2S$. This energy density is 3–5 times higher than that for the commercial lithium-ion batteries [9]. However, commercialization of lithium–sulfur batteries is still hindered by unsolved problems, including low utilization of sulfur, poor coulombic efficiency, and rapid capacity fading. One of the major problems in lithium–sulfur batteries is that sulfur is an electrically insulating material (5×10^{-30} S cm^{-1} at 25 °C), so that sulfur cannot be converted into the final discharge product (Li_2S). For more efficient utilization, sulfur powder must be in intimate contact with a conductive matrix that can reinforce the electrically conductive network within the composite positive electrode layer. For this purpose, various types of sulfur–carbon composites have been suggested, with partial success.

Another problem is polysulfide dissolution in electrolytes. During discharge, sulfur undergoes a sequential reduction reaction to form Li_2S via the formation of a series of polysulfides, Li_2S_n ($n = 2$–8). The long-chain polysulfide ions S_n^{2-} ($n > 2$) are soluble in common nonaqueous electrolytes. The dissolved polysulfides diffuse away from the positive electrode and cross over to the negative Li metal electrode, where they are reduced to insoluble Li_2S_2 or Li_2S. The reduced polysulfides can also diffuse back to the positive electrode and are reoxidized during charge. This is an internal short process, which is known as the shuttle effect. The poor electrochemical performance, including severe capacity fading and low coulombic efficiency, is accounted for by this shuttle effect. The dissolved polysulfides lose electrical contact with the current collector, resulting in a loss of active materials. Moreover, Li metal is passivated by the deposited polysulfides, leading to an increase in polarization for the electrochemical deposition/dissolution of Li. Protection of the Li metal surface could be a way to improve the electrochemical performance, and it was recently reported [10] that the additives in electrolytes such as $LiNO_3$ are effective in protecting Li metal by forming a stable solid electrolyte interphase.

The dendritic growth of Li metal and formation of dead lithium are inevitable, since Li metal is used as the negative electrode. Hence countermeasures should be taken to solve the safety problem and poor cycle performance.

28.2.5
Lithium–Air Batteries

The lithium–air battery is the lightest among metal–air battery systems (Figure 28.5). Gaseous oxygen from ambient air is utilized as the positive active material and lithium, the lightest metallic element, is used directly as the negative electrode without intercalation or insertion reactions. The absence of host material gives a high theoretical specific energy (~3500 Wh kg^{-1}) that is nearly 10 times higher than that of current lithium-ion batteries. Also, the practical specific energy is estimated to be well over 1000 Wh kg^{-1}, which allows more than 800 km of driving with a single charge [11]. Hence, if this is realized, lithium–air cells will be competitive with gasoline when employed in EVs.

Two different chemistries occur, depending on the electrolyte. An aqueous electrolyte system provides a higher operating voltage and much faster kinetics:

$$4\,Li + O_2 + 2\,H_2O \rightarrow 4\,LiOH \qquad E° = 3.45\,V$$

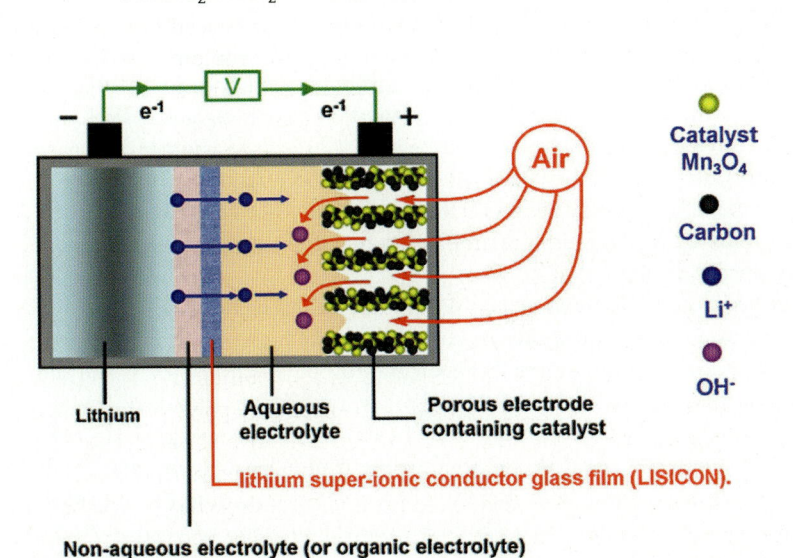

Figure 28.5 Schematic diagram of a lithium–air battery.

However, as the water consumption in the oxygen reduction reaction is so significant, an excessive amount of water has to be loaded to keep the discharge products such as lithium hydroxide dissolved, limiting the energy density of the cells [12]. Accordingly, the majority of research activities are focused on nonaqueous electrolyte systems, in spite of the slower kinetics due to the insolubility and low electronic conductivity of the discharged product (lithium peroxide):

$$2\,Li + O_2 \rightarrow Li_2O_2 \qquad E° = 3.10\,V$$

The positive electrode is typically composed of porous carbon, optionally with supporting catalyst particles on its surface. The conductive surface of carbon or catalyst provides the reaction sites for oxygen reduction (discharge) and oxygen evolution (charge). The positive electrode is often compared with that of the proton-exchange membrane fuel cell (PEMFC) owing to the similar constitutional elements: carbon, catalyst, and binder. In lithium–air batteries, however, the oxygen reduction products are accumulated on the positive electrode, not consumed by a reaction with protons as in PEMFCs. When nonaqueous electrolytes are used, the build-up of discharge products such as lithium peroxide (Li_2O_2) reduces the effective electrode area continuously, and retards the oxygen and ion transport inside the positive electrode, causing the overall kinetics to deteriorate. It is not usual to apply a current density higher than 1 mA cm^{-2} in lithium–air cells, whereas even 1 A cm^{-2} can be used in PEMFCs [13]. This means that much larger amounts of carbon and catalyst are required compared with PEMFCs to achieve a similar performance for EV application.

The capacity and rate capability of positive electrodes are known to be largely dependent on the morphology of carbon materials. Carbons with a large surface area can provide more reaction sites and consequently greater discharge capacity. However, the surface associated with micropores (diameter < 2 nm) is not effective since there is no space for storing discharge products and oxygen transport. It is widely accepted that carbons with mesopores and larger pores are more effective [14]. The pore volume is also considered an important parameter since discharge products are stored there. A larger pore volume generally leads to a higher discharge capacity. Nanostructured carbon materials such as carbon nanotubes (CNTs) and graphene are beneficial in this respect. However, it has to be considered that the increase in pore volume requires a larger amount of electrolyte, resulting in a decrease in energy density.

Catalysts have been the most extensively studied area in lithium–air battery research. The importance of the catalyst comes from the high polarization experienced in nonaqueous lithium–air cells. The charging potential is > 4.0 V and the average discharge potential is about 2.8 V or below, making the round-trip energy efficiency 70% at best. Since low energy efficiency implies more energy waste, considerable efforts have been made to reduce the polarization by applying catalysts in the positive electrodes. Candidates include noble metals, transition metal oxides, and heteroatom-doped carbon materials. Noble metal nanoparticles such as Pt–Au alloy and Pd show some improvement in oxygen evolution kinetics and there is a report [15] that gold without carbon can lead to a better cycle life. Although noble metals have a better catalytic activity than the others, the cost is an important issue since a very large amount of catalyst will be required in an EV. Mn-based oxides seem to be advantageous in terms of cost, but Mn dissolution has to be suppressed [16]. Perovskite-structured oxides are also being widely studied, as their catalytic activity in aqueous electrolytes is well proven, but not in nonaqueous electrolytes [17]. Nitrogen-doped carbon materials are also being actively investigated. Despite extensive catalyst research, there is a claim that no catalyst is actually required, and it is speculated that the polarization is due mainly to mass transport rather than the intrinsic activity of the catalysts [18].

The discharge product in the positive electrode has been identified as lithium peroxide, but its formation mechanism has not yet been clarified. Decomposition of solvent molecules by superoxide and peroxide has become one of the most important issues in lithium–air battery research. It was found that organic carbonates are vulnerable to nucleophilic substitution reactions and easily decomposed by an attack of superoxide ions [19]. Ether-based solvents were found to be much more stable, but long-term stability has yet to be established [20, 21].

The Li metal negative electrode is another important part of the lithium–air battery and there are many technical issues to be resolved. It has to be protected from oxygen or other oxygen reduction products formed in the positive electrode. Otherwise, Li metal will be oxidized, worsening not only the coulombic efficiency but also the rechargeability. To avoid direct contact between Li and oxygen, Li^+ ion-conducting ceramic membranes are often positioned on the Li surface. Na superionic conductor (NASICON) structured materials are most commonly employed to this end [22]. Since these NASICON-based materials are not stable against Li metal, a protecting layer (polymer, liquid or another solid electrolyte) is inserted between the Li metal and the ceramic membrane.

There have been attempts to estimate the practical specific energy of lithium–air batteries. Researchers at Bosch suggested that 1000 Wh kg^{-1} can be achieved with 200 μm thick positive electrodes. Another calculation by researchers at Ford indicated that more than 600 Wh kg^{-1} is probable in their bipolar cell design [23]. However, both of them assumed an areal capacity of > 30 mAh cm^{-2}, which is extremely difficult to achieve in nonaqueous rechargeable batteries.

28.3
Secondary Batteries for Electric Vehicles

EVs include hybrid electric vehicles (HEVs), plug-in hybrid electric vehicles (PHEVs), and battery electric vehicles (BEVs), in order of increasing electrical energy usage (Figure 28.6). EVs have emerged as environmentally friendly vehicles according to the ZEV (zero-emission vehicle) regulation in California in the early 1990s. Immature technologies, however, did not satisfy the requirements of energy and power density, such that R&D interest and investment have been declining since the mid-2000s. Recently, however, R&D activities for EVs have been revisited since LIBs with high energy and power were adopted as the main power source for HEVs and EVs.

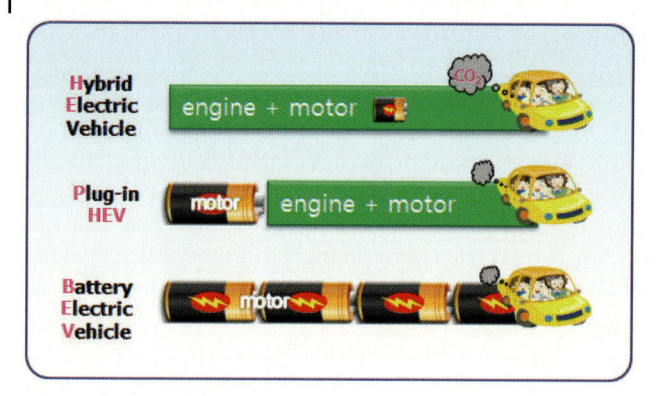

Figure 28.6 Types of EVs.

Table 28.2 Components of electric vehicles and their functions.

System	Component	Function
Electrical energy Drive system	Motor Transmission	Driving (by motor or by motor with engine)
Electrical energy Conversion system	Inverter Converter Charger	Energy supply to the motor Energy storage to the batteries
Electrical energy Storage system	Batteries Supercapacitors	Energy storage and supply

The energy storage system for EVs works by: (i) storing/releasing energy from/into the battery, (ii) controlling the input/output energy, (iii) managing for the optimization of battery operation through a feedback by other control systems such as driving/charging/cooperation systems, and (iv) protecting the battery and its components. Table 28.2 lists the components in EVs and their functions.

The energy storage system can be divided into two components: (i) hardware such as cells, battery packs, structure, cooling system, electrical connection systems for high voltages, and control devices, and (ii) software related to control algorithms for driving and charging. For the development of the battery system, not only the battery or cell technology but also the technologies for charging, interfaces between the systems, reliability for high voltage and high current, and safety are required. Additionally, other aspects including the characteristics against vibration, mechanical shock, and thermal changes should be considered for design, control, and manufacture. Overall, the battery system is an interdisciplinary technology that encompasses mechanical, electrical, electronic, chemical, and materials science and engineering.

The performance requirements for EVs battery systems are listed in order.

1. *Energy density.* Since restrictions on space and weight are stringent for EVs, the battery systems should have a higher energy density (more than 10 times higher) than that of the current secondary batteries. In this regard, LIBs with superior energy density to the others are one of the most promising energy storage devices for EVs and have driven the third boom in the development of EVs. The specific energy or energy density (Wh kg^{-1}, Wh l^{-1}) can be increased by either increasing energy (Wh) or by decreasing weight (kg)/volume (l). As energy (Wh) is the capacity (Ah) times voltage (V), an increment in either capacity or voltage works to increase the specific energy or energy density. The previous approach has been to aim for an increase in either capacity or packing density of the electrode materials. In this approach, a loss of power and cell life is inevitable. Hence the main research direction now is to develop high-voltage positive electrodes and electrolytes with sufficient stability against oxidation.
2. *Safety.* In general, there is a conflict between energy density and safety of batteries. The safety issue is not severe for HEVs, in which the battery loading is not large and the usage of batteries is relatively limited with a shallow charging and discharging. In contrast, PHEVs or BEVs must meet severe safety requirements because of the high energy density and wide charging ranges. In this regard, there should be a compromise between energy density and safety.
3. *Cost.* In practice, cost is the most demanding feature for the widespread adoption of EVs. As the cost of the battery system represents more than half the overall electric power component costs in PHEVs and BEVs, a cost reduction of batteries is required to increase the competitiveness over gasoline-engine vehicles and to expand the EV market. It cannot be overemphasized that battery development roadmaps focus on both cost reductions and energy and power density increases.
4. *Durability and reliability in harsh conditions.* Durability and reliability in harsh conditions such as low or high temperatures, high humidity, and vibration are another important issue for EV batteries. It should be noted that the batteries in EVs are exposed to severer environments or conditions than those for portable electronic devices. Durability also affects the overall TCO (total cost of ownership), which includes driving distance, driving performance, driving range, and after-service cost.
5. *Development of new materials.* Commonly, the performance of batteries is limited by the performance of the constituent materials, such that incremental evolutionary improvements in material technologies are desperately needed. Moreover, it takes time for the development of new materials, including characterization, evaluation, and verification. Hence the material development cannot be achieved without patience and strenuous efforts.

Table 28.3 compares the performance characteristics of secondary batteries for EV application. LIBs are the present power sources for EVs, but their energy density should be increased in order to extend the driving distance. Also, the price should be reduced for the widespread take-up of EVs. Even though lithium–air and lithium–sulfur batteries have attracted a great deal of interest because of their higher theoretical capacities, many technical issues need to be resolved before commercialization.

Table 28.3 Performance characteristics of secondary batteries.

Property	For electric vehicles			For stationary energy storage		
	LIB	Li–S	Li–air	LIB	Redox	Na–S
Theoretical energy density (Wh kg^{-1})	400–600	~2500	~3500	400–600	80–100	700–800
Efficiency (%)	> 90	80–90	70–85	> 90	75–85	85–90
Practical energy density (Wh kg^{-1})	~300 (electrode)	~500 (electrode)	~1000 (electrode)	~250 (electrode)	– (electrode)	~500 (electrode)
	~160 (cell)	~350 (cell)	– (cell)	~150 (cell)	~30 (cell)	~200 (cell)
	< 120 (system)	– (system)	– (system)	< 100 (system)	< 20 (system)	< 150 (system)
Cycle life (year)	~5	–	–	~10	> 10	> 10
Capital cost (US$ kWh^{-1})	> 1000	–	–	700–2000	200–500	300–600
Prospective	–	High theoretical energy density	High theoretical energy density	–	Easy control of capacity and power	Environmental problem
Constraint	High cost Limited driving distance	Low conductivity Polysulfide	Low energy efficiency	High cost	Low efficiency High maintenance cost	High operating temperature Safety

28.4
Secondary Batteries For Energy Storage Systems

The current grid system has limitations on accommodating fluctuations in renewable energy generation. In order to manage and control such unstable energy supplies, an ESS is required, which can stabilize the energy supply by storing the excess energy when production is higher than the need, and supplying extra energy when production is insufficient for the need. An ESS can be utilized for the following purposes:

1. *Load leveling:* Storage of extra energy in the ESS when the load level is low, and release from the ESS when the load level is high and the energy of the main source is insufficient.
2. *Peak shaving:* Reducing the maximum load level of a grid and securing extra energy supply when energy consumption is high.
3. *Emergency power supply:* Protecting expensive manufacturing facilities and supercomputers against any sudden blackout accidents.
4. *Storage of renewable energy:* Stabilizing the energy supply or frequency from fluctuating renewable energy sources such as wind and solar energy.

28.4.1
Lithium-Ion Batteries for ESS

LIBs are now evaluated for megawatt-scale ESSs for frequency control, 3–10 kWh-scale ESSs for houses/industry, and blackout-free energy supply devices for Internet Data Centers (IDCs).

The ESS for frequency control is a high-power system that can charge and discharge in a short period, and can operate without rest. Figure 28.7 illustrates an ESS evaluated by LG Chem. (South Korea) and EKZ (Switzerland). It operates by charge and discharge in ~15 min. As the charge and discharge are made at a similar rate, $LiFePO_4$ is considered to be an appropriate positive electrode since its charge and discharge voltages are comparable.

Some 3–10 kWh house-use energy storage devices have been installed in Germany and the United States (Figure 28.8). The ESS combined with a house-use solar power source can store the extra energy in the daytime and release the stored energy at night. The ESS can also purchase electricity at a time of lower price and consume it at a time of higher price. For this application, not only the price but also the size and life are important, which inevitably require advanced LIBs with a smaller size and longer life.

Figure 28.7 ESS evaluated by LG Chem. (South Korea) and EKZ (Switzerland): 500 kWh LIBs and the cooperation system are placed in the left container, and the energy conversion system in the right container.

Figure 28.8 A prototype house-use ESS developed by LG Chem.

28.4.2
Redox-Flow Batteries for ESS

Since NASA in the United States started a project on Fe–Cr-based redox-flow batteries in 1970s, many efforts have been made to develop redox-flow batteries for ESS. Lately, the US Department of Energy (DOE) has been targeting $ 200–2.00 kW^{-1}, 5 kW vanadium redox-flow battery that is connected to solar power of several kilowatts. Also, an ESS with a $ 2.5 kW^{-1} target is under operation by the Self Generation Incentive Program (SGIP) and California Solar Initiative. ZBB Energy in the United States is developing a Zn–Br redox couple. A prototype Zn–Br redox-flow battery, the ZESS 50, was exhibited during the Beijing Olympics in 2008. In Canada, a vanadium redox-flow battery is being tested by the Distributed Energy Production (DEP) plan under the Technology and Innovation (T&I) Program. Plurion in the United Kingdom first developed a cell that is based on a Zn–Ce redox couple and methanesulfonic acid (MSA).

In Japan, the project Stabilization of Wind Power was launched by NEDO (New Energy and Industrial Technology Development Organization). A vanadium redox-flow battery combined with wind power was built in Hokkaido in 2003 and has been under operation since then. Sumitomo has developed a 5 kW vanadium redox-flow battery stack. VANaSAVER, an electrolyte for the vanadium redox-flow battery, was developed by Ryukyu and is now on the market. Prudent Energy in China reported a prototype 4 kW vanadium redox-flow battery in 2007 and provided 5 kW × 4 h vanadium redox-flow ESS for Kenya, which was followed by building a 20 kW × 9 h vanadium redox-flow ESS at the Telecoms Site in California.

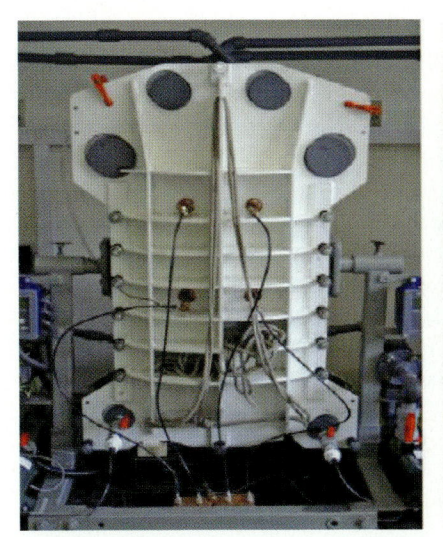

Figure 28.9 The Zn–Ce redox-flow battery developed by Plurion.

28.4.3
Sodium–Sulfur Batteries for ESS

In the 1980s, Yuasa in Japan developed a 300 Ah sodium–sulfur cell for ESS. FACC (Ford Aerospace and Communications Corporation) and GE (General Electric) developed 150, 450, and 1250 Wh sodium–sulfur batteries for load leveling. TEPCO and NGK Insulators developed large-scale sodium–sulfur cell batteries for ESS. In 2003, AEP (America Electronic Power) started a demonstration project where the power system was provided by ABB and sodium–sulfur cell batteries by NGK

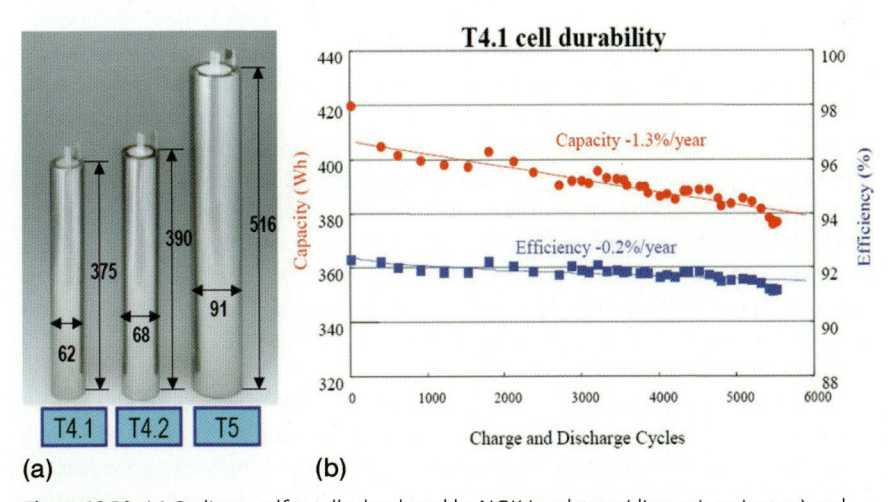

Figure 28.10 (a) Sodium–sulfur cells developed by NGK Insulators (dimensions in mm) and (b) a typical cycle performance.

Insulators. In 2008, a sodium–sulfur battery was employed for stabilizing 34 MW wind power. NGK Insulators, which is one of the leading companies in the field of sodium–sulfur batteries, has a capability to produce 65 MW batteries for ESS at its Komaki factory, where the production of alumina using a kiln furnace, welding by robots, and assembly of the modules are available (Figure 28.10). A 400 Wh sodium–sulfur unit cell fabricated by NGK Insulators exhibited excellent performance: retention of 90% of its initial capacity after 5000 charge–discharge cycles, 1.3% capacity loss per year, and 0.2% efficiency loss per year.

The performance characteristics of secondary batteries for ESS application are compared in Table 28.3. It is expected that LIBs are suitable for small-scale ESSs that require high power. Redox-flow cells and sodium–sulfur batteries would be the right choice for use in large-scale ESSs with a capacity of a few megawatts.

28.5
Conclusion

For small mobile IT device applications, just a few lithium-ion batteries connected in series or in parallel are used. However, EVs and stationary ESSs need large-scale energy storage on the scale from tens of kilowatts to several megawatts. The size of the cells is 20–200 times larger than that used in mobile IT devices, and hundreds or thousands of cells are connected in series or in parallel for typically 300–500 V, 10 kW–2 MW battery pack systems.

Even though lithium-ion batteries have sparked the third boom for EVs offering a higher energy density than other commercial secondary batteries, the existing technology cannot satisfy the requirements for this application. For instance, the EV driving distance is limited to 150–200 km per charge using the common battery packs fabricated with lithium-ion cells, which is far below the level of internal combustion engines that can drive up to 500–1000 km on one fueling. To increase the driving distance, advanced battery technologies are desperately needed, which should have at least 10 times higher energy density than that for the current types. The approach to increase the energy density of lithium-ion cells by either increasing the capacity or packing density shows limitations. The most promising approach seems to be to increase the working voltage of positive electrodes.

Ultimately, lithium–air and lithium–sulfur batteries should be commercialized, in which high-capacity lithium metal negative electrodes and high-capacity positive electrodes are employed. Challenges remain for the sulfur positive electrode and oxygen electrode. The problems associated with dendritic growth of lithium and formation of dead lithium need to be overcome to realize high energy-density secondary battery systems.

From a practical point of view, price reduction is the most demanding requirement for EV and ESS applications. In addition, considering that EV and ESS are a heavy-duty application and the safety characteristics are generally inversely proportional to the energy density, technologies on materials, cell design, pack fabrication and battery management systems need to be developed.

Acknowledgments

S. M. Oh thanks to Professor Kyu T. Lee (UNIST) and Professor Yoon S. Jung (UNIST) for their assistance with preparation of the manuscript.

References

1. IEA (2009) *Technology Roadmap, Electric and Plug-in Hybrid Electric Vehicles (EV/PHEV)*, International Energy Agency, Paris.
2. Ponce de Leon, C., Frias-Ferrer, A, Gonzalez-Garcia, J., Szanto, D. A., and Walsh, F. C. (2006) Redox flow cells for energy conversion. *J. Power Sources*, **160**, 716.
3. State of California Air Resources Board (2009) Attachment A: Status of ZEV Technology Commercialization (Technical Support Document), in *White Paper: Summary of Staff's Preliminary Assessment of the Need for Revisions to the Zero Emission Vehicle Regulation*, State of California Air Resources Board, Sacramento, CA.
4. Bandivadekar, A., Bodek, K., Cheah, L., Evans, C., Groode, T., Heywood, J., Kasseris, E., Kromer, M., and Weiss, M. (2008) *On the Road in 2035: Reducing Transportation's Petroleum Consumption and GHG Emissions*, Report No. LFEE 2008-05, MIT Energy Initiative, Massachusetts Institute of Technology, Cambridge, MA.
5. Lache, R., Galves, D., and Nolan, P. (2008) *Electric Cars: Plugged In, Batteries Must Be Included*, Deutsche Bank, Frankfurt am Main.
6. Kummer, J. T. and Weber, N. (1968) Battery having a molten alkali metal anode and molten sulfur cathode, US Patent 3,413,150.
7. Yu Yao, Y.-F. and Kummer, J. T. (1967) Ion exchange properties of and rates of ionic diffusion in beta-alumina. *J. Inorg. Nucl. Chem.*, **29**, 2453–2475.
8. Christensen, J., Albertus, P., Sanchez-Carrera, R. S., Lohmann, T., Kozinsky, B., Liedtke, R., Ahmed, J., and Kojic, A. (2012) A critical review of Li/air batteries, *J. Electrochem. Soc.*, **159**, R1–R30.
9. Bullis, K. (2009) Revisiting lithium sulfur batteries. advances could at last make the high-energy batteries practical. *MIT Technol. Rev.*, 22 May, http://www.technologyreview.com/energy/22689/?a=f (last accessed 31 January 2013).
10. Aurbach, D., Pollak, E., Elazari, R., Salitra, G., Kelley, C. S., Affinito, J. (2009) On the Surface Chemical Aspects of Very High Energy Density, Rechargeable Li–Sulfur Batteries, *J. Electrochem. Soc.*, **156**, A694–A702.
11. Girishkumar, G., McCloskey, B., Luntz, A. C., Swanson, S., Wilcke, W. (2010) Lithium-air battery: Promise and challenges, *J. Phys. Chem. Lett.*, **1**, 2193–2203.
12. Christensen, J., Albertus, P., Sanchez-Carrera, R. S., Lohmann, T., Kozinsky, B., Liedtke, R., Ahmed, J., Kojic, A. (2012) A critical review of Li/air batteries, *J. Electrochem. Soc.*, **159**, R1–R30.
13. Ahn, M., Cho, Y.-H., Jung, N., Lim, J. W., Kang, Y. S., Sung, Y.-E. (2012) Structural modification of a membrane electrode assembly via a spray coating in PEMFCs, *J. Electrochem. Soc.*, **159**, B145–B149.
14. Shitta-Bey, G. O., Mirzaeian, M., Hall, P. J. (2012) The electrochemical performance of phenol-formaldehyde based activated carbon electrodes for lithium/oxygen batteries, *J. Electrochem. Soc.*, **159**, A315–A320.
15. Peng, Z., Freunberger, S. A., Chen, Y., and Bruce, P. G. (2012) A reversible and higher-rate Li–O_2 battery. *Science*, **337**, 563–566.
16. Ogasawara, T., Débart, A., Holzapfel, M., Novák, P., and Bruce, P. G. (2006) Rechargeable Li_2O_2 electrode for lithium batteries. *J. Am. Chem. Soc.*, **128**, 1390–1393.

17 Jung, K.-N., Lee, J.-I., Im, W. B., Yoon, S., Shin, K.-H., and Lee, J.-W. (2012) Promoting Li_2O_2 oxidation by an $La_{1.7}Ca_{0.3}Ni_{0.75}Cu_{0.25}O_4$ layered perovskite in lithium–oxygen batteries. *Chem. Commun.*, **48**, 9406–9408.

18 McCloskey, B. D., Scheffler, R., Speidel, A., Bethune, D. S., Shelby, R. M., and Luntz, A. C. (2011) On the efficacy of electrocatalysis in non-aqueous $Li–O_2$ batteries. *J. Am. Chem. Soc.*, **133**, 18038–18041.

19 Bryantsev, V. S. and Blanco, M. (2011) Computational study of the mechanisms of superoxide-induced decomposition of organic carbonate-based electrolytes. *J. Phys. Chem. Lett.*, **2**, 379–383.

20 Ryan, K. R., Trahey, L., Ingram, B. J., and Burrell, A. K. (2012) Limited stability of ether-based solvents in lithium–oxygen batteries. *J. Phys. Chem. C.*, **116**, 19724–19728.

21 Lim, H.-D., Park, K.-Y., Gwon, H., Hong, J., Kim, H., and Kang, K. (2012) The potential for long-term operation of a lithium–oxygen battery using a non-carbonate-based electrolyte. *Chem. Commun.*, **48**, 8374–8376.

22 Shimonishi, Y., Zhang, T., Imanishi, N., Im, D., Lee, D. J., Hirano, A., Takeda, Y., Yamamoto, O., and Sammes, N. (2011) A study on lithium/air secondary batteries – stability of the NASICON-type lithium ion conducting solid electrolyte in alkaline aqueous solutions. *J. Power Sources*, **196**, 5128–5132.

23 Adams, J. and Karulkar, M. (2012) Bipolar plate cell design for a lithium air battery. *J. Power Sources*, **199**, 247–255.

29
Pumped Storage Hydropower

Atle Harby, Julian Sauterleute, Magnus Korpås, Ånund Killingtveit, Eivind Solvang, and Torbjørn Nielsen

29.1
Introduction

Hydropower with reservoirs is the only form of renewable energy storage that is well developed and in wide commercial use today. Storing potential energy in water in a reservoir behind a hydropower plant is used for storing energy at multiple time horizons, ranging from hours to several years. Reservoirs for hydropower are very often multi-purpose reservoirs also providing other services such as domestic and industrial water supply, irrigation for agriculture, flood protection, fish farming, and recreational use. As hydropower technology is described separately (Chapter 20), this chapter focuses on the use of pumped storage hydropower plants for multiple time scales of energy storage and balancing services.

29.1.1
Principle and Purpose of Pumped Storage Hydropower

Pumped storage hydropower uses the potential energy in water to produce electricity in *turbine mode*. In *pump mode*, electricity is used to pump water to a higher elevation to store energy as potential energy in water. A pumped storage hydropower plant is attached to an upper reservoir and a lower reservoir by a conduit system consisting of a headrace tunnel and pressure shaft, draft tube, and tailrace tunnel, in the same way as in conventional hydropower (see Chapter 20). One may either install two separate aggregates, one pump and one turbine, or use a machine that runs both ways. These types of hydraulic machinery are called reversible pump turbines (RPTs). The basic principle of pumped storage hydropower plants (often abbreviated to PSH or PSP) is illustrated in Figure 29.1.

The main purpose of PSH is to allow efficient base load generation by covering periods of peak demand and absorbing energy during hours of low demand, in addition to providing ancillary services, such as black start capability and stabilization

Transition to Renewable Energy Systems, 1st Edition. Edited by Detlef Stolten and Viktor Scherer.
© 2013 Wiley-VCH Verlag GmbH & Co. KGaA. Published 2013 by Wiley-VCH Verlag GmbH & Co. KGaA.

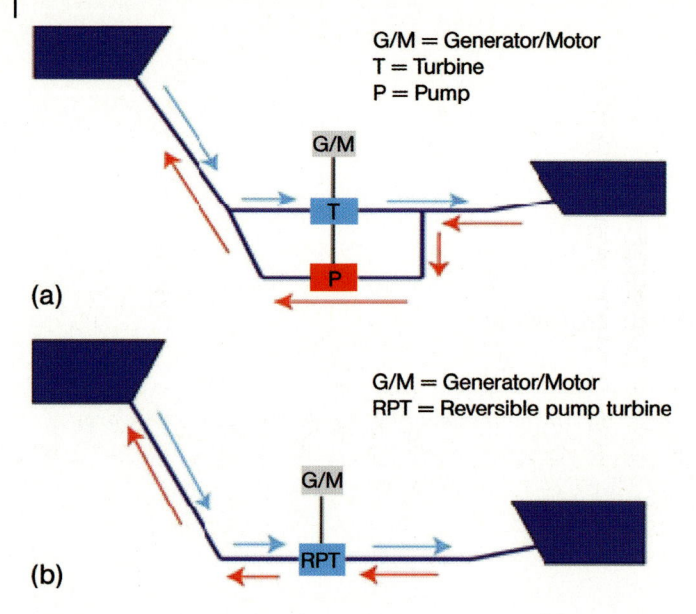

Figure 29.1 Basic principles of pumped storage plant with separate turbine and pump (a) and with reversible pump turbine (RPT) (b).

of the network frequency and voltage level [1]. New commercial and technical interest in PSH has been rising in recent years with political targets on the development of renewable energy sources [2], expected increasing demand for electricity, growing interconnected markets across Europe [3], security of supply, and upgrading of existing plants being the main driving forces [1].

29.1.2
Deployment of Pumped Storage Hydropower

As of today, the world-wide installed PSH capacity is ~130 GW [4], of which ~45 GW is installed in Europe [5], 30 GW in Japan, 24 GW in China and 22 GW in the United States [1]. About 24 GW has been added since 2005, and projections indicate installation of a further 500–600 GW PSH by 2050. The countries with the largest capacities in Europe are Germany, Italy, Spain, France, the United Kingdom, Austria, and Ukraine (Figure 29.2). A typical PSH plant in the current European power system has an installed capacity of 200–300 MW and a relatively short duration for one storage cycle, with 4–9 h generation and 6–12 h pumping. In many PSH stations, the installed capacity for generating electricity is greater than that for pumping the water back into the upper reservoir. In Germany, for instance, the total installed capacity by 2012 was 6.8 GW for generation and 6.45 GW for pumping, and in Switzerland 1.9 and 1.4 GW, respectively. Mostly, the volume of the lower reservoir is smaller than that of the upper reservoir, meaning that the lower reservoir usually limits the amount of water or energy that can be used in the storage cycles.

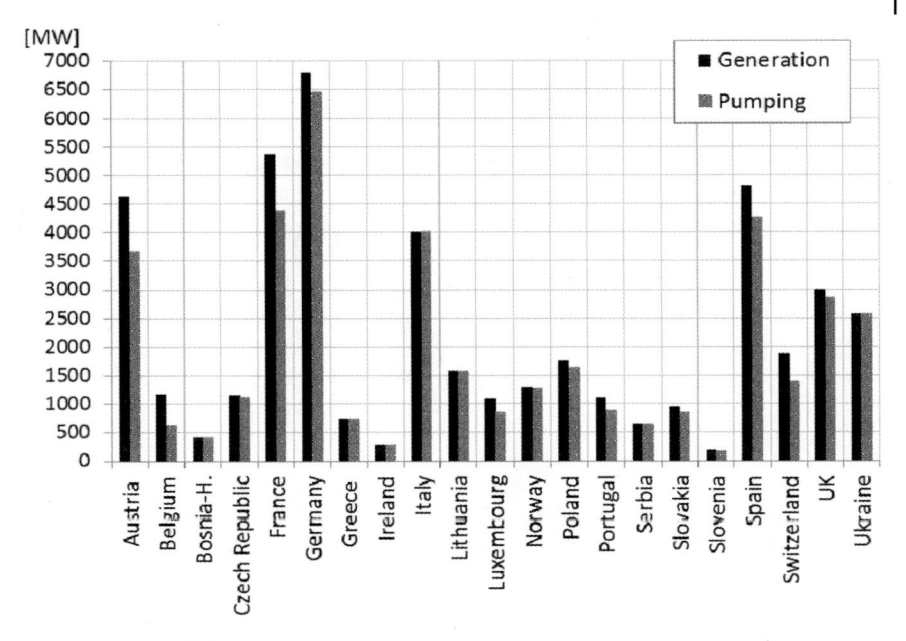

Figure 29.2 Installed PSH capacity in Europe.

The first PSH plants were built in the Alpine regions of Switzerland, Italy, and Austria [1, 6] and in Germany [5]. Most of the plants were constructed in the period between 1960 and 1990, the time when large capacities of conventional power plants were integrated into the energy system.

29.2
Pumped Storage Technology

PSH are characterized by long lifetime expectancy, typically between 50 and 100 years, a round-trip efficiency of 70–85% and a fast response time, usually in the order of seconds or minutes. Pumping and generating can follow a daily cycle, or weekly or even seasonal cycling in large systems. PSH has site-specific and high capital costs and low operating and maintenance costs. The construction time is normally long, up to 10 years. PSH may use existing lakes as reservoirs, but in many cases artificial reservoirs are built.

Reservoirs, waterways, turbines, and generators used in PSH are similar to the technology used for hydropower, described in Chapter 20. Special issues related to pumped storage are described here.

Pumped storage plants can be divided into two main categories (see Figure 29.1): (i) separate turbine and pump and (ii) reversible pump turbine (RPT). The design of PSH is closely linked to the planned operational strategy for each plant. In addition to providing energy storage, PSH also provides services to the grid to maintain frequency and voltage. The frequency alters because the demand for electric power varies. If there is a surplus of generated power, the frequency will increase, and with a lack of generated power, the frequency will decrease. The voltage varies according to the ratio of active and reactive power. The generator must be able to produce exactly the same ratio. Voltage regulation is achieved by adjusting the magnetic field of the generator. In order to regulate, one has to have available generators connected to the grid, even if they are only idling. This is the condensing mode of operation. Many PSH are used both for these ancillary services and to store energy.

The design of PSH is based on more starts and stops and alternating electricity production than conventional hydropower plants. Therefore, it is very important to ensure safe dynamic behavior of the whole system, including waterways, turbine/pump, and generator. On the one hand, the dynamic behavior is connected to the conduit system design and the performance characteristics of the plant. On the other hand, there is a demand to have a machine with stable operation at both low and high loads (i.e., low and high production or low and high flow). Noise, vibrations, and pressure pulsations must be controlled. Instability can cause failure in operation, but also fatigue breakage, with catastrophic consequences.

A reversible pump turbine (Figure 29.3) is a compromise between an optimal pump and an optimal turbine. The design challenge is to achieve a stable pump and an efficient turbine at different loads. Starting up a turbine is fairly conventional.

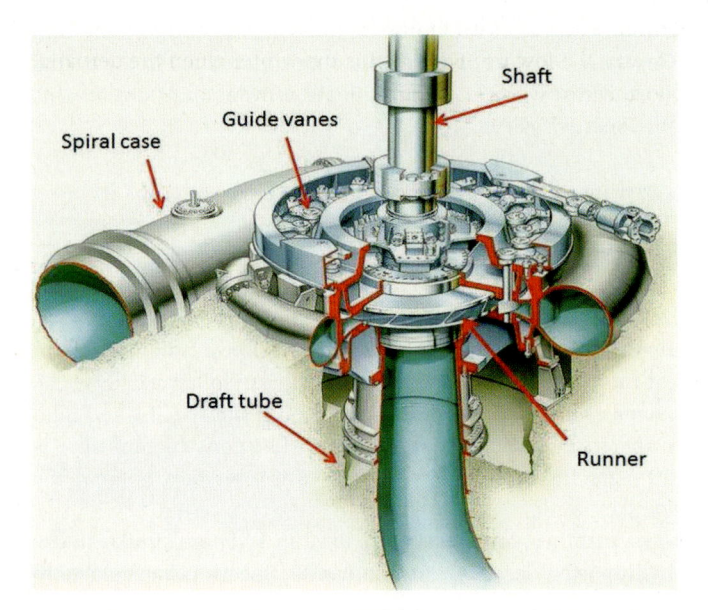

Figure 29.3 A reversible pump turbine where water comes into the machine from the spiral case in the turbine mode and is guided by the guide vanes through the runner, before being discharged through the draft tube. Potential and kinetic energy in the water makes the shaft rotate and is converted to electric energy in a generator connected to the shaft. In the pump mode, this operation is reversed.

By using the wicket gate for regulating power, the machine can be attached to the grid, when stable operation at synchronous speed of rotation is achieved. Starting up a pump is more complicated, and a synchronous machine cannot be connected immediately to the grid because of a tremendous torque. The traditional method to start a pump is then some sort of back-to-back start, that is, using a turbine electrically or mechanically connected to the pump shaft. By starting the turbine, the pump starts to rotate, and when synchronous speed of rotation is achieved, the pump motor is connected to the grid and the turbine is disconnected. In order to reduce the starting torque, it is common to evacuate the water from the pump runner using a compressor and blowing out the water.

29.2.1
Operational Strategies

Most PSH are built to operate for some hours, but they can also participate in regulation at shorter time scales, down to tertiary and secondary regulation providing ancillary services to the grid. However, there are also some PHP or pumps that operate on a seasonal scale, that is, pumping water for weeks and months to higher elevation where larger storage facilities are located. These "seasonal" turbines may also participate in the market for short-term regulation.

In systems such as the Norwegian electricity supply, largely dominated by hydropower (~99% hydro) and with a strong seasonality in both consumption and inflow, there is a strong need to store energy as water from periods of high inflow to periods with low inflow. In Norway, the low inflow period is the winter when the demand for energy is also highest. Figure 29.4 shows the energy content in Norwegian reservoirs during the last 10 years. The seasonal variation is evident, with a maximum

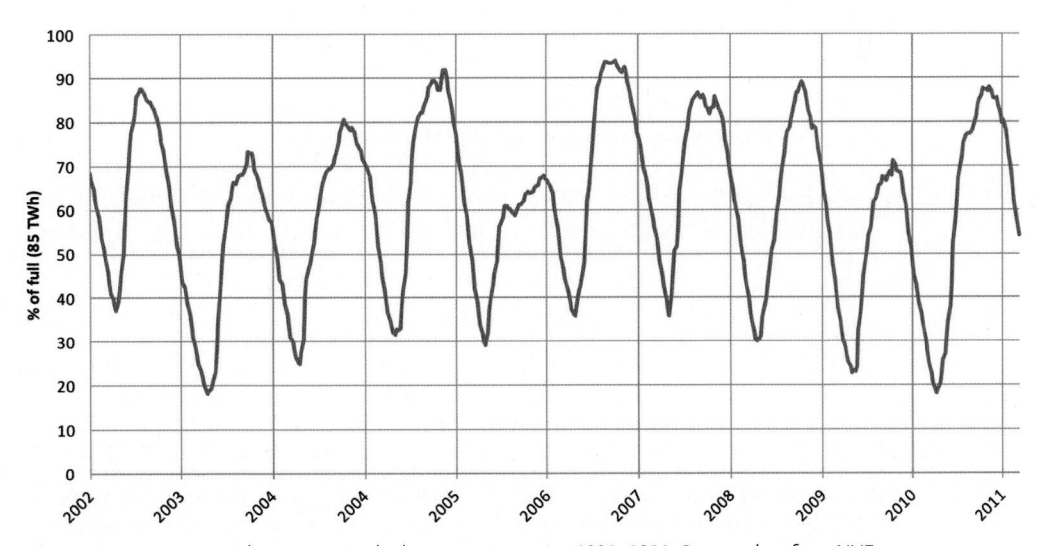

Figure 29.4 Energy stored in Norwegian hydropower reservoirs, 2002–2012. Source: data from NVE.

in late summer and a minimum at the end of winter. It also shows that the sum of all reservoirs always has some free capacity, especially during autumn (fall) and winter. Single reservoirs may have limited capacity, but overall there is always available storage capacity. This capacity may be used by increasing the capacity in existing power plants and by installing PSH connected to these reservoirs (see the case-study in Section 29.5).

29.2.2
Future Pumped Storage Plants

Over the next few years, Europe is likely to experience a strong rise in installed PSH capacity. According to a market analysis by Ecoprog [5], about 27 GW will be installed throughout Europe in a 10 year perspective. Most of this capacity will be developed in Germany, Switzerland, Portugal, Spain, Austria, and the United Kingdom. In these countries, the capacity of projects under construction or undergoing the licensing process is around 4.9, 4.8, 4.0, 2.6, 1.8, and 1.2 GW, respectively. Regarding projects at the early stage of the planning process and project ideas, considerable capacity is likely to be developed in Turkey (3.2 GW), Austria (3 GW), France (3 GW), Switzerland (1.9 GW), Spain (1.7 GW), and Portugal (1.6 GW) [7]. New PSH facilities are mostly upgrades and extensions of current power stations using existing dams and reservoirs. In some cases, dams are enlarged to increase the storage capacity, and in a few cases new facilities including the construction of reservoirs are built (e.g., Atdorf in Germany).

Studies in the United States in the early 1980s indicated a tremendous potential for new PSH, identifying potential sites of more than a total of 1000 GW [8]. Very few sites have been developed, but due to proven technology and low costs compared with alternatives, about 30 GW of new pumped hydro capacity has been proposed in the United States between 2006 and 2009 [8].

29.3
Environmental Impacts of Pumped Storage Hydropower

Regarding environmental impacts, there are two fundamental differences in the magnitude of the environmental effects of PSH projects, depending on the project type:

- creation of new dams and reservoirs for PSH
- PSH construction using existing reservoirs.

For both types, there are environmental impacts related to the construction of power houses, water tunnels, access roads, and power grid connection, in addition to environmental impacts related to the operation of reservoirs and downstream rivers, lakes, or estuaries/fjords/sea. Creation of new dams and reservoirs implies serious interference with Nature in the form of flooding terrestrial areas, land use change, modification of natural stream flow regimes, disruption of the river continuum, and change of terrestrial and aquatic ecosystems. Flooding may involve resettlement of people, loss of biodiversity, and increased greenhouse gas emissions from the new reservoirs. Modified stream flow regimes downstream of dams affect aquatic ecosystems by hydro-morphological changes, alterations in water temperature patterns, habitat changes, disruption of the lateral connectivity, and changes in water quality. These impacts are the same as those occurring during the construction of reservoirs for hydropower or other purposes. This section focuses on impacts related to water bodies when using existing reservoirs to deploy PSH.

Today, state-of-the-art design of PSH facilities is to build many of the structures underground. This requires drilling, blasting, and excavation of tunnels and rock caverns. Access to construction sites and transport of infrastructure have to be ensured by establishing roads. Excavated rock masses need to be deposited. These activities affect the terrestrial ecosystem and biodiversity by area use and landscape fragmentation.

The construction of PSH using existing reservoirs implies both direct/immediate and indirect/long-term abiotic and biotic impacts on the affected reservoirs and downstream rivers. Abiotic factors include changes in water level, water temperature, erosion, circulation patterns, and ice cover. Compared with operational regimes for traditional hydropower, PSH operation introduces greater magnitude, frequency, and rates of variations in water level, water volume, and inundated area in addition to regular emptying and filling. This leads to changes in the hydrodynamics and water quality of reservoirs. Modifications of circulation patterns may occur in the long term, induced by the occurrence of strong currents, particularly in the proximity of turbine outlets [9, 10]. The water column may be less stable, meaning more vertical mixing and alterations in thermal stratification [11, 12], in addition to dissolved oxygen concentration and turbidity [13]. Hence the water quality and temperature in downstream water bodies may be modified. Shoreline erosion may be induced by strong water level fluctuations affecting substrate characteristics and amount, and also re-suspension of nutrients [14]. Moreover, ice formation, ice cover stability, and ice break-up may be altered in lakes with ice cover during winter [15].

These abiotic factors are expected to influence lake organisms and interactions between species [16, 17] by affecting the food web, habitat area, population dynamics, and nutrient level. The timing of the water level fluctuations and also the extent to which the littoral and pelagic zones are reduced is of vital importance for the biological communities in reservoirs. A direct impact is stranding of juvenile fish in littoral zones with low slope during dewatering [18]. Water level fluctuations of tens of meters annually lead to reservoirs with a typically barren shoreline, where both terrestrial and aquatic organisms have little chance of surviving [14], in contrast to the species richness of shorelines of lakes with natural fluctuations. Jonsson and Jonsson [19] studied salmonids and found that water level fluctuations reduce the connectivity of lakes to their tributaries and consequently, since the fish may spawn in tributaries, the accessibility to vital habitats in adjacent streams. Changes in water temperature and water current patterns may lead to a reduction in ice cover, which may affect fish population dynamics [17]. Furthermore, pumping water from a downstream system may increase the nutrient level in upstream reservoirs and change the population dynamics of plankton communities and also, in some cases, increase the fish production [20]. Another significant factor is the possibility of transferring alien species from a downstream reservoir to an upstream reservoir and catchment or to a neighboring catchment if the upstream reservoir is located there.

Reservoirs are most often multi-purpose projects providing water for purposes other than hydroelectric power generation, such as flood protection, irrigation for agriculture, water supply for domestic or industrial use, and recreation. Negative effects of dam constructions have to be seen in relation to positive and desired effects and services.

29.4
Challenges for Research and Development

Recent developments in PSH technology are mainly related to two areas. First, the use of a double-stage regulated pump–turbine allows a very high head to be used for pumped storage, which provides higher energy output and better efficiency than previously. Second, the use of a variable-speed drive allows for frequency regulation in pumping mode, better efficiency, more flexibility, and improved reliability. Research and development to improve the technology will probably still be concentrated around these topics. To improve the operation of PSH, better tools for understanding the energy system and grid are important, in addition to dynamic simulation models for analyzing details in the operation.

The development of PSH in the future is largely connected with the integration of intermittent renewable energy sources such as wind and solar power in the grid. There are great uncertainties about how the differences between electricity consumption and production will be handled. There are many options for balancing the gap between consumption and production, and PSH will certainly contribute to this storage challenge. The main challenges for research and development of PSH are therefore connected with issues about future energy system development.

As PSH have relatively large capital costs (and very low operating costs), the development of business models that include PSH, grid connections, and market models are probably crucial for the future development of PSH. Today's market for balancing the production and demand for energy is concentrated only on short time scales such as within-hour, intra-day, or at most looking one day ahead. To utilize the potential for PSH fully, it is also important to develop business models and markets for longer time horizons, as many PSH will able to participate in multiple markets. A regulatory framework to handle multiple time horizons and multiple markets may be needed.

Other issues of high importance in research and development are connected with environmental impacts of PSH and necessary grid connections, and also societal acceptance. The siting of new PSH is then challenging, and we may see more use of seawater in PSH and underground storage in special cases.

29.5
Case Study: Large-Scale Energy Storage and Balancing from Norwegian Hydropower

In countries with few natural lakes and no available existing reservoirs for PSH, artificial reservoirs are built to serve the PSH. Some countries have large reservoirs and/or lakes used for traditional hydropower production today, and it might be possible to increase the PSH capacity by using existing reservoirs. A case study from Norway shows examples of this, suggesting PSH with storage volumes that could serve storage and balancing for several weeks.

Norway's electricity supply is almost completely based on hydropower. The hydropower system was mainly built in the 1960s, 1970s, and 1980s, and was designed for securing the national electricity supply. Water is stored during the period with high precipitation and run-off, but low electricity demand in spring and summer, to be used during winter, the period with accumulation of snow and little run-off, but high demand. Norwegian hydropower is characterized by a large number of reservoirs, many of which are located in natural lakes in mountainous regions, high storage capacity, and high head. As of 2012, Norway has 1250 hydropower stations with a total of 30.14 GW of installed capacity, a yearly production of 130 TWh, and a storage potential of 84 TWh, which makes up 50% of the total storage capacity in Europe.

In the future, the large storage potential existing in Norway could be used to balance fluctuations in power generation of intermittent renewable energy sources in the European power grid. The intention is to extend existing hydropower stations for the use of pumped storage without constructing new dams or reservoirs. These PSH facilities could be used to back up the electricity production in times with little generation from wind and solar power, and to absorb energy and pump back water when wind and solar power production is high.

The research center CEDREN (www.cedren.no) studied the potential for the deployment of Norway's hydropower for large-scale balancing of intermittent renewable energy sources in a preliminary study focusing on reservoir pairs in south-west Norway as potential sites for PSH development [21]. This study was followed by a more detailed analysis of three cases, aiming at analyzing implications for the operational schemes of the affected reservoirs in addition to current operation, when balancing wind power from the North Sea area. A simulated wind power time series for the North Sea area from the TradeWind project was used for determining the daily required amount of balancing power [22]. Based on time series of stage and live storage volume of the reservoirs, balancing power on a daily basis was simulated on top of the current operation. This was assumed to be realized by installing reversible turbines in addition to the existing power stations. The objectives of this study were to compare the current patterns of water level fluctuations with the simulated patterns (season, frequency, rate of change) and to analyze which factors determine how much power can actually be balanced compared with how much is required to be balanced (turbine capacity, free reservoir volumes). The characteristics of these patterns may be important when studying environmental consequences of providing balancing power and could serve as parameters related to impacts on the ecosystem.

This section presents the potential and limits of balancing power operation by use of existing reservoirs in Norway and points to upcoming environmental challenges related to future reservoir operation.

The case studies do not include topics related to grid connections, costs, business model, regulatory framework, or societal acceptance in details, and these topics were briefly reported by Solvang et al. [21].

29.5.1
Demand for Energy Storage and Balancing Power

The electricity balancing needs as seen from Norwegian hydropower's point of view are expected to be closely related to the variation in wind power production and the demand for electricity in northern and western Europe.

Scenarios for wind power production in Europe are developed in the EU-funded project TradeWind. The variation in wind power production is illustrated here based on TradeWind's medium wind power capacity scenario year 2030. The offshore wind power part of this scenario in 2030 consists of 94.6 GW (94 600 MW) installed capacity in the North Sea in Belgium (3.0 GW), Denmark (5.6 GW), Germany (25.4 MW), the United Kingdom (43.3 MW), The Netherlands (12.0 GW), and Norway (5.4 MW).

Wind speed data from the Reanalysis global weather model, combined with regional wind power curves and wind speed adjustment factors, are used for constructing synthetic wind power time series for specific grid model zones in the TradeWind project. The calculations presented in Section 29.5.4 are based on time series of hourly electricity generation from the 94.6 GW installed capacity in the North Sea in 2030 from TradeWind with weather data for the period 2000–2006.

Figure 29.5 shows the variation in wind power production (megawatts) during three months (January–March) based on average simulated hourly electricity generation for 2001, and demonstrates the need for power balancing. From an average winter output of around 45 000 MW, we can see both rapid and more long-term variability in power output. During two calm periods in February and March, the power output was down to only ~15 000 and 25 000 MW as an average for an entire week. In another one week period in early February, the output was ~75 000 MW for one full week. In some extreme cases, the total output can be well below 10 000 MW for a few days and as low as < 2000 MW for a few hours. In order to balance this system and maintain a steady supply, of for example 45 000 MW, one will need a technology that can provide 30 000 MW extra for a full week, or create a demand of 30 000 MW for another week. The energy storage needed for this balancing will be in the order of 5000 GWh (5 TWh) for each event of 1 week.

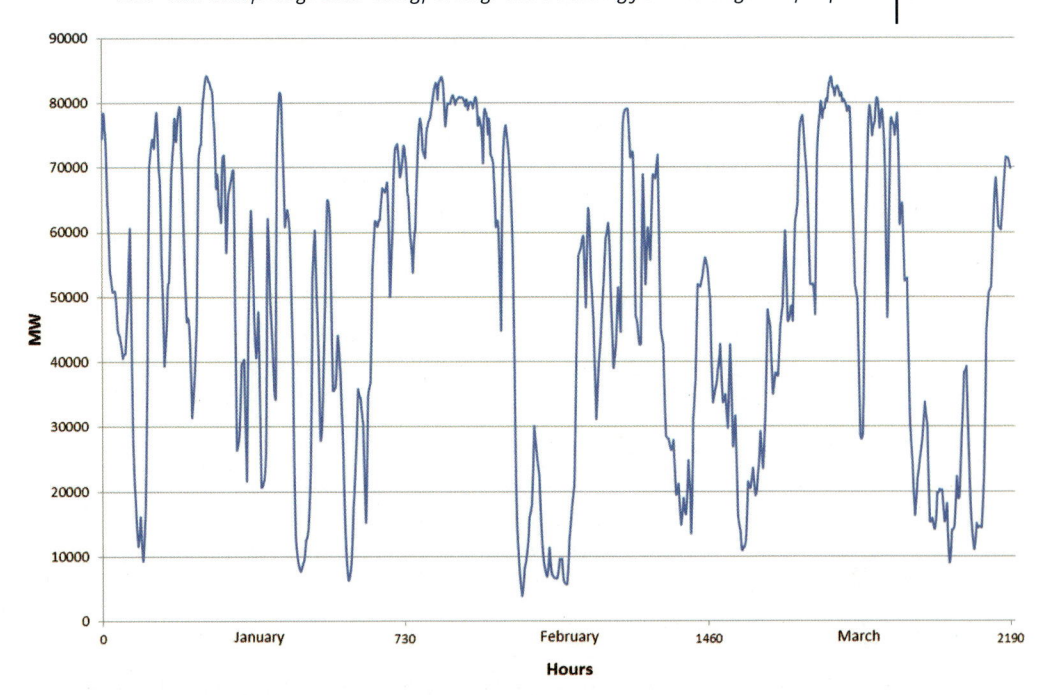

Figure 29.5 Simulated wind power production (MW) in the North Sea area, January–March 2001.

29.5.2
Technical Potential

Table 29.1 shows results from a preliminary study [21] relating to increasing the power output of existing hydroelectric reservoir plants in southern Norway subject to the constraints of current regulations relating to maximum and minimum regulated water levels (HRWL and LRWL). The main scenario involves 12 new power stations with a combined power output of 11 200 MW. It is envisaged that these power stations would be constructed with new tunnels to an upstream reservoir and to the downstream outflow into a reservoir or to the sea. Five of the power stations are pumped storage power stations with a combined output of 5200 MW, while the remainder are conventional hydroelectric power stations with a combined output of 6000 MW, all but one of which (Tyin) discharge into the sea. The pumped storage power stations have reversible pump turbines, pumping water between two reservoirs, while the conventional power stations are not fitted with such pump turbines.

The water level variations in the upper and lower reservoirs include any inflow and discharge resulting from maximum power generation in other power stations associated with the reservoirs in each case.

The power generation outputs (design) were chosen mainly so that the water level change in the upper and lower reservoirs does not exceed 13 cm h^{-1}. For two of the reservoirs (Nesjen and Juklavatn) the rate is 14 cm h^{-1}. According to research into the stranding of salmon in rivers, the water level should not fall by more than 13 cm h^{-1} [23]. Although this is not directly applicable to lakes, this was used as a rule of thumb for acceptable water level reduction in reservoirs.

The output of the 12 power stations in the main scenario can be increased by 18 200 MW without the water level changes in the upper and lower reservoirs exceeding 14 cm h^{-1}. How long the power stations are able to deliver this power output will depend on, among other things, the current regulations regarding HRWL and LRWL, and also what strategies are adopted with regard to pumping in the case of PSH. By including more cases in southern Norway in addition to some in northern Norway, it will be possible to increase the output of existing hydroelectric reservoirs by a further 1800 MW to give a total of 20 000 MW for the whole country.

Table 29.1 New power generation and pump storage hydropower plant (PSH) installations.

Case	Output (MW)	Upper reservoir[a]	Lower reservoir[a]
Tonstad PSP	1400	Nesjen (14 cm h^{-1})	Sirdalsvatn (3 cm h^{-1})
Holen PSP	700	Urarvatn (8 cm h^{-1})	Bossvatn (8 cm h^{-1})
Kvilldal PSP	1400	Blåsjø (7 cm h^{-1})	Suldalsvatn (4 cm h^{-1})
Jøsenfjorden conventional power station	1400	Blåsjø (7 cm h^{-1})	Jøsenfjorden (sea)
Tinnsjø PSP	1000	Møsvatn (2 cm h^{-1})	Tinnsjø (1 cm h^{-1})
Lysebotn conventional power station	1400	Lyngsvatn (9 cm h^{-1})	Lysefjorden (sea)
Mauranger conventional power station	400	Juklavatn (14 cm h^{-1})	Hardangerfjorden (sea)
Oksla conventional power station	700	Ringedalsvatn (12 cm h^{-1})	Hardangerfjorden (sea)
Tysso PSP	700	Langevatn (9 cm h^{-1})	Ringedalsvatn (7 cm h^{-1})
Sy-Sima conventional power station	700	Sysenvatn (9 cm h^{-1})	Hardangerfjorden (sea)
Aurland conventional power station	700	Viddalsvatn (12 cm h^{-1})	Aurlandsfjorden (sea)
Tyin conventional power station	700	Tyin (1 cm h^{-1})	Årdalsvatnet (unknown)

a Water level decrease/increase in parentheses.

29.5.3
Water Level Fluctuations in Reservoirs

The magnitude, frequency, and rate of change of water level fluctuations are case specific, depending on the installed capacity, load scenario, and characteristics of the reservoirs, that is, the live storage volume, how steep or gentle the bank slopes, are and the size of the lower reservoir in proportion to the upper reservoir. Table 29.2 shows the characteristics of the two cases presented in this section.

The seasonal cycle is changed in some cases, but the simulated water levels roughly follow the current course over the years (Figure 29.6). Regarding the upper reservoirs, Rjukan shows only a slight shift with periods of both high stage and low stage having a lower stage in all studied years. In the case of Holen, the seasonal cycle is modified. High stage periods show lower stages than in the current situation and the emptying phase often starts earlier. Regarding the lower reservoirs, Rjukan has a strongly modified pattern with fluctuations occurring constantly. The water level reaches the HRWL more often than in the current situation. In the lower reservoir of Holen, the simulated water level follows approximately the current seasonal course; only during the period of high stage does it deviate, showing more fluctuations up to the HRWL.

Obviously, PSH operation introduces strong daily water level fluctuations in both the upper and lower reservoirs. Compared with the current pattern, they occur more frequently, are greater, and have higher rates of change. In all cases studied here, the percentage of days on which the water level changes in the opposite direction to the day before increases to around 40% (Table 29.3). Rates of change in water level are higher, especially the case Holen, showing a simulated median rate of change in the upper reservoir of 1.17 m d^{-1} compared with 0.08 m d^{-1} under

Table 29.2 Characteristics of the reservoirs involved in the two selected cases.

Characteristic[a]	Case			
	Rjukan: installed capacity 2800 MW		Holen: installed capacity 1400 MW	
	Upper reservoir	Lower reservoir	Upper reservoir	Lower reservoir
Volume (Mm^3)	1064	204	253	296
Area at HRWL (km^2)	78.4	51.4	13.2	7.7
LRWL (m a.s.l.)	919	191	1175	551
HRWL (m a.s.l.)	900	187	1141	495
Difference: HRWL − LRWL (m)	19	4	34	56

a LRWL = lowest regulated water level; HRWL = highest regulated water level; a.s.l. + above sea level.

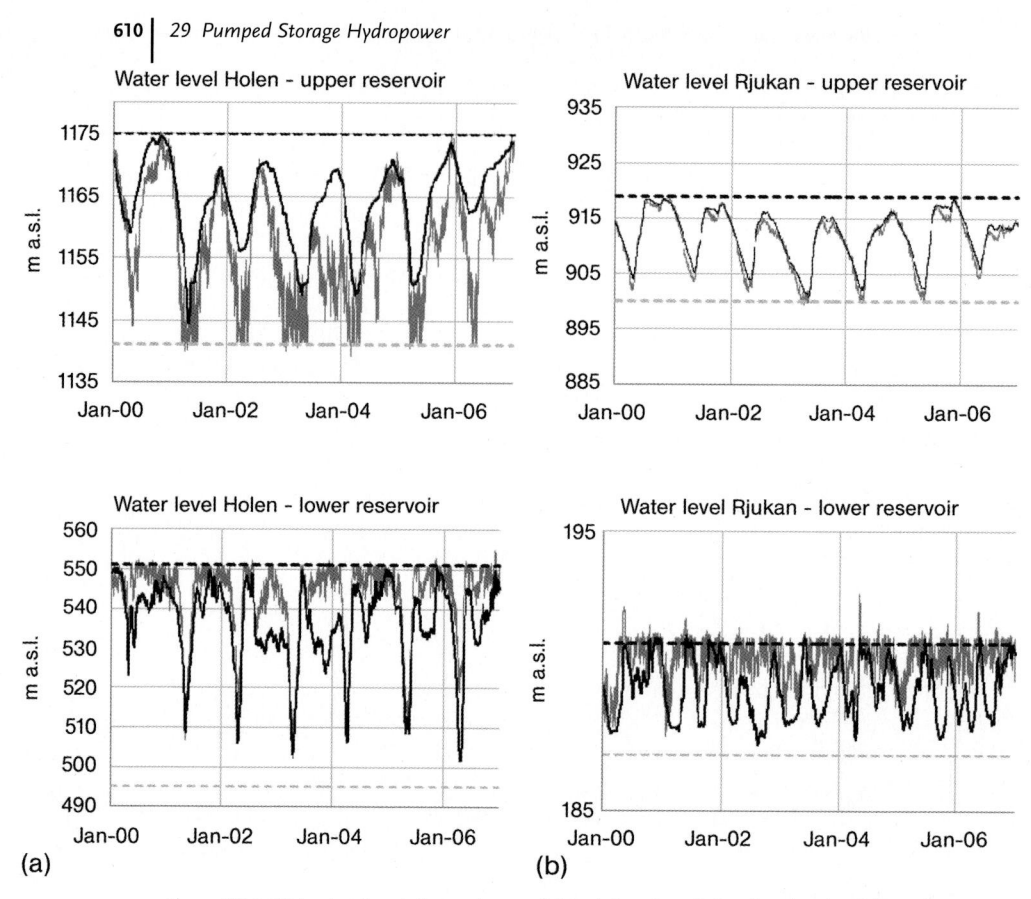

Figure 29.6 Water level variations observed (black lines) and simulated with additional PSH installation (gray lines) in the (a) Holen and (b) Rjukan case studies (a.s.l., above sea level).

current conditions and 1.20 m d^{-1} compared with 0.28 m d^{-1} in the lower reservoir (Table 29.4), while the rise is more moderate in the Rjukan case. This is related to the size of the reservoirs. The Holen case has an upper and a lower reservoir with equal volumes, which allows the transfer of the same amount of water between the reservoirs without any volume limitations, whereas Rjukan has a five times larger upper than lower reservoir. Even though a larger volume of water is transferred in the Rjukan case (double installed capacity compared with Holen), the water level changes are more moderate. This can be explained by the shape of the reservoirs. The lower reservoir of Rjukan has a much larger surface area than the lower Holen reservoir. Analysis of the seasonal pattern in the magnitude of rates of change of water level shows that PSH operation reverses the pattern in the case of Rjukan. Under the current pattern, rates of change are higher during the summer months, whereas they are greater during winter in the simulated pattern.

Table 29.3 Percentage of days on which the water level changes in the opposite direction to the day before in the upper and lower reservoirs of the two cases.

Case	Percentage of days			
	Upper reservoir		Lower reservoir	
	Current	Simulated	Current	Simulated
Rjukan	8.4	38.5	15.5	40.3
Holen	3.9	39.8	17.7	39.9

Table 29.4 Rate of change in water level in the upper and lower reservoirs of the two cases.

Case	Value[a]	Rate of change (m d^{-1})			
		Upper reservoir		Lower reservoir	
		Current	Simulated	Current	Simulated
Rjukan	Median	0.07	0.22	0.04	0.26
	P90	0.15	0.58	0.10	0.62
Holen	Median	0.08	1.17	0.28	1.20
	P90	0.24	3.48	0.94	2.87

a P90 = The 90th percentile of a duration curve of all rates of change.

29.5.4
Environmental Impacts

During the construction phase, interference with Nature will occur to the same extent as it does related to the construction of other hydroelectric facilities (see Chapter 20), except from impacts arising from flooding and the creation of dams, as only existing dams and reservoirs are considered for use. For instance, it will be necessary to deposit the excavated rock masses from tunnel drilling and blasting. These activities affect the terrestrial ecosystem and biodiversity by area use and landscape fragmentation, making it important to pay attention to vulnerable species, such as reindeer and their predators, and landscape types deserving protection.

The strong daily water level fluctuations that PSH operation introduces are likely to change the mixing and current conditions in the affected reservoirs. This may result in altered stratification patterns, that is, the water temperature pattern, affecting the growth of species, lifecycles of organisms, water quality, and ice cover. Reduced ice cover may have effects on fish behavior, leading to increased energy expenditure and reduced winter survival, but this has to be studied in further detail.

More unstable ice will limit recreational use of the reservoirs, for example, for game, fishing, or skiing. Another consequence of water level fluctuations is an increased risk of bank erosion, caused by relatively rapid changes in pore water pressure.

In the studied cases, the average rates of change in water level are obviously higher than during the current operation, but they are still below the range of critical rates as defined related to the stranding of fish in running waters [23, 24]. The change in the seasonal pattern of the water level fluctuations suggests the potential necessity to restrict the rate of changes according to seasonal requirements, such as limiting the magnitude of the water level changes during winter/spring for the purpose of recreation or adjusting thresholds in accordance with seasonal requirements of aquatic and terrestrial species.

29.6
System Analysis of Linking Wind and Flexible Hydropower

Several studies have shown that hydro reservoirs are suitable as a means for storing wind energy in areas with limited power transfer capability [25–28]. This is especially relevant in mountain regions where reservoir hydro plants have already been developed to some extent, and there is high potential for wind power due to the large available areas and usually excellent wind conditions. The coordination idea is simple: when wind speeds are high, the hydropower production is reduced or, if possible, reversed to pumping, in order to keep the power transfer out of the area within acceptable operational limits. At a later stage, when the wind has calmed, more water is released from the reservoir. The flexibility of the hydro reservoirs can be utilized either by direct coordination or by a market-based approach. In this section, we present a coordination approach for wind and hydropower control, based on AGC (Automatic Generation Control), and applied to a simulation case study for northern Norway.

29.6.1
Method

The studied system shown in Figure 29.7 is part of the 132 kV regional power system in northern Norway with an assumed total export capacity southwards of 270 MW. The grid capacity is represented for simplicity as a constant value for transmission of active power, although the actual capacity depends on several other factors such as reactive power transfer, voltage stability margins, and steady-state voltage limits [25]. Several hydropower plants are connected to the regional grid; one of them is the Goulas power plant, which is considered in this study for wind and hydro coordination.

Currently, two wind farms with a total installed capacity of 80 MW are connected to the regional grid. However, many more wind farms are being planned, and the potential is huge. Because of the relatively weak connection to the rest of the national grid, the main issue is to establish how much wind power it is viable to

29.6 System Analysis of Linking Wind and Flexible Hydropower | 613

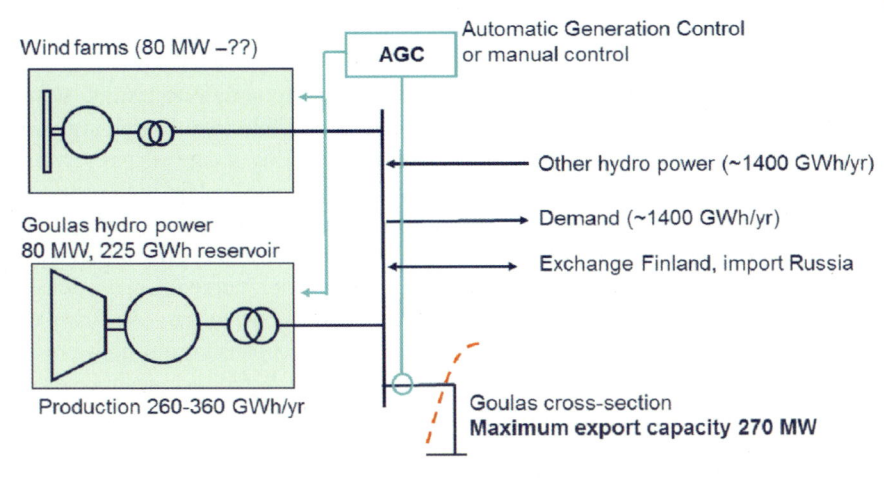

Figure 29.7 Overview of the case study power system. The regional grid is connected to the 420 kV national grid via a corridor of several 132 kV lines. Automatic Generation Control (AGC) is utilized for keeping the power transmission below the maximum export capacity of 270 MW.

install without extensive grid expansions, when taking into account the flexibility of one of the hydropower plants that are installed in the region.

As shown in Figure 29.7, AGC [25] is considered for keeping the power transmission below the maximum export capacity of 270 MW. The AGC is assumed to be applied to two different control strategies [25]:

- "Control wind": the power output of the wind farms is constrained if required. The hydropower plant is operated according to a generation schedule which is unaffected by the wind power output.
- "Control hydro": first, the output of the hydropower plant is decreased as much as possible to prevent overloading of the grid. If this is not sufficient, the wind power output is constrained as for "control wind." The hydropower is increased above the generation schedule at a later stage to keep the annual hydro generation as close to the schedule as possible.

29.6.2
Results

The system shown in Figure 29.7 was simulated using a power system model implemented in MATLAB [26] with hourly resolution and time-series input from the hydrological and meteorological years 2003–2007.

The first simulated example is an approximation of the regional system as it is today with 80 MW wind power installed. As can be seen from the duration curve in Figure 29.8a, the power export is always below the maximum limit of 270 MW, meaning that no additional power control is needed. In Figure 29.8b, the same system is simulated with 220 MW new wind power installed. The power export now reaches the maximum limit about 25% of the time, forcing wind turbine shut-off or wind production reduction to prevent line overloading. This results in about 15% lost wind energy compared with a situation without grid constraints. With coordinated hydro control, on the other hand, it is possible to store the surplus wind energy with negligible wind energy losses and flooding of the hydro reservoir. As a consequence, the full operating range of the hydropower plant is used more extensively, with faster variations in power output, more production at rated power, and more often start–stop operation in order to follow the wind power variations. This is visualized in Figure 29.9, which shows the duration curve for hydropower output with and without coordinated control, for a simulation setup with 300 MW wind power installed. With coordinated control, the production is at rated power over 20% of the time, while the total time with full stop of the hydro turbines increases from 18 to 30% compared with the original production plan.

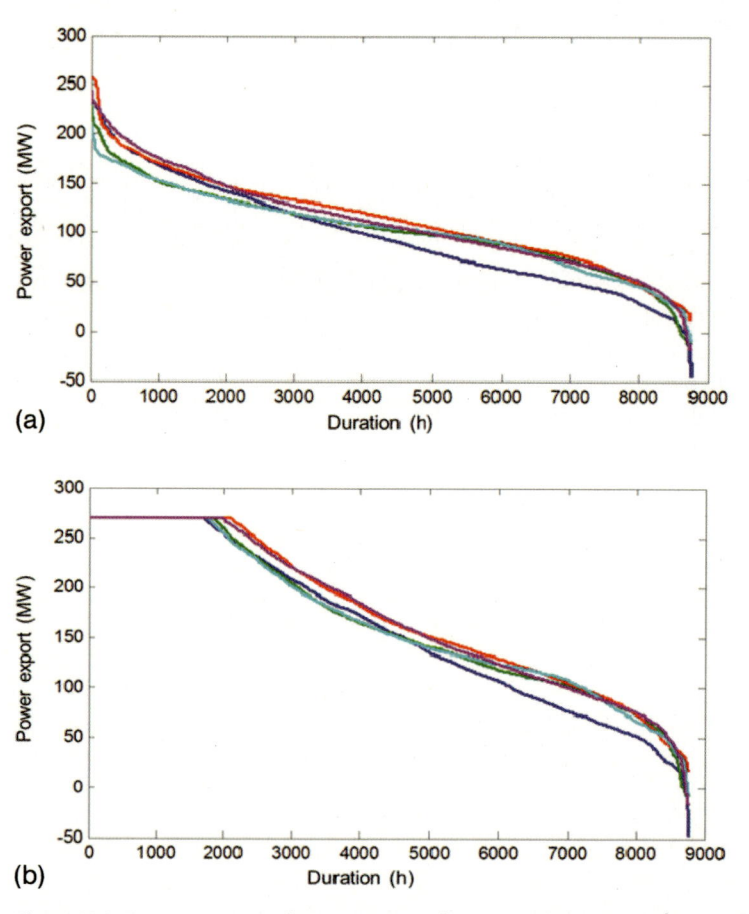

Figure 29.8 Duration curves of power export with (a) 80 and (b) 300 MW installed wind power. The different graphs represent the different simulated hydrological and meteorological years 2003–2007, which are used for inflow and wind speed.

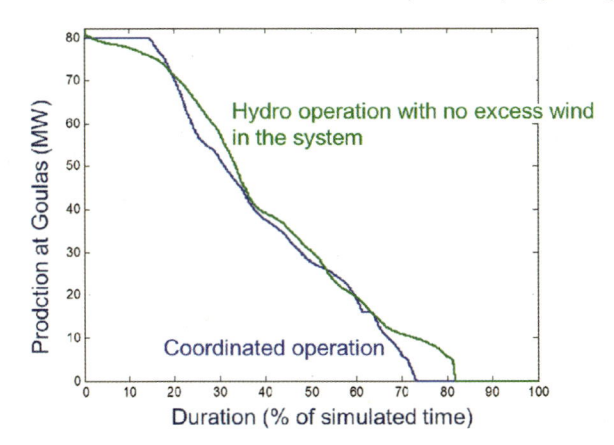

29.6 System Analysis of Linking Wind and Flexible Hydropower | 615

Figure 29.9 Duration curve for simulated hydropower production for a case with 300 MW wind power installed in the region.

To quantify the benefits of operating the Goulas hydropower plant in a flexible way, subsequent simulations were run for increasing wind power capacity up to 400 MW, in steps of 80 MW. Figure 29.10 shows the average power export as a function of installed wind power, with and without coordinated control. In the ideal case, without any power transfer limits (black line), the relation between power export and installed wind power is linear, that is, a steady increase in power export as more wind power is installed. The red curve starts to deviate from the black curve at around 200 MW wind power. This is where energy losses due to grid limitations start to become evident with no additional hydro control. This situation occurs first at 250 MW with hydro control (blue curve), which shows the significant benefits of using the flexibility of the hydropower plant to increase the utilization of existing grid capacity in areas with high wind potential.

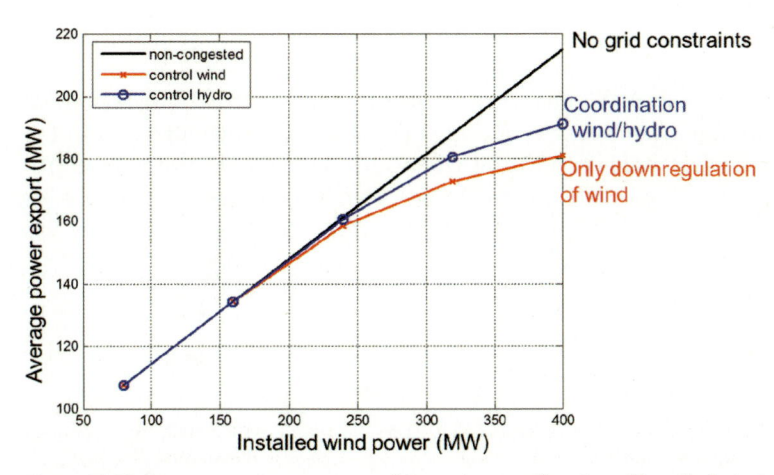

Figure 29.10 Average power export out of the region as a function of installed wind power.

29.7
Conclusion

Hydropower is the major renewable source for electricity generation worldwide and will remain so for a long time [29]. PSH are also currently the only emission-free available technology to store energy for large power output over time horizons longer than minutes. Currently, the total worldwide installed PSH capacity is ~130 GW. While we can foresee a doubling of global hydropower capacity up to almost 2000 GW and 7000 TWh by 2050, we may see PSH capacities multiplied by a factor of 3–5 [28].

The technology for PSH is the same as for conventional hydropower, but it includes either an RPT or a combination of turbine and pump, to be able to pump water to higher elevation for later electricity generation. PSH may provide balancing and storage at multiple time horizons, ranging from minutes to months.

PSH is a well-known technology, but there is a need for improvements related to improving the efficiency and flexibility, which has increased importance in the future when large amounts of intermittent renewable energy are going to be integrated in the grid. There are also research needs connected with the development of business models, regulatory frameworks, grid integration, environmental impacts, and societal acceptance of PSH.

Environmental impacts of PSH are similar to the impacts from conventional hydropower. In addition, PSH may threaten biodiversity by transferring water upstream and to neighboring catchments and may also have special effects in reservoirs and partly in rivers as water levels will fluctuate more frequently and potentially at a higher rate. This may impact both physical processes and ecosystem responses.

The results of the case study on Norwegian hydropower for large-scale balancing and storage simulations show that the reservoirs are not often completely filled or emptied, that is, the available storage volume is not completely used. It is possible to build new hydropower and pumped storage plants with a total capacity of nearly 20 000 MW in southern Norway. This will have small environmental impacts, as only existing reservoirs are used. However, the local impacts may be large in some cases, and detailed analysis must be carried out to ensure optimum environmental design.

To utilize fully the wind power production from remote areas, a strong grid connection is necessary. Simulations in a test case in northern Norway showed that there are significant benefits of using the flexibility of a hydropower plant to increase the utilization of existing grid capacity in areas with high wind potential. Less energy will be lost due to overproduction of wind power, grid constraints, or flooding of the hydro reservoir. This also shows how important it is to study and implement renewable energy production in an integrated way.

References

1. Deane, J. P., Gallachoir, B. P., and McKeogh, E. J. (2010) Techno-economic review of existing and new pumped hydro energy storage plant. *Renew. Sustain. Energy Rev.*, **12**, 1293–1302.
2. European Commission (2009) *Directive 2009/28/EC of the European Parliament and of the Council of 23 April 2009 on the promotion of the use of energy from renewable sources and subsequently repealing Directives 2001/77/EC and 2003/30/EC*, European Commission, Brussels.
3. UCTE (2007) *Final Report: System Disturbance on 4 November 2006*, Union for the Coordination of Transmission of Electricity, Brussels.
4. Ingram, E. A. (2010) Worldwide pumped storage activity, *Renewable Energy World*, http://www.renewableenergyworld.com/rea/news/article/2010/10/worldwide-pumped-storage-activity (accessed 12/03/2013)
5. Ecoprog (2011) *The European Market for Pumped-Storage Power Plants*, Ecoprog, Cologne, http://www.ecoprog.com/en/publications/energy-industry/pumped-storage-power-plants.htm (last accessed 31 January 2013).
6. Chen, H., Cong, T. N., Yang, W., Tan, C., Li, Y., and Ding, Y. (2009) Progress in electrical energy storage system: a critical review. *Prog. Nat. Sci.*, **19** (3), 291–312.
7. Eurelectric (2012) *Hydro in Europe: Powering Renewables. Full Report*, Eurelectric Renewables Action Plan, Eurelectric, Brussels.
8. Yang, C. and Williams, E. (2009) *Energy Storage for Low-Carbon Electricity*, Policy Brief, Climate Change Policy Partnership, Duke University, Durham, NC, USA.
9. Anderson, M. A. (2006) *Analysis of the Potential Water Quality Impacts of LEAPS on Lake Elsinore*, Department of Environmental Sciences, University of California Riverside, Riverside, CA.
10. Gailiusis, B. (2003) Modelling the effect of the hydroelectric pumped storage plant on hydrodynamic regime of the Kaunas Reservoir in Lithuania. *Nordic Hydrol.*, 34 (9), 507–518.
11. Anderson, M. A. (2010) Influence of pumped-storage hydroelectric plant operation on a shallow polymictic lake: predictions from 3-D hydrodynamic modeling. *Lake Reserv. Manage.*, **26**, 1–13.
12. Potter, D. U., Stevens, M. P., and Meyer, J. L. (1982) Changes in physical and chemical variables in a new reservoir due to pumped storage operations. *J. Am. Water Resour. Assoc.*, **18**, 627–633.
13. Bonalumi, M., Anselmetti, F. S., Kägi, R., and Wüest, A. (2011) Particle dynamics in high-Alpine proglacial reservoirs modified by pumped-storage operation. *Water Resour. Res.*, **47**, W09523.
14. Zohary, T. and Ostrovsky, I. (2011) Ecological impacts of excessive water level fluctuations in stratified freshwater lakes. *Inland Waters*, **1**, 47–59.
15. Liu, L. X. and Wu, J. C. (1999) Research on ice formation during winter operation for a pumped storage station. *Ice Surf. Waters*, **2**, 753–759.
16. Stanford, J. A. and Hauer, F. R. (1992) Mitigating the impacts of stream and lake regulation in the Flathead River catchment, Montana, USA – an ecosystem perspective. *Aquat. Cons. Mar. Freshwater Ecosyst.*, **2** (1), 35–63.
17. Helland, I., Finstad, A. G., Forseth, T., Hesthagen, T. and Ugedal, O. (2011) Ice-cover effects on competitive interactions between two fish species. *J. Anim. Ecol.*, **80**, 539–547.
18. Bell, E., Kramer, S., Zajanc, D., and Aspittle, A. (2008) Salmonid fry stranding mortality associated with daily water level fluctuations in Trail Bridge Reservoir, Oregon. *N. Am. J. Fish. Manage.*, **28**, 1515–1528.
19. Jonsson, B. and Jonsson, N (2011) *Ecology of Atlantic Salmon and Brown Trout: Habitat as a Template for Life Histories*, Fish & Fisheries Series 33, Springer, Dordrecht.

20 Stockner, J. G. and Macisaac, E. A. (1996) British Columbia lake enrichment programme: two decades of habitat enhancement for Sockeye salmon. *Regul. Rivers Res. Manage.*, **12**, 547–561.

21 Solvang, E., Harby, A., and Killingtveit, Å. (2012) *Increasing Balance Power Capacity in Norwegian Hydroelectric Power Stations. A Preliminary Study of Specific Cases in Southern Norway*, CEDREN, SINTEF Energy Report TR A7126, SINTEF Energi, Trondheim.

22 Tande, J. O.G, Korpås, M., Warland, L, Uhlen, K., and Van Hulle, F. (2008) Impact of TradeWind offshore wind power capacity scenarios on power flows in the European HV network, presented at the 7th International Workshop on Large Scale Integration of Wind Power and on Transmission Networks for Offshore Wind Farms, Madrid, 26–27 May 2008.

23 Saltveit, S. J., Halleraker, J. H., Arnekleiv, J. V., and Harby, A. (2001) Field experiments on stranding in juvenile Atlantic salmon (*Salmo salar*) and brown trout (*Salmo trutta*) during rapid flow decreases caused by hydropeaking. *Regul. Rivers Res. Manage.*, **17**, 609–622.

24 Halleraker, J. H., Saltveit, S. J., Harby, A., Arnekleiv, J. V., Fjeldstad, H.-P., and Kohler, B. (2003) Factors influencing stranding of wild juvenile brown trout (*Salmo trutta*) during rapid and frequent flow decreases in an artificial stream. *River Res. Appl.*, **19**, 589–603.

25 Tande, J. O. G. and Uhlen, K. (2004) Cost analysis case study of grid integration of larger wind farms. *Wind Eng.*, **28** (3), 265–273.

26 Tande, J. O. G., Korpås, M., and Uhlen, K. (2012) Planning and operation of large offshore wind farms in areas with limited power transfer capacity. *Wind Eng.*, **36** (1), 69–80.

27 Matevosyan, J. and Söder, L. (2007) Short-term hydropower planning coordinated with wind power in areas with congestion problems. *Wind Energy*, **10** (3), 195–208.

28 Acker, T., Robitaille, A., Holttinen, H., Piekutowski, M., and Tande, J. O. G. (2012) Integration of wind and hydropower systems: results of IEA Wind Task 24. *Wind Eng.*, **36** (1), 1–18.

29 IEA (2012) *Technology Roadmap Hydropower*, International Energy Agency, Paris.

30
Chemical Storage of Renewable Electricity via Hydrogen – Principles and Hydrocarbon Fuels as an Example

Georg Schaub, Hilko Eilers, and Maria Iglesias González

30.1
Integration of Electricity in Chemical Fuel Production

Renewable electricity from fluctuating sources (wind, solar) has shown significant increases in the recent past worldwide (e.g., in Germany 19.9% of total electricity generation in 2011 versus 6.6% in 2001 [1]). Fluctuations include day–night cycles for solar energy and noncyclic changes of wind speed, the latter with high rates of change and potentially longer periods of zero input [2]. In connection with the limited capacities of electrical distribution grids, this will lead to low-value or cheap electricity during time periods of high generation and low demand. In these excess situations, the electric energy should be converted into storable forms of energy. A variety of storage technologies are either proven (e.g., conversion to gravitational energy such as pumped hydropower) or at some stage of development (pressure–volume energy of compressed air, electrochemical storage, conversion to chemical energy, H_2 via electrolysis, and others). Chemical fuels in general are seen as important options for medium- and long-term storage systems [3]. A general flow diagram for electricity-to-fuel conversion is shown in Figure 30.1.

Today, chemical energy carriers or fuels are being produced based on various organic raw materials,, for example, via refining of petroleum, cleaning of natural gas, and conversion of coal or biomass into solid, liquid, or gaseous fuels. The respective infrastructures and utilization are highly developed. With respect to

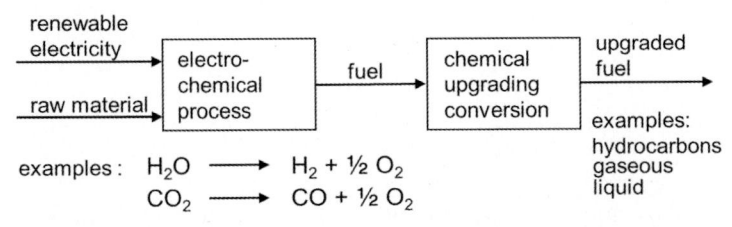

Figure 30.1 General flow diagram of electricity conversion to chemical fuels.

Transition to Renewable Energy Systems, 1st Edition. Edited by Detlef Stolten and Viktor Scherer.
© 2013 Wiley-VCH Verlag GmbH & Co. KGaA. Published 2013 by Wiley-VCH Verlag GmbH & Co. KGaA.

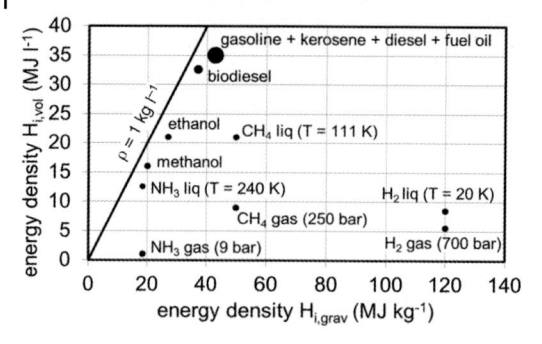

Figure 30.2 Energy density of selected compounds, based on reaction with O_2 (298 K if not specified otherwise) [6].

potential future fuels made from renewable electricity, the selection of preferred energy carriers will be based on criteria such as (i) fuel properties with respect to present infrastructures [e.g., energy density (Figure 30.2), combustion, and handling properties], (ii) production cost, depending on availability and price of raw materials and the required process cost for transformation/conversion, and (iii) environmental aspects (e.g., generation of pollutants and greenhouse gases during transformation and combustion). It is currently an open question which kind of fuels may become most significant in the future [4, 5].

Electrolysis of water is a proven electrochemical technology for the production of hydrogen on a small scale. There are different kinds of electrochemical cells, based on different kinds of electrolyte and electrodes [alkaline, proton exchange membrane (PEM), solid oxide]. They have different characteristics with respect to module capacity, efficiency, stability, and dynamic operation. Although all of these currently have high specific investment figures, there are positive expectations for cost reduction, high rates of load change, and operability [7]. Upgrading or conversion of hydrogen to upgraded fuels will be required according to the criteria mentioned earlier, for example, for achieving desired fuel properties. Liquid or gaseous hydrocarbons are examples where hydrogen can be integrated in chemical upgrading processes leading to an overall electricity-to-fuel process (Figure 30.3).

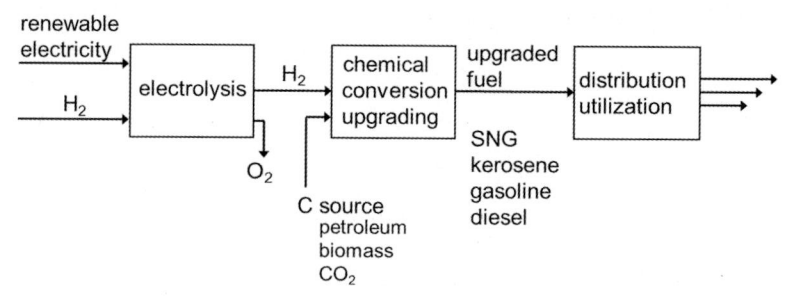

Figure 30.3 Example: flow diagram for hydrocarbon fuels (gaseous or liquid) from renewable electricity via hydrogen. SNG, substitute natural gas.

In the case of substitute natural gas (SNG) (based on methane) or liquid hydrocarbon fuels (kerosene, gasoline, diesel), integration in present infrastructures would be easy. However, any upgrading process has internal energy demands, so for energy efficiency reasons direct utilization of hydrogen would be preferable.

30.2
Example: Hydrocarbon Fuels

30.2.1
Hydrocarbon Fuels Today

Hydrocarbon fuels such as petroleum products or natural gas currently contribute significant amounts to the secondary energy supply in industrialized countries (in Germany about 60%; see Table 30.1, right). Liquid hydrocarbon fuels exhibit the highest energy densities in the form of gasoline, kerosene, diesel fuel, and fuel oil. This is significant in particular for mobile applications, for example, in automobiles and airplanes. In the case of natural gas, there are two major advantages: no significant processing of the natural feedstock is required before distribution and utilization, and the distribution infrastructures in gas grids help in transport to final consumers and in storage (via variations in pressure). Figure 30.2 shows energy densities related to volume and weight as an advantage of chemical energy carriers in general and of hydrocarbons in particular, which is due to the high reaction enthalpy of combustion. Electric batteries currently exhibit lower volume densities by a factor of about 40.

The following discussion presents the most prominent hydrocarbon fuel upgrading/production pathways and the stoichiometric potential to include hydrogen made from excess electricity, using different carbon sources. The integration of renewable H_2 may have two main advantages: replacing fossil resources and/or increasing hydrocarbon yield per carbon feedstock (both leading to lower fossil CO_2 emissions).

Table 30.1 Structure of primary energy supply and end energy carriers, Germany 2010 [1].

Primary energy[a]	Proportion (%)	Secondary energy	Proportion (%)
Petroleum	33	Petroleum products	37[b]
Natural gas	23	Natural gas	23
Coal	23	Electricity	21
Nuclear	11	Renewables, coal products, district heat	17
Renewables	10		

a ~5.4 kW per capita.
b 26% for mobility/transport.

30.2.2
Hydrogen Demand in Hydrocarbon Fuel Upgrading/Production

An overview of H_2 demand values for process pathways/raw materials to produce liquid hydrocarbon fuels by hydrogenating various carbon sources is given in Figure 30.4. Calculation of stoichiometric H_2 demands for the production of $-(CH_2)-$ is based on Eq. 30.1, which implies complete conversion of the feedstock carbon into high-value hydrocarbons and complete conversion of oxygen into water. This is equivalent to maximizing the carbon yield (per feedstock). There may be strategies with lower H_2 demand, converting the feedstock oxygen into CO_2 (Eq. 30.2). The stoichiometric figures shown are considered as characteristic values that may differ in industrial applications. In the case of petroleum refining, there is always the need for hydrodesulfurization (Eq. 30.3), given the strict environmental regulations with respect to emission control during utilization.

$$CH_xO_y + (1 + y - x/2) H_2 \rightarrow -(CH_2)- + y\, H_2O \qquad (30.1)$$

$$CH_xO_y + (1 - y/2 - x/2) H_2 \rightarrow (1 - y/2) - (CH_2)- + y/2\, CO_2 \qquad (30.2)$$

$$CH_xS_z + (1 + z - x/2) H_2 \rightarrow -(CH_2)- + z\, H_2S \qquad (30.3)$$

It can be seen from Figure 30.4 that, with the exception of natural gas as feedstock, hydrogen is always needed to produce $-(CH_2)-$ hydrocarbons. H_2 demand increases with increasing oxidation value of the carbon source, defined as $0.5\,y - 0.25\,x$ for CH_xO_y. It is evident that petroleum and vegetable oil have the lowest stoichiometric H_2 demand. Molecular compositions are close to the desired $-(CH_2)-$ hydrocarbon product, hence they are the most advantageous feedstocks for liquid hydrocarbon production from a stoichiometric viewpoint.

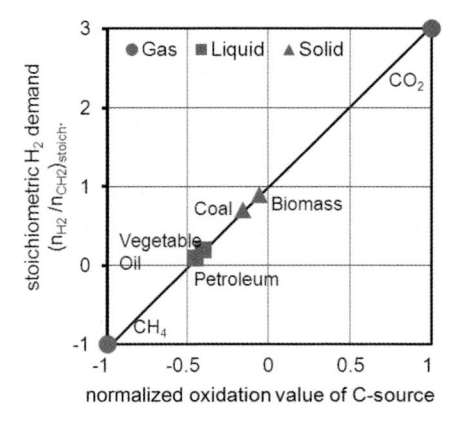

Figure 30.4 Stoichiometric H_2 demand for the production of liquid hydrocarbons (as $-CH_2-$), depending on carbon source (elementary composition: CH_xO_y), normalized oxidation value of carbon source ($OV = +0.5\,y - 0.25\,x$).

30.2.3
Hydrogen in Petroleum Refining

Petroleum refining has developed in the past from simple fractionation according to boiling temperature to chemical upgrading of individual boiling fractions [8, 9]. Some of these upgrading processes require the addition of hydrogen (hydrodesulfurization, hydrocracking). Liquid and gaseous distillation products, which constitute about 65–85% of crude oil, need to be desulfurized to reach today's allowed sulfur contents. The H_2 generated in the refinery during catalytic reforming of straight-run gasoline (Figure 30.5), which constitutes only about 20–30% of crude oil, is not sufficient to remove the sulfur present in the distillation products. Conventional refining can therefore be seen as a redistribution of H_2 available in the raw material crude oil into the different product fractions. Today, in spite of efficient internal H_2 management, there is need to generate additional H_2 [10], commonly done by steam reforming of imported natural gas or gasification of distillation residues (Eqs. 30.4 and 30.5).

$$CH_4 + 2\,H_2O \rightarrow 4\,H_2 + CO_2 \tag{30.4}$$

$$CH_{\leq 1.4} + O_2/H_2O \rightarrow <2.7\,H_2 + CO_2 \tag{30.5}$$

In the case of heavy crude oil or bitumen from tar sands, there is less hydrogen and generally more sulfur and oxygen present in the feedstock. Increased H_2 content in the products can be achieved with two different strategies [11]: carbon removal via a coking process or hydrogen addition via a hydrogenation process (Eqs. 30.6 and 30.7, example stoichiometries). As an advantage of hydrogenation, higher yields per feedstock can be achieved.

C removal:

$$CH_{1.4}O_{0.05}S_{0.02} + 0.01\,H_2 \rightarrow 0.71\,CH_{1.8} + 0.29\,C + 0.05\,H_2O + 0.02\,H_2S \tag{30.6}$$

H_2 addition:

$$CH_{1.4}O_{0.05}S_{0.02} + 0.23\,H_2 \rightarrow 0.95\,CH_{1.8} + 0.05\,C + 0.05\,H_2O + 0.02\,H_2S \tag{30.7}$$

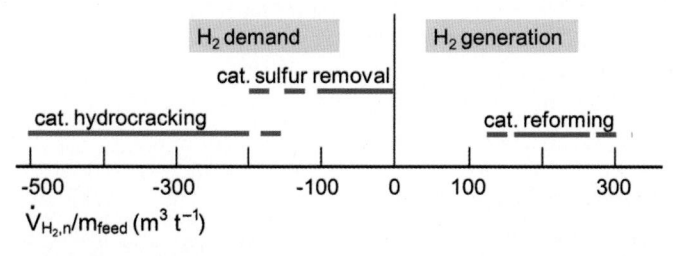

Figure 30.5 Hydrogen demand and generation in conventional petroleum refining processes per unit mass of process feedstock.

Table 30.2 Ranges of hydrogen demand values in refining/upgrading of petroleum and hydrogenation of vegetable oil.

Material	$\dfrac{m_{H_2}}{m_{HC\,product}}$ (kg t^{-1})
Petroleum, conventional[a]	2–10
Petroleum, heavy[b]	4–35
Vegetable oil[c]	15–25

a $CH_{1.8}S_{0.01}$.
b $CH_{1.4}O_{0.05}S_{0.02}$, based on Eqs. 30.6 and 30.7.
c $CH_{1.8}O_{0.1}$.

During the last decade, hydrogenation of vegetable oil has been established as a proven technology. It is based on experience with desulfurization of petroleum fractions [12–14]. The resulting hydrocarbon product can be used as high-value diesel or kerosene fuels, integrated into present infrastructures for blending with petroleum products without limitation or for separate use.

Demand figures for hydrogen in petroleum and vegetable oil refining are listed in Table 30.2, based on stoichiometric estimates analogous to Figure 30.4. In vegetable oil hydrogenation, the overall reaction is a combination of Eqs. 30.1 and 30.2. Resulting hydrogen demand flows in typical petroleum refineries are in the range 20 000–200 000 m^3 h^{-1} and in vegetable oil hydrogenation plants (assumed 200 000 t a^{-1} of product) in the range 8000–12 000 m^3 h^{-1}.

30.2.4
Hydrogen in Synfuel Production

Whenever solid raw materials (biomass, coal, etc.) are to be converted to hydrocarbon fuels, major chemical changes of feedstock molecules are needed. Since the feedstock typically is low in hydrogen and/or high in oxygen content, H_2 addition will help to increase the yield of hydrocarbons per unit mass of raw material. The pathway preferred is via gasification to synthesis gas (syngas) (H_2–CO), to which H_2 can be added to reach the required stoichiometric ratio in the syngas. If CO_2 is used as carbon source, no gasification is needed; instead, it has to be reduced with H_2 to CO as syngas constituent.

30.2.5
Example: Substitute Natural Gas (SNG) from H_2–CO_2

Synthesis of methane from H_2–CO_2 mixtures was recently proposed as a pathway to produce nonfossil SNG, using H_2 generated via electrolysis with renewable electricity

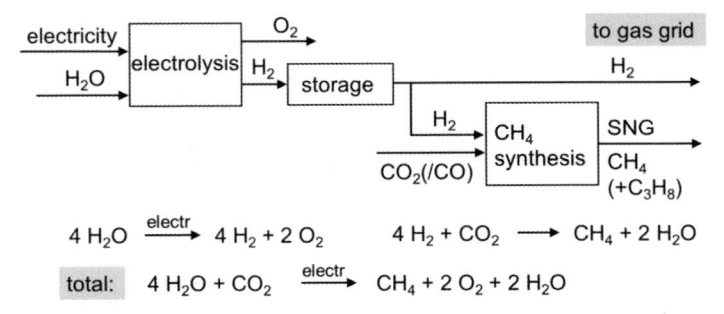

```
electricity ──→ ┌───────────┐  O₂ ──→
                │electrolysis│ H₂ ──→ ┌──────┐
H₂O ──────────→ └───────────┘        │storage│ ──→                                    to gas grid
                                     └──────┘    H₂                                   H₂ ──→
                                                 ├────────→ ┌─────────┐  SNG
                                                            │  CH₄    │ ──→
                                                 CO₂(/CO) ─→│synthesis│  CH₄
                                                            └─────────┘ (+C₃H₈)
```

$$4\,H_2O \xrightarrow{electr} 4\,H_2 + 2\,O_2 \qquad 4\,H_2 + CO_2 \longrightarrow CH_4 + 2\,H_2O$$

total: $\quad 4\,H_2O + CO_2 \xrightarrow{electr} CH_4 + 2\,O_2 + 2\,H_2O$

Figure 30.6 Hydrogen from fluctuating electric energy in synfuel production – example flow diagram for SNG from H_2–CO_2.

(Figure 30.6) [15]. Potential CO_2 sources are biogas plants, biomass combustion or gasification (leading to CO–CO_2 mixtures), and industrial production (e.g., fertilizer). Addition of hydrogen from electrolysis directly to the distributed natural gas is limited according to present standards (5 vol.%). Methanation therefore allows the amount of nonfossil SNG to be increased. Overall efficiencies are envisaged to be around 60% (as heating value of product gas per unit electricity). Losses could be decreased by using part of the thermal energy from the process in some suitable way. The advantage is that the product gas is fully compatible with today's natural gas. Both transport and distribution grids (and additional underground storage capacities) can be used. Hydrogen flows corresponding to typical CO_2 flows of 500 m³ h⁻¹ (biogas plants) and 10 000 m³ h⁻¹ (industrial biomass combustion) will be around 2000 and 40 000 m³ h⁻¹, respectively. There are several R&D projects ongoing in Germany, and the current projection foresees commercial application from 2020 on [15].

30.2.6
Example: Liquid Fuels from Biomass

Integration of H_2 in the production of liquid hydrocarbon synfuels from biomass has as the main incentive the significant increase in hydrocarbon product yield per unit biomass (or per hectare of agricultural or forest land). Given the shortage of arable land and water in many countries, there is a need to achieve maximum fuel yields per unit of land (while at the same time ensuring sustainable protection of biotic and abiotic resources, namely flora and fauna, water, and soil). Chemical conversion of biomass to liquid hydrocarbons includes as process steps after gasification the cleaning and adjustment of syngas, Fischer–Tropsch synthesis, and product hydrocarbon upgrading (Figure 30.7). Given the limited amount of hydrogen available in the feedstock, the resulting H_2/CO ratio in syngas from gasification is < 2. Without H_2 addition, part of the CO must be shifted to CO_2 in order to generate additional hydrogen and achieve the required stoichiometry for Fischer–Tropsch synthesis (H_2/CO ≈ 2). Less than 50% of the feedstock carbon is ultimately converted to hydrocarbons if no H_2 is added.

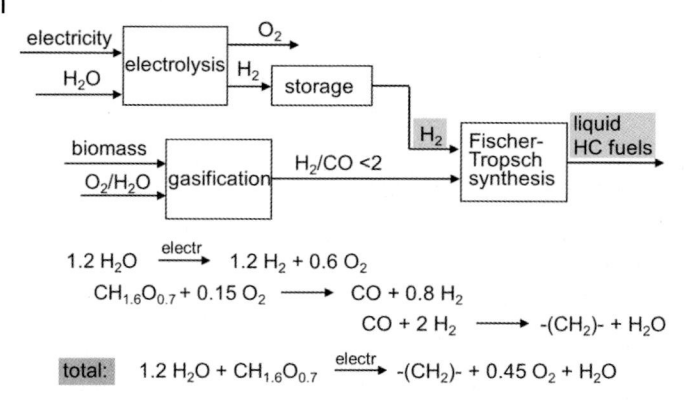

Figure 30.7 Hydrogen from fluctuating electric energy in synfuel production – example flow diagram for liquid hydrocarbons from biomass.

$$1.2\ H_2O \xrightarrow{electr} 1.2\ H_2 + 0.6\ O_2$$
$$CH_{1.6}O_{0.7} + 0.15\ O_2 \longrightarrow CO + 0.8\ H_2$$
$$CO + 2\ H_2 \longrightarrow -(CH_2)- + H_2O$$

total: $1.2\ H_2O + CH_{1.6}O_{0.7} \xrightarrow{electr} -(CH_2)- + 0.45\ O_2 + H_2O$

With H_2 addition, hydrocarbon yields can be increased significantly such that, in principle, all carbon present in the feedstock biomass would be converted to hydrocarbons. This reflects the situation with a maximum use of biomass carbon for the production of liquid hydrocarbon synfuels. Hydrogen addition could more than double the yield per unit biomass (or per unit arable land), based on a stoichiometric estimate. Hydrogen flows for achieving this increase in hydrocarbon yield would amount to about 40 000 m³ h⁻¹ in a biomass-to-liquid plant with a product capacity of 200 000 t a⁻¹. There are R&D projects ongoing, with improving the economics of the process being a significant challenge owing to the high investment required [16,17].

30.2.7
Cost of Hydrogen Production

The economics of H_2 integration in upgrading/production of hydrocarbon fuels depend on the cost of hydrogen production. The cost of H_2 produced from fluctuating renewable electricity is dominated by (i) capital expenditure for electrolysis, (ii) size or production capacity, and (iii) operation time, if production is not continuous but only in excess situations. Capital expenditure today is relatively high (€1300–1600 kW⁻¹) [7], with a strong incentive for significant reductions (to €500–900 kW⁻¹). Upscaling of present module capacities (up to about 4 MW for 1000 m³ h⁻¹ H_2) to the required capacities for H_2 integration is under way. As for operation time, load change patterns of fluctuating renewable electricity generation and economic boundary conditions are not yet known. It is clear that a high-investment plant can only operate with reasonable economics if the operation time is high. From preliminary cost studies, it can be seen that the H_2 production cost for electrolysis may reach current values for natural gas steam reforming, if electrolysis becomes cheaper and input electricity has a low value.

30.3
Conclusion

Conversion of electric energy into fuels may become an option for energy storage with increasing renewable electricity generation in the future. With liquid hydrocarbons probably remaining important for mobile applications and natural gas grids offering possibilities for storing fuel gases, integration of H_2 from electrolysis in hydrocarbon fuel production offers a potential route, in addition to the direct use of H_2. Based on the discussion presented here, some conclusions may be drawn.

- Electrolysis of water to produce H_2 is a proven process today for small-scale applications and will be extended to a larger scale in the future.
- If H_2 is to be converted to hydrocarbons, different carbon sources can be used. (i) In the case of fossil raw materials, yields per feedstock can be increased and fossil CO_2 emissions decreased since fossil resources for H_2 production can be avoided. (ii) In the case of biomass conversion to synfuels, product yields per biomass/hectare can be increased, and in the case of vegetable oil hydrogenation fossil resources for H_2 production can be avoided. (iii) With CO_2 as feedstock, no gasification step is needed; however, the lowest hydrocarbon product yields per H_2 are achieved, owing to the high oxidation value of CO_2.
- Most critical aspects of H_2 integration in hydrocarbon fuel production are (i) the currently high capital cost of electrolysis resulting in high H_2 generation cost, and (ii) medium to higher capacities of H_2 generation are generally needed in present fuel production processes, so upscaling of electrolysis will be required.
- Related research topics are cost reduction and upscaling of electrolysis, system development with respect to availability to fluctuating electricity (scale of flow, time pattern, location), and dynamic behavior resulting from electric power fluctuations.
- Research into new production and utilization processes may result in innovative chemical energy carriers besides hydrocarbons.

30.4
Nomenclature

grav	gravimetric
HC	hydrocarbons
liq	liquid
n	number of moles
n	1 bar, 0 °C
OV	oxidation value
V	volume
x	molar ratio H/C
y	molar ratio O/C

Acknowledgments

Financial support from the Bundesministerium für Bildung und Forschung (BMBF, FKZ 01RC1010C) and the Fachagentur Nachwachsende Rohstoffe (FNR, FKZ 22403711) for parts of the present study is gratefully acknowledged.

References

1. AG Energiebilanzen (2012) *AGEB Home Page*, www.ag-energiebilanzen.de (last accessed 8 February 2013).
2. Jarass, L., Obermai, G. M., and Voigt, W. (2009) *Windenergie: Zuverlässige Integration in die Energieversorgung*, 2nd edn., Springer, Berlin.
3. Droste-Franke, B., Paal, B. P., Rehtanz, C., Sauer, D. U., Schneider, J. P., Schreurs, M., and Ziesemer, T. (2012) *Balancing Renewable Electricity – Energy Storage, Demand Size Management, and Network Extension*, Springer, Berlin.
4. Schüth, F. (2011) Chemical compounds for energy storage. *Chem.-Ing.-Tech.*, **83**, 1994–2001.
5. Nitsch, J., et al. (2012) *Langfristszenarien und Strategien für den Ausbau der erneuerbaren Energien in Deutschland bei Berücksichtigung der Entwicklung in Europa und global*, "Leitstudie 2010", Report BMU-FKZ 03MAP146, http://www.bmu.de/fileadmin/bmu-import/files/pdfs/allgemein/application/pdf/leitstudie2010_bf.pdf (last accessed 8 February 2013).
6. Eilers, H., Iglesias González, M., and Schaub, G. (2012) Chemical storage of renewable electricity in hydrocarbon fuels via H_2, presented at Tagung "Reducing the Carbon Footprint of Fuels and Petrochemicals", Berlin, October 2012, *DGMK-Tagungsbericht*, 2012-3, pp. 83–90, ISBN 978-3-941721-26-5.
7. Smolinka, T., Günther, M., and Garche, J. (2011) *NOW-Studie: Stand und Entwicklungspotenzial der Wasserelektrolyse zur Herstellung von Wasserstoff aus regenerativen Energien*, 2011, NOW.
8. Lucas, A. G. (ed.) (2000) *Modern Petroleum Technology. Vol. 2. Downstream*, John Wiley & Sons, Ltd., Chichester.
9. Alfke, G. et al. (2008) Oil refining, in *Handbook of Fuels – Energy Sources for Transportation* (ed. B. Elvers), Wiley-VCH Verlag GmbH, Weinheim.
10. Neumann, G. (2003) Hydrogen management in the MiRO refinery, in *Proceedings 2003-2 of the DGMK Conference Innovation in the Manufacture and Use of Hydrogen*, pp. 103–107, ISBN 3-936418-04-7.
11. Berkowitz, N. (1997) *Fossil Hydrocarbons – Chemstry and Technology*, Academic Press, San Diego.
12. Neste Oil (2012) *NExtBTL Renewable Diesel, Product Information Brochure*, www.nesteoil.com/default.asp?path=1,41,11991,12243,12335 (last accessed 12 March 2013).
13. UOP (2012) *Green Diesel*, www.uop.com/processing-solutions/biofuels/green-diesel/ (last accessed 12 March 2013).
14. Endisch, M., Balfanz, U., Olschar, M., and Kuchling, Th. (2008) Hydrierung von Pflanzenölen zu hochwertigen Dieselkraftstoffkomponenten, in *Proceedings 2008-2 of the DGMK Conference Innovation in the Manufacture and Use of Hydrogen*, pp. 261–270, ISBN 978-3-936418-80-4.
15. Dena (2012) *Integration erneuerbaren Stroms in das Erdgasnetz*, http://www.powertogas.info/power-to-gas/strom-in-gas-umwandeln.html (last accessed 12 March 2013)
16. Iglesias González, M., Kraushaar-Czarnetzki, B., and Schaub, G. (2011) Process comparison of biomass-to-liquid (BtL) routes – Fischer–Tropsch synthesis and methanol-to-gasoline. *Biomass Conversion Biorefinery*, **1** (4), 229–243.
17. Schaub, G. and Edzang, R. (2011) Erzeugung synthetischer Kraftstoffe aus Erdgas und Biomasse – Stand und Perspektiven, *Chem.-Ing.-Tech.*, **83** (11), 1912–1924.

31
Geological Storage for the Transition from Natural to Hydrogen Gas

Jürgen Wackerl, Martin Streibel, Axel Liebscher, and Detlef Stolten

31.1
Current Situation

A continuous and reliable power and fuel supply is one of the key elements for a stable and growing economy. In the past, this supply was mainly governed by the fuel-producing countries and their price politics. To ensure continuity despite short-term shortages, strategic reserves were introduced. In the European Community (EC) this became so important that a directive was introduced to maintain a reserve for each EC country of at least 90 days of the average daily consumption of hydrocarbons [1]. The reserves were also used to buffer and average the daytime-dependent consumption. This worked well in the past, but in recent years an additional component introduced minor instabilities, and not only from the former suppliers. The increased amount of available electricity derived from alternative, regenerative sources such as solar and wind power make it difficult to predict fluctuations of both the electrical power grid and fuel consumption. Therefore, these reserve storages were increasingly used not only as seasonal but also as sporadic buffers. As long as the fraction of energy produced by the regenerative sources covered only small parts of the overall consumption, the interferences thereof could be easily handled. However, the peak power from these sources combined with the "traditional," hydrocarbon-based power now sometimes exceeds the overall electricity consumption and adds serious problems to the control of the electrical power grid [2]. In 2011, already an average of 11% of the primary energy in Germany was covered by renewables [3]. For the 6.15×10^{11} kWh of electricity produced in Germany, the fraction of the renewables is 20% [4], which is a significant amount.

Especially for Germany, there is an additional directive to reduce significantly the overall CO_2 emissions. This is also on the roadmap of the EC [5]. Therefore, a new concept to handle the energy surplus and not to waste it is needed. One of the promising approaches is to convert the surplus electrical energy into hydrogen gas. For this, a new infrastructure including not only an electrolyzer, gas distribution grid and points of end use but also different types of storage options is needed.

Transition to Renewable Energy Systems, 1st Edition. Edited by Detlef Stolten and Viktor Scherer.
© 2013 Wiley-VCH Verlag GmbH & Co. KGaA. Published 2013 by Wiley-VCH Verlag GmbH & Co. KGaA.

Since the amount of power involved is huge, the total storage capacity must also be of large dimensions. In 2011, a total working gas volume of 2.04×10^{10} scm (standard cubic meters) was available for natural gas at 48 different storage sites in Germany [4]. For 2012, this volume increased to 3.3×10^{10} scm including all planned sites [6]. This is about the strategic reserve required for an actual natural gas consumption of 1.26×10^{11} scm (equivalent to 1.06×10^{12} kWh) in 2011 [7]. However, for the future gas storage volume requirement, the total primary energy consumption of Germany in 2011 of about 3.7×10^{12} kWh [4] is used as a reference in this chapter. This value is believed to be stable for the forthcoming years [8]. Since electricity from nuclear power is no longer an option and an overall CO_2 reduction of 90% compared with the emissions of 1990 is demanded [9], a switch-over to hydrogen gas as an energy carrier is assumed. With the lower heating value of hydrogen gas of 3 kWh scm^{-1}, a total amount of hydrogen gas of about 1.1×10^{12} scm would be consumed per year. For the strategic reserves this would involve a working gas volume of about 2.7×10^{11} scm (2.5×10^7 t of H_2), which is in agreement with other estimates [10]. To visualize this value, a nominal compression of hydrogen gas to 150 bar is assumed. At that pressure, a single sphere roughly 750 m in diameter would be needed. For economic reasons, only geological storage options are suitable to reach such volumes. This volume is, however, without any buffer, for example, when the complete pressure difference between the full and empty states could be used. A decrease in this volume could be achieved in theory by liquefying the hydrogen (liquid hydrogen, LH_2) since 1 l of LH_2 expands to 0.79 scm of hydrogen gas. The largest state-of-the-art surface hydrogen storage for LH_2 is located at the Kennedy Space Center (KSC) in the United States. It has a diameter of about 20 m and a storage capacity of 230 t [11]. The use of LH_2 for large scale storage is not economic because every storage site must be facilitated with a liquefying station and handling of the extremely cold gas is demanding concerning materials and costs. Moreover, the losses during storage are too high, although the relative losses become smaller when using larger storage tanks. As an example, the LH_2 storage at the KSC currently has a leak rate of about 1200 liters of LH_2 per day – a value that has been lowered and reached after 40 years of operation. This still represents around 85 kg of hydrogen per day. Since about 110 000 of these storage spheres would be needed if all hydrogen gas were to be stored liquefied, the losses would amount to about 10^4 t per day. Aside from the space needed to install that quantity of storage units, this amount of gas loss is uneconomic. In addition, there is the high security risk due to the high visibility and surface of such storages sites. Another option for storing hydrogen above ground is to use high-pressure gas storage units. These are considered safe even for long-term operation and can store up to 2 scm at 185 bar [12]. However, they can only be used for local buffering or peak smoothing as large volumes would be highly space and materials consuming and costly.

It is evident from these considerations that storage of the required large amounts of gas for energy supply purposes becomes feasible only if one considers geological, subsurface storage sites. Several state-of-the-art options are already available from natural gas storage. However, some of them are more advantageous than others or are not feasible in the worst cases, depending on the geological situation at the desired location. Therefore, a trade-off between storage type and location has to be made. Additional factors can further complicate the choice.

As will be shown later in detail, the dynamics of geological storage sites can be very different. Some types can be used only for slow and continuous, static fluxes, whereas others can also be used for high-rate, dynamic, peak demands. However, the storage sites are only one part of a more complex system. In the most ideal case, geological storage units with high throughput ability can be installed close to major points of end use. These points include the major cities of a country and large industrial complexes. This lowers the requirements and possible pressure losses of the gas distribution pipeline grid, thus making the complete infrastructure less costly and complex. However, it is also advantageous to have slow gas storage units close to the feed points of the gas distribution grid, for example, where the gas is produced or delivered from other countries. In the case of hydrogen gas, the storage units on the production sites, for example, where the electrolyzers are installed, must be partly of a dynamic type. The reason is the on-demand operation of the electrolyzers, depending on the electrical energy available from renewable sources. Therefore, when discussing the "how and where" of the strategic reserves, one also has to consider distributed gas storage sites all over the country, especially for a large country such as Germany.

Another issue to consider is the ability of a storage site to enable at least one complete drain and refill cycle each year over an assumed lifetime of 50 years. This annual cycle is attributable to the seasonally dependent electricity production rate and consumption. For Germany in particular, this kind of operation goes back to 1955. The natural gas storage units are mainly used to buffer the seasonal changes. Additionally, short-term buffering on a daily basis between the gas provider, for example, the operator of the gas distribution grid, and the consumer is of interest. This became more and more important in terms of strategic concerns in recent years due to shortages in the natural gas supply chain [13].

In the case of hydrogen gas, considerable amounts are currently stored at geological sites in only a few places, such as Teesside in England operated by Sabic UK Petrochemicals and Spindletop, Texas, United States operated by Air Liquid Phillips and Praxair, hence the practical experience in operating and long-term effects with pure hydrogen gas is limited. Some of the experience with natural gas geological storage is described in this chapter to allow an evaluation of the different storage types.

31.2
Natural Gas Storage

Natural gas storage units can be divided into two types: hollow, such as mines and caverns, and porous. The porous type are mainly used to cover the base demand over the long term. This can be attributed to the lag properties. Caverns, on the other hand, mainly cover daily fluctuations. However, independent of the type of storage, the capacity stated by the operator must be treated with caution.

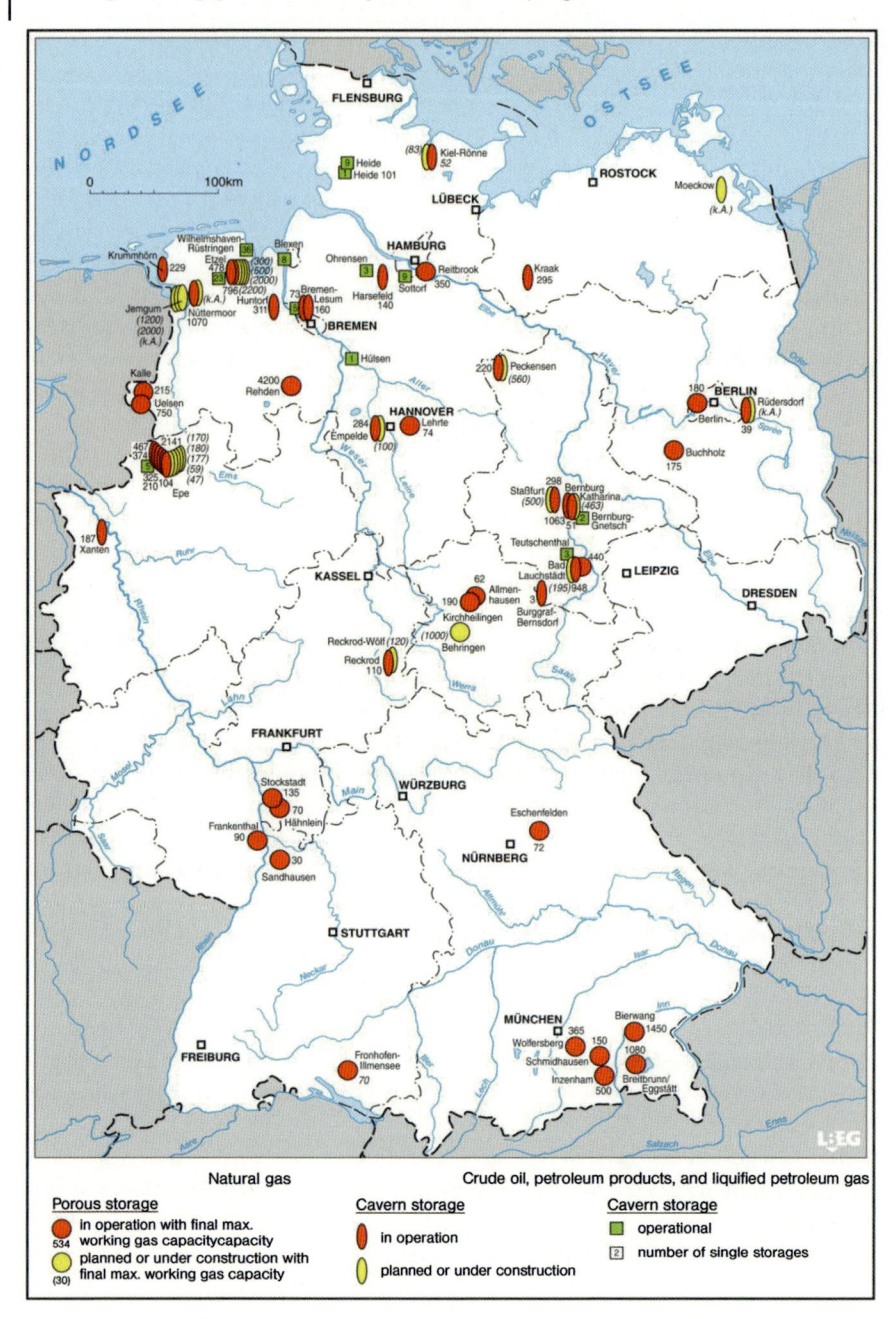

Figure 31.1 Subsurface storage sites in Germany for natural gas, crude oil, mineral oil products, and liquefied gas [14].

Table 31.1 Segmentation of natural gas storage sites in Germany for 2009.

Parameter	Porous storage	Cavern storage	Total
Working gas volume in operation (10^9 scm)	12.7	8.1	20.8
Maximum steady drain rate (10^9 scm per day)	197.9	296.4	494.3
Number of storage sites in operation	23	24	47

There are several types of volumes. The first is the geometric volume, which for a cavern represents the total usable volume under standard conditions. The second is the working volume, which is the nominal gas volume that is usable when operating the storage unit between its nominal upper and lower conditions (e.g., pressure). The third is the cushion gas volume, which is either the volume needed to keep the storage site in a stable mechanical condition (structural integrity) or the amount of gas that is lost on the first flush to initialize the storage site for operation. The cushion gas volume can normally not be used, but might be of interest in the case of a fatal failure scenario or when the storage unit has to be drained completely. For salt caverns the residing cushion gas volume is about 25% and for porous storage units it can be up to 50%.

In Figure 31.1, the geographic locations of subsurface gas storage units including those for liquid hydrocarbons are given. Owing to the geological situation in Germany, cavern-type storages units are located with very few exceptions only in the northern region up to the Mittelgebirge [14].

Some of the most important data concerning natural gas storage are given in Table 31.1 [14]. As can be seen, the predominant amount of the natural gas is stored in porous storage units. For planned storage units or those already in construction, this kind of storage does not have a significant role any longer. Instead, the total working volume of the cavern storage units will increase by up to a factor of three until 2018 according to current plans [6, 15].

31.3
Requirements for Subsurface Storage

The lifecycle of a geological storage site can be divided into four phases:

1. Exploration, where the possible location is examined and its suitability is determined.
2. Drilling: in the case of caverns and mines also the excavation of the geometric volume.
3. Operation of the storage site.
4. Shut-down and sealing of the storage site.

Access to porous storage units and salt caverns is via the wellbores only. Therefore, the required data for each step can be acquired only using remote sensing techniques and not using human accessible inspection. Hence the geological formations must first be examined thoroughly to establish whether they meet the requirements of German mining laws. The requirements for a geological storage unit that concern all phases of the lifecycle include the following [16]:

- geological safety – on a local, limited area around the site and a more global scale
- sustained and reliable integrity of the surrounding rock and also the drill hole
- acceptable elevation or lowering of the affected surface area due to changing of the underground
- environmentally friendly closing of the storage site at the end of its lifetime.

These evaluations are accompanied by numerical simulation to model and predict the geological behavior for each step. Geological fault zones are especially demanding and must be included in this process. They are the main reason for possible leakages or complications during the operation of the storage site. Simultaneous rock extractions from different horizons (depths) of test drillings of the possible storage site complete the picture, since they reveal the actual geological situation [16].

Additional requirements arise in the case of hydrogen gas storage units:

1. *Pressure level of the storage site.* A future hydrogen gas grid will be operated at the same pressure level as the existing natural gas grid, which is in the range 20–60 bar. The cushion gas pressure should therefore amount to at least 60 bar. This keeps the costs of compressors low. Hence the lower pressure limit given above should allow draining from the storage to the grid without the use of additional compressors and thus having a less complex infrastructure. The upper pressure limit is set to about 120 bar owing to the high-volume compressor technology currently available [12]. If an average increase of the lithostatic, rock pressure of 22.6 bar per 100 m depth increase is assumed [17], the maximum depth of such a cavern would be about 500 m below the surface. At greater depths, the pressure of the surrounding rock becomes higher than the gas pressure, resulting in possible compression of the cavern and also increasing the risk of a loss of structural integrity and possible collapse. This implies also that a storage unit aligned vertically might be more limited in the pressure levels allowed compared with one aligned horizontally. Technically reachable depths for excavation can go down to 3000 m. To use caverns at such depths would,, however, require first a higher amount of cushion gas to maintain a higher pressure level and second higher gas compression levels, increasing the costs of the infrastructure and operation, although it adds to the storage capacity.
2. *Cycling.* Hydrogen gas storage systems are supposed to be used in the same way as the natural gas ones currently operated. This means that there is the need for slow buffers which are part of the strategic reserve but also cover the annual and seasonal fluctuations. An annual full drain and refill cycle during the lifetime must therefore be acceptable for such a storage system. Such usage will have long-lasting moderate drain and fill rates but will stress the storage with large pressure differences with respect to the swings. In addition, there will be partial

cycles on a daily basis or even shorter which require high drain and fill rates. This will cause only small pressure changes and the effect will be less pronounced in larger storage sites. For a reasonable lifetime of about 50 years, this would imply at least 50 full cycles that a storage site would have to withstand but also roughly 20 000 partial cycles.

3. *Contamination.* One of the very important economic aspects is the preservation of the quality, mainly the purity, of the stored hydrogen gas. In the most ideal case the drained hydrogen gas has the same quality as that used for filling. Enrichment of the gas with chlorine- or sulfur-based compounds, hydrocarbons, and water should be avoided. Since the value of hydrogen gas and its possible use are strongly related to the amount of these contaminants, cleaning and filtering of the drained gas might become important and can add significant costs to operation. If the storage unit itself does not have any significant influence on the quality of the gas, such infrastructure considerations can be neglected.

4. *Safety.* As with natural gas and other fuels, burning and explosion limits have to be considered. For hydrogen – under standard conditions and in air – the flame limits are between 4 and 74 vol.% and the explosion limits are between 18 and 59 vol.% [14, 16, 18]. These ranges narrow with increasing gas pressure. In the case of an explosion, hydrogen can be considered no more hazardous than natural gas. However, there are major differences between these two gases in the case of burning and leakage. Since hydrogen has a far higher diffusivity and a lower density than other gaseous fuels, it burns fast and with a mostly upright flame. Effective heat shielding of the flame due to the water produced takes place that is not interfered with by the presence of carbon, CO, and CO_2 in the exhaust gases. Therefore the damage caused by burning hydrogen is more localized and is significantly lower apart from the actual flame than for other fuels, despite the extremely high flame temperature itself. Moreover, documented hydrogen gas accidents with significant volumes have shown no explosion risk so far. This is attributed to the fact that hydrogen self-ignites in contact with air even at very small leak holes, and in most cases the hydrogen will therefore burn before the critical limit for an explosion is reached [19].

5. *Long-term stability of the storage site.* Many materials become brittle or react in the presence of hydrogen gas, especially if high pressure is applied or if water is present. This has to be considered not only for the surrounding rock materials of the storage site but also for the rock formations used in porous storage systems. Mechanical stability issues may arise which can lead to a fatal failure scenario. Unfortunately, data about the mechanical issues of rocks in contact with hydrogen are scarce.

6. *Gas losses.* Hydrogen gas diffuses through most materials fairly easily owing to its low density and mass. Whereas data exist for iron and various metal alloys and steels, only limited data are available for construction materials. The diffusion coefficient of hydrogen in concrete is 2.34×10^{-3} cm^2 s^{-1}, which is rather fast, whereas for tar the value is 9.9×10^{-7} cm^2 s^{-1} [20]. The situation is worse for rock, for which almost no data are available. To evaluate the losses during operation, the losses due to reaction also have to be considered. Nevertheless, experience

with geological caverns filled with hydrogen gas show a loss rate of only about 0.01% per year, which is considered acceptable. The overall losses of the currently operated natural gas storage sites amount to about 0.2% [21].

31.4
Geological Situation in Central Europe and Especially Germany

Since not every location is suitable as a geological storage site, the geological situation is discussed first. It is important to consider the geological formations down to a depth of several thousand meters from the surface. Not only technical but also economic aspects make the use of great depths less attractive.

One not very obvious factor to consider is the subsurface temperature, which increases on average by 3 °C with every 100 m increase in depth. For the operation of a gas storage unit, this means that the effective working volume (amount of gas molecules, n) for a given pressure (p) and geometric site volume (V) decreases with increasing depth according to the ideal gas law (p multiplied by V is directly proportional to n and temperature!) in a first approach. At high pressures and higher temperatures, the pressure–volume relationship becomes significantly nonlinear,

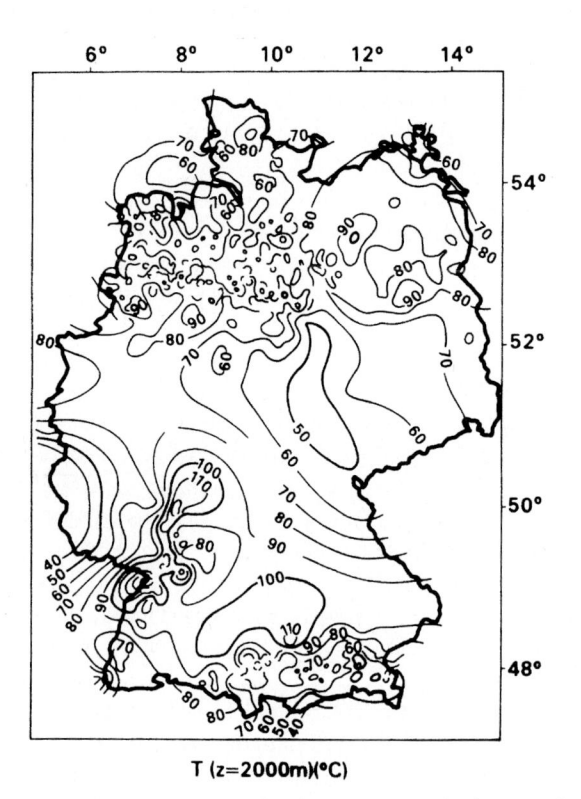

Figure 31.2 Temperature distribution at a depth of 2000 m for Germany [22].

whereas the effective volume reduction becomes smaller with increasing temperature and pressure. For methane (CH_4), as the major component of natural gas, this is more pronounced than for pure H_2 gas, which can be treated as ideal for pressures up to 1000 bar. This means for the operator of a geological storage site that at high degrees of filling the storage can store less gas for a given pressure difference than at low degrees of filling. In addition to the average increase in temperature, areas of increased geothermal gradients have to be avoided. This is an issue not only for the effective working volume available at a specific storage site but also because in these warmer areas the rock can be more prone to creep [23, 24] and therefore significant shape changes of the storage unit can occur during pressure cycling. Cooler areas are therefore more favorable for large-sized storage units. An overview of the temperature distribution in the German underground is shown in Figure 31.2.

Another important aspect to consider is the local seismic activity. A smaller amount of activity at a chosen site in the past implies a lower risk of future events – at least on a short geological timescale of several hundred years. Moreover, fewer safety measures are necessary and there will be a greater acceptance by the population Figure 31.3 shows an overview of the documented seismic events in Germany for the past 1000 years.

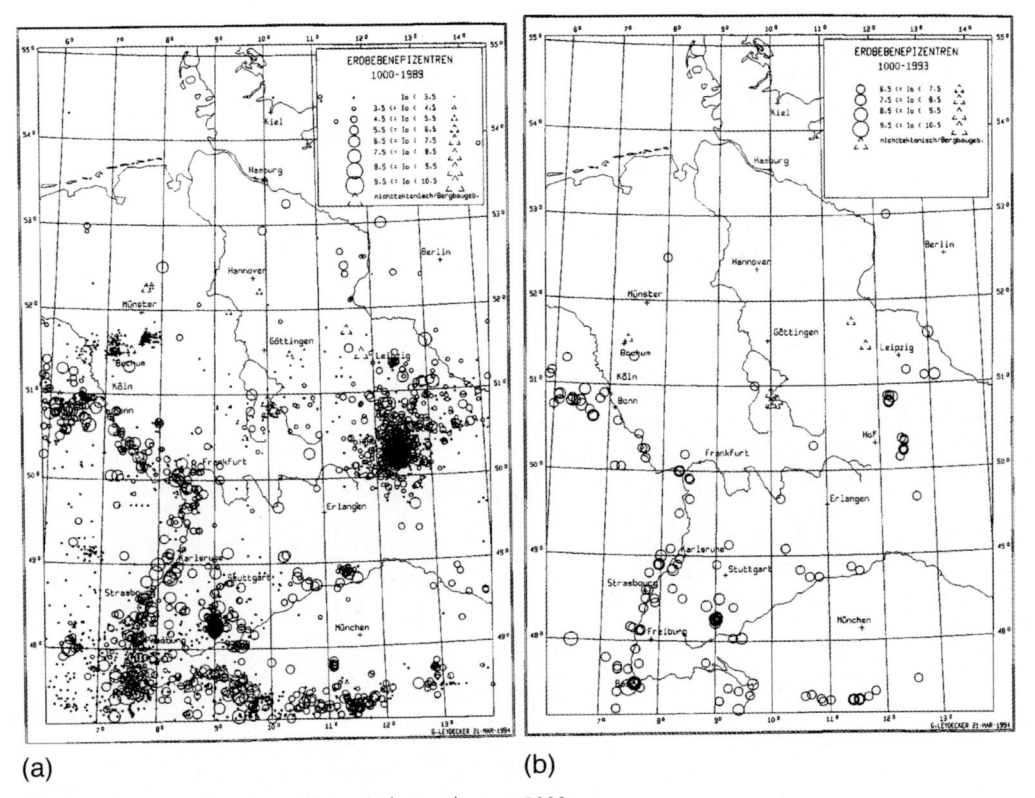

Figure 31.3 Seismic activity in Germany during the past 1000 years.
(a) All recorded events and (b) only the major events. Source: taken from [25].

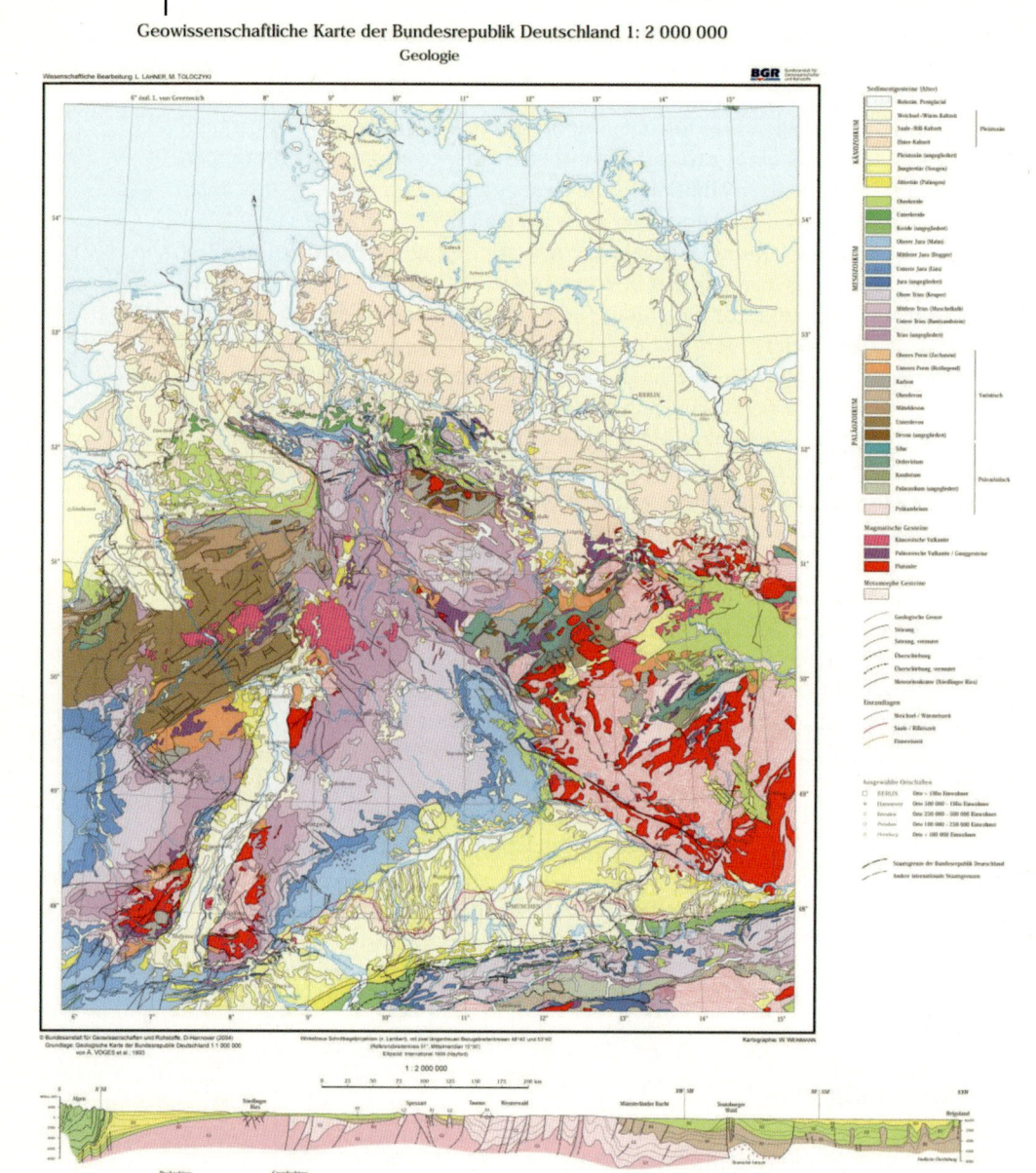

Figure 31.4 The near-surface geological rocks in the German underground [26].

It becomes evident from an examination of the maps that the areas close to the Rhine, the Alps, in the Erzgebirge (area around Hof), and in most parts of the Swabian Alb (area north of the river Danube) show a lot of seismic activity and are therefore prone to stability issues and should be considered carefully when choosing possible sites for geological storage.

A further factor for large-sized storage sites is the type and homogeneity of the underground. An overview of the uppermost formations is given in Figure 31.4. As can be seen, different types of rocks can be found. Some are magmatic, marked in red on the map, where no other rocks can be expected below, very old rocks such as those marked in brown ones, and also rather newer rocks such as in the northern part of Germany, where different rocks can be expected in the underground. Owing to this diversity, Germany can be used as an example for other countries.

31.5
Types of Geological Gas Storage Sites

Depending on the geological situation, the types of gas storage that can be applied differ. The type of rock dominates the decision, but technical and economic factors are also important. There are several "traditional" storage types, where the technology is already known and used such as for porous storage sites, and new, more experimental ones such as aquifers. In this section the four most promising types for Germany are discussed. The main aspects used for the selection are availability, setup and maintenance costs, safety, and time to operation. Especially the last point is critical since the timeline for the CO_2 reduction strategy is very short and demanding. This means that within the next 10 years several additional gas storage sites, especially for hydrogen as alternative to natural gas, have to be established and ready for use. However, these sites should be operable not on an intermediate timescale but for a predicted lifetime of about 50 years or more. Since the future hydrogen storage sites will use the same geological settings as for the natural gas ones already in use, the storage types discussed here are mostly the same as those already in use for natural gas storage in Germany.

31.5.1
Pore-Space Storage Sites

Storage sites of this kind consist of open porous rock, for example, where gas can permeate through, embedded in dense cover rock which is impermeable to the stored gas. The porous rock must have a high porosity to provide sufficient space for storage. The pores in the rock must be connected, so-called open porosity, to provide a reasonable permeation rate. Hence the fill and drain rates are high enough for economic use, which means that a reasonable amount of gas can be stored at a site and at least one filling and draining cycle can be applied once per year. However, the mechanical stability has to be sufficient to withstand the various pressure swings. The rock formations found suitable for this purpose are mainly of sandstone. The sandstones consist of single sand grains with diameters from 60 μm to 2 mm and the required porosity is about 20% [27]. Since a lot of these formations exist in Germany, this kind of storage site has high potential. However, a major problem arises from the rocks surrounding the sandstone. For gas storage purposes these have to be, in the ideal case, completely impermeable to gas [28], and indeed there are many locations where the natural sealing for gas seems to be satisfactory, since this kind of storage is still of high interest, as Table 31.1 and Figure 31.1 have already revealed.

Typical porous natural gas storage sites have a working volume between 10^8 and 10^9 scm. Some former natural gas production sites now used as storage sites have a working gas capacity of over 2×10^9 scm. However, to maintain the structural integrity, a cushion gas level of up to 50% is needed. The major drawback – especially for the very large-sized sites – is the low drain and fill rates due to the permeation of the gas through the porous structures. For most storage sites of this kind the maximum drain rate is so low that a complete filling–draining cycle down to the cushion gas level takes around 60 days. Therefore, the porous storage sites in Germany are only used for buffering the seasonal fluctuations and to provide a basic supply [4]. The operational losses for this storage type are assumed to be about 1% per year, whereas a significant amount is attributed to leakages at the bore hole and infrastructure. When hydrogen is used in such storage units, the losses might increase. Since the porous structure also provides a high surface area, losses due to adsorption, filling of closed pores due to diffusion, and saturation of residing water with hydrogen can amount to up to 0.4% of the total volume. However, there is great uncertainty concerning the actual losses since not only chemical but also biological processes that consume hydrogen exist. These processes are known and sometimes induce a significant heat production of several tens of degrees Celsius in the storage site.

31.5.2
Oil and Gas Fields

As can be seen from Figure 31.5, most of the natural gas and crude oil production sites in Germany are located in the North German Basin and the Molasse Basin. The main advantage of these kinds of storage sites is the already existing infrastructure and the well-known geology below the surface. Most of these sites are of the porous type. A benefit is that the site can be considered to be gas-tight on the human timescale because the former gas has been enclosed. Therefore, depleted fields seem to be an obvious choice for the future storage of natural gas and hydrogen gas. This is based on the assumption that hydrogen has about the same properties and thus behaves in almost the same way as methane [30], which is a major component of natural gas. However, the following comparison shows that this assumption might be not correct. First, the van der Waals radii are significantly different – the size of an H_2 molecule is about 120 pm whereas that of CH_4 is about 170 pm [31]. Hydrogen not only can penetrate smaller pores than methane but can also propagate through smaller cracks. A higher leakage rate can therefore be assumed for hydrogen for

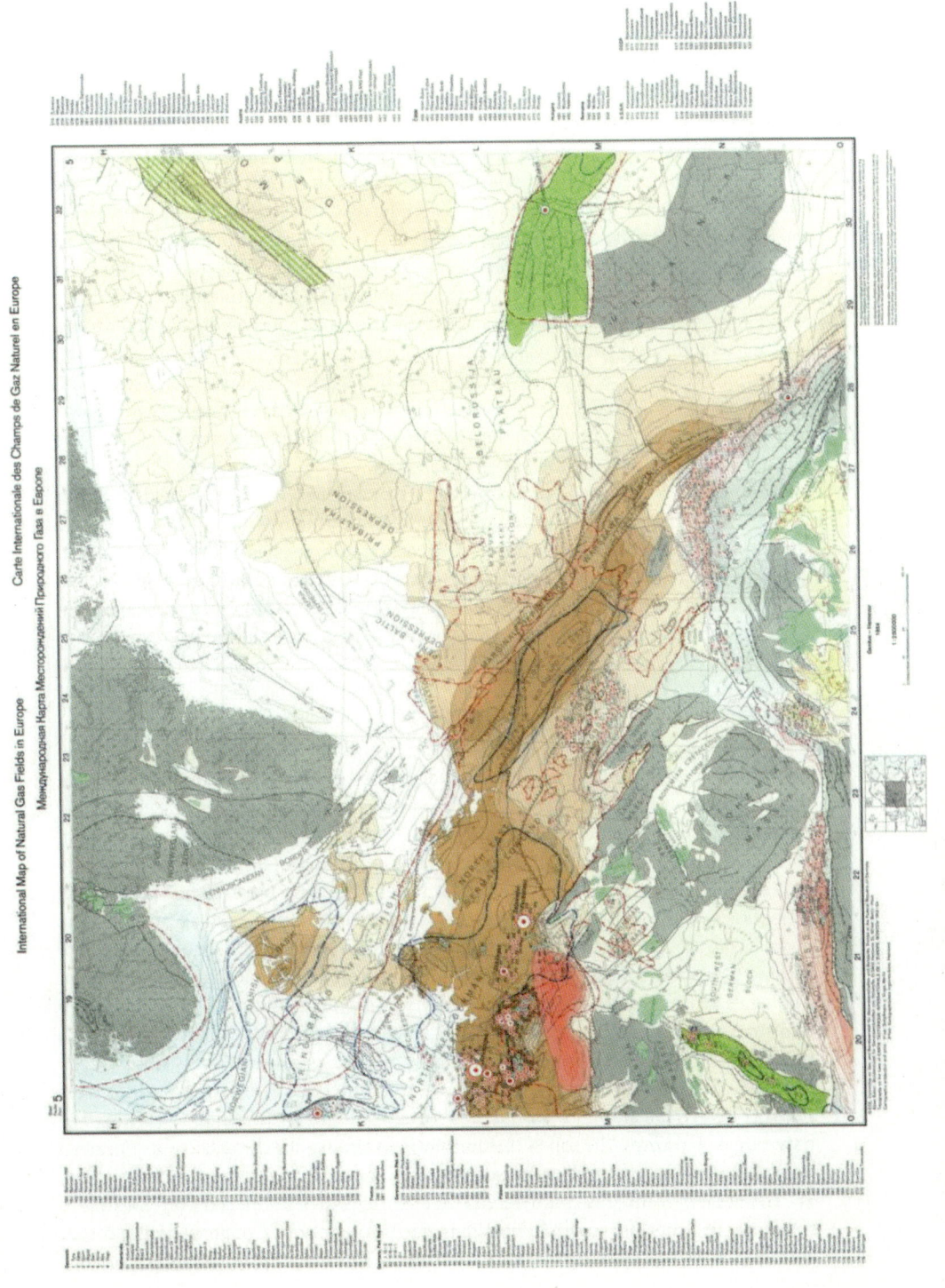

Figure 31.5 Crude oil and natural gas production sites in Germany [29].

the same storage site. Second, the free mean path of gaseous H_2 of about 110 nm is substantially higher than that of CH_4 of about 48 nm at 0 °C and 1 bar [32]. This translates to a much faster penetration of H_2 than methane along narrow paths since the velocity of a hydrogen molecule is also far higher. In a porous medium, the pressure gradients will therefore be less pronounced or equilibrated faster.

Most of the depleted or nearly depleted gas and oil fields in Germany are already in use or are reserved for future use for natural gas storage. These are porous-type storage sites which, as already stated earlier, have long cycling periods. Therefore, the interest in this type of storage is decreasing. Especially for those fields close to depletion, fracturing is considered to exploit more natural gas than solely with traditional methods [33]. This will provide problems for the future use of such sites for storage. The rock is artificially fractured and disintegrates during fracturing, thus releasing the rest of the gas residing there but also mechanically disintegrating the structure. For several already used natural gas storage sites, partial structural disintegration of the open porous, consolidated sandstone is observed. Depressions of the land surface and in the worst case damage to buildings due to subsidence are already well-known issues [34].

Another issue associated with the potential storage of H_2 in depleted fields is purity. Depletion means in general reaching the part of production cycle that is no longer economic, but a significant amount of hydrocarbon species still remains. When used later as a storage site, with every draining a certain amount of the remnants will be mixed with the hydrogen [35]. Cleaning either before or during operation (gas separation) will add significant costs to this type of storage site.

31.5.3
Aquifers

An aquifer in its simplest form is a single stream or path of ground water, and the water permeates through suitable rock formations. This does not necessarily imply porous structures. Water-soaked minerals or connected cave systems (so-called karst) are other possibilities. In Germany, the definition of an aquifer according to DIN 4049-3 also includes the water-unsaturated boundary zones, while the ground water horizon only includes the water-saturated zone. Therefore, the aquifer zones as given in Figure 31.6 include more than the ground water horizons. Despite this, the area with aquifers is large for Germany. Aquifer storage sites can be highly interesting, since a single site can extend to several kilometers in width, thus providing an enormous volume. Moreover, small cavern-like open structures can be found which hold a significantly greater gas volume than a porous structure of the same size.

To use an aquifer for storage, a structural trap is needed to keep the gas localized. Unfortunately, the aquifer systems in Germany are not as thoroughly examined as those in, for example, Texas [37]. Therefore, extensive and expensive exploration would first be needed to find suitable storage sites. Since many aquifer systems are connected, it must be assured that the storage sites themselves are located in unconnected ones. It is already known from the operation of oil and gas fields that multiple drainings from connected systems may seriously affect the performance of

31.5 Types of Geological Gas Storage Sites | **643**

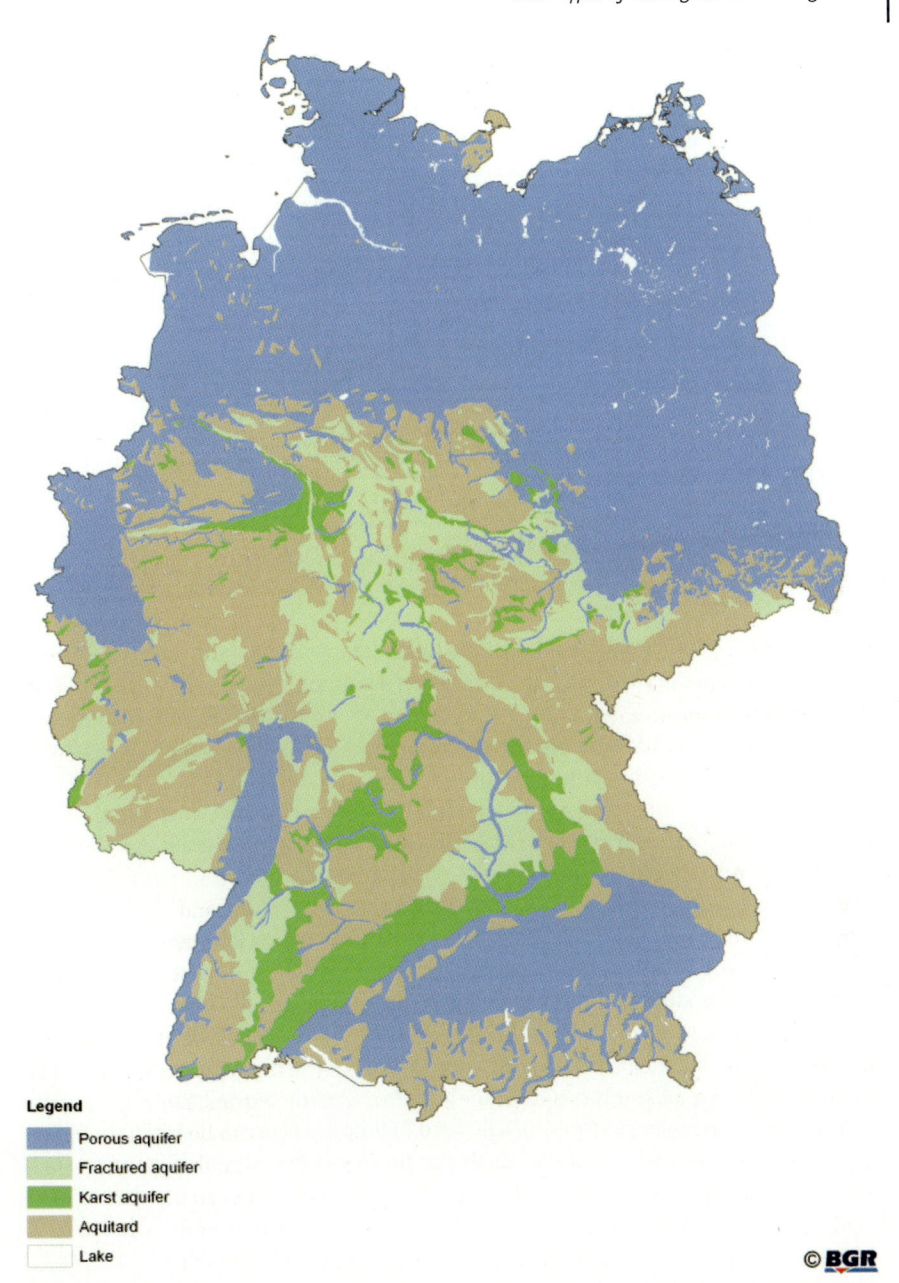

Legend
- Porous aquifer
- Fractured aquifer
- Karst aquifer
- Aquitard
- Lake

© **BGR**

Figure 31.6 Overview of the aquifer systems in Germany according to DIN 4049-3 [36].

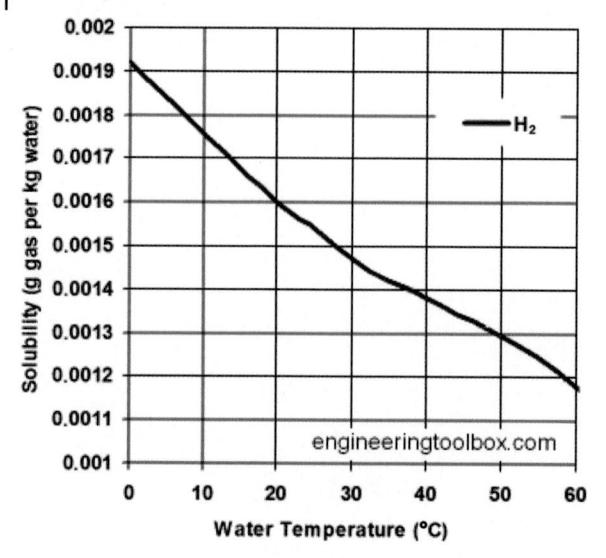

Figure 31.7 Solubility of hydrogen in water for ambient pressure as a function of temperature.

the related sites [34]. Such effects include pressure loss at one or more of the drain sites or overall reduced drain rates. From an operator's standpoint, the aquifers need a large amount of cushion gas [38] owing to the mainly porous structure of the aquifer systems. Additionally, the operating pressure of the storage site is largely determined by the hydrostatic pressure of the aquifer, which limits the working volume as the gas must be kept in the structural trap to avoid permanent losses due to spill-out. Losses also occur from the nature of the aquifer itself. As gases are soluble in water [18, 39], as shown for the particular case of H_2 in Figure 31.7, and the water is flowing in most cases, operating losses can be expected. A high uncertainty factor for loss is the gas permeability of the covering rock layer for natural gas or hydrogen. For aquifers in general, it is sufficient for the covering rocks to repel water, but for gas storage the covering layer must impermeable for the gas to keep back the stored gas even when the rock dries due to the gas storing. Research and field studies are still needed to quantify the losses arising from these uncertainties. Therefore, the use of aquifers as storage sites is currently not of great interest [40].

As can be seen from Figure 31.6, this kind of geological gas storage might nevertheless be suitable for most of the north and north-eastern parts of Germany, such as the area along the river Danube.

31.5.4
Abandoned Mining Sites

Another possible way to store large amounts of gas could be to use abandoned mining sites or those no longer of economic interest. Although many mining sites exist in Germany, as can been seen from Figure 31.8, only two sites are currently used for natural gas storage, one close to Halle and the other close to Treuchtlingen.

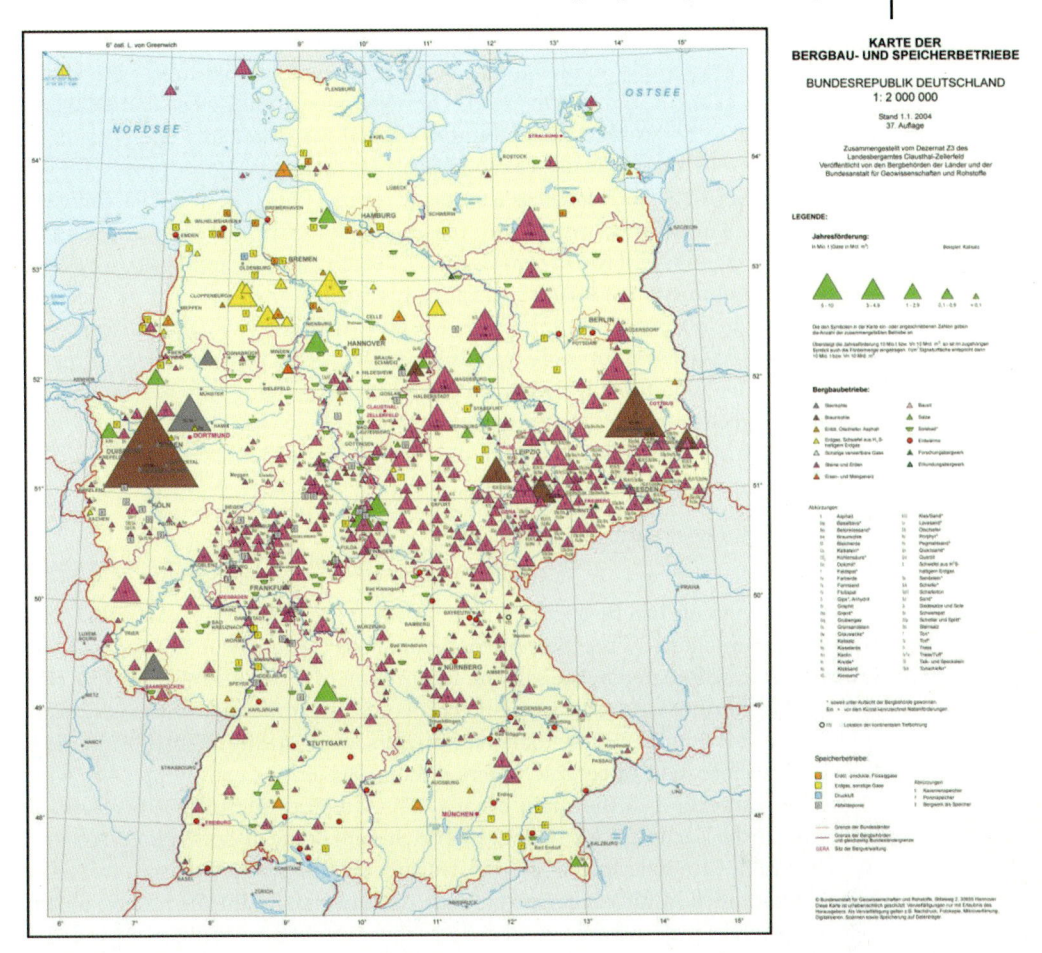

Figure 31.8 Official map showing the position, type, size, and use of abandoned mining sites in Germany [41].

Apart from the information that about 0.01% of the total natural gas storage is covered with this kind of storage [42], unfortunately no public data about these sites are available.

The small number of sites also indirectly shows that there are major problems such as mechanical stability and the gas tightness of the site. Most of the mining sites were formerly used to mine ore or other solids, so the mine walls are commonly brittle and therefore prone to cracks when a pressure cycle is applied. Covering the mine walls with asphalt, concrete, or steel might solve this issue and also the gas and water tightness. Another option is the so-called water curtain, where water is pressed into the surrounding rocks in order to prevent gas from permeating through cracks and fissures. However, depending on the geology, this might be costly. Nevertheless, this kind of storage site might cover the needs of middle Germany, where other types of geological storages are not feasible, especially when using storage tubes.

A smaller facility of this type is already available and in operation at various locations in Germany. For example, in Bietigheim-Bissingen, a bundle of six tubes each with a length of 142 m and a diameter to 1.4 m is buried underground, close to the surface [43]. Operated at 22 bar, the total amount of natural gas stored to smooth the daily peaks is ~2.5×10^4 scm. It is used to cover peak gas demands. Nevertheless, for the large total amount of gas needed as a strategic reserve, this type of small storage is not favorable. For other countries with massive rock formations such as gneiss or granite, this option might worth considering. In South Korea, natural gas with a total working gas volume to 2.2×10^6 scm is already stored in artificial caverns specially built for storage purposes [44, 45], and in Norway experience with operating mines with propane has existed for more than 30 years [46]. Often so-called water curtains have to installed additionally to reduce gas losses [47]. Using this technique, the currently largest storage of this kind worldwide is installed in Japan and has a total volume of 8.6×10^5 scm [48].

31.5.5
Salt Caverns

This kind of storage can be operated in areas where large amounts of soluble salts with high homogeneity are present. Homogeneity means that the mechanical properties are dominated by the salt and the amount of insoluble rocks such as clay are low. Furthermore, the salt is not brecciated. Such formations with significant thicknesses, for example, more than a few meters, were mostly formed by the slow but steady evaporation of oceanic waters in huge flat seas of continental size with only little connection to the main oceans [23]. A typical member of this type of formation is the Zechstein formation. By enrichment of the dissolved ions, first the most insoluble minerals such as carbonates and sulfates precipitated and formed a stable bottom rock. With further evaporation of the water, finally the highly soluble minerals such as rock salt precipitated. Since this was not a one-time event and this evaporation took place several times, huge salt layers with a total thickness of up to 1000 m could be formed [49, 50]. After the final evaporation of the sea water, the remaining salt basin was covered with dust and rocks from surrounding mountains by wind and rain. The meteoric water, for example, originating from rainfall and forming groundwater, later also separated the minerals in the salt layer carrying the more highly soluble fractions such as the rock salt deeper and leaving the less soluble minerals on top. With time, a cap layer of mainly sulfates such as anhydrite formed, protecting the underlying remaining salt from further erosion since less water could permeate through that layer. As more and more rocks formed on top of the salt layers with time, the salt layers started to creep significantly owing to the increasing lithostatic pressure on the geological timescale and were extruded at fault zones or weak cover parts [51, 52]. There the salt moved upwards and formed geological term-, cushion-, or dome-like structures called diapir [53], as shown in Figure 31.9. In Europe, these structures are mainly found in Germany, Poland, The Netherlands, and the United Kingdom.

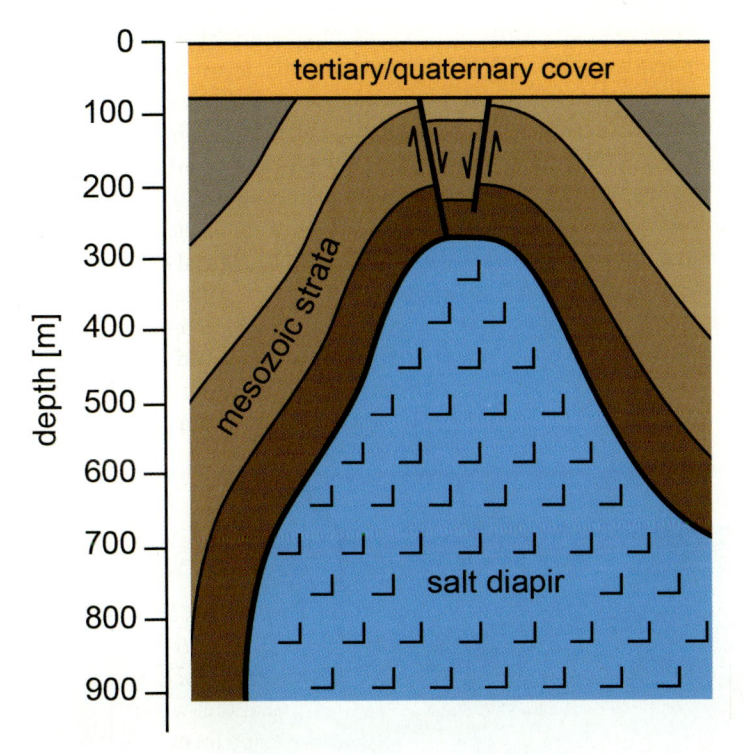

Figure 31.9 Scheme of the Asse diapir close to Braunschweig. The dome-like structure of the salt can be clearly seen, freely adapted from [34].

With time, these dome-like structures evolved to very great thicknesses of up to 9 km and reach close to the surface, as can be seen in Figure 31.10. In some areas such as Turkey, these diapirs broke through the surface and formed hills. Owing to the vertical alignment, these structures are of great interest for solution mining techniques and therefore also as storage sites.

An advantage of using salt formations is the gas tightness. As this kind of geological storage site has already been in use for natural gas for decades, experience shows that these storage sites are gas-tight for methane, and the same is considered to be true for H_2. Although laboratory-scale experiments are difficult, the diffusion coefficient of hydrogen in rock salt can be estimated as 2×10^{-13} m^2 s^{-1} (interpolated to 40 °C) [55]. Experimental data show an effective permeability for unscratched salt between $K = 10^{-21}$ and 10^{-23} m^2. Estimations with these data lead to an expected loss rate of about 0.015% per annum for a cavern with a geometric size of 5×10^5 m^3 [56]. However, the real values can only be calculated from core samples from the actual location and with more experimental data.

A positive aspect of using salt formations for gas storage sites arises from the high ductility and creep tendency [25, 57]. If operated correctly, small fractures of the cavern walls due to pressure cycling can be closed again using the creep behavior of the salt by adjusting the operating pressure of the cavern itself [58]. However, the ductility of the salt also results in a disadvantage already observed for natural gas storage sites. A convergence of the cavern is observed, for example, the cavern itself becomes smaller with time. Creep is one reason for convergence, and another is the continual reduction of the rock pressure on the salt horizons as the additional weight of the former glaciers apparent on the surface during the last Ice Age are still leading to an increased upward flow of the salt at the fault zones and diapirs. However, this effect can be mitigated by repeating the solution mining process from time to time.

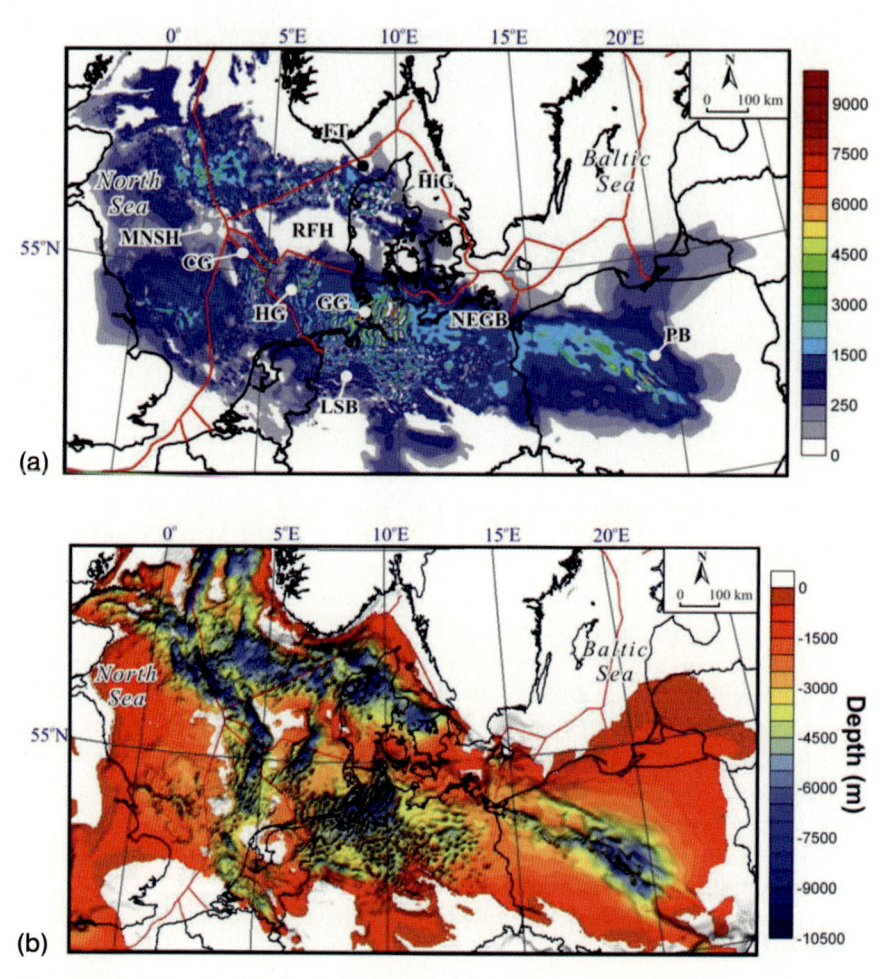

Figure 31.10 (a) Depth of the formation and (b) depth from the surface of the northern German salt deposit [54].

31.5 Types of Geological Gas Storage Sites

For this type of storage, highly water-soluble rock formations – mainly rock salt (NaCl) – are used. To create the storage, the caverns are produced by solution mining. Vertically aligned cylindrical hollow bodies are formed, as shown in Figure 31.11. In general, fresh water is pumped into a stabilized bore hole. In coastal areas, sea water is preferred as the feed, since it is less costly and the solution process is faster than with fresh water. The water resides in the cavern and is enriched with dissolved salt until a saturated salt solution or brine is formed. Since the density of the brine is higher than that of the feed water, the brine sinks and accumulates at the bottom. Depending on the local geology and preferred shape of the cavern, the brine is pumped out on the bottom of the cavern and replaced with feed water close to the top, as shown in Figure 31.11, or vice versa. Hence a steady circulation and a fast excavation are achieved. By adjusting the position of the feed and the drain for the brine, the subsequent shape of the cavern can be controlled. However, this technique also defines the vertical shape. The initial phase is slow until saturation of the brine is achieved, but the mining speed increases asymptotically in the first 100 days [59] and then an almost constant mining rate is reached.

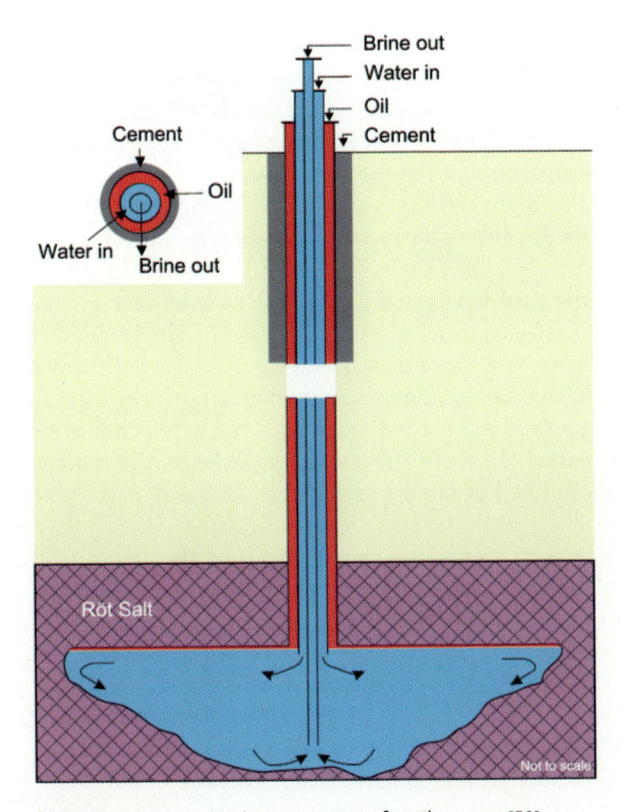

Figure 31.11 Typical solution mining of a salt cavern [53].

Depending on the composition and type of brine, for example if the brine contains a high concentration of potassium, it is even processed during excavation for further use in the chemical industry. Otherwise the brine pumped off is either dumped in rivers or the sea. When the salt cavern is operated later with a nearly constant internal pressure, part of the brine is collected in a large temporary brine storage pond. There are two different ways to operate salt cavern storage sites: either dry or in exchange with brine. In the first case, the salt cavern is dried after excavation by replacing the brine with, for example, natural gas or hydrogen. Several fill and drain cycles might be needed to reach an acceptable humidity level of the drained gas. In that stage, the cushion gas volume that normally resides in the cavern for its lifetime is also inserted. For the second case, the cavern can be operated directly after excavation. The brine is pushed out by the gas filling gas of the cavern and collected in an above-ground reservoir. The brine is reused to keep the pressure in the cavern stable when gas is drained from it. On increasing the pressure of the filled in brine, the drain rate can be enhanced. Additionally, the cushion gas volume used for this kind of operation is lower than that for a dry cavern. However, a more complex infrastructure is needed. Wetting of the gas can be an issue, so drying stages for the draining have to be considered. Additional operating costs are also likely since losses due to the solubility of the gas in the brine occur.

The currently operated salt caverns have an average diameter of several tens of meters. Those planned for hydrogen may have a height of up to 300 m and a diameter between 40 and 50 m [60]. This diameter is not constant over the height since the salt formations are not fully homogeneous. Therefore, a demanding part of the excavation process is the control of the geometry to produce mechanically stable caverns. Since more and more caverns are being built using the solution mining technique, this is now becoming an increasingly reliable state-of-the-art process.

In Germany, the salt caverns used for natural gas storage have an average size of about 6×10^7 scm each. At selected locations, up to 30 caverns are operated at the same site, as shown in Figure 31.1. Since these storage units are hollow bodies and have few problems with pressure gradients inside, they are mainly used to smooth the short-term gas supply and demand fluctuations due to peak demands and shortages on the gas market [4]. From operational experience, the average time for draining a full storage unit down to the cushion gas level is 27 days at the maximum drain rate.

However, this extreme operation scenario is very stressful on the cavern, and micro-fracturing and partial disintegration of the cavern wall can occur. This can be accepted to a certain extent when it is already included in the design of the cavern itself. Steady and thorough monitoring of the cavern is nevertheless mandatory to prevent a premature end of life of the cavern due to weakening of the salt rocks [16].

At the end of the operation time of the cavern, there are two main options for a safe shut-down and closing of the storage site. Either the cavern is sealed with the cushion gas still included to sustain the mechanical stability further or the cavern is refilled with brine or water. As there is currently no practical experience with salt cavern shut-down, the options are still under consideration [16].

Especially for Germany, another issue with the salt caverns exists. Although they seem to be the best choice in terms of short- and also long-term cycling, variable drain and filling rates, and low gas losses, the favorable sites are located mainly in the northern part since that had been the basin of the Zechstein Sea. For the larger cities in the southern part of Germany, only a few useable salt formations are present, located in parts of Baden-Württemberg and south of Munich, as can be seen from Figure 31.12 and Figure 31.13. Since these salt layers are thin, for example only up to 40 m in height and horizontally aligned [63], no expressed diapir formation is observed. Currently, solution mining techniques to use these formations are being evaluated. A main drawback, however, will be the huge ceiling area of such caverns that will lead to high mechanical stresses and make the operating conditions harsher. Such horizontally aligned caverns are more prone to failure via collapse as in traditional mining sites.

Concerning the operation of salt caverns with hydrogen, one exists in Germany that is operated with town gas containing ~60% hydrogen. It is located in Kiel and has a geometric volume of ~3.2×10^4 m^3, and has been operated at 80–160 bar since 1971 without any documented severe problems [64].

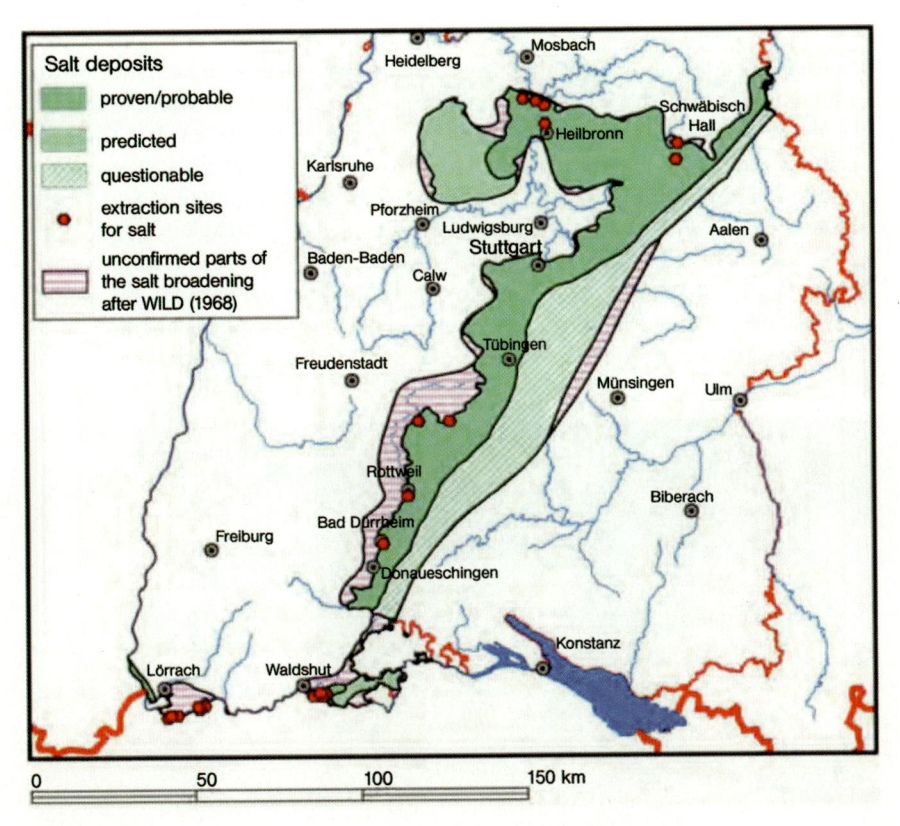

Figure 31.12 Salt deposits in southern Germany: Baden-Württemberg [61].

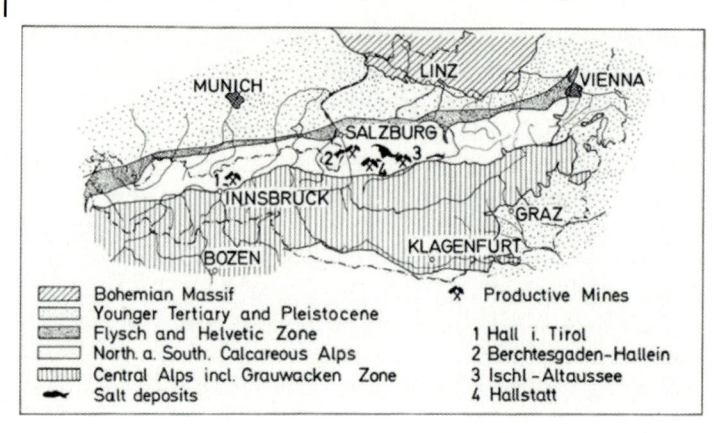

Figure 31.13 Salt deposits in southern Germany: Bavaria [62].

31.6
Comparisons with Other Locations and Further Considerations with Focus on Hydrogen Gas

Since the considerations and options discussed so far are not just valid for German locations, and many discussions are taking place regarding hydrogen, it is worth looking at current installations of hydrogen storage sites worldwide. Of special interest is Teesside, United Kingdom, with three caverns each with a 1.5×10^5 scm working gas capacity (total of 600 t, 5 t h^{-1} drain rate) [58, 65] and Spindletop, Texas, United States (Figure 31.14) with a working volume of 8.5×10^7 scm and a cushion gas amount of 5.8×10^7 scm. The site in Teesside, operated by Sabic UK Petrochemicals, is salt cavern operated at a constant pressure of 45 bar with a brine storage pond used to equilibrate short-term demands [58]. The one in Spindletop is a dry salt cavern. However, a direct comparison to the situation in Germany is inappro-

Figure 31.14 Salt cavern hydrogen storage site run by Air Liquide in Spindletop, Texas [66].

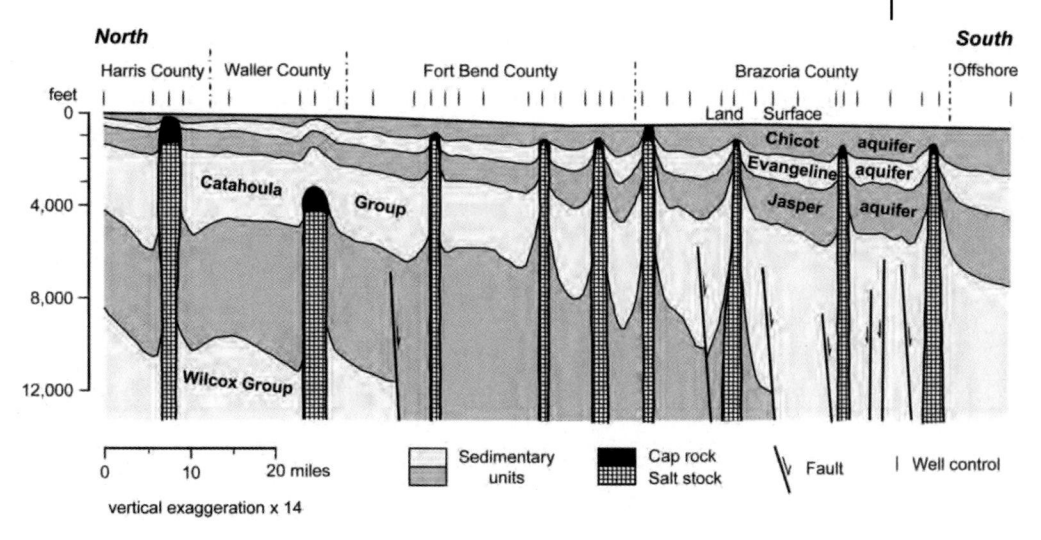

Figure 31.15 North–south cut of the geological situation in Texas. The Spindletop cavern resides in one of the outermost right diapirs at the border between Brazoria County and offshore (not shown here) [37].

priate, mainly owing to the geological situation. In Teesside, a flat salt layer with a low height of the salt layer is found, requiring operation with constant pressure and therefore of pendulum type. In Spindletop, the cavern is located in a diapir as shown in Figure 31.15. The main difference between there and the situation in Germany is the geometry of the diapir. In Spindletop it has a smaller diameter since the salt rock forming these diapirs originates from greater depths than in Germany. Additionally, the diapirs of this area are surrounded with huge aquifer systems with fairly high flow rates [60]. The operating conditions are therefore limited and a high cushion gas level is needed to mechanically stabilize the cavern.

Another aspect that must to be handled in addition to the technical issues is public acceptance. This can be very controversial, as studies from the Teesside location show [67]. In contrast, such problems are yet unknown for Germany.

31.7 Conclusion

There is not a specific geological storage option that will work for all of Germany owing to the various geological situations found in the German underground. Salt caverns have the greatest potential as the technology to produce and operate them is well known. There are several locations still available to build new ones, so little or no interference with existing natural gas storage sites will arise. Size seems to be of little concern and the production time is fairly short. Additionally, the requirements for purity in the case of hydrogen as operating gas can easily be fulfilled and the infrastructure can be kept relatively simple. For the northern part of Germany, dry storage sites will be preferred, whereas for the southern part the pendulum type might be a reasonable option.

In contrast, aquifer storage sites are more costly to operate but might be of interest for the southern part of Germany since huge working volumes can be achieved. The drawbacks are problems due to humidified gas and the low drain and fill rates. Unresolved issues include not only the lack of exact knowledge about connections of the aquifers [68] but also the unclear situation about the gas tightness and losses.

Finally, to cover the middle part of Germany and the short-term fluctuations in the southern part, the modification of abandoned mining sites including tubular subsurface storage sites might be an option. As a last alternative, above-ground storage sites of the KSC/NASA type should be considered.

The overall situation can be therefore been seen as positive, especially in terms of further geological storage sites. This is also valid for hydrogen gas, since large and safe storage sites can be built using existing geological storage technologies on reasonable and predictable timescales.

References

1 Council of the European Union (2009) Council Directive 2009/119/EC of 14 September 2009 Imposing an Obligation on Member States to Maintain Minimum Stocks of Crude Oil and/or Petroleum Products. *Off. J. Eur. Union*, **L 265**, 9–23.
2 Biegel, B., *et al.* (2012) Model predictive control for power flows in networks with limited capacity, presented at the American Control Conference (ACC).
3 Flottenkommando Dezernat Handelsschifffahrt und Marineschifffahrtleitung (2012) *Kennzahlen zur maritimen Abhängigkeit der Bundesrepublik Deutschland*, Bundeswehr Marine, Berlin.
4 Flottenkommando Deutsche Marine (2012) *Jahresbericht 2012 – Fakten und Zahlen zur maritimen Abhängigkeit der Bundesrepublik Deutschland*, Bundeswehr Marine, Berlin.
5 European Commission (2011) *Communication from the Commission to the European Parliament, the Council, the European Economic and Social Committee and the Committee of the Regions: a Roadmap for Moving to a Competitive Low Carbon Economy in 2050*, Publications Office of the European Union: EUR-Lex, Brussels.
6 LBEG: Landesamt für Bergbau Energie und Geologie Niedersachsen (2011) *Erdöl und Erdgas in der Bundesrepublik Deutschland 2011*, Landesamt für Bergbau, Energie und Geologie Referat Energiewirtschaft Erdöl und Erdgas, Bergbauberechtigungen, Hannover.
7 Müller-Syring, G., *et al.* (2011) Power to gas: Untersuchungen im Rahmen der DVGW-Innovationsoffensive zur Energiespeicherung. *bbr*, **2011** (10), 14–21.
8 Flottenkommando Deutsche Marine (2011) *Jahresbericht 2011 – Fakten und Zahlen zur maritimen Abhängigkeit der Bundesrepublik Deutschland*, Bundeswehr Marine, Berlin.
9 BMWi and BMU (2010) *Das Energiekonzept der Bundesregierung 2010 und die Energiewende 2011*, Bundesministerium für Wirtschaft und Technologie (BMWi), Öffentlichkeitsarbeit und Bundesministerium für Umwelt Naturschutz und Reaktorsicherheit (BMU), Abteilung K I.
10 Schmitz, S. (2012) Gasspeicherung – Voraussetzung für die Energiewende, in *6. Dow Jones Konferenz – Gasmarkt 2012*, DBI – Gastechnologisches Institut, Freiberg, and Aninstitut der TU Bergakademie Freiberg, Frankfurt, pp. 1–44.

11 Gu, L., et al. (2005) *LH₂ Storage Tanks at KSC*, https://securedb.fsec.ucf.edu/hr/document_api.download?file=F856558843/Gu%20-%20LH2%20Storage%20Tanks%20at%20KSC.pdf (last accessed 18 March 2013).

12 Tzimas, E., et al., *Hydrogen Storage: State-of-the-Art and Future Perspective*, European Commission Joint Research Centre, Petten, ISBN 92-894-6950-1.

13 Flottenkommando Deutsche Marine (2009) *Jahresbericht 2009 – Fakten und Zahlen zur maritimen Abhängigkeit der Bundesrepublik Deutschland*, Bundeswehr Marine, Berlin.

14 Helmholtz-Zentrum Potsdam Deutsches GeoForschungsZentrum GFZ (2011) Untertage-Gasspeicherung in Deutschland. *Erdöl Erdgas Kohle*, **126** (11), 394–403.

15 LBEG: Landesamt für Bergbau Energie und Geologie Niedersachsen (2012) Untertage-Gasspeicherung in Deutschland/Underground gas storage in Germany. *Erdöl Erdgas Kohle*, **128** (11), 412–421.

16 Lux, K.-H. (2009) Design of salt caverns for the storage of natural gas, crude oil and compressed air: geomechanical aspects of construction, operation and abandonment. *Geol. Soc. London Spec. Publ.*, **313** (1), 93–128.

17 Yardley, G. S. and Swarbrick, R. E. (2000) Lateral transfer: a source of additional overpressure? *Mar. Pet. Geol.*, **17** (4), 523–537.

18 The Engineering ToolBox (2012) *Solubility of Gases in Water*, http://www.engineeringtoolbox.com/gases-solubility-water-d_1148.html (last accessed 16 January 2012).

19 Molkov, V. (2008) Hydrogen safety research: state-of-the-art, in *5th International Seminar on Fire and Explosion Hazards*, University of Edinburgh, p. 16.

20 Meacham, J. E. (2003) *Flammable Gas Diffusion Through Single-Shell Tank Domes*, US Department of Energy, Washington, DC.

21 Buchholz, D. (2012) Verfügbarkeit Infrastruktur: Speicherfähigkeit als Schlüssel, in *Biogaspartner – die Konferenz: Biomethan – der Joker der Energiewende*, Deutsche Energie-Agentur (dena), Berlin, pp. 1–14.

22 Hänel, R. and Homilius, J. *Geophysik bei den Geowissenschaftlichen Gemeinschaftsaufgaben in Hannover*, http://www.dgg-online.de/geschichte/kapitel_pdf/kapitel_2_7.pdf (last accessed 16 January 2012).

23 Warren, J. K., et al. (2008) Salt dynamics, in *Dynamics of Complex Intracontinental Basins* (eds. R. Littke et al.), Springer, Berlin, pp. 248–344.

24 Cosenza, P. and Ghoreychi, M. (1999) Effects of very low permeability on the long-term evolution of a storage cavern in rock salt. *Int. J. Rock Mech. Min. Sci.*, **36** (4), 527–533.

25 BGR (1995) *Untersuchung und Bewertung von Salzformationen*, Bundesanstalt für Geowissenschaften und Rohstoffe, Hannover, http://www.bgr.bund.de/DE/Themen/Endlagerung/Downloads/Schriften/3_Wirtsgesteine_Salz_Ton_Granit/BGR_salzstudie.html (last accessed 16 January 2012).

26 BGR (2004) *Geowissenschaftliche Karte der Bundesrepublik Deutschland 1 : 2 000 000 (GK2000)*, Bundesanstalt für Geowissenschaften und Rohstoffe, Hannover, http://www.bgr.bund.de/DE/Themen/Sammlungen-Grundlagen/GG_geol_Info/Karten/Deutschland/GK2000/gk2000_node.html (last accessed 13 February 2012).

27 Sedlacek, R. (1999) Untertage Erdgasspeicherung in Europa/Underground gas storage in Europe. *Erdöl Erdgas Kohle*, **115** (11), 537–541.

28 Kühn, M. (2011) CO_2-Speicherung. Chancen und Risiken. *Chem. Unserer Zeit*, **45** (2), 126–138.

29 BGR (1984) *International Map of Natural Gas Fields in Europe 1 : 2 500 000 (IGasFE2500)*, Bundesanstalt für Geowissenschaften und Rohstoffe, http://www.bgr.bund.de/DE/Themen/Sammlungen-Grundlagen/GG_geol_Info/Karten/International/Europa/%C3%96l-%20und%20Gasfelder/oel_gasfelder_node.html (last accessed 16 January 2012).

30 NATURALHY (2008) *Workshop: Hydrogen – Does it Have a Future in Natural Gas Networks?*,

http://www.naturalhy.net/docs/3rd_workshop/Naturalhy%20Workshop%20IGRC%20Rapporteur%20feedback.pdf (last accessed 16 January 2012)

31 Bondi, A. (1964) van der Waals volumes and radii. *J. Phys. Chem.*, **68** (3), 441–451.

32 Hirschfelder, J. O., et al. (1954) *Molecular Theory of Gases and Liquids*, John Wiley & Sons, Inc., New York.

33 Gaupp, R., Liermann, N., and Pusch, G. (2005) Adding value through integrated research to unlock the tight gas potential in the Rotliegendes Formation of northern Germany, in *SPE Europec/EAGE Annual Conference*, Society of Petroleum Engineers, Madrid, p. 9.

34 Bentz, A. and Martini, H.-J. (eds.) (1968) *Lehrbuch der Angewandten Geologie, Zweiter Band, Erster Teil, Geowissenschaftliche Methoden*, Enke, Stuttgart, p. 1071.

35 Schmitz, S. (2011) Einfluss von Wasserstoff als Gasbegleitstoff auf Untergrundspeicher, in *DBI-Fachforum Energiespeicherkonzepte und Wasserstoff*, DBI Gas- und Umwelttechnik, Berlin.

36 BGR. *Distribution of Aquifer Types in Germany*, Bundesanstalt für Geowissenschaften und Rohstoffe, http://www.bgr.bund.de/EN/Themen/Wasser/Bilder/Was_wasser_startseite_gwleiter_g_en.html (last accessed 13 February 2012).

37 Texas Water Development Board (2006) *Report 365 – Aquifers of the Gulf Coast of Texas*, http://www.twdb.state.tx.us/publications/reports/numbered_reports/doc/R365/Report365.asp (last accessed 9 February 2013).

38 Lord, A. S. (2009) *FY 2009 Annual Progress Report – III. Hydrogen Delivery – 22. Geological Storage of Hydrogen*, http://www.hydrogen.energy.gov/annual_progress09_delivery.html (last accessed 9 February 2013).

39 Pray, H. A., Schweickert, C. E., and Minnich, B. H. (1952) Solubility of hydrogen, oxygen, nitrogen, and helium in water at elevated temperatures. *Ind. Eng. Chem.*, **44** (5), 1146–1151.

40 Bayerisches Staatsministerium für Wirtschaft Verkehr und Technologie (2001) *Rohstoffe in Bayern*, BayStMWV, Munich.

41 BGR. *Bergbau Deutschland*, Bundesanstalt für Geowissenschaften und Rohstoffe, http://www.bgr.de/karten/bergbau_spbetriebe/bergbau_2004.pdf (last accessed 16 January 2012).

42 WEG (2012) *Jahresbericht 2011: Zahlen & Fakten*, Wirtschaftsverband Erdöl- und Erdgasgewinnung, Hannover, pp. 1–64.

43 Stadtwerke Bietigheim-Bissingen (2001) *Gasspeicher*, http://www.sw-bb.de/de/privatkunden/erdgas/service-erdgas/kundeninformationen/gasspeicher/ (last accessed 13 February 2012).

44 Lee, C.-I. and Song, J.-J. (2003) Rock engineering in underground energy storage in Korea. *Tunnelling Underground Space Technol.*, **18** (5), 467–483.

45 Park, J. J., Jeon, S., and Chung, Y. S. (2005) Design of Pyongtaek LPG storage terminal underneath Lake Namyang: a case study. *Tunnelling Underground Space Technol.*, **20** (5), 463–478.

46 Blindheim, O. T., Broch, E., and Grøv, E. (2004) Gas storage in unlined caverns – Norwegian experience over 25 years. *Tunnelling Underground Space Technol.*, **19** (4–5), 367.

47 Kjørholt, H. and Broch, E. (1992) The water curtain – a successful means of preventing gas leakage from high-pressure, unlined rock caverns. *Tunnelling Underground Space Technol.*, **7** (2), 127–132.

48 Yoshida, H., et al. (2013) Features of fractures forming flow paths in granitic rock at an LPG storage site in the orogenic field of Japan. *Eng. Geol.*, **152** (1), 77–86.

49 Geluk, M. C. (2007) Permian, in *Geology of The Netherlands* (eds. T. E. Wong, D. A. J. Batjes, and J. de Jager), Royal Netherlands Academy of Arts and Sciences, Amsterdam, pp. 63–83.

50 Kucha, H. and Pawlikowski, M. (1986) Two-brine model of the genesis of strata-bound Zechstein deposits (Kupferschiefer type), Poland. *Mineral. Deposita*, **21** (1), 70–80.

51 Mohr, M., et al. (2005) Multiphase salt tectonic evolution in NW Germany: seismic interpretation and retro-deformation. *Int. J. Earth Sci.*, **94** (5–6), 917–940.

52 Warsitzka, M., Kley, J., and Kukowski, N. (2012) Salt diapirism driven by differential loading – some insights from analogue modelling. *Tectonophysics*, in press.

53 Geluk, M. C., Paar, W. A., and Fokker, P. A. (2007) Salt, in *Geology of The Netherlands* (eds. T. E. Wong, D. A. J. Batjes, and J. de Jager), Royal Netherlands Academy of Arts and Sciences, Amsterdam, pp. 283–294.

54 Maystrenko, Y., Bayer, U., and Scheck-Wenderoth, M. (2010) *Structure and Evolution of the Central European Basin System According to 3D Modeling*, DGMK Research Report 577-2/2, DGMK, Hamburg, 90 pages.

55 Sutherland, H. J. and Cave, S. P. (1980) Argon gas permeability of New Mexico rock salt under hydrostatic compression. *Int. J. Rock Mech. Min. Sci., Geomech. Abstr.*, **17** (5), 281–288.

56 Jockwer, N. and Wieczorek, K. (2008) Excavation of damaged zones in rock salt formations, in *Waste Management Conference 2008*, Phoenix, AZ, p. 8172.

57 Yang, C., Daemen, J. J. K., and Yin, J.-H. (1999) Experimental investigation of creep behavior of salt rock. *Int. J. Rock Mech. Min. Sci.*, **36** (2), 233–242.

58 Ozarslan, A. (2012) Large-scale hydrogen energy storage in salt caverns. *Int. J. Hydrogen Energy*, **37** (19), 14265–14277.

59 Crotogino, F. (2011) Wasserstoffspeicherung im geologischen Untergrund – Stand der Technik und Potential, in *Fachkonferenz Energiespeicher für Deutschland*, Süddeutscher Verlag Veranstaltungen, Cologne.

60 Acht, A. (2012) Gasspeicher – Technologie und Herausforderungen, in *Leuphana Energieforum 2012*, DEEP Underground Engineering, Bad Zwischenahn.

61 Regierungspräsidium Freiburg (2009) *Die Verbreitung der steinsalzführenden Schichten in Baden-Württemberg – eine Aktualisierung des Wissenstandes*, LGRB-Nachrichten, Regierungspräsidium Freiburg Landesamt für Geologie Rohstoffe und Bergbau, Freiburg, vol. 8, p. 2.

62 Ambatiello, P. and Ney, P. (1983) *The Berchtesgaden Salt Mine, in Mineral Deposits of the Alps and of the Alpine Epoch in Europe* (ed. H.-J. Schneider), Springer, Berlin, Heidelberg, p. 146–154.

63 Gillhaus, A., et al. (2006) *Compilation and Evaluation of Bedded Salt Deposit and Bedded Salt Cavern Characteristics Important to Successful Cavern Sealing and Abandonment*, Research Project Report 2006-2-SMRI, Solution Mining Research Institute, Clarks Summit, PA.

64 Sherif, S. A., Barbir, F., and Veziroglu, T. N. (2005) Wind energy and the hydrogen economy – review of the technology. *Solar Energy*, **78** (5), 647–660.

65 Pritchard, D. (2005) Blowing bubbles: are wind and hydrogen electrolysis the ultimate couple? *Power Eng.*, **19** (1), 32–33.

66 Parsons Brinckerhoff (2012) *Underground Storage*, http://www.pbworld.com/capabilities_projects/power_energy/underground_storage.aspx (last accessed 16 January 2012).

67 Ricci, M., Bellaby, P., and Flynn, R. (2008) What do we know about public perceptions and acceptance of hydrogen? A critical review and new case study evidence. *Int. J. Hydrogen Energy*, **33** (21), 5868–5880.

68 Hebig, K. H., et al. (2012) Review: deep groundwater research with focus on Germany. *Hydrogeol. J.*, **20** (2), 227–243.

32
Near-Surface Bulk Storage of Hydrogen

Vanessa Tietze, Sebastian Luhr and Detlef Stolten

32.1
Introduction

Future sustainable energy systems face the challenging task of simultaneously providing energy supply security, economic competitiveness, environmental friendliness, and safety. Strict and consistent protection of people and nature will eventually demand the abandonment of exploitation and combustion of fossil energy sources and also phasing out of nuclear power usage. Therefore, the only option remaining in the long term is the transition to renewable energy sources and the replacement of current fossil energy carriers.

Hydrogen offers many advantages as an energy carrier for future sustainable energy systems. It can play an important complementary role to electricity since it can be produced from and converted into electricity with relatively high efficiency [1] and it can easily be stored in large quantities. These properties allow an increased amount of intermittent renewable energy sources to be integrated into the power generation mix. In addition to converting hydrogen back to electrical energy, it can also be employed as a carbon-free fuel in the transport sector. This makes hydrogen one of the few energy carriers that enable renewable energy also to be introduced into transport systems [2]. Hydrogen use as a transportation fuel is in economic and environmental terms with respect to CO_2 reduction the more efficient option [3]. Therefore, it is chosen as the main final utilization application in this chapter.

Currently, most scientific interest is concentrated on the challenging task of developing hydrogen vehicles with similar performance and cost levels to today's internal combustion engine vehicles. Within the framework of the FreedomCAR and Fuel Partnership, performance requirements for passenger vehicles were converted into concrete values for the needs of storage systems, resulting in the US Department of Energy (DOE) hydrogen storage system targets [4]. Supported research areas include both reversible on-board storage and regenerable off-board storage approaches, but none of them is yet able to meet all of the DOE performance targets [5]. Reversible on-board storage includes all methods that allow storage of hydrogen in molecular

Transition to Renewable Energy Systems, 1st Edition. Edited by Detlef Stolten and Viktor Scherer.
© 2013 Wiley-VCH Verlag GmbH & Co. KGaA. Published 2013 by Wiley-VCH Verlag GmbH & Co. KGaA.

or in bonded form with low binding energies. In the former case, densities sufficient for storage can be achieved by physical methods such as compression and liquefaction. In the latter, hydrogen is stored in molecular or atomic form absorbed in or adsorbed on solid-state materials [6]. Examples are metal hydrides, high surface area sorbents, and carbon-based materials [5]. In regenerable off-board storage approaches, the hydrogen is stored in irreversible chemical form in hydrogen-rich liquids and solids, which results in much higher binding energies [5, 6]. Since most of the reactions involved are not reversible under operating conditions convenient at refueling stations, the spent material has to be regenerated off-board the vehicle [5, 7]. Examples are chemical hydrides, organic liquids, ammonia borane, and other boron hydrides [5].

In order to introduce hydrogen widely as a fuel for the transport sector, in addition to the question of the best automotive on-board storage solution, the question of a feasible and cost-effective means for its delivery also has to be addressed. Here delivery means the process of transporting and distributing hydrogen from the production site to the final utilization application [8]. This can be done by gaseous, liquid, and carrier-based delivery pathways. Hydrogen carriers store hydrogen in any chemical state other than free molecules [9]. Especially which approach, reversible on-board storage or regenerable off-board storage, will be successful would have a huge impact on the delivery infrastructure. Currently, the state-of-the-art for onboard hydrogen storage is compressed gaseous tanks with pressures of 350 and 700 bar and cryogenic liquid tanks [4]. An analysis [10] came to the conclusion that using hydrogen carriers in a pathway that discharges hydrogen at the station and supplies compressed hydrogen to vehicles will hardly or not at all reduce the costs for fueling stations because there is still a need for compressed hydrogen equipment at the fueling station. Therefore, this chapter is limited to a delivery infrastructure that is designed for the transport of hydrogen in its molecular form.

Stationary storage facilities are a fundamental part of every hydrogen delivery system. Small-scale storage facilities are typically utilized at the distribution or final-user level, such as refueling stations, whereas large-scale storage facilities are typically applied at the production or transport level, including the production sites, the terminals of pipelines, and other transportation paths [11, 12]. Generally, bulk storage vessels located above or buried a few meters below the surface and underground storage are of relevance here. Compressed hydrogen storage in salt caverns is especially suitable for very large quantities [11, 13], but good geological conditions are limited to a few sites. For example, in the case of Germany they appear mainly in the northern part. Consequently, other options for the storage of large quantities of hydrogen could be of major importance in the future.

This chapter therefore concentrates on all storage mechanisms suitable for bulk storage at or near ground level. Principally, the hydrogen storage mechanisms for stationary applications are the same as for mobile applications, but the requirements differ and are more stringent in the latter case. The most common methods for stationary hydrogen storage are compressed gaseous hydrogen, cryogenic liquid hydrogen, and metal hydrides [14]. These could be applied in a future delivery infrastructure with gaseous- and/or liquid-based pathways and that is the focus here.

Emphasis is especially set on the description of the storage vessel: storage vessel types that can be considered as being readily available are described, and both technical and economic data are given. In addition, the storage capacity and efficiency of selected storage technologies are assessed in relation to the daily hydrogen demand of a future fueling station and with respect to the boundary conditions of a chosen delivery scenario. However, first some important storage parameters are defined and explained for the sake of clarity in the subsequent sections.

32.2
Storage Parameters

Assessing the suitability of different storage methods for a certain application field is best addressed by using appropriate storage parameters. It is common usage to compare storage technologies on the basis of their gravimetric and volumetric capacities. Gravimetric and volumetric capacities are defined as the amount of hydrogen stored per unit weight and volume, respectively [6, 15, 16]. However, these terms can be misleading since the amount of hydrogen stored can be given in terms of thermal energy, mass, or volume. To avoid ambiguity, the terms gravimetric and volumetric capacity are applied here only with respect to the mass of hydrogen stored per unit weight or volume. The terms specific energy and energy density are used here to define the amount of thermal energy per unit weight and volume, respectively. To express the amount of hydrogen stored in terms of thermal energy, mass, or volume, the terms energy, mass, or volume capacity are chosen. Thereby a difference is made between the total and usable amounts of hydrogen stored, which are defined here as gross and net storage capacity, respectively.

Furthermore, care has to be taken concerning the basis to which the specific energy, energy density, and gravimetric and volumetric capacity are referred. This can be the storage material alone, or including the vessel and also the total storage system with all the necessary components. The material-only value is a measure of the absolute capacity whereas the total system-based values represent an upper limit to the amount of usable hydrogen [15]. In the following, the term physical is used when reference is made solely to the storage material, and the term technical is used when reference is made to the storage vessel or the total system.

When hydrogen is stored in molecular form, the corresponding thermal energy can be calculated on the basis of the lower heating value (LHV). The physical specific energy is 33.33 kWh kg^{-1} (120 MJ kg^{-1}) and the energy density is 2.99 kWh m^{-3} (10.78 MJ m^{-3}) at standard temperature and pressure (STP) conditions. The former value remains constant and only the latter changes with density. Consequently, in this case only the physical energy density is the crucial storage parameter. It should be noted that for hydrogen carriers both the physical gravimetric capacity and the physical volumetric capacity are material-specific values and have to be determined empirically.

32.3
Compressed Gaseous Hydrogen Storage

32.3.1
Thermodynamic Fundamentals

At STP conditions, which means 0 °C and 1.01325 bar, the density of hydrogen is 0.09 kg m^{-3} [17]. When hydrogen is compressed to pressures of 250 and 1000 bar, the densities are 19.02 and 52.12 kg m^{-3}, respectively, at standard temperature [18]. From these data, it can be seen that there is no linear correlation between the density and the pressure. The thermodynamic properties of hydrogen can be calculated with only minor errors by using the ideal gas equation up to a pressure of 100 bar. However, for higher pressures the real gas behavior has to be considered, since at a pressure of 700 bar a large error of ~44% occurs when using the ideal gas equation [19].

For the calculation of the compression work, two different processes have to be distinguished. In the case of adiabatic compression, no heat exchange with the environment is considered. In contrast, for isothermal compression, cooling is assumed in such a way that the temperature remains constant. The power demand for real multistage compression processes lies between these two borderline cases [20]. Thereby, isothermal compression at ambient temperature represents the theoretical optimum, because work only decreases logarithmically with pressure. For example, the thermodynamic minimum work for the compression of hydrogen to pressures of 253 and 1013 bar (250 and 1000 atm) amounts to 1.5 and 2.0 kWh kg^{-1}, respectively [21]. Since in practice losses such as parasitic compression losses and the general energy consumption of the compressor station also have to be considered [20], the power demand is 2.5 and 4.0 kWh kg^{-1}, respectively [21], which correspond to 7.5 and 12% of the LHV of hydrogen. Other sources state a consumption of ~5 and ~20% of the LHV when compressing to 350 and 700 bar, respectively [22].

32.3.2
Hydrogen Compressors

Hydrogen compressors are employed satisfactorily at relatively large numbers in industrial gas handling and also in the chemical and oil industries. They can be regarded today as state-of-the-art [11]. Different compressor types such as reciprocating and rotary displacement compressors and also centrifugal compressors are generally used for the compression of gases. However, because of the special characteristics of hydrogen, only reciprocating displacement compressors are of practical use. In this category, piston, diaphragm, and ionic compressors can be discerned. The state-of-the-art type for high-volume applications is reciprocating pistons, whereas for small-volume applications pistons or diaphragms are used [9]. Another interesting approach is nonmechanical hydrogen compressors such as high-pressure electrolyzers, metal hydrides, and electrochemical compressors [20].

In a hydrogen delivery system, which is based on gaseous hydrogen, there are many different application areas for compressors. For example, in transmission pipelines compressors are needed that can manage a high throughput of 50 000–2 000 000 kg d^{-1} with a modest compression ratio, since the pressure typically has to be increased from ~5 to ~70 bar. However, in refueling stations a pressure of 350–700 bar has to be provided for high-pressure hydrogen onboard storage, but with lower flow rates of 50–3000 kg d^{-1} [9].

32.3.3
Hydrogen Pressure Vessels

The technique of storing hydrogen under pressure has been applied for many years and gaseous pressure vessels are currently the most common means of storing hydrogen [9]. Typical bulk gas storage systems commercially available today consist of a number of steel pressure tubes that can be manifolded together to give extended storage volumes (Figure 32.1). The systems are modular in design and hence can be sized according to customer use rates [23]. Depending on the purpose, the tubes can be erected horizontally or vertically and can be designed for different pressure ranges. Large cylindrical storage vessels for industrial applications typically are in the range 50–70 bar [24, 25] with a water volume of ~100–150 m^3 [26]. Larger volumes require operation at lower pressure [26]. However, some applications, for example fueling stations, require very high pressure tubes which can reach up to 1000 bar [9, 27]. There are many international and national standards and technical regulations for bulk gas storage vessels available. However, for the very high pressure range, additional testing and certification are required [27, 28].

Generally, four different types of high-pressure tubes can be discerned based on their structural element and their permeation barrier according to the classification of the European Integrated Hydrogen Project [9, 29]. These types have different main features, which are summarized in Table 32.1. For stationary gaseous storage, commonly pressure vessels of Type I are applied, which have the highest weight but are the cheapest [9, 27, 29]. For stationary applications with higher pressures, Type II is preferred. Type III and IV are much more expensive, but their lower weight makes transport easier. Consequently, a compromise between technical performance and cost-competitiveness has to be made with respect to the final application [29].

Figure 32.1 Different ground storage assemblies [30].

Table 32.1 Classification and main features of hydrogen pressure vessel types in 2006 [9, 29].

Type I	Type II	Type III	Type IV
All-metal cylinder	Load-bearing metal liner hoop wrapped with resin-impregnated continuous filament	Non-load-bearing metal liner axial and hoop wrapped with resin-impregnated continuous filament	Non-load-bearing, non-metal liner axial and hoop wrapped with resin-impregnated continuous filament
Technology mature: ++ Pressure limited to 300 bar ($\rightarrow$ density: –)	Technology mature: + Pressure not limited ($\rightarrow$ density: +)	Technology mature for $p \leq 350$ bar; 700 bar under development	Technology mature for $p \leq 350$ bar; 700 bar under development
Cost performance: ++	Cost performance: +	Cost performance: –	Cost performance: –
Weight performance: –	Weight performance: 0	Weight performance: +	Weight performance: +

The diameter, length, and pressure rating of each tube determine the quantity of hydrogen stored. On the one hand, a higher pressure level enables more hydrogen per unit volume to be stored, but on the other hand, thicker walls are necessary and the cost increases. Therefore, larger storage capacities are commonly realized on a modular basis. Typical industrial pressure tubes of Type I have a volume of 54 l and contain ~0.61 kg of hydrogen at a pressure of 156 bar at 21 °C. Modules are available in configurations of 3–18 tubes and can reach a hydrogen storage capacity of ~700 kg at 165 bar [9].

It is considered that pressure vessels which are typically used for the storage of natural gas or town gas are basically also suitable for decentralized hydrogen storage [11, 31]. These vessels exist in different sizes and shapes and are explained in the following.

Gas holders store gas at low pressure of only a few millibars above atmospheric pressure. There are many different ways of construction. Generally, they can be divided according to the mechanism of sealing in water-sealed gas holders and dry-type gas holders. At operating pressures of up to 1.5 bar, a gas volume of 600 000–700 000 m^3 can be stored in some types [32]. However, the space requirements are huge owing to the low pressure (Figure 32.2).

Gas holders were widely applied for town gas at the beginning of the twentieth century. However, some were even used for hydrogen, for example, a gas holder with a capacity of 56 000 m^3 in Neustadt/Coburg, Germany, which was built to fuel airships in 1914 [33]. Although this technique is now considered obsolete, there are still a few hydrogen gas holders in operation today (Table 32.2). One recent example is a gasholder at the Höchst Industrial Park. It collects hydrogen which is a chemical by-product for use at a multi-fuel filling station [34].

Another possibility is to use spherical pressure vessels. These allow relatively large amounts of gas to be stored in a small space. In comparison with low-pressure gas holders they offer the benefit of reduced procurement and operating costs [35].

32.3 Compressed Gaseous Hydrogen Storage

Figure 32.2 (a) Hydrogen gas holder of Infraserv at Höchst/Frankfurt, Germany [38] and (b) gas holders of CABB GmbH at Gersthofen, Germany [39].

Table 32.2 Technical data of hydrogen gas holders.

Operator		Infraserv Höchst	BASF	CABB GmbH
Site/year built		Höchst/1997[a]	Ludwigshafen/N. A.[a]	Gersthofen/1929[c]
Type/material		Dry type gas holder/ S235JRG2[a]	Water-sealed gas holder/ N. A.[a]	Water-sealed gas holder/ black steel[c]
Parameter	Unit			
Diameter	m	26[a]	N. A.	N. A.
Height	m	30[a]	N. A.	N. A.
Pressure	bar	1.07[b]	N. A.	1.025[c]
Net volume	m³	10,000[a,b]	30,000[a]	1,000[c]

N. A.: not available.
Source: [a][37], [b][34], [c][39].

Spherical pressure vessels can store gas volumes of up to 300 000 m³ and can be operated at pressures of up to 20 bar [36].

The first spherical pressure vessels were built in 1923 by the Chicago Bridge and Iron Company [40]. As in the case of gas holders, they were usually employed to store town gas. Table 32.3 provides some technical data. Today, spherical pressure vessels are a widespread technology. They are employed, for example, in the industrial sector for hydrocarbons and in public utilities for natural gas (Figure 32.3).

Table 32.3 Technical data of spherical pressure vessels for town gas [35, 37].

Parameter	Operator			
	STAWAG	GASAG	Public utility	Public utility
Site	Aachen	Berlin	Lahr	Stockholm
Form/material	Spherical/ STE 29	Spherical/ STE 51	Spherical/ STE 47	Spherical/ FG 51
Diameter (m)	27.16	39.5	26.74	22.4
Pressure (bar)	5.5	10.4	10	17
Water volume (m^3)	10 490	32 057	10 000	5885
Wall thickness (mm)	21	34	25.3	32–34

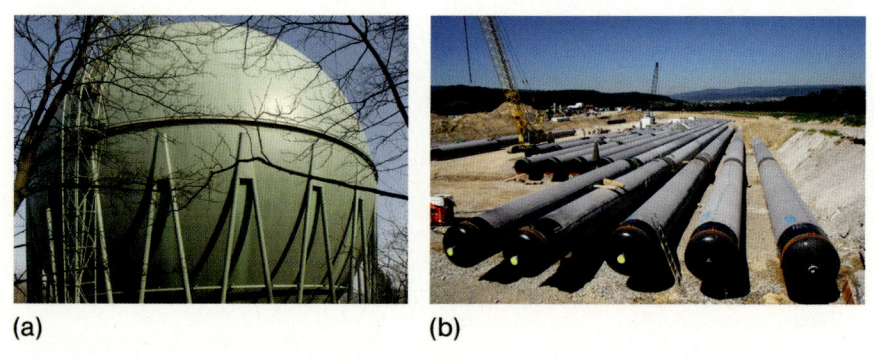

Figure 32.3 (a) Spherical pressure vessel of Wuppertaler Stadtwerke at Möbeck/Sonnborn, Germany [49] and (b) pipe storage facility of Erdgas Zürich at Urdorf, Switzerland [50].

A spherical pressure vessel for hydrogen with a volume of about 15 000 m^3 and an operating pressure between 12 and 16 bar has been reported [41–46], but despite intensive efforts no proof for the actual existence of this storage tank was found. Instead, back-tracing of the literature cited lead to a book chapter [11] published in 1988 with the statement that the application range of typical low-pressure spherical containers (> 15 000 m^3 at 12–16 bar) in use by the gas sector might be similar for hydrogen storage. Since no other hint of hydrogen-containing spherical pressure vessels could be found, Table 32.4 uses technical data from natural gas spherical pressure vessels to calculate hydrogen storage parameters under the assumption that the vessel would be equal. Consequently, these values can only be considered as an estimate.

In addition to spherical pressure vessels, gas supply companies also use pipe storage facilities. This kind of storage vessel is constructed from standard tubes customary in the trade with a high nominal diameter of about DN 1400 and is able to withstand high operating pressures of 64–100 bar [48]. Table 32.5 summarizes technical data for different storage facilities for natural gas in operation by public utilities and gas supply companies.

Table 32.4 Calculated hydrogen storage parameters based on technical data for spherical pressure vessels for natural gas of different public utility companies.

Site/year built		Heilbronn am Neckar/1964[h]	Reutlingen/ 1965[i]	Gießen/ 1961[j]	Wuppertal/ 1956[k]
Parameter	Unit				
Diameter[a]	m	34[h]	34.11[i]	16.88[j]	47.3[k]
Wall thickness	mm	30[h]	30[i]	20[j]	30[k]
Pressure[b]	bar	7[h]	9.3[i]	8[j]	5.05[k]
Weight[c]	t	759[h]	965[l]	175[j]	1 944[k]
Parameter	Unit	Calculated hydrogen storage parameters			
Geometric volume[d]	m^3	20 471	20 670	2 500	55 199
Gross volume capacity[e]	m$^3_{(STP)}$	141 422	189 721	19 742	275 108
Net volume capacity[f]	m$^3_{(STP)}$	101 015	148 921	14 807	166 154
Gross mass capacity[e]	kg	12 711	17 052	1 774	24 727
Net mass capacity[f]	kg	9 079	13 385	1 331	14 934
Gross energy capacity[e]	MWh	424	568	59	824
Net energy capacity[f]	MWh	303	446	44	498
Gross specific energy[e,g]	kWh t^{-1}	524.39	578.60	334.48	418.52
Net specific energy[f,g]	kWh t^{-1}	374.57	454.17	250.86	252.77
Gross energy density[e,g]	kWh m^{-3}	20.58	27.34	23.48	14.87
Net energy density[f,g]	kWh m^{-3}	14.70	21.46	17.61	8.98

Notes: *own calculation.*
a   is assumed to be external diameter,
b   is assumed to be maximum operating pressure,
c   is assumed to be weight of empty vessel,
d   calculated from inner radius (external radius minus wall thickness),
e   gross capacity is calculated at maximum operating pressure,
f   net capacity is calculated assuming a minimum operating pressure of 2 bar,
g   technical value (on basis of storage vessel including hydrogen),
Source: **h** [88], [89], **i** [90], **j** [91], **k** [92], **l** [47].

Table 32.5 Technical data for pipe storage facilities for natural gas [48].

Operator	Length (m)	Diameter (DN)	Pressure (bar)	Water volume (m³)	Net volume (m³)
SBL, Linz	1720	1600	5–22	5000	85 000
SW Bietigheim-Bissingen	852	1400	22	1330	25 000
EV Hildesheim	6000	1400	75	9200	700 000
Gas- und E-Werk Singen	1200	1400	16–80	1800	144 000
Erdgas Zürich	5500	1500	7–70	9540	714 000

DN = Diameter Nominal

One of the most recent projects is the construction of a pipe storage facility for natural gas in Urdorf, Switzerland, by Erdgas Zürich Transport, which will be one of the largest in Europe (Figure 32.3). The investment costs amount to CHF21 million (~€17 million) and completion is planned for July 2013. The entire construction pit has dimensions of 210 × 50 m. A total of 260 single tubes with a weight of 14 t each are welded together, resulting in 20 pipe stings and an overall length of 4140 m. The tube diameter is 1422 mm with a wall thickness of 25.5 mm. The total geometric volume is 6112 m³ and offers a net volume capacity of 720 000 m³ between upper and lower operating pressures of 100 and 7 bar, respectively [50–53]. Transferring this net volume capacity to hydrogen results in a net mass of 64 714 kg and a thermal energy equivalent of 2156 MWh stored. This demonstrates that with this type of facility, storage at higher pressures in the GWh range is possible.

In comparison with spherical pressure vessels, buried pipe storage facilities offer technical and economic advantages [51, 54, 55]. The geometric shape of the ball permits minimal surface areas and small wall thicknesses, leading to reduced material expenditure per unit gas volume [48]. However, the maximum pressure is limited since the wall thickness has to increase with higher stress levels. In contrast, pipe storage facilities can be operated at much higher pressure ranges, require comparatively small storage volumes, and have an unproblematic technical construction, which altogether results in low investment costs [48]. Furthermore, pipe storage facilities are typically buried a few meters below ground, which has the benefit of protection against adverse weather conditions and external mechanical impacts [54]. In addition, the land area can still be used for other purposes. For example, in the case of the pipe storage facility in Urdorf, the area will be made agriculturally usable again [51].

In a future hydrogen infrastructure, storage facilities similar to the present ones for natural gas could fulfill useful tasks. However, it should be noted that hydrogen has a different impact on materials than natural gas and consequently requires special attention. Vessels have to be constructed with materials resistant to hydrogen embrittlement and fatigue. In addition, they have to maintain structural integrity under high-pressure cycling environments [9]. Since hydrogen storage adds significantly to the cost of a delivery infrastructure, research into new cost-effective materials, coatings, and fiber or other composite structures is needed [8, 9].

32.4
Cryogenic Liquid Hydrogen Storage

32.4.1
Thermodynamic Fundamentals

Cryogenic liquid hydrogen has the advantage of a higher physical energy density than compressed gaseous hydrogen. For example, liquid hydrogen at normal pressure has a physical energy density of 8.49 MJ l^{-1} which is about four and three times the energy per unit volume compared with gaseous hydrogen at 250 and 350 bar, respectively [24, 56, 57]. The reason for this is the difference in the densities. Liquid hydrogen has a density of 70.8 kg m^{-3} at the normal boiling point, which is –252.85 °C (20.3 K) at normal or standard pressure (Table 32.6). In contrast, the volumetric density of gaseous hydrogen is only 17.9 and 23.7 kg m^{-3} [18] at an ambient temperature of 293.15 K and pressures of 250 and 350 bar, respectively. This volume reduction makes liquid hydrogen especially attractive for bulk hydrogen delivery by truck, train, or ship since the required number of runs for transporting the same energy content is drastically reduced.

However, to liquefy hydrogen a considerable amount of energy is required. Various parameters have an influence on the minimum theoretical liquefaction power of hydrogen [57]. One important feature is the existence of two different forms of the hydrogen molecule, namely ortho- and para-hydrogen. Their difference is caused by the electron configuration, or more precisely, by the orientation of the nuclear spin, which results in slightly different properties. Para-hydrogen, in which electrons rotate in opposite directions, has a lower energy level than ortho-hydrogen, in which electrons rotate in the same direction. Under normal conditions, molecular hydrogen consists of about 75% ortho- and 25% para-hydrogen, which is commonly defined as normal hydrogen [17, 58]. The equilibrium between ortho- and para-hydrogen is temperature dependent and the amount of para-hydrogen increases with decrease in temperature. For example, at –254.15 °C (19 K) an equilibrium sample of hydrogen contains 99.75% para-hydrogen [58].

Table 32.6 Physical and chemical properties relevant for hydrogen liquefaction.

Property	Value	Conditions	Source
Normal boiling temperature	20.268 K	$p_{(NTP)} = p_{(STP)} = 1.01325$ bar	[59]
Density of liquid hydrogen	70.78 kg m^{-3}	At normal boiling point	[59]
Density of gaseous hydrogen	1.338 kg m^{-3}	At normal boiling point	[59]
Physical energy density	8.49 MJ l^{-1}	$p_{(NTP)} = p_{(STP)} = 1.01325$ bar	[24]
Heat of ortho-to-para conversion	555 kJ kg^{-1}	At normal boiling point	[60]
Heat of vaporization	445.6 kJ kg^{-1}	At normal boiling point	[59]

In this context, two factors are relevant for long-term liquid hydrogen storage. First, the exothermic conversion of ortho- to para-hydrogen is very slow, and second, the heat of hydrogen vaporization is lower than the heat of ortho-to-para-conversion (Table 32.6). If liquid hydrogen that has not yet attained its equilibrium concentration is stored over a longer period, the ortho-hydrogen present would finally transform to para-hydrogen. This conversion produces a non-negligible amount of heat leading to vaporization and loss of the liquid stored hydrogen in the long term, and consequently contributes to the boil-off loss of the storage tank. Therefore, hydrogen has to be converted into nearly pure para-hydrogen with the help of catalyzed reactions during the liquefaction process [48, 50–52].

Commercial liquid hydrogen production normally operates with a product para-hydrogen fraction of at least 95%. The conversion from normal hydrogen to a very high proportion of para-hydrogen makes up a significant proportion of the liquid hydrogen exergy content. For example, the minimum specific power amounts to 3.94 kWh kg^{-1} for the transformation of normal hydrogen at 26.85 °C (300 K) to saturated liquid hydrogen of equilibrium ortho–para composition with both feed and product stream at a pressure of 1 bar. The conversion of ortho- to para-hydrogen thereby makes up 0.59 kWh kg^{-1}, which is about 15% of the total reversible work [57, 61].

In addition to the ortho-to-para conversion, other process parameters also have an influence on the minimum specific liquefaction power, which makes it difficult to compare the efficiencies of different liquefaction processes. For example, the minimum specific liquefaction power reacts much more sensitively to a change in feed pressure than to product para-hydrogen fraction within defined intervals of 1–60 bar and 95.0–99.8%, respectively. Furthermore, within these intervals, variation of the feed stream temperature between 283 and 303 K has a greater influence than variation of the liquid product saturation pressure between 1 and 4 bar [57].

32.4.2
Liquefaction Plants

Several liquefaction processes for hydrogen exist, but only a few are actually applied at liquefaction plants. For example, Linde Kryotechnik uses a helium precooled Joule–Thomson process for small-capacity liquefaction plants of up to 1000 l h^{-1} and a liquid nitrogen precooled Claude process for higher capacities [26, 62]. The precooled Claude process employs additional expansion turbines, because the isenthalpic expansion is more effective than the Joule–Thomson expansion. However, liquid formation damages the expansion engine [20]. The approximate sequence of the main process steps is nitrogen precooling to 80 K, expansion in turbines to 80–30 K and Joule–Thomson expansion to 30–20 K [31]. Common to all hydrogen liquefaction processes is the need for an upstream purification step, which lowers the level of impurities to below 1 ppm. This is necessary, since all impurities except helium are solid at 20 K and would cause clogging of the different components in a liquefier [20, 63].

The typical energy demand of existing hydrogen liquefaction plants is in the range 36–54 MJ kg^{-1} [20], which corresponds to 30–45% of the LHV of hydrogen. The cost and the energy requirements per kilogram of hydrogen liquefied decrease as the plant capacity increases [64]. Currently, only a few plants exist worldwide, with more than nine plants located in the United States, four plants in Europe, and eleven plants in Asia. Their production capacities are in the range 5–34, 5–10, and 0.3–11.3 t d^{-1}, respectively. The exergy efficiencies of these plants are 20–30%, but there are some proposed conceptual plants with efficiencies of 40–50% [65]. To increase the liquefaction capacity and to lower the specific power requirements by process optimization and new techniques are especially important for any future large-scale hydrogen production [31].

32.4.3
Liquid Hydrogen Storage Tanks

Storage and transportation of industrial and medical gases have been accomplished with cryogenic tanks, also called dewars, for more than 40 years [29]. The storage of liquid hydrogen is similar to that of liquid helium and is state-of-the-art today, especially due to intensive applications in space flight [31].

Liquid storage tanks can have either a spherical or a cylindrical form. The latter is available in horizontal and vertical configurations. Larger tanks for long-term storage usually have a spherical form to ensure the smallest surface-to-volume ratio and hence to decrease evaporative losses [26]. In addition to stationary applications, these vessels can also be applied in the bulk transportation sector. The spherical vessel is more advantageous for ships, whereas the cylindrical form is better for transportation by railway or truck.

A typical installation with subsequent gaseous hydrogen use normally consists of a cryogenic storage tank, ambient air vaporizer, and controls (Figure 32.4). Generally, bulk liquid storage systems can be selected based on customer volume, desired pressure, purity level, flow rate, and operating pattern [23]. Common tank capacities range from 1500 l or 100 kg [42] to 95 000 l or 6650 kg [9] of hydrogen. The pressure in the tank is usually below 5 bar [9].

There are only a few manufacturers of cryogenic vessels throughout the world. Well-known companies include Linde in Germany, Air Liquide in France, Air Products and Praxair in the United States, British Oxygen Company in the United Kingdom, Kobe Steel in Japan, and JSC Cryogenmash in Russia [26]. Table 32.7 shows typical technical data for cylindrical storage tanks in horizontal and vertical constructions and various sizes commercially available today. As in the case of bulk ground storage modules for compressed gaseous hydrogen, liquid storage tanks can also be used in modular format. Table 32.8 gives technical data provided by DLR (the German Aerospace Center) [26] assuming that local storage ("Local") at filling stations or other consumer sites requires storage volumes of ~60 m^3 and that future storage tanks (Large/"long-term") can be built larger than the largest tanks of today (Large/"today").

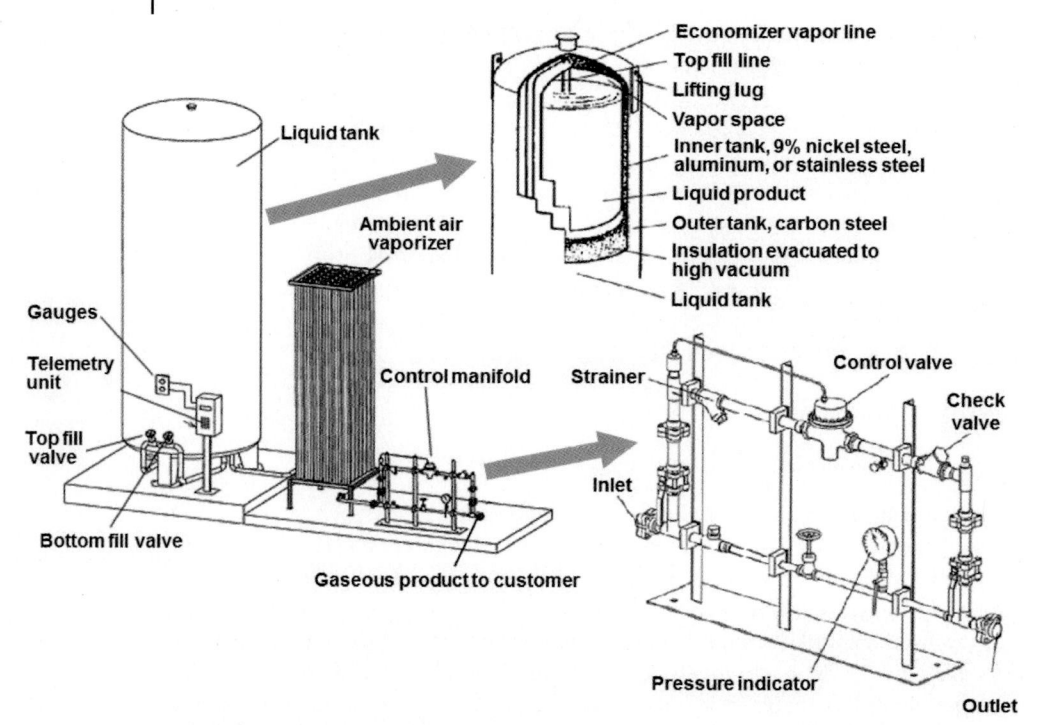

Figure 32.4 Typical bulk liquid storage system with cryogenic storage tank, ambient air vaporizer, and control manifold [23].

Table 32.7 Technical data for horizontal and vertical tanks for liquid hydrogen storage [66].

Parameter	Type specification					
	TLH-1500	VLH-4500	TLH-4500	TLH-9000	VLH-15000	TLH-18150
Diameter (ft-in)	7′-3″	8′-0″	8′-0″	8′-8″	10′-8″	10′-8″
Height (ft-in)	17′-5″	23′-5″	23′-$\frac{1}{8}$″	38′-8″	41′-4″	43′-3″
Weight (empty vessel) (t)	3.86	9.25	9.07	21.05	28.89	34.84
Gross volume capacity (m^3$_{(liquid)}$)	5.7	16.3	16.4	33.3	53.8	64.9
Gross volume capacity (m^3$_{(STP)}$)	4517	12 813	13 000	26 377	42 119	51 404
Gross mass capacity (kg)	408	1 179	1 247	2 363	3 765	4 604
Gross specific energy (MWh kg^{-1})[a]	13.6	39.28	41.54	78.70	125.38	153.32

Values in italics: own calculation.
a Technical value (on the basis of the storage vessel including hydrogen).

32.4 Cryogenic Liquid Hydrogen Storage

Table 32.8 Technical data for liquid hydrogen storage tanks depending on their size [26].

Parameter	"Local"	Large/"today"	Large/"long-term"
Water volume (m^3)	60	3000	100 000
Net mass capacity (t)	3.8	191	6 371
Net energy capacity (MWh)	127	6370	212 349
Technical system lifetime (a)	30	30	30
Utilization by time[a] (h a^{-1})	8,400	8400	N. A.
Utilization by volume (%)	90	90	90
Loss rate due to boil-off (%)	0.4	0.07	0.01

a According to [45].

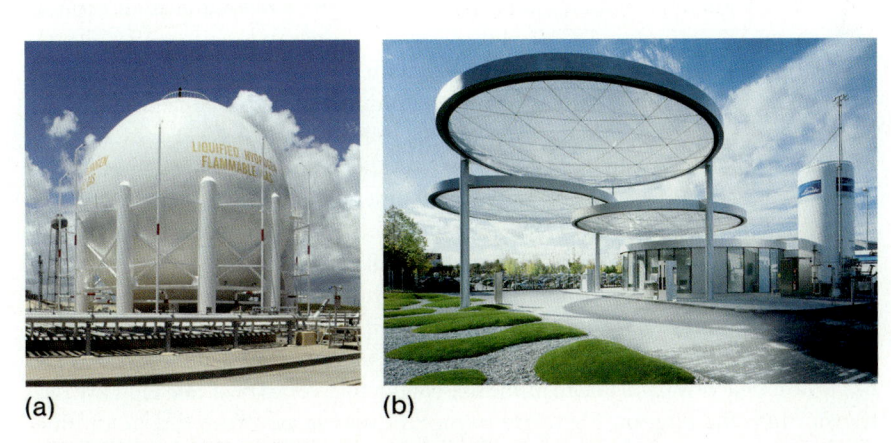

Figure 32.5 (a) Spherical storage tank of NASA [67] and (b) horizontal storage tank at the Linde Hydrogen Center, Unterschleissheim, Germany [68].

Stationary large-scale tanks are required for space applications, since the space shuttles have to be loaded with huge amounts of liquid hydrogen and oxygen just a few hours before take-off. At present, NASA has the largest spherical storage tank (Figure 32.5) [26, 31, 42]. The tank is located at the north-east corner of Launch Pad 39 A at Cape Canaveral in Florida and stores about 3400 m^3 (850 000 gal) of liquid hydrogen at −253 °C (−423 °F) [67]. The cited data for its diameter and its mass capacity range from 20 m [9, 31, 42] to 22 m [26] and from 230 t [26] to 270 t [31, 42], respectively.

In addition to launch sites for space flights, liquid hydrogen storage tanks are also applied at hydrogen fueling stations. An example is the Linde Hydrogen Center near Munich, Germany, which started operation in 2007. The fueling station has a vertical above-ground liquid hydrogen storage tank, which is super-insulated and provides a storage volume of 17 600 l. The fueling process for both liquid and compressed gaseous hydrogen can be accomplished in a few minutes, whereby the liquid hydrogen is kept at a cryogenic temperature of −253 °C and the compressed gaseous hydrogen is provided at 350 bar through corresponding dispenser systems. The boil-off is utilized as the supply for compressed hydrogen [69]. A similar fueling station was erected in Berlin, Germany, in 2006 [70].

As in the case of compressed gaseous storage vessels, liquid hydrogen tanks can be placed above and below ground. Although the costs for underground liquid hydrogen storage would likely be higher than for a traditional above-ground pressurized hydrogen system, this approach could combine advantages from both underground and liquid storage. Instead of the storage tank, the area above ground is saved for other purposes. This is especially advantageous where space is limited as in filling stations in cities. Furthermore, because of the higher density of liquid hydrogen, less space is needed compared with storage of gaseous hydrogen. In addition, this concept makes a fueling station inherently safer [9]. Currently, there are only a few fueling stations with underground liquid hydrogen tanks, situated in London, Munich, and Washington, DC [41, 42].

Special attention has to be paid to the selection of materials for the construction of liquid hydrogen storage tanks. Hydrogen embrittlement can be neglected at the boiling temperature because the hydrogen solubility is low [20, 29]. For example, in the case of unstable austenitic stainless steels, the hydrogen embrittlement effect is maximum at −100 °C, but negligible for temperatures below −150 °C. However, the brittleness of metals at cryogenic temperatures limits the choice of materials. Therefore, nickel ferritic steels, which can be applied down to −200 °C, or stabilized austenitic stainless steels and aluminum alloys, which are adaptable down to absolute zero, are commonly used for cryogenic tanks [29].

In order to avoid undesired evaporation of the stored liquid hydrogen, the vessel has to be excellently insulated to minimize conductive, convective, and radiative heat transfer into the interior. Therefore, cryogenic vessels are composed of an inner pressure vessel and an external protective container with an insulating vacuum layer between them. There are different options for insulation. Multi-layer insulation, which consists of several layers of aluminum foil alternating with glass-fiber matting to avoid heat radiation, or perlite vacuum insulation is often applied [17, 26, 29, 31, 42].

Although the vessel is excellently insulated, evaporation occurs to a certain extent, which causes a pressure increase over longer storage periods [17]. Therefore, a liquid hydrogen storage tank has to be designed as an open system and equipped with a pressure-relief system, leading to hydrogen losses, which are commonly called boil-off [17]. The released gaseous hydrogen can either be utilized directly or stored in an auxiliary system, or it can be returned to the liquefaction plant when the storage tank is situated next to it [20, 26].

The heat transfer from the environment to the inside of the storage tank depends on the thermal insulation, tank size, and shape [20, 26]. As already explained in

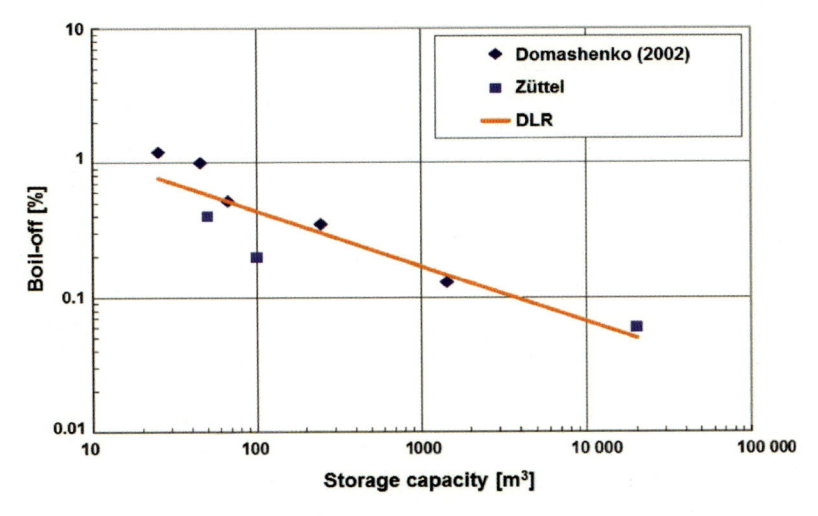

Figure 32.6 Boil-off rate of large liquid hydrogen storage tanks [26].

Section 32.4.1, the ortho-to-para ratio of the liquid hydrogen also has an influence on the boil-off rate. The spherical form is advantageous because it offers a minimal surface area-to-volume ratio. With increasing storage capacity or volume, respectively, this ratio increases so that larger vessels have a lower evaporation rate as smaller vessels assuming equal insulation (Figure 32.6) [26]. Typical values for boil-off rates are below 0.03% per day for large storage spherical tanks with perlite vacuum insulation such as the NASA tank, 0.4% per day for vacuum-super-insulated tanks and 1–2% per day for large tanks with vacuum powder insulation depending on their geometry [31].

32.5
Metal Hydrides

32.5.1
Characteristics of Materials

Metal hydrides are single-phase compounds of one or more metal cathodes and one or more hydride anions [64], where hydrogen can be locked in interstitial sites of the metal lattice [26, 71]. Based on their structure and the elements involved, simple and complex hydrides can be discerned [72]. Simple metal hydrides can be operated at temperatures just above or around ambient, whereas complex hydrides need temperatures in the range 100–300 °C or even higher [5, 6]. For low-temperature metal hydrides, a heat input of ~12% of the energy content of the delivered hydrogen is required [14]. Generally, operating conditions and performance characteristics, such as cycle life and heat of reaction, can be very different depending on the material [45].

Table 32.9 Characteristics of different metal hydrides [73].

Parameter	Mg_2NiH_4	$FeTiH_{1.95}$	$LaNi_5H_{7.0}$	$R.E.Ni_5H_{6.5}$[b]
Gravimetric capacity(kg kg^{-1})[a]	0.0316	0.0175	0.0137	0.0135
Volumetric capacity(kg l^{-1})[a]	0.081	0.096	0.089	0.090
Specific energy(kJ kg^{-1})[a]	4 484	2 483	1 944	1 915
Energy density(kJ l^{-1})[a]	11 494	13 620	12 630	12 770
Reaction heat (kJ mol^{-1} H$_2$)	−64.4	−23.0	−30.1	−38.1
Dissociation pressure (MPa)	0.1 (523 K)	1.0 (323 K)	0.4 (323 K)	3.4 (323 K)

a Value is assumed to be gross physical capacity (on a material-only basis).
b R. E.: rare earth.

In Table 32.9, characteristics of different metal hydrides are presented. In addition to storage, consideration is being given to using the special properties of metal hydrides for refrigeration, pumping, purification, and other purposes. However, these technologies are still far from commercialization [1].

32.5.2 Metal Hydride Tanks

The hydrogen storage system is composed of the metal hydride itself, a pressure vessel, and an integrated heat exchanger for cooling and heating during the charging and discharging processes [26, 45, 73]. The attendant temperature and pressure changes result in high requirements for the storage unit regarding structural and thermal stability in order to withstand a series of process cycles and also the capability of rapid heat transfer [45].

Metal hydride tanks are available in different sizes and shapes such as tubular or rectangular. The design of the hydrogen charging pressure is flexible and can vary from 1 to 100 bar or more. The heat exchange can be based on air and also on water, resulting in different achievable charging and discharging rates. The gravimetric density of the storage system, including metal hydride and tank, depends on the operating pressure and overall tank design, and lies in the range 0.7–1.1 wt% [20]. Metal hydrides have been put to practical use for hydrogen storage to only a limited extent [73]. Furthermore, they are usually applied for stationary small-scale storage [14].

32.6
Cost Estimates and Economic Targets

Costs for bulk stationary hydrogen storage are a decisive factor [9, 74]. This section presents economic data for different storage systems together with cost targets of the DOE [8] and the European Commission (EC) [75]. A literature search revealed that several detailed assessments and review articles [11, 12, 44, 45, 76–79] were published before 2000, but only two sources [14, 80] with newer data, published in 2004–2005, could be found. In a DLR report [26], investment costs for bulk liquid and gaseous hydrogen storage taken from [45, 76] were adjusted to €2000, but on comparing these data with newer data from [14, 80], no substantial match could be found. Consequently, the economic data given before 2000 are considered outdated and are not included here.

In Figure 32.7, investment costs of different hydrogen storage options are depicted as a function of the gross storage volume capacity. The economic data in Figure 32.7, which were obtained from an EC report [80], are estimates on the basis of data requested from industry [81]. In addition, economic data listed in Table 32.10 and EC targets listed in Table 32.11 are included in Figure 32.7 for comparative purposes. It should be noted that the EC targets are on a fully installed basis, whereas the specific investment given in [14, 80] are not specified in detail. The different cost basis also has to be considered with respect to Table 32.12. Consequently, direct comparison and assessment of these data is not possible.

In Figure 32.7 and Table 32.11, the high cost level of solid-state storage is clearly visible. It is also clear that gaseous storage costs are influenced by the size, the pressure range, and the shape of the vessel. In Figure 32.7, it can be recognized that liquid storage tanks are cheaper than gaseous storage vessels. However, in an overall analysis the costly and energy-intensive liquefaction process also has an influence. Therefore, an economic assessment of the different storage alternatives is

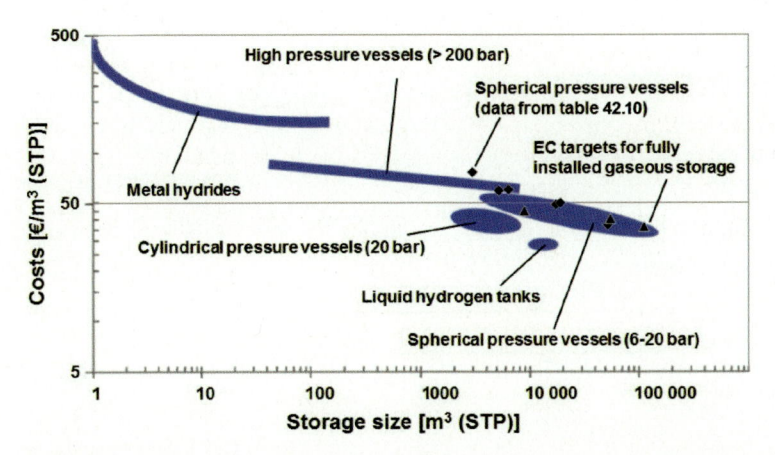

Figure 32.7 Investment cost estimates of different hydrogen storage options depending on the storage size and targets of the EC for fully installed gaseous storage [80] (modified).

only reasonable on the basis of a detailed application scenario including all relevant aspects such as the charging-discharging schedule [76, 78].

Table 32.10 Technical and economic data of spherical pressure vessels [14].

Volume (m^3)	Maximum pressure (MPa)	Minimum pressure (MPa)	Volume capacity (m$^3_{(STP)}$)	Investment (€)	Specific investment (€ m$^{-3}_{(STP)}$)
300	1.2	0.1	3 000	230 000	76
	2	0.1	5 200	307 000	59
1000	0.8	0.1	6 400	383 000	60
	2	0.1	17 400	844 000	49
3000	0.8	0.1	19 300	971 000	50
	2	0.1	52 100	1 917 000	37

Table 32.11 Economic data and targets of the EC [75].

Parameter	Units	Current 2010	Target 2015	Target 2020
Distributed storage of gaseous hydrogen				
Capacity per site (high end of size range)	t	0.8	5	10
	m$^3_{(STP)}$	8 899	55 617	111 235
Capital expenditure (fully installed) per capacity	M€ t	0.5	0.45	0.4
	€ m$^{-3}_{(STP)}$	45	40	36
Storage of hydrogen in solid materials				
Capacity per site (high end of size range)	t	3	5	10
	m$^3_{(STP)}$	33 370	55 617	111 235
Capital expenditure (fully installed) per capacity	M€ t	5	1.5	0.83^a/0.85^b
	€ m$^{-3}_{(STP)}$	450	135	75/76

Values in italics: own calculation.
a Value given in Annex on p. 37 of [75].
b Value given on p. 7 of [75].

Table 32.12 Economic data and targets of the DOE for stationary gaseous hydrogen storage [8].

Category	Units	2005 status	FY 2011 status	FY 2015 target	FY 2020 target
Low pressure (160 bar) purchased capital cost	$ kg^{-1}	1000	1000	850	700
	€ m^{-3}$_{(STP)}$ [a]	68	68	58	48
Moderate pressure (430 bar) purchased capital cost	$ kg^{-1}	1100	1100	900	750
	€ m^{-3}$_{(STP)}$ [a]	75	75	61	51
High pressure (860 bar) purchased capital cost	$ kg^{-1}	N. A.	1450	1200	1000
	€ m^{-3}$_{(STP)}$ [a]	N. A.	82	68	99

FY: fiscal year. N. A.: not available. Values in italics: own calculation.
a US$ 1 is assumed to be equivalent to €0.76.

32.7
Technical Assessment

Depending on the application area, different solutions can be beneficial since each of the storage methods has its own advantages and disadvantages and consequently can fulfill distinct requirements. However, for an evaluation of the suitability of a technology the whole storage system, including charging, storage, and release of the hydrogen, has to be considered with respect to the application purpose and the boundary conditions. In this context, aspects concerning the transport of hydrogen to and from the storage system are also important.

The technical assessment, which is presented in the following and in Table 32.13, is based on two different cases. Both assume that gaseous hydrogen is delivered to the storage system via pipeline at 30 bar. According to reports [8, 82], this value represents the lower operating pressure of a transmission pipeline. In the first case, no special application purpose or means of transport is considered for the hydrogen after storage. Instead, the lower storage pressure, which is set to 1.5 bar for all storage systems, is taken as the reference point. This leaves room for other delivery options such as transport by truck in either the liquid or gaseous state. In the second case, it is assumed that the hydrogen has to be transported again by pipeline, for which a pressure of 30 bar was also chosen. This value lies in the middle of the pressure range of a hydrogen distribution pipeline [8]. Furthermore, this assumption allows the calculation of the storage efficiency from the same reference point before and after storage.

Table 32.13 Technical assessment of different bulk hydrogen storage technologies.

Parameter	Storage technology			
	Gas holder	Spherical pressure vessel	Pipe storage facility	Spherical dewar
Average operating temperature (°C)	10	10	10	−253
Minimum operating pressure (bar)	1.5[d]	1.5	1.5	1.5
Maximum operating pressure (bar)	1.5[d]	20[g]	100[i]	1.5
Inner diameter (m)	23.2	31.1	1.4[j]	19.4
Height/length (m)	100	–	6 496	–
Footprint (m^2)	420	760	15 391[k]	295
Geometric volume/water volume (m^3)	42 014	15 755	10 000[j]	3 814
Gross volume capacity (m$^3_{(STP)}$)	60 000[e]	300 000[g]	952 068	3 004 005
Net volume capacity (m$^3_{(STP)}$)	54 000	277 500	937 787	2 703 605
Gross mass capacity (kg)	5 393	26 964	85 572	270 000[m]
Net mass capacity (kg)	4 854	24 942	84 288	243 000
Gross energy capacity (MWh)	180	899	2 851	8 997
Net energy capacity (MWh)	162	831	2 809	8 097
Utilization by volume (%)	90	93	99	90[n]
Supply equivalent (d)[a]	3	17	56	162
Specific power demand for loading (kWh kg^{-1})	–	–	< 0.7[h,l]	9.2[o]
Total power demand for loading (MWh)	–	–	< 40[h]	2236
Storage efficiency at 1.5 bar (%)[b]	100	100	> 98.6[h]	78.4
Specific power demand for unloading (kWh kg^{-1})	1.8[f]	< 1.8[f,h]	< 1.8[f,h]	1.8[f]
Total power demand for unloading (MWh)	9	< 45[h]	< 44[h]	435
Storage efficiency at 30 bar (%)[c]	94.9	> 94.9[h]	> 97.1[h]	75.2

Legend for Table 32.13

Values in bold: design value/assumption. Values in italics: own calculation.

a Represents the number of days a fueling station with a hydrogen consumption of 1500 kg d^{-1} could be supplied.
b Reference point is 1.5 bar and only the energy required for loading has to be considered.
c Reference point is 30 bar and the energy required for loading and unloading has to be considered.
d According to [32].
e According to [33].
f Compression from 1.5 to 30 bar with isentropic efficiency of 70%, motor efficiency of 94%, three stages with intercooling to 30 °C, and calculation according to ideal gas law.
g According to [36].
h In reality, less power is required for compression and the storage efficiency is higher, because of the dynamic nature of the storage system.
o According to [53].
j Selected according to Table 32.5.
k A length of 200 m for the pipe stings and space of 1 m between them is assumed.
l Compression from 30 to 100 bar with isentropic efficiency of 70%, motor efficiency of 94%, two stages with intercooling to 30 °C, and calculation according to ideal gas law.
m According to [31, 42].
n According to [26].
o According to [57, 86] considering 30 bar feed pressure.

The storage efficiency describes the ratio of useful energy output to the required energy input of the storage system. Thereby, the power demand for loading and unloading is also considered. The difference between the two cases, explained above, is that for the first one no energy is required for unloading. For the sake of clarity, the terms storage efficiency at 1.5 bar and storage efficiency at 30 bar are used in the following. In addition, another performance indicator was defined here, termed supply equivalent, which describes the number of days a fueling station with a hydrogen consumption of 1500 kg d^{-1} [82] could be supplied, and thus gives a realistic impression of the net mass capacity of the storage system. To ensure traceability, the assessment is kept deliberately simple, which means that the dynamic behavior of the storage systems is largely neglected and the ideal gas equation is used. In addition, losses caused by leakage and boil-off are not considered.

Selected storage types of this assessment are gas holders, spherical pressure vessels, pipe storage facilities, and spherical dewars. Metal hydrides are not included, because even with a drastic cost reduction in the future they will be double the price compared with gaseous storage vessels, which can be seen in Table 32.11. For the calculation, the different containments were designed in such a way that they represent the upper range in terms of size. Since large spherical pressure vessels and pipe storage facilities for hydrogen have not yet been built, the design of the vessel is aligned with existing ones for natural gas and values found in the literature and thus describe potential future technologies. The storage capacity of the gas holder and the spherical dewar are chosen according to the gas holder which existed in Neustadt/Coburg to fuel airships and the spherical dewar of NASA, respectively.

For the calculation of the latter, a tank temperature near to the normal boiling point is assumed. According to the literature[83–85], the annual average temperature of the air and the soil do not show much difference and therefore for simplification a temperature of 10 °C is chosen and it is assumed that the containments are always in thermal equilibrium with their surroundings.

The operating pressure of gas holders remains approximately constant, because the geometric volume changes with storage capacity. Therefore, here a constant operating pressure of 1.5 bar was selected according to Rummich [32]. Since the minimum and maximum operating pressures are the same, for the calculation of the net volume capacity a utilization by volume of 90% is assumed, which represents the geometric dead volume. With this, the net volume capacity amounts to 54 000 $m^3_{(STP)}$ and the net mass capacity to 4854 kg. This amount is sufficient for a supply equivalent of 3 d, which means that one fueling station with a hydrogen consumption of 1500 kg d^{-1} can be supplied for 3 d. During loading no energy is required, because the hydrogen is delivered at a pressure of 30 bar and filled into the gas holder after expansion. Consequently, the storage efficiency at 1.5 bar is 100%. However, in the second case the hydrogen has to be fed into the distribution grid at a pressure of 30 bar. With a total power demand of 9 MWh for compression of the net mass capacity, which is taken from the storage system, the storage efficiency at 30 bar is 94.9%.

According to Arenz [36], the gross volume capacity and the maximum operating pressure of the spherical pressure vessel are set to 300 000 $m^3_{(STP)}$ and 20 bar, respectively. The minimum operating pressure for all storage types is fixed at 1.5 bar for comparability reasons. Consequently, the utilization by volume amounts to 93% with a net volume capacity of 277 500 $m^3_{(STP)}$, which equals a net mass capacity of 24 942 kg or a supply equivalent of 17 d. As in the case of the gas holder, the storage efficiency at 1.5 bar is 100%. For the calculation of the storage efficiency at 30 bar, it was assumed that the total net mass capacity has to be compressed from a constant outlet pressure of 1.5 bar to the distribution pipeline pressure of 30 bar. However, in reality, the outlet pressure depends on the amount of hydrogen stored and decreases from 20 to 1.5 bar during unloading. Therefore, in practice, the total power demand for compression is less than 45 MWh and accordingly the storage efficiency at 30 bar is > 94.9%.

Based on Table 32.5, a water volume of 10 000 m^3 and an inner tube diameter of 1.4 m were selected for the design of the pipe storage facility. The maximum operating pressure was assumed to be 100 bar, according to Erdgas Zürich [53]. Since the minimum operating pressure of 1.5 bar is fairly low, the storage range with a utilization by volume of 99% is nearly fully exploited. Accordingly, the net volume capacity, the net mass capacity, and the supply equivalent of 937 787 $m^3_{(STP)}$, 84 288 kg and 56 d, respectively, are the highest of all storage types for gaseous hydrogen. To calculate the power demand for loading, it is assumed that no energy is required to fill the pipe storage facility up to a pressure of 30 bar. Then the power demand is calculated as if the remaining mass of 69 900 kg had to be compressed from 30 to 100 bar. Again, this is not the case in reality, because during loading the pressure rises and a specific power demand of 0.7 kWh kg^{-1} is only necessary in the final part of the loading process. Therefore, the storage efficiency at 1.5 bar is > 98.6%. Similarly, to calculate the power demand for unloading, it is assumed that no energy is required until an equal vessel pressure of 30 bar is reached. Subsequently, the total power demand was calculated as was done for the spherical pressure vessel. Consequently, also the storage efficiency at 30 bar has to be > 97.1% in practice. A change of the compression efficiency produces only minor deviations. For example, with an isentropic efficiency of 56% and 85% and a motor efficiency of 92% and 96%, selected according to [87] and [57, 61], respectively, the storage efficiency at 30 bar amounts to 96.3% and 97.7%, respectively.

For the spherical dewar, a gross mass capacity of 270 t corresponding to literature data [31, 42] for the NASA tank was taken. Assuming a utilization by volume of 90%, the net volume capacity and the net mass capacity amount to 2 703 605 $m^3_{(STP)}$ and 243 000 kg, respectively. These are nearly three times the net capacity of the pipe storage facility and equal a supply equivalent of 162 d. With respect to the amount of hydrogen stored, the spherical dewar exceeds the other storage types. However, the power demand is also the highest. With an exergy efficiency of 30%, which represents good state-of-the-art liquefaction processes [65], and an inlet feed pressure of 30 bar, the specific power demand for liquefaction is ~9.2 kWh kg^{-1} according to the literature [57, 86]. In the calculation of the storage efficiency, only the energy demand that is required to load and unload the same amount of hydrogen is considered. This means that although the entire spherical dewar has to be filled with liquid hydrogen the first time, only the energy demand for liquefaction of the net mass capacity, which is 2236 MWh, has to be used here. With this, the storage efficiency at 1.5 bar amounts to 78.4%. To compress the entire usable mass of 243 000 kg, a power demand of 435 MWh is needed and the specific efficiency at 30 bar decreases to 75.2%. Improving the exergy efficiency of the liquefaction process to 50%, which is postulated in some theoretical concepts [65], results in a much higher storage efficiency at 1.5 and 30 bar of 84.3% and 81.8%, respectively.

In the following, the results of the technical assessment are briefly summarized. With respect to the usable mass of hydrogen, the supply equivalent, and the footprint, the spherical dewar performs best. However, the power demand for loading and unloading are fairly high under the chosen boundary conditions and consequently also the storage efficiency for the two reference points set at 1.5 and 30 bar are the lowest. In this regard, the gas holder and the spherical pressure vessel achieve the highest results in the first case and the pipe storage facility in the second case. However, it has to be considered that finally at a fueling station a very high pressure is required to load the compressed gaseous on-board storage tank. Since the compression work depends on the pressure ratio, especially a drop near to the atmospheric pressure level should be avoided.

32.8
Conclusion

Currently, only very few near-surface bulk hydrogen storage facilities exist. They store hydrogen in either the gaseous or liquid state mainly for further utilization in the chemical industry or in the space flight sector. In the future, however, when hydrogen is employed in addition to its current chemical usage as a versatile energy carrier, a greater number of storage facilities will be needed. One promising application field is to employ hydrogen as a carbon-free fuel, which allows integration of renewable energy also into the transport sector. In a delivery infrastructure with gaseous- and/or liquid-based pathways, hydrogen could be transported by truck or by pipeline and stored in pressure vessels, in cryogenic liquid tanks, and as metal hydrides.

One challenge for future pipeline grids will be to balance fluctuations resulting from irregular injection from renewable energies and inconsistent withdrawal at fueling stations. Pipeline grids are able to provide some built-in buffering storage, but only to a small extent. Therefore, further management measures for compensating fluctuations might be required. It is conceivable that for this purpose, pressure vessels similar to those which are already used for the storage of natural gas or town gas are applied. Vessel types in this category include gas holders, spherical pressure vessels, and pipe storage facilities.

Gas holders are able to store large gas volumes but only at pressures slightly above the atmospheric pressure level. Since electrolyzers produce hydrogen already at elevated pressure, which will even increase in the future, and also a hydrogen grid would be operated with higher pressures, gas holders are unsuitable in this case. However, they could be an interesting option for other applications, where hydrogen is produced at low pressure. Characteristics of spherical pressure vessels are medium operating pressures of up to 20 bar and gas volumes of up to 300 000 m^3. The advantages of the spherical shape are a minimal surface area and a small wall thickness, leading to reduced material expenditure per unit gas volume. Furthermore, at the same pressure it permits the storage of the largest gas volume with the smallest footprint. Pipe storage facilities can achieve net volume capacities of > 700 000 m^3 and operating pressures up to 100 bar, which makes storage in the GWh range possible. They are typically buried below ground, where they are protected against adverse weather conditions and external mechanical impacts. In addition this permits the land area to be used for other purposes.

Storage capacities exceeding those for gaseous pressure storage can be achieved when hydrogen is stored in liquid form. Currently, the largest existing spherical dewar is that of NASA, which has a mass storage capacity between 230 and 270 t of hydrogen and a boil-off rate of 0.03% d^{-1}. However, the disadvantages of this technology are the energy- and cost-intensive liquefaction process and the losses due to boil-off.

Hydrogen can also be stored as metal hydrides. This technology has only recently become commercially available, and it has been used only to a limited extent and for the storage of small quantities. Metal hydrides release hydrogen only upon heat input, which is a benefit with regard to safety, but requires special heat management.

At present, the costs for metal hydrides are still fairly high, but they are expected to decrease in the future.

Investment cost estimates for gaseous and liquid hydrogen storage vessels were also considered. The costs for gaseous storage vessels clearly depend on the size, the pressure range, and the shape of the vessel. For the same storage size, liquid storage tanks were estimated to be cheaper than gaseous storage vessels. However, considering only the storage vessels can be misleading, since this neglects other important aspects such as the cost and energy requirements for charging and discharging. Other questions such as safety aspects or the utilization of hydrogen after storage are also of relevance here. Therefore, the different storage technologies have to be assessed in close relationship with their application purpose within an overall analysis.

References

1 Sherif, S. A., Barbir, F., and Veziroglu, T. N. (2005) Wind energy and the hydrogen economy – review of the technology. *Solar Energy*, **78** (5), 647–660.
2 European Commission (2003) *Hydrogen Energy and Fuel Cells: a Vision of Our Future*, Final Report of the High Level Group, EUR 20719 EN, http://ec.europa.eu/research/energy/pdf/hydrogen-report_en.pdf (last accessed 16 February 2013).
3 Stolten, D., Grube, T., and Mergel, J. (2012) *Beitrag elektrochemischer Energietechnik zur Energiewende*, Innovative Fahrzeugantriebe 2012, VDI-Berichte 2183, 8. VDI-Tagung.
4 US Department of Energy – Energy Efficiency and Renewable Energy (2011) *Hydrogen Storage (updated September 2011)*. Hydrogen, Fuel Cells and Infrastructure Technologies Program. Multi-Year Research, Development and Demonstration Plan. Planned Program Activities for 2005–2015, https://www1.eere.energy.gov/hydrogenandfuelcells/mypp/pdfs/storage.pdf (last accessed 16 February 2013).
5 Satyapal, S., Petrovic, J., Read, C., *et al.* (2007) The U.S. Department of Energy's National Hydrogen Storage Project: progress towards meeting hydrogen-powered vehicle requirements. *Catal. Today*, **120** (3–4), 246–256.
6 Broom, D. P. (2011) *Hydrogen Storage Materials: the Characterisation of Their Storage Properties*. Springer, London.
7 Ott, K. C., Autrey, T., andStephens, F. (2012) *Final Report for the DOE Chemical Hydrogen Storage Center of Excellence*, Los Alamos National Laboratory, http://www1.eere.energy.gov/hydrogenandfuelcells/pdfs/chemical_hydrogen_storage_coe_final_report.pdf (last accessed 16 February 2013).
8 US Department of Energy – Energy Efficiency and Renewable Energy (2012) *Hydrogen Delivery (updated September 2012)*. Hydrogen, Fuel Cells and Infrastructure Technologies Program. Multi-Year Research, Development and Demonstration Plan. Planned Program Activities for 2005–2015, https://www1.eere.energy.gov/hydrogenandfuelcells/mypp/pdfs/delivery.pdf. (last accessed 16 February 2013).
9 US Department of Energy – Energy Efficiency and Renewable Energy (2007) *Hydrogen Delivery Technology Roadmap*, http://www1.eere.energy.gov/vehiclesandfuels/pdfs/program/delivery_tech_team_roadmap.pdf (last accessed 16 February 2013).
10 Nexant (2008) *Hydrogen Delivery Infrastructure Options Analysis*, Final Report, DOE Award Number DE-FG36-05GO15032, http://www1.eere.energy.gov/

hydrogenandfuelcells/pdfs/delivery_infrastructure_analysis.pdf (last accessed 16 February 2013).
11 Carpetis, C. (1988) Storage, transport and distribution of hydrogen, in *Hydrogen as an Energy Carrier: Technologies, Systems, Economy* (ed. C. J. Winter and J. Nitsch) Springer, Berlin, pp. 249–290.
12 Carpetis, C. (1994) Technology and cost of hydrogen storage. *TERI Inf. Dig. Energy Environ.*, **4** (1), 1–13.
13 Detlef Stolten, Thomas Grube (Eds.): *18th World Hydrogen Energy Conference 2010 – WHEC 2010*, Parallel Sessions Book 4: Storage Systems/Policy Perspectives, Initiatives and Co-operations, Proceedings of the WHEC, May 16.–21. 2010, Essen, Schriften des Forschungszentrums Jülich/Energy & Environment, Vol. 78-4, Institute of Energy Research – Fuel Cells (IEF-3), Forschungszentrum Jülich GmbH, Zentralbibliothek, Verlag, 2010.
14 Blandow, V., Schmidt, P., Weindorf, W., et al. (2005) *Earth and Space-Based Power Generation Systems – a Comparison Study*, A Study for ESA Advanced Concepts Team, Final Report, LBST, http://www.lbst.de/ressources/docs2005/Final-Report_LBST_SPS-Comparison-Study_Chap0-3_050318.pdf (last accessed 16 February 2013).
15 McWhorter, S., Read, C., Ordaz, G., et al. (2011) Materials-based hydrogen storage: attributes for near-term, early market PEM fuel cells. *Curr. Opin Solid State Mater. Sci.*, **15** (2), 29–38.
16 Marrero-Alfonso, E. Y., Beaird, A. M., Davis, T. A., et al. (2009) hydrogen Generation from Chemical Hydrides. *Ind. Eng. Chem. Res.*, **48** (8), 3703–3712.
17 Klell, M. (2010) Storage of hydrogen in the pure form, in *Handbook of Hydrogen Storage: New Materials for Future Energy Storage* (ed. M. Hirscher), Wiley-VCH Verlag GmbH, Weinheim, pp. 1–38.
18 National Institute of Standards and Technology (2011) *Thermophysical Properties of Fluid Systems*, http://webbook.nist.gov/chemistry/fluid/ (last accessed 20 December 2012).
19 Hobein, B. and Krüger, R. (2010) Physical hydrogen storage technologies – a current overview, in *Hydrogen and Fuel Cells: Fundamentals, Technologies and Applications* (ed. D. Stolten), Wiley-VCH Verlag GmbH, Weinheim, pp. 377–394.
20 Godula-Jopek, A., Jehle, W., and Wellnitz, J. (2012) *Hydrogen Storage Technologies*. Wiley-VCH Verlag GmbH, Weinheim.
21 Berry, G. D., Martinez-Frias, J., Espinosa-Loza, F., et al. (2004) Hydrogen storage and transportation, in *Encyclopedia of Energy* (ed. J. C. Cutler), Elsevier, New York, pp. 267–281.
22 Zheng, J., Liu, X., Xu, P., et al. (2012) Development of high pressure gaseous hydrogen storage technologies. *Int. J. Hydrogen Energy*, **37** (1), 1048–1057.
23 Air Products (2012) *Typical Bulk Liquid Storage Systems*, http://www.airproducts.com/products/gases/supply-options/typical-bulk-liquid-storage-systems.aspx (last accessed 22 December 2012).
24 Wolf, J. (2003) Die neuen Entwicklungen der Technik – Wasserstoff-Infrastruktur: von der Herstellung zum Tank, in *Linde Technology 2/2003* (ed. A. Belloni), Linde, Wiesbaden, pp. 20–25.
25 Reitzle, W. (2003) Eine Vision für die Wirtschaft 4 – Potenziale und Marktchancen von Wasserstoff, in *Linde Technology 2/2003* (ed. A. Belloni), Linde, Wiesbaden, pp. 4–10.
26 Krewitt, W. and Schmid, S. (2005) *CASCADE Mints, WP 1.5 Common Information Database, D 1.1 Fuel Cell Technologies and Hydrogen Production/Distribution Option*, Final Report, DLR, Stuttgart, http://www.dlr.de/fk/Portaldata/40/Resources/dokumente/publikationen/2005-09-02_CASCADE_D1.1_fin.pdf (last accessed 16 February 2013).
27 ASME (2005) *Hydrogen Standardization Interim Report for Tanks, Piping, and Pipelines*, ASME Standards Technology, New York.
28 Lipman, T. (2011) *An Overview of Hydrogen Production and Storage Systems with Renewable Hydrogen Case Studies. A Clean Energy States Alliance Report*, Office of Energy Efficiency and Renewable Energy Fuel Cell Technologies Program,

http://www.cleanenergystates.org/assets/2011-Files/Hydrogen-and-Fuel-Cells/CESA-Lipman-H2-prod-storage-050311.pdf (last accessed 16 February 2013).

29 Barthélémy, H. (2012) Hydrogen storage – industrial prospectives. *Int. J. Hydrogen Energy*, **37** (22), 17364–17372.

30 Weldship Corporation (2012) *Inventory: Ground Storage*, http://www.weldship.com/tools/inventory.html (last accessed 14 December 2012).

31 Hydrogen Strategy Group of the Federal Ministry of Economics and Labour (2005) *Strategy Report on Research Needs in the Field of Hydrogen Energy Technology*, http://www.wiba.de/download/SKH2_english_01022005.pdf (last accessed 16 February 2013).

32 Rummich, E. (2011) *Energiespeicher: Grundlagen – Komponenten – Systeme und Anwendungen*, Expert-Verlag, Renningen.

33 Bardua, S. (2011) *Gasbehälter und Gaswerke in Deutschland*, http://www.industrie-kultur.de/index.php?module=html01pages&func=display&pid=90 (last accessed 20 November 2012).

34 Höchst (2012) *Innovation: Visionary Ideas from Industriepark Höchst*, http://www.industriepark-hoechst.com/en/index/industriepark/innovation.htm (last accessed 17 December 2012).

35 Ackermann, W., Gutsme, H., Jerusalem, E. G., et al. (1975) Bau von Hochdruckkugelgasbehältern in Berlin·Charlottenburg. *Stahlbau*, **44** (11), 321–330.

36 Arenz, B. (2008) *Netzmeister: Technisches Grundwissen Gas Wasser Fernwärme Bereich Umwelt*, Oldenbourg Industrieverlag, Munich.

37 Küchler, L. (2012) F. A. Neuman Anlagentechnik GmbH, Eschweiler, personal communication, 14 December 2012, http://www.neuman-eschweiler.de/ (last accessed 16 February 2013).

38 Infraserv Höchst KG (2012) *Pressebild: Anlagen im Industriepark Höchst, Wasserstoffgasometer*, http://www.infraserv.com/de/aktuelles/pressefotos/index.html (last accessed 20 December 2012).

39 Borucinski, T. (2013) CABB GmbH, Gersthofen, personal communication, 20.02.2013.

40 Chicago Bridge and Iron Company (2012) *History*, http://www.cbi.com/about-cbi/history/(last accessed 23 November 2012).

41 Weber, M., and Perrin, J. (2008) Hydrogen transport and distribution, in *Hydrogen Technology* (ed. A. Léon), Springer, pp. 129–149.

42 Barbier, F. (2010) Hydrogen distribution infrastructure for an energy system: present status and perspectives of technologies, in *Hydrogen and Fuel Cells: Fundamentals, Technologies and Applications* (ed. D. Stolten), Wiley-VCH Verlag GmbH, Weinheim, pp. 121–148.

43 Häussinger, P., Lohmüller, R., and Watson, A. M. (2000) Hydrogen, in *Ullmann's Encyclopedia of Industrial Chemistry*, Wiley-VCH Verlag GmbH, Weinheim, pp. 1–155.

44 Padró, C. E. G. and Putsche, V. (1999) *Survey of the Economics of Hydrogen Technologies*, NREL/TP-570-27079, National Renewable Energy Laboratory, http://www1.eere.energy.gov/hydrogenandfuelcells/pdfs/27079.pdf (last accessed 16 February 2013).

45 Amos, W. A. (1998) *Hydrogen Technical Publications: Costs of Storing and Transporting Hydrogen*, NREL/TP-570-25106, National Renewable Energy Laboratory, http://www1.eere.energy.gov/hydrogenandfuelcells/hydrogen_publications.html (last accessed 16 February 2013).

46 Hart, D. (1997) *Hydrogen Power: the Commercial Future of the Ultimate Fuel*, Financial Times Energy Publishing, London.

47 Gaswerk Augsburg (2012) *Kugel Gasbehälter in Deutschland und Europa*, http://www.gaswerk-augsburg.de/kugeleuropa.html (last accessed 20 December 2012).

48 Möller, A. and Niehörster, C. (2003) *Optimierung des Gasbezugs durch Röhrenspeicher*, ET Energiewirtschaftliche Tagesfragen 6/2003, Büro für Energiewirtschaft und technische Planung GmbH, http://www.bet-aachen.de/

fileadmin/redaktion/PDF/
Veroeffentlichungen/2003/BET-Artikel_
Gasbezug_0306.pdf (last accessed
16 February 2013).

49 Wuppertaler Stadtwerke (2012)
*Pressebilder, Energie & Wasser: Gaskugel
in Sonnborn*, http://www.wsw-online.de/
unternehmen/presse/Fotoarchiv/
Energie_Wasser/Startseite_
Energie_Wasser (last accessed
20 December 2012).

50 Erdgas Zürich (2012) *Röhrenspeicher
Urdorf, Tag der offenen
Baustelle*, Medienmitteilungen,
http://www.erdgaszuerich.ch/de/
netzbau-planauskunft/urdorf/tag-der-
offenen-baustelle.html (last accessed
20.12.2012).

51 Fuoli, F. (2012) *Für 21 Millionen
entsteht ein riesiger Erdgasspeicher*,
Limmattaler Zeitung,
http://www.limmattalerzeitung.ch/
limmattal/region-limmattal/fuer-
21-millionen-entsteht-ein-riesiger-
erdgasspeicher-125016057 (last accessed
16 February 2013).

52 Erdgas Zürich (2012) *Baustart des
Erdgas-Röhrenspeichers in Urdorf*,
http://www.erdgaszuerich.ch/medien/
archiv/mm-1142012-baustart-des-
erdgas-roehrenspeichers-in-urdorf.html
(last accessed 19 September 2012).

53 Erdgas Zürich (2012) *Erdgas Röhrenspeicher in Urdorf: Technische Daten*,
http://www.erdgaszuerich.ch/fileadmin/
media/ueber_uns/publikationen/
produktebroschueren/Roehrenspei-
cher-Urdorf_Technische-daten.pdf
(last accessed 19 September 2012).

54 Kühn, U. (2008) *Safe and
Cost-Effective Pipe Storage
Facility in Bocholt (Germany)*,
http://www.tuev-sued.de/uploads/
images/1207896537119176350047/
Pipe_storage_facility_Bocholt.pdf
(last accessed 16 February 2013).

55 Erdgas Zürich 2012) *Erdgas erfolgreich
speichern*, http://www.erdgaszuerich-
transport.ch/de/medien/mm-1482012-
erdgas-erfolgreich-speichern.html
(last accessed 16 February 2013)

56 Wolf, J. (2003) Flüssigwasserstoff-
technologie für Kraftfahrzeuge.
LH2 macht mobil, in *Linde Technology

1/2003* (ed. A. Belloni), Linde,
Wiesbaden, pp. 20–25.

57 Berstad, D. O., Stang, J. H., and
Nekså, P. (2009) Comparison criteria
for large-scale hydrogen liquefaction
processes. *Int. J. Hydrogen Energy*, 34 (3),
1560–1568.

58 Leachman, J. W., Jacobsen, R. T.,
Penoncello, S. G., *et al.* (2009)
Fundamental equations of state for
parahydrogen, normal hydrogen, and
orthohydrogen. *J. Phys. Chem. Ref. Data*,
38 (3), 721–748.

59 HySafe (2007) *Biennial Report on
Hydrogen Safety. Chapter I: Hydrogen
Fundamentals*, http://www.hysafe.org/
download/1196/BRHS_Chap1_V1p2.pdf
(last accessed 16 February 2013).

60 Leachman, J. (2007) Fundamental
equations of state for parahydrogen,
normal hydrogen, and orthohydrogen,
Master's thesis, College of Graduate
Studies, University of Idaho.

61 Quack, H. (2001) Conceptual design
of a high efficiency large capacity
hydrogen liquefier. *Adv. Cryog. Eng.*, 47,
255–263.

62 Linde Kryotechnik (2005)
Hydrogen Liquefiers,
http://www.linde-kryotechnik.ch/
1259/1260/1308/1309.asp (last accessed
28 December 2012).

63 Léon, A. (2008) Hydrogen storage, in
*Hydrogen Technology: Mobile and Portable
Applications* (ed. A. Léon), Springer,
Berlin, pp. 82–128.

64 Kevin, M., and Warren, V. (2012)
Hydrogen infrastructure: production,
storage, and transportation, in *Hydrogen
Energy and Vehicle Systems* (ed. S. E.
Grasman), Green Chemistry and
Chemical Engineering, CRC Press, Boca
Raton, FL, pp. 23–44.

65 Krasae-in, S., Stang, J. H., and Neksa, P.
(2010) Development of large-scale
hydrogen liquefaction processes from
1898 to 2009. *Int. J. Hydrogen Energy*,
35 (10), 4524–4533.

66 Praxair (1997) *Bulk Supply Systems for
Nitrogen, Oxygen, Argon and Hydrogen*.
http://www.praxair.com/praxair.nsf/0/
6c0316bc10aa86518525654b00583771/
$FILE/BulkSupply.pdf (last accessed
20 December 2012).

67 National Aeronautics and Space Administration (2009) *Kennedy Media Gallery*, http://mediaarchive.ksc.nasa.gov/detail.cfm?mediaid=48266 (last accessed 12 December 2012).
68 Linde (2012) *Linde Hydrogen Center*, http://www.the-linde-group.com/de/news_and_media/image_library/index.html (last accessed 20.12.2012).
69 Linde (2012) *Linde Hydrogen Center in Munich, Germany. Liquid and Gaseous Hydrogen Fuelling Station*, http://www.linde-gas.com/internet.global.lindegas.global/en/images/Linde%20Hydrogen%20Center%20in%20Munich,%20Germany17_15303.pdf (last accessed 28 December 2012).
70 Linde (2012) *CEP/TOTAL at Heerstrasse, Berlin, Germany. Liquid and Gaseous Hydrogen Fuelling Station*, http://www.linde-gas.com/internet.global.lindegas.global/en/images/CEP-TOTAL%20hydrogen%20fuelling%20stations%20at%20Heerstrasse,%20Berlin,%20Germany.17_15299.pdf (last accessed 18 December 2012).
71 Hottinen, T. (2001) Technical review and economic aspects of hydrogen storage technologies, Master's thesis, Helsinki University of Technology.
72 Sullivan, E. A. (2000) Hydrides, in *Kirk-Othmer Encyclopedia of Chemical Technology*, Wiley-VCH Verlag GmbH, Weinheim, pp. 607–631.
73 Takahashi, K. (2009) Hydrogen storage, in *Energy Carriers and Conversion Systems* (ed. T Ohta), EOLSS, Oxford, pp. 52–70.
74 Shibata, T., Yamachi, H., Ohmura, R., et al. (2012) Engineering investigation of hydrogen storage in the form of a clathrate hydrate: conceptual designs of underground hydrate-storage silos. *Int. J. Hydrogen Energy*, **37**, 7612–7623.
75 European Comission (2011) *Fuel Cells and Hydrogen Joint Undertaking (FCH-JU). Multi-Annual Implementation Plan 2008–2013, Document FCH JU 2011 D708*, http://www.fch-ju.eu/sites/default/files/MAIP%20FCH-JU%20revision%202011%20final.pdf (last accessed 16 February 2013).
76 Taylor, J. B., Alderson, J. E. A., Kalyanam, K. M., et al. (1986) Technical and economic assessment of methods for the storage of large quantities of hydrogen. *Int. J. Hydrogen Energy*, **11** (1), 5–22.
77 Carpetis, C. (1980) A System consideration of alternative hydrogen storage facilities for estimation of storage costs. *Int. J. Hydrogen Energy*, **5** (4), 423–437.
78 Carpetis, C. (1982) Estimation of storage cost for large hydrogen storage facilities. *Int. J. Hydrogen Energy*, **7** (2), 191–203.
79 Venter, R. D., Pucher, G. (1997) Modelling of stationary bulk hydrogen storage systems. *Int. J. Hydrogen Energy*, **22** (8), 791–798.
80 Altmann, M., Schmidt, P., Wurster, R., et al. (2004) *Potential for Hydrogen as a Fuel for Transport in the Long Term (2020–2030)*, Full Background Report, EUR 21090 EN, European Commission, http://www.lbst.de/ressources/docs2004/LBST-study-IPTS_2004_eur21090en.pdf (last accessed 16 February 2013).
81 Altmann, M., Ludwig-Bölkow-Systemtechnik GmbH, personal communication, 17 December 2012, http://www.lbst.de/ (last accessed 16 February 2013).
82 Krieg, D. (2012) *Konzept und Kosten eines Pipelinesystems zur Versorgung des deutschen Strassenverkehrs mit Wasserstoff Dennis Krieg [E-Book]*, Forschungszentrum Jülich, Zentralbibliothek, Jülich.
83 Umweltbundesamt (2012) *Trends der Lufttemperatur*, http://www.umweltbundesamt-daten-zur-umwelt.de/umweltdaten/public/theme.do?nodeIdent=2355 (last accessed 16 February 2013).
84 Umweltbundesamt (2013) *Klimadaten und Bodentemperatur*, http://www.umweltbundesamt.de/boden-und-altlasten/altlast/web1/berichte/langzeit/langzeit-3.5.html (last accessed 14 January 2013).
85 ENREGIS (2012) *Temperatur des Bodens*, http://www.enregis.de/pop_popupshowimg.htm?img=auftritt/daten/bilder/popup/Waerme_Temperaturverteilung_gr.jpg&x=698&y=750&print=0 (last accessed 14 January 2012).
86 Berstad, D., Stang, J. and Nekså, P. (2009) *A Future Energy Chain Based on*

Liquefied Hydrogen, SINTEF Energy Research, http://www.sintef.no/upload/lh2%20pres%20berstad.pdf (last accessed 16 February 2013).

87 Nexant (2008) *H2 A Hydrogen Delivery Infrastructure Analysis Models and Conventional Pathway Options Analysis Results*, DE-FG36-05GO15032, Interim Report, http://www1.eere.energy.gov/hydrogenandfuelcells/pdfs/nexant_h2a.pdf (last accessed 16 February 2013).

88 STIMME.de (2010): Heilbronner Gaskugel wird im November demontiert, http://www.stimme.de/heilbronn/nachrichten/stadt/sonstige-Heilbronner-Gaskugel-wird-im-November-demontiert;art1925,1948974 (last accessed 30 January 2013).

89 STIMME.de (2011): Heilbronner Gaskugel vor Abriss im Frühjahr?, http://www.stimme.de/heilbronn/nachrichten/stadt/Heilbronner-Gaskugel-vor-Abriss-im-Fruehjahr;art1925,2026437 (last accessed 30 January 2013).

90 Leibfritz, K. (2013) FairEnergie GmbH, Reutlingen, personal communication, 30.01.2013.

91 Althaus, D. (2013) Stadtwerke Gießen AG, Gießen, personal communication, 31.01.2013.

92 Seipenbusch, B. (2013) WSW Energie & Wasser AG, Wuppertal, personal communication, 31.01.2013.

33
Energy Storage Based on Electrochemical Conversion of Ammonia

Jürgen Fuhrmann, Marlene Hülsebrock, and Ulrike Krewer

33.1
Introduction

The restructuring of the energy supply in favor of renewable sources faces significant challenges owing to the intermittency of wind and solar energy input. For Germany alone, a reliable electricity supply based solely on renewable sources from within the country would require an increase in the electricity storage capacity by two orders of magnitude [1]. In this context, an advantage of storage schemes based on the energy of chemical bonds is the high energy density [2].

Hydrogen is widely discussed as an intermediate energy carrier. However, its thermodynamic properties complicate its handling on a large scale. At atmospheric pressure it becomes liquid at −253 °C. Even in pressurized form its energy density is significantly lower than that of most competing molecules (see Table 33.1).

Methane has been proposed as an alternative to hydrogen in the context of a buffer storage system for fluctuating renewable energy [1, 3]. It is attractive owing to its easy integration into the existing natural gas infrastructure and efficient means of recovering the stored energy. However, at the predicted scale, the source of the carbon dioxide necessary for the production of methane from hydrogen remains uncertain [4]. In order to become carbon neutral, ultimately this scheme probably would have to include additional carbon dioxide recovery and storage facilities at a scale comparable to the methane infrastructure.

Methanol, owing to its energy content, thermodynamic properties, and synthesis options, appears to be another suitable option [5]. With regard to the availability of carbon dioxide during the synthesis process, it shares the same concerns as methane.

Similarly to carbon, nitrogen can serve as a chemical carrier for hydrogen. Among the nitrogen–hydrogen compounds, ammonia is a sufficiently stable option. Nitrogen in large quantities is available from the air. It therefore appears justified to investigate the feasibility of ammonia as an energy carrier in the context of storage of energy from fluctuating renewable sources. This chapter attempts to give an overview of contemporary research activities on this issue.

Transition to Renewable Energy Systems, 1st Edition. Edited by Detlef Stolten and Viktor Scherer.
© 2013 Wiley-VCH Verlag GmbH & Co. KGaA. Published 2013 by Wiley-VCH Verlag GmbH & Co. KGaA.

In Section 33.2, the thermodynamic and chemical properties of ammonia that would allow it to be used as a medium for energy storage are discussed. A number of historical investigations and uses in this context are also reported. In Section 33.3, known processes for ammonia synthesis are reviewed. The discussion mainly focuses on options for a dynamic operation mode of the Haber–Bosch process, which is responsible for the overwhelming share of present ammonia production, and on contemporary, laboratory-scale research on electrochemical methods for ammonia synthesis that promise high flexibility with respect to fluctuating energy input. Recovery of the energy contained in ammonia is briefly discussed in Section 33.4, focusing on the historically well-known option of combustion, and possible electrochemical methods. A number of possible pathways emerges from these different options, which are compared in Section 33.5 based on rough efficiency estimates.

33.2
Ammonia Properties and Historical Uses as an Energy Carrier

The thermodynamic properties of ammonia are close to those of propane (Table 33.1), which is successfully handled even at the household scale. At atmospheric pressure it becomes liquid at −33 °C, and at a pressure of 10 bar it is liquid at room temperature. These properties make it easier to handle and to store ammonia than pure hydrogen in liquid or compressed form. Based on the Haber–Bosch synthesis process, ammonia is one of the chemicals with the highest production numbers. It is a core chemical for fertilizer production.

Similarly to other fuels, such as petrol, methanol, and even hydrogen, ammonia needs to be handled with care to prevent any harm to the environment. Whereas hydrogen is nontoxic but has a high fugacity, petrol, methanol, and ammonia are toxic and hazardous to the environment. As such, special precautions during handling and packaging need to be taken to prevent any possible damage. For the case of petrol and methanol, such precautions have been successfully implemented and these fuels are integrated in commercially available mobile and portable energy systems. The scale and structure of the necessary storage capacities discussed in this chapter suggest that an energy storage system based on ammonia could be based on a smaller number of larger, stationary units, for which high safety levels can be guaranteed. Already today, ammonia is being transported in pipelines over large distances [6] and stored in large amounts, for example, in port terminals [7]. In some countries, it is even directly used as a fertilizer [8].

The higher heating value is close to that of methanol, and half of that of diesel fuel. The volumetric energy density of ammonia in the liquid state is slightly less than that of methanol, and it is between 35 and 40% of the value for diesel fuel. Therefore, it is not surprising that ammonia was identified as an energy carrier already in the middle of the last century. It has been investigated as a fuel for combustion engines [12, 13] and gas turbines [14]. Practical uses included civilian buses [15] and the XLR99 rocket engine for the X15 spacecraft [16].

Table 33.1 Higher heating values (HHV) calculated from enthalpies of formation of pure substances [9], boiling points $\left(T_{1\,\text{bar}}^B \text{ and } T_{10\,\text{bar}}^B\right)$ [10], HHV-based volumetric energy densities (E_V) using density data from [10], and volumetric energy densities relative to NH_3 $\left(E_V^R = E_V^*/E_V^{NH_3}\right)$ for selected chemical energy carriers.

Parameter	p (bar)	T (°C)	NH_3	H_2	CH_4	C_3H_8	CH_3OH	Diesel
HHV (kWh kg^{-1})			6.25	39.39	15.42	13.89	6.11	12.5
$T_{1\,\text{bar}}^B$ (°C)	1		−33	−253	−161	−42	64	
$T_{10\,\text{bar}}^B$ (°C)	10		25	−242	−124	27	136	
E_V (kWh m^{-3})	1	$T_{1\,\text{bar}}$	4258.30 (l)	2792.40 (l)	6515.71 (l)	8128.90 (l)	4715.45 (l)	10759.58 (l)
	1	0	4.68 (g)	3.49 (g)	10.91 (g)	27.73 (g)	5099.90 (l)	10759.58 (l)
	10	0	3985.35 (l)	34.70 (g)	111.29 (g)	7408.02 (l)	5104.74 (l)	10759.58 (l)
	200	0	4064.43 (l)	615.17 (g)	2867.71 (c)	7813.17 (l)	5198.78 (l)	10759.58 (l)
E_V^R	1	$T_{1\,\text{bar}}$	1	0.65 (l)	1.52 (l)	1.91 (l)	1.10 (l)	2.52 (l)
	1	0	1	0.74 (g)	2.33 (g)	5.93 (g)	1090.50 (l)	2368.37 (l)
	10	0	1	0.0087 (g)	0.028 (g)	1.86 (l)	1.28 (l)	2.70 (l)
	200	0	1	0.15 (g)	0.70 (c)	1.92 (l)	1.28 (l)	2.65 (l)

Data for diesel fuel taken from [11]. Letters in parentheses indicate the aggregate state.

These historical aspects of ammonia as an energy carrier had to consider the substance in competition with other energy carriers within the classical, carbon-bound energy system. The ready availability of hydrocarbons made it attractive only in very narrow niches. The restructuring of the energy system in favor of renewables brings new facets into this discussion.

33.3
Pathways for Ammonia Conversion: Synthesis

The key element of the synthesis of ammonia from molecular nitrogen available from the air is nitrogen fixation – the splitting of the triple bond in the nitrogen molecule.

Biological nitrogen fixation is performed, for example, by nodule bacteria in the roots of legumes using nitrogenase enzymes containing transition metal centers [17, 18]. The energy for this reaction is supplied by the oxidation of ATP to ADP. The amount of reactive nitrogen generated by this natural process at present is not able to deliver sufficient amounts for human nutrition [19].

Another natural process is the direct oxidation of nitrogen in air, which is observed at very high temperatures in lightning. Based on this fact, the "Norwegian arc process" (also called the Birkeland–Eyde process) running at temperatures around 3000 °C, was the first candidate for industrial nitrogen fixation [20, 21]. The resulting nitrogen oxides provide only the first step for ammonia synthesis.

When the importance of ammonia as a carrier of reactive nitrogen (for producing both fertilizers and explosives) became evident, significant research efforts were made in order to establish possibilities of producing ammonia from nitrogen as a commodity.

Under appropriate conditions, nitrogen reacts with metals and other elements to form nitrides, which react with water to form hydroxides and ammonia. The aluminum-based historical Serpek process [20, 21] attempted to utilize this route. At sufficiently high temperatures, nitrogen reacts with carbides, forming cyanides that can be used to produce ammonia. This rather complex route had been attempted in the Frank–Caro process [20, 21]. However, the main result of the historical research efforts was the catalytic Haber–Bosch process, which at present is used to produce most of todays ammonia, and which is discussed in more detail in Section 33.3.1.

Under the premise of synthesis based on surplus energy from renewable sources, it would be interesting to investigate other possible pathways. Of particular interest are newly emerging electrochemical methods, which are discussed in Section 33.3.2.

33.3.1
Haber–Bosch Process

More than 90% of ammonia is presently produced by the Haber–Bosch process. Owing to the importance of fertilizer production for society, it has been extremely thoroughly studied.

The process involves the conversion of H_2 with N_2 to NH_3 at high pressure (> 100 bar) and high temperature (300–500 °C) in a heterogeneously catalyzed reaction. The reaction is limited by equilibrium and has a low conversion of ~20%. Therefore, the reactants and product are separated after the reactor by condensation of NH_3, and the unused reactants are fed back into the reactor. The reaction is exothermic (–46 kJ mol^{-1} NH_3); the released reaction enthalpy is used to heat the recycled reactants and, in current systems, to produce the highly pure reactant gases H_2 and N_2 from steam reforming and cryogenic air separation. A detailed discussion of contemporary Haber–Bosch processes was given recently by Appl [22].

Modern ammonia plants have been optimized for economical mass production, are operated at steady state, and can generate up to 3300 t d^{-1} [23] in a single line. Dynamic ammonia production would require the Haber–Bosch process to operate dynamically. Existing large-scale plants such as that operated by SKW Piesteritz allow a change in ammonia production rate between 65 and 100% [24]. In the event of complete shut-down of this plant, however, a subsequent start-up time of several days is required. This long start-up time is due to the to complex start-up procedure to generate purified reactants via reforming and cryogenic air separation and to regenerate the catalyst employed in the Haber–Bosch process: the usual Fe-

or Ru-based catalysts for ammonia production [25] are oxidized during shut-down and need to be reduced with the help of hydrogen or synthesis gas to become active again. This activation process can take 30 h and more; already during this process the reactor can produce ammonia [22]. Most catalysts employed are sensitive to oxygen-containing compounds up to the parts per million level [22], which necessitates purification of the reactants.

The focus of this section is a first rough evaluation of the Haber–Bosch process for its suitability as a component in a dynamically operating energy storage system, based on NH_3. Currently operated modern plants are highly complex and optimized for steady-state operation. A smaller scale plant with a capacity of ~300 t d^{-1} could be operated more flexibly and would fit better into such a system [26]. As we do not know of any small, contemporary, dynamic Haber–Bosch process plants, the discussion will be based on the second ammonia plant built worldwide, which was located in Leuna and produced up to 350 t d^{-1} in 1918 [27]. The reactor itself was small (~24 m^3) and could be brought to the operating conditions within 15 min by electrical heating.

Figure 33.1 shows the general scheme of the Haber–Bosch process of the historical Leuna plant [27], and an extension by compression and expansion of entering or exiting reactants as it might be feasible for a chemical energy storage system.

The large size of contemporary production sites is attributed mostly to the facilities for producing pure reactants: N_2 is obtained by cryogenic air separation and H_2 by reforming of natural gas and cleaning of the resulting synthesis gas.

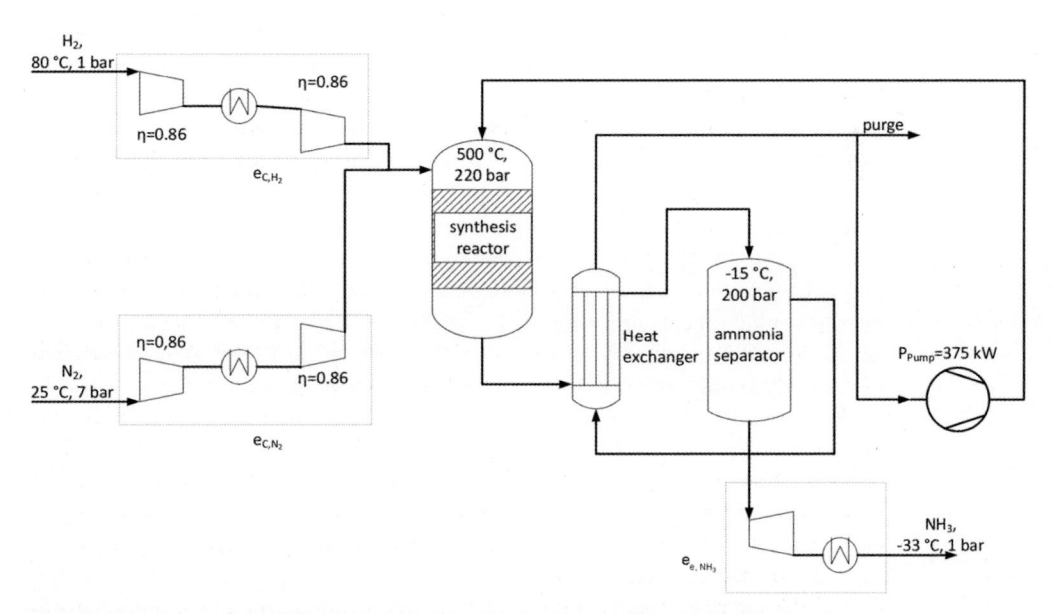

Figure 33.1 Scheme of the Haber–Bosch process with process conditions of the historical Leuna plant [27] and assuming reactant supply and withdrawal conditions feasible for chemical energy storage.

When using ammonia to buffer intermittent electricity generated from renewables, alternative, small scale processes to generate the reactant streams for the Haber–Bosch process may be employed: hydrogen may be supplied by a water electrolysis facility which is powered by intermittent energy; pure nitrogen may be supplied by pressure swing adsorption (PSA).

The efficiencies for converting electricity into hydrogen can reach more than 80% depending on the type of electrolyzer and operating point; the most common types are alkaline electrolyzers and PEM electrolyzers. Both are excellently scalable, produce highly pure H_2 and allow dynamic operation [28].

PSA is an attractive alternative to cryogenic air separation by rectification as it can operate at ambient temperature. The use of carbon molecular sieves allows high purities of up to 99.9999% to be achieved [29]. Existing PSA plants start up and shut down within minutes and permit a control of flow rate following demand [29]; as they can be operated dynamically and are also scalable, they seem to be suitable for supplying pure nitrogen for dynamic ammonia production for energy storage. Alternative nitrogen separation processes using membranes are also excellently scalable but currently do not seem to supply the required high purity of reactant gases.

Taking into account the proposed reactant supply techniques, namely water electrolysis to generate H_2 and PSA to generate N_2, setting up smaller and decentralized Haber–Bosch plants for dynamic ammonia production may be feasible.

In the following, a rough analysis of the energy efficiency of such a setup is given based on available data for the historical Leuna plant [27]. The pump in the reactant cycle had a power consumption of P_{pump} = 375 kW. Furthermore, it is assumed that H_2 is supplied by water electrolysis at ambient pressure and 80 °C and N_2 by PSA at 7 bar and ambient temperature.

The gases need to be compressed to the operating conditions of the Haber–Bosch reactor (500 °C, 220 bar). To guarantee these conditions, H_2 and N_2 compression are assumed to take place in two steps with intermittent cooling (Fig. 33.1). The respective energy demand was calculated with the in-house software Enbipro [30], taking into account typical compression efficiencies of 86% for each compressor; the resulting specific energies for compression of H_2 and N_2 are e_{c,H_2} = 13.46 MJ kg^{-1} and e_{c,N_2} = 0.89 MJ kg^{-1}. The NH_3 produced is expanded in an adiabatic turbine and subsequently cooled to the storage conditions of 1 bar, −33 °C; the specific energy needed for this process is e_{e,NH_3} = 0.021 MJ kg^{-1}, which is negligibly small compared with the other processes. The efficiency of the Haber–Bosch process is then calculated as follows ($\dot{m}_*$ denoting the respective molar masses):

$$\eta_{\text{Haber-Bosch}} = \frac{\dot{m}_{NH_3} \, LHV_{NH_3}}{\dot{m}_{H_2} \, LHV_{H_2} + \dot{m}_{H_2} \, e_{c,H_2} + \dot{m}_{N_2} \, e_{c,N_2} + P_{pump} + \dot{m}_{NH_3} \, e_{e,NH_3}} \quad (33.1)$$

Taking into account the stoichiometric feed of reactants, a NH_3 mass flow rate of 4.05 kg s^{-1} NH_3, and lower heating values of LHV_{NH_3} = 18.6 MJ kg^{-1} and LHV_{H_2} = 120 MJ kg^{-1}, this yields an encouraging energy efficiency of 76% for the non-optimized Leuna process.

33.3.2
Electrochemical Synthesis

In many cases, electrochemical methods have the potential to circumvent the thermodynamic restrictions of purely chemical methods. Along with high pressure, a properly applied voltage difference can shift the reaction equilibrium in favor of the reaction products.

A number of recent laboratory results suggest that there is some potential in electrochemical ammonia synthesis that may be worth investigating further. A fairly comprehensive overview of experimental results was presented recently by Amar et al. [31]. Most of the methods investigated can proceed at atmospheric pressure; however, a higher nitrogen pressure is beneficial for some of them. Depending on the electrolyte, the processes differ significantly in their temperature range.

Following the general idea of electrochemical promotion, Panagos et al. [32] performed a model-based investigation of hydrogenation reactions in proton-conducting solid oxide membrane reactors. Based on their predictions, the synthesis of ammonia at atmospheric pressure using a solid-state proton conductor acting at temperatures above 500 °C was reported [33, 34]. Under these conditions, however significant rates of product decomposition are observed. A solid-state proton conductor and an industrial Fe catalyst were investigated at atmospheric pressure and elevated temperatures, ensuring the proton conductivity of the solid electrolyte [35]. It was concluded that it is possible, depending on the surrounding conditions, to promote the synthesis reaction by electrochemical supply of hydrogen in the form of protons. A proton-conducting carbonate–$LiAlO_2$ electrolyte was used by Amar et al. [36].

Tsuneto et al. [37, 38] observed that Li ions may mediate nitrogen fixation because the Li can "burn" in an N_2 atmosphere, creating Li_3N, which in contact with water produces ammonia and LiOH [20]. If electrochemical recovery of Li from LiOH could be established, this pathway could be another route for ammonia synthesis. However, in an aqueous solution, hydrogen evolution sets in at potential levels much earlier than Li precipitation from LiOH [38]. As a consequence, experimental investigations of Li-mediated ammonia synthesis focused on molten salt electrolytes [38–41]. Room temperature ionic liquids have been investigated for this purpose [42]. Multi-chamber cells with Li-conducting membranes separating the aqueous solution from the region of Li precipitation could circumvent the problem of early set-in of hydrogen evolution. Such cells have been investigated in the framework of lithium battery research [43]. This seems to be another option to close the Li-mediated synthesis process.

Following the work of Marnellos et al. [34], another body of work started with solid-state proton conductors [44, 45] and then switched to low-temperature polymer membrane-based synthesis cells that use oxidic catalysts containing rare earth elements [46–50]. The device with the best evolution rate [48] used a Nafion membrane, an SFCN ($SmFe_{0.7}Cu_{0.1}Ni_{0.2}O_3$) cathode catalyst, and nickel–samaria-doped ceria (NiO–$Ce_{0.8}Sm_{0.2}O_{2-\delta}$)-based anode catalyst.

Conducting polymers (polyaniline) [51] and fullerene–γ-cyclodextrin complexes [52] have been reported as well.

Hellman et al. [25] surveyed theoretical work based on molecular-scale calculations. They considered ammonia synthesis with the Haber–Bosch-method to be fairly well understood, and to be based on an associative Langmuir–Hinshelwood mechanism starting with N_2 adsorption on the catalyst surface. On the other hand, biocatalytic ammonia synthesis follows a different route starting with N_2 hydrogenation [17], suggesting that a mild temperature process should follow a dissociative, biocatalytic route.

A possible explanation of the low electrochemical formation rate of NH_3 and the strong competition of the H_2 evolution reactions in an aqueous environment has been proposed [18]. Using DFT calculations, it was suggested that catalysts based on early transition metals such as Sc, Y, Ti, and Zr should be considered, where NH_3 formation is the preferred process in comparison with H_2 formation [18].

33.4
Pathways for Ammonia Conversion: Energy Recovery

Recovery of the energy contained in ammonia is possible through a number of pathways, which are briefly discussed in this section.

33.4.1
Combustion

As already stated in Section 33.2, combustion is a possible pathway for the recovery of the energy stored in ammonia. It has been successfully tested in combustion engines [13, 15], in gas turbines [14], and in rocket engines [16]. A particular advantage of ammonia in this context is the absence of soot emissions. On the other hand, it is necessary to deal with possible NO_x emissions and ammonia slip in the combustion devices. Early experiments on NO_x concentrations in exhaust gases from combustion engines [13] indicated that NO_x concentrations are in the region of 500 ppm, which is the same order of magnitude as prescribed by the EURO3 norm for diesel engines. The authors did not find comparable data for gas turbines. It should be noted, however, that reportedly, the combustion temperature of ammonia is lower than for hydrocarbons and H_2, and therefore it is likely that the direct oxidation of N_2 generates only a small amount of NO_x emissions. On the other hand, the splitting reaction

$$2\,NH_3 \rightarrow 2\,N_2 + 3\,H_2 \tag{33.2}$$

may take place before oxidation at a significant rate, thus making reactive nitrogen unavailable for direct oxidation.

In contemporary diesel engine technology, suppression of NO_x is achieved by admixture of ammonia (via an aqueous solution of urea, "diesel exhaust fluid") into the exhaust gas and subsequent selective catalytic reduction (SCR) [53, 54]. This technology also exhibits ammonia slip.

Combustion chambers for ammonia would have to be optimized for the comparably low combustion rate of ammonia. At the same time, this optimization would be the primary way to reduce ammonia slip, and possibly NO_x.

The reduction of ammonia slip in the exhaust is currently under discussion in connection with the SCR–urea technology and can be performed using a second catalytic oxidation step [54] or catalysts that are able to reduce both NO_x and NH_3 [55].

Widespread usage of ammonia combustion would have to be backed by basic research to resolve the issues discussed. If successful, combustion could be a promising pathway for energy recovery from ammonia. In particular, gas turbines are well established and highly efficient. The process of energy recovery from stored H_2 and CH_4 is assumed to be based on gas turbines [1]. Modern gas turbines include low-load operation capabilities in order to fit into an energy system with significant contributions from renewables [56].

33.4.2
Direct Ammonia Fuel Cells

Complementary to electrochemical methods of ammonia synthesis, fuel cells are attractive devices to recover the energy stored in ammonia.

According to a number of workers, the use of alkaline fuel cells to generate electricity from the oxidation of ammonia is feasible [57–60]. At the same time, the alkaline character of aqueous ammonia solutions renders acidic fuel cells impractical [61].

High-temperature cells based on oxygen-conducting YSZ ceramic electrolytes have been investigated [62], studying the co-generation of electricity and NO from ammonia. More recent studies have involved ammonia-fed high-temperature fuel cells with oxygen-conducting electrolytes [60, 63–67]. Staniforth and Ormerod [63] stated that NO_x generation is not a serious problem.

Medium-temperature cells with proton-conducting membranes based on doped barium cerate have been investigated with respect to the utilization of ammonia [68–72]. It seems that in these cells, the efficiency of ammonia conversion can be comparable to that of hydrogen.

Ammonia has been used to produce electricity in a molten hydroxide high-temperature cell [73].

33.4.3
Energy Recovery via Hydrogen

At low pressure, the endothermic reaction in Eq. 33.2 in the presence of a suitable catalyst [74, 75] allows ammonia to be split into nitrogen and hydrogen. The hydrogen can then be utilized, for example, in a hydrogen PEM fuel cell. This hydrogen-based pathway of electrochemical conversion of ammonia into energy is attractive because it allows to use the significant amount of accumulated knowledge of using hydrogen as a fuel in electrochemical cells to be utilized. A particular advantage of this approach in comparison with other possible chemical sources of hydrogen is the absence of CO poisoning. Residual ammonia, however, negatively influences the conductivity of a Nafion membrane [61]. This effect is significantly less evident in alkaline cells [76, 77].

33.5
Comparison of Pathways

Figure 33.2 gives an overview of the different pathways for chemical energy storage based on ammonia.

The Haber–Bosch process and pressure swing adsorption (PSA) are already established and commercially available on a large scale. Water electrolysis and hydrogen fuel cells seem to be at the threshold of large-scale commercialization. Except for the Haber–Bosch process, these processes can already now be operated dynamically. The remaining processes are either at the research stage, such as electrochemical synthesis or decomposition of ammonia, or they might already be technically feasible but have no attractive application area yet, for example, ammonia combustion. As such, less is known about how to design and operate them and on their achievable performance.

To the knowledge of the authors, no complete cycle based on the pathways for synthesis and conversion of ammonia presented in Figure 33.2 has yet been established, which indicates the need for significant research into the individual pathways and processes and a systematic comparison between them.

Because the theoretical energy efficiencies of electrochemical processes are often close to 100%, pathways based on such processes may also yield high total energy efficiencies. Currently, however, many electrochemical low-temperature processes are often hampered by slow reaction kinetics or side reactions that lead to significant losses in efficiency; they can be improved by using highly active catalysts and highly conductive, inert electrolytes, as has been done for hydrogen PEM fuel cells.

Comparison based on currently achievable performances means that routes with already established processes such as Haber–Bosch, electrolysis, and combustion may have practical efficiencies that are closer to theoretical efficiencies than the direct electrochemical synthesis or conversion processes for ammonia.

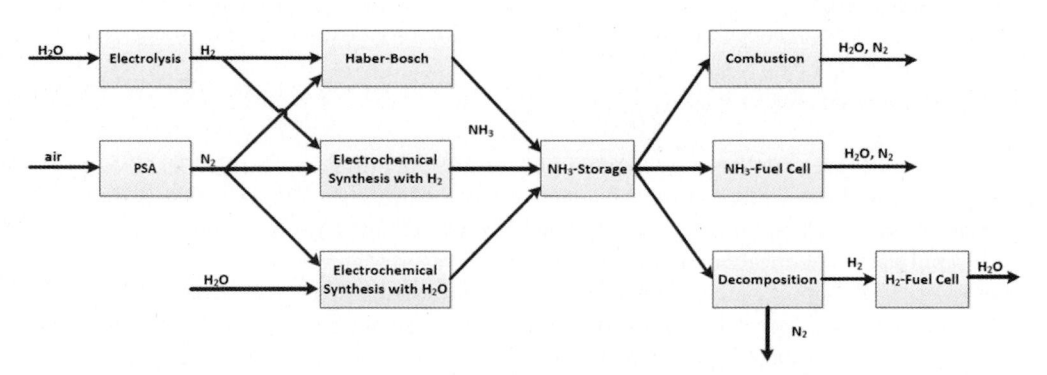

Figure 33.2 Pathways of a chemical energy storage system based on NH_3.

Nonetheless, to get a first idea about the currently achievable practical efficiencies of the pathways, the ammonia production pathways are discussed in the following.

Hydrogen production by water electrolysis is needed for ammonia generation by Haber–Bosch or electrochemical synthesis. It is assumed to take place at 80 °C and 1 bar; under these conditions, alkaline and PEM electrolysis both can yield cell efficiencies between 86% and 76% depending on the applied current density. These values have been calculated based on an efficiency equation [78], an equation for the current–voltage curve for an alkaline electrolyzer [79], and for a PEM electrolyzer [80] using the assumption of negligible system and faradaic losses. With respect to the energy needed for the compression of hydrogen at the input of the Haber–Bosch synthesis, these estimates are conservative, as some modern electrolyzers are able to deliver high-pressure hydrogen at the output, so for simplicity an efficiency of 80% is assumed.

PSA takes place in all synthesis routes and therefore its energy efficiency should be considered for estimating the full energy efficiency of each pathway; it is omitted for direct comparison of the synthesis routes here. The efficiency of the Haber–Bosch process has already been estimated in Section 33.3.1 to be ~76%; coupling of electrolysis and the Haber–Bosch process yields an overall efficiency of 60%. The efficiency of the electrochemical synthesis pathways can only be estimated roughly because complete data sets for the processes are lacking. Murakami and co-workers presented faradaic efficiencies of 23% for an electrochemical process using H_2O and molten salts as electrolyte [39], and up to 72% for an electrochemical process using H_2 and molten salts as electrolyte [81]; the efficiency of the electrochemical ammonia production route based on H_2 as reactant also needs to account for water electrolysis, yielding efficiencies of up to 58%. Both electrochemical processes take place at 400 °C, so one needs to account for additional heating of reactants and cooling of the product ammonia; the exit streams may be used to heat the inlet streams, however; as such, only little further energy loss may occur due to heating and cooling. Comparison of the above efficiencies for ammonia production suggests that Haber–Bosch and electrochemical conversion with H_2 may yield comparable efficiencies of up to 60% (without taking into account the energy used for PSA and based on the historical Leuna process), whereas electrochemical synthesis via H_2O needs to be improved significantly by preventing side reactions.

Assuming the energy recovery in gas turbines to have a similar efficiency of 57% as for H_2 and CH_4 [1], we arrive – without taking into account the energy necessary for PSA – at an overall system efficiency of 34.2%. Klaus et al. [1] gave a value of 42% for an H_2-based storage system and 35% for a system based on CH_4. Taking into account the high availability of nitrogen and the ease of large-capacity storage for ammonia, the slightly lower energy efficiency for ammonia may be acceptable. It should be mentioned that the presented absolute values of theoretical system efficiencies need to be treated with caution owing to the underlying assumptions or system boundaries. This is illustrated by a different model [5] which estimates an overall efficiency of ~36% for an H_2-based system and of 20% for a methanol-based strategy.

Further aspects influencing the decision on a pathway are scalability, which may be excellent especially for electrochemical pathways, as electrochemical processes allow for stacking of cells, and the possibility of dynamic operation, which should be excellent for electrochemical pathways also. Both aspects need to be investigated in greater depth for the pathways including the Haber–Bosch process.

33.6
Conclusions

Owing to its thermodynamic properties, its energy content, the good availability of its precursors, and a number of possibilities for energy recovery, ammonia is a possible candidate for large-scale chemical energy storage in the framework of a renewable energy system.

The main problem in the utilization of ammonia as buffer storage in a renewable energy system is the dynamic operation of ammonia production from its elements on a large scale. Current ammonia production is mainly based on the Haber–Bosch synthesis optimized for large-throughput, steady-state operation. Optimization of this process for production following energy supply in a renewable energy system may be feasible, but to the knowledge of the authors has not yet been investigated. At the same time, the operation of such a system has to rely on large-scale hydrogen production from renewables.

A number of other, electrochemical pathways for ammonia production are under investigation. Their technological feasibility at the necessary scale is an open question. Given currently available experimental and theoretical results, significant basic research appears to be necessary in order to be able to develop them into a feasible option for ammonia production. Advantages of this class of processes may be the possibility of closely following fluctuating energy input and operation at lower pressure.

From the very coarse estimates of the overall storage efficiency, one can draw a first conclusion that electrical energy storage via ammonia has an efficiency similar to that of currently proposed methanol- and methane-based systems. The energy efficiency of an H_2-based system is better, but would have to be paid for by a significant increase in effort to store it.

Obviously, at this stage, it is too early to carry out an economic analysis. However, it is worth mentioning that ammonia production based on renewable energy could provide a carbon-free alternative to current ammonia production facilities, which mainly use natural gas as a hydrogen source and with CO_2 as by-product. This might be an advantage as it would provide another, carbon-neutral source for nitrogen fertilizers. On the other hand, ammonia as an energy carrier is connected with the potential danger of competing with fertilizer use, similar to biofuels competing with food production. A well-planned and balanced approach would therefore be necessary, should an ammonia economy be implemented on a large scale.

Based on the analysis presented here, the authors consider that it is worth carrying out basic research on chemical energy storage in the form of ammonia alongside similar investigations on hydrogen, methane, and other substances.

Acknowledgment

The authors thank Lasse Nielsen for his support of the investigations presented in this chapter, especially in computing the energy for compression and expansion of the gases entering and exiting the Haber–Bosch process with Enbipro, the in-house software of the Institute of Energy and Process Systems Engineering.

References

1 Klaus, T., Vollmer, C., Werner, K., Lehmann, H., and Müschen, K. (eds.) (2010) *Energieziel 2050: 100% Strom aus erneuerbaren Quellen*, Umweltbundesamt, http://bit.ly/lrZLRW (last accessed 25 June 2011).

2 Behrens, M. and Schlögl, R. (2011) Energie ist Chemie – Katalyse als Schlüsseltechnik, in *Herausforderung Energie. Ausgewählte Vorträge der 126. Versammlung der Gesellschaft Deutscher Naturforscher und Ärzte e. V.* (eds. J. Renn, R. Schlögl, and H. P. Zenner), Max Planck Research Library for the History and Development of Knowledge Proceedings, No. 1, http://bit.ly/UDeEx1 (last accessed 10 January 2013).

3 Sterner, M. (2009) Bioenergy and renewable power methane in integrated 100% renewable energy system, Dissertation, University of Kassel, University of Kassel Press.

4 Sterner, M., Jentsch, M., and Holzhammer, U. (2011) *Energiewirtschaftliche und ökologische Bewertung eines Windgas-Angebotes*, Fraunhofer IWES, Gutachten für Greenpeace Energy e.G., http://bit.ly/kGwkIc (last accessed 27 June, 2011).

5 Rihko-Struckmann, L. K., Peschel, A., Hanke-Rauschenbach, R., and Sundmacher, K. (2010) Assessment of methanol synthesis utilizing exhaust CO_2 for chemical storage of electrical energy. *Ind. Eng. Chem. Res.*, **49**, 11073–11078.

6 Wikipedia (2013) *TogliattiAzot*, http://en.wikipedia.org/wiki/TogliattiAzot (last accessed 25 January 2013).

7 Odessa Port Plant (2013) *Ammonia Terminal*, http://bit.ly/11YSp7e (last accessed 25 January 2013).

8 Appl, M. (2012) Ammonia, in *Ullmann's Encyclopedia of Industrial Chemistry*, Wiley-VCH Verlag GmbH, Weinheim, pp. 107–137.

9 Atkins, P. and de Paula, J. (2006) *Atkins Physical Chemistry*, Oxford University Press, Oxford.

10 NIST (2012) *NIST Chemistry Webbook*, National Institute of Standards and Technology, http://webbook.nist.gov/chemistry (last accessed 12 December 2012).

11 US Department of Energy (2012) *Biomass Energy Data Book*, http://cta.ornl.gov/bedb (last accessed 12 December 2012).

12 Halvorsen, B. F. (1934) Patentschrift, N. P. 55384.

13 Tanner R. (1945) Über die Verwendung von Ammoniak als Treibstoff, PhD thesis, ETH Zürich, http://dx.doi.org/10.3929/ethz-a-000097159 (last accessed 10 February 2013).

14 Pratt, D. T. (1967) *Performance of Ammonia-Fired Gas-Turbine Combustors*, Technical Report 9, College of Engineering, University of California Berkeley, http://1.usa.gov/W84Oi3 (last accessed 10 January 2013).

15 Kroch, E. (1958) Ammonia – a fuel for motor buses. *J. Inst. Pet.*, **31**, 213, http://bit.ly/T4yaxI (last accessed 10 January 2013).

16 Gibb, J. W. (1958) *X-15 Propellant System Description*, http://1.usa.gov/UDeYMf (last accessed 10 January 2013).

17 Rod, T. H., Logadottir, A., and Nørskov, J. K. (2000) Ammonia synthesis at low temperatures. *J. Chem. Phys.*, **112**, 5343.

18 Skúlason, E., Bligaard, T., Gudmundsdóttir, S., Studt, F., Rossmeisl, J., Abild-Pedersen, F.,

Vegge, T., Jónsson, H., and Nørskov, J. K. (2012) A theoretical evaluation of possible transition metal electro-catalysts for N_2 reduction. *Phys. Chem. Chem. Phys.*, **14** (3), 1235–1245.

19 Smil. V. (1999) Detonator of the population explosion. *Nature*, **400** (6743), 415.

20 Mellor, J. W. (1935) *A Comprehensive Treatise on Inorganic and Theoretical Chemistry*, vol. 8, Longmans, Green, London.

21 Leigh, G. J. (2004) Haber–Bosch and other industrial processes, in *Nitrogen Fixation: Origins, Applications, and Research Progress, Volume 1: Catalysts for Nitrogen Fixation: Nitrogenases, Relevant Chemical Models and Commercial Processes* (eds. B. E. Smith, L. R. Richards, and W. E. Newton), Springer, Berlin.

22 Appl, M. (2012) Ammonia, in *Ullmann's Encyclopedia of Industrial Chemistry*, Wiley-VCH Verlag GmbH, Weinheim, pp. 107–137.

23 ThyssenKrupp Uhde (2012) *Fertiliser Maaden Project*, http://bit.ly/UNolaR (last accessed 4 January 2013).

24 SKW Piesteritz (2012) *Home Page*, http://www.skwp.de/de/home.html (last accessed 10 February 2013), and personal communication.

25 Hellman, A., Baerends, E. J., Biczysko, M., Bligaard, T., Christensen, C. H., Clary, D. C., Dahl, S., van Harrevelt, R., Honkala, O.K, Jonsson, H., *et al.* (2006) Predicting catalysis: understanding ammonia synthesis from first-principles calculations. *J. Phys. Chem. B*, **110** (36), 17719–17735.

26 Morgan, E., McGowan, J., and Manwell, J. (2011) Offshore wind production of ammonia: a technical and economic analysis, presented at the 8th Annual NH_3 Alternative Fuel Conference, Portland, OR.

27 Deutsches Chemiemuseum Merseburg (2012) *Home Page*, http://bit.ly/ZKz60L (last accessed 10 January 2013).

28 Smolinka, T., Günther, M., and Garche, J. (2011) *NOW-Studie: Stand und Entwicklungspotenzial der Wasserelektrolyse aus regenerativen Energien*, Technical Report, Nationale Organisation Wasserstoff- und Brennstoffzellentechnologie, http://bit.ly/XmQRig (last accessed 10 January 2013).

29 Linde (2009) *Nitrogen Generation by Pressure Swing Adsorption*, http://bit.ly/128Sl3y (last accessed 12 December 2012).

30 Apascaritei, B., Hauschke, A., Leithner, R., Schlitzberger, C., and Zindler, H. (2009) Stationary design calculation and part load and dynamic simulation as well as validation of energy conversion and power plant cyclesa. *VGB PowerTech*, **89** (4), 82–88.

31 Amar, I. A., Lan, R., Petit, C. T. G., and Tao, S. (2011) Solid-state electrochemical synthesis of ammonia: a review. *J. Solid State Electrochem.*, **15** (9), 1845–1860.

32 Panagos, E., Voudouris, I., and Stoukides, M. (1996) Modeling of equilibrium limited hydrogenation reactions carried out in H^+ conducting solid oxide membrane reactors. *Chem. Eng. Sci.*, **51** (11), 3175–3180.

33 Marnellos, G. and Stoukides, M. (1998) Ammonia synthesis at atmospheric pressure. *Science*, **282**, 98–100.

34 Marnellos, G., Zisekas, S., and Stoukides, M. (2000) Synthesis of ammonia at atmospheric pressure with the use of solid state proton conductors. *J. Catal.*, **193** (1), 80–87.

35 Ouzounidou, M., Skodra, A., Kokkofitis, C., and Stoukides, M. (2007) Catalytic and electrocatalytic synthesis of NH_3 in a H^+ conducting cell by using an industrial Fe catalyst. *Solid State Ionics*, **178** (1–2), 153–159.

36 Amar, I. A., Lan, R., Petit, C. T. G., Arrighi, V., and Tao, S. (2011) Electrochemical synthesis of ammonia based on a carbonate–oxide composite electrolyte. *Solid State Ionics*, **182** (1), 133–138.

37 Tsuneto, A., Kudo, A., and Sakata, T. (1993) Efficient electrochemical reduction of N_2 to NH_3 catalyzed by lithium. *Chem. Lett.*, **22** (5), 851–854.

38 Tsuneto, A., Kudo, A., and Sakata, T. (1994) Lithium-mediated electrochemical reduction of high pressure N_2 to NH_3. *J. Electroanal. Chem.*, **367** (1–2), 183–188.

39 Murakami, T., Nohira, T., Goto, T., Ogata, Y. H., and Ito, Y. (2005) Electrolytic ammonia synthesis from

water and nitrogen gas in molten salt under atmospheric pressure. *Electrochim. Acta*, **50** (27), 5423–5426.
40. Murakami, T., Nishikiori, T., Nohira, T., and Ito, Y. (2005) Investigation of anodic reaction of electrolytic ammonia synthesis in molten salts under atmospheric pressure. *J. Electrochem. Soc.*, **152** (5), D75–D78.
41. Murakami, T., Nohira, T., Araki, Y., Goto, T., Hagiwara, R., and Ogata, Y. H. (2007) Electrolytic synthesis of ammonia from water and nitrogen under atmospheric pressure using a boron-doped diamond electrode as a nonconsumable anode. *Electrochem. Solid-State Lett.*, **10** (4), E4–E6.
42. Pappenfus, T. M., Lee, K., Thoma, L. M., and Dukart, C. R. (2009) Wind to ammonia: electrochemical processes in room temperature ionic liquids. *ECS Trans.*, **16** (49), 89–939.
43. Kraytsberg, A. and Ein-Eli, Y. (2011) Review on Li–air batteries: opportunities, limitations and perspective. *J. Power Sources*, **196** (3), 886–893.
44. Xie, Y. H., Wang, J. D., Liu, R. Q., Su, X. T., Sun, Z. P., and Li, Z. J. (2004) Preparation of $La_{1.9}Ca_{0.1}Zr_2O_{6.95}$ with pyrochlore structure and its application in synthesis of ammonia at atmospheric pressure. *Solid State Ionics*, **168** (1–2), 117–121.
45. Liu, R. Q., Xie, Y. H., Wang, J. D., Li, Z. J., and Wang, B. H. (2006) Synthesis of ammonia at atmospheric pressure with $Ce_{0.8}M_{0.2}O_{2-\delta}$ (M = La, Y, Gd, Sm) and their proton conduction at intermediate temperature. *Solid State Ionics*, **177** (1–2), 73–76.
46. Wang, J. and Liu, R. Q.(2008) Property research of SDC and SSC in ammonia synthesis at atmospheric pressure and low temperature. *Acta Chim. Sin.*, **66** (7), 717–721.
47. Xu, G. C., Liu, R. Q., and Wang, J. (2009) Electrochemical synthesis of ammonia using a cell with a Nafion membrane and $SmFe_{0.7}Cu_{0.3-x}Ni_xO_3$ (x = 0–0.3) cathode at atmospheric pressure and lower temperature. *Sci. China Ser. B Chem.*, **52**, 1171–1175.
48. Xu, G. and Liu, R. (2009) $Sm_{1.5}Sr_{0.5}MO_4$ (M = Ni, Co, Fe) cathode catalysts for ammonia synthesis at atmospheric pressure and low temperature. *Chin. J. Chem.*, **27** (4), 677–680.
49. Zhang, Z. F., Zhong, Z. P., and Liu, R. Q. (2010) Cathode catalysis performance of $SmBaCuMO_{5+\delta}$ (M = Fe, Co, Ni) in ammonia synthesis. *J. Rare Earths*, **28** (4), 556–559.
50. Liu, R. and Xu, G. (2010) Comparison of electrochemical synthesis of ammonia by using sulfonated polysulfone and Nafion membrane with $Sm_{1.5}Sr_{0.5}NiO_4$. *Chin. J. Chem.*, **28** (2), 139–142.
51. Köleli, F. and Röpke, T. (2006) Electrochemical hydrogenation of dinitrogen to ammonia on a polyaniline electrode. *Appl. Catal. B Environ.*, **62** (3–4), 306–310.
52. Pospíšil, L., Bulíčková, J., Hromadová, M., Gál, M., Civiš, S., Cihelka, J., and Tarábek, J. (2007) Electrochemical conversion of dinitrogen to ammonia mediated by a complex of fullerene C60 and γ-cyclodextrin. *Chem. Commun.*, (22), 2270–2272.
53. Gieshoff, J., Pfeifer, M., Schafer-Sindlinger, A., Spurk, P. C., Garr, G., Leprince, T., and Crocker, M. (2001) Advanced urea SCR catalysts for automotive applications. *SAE Tech. Pap.*, 2001010514.
54. Koebel, M., Elsener, M., and Kleemann, M. (2000) Urea-SCR: a promising technique to reduce NO_x emissions from automotive diesel engines. *Catal. Today*, **59** (34), 335 – 345.
55. Ito, E., Hultermans, R. J., Lugt, P. M., Burgers, M. H. W., van Bekkum, H., and van den Bleek, C. M. (1995) Selective reduction of NO_x with ammonia over cerium exchanged zeolite catalysts: towards a solution for an ammonia slip problem. *Stud. Surf. Sci. Catal.*, **96**, 661–673.
56. Varley, J., Hiddemann, M., Hummel, F., Schmidli, J., and Arguelles, P. (2011) KA26 with new GT26: over 61% efficiency, plus additional flexibility. *Mod. Power Syst.*, **31** (7), 10–12.
57. Lan, R. and Tao, S. (2010) Direct ammonia alkaline anion-exchange membrane fuel cells. *Electrochem. Solid-State Lett.*, **13** (8), B83–B86.
58. Kuppinger, R. E. (1964) *Direct Ammonia–Air Fuel Cell*, Technical Report 4, Electrochimica Corp., Menlo Park, CA.

59 Cairns, E. J., Simons, E. L., and Tevebaugh, A. D. (1968) Ammonia–oxygen fuel cell. *Nature*, **217** (5130), 780.

60 Wojcik, A., Middleton, H., Damopoulos, I., and Van Herle, J. (2003) Ammonia as a fuel in solid oxide fuel cells. *J. Power Sources*, **118** (1–2), 342–348.

61 Halseid, R. (2004) Ammonia as hydrogen carrier. Effects of ammonia on polymer electrolyte membrane fuel cells, PhD thesis, Norwegian University of Science and Technology.

62 Farr, R. D. and Vayenas, C. G. (1980) Ammonia high temperature solid electrolyte fuel cell. *J. Electrochem. Soc.*, **127**, 1478–1483.

63 Staniforth, J. and Ormerod, R. (2003) Running solid oxide fuel cells on biogas. *Ionics*, **9**, 336–341.

64 Fournier, G. G. M., Cumming, I. W., and Hellgardt, K. (2006) High performance direct ammonia solid oxide fuel cell. *J. Power Sources*, **162** (1), 198–206.

65 Dekker, N. J. J. and Rietveld, G. (2006) Highly efficient conversion of ammonia in electricity by solid oxide fuel cells. *J. Fuel Cell Science Technol.*, **3** (4), 499–502.

66 Ma, Q., Ma, J., Zhou, S., Yan, R., Gao, J., and Meng, G. (2007) A high-performance ammonia-fueled SOFC based on a YSZ thin-film electrolyte. *J. Power Sources*, **164** (1), 86–89.

67 Fuerte, A., Valenzuela, R. X., Escudero, M. J., and Daza, L. (2009) Ammonia as efficient fuel for SOFC. *J. Power Sources*, **192** (1), 170–174.

68 Xie, K., Ma, Q., Lin, B., Jiang, Y., Gao, J., Liu, X., and Meng, G. (2007) An ammonia fuelled SOFC with a $BaCe_{0.9}Nd_{0.1}O_{3-\delta}$ thin electrolyte prepared with a suspension spray. *J. Power Sources*, **170** (1), 38–41.

69 Q. Ma, R. Peng, Y. Lin, J. Gao, and G. Meng. A high-performance ammonia-fueled solid oxide fuel cell. *J. Power Sources*, **161** (1), 95–98, 2006.

70 Meng, G., Jiang, C., Ma, J., Ma, Q., and Liu, X. (2007) Comparative study on the performance of a SDC-based SOFC fueled by ammonia and hydrogen. *J. Power Sources*, **173** (1), 189–193.

71 Zhang, L. and Yang, W. (2008) Direct ammonia solid oxide fuel cell based on thin proton-conducting electrolyte. *J. Power Sources*, **179** (1), 92–95.

72 Lin, Y., Ran, R., Guo, Y., Zhou, W., Cai, R., Wang, J., and Shao, Z. (2010) Proton-conducting fuel cells operating on hydrogen, ammonia and hydrazine at intermediate temperatures. *Int. J. Hydrogen Energy*, **35** (7), 2637–2642.

73 Ganley J. C. (2008) An intermediate-temperature direct ammonia fuel cell with a molten alkaline hydroxide electrolyte. *J. Power Sources*, **178** (1), 44–47.

74 Klerke, A., Christensen, C. H., Nørskov, J. K., and Vegge, T. (2008) Ammonia for hydrogen storage: challenges and opportunities. *J. Mater. Chem.*, **18** (20), 2304–2310.

75 Schüth, F., Palkovits, R., Schlögl, R., and Su, D. S. (2012) Ammonia as a possible element in an energy infrastructure: catalysts for ammonia decomposition. *Energy Environ. Sci.*, **5** (4), 6278–6289.

76 Metkemeijer, R. and Achard, P. (1994) Ammonia as a feedstock for a hydrogen fuel cell; reformer and fuel cell behaviour. *J. Power Sources*, **49** (1–3), 271–282.

77 Hejze, T., Besenhard, J. O., Kordesch, K., Cifrain, M., and Aronsson, R. R. (2008) Current status of combined systems using alkaline fuel cells and ammonia as a hydrogen carrier. *J. Power Sources*, **176** (2), 490–493.

78 Smolinka, T. (2009) Water electrolysis, in *Encyclopedia of Electrochemical Power Sources* (eds. C. K. Dyer et al.), Elsevier, Amsterdam, pp. 394–413.

79 Artuso, P., Gammon, R., Orecchini, F., and Watson, S. J. (2011) Alkaline electrolysers: model and real data analysis. *Intl. J. Hydrogen Energy*, **36** (13), 7956–7962.

80 Garcia-Valverde, R., Espinosa, N., and Urbina, A. (2012) Simple PEM water electrolyser model and experimental validation. *Int. J. Hydrogen Energy*, **37** (2), 1927–2938.

81 Murakami, T., Nishikiori, T., Nohira, T., and Ito, Y. (2003) Ammonia synthesis by molten salt electrochemical process under atmospheric pressure, presented at the 204th Meeting of the Electrochemical Society, October 2003.

Part VI
Distribution

34
Introduction to Transmission Grid Components

Armin Schnettler

34.1
Introduction

Worldwide power systems are facing a significant change. In developing and newly industrialized countries, power systems are developing rapidly to match the fast-growing demand and to guarantee higher system security and safety. Here, the installed power of renewable energy sources is increasing but not by as much as thermal power units (based on fossil power generation) are being built. In industrialized countries, the focus of changing the energy supply is mainly driven by national environmental targets, such as the "Energiewende" in Germany, and also by long-term targets to step out of the nuclear energy supply. Worldwide, by the end of 2010 the installed capacity of renewable energy sources (excluding water power) had reached more than 310 GW (EU-wide 135 GW; Germany 50 GW). It is expected that the strong growth of renewables will continue, mainly driven by wind power (additional installed power in 2011: 40 GW) and solar power (PV) (additional capacity in 2011: 22 GW worldwide, mainly in Italy and Germany) [1].

In many countries, these developments will create a significantly different power generation pattern, demonstrating the need for

- higher average transmission distance of electricity
- more flexible (thermal) power generation units
- more flexible transmission ("super grids," "overlay grids") and distribution systems ("smart grids")
- more flexible demand ("fossil generation follows demand follows renewable generation" instead of "generation follows demand").

The role of electricity power grids can be described by (i) integrating and interconnecting generators and consumers, (ii) securing a reliable, stable, safe, and cost-efficient system operation, and (iii) securing international electricity trading and energy balancing for both normal and emergency conditions. Owing to their

Transition to Renewable Energy Systems, 1st Edition. Edited by Detlef Stolten and Viktor Scherer.
© 2013 Wiley-VCH Verlag GmbH & Co. KGaA. Published 2013 by Wiley-VCH Verlag GmbH & Co. KGaA.

very high importance (backbone of the energy grid), transmission grids have to be designed, planned, and operated with extremely high availability and reliability. By whatever means, any malfunction of any component under any environmental conditions must not yield a regional or national/international blackout (e.g., [2]).

34.2
Classification of Transmission System Components

Today, electricity grids of rated voltage levels ≤ 170 kV are often perceived as distribution grids; rated voltage levels > 170 kV (e.g. 245, 420, 550, 800, 1200) are assigned to transmission system voltages.

The most commonly used AC transmission system voltages are 420 and 550 kV with typical transmission capacities of 2–3 GW per system. Typical voltage levels for high-voltage DC (HVDC) transmission systems are ±320, ±500 up to ±800 kV with transmission capacities of up to 7 GW per system.

In general, transmission system components can be classified into two major categories [3]:

- transmission technologies such as overhead lines, cables
- conversion technologies, including transformers, power electronic converters.

34.2.1
Transmission Technologies

In general, transmission technologies can be subdivided into overhead lines and underground lines and are mainly used for interconnecting power system nodes (substations, converter stations).

34.2.1.1 Overhead Lines
Overhead line technologies have been used for more than 100 years and are perceived as a well-known, cost-efficient, and reliable technology. Since nearly all transmission grids are based on overhead line technologies, any grid extension today is mainly realized by these techniques. As overhead lines (and the towers) are affected by environmental conditions (e.g., storms) and their public acceptance in densely populated areas is steadily decreasing, there is a trend towards the use of underground technologies.

Today, owing to their excellent overall performance, overhead lines are state-of-the-art for all voltage levels and both AC and DC transmission; recent developments are related to increases in transmission capacity, for example, by use of high-temperature, low-sag conductors or even more compact designs.

34.2.1.2 Underground Lines

With the increasing trend towards a reduced environmental and visible impact, and with steadily growing electricity consumption in cities, underground technologies have continuously increased in importance. Today, three classes of underground transmission technologies are used: (i) AC and DC underground (submarine) cables [mainly with oil-paper or XLPE (cross-linked polyethylene) insulation], (ii) gas insulated lines (GILs) (N_2–SF_6 insulated), and (iii) high-temperature superconducting (HTS) cables (Figure 34.1).

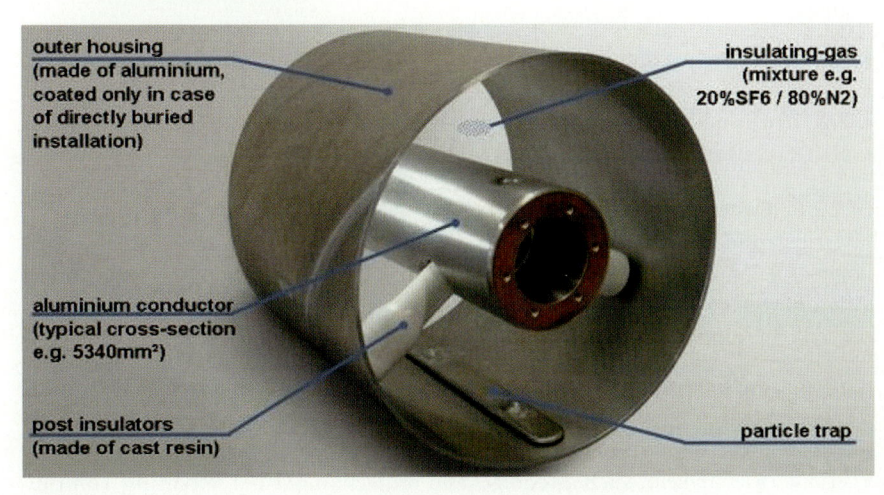

Figure 34.1 (a) High-voltage XLPE cable (Nexans); (b) GIL (Siemens).

Recent underground line technologies are used for the following:
- AC cables: up to 550 kV
- DC cables: up to ±320 kV (XLPE insulation)
               up to ±600 kV (oil-paper insulation)
- GILs: up to 550 kV
- HTS cables: up to 138 kV AC

with typical transmission capacities for cables of up to 1.5 GW and for GILs of more than 4 GW. Depending on the insulation level and the transmission capacity (cross-section of the conductor), typical manufacturing lengths of high-voltage (HV) cables are limited to 400–800 m (for transport reasons). Therefore, all cables have to be connected to each other by prefabricated cable joints. GILs are prefabricated in single units of ~10 m, which are connected on-site to give GIL lengths of up to several kilometers.

For AC applications, underground cables normally require reactive power compensation if the cable length exceeds a total length of more than 20–30 km. Here, GILs offer some advantages since the capacitive load of the GIL is significantly reduced compared with HV cables. Underground transmission line technologies normally do not require maintenance; however, owing to their high importance, monitoring systems are often applied.

So far, HTS cables do not belong to state-of-the-art transmission technologies. However, assuming a significant growth in product and system integration experience with this technique, in the future HTS cables may take over transmission system tasks (owing to their very high transmission capacity).

34.2.2
Conversion Technologies

Generally, conversion technologies can be subdivided into several subsystems, such as:

- switchgears/substations, including breakers, switches, and measurement devices
- power transformers
- FACTS (flexible AC transmission system) devices, such as load flow control devices
- HVDC converters.

34.2.2.1 Switchgears/Substations
HV substations operate as connecting stations between several transmission technologies (e.g., connecting different regions via AC or DC lines/cables) and connect different voltage levels by power transformers. Transmission grid substations may consist of a huge number of

- *primary equipment/components,* such as incoming/outgoing transmission lines/cables, busbars, power transformers, switching devices (e.g., circuit breakers, disconnectors), instrument transformers, towers, earthing systems, and
- *secondary equipment,* such as control and protection devices, metering, communication technologies.

The higher the importance of a substation for the operation of the power system, the higher the design redundancy will be, for example busbar layout with up to three busbar systems, and so on. For transmission voltage levels, substation are normally air insulated [AIS (air insulated substation); Figure 34.2], with significant space requirements (up to several tens of hectares) but at relatively low cost and with standardized components being used.

HV substations in load centers are often built in a gas-insulated system (GIS) using SF_6 as insulating gas within a metal enclosure (Figure 34.3). This technique requires significantly less space than the AIS technology. Today, substations of this type are built for voltage levels of up to 800 kV; typical applications, however, are for voltage levels of up to 550 kV, mainly in densely populated areas, or with severe environmental conditions (e.g., the Middle East) or special space requirements (e.g., hydro power stations). In addition to these air-insulated and gas-insulated substations, hybrid substation also exist, combining both techniques into one substation, for example, busbars in AIS technology and switchbays (breakers, disconnectors, instrument transformers) in GIS technology. Hybrid technologies are often used for retrofitting of existing substations or for upgrading (extensions) of substations with space limitations.

Figure 34.2 AIS substation Basslink-interconnector (Siemens).

Figure 34.3 Overview of a gas-insulated transmission system substation (ABB).

Substations and their main components are normally used for AC systems. Since DC transmission is used for point-to-point transmission, HVDC substations today mainly consist of an AC part including AC filter circuits, transformers with special requirements for application close to power electronic converters, and the DC converters connected to the DC transmission line.

34.2.2.2 Power Transformers

Power transformers connect different voltage levels and provide this voltage transformation with excellent performance, high reliability, and very high efficiency (above 99%). Transformers are normally used in public power systems to interconnect different voltage levels and in power generation stations as a step-up transformer to connect the generator to the grid. Depending on the power rating, transformers are built in one unit (three phases) or in three units (three single-phase transformers will form one three-phase transformer) (Figure 34.4).

In addition to transforming the voltage level, power transformers significantly affect the short-circuit power and voltage quality at lower voltage levels and therefore play an important role in the power system stability and safety and power quality. Owing to their high importance, cost, and delivery time, power transformers are often monitored by on-line monitoring systems and protected against any kind of external influence (overvoltages, environmental and human impact).

Special applications require different designs of power transformers, for example, for traction, electro arc furnaces, HV testing, and load flow control (phase-shifting transformers).

Figure 34.4 (a) Power transformer in an HV laboratory (Siemens); (b) transportation of a power transformer (ABB).

34.2.2.3 FACTS Devices

FACTS are power system components composed of devices to ensure more flexible operation of electrical transmission systems. Normally, FACTS are based on power electronic devices. Generally, FACTS improve AC power system operation by provision of control capabilities in the load flow and therefore increase the power transmission capacity [4]. These devices change the complex power line impedance by means of series compensation or shunt compensation.

Figure 34.5 SVC installation at Statnett, Norway (ABB).

FACTS devices are used in long transmission lines for compensating voltage drops (e.g., due to the demand of reactive power) or compensating capacitive voltage increases (e.g., Ferranti effect). Owing to their fast control capability, FACTS devices are used for power quality control applications, for example,. due to fast changing load conditions [e.g., SVC (static VAR compensator)-supported operation of an electric arc furnace (EAF) (Figure 34.5)] in areas with low short-circuit power ratings. Today, FACTS devices are available for all voltage levels.

In addition to power electronics-based FACTS devices, phase-shifting transformers are also used for power flow control in electricity transmission grids. Phase-shifting transformers are installed in series to power lines and change (by using tapped windings) the total impedance of a power line between two substations. These units are usually used in meshed power networks in order to control the load flow in the region (e.g., in the case of parallel lines with different impedances). Phase-shifting transformers require less space than power electronic-based FACTS devices, are less expensive (at given power ratings), but cannot be used for power quality regulation (only steady-state regulation instead of control in the millisecond range).

34.2.2.4 HVDC Converters

HVDC technology is used for bulk power transmission over long distances. Mainly for long lines (several hundred kilometers), AC transmission of high power faces significant power losses, strong reactive power demand, and voltage stability issues (see Section 34.2.3). Therefore, today HVDC transmission is used for

- bulk power transmission for distances > 600–800 km
- underwater power transmission connections (with unit cable lengths of so far > 600 km realized)
- interconnection of unsynchronized electricity systems (e.g., with different frequencies or power system conditions).

Currently, voltage levels of up to ±800 kV and power ratings of up to 7 GW have been realized, with transmission distances of > 2000 km and total losses < 10% overall. In general, for point-to-point bulk power transmission, HVDC technologies

Figure 34.6 HVDC power electronic converter (Siemens).

are the preferred solution, especially considering that there is no specific difference if either overhead lines or underground cables are used (Figure 34.6). Today, high power-rated HVDC converters are based on thyristor valve technology. Since thyristors require an external circuit (which is provided by the AC system) to turn them off, this technique is often termed the line-commutated converter (LCC) technique.

However, so far mainly point-to-point DC connections have been realized. Compared with AC transmission systems, operation of multi-terminal connections or even a meshed DC grid is much more complex:

- Power flow control requires high-speed and safe communication between the terminals.
- Currently less standardized components (primary, secondary) of HVDC systems between the main manufacturers exist.
- Interruption of short-circuit currents on the DC side is complex since traditional circuit breaker techniques require a natural current zero phase.

For about 10 years, more modern HVDC technologies based on insulated gate bipolar transistors (IGBTs), gate turn-off (GTO) thyristors, and integrated gate-commutated thyristors (IGCTs) are used. These so-called voltage-source converter (VSC) technologies offer the advantage of turning off the current through the valve even before a natural current zero crossing is reached. Since the distortion of the power system voltage by harmonics is reduced, HVDC systems based on these technologies have become smaller (lower ratings), less complex, and therefore less expensive (Figure 34.7).

Since these systems do not necessarily require a power system connection with a high short-circuit power rating, interconnections to island grids or offshore farms (oil and gas; wind power) by VSC technologies have been realized. At present, VSC technologies are operated at voltage levels of up to 320 kV and power ratings of up to 1200 MW, mainly connected by XLPE DC cables, and offering another environmental advantage since oil-paper insulations can be replaced by XLPE (Figure 34.8).

34.2 Classification of Transmission System Components

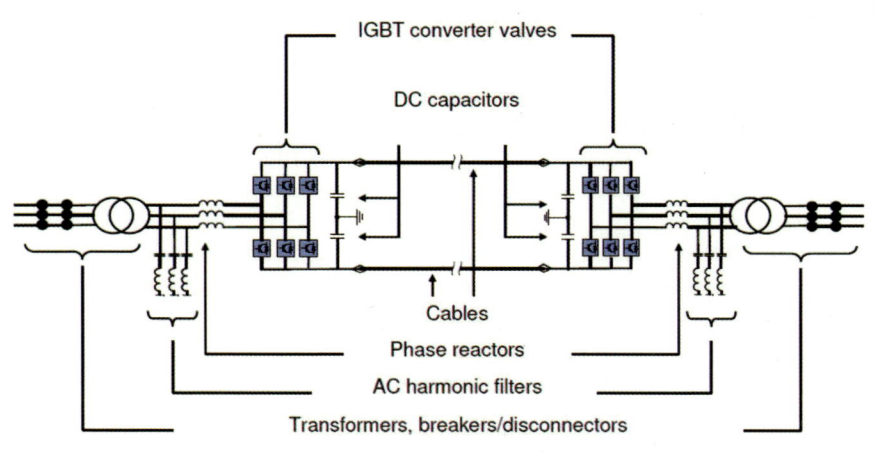

Figure 34.7 Structure of a VSC HVDC transmission system (ABB).

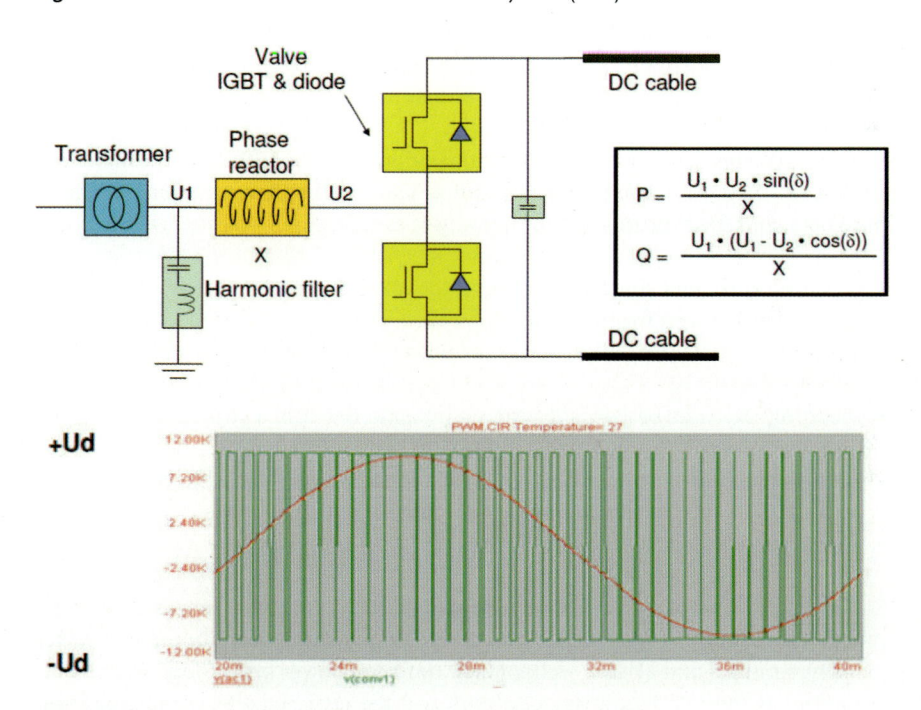

Figure 34.8 Basic principle of VSC HVDC converter technology (ABB).

$$P = \frac{U_1 \cdot U_2 \cdot \sin(\delta)}{X}$$

$$Q = \frac{U_1 \cdot (U_1 - U_2 \cdot \cos(\delta))}{X}$$

34.2.3
System Integration of Transmission Technologies

With the growing demand for electricity, stronger transmission grids have been planned, designed, and implemented worldwide. Starting in the 1950s, in Europe a 380 kV AC power system was installed, mainly using overhead lines for transmission.

Today, the European transmission grid is highly meshed, interconnected throughout all western and central European countries and operated very reliably (so far, no complete blackout has been reported). HVDC systems (LCC technology) are in use for undersea transmission/interconnectors, for example, between Sweden and Poland, Sweden and Germany, Norway and The Netherlands (with power ratings of up to 600 MW). In addition, some first VSC converter systems have been installed and operated, mainly to integrate offshore windfarms into the transmission grid or to provide bulk power transmission capacity with ancillary service to the AC grid (a ±320 kV/2000 MW DC-interconnector France–Spain; operation is expected by 2014).

With increasing demand for bulk power transmission in Europe, especially in Germany, there is a strong need for power system extensions [5, 6)]. At the end of 2012, the German government decided on one of the most intensive power system investment plans for more than 50 years, considering an extension and upgrading/uprating of the existing AC transmission grid and the integration of a new HVDC-overlay system based on up to four DC bulk power corridors with corridor ratings of up to 12 GW so far (preferably VSC technology) (Figure 34.9) [7]. The total investment in the German transmission grid expansion is expected to exceed €25 billion within the next 20 years.

Major challenges for the power system operation in the next 10–20 years, including significant changes in the power generation pattern, the power system extension measures, and the expected consumption, are as follows:

- providing sufficient short-circuit power for voltage control and mass moment of inertia/active power for frequency control
- providing sufficient reactive power for power system stability (avoiding voltage collapse situations)
- operating DC links or meshed systems integrated within a strong AC transmission system
- electromagnetic interference between AC and DC power lines which are operated partly in the same transmission corridor and on the same tower
- integration of ancillary services by regional distribution grids.

So far, in the European power system operation DC interconnectors have been considered as active power generators (which is a valid assumption considering the power ratings and HVDC technology). In future power system operation, both the rating of the HVDC overlay corridors and the expected VSC technology will play a very important role in controllable active power transmission, reactive power provision, voltage regulation, and power quality control. Therefore, new technologies and concepts for system operation, system protection, and emergency operations need to be developed. These new technologies (e.g., DC circuit breakers) may offer new operational concepts, such as meshed DC grids, although it has not yet been proved that meshed DC overlay grids will provide greater reliability and system stability compared with separate DC bulk power transmission corridors within strong AC transmission systems.

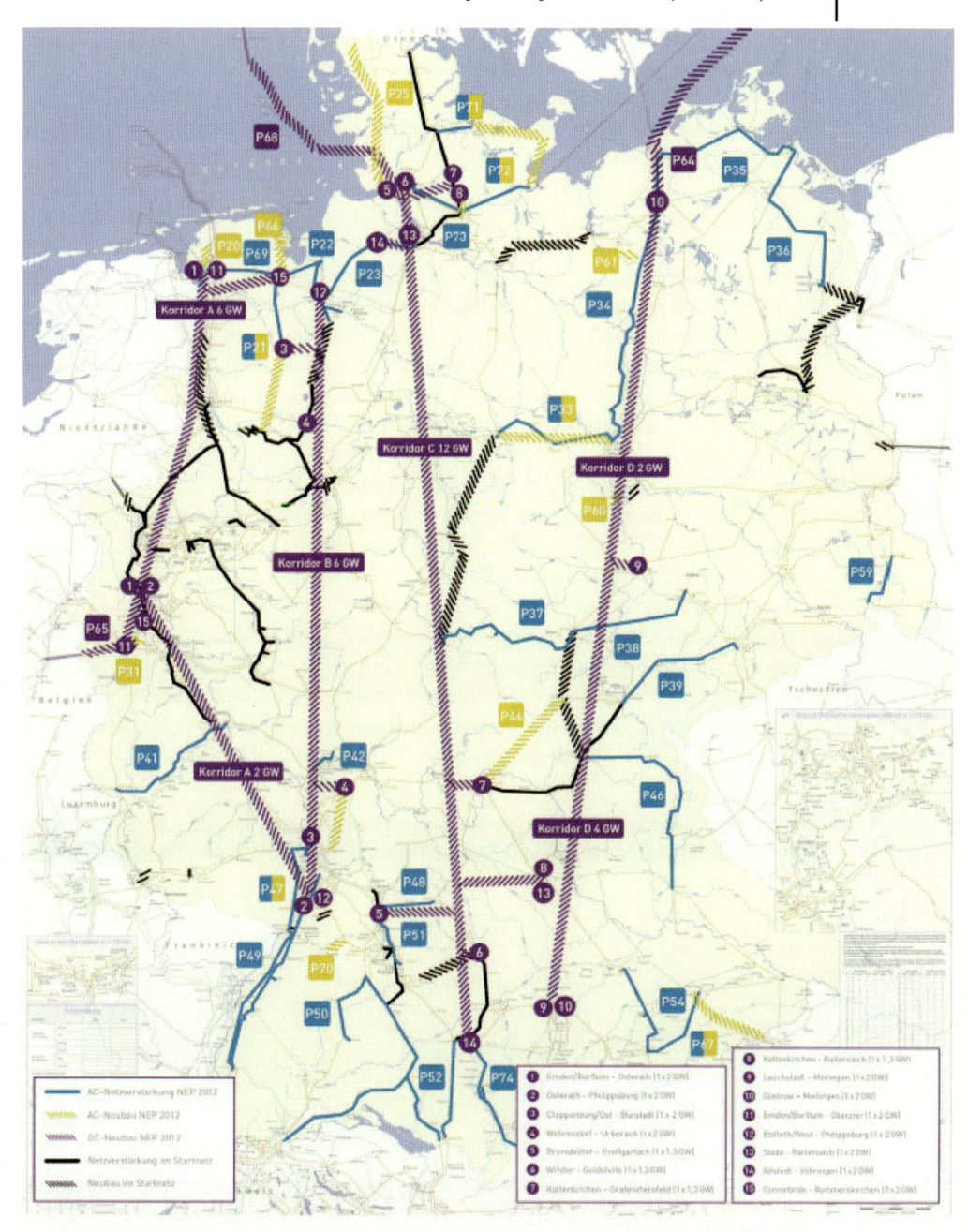

Figure 34.9 Power transmission extensions and overlay grid in Germany (Scenario B2032) [7].

34.3
Recent Developments of Transmission System Components

With respect to power system developments in Europe, it is expected that three major topics will become essential:

- Strong development of insulation strength and current-carrying capacity of power electronic blocks/modules combined with innovative control mechanisms, such as multilevel converters for high voltage levels and power ratings (> ±500 kV; > 4000 MW) (Figure 34.10). These technologies will offer additional system services that are today provided by large power generators/stations and may support the securing of power system stability even with higher shares of power electronic-fed generation (such as wind power and solar power).
- Higher transmission capacity of existing (mature) transmission system components, mainly driven by AC and DC power cable system developments (plastic insulations as well as HTS) and more compact overhead lines. These developments will reduce the visual impact of electrical power generation especially in fast-developing countries with strong load growth and demand for an extensive power system expansion at the lowest cost and increasing importance for a sustainable power supply.
- Highly standardized HVDC components and systems, mainly driven by increased competition and expectations by power system operators worldwide, resulting in significant cost reductions. Thereby the market potential for power electronic devices in transmission and later (with lower costs in euros per kilowatt hour or per kilowatt) in distribution grids will increase significantly. Hence even the

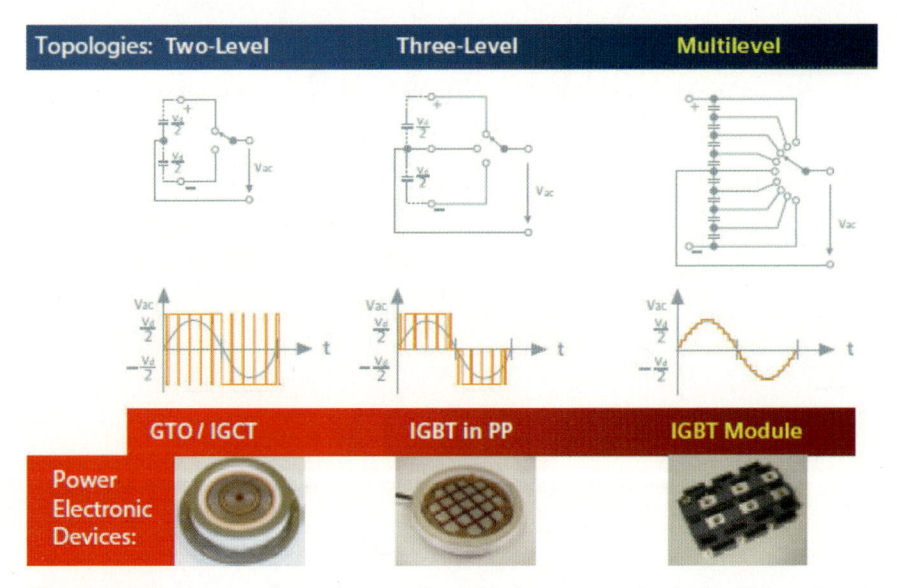

Figure 34.10 Comparison of two-, three-, and multilevel converter technologies (Siemens).

application of DC technologies on a broader, more cost-sensitive power distribution scale may become realistic. Following the trend towards an underground (invisible) power transmission and distribution system, it may be expected that even higher voltage levels will go underground.

References

1 *Smart Grid Strategic Research Agenda 2035*, http://www.smartgrids.eu/documents/sra2035.pdf.
2 Ministry of Power, Government of India (2012) *Report of the Enquiry Committee on Grid Disturbance in Northern Region on 30th July 2012 and in Northern, Eastern & North-Eastern Region on 31st July 2012*, http://www.powermin.nic.in/pdf/GRID_ENQ_REP_16_8_12.pdf (last accessed 28 August 2012).
3 European Union (2012) *Infrastructure Roadmap for Energy Networks in Europe (IRENE-40)*, www.irene-40.eu (last accessed 8 February 2013).
4 Edriss, A. A., Aapa, R., Baker, M. H., et al. (1997) Proposed terms and definitions for flexible AC transmission system (FACTS), *IEEE Trans. Power Deliv.*, **12** (4), 1848–1853.
5 European Network of Transmission System Operators for Electricity (ENTSO-E) (2012) *Ten-Year Network Development Plan 2012*, www.entsoe.eu/major-projects/ten-year-network-development-plan/tyndp-2012/ (last accessed 8 February 2013).
6 Deutsche Energie-Agentur (dena) (2010) *dena-Netzstudie II – Integration erneuerbarer Energien in die deutsche Stromversorgung im Zeitraum 2015–2020 mit Ausblick auf 2025*, www.dena.de/fileadmin/user_upload/Publikationen/Erneuerbare/Dokumente/Endbericht_dena-Netzstudie_II.PDF (last accessed 8 February 2013).
7 Netzentwicklungsplan Strom (2012), *Neue Netze für neue Energien*, www.netzentwicklungsplan.de (last accessed 8 February 2013).

35
Introduction to the Transmission Networks

Göran Andersson, Thilo Krause, and Wil Kling

35.1
Introduction

Typically, the value creation chain in electricity markets comprises three constituents: the production side (generation of electric power), the demand side ("consumption" of electricity), interlinked by transmission and distribution grids. The latter play a crucial role as they "transport" electrical energy from the power plants to the customer. The transmission system operates with high voltage levels over long distances, similar to a system of motorways or highways. From the transmission level, the power "flows" into the sub-transmission grid and eventually into the distribution grid comparable to a system of local roads. The transmission and distribution of electrical energy can be seen as different functions having different technical characteristics, that is, in terms of voltage level, grid topology, and assets used. This chapter deals primarily with the transmission system, targeting questions of bulk power transport, namely the transmission of "massive amounts" of electricity. First, an overview is presented of how the power systems and transmission grids have developed in industrialized countries over the last century. The role of the transmission grid in the operation of the electric power system is discussed and the most important technical limitations with regard to power transmission are deliberated. Second, a more specific description is given of how the transmission system in Europe has evolved. Current and future challenges for the transmission system are described, including the integration of high shares of fluctuating in-feeds from renewables, the creation of the internal European electricity market, together with related questions in terms of cross-border congestion management. The chapter concludes with prospective economic and technological options in order to ensure the secure functioning of transmission grids in Europe.

Transition to Renewable Energy Systems, 1st Edition. Edited by Detlef Stolten and Viktor Scherer.
© 2013 Wiley-VCH Verlag GmbH & Co. KGaA. Published 2013 by Wiley-VCH Verlag GmbH & Co. KGaA.

35.2
The Transmission System – Development, Role, and Technical Limitations

The electric power system has during its existence developed into a structure consisting of in essence three different parts: generation, transmission, and distribution. In addition there are, of course, the consumers, or loads as they often are referred to, which also must be considered in the planning, design, and operation of the power system. The current structure is a consequence of historical and technical factors during the last 100 years. Dominating the electric power generation are still large power plants converting different primary energy sources into electric power. Parts of this electric power can be consumed locally but often the power is transported in the transmission grid, which consists of high-voltage lines and cables, to remotely located loads. The transmission grid is normally meshed and is designed for bulk power transport, and very large loads, for example industries consuming much electric power, might also be connected to the transmission system. For the final transport of the power to most consumers, the distribution system is used. Sometimes an intermediate level, sub-transmission, is used in a region or part of a country. The sub-transmission system is usually not as meshed as the transmission system and the distribution grids are normally operated as radial systems. The different systems are interconnected by transformers, which transform the electric power between different voltage levels. In the past, the electric power was flowing from the higher to the lower voltage levels since the large generators dominated the power production. With the advent of distribution generation, that is, wind and photovoltaic (PV), this will drastically change the picture when the power production of these latter power sources increases.

This chapter focuses on the transmission system, but the developments in the other parts of the electric power system must also be taken into account when discussing the future development of the transmission system.

35.2.1
The Development Stages of the Transmission System

Most of the first electric power systems were based on direct current (DC). Prominent examples are the systems designed, built, and operated by Thomas Alva Edison in North America at the end of the nineteenth century, but similar systems were also functioning at this time in Europe and other places [1–3]. The first larger system in commercial operation was the Pearl Street system in lower Manhattan in New York City. It started its operation in 1882 and served initially 85 customers with a total load consisting of 400 incandescent light bulbs. The system grew rapidly and by 1884 there were 508 customer connected with a total of more than 10 000 light bulbs. Coal was the primary energy source and the electric power was generated by DC generators, which usually were called "dynamos," with a power rating of 100 kW, enough to power 1200 light bulbs, and the operation voltage was ±110 V. DC voltage was very appropriate for the incandescent light bulbs, which were by far the dominant load of the system, since for this kind of voltage the bulb emits a constant light that is pleasant for the human eye. The competing technology based on alternating current (AC) gave at lower frequencies a flickering light, which was annoying. However, as the DC systems grew, the disadvantages of this technology became more and more obvious.

It is clear that the ohmic losses in a system increase with the distance the electric power has to be transported. This means that for a given voltage level there is a maximum distance for which the energy transport is economical. Furthermore, the voltage drop at the boundaries of the system will so large that the light from the bulbs might be unsatisfactory. For the voltage levels selected by Edison and others, this meant that the size of the system was limited to 1–2 km. If a larger system is to be built, a higher voltage must be used for the transport of the power. The electric power engineers were faced with two conflicting requirements. On the one hand, for the end use of power the voltage should not be too high for safety reasons. On the other hand, for efficient transport of power over longer distances, a high voltage is desirable. The obvious solution is to have different voltage levels for the two purposes, but it was not possible to achieve this for DC systems in an effective way with the technologies available at that time. However, it was fairly easy to achieve with a transformer if the voltage was alternating periodically, which is referred to as AC voltage. In the power transformer, two or more windings are magnetically coupled through an iron core and this enables a transformation with low losses. The principle of the transformer was already known by Michael Faraday through his work on electromagnetic induction, and around 1880 the first power transformers for practical use were manufactured. For the AC systems, it was thus possible to have different voltage levels in one system and the voltages could be optimized taken into account economy and safety constraints.

The AC system could therefore cover larger areas than the DC systems and could benefit from the economies of scale that were applicable for the generators. Furthermore, the induction motor designed by Nikolai Tesla was a very simple and cheap electric motor, which can only be operated with AC voltage, and opened up applications of electric power besides the hitherto dominant lighting. Particularly Edison refused to accept the advantages of AC and tried by different means to discredit it by claiming that it was dangerous for humans and animals. He and his protagonists wrote articles and arranged lectures in which the AC technology was denigrated. Tesla, who originally had worked for Edison, joined the AC side and started to work together with the industrialist George Westinghouse, who was the main champion of AC systems in North America. For a couple of year a fierce combat, not always fought by decent methods, took place between the AC and DC camps and this period is often referred to as the battle of the currents. Eventually the advantages of the AC system prevailed and from the mid-1890s it became dominant. The AC technology was further improved by the introduction of the use of several phases, so called multi-phase systems, and very soon the three-phase AC voltage was the leading technology for electric power transmission, and this is still the case today.

For AC systems, the frequency is a crucial parameter and it was optimized for each system. As already indicated above, the frequency needed to be high enough to give a pleasant illumination from the incandescent bulb. Also, transformers and

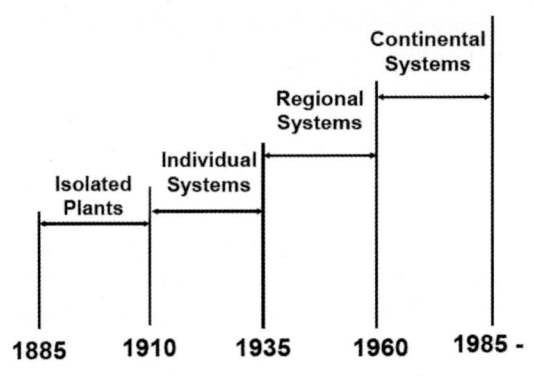

Figure 35.1 Development stages of the electric power systems in Europe and North America.

generators require less iron in the magnetic circuits for higher frequencies, while the possibility of transporting power decreases with increase in frequency. Hence there is a certain power frequency that would optimize the system cost and performance, and this optimal frequency is system specific. For the early AC systems, the frequencies were therefore different, typically in the range 40–70 Hz. However, it was soon realized that there were economic and operational advantages to interconnecting the individual AC systems and this required a standardization of the power frequency, which resulted in 50 Hz being selected in Europe and 60 Hz in North America. For other parts of the world, one of these frequencies was usually chosen, depending on whether the companies involved in the electrification were European or North American.

In Europe and North America, the power systems grew larger and larger. In this process, one can distinguish different stages, as shown in Figure 35.1. It is outwith the scope of this chapter to discuss all the engineering and technical developments that made this possible, but a few will be mentioned. A more detailed description of the European system and its development is given in subsequent sections.

The process shown in Figure 35.1 was made possible through technical developments on different levels. First, the voltage levels[1] used for the transmission lines have been increased from originally tens of kilovolts to 400 kV in Europe and maximum 750 kV in North America. The increased voltage levels made power transmission feasible over longer distances and could be accomplished by advances in insulation technology for power lines, transformers, and other equipment in the high-voltage grid. Another very important enabling technology was automatic control, and in recent decades fast communication and IT have become ubiquitous in transmission systems. Today, these so called secondary systems are as important as the actual high-voltage equipment for an economical, efficient, and stable operation.

As mentioned above, the electric power systems today are almost exclusively based on three-phase AC voltage. However, the DC technology, which lost the competition against AC in the battle of currents at the end of the nineteenth century, offers

---

1) The voltage level for a three-phase system is given as the root mean square (r.m.s.) value of the phase-to-phase voltage, which is $\sqrt{3}$ times the phase-to-neutral (phase-to-ground) r.m.s. voltage.

some attractive advantages for special applications. With the development of power electronics for high voltages and currents, the DC technology has experienced a renaissance and today is regarded as an important complement to AC technology and an essential part in many transmission grids. In the most basic form, the DC transmission system, which for high voltages is called high-voltage direct current (HVDC) transmission, consists of a rectifier, a transmission line or cable, and an inverter. The rectifier converts the AC voltage and current to DC voltage and current. The electric power is then transported as DC on the transmission line and converted to AC in the inverter and fed into the AC power system. One advantage with HVDC transmission is that power systems with different frequencies can be connected with each other or systems with nominally the same frequency but where the frequency controls are not coordinated. Another advantage with HVDC is that the power line or cable does not consume or produce reactive power. Reactive power is a measure of the oscillating power stored in the electric and magnetic fields surrounding an AC power line and gives rise to reactive currents in the power line. This current does not transport any energy but it contributes to ohmic losses and must be considered when calculating the current density, which must be taken into account when designing the conductors (see Section 35.2.3). Despite the fact that the converter stations are costly, power transmission with cables longer that around 100 km are more cost-effective with HVDC than with AC. For overhead transmission lines, the breakeven distance is typically 400–600 km, but can vary considerably because of special circumstances. Another important feature of HVDC transmission is that the transmitted power can be controlled accurately by the control system. In an AC system, one usually has limited possibilities to control the power on individual power lines and new lines do not always increase the transmission capacity as much as required. HVDC can in many cases offer powerful solutions, which enhance the performance of the transmission grid.

35.2.2
Tasks of the Transmission System

During the development of the transmission grid, its role and tasks have changed over time and with the new advances in the power industry it will most certainly adapt to new conditions. The most important tasks can summarized as

- transport of bulk power from remote energy sources;
- increased security by providing redundancy;
- improved efficiency by optimized operation;
- enabling power markets.

Which of these is most important is system dependent, but the last three are fundamental in most large power systems today. The first point was the motivation for the first high-voltage transmission lines, which in most cases were used to transport the energy from remote hydro power plants to load centers. The second and third points are related to each other and were the driving forces behind the high-voltage interconnections in Europe from the 1950s and later. Even if each country was self-sufficient concerning electric power, it was soon realized that during disturbances, such as power plant outages, interconnections could provide reserve power so that the operation was not interrupted. The reserves needed for secure operation of the system could therefore be shared between several countries, which led to improved security and/or reduced costs. Also, temporal differences in power consumption and production, for example, from hydro resources, give the impetus for interconnections that made more optimized operation possible. A prerequisite for the liberalization of the power industry during recent decades is that well-functioning power markets can be formed. In order to increase the efficiency of the markets, liquid markets with many competing actors need to be established. One country or region is often too small for a market and several countries are merged into one common market. For such a merger to meaningful, it is apparent that sufficient transmission capacity must be available to avoid market distortions and prevent some power producers from executing market power.

35.2.3
Technical Limitations of Power Transmission

In this section, we briefly review the physical and technical limits concerning the transport of electric power. Also, some recent technical advances that can increase these limits are presented.

A fundamental limit of a power line or cable is the current capacity of the conductors. Currents lead to ohmic losses, which implies that the temperature of the conductor will reach a value for which the conductor has achieved a heat balance with the surroundings, namely the air for overhead lines and the insulation medium for cables. This temperature cannot be too high since that can lead to destruction of the conductor itself or the insulators, or to accelerated aging. For overhead lines, higher temperatures mean also that the conductor sag between the line towers increases owing to the length expansion of the conductors. If these sags are too great, the distance between the conductor and objects on the ground, such as trees, can become too small and might lead to flash-overs and short-circuits. The limit set by these considerations is called the thermal limit of the power line or cable. For overhead lines this limit is normally the deciding limit for lines shorter than 80–100 km. For cables, this limit is in most cases the dominant one.

For stable and secure operation of the power system, the voltages of different nodes must also be kept within certain limits. High power transfers lead to voltage drops, which cannot be too large. Therefore, a maximum power is allowed and this limit is normally lower than the thermal limit for lines longer than 80–100 km. A third limit imposed on the power transfer is due to the dynamic angle stability, which must be maintained in order to maintain the synchronism between the generators in the system. This limit is typically binding for very long power lines, that is, lines longer than around 300 km.

As indicated above, a meshed transmission grid will enhance the redundancy and security of the system. However, a more meshed system implies also higher short-circuit currents. In case of short-circuits, for example as a consequence of a ground fault, these currents must be interrupted and the circuit breakers used for this task must have the required capability. These deliberations might impose restrictions on the expansion and operation of the transmission grid.

The various limits described above are all of fundamental physical nature and a number of engineering solutions and inventions have been implemented in the power system to enhance these limits in order to allow higher power transfers without endangering the system stability or jeopardizing the equipment. This is an ongoing improvement and two lines of development are today particularly important. First, power electronics is becoming less costly and associated with lower losses, and offers many new interesting solutions. HVDC was mentioned above, but a whole new class of equipment is also being developed. AC systems with a high penetration of power electronics-based devices are often referred to as flexible AC transmission systems (FACTS). Also concerning the deployment of ICT in power systems, several new technologies and systems are being introduced. The term "smart grid" is well known and established today. The transmission system engineers were in many cases pioneers when it came to the use of modern computer and communication solutions. There are examples of control systems from the 1970s that used basically the same principles as the Internet. The transmission system has already been "smart" for several decades, but needs to be smarter to meet the future challenges. Among the most recent advances, the Phasor Measurement Unit (PMU) can be mentioned. This device is based on GPS, and PMUs are used to monitor and control the stability of very large power systems, such as the European one.

35.3
The Transmission Grid in Europe – Current Situation and Challenges

After this more general discussion of the transmission system, the focus will be on the interconnected power grid in continental Europe.

35.3.1
Historical Evolution of the UCTE/ENTSO-E Grid

The existing power grid in Europe is a highly interconnected system, spanning the whole of continental Europe with connections to neighboring systems such as in Scandinavia (Nordel), the United Kingdom, and Russia. The current structure of this meshed, supra-national system was largely influenced by available generation technologies. Conventional thermal power plants and hydro power plants have been characterized in economic terms by significant economies of scale, that is, the larger the plant, the more cost-efficient is electricity generation. Hence it was reasonable to build plants with a rating of several hundred megawatts and to transport electricity over long distances to the load centers, that is, cities. In the 1950s, efforts started to couple the different national networks to a supra-national system of networks. This was achieved through the foundation of the UCTE (Union for the Coordination of

Transmission of Electricity). In 1958, the transmission grids of Switzerland, France, and Germany were coupled. During the following decades, other national networks joined the UCTE. In 2008, the UCTE was replaced by its successor organization, ENTSO-E (European Network of Transmission System Operators for Electricity), with 41 members. Technically, this means that all transmission systems of the member countries are operated together as one so-called synchronous zone. The objectives of creating such a "compound" system were (among others) to benefit from the possibilities of helping each other in emergency situations, providing back-up power, and exchanging "excess" power. To fulfill these functions, the cross-border interconnections between the countries gained more and more importance. However, individual countries operated to a great extent as autarkic zones, that is, generation capacities were in most situations sufficient to cover domestic demand. This resulted in limited cross-border exchanges.

35.3.2
Transmission Challenges Driven by Electricity Trade

For optimal coordination of power plant and network expansion and also operational requirements, it appeared natural that one vertically integrated company would serve all elements of the value creation chain. Through this integrated view, utilities were able to optimize the operation, maintenance, and reinforcement of the whole system. This structure persisted until the early 1990s, when the power supply industries worldwide started to undergo extensive changes. Electricity markets moved away from vertically integrated monopolies towards liberalized structures. The value creation chain was unbundled with generation, transmission, and distribution being separate services, no longer offered by one large utility but by several distinct providers.

The national liberalization processes were supported by legislation from the European Union targeting the creation of a single European electricity market. In regulation 1228/2003, it is stated that, "The creation of a real internal electricity market should be promoted through an intensification of trade in electricity, which is currently underdeveloped compared with other sectors of the economy." Generally, the liberalization efforts led to a significant increase in cross-border trading activities and, in turn, to an increase in cross-border power flows. As the interconnectors usually served purposes in terms of handling emergency situations and sharing system services with only very limited power exchanges, strong cross-border congestion can be observed nowadays. Interconnectors are more often driven to their thermal and stability limits. Challenges are also posed by fluctuating flows induced by very short-term (intra-day) trading activities. Potential remedies include transmission investments, that is, the reinforcement of congested corridors, in addition to the further development and implementation of so-called efficient congestion management schemes such as market coupling. These options are described in the following sections.

35.3.3
Transmission Challenges Driven by the Production Side

In the period between the 1950s and the 1980s, investments in large "conventional" thermal or hydro power plants prevailed. However, the introduction of the combined cycle gas turbine (CCGT) provided a technological justification for competition. The CCGT technology allowed for smaller plant sizes, being at least as economical as conventional thermal and hydro plants with their large economies of scale. This trend was continued by advances in technologies concerning electricity production by means of wind turbines and PV and solar thermal facilities. Electricity in-feed from these technologies fluctuates due to geographically and temporarily changing weather conditions. Fluctuations can be predicted with only limited accuracy. As in power systems, supply and demand always have to be balanced, and appropriate measures must be deployed in order to react to variations on both the generation and load sides. In case of imbalances, primary reserves stop any frequency change in the system due to a mismatch of generation and demand. This automatic control stabilizes the grid frequency. Secondary reserves are activated to bring the grid frequency back to its nominal value (50 Hz in Europe). Finally, tertiary reserves are used to relieve primary and secondary reserves and to guarantee that new disturbances in the system can be faced. Primary, secondary, and tertiary control are standard concepts that have been in operation for decades. However, the change of the production mix towards higher shares of in-feed from fluctuating renewable sources leads also to an increased demand for reserve power.

Another challenge for the transmission system is the change in the spatial distribution of generation facilities. Investments in offshore and onshore wind farms led to substantial in-feeds away from the traditional load centers. Thus, power has to be transported over long distances from, for example, offshore sites to the coast and then further inland. Additionally, the so-called DESERTEC initiative foresees the building of solar-thermal power plants in North Africa. If this plan materializes, the new generation areas have to be connected to the existing grids in Africa and Europe. It is also likely that additional investments will need to be made in order to reinforce the existing system and adapt to the changing power flow patterns in the network. Technological options are described in the following section.

35.3.4
Transmission Challenges Driven by the Demand Side and Developments in the Distribution Grid

Climate change, fossil resource depletion, policy incentives, and higher public awareness in term of sustainability have promoted the deployment of small decentralized and renewable generation technologies, typically including photovoltaics, microturbines, combined heat and power (CHP), and so on. Together with the implementation of distributed storage technologies or the prospective integration of plug-in hybrid electric vehicles (PHEVs), complex interactions in distribution systems arise. The traditional "setup" of the power system with the typical power flow from higher to lower voltage levels may be altered. In-feeds from lower voltage levels are becoming increasingly common. Additional developments on the demand side concern the transformation of formerly "passive" consumers to loads, which can be integrated actively by means of demand-side participation. In doing so, loads (consumers) in conjunction with storage devices and small renewable generation units may contribute to traditional concepts, such as secondary control. Such developments are likely also to influence higher network levels. There may be a trade-off between supra-national investments and regional or local "solutions," such as by investing into micro-grids.

35.3.5
Conclusion

A short introduction was given to the historical evolution of the UCTE/ENTSO-E grid, and a number of prevailing challenges for the transmission system, and thus for bulk power transport, were identified. The following sections try to identify market-based and technological solutions to respond to the identified challenges and to ensure the reliable operation of the power system in the long run.

35.4
Market Options for the Facilitation of Future Bulk Power Transport

35.4.1
Cross-Border Trading and Market Coupling

It is expected that large-scale integration of renewable sources will put stress on power balancing mechanisms even in large systems. For the efficient integration of renewable sources (e.g., wind and solar generation), new market structures are required. One of the new approaches is to combine the different national market places in order to use trading and balancing procedures across borders. This has been proposed by policy makers and regulators. With so-called market coupling, the daily cross-border transmission capacity between the various areas is not explicitly auctioned among the market participants, but is implicitly made available via energy auctions on the power exchanges on either side of the border (hence the term implicit auction).

Spot markets of different countries are already merged. Currently, Germany, Belgium, France, Luxembourg, and The Netherlands share their day-ahead markets [4]. Even control areas outside the Central West European system are coupled to this market. For example, the NorNed cable (HVDC) connects the Dutch grid with the Norwegian grid.

Furthermore, the integration of more renewable generation demands trading closer to real-time operation. This ensures that positions are kept balanced between the day-ahead stage and real-time operation. In addition, the intra-day spot market coupling would be a further step towards market integration. Belgium and The Netherlands are the first countries already performing such an intra-day coupling.

35.4.2
Cross-Border Balancing

Another step forward is the coupling of balancing markets for real-time operation to deploy balancing resources (secondary reserves) from other control areas. When a certain control area monitors an area control error (ACE), balancing services could be deployed from transmission system operators (TSOs) across the border. The first step towards cross-border balancing is the share of ACEs among participating control areas. If a certain area is long (more export as scheduled) and a certain area is short (more import as scheduled), both TSOs will not deploy balancing services but share their opposite imbalance. In general, ACE netting decreases the total amount of requested balancing services. Since the beginning of 2012, the four TSOs of Germany, together with TSOs from Denmark, The Netherlands and Switzerland, have applied ACE netting [5]. In addition to this type of balancing services via market coupling, there are other market designs to balance across the border, defined by ENTSO-E and EER (European Energy Regulators). The ultimate objective is to implement one common merit order list for the entire Central West European system to deploy the most cost-efficient balancing services.

Nevertheless, these new types of cross-border deployment of energy and capacity will affect system operation and power transmission. Frequency deviations due to cross-border trading are already noticeable on hourly transitions [6]. In addition, the future will reveal if cross-border activities will touch congestion areas. It has to be taken into consideration that cross-border trading and balancing must not violate the N-1 criterion[2] at any time.

35.4.3
Technological Options for the Facilitation of Future Bulk Power Transport

When it comes to network investments, different technological choices exist that depend on the specific project. Generally, a cross-border interconnector can be regarded as a coupling between two neighboring grids. The network parameters and the method of operation of either system influence to a large extent the final investment decision. Most networks around the world are based on AC. Therefore, if there are no specific reasons of technical or economic nature, these networks will be coupled AC synchronously.

In some cases, synchronous coupling is impossible or economically not desirable, namely:

- with coupling between systems with different nominal frequency or a different mode of frequency and voltage regulation;
- with long cable connections, as in the case of sea crossings (distances greater than 30–40 km) due to the capacitive effect of AC cables;
- with a weak coupling between two grids because of stability problems.

2) The N-1 criterion expresses the ability of the transmission system to lose any component, e.g. generator, line, or transformer, without causing overloads elsewhere or bring the system into unstable operation.

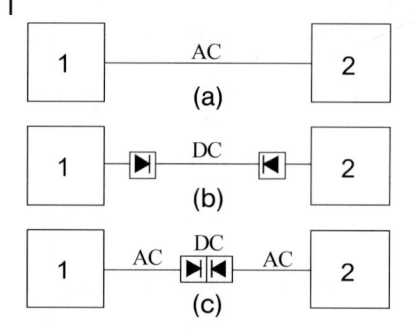

Figure 35.2 Connection of systems.

In these cases, instead of using AC, energy may be transmitted by HVDC using an overhead line, an underground or submarine cable, or a combination of these [7]. Converter stations to switch between AC and DC are needed at both ends of an HVDC connection. Distinction can be made in true DC connections and the so-called back-to-back installations (see Figure 35.2).

In the case of synchronous connection, the choice of the voltage level of the links is also relevant. Nowadays, 380 kV in Europe and 550 kV in North America are the predominantly applied highest voltage levels, but there may be a need for a higher voltage level in the future, as studies show, ref. [8]. Voltage levels of 750, 1000, and 1200 kV are conceivable. In North and South America, 765 kV lines are already present. As in Eastern Europe a 750 kV line was built, it is possible that this trend will continue there, and perhaps spread also towards Western Europe. In Russia, from the Urals to Kazakhstan, an 1150 kV line is in service. Also in other places around the world there are experiments with voltages higher than 1000 kV. The main reason for using a higher voltage is the ability to increase the transmission capacities, and in turn decrease network losses, which is important for transport over longer distances.

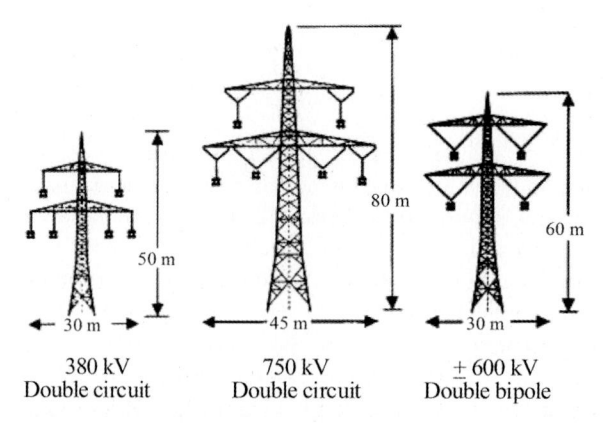

Figure 35.3 Tower images of AC and DC lines.

Figure 35.3 shows what a higher voltage means for the tower image. The DC solution is also drawn. The capacity of the 750 kV line is twice that of the 380 kV line; the DC line has the same capacity as the 750 kV line, but can be built in the same trace width as the 380 kV line.

For AC coupling, the line costs are dominant over the substation cost. With DC, the substation costs (converters) are higher than with AC, but the cost of an overhead DC line is lower than that of an AC line. Hence there is a break-even point where the total costs are equal to each other (see Section 35.2.1).

AC cables are most often used in heavily populated areas such as large cities or under water. The cost of manufacturing and installing such cables is higher than the corresponding costs of overhead lines for a given length. The most serious limitation for the use of AC cables is t that they cannot be used for long connections without extra devices, compensation reactors, which are needed at intervals along the cable to limit the reductions in transmission capacity arising from the reactive currents in AC cables.

The transmission capacity of an AC cable is rather limited compared with overhead lines. It is constrained by heating of the cables and is strongly restricted by the length of the connection. The major disadvantages of AC cables are therefore their high costs and short length. Their main advantage compared with overhead AC lines is the reduced visual impact. HVDC cables are not limited in length.

35.5
Case Study

This case study is intended to give an indication of how network investments will influence total operating costs for the power system in Europe. The research was performed within the FP7 project Infrastructure Roadmap for Energy Networks in Europe (IRENE-40) financed under grant agreement 218903 [8]. IRENE-40 aims at identifying appropriate transmission expansion measures in order to achieve a more secure, sustainable, and competitive European power system. Here we focus on the comparison of different technological options for transmission expansion and the influence of such measures on the operating costs of the overall system, that is, including generation and transmission costs.

The results were computed for a scenario in 2050 with a high share of in-feed from renewable sources (~80%). Generally, network congestion leads to higher generation costs as cheap generation units are constrained-off by the bottlenecks in the transmission system. Congestion relief will lead to lower operational costs as generation can be utilized more efficiently. The studies were performed with a single node per country model, comprising 32 nodes and 104 branches. The model includes the former UCTE area, Nordel, United Kingdom and Ireland, and the Baltic States. Each interconnection (except for the submarine cables) was modeled as two identical AC lines. The line data were aggregations of real data provided by UCTE (now ENTSO-E). The network model used is illustrated in Figure 35.4, Figure 35.5, and Figure 35.6.

In the following we compare three expansion technologies: (1) double-circuit AC–400 kV lines with a rating of 3000 MVA, (2) a single-circuit AC–750 kV line with a rating of 3900 MVA, and (3) voltage source converter high-voltage DC transmission with a rating of 3000 MVA. In all cases we assume that we add one parallel line along interconnections which are congested (i.e., 100% loaded) over 50% of the time during the year in the base-case scenario. For all three technological options we calculate the operational costs together with potential cost savings.

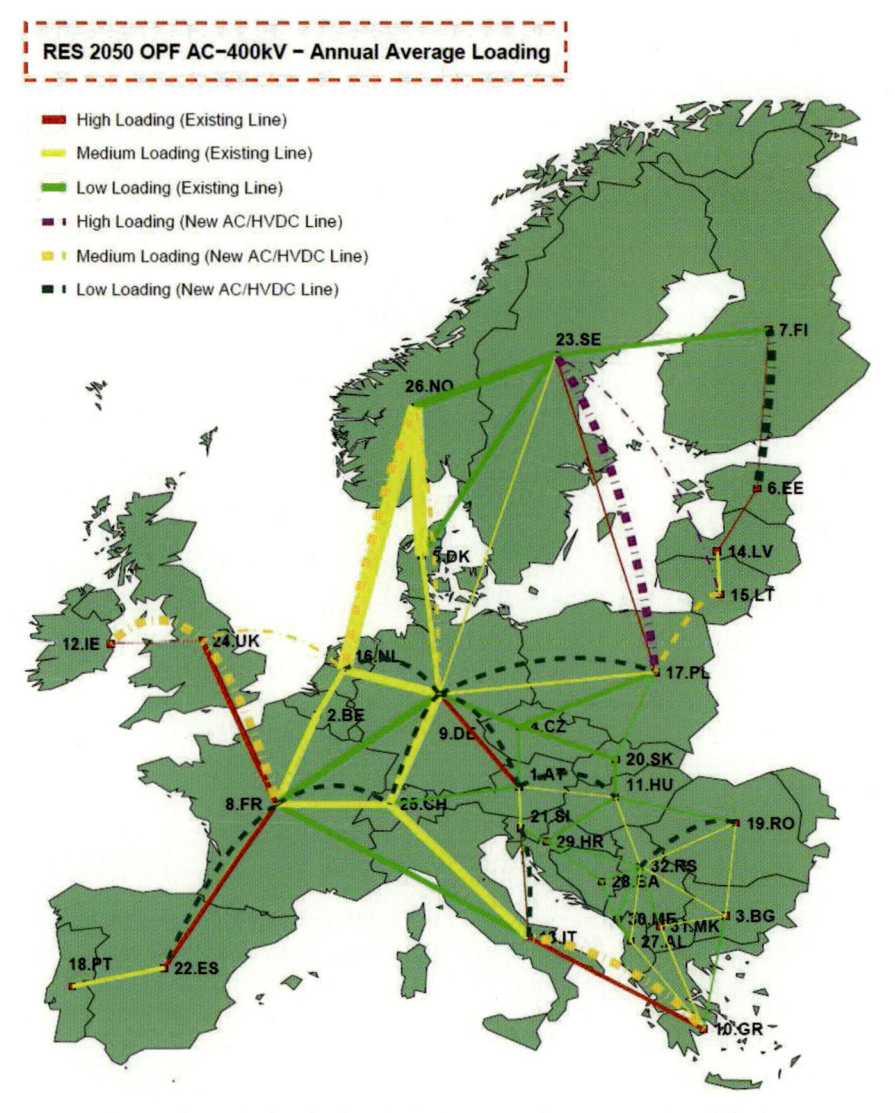

Figure 35.4 Annual average line loading in 2050 (AC–400 kV expansion). High loading, > 85%; medium loading, 50–85%; low loading, < 50%. The line width is proportional to the line capacity. All submarine cables are HVDC lines.

35.5 Case Study

Figure 35.4 shows the annual average of line loadings when we rely on double-circuit AC–400 kV lines with a rating of 3000 MVA for network expansion. Eleven new lines are built. Furthermore, three HVDC lines are installed representing projects that are already planned today (2012). Red lines indicate congestion. The operating costs (generation and network) for such a scenario amount to €95.29 billion. Compared with the base-case scenario (no network expansion), we save about 10.3% of operating costs.

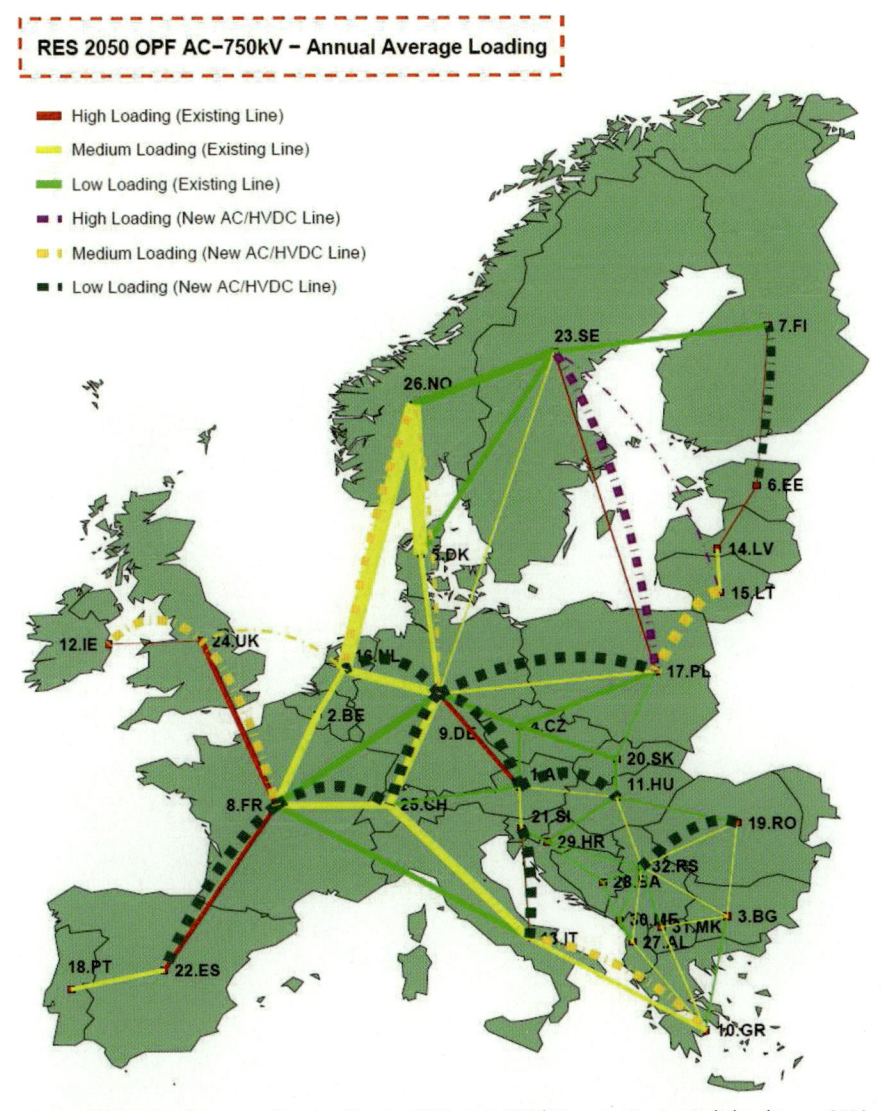

Figure 35.5 Annual average line loading in 2050 (AC–750 kV expansion). High loading, > 85%; medium loading, 50–85%; low loading, < 50%. The line width is proportional to the line capacity. All submarine cables are HVDC lines.

Figure 35.5 shows an investment scenario where 11 new 750 kV AC lines are built as indicated by the different colors. As in the previous case, the three already existing HVDC investment projects are considered. In this case, operating costs amount to €94.01 billion. This means a decrease of 11.5% compared with a situation without any network investments.

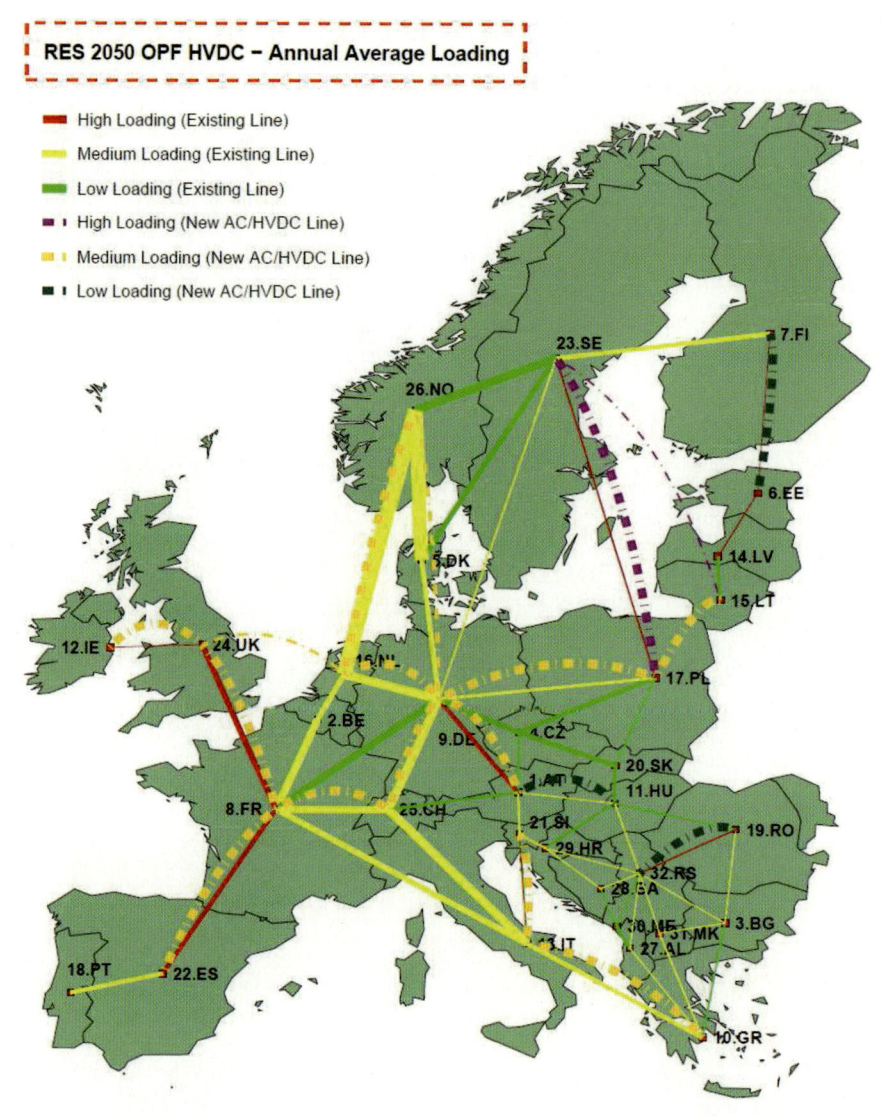

Figure 35.6 Annual average line loading in 2050 (HVDC expansion). High loading, > 85%; medium loading, 50–85%; low loading, < 50%. The line width is proportional to the line capacity. All submarine cables are HVDC lines.

Figure 35.6 shows the results for the case when the network is reinforced, relying completely on VSC (voltage source converter)–HVDC technology. Fourteen new HVDC lines are built. This investment scheme leads to a 16.5% decrease in operating costs to €88.74 billion per year.

The results are dependent on several parameters assumed for the development of generation and load and also the spatial distribution of generation. However, it becomes obvious that the choice of technology and investment location will have an influence on operating costs and congestion. This mutual dependency indicates that different solutions are conceivable, that is, prospective investment plans (preferred technologies and locations) might change with respect to the scenario assumptions. However, from our analysis, a VSC–HVDC backbone network is the preferred solution as it leads to the highest savings in operating costs. This effect is due to the "controllability" of power flows by means of HVDC technology, that is, HVDC lines can be actively used to optimize network utilization.

References

1 Hughes, T. P. (1993) *Networks of Power: Electrification in Western Society, 1880–1930*, Johns Hopkins University Press, Baltimore.
2 Jonnes, J. (2003) *Empires of Light*, Random House, New York.
3 IEEE (2012) *IEEE Global History Network*, http://www.ieeeghn.org (last accessed November 2012).
4 ApxEndex (2012) *CWE Market Coupling*, http://www.apxgroup.com/services/market-coupling/cwe/ (last accessed March 2013).
5 Regelleistung.net (2012) *Grid Control Cooperation*, https://www.regelleistung.net/ip/action/static/gcc (last accessed March 2013).
6 UCTE Ad-Hoc Group (2012) *Frequency Quality investigation*, excerpt from the final report, https://www.entsoe.eu/resources/publications/former-associations/ucte/other-reports/ (last accessed November 2012).
7 EASAC (2009) *Transforming Europe's Electricity Supply – an Infrastructure Strategy for a Reliable, Renewable and Secure Power System*, EASAC Policy Report, ISBN 978-0-85403-747-6, EASAC, Royal Society, London, easac.eu/fileadmin/PDF_s/reports_statements/Transforming.pdf (last accessed March 2013).
8 IRENE-40.eu (2012) *Infrastructure Roadmap for Energy Networks in Europe*, www.irene-40.eu (last accessed 9 January 2013).

36
Smart Grid:
Facilitating Cost-Effective Evolution to a Low-Carbon Future

Goran Strbac, Marko Aunedi, Danny Pudjianto, and Vladimir Stanojevic

36.1
Overview of the Present Electricity System Structure and Its Design and Operation Philosophy

In most industrialized countries, the present electricity system was designed to support the post-World War II economic growth and the developments in electricity generation technology. The system is characterized by small numbers of large generators, mainly coal-, oil-, hydro-, nuclear-, and gas-based generation. Typical power station ratings would be from a few hundred to a couple of thousand megawatts. These stations are connected to a very high voltage transmission network operating at 400 kV. The role of the transmission system is to provide bulk transport of electricity from these large stations to demands centers. The electricity is then taken over by the distribution networks that, through a number of voltage transformations,[1] provide the final delivery of electricity to consumers. The flow is unidirectional from higher to lower voltage levels, as illustrated by the structure presented in Figure 36.1.

The operation and design of the electricity system were driven by an overall "predict and provide" design philosophy: electricity demand is predicted and then electricity generation and network infrastructure are delivered to supply this demand securely.[2]

In order to supply demand that varies daily and seasonally, and given that demand is uncontrollable and any supply interruptions are very costly, installed generation capacity should be able to meet maximum (peak) demand.[3] In addition, there needs to be sufficient capacity available to deal with the uncertainty in generation availability and unforeseen demand increases. Historically, a capacity margin of around 20% was considered to be sufficient to provide adequate generation security.

1) Typical voltage levels include 132, 110, and 33 kV for high voltage, 20 and 10 kV for medium voltage, and 400 and 230 V for low voltage.
2) It is important to stress that the key function of a network is to provide secure and efficient transport of electricity from generation (production) to demand (consumption). Hence the position of generation relative to demand and the amount of power to be transported are the key factors driving the design and operation of electricity networks.
3) In northern Europe, the peak demand occurs in evening hours during winter, whereas in southern Europe the peak demand generally occurs around midday in summer

Transition to Renewable Energy Systems, 1st Edition. Edited by Detlef Stolten and Viktor Scherer.
© 2013 Wiley-VCH Verlag GmbH & Co. KGaA. Published 2013 by Wiley-VCH Verlag GmbH & Co. KGaA.

```
                    ┌──────────────────┐ ┐
                    │  Large Central   │ ├  Source of energy and system
                    │   Generation     │ │  control
                    └────────┬─────────┘ ┘
                             │
                    ┌────────▼─────────┐ ┐
                    │   Transmission   │ ├  Bulk transport of electricity,
                    └────────┬─────────┘ │  coordination of control
                             │           ┘
                    ┌────────▼─────────┐
                    │  EHV Distribution│
                    └────────┬─────────┘ ┐
                   D         │           │
                   E ┌───────▼──────────┐│
        Passive,   M │  HV Distribution │├ Delivery system,
     uncontrollable A└────────┬─────────┘│ passive, radial networks
                   N          │          │
                   D ┌────────▼─────────┐│
                     │  LV Distribution ││
                     └──────────────────┘┘
```

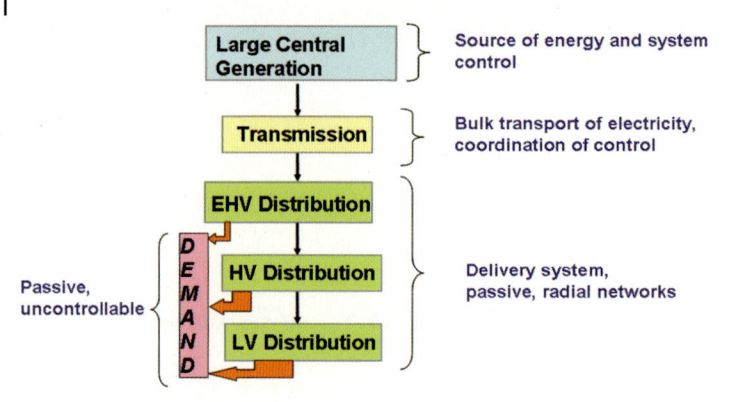

Figure 36.1 Structure of the present electricity system.

Given the average demand across the year, the average utilization of the generation capacity is below 55%.[4]

One of the key distinguishing features of the electricity system is that the balance between demand and supply must be maintained at all times. Given that the demand is not controllable (or not responsive), the only source of control is the generation system. Any changes in demand are met by almost instantaneous changes in generation. In order to deal with unpredicted changes in demand and generation availabilities, various forms of generation reserve services are made available to ensure that demand and supply can be matched on a second-by-second basis. Hence the generation system not only provides energy to supply the demand but is also the key source of system control.

Flows in the transmission network need to be maintained within given limits to ensure that the equipment is not overloaded and that the system is operated stably and securely. The primary means of controlling transmission network flows are the large central generators themselves, that is, the transmission network flows are controlled by varying the output of generators in different locations. This is because the transmission network operates with fixed topology and fixed impedance and the power flows are therefore determined by the magnitudes of generation and demand.

Historically, the design and structure of electricity transmission and distribution networks were driven by an overall design philosophy developed to support large-scale generation technologies. The network is required to continue to function after a loss of a single circuit or a double circuit on the same tower. Given the present operation philosophy, this means that under normal operation (during the peak load conditions), circuits in the interconnected transmission network are generally loaded below 50%. Given that very little generation is connected to distribution networks, these networks are not controlled in real time. According to the historical principles of electricity distribution network design, the real-time control of distri-

---

4) There is a significant spread in utilization among different generators. The lowest marginal cost plant would operate at high load factors [say more than 85% load factor (e.g., nuclear)], whereas plant with high fuel costs would operate at low load factors [less than 5% load factor (e.g., open-cycle gas turbines)].

bution network is resolved through the robust specification of primary network infrastructure, hence these networks traditionally operate as passive systems (i.e., the network control problem is resolved at the planning stage).

36.2
System Integration Challenges of Low-Carbon Electricity Systems

Worldwide, future electricity systems face challenges of unprecedented proportions. By 2020, 20% of European electricity demand will be met by renewable generation. In the context of the targets proposed by the EU governments (greenhouse gas emission reductions of at least 80% by 2050), it is expected that the electricity sector would become largely decarbonized by 2030, with potentially significantly increased levels of electricity production and demand driven by the incorporation of heat and transport sectors into the electricity system [1].

There are two key concerns with the future low-carbon electricity systems:

1. *Degradation in generation and network asset utilization.* Wind generation and other low-carbon distributed generation will displace energy produced by conventional plant, but their ability to displace capacity will be very limited. If the security of supply is to be maintained (and demand supplied through periods of low renewable generation output), this will require that a significant amount of conventional plant is maintained on the system, leading to asset utilization degradation. Our analysis of the British system[5] suggests that the utilization of generation capacity will reduce from the present level of 55% to 35% by 2020, with more than 25 GW of conventional plant operating at less than 10% load factor. Similarly, new low-carbon distributed generation (DG) technologies have been connected to the systems under the current policy based on the "fit and forget" approach. Under this historical passive operation paradigm, the DG will not be able to displace the capacity of conventional generation owing to the lack of controllability of DG as the system control and security services may continue to be provided by central generation. Furthermore, the incorporation of the heat and transport sectors into the electricity system will lead to a very significant increase in peak demand that is disproportionately higher than the increase in energy. If the present ("predict and provide") network operation and design philosophy is maintained, massive electricity infrastructure reinforcements will be required, leading to high investments and low utilization of generation, transmission, and distribution network capacity.
2. *Reduced ability to incorporate increased amounts of intermittent generation.* Efficient real-time demand–supply balancing with a significant penetration of intermittent wind power and increased contribution from less flexible low-carbon generation will become a major challenge. This will be primarily driven by the increased need for various reserve and frequency regulation services to deal with wind

5) Current British electricity system is characterised by approximately 60 GW peak demand, and 72 GW of conventional plant capacity generating about 350 TWh annually.

output uncertainty [2]. Provision of a significant part of these services will be accompanied by energy production, given the involvement of part-loaded fossil fuel generation. The increased need for system management services will not only reduce the efficiency of operation of conventional generation in the presence of intermittent generation, but also may limit the ability of the system to absorb renewable output, potentially leading to renewable generation curtailments. In systems with significant contributions of inflexible nuclear generation technologies, this is particularly likely to occur during periods when high wind generation output coincides with relatively low demand, given the need to accommodate not only must-run nuclear generation but also a significant amount of part-loaded generation required to maintain system integrity. Any curtailment of renewable energy would most likely be compensated by fossil fuel plant (whose output was intended to be displaced by intermittent generation), which will also deteriorate the emissions-related performance of the system as the full emission-saving potential of renewables will not be realized. Our analysis shows that in 2020 scenario with a 25% contribution from wind generation together with 25% of less flexible nuclear in combination with energy efficiency measures, more than 20% of wind production may need to be curtailed. This would also lead to an under-delivery in emissions reductions.

36.3
Smart Grid: Changing the System Operation Paradigm

As discussed above, the historical approach to grid control based on redundancy in assets will lead to inefficient and costly overinvestment, and an alternative solution, involving innovations in ICT technology and active system control, would be potentially more cost-effective and more commercially adaptable than simply building more primary assets. The core issue facing the network in the future is not only to make grids larger but also, more importantly, to make them more intelligent. This is illustrated in Figure 36.2.

As the future cost of system infrastructure reinforcements will be driven by the control and design concepts, we contrast two approaches:

- First, following the present "predict and provide" philosophy with only conventional generation resources providing control, while transmission and distribution network control problems are being resolved in the planning stage (through asset redundancy), that is, a Business as Usual (BaU) approach that will lead to massive degradation in asset utilization, as indicated in Figure 36.2.
- Second (smart grid), involving real-time demand–supply and network management through the application of alternative smart-grid technologies, involving a paradigm shift in network control philosophy from preventive to corrective control, making use of advanced grid control functionality facilitated by appropriate communication infrastructure that can increase asset utilization and avoid or postpone generation and network reinforcements [3, 4].

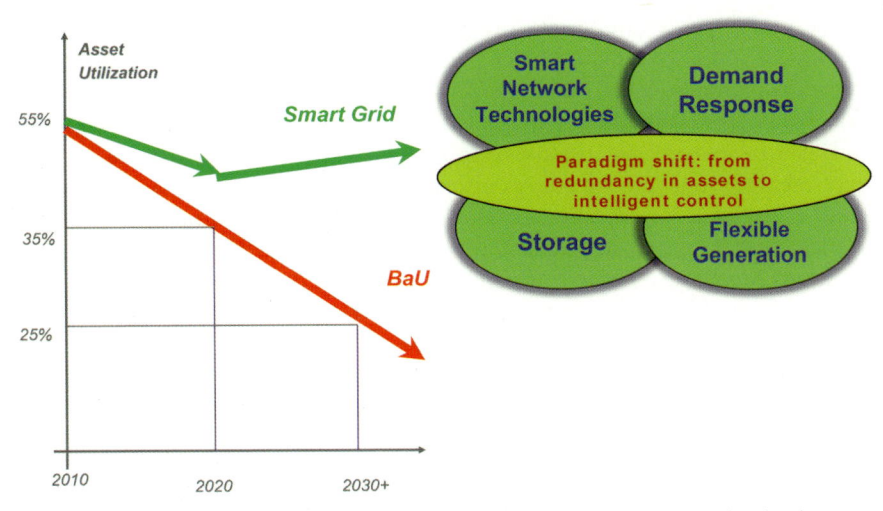

Figure 36.2 Reversing the trend of degradation of asset utilization by smart grid technologies supported by ICT.

If the asset utilization is not to degrade but rather potentially to become enhanced, the system flexibility that has been traditionally delivered through asset redundancy would need to be provided through more sophisticated control that incorporates advanced technologies (supported by appropriate communication and information technologies):

- Network technologies, such as phasor measurement units and network sensors, dynamic line rating, advanced power electronics technologies [e.g., various types of distribution and transmission network flexible AC transmission systems (FACTS)], various novel control and protection schemes, that all enhance the utilization of the network and generation assets through facilitating more sophisticated real-time control of the system [5, 6].
- Demand-side response (DSR), through utilizing the inherent demand-side flexibility, particularly demand associated with heat and transport; DSR can be used for real-time system management, while ensuring that the intended service quality is not adversely affected [7].
- Energy storage technologies, that can be used to support demand–supply balancing or control of network flows, and hence increase the utilization of electricity infrastructure assets [8, 9].
- Enhancing the flexibility of central and distributed generation, which can be used to facilitate more cost-effective demand–supply balancing and control of network flows, hence enhancing the ability of the system to absorb intermittent generation.

Our recent work suggests that the volume of the market for these technologies in the UK system could be very significant, exceeding £10 billion per year by 2050 [1].

The proliferation of energy storage, distributed generation, and solid-state equipment and greater demand-side participation are at present not appropriately integrated, for a variety of reasons (such as market, regulatory, and policy barriers discussed later). Furthermore, information management, wide-area measurement, and disturbance recognition and visualization tools are yet to be fully developed and implemented to enhance the processing of real-time information, accelerate response times to various system disturbances (such as power imbalance of voltage deviations) and achieve compliance with reliability criteria at lower cost [10]. This also includes the development of interface technologies and standards to permit the seamless integration of distributed energy and loads with the local distribution system.

Although the key ingredients of these technologies exist, the key unresolved challenge is in the development and demonstration of effective energy system integration, showing that smart grid can deliver the functionality and performance needed for real-time control of the power system. This represents a shift from a traditional central control philosophy, currently used to control typically hundreds of generators, to a new distributed control paradigm applicable to the operation of hundreds of thousands or millions of generators and controllable loads [11, 12].

Hence the implementation of ICT for monitoring and control of the electricity system, including the application of new technologies, will lead to the development of an integrated energy and information and communication system architecture that is intended to bring together two elements of the power industry: the electrical delivery system and the information system that controls it [13]. Maximizing the utilization of primary electricity assets and infrastructure, by deploying and utilizing the information and communication technologies and developing effective energy system integration strategies, is the core objective of the concept of smart grids.

36.4
Quantifying the Benefits of Smart Grid Technologies in a Low-Carbon future

When considering system benefits of smart grid technologies, such as storage, demand-side response, network and flexible generation technologies, it is important to consider two key aspects:

- *Different time horizons:* From long-term investment-related time horizon to real-time balancing on a second-by-second scale (Figure 36.3). This is important as the alternative balancing technologies can both contribute to savings in generation and network investment and also increase the efficiency of system operation.
- *Different assets in the electricity system:* Generation assets (from large scale to distributed small scale), transmission network (national and interconnections), and local distribution network operating at various voltage levels. This is important as alternative technologies may be placed at different locations in the system and at different scales. For example, bulk storage is normally connected to the national transmission network, whereas highly distributed technologies may be connected to local low-voltage distribution networks.

36.4 Quantifying the Benefits of Smart Grid Technologies in a Low-Carbon future

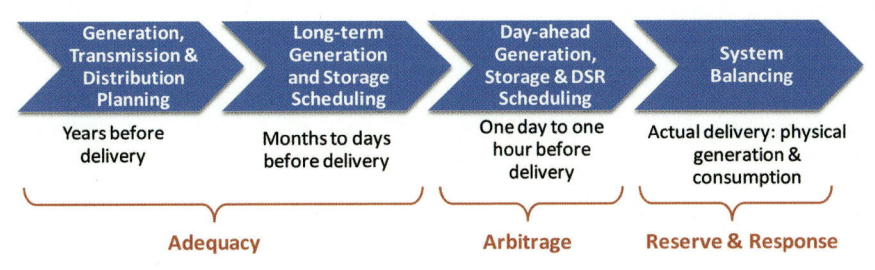

Generation, Transmission & Distribution Planning → **Long-term Generation and Storage Scheduling** → **Day-ahead Generation, Storage & DSR Scheduling** → **System Balancing**

Years before delivery | Months to days before delivery | One day to one hour before delivery | Actual delivery: physical generation & consumption

Adequacy | Arbitrage | Reserve & Response

Figure 36.3 Balancing electricity supply and demand across different time horizons.

Capturing the interactions across different timescales and across different asset types is essential for the analysis of future low-carbon electricity systems that include alternative balancing technologies such as storage and demand-side response. Clearly, applications of those technologies not only may improve the economics of real-time system operation, but can also reduce the investment into generation and network capacity in the long term.

In order to capture these effects and in particular tradeoffs between different flexible technologies, it is important that they are all considered in a single integrated modeling framework, so that long-term investment decisions against short-term operation decisions can be simultaneously balanced across generation, transmission, and distribution systems, in an integrated fashion.

Such a holistic approach would provide optimal decisions for investing in generation, network, and/or storage capacity (in terms of both volume and location), in order to satisfy the real-time supply–demand balance in an economically optimal way, while at the same time ensuring efficient levels of security of supply.

Such approach has been used for studying the interconnected European electricity systems,[6] enabling simultaneous consideration of system operation decisions and capacity additions, with the ability to quantify the value of alternative mitigation measures, such as DSR and storage, for real-time balancing and transmission and distribution network and/or generation reinforcement management. For example, the model captures potential conflicts and synergies between different applications of distributed storage in supporting intermittency management at the national level and reducing reinforcements in the local distribution network.

Figure 36.4 provides a summary of system benefits (through reduction of investment and operating costs) provided by DSR and storage technologies in the future low-carbon UK system. This also includes the application of advanced coordinated voltage control and power flow control in transmission and distribution networks. Clearly, the benefits of these technologies will already be very significant by 2030, with a growing trend as the decarbonization of the system accelerates.

---

6) This model has been used in recent European projects aimed at quantifying the system infrastructure requirements and operation cost of integrating large amounts of renewable electricity in Europe, including (i) "Roadmap 2050: A Practical Guide to a Prosperous, Low Carbon Europe," (ii) "Power Perspective 2030: On the Road to a Decarbonised Power Sector," (iii) "The Revision of the Trans-European Energy Network Policy (TEN-E)," and (iv) "Infrastructure Roadmap for Energy Networks in Europe (IRENE-40)."

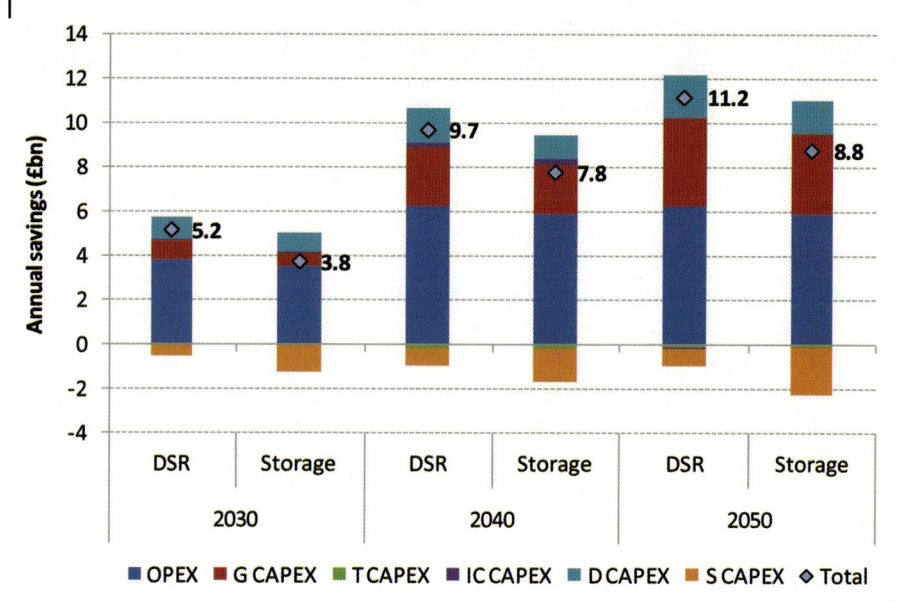

Figure 36.4 Annual system integration savings in the UK system from optimizing demand-side response and distributed storage in transition to a low-carbon future.

The savings are calculated as the difference in total annuitized investment and annual operating costs between (a) a counterfactual system, with energy storage not being available, and (b) the system with DSR and energy storage, given its cost, being optimally used to minimize the total system cost. Optimal levels of annuitized investment in new storage capacity are plotted as negative benefit (S CAPEX), and the resulting net system benefit is also depicted in Figure 36.4 as the difference between storage expenditure and the resulting savings. The benefits of energy storage are grouped in the following categories: generation investment (G CAPEX), interconnection investment (IC CAPEX), transmission investment (T CAPEX), distribution investment (D CAPEX), and operating cost (OPEX).

We observe that DSR and energy storage technologies deliver multiple benefits: DSR and storage technologies (i) reduce the generation operating cost by enhancing the ability of the system to absorb renewable generation and displace low load factor backup generation with low efficiencies, (ii) reduce generation investment costs by contributing to delivery of adequacy/security of supply, (iii) offset the need for interconnection and transmission investment, and potentially (iv) reduce the need for distribution network reinforcement driven by electrification of transport and heat sectors. In some cases, the objective of overall cost minimization may lead to an increase in cost in particular assets, such as expenditure in the transmission network between Scotland and England that may slightly increase with the application of DSR and distributed energy storage.

We notice that the savings in generation capacity increase significantly after 2030, when emission constraints necessitate the investment in of abated plants in the absence of DSR and energy storage. Clearly, the system is unable to accommodate the high penetration of renewable generation (we observe a significant curtailment of renewable energy). In order to comply with the carbon emission targets of 100 and 50 g CO_2 kWh^{-1} in 2040 and 2050 respectively, if DSR or energy storage are not available then significant additional capacity of low-carbon generation such as carbon capture and storage (CCS) or nuclear would need to be built. On the other hand, DSR and energy storage increases the ability of the system to absorb intermittent generation and hence costly CCS or nuclear plant can be displaced, which leads to very significant savings.

These savings represent the overall budget that would be available for implementing the shift from traditional central control philosophy to a smart grid distributed control paradigm. The savings are an order of magnitude larger than the cost associated with ICT infrastructure that will be needed for the full implementation of smart grid, which clearly shows that smart grid will provide a cost-effective transition to low-carbon futures.

36.5
Integration of Demand-Side Response in System Operation and Planning

As indicated, incorporating flexible demand into the electricity system can benefit both its operation and design [14, 15]. These benefits include increased generation and network asset use, increased capability to accommodate low-carbon generation and load growth [16], and enhanced network flexibility and resilience [17].

Understanding the characteristics of flexible demand is vital to establishing its economic value. For there to be regular flexible demand, controlled devices (or appliances) must have access to certain type of storage when rescheduling their operation or when needing to function during interruptions. Storage may be in the form of thermal, chemical, or mechanical energy, or intermediate products. Flexible demand redistributes the load but may not reduce the total energy that a flexible device uses. Load reduction periods are therefore followed or preceded by load recovery. The duration of load recovery will depend on the type of interrupted process and the type of storage. The amount of energy recovered may even exceed the amount of load that is restricted through flexible operation due to losses in the storage or energy conversion process.

Achieving flexible demand means carefully managing this process of load reduction and load recovery. For example, let us assume we have 1000 refrigerators and that each refrigerator, when operating (using electricity), has a load of 200 W. Next, assume a coincidence factor of 25%, as only one-quarter of the devices will use electricity simultaneously. This means that the diversified load of 1000 refrigerators is 50 kW (1000 × 0.2 × 0.25). Therefore, if all 1000 refrigerators are switched off, the expected load reduction will be only 50 kW. If these refrigerators are then reconnected back to the grid after say 1 h, the load is likely to be close to 200 kW (1000 × 0.2). This is because the temperature in every refrigerator will be above the set levels and thermostat control will be trying to reduce the temperature by having all refrigerators use electricity at the same time.

Hence the total load of a group of controlled devices will increase during the load recovery period. To counteract this load increase, some other appliances must be switched off. This would reduce load control efficiency. A key technical challenge is to design ways to maximize both the efficiency and use of controlled loads, while at the same time not compromising consumers' comfort levels.

This section describes the flexibility of various forms of demand. The modeling framework sets out to minimize the cost of system reinforcement and operation. To do so, it optimizes the aggregate effect of various types of flexible demand technologies examined in this study, including water heaters, smart wet appliances, smart refrigerators, smart control of electric vehicles (EVs), and smart control of heat pumps (HPs).

36.5.1
Control of Domestic Appliances

In this section, we discuss the effects of control of various domestic appliances, including water heaters, wet appliances, and refrigerators.

The diversified water heater demand model used to illustrate the key concepts is shown in Figure 36.5. This diversified curve shows the average uncontrolled water heater use from a large sample of water heaters. The diagram shows the use of electricity peaking at 8 p.m., with a smaller peak occurring at 8 a.m. These two peaks are the result of water heaters being used mostly in the morning and in the evening (as shown by the solid line).

Reconnecting a group of water heaters to the system after the assumed disconnection due to DSR actions between 6 p.m. and 8 p.m. affects their subsequent operation, and we observe a spike in demand after 8 p.m. that is commonly referred to as payback. This effect is represented by the dashed line in Figure 36.5. The diversified demand of controlled water heaters during the load reduction period (6–8 p.m.) is zero. During payback, additional energy to restore the demand is needed on top of the uncontrolled diversified demand.

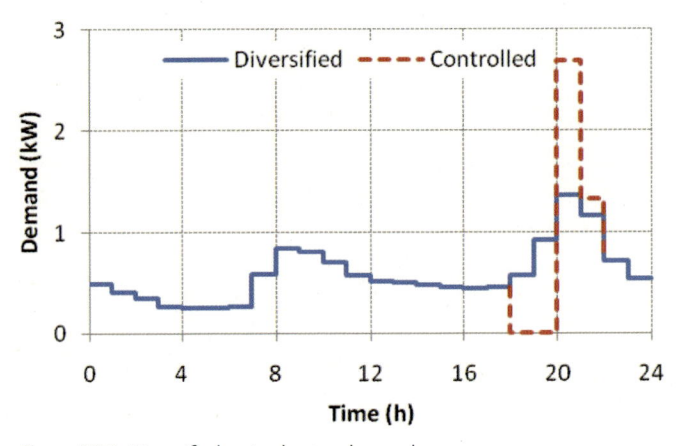

Figure 36.5 Diversified water heater demand.

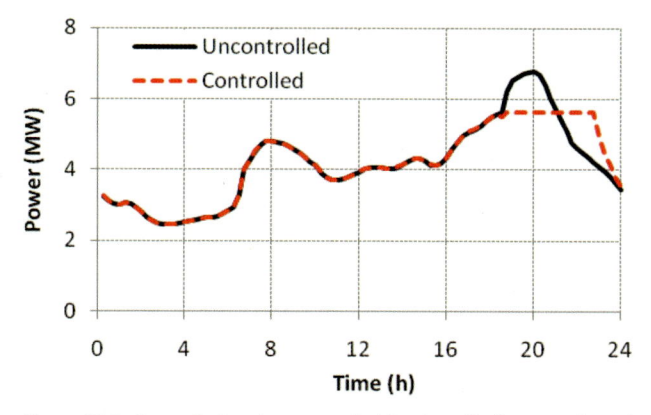

Figure 36.6 Controlled and uncontrolled load profile for a residential area.

The models developed for this analysis are able to schedule optimally the control of groups of water heaters to minimize peak demand (or generation cost), while considering the load reduction and load recovery effects shown in Figure 36.5. The model was calibrated using data obtained from field trials. Figure 36.6 shows examples of controlled and uncontrolled load profiles for a typical residential area, where we observe a peak demand reduction from 7 to 5.5 MW.

Figure 36.7 details the water heater control schemes applied in the residential area. Water heater control in this example occurs only in the evening, around the time of peak demand. The position of each horizontal bar shows the control period during which the water heaters are switched off and the number indicated next to that control scheme is the number of controlled water heaters following the scheme. For example, 120 water heaters are switched off for 2 h between 6.15 p.m. and 8.15 p.m. It is interesting that the control of water heaters continues well beyond the duration of peak demand, and finishes just before 11 p.m. This is necessary in order to manage load recovery adequately, that is, avoid causing a new peak after reducing the original one.

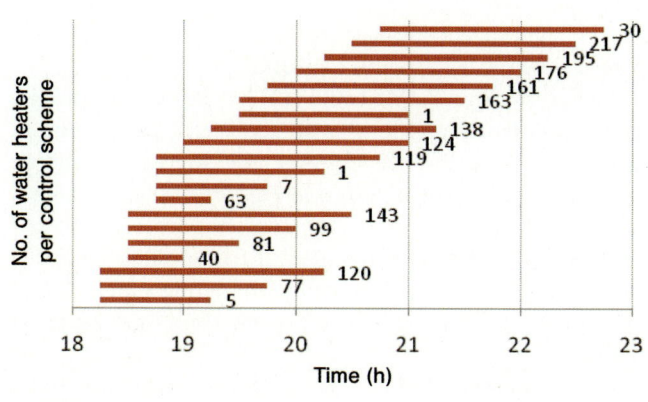

Figure 36.7 Water heater control schemes for the residential area.

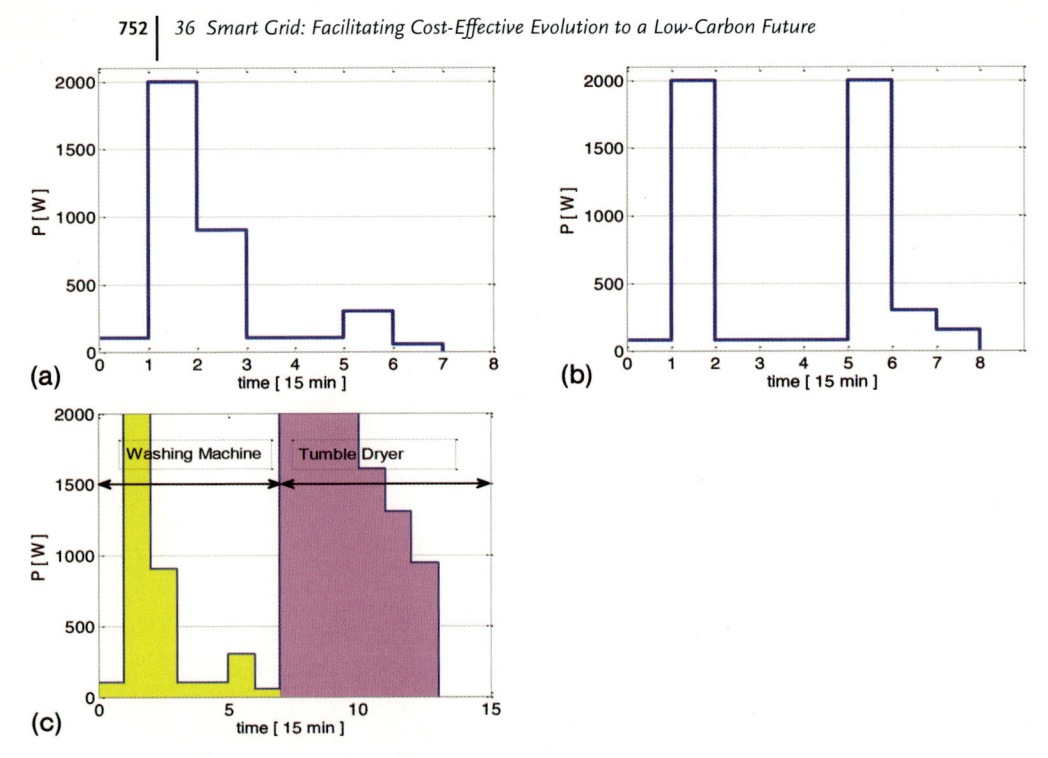

Figure 36.8 Typical operation cycles for (a) WM, (b) DW, and (c) WM+TD.

Similarly, control of wet appliances, including washing machines (WM), dishwashers (DW), and washing machines equipped with tumble dryers (WM+TD) could contribute to the delivery of savings in system operation and investment. The operation cycle defines the duration and power consumption at each time instant when the appliance is in use (typical operation cycle patterns of these appliances have been sourced from the Intelligent Energy Europe Smart-A project [18], and are presented in Figure 36.8).

Assuming that operation of smart appliances (SAs) can be postponed for 1–3 h, their use can be optimized to minimize the system peak as presented in Figure 36.9 as an example of a typical residential distribution network.

In addition to minimizing system peak, SAs could also be used to provide flexibility, which is particularly relevant for systems with significant penetration of wind power. The monetary value of flexible use of SAs in order to provide standing reserve and contribute to system balancing is shown in Figure 36.10. This value is expressed as annual fuel cost savings obtained per appliance for different levels of wind penetration. Figure 36.10 also shows the fractions of the SA value generated by different types of appliances. The largest contribution among different SAs expectedly originates from DW, since this device is characterized by the highest demand shifting flexibility.

In electricity systems, the automatic frequency regulation services maintain the tight balance between generation and demand on a second-by-second basis.

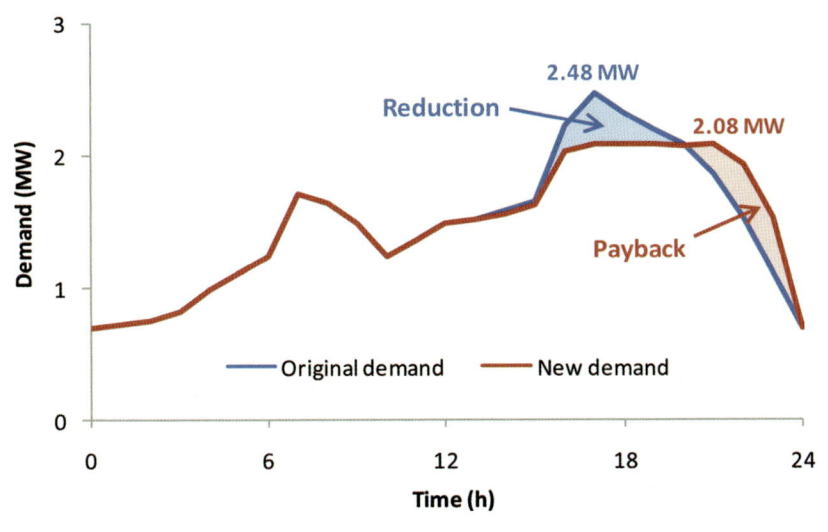

Figure 36.9 Smart wet appliances reducing the peak load of a distribution network.

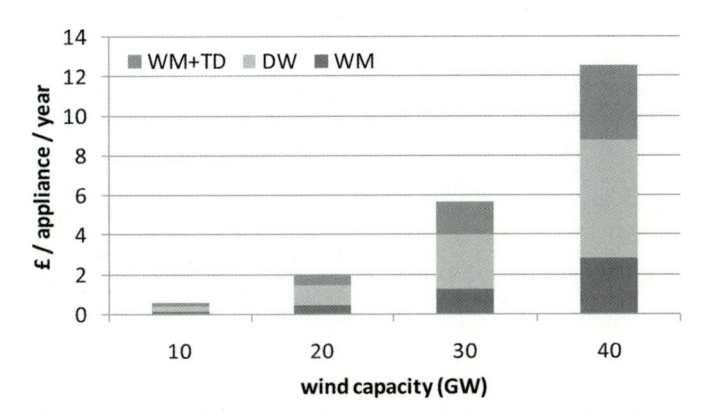

Figure 36.10 Value of smart appliances for different wind penetration levels.

In particular, system frequency needs to be carefully managed following a sudden loss of a large generator (due to, e.g., an unforeseen outage), and restored to its nominal value within a relatively short time frame. Frequency regulation services are usually provided by synchronized generators, running part loaded, and by frequency-sensitive load control actions involving certain industrial customers. These fast-responding services are used to manage system frequency fluctuations over timescales ranging from seconds to tens of minutes.

Smart refrigeration (SR) could potentially contribute to frequency regulation on the demand side [19, 20]. The value of SR providing frequency regulation will be reflected in reduced system operation costs, lower carbon emissions, and an increased ability of the system to absorb intermittent generation [21].

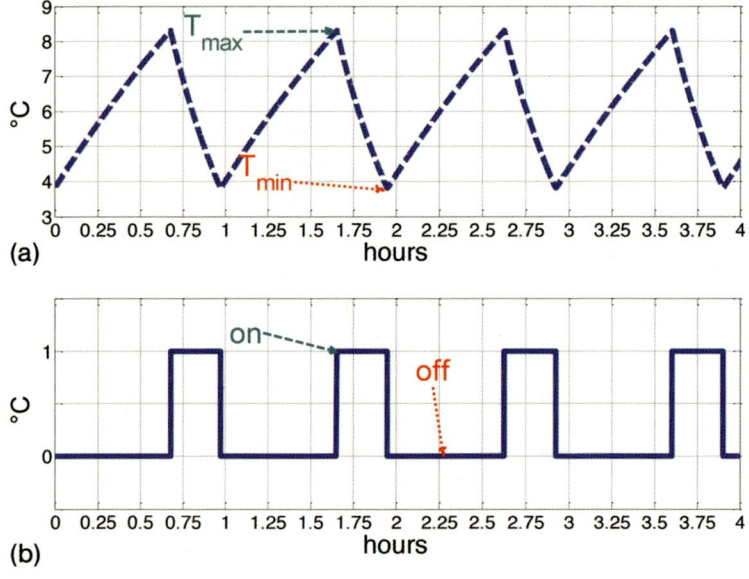

Figure 36.11 Standard refrigerator operating cycles with temperature variations (a) and compressor switching (b).

A domestic refrigerator generally maintains its internal temperature between two set points. Once the internal temperature has reached the set point T_{max}, the compressor switches on and the refrigerator starts to cool. Once the refrigerator's internal temperature reaches the minimum temperature set point T_{min}, the compressor stops. The cycle then repeats, as demonstrated in Figure 36.11. The compressor has a duty cycle (that is, the fraction of time when it consumes electricity) of 20–30%.

Assuming a typical duty cycle of about 30% and a rating between 120 and 150 W for individual appliances, the diversified (average) load of an individual refrigerator is between 40 and 50 W.

The inclusion of SR control in a domestic refrigerator modifies its standard operation pattern by adjusting its duty cycle length as a function of frequency deviations, so that the average energy demand of all appliances should decrease as the system frequency decreases.

The economic value of SR in a particular system is determined by the reduction in system operating costs compared with the appropriate base-case scenario without SR. Environmental benefits of SR include the avoided carbon emissions from the electricity system and the reduced need to curtail wind output.

The benefits of SR will be system-specific and will depend significantly on the contribution of wind generation to system inertia and frequency regulation, as shown in Figure 36.12. In the case that wind generation contributes to inertia and frequency regulation (active wind), the value of SR from 2030 onwards would be around £50 per SR unit whereas in the case that wind generation does not contribute to the provision of inertia and frequency regulation (passive wind), this would increase to around £300 per SR unit in 2030 and around £500 per SR unit in 2050.

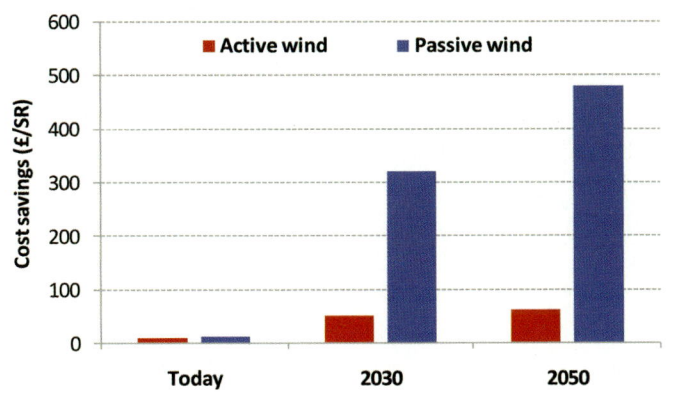

Figure 36.12 Value of SR in today's and future systems for different contributions of wind to system management.

36.5.2
Integration of EVs

EV loads are particularly well placed to support power system operation, for several reasons: (i) their energy requirements are relatively modest; (ii) light passenger vehicles are generally associated with short driving times (relevant databases indicate that vehicles are stationary on average for 90% of the time); and (iii) the relatively high power ratings of EV batteries. Clearly, there is considerable flexibility regarding the time when the vehicles can be charged (provided that an adequate charging infrastructure is available) and this can provide significant benefits both to the operation of distribution and transmission networks and to the efficient dispatch and utilization of generation [22, 23].

Figure 36.13 shows the hourly demand profiles in a system with a high penetration of EVs: (a) for the case of uncontrolled charging with very high demand peaks and (b) the effect of smart charging with the objective of reducing system peak demand. The net demand in Figure 36.13 refers to the difference between system demand and wind output, representing the part of the demand that needs to be supplied by conventional generators. Minimization of peak demand results in the charging demand being spread across periods of low demand, thus effectively filling the demand valleys during the night hours.

Optimized EV charging also improves the ability of the system to absorb intermittent (renewable) generation, as charging is shifted towards periods of low net demand (system demand minus intermittent wind generation [24, 25]).

At the local distribution network level, the effects are similar, although the magnitudes may be different [26]. The impact of EV charging patterns associated with commuting to a town/business park area in the case of uncontrolled and smart charging are shown in Figure 36.14. This study analyzed a commercial district area considering both uncontrolled and smart modes of EV charging. As expected, a significant increase in morning peak demand occurs under the uncontrolled charging regime, driven by concentrated EV charging around the time of arrival

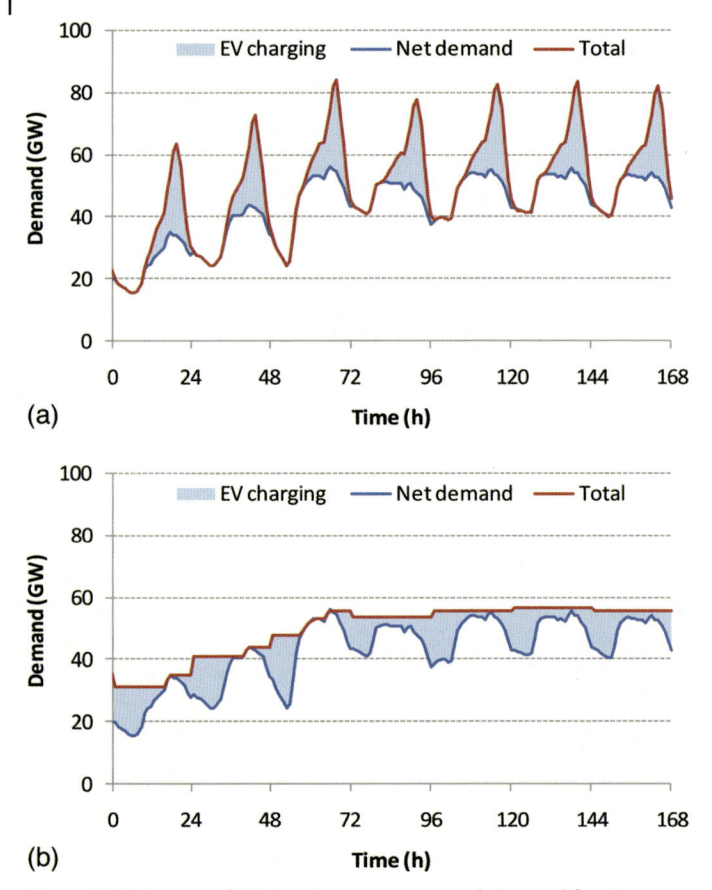

Figure 36.13 Impact of EV charging on system peak demand for (a) uncontrolled and (b) optimized charging.

at work, as illustrated in Figure 36.14. On the other hand, a very flat profile can be obtained if charging is optimized, while still allowing batteries to be fully charged before vehicle owners leave work.

Figure 36.15 contrasts the increases in network peak demand for uncontrolled and smart charging modes. Clearly, not incorporating demand side in network real-time operation will result in a massive degradation of network asset utilization.

Figure 36.16 shows the changes in the demand profile for the case of a residential area, driven by EV charging when people return home from work, for both uncontrolled and smart charging modes. As expected, peak demand occurs in the evening, following the assumption that charging starts upon returning home from work. We assume that evening charging will recover the energy of the return journey only, whereas the energy associated with the journey to work is recovered through charging during working hours at the workplace. Under a smart operating regime, this demand peak (and hence a massive network reinforcement cost) can be avoided, as shown in Figure 36.16.

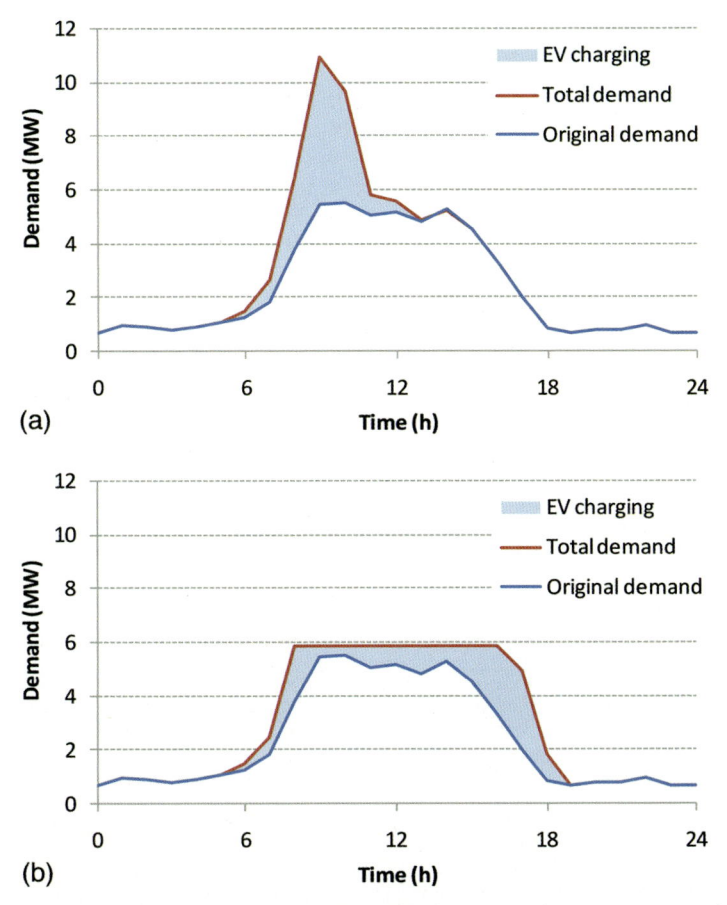

Figure 36.14 (a) Uncontrolled and (b) smart charging profiles in a commercial district (1 km^2) driven by charging of 5000 EVs following arrivals at work.

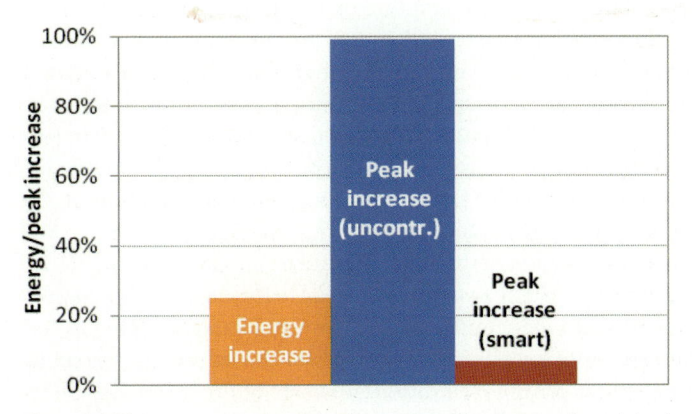

Figure 36.15 Increases in electricity demand and local network peak load.

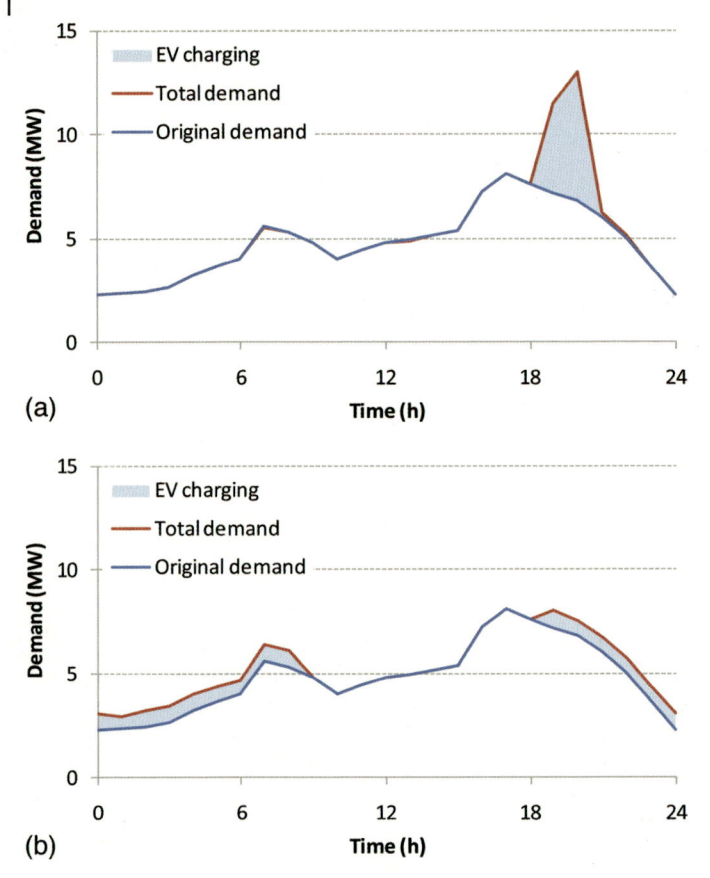

Figure 36.16 (a) Uncontrolled and (b) smart charging in a residential area (8000 properties) driven by charging of 5000 EVs when people return from work.

Figure 36.17 provides an estimate of the necessary reinforcement cost for UK distribution networks in the 2050 horizon for three different approaches to network control: (i) passive control (business as usual approach), (ii) smart voltage control, and (iii) smart charging of EVs. The figure suggests that although smart EV control can deliver massive savings in necessary reinforcement cost (with up to seven times less cumulative investment), significant cost reductions can also be achieved by applying smart voltage control in distribution networks, effectively halving the necessary reinforcement cost (in the case when a significant proportion of the network reinforcement is driven by voltage constraint violations).

For a comprehensive assessment of the value that smart EV charging can deliver to various segments of the electricity system, one needs to look at services provided to these segments by EVs simultaneously. In that context, Figure 36.18 quantifies the annual economic benefits for different segments of the future UK electricity system from applying smart control over different shares of the national EV fleet (which is assumed to cover 30% of the total national vehicle fleet including around

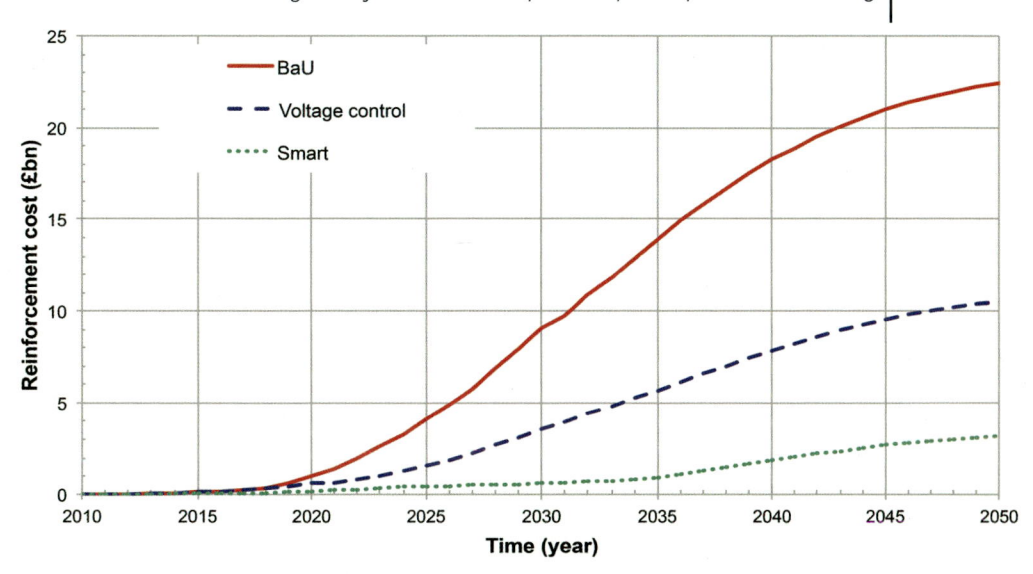

Figure 36.17 Cumulative distribution network reinforcement cost with current practice of business as usual (BaU), with smart voltage control, and with smart EV charging.

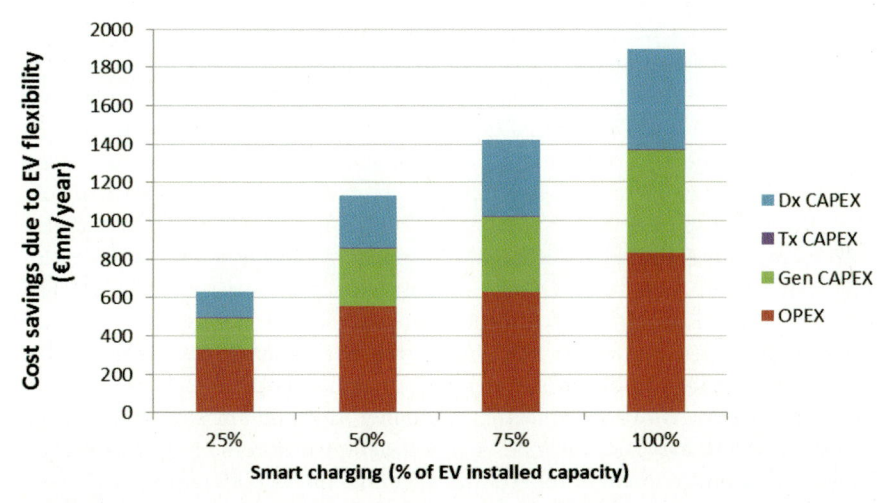

Figure 36.18 Annual cost savings from flexible EV charging for different shares of smart EVs.

35 million vehicles). The values shown were obtained using an advanced whole-system analytical model capable of optimally utilizing smart EV charging to reduce both operation cost and investment into generation and network capacity. When all EVs are controlled in a smart manner (i.e., for 100% smart charging), the value generated to the system amounts to around €180 per vehicle annually. As indicated in Figure 36.18, almost half of this value is created by savings in operating cost, and avoided investment in generation and distribution reinforcements (in similar proportions) accounts for the other half.

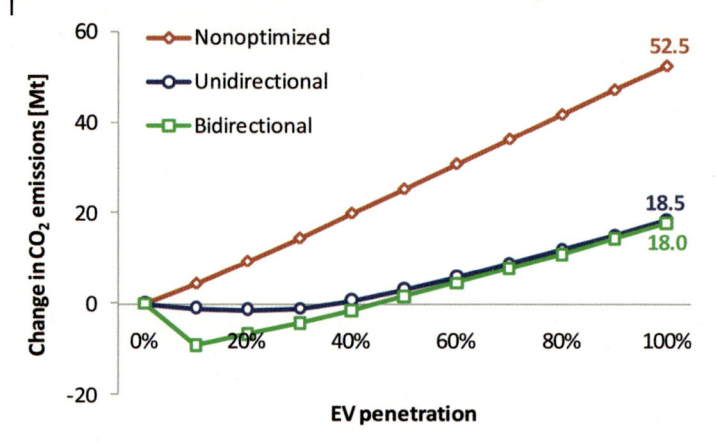

Figure 36.19 Change in annual carbon emissions from the UK electricity system for different EC charging control regimes.

The transport sector is a major contributor to environmental emissions. For instance, road-based transport accounted for almost one-quarter of the UK CO_2 emissions in 2010.[7] Therefore, reducing the reliance on carbon-based fuels in this sector is seen as a key contributor to reducing the overall CO_2 emissions.

As an illustrative example, the incremental annual CO_2 emissions from electricity generation for the 2030 UK system are presented in Figure 36.19 for various levels of EV penetration and for three different charging strategies, using the 0% EV penetration as reference. Charging strategies include uncontrolled (nonoptimized) charging, optimization of charging only (unidirectional), and optimization of both charging and discharging (bidirectional) [27]. The emissions generally increase with increase in EV penetration for all charging strategies. However, at lower EV penetrations (of up to 40%), CO_2 emissions actually decrease in the controlled charging cases. The initial reduction in CO_2 emissions in the case of controlled charging strategies, observed at 10% EV penetration, amounts to about 2% and 10% for unidirectional and bidirectional charging strategies, respectively. This is driven by a combination of two main factors: (i) higher utilization of zero-carbon wind energy due to the avoidance of wind energy curtailment in both of these cases, and (ii) increased utilization of lower-emitting conventional technologies.

In order to assess the overall impact of transport electrification on the emissions from the whole energy sector, the incremental emissions quantified in Figure 36.19 would need to be offset by carbon emissions potentially avoided in the transport sector through replacing fossil fuels with electricity.

---

7) Source: Office for National Statistics, *Road Transport Emissions by Industry and Type, 1990–2010*, December 2012, http://www.ons.gov.uk/ons/datasets-and-tables/index.html (last accessed 8 February 2013).

36.5.3
Smart Heat Pump Operation

The residential heating sector is another area that has significant potential for decarbonization. There has been significant interest in various electrical heat pump (HP) technologies, in the form of domestic-scale installations or large-scale installations supporting heat networks. This concept relies on the assumption that future electricity systems will be largely carbon neutral as a result of adopting renewable, nuclear, and other low-carbon generation technologies.

Regarding the domestic scale, two main types of HPs include air-source HPs (ASHPs) and closed-loop ground-source HPs (GSHPs). The key parameter of HP performance is the coefficient of performance (COP). When HPs are used for heating, COP is defined as the ratio of the heat supplied to the energy carrier medium, and the electric input into the compressor, with typical values varying between 2 and 4. The COP generally decreases (i.e., the HP becomes less efficient) at lower temperatures of the heat source (air or ground). In that sense, GSHPs generally provide a better energy performance as the ground or underground water provides a more stable temperature source than air, but on the other hand their installation costs are higher.

Based on representative data for the operation of HPs in the UK system, derived from empirical studies and field trials, an aggregate HP demand profile is constructed, while assuming high standards of building insulation levels to account for the expected future improvements in energy efficiency. The resulting uncontrolled electricity demand profile for heating during a cold winter day is presented in Figure 36.20 as the dashed line, assuming a full penetration of HPs, that is, 100% of residential dwellings heated by HPs. The additional instantaneous electricity demand could reach up to 45 GW, with the maximum broadly coinciding with the existing system peak, which significantly adds to the challenge of supplying peak system demand. The energy required by all heat pumps on the same (cold) day is around 460 GWh, representing more than 40% of the existing winter daily demand.

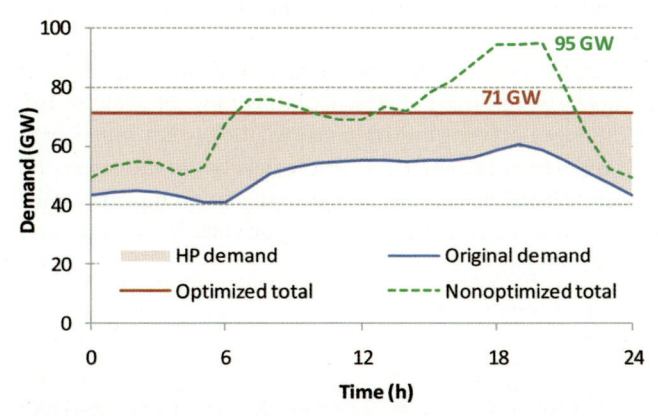

Figure 36.20 National load profile with uncontrolled and optimized HP operation using heat storage.

Given the characteristics and constraints of HPs (low-temperature operation and reduced rate of heat delivery), it might be beneficial to equip an HP-based system with thermal storage (such as a hot water tank) in order to enable the HP to follow the same heat requirements with a lower electrical rating. This would potentially lead to a more uniform operation of HPs, but will also provide an opportunity to optimize HP operation not only to meet local heat requirements, but also to contribute to grid management and support the integration of low-carbon generation. An analysis conducted by the authors has shown that heat storage with the capacity of less than 25% of daily heat demand would be sufficient for fully flattening the national daily demand profile in the case of a 100% rollout of HPs.[8] In that context, Figure 36.20 also shows how the total peak demand can be significantly reduced using the flexibility of HPs equipped with heat storage; the resulting increase in peak demand, compared with the original demand profile, is reduced from 56% to 17%.

36.5.4
Role and Value of Energy Storage in Smart Grid

As indicated, the considerable increase in system integration costs associated with intermittent generation and electrification of transport and heating sectors may be mitigated with cost-effective energy storage technologies [8, 28]. Using an example of the future UK system, Figure 36.21 illustrates that the potential system savings, enabled by energy storage, increase markedly as the system decarbonizes towards 2050. The composition of the annual system benefits is given in billions of pounds per year, for a range of assumed energy storage costs (top horizontal axis) also corresponding to different optimal volumes of energy storage deployed by the model (bottom horizontal axis).[9]

It is critical to stress the importance of the assumptions associated with the cost of capital and lifetime when assessing the capitalized cost of energy storage technologies. For a given level of annualized cost of storage (top horizontal axis), a more risky and short-lived technology needs to be available at a much lower cost than if the same storage technology had a longer economic life and was considered to be technically mature.[10]

The savings presented in Figure 36.21 are calculated as the difference in total system annuitized investment and annual operating costs between (a) a counterfactual system, with energy storage not being available, and (b) the system with energy storage, given its cost, being optimally placed and operated so as to minimize the total system cost. We again observe that energy storage delivers multiple benefits: reduction in generation operating cost, reduction in low load factor backup genera-

---

8) The analysis also took into account efficiency losses that might accompany the process of storing heat and releasing it back in to the heating system.
9) These results are presented for storage with 75% efficiency and 6 h duration.
10) For a mature and long-lived technology, similar to network assets with the assumed life of 40 years and WACC (weighted average cost of capital) of 5.7%, the annuitized cost of storage of £100 kW^{-1} per year would correspond to a capitalized value of storage of £1563 kW^{-1}. If, however, we assume a more risky technology, with an economic life of only 25 years and an appropriately higher WACC of 14.5% (similar to CCS technology), we obtain a capitalized value of storage of only £666 kW^{-1}.

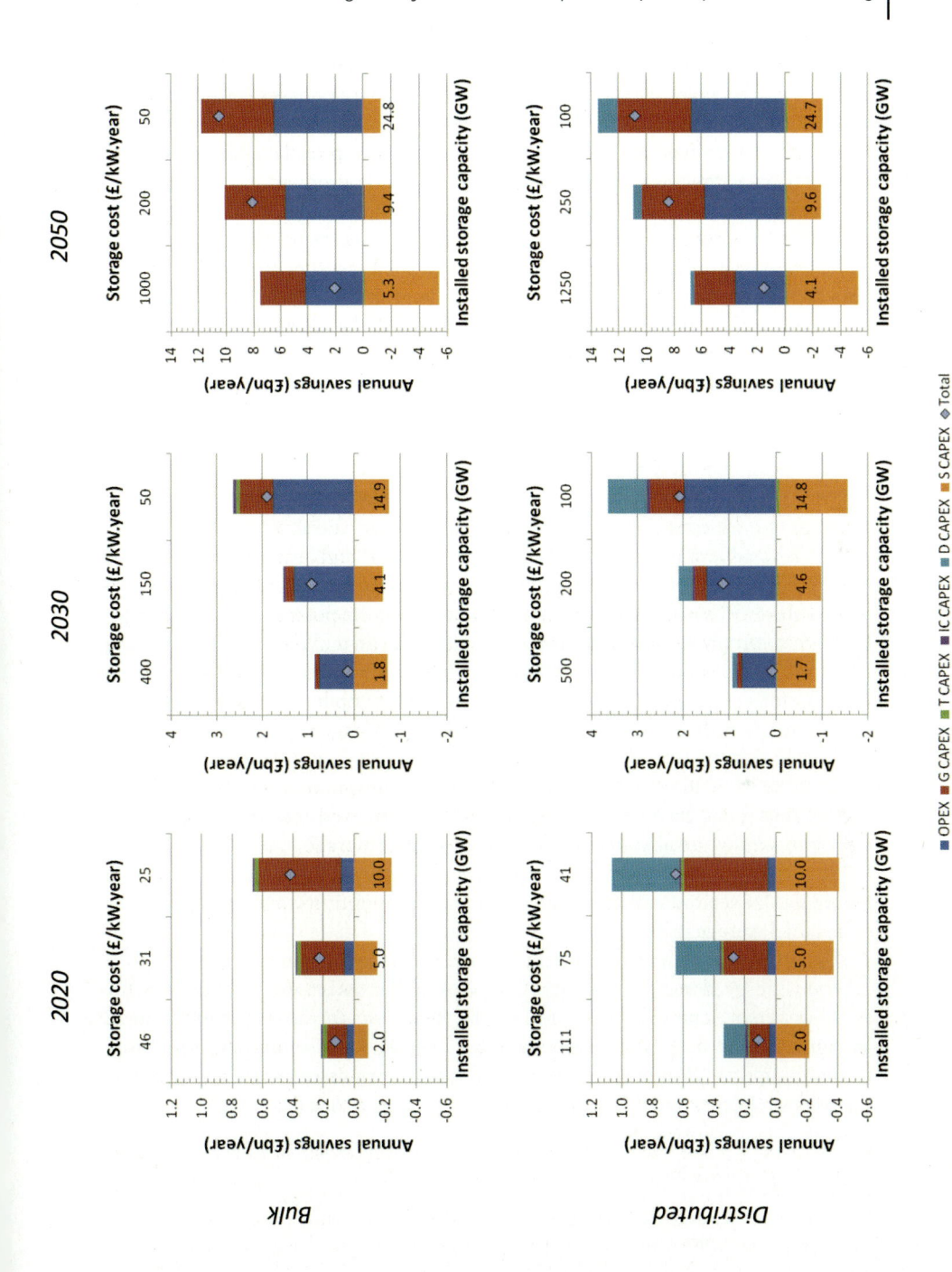

36.5 Integration of Demand-Side Response in System Operation and Planning

Figure 36.21 Net benefits of storage in the future UK system with a significant contribution of renewable generation and electrified transport and heat.

tion, and reduction in the need for interconnection, transmission, and distribution network investment. In some cases the objective of overall cost minimization may lead to an increase in cost in particular assets, such as expenditure in transmission network reinforcement between Scotland and England (note that, for example, in 2030 bulk storage reduces the need for transmission whereas distributed storage increases transmission investment).

The magnitude of these savings will be system specific and is driven by the characteristics of generation and demand, including their location. Net system savings in the UK system will increase radically between 2020 and 2050, given the expected growth in intermittent renewable generation and the levels of electrification of transport and heat demand. For instance, in the case of bulk storage, with the cost of £50 kW^{-1} per year, the achievable net annual system benefits are £0.12 billion in 2020, around £2 billion in 2030, and over £10 billion in 2050. Similar trends are observed for distributed storage.

Relative shares of savings across different sectors vary depending on the assumed system background and the deployment levels of storage.

Savings in generation capacity will increase significantly after 2030, when emission constraints necessitate investment in abated plants in the absence of storage, as storage is capable of displacing this expensive capacity, generating a very high value to the system. Clearly, the system is unable to accommodate the high penetration of renewable generation (we observe a significant curtailment of renewable energy). In order to comply with the carbon emission target of 50 g kWh^{-1} in 2050, if storage is not available then significant additional capacity of CCS would need to be built. Adding storage increases the ability of the system to absorb intermittent sources and hence costly CCS plant can be displaced, which leads to very significant savings.

We also observe some savings in transmission and interconnection (T and IC CAPEX) in 2030 as a result of adding bulk storage, whereas for distributed storage these may even become negative. This is the consequence of the location of energy storage relative to the location of wind generation: most bulk storage capacity is located in Scotland where it is able to support balancing of wind output locally and hence reduce the need for transmission reinforcement, whereas most distributed storage is located closer to large demand centers in England, driven by the opportunity to reduce the distribution network reinforcement cost.

When energy storage faces competition from other smart grid technologies, the volumes deployed and benefits generated by energy storage decrease. Figure 36.22 presents the net benefits of energy storage in 2030 when competing with other flexibility options. We observe that in the majority of cases, when storage faces competition from other technologies, the net benefit generated by storage decreases, and the optimal storage capacity drops.

The key impact of competing technologies is that OPEX savings of storage decrease compared with the storage-only case (OPEX savings predominantly result from avoided renewable curtailment). Competing technologies can deliver these savings and significantly reduce the scope for storage to deliver further avoided renewable curtailment. Although the presence of interconnection and flexible generation reduces the value of storage, the volume of storage deployed is not affected.

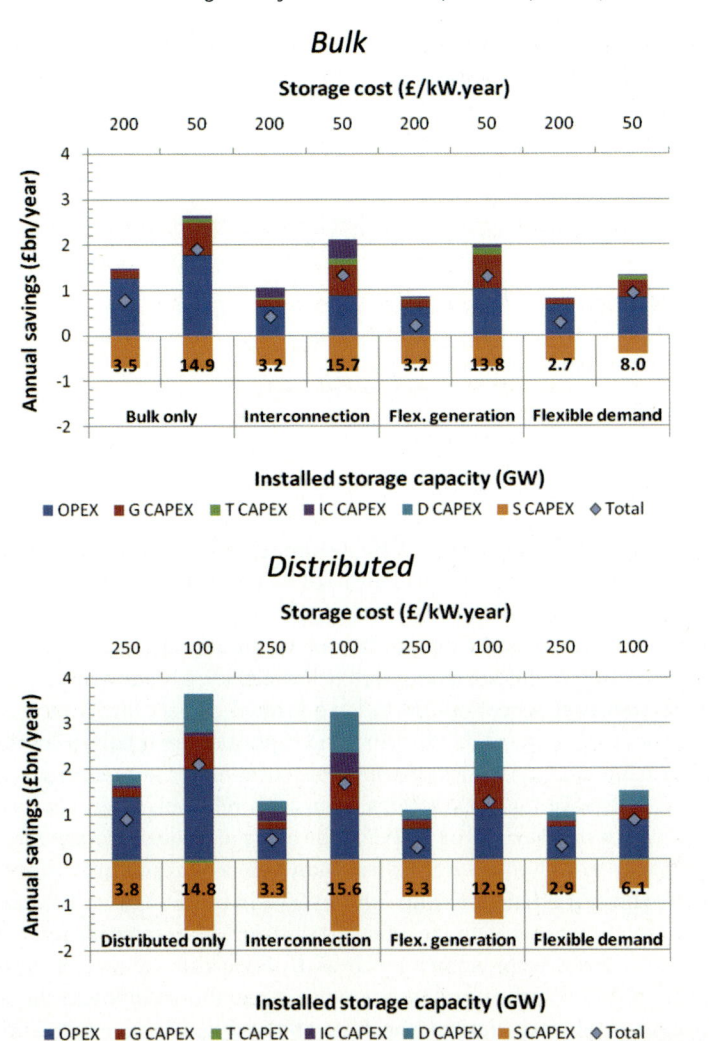

Figure 36.22 Net benefit and value of storage in 2030 in the presence of competing flexibility options. Storage duration is 24 h. Installed capacity is given beneath each bar.

A highly flexible demand side, however, is the most direct competitor to energy storage. Both offer very similar services (deferring or avoiding distribution network reinforcement in particular) and are therefore not complementary. High levels of demand flexibility may reduce the market size for storage in 2030 by more than 50%.

Given the shape of the peak demand, the value of storage in the future UK is not strongly affected by increases in storage duration beyond 6 h. The capacity value of both bulk and distributed technologies has been found to be high throughout. Not only are these storage technologies able to displace (back up) generation capacity roughly on a megawatt per megawatt basis in maintaining capacity adequacy re-

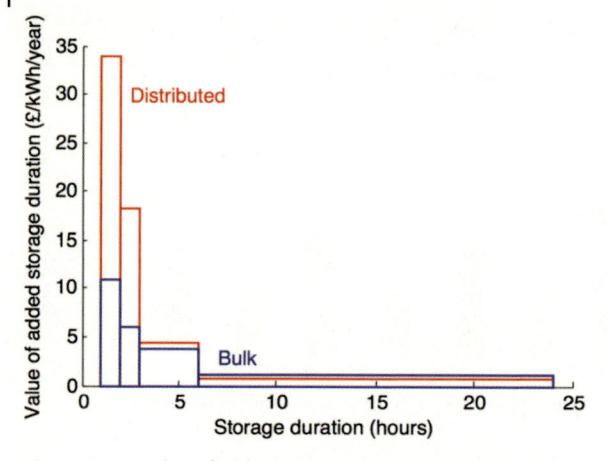

Figure 36.23 Value of adding energy capacity (duration) to a 10 GW storage in 2030.

quirements,[11] but they can further facilitate a more effective use (i.e., higher load factors) of more efficient plant, or avoid the need for costly abated plants in providing peaking services in low-carbon systems.

As illustrated in Figure 36.23, additional storage duration leads to rapidly diminishing value per unit of energy falling well below £20 kWh^{-1} per year.

Storage efficiency has been found to have limited impact on its value, in the systems considered; provided that the installed capacities are relatively low and the potential for arbitrage, that is, saving renewable curtailment remains high, storage can effectively displace high-cost energy, even with relatively low round-trip efficiency, while achieving savings in CAPEX. The average value of storage for 10 GW installed capacity in 2030 increases by less than 10% as a result of improvements in storage round-trip efficiency from 50% to 90%. The change in the accessible market size, as a function of storage efficiency, is much more significant, however.

Fast energy storage technologies (such as flywheels or super-capacitors) can contribute to primary and secondary frequency regulation. Although the market for fast storage is relatively limited, the value and savings are very substantial and could reach £700 kW^{-1} per year and come from a significantly reduced need to run conventional generation part-loaded and hence enhanced capability of the system to absorb renewable generation. However, the presence of flexible generation with significantly enhanced capability to provide fast frequency regulation services could materially reduce the value of fast storage to around £200 kW^{-1} per year. In addition to more flexible conventional generation, there may be a spectrum of other technologies that may contribute to this market and reduce the value of fast storage, such as refrigeration load or suitably controlled wind turbines.

Uncertainty over changes in wind output over several hours necessitates the presence of high reserve capacity and this is a major source of value for energy storage.

---

11) Resource adequacy requires several hours of storage duration, if peaking generation is to be displaced securely, based on the shape of the demand profile derived for 2030.

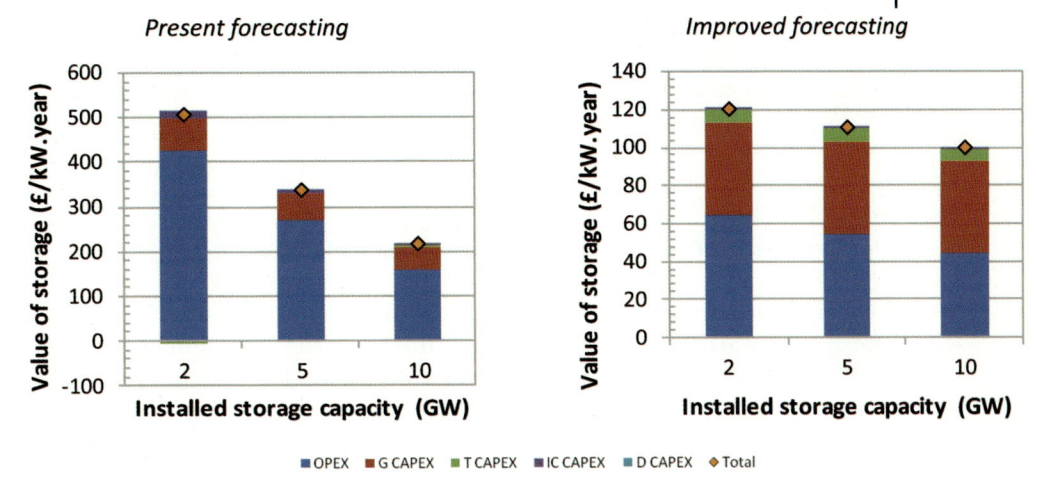

Figure 36.24 Impact of improved wind forecasting on the value of storage.

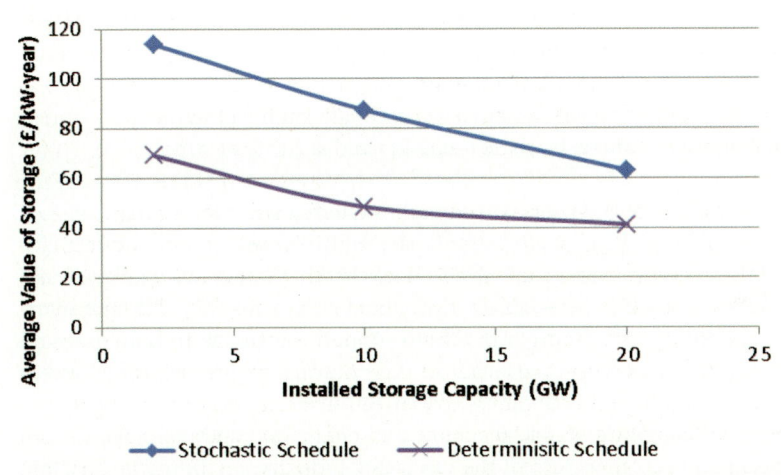

Figure 36.25 Value of storage under stochastic and deterministic scheduling.

Improved wind forecasting (50% improvement in the root mean square error) will result in reduced reserve requirements and will hence reduce the value of storage, as illustrated in Figure 36.24.

Furthermore, the approach to allocating storage resource between energy arbitrage and reserve provision will be critical. Figure 36.25 presents the difference in the value of storage being evaluated using conventional deterministic scheduling and the stochastic scheduling approach.

It will clearly be very important to allocate optimally the storage resource between providing reserve and conducting energy arbitrage, which only stochastic scheduling can facilitate. Stochastic scheduling is therefore superior to its deterministic counterpart, because the allocation of storage resources between energy arbitrage and reserve varies dynamically depending on the system conditions. We observe that with 2 GW of storage when considering a particular scenario, stochastic scheduling increases the value of storage by more than 75%, whereas for the installed capacity of 20 GW of storage this would be around 50%.

36.6
Implementation of Smart Grid: Distributed Energy Marketplace

The smart grid presents a technical vision for electricity networks that intelligently integrates the actions of all users connected to it – generators, consumers, and those that do both – to allow an efficient delivery of sustainable, economic, and secure electricity supplies. The smart grid will employ innovative products and services together with intelligent monitoring, control, and communication technologies to:

- facilitate the connection, operation, and competition of generators of all sizes and technologies, advanced network and storage technologies
- provide consumers with greater information and choice in the way they secure their electricity supplies, allowing them to play a part in optimizing the operation of the system and its future development
- significantly reduce the environmental impact of the total electricity supply system while delivering enhanced levels of reliability and security of supply.

Clearly, vast numbers of end users will enter the system under the smart grid vision, bringing with them new possibilities for smart solutions and system interactivity. To support this technical vision, an advanced market that integrates wholesale and retail markets will need to be established. Without such a market, although many intelligent and effective tools are available to support the smart grid, there would be no framework for end users to establish their optimal responses or for system operators to access efficiently the most cost-effective resources.

As demonstrated, demand-side response and energy storage technologies can bring benefits to several sectors in the electricity industry, including generation, transmission, and distribution, while providing services to support real-time balancing of demand and supply and network congestion management and reduce the need for investment in system reinforcement [1]. These "split benefits" of demand-side response and energy storage pose significant challenges for the present markets as these would prevent investors in smart grid technologies from being adequately rewarded for delivering the diverse sources of value that these technologies can provide.

This will require the development of the concept of a distributed energy marketplace, which will put end users at the center of the development and evolution of the power system and provide system operators with access to the most cost-effective solutions for system management [29]. Such a marketplace is vital to the development of the least-cost, secure, and efficient power system of the future and, as such, its successful implementation occupies the critical path to realization of the smart grid.

The distributed energy marketplace will bring together all energy system end users (both generation and demand) from all levels of the power system to interact with each other and the system operators in a competitive market-based environment, buying and selling energy and ancillary services [30].

Crucially, the marketplace will link all market participants in a single real-time marketplace and remove the disconnection caused by the separation of wholesale and retail energy markets. It will identify the value of end-user response according to time and location (unlocking the potential for demand response) and allow the system operator to access all resources suitable for system management activities. The distributed energy marketplace will be capable of integrating all end-use technologies into a competitive market framework and provides an accessible and flexible market framework to incorporate these and future technological innovations [31]. The marketplace will open a market gateway so that end users may be exposed to wholesale energy prices and that these resources would be made visible to the wider energy market in real time. This will be important for making optimal decisions and for fulfilling the energy services vision that would allow fuel switching and optimal decision-making in response to prices.

The realization of the demand-side potential needs to be coupled with integration schemes driven by competitive market dynamics and individual consumers' interests, and enabling the latter to access the price-setting process. Such approaches will close the gap between wholesale and retail electricity market segments, enable a more active demand participation in the market setting, and consequently lead to more efficient and competitive markets. This paradigm change necessitates suitable modifications in one-sided market mechanisms which were traditionally designed to treat the demand as a fixed, forecast load, needing to be served under all conditions, and despite the recently attained research interest it still exhibits outstanding challenges [32]. In this context, the concerns about cyber security are attracting increasing attention [13, 33].

We also need to stress that the present network operation and design standards require that network security is provided through asset redundancy, through so-called N-1/N-2 (historical asset-heavy paradigm). This approach fundamentally contradicts the concepts of the smart grid that focuses on non-network solutions to network problems. In other words, the network design standards may impose a barrier for innovation in network operation and design and prevent the implementation of technically effective and economically efficient solutions that enhance the utilization of the existing network assets and maximize network users' benefits. There is therefore significant interest in reviewing the historical network design standards, which will be important for the development of smart grid concepts and solutions. Specifically, there has recently been growing interest in connecting various forms of distributed generation (DG) to traditionally passive distribution networks. This gave rise to the question of the treatment of different generation technologies in distribution network planning and design, particularly from the network security perspective, which led to the updating of distribution network security standards in the United Kingdom [34].

The timescales and rate at which the value of demand-side response and energy storage increase pose a strategic challenge. The long-term value of the smart grid is unlikely to be very tangible to market participants in 2020, yet a failure to deploy these technologies in a timely manner may lead to higher system costs in 2030 and beyond. Strategic policies will be needed to ensure that markets can deliver long-term system benefits.

References

1 Imperial College London and NERA Economic Consulting (2012) *Understanding the Balancing Challenge*, Report for the UK Department of Energy and Climate Change, http://www.decc.gov.uk/assets/decc/11/meeting-energy-demand/future-elec-network/5767-understanding-the-balancing-challenge.pdf (last accessed 8 February 2013)

2 Kabouris, J. and Kanellos, F. D. (2010) Impacts of large-scale wind penetration on designing and operation of electric power systems. *IEEE Trans. Sustain. Energy*, **1** (2), 107–114.

3 Amin, M. and Wollenberg, B. (2005) Toward a smart grid: power delivery for the 21st century. *IEEE Power Energy Mag.*, **3** (5), 34–41.

4 Farhangi, H. (2010) The path of the smart grid. *IEEE Power Energy Mag.*, **8** (1), 18–28.

5 Bose, A. (2010) Smart transmission grid applications and their supporting infrastructure. *IEEE Trans. Smart Grid*, **1** (1), 11–19.

6 Li, F., Qiao, W., Sun, H., Wan, H., Wang, J., Xia, Y., Xu, Z., and Zhang, P. (2010) Smart transmission grid: vision and framework' *IEEE Trans. Smart Grid*, **1** (2), 168–177.

7 Strbac, G. (2008) Demand side management: benefits and challenges' *Energy Policy*, **36**, 4419–4426.

8 Imperial College London (2012) *Strategic Assessment of the Role and Value of Energy Storage Systems in the UK Low Carbon Energy Future*, Report for Carbon Trust, http://www.carbontrust.com/resources/reports/technology/energy-storage-systems-strategic-assessment-role-and-value (last accessed 8 February 2013).

9 Roberts, B. P. and Standberg, C. (2011) The role of energy storage in development of smart grids' *IEEE Proc.*, **99** (6), 1139–1144.

10 Liu, H., Chen, X., Yu, K., and Hou, Y. (2012) The control and analysis of self-healing urban power grid. *IEEE Trans. Smart Grid*, **3** (3), 1119–1129.

11 Ramchurn, S. D., Vytelingum, P., Rogers, A., and Jennings, N. R. (2011) Agent-based control for decentralised demand side management in the smart grid, in *Proceedings of the 10th International Conference on Autonomous Agents and Multiagent Systems*, pp. 5–12.

12 Vytelingum, P., Voice, T. D., Ramchurn, S. D., Rogers, A., and Jennings, N. R. (2011) Theoretical and practical foundations of large-scale agent-based micro-storage in the smart grid. *J. Artif. Intell. Res.*, **42**, 765–813.

13 Kirschen, D. and Bouffard, F. (2009) Keep the lights on and the information flowing. *IEEE Power Energy Mag.*, **7** (1), 50–60.

14 Schweppe, F. C., Tabors, R. D., Kirtley, J. L., Outhred, H. R., Pickel, F. H., and Cox, A. J. (1980) Homeostatic utility control. *IEEE Trans. Power Apparatus Syst.*, **PAS-99**, 1151–1163.

15 Kirby, B. and Hirst, E. (1999) *Load as a Resource in Providing Ancillary Services*, http://www.consultkirby.com/files/Load_as_a_Resource_-_APC_99.pdf (last accessed 8 February 2013).

16 Cecati, C., Citro, C., and Siano, P. (2011) Combined operations of renewable energy systems and responsive demand in a smart grid. *IEEE Trans. Sustain. Energy*, **2**, 468–476.

17 Medina, J., Muller, N., and Roytelman, I. (2010) Demand response and distribution grid operations: opportunities and challenges. *IEEE Trans. Smart Grid*, **1** (2), 193–198.

18 Silva, V., Stanojevic, V., Pudjianto, D., and Strbac, G. (2009) *Value of Smart Appliances in System Balancing, Part I of Deliverable 4.4 of Smart-A Project (No. EIE/06/185//S12.447477)*, http://www.smart-a.org/W_P_4_D_4_4_Energy_Networks_Report_final.pdf (last accessed 8 February 2013).

19 Short, J. A., Infield, D. G., and Freris, L. L., Stabilization of grid frequency through dynamic demand control. *IEEE Trans. Power Syst.*, **22**, 1284–1293.

20 Molina-García, A., Bouffard, F., and Kirschen, D. S. (2011) Decentralized demand-side contribution to primary frequency control. *IEEE Trans. Power Syst.*, **26**, 411–419.

21 Aunedi, M., Ortega Calderon, J. E., Silva, V., Mitcheson, P., and Strbac, G. (2008) *Economic and Environmental Impact of Dynamic Demand*, Report for the UK Department of Energy and Climate Change, http://www.supergen-networks.org.uk/filebyid/50/file.pdf.

22 Kempton, W. and Tomić, J. (2005) Vehicle-to-grid power fundamentals: calculating capacity and net revenue. *J. Power Sources*, **144**, 268–279.

23 Kempton, W. and Tomić, J. (2005) Vehicle-to-grid power implementation: from stabilizing the grid to supporting large-scale renewable energy. *J. Power Sources*, **144**, 280–294.

24 Shakoor, A. and Aunedi, M. (2011) *D3.1: Report on the Economic and Environmental Impacts of Large-Scale Introduction of EV/PHEV Including the Analysis of Alternative Market and Regulatory Structures, Deliverable 3.1 of the Grid-for-Vehicles (G4 V) Project (FP7 No. 241295)*, http://www.g4v.eu/datas/reports/G4 V_WP3_D3_1_economic_and_environmental_impact.pdf (last accessed 8 February 2013).

25 Lund, H. and Kempton, W. (2008) Integration of renewable energy into the transport and electricity sectors through V2G. *Energy Policy*, **36** (9), 3578–3587.

26 Imperial College London and Energy Networks Association (2010) *Benefits of Advanced Smart Metering for Demand Response-Based Control of Distribution Networks*, http://www.energynetworks.org/modx/assets/files/electricity/futures/smart_meters/Smart_Metering_Benerfits_Summary_ENASEDGImperial_100409.pdf (last accessed 8 February 2013).

27 Guille, C. and Gross, G. (2009) A conceptual framework for the vehicle-to-grid (V2G) implementation. *Energy Policy*, **37**, 4379–4390.

28 Black, M. and Strbac, G. (2007) Value of bulk energy storage for managing wind power fluctuations. *IEEE Trans. Energy Convers.*, **22**, 197–205.

29 Pudjianto, D., Mancarella, P., Gan, C. K. and Strbac, G. (2011) Closed loop price signal based market operation within microgrids. *Eur. Trans. Electr. Power*, **21**, 1310–1326.

30 Rahimi, F. and Ipakchi, A., Demand response as a market resource under the smart grid paradigm. *IEEE Trans. Smart Grid*, **1** (1), 82–88.

31 Nyeng, P. and Ostergaard, J. (2011) Information and communications systems for control-by-price of distributed energy resources and flexible demand. *IEEE Trans. Smart Grid*, **2** (2), 334–341.

32 Papadaskalopoulos, D. and Strbac, G. (2012) Decentralized participation of electric vehicles in network-constrained market operation, presented at the IEEE PES Innovative Smart Grid Technologies Conference, Berlin, October 2012.

33 Hahn, A. and Govindarasu, M. (2011) Cyber attack exposure evaluation framework for the smart grid. *IEEE Trans. Smart Grid*, **2** (4), 835–843.

34 Allan, R. N., Djapic, P., and Strbac, G. (2006) Assessing the contribution of distributed generation to system security, presented at the International Conference on Probabilistic Methods Applied to Power Systems (PMAPS), Stockholm, June 2006.

37
Natural Gas Pipeline Systems

Gerald Linke

37.1
Physical and Chemical Fundamentals

At first glance, the transport of natural gas within pipeline systems appears to be a physical phenomenon, based on the compressibility of the medium, and less on chemical parameters. However, this is not the case, for several reasons: On the one hand, the transport capacity depends on the properties of the medium such as density and calorific value, as will be seen below when the transport equation is derived from the Darcy–Weissbach equation. On the other hand, the transported medium is used as a fuel gas for the compressors that boost the pressure in the transport system to a higher level – a premise for the further transport in the downstream pipeline section.

Natural gas is the fuel with the lowest carbon dioxide emissions compared with other fossil fuels and it can even compete with the emission profile of some renewable gases. Its main component is methane, and other hydrocarbons and compounds such as nitrogen and carbon dioxide are only minor components. An example of two different but typical composition, denoted a low caloric gas (or L gas) and high caloric gas (or H gas), is given in Table 37.1.

The density of natural gas is ~80% of that of air, and it ascends if there is an unintentional gas release, which is important for the assessment of the risks of natural gas transport systems. One of the most important parameters is the Wobbe index: the ratio of the calorific value to the square root of the relative density. This index characterizes the quality of the gas. Different gas compositions with the same Wobbe index show the same behavior during combustion in gas appliances. Since natural gas is not an ideal gas, the precise value of the compressibility factor is important. There is a very simple ansatz in the German Standard DVGW-G 2000 where the compressibility is approximated via

$$K = 1 - \frac{p}{450} \qquad (37.1)$$

Transition to Renewable Energy Systems, 1st Edition. Edited by Detlef Stolten and Viktor Scherer.
© 2013 Wiley-VCH Verlag GmbH & Co. KGaA. Published 2013 by Wiley-VCH Verlag GmbH & Co. KGaA.

Table 37.1 Typical compositions of low calorific value gas (or L gas) and high calorific value gas (or H gas).

Component		High Caloric Gas			Low Caloric Gas		
		Russian Federation	North Sea	Blended gas	The Netherlands	"Verbundgas" area	Weser/Ems area
Methane	CH_4	98.09	86.90	87.52	83.49	85.30	87.73
Ethane	C_2H_6	0.66	8.03	7.30	3.80	3.01	0.65
Propane	C_3H_8	0.22	1.91	1.17	0.76	0.56	0.04
Butane	C_4H_{10}	0.08	0.50	0.30	0.24	0.18	0.02
Pentane	C_5H_{12}	0.02	0.08	0.05	0.07	0.04	–
Hexane	C_6H_{14}	0.01	0.05	0.05	0.05	0.05	–
Nitrogen	N_2	0.84	1.05	2.17	10.07	9.27	9.04
Carbon dioxide	CO_2	0.08	1.48	1.44	1.52	1.59	2.52
Oxygen	O_2	–	–	–	–	–	–

However, it is recommended that this equation is used for pressures below 70 bar (numerical values are correct only if the pressure is given in bar) and for temperatures between 10 and 15 °C. More precise analytical descriptions exist and have been comprehensively studied and compared [1].

To compare different energies (e.g., environmental energy, district heating, power, and natural gas) it is common to use a factor, the so-called primary energy factor, that describes the effort to provide a certain amount of energy to the ultimate consumer. To deliver natural gas to the German market, ~10% of the final energy is needed for the entire supply chain, especially for transport over the average distance between the producer and the consumer. Therefore, the primary energy factor of natural gas is 1.1, whereas it is 3 for the current power mix, based on an evaluation in accordance with DIN V 4701-10.

Natural gas emits ~200 g of CO_2 per kilowatt hour thermal energy content. The total emissions along the entire energy chain cumulate to 244 g of so-called "CO_2 equivalent emissions," whereas the power mix in Germany has an emission value of 633 g kWh^{-1} (CO_2 equivalent). A survey of the most interesting properties natural gas and its main component methane is given in Table 37.2.

Natural gas is transported by a pressure gradient along a gas pipeline. For reasons of simplicity, we assume that the whole pipeline has been laid at the same altitude. There is a physical relation between the flow rate and the inlet and outlet pressures, which can be derived from the Darcy–Weisbach equation for the relation between the differential pressure drop dp along an infinitesimal length dx:

Table 37.2 Gas properties of a typical high calorific value Russian natural gas and of methane (main component).

Symbol	Property	Equation	Value
ρ	Density of methane	$\rho = \dfrac{M}{V}$	0.717 kg m^{-3}
W_S	Wobbe index of methane (based on upper calorific value)	$W_S = H_S \left(\dfrac{\rho}{\rho_{air}}\right)^{-\frac{1}{2}}$	55.45 MJ m^{-3}
–	Primary energy factor of natural gas	–	1.1
–	CO_2 emissions (of an ideal combustion of natural gas)	–	200 g kWh^{-1}

$$dp = -\lambda \cdot \frac{1}{D} \cdot \frac{\rho(p)}{2} \cdot w(p)^2 \cdot dx$$

$$= -\lambda \cdot \frac{1}{D} \cdot \frac{\rho_{id}(p)}{2} \cdot w_{id}(p)^2 \cdot K \cdot dx \qquad (37.2)$$

where p is the pressure, λ the integral friction coefficient, D the pipe diameter, ρ the gas density, ρ_{id} the gas density of an ideal gas, w the flow velocity, w_{id} the flow velocity of an ideal gas, K the compressibility, x the line element (section length), and L the total pipeline length. The second half of the equation is a manipulation making use of the ideal gas equation. Through this substitution, the compressibility enters the equation. After integration, this leads to

$$\int_{p_1}^{p_2} p \, dp = -\frac{1}{D} \cdot \frac{\rho_{1,id}}{2} \, w_{1,id}^2 \, \frac{p_1}{T_1} \int_0^L \lambda \, T \, K \, dx \qquad (37.3)$$

where the subscript 1 stands for the inlet position of the pipeline section and 2 for the outlet position at a distance L (path length). With certain assumptions, the integral on the right can be solved: if we assume that the integral flow friction, the temperature, and the compressibility do not change significantly with increasing distance from the inlet point, they can be drawn put of the integral, which becomes trivial. The three functions are replaced by their mean values (λ_m, T_m, K_m). Then the density and the flow velocity are taken at normal conditions (utilizing the ideal gas law a second time). This gives us the transport equation in the desired format:

$$p_1^2 - p_2^2 = \lambda \frac{L}{D^5} \frac{16}{\pi^2} \rho_n \frac{T_m}{T_n} p_n \dot{V}_n^2 K_m \qquad (37.4)$$

Rearrangement of Eq. 37.4 leads to the flow rate as a function of the inlet and outlet pressure and several other transport system parameters such as pipeline diameter, pipeline length, integral pipeline friction, and average conditions (such as the temperature):

$$\dot{V}_n = \sqrt{\left[1-\left(\frac{p_2}{p_1}\right)^2\right]\frac{1}{\lambda}\frac{D^5}{L}\frac{\pi^2}{16}\frac{T_n}{T_m}\frac{p_1^2}{p_n}\frac{1}{\rho_n}\frac{1}{K_m}} \qquad (37.5)$$

It should be noted that the transport capacity of a pipeline (measured in cubic meters per second) is proportional to $D^{2.5}$ and not, as one might expect, to the cross-sectional area of the pipe or to the square of the diameter.

Furthermore, it can be shown that the pressure drop along the line is a simple function of the path length according to

$$p(x) = p_1 \sqrt{1-\frac{x}{L}\left[1-\left(\frac{p_2}{p_1}\right)^2\right]} \qquad (37.6)$$

At certain distances, typically between 50 and 100 km, compressor stations are installed to raise the pressure again. These compressors accelerate the gas molecules or, in other words, they increase the dynamic pressure. The gas is released into the subsequent section and the next pressure drop along this downstream section generates the transport successively over the following kilometers.

37.2
Technological Design

How does one design a gas transport system sufficiently? The process starts with the collection of the basic points of the scenario. We will take as an example the planning of the supply of a large customer such as a gas-fired power plant. It is intended to build a new pipeline over a certain distance to link the customer into an existing gas source. This could be an exploration site but here we assume that it is a connection to a larger supply system – a large-diameter pipeline – several kilometers away. The customer defines its maximum energy demand per hour or, in other words, the maximum flow rate. In addition, all gas installations need a minimum feed-in pressure which corresponds to p_2 in Eq. 37.4. Since the spur line to the power plant is connected to an existing large-diameter pipeline, the inlet pressure p_1 is also known. The spur line length depends on the route that has to be followed from the inlet point to the outlet location. This will vary from local specifications if, for example, the shortest or straight connection is not feasible or if the spur line needs to be laid in an "energy highway" together with other pipes or cables to minimize the environmental impact. However, once the pipeline route planning has been completed, the length L is known and fixed as well. Therefore, the remaining parameter, the pipeline diameter D, can be calculated from Eq. 37.4.

What happens next? Since a pipeline is a large pressure vessel, it has to be constructed in accordance with proven rules that ensure sufficient provision against bursting. Two parameters have an influence on the stability of the vessel: the wall thickness and the yield strength of the chosen material. The Barlow law quantifies this interdependence between the operational pressure in the pipe, its diameter, and these two terms:

$$t_{min} \, R_{t0.5} = \frac{P_D \, D}{20 \, f} \quad \text{or} \quad t_{min} = \frac{P_D \, D}{20 \, f \, R_{t0.5}} \tag{37.7}$$

where t_{min} is the calculated minimum wall thickness (mm), f the design factor, P_D the design pressure (bar), $R_{t0.5}$ the specified minimum yield strength (N mm^{-2}), D the outside pipe diameter (mm), and $S = 1/f$ the safety factor. The design pressure in Eq. 37.7 is at least the maximum operational pressure and the design factor has a value < 1. Mandatory rules for a proper choice are given in [2], a European standard for the design of pipelines. A design factor of < 1 ensures that the stress load on the pipeline remains below a state when the steel deforms plastically (see Figure 37.1). It is therefore sometimes called the "degree of utilization". In Germany, f is typically only 62%. There remains a high safety margin which is beneficial for several reasons: on the one hand, unknown loads might generate additional stresses, and on the other, aging and corrosion might cause a decrease in wall thickness over time. We will come back to such aspects of aging and suitable precautionary measures later.

Once the steel type has been selected the minimum wall thickness can be calculated from [2]. As an example, we calculate the minimum wall thickness for a pipeline with a diameter of 600 mm and an operational pressure of 80 bar, built out of steel with a minimum yield strength of 560 N mm^{-2} and with an degree of utilization of 62%:

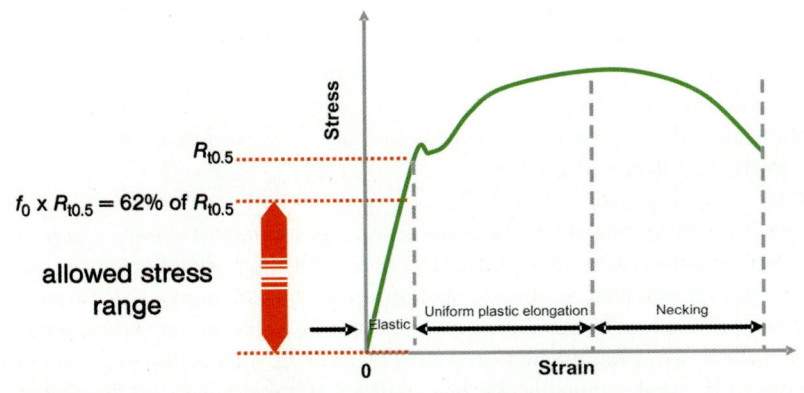

Figure 37.1 Strain–stress diagram for a mild steel. The internal pressure inside a pipeline is kept in the "allowed stress range" where the degree of utilization is below 62%. The maximum or 100% utilization characterizes a stress where this steel would deform elastically. Even at the borderline of this elastic or linear area, the pipe has additional loading capacities and a safety margin to the ultimate yield strength.

$$t_{\min} = \frac{P_D \, D}{20 \, f \, R_{t0.5}} = \frac{600 \times 80}{20 \times 0.62 \times 560} = 6.9 \tag{37.9}$$

A wall thickness of ~7 mm would be sufficient in this case.

During its lifetime, a pipeline is not only exposed to stresses from internal pressure but also has to cope with other influences. However, a set of suitable protection and counter-measures has been developed and specified in an international standard[1]:

- *Threat from corrosion:* The pipeline is protected against corrosion via a coating (passive protection) and via a cathodic protection system – a protection current that isolates bare parts of metal from corrosion by prohibiting the electrochemical process of oxidation.
- *Threat from mechanical impact:* The pipeline is buried and the depth of cover serves as sufficient protection against damage by excavators. Sometimes a warning tape is buried above the vertex of the pipe. Marker posts provide visible hints to where the pipeline is buried. In addition, the pipe route is survey and patrolled regularly.
- *Threat from hot tapping:* This type of threat was a "teething problem" of the early "drilling during operation" activities in the 1950s and 1960s when it happened that the wrong pipeline was tapped. The cause of failure has been completely removed owing to the much higher quality of modern documentation, especially since pipelines are recorded in geographic information systems (GIS).
- *Threat from ground movement:* In general, at locations where ground movement is a potential risk, a suitable monitoring technique is installed. As an example, in areas with mining subsidence the pipes are equipped with strain gauges to measure additional stresses before a critical threshold is exceeded.
- *Threat from additional loads:* Additional loads (temperature-induced stresses, loads from traffic, etc.) are anticipated from the very beginning and can be borne due to the sufficient safety margin of the design.
- *Threat from pressure cycles or alternating loads or shock waves:* Natural gas pipelines are not exposed to this risk since they are operated quasi-static. The amplitudes of the pressure changes are too low to cause alternating loads (according to Wöhler's law). In addition, the compressibility of gas is beneficial to limiting shock wave effects from any local, spontaneous pressure peak (e.g., generated from a fast shut-down of a valve).

In Section 37.1, it was mentioned that long-distance gas transport would be impossible without interjacent compression. Here we make a short excursion to discuss the proper design of such a compressor station and specify the necessary compression power. For reasons of reliability, it has proven its value in the past to work with a so-called ($n + 1$) redundancy on a station site. This means that even on the coldest day (or at the design point for the maximum transport load and the largest demand for intermediate compression) there should always be one unit in reserve.

---

[1] In Germany, DVGW-G 466-1 describes the requirements for operation of a pipeline system above 16 bar and DVGW-G 463 explains the design principles.

But how many cubic meters per hour can a single compressor raise to the required pressure level? This depends on the amount of energy for the thermodynamic process of boosting a cubic meter of the relevant gas to the outlet pressure and on the installed power of the prime engine. A derivation can be found in [1]; we just show the results:

$$P_{engine} = P_{specific}(1-\alpha)\dot{V}_N$$

$$P_{specific} = 1.0304 \times 10^{-4} \times \left(\frac{k}{k-1}\right) K_1 T_1 \left(\frac{1}{\eta_S \eta_m}\right)\left(\pi^{\frac{k-1}{k}} - 1\right) \qquad (37.10)$$

where P_{engine} is the power of prime engine (kW), $P_{specific}$ the specific power (per cubic meter) [kW (m^3 h^{-1})$^{-1}$], α the fraction of gas flow that is used as fuel gas, $\dot{V}_N$ the normal volume flow (m^3 h^{-1}), k an isentropic exponent, K_1 the compressibility (at the inlet point), T_1 the gas temperature (at the inlet point) (K), π the pressure ratio (p_2/p_1), p_1 the inlet pressure (bar), p_2 the outlet pressure (bar), η_S the isentropic efficiency, and η_m the mechanical efficiency. For natural gas, and assuming that $\eta_S \eta_m \approx 0.88$, this can be simplified to

$$P_{engine} = 4.33 \times 10^{-7} \times T_1 \dot{V}_N \left(\pi^{0.26} - 1\right) \qquad (37.11)$$

giving us the prime engine power in megawatts if the normal volume flow is given in cubic meters per hour and the temperature in kelvin.

Since we are now able to design an entire transport system – a pipeline and a compressor – there remains the question of what costs will arise in the construction of such a system. We give an answer to this with two rules of thumb:

$$\text{Invest}_{Pipeline}\left[10^6 \text{ € km}^{-1}\right] = 0.17 \, D\,[m]\sqrt{P_D\,[bar]}$$

$$\text{Invest}_{Compressor\,Station}\left[10^6 \text{ €/unit}\right] = 1.3 \times \frac{P_{engine}\,[MW] + 10}{2} \qquad (37.12)$$

However, a natural gas transport system comprises many additional components. A vital aspect is storage. Transmission systems are designed for an average load, and when the demand deviates from this value, gas is sent to or extracted from storage. Nearly all storage facilities are underground units, either aquifer reservoirs or cavity storage operated at depths between 1000 and 1500 m. They add up to a giant capacity – usually in the region of 20% of the annual gas consumption of a country. Table 37.3 shows a comparison of the power and gas industries in Germany. The gas system can provide a range of coverage of more than 2000 h!

Another impressive figure is the energy rate of a typical underground storage facility. Storage sites such as Epe in the north-west of Germany can generate a gas withdrawal rate of ~2×10^6 m^3 h^{-1}, which represents a supply of 20 GW in 1 h!

A natural gas system is composed of the pipelines, the compressor, underground storage, and other stations, such as valve stations (to isolate individual pipe sections from the rest of the infrastructure if repair work needs to take place), metering

Table 37.3 Comparison of the power and gas industries in Germany.

Parameter	Power	Natural gas
Demand (in 2011) (TWh a^{-1})	619	930
Average capacity (GW)	70	106
Storage capacity (TWh)	0.04	217
Range of coverage (h)	0.6	2000

stations (to measure the delivered gas quantities and the compositions for billing purposes), and blending stations (to prepare gases with different calorific values).

The efficiency of the overall natural gas transmission system is extremely high. Since the gas is clean and dry and since modern pipeline systems have an internal flow coating, the losses due to friction are very low. The integral friction factor λ in Eq. 37.4 is, in general, less than 0.01 mm. In Germany, on average, only less than 0.4% of the transported energy is used as fuel gas.

The lifetime of natural gas systems depends on several conditions. Systems built in permafrost areas are very different to those with pipes buried in relatively dry soil. If the functionality of the cathodic protection system (CPS) is suboptimal or disturbed (for example, by alternating currents of cable car lines or other external electromagnetic fields), failures of the pipe coating become more serious and can lead to corrosion defects. However, no incident is known, worldwide,.. where corrosion was the cause of a total pipeline failure, that is, a rupture. The average corrosion rate of well-protected piping with a proper CPS is less than 0.01 mm per year. Also, real cases demonstrate impressively that pipelines have been operated for 100 years even though a CPS was not installed during the early decades. The codes and standards that need to be applied for prudent operation of a natural gas system aim for high reliability and safety of the system. This is an essential requirement to achieve public acceptance and to deliver energy into the heart of populated areas. No other means of transport is capable of managing such high quantities of energy at a comparable safety level and with less emissions.

37.3
Cutting Edge Technology of Today

The design and construction of natural gas pipeline systems have been improved over decades and further increases will be marginal. Nowadays, the focus lies on further improvement of operation, increased efficiency of surveillance and maintenance, and more sophisticated diagnostics. Most pipelines are "piggable," that is, they can be inspected by inline tools, called pigs or intelligent pigs, which are driven by a difference in pressure and pass through the pipeline over distances of several

Figure 37.2 Magnetic flux line (MFL) pipeline inspection tool.

hundred kilometers. The pigs deliver very reliable diagnostics of the remaining wall thickness of the pipe. In many countries, pigging of gas pipelines has become mandatory. However, not every pipeline is piggable owing to installations present in some of the pipe section, or to too tight curves, or for some other reasons. Then alternatives need to be applied and the state of the system has to be assessed via other methods (for example, control sample excavations or intensive measurement for the identification of coat defects). Whereas pipelines for liquid substances can be inspected with pigs fitted with ultrasonic sensors, where the liquid ensures acoustic coupling with the steel wall, this technique cannot be applied to gas pipelines. Instead magnetic flux line (MFL) technology is used (Figure 37.2). Further, a new concept, electromagnetic induced ultrasound (EMUS), seems to have the potential to achieve a leap in resolution of future inline inspections.

For the assessment of the surface defects of a pipe, advanced tools are available. With the finite element method (FEM) it can be checked numerically whether a defect remains stable even if additional stresses occur. Defects found during excavations can be scanned with systems such as Optocam®, which make the data available for a 3D model of the defect that again is interpreted by FEM.

Also in gas detection technology one can find new appliances utilizing optical methods. One is know as CHARM® (CH_4 airborne remote monitoring). CHARM® comprises a laser system that generates two successive rays with different, distinct wavelengths in a double pulse (Figure 37.3). On of them is methane-sensitive (the "on" signal) and the other is not and serves as a reference signal (the "off" signal); 100 double pulses are emitted per second. The laser is mounted in a box together with other measurement and control equipment and is carried by a helicopter. The helicopter is used for a visual inspection of the pipeline route but with the CHARM® system operational methane can also be detected. The laser double pulses are directed automatically towards the pipeline sections and cover a corridor with a width of up to 18 m. The light reflected from the observed surface is collected by a lens system (an integral part of CHARM®) and undergoes a comparison of the intensities of the two pulses ("on" and "off"). CHARM® is able to detect methane with an integral concentration of less than 50 ppm m. The measuring principle has received the formal approval of the German Gas and Water Association.

Figure 37.3 CHARM® (CH$_4$ airborne remote monitoring).

A similar gas detection system is GasCam, a camera system that visualizes the smallest gas releases and provides pictures comparable to those of a thermo camera (Figure 37.4). The application was developed for the inspection of gas plants. Compared with CHARM®, the GasCam is more sensitive but not appropriate for mobile operation in a helicopter.

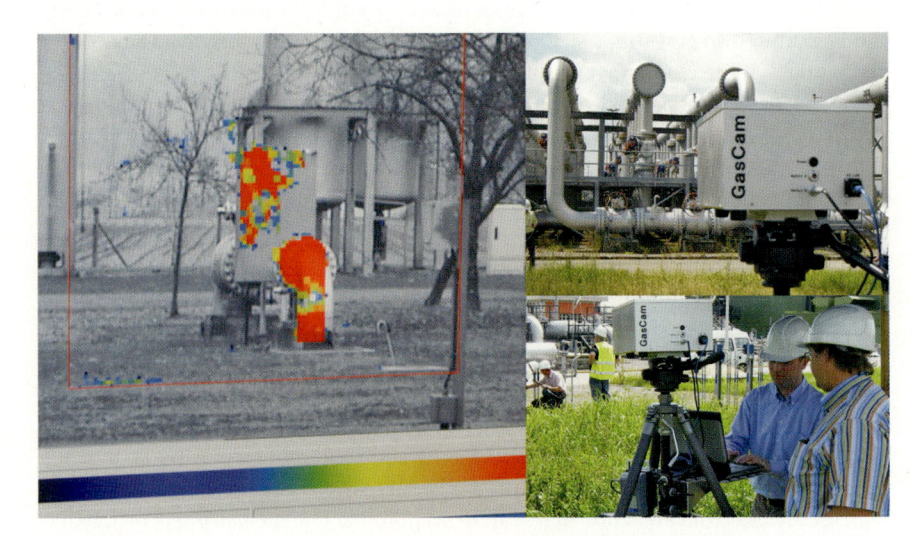

Figure 37.4 GasCam in operation to identify the smallest gas release in gas plants.

37.3 Cutting Edge Technology of Today

Statistics show that unintended and undetected damage to a pipeline system, for example, a cut in the coating or dents and gouges from excavators, is the most common failure. This can lead to a corrosion defect years later when its initiator can no longer be discovered. If the impact takes place shortly after the last inline inspection, the defect might grow over several years before the next pig run is performed. Especially combinations of a dent and a gouge can cause a dangerous configuration near to a stability threshold. Historical examples, such as a pipeline accident caused by a moulding cutter in 2004 in Ghislenghien, Belgium, prove that undetected damage can emerge as a collapse of the pipe section.

However, cutting-edge technologies in pipeline integrity monitoring now exist. Several companies have developed competing systems that claim to detect and locate an impact. They trust in different physical phenomena by searching for acoustic signals that can be interpreted as an impact from an excavator or by observing interference signals in fiber-optic cables laid parallel to the gas pipeline. It is stated that vibrations from soil displacements or temperature changes due to spilled gas will change the optical length of a laser signal that is send through the fiber.

Another promising technology is a system developed by Open Grid Europe, Germany's largest transmission grid operator. The system compares the currents registered from two neighboring marker posts which are equipped with a CPS amperemeter. Any pipe contact of an excavator – even a light touch – causes a short cut of the cathodic protection current and will be identified by the two posts. The system has been tested successfully, received a recommendation from the German Gas and Water Association, and has the advantage that old pipeline systems can easily be retrofitted (Figure 37.5).

This section on cutting-edge technologies of today would be incomplete if the change in gas composition and the related challenges were not mentioned. The steady growth in the number of biomethane-generating plants and the amount of renewable gas fed into the natural gas system reveals two sides: on the one hand, gas becomes "greener" and the overall environmental footprint of the blended gas (a mixture of natural and green gas) improves; on the other hand, since biomethane is generally fed into low-pressure distribution grids, the gas composition and quality fluctuate depending on the daily ratio of the two flows of biomethane and natural gas. This causes difficulties in correct billing of the energy consumption (which

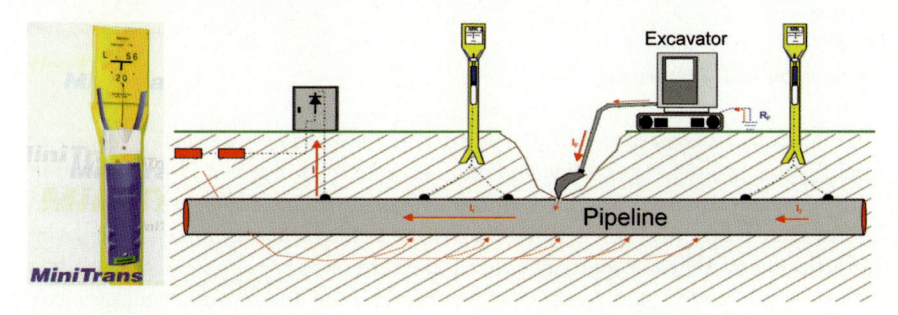

Figure 37.5 The "Potential Fernüberwachung" is used as a pipe contact detection system.

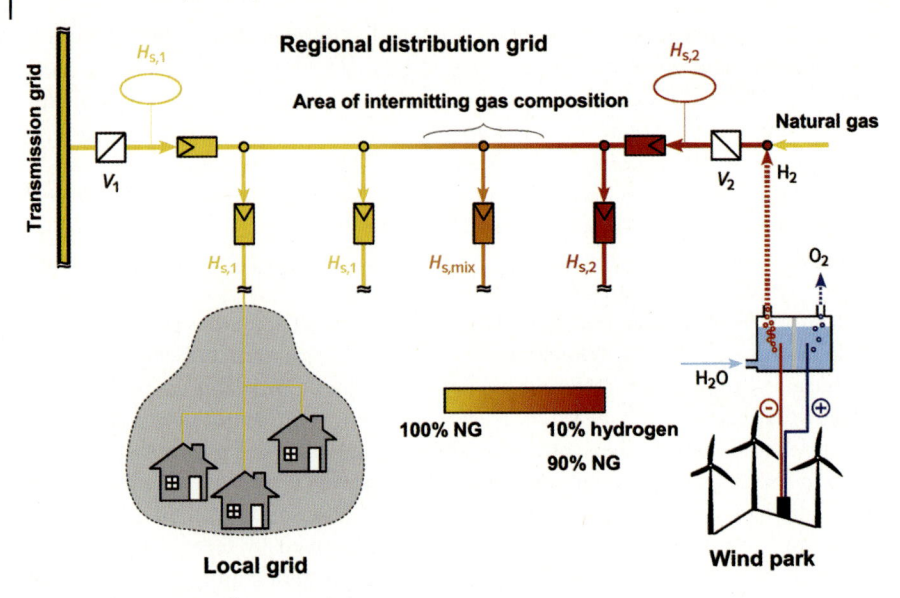

Figure 37.6 SmartSim® is used to reconstruct the correct calorific value in distribution areas with changes in gas composition. Here it is assumed that a section is fed in bi-directionally from a transmission system providing natural gas and from a power-to-gas plant injecting hydrogen.

needs to be determined from the gas quantity and its specific energy content or calorific value) and can make it necessary to equip the biomethane plant with an expensive gas conditioning system. The additional installation of chromatographs is the most uneconomic approach. However, E.ON New Build & Technology has developed a flow simulation software package that has proven its capability to reconstruct the local calorific value of the gas mixture with the highest accuracy. The software, known as SmartSim®, has received formal approval from the national body of metrology (PTB, Physikalisch-Technische Bundesanstalt) for billing purposes. Possible relative deviations between real and reconstructed calorific data are below 0.06% (Figure 37.6). Its implementation turns the pipeline grid into a smart grid.

37.4
Outlook on R&D Challenges

The steady increase in renewables in power generation leads to a substantially more volatile load profile. Natural gas systems can help to compensate for "generation valleys" by providing the necessary amount of energy for gas-fired power plants that produce the power when the sun does not shine or the wind does not blow. They are fairly flexible and suitable for quick start-up and for a rapid boost in power generation. However, the natural gas system – the pipeline and the underground storage – can also serve as a giant reservoir to store excess power (Figure 37.7 and Figure 37.8).

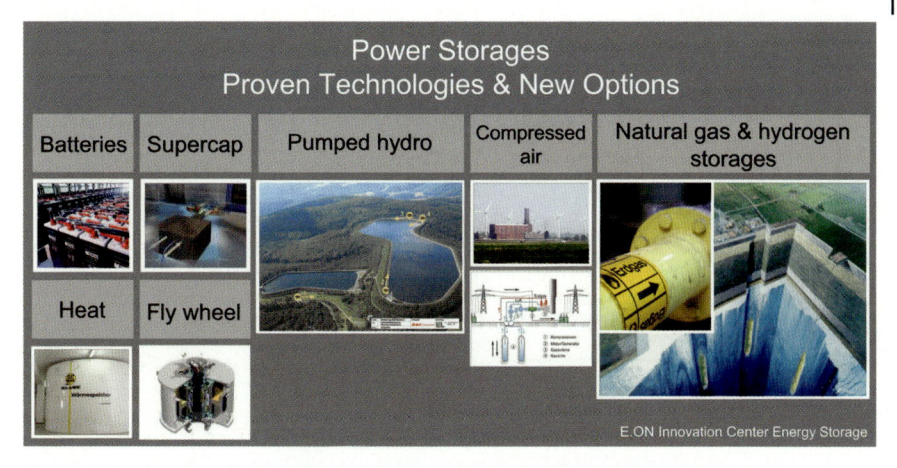

Figure 37.7 The natural gas system can store wind or solar excess power.

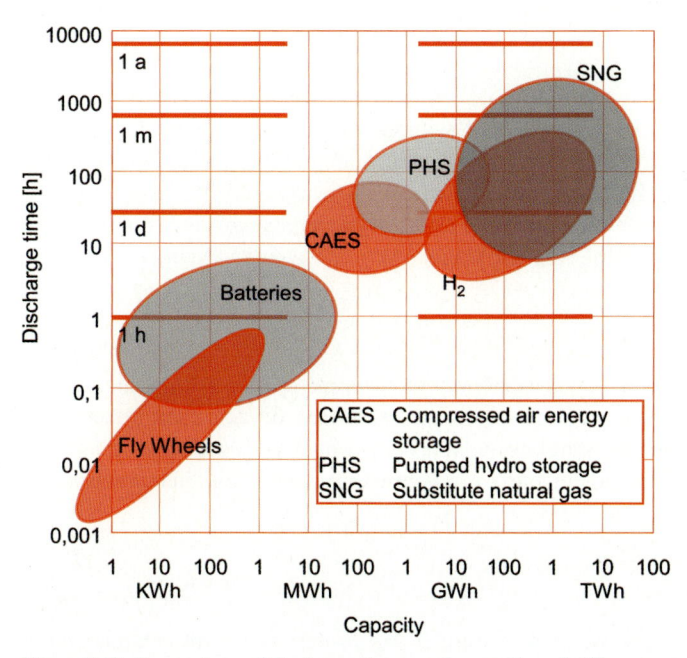

Figure 37.8 Comparison of discharge times and capacities of different power storage systems.

Several studies have come to the conclusion that power grids will need large storage facilities to manage the growth of renewable generation [3] and the solution could be to produce hydrogen from excess power and to inject it into the natural gas grid. Figure 37.8 shows the order of magnitude of the storage capacity that the natural gas system can provide. But is the blending of hydrogen and natural gas inoffensive? Experience from the past – when at the beginning of the European gas industry coke-oven gas was shipped in pipelines with hydrogen concentrations of up

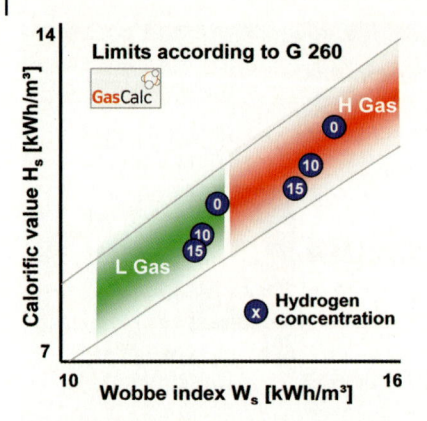

Figure 37.9 Schematic view of the alteration of gas quality if the hydrogen concentration is increased stepwise. The scheme differentiates between low and high calorific value gases and refers to the boundary values laid down in the quality code DVGW-G 260 (which serves as a reference for European gas quality standardization). More precise calculations can be found in [4] based on the software package GasCalc®.

to 60% – and calculation of the effect of hydrogen injection on the most important quality parameters (Wobbe index and calorific value) suggests that even modern gas appliances should function properly with a 10% hydrogen content or even more (Figure 37.9). The power of a burner – a classical gas appliance – is proportional to the gas volume flow and the calorific value:

$$P = \dot{V} H_S = A w H_S \tag{37.13}$$

whereas the second conversion considers the average cross-sectional area of the burner tip, A, and the flow velocity of the gas, w. Since in distribution systems the gas streams at low velocities and nearly as an incompressible medium through the final centimeters of the burner, one can assume that the Bernoulli equation:

$$\frac{p_1}{\rho_1} + \frac{w_1^2}{2} = \frac{p_2}{\rho_2} + \frac{w_2^2}{2} \tag{37.14}$$

is still valid for this case if we set $w_1 \approx 0$ and $\rho_1 \approx \rho_2$, where subscript 1 stands for the flow before the gas passes the burner tip and 2 for the conditions in the release zone. We drop the subscript 2, rearrange the Bernoulli equation to get another expression for w, and insert this new term in Eq. 37.13, which leads us to the following result:

$$P = A \sqrt{\frac{2(p_1 - p_2)}{\rho}} H_S \tag{37.15}$$

$$= A \sqrt{\frac{2 \Delta p}{\rho_{air}}} \frac{H_S}{\sqrt{d}} = A \sqrt{\frac{2 \Delta p}{\rho_{air}}} W_S$$

where d is the relative density of the gas, the ratio between the densities of the gas and air. Therefore, the heating power of a burner is proportional to the Wobbe index and should not change if the index remains constant. However, calculations by Altfeld and Schley have shown that this is the case if the hydrogen admixed with natural gas is kept below ~10% (see [4], but also [5], where a detailed analysis of the functional relationship between Wobbe index and hydrogen concentration can be found).

However, some constraints on the proportion of hydrogen in the natural gas system have been identified. They result from warranty limits from turbine manufacturers, from security restrictions on tanks in compressed natural gas (CNG) vehicles, from the inability of chromatographs to register hydrogen, and from possible lifetime reductions of underground storage formations. This is the reason why these constraints need to be studied, understood, and assessed in more detail (see, e.g., [5]). Some publications suggest that hydrogen feed-in would reduce the transport capacity of a pipeline system significantly. We evaluate this effect by rewriting the transport equation, Eq. 37.4. If the volume flow is replaced by the more meaningful energy flow

$$\dot{Q} = H_S \dot{V}$$

we obtain a new expression for the relation between the pressure drop along a transmission pipeline system and the amount of energy transported per unit time:

$$p_1^2 - p_2^2 \sim \frac{\rho_n}{H_S^2} \dot{Q}_n^2 \tag{37.16}$$

However, with respect to the definition of the Wobbe index, this means that the capacity of the system – in terms of thermal energy per second – is

$$\dot{Q}_n \sim W_S \sqrt{p_1^2 - p_2^2} \tag{37.17}$$

If the Wobbe index does not change significantly by altering the composition of the gas, this term remains stable and the capacity of a pipeline system remains almost unchanged.

However, the injection of hydrogen into the gas grid will reach its limit, but there is a way out even if one intends to produce more gas from excess or green power, namely that hydrogen from these sources can be converted into synthetic methane (Figure 37.10).

The challenges for future research on the process of methanization include the following:

- downscaling of catalytic methanization technology to the power range of electrolyzers for a typical wind park;
- enhancements of today's biological methanization process operation from the laboratory scale to the industrial scale (Figure 37.11);
- future reduction of operating costs and increase in flexibility of the operating regime.

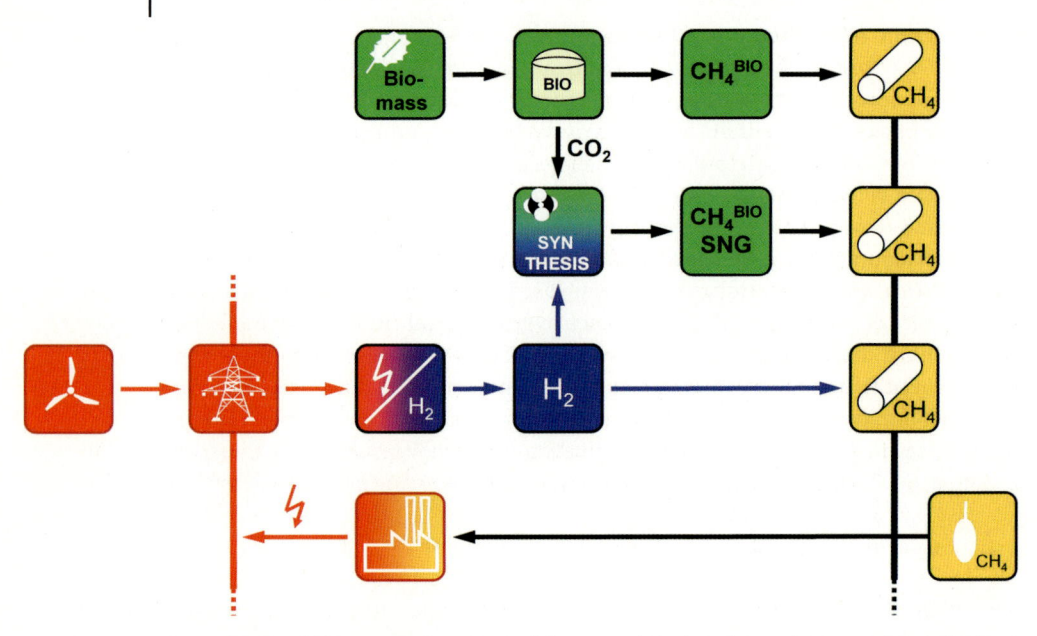

Figure 37.10 Schematic representation of the utilization paths of excess power. In addition to direct hydrogen injection into the natural gas grid, another option exists: hydrogen and carbon dioxide – in this case a biomethane plant serves as the source of CO_2 – can be used to produce (synthetic) methane, which is 100% compatible with natural gas.

Figure 37.11 Biological methanization. (a) 100 m^3 production plant designed by MicrobEnergy. (b) Reactor prototype of Krajete GmbH.

Methanization will lead to further reductions of the entire emissions of our energy system. It will become necessary to undertake also investigations of the utilization of CO_2 from industrial processes. The CO_2 – together with the renewable hydrogen bound in the CH_4 molecules – is recycled and thus helps to reduce the overall carbon freight of the total consumed power. Methanization could be the dominant carbon capture/cycling (CCC) technology of the future, reducing or eliminating the demand for carbon storage as in the case of carbon capture and storage (CCS).

Both the injection of wind power-generated hydrogen and of synthetic natural gas – the product of methanization of green hydrogen and carbon dioxide – are options to provide energy storage to the power grid. Hence they are the keys for a convergence of the power grid and the natural gas grid to form a more flexible and powerful energy grid of the future.

We want to estimate a reasonable and likely amount of renewable gas that can be expected to be part of the future natural gas value chain. Two main sources need to be considered: (i) hydrogen or synthetic methane from surplus energy and (ii) biomethane. Of course, the total amount to be expected in the coming decades depends on the growth rate of power generation from renewables and the storage capacity demand for excess power, and also on the agricultural area released for energy plant cropping. An example of an industrial country with limited agricultural areas is Germany: the Fraunhofer-Institut für Windenergie und Energiesystemtechnik (IWES) predicted a lack of 20–40 TWh of power storage capacity for the next decade and growth with a 4.5-fold higher storage demand by 2050. As we have seen before, this could be provided easily by power-to-gas storage technologies since the annual German gas consumption is ~930 TWh (see Table 37.3). However, it is very unlikely that by 2020 even 10% of it or 9 TWh will originate from renewable wind or solar power due to the higher generation costs of renewable gases caused by the price for electrolysis. However, adequate compensation of energy storage allocation and of related services can stimulate this business enormously and provide the chance to replace far more than 10% of the natural gas demand by hythane or synthetic natural gas. Nowadays, biomethane is even more important: ~70 biomethane plants are operated in Germany with an annual production of 5 TWh. Compared with the 58 TWh generated by more than 7000 biogas plants, this figure is small. This is due to the fact that biogas production is a simple and proven technology and cost reductions of gas generation techniques have already been achieved. Biomethane production requires an additional purification (additional costs) to upgrade the raw gas to natural gas quality and compression to the pressure regime of the gas distribution or transmission system. The competitive advantage of biomethane plants results from their size and optimization of the feedstock logistics. Also, biomethane can be fed into the natural gas grid almost without any limitations where more efficient utilization [e.g., via CHP (combined heat and power)] can be achieved. Biomethane can be generated via conventional "wet digestion" or from residual biomass and via thermochemical gasification of wood and straw. Both feedstock diversification and the capability of thermochemical gas generation indicate that the political goal to increase German biomethane production to 60 TWh by 2020 and to 100 TWh by 2030 can still be achieved. Therefore, one can predict that by 2030 at least 10% of the gas utilized will have a renewable origin either from biomass congestion or thermochemical processes or from power-to-gas technologies.

In addition to power-to-gas, a second and similar interaction of the power and the natural gas grid will also gain increased importance: distributed generation. CHP appliances for small- and medium-sized premises will be the key technology to

- reduce the consumption of primary energy
- increase the energy efficiency (Figure 37.12)
- relieve the local power distribution grids
- make the construction of new, large power plants dispensable on the long run once a significant number of CHP plants together add up to a so-called virtual power plant of comparable size.

The outlook on future R&D challenges could be extended to the entire natural gas chain. This would bring us to questions such as:

- How do we safeguard long-term supply using natural gas?
- How do we finance future investment in new exploration technologies to gain access to Arctic gas or to deep-water gas hydrates?
- How do we produce gas from other renewable sources (e.g., via thermochemical synthesis from wood or straw) to solve the "table versus tank" issue?

However, these questions are beyond the scope of this chapter, which is focused on the future role of the natural gas backbone: the pipeline system.

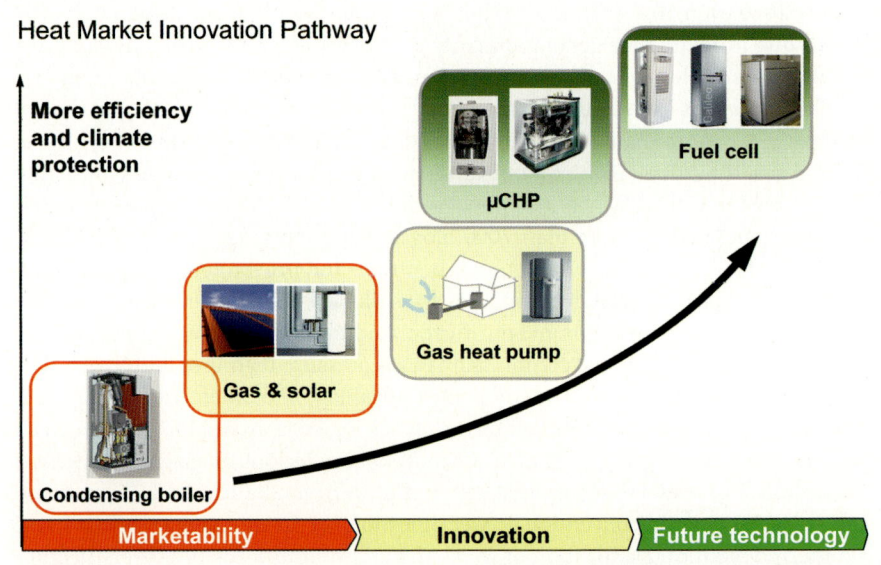

Figure 37.12 Increase in efficiency of natural gas appliances. CHP units such as a Stirling or Otto engine are already available. However, further increases in efficiency will come with the commercial launch of the fuel cell heating system. The R&D challenges are to achieve (i) an increase in electrical efficiency due to a decreased heat demand in new premises and a greater need for power output, (ii) an improved ability for intermittent operation, and (iii) further cost reductions and integration into a smart home solution.

37.5
System Analysis

Driven by market mechanisms, the natural gas grid has grown rapidly during recent decades. Figure 37.13 illustrates this boost impressively by comparing the European gas infrastructure fin 1970 and 2011.

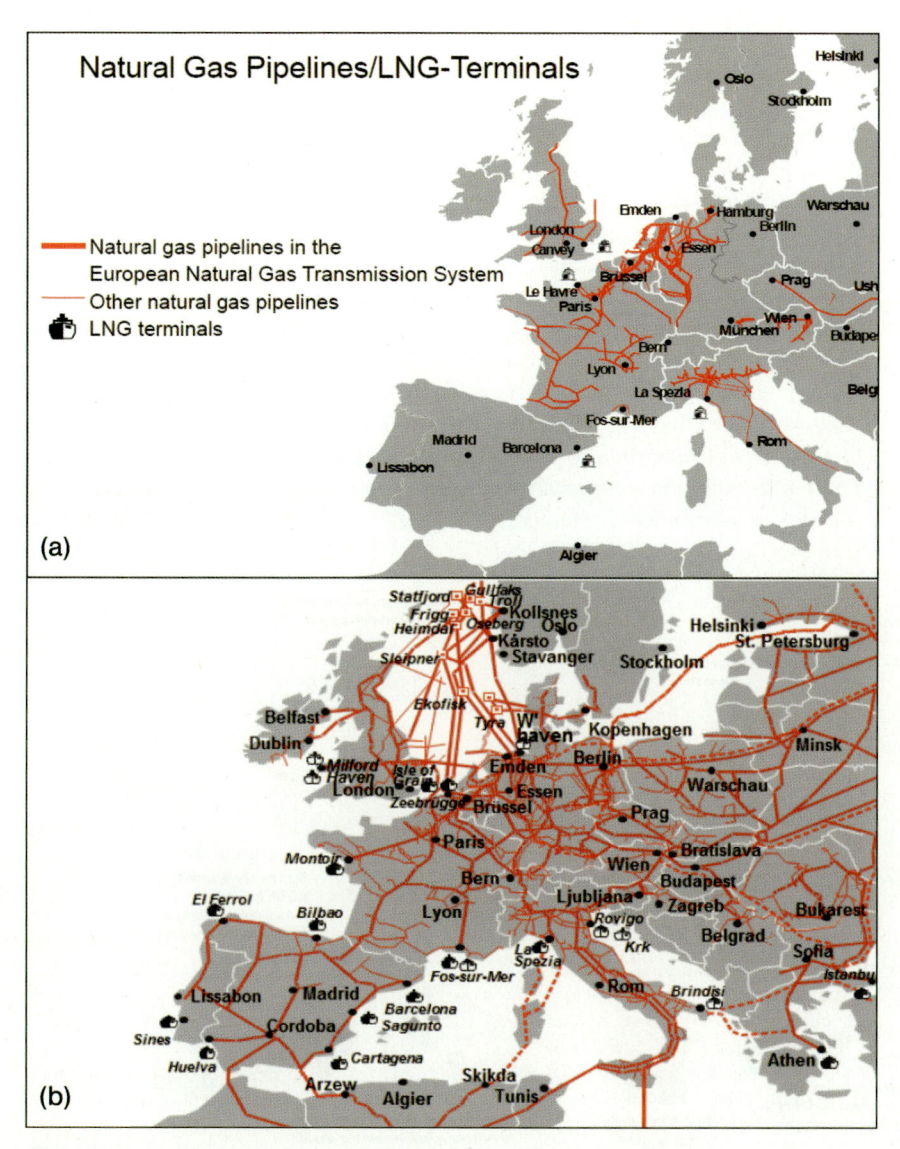

Figure 37.13 Development of the natural gas infrastructure in recent decades: (a) 1970 and (b) 2011. Source: E.ON Ruhrgas AG.

These infrastructure investments will reveal their benefits in the decades to come: The pipeline system is the backbone for

- the gas supply of distributed generation units such as micro-CHPs or fuel cell heating systems;
- the integration of renewable gases such as biomethane or of excess power from wind and solar via electrolysis or methanization;
- extended clean mobility based on proven CNG technology or mobile LNG units (trucks or ships);
- better convergence of the power and the natural gas industries to combine the strengths of both (Figure 37.14).

However, future outlooks exist according to which the importance of the natural gas pipeline system is declining. In general, two main arguments are stressed:

- natural gas is a fossil fuel and its consumption should be reduced over time to match carbon reduction targets;

However, natural gas has the lowest emissions among all fossil fuels and it should not be overlooked that the activation of wind and solar plants also causes emissions. The natural gas infrastructure provides the benefits shown in Figure 37.14 and will be part of the solution – especially where other techniques fail – as in case of large scale and long-term energy storage.

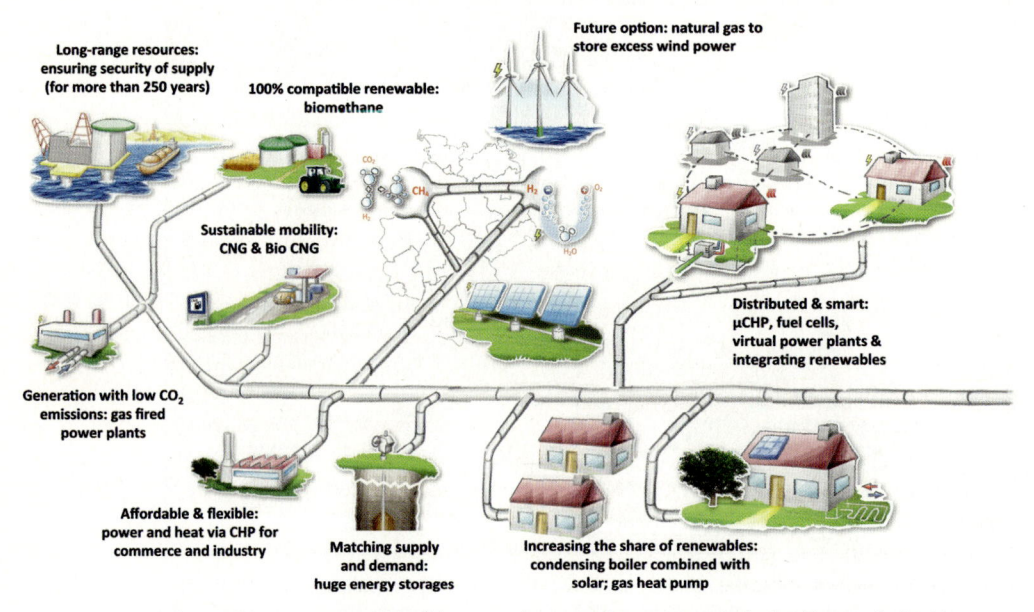

Figure 37.14 Future model of the natural gas pipeline system as the backbone of a highly efficient energy supply combining the strengths of power and gas and renewables.

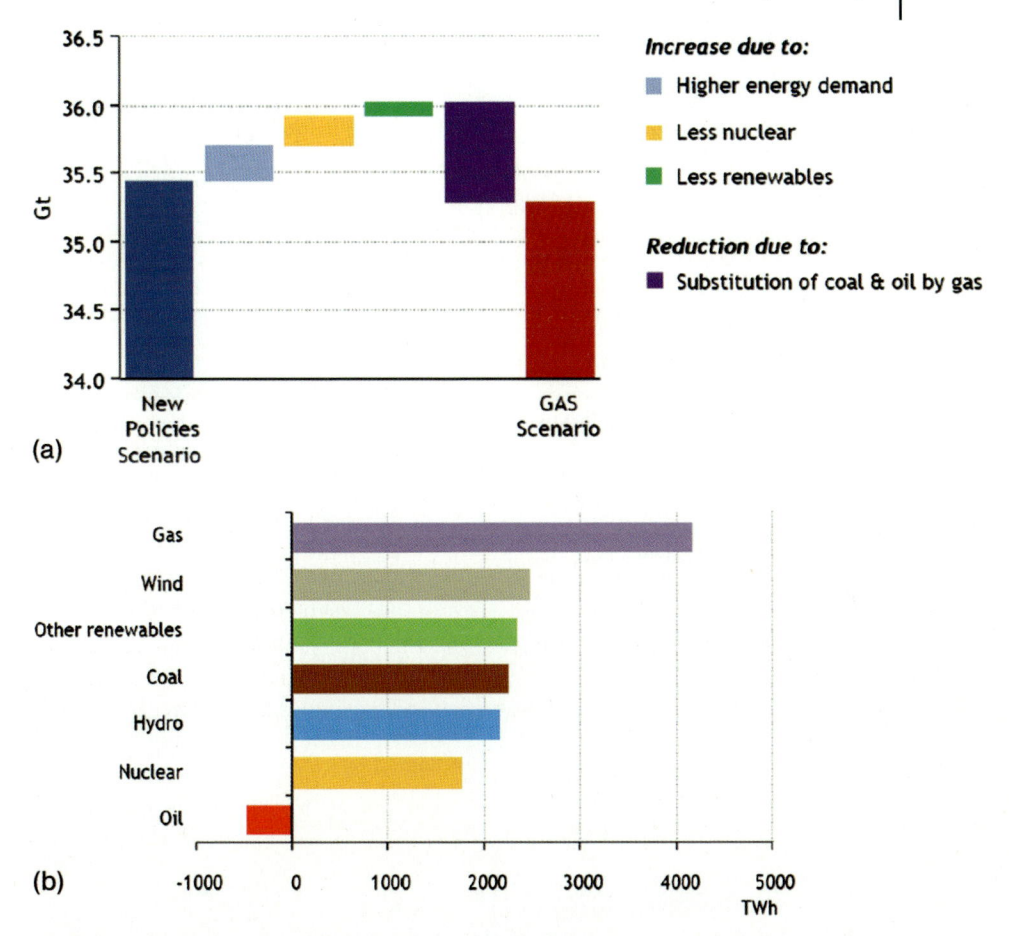

Figure 37.15 (a) Prediction of the change in power generation by fuel in the Gas Scenario (2010–2035). Total electricity demand increases by 70% by 2035, underpinned by a near doubling of gas-fired generation. (b) The CO_2 emissions of the Gas Scenario compared with the New Policy Scenario 2035. Source: *WEO Special Report 2011: Are We Entering the Golden Age of Gas?* [6].

The IEA has studied how increased utilization of natural gas and of the existing capable infrastructure can lead to a reduction of emissions [6] (Figure 5.3).

Other researchers have also proved that the cost for saving 1 t of CO_2 is significantly lower when the utilization of modern gas appliances is pushed forward, and when power-to-gas technologies are fostered to design the emission profile of the gas instead of investment in insulation of premises [7].

In conclusion, the natural gas system still guarantees a safe and reliable, an environmentally sustainable, and – last but not least – an affordable future energy supply – an option that without doubt can display its strengths in a sensible interaction with power and renewables.

References

1 Mischner, J., Fasold, H.-G., and Kadner, K. (2011) *gas2energy.net – Systemplanerische Grundlagen der Gasversorgung*, ISBN 978-3-8356-3205-9, Oldenbourg Industrieverlag, Munich.

2 DIN (2009) *Gas Supply Systems – Pipelines for Maximum Operating Pressure over 16 bar – Functional requirements*, DIN EN 1594, Deutsches Institut für Normung, Berlin.

3 Pieper, C. and Rubel, H. (2011) *Revisiting Energy Storage – There is a Business Case*, Boston Consulting Group, Boston.

4 Altfeld, K. and Schley, P. (2011) Entwicklung der Gasbeschaffenheiten in Europa, *gwf Das Gas- und Wasserfach, Gas/Erdgas*, September.

5 Müller-Syring, G., Henel, M., Köppel, W., Mlaker, H., Sterner, M., Höcher, Th. (2013) *Entwicklung von modularen Konzepten zur Erzeugung, Speicherung und Einspeisung von Wasserstoff und Methan ins Erdgasnetz*, Abschlussbericht des DVGW-Forschungsvorhabens G01-07-10 (http://www.dvgw-innovation.de/die-projekte/archiv/energiespeicherkonzepte).

6 International Energy Agency (2011) *World Energy Outlook 2011 Special Report: Are We Entering the Golden Age of Gas?*, IEA, Paris

7 Krause, H., Erler, F., Köppel, F., Fischer, M., Hansen, P., Markewitz, P., Kuckshinrichs, W., and Hake, J. (2011) *Systemanalyse – Teil II: Bewertung der Energieversorgung mit leitungsgebundenen gasförmigen Brennstoffen im Vergleich zu anderen Energieträgern – Einfluss moderner Gastechnologien in der häuslichen Energieversorgung auf Effizienz und Umwelt*, Abschlussbericht des DVGW-Forschungsberichtvorhabens G 5/04/09-TP2.

38
Introduction to a Future Hydrogen Infrastructure

Joan Ogden

38.1
Introduction

Hydrogen has been widely proposed as a future energy carrier to address environmental and energy security problems posed by current fuels. There is growing interest in using hydrogen as a transport fuel, and fuel-cell vehicle (FCV) demonstrations are proceeding in Europe, Asia, and North America. Hydrogen fuel-cell cars are 1.5–2.5 times more efficient than advanced gasoline cars (on a tank to wheels basis), and produce zero tailpipe emissions. They offer good performance, a range of 500 km or more and can be refueled in a few minutes [1] Hydrogen can be made with zero or near-zero emissions from widely available resources, including renewables (such as biomass, solar, wind, hydropower, and geothermal), fossil fuels (such as natural gas or coal with carbon capture and sequestration), and nuclear energy. In principle, it should be possible to produce and use hydrogen transportation fuel with near-zero lifecycle emissions of greenhouse gases and greatly reduced emissions of air pollutants while simultaneously diversifying away from our current dependence on petroleum [2–9]. Moreover, hydrogen can help enable the use of vast intermittent renewable sources such as wind and solar for applications now served by liquid fuels and as a means of electricity storage.

To reach stringent long term goals for reducing greenhouse gas emissions from the transport sector, it appears likely that the light-duty fleet will be substantially electrified by 2050 [10]. Hydrogen fuel cells are an important enabling technology for this vision. Automakers are actively developing both batteries and fuel cells and foresee a future electrified light-duty fleet with batteries powering smaller, shorter range cars and hydrogen fuel cells powering larger vehicles with longer range. Hydrogen is also widely seen as a possible electricity storage medium for intermittent renewable energy sources such as wind and solar. Hydrogen and electricity could be complementary energy carriers in a renewable-intensive future [11, 12]. Hydrogen is a key enabling technology for using intermittent renewable sources in applications such as transport where high energy density and fast refueling are desirable.

Transition to Renewable Energy Systems, 1st Edition. Edited by Detlef Stolten and Viktor Scherer.
© 2013 Wiley-VCH Verlag GmbH & Co. KGaA. Published 2013 by Wiley-VCH Verlag GmbH & Co. KGaA.

However, hydrogen faces significant technical, economic, infrastructure, and societal challenges before it could be implemented as a transportation fuel on a large scale or used for the storage of intermittent renewable sources. In particular, the need to develop a hydrogen infrastructure is a key issue. This chapter examines the current and projected future status of hydrogen infrastructure technologies, including hydrogen production and delivery systems, and infrastructure design issues associated with a transition towards large-scale use of hydrogen. We compare hydrogen with other alternative transportation fuels in terms of technical readiness, timing, cost, and reductions in oil use and greenhouse gas (GHG) emissions. We also present various strategies for building up infrastructure over time.

38.2
Technical Options for Hydrogen Production, Delivery, and Use in Vehicles

38.2.1
Hydrogen Vehicles

Although internal combustion engines can run on hydrogen, it is the higher efficiency, zero-emission hydrogen fuel cell that has largely captured the attention of automakers. Several automakers have embraced fuel cells as a promising zero-emission technology and have large development and commercialization programs. Honda, Toyota, Daimler, General Motors, and Hyundai have announced plans to commercialize FCVs sometime between 2015 and 2020 [13]. Hydrogen and fuel cells represent a logical progression beyond efficiency and increasing electrification of cars with hybrid and electric drive trains. Many automakers see complementary roles for hydrogen fuel cells and battery electric vehicles and are pursuing both technologies.

As indicated in Table 38.1, hydrogen FCVs have already met 2015 goals for fuel economy and range. However, further development is needed for key hydrogen vehicle issues such as the proton exchange membrane (PEM) fuel-cell cost and durability, hydrogen storage on vehicles, and technologies for zero-carbon hydrogen production, as discussed in the next section.

Estimates of the price of mass-produced FCVs based upon projections for 2015 technology are within a few thousand dollars of conventional vehicles [3, 6]. According to a study by the National Research Council (NRC), mass-produced, mature technology mid-sized FCV passenger cars are estimated to have a retail price $ 3600–6000 higher than that of a comparable gasoline internal combustion engine vehicle (ICEV) [3]. However initial FCV models will not be produced in such high volumes and as a result will have a higher price premium. At a scale of 50 000 FCVs being produced worldwide, the NRC model gave estimated prices of around $ 75 000 per vehicle. Prices can drop quickly as manufacturing volume increases. In 2012, Daimler announced its intention to produce thousands of hydrogen FCVs in 2015 at a selling price of $ 50 000.

38.2 Technical Options for Hydrogen Production, Delivery, and Use in Vehicles

Table 38.1 Current status and 2015 goals for hydrogen fuel-cell vehicles [14].

Parameter	Today	2015 goals
Fuel cell in-use durability (h)	2500 (4000 in laboratory)	5000
Vehicle range (miles per tank)	280–400	300
Fuel economy (miles kg^{-1} H$_2$)	72	60
Fuel cell efficiency (%)	53–58	60
Fuel cell system cost ($ kW^{-1})	49	30
H2 storage cost ($ kWh^{-1})	15–23	10–15 (NRC) 2–4 (USDOE)

38.2.2
Hydrogen Production Methods

Like electricity, hydrogen is an energy carrier that can be produced from diverse primary energy resources (Figure 38.1). A variety of hydrogen production processes are commercially available today, such as thermochemical methods, which are used to generate hydrogen from hydrocarbons, and electrolysis of water, where electricity is used to split water into its constituent elements hydrogen and oxygen. Potential future hydrogen production methods such as thermochemical water splitting processes involving a series of coupled chemical reactions at high temperature, direct conversion of sunlight to hydrogen in electrochemical cells, and biologically hydrogen production are being researched at a fundamental science level. In this section, we describe methods of hydrogen production, the current status and projections for technical progress, and economics.

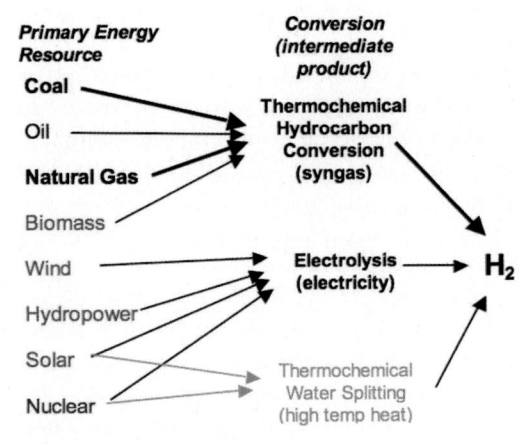

Figure 38.1 Hydrogen production methods.

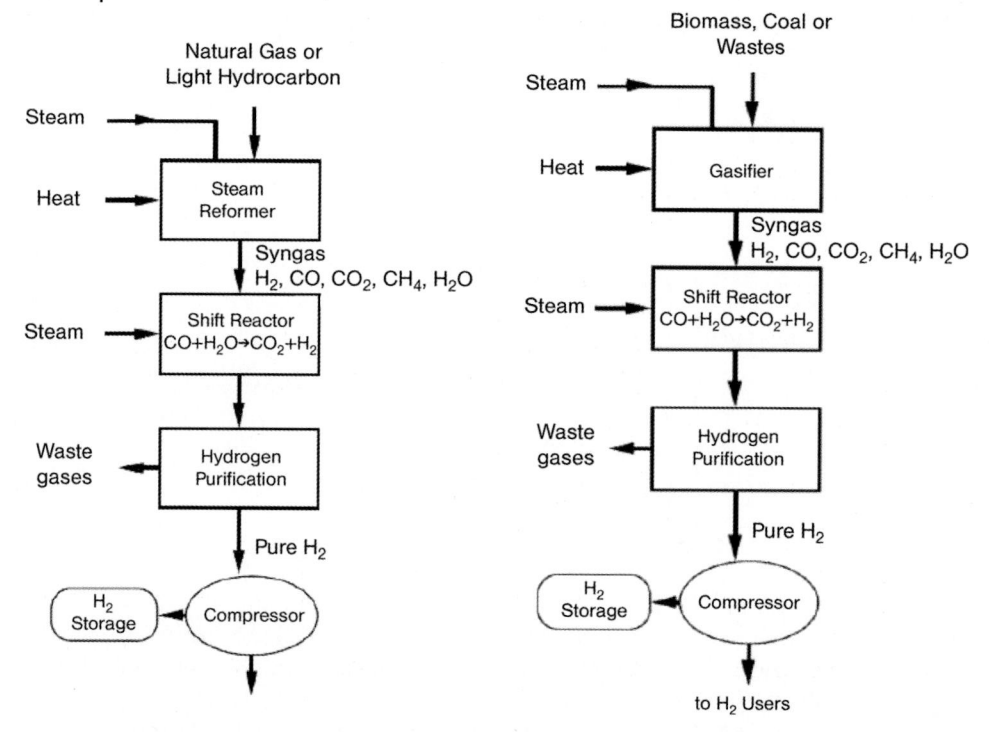

Figure 38.2 Examples of thermochemical hydrogen production methods [15].

Hydrogen is made *thermochemically* by processing hydrocarbons (such as natural gas, coal, biomass, or wastes) in high-temperature chemical reactors to make a synthetic gas or "syngas," comprised of H_2, CO, CO_2, H_2O, and CH_4. The syngas is further processed to increase the hydrogen content via the water gas shift reaction, and hydrogen is separated out of the mixture at the desired purity. Figure 38.2 shows process steps for typical thermochemical hydrogen production plants based on steam methane reforming and gasification of coal or biomass. Energy conversion efficiencies of 75–80% are typical for large steam methane reformers (on a higher heating value basis). Coal or biomass gasification systems have energy conversion efficiencies of about 60–65%. During thermochemical hydrogen production, it is possible to capture much of the carbon in the fuel as a stream of CO_2, which can then be compressed, transported by pipeline, and injected into secure underground geological formations for permanent storage. Carbon capture and sequestration (CCS) allows the production of hydrogen from fossil fuels with near-zero emissions of CO_2 to the atmosphere. Net negative emissions of CO_2 may be possible with biomass hydrogen production plus CCS, since growing biomass removes atmospheric carbon, which is then captured and sequestered during hydrogen production.

In water electrolysis, electricity is passed through a conducting aqueous electrolyte, breaking down water into its constituent elements hydrogen and oxygen (Figure 38.3) via the reaction

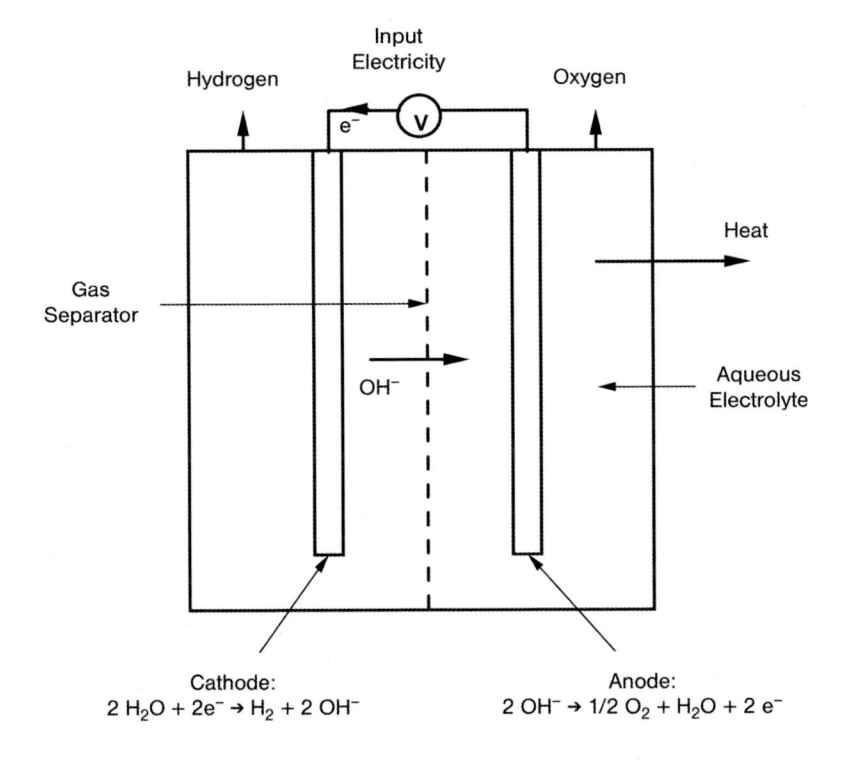

38.2 Technical Options for Hydrogen Production, Delivery, and Use in Vehicles | 799

Figure 38.3 Electrolytic hydrogen production [15].

Diagram labels:
- Input Electricity
- Hydrogen
- Oxygen
- Gas Separator
- Heat
- Aqueous Electrolyte
- OH$^-$
- Cathode: $2\,H_2O + 2e^- \rightarrow H_2 + 2\,OH^-$
- Anode: $2\,OH^- \rightarrow 1/2\,O_2 + H_2O + 2\,e^-$
- Overall Reaction: $H_2O \rightarrow H_2 + 1/2\,O_2$

$$2\,H_2O \rightarrow 2\,H_2 + O_2 \tag{38.1}$$

Any source of electricity can be used, including intermittent (time varying) sources such as off-peak power, solar, or wind. Various types of electrolyzers are in use. Commercially available systems today are based on alkaline or PEM technologies. PEM electrolyzers also have advantages of quick start-up and shut-down and the ability to handle transients well. Electrolyzers have electricity to hydrogen conversion efficiencies of 70–85% (on a higher heating value basis). Electrolyzers are modular in design and production plants range in size from a few kilowatts to many megawatts. Electrolyzers have been developed using solid oxide electrolytes and operating at temperatures of 700–900 °C. High-temperature electrolysis systems offer higher efficiency of converting electricity to hydrogen, as some of the work required to split water is done by heat, but materials requirements are more stringent and capital costs are generally higher.

38.2.3
Options for Producing Hydrogen with Near-Zero Emission

To realize the full environmental benefits of hydrogen, it should be produced via pathways with zero or near-zero net emissions of carbon. There are several zero emission options, all of which face technical issues.

For H_2 from renewable sources (wind or solar electrolysis and biomass gasification), the issue is primarily cost rather than technical feasibility. There are ample wind and solar resources. Hydrogen via biomass gasification is a promising and relatively low-cost option for 2020 and beyond.

For nuclear H_2, the issues are cost (for electrolytic H_2), technical feasibility (for water splitting systems powered by high temperature nuclear heat). Nuclear hydrogen faces the same waste and proliferation issues as nuclear power.

Fossil H_2 production with CO_2 capture and sequestration offers nearly zero emissions and relatively low cost, assuming that suitable CO_2 disposal sites are available nearby and hydrogen is produced at large scale. Biomass hydrogen with CCS could have net negative emissions. Much remains unknown about the potential environmental impacts and feasibility of CO_2 sequestration.

38.2.4
Hydrogen Delivery Options

Once hydrogen has been produced, there are several ways to deliver it to vehicles (Figure 38.4). It can be produced regionally in large plants, stored as a compressed gas at 70–700 bar or as a cryogenic liquid (at −253 °C), and distributed by truck or gas pipeline. Alternatively, hydrogen can be produced on-site at refueling stations (or even homes) from natural gas, alcohol (methanol or ethanol), or electricity.

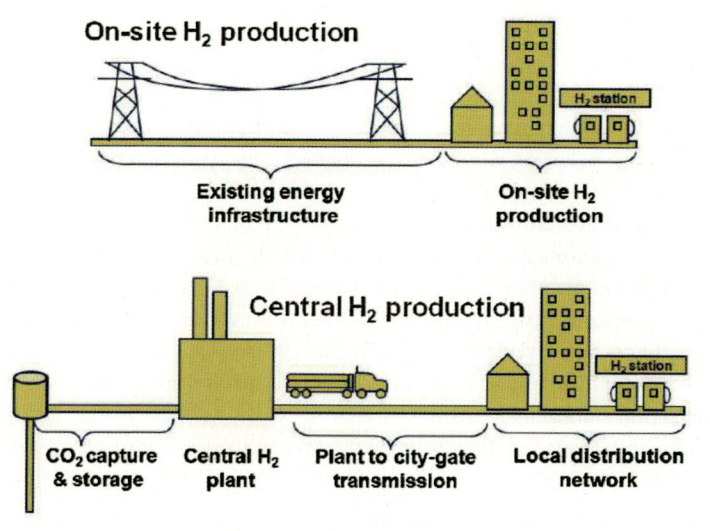

Figure 38.4 Options for hydrogen delivery [16].

Hydrogen delivery technologies are well established in the merchant hydrogen and chemical industries today. While most industrial hydrogen is produced and used on-site, a significant fraction is delivered by pipeline or truck to more distant users. No one hydrogen supply pathway is preferred in all situations, so, like electricity, it is likely that diverse primary sources will be used to make hydrogen in different regions.

38.2.5
Hydrogen Refueling Stations

Hydrogen can be dispensed to vehicles in refueling stations as a compressed gas at 350–700 bar. The configuration of the hydrogen station depends on the details of the supply pathway and the demand served [17, 18]. Hydrogen refueling systems can be co-located with conventional fuels at existing gasoline stations or developed as stand-alone hydrogen stations.

Another refueling option is a so-called tri-generation (Figure 38.5) system that reforms natural gas to hydrogen to produce three energy co-products, heat and power from a fuel cell for a building and hydrogen fuel for vehicles. Tri-generation systems can be sited at homes, neighborhood, or commercial buildings [19, 20]. The economics of hydrogen refueling are improved by credits for coproducing electricity and heat. Home and neighborhood refueling both potentially offer convenience along with early availability of hydrogen fuel with less investment than a dedicated hydrogen station network. Based on near-term projections for system cost and performance, modeling by researchers at the University of California, Davis (UC Davis) shows that tri-generation systems in residential, neighborhood, and commercial building applications can become economically competitive, especially in regions with low natural gas prices and high electricity prices.

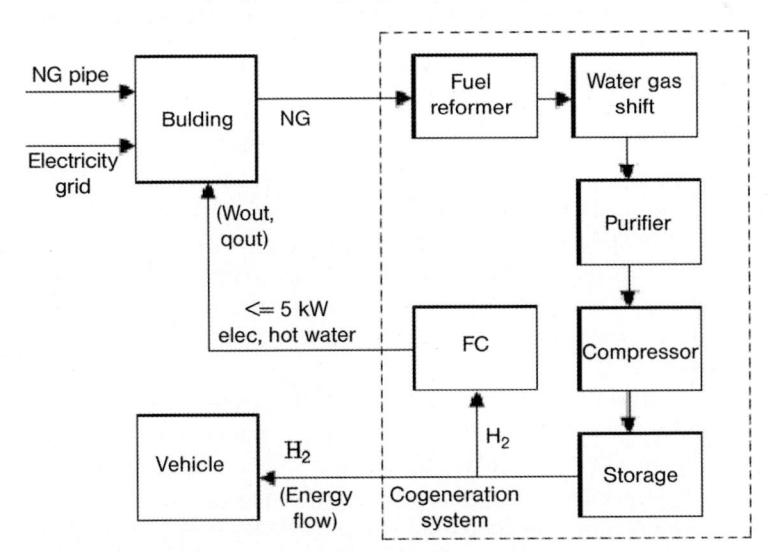

Figure 38.5 A typical tri-generation system simultaneously provides electricity and heat for a building along with hydrogen for a vehicle [20].

38.3
Economic and Environmental Characteristics of Hydrogen Supply Pathways

Clearly, there are many different options for producing and supplying hydrogen to users (Table 38.2). Each has differing costs, performance, and environmental characteristics with respect to emissions, primary energy use, land, water, and materials use. In this section, we compare a range of near- and long-term hydrogen pathways.

Table 38.2 Hydrogen supply pathways considered in this analysis.

Resource	H_2 production technology	H_2 delivery method to station (for central plants)
Central production		
Natural gas	Steam methane reforming	Liquid H_2 truck
Coal	Coal gasification with CCS	Compressed gas truck
		H_2 gas pipeline
Biomass (agricultural, forest and urban wastes)	Biomass gasification	
On-site production (at refueling station)		
Natural gas	Steam methane reforming	n/a
Electricity (from various electricity generation resources)	Water electrolysis	

38.3.1
Economics of Hydrogen Supply

The projected capital cost of centralized hydrogen production plants is shown in Table 38.3 for a variety of technologies and plant sizes. For reference, a hydrogen production capacity of about 1 kg d^{-1} is needed to support a single mid-sized hydrogen fuel-cell car. A plant producing 50 t of hydrogen per day could support a fleet of perhaps 50 000 such cars. In addition, hydrogen storage, delivery, and refueling contribute to capital costs for the system. Storage capacity and cost depend on the time variation of the hydrogen demand and how much hydrogen must be stored to meet these variations.

The best choice for delivering hydrogen from the plant to users depends on the size and type of demand and the distances involved to reach consumers [16]. For short distances (< 50 km) and small amounts (< 500 kg d^{-1}), compressed gas trucks are preferred. For medium amounts of hydrogen (thousands of kilograms per day) and long distances (> 50 km), liquid hydrogen (LH$_2$) truck delivery is preferred. There are significant scale economies associated with liquefaction, and large electrical energy input, equal to 33% of the energy value of the hydrogen. The delivery cost

Table 38.3 Estimated capital costs for future hydrogen production systems [3].

	Plant size: t H_2 d^{-1} (MW LHV)	2015 technologies: capital cost ($million MW^{-1} H_2)
Central H_2 production plants		
Central natural gas steam methane reformer	50 (70)	0.45
	300 (417)	0.29
	400 (556)	0.27
Central natural gas steam methane reformer with CCS (capital cost assumed to be 50% higher with CCS) [21]	50 (70)	0.67
	300 (417)	0.43
	400 (556)	0.40
Central coal	250 (348)	0.92
	400 (556)	0.84
	1200 (1669)	0.68
Central coal with CCS (capital cost assumed to be 10% higher with CCS) [21]	250 (348)	1.01
	400 (556)	0.92
	1200 (1669)	0.75
Central biomass gasifier	30 (42)	0.91
	155 (216)	0.62
	200 (278)	0.58
On-site production at refueling station		
Electrolysis on-site (station)	0.1 (0.14)	3.1
	0.1 (0.7)	1.51
	1.1 (2.1)	1.2
Steam methane reformer on-site (station)	0.1 (0.14)	2.9
	0.1 (0.7)	1.3
	1.51 (2.1)	1.0

for LH_2 is sensitive to the cost of electricity. For very large amounts of hydrogen (tens of thousands of kilograms per day), pipeline transmission is preferred. The pipeline capital cost is the largest single factor. Pipeline costs scale strongly with both distance and flow rate.

The layout and cost of hydrogen distribution within a city depend on the city population, the city radius (or equivalently the population density), the market fraction of hydrogen vehicles, and the station size. Although truck delivery or on-site production are favored early on, pipeline distribution can yield the lowest delivery costs for dense cities with a large population, high penetration of hydrogen vehicles, and large refueling stations. Pipelines are expected to become the lowest cost delivery system in most cities, once market penetration of hydrogen vehicles exceeds 25–50%.

Studies by the NRC [3] suggest that the total capital investment for mature hydrogen infrastructure in the United States would be $ 1400–2000 per light-duty vehicle served, depending on the pathway. Early infrastructure investment costs per car (to serve the first million vehicles) would be higher ($ 5000–10 000 per car).

Figure 38.6 shows the levelized cost of hydrogen delivered to vehicles including production, delivery, and refueling, for the supply pathways listed in Table 38.2. We compare the delivered cost of hydrogen transportation fuel for "near term" (scaled-up infrastructure with current technology) and "future" (full-scale infrastructure with advanced technologies beyond 2015). Costs are projected to decrease as technology advances. We also find that hydrogen from hydrocarbons generally costs less than electrolytic hydrogen production. All central alternatives assume that hydrogen is deployed at large scale. On-site alternatives use stations serving numbers of cars similar to today's gasoline stations (at 1500 kg of H_2 per day). We also show estimated H_2 costs for smaller size stations (100 kg of H_2 per day), typical of near-term demonstration H_2 stations which serve a relatively small number of early FCVs. These small stations would have significantly higher hydrogen cost because of scale economies. The range for hydrogen fuel costs to compete with gasoline on a cents-per-mile basis is shown, based on an efficient gasoline hybrid competing with an FCV. If hydrogen costs $ 3–6 kg^{-1}, the fuel cost per mile for an FCV is about the same as for an efficient gasoline hybrid using gasoline at $ 2–4 gal^{-1}, assuming that the fuel economy of a fuel cell vehicle is 1.5 times higher than that of a comparable gasoline hybrid.

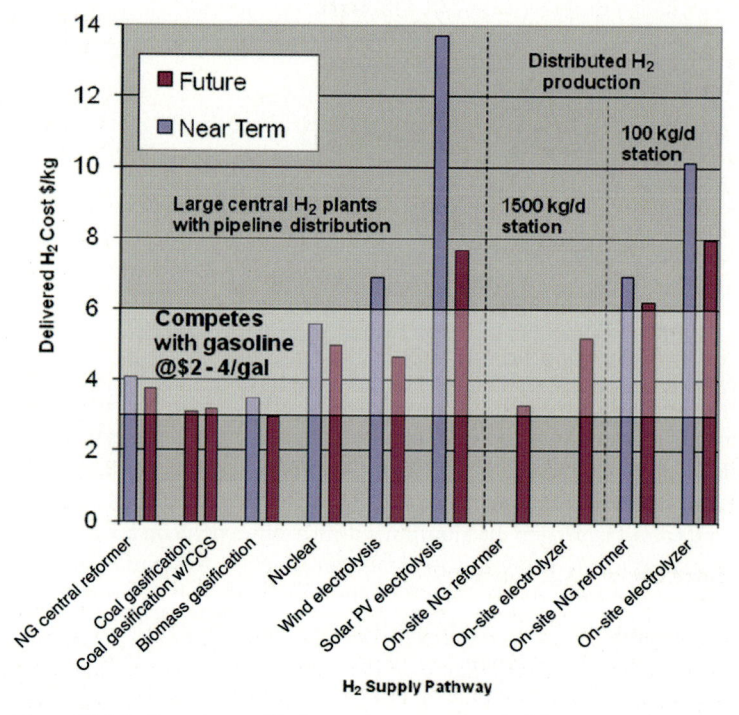

Figure 38.6 Delivered cost of hydrogen transportation fuel [14].

38.3.2
Environmental Impacts of Hydrogen Pathways

38.3.2.1 Well-to-Wheels Greenhouse Gas Emissions, Air Pollution, and Energy Use

Most hydrogen production today is from fossil fuels, which releases CO_2, the major GHG linked to climate change. For the near term, FCVs using hydrogen produced from natural gas would reduce well-to-wheels (WTW) GHG emissions by about half compared with current gasoline vehicles. Production of hydrogen from renewable biomass is a promising mid-term option with very low net carbon emissions. For large central plants producing hydrogen from hydrocarbons (natural gas, coal or biomass), it is technically feasible to capture the CO_2 and permanently sequester it in deep geological formations, although sequestration technology will not be in widespread use before 2020 at the earliest. In the longer term, carbon-free renewables such as wind and solar energy might be harnessed for hydrogen production via electrolysis of water.

In Figure 38.7, we compare the WTW emissions of GHGs for a variety of alternative fuels including fuel cells, based on analysis by the US Department of Energy.

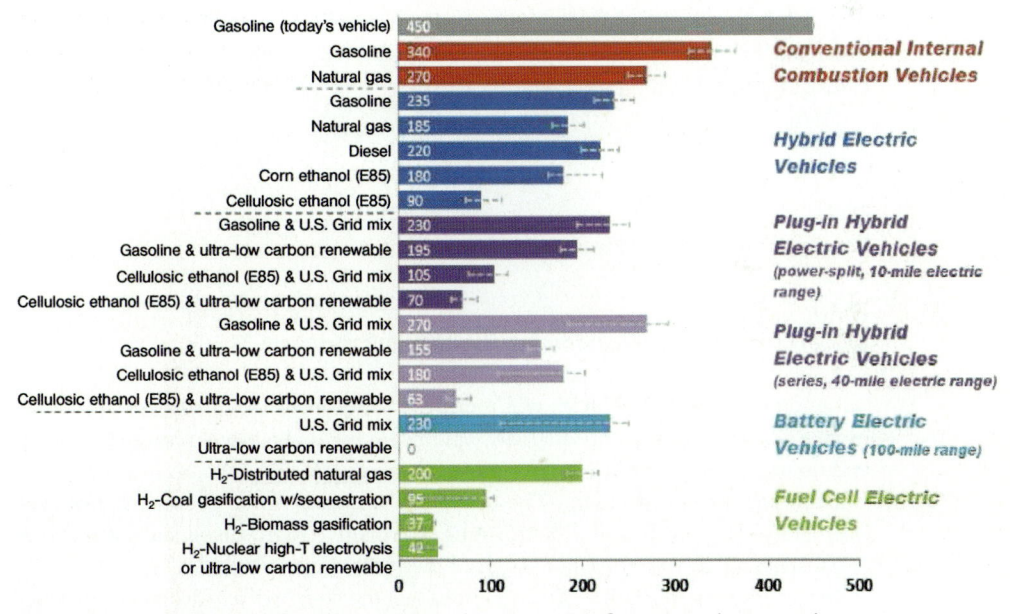

Figure 38.7 WTW emissions for a future mid-sized car, in grams of CO_2-equivalent per mile [22].

38.3.2.2 Resource Use and Sustainability

Hydrogen could access a wide primary resource base, including low-carbon options such as fossil fuels with CCS, renewables (solar, wind, biomass, hydro, geothermal), and nuclear. In theory, the availability of low-carbon resources should not be a limiting factor for hydrogen, although the higher cost of zero-carbon pathways could increase fuel costs.

With hydrogen FCVs, the amount of primary energy required is similar to that for gasoline hybrids and considerably less than that for conventional gasoline cars. There are plentiful near-zero carbon resources for hydrogen production in the United States. For example, a mix of low-carbon resources including natural gas, coal (with carbon sequestration), biomass, and wind power could supply ample hydrogen for vehicles. With 20% of the biomass resource, plus 15% of the wind resource, plus 25% added use of coal (with sequestration), 300 million hydrogen vehicles (approximately the entire US fleet projected in 2030) could be served with near-zero WTW GHG emissions.

38.3.2.3 Infrastructure Compatibility

There appears to be relatively little opportunity to use hydrogen directly in existing energy systems, and a new dedicated infrastructure would be needed if hydrogen becomes a major energy carrier. It has been suggested that hydrogen could be blended at up to 15% by volume with natural gas without infrastructure changes, but there would be only a modest environmental benefit to this approach. Like electricity, hydrogen would rely on other underlying infrastructures that deliver feedstocks to production plants. Expanding use of hydrogen could also require an expansion of the underlying feedstock infrastructure (for example, to make large quantities of hydrogen from coal would require extra rail and barge capacity to deliver coal to hydrogen production plants.)

38.4
Strategies for Building a Hydrogen Infrastructure

38.4.1
Design Considerations for Hydrogen Refueling Infrastructure

Adoption of hydrogen vehicles will require a widespread new hydrogen refueling infrastructure. Because there are many options for hydrogen production and delivery, and no one supply option is preferred in all cases, creating such an infrastructure is a complex design problem. The challenge is not so much producing low-cost hydrogen at large scale as it is distributing hydrogen to many dispersed users at low cost, especially during the early stages of the transition.

Recent studies [14, 16, 23–26] have found that the design of a hydrogen infrastructure depends on many factors, including the following:

- *Scale*. Hydrogen production, storage, and delivery systems exhibit economies of scale, and costs generally decrease as demand grows.
- *Geography/regional factors*. The location, size, and density of demand, the location and size of resources for hydrogen production, the availability of sequestration sites, and the layout of existing infrastructure can all influence hydrogen infrastructure design.

- *Feedstocks.* The price and availability of feedstocks for hydrogen production, and energy prices for competing technologies (e.g., gasoline prices), must be taken into account.
- *Technology status.* Assumptions about hydrogen technology cost and performance determine the best supply option.
- *Supply and demand.* The characteristics of the hydrogen demand and how well it matches supply must be considered. Time variations in demand (refueling tends to happen during the daytime, with peaks in the morning and early evening) and in the availability of supply (e.g., wind power is intermittent) can help determine the best supply and how much hydrogen storage is needed in the system.
- *Policy.* Requirements for low-carbon or renewable hydrogen influence which hydrogen pathways are used.
- *Transition issues/coordination of stakeholders.* A hydrogen transition means many major changes at once: adoption of new types of cars, building a new fuel infrastructure, and development of new low-carbon primary energy resources. These changes will require coordination among diverse stakeholders with differing motivations (fuel suppliers, vehicle manufacturers, and policy makers), especially in the early stages when costs for vehicles are high and infrastructure is sparse. Factors that could ease transitions, such as compatibility with the existing fuel infrastructure, are more problematic for hydrogen than for electricity or liquid synthetic fuels.

38.4.2
Hydrogen Transition Scenario for the United States

In this section, we discuss a transition strategy for introducing hydrogen for the United States, considering early market issues for launching initial hydrogen infrastructure in "lighthouse" regions such as southern California and developing of a mature large national infrastructure [28].

In the early stages of infrastructure development, hydrogen might rely on truck delivery of small quantities of "merchant" hydrogen produced from natural gas, moving towards on-site hydrogen production at stations and eventually towards centralized production of low-carbon hydrogen with pipeline delivery.

The early stages of hydrogen infrastructure development pose special problems. Consumers will not buy the first hydrogen cars unless they can refuel them conveniently and travel to key destinations, and fuel providers will not build an early network of stations unless there are cars to use them. Major questions include how many stations to build, what type of stations to build, and where to locate them. Key concerns are cost, fuel accessibility, customer convenience, the quality of the refueling experience, network reliability, and technology choice.

Stakeholder coordination is key for launching early hydrogen infrastructure. Automakers seek a convenient, reliable refueling network, recognizing that a positive customer experience is largely dependent on making hydrogen refueling just as convenient as refueling gasoline vehicles. Energy suppliers are concerned about the cost of building the first stages of hydrogen infrastructure when stations are small and under-utilized. Installing a large number of stations for a small number of vehicles might solve the problem of convenience but would be prohibitively expensive.

A series of studies by UC Davis researchers [17, 18, 28, 29] analyzed how many stations would be needed for consumer convenience (defined as travel time to the station), and used spatial analysis tools to estimate where stations would be located. Based on studies of four urban areas in California, Nicholas et al. found that a strategically sited hydrogen network could provide an acceptable level of convenience if only 10–30% of gas stations offered hydrogen [28]. Refining this analysis, Ogden and Nicholas analyzed a "cluster strategy" for co-locating the first thousands of vehicles and tens of stations in 4–12 communities within the southern California region [18]. They found that this strategy gave acceptable travel times of less than 4 min from home to station even with a very sparse initial regional network of 20–40 stations (< 1% of gasoline stations). We find that hydrogen costs drop over a period of about 5–10 years to \$ 5–8 kg^{-1}, competitive with gasoline on a cent per kilometer basis, considering the higher fuel economy of hydrogen FCVs compared with gasoline cars.

Moving beyond early markets, building a national hydrogen refueling infrastructure in a large, diverse country such as the United States is a complex design problem involving regional considerations. Researchers at UC Davis developed models to determine the least-cost method for supplying hydrogen to a particular city at a given market penetration [3, 16]. In our scenario, we assume that the first few thousand FCVs are successfully introduced in 2012, with tens of thousands of FCVs by 2015, 2 million by 2020, 10 million by 2025, and about 200 million (60% of the fleet) by 2050. Because of the need to locate infrastructure and vehicles together, hydrogen is introduced in a succession of "lighthouse" cities, starting with the Los Angeles area.

Looking to the long term in the United States, the lowest cost low-carbon hydrogen supply pathways appear to be biomass gasification and hydrogen from coal with CCS. If low-cost shale gas is available, this could lead to a larger role for central steam methane reforming with CCS. Each could contribute significantly to the long-term hydrogen supply. The lowest cost option depends on the market penetration of FCVs, the local feedstock and energy prices, ands geographic factors such as city size and density of demand. Detailed regional studies reveal possibilities for further optimizing the hydrogen supply system at the regional level. It appears that hydrogen could be delivered to consumers for about \$ 3–4 kg^{-1}, with near-zero emissions of GHGs, on a WTW basis, which leads to a reduction in fuel cost per mile compared with gasoline vehicles, given the increased efficiency of FCVs. GHG emissions could be reduced further with renewable sources such as wind and solar, but the fuel cost would be higher.

Studies by the NRC [3] suggest that the capital investment for mature infrastructure would be \$ 1400–2000 per light-duty vehicle served, depending on the pathway. The NRC found that building a fully developed hydrogen infrastructure serving 220 million vehicles in the United States in 2050 would cost about \$ 400 billion over a period of about 40 years. (The NRC scenario is based mostly on fossil-fueled hydrogen with CCS and biomass hydrogen. Electrolysis-based pathways could cost more to build.) Early infrastructure investment costs per car (to serve the first million vehicles) would be higher (\$ 5000–10 000 per car).

38.5 Conclusion

Hydrogen FCVs are making rapid progress; it appears likely that they will meet their technical and cost goals and could be commercially ready by 2015. Hydrogen infrastructure technologies are also progressing, and the technology to produce natural gas-based hydrogen is commercial today. In the near term (up to 2025), hydrogen fuel will likely be produced from natural gas, via distributed production at refueling stations or, where available, excess industrial or refinery hydrogen. Beyond 2025, central production plants with pipeline delivery will become economically viable in urban areas and regionally, and low-carbon hydrogen sources such as renewables and fossil with CCS, will be phased in.

The environmental impacts of hydrogen fuel vary with the production pathway. For the near term, FCVs using hydrogen made from natural gas would reduce WTW GHG emissions by about half compared with current gasoline vehicles. Future hydrogen production technologies could virtually eliminate GHG emissions. On the other hand, important constraints on the use of land, water, and materials required by the hydrogen pathway are not well understood. This is a key area for future research.

Building a hydrogen infrastructure will be a decades-long process in concert with growing vehicle markets. We have modeled infrastructure deployment in individual "lighthouse" cities as well as at the regional level. Since it is likely that hydrogen will be produced from a variety of feedstocks, optimal supply strategies will differ between geographic regions.

When FCVs are mass marketed and sold to consumers in 2015 or soon after, hydrogen must make a major leap to a commercial fuel available initially at a small network of refueling stations and must be offered at a competitive price. The first steps are providing hydrogen to test fleets and demonstrating refueling technologies in mini-networks Several such projects are now under way in Europe, Asia, and North America. Learning from these programs will include the development of safety codes and standards. If strategically placed, these early sparse networks could provide good fuel accessibility for early users, while forming a seedbed for a large-scale hydrogen infrastructure rollout after 2015.

Getting through the transition to hydrogen will involve costs and some technological and investment risks. Concentrating hydrogen projects in key regions such as southern California will focus efforts, lower investment costs to make refueling available to consumers, and hasten infrastructure cost reductions through faster market growth and economies of scale.

Even under optimistic assumptions, it will be several decades before hydrogen FCV technologies can significantly reduce emissions and oil use globally, because of the time needed for new vehicle technology to gain major fleet share. Beyond this, hydrogen can yield significant benefits, greater than those possible with efficiency alone. Hydrogen should be seen as an important enabling technology for electrifying the light-duty vehicle fleet. This underscores the importance of providing consistent support for hydrogen and fuel-cell vehicle technologies as they approach commercial introduction, so they can progress more quickly to scale, yielding competitive costs and greater societal benefits.

Acknowledgments

The author would like to acknowledge her colleagues at UC Davis Dr Christopher Yang, Dr Michael Nicholas, Dr Nils Johnson, Dr Nathan Parker, Dr Mark Delucchi, Dr Yongling Sun, Dr Xuping Li, Prof. Yueyue Fan, and Prof. Daniel Sperling and the sponsors of the Sustainable Transportation Energy Pathways research program at UC Davis for research support.

References

1 Wipke, K., Anton, D., and Sprik, S. (2009). *Evaluation of Range Estimates for Toyota FCHV-adv Under Open Road Driving Conditions*, Report SRNS-STI-2009-00446, Savannah River National Laboratory, Aiken, SC.

2 National Research Council, National Academy of Engineering, Committee on Alternatives and Strategies for Future Hydrogen Production and Use (2004) *The Hydrogen Economy: Opportunities, Costs, Barriers, and R&D Needs*, National Academies Press, Washington, DC.

3 National Research Council, Committee on Assessment of Resource Needs for Fuel Cell and Hydrogen Technologies (2008) *Transitions to Alternative Transportation Technologies: a Focus on Hydrogen*, National Academies Press, Washington, DC.

4 Gielen D. and Simbolotti, G. (2005) *Prospects for Hydrogen and Fuel Cells*, International Energy Agency, Paris.

5 Ball, M. and M. Wietschel, M. (2009) *The Hydrogen Economy: Opportunities and Challenges*, Cambridge University Press, Cambridge.

6 Kromer, M. A. and Heywood, J. B. (2007) *Electric Powertrains: Opportunities and Challenges in the U. S. Light-Duty Vehicle Fleet*, LEFF 2007-02 RP, Sloan Automotive Laboratory, Massachusetts Institute of Technology, Cambridge, MA.

7 Bandivadekar, A., Bodek, K., Cheah, I., Evans, C., Groode, T., Heywood, J., Kasseris, E., Kromer, M., and Weiss, M. (2008) *On the Road in 2035: Reducing Transportation's Petroleum Consumption and GHG Emissions*, MIT Laboratory for Energy and the Environment, Massachusetts Institute of Technology, Cambridge, MA.

8 Plotkin, S. and Singh, M. (2009) *Multi-Path Transportation Futures Study: Vehicle Characterization and Scenario Analyses (Draft)*, Argonne National Laboratory, Argonne, IL.

9 EUCAR (European Council for Automotive Research and Development), CONCAWE, and ECJRC (European Commission Joint Research Centre) (2007) *Well-to-Wheels Analysis of Future Automotive Fuels and Powertrains in the European Context*, Well-to-Wheels Report, Version 2c, European Commission Joint Research Centre, Brussels.

10 IEA (2009). *Transport, energy, and CO_2: Moving toward sustainability*. Paris: International Energy Agency, IEA/OECD. Retrieved from www.iea.org.

11 Jacobson, M. Z. and Delucchi, M. A. (2011) Providing all global energy with wind, water, and solar power. Part I: technologies, energy resources, quantities and areas of infrastructure, and materials. *Energy Policy*, **39**, 1154–1169.

12 Yang, C. (2008) Hydrogen and electricity: parallels, interactions, and convergence. *Int. J. Hydrogen Energy*, **33** (8), 1977–1994.

13 For statements from automakers about their HFCV commercialization timelines, see http://www.fuelcells.org/automaker_quotes.pdf (last accessed 10 January 2013); see also the 2009 USCAR (United States Council for Automotive Research) white paper *Hydrogen Research for Transportation: the USCAR Perspective*, http://www.uscar.org/guest/article_view.php?articles_id=312 (last accessed 10 January 2013).

14 Ogden, J. M. and Anderson, L. (2011) *Sustainable Transportation Energy Pathways*, Institute of Transportation Studies. University of California, Davis, Regents of the University of California, Davis Campus. Available under a Creative Commons BY-NC-ND, 3.0 license.

15 Ogden, J. (1999) Prospects for Building a hydrogen energy infrastructure. *Annu. Rev. Energy Environ.*, **24**, 227–791.

16 Yang C. and Ogden, J. (2007) Determining the lowest-cost hydrogen delivery mode. *Int. J. Hydrogen Energy*, **32**, 268–286.

17 Nicholas, M. A. and Ogden, J. M. (2007) Detailed analysis of urban station siting for california hydrogen highway network. *Transport. Res. Rec.*, **1983**, 121–128.

18 Ogden, J. and Nicholas, M. (2011) Analysis of a cluster strategy for introducing hydrogen vehicles in southern California. *Energy Policy*, **39**, 1923–1938.

19 Li, X. and Ogden, J. M. (2011) Understanding the design and economics of distributed tri-generation systems for home and neighborhood refueling – Part I: single family residence case studies. *J. Power Sources*, **196**, 2098–2108.

20 Li, X. and Ogden, J. M. (2012) Understanding the design and economics of distributed tri-generation systems for home and neighborhood refueling – Part II: neighborhood system case studies. *J. Power Sources*, **197**, 186–195.

21 Intergovernmental Panel on Climate Change (2005) *Carbon Dioxide Capture and Storage*, Special Report prepared by Working Group III of the IPCC, Cambridge University Press, Cambridge.

22 United States Department of Energy (2012) *Well-to-Wheels Greenhouse Gas Emissions and Petroleum Use for Mid-Size Light-Duty Vehicles*, http://www.hydrogen.energy.gov/pdfs/10001_well_to_wheels_gge_petroleum_use.pdf (last accessed 10 January 2013).

23 Mintz, M. (2008) *Hydrogen Delivery Infrastructure Analysis*, US Department of Energy Hydrogen Program FY 2008 Annual Progress Report, US Department of Energy, Washington, DC, pp. 368–371.

24 Schindler, J. (2005) E3 Database – a tool for the evaluation of hydrogen chains, presented at the IEA Workshop.

25 Greene, D. L., Leiby, P. N., James, B., Perez, J., Melendez, M., Milbrandt, A., Unnasch, S., and Hooks, M. (2008) *Analysis of the Transition to Hydrogen Fuel Cell Vehicles and the Potential Hydrogen Energy Infrastructure Requirements*, ORNL/TM-2008/30, Oak Ridge National Laboratory, Oak Ridge, TN.

26 California Fuel Cell Partnership (2010) *Hydrogen Fuel Cell Vehicle and Station Deployment Plan, Action Plan. Hydrogen Fuel Cell Vehicle and Station Deployment Plan, Progress and Next Steps*, California Fuel Cell Partnership, West Sacramento, CA.

27 Ogden, J. and Yang, C. (2009) Build-up of a hydrogen infrastructure in the U. S., in *The Hydrogen Economy: Opportunities and Challenges* (eds. M. Ball and M. Wietschel), Cambridge University Press, Cambridge, pp. 454–482.

28 Nicholas, M., Handy, S., and Sperling, D. (2004) Using geographic information systems to evaluate siting and networks of hydrogen stations. *Transport. Res. Rec.*, **1880**, 126–134.

29 Nicholas, M. (2009) The *Importance of Interregional Refueling Availability to the Purchase Decision*, UCD-ITS-WP-09-01, Institute of Transportation Studies, University of California, Davis, Davis, CA.

39
Power to Gas

Sebastian Schiebahn, Thomas Grube, Martin Robinius, Li Zhao, Alexander Otto, Bhunesh Kumar, Michael Weber, and Detlef Stolten

39.1
Introduction

Since renewable energies have been introduced in greater quantities, it is well known that there are very strong fluctuations of the energy input over time and that there are even periods stretching out over weeks where there is little or no renewable power input. This can be alleviated by a mixture of different renewable energies, but major fluctuations will remain. Smart grids represent another approach to alleviating this problem by introducing flexibility on the demand side, which generally is not the case yet. Nonetheless, there is strong evidence that these measures will not be sufficient to compensate for short-term fluctuations, or for the season or differences in the level of pod generation.

Hence energy storage is major issue being investigated these days. Gas storage offers the opportunity for storing great quantities in geological formations. Additionally, it provides the advantage of high energy density of chemical storage in general, compared with mechanical storage. Thus hydrogen and natural gas need just 1% of the storage volume of compressed air for the same energy.

Compared with batteries, there is the disadvantage of a lower turnaround efficiency with hydrogen production through water electrolysis as the first step and subsequent reconversion to electric power or the use of the hydrogen as fuel. On the other hand, gas storage provides capacities that are out of reach of battery storage. Furthermore, it is suitable for long-term storage for weeks and even for seasonal storage. If CO_2 is available, be it from carbon capture, biomass, or air, methanation can be considered owing to its better compatibility with the existing natural gas grid and end-use technologies (see Figure 39.1).

This chapter provides an overview of the technologies of and options for gas storage. In addition, some pivotal economic points are raised. Particularly the end use of the produced gas in transportation as a fuel and the feed-in to the natural gas grid are compared.

Transition to Renewable Energy Systems, 1st Edition. Edited by Detlef Stolten and Viktor Scherer.
© 2013 Wiley-VCH Verlag GmbH & Co. KGaA. Published 2013 by Wiley-VCH Verlag GmbH & Co. KGaA.

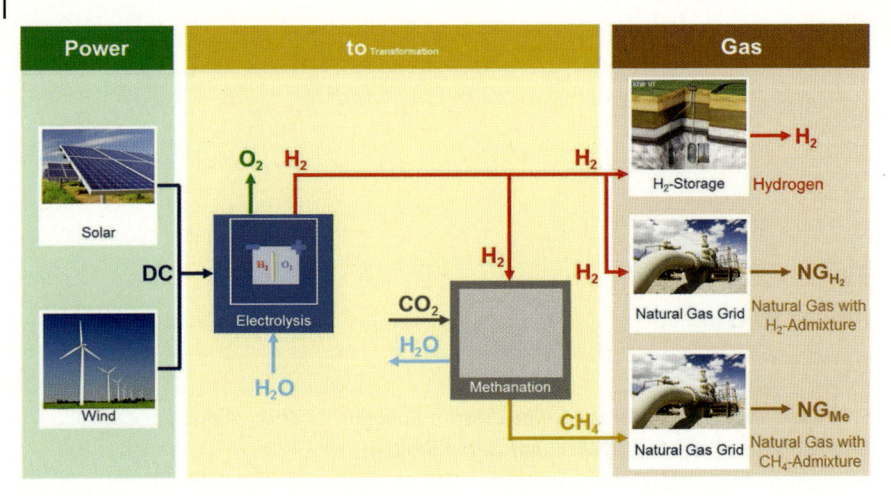

Figure 39.1 Principle of power to gas concept.

39.2
Electrolysis

In the 1800s, Nicholson and Carlisle were first to demonstrate the process of water electrolysis. Faraday clarified the principle in 1820 and introduced the word electrolysis in 1834. It involves two porous graphite electrodes and an electrolyte, and the system is the most important method for producing hydrogen from water by using electricity [1, 2]. The basic chemical reaction of water electrolysis is

$$H_2O_{(l)} + \text{electrical energy} \rightarrow H_{2(g)} + \tfrac{1}{2} O_{2(g)} \tag{39.1}$$

Regarding the current worldwide hydrogen production, 77% is produced from oil and gas, 18% comes from coal and the other 4% via electrolysis [1]. Over the last decade hydrogen production via electrolysis process has begun to be employed, and this approach has been attracting more attention in recent years. A more detailed description of this electrolysis approach is given in Chapter 20 by Mergel *et al*. [3].

The main types of electrolysis techniques are alkaline water electrolysis (AWE), proton exchange membrane electrolysis (PME) and high-temperature water electrolysis (HTE), which are described in the following. Subsequently, integration of electrolysis with renewable energies is discussed.

39.2.1
Alkaline Water Electrolysis

AWE is a mature and commercialized technology. Since the 1920s, numerous 100 MW plants have been developed worldwide. The design of the electrolysis cell consists of electrodes, namely the cathode (+) and anode (−), and the separator or diaphragm (Figure 39.2) [3, 4]. In the zero gap configurations, a highly insulating diaphragm is placed between the electrodes, acting as a barrier to keep the gases apart and to avoid short-circuiting. In the electrolysis module, cells are connected in series or in parallel, called bipolar and monopolar, respectively. In the monopolar configuration, the power supply is connected to the corresponding electrodes of each cell. This configuration is known as a conventional tank or monopolar electrolyzer. This type of configuration has advantages of reliability, flexibility, and simplicity. In the bipolar configuration, the same current flows through all cells while the voltages of each cell are summed. This configuration is denoted a filter press or bipolar electrolyzer. Bipolar-type electrolyzers have the advantages of lower ohmic losses and a compact design that permits a reduction of the space and also the electrical wire or mesh.

In AWE, water splits into hydrogen and oxygen at the cathode and anode, respectively. This is achieved by applying a direct current across the two electrodes, separated by the electrolyte. Owing to optimal conductivity, AWE generally uses aqueous potassium hydroxide (KOH) solution, usually at a concentration of 20–40 wt%. Water electrolyzers are operated at a temperature of about 80 °C with current densities in the range 0.2–0.4 A cm^{-2}. The voltage stack efficiency of AWE is 62–82% [3]. The following electrochemical reactions take place when an electric current is applied to the electrodes:

$$\text{Anode} \quad 2\,OH^-_{(aq)} \rightarrow \tfrac{1}{2}O_{2(g)} + H_2O_{(l)} + 2\,e^- \tag{39.2}$$

$$\text{Cathode} \quad 2\,H_2O_{(l)} + 2\,e^- \rightarrow H_{2(g)} + 2\,OH^-_{(aq)} \tag{39.3}$$

$$\text{Total} \quad H_2O_{(l)} \rightarrow H_{2(g)} + \tfrac{1}{2}O_{2(g)} \tag{39.4}$$

The main manufacturers and performance data are given in Table 39.1 and advantages and targeted improvements of alkaline water electrolyzers are summarized in Table 39.2.

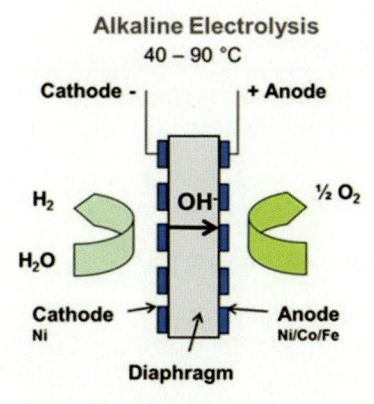

Figure 39.2 Schematic working principle of alkaline water electrolysis [3].

Table 39.1 Main manufacturers of and performance data for currently available alkaline water electrolyzers [3, 5].

Manufacturer	Technology	Rated production [m^3 (STP) h^{-1}]	Maximum pressure (bar)	Energy consumption [kWh m^{-3} (STP)]	Location
Hydrogenics	Bipolar	10–60	10–25	5.2–5.4 (system)	Canada
H2 Logic	Bipolar	0.66–1.33	1.0/2.0/atmospheric	5.4 (system)	Denmark
H2 Logic	Bipolar	32–64	4–12	4.9–5.0	Denmark
NEL Hydrogen	Bipolar	10–500	Atmospheric	4–4.35	Norway
NEL Hydrogen	Bipolar	60	15	4.9	Norway
Sagim SA	BP-MP 100/1000	0.1–1	10	5	France
Sagim SA	BP-MP 100/5000	1–5	10	5	France
Sagim SA	MP8	0.5	8	5	France
Teledyne Energy System	Bipolar	2.8–56	10	NA	USA
ELT	Bipolar	3–330	Atmospheric	4.3–4.6	Germany
ELT	Bipolar	100–760	30	4.3–4.65	Germany
IHT	Bipolar	380–760	32	4.3–4.65	Switzerland

Table 39.2 Advantages and targeted improvement of alkaline water electrolyzers [3, 5].

Advantage	Targeted improvement
Mature technology	Reduce the ohmic losses by minimization of space between electrodes
Long-term stability	Replace the previous asbestos/diaphragms with new advanced material
Module up to 760 m^3 (STP) h^{-1} (3.4 MW)	Working temperature up to 150 °C
No noble metal catalysts	New advanced electrocatalytic materials, to reduce the electrode overvoltage
Reduced operating costs, increased efficiency	Reduce system size and complexity

The cost of an alkaline electrolyzer on the megawatts scale has been estimated to be ~€1000 kW^{-1} [3]. Moreover, the investment costs of large units of alkaline water electrolyzers are approximately proportional to the electrolysis cell surface area [5]. The US Department of Energy (DOE) National Renewable Energy Laboratory (NREL) has performed several analyses on forecourt and electrolysis costs based on a thorough suppliers' questionnaire covering all cost contributors in recent years. The 2009 update gives the estimated investment cost of a central electrolysis plant producing 50 000 kg d^{-1} of hydrogen as US$ 50 million [6].

39.2.2
Proton Exchange Membrane Electrolysis

Since the 1950s, PEM technology (Table 39.3) has undergone continued development for space program applications. The first commercial water electrolyzer was developed by General Electric in 1966 based on the proton conducting method using a polymer membrane as the electrolyte [5, 7]. The membrane electrode assembly (MEA) is composed of anode, cathode, and membrane [3].

The electrocatalyst consists of noble metals such as platinum and iridium. For water electrolysis in the PEM fuel cell, Nafion, developed by DuPont, is used as the membrane material [5, 7]. The conventional liquid electrolyte is replaced with a solid polymer membrane in PEM electrolysis (Figure 39.3). Compared with PEM electrolysis, AWE is a more mature technology and is still considered as the first priority for medium- and large-scale units [5].

Water is oxidized electrochemically at the anode to form oxygen, hydrogen ions, and electrons (Eq. 39.5). The hydrogen ions then migrate through the membrane and recombine with electrons, via passing through an external circuit to form hydrogen gas at the cathode (Eq. 39.6). The total electrochemical reaction is represented by Eq. 39.7 [3].

Table 39.3 Main manufacturers and operating parameters of PEM electrolyzers [3, 5].

Manufacturer	Technology	Rated production [m^3 (STP) h^{-1}]	Maximum pressure (bar)	Energy consumption [kWh m^{-3} (STP)]	Location
Hydrogenics	Bipolar	1.0	7.9	7.2	Canada
Giner	Bipolar	3.7	85	20	USA
Proton onSite	Bipolar	0.265–1.0	14	6.7	USA
Proton onSite	Bipolar	2–6	15	6.8–7.3	USA
Proton onSite	Bipolar	10–30	30	5.8–6.2	USA
Siemens	No details	~20–50	50	No details	Germany
H-TEC Systems	No details	0.3–40	30	5.0–5.5	Germany

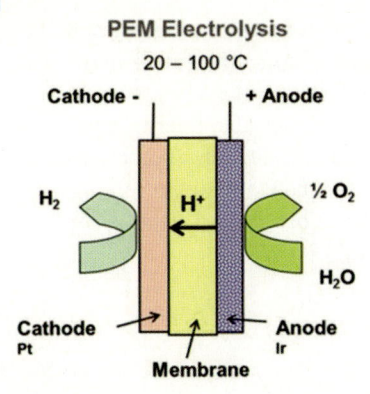

Figure 39.3 Schematic working principle of PEM electrolysis [3].

Table 39.4 Advantages and targeted improvements of PEM electrolyzers [3, 5].

Advantage	Targeted improvement
High power density	Increase lifetime
Higher efficiency	Reduce investment cost
Fast response under fluctuating power regimes	Increase hydrogen throughput capacity
Compact stack design permits high-pressure operation	Scale-up of the stack and hardware in the MW range

$$\text{Anode} \quad H_2O_{(l)} \rightarrow 2\,H^+_{(aq)} + \tfrac{1}{2} O_{2(g)} + 2\,e^- \tag{39.5}$$

$$\text{Cathode} \quad 2\,H^+_{(aq)} + 2\,e^- \rightarrow H_{2(g)} \tag{39.6}$$

$$\text{Total} \quad H_2O_{(l)} \rightarrow H_{2(g)} + \tfrac{1}{2} O_{2(g)} \tag{39.7}$$

The advantages and targeted improvements of PEM electrolyzers are summarized in Table 39.4.

39.2.3
High-Temperature Water Electrolysis

In the 1970s and 1980s, HTE technology attracted considerable interest. It was developed by Dornier System and Lurgi (HOT ELLY) in Germany, using an electrolyte-supported tubular concept for a solid oxide electrolysis cell (SOEC). Typically solid ceramic material is used for the electrolyte that conducts oxygen ions. High-temperature electrolyzers have not yet been commercialized and are still in research

High-Temperature Electrolysis
700 – 1000 °C

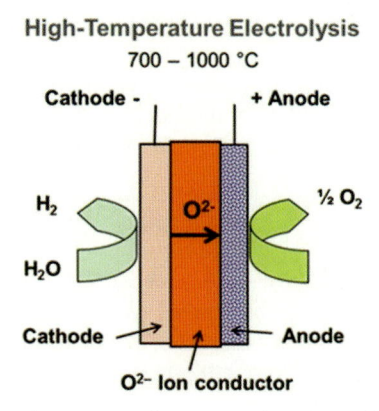

Figure 39.4 Schematic working principle of high-temperature water electrolysis [3].

and development phase. HTE typically operates at much higher temperatures, in the range 900–950 °C, and feeding water is replace with steam. During continuous tests on a single cell, voltages of < 1.07 V were achieved with a current density of 0.3 A cm^{-2} [3]. In HTE, water is fed as steam towards the cathode, where it acts as a reactant with electrons to split water into hydrogen and oxygen ions. Oxygen ions move towards the anode, then discharge electrons and make pairs to produce oxygen gas. The principle is illustrated in Figure 39.4 [2, 3].

$$\text{Anode} \qquad H_2O_{(l)} \rightarrow 2\,H^+_{(aq)} + \tfrac{1}{2}O_{2(g)} + 2\,e^- \qquad (39.8)$$

$$\text{Cathode} \qquad 2\,H^+_{(aq)} + 2\,e^- \rightarrow H_{2(g)} \qquad (39.9)$$

$$\text{Total} \qquad H_2O_{(l)} \rightarrow H_{2(g)} + \tfrac{1}{2}O_{2(g)} \qquad (39.10)$$

39.2.4
Integration of Renewable Energies with Electrolyzers

Hydrogen can be produced from both on-grid and off-grid systems. Typically, two major scenarios are promising for integrating electrolyzers in off-grid applications. In the first arrangement, electrolyzers are directly coupled with wind power or photovoltaics (PV). Due to the intermittency of wind and PV, the electrolyzers are always oscillating. The oscillation of electrolysis can be removed by using AC/DC or DC/DC converter technology as an intermediate connection source. The second arrangement uses surplus wind energy from a remote-area grid where this grid is not connected with the national grid. This is known as the energy management concept where the source and load could be balanced, using the excess amount of energy when the load is not high. The stored energy in the form of hydrogen can be used via a fuel cell to produce electricity and feed this energy into the grid while the wind energy production is low with respect to the load. The integration technique is known as closely coupled. The hydrogen produced through the off-grid system is called renewable hydrogen, where the electrolyzers are directly integrated with the renewable energy sources [5]. In on-grid applications, the surplus amount of energy can be used because the transmission grid is connected with the national grid and the electrolyzers are directly integrated with the grid for hydrogen production.

39.3
Methanation

To produce renewable power methane, hydrogen has to be reacted with carbon dioxide via the methanation process. Here the Sabatier process is described in detail and a possible plant configuration is presented. In addition, different possibilities for CO_2 sources are considered.

39.3.1
Catalytic Hydrogenation of CO_2 to Methane

The catalytic hydrogenation of carbon dioxide to methane (Eq. 39.11), named after its discoverer Paul Sabatier (1902) as the Sabatier reaction [8], is a combination of a reversed endothermic water-gas shift reaction (Eq. 39.12) and an exothermic methanation of carbon monoxide (Eq. 39.13). The methanation of CO_2 is thermodynamically favorable (exergonic), but the full reduction of the completely oxidized carbon atom needs a catalyst to achieve acceptable reaction rates and selectivity. Different metal catalytic systems based on group VIII of the periodic table (e.g., Ni, Ru, Rh, Fe), which are supported on various metal oxides (Al_2O_3, TiO_2, SiO_2, ZrO_2), have been investigated, but nickel- and ruthenium-based catalysts are most effective for methanation [9, 10]. Hu [11] presented a detailed review of the advances in catalyst systems. Generally, the catalysts are sensitive to sulfur and carbonyl compounds and also coke formation [12].

	$\Delta H°_{298\,K}$ (kJ mol^{-1})	$\Delta G°_{298\,K}$ (kJ mol^{-1})	
$CO_{2(g)} + 4\,H_{2(g)} \rightleftharpoons CH_{4(g)} + 2\,H_2O_{(g)}$	−164.94	−113.50	(39.11)
$H_{2(g)} + CO_{2(g)} \rightleftharpoons CO_{(g)} + H_2O_{(g)}$	41.19	28.59	(39.12)
$3\,H_{2(g)} + CO_{(g)} \rightleftharpoons CH_{4(g)} + H_2O_{(g)}$	−204.13	−370.66	(39.13)

Depending on the catalyst type, the methanation reaction is typically conducted at temperatures between 250 and 400 °C and pressures between 1 and 80 bar [13]. For the kinetics, high temperatures are preferred, but the thermodynamic equilibrium shifts to the product side at low temperatures and higher pressures [14] (Le Chatelier principle).

39.3.2
Methanation Plants

There are different reactor designs for the methanation process, such as fixed-bed, fluidized-bed, and three-phase reactors. Because of the exothermic reaction, the reactor should have good heat removal and a homogeneous distribution for better control of the reaction. In a fixed-bed reactor, this is only possible with high technical investment, such as additional cooling elements inside the reactor; however, they are often used because of the low investment cost and simple handling. In a fluidized-bed reactor, the fluctuation of the catalyst allows better heat release and distribution, but a disadvantage is the relatively high abrasion effects of the catalyst [12]. The three-phase reactor uses a temperature-stable heat-transfer fluid, in which the solid catalyst is suspended. This reactor type is expected to be suitable for dynamic operation of a methanation plant owing to the excellent control of temperature by the heat-transfer fluid [15].

SolarFuel described in a patent [16] the design and operational parameters for a methanation plant operated at 6 bar (Figure 39.5). The plant consists of two fixed-bed reactors with a nickel catalyst in line with an intercalated cooling unit with water condensation. The condensation of water and lower operating temperature in the second reactor lead to a shift of the thermodynamic equilibrium towards the product, which is necessary to achieve a high conversion to methane (99%). An alternative is methanation in one reactor at higher pressures (e.g., 20 bar), but from the technical point of view a higher pressure could lead to leakage (mainly H_2, because of the small molecule size) and higher investment costs.

The conversion of CO_2 into methane is not yet state-of-the-art and is currently under research and testing in pilot plants. Since October 2012, the Zentrum für Sonnenenergie- und Wasserstoff-Forschung Baden-Württemberg (ZSW) operated, in cooperation with SolarFuel, a 250 kW_{el} methanation plant [300 m^3 (STP) d^{-1}]. This pilot plant is the world's largest of its type and should be the pre-stage to a 20 MW_{el} industrial methanation plant [17].

For complete conversion, the maximum achievable efficiency is ~83%, because 17% of the chemical energy of the hydrogen is converted into heat during the reaction. Additionally, the methanation process needs electric power, mainly for compression of the incoming gases to the desired operating pressure and, if applicable, compression of the produced methane to the pipeline pressure level. As a reason for this, the efficiency is strongly dependent on the pre-pressure of the incoming hydrogen and carbon dioxide. Depending on the operating conditions, the overall efficiency is estimated to be between ~75 and ~80%. Usually, in the literature a mean efficiency of 80% for methanation is reported [18, 19].

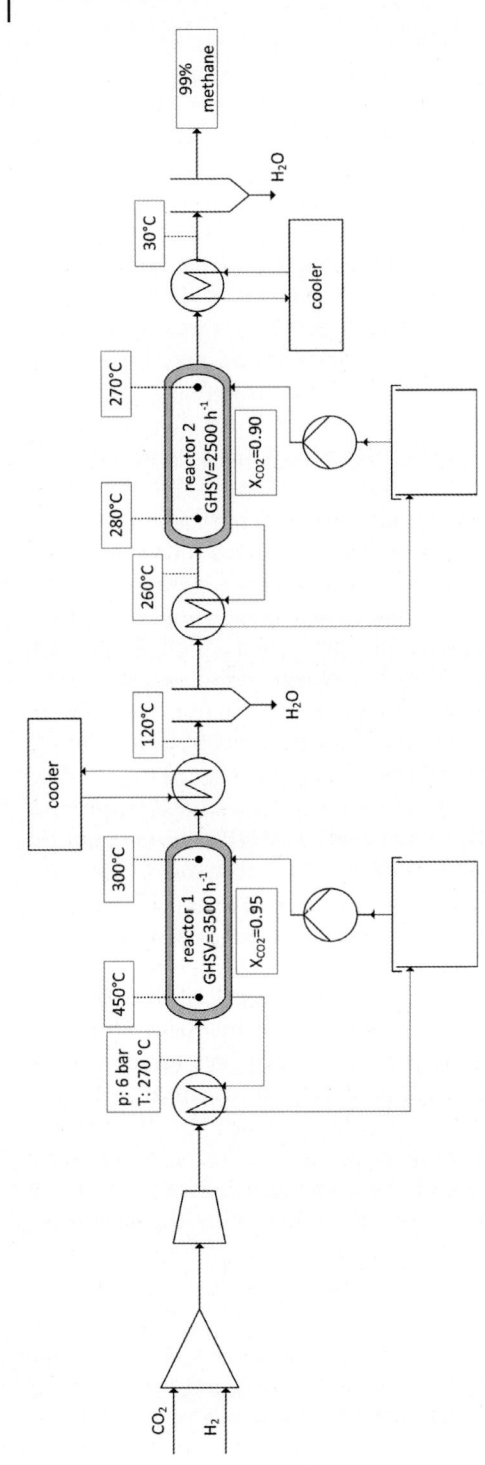

Figure 39.5 Process flowsheet of a methanation plant patented by SolarFuel [16]. GHSV = gas hourly space velocity.

39.3.3
CO$_2$ Sources

In addition to H$_2$, CO$_2$ is the other important component of the Sabatier process, so obtaining CO$_2$ efficiently and economically is a key factor in the renewable power methane (RPM), as mentioned in Section 39.3.1. An investigation of possible CO$_2$ sources is explored here. In order to ensure the downstream product synthetic natural gas (SNG) with high quality and at low cost, the CO$_2$ should be produced economically with high purity.

The quality issue of SNG is important. If the SNG produced is to be integrated with the existing natural gas network, then the CH$_4$ content must be > 90 mol% [20, 21].

39.3.3.1 CO$_2$ via Carbon Capture and Storage

Considering the climate issue, the energy structure should be adjusted to increase the renewable energy aspect and decrease the dependence on fossil fuels. During this transition phase, CO$_2$ sequestration and utilization must be considered. A large CO$_2$ source can be found in world power sector emissions, which reached 12.5 Gt in 2010 [22]. Carbon capture and storage (CCS) technology can help to achieve CO$_2$ as a byproduct of electricity production. CCS consists of four major parts: capture, compression, transport, and storage of CO$_2$.

There are three main routes to capture the CO$_2$ from power processes: post-combustion, pre-combustion, and oxy-fuel combustion. *Post-combustion* capture is aimed at the separation of CO$_2$ from the flue gases (13–15 mol% CO$_2$, 70 mol% N$_2$, 3–6 mol% O$_2$, water steam saturated at 50 or 70 °C [23]) generated in a large-scale combustion process fired by fossil fuels. *Pre-combustion* systems process the primary fuel in a reactor with steam and air or oxygen to produce a mixture consisting mainly of CO and H$_2$. Additional H$_2$, together with CO$_2$, is produced by reacting the CO with steam in a second reactor (a "shift reactor"). The resulting mixture of H$_2$ and CO$_2$ can then be separated into a CO$_2$ gas stream, and a stream of H$_2$, prior to a combustion process. The converted syngas is composed mainly of 35–40 mol% CO$_2$ and 55–60 mol% H$_2$; minor components are N$_2$, CO, and Ar [23–25]. When oxygen is used instead of air for combustion, the process is called *oxy-combustion*. This results in a flue gas that consists mainly of CO$_2$ (> 80 mol%) the minor components being 8–10 mol% N$_2$ and 5 mol% O$_2$ and saturated H$_2$O [23, 26]; consequently, the CO$_2$-rich stream should be treated to meet purity requirements.

Absorption, adsorption, membrane, and cryogenic fractionation are the common methods for separating CO$_2$ from flue gas or syngas. A CO$_2$ purity of 95–99 mol% can be reached by different capture methods with relevant energy consumption [27]. On the basis of different capture methods, the energy consumption varies in the range 100–240 kWh$_{el}$ t$_{CO_2}^{-1}$ [27–30], the CO$_2$ capture cost is €20–60 t$_{CO_2}^{-1}$ net captured [27, 29–32], and the levelized cost of electricity (LCOE) amounts to €120–150 MWh$_{el}^{-1}$ [28, 30]. Several points should be highlighted here: the energy consumption is exchanged to CO$_2$ capture, through subtracting 100 kWh$_{el}$ t$_{CO_2}^{-1}$ compression energy from the original literature data; avoided emissions are less than the amount of CO$_2$ captured; avoidance costs are greater than capture costs;

in the LCOE evaluation the avoidance cost is considered; and the exchange rate between the US dollar and euro is assumed to be 1, and this will be done in the following discussion also.

39.3.3.2 CO_2 Obtained from Biomass

The most direct source for SNG is to use biomass as feedstock, which is available worldwide, such as agricultural residues, forestry biomass, energy crops, food and food processing wastes, and some novel feedstocks such as algae. The *World Energy Outlook 2012* [22] reports that bioenergy contributed 1277 Mtoe (million tons of oil equivalent), ~10%, to the total primary energy demand (TPED) in 2010.

Three basic concepts are commonly used for bioenergy conversion from biomass: fermentation, gasification, and combustion. The relevant products from fermentation and gasification are *biogas* and *biosyngas*, respectively. Biogas is produced as landfill gas (LFG) or anaerobically digested gas. Depending on the different sources, biogas is mainly composed of 50–70 mol% CH_4 and 30–50 mol% CO_2 [33], together with water vapor and minor components H_2S, NH_3, N_2, and O_2 [33]. If biomass reacts with air, or oxygen and steam, a raw syngas can be produced. The major components of biosyngas are H_2 and CO, with minor components CO_2, CH_4, and H_2O depending on the gasification temperature [33, 34]; if air is used for the reaction, N_2 is also one of the main components [13]. Using circulated fluidized-bed (CFB) technology, 100% biomass combustion can be realized in combined heat and power (CHP) plants [35, 36] in the size range from a few kilowatts up to more than 100 MW [37]; 100% biomass combustion is related to significant pollutant formation [37], hence co-firing with coal is promising approach [35, 36].

In view of the different gas components from the aforementioned concepts, how these processes can be integrated with methanation process efficiently should be investigated in detail.

- *Fermentation concept, Scenario 1 (biogas):* After dewatering, using pressure swing absorption (PSA) or temperature swing absorption (TSA), membrane, or chemical absorption, CH_4 can be stripped from the biogas to a fairly high purity of 97 mol% [33, 38], and CO_2 can be obtained almost freely as by-product (99 mol% purity [38]) for further SNG production.
- *Fermentation concept, Scenario 2 (biogas):* The mixture of CH_4 and CO_2 can be directly fed to a methanation unit, upgraded to SNG with > 90 mol% CH_4 [34]. It is very attractive as an *in situ* methanation process, because an extra separation of CO_2 can be avoided.
- *Gasification concept, Scenario 3 (biosyngas, composed of $H_2 + CO + CO_2 + CH_4$, using oxygen as gasification medium):* The syngas is converted by an absorption-enhanced reforming (AER) reaction, and the biosyngas can reach a fairly high H_2 content of > 60 mol%. This converted biosyngas can be directly upgraded to SNG with > 90 mol% CH_4 [33]. Although such an extra shift reaction is very energy consuming, the high H_2 content and the available CH_4 content can assure a quantitative product volume and save on relevant gas treatment processes [33].

- *Gasification concept, Scenario 4 (biosyngas, composed of $H_2 + CO + N_2 + CO_2 + CH_4$, using air as gasification medium):* The gas mixture is composed of 15–20 mol% CO, 10–12 mol% H_2, up to 4 mol% CH_4, 45–55 mol% N_2, and 8–12 mol% CO_2 [13], and direct combustion for CHP plants is a feasible option [39]. The CO_2 content in the flue gas is fairly low, < 10 mol%. The subsequent CO_2 separation is similar to a natural gas combined cycle (NGCC) plant in CCS, using chemical absorption and so on [40].
- *Combustion concept, Scenario 5 (for biomass combustion):* CO_2 can be separated after combustion. If 30% biomass is co-fired with coal, the flue gas composition will be the same as that from post-combustion in CCS. Hence a similar CO_2 capture using chemical absorption or other methods can be carried out [35, 36].

At present, it is still difficult to obtain detailed information about energy consumption and costs for each scenario. For Scenarios 5 and 6, it is known [35] that the LCOE from firing with biomass is generally higher than for fossil fuels owing to the difference in fuel costs. As an example, the Intergovernmental Panel on Climate Change (IPCC) [23] states that the capturing of $0.19\ Mt_{CO_2}$ per year in a 24 MW_{el} biomass integrated gasification combined cycle (IGCC) plant is estimated to be about €80 $t_{CO_2}^{-1}$ net captured. An LCOE value of biomass of €80–140 MWh^{-1} has been reported [20].

39.3.3.3 CO_2 from Other Industrial Processes

In addition to the energy sector, CO_2 is emitted in many other industrial processes, for example, the steel and cement industries, petrochemical processes, the chemical industry, and the food sector. During iron and steel production, the largest CO_2 source occurs as the reducing agent is consumed during iron production, that is, coal, coke produced on-site, or purchased coke is used as the reducing agent in stationary combustion processes [41]. In the cement industry, CO_2 is produced in the decomposition of $CaCO_3$ to CaO and CO_2 that takes place in the kiln. CO_2 emissions from clinker production amount to about $0.5\ kg_{CO_2}\ kg_{CaCO_3}^{-1}$. The specific process CO_2 emission per ton of cement depends on the ratio of clinker to cement, which normally varies from 0.5 to 0.95 [42].

The CO_2 concentration in these processing gases is normally higher than that in flue gas in power plants [23]. Table 39.5 shows the worldwide CO_2 emission and CO_2 concentration from these sources in 2005 [43, 44].

CO_2 has been captured from industrial process streams for about 80 years [45], although most of the CO_2 captured is vented to the atmosphere because there is no incentive or requirement to store it. CO_2 could be captured from these streams using techniques that are commonly used in CCS [23]. The specific energy for different processes varies widely according to the different CO_2 compositions and different capture processes. The capture cost is €25–115 $t_{CO_2}^{-1}$ net captured [23].

Table 39.5 Worldwide CO_2 emissions from different industrial processes in 2005 [43, 44].

Sector	CO_2 emissions (Gt)	Percentage of total amount	CO_2 concentration in flue gas (mol%)
Steel	1.5	6	15–27
Cement	0.93	4	14–33
Refinery	0.8	3	3–13
Chemical industry	0.41	2	100 in some processes

39.3.3.4 CO_2 Recovery from Air

The capture of CO_2 from the atmosphere (air capture) means extracting CO_2 at a very low concentration (~390 ppm) to produce a highly concentrated stream of CO_2.

In thermodynamic theory, the minimum energy demand to extract CO_2 with a partial pressure p_0 in a gas mixture to generate a pure CO_2 stream is the Gibbs free energy $\Delta G = R\,T \ln(p/p_0)$, which is proportional to the chemical potential μ, where R is the specific gas constant (8.31 J mol^{-1} K^{-1}), T the ambient temperature in kelvin, p the ambient pressure and $p_0 = 0.39$ mbar. This energy demand is only 0.45 GJ $t_{CO_2}^{-1}$ or 20 kJ mol^{-1}. Nevertheless, in real process technology, the energy demand is much higher since activation energy is required [34, 46].

Nowadays, various technologies exist for the extraction of CO_2 from the atmosphere (air capture), for example, adsorption, absorption, condensation, and membrane. Adsorption processes such as PSA are standard technology for gas cleaning and conditioning. The drawback of CO_2 recovery via PSA is the mutual adsorption of CO_2 and water vapor. To recover the absorbent and separate CO_2 from water vapor is very energy intensive. Recovering a CO_2–water mixture consumes 20 times more energy than recovering pure CO_2. Further, membrane sieves are sensitive to air impurities such as dust [34]. Therefore, the only promising pathway for atmospheric CO_2 concepts is absorption processes, for example, reacting CO_2 with a solution of a strong alkali, such as NaOH and KOH [34, 47].

In an absorption process, CO_2 forms a sodium carbonate (Na_2CO_3) solution, which is highly soluble in water. However, this property prevents the precipitation of Na_2CO_3 from the aqueous solution and easy collection for regeneration. Hence establishing how to regenerate Na_2CO_3 efficiently is a key factor and has been investigated intensively [48–53]. The process usually proposed for the regeneration of NaOH is known as caustic recovery. An extra "causticization" process is applied for the regeneration of NaOH, where Na_2CO_3 is reacted with $Ca(OH)_2$. The resulting solid $CaCO_3$ precipitates out of the solution whereas the regenerated NaOH solution is sent back to the CO_2 contactor. $CaCO_3$ is then dried and calcined in a kiln to drive off concentrated CO_2 and obtain CaO (over 700 °C), which is hydrated with water (slaked) to form $Ca(OH)_2$ to close the cycle [47–50]. The most energy-intensive step in this cycle is the regeneration of $Ca(OH)_2$ by calcination, which requires high temperatures and pure oxygen [47]. In order to reduce the carbon footprint in air capture, the use of non-fossil energy to run the process has been investigated [47]. On the other hand, non-conventional causticization techniques are being developed. Most of them are based on the addition of a metal oxide (Me_xO_y) to convert Na_2CO_3 directly to $Na_2Me_xO_{y+1}$ and CO_2. NaOH is then regenerated by dissolving $Na_2Me_xO_{y+1}$ in water [47]. Furthermore, owing to the corrosiveness of strongly alkaline solutions, amines immobilized either physically or chemically on solid supports have emerged as possible alternatives. An important advantage is the relatively low regeneration temperatures. However, the long-term stability of amine-based absorbents under air capture conditions remains to be determined [47].

House *et al.* [46] made a detailed energetic and economic evaluation of different air capture methods. The energy requirements of those for NaOH scrubbing/lime causticization systems vary widely, from around 500 to 800 $kJ_{primary\ energy}\ mol_{CO_2}^{-1}$, whereas the primary energy required to strip CO_2 from the rich amine stream (115–140 $kJ\ mol_{CO_2}^{-1}$) dominates the CO_2 capture process. Regardless of the technology used, the overall cost of CO_2 capture from the air remains a highly debated question which will probably receive clearer answers only after the construction of demonstration and pilot plants [47]. According to House *et al.*'s estimation [46], the total system costs of an air capture system will be on the order of €1000 $t_{CO_2}^{-1}$.

A summary of the different CO_2 sources is listed in Table 39.6. One important aspect to be pointed out is that the heat value of methane is ~ 800 $kJ\ mol_{CO_2}^{-1}$; the maximum methanation efficiency is 83% (see Section 39.3.2) without considering CO_2 sources. If CO_2 as a key component of the methanation process must be additionally separated from the listed sources, the relevant efficiency must be decreased in view of the CO_2 yield; for example, if using 800 $kJ\ mol_{CO_2}^{-1}$ energy to separate CO_2, then the efficiency of the methanation process will be decreased to 45%. Therefore, biogas and biosyngas upgrading (Senarios 2 and 3) should be attractive options for the present purpose.

Table 39.6 Specific energy consumption and cost of different CO_2 sources.

Source	Energy consumption	Capture cost (€ t_{CO2}^{-1})	LCOE (€ MWh^{-1})	Ref.
CCS	100–240 $kWh_{el}\ t_{CO_2}^{-1}$ Post-combustion (amine): 115–140 $kJ_{primary\ energy}\ mol_{CO_2}^{-1}$	20–60	120–150	[27–32, 46]
Biomass	–	–	80–140	[20]
Other Industries	–	25–115	–	[23]
Air capture	500–800 $kJ_{primary\ energy}\ mol_{CO_2}^{-1}$	1000	–	[46]

39.4
Gas Storage

An increasing share of renewable energy, particularly by wind and PV sites, to the total power supply will lead to increased fluctuation of the power input into the grid. In the case of predominant power supply by renewables, temporary excess and insufficient power generation, which can last for either very short periods or over weeks, are the consequences. The storage of large amount of energy will become mandatory. The Association for Electrical, Electronic, and Information Technologies (VDE) estimates the demand for short-term (5 h) and long-term (17 days) energy storage capacity in Germany for an 80% renewable power supply scenario to be 70 GWh and 7.5 TWh, respectively. In case of a 100% renewable power supply scenario, the demand will be more than tripled and the long-term storage demand rises to 26 TWh [54].

Long-term storage for weekly to seasonal balancing can only be accomplished on a chemical basis via large-scale underground storage. The current storage capacity for natural gas in Germany is about 200 TWh_{th} [55], which corresponds to about 20% of the annual natural gas consumption. If hydrogen is to be stored in the same facilities, this value would shrink to about 30% because of the lower volumetric energy content [56].

When planning underground storage for hydrogen, either pure or as an addition to natural gas, several aspects have to be considered [57]:

- Containment of the reservoir.
- High ratio of working to cushion gas. The cushion gas is the amount of gas that has to remain inside the storage in order to maintain the minimum allowable pressure, whereas the working gas is the part that can be utilized for balancing.
- High annual cycling capability in addition to a high rate of injection and withdrawal of storage gas.
- Chemical inertness of the reservoir, in order to avoid reactions with the gas.
- Avoid/limit contamination of the stored gas to maintain purity and avoid additional cleaning steps.

Suitable formations for underground gas storage can be found in porous rock structures and also caverns mined in salt rocks, which must be covered with impermeable caprock to prevent gas from escaping. Worldwide, the most relevant underground gas storage facilities are depleted gas and oil fields, aquifer storage, and artificially made caverns in salt rocks (Figure 39.6). To a lesser extent, inoperative mining plants and rock caverns can also be used for underground storage [57].

In Germany, porous rock structures and caverns contribute in equal parts to the total storage capacity (cf. Table 39.7). For 2050 it has been estimated that there will be additional construction of 400 new salt caverns, whereas no new porous rock storage projects are planned. [59] For this scenario, an increase in storage capacity to 514 TWh_{CH_4} will be achieved. Assuming that only salt caverns can be used for hydrogen storage, the corresponding storage capacity amounts to 110.4 TWh_{H_2}.

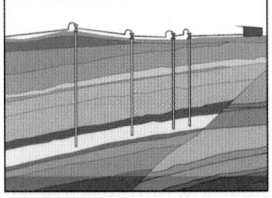

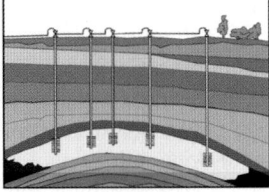

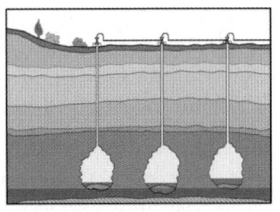

| Depleted oil and gas fields | Aquifers | Salt caverns |

Figure 39.6 Underground storage types [58].

Table 39.7 Underground gas storage capacity in Germany [59, 60].

Date	Storage capacity (volumetric) [10^9 m^3 (STP)]^a			Storage capacity (energetic) (TWh$_{th}$)	
	Porous rock	Caverns	Total	Methane^b	Hydrogen^c
Status end of 2011	10.0	10.4	20.4	204	31.2
2011 including planned	11.7	21.5	33.2	332	64.5
Expected in 2050	13.6	36.8	51.4	514	110.4

a Corresponds to working gas.
b Calculated with lower heating value LHV(CH_4) = 10 kWh m^{-3}.
c Calculated with LHV(H_2) = 3 kWh m^{-3} and cavern storage only.

39.4.1
Porous Rock Storage

Rock pores below ground can be utilized to store gas. Therefore, the rock structure should possess sufficient porosity and permeability and should be covered by a tight caprock structure, which can seal the entrapped gas and guarantee containment [61]. Structures used for porous rock storage can be found in depleted oil and gas fields and in aquifer structures [62]. Typical features of depleted gas and oil reservoirs and aquifers are their large storage capacities and high cushion gas requirements, which total 50–67% of the total amount of stored gas [63].

Especially depleted gas fields offer an economical way to store methane, since the technical equipment is already installed and the required information about rock properties and tightness of the cap rock are already available. The operating parameters, such as the working gas capacity, the maximum reservoir pressure, and the delivery rates, are known from earlier gas extraction [61].

In aquifer storage reservoirs, gas is injected at high pressures in order to force out the interstitial water in the pores. For evaluating whether an aquifers structure is suitable for underground storage, extensive geological and reservoir engineering effort is required. Critical issues for exploration are the thickness, porosity, and permeability of the storage rock and the thickness of the caprock [61].

Because of the high porosity, internal pressure losses in porous rock storages are high, leading to lower achievable extraction rates and making the storage feasible only for long-term storage ranging from weeks to months. In terms of hydrogen storage in depleted fields and particularly aquifer formations, reactions between hydrogen and microorganisms, and also between hydrogen and mineral constituents, may occur. Deterioration or depletion of the hydrogen storage and plugging of the microporous pore spaces by the reaction products are possible consequences [64, 65]. Further investigations are required, but the maximum H_2 concentration is estimated to be just a few percent.

39.4.2
Salt Cavern Storage

Salt caverns are man-made caverns in deep salt beds and domes produced by solution mining, where water is injected into wells, dissolves the salt, and forms brine, then the brine is leached out again. Thereby, the sizes and shape of caverns can be specified precisely. Special requirements for the application of the solution mining technique are deliverability of large volumes of freshwater and an available disposal system for discharging the resulting brine [62].

Salt caverns provide very high withdrawal and injection rates compared with their working gas capacity. Typical cavern volumes are 500 000 m^3 with operating pressures between 60 and 180 bar [57] Salt caverns for gas storage purposes have the ability to perform several withdrawal and injection cycles each year and can be utilized to meet peak load demands [62]. Cushion gas requirements are relatively low and typically lie between 20 and 35% of the total amount of stored gas [55, 61, 63].

In salt caverns, the surrounding rock salt is chemically inert regarding hydrogen and additionally very gas-tight [64]. Hence underground salt caverns are considered to be very favorable facilities for storing compressed hydrogen [62], as demonstrated decades ago. A plant with three relatively small single caverns with a volume of 70 000 m^3 each is operated in the United Kingdom (Teesside) with a constant operating pressure of 45 bar using a separate brine storage to extract the hydrogen [66]. Two much larger caverns are in operation in the United States (Texas), operated by ConocoPhillips and Praxair, with volumes of 580 000 and 566 000 m^3, respectively [66].

Regarding the economics, the investment costs for a new cavern with a volume of 750,000 m^3 has been estimated to be between €20 and 30 million, depending on the particular site and the necessary exploration and investigation efforts [66]. The VDE calculated that for long-term storage the costs amount to €0.23 $kWh_{H_2}^{-1}$ for storing hydrogen in salt caverns [67]. It is expected that these costs can be reduced to below €0.10 $kWh_{H_2}^{-1}$ in the future (Figure 39.7).

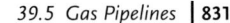

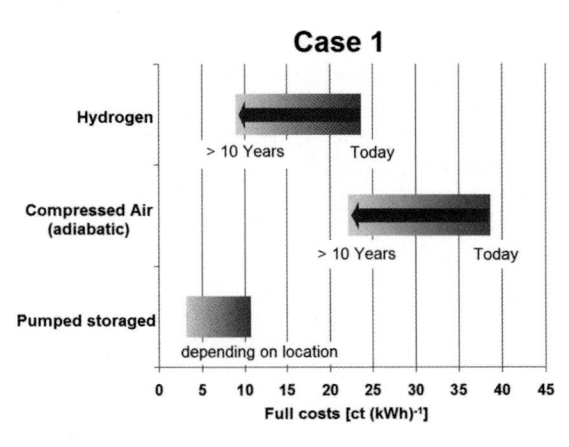

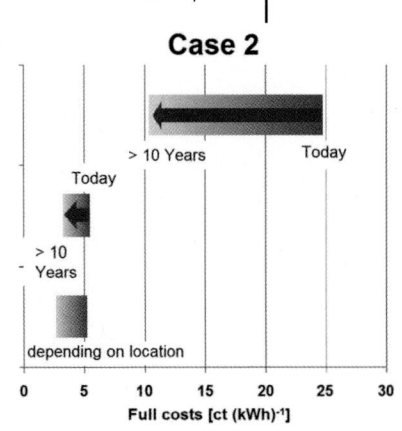

Figure 39.7 Full costs of different storage systems in two cases
(Case 1, long-term storage, 500 MW, 100 GWh, 1.83 cycles per month;
Case 2, load-leveling high-voltage network, 1 GW, 8 GWh, one cycle each day) [67].

39.5
Gas Pipelines

A gas transport system in principle has the task of transporting gas from the source of supply to the demand sink. Gas transportations are based on compressed gas transport, direct pipeline transfer, gas-to-power, gas-to-solid, and gas-to-liquid conversion. The most common and oldest transportation technology is pipeline transfer [68]. The highest gas consumption in Germany in 2011 was that of natural gas with 842 TWh. Germany imports 943 TWh and 90% of this import was from Russia, Norway, and The Netherlands [69]. From these perspectives the importance of natural gas is obvious; also, in addition to the classical uses such as gas combustion for cooking and heating, there are many other applications for natural gas such as NGCC power plants.

39.5.1
Natural Gas Pipeline System

The major elements of the gas pipeline grid are the pipes, storage facilities, and compressor, measuring, and regulating stations. The pipes are not only used as a transport medium, but also as a dynamic memory in case of natural gas in the gaseous state. However, this dynamic memory cannot balance supply and demand at any time. In particular, the seasonal adjustment scheme has to be balanced by underground storage. In a long-distance pipeline, the pressure decreases owing to friction during the transport by 10 bar per 100 km. Therefore, every 100–200 km a compressor station is located to compensate for the pressure drop in the pipes [70]. The measurement and control systems analyze and correct the pressure and the amount within one supply level or on different supply levels. Basically, the natural

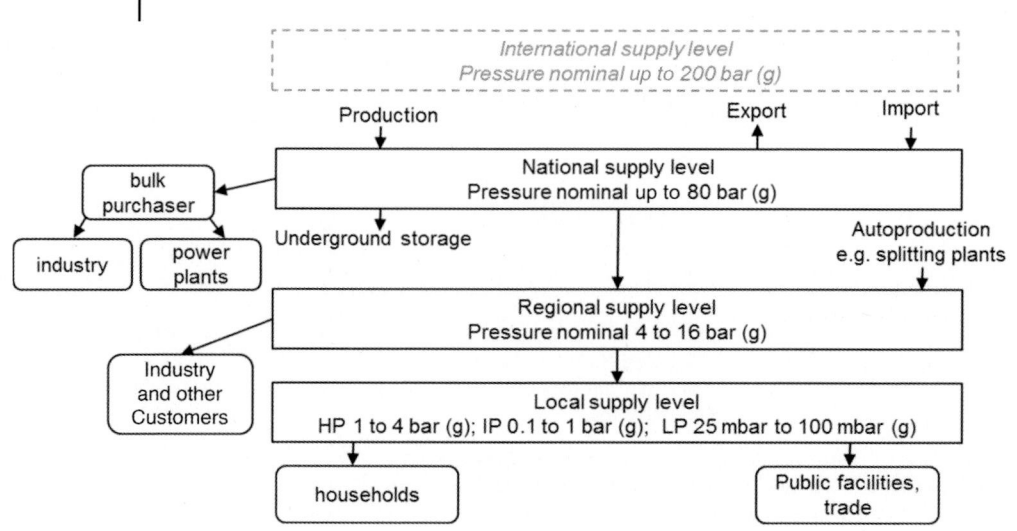

Figure 39.8 Structure of the natural gas grid in Germany [70, 75]. HP, high pressure; IP, intermediate pressure; LP, low pressure.

gas grid is divided into three supply levels, the national, the regional, and the local supply levels, which are operated at different pressure levels (see Figure 39.8). The total length of the German pipeline gas grid amounts to 524 000 km [71]. Applying the Sabatier process, where hydrogen reacts with CO_2 to give methane, SNG is produced that is in accordance with the standard quality definition from the German Technical and Scientific Association for Gas and Water (DVGW) [72]. Hence it can be added in any amount to the natural gas grid. Among other benefits are that today about 50% of households are connected to the natural gas grid. In 2009, the grid charges for households, industry, and trade amounted to €0.014, 0.003, and 0.012 kWh^{-1}, respectively [73]. In particular, the transmission losses in the gas grid add up to 0.5% per 1000 km, which is low compared with the electricity grid with 3–10% per 1000 km [74].

For the transport of hydrogen, there are basically two options: first, direct admixing of hydrogen into the existing natural gas grid, and second, the construction and use of a dedicated hydrogen pipeline system. Both options must be examined and compared in terms of their technical and economic feasibility. The calorific value of hydrogen is about one-third of that of natural gas. with admixture of 20 vol.% hydrogen, the energy content and the Wobbe index of the gas mixture would decrease by about 15% and 5%, respectively [76]. According to this circumstance and regulations [72], a maximum of 5 vol.% hydrogen can be fed as additional gas into the gas grid [77]. Nonetheless, further restrictions could also apply from the end users. Hence detailed investigation of the whole process chain is mandatory [76]. For example, in order to increase the hydrogen feed to 10 vol.%, changes in the compressor, measurement, and control stations are necessary. These technical measures are designed to adapt the increased hydrogen concentration and calculated according to [77] to a total of

Table 39.8 Costs [77] and potential feed [78] of 10 vol.% H_2 in the natural gas grid.

Measure, equipment	Cost (10^6 €)	Import point	Maximum (MWh$_{th}$ h^{-1})
Measuring and control station	30	Emden	769
Converting the gas turbines for gas processing	75	Dornum	1035
Modification/replacement of the gas turbine and compressor converting	3625 (replacement)	Lubmin	2345
Sum of feed, conversion and connection	3730		

€3730 million. Furthermore, in a hydrogen feed, local concentration jumps have to be avoided. Therefore, hydrogen should be injected into natural gas carrier streams. In Germany, potential entry points for hydrogen (produced from offshore wind power) according to [78] could be, for example, Emden, Dornum, and Lubmin (Table 39.8). It has been determined, assuming an efficiency of 75% for electrolysis plants, that a maximum of 4149 MWh$_{th}$ h^{-1} of hydrogen can be fed into the natural gas grid at 10 vol.% H_2 [78].

39.5.2
Hydrogen Pipeline System

It has been reported [78] that at 2.7×10^6 m^3 (STP) hydrogen in large parts of the gas transmission grid, a hydrogen concentration of about 60 vol.% is reached. At such high concentrations, a dedicated hydrogen pipeline system would be reasonable. For comparison, 2.7×10^6 m^3 (STP) hydrogen could be generated in 2022 by wind power (offshore) and electrolysis with an efficiency of 75% in Scenario B, according to [78] with data from [79]. A hydrogen pipeline system can be realized, for example, with 14 sources (Figure 39.9). These sources are according to Krieg [80] based on data from [81] feeding in equal parts of electrolysis and coal gasification. They produce in a total of 5.4×10^6 t of H_2 per year. The same hydrogen production for the transport sector is one key element of the scenario in [82], which can reduce CO_2 emissions by 55% compared with 1990. According to Krieg [80], the necessary pipeline system would have a length of 48 000 km. The system would be able to supply 9860 of the total of 12 000 petrol stations in Germany. Hence the collection rate amounts to about 80%; based on the sales volume, the rate is almost 90%. The associated cost for this scenario would be ~€23 billion for the pipeline system and €0.79 kg$_{H_2}^{-1}$ [80]. Other literature values range between €0.13 and 0.84 kg$_{H_2}^{-1}$ [80, 83–86], depending on the particular study and the assumed conditions.

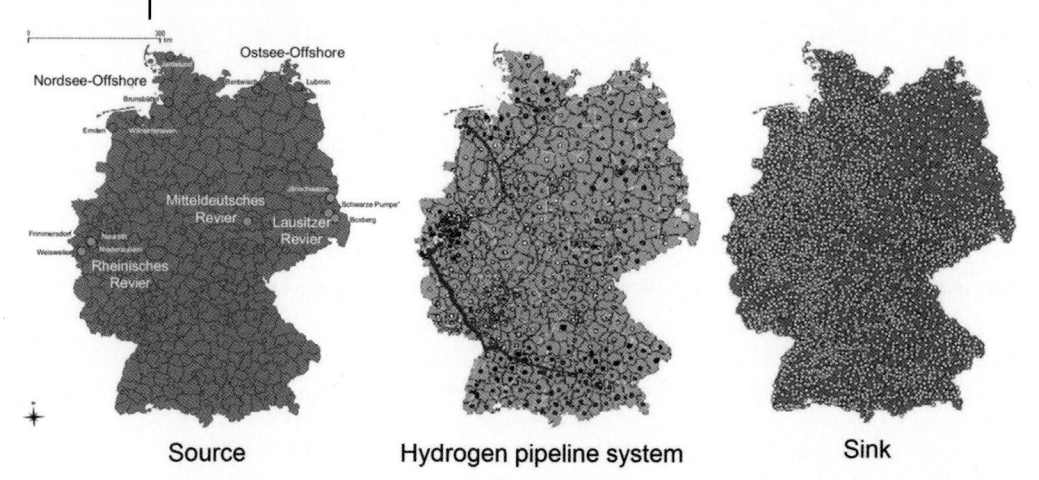

Figure 39.9 Example of hydrogen pipeline system from source (electrolysis and coal gasification) to sink (petrol stations) [80, 81].

39.6
End-Use Technologies

The various hydrogen use paths encounter specific requirements from end-use technologies. Uses for SNG are the same as those for (fossil) natural gas; end use does not impose additional requirements. The German natural gas consumption in 2010 was 880 TWh, of which 35% was used for residential heating, 14% in commerce and services, mostly heating, 25% for power generation and CHP, and 9% in the chemical industry and refineries [87]. Other industrial uses are highly diversified.

Relevant existing end uses for "pure" hydrogen are ammonium/fertilizer, steel, and glass production and petroleum refining. Today, most end uses are served with reformed natural gas. The world ammonium demand was estimated to be 152×10^6 t a^{-1} in 2010 [88]; the bound hydrogen amounted to 900 TWh$_{LHV}$. The production is shifting towards countries with low-cost natural gas. The use of hydrogen for fuel cell-driven vehicles and reconversion to electricity offers an even larger *potential* market. The current energy consumption for road transportation in Germany alone is 586 TWh [87]. A market of this order of magnitude easily justifies the investment in an energy-efficient pipeline system; this system could also serve existing applications, but the transport sector determines the required purity level, because of its volume and the wide distribution of the refueling stations.

For the easiest path, feeding the hydrogen into the natural gas grid, the strictest requirement of today's prevailing end-use technologies defines the acceptable hydrogen limit at least for the next two decades. Dehydrogenation would be affordable only for special applications and a tiny volume share. An increase in the limit would require a detailed analysis of the extent of replacement needed. We focus in the following on limits for gas turbines, piston engines, and heating vessels; for end-use in chemistry, we do not see relevant limitations.

39.6.1
Stationary End Use

39.6.1.1 Central Conversion of Natural Gas Mixed with Hydrogen in Combustion Turbines

According to the DVGW, turbine developers have no experience with the combustion of natural gas with hydrogen proportions above 3–4% [89]. Nevertheless, there have been theoretical investigations and experiments in model combustors. Hydrogen addition has several effects on premixed combustion: a slight increase in the flame temperature, a reduction in ignition delay times, and an increase in the laminar flame speed. Combustion modeling for relevant conditions shows that the laminar flame speed doubles with a hydrogen content of 60%, but the ignition delay time is shortened by a factor of 10 [90]. As reheat combustors allow the inlet temperature of the second burner to be set independently of its flame temperature, reactivity can easily be kept within safe margins [90]. Specialists estimate for such systems that a hydrogen content of 5 vol.% would not cause problems. An important side effect of a minor hydrogen addition would be an NO_x reduction and an extension of the part load capability; for example, a hydrogen content of 10% would reduce the lean limit fuel equivalence ratio by 4% [91].

39.6.1.2 Decentralized Conversion of Natural Gas Mixed with Hydrogen in Gas Engines

Today, many small- and medium-sized CHP plants use gas piston engines. For this application, the limiting hydrogen content of the gas is determined by the mixture's methane number, which describes the knock resistance: 100 stands for pure methane, 0 for pure hydrogen. Methane numbers of different gas sources vary significantly and the values depend slightly on the investigation method. The algorithm of Andersen gives 96.6 for the CIS, 93.3 for The Netherlands, 75.7 for Norway/North Sea, and 84.1 for the EU [92]. An important engine supplier specifies a methane number of 80 as general limit for its piston engines. This limit allows the dilution of natural gas from Russia, which makes up 41% of German imports [93], with 17 vol.% of hydrogen, whereas gas from Norway, which makes up 25% of the imports, would not allow any hydrogen addition and even requires treatment or a power limitation measure.

39.6.1.3 Conversion of Hydrogen Mixed with Natural Gas in Combustion Heating Systems

The DVGW states that typical residential end-use applications do not show any problems with hydrogen contents up to 20% [89]. Our enquiries with important manufacturers do not fully confirm this: although heating vessels successfully pass a so-called limit test procedure according to EN437 with 23% hydrogen content, for normal operation a hydrogen content of only 10% is confirmed as nonhazardous; an increase in this limit would require thorough investigations [94].

39.6.2
Passenger Car Powertrains with Fuel Cells and Internal Combustion Engines

With the availability of the power-to-gas products hydrogen and methane, new options for their efficient and low- or even zero-emission usage in transport arise. Hydrogen can be utilized as a neat fuel in fuel cell systems with zero tailpipe emissions and superior efficiency compared with the conversion of pure hydrogen, hydrogen–natural gas mixtures, or methane–natural gas mixtures in internal combustion engines (ICEs). This section describes the state-of-the-art of the conversion technologies to be addressed in the context of power-to-gas products as fuels for transportation. Criteria are tank-to-wheel (TTW) fuel consumption, CO_2 emissions, and costs.

39.6.2.1 Direct-Hydrogen Fuel Cell Systems

Passenger cars with direct-hydrogen fuel cell systems [fuel cell electric vehicles (FCVs)] have been developed for many years. Since 1994, a total of nearly 800 vehicles has been demonstrating the technology and proving progress in commercialization [95]. The development of performance indicators, such as fuel consumption, cruising range, and stack power, have been documented [95].

Regarding TTW fuel consumption, 33 kWh per 100 km has been achieved for the Mercedes-Benz B-Class F-Cell, 35 kWh per 100 km for the Honda FCX Clarity, and 44 kWh per 100 km for the Opel Hydrogen4 [96–98]. As a sports utility vehicle (SU) V, the Hydrogen4 shows comparatively high fuel consumption due to higher driving resistance. Cruising ranges are 380 km for the F-Cell, 320 km for the Hydrogen4 and 380 km for the FCX Clarity. These figures are well within the range of ICE-based passenger cars with compressed natural gas (CNG) as the fuel (cf., Section 6.2.2).

The documentation of the progress achieved towards cost reduction is difficult. The number of cars produced in series is at least three orders of magnitude smaller compared with today's passenger car mass production volumes. Moreover, critical components, such as fuel cell stacks and hydrogen storage tanks, still require substantial progress in cost reduction. The platinum requirement, which is one of the key issues regarding fuel cell cost reduction, is around 30 g per vehicle [99] at present. After 2020, a further reduction to below 10 g per vehicle has been estimated [98], approaching the levels of today's ICE exhaust gas treatment systems. The current status of fuel cell system costs without hydrogen storage, power electronics, and electric drives is estimated by the US DOE at US\$ 49 kW_{el}^{-1} for an 80 kW_{el} system. The DOE target value for 2017 and beyond is US\$ 30 kW_{el}^{-1} [100]. All DOE figures were based on high-volume production, namely 500 000 units per year. The European Commission (EC) approximates the 2010 status of specific fuel cell system costs at €1000 kW_{el}^{-1} at the minimum. EC targets for the 2015 and 2020 are €100 and €50 kW_{el}^{-1}, respectively [101]. However, it is not clear what production volume is assumed and whether hydrogen storage costs were considered in the values cited. A cost estimation for hydrogen storage tanks of US\$ 10 $kWh_{H_2}^{-1}$ (2008) was given by Kromer and Heywood [102]. If the more optimistic DOE values are used and an exchange rate of €1 = US\$ 1 is assumed, a complete fuel cell power generation system with 80 kW$_{el}$ rated power output and a typical storage capacity of 5 kg of H$_2$ corresponding to 170 kWh$_{H_2}$ would cost €4100.

39.6.2.2 Internal Combustion Engines

Today, passenger cars with ICEs [internal combustion engine vehicles (ICVs)] running on CNG are commercially available from a variety of car manufacturers such as Fiat, Opel, and Volkswagen. As of December 2011, the CNG vehicle population in the EU countries was 1 million (96 000 for Germany), and 2800 CNG refueling stations were available (903 for Germany) [103]. CNG engines are derived from gasoline ICEs. Tank pressure vessels are operated at up to 200 bar. CNG cylinders can be all-metallic, composite with metallic liners, or composite with nonmetallic liners [104].

Assuming comparable driving resistances and powertrain efficiencies, CNG powertrains show 23% lower CO$_2$ emissions than powertrains with gasoline, assuming CO$_2$ emission factors of 56.4 g$_{CO_2}$ MJ^{-1} for natural gas and 73.2 g$_{CO_2}$ MJ^{-1} for gasoline according to Appendix 1 in [105]. However, as Figure 39.10 shows, higher fuel consumption values reduce the CO$_2$ emissions advantage for some of the vehicles selected. Vehicles with the highest CO$_2$ emissions advantage also have the highest premium to be paid. Cruising ranges of the vehicles in natural gas operation are between 310 km (Fiat Punto) and 456 km (Volkswagen Caddy). The Audi A3 TCNG which is part of Audi's e-gas project will be available in late 2013. Fuel consumption is announced to be below 4 kg$_{CNG}$ per 100 km or 56 kWh$_{CNG}$ per 100 km, resulting in less than 113 g$_{CO_2}$ km^{-1} [106].

Regarding the compatibility of CNG vehicles with natural gas–hydrogen blends, Klell and Sartory [107] concluded that fuel systems and combustion engines can be operated reliably on such blends with up to 30% hydrogen. However, the application of storage vessels for today's CNG vehicles is limited to blends with up to 2% hydrogen, if cylinders made of steel with a minimum ultimate tensile strength of 950 MPa are used [108].

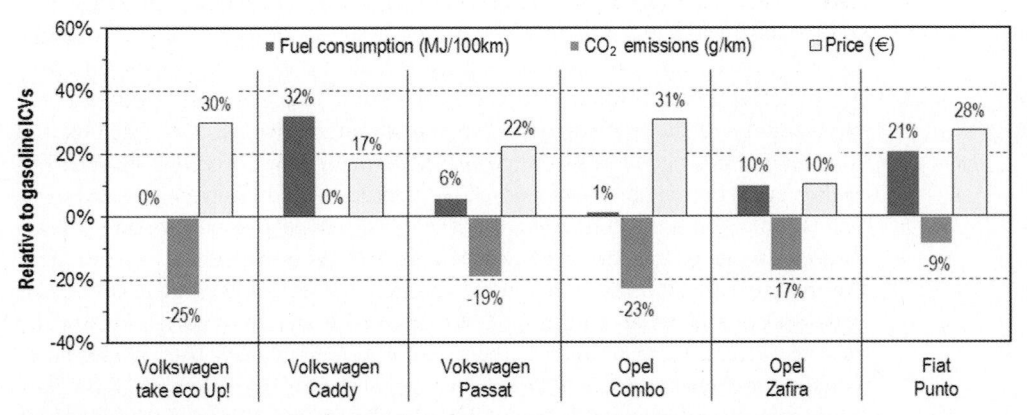

Figure 39.10 Comparison of fuel economy and CO$_2$ emissions of ICVs, model year 2012. Values for gasoline ICVs relate to 100%; basic car configurations have been selected, engine power in kW (CNG/gasoline): Fiat Punto 51/51, Opel Combo 88/70, Volkswagen Caddy 80/77.

ICEs for operation with hydrogen have been developed and demonstrated for many years. BMW demonstrated this technology with one hundred 7-Series Hydrogen 7 between 2006 and 2008. Other car manufacturers with notable developments in that area include Ford (passenger cars and buses), MAN (buses) [109], and Mazda (battery-electric passenger car with hydrogen-fueled rotary engine as range extender) [110]. Regarding fuel efficiency, BMW reported a peak value of 42% for its 12-cylinder ICE, which is between the peak efficiencies of gasoline and diesel engines. For hydrogen ICE powertrains, fuel consumption can be assumed to be comparable to that of conventional cars with gasoline or diesel ICEs. More recently, reported development and demonstration activities of passenger cars with hydrogen-fueled ICEs have decreased. This option is not considered further in this section. Electric drives with direct-hydrogen fuel cells offer considerably higher fuel economies, allowing for reduced fuel costs and reduced on-board hydrogen storage capacity (cf., Section 39.6.2.1).

39.7
Evaluation of Process Chain Alternatives

A comparative evaluation of power-to-gas process chain alternatives can be based on energy use and costs of fuel supply. Whereas the energy use can be analyzed using component efficiencies, the economic assessment must include a component scaling for the estimation of investment volumes. In this section, results of a hydrogen infrastructure assessment for Germany are used [111]. Infrastructure components have been scaled assuming an extensive use of renewable electric power in Germany and abandonment of fossil and nuclear energy-based electric power production. In a scenario based on these assumptions, all excess renewable power is used for hydrogen production [renewable power hydrogen (RPH)] via large-scale electrolysis. The hydrogen is then utilized in the transportation sector assuming highly efficient conversion in FCVs. Grid power is balanced using natural gas-fired power plants. An overview of investment volumes of electrolysis, geological storage, transmission and distribution pipeline grids, refueling stations, and balancing power production is given in Table 39.9. Values according to [111] are supplemented by specifications of the methanation plants as an alternative process chain for the utilization of hydrogen from wind power for methane production (RPM). Whereas the RPH path for transportation comprises all processes between the electric power generation and the end consumer, the RPH and RPM feed-in pathways terminate at the hydrogen or methane feed-in to the natural gas transmission pipeline. It is assumed that the specific investment for methanation plants is equal to that of electrolysis plants on an energy output basis. Infrastructure requirements of the CO_2 capture and transportation to the methanation plants have been neglected as reliable cost data are not available.

Further assumptions for the economic assessment are depreciation period and interest rate. Regarding the depreciation period, 10 years was selected for large conversion plants, 20 years for gas distribution grids, and 40 years for gas transmission

Table 39.9 Investment volumes for power-to-gas process chains according to [111], supplemented by assumptions for large-scale methanation plants[a].

Process chain component	Installed capacity (size or number)	Investment (10^9 €)
Hydrogen for transportation, according to [111]		
Electrolyzers (70% efficiency; €720 kWh$_{H_2}^{-1}$)	59 GW$_{H2}$ (84 GW$_{el}$)	42
Pipeline grid[b]	5.4×10^6 t a^{-1}	19–25
Geological storages	Seasonal balancing only	5
	60 day reserve	15
Refueling stations	9800 units	20
Balancing power generation[c]	42 GW$_{el}$	24
Total investment		**110–126**
Methane feed-in to natural gas pipeline		
Electrolyzers (70% efficiency; €720 kWh$_{H_2}^{-1}$)	59 GW$_{H2}$ (84 GW$_{el}$)	42
Methanation plants (80% efficiency; €720 kWh$_{CH_4}^{-1}$)	47 GW$_{CH4}$	34
Balancing power generation[c]	42 GW$_{el}$	24
Total investment		**100**
Hydrogen, direct feed-in to natural gas pipeline		
Electrolyzers (70% efficiency; €720 kWh$_{H_2}^{-1}$)	59 GW$_{H2}$ (84 GW$_{el}$)	42
Balancing power generation[c]	42 GW$_{el}$	24
Total investment		**66**

a   Lower heating values have been used.
b   Comprising transmission (12 000 km) and distribution (39 000 km) networks.
c   Open gas turbines (< 750 operational hours per year) and combined cycle power plants.

network and geological storage. The interest rate was assumed to be 8%. Operation and maintenance costs were assumed to be 3% of the investment. Results of this comparative assessment are displayed in Figure 39.11.

Figure 39.11a compares RPH supply costs including retail at the refueling stations with today's gasoline cost of €0.08 kWh^{-1} or €0.70 l^{-1} before tax. With the assumption that FCVs have only half of the fuel consumption compared with ICVs, the allowable hydrogen costs are twice the gasoline cost, €0.16 kWh^{-1}. Figure 39.11a shows that RPH supply costs could be 25 and 53%, respectively, higher than the allowable costs if wind power were to be available at €0.06 and €0.09 kWh^{-1}, respectively.

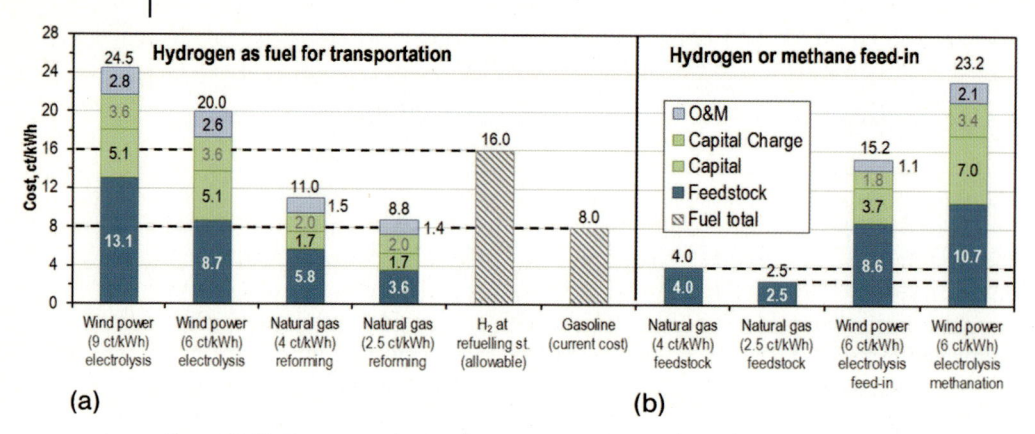

Figure 39.11 Cost comparisons of power-to-gas process chain alternatives.
(a) Hydrogen as fuel for transportation; (b) hydrogen or methane feed-in. All costs before tax.
O&M, operation and maintenance; refueling st., refueling station; ct, euro cents.

For comparison, values regarding natural gas conversion to hydrogen in large plants based on [105] are displayed that show that hydrogen supply can already today be cost-competitive with gasoline at the refueling station. However, CO_2 emissions must be considered in this case.

Figure 39.11b shows the results of comparing natural gas feedstock costs and the costs of the RPH or RPM feed-in alternatives. In relation to today's natural gas feedstock costs of €0.04 kWh^{-1}, RPM and RPH feed-in costs would be about four and six times higher, respectively, if the cost of wind power was €0.06 kWh^{-1}.

If RPM admixed with natural gas were to be used in transportation, the cost comparison must also include the costs of transport, distribution, storage, and refueling station costs. For business customers, these costs were €0.012 kWh^{-1} for natural gas [112], not including refueling station costs. Based on values given in [113], refueling station costs could be estimated as €0.02 kWh^{-1} of CNG. In total, the cost of supplying RPM to the fuel tank of a vehicle would be €0.26 kWh^{-1}. Assuming an ICV with a fuel consumption of 56 kWh per 100 km or 4 kg per 100 km of natural gas, the specific fuel cost would be €0.15 km^{-1}. In comparison, an FCV with a fuel consumption of 33 kWh per 100 km as for the Mercedes-Benz B-Class F-Cell, the specific fuel cost could be estimated as €0.07 km^{-1}, based on the hydrogen costs according to Figure 39.11a.

As mentioned above, the energetic process chain assessment basically relies on the efficiencies of the components employed. In the following, RPH and RPM are analyzed with regard to their utilization as fuels for passenger cars. All electric energy inputs are assumed to be renewable. First, a well-to-tank (WTT) analysis compares the process chain efficiencies of RPH and RPM for transportation. In the RPH case, in addition to the 70% electrolysis efficiency (all values are based on the lower heating value), it is assumed that the hydrogen produced is compressed twofold from 30 to 100 bar. The pressure level of 30 bar is assumed to be the delivery pressure of the electrolysis, but also the lower pressure limit of the hydrogen pipeline grid.

The compressors are operated on renewable electric power. At the refueling station, a three-stage compression from 30 to 870 bar high-pressure storage is assumed. Again, renewable electric power is used. In total, the WTT efficiency of RPH for transportation is 65%.

The RPM case employs a methanation step with an assumed efficiency of 80%, subsequent to the electrolysis. RPM is then fed into the natural gas pipeline grid and delivered to refueling stations. According to the National Inventory Report [114], the feedstock loss in the German pipeline grid is 0.7%, which is assumed to cover all energy expenses for gas transport and storage. At the refueling station, the delivered gas is then compressed to 250 bar utilizing renewable electric power. The resulting WTT efficiency can be determined as 55%.

It can be seen that owing to the additional process step of hydrogen methanation, the WTT efficiency of RPM supply is 10% lower compared with the RPH case. If TTW fuel consumption is assumed to have the values cited above, 33 kWh per 100 km for the FCV and 56 kWh per 100 km for the CNG vehicle, the driven distance per unit energy delivered is twofold greater for the FCV. This means, for an example renewable electric energy input of 100 kWh_{el}, that the FCV could travel 196 km. For the CNG vehicle, this value would decrease to 100 km.

It can be concluded that regarding energy use, the utilization of RPH in the transportation sector would be the more beneficial option compared with the RPM alternative. Reasons are the more efficient process chain and the considerably higher efficiency of the end-use technology FCV compared with the CNG-powered ICV. Moreover, a precondition for the technical and economic feasibility of both options is the commercial availability of large-scale electrolysis at the investment costs and efficiency levels assumed here. Specifically for the RPM path, large-scale methanation of hydrogen and CO_2 must be demonstrated and developed to a commercial level. Likewise, separation processes for various CO_2 sources such as biogas and air and the subsequent CO_2 delivery have yet to be verified. These values have not been part of the analysis presented here.

39.8
Conclusion

A rising share of renewable power sources, particularly in the form of wind and solar electric power, will lead to an increase in fluctuating electric power production. As a consequence, temporary situations of excess power production will occur more frequently, causing increased imbalances of supply and demand.

In order to avoid wasting large amounts of energy, excess energy needs to be stored. Converting electric power into chemically bound energy is a feasible way of storing large amounts of energy in the range of several terawatt hours. This concept, known as power to gas, is based on the utilization of excess energy for producing hydrogen via electrolysis.

The chemically bound energy can be used in multiple ways. Reconversion into electricity in times of insufficient renewable power production is reasonable in order to homogenize the power production. Another application can be use as a heat source for households and industry. Moreover, each chemical energy carrier can be applied as fuel in the transportation sector.

Subsequent to the production of hydrogen in the electrolyzer, it can be processed in several ways. The first and easiest path is to feed the hydrogen directly into the natural gas grid. The existing infrastructure, including pipelines, storage facilities, and end users, can be used for small amounts of injected hydrogen. Additionally, no further processing steps are required, which saves costs and ensures the maximum energetic efficiency. The drawback of this concept can be found in the limited amount of injectable gas, since pipelines, storage facilities, and end users are designed for natural gas with only a limited proportion of hydrogen. The maximum proportion is not yet fully known and depends on the particular component. Gas turbines and porous rock storage are expected to be the most sensitive parts, tolerating maximum hydrogen contents in the range 1–5%. Hence this path can only serve as a temporary solution. For higher hydrogen contents, extensive adaptation of the pipeline grid and also of the end users will probably be necessary.

The second alternative is to utilize hydrogen with carbon dioxide via the Sabatier reaction in order to produce methane, which as SNG can be fed into the natural gas grid in any amount without further restrictions, since no issues occur regarding storage, gas power plants, or CNG vehicles. Additionally, the volumetric energy density of methane is over three times higher than that of hydrogen, leading to a lower storage demand. On the other hand, the efficiency of the whole process chain is reduced by 20%, due to the exothermic methanation process. In combination with the additional investment costs for the process unit, this leads to an increase in price per kilowatt hour of about 50% in comparison with renewably produced hydrogen. Another issue is the supply of CO_2 for the reaction. Whereas CO_2 from biogas is neutral in terms of CO_2 emissions and, as a by-product, is available free, the effort in separating CO_2 from fossil power plants or from air may have a major impact on the overall efficiency, making the whole process chain unattractive.

The third way is to distribute the hydrogen directly via a dedicated hydrogen infrastructure. Thereby, similarly to the first case, no further process steps are required. The maximum amount of usable hydrogen is dependent only on the size of the infrastructure, which includes dedicated salt caverns as underground storage. Another advantage in comparison with the previous options is the avoidance of mixing the hydrogen with natural gas. Thereby, the hydrogen can be utilized in a pure form. This is especially relevant for the transport sector. Here, FCVs need pure hydrogen as fuel but have a lower energetic fuel consumption of about 60% in comparison with cars based on ICEs. The disadvantage of a dedicated hydrogen infrastructure is a higher barrier to starting this project, since the whole infrastructure has to be built completely new. Also, the required end users are not yet available or widely spread in the existing economy. The adaptation would need to take place stepwise.

The different scenarios can be assessed energetically and economically. From an energetic point of view, the additional process step of methanation should be avoided, since it decreases the overall efficiency by about 20%. Evaluating the economic potential, the application of the chemical energy also has to be taken into account.

If the renewably produced hydrogen or methane is to be used as a substitute for natural gas, the current production costs calculated at the grid feed-in point exceed the costs for natural gas by four- and six-fold, respectively. Hence, under present conditions this will not be economically viable. Regarding the substitution of fuels for the transportation sector, especially the use of hydrogen may become economically reasonable, since the higher efficiency of fuel cells in comparison with conventional ICEs partially compensates for the higher production costs. Depending on the costs of the excess wind power, the price per kilometer before tax exceeds that for gasoline by 25–50%.

In conclusion, power to gas is a feasible way of storing excess energy for the long term. The different process chains presented come with their particular advantages and drawbacks, which have to be considered with regard to the targeted aim when applying power to gas. Nevertheless, economic analysis reveals that, although ecologically expedient, the reconversion of a renewably produced chemical energy carrier in order to produce electricity or to utilize it as a heat source is not yet economically feasible. On the other hand, the use of hydrogen as a fuel for FCVs has the potential to become an economically sound business case.

References

1 Stolten, D. and Krieg, D. (2010) Alkaline electrolysis–introduction and overview, in *Hydrogen and Fuel Cells* (ed. D. Stolten, Wiley-VCH Verlag GmbH, Weinheim.

2 Laguna-Bercero, M. A. (2012) Recent advances in high temperature electrolysis using solid oxide fuel cells: a review. *J. Power Sources*, **203**, 4–16.

3 Mergel, J., Carmo, M., and Fritz, D. (2013) Status on technologies for hydrogen production by water electrolysis, in *Transition to Renewable Energy Systems* (eds D. Stolten and V. Scherer), Wiley-VCH Verlag GmbH, Chapter 20.

4 Zeng, K. and Zhang, D. (2010) Recent progress in alkaline water electrolysis for hydrogen production and applications. *Prog. Energy Combust. Sci.*, **36**, 307–326.

5 Ursúa, A., Gandía, L. M., and Sanchis, P. (2012) Hydrogen production from water electrolysis: current status and future trends. Proc. IEEE, **100**, 410–426.

6 US DOE NREL (2009) *Current (2009) State-of-the-Art Hydrogen Production Cost Estimate Using Water Electrolysis*, National Renewable Energy Laboratory, Golden, CO.

7 Smolinka, T., Rau, S., and Hebling, C. (2010) Polymer electrolyte membrane (PEM) water electrolysis, in *Hydrogen and Fuel Cells* (ed. D. Stolten), Wiley-VCH Verlag GmbH.

8 Sabatier, P. and Senderen, J.-B. (1902) New synthesis of methane. *J. Chem. Soc.*, **82**, 333.

9 Wang, W. and Gong, J. (2011) Methanation of carbon dioxide: an overview. *Front. Chem. Sci. Eng.*, **5**, 2–10.

10 Yaccato, K. and Cahart, R. (2005) Competitive CO and CO_2 methanation over supported noble metal catalysts in high throughput scanning mass spectrometer. *Appl. Catal. A: Gen.*, **296**, 30–48.

11 Hu, Y. H. (ed) (2010) *Advances in CO_2 Conversion and Utilization*, ACS Symposium Series, vol. 1056, American Chemical Socety, Washington, DC.

12 Rönsch, S. (2011) *Optimierung und bewertung von Anlagen zur Erzeugung von Methan, Strom und Wärme aus biogenen Festbrennstoffen*, DBFZ Report No. 5, Dissertation, Technische Universität Hamburg-Harburg: Deutsches BiomasseForschungsZentrum, Leipzig.

13 Hoekman, S. K. and Broch, A. (2010) CO_2 recycling by reaction with renewably-generated hydrogen. *Int. J. Greenhouse Gas Control*, **4**, 44–50.
14 Lunde, P. J. and F. L. Kester (1974) Kinetics of carbon dioxide methanation on a ruthenium catalyst. *Ind. Eng. Chem. Process Des. Dev.*, **13**, 27–33.
15 Bajohr, S., Götz, M., Graf, F., and Kolb, T. (2012) Dreiphasen-Methanisierung als innovatives Element der PtG-Prozesskette. *gwf-Gas/Erdgas*, (5), 328–325.
16 Solar Fuel GmbH (2011) Hocheffiziente Verfahren zur katalytischen Methanisierung von Kohlendioxid und Wasserstoff enthaltenden Gasgemischen. German Patent Application, DE 10 2009 059 310 A1.
17 Zentrum für Sonnenenergie- und Wasserstoff-Forschung Baden-Württemberg (ZSW) (2012) *Weltweit größte Power-to-Gas-Anlage zur Methan-Erzeugung geht in Betrieb*, http://www.zsw-bw.de/infoportal/presseinformationen/presse-detail/weltweit-groesste-power-to-gas-anlage-zur-methan-erzeugung-geht-in-betrieb.html (last accessed 22 January 2013).
18 Specht, M. and Brellochs, J. (2010) Speicherung von Bioenergie und erneuerbarem Strom im Erdgasnetz. *Erdöl Erdgas Kohle*, **126**, 342–346.
19 Sterner, M. (2009) Bioenergy and renewable power methane in integrated 100% renewable energy systems. Dr.-Ing. thesis, University of Kassel.
20 Kavalov, B., Petrić H., and Georgakaki, A. (2009) *Liquefied Natural Gas for Europe—Some Important Issues for Consideration*, JRC Reference Report, European Commission Joint Research Centre, Petten, http://ec.europa.eu/dgs/jrc/downloads/jrc_reference_report_200907_liquefied_natural_gas.pdf (last accessed 22 January 20130.
21 Brown, M., Bryant, N., and Haynes, D., *Study on LNG Quality Issues, a Study for the European Commission – JRC Institute for Energy*, Advantica, Loughborough.
22 International Energy Agency (2012) *World Energy Outlook 2012*, IEA, Paris.
23 Metz, B., Davidson, O., de Coninck, H., Loos, M., and Meyer, L. (eds) (2005), *IPCC Special Report on Carbon Dioxide Capture and Storage*, Cambridge University Press, Cambridge, http://www.ipcc.ch/pdf/special-reports/srccs/srccs_wholereport.pdf (last accessed 22 January 2013).
24 Franz, J. and Scherer, V. (2010) An evaluation of CO_2 and H_2 selective polymeric membranes for CO_2 separation in IGCC processes. *J. Membr. Sci.*, **359**, 173–183.
25 Göttlicher, G. (2004) *The Energetics of Carbon Dioxide Capture in Power Plants*, National Energy Technology Laboratory (NETL), www.netl.doe.gov/technologies/carbon_seq/refshelf/refshelf.html (last accessed 22 January 2013).
26 Pfaff, I. and Kather, A. (2009) Comparative thermodynamic analysis and integration issues of CCS steam power plants based on oxy-combustion with cryogenic or membrane based air separation. *Energy Procedia*, **1**, 495–502.
27 Riensche, E., et al. (2011) Capture options for coal power plants, in *Efficient Carbon Capture for Coal Power Plants* (eds D. Stolten and V. Scherer), Wiley-VCH Verlag GmbH, Weinheim.
28 Finkenrath, M. (2012) Carbon dioxide capture from power generation–status of cost and performance. *Chem. Eng. Technol.*, **35** (3). 482–488.
29 Wall, T. F. (2007) Combustion processes for carbon capture. *Proc. Combust. Inst.*, **31** (1), 31–47.
30 Finkenrath, M. (2011) *Cost and Performance of Carbon Dioxide Capture from Power Generation*, IEA, Paris.
31 Singh, D., et al. (2003) Techno-economic study of CO_2 capture from an existing coal-fired power plant: MEA scrubbing vs. O_2/CO_2 recycle combustion. *Energy Convers. Manage.*, **44**, 3073–3091.
32 Abu-Zahra, M. R. M., Niederer, J. P. M., and Feron, P. H. M., CO_2 capture from power plants. Part II. A parametric study of the economical performance based on mono-ethanolamine. *Int. J. Greenhouse Gas Control*, **1**, 135–142.
33 Specht, M., et al. (2010) Speicherung von Bioenergie und erneuerbarem Strom im Erdgasnetz, *Erdöl Erdgas Kohle*, **126** (10), 342–346.
34 Sterner, M. (2009) Bioenergy and renewable power methane in integrated

100% renewable energy systems, Dissertation of Fraunhofer IWES, University of Kassel.

35 European Biofuels Technology Platform: Zero Emissions Platform (ZEP) (2011) *Biomass with CO_2 Capture and Storage (Bio-CCS)*, ZEP and EBTP.

36 IEAGHG (2011) *Potential for Biomass and Carbon Dioxide Capture and Storage*, Report 2011/06, IEAGHG, Cheltenham, http://www.eenews.net/assets/2011/08/04/document_cw_01.pdf (last accessed 22 January 2011).

37 Nussbaumer, T. (2003) Combustion and co-combustion of biomass: fundamentals, technologies, and primary measures for emission reduction. *Energy Fuels*, **17**, 1510–1521.

38 Shao, P., et al. (2012) Design and economics of a hybrid membrane–temperature swing adsorption process for upgrading biogas. *J. Membr. Sci.*, **413–414**, 17–28.

39 Wahlund, B., Yan, J., and Westermark, M. (2004) Increasing biomass utilisation in energy systems: a comparative study of CO_2 reduction and cost for different bioenergy processing options. *Biomass Bioenergy*, **26** (6), 531–544.

40 IEA (2012) *Technology Roadmap: Bioenergy for Heat and Power*, http://www.iea.org/publications/freepublications/publication/bioenergy.pdf (last accessed 22 January 2013).

41 US Environmental Protection Agency, Climate Leaders Greenhouse Gas Inventory Protocol (2003) *Direct Emissions from Iron & Steel Production*, http://www.epa.gov/climateleadership/documents/resources/ironsteel.pdf (last accessed 22 January 2013).

42 Worrell, E., et al. (2001) Carbon dioxide emissions from the global cement industry. *Annu. Rev. Energy Environ.*, **26**, 303–329.

43 Ausfelder, F. and Bazzanella, A. (2008) *Diskussionspapier: Verwertung und Speicherung von CO_2*, Dechema, Frankfurt.

44 Sterner, M., Jentsch, M., and Holzhammer, U. (2011) *Energiewirtschaftliche und ökologische Bewertung eines Windgas-Angebotes*, Fraunhofer IWES, Kassel, http://michaelwenzl.de/wiki/_media/ee:greenpeace_energy_gutachten_windgas_fraunhofer_sterner.pdf (last accessed 22 January 2013).

45 Kohl, A. and Nielsen, R. (1997) *Gas Purification*, 5th edn, Gulf Publishing, Houston, TX.

46 House, K. Z., et al. (2011) Economic and energetic analysis of capturing CO_2 from ambient air. *Proc. Natl. Acad. Sci. U.S.A.*, **108** (51), 20428–20433.

47 Goeppert, A., et al. (2012) Air as the renewable carbon source of the future: an overview of CO_2 capture from the atmosphere. *Energy Environ. Sci.*, **5** (7), 7833.

48 Keith, D., Ha-Duong, M., and Stolaroff, J. (2005) Climate strategy with CO_2 capture from the air. *Climate Change*, **74**, 17–45.

49 Baciocchi, R., Storti, G., and Mazzotti, M. (2006) Process Design and Energy Requirements for the Capture of Carbon Dioxide from Air. *Chem. Engin. Process.*, **45**, 1047–1058.

50 Zeman, F. (2007) Energy and material balance of CO_2 capture from ambient air. *Environ. Sci. Technol.*, **41**, 7758–7563.

51 Stolaroff, J., Lowry, G., and Keith, D. (2008) Carbon dioxide capture from atmospheric air using sodium hydroxide spray. *Environ. Sci. Technol.*, **42**, 2728–2735.

52 Mahmoudkhani, M. and Keith, D. (2009) Low-energy sodium hydroxide recovery from CO_2 capture from atmospheric air – thermodynamic analysis. *Int. J. Greenhouse Gas Control*, **3**, 376–384.

53 Specht, M., et al. (2000) *CO_2-Recycling zu Herstellung von Methanol*, ZSW – Zentrum fuer Sonnenenergie- und Wasserstoffforschung Baden-Wuerttemberg, Stuttgart.

54 Adamek, F. and Aundrup, T. (2012) *Energiespeicher für die Energiewende*. http://www.chemieingenieurwesen.de/VDE-Studie_Energiespeicher_Kurzfassung.pdf.

55 Crotogino, F. and Donadei, S. (2012) *Power-2-Gas – Zukünftige Nutzung der bestehenden Gasinfrastruktur zur Speicherung erneuerbarer Energien*. European Hydrogen Road Tour, KBB Undergroud Technologies, Hannover, http://www.kbbnet.de/wp-content/

uploads/2011/05/120917-Power-2-Gas-European-Hydrogen-Road-Tour.pdf (last accessed 22 January 2013).

56 Müller, K., et al. (2012) Energetische Betrachtung der Wasserstoffeinspeisung ins Erdgasnetz. *Chem.-Ing.-Tech.*, **84** (9), 1513–1519.

57 Crotogino, F. (2011) Wasserstoff-Speicherung in Kavernen, presented at the PRO H2 Technologie Forum, http://www.kbbnet.de/wp-content/uploads/2011/09/Wasserstoffspeicherung-in-Kavernen.pdf.

58 Schmitz, S. (2011) Einfluss von Wasserstoff als Gasbegleitstoff auf Untergrundspeicher, presented at DBI-Fachforum Energiespeicherkonzepte und Wasserstoff, http://www.dbi-gti.de/fileadmin/downloads/5_Veroeffentlichungen/Tagungen_Workshops/2011/H2-FF/13_Schmitz_DBI_GUT.pdf.

59 Klaus, T., Vollmer, C., Werner, K., Lehmann, H., and Müschen, K. (2010) Energieziel 2050: 100% Strom aus erneuerbaren Quellen, Preprint for Federal Press Conference, 7 July 2010, http://www.umweltdaten.de/publikationen/fpdf-l/3997.pdf.

60 Sedlacek, R. (2012) Untertage-Gasspeicherung in Deutschland. *Erdöl Erdgas Kohle*, **126** (11), 453–465.

61 Hammer, G., Lübcke, T., Kettner, R., Pillarella, M. R., Recknagel, H., Commichau, A., Neumann, H.-J., Paczynska-Lahme, B. (206) Natural gas, in *Ullmann's Encyclopedia of Industrial Chemistry*, Wiley-VCH Verlag GmbH, Weinheim.

62 Ozarslan, A. (2012) Large-scale hydrogen energy storage in salt caverns. *Int. J. Hydrogen Energy*, **37** (19), 14265–14277.

63 Civan, F. (2004) Natural gas transportation and storage, in *Encyclopedia of Energy* (ed. C. Cleveland), Elsevier, Oxford, vol. 4, pp. 273–282.

64 Crotogino, F., Donadei, S., Bünger, U., and Landinger, H. (2010) Large-scale hydrogen underground storage for securing future energy supplies, in *Proceedings of the 18th World Hydrogen Energy Conference – WHEC 2010. Parallel Sessions Book 4: Storage Systems/Policy Perspectives, Initiatives and Cooperations* (eds D. Stolten and T. Grube), Forschungszentrum Jülich, Jülich.

65 Panfilov, M. (2010) Underground storage of hydrogen: in situ self-organisation and methane generation. *Transport Porous Media*, **85** (3), 841–865.

66 Friedrich, K. A. (2012) Wasserstoff als Chemischer Speicher: Erzeugung, Verteilung und Speicherung, presented at Energiespeichersymposium, Stuttgart, http://www.dlr.de/tt/Portaldata/41/Resources/dokumente/ess_2012/Friedrich_Wasserstoff_Chemische_Speicher.pdf.

67 Bünger, U. and Crotogino, F. (2009) in *Energiespeicher in Stromversorgungssystemen mit hohem Anteil erneuerbarer Energieträger*, VDE, Frankfurt. http://www.vde.com/de/fg/ETG/Arbeitsgebiete/V1/Aktuelles/Oeffentlich/Seiten/Studie-Energiespeicher.aspx

68 Civan, F. (2004) Natural gas transportation and storage, in *Encyclopedia of Energy* (ed. C. Cleveland), Elsevier, Oxford, vol. 4, pp. 273–282.

69 AGEB (2012) *Energieverbrauch in Deutschland im Jahr 2011*, AGEB: Arbeitsgemeinschaft Energiebilanzen, Berlin.

70 Konstantin, P. (2009) *Praxisbuch Energiewirtschaft: Energieumwandlung, -transport und -beschaffung im liberalisierten Markt*, Springer, Berlin.

71 Franke, P. (2012) Strom- und Gasnetze: Zwei ungleiche Partner auf gemeinsamem Weg?, presented at dena Konferenz der Strategieplattform Power to Gas.

72 DVGW (Deutscher Verein des Gas- und Wasserfaches e. V.) (2008) *Technische Regel*, Arbeitsblatt G 260 – Gasbeschaffenheit.

73 Frontier Economics (2010) *Energiekosten in Deutschland – Entwicklungen, Ursachen und internationaler Vergleich (Projekt 43/09)*, Frontier Economics, London.

74 Schmid, J., Specht, M., Sterner, M., et al. (2011) *Welche Rolle spielt die Speicherung erneuerbarer Energien im zukünftigen Energiesystem?*, Fraunhofer IWES, Kassel.

75 Fasold, H. G. (1995) *Erdgastransport, Erdgasspeicherung und Erdgasverteilung*, Forschungszentrum Jülich, Jülich.

76 Hüttenrauch, J. and Müller-Syring, G. (2010) Zumischung von Wasserstoff zum Erdgas. *Energie Wasser-Praxis*, **61** (10), 68–71.
77 Sterner, M, Jentsch, M., and Holzhammer, U. (2011) *Energiewirtschaftliche und ökologische Bewertung eines Windgas-Angebotes*, Fraunhofer IWES, Kassel.
78 Deutsche Fernleitungsnetzbetreiber (2012) *Netzentwicklungsplan Gas 2012 – Entwurf* (ed. M. Wild), http://www.netzentwicklungsplan-gas.de/files/130310_netzentwicklungsplan_gas_2012.pdf.
79 BNetzA (2011) Genehmigung des Szenariorahmens zur energiewirtschaftlichen Entwicklung nach § 12a EnWG, presented at a Press Conference, 7 December 2011, Bundesnetzagentur, Bonn.
80 Krieg, D. (2012) *Konzept und Kosten eines Pipelinesystems zur Versorgung des deutschen Strassenverkehrs mit Wasserstoff*, Forschungszentrums Jülich, Jülich, http://juwel.fz-juelich.de:8080/dspace/bitstream/2128/4608/3/Energie%26Umwelt_144.pdf.
81 GermanHy (2009) *Woher kommt der Wasserstoff in Deutschland bis 2050?*, Deutsche Energie-Agentur (dena), Berlin.
82 Stolten, D., Grube, T., and Mergel, J. (2012) Beitrag elektrochemischer Energietechnik zur Energiewende, VDI-Berichte Nr. 2183, pp. 199–215.
83 Ball, M. (2006) *Integration einer Wasserstoffwirtschaft in ein nationales Energiesystem am Beispiel Deutschlands: Optionen der Bereitstellung von Wasserstoff als Kraftstoff im Straßenverkehr bis zum Jahr 2030*, Fortschritt-Berichte VDI No. 177, VDI, Düsseldorf.
84 Wietschel, M., Bünger, U., and Weindorf, W. (2010) *Vergleich von Strom und Wasserstoff als CO_2-freie Endenergieträger: Endbericht; Studie im Autrag der RWE AG*, Fraunhofer ISI, Karlsruhe.
85 Yang, C. and Ogden, J. (2007) Determining the lowest-cost hydrogen delivery mode. *Int. J. Hydrogen Energy*, **32** (2), 268–286.
86 Johnson, N. and Ogden J. (2012) A spatially-explicit optimization model for long-term hydrogen pipeline planning. *Int. J. Hydrogen Energy*, **37** (6), 5421–5433.
87 AG Energiebilanzen (2012) *Energy Balance 2010*, www.ag-energiebilanzen.de (last accessed 23 January 2013).
88 Appl, M. (2006) Ammonia, in *Ullmann's Encyclopedia of Industrial Chemistry*, Wiley-VCH Verlag GmbH, Weinheim, p. 122.
89 Müller-Syring, G., Henel, M., and Rasmusson, H. (2011) *Power to Gas: Untersuchungen im Rahmen der DVGW-Innovationsoffensive zur Energiespeicherung*, 2011, Deutscher Verein des Gas- und Wasserfaches (DVGW), Bonn.
90 Brower, M., et al. (2012) Ignition delay time and laminar flame speed calculations for natural gas/hydrogen blends at elevated pressures, presented at ASME Turbo Expo GT2012, Copenhagen, 11–15 June 2012.
91 Giesecke, J. and Mosonyi, E. (2009) *Wasserkraftanlagen*, Springer, Berlin.
92 Edwards, R., Larivé, J.-F., and Beziat, J.-C. (2011) *Well-to-Wheels Analysis of Future Automotive Fuels and Powertrains in the European Context. WTT Appendix 1: Description of Individual Processes and Detailed Input Data*, EUR 24952 EN, European Comission Joint Research Center, Institute for Energy and Transport, Ispra.
93 Bundesnetzagentur and Bundeskartellamt (2012) *Monitoringbericht Strom und Gas 2012*, Bundesnetzagentur and Bundeskartellamt, Bonn.
94 Rogatty, W. (2012) *Personal Communication*, Viessmann Werke, Allendorf, Germany.
95 Stolten, D. and Bernd, E. (eds) (2012) *Fuel Cell Science and Engineering*, Wiley-VCH Verlag GmbH, Weinheim.
96 Mercedes-Benz (2012) *Presseinformation – Mercedes-Benz B-Klasse F-CELL: Mercedes-Benz übergibt erste Elektrofahrzeuge mit Brennstoffzelle in Kundenhand*, Press Release, Daimler, Stuttgart.
97 Honda (2012) *FCX Clarity Specifications*, http://automobiles.honda.com/fcx-clarity/specifications.aspx (last accessed 22 January 2013).

98 Thiesen, L. P., von Helmolt, R., and Berger, S. (2010) *HydroGen4 – The First Year of Operation in Europe*, presented at the 18th World Hydrogen Energy Conference 2010.

99 Rees, J. (2010) *Brennstoffzelle Reloaded*, http://www.wiwo.de/technologie/auto/wasserstoffautos-brennstoffzelle-reloaded/5666152.html (last accessed 22 January 2013).

100 US Department of Energy (2012) *Multi-Year Research, Development and Demonstration Plan, Hydrogen, Fuel Cells and Infrastructure Technologies Program*, US Department of Energy, Washington, DC.

101 European Commission (2011) *2011 Technology Map of the European Strategic Energy Technology Plan (SET-Plan)*, EUR 24979 EN European Commission, Luxemburg.

102 Kromer, M. A. and Heywood, J. B. (2008) A comparative assessment of electric propulsion systems in the 2030 US light-duty vehicle fleet, presented at the SAE 2008 World Congress, Detroit, MI.

103 Boisen, P. (2012) *NGVs and Refuelling Stations in Europe*, Natural and Bio Gas Vehicle Association Europe, Madrid.

104 Backhaus, R. (2012) Entwicklungen bei Tanksystemen, *ATZ Automobiltech. Z.*, **114** (2), 132–135.

105 Edwards, R., Larivé, J.-F., Mahieu, V., and Rouveirolles, P. (2008) *Well-to-Wheels Analysis of Future Automotive Fuels and Powertrains in the European Context. Well-to-Tank Report, Version 3, October 2008*, European Comission Joint Research Center, Institute for Environment and Sustainability, Ispra.

106 Mangold, R. (2011) Das Projekt e-Gas: Power-to-Gas im Verkehrssektor, in *Power-to-Gas – Erdgasinfrastruktur als Energiespeicher*. Bundesnetzagentur, Berlin.

107 Klell, M. and Sartory, M. (2007) *Wasserstofferdgasgemische in Verbrennungsmotoren*, Hydrogen Center Austria, Graz.

108 ISO (2000) *Gas Cylinders – High Pressure Cylinders for the On-Board Storage of Natural Gas as a Fuel for Automotive Vehicles*, ISO 11439:2000, ISO, Geneva.

109 Åkermann, K., Hutton, D., Keenan, M., Owen, N., and Patterson, J. (2008) *Deliverable 1.9: A Fuel Cell and Hydrogen Technology Watch Based on Emerging Products – Final Report*, Document No. R2H1021PU.2, Roads2HyCom, http://www.roads2hy.com/r2h_Downloads/Roads2HyCom%20R2H1021PUv2%20-%20Technology%20Watch%20Report.pdf (last accessed 22 January 2013).

110 H2Mobility (2012) *H2Mobility: Hydrogen Vehicles. Timeline of all Hydrogen Vehicles Worldwide*, http://www.netinform.net/h2/H2Mobility/Default.aspx (last accessed 22 January 2013).

111 Stolten, D., Grube, T., and Mergel, J. (2012) Beitrag elektrochemischer Energietechnik zur Energiewende, in *VDI-Tagung Innovative Fahrzeugantriebe*, VDI-Verlag, Dresden.

112 Frontier Economics (2010) *Energiekosten in Deutschland – Entwicklungen, Ursachen und internationaler Vergleich (Projekt 43/09). Endbericht für das Bundesministerium für Wirtschaft und Technologie*, Frontier Economics, London.

113 Selke, J. and Haas, A. (2011) *National Report on State of CNG/Biomethane Filling Station – Germany*, GasHighway Project, WP3, Deliverables 3.1, GERBIO, German Society for Sustainable Biogas and Bioenergy Utilization, Kirchberg.

114 Umweltbundesamt (2011) *National Inventory Report for the German Greenhouse Gas Inventory 1990–2009. Submission Under the United Nations Framework Convention on Climate Change and the Kyoto Protocol 2011*, Climate Change 12/2011, Umweltbundesamt, Dessau-Rosslau, http://www.umweltdaten.de/publikationen/fpdf-l/4127.pdf (last accessed 23 January 2013).

Part VII
Applications

40
Transition from Petro-Mobility to Electro-Mobility

David L. Greene, Changzheng Liu, and Sangsoo Park

40.1
Introduction

From the perspective of physics, transportation is work: force must be applied to overcome inertia and friction in order to move mass over a distance. Work cannot be accomplished without energy; energy is the ability to do work. Hence energy is and always will be essential for transportation. Securing adequate energy for transportation while protecting the environment and allowing continued economic development poses an enormous challenge for society. The recent Global Energy Assessment (GEA) report concluded that solving the world's energy problems would require an urgent transition to alternative forms of energy:

> *Without question a radical transformation of the present energy system will be required over the coming decades ... An effective transformation requires immediate action.* [1]

Transporting people and materials is essential to human society, yet the global transportation system has grown to such an extent that it is now large relative to the resources that it requires and the environmental systems on which humanity depends. Modern economies move ~15 trillion ton-kilometers of freight each year, and passengers travel 40 trillion kilometers by motorized transport [2]. Today, over 93% of the energy used globally in motorized transport is obtained from one fossil form of energy: petroleum [3]. The economic and social importance of transportation and its near total dependence on petroleum, combined with the geographical concentration of petroleum resources and their control by nation states, has led to serious concerns about energy security among many oil-importing economies. Combustion of petroleum fuels to power the world's transportation systems also causes significant damage to the global environment. Carbon dioxide (CO_2) accounts for nearly all of the global warming potential of greenhouse gases (GHGs) emitted by transportation vehicles, which comprise ~15% of global anthropogenic GHG emissions and 25% of global emissions from energy use [2]. Transportation's GHG

Transition to Renewable Energy Systems, 1st Edition. Edited by Detlef Stolten and Viktor Scherer.
© 2013 Wiley-VCH Verlag GmbH & Co. KGaA. Published 2013 by Wiley-VCH Verlag GmbH & Co. KGaA.

emissions are almost entirely comprised of CO_2 produced by the combustion of fossil petroleum.

Global resources of conventional petroleum are finite and the rate of global consumption is now large (1.5–3%) relative to best estimates of annual ultimately recoverable conventional petroleum resources (1–2 trillion barrels). Unconventional fossil resources that can be converted into liquid hydrocarbon fuels at costs that the world's transportation system has already demonstrated it is willing to pay are vast. Indeed, a transition to unconventional petroleum resources such as oil sands and extra-heavy oil has already begun. Combustion of unconventional petroleum emits more CO_2 per unit of final energy than conventional petroleum but, more importantly, it begins the exploitation of more than three times as much fossil carbon [4]. Over the past 40 years, energy security and environmental concerns have induced the world's governments to promote biofuels, electricity, and hydrogen as alternatives to petroleum. However, the energy density, relatively low cost, and convenience of petroleum fuels, in addition to the technology "lock-in" of the internal combustion engine (ICE), have frustrated these efforts. To date, increasing energy efficiency has been the most effective strategy for restraining the growth of transport petroleum consumption and it will continue to be so for the next decade or two.

Road vehicles produce the most transportation and use the greatest quantities of energy (Figure 40.1). Of the 92 EJ (92×10^{18} J) of final energy used by transport each year, cars, trucks, and buses account for about 75% [4]. Air transport is second in passenger-kilometers and energy use, using 10 EJ, nearly all of which is kerosene for subsonic turbine engine aircraft. About 90% of the 9 EJ of energy used for waterborne transport is used in international shipping. All modes of transport are important energy users and GHG emitters, but the focus of this chapter will be on road vehicles and especially light vehicles (< 3.5 t).

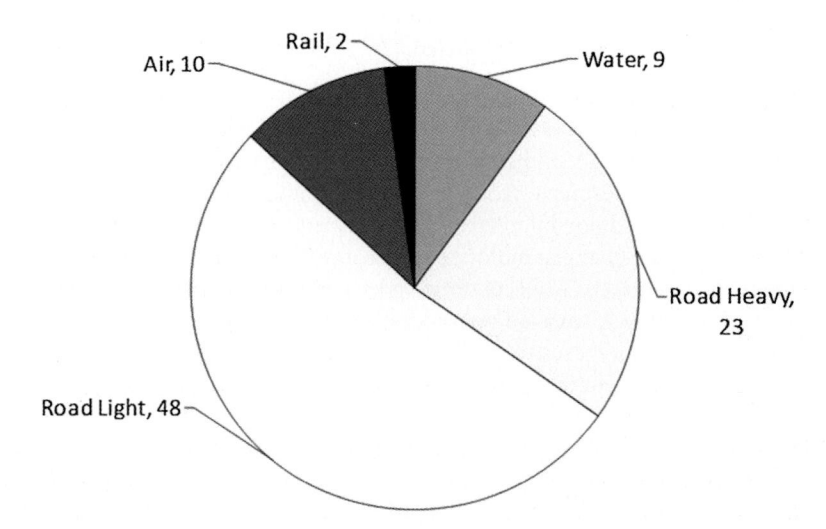

Figure 40.1 World transport energy use, 2009 (final use, 92 EJ) [2].

Transport energy use has more than doubled since 1970, increasing at an average rate of about 2% per year. Recent "business as usual" projections anticipate slower growth in the future, with nearly all the growth occurring in the developing economies, but still foresee a near doubling of global energy use for transport by 2050 [2].

This chapter briefly reviews recent progress in electric drive technologies (Section 40.2), discusses the importance of increasing energy efficiency to the transition to sustainable energy for motor vehicles (Section 40.3), describes the challenges a transition poses for public policy and outlines a framework for decision-making (Section 40.4), briefly reviews the status of electric drive vehicles and transition plans in several countries (Section 40.5), and presents some insights from modeling of the transition process (Section 40.6).

40.2
Recent Progress in Electric Drive Technologies

The rate of progress of electric drive technologies over the past 10–15 years is consistent with what will be necessary to compete successfully with conventional ICE vehicles some time in the next decade. Still, further cost reductions and performance improvements are needed and it is not certain that they can be achieved.

Lithium-ion battery cells for all uses sold for an average of US$ 2600 kWh^{-1} as recently as 1999; in 2011 their price was an order of magnitude lower: $ 240 kWh^{-1} [5]. Yet there is still a long way to go to reach competitive costs. The $ 725 kWh^{-1} cost of a full battery pack (in contrast to cell cost) in 2011 decreased to $ 500 kWh^{-1} in 2012 [2] and is projected to fall to $ 320 kWh^{-1} by 2020 and $ 215 kWh^{-1} by 2030 (although some reports put the cost already at $ 400 kWh^{-1} in 2013). Even at the 2030 cost, the battery pack for a battery electric vehicle (BEV) with 30 kWh of on-board storage would cost $ 6400, more than a complete ICE powertrain [5]. Costs of electric motors and controllers have also been reduced significantly [6].

Estimated costs of automotive fuel cell systems in high-volume production have decreased from over $ 250 kWh^{-1} in 2002 to $ 48 kWh^{-1} in 2011 [7, 8] (Figure 40.2). Although the cost estimates are hypothetical, most original equipment manufacturers (OEMs) developing fuel cell technology consider them a reasonable description of technological progress (this assertion is based on the author's private discussions with seven major OEMs during 2012). Still, stack costs must reach $ 20–25 kWh^{-1} to compete effectively with ICE powertrains. Over the same time period, the durability of automotive cells increased from 1500 to 5500 h, platinum content was greatly reduced, and stack size was halved [9, 10]. The reductions in platinum loadings together with the ability to recycle platinum at the rate of 98% have allayed early concerns about the adequacy of world platinum resources to support large numbers of hydrogen fuel cell vehicles (FCVs). The range of FCVs has increased to over 400 km and the time required to refuel has decreased to 5 min.

Standardization, increased production volumes, and competition among suppliers have already reduced the cost of recharging and hydrogen refueling infrastructure by half and further cost reductions are expected in the future.

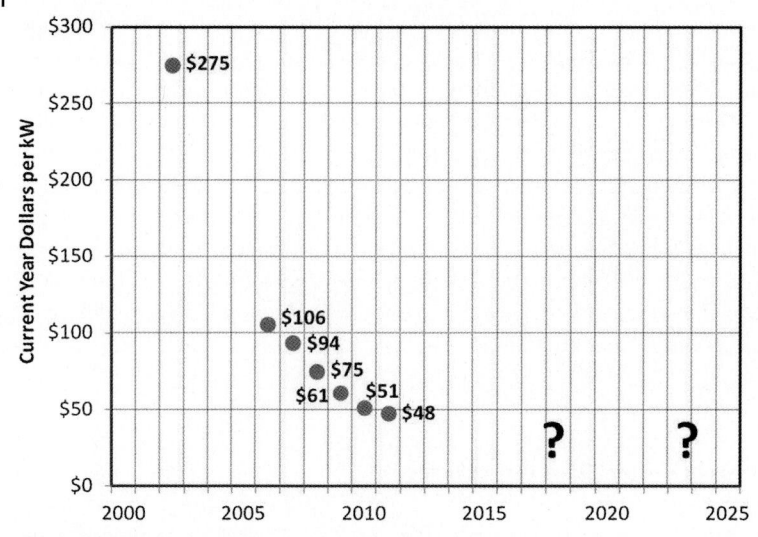

Figure 40.2 Estimated costs of automotive fuel cell systems assuming current year technology and production of 5 000 000 units per year [7].

40.3
Energy Efficiency

In the next one to two decades, increasing energy efficiency will be the most effective strategy for improving the sustainability of transportation. Even after a century of technological advances, it is still possible to double, triple, or even quadruple the energy efficiency of motor vehicles. On the US combined drive cycle, a modern gasoline ICE delivers only about one-sixth to one-seventh of the energy content of the fuel to the vehicle's wheels (Figure 40.3). Three-quarters is lost in the conversion of chemical energy to mechanical power by the ICE. Energy conversion losses in a hydrogen fuel cell vehicle (HFCV) are only half as large, while the energy losses of a BEV from plug to wheels are less than one-quarter of that of an ICE vehicle. While the energy efficiency of ICEs will continue to improve, even greater energy efficiencies will be attainable by switching to electric drive.

Reducing load by reducing vehicle mass, aerodynamic drag, and rolling resistance directly reduces energy consumption and benefits all powertrain technologies. In addition, it is likely that reductions in vehicle load will favor electric drive vehicles because the most costly components, namely battery packs and fuel cell stacks, will scale directly with power requirements. If loads can be significantly reduced, it is possible that BEVs and fuel cell vehicles (FCVs) may eventually cost less to produce than ICE vehicles [11].

Improving energy efficiency produces immediate reductions in petroleum use and GHG emissions, buying time for the development of electro-mobility technology and markets. However, it can also dramatically reduce the amount of clean energy required to power future vehicles, as the following simple calculation illustrates.

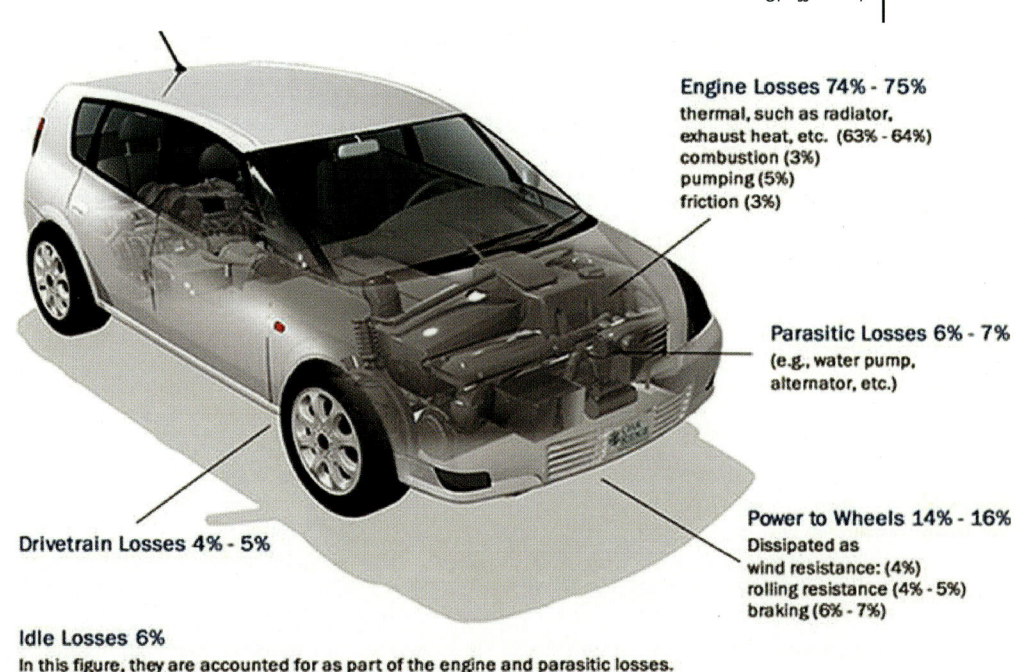

Engine Losses 74% - 75%
thermal, such as radiator, exhaust heat, etc. (63% - 64%)
combustion (3%)
pumping (5%)
friction (3%)

Parasitic Losses 6% - 7%
(e.g., water pump, alternator, etc.)

Drivetrain Losses 4% - 5%

Power to Wheels 14% - 16%
Dissipated as
wind resistance: (4%)
rolling resistance (4% - 5%)
braking (6% - 7%)

Idle Losses 6%
In this figure, they are accounted for as part of the engine and parasitic losses.

Figure 40.3 Energy losses in a modern ICE vehicle: combined U. S. test cycles. Source: www.fueleconomy.gov [12].

The importance of increased energy efficiency to reducing global transportation GHG emissions to 50% of their current level can be illustrated by a simple calculation. A very simple extrapolation of 2009 transportation energy use by mode leads to the approximate doubling of energy use in 2050, as expected by the International Energy Agency (IEA). The IEA has estimated that if light-duty vehicles achieved the target fuel consumption rate of the Global Fuel Economy Initiative (5.6 l per 100 km by 2020 and 4.1 l per 100 km by 2030), fuel consumption in 2050 would be 50% lower than it would have been with no fuel economy improvement [2]. Other modes are capable of similar or possibly smaller efficiency improvements [2, 13]. If we assume that road and air energy intensity can be halved by 2050 and that waterborne and rail energy use can be reduced by one-third, transport energy use in 2050 would be just slightly lower than its 2009 level. Further, if we assume that every 10% improvement in energy efficiency results in a increase of 1% in activity (rebound effect), we are almost exactly back to the level of 2009 energy use in 2050. First, this illustration shows that even the ambitious energy efficiency improvements assumed in Table 40.1 are not enough. Second, it illustrates the importance of energy efficiency to enabling low GHG energy. Without any efficiency improvement, 145 EJ of zero GHG energy would be required to reduce GHG emissions to 50% of the 2009 level in 2050. With the assumed energy efficiency improvements, only one-third as much zero emission energy is required.

Table 40.1 Illustration of the effect of energy.

Mode	Energy use 2009 (EJ)	Growth rate (%)	Energy use 2050 (EJ)	Energy intensity (%)	Energy use with rebound	
					2050	2050
Light-duty	48	1.3	81.5	−50	40.8	43.7
Heavy-duty	23	1.5	42.3	−50	21.2	22.7
Air	10	2.0	22.5	−50	11.3	12.1
Water	9	1.8	18.7	−33	2.0	2.1
Rail	2	1.0	3.0	−33	2.0	2.1
Total	92		168		88	94

40.4 The Challenge of Energy Transition

Achieving sustainability, assuring that the world we leave to future generations allows a quality of life at least as good as our own, may be humanity's greatest task in the twenty-first century. Transitioning to sustainable energy systems is an essential element of a sustainable global society [3, 14]. As the world's vehicle population grows towards 2 billion, motorized transportation is challenged to contribute to protection of the global climate system, energy security, and the elimination of the adverse health effects of local air pollution [15]. There is no single solution to these problems [13, 16, 17]. However, it is becoming increasingly clear that achieving sustainable transportation implies a transition from petroleum-based ICEs to zero emission electric drives.

Bringing about a large-scale energy transition for transportation to achieve public goods requires a new paradigm for public policy. The concept of externalities enabled a new understanding of the causes of environmental pollution and provided a paradigm for formulating efficient policy responses. Accomplishing major energy transitions for the public good poses a new and different challenge and calls for a new paradigm. It is proposed here that the new economic paradigm for efficient energy transitions should be built on the concepts of net social value, network external benefits, and adaptation to an uncertain future [18]. Net social benefits must be the key metric because a broadly based cost–benefit framework is necessary to compare very different future states of the world; marginal analysis is inadequate. Network external benefits are key concepts because the process of breaking down the natural economic barriers to transition is comprised of actions that provide future benefits to others with costs incurred in the present [19]. Uncertainty is an inescapable dimension because the future technologies and markets will undoubtedly surprise us, and our knowledge even of current market behaviors is in many ways inadequate.

The transition to electric drive vehicles faces six major economic barriers that help lock in petroleum powered ICE vehicles:

1. current technological limitations of alternative powertrains and fuels;
2. learning by doing;
3. scale economies;
4. consumers' aversion to the risk of novel products;
5. lack of diversity of choice in the early market for alternatives;
6. lack of an energy supply infrastructure for alternatives.

Market researchers add lack of awareness, social exposure, and willingness to consider to this list [20]. Each of these barriers is difficult to quantify and uncertain in the future. However, it is possible to quantify them approximately using the best available information with the understanding that as knowledge improves and conditions change, assessments must be revised and policies adapted. By quantifying the transition barriers, the costs of overcoming them can be measured. Each of the six barriers above can be viewed either as a *transition cost* or as an *external benefit* produced for subsequent customers as new vehicles are purchased and new fuels are deployed, provided that the transition leads to a better, sustainable future.

Modern economics recognizes "network externalities," positive external benefits that one user of a commodity can produce for another [21]. Network external benefits have been extensively studied in the development of personal computer operating systems and cell phones, for example. Every consumer who purchases a given computer operating system increases the value of that system to other users by expanding the network of users with whom files can be exchanged seamlessly and by increasing the size of the market for software developers. These are called direct network external benefits. When software developers produce a new product for the operating system, it increases the value of the systems to those who have already purchased them; these are referred to as indirect network external benefits. Analogously, when a city installs a public recharging station, it increases the value of owning an EV for every EV owner, an indirect network externality. When an innovator purchases one of the first HFCVs, direct external benefits are produced for other buyers in the form of reduced risk of buying a now slightly less novel technology, as well as via scale economies and learning by doing. The last two benefits are referred to as pecuniary network external benefits because they are reflected in the prices of the vehicles. Network external benefits will be key drivers of the transition to electric drive vehicles. They create important positive feedbacks that can lead to a self-sustaining transition. Although it is difficult, it is not impossible to measure these external benefits and to use them in formulating efficient public policies. Measuring the value of early deployment of vehicles and infrastructure at a time when it might seem uneconomical is an important part of understanding the new paradigm for energy transitions.

40.5
A New Environmental Paradigm: Sustainable Energy Transitions

If every path to the future could be assigned a value reflecting its overall worth to society, then alternative paths with different policies inducing different combinations of vehicles and fuels could be compared. The net social value (NSV) of a path is the sum over all future years ($t = 0, \infty$) of its full benefits minus its full costs. To simplify, assume both benefits (B) and costs (C) depend only the numbers of vehicles of each type (N_i), the supporting energy infrastructure (K_i), the status of technology (T), and economic conditions, such as energy prices (E), for all years ($\forall$). This equation of net social value can be written as follows:

$$\text{NSV} = \sum_{t=s}^{\infty} \sum_{i=1}^{n} \left[B(N_{it}, K_{it}, T_{it}, E_{it}) - C(N_{it}, K_{it}, T_{it}, E_{it}) \right] \forall\ s = 0, \infty \qquad (40.1)$$

In any given year, increasing the number of EVs or expanding their refueling infrastructure will change the costs and benefits not only in that year but also in succeeding years. The costs and benefits of adding EVs and infrastructure in any year will also depend on what has happened in previous years. The optimal number of alternative vehicles to deploy in any given year can be calculated but is conditional on the quantities in previous and succeeding years also having been chosen in the best possible way. There will be a change in NSV for the first 100 EVs sold in year t, and also for the second, third and fourth 100 EVs, and so on. A similar set of calculations could be made for infrastructure, but for simplicity it is assumed that the optimal quantity of infrastructure has been constructed for year t.

Plotting the change in net present value (NPV) against the number of vehicles produces a downward-sloping curve describing the value to society (or the maximum amount society would be willing to pay) for producing and selling different quantities of EVs in year t (this is the marginal net present social value curve in Figure 40.4). The subsidy per vehicle required to produce a given level of sales in year t is the marginal cost to society of (or the market's willingness to accept) that level of sales. Where the two curves intersect, society's willingness to pay to sell one more EV exactly equals the market's willingness to buy the vehicle. If these points could be discovered for both vehicles and infrastructure, *for every year in the future, the resulting transition pathway would be efficient.*

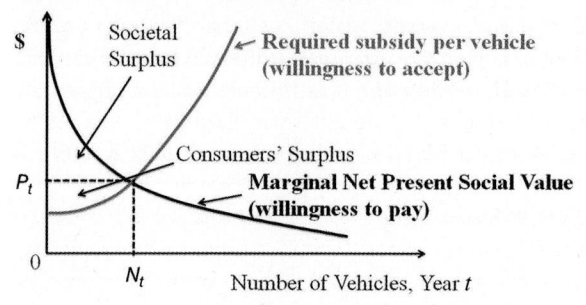

Figure 40.4 Societal willingness to pay and market willingness to accept alternative energy vehicles.

However, energy transitions take decades [22] and, because of this, deep uncertainty about technology and markets is unavoidable. As a consequence, risk-averse decision-makers may further discount expected future benefits, lowering the marginal NPV curve and reducing the efficient number of vehicles to be deployed in year t.

40.6
Status of Transition Plans

Nearly all countries have PEVs on their roads and many countries have hydrogen vehicle demonstration programs in operation, in addition to plans to expand the market for electric drive vehicles. Owing to limitations of space, this section focuses on a few cities and countries that have been at the forefront of electro-mobility. Japan, Germany and the United States (California) also have major automobile manufacturers with mass market vehicles in the market or undergoing premarket development. Selling 40 000 BEVs worldwide in 2011 was a disappointment in comparison with analysts' and automakers' projections despite breaking all previous sales records [23]. The 2011 sales volume fell short of government targets that call for cumulative sales of 20 million vehicles by 2020, but it is clearly too early to pass judgment [2]. For the first 10 months of 2012, US sales of plug-in hybrid electric vehicles (PHEVs) totaled 29 000 units, more than triple the 2011 volume [24]. US sales of BEVs up to October 2012 were only 15% higher than for the comparable period in 2011, despite somewhat improved economic conditions. The cost of batteries and the combination of limited range and long recharging times are generally cited as the reasons why BEV sales lag behind expectations.

Most manufacturers consider BEVs to be best suited to urban markets where typical daily driving distances are well within the range of an EV's battery storage capacity [6]. A sampling of the status and plans for BEV deployment in five world cities is shown in Table 40.2. The early stage of market development is clear from the fact that in none of the cities do EVs comprise as much as 1% of the vehicle park. The ambitiousness of plans for the near future varies enormously, with Rotterdam's vision of replacing nearly all vehicles with EVs by 2025 being by far the most ambitious. Policies also vary widely in nature and intensity. Substantial subsidies are offered in Shanghai, Kanagawa, and Los Angeles (and probably Rotterdam, although the IEA report [23] does not provide specifics). The US government offers a tax credit of $ 7500 per EV and the State of California will add up to $ 2500 more in the form of a rebate [23]. The Japanese government subsidizes 50% of the purchase price difference between an EV and a conventional gasoline vehicle, which is matched by the Kanagawa Prefecture. The Chinese government subsidizes each EV with 60 000 RMB (50 000 RMB per PHEV) and the city of Shanghai adds another 40 000 RMB (20 000 RMB per PHEV). EV purchasers in Berlin are exempt from annual registration fees (about €200 per year) for 10 years. Numerous nonmonetary incentives that vary from city to city are also provided, such as priority lane access, toll reductions, free parking, and free recharging.

Table 40.2 Current status and future plans for BEV market development in five cities.

City	Population (millions)	Vehicles (millions)	Average km per vehicle per day	EVs 2012	EVs Future	Stations 2012	Stations Future
Berlin	3.5	1.3	20	350	15 000 (2015)	220	1 400
Kanagawa	9.1	3.1	n.a.	2 183	3 000 (2013)	450	1 000 (2014)
Los Angeles	4.1	2.5	23	2 000	80 000 (2015)	106	To be determined
Rotterdam	1.2	0.2	20	1 124	200 000 (2025)	100	1 000 (2014)
Shanghai	23.0	1.7	39	1 128	30 000–50 000 (2015)	696	5 000 (2015)

Source: [23].

In Japan, Toyota, Honda, and Nissan all have PEVs in the market and plan to introduce HFCVs for limited mass market sales in 2015. There were 9000 BEVs and 400 PHEVs in use on Japanese roads at the beginning of 2011, with numbers growing rapidly, and 833 charging stations, nearly all of which were open to the public [25]. The Research Association of Hydrogen Supply/Utilization Technology (HySUT) is a private sector organization that runs Japan's 12 demonstration hydrogen refueling stations, serving 60 FCVs, for the Fuel Cell Commercialization Conference of Japan (FCCJ). The goal of the FCCJ is to put 100 publicly accessible hydrogen stations in operation by 2015. All stations and vehicles will operate at 700 bar pressure. By 2025, they expect to have 1000 stations in operation, at which point they expect the FCV market to become self-sustaining (there are currently 47 000 refueling stations in Japan).

The Japanese Ministry of Economy, Trade, and Industry (METI) has authorized the FCCJ's 100 station hydrogen infrastructure plan and has requested 5 billion yen in its 2013 budget to subsidize half of the construction cost of the stations. The other half will be paid by the FCCJ members. The government and manufacturers expect FCVs to be priced at about 5 million yen when they are introduced in 2015, about twice the cost of a conventional vehicle. FCVs will qualify for a clean energy vehicle subsidy that will amount to about half the difference in price compared with a conventional ICE vehicle.

Although virtually all European Union (EU) nations have active programs to create viable markets for electric drive vehicles, this review will focus on Germany, which has been a leader with respect to both electric drive vehicles and developing the renewable energy resources to power them. Germany's National Innovation Program supports both BEV and HFCV programs.

The German Ministry of Transport, Building, and Urban Development (BMVBS) is supporting the creation of an EV market within eight Electromobility Model Regions under the management of the Nationale Organisation Wasserstoff (NOW). Some 2000 EVs and associated recharging infrastructure participate in the project, supported by €500 million of government funding. In 2011, NOW reported there were over 1000 public and semipublic recharging stations in Germany [9, 13].

There are currently 15 hydrogen refueling stations in Germany, all of which are part of a demonstration program consisting of 150 HFCVs supported by the Clean Energy Partnership [13]. In June 2012, the BMVBS together with a group of industrial companies formally agreed to construct and operate at least 50 public hydrogen refueling stations by 2015 [26]. The stations will be concentrated in seven metropolitan areas. Like Japan, the German government, vehicle manufacturers, and energy providers have settled on 700 bar as the standard pressure for on-board storage and dispensing. Also like the Japanese plan, government and industry will share station construction costs 50:50.

The original Clean Energy Partnership plan for HFCVs called for 100 refueling stations in Germany by 2015 to support ~5000 HFCVs. At this time, it seems unlikely that there will be more than hundreds of FCVs on German roads in 2015. The plan also envisioned 400 stations with increased density in urban areas and stations along major intercity routes serving 150 000 vehicles by 2010, and 1000 stations and 1.8 million vehicles in a fully connected highway network by 2025 [27].

In the United States, California is leading the market development for electric drive vehicles by means of its Zero Emission Vehicle (ZEV) standard, and other policies. The California Fuel Cell Partnership (CaFCP) has developed a plan for early market development that calls for 68 public hydrogen refueling stations to be in operation by 2016, mostly concentrated in the Los Angeles area (Table 40.3).

Table 40.3 Station deployment and expected vehicles sales in California.

Year	Start of year station total	Added stations	CaFCP No. of vehicles on the road	CaFC sales	Estimated minimum ZEV sales requirement
2012	4	4	312	100	0
2013	8	9	430	118	0
2014	17	20	1 389	959	0
2015	37	31	10 000	8 611	2 134
2016	68	Market needs	20 000	10 000	2 269
2017	84	Market needs	53 000	33 000	2 297
2018	100	Market needs	95 000	42 000	2 943

Numbers in italics have been approximated based on lower bounds given in CaFCP Table 5.
Source: Table 5 in [28].

The CaFCP's estimates of FCV sales are much higher than estimates of the number of FCVs that are likely to be induced by the state's ZEV requirements. Only one manufacturer, Hyundai, has announced plans to sell FCVs to the public before 2015, and they have estimated total sales by 2015 of 1000 vehicles. The CaFCP's estimates imply that essentially all those vehicles would go to California, an unlikely allocation given Germany's plans to have 5000 FCVs on the road in 2015, as well as the plans of other EU countries and South Korea itself. It is also unlikely that California will receive over 8000 FCVs in 2015, even if all the manufacturers who have announced their intent to begin limited commercial sales in that year follow through on schedule. More likely, the early market development will proceed more slowly.

All three countries' hydrogen infrastructure plans call for concentrating refueling stations in clusters in order to achieve a critical mass of availability in at least a few places with very limited investment [29].

40.7
Modeling and Analysis

Accomplishing a transition to electric mobility requires a new public policy paradigm. Internalizing external costs and benefits and allowing the market to work is not necessarily a sufficient solution. To achieve a transition to clean energy, public policy must break through the barriers that lock in conventional technology and must cope with the uncertainties of future technological progress and market acceptance. At present, there is not a consensus on what the appropriate policy paradigm should be, nor are there accepted tools for modeling and analysis of alternative policies. This section briefly describes one modeling tool, the Light-duty Alternative Vehicle Energy Transitions (LAVE-Trans) model, developed for the International Council on Clean Transportation (ICCT) to analyze the transition to electro-mobility in California, and used in a forthcoming US National Research Council study of the transition to alternative vehicles and fuels [30]. Space does not permit a full description of the model or the analysis, which are available elsewhere [31]. The description that follows aims to convey a general impression of how the model works and of the kind of insights modeling and analysis can produce.

The structure of the LAVE-Trans model is illustrated in Figure 40.5. Its core is a combination of a model of consumers' choices among competing automotive technologies, linked to models that age and scrap vehicles and predict their usage. Feedback loops (shown as dashed lines) represent the network external benefits of scale, learning, infrastructure development, and the reduction of the risk aversion of majority consumers. Costs and benefits are calculated by comparing policy-driven transition scenarios to a base-case scenario with identical assumptions about market behavior and technological progress but without policy interventions.

The progress of electric drive technology is also critical to its ability to succeed in the marketplace. The estimated retail prices for average US passenger cars shown in Figure 40.6 and used in the scenarios described below assume that industry and government targets will be met. Figure 40.6 shows long-run prices at high-volume

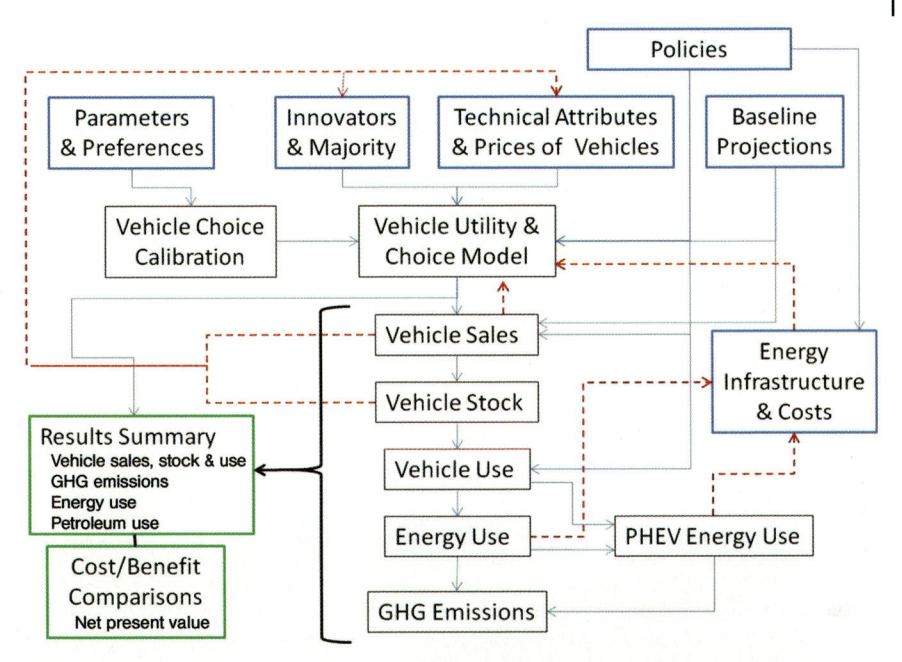

Figure 40.5 Diagrammatic representation of light-duty alternative vehicle energy transition model.

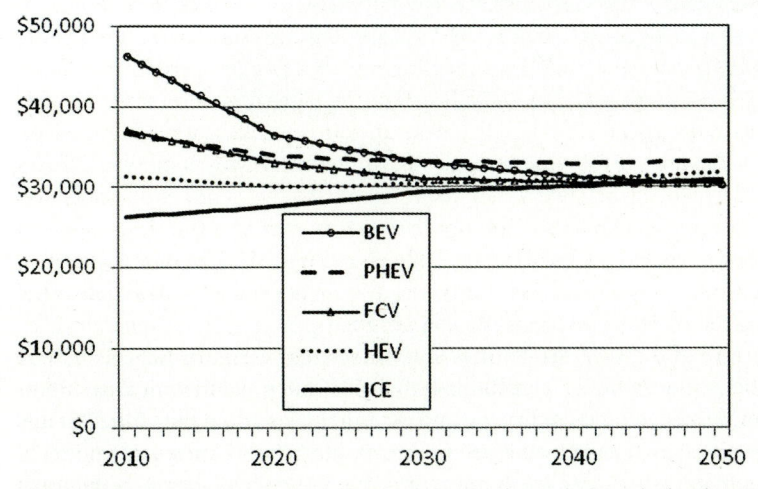

Figure 40.6 Retail price equivalents of advanced technology vehicles at high volume, fully learned.

production and with substantial cumulative learning. In the LAVE-Trans model, prices are far higher during the initial periods of market development owing to lack of scale economies and the immature state of learning-by-doing. Energy efficiencies are also expected to increase several-fold, driven by increasingly strict fuel economy and GHG emissions standards (Figure 40.7).

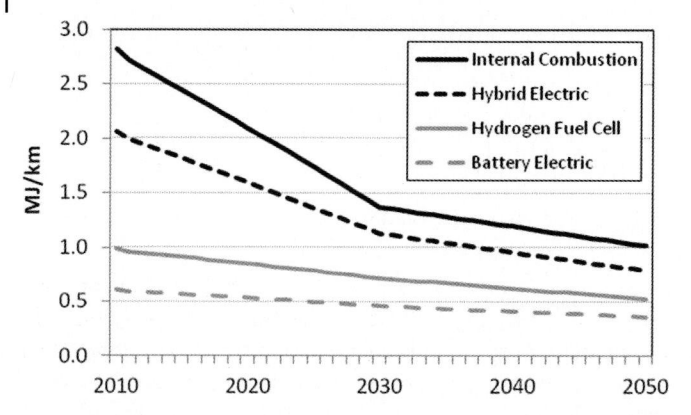

Figure 40.7 Energy efficiencies of alternative power train technologies to 2050.

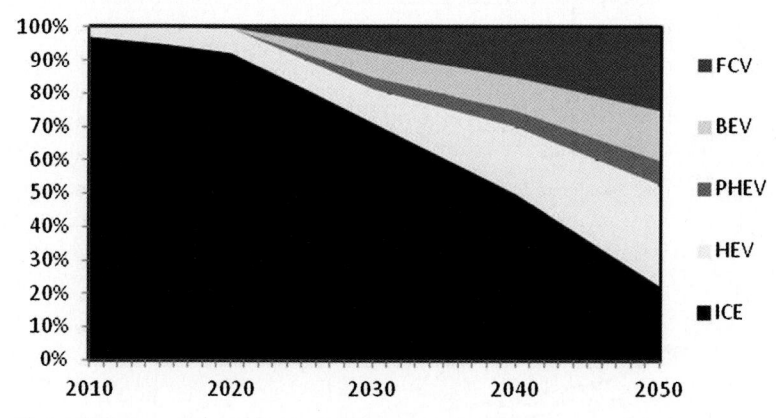

Figure 40.8 Assumed world new passenger car sales technology market shares.

The example scenario presented in the following is one of six scenarios analyzed in the ICCT study. It assumes that California and the subset of 14 US States that have adopted California's emissions regulations attempt a transition to electric drive vehicles but that the rest of the United States does not. Other countries such as Germany, Japan, South Korea, and China are assumed to proceed with a transition to electro-mobility as illustrated in Figure 40.8, and sales outside the United States are assumed to be unaffected by the US market. By 2050, 80% of new passenger car sales are assumed to be FCVs, PEVs, or hybrid EVs (HEVs). To achieve reductions in GHG emissions, the transition to electric drive vehicles must be accompanied by a transition to renewable and low-carbon electricity production and the creation of a low-carbon hydrogen supply infrastructure.

For the purpose of calculating costs and benefits, the scenario is compared to a "base case" with identical assumptions about market conditions and technological progress but without transition policies. The base case does assume that fuel economy and emissions standards consistent with the US Environmental Protection

Agency's (EPA's) rules for 2025 continue to tighten until 2050 and that they induce pricing of the alternative technologies that reflects the social costs of their GHGs and petroleum use.[1] The current tax credits for advanced technology vehicles are replaced in 2015 by policies estimated to be necessary to achieve a transition. Decarbonization of the electricity grid is assumed, as is low-carbon production of hydrogen. By 2050, each kilowatt hour is responsible for 150 g of CO_2, and each kilogram of hydrogen consumed is associated with 2.7 kg of CO_2. This compares with 11.2 kg per gallon of gasoline today.

The production and sale of electric drive vehicles outside the United States will generate a subset of network benefits for the US market, namely scale economies and learning-by-doing throughout the supply chain, in addition to breaking down the risk aversion of majority consumers, to some extent. It will not, however, create diversity of choice within the US market or create infrastructure for refueling and recharging. That requires selling vehicles and building infrastructure in the United States. In scenario 3, the world market evolution illustrated in Figure 40.8 is connected to the US market. The ZEV program proceeds in California and the Section 14 States as usual but the rest of the United States takes no action to induce a transition to electric drive vehicles.

The California ZEV standards are assumed to continue to 2025. However, after 2021 the standards are no longer binding as the market begins to develop under its own momentum. The market "takes off" between 2020 and 2025, at which time transition barriers have been greatly reduced. Sales increase rapidly, reaching sustainable levels around 2040. FCVs are estimated to claim about half the passenger car market, with BEVs taking almost 20% and PHEVs just over 5% (Figure 40.9). The relatively modest market share of PHEVs relative to BEVs is due to their higher cost even in the long run (Figure 40.8).

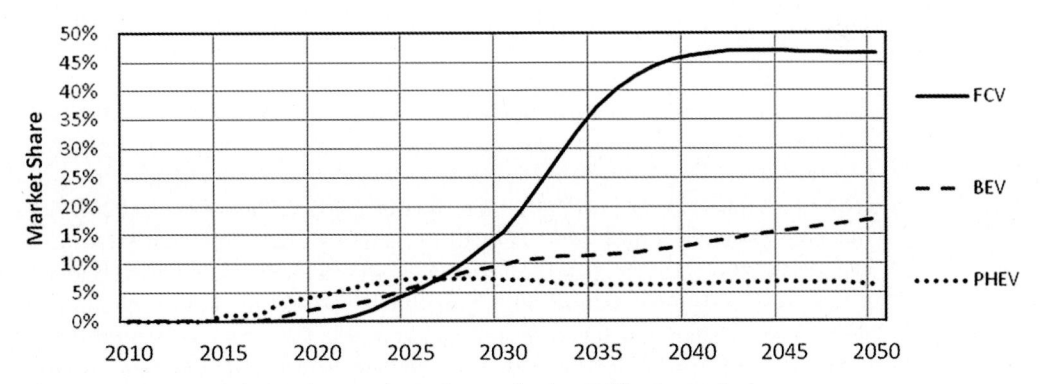

Figure 40.9 Estimated electric drive market in States adopting California standards.

---

[1] The strict fuel economy and emissions standards are assumed to induce a "shadow price" on a vehicle's expected GHG emissions and petroleum use, reflecting the vehicle's value to a manufacturer constrained to meet strict fuel economy and emissions standards. The value is determined by the value of reducing GHGs and petroleum.

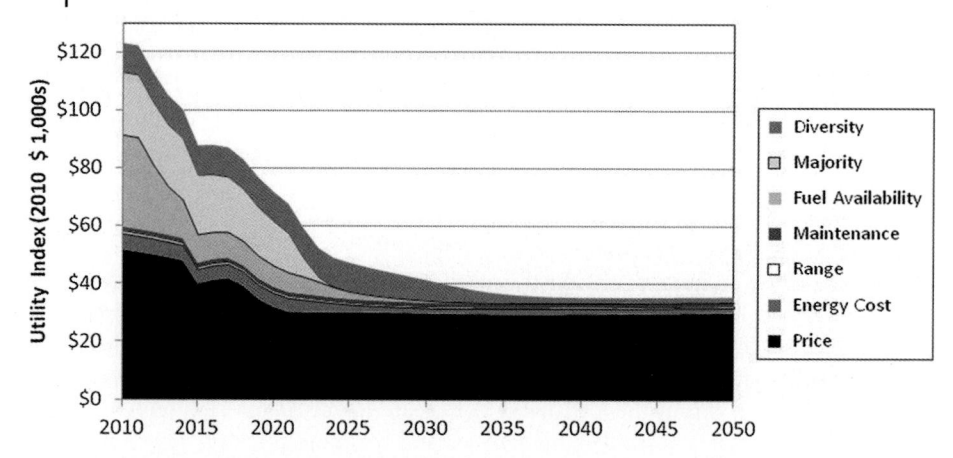

Figure 40.10 Dollar equivalent utility index for hydrogen fuel cell passenger cars in States adopting California standards: majority consumers.

The effect of the ZEV program and subsequent federal policies on the marketability of FCVs is illustrated in Figure 40.10. The costs shown pertain to majority consumers who, unlike innovators and early adopters, are averse to the risk of new technology. Initially priced at over $ 45 000 per vehicle (after subtracting a $ 7500 per vehicle tax credit), the HFCV is also burdened by lack of fuel availability, lack of a diverse array of makes and models to choose from, and, for the majority of consumers, the risks of being among the first to purchase a novel technology. Together, these factors raise the FCV's utility index cost to over $ 120 000 per vehicle. The early placement of hydrogen refueling stations greatly reduces the estimated cost of lack of fuel availability from over $ 20 000 to less than a $ 5000 per vehicle penalty. Subsidies by manufacturers required to sell ZEVs further reduce the price to potential buyers.

The attractiveness of BEVs to future buyers is also enhanced by policy-induced sales between 2015 and 2025. Unlike hydrogen vehicles, fuel availability is a relatively minor issue. The value of public recharging infrastructure is unfortunately invisible in Figure 40.11 because it is a negative cost (benefit); it grows from about $ 200 per vehicle in 2025 to more than $ 400 per vehicle in 2040. The limited range and longer recharging time of EVs remain significant cost factors through 2050. The greater early sales volumes for BEVs erode the majority's risk aversion more quickly than for FCVs, so that by 2025 BEVs are no longer seen as a risky new technology by majority consumers. Because BEVs' market share does not exceed 20%, limited diversity of make and model choice remains somewhat of an issue through 2050.

Still, sales to majority consumers are essentially zero during the early years of market development; innovators and early adopters are virtually the only purchasers (Figure 40.12), according to the assumptions used in the LAVE-Trans model.[2]

2) In fact, the number of innovators and early adopters, their willingness to pay for electric drive technologies, and how that willingness to pay will decline with increasing sales are not well understood. This is just one of many areas where a better quantification of market behavior would permit better analysis for decision-making.

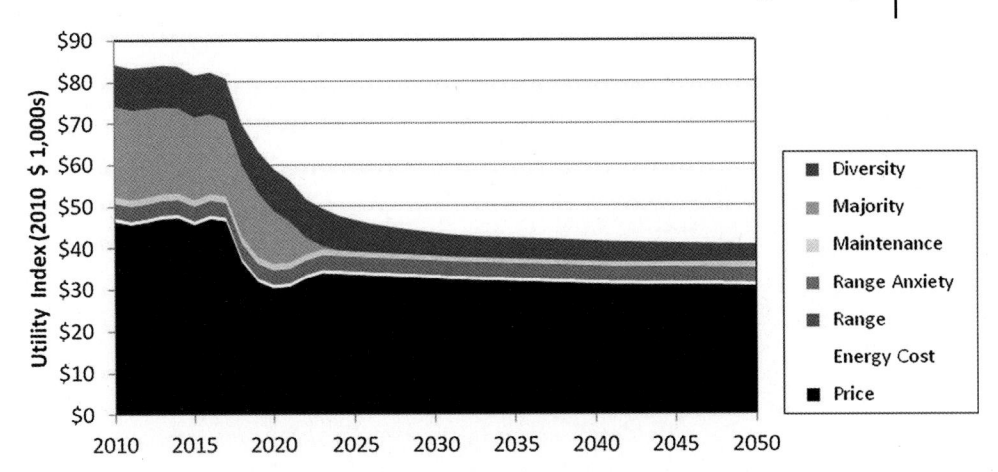

Figure 40.11 Dollar equivalent utility index for battery electric passenger cars in States adopting California standards: majority consumers.

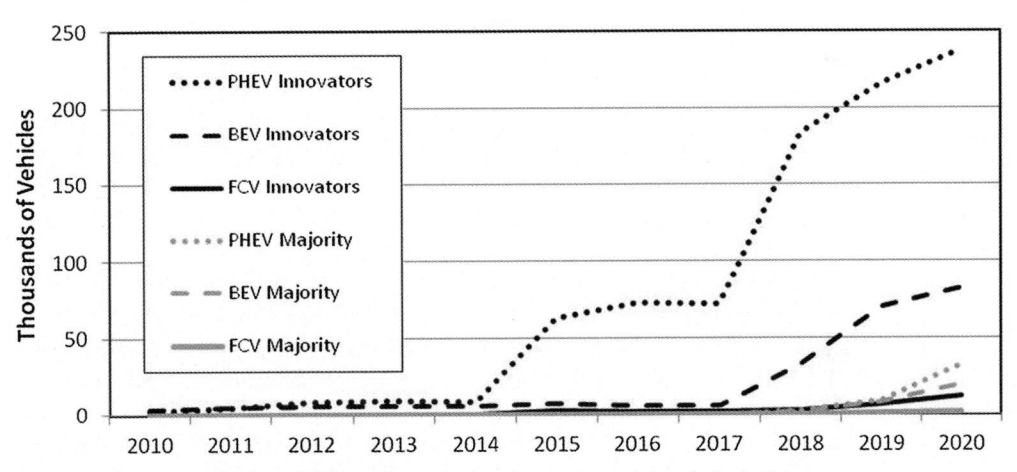

Figure 40.12 Estimated sales of electric drive vehicles to innovators and majority in States adopting California standards 2010–2020.

Innovators and early adopters generate network external benefits not only for future California and Section 177 State buyers but also spillover benefits for the rest of the United States. The spillover benefits allow the transition to proceed more rapidly and permit a less intense policy effort. Early installation of refueling infrastructure and sales of the first FCVs to the earliest adopters between 2015 and 2020 reduce fuel availability costs by another estimated $ 5000 per vehicle. Sales required by the ZEV mandates and by the federal program also reduce costs by scale economies and learning-by-doing by an estimated $ 25 000 per vehicle by 2025.[3]

3) Unlike indirect network external benefits, the benefits generated via scale economies and learning are pecuniary external benefits, that is, they are reflected in market prices in a competitive market.

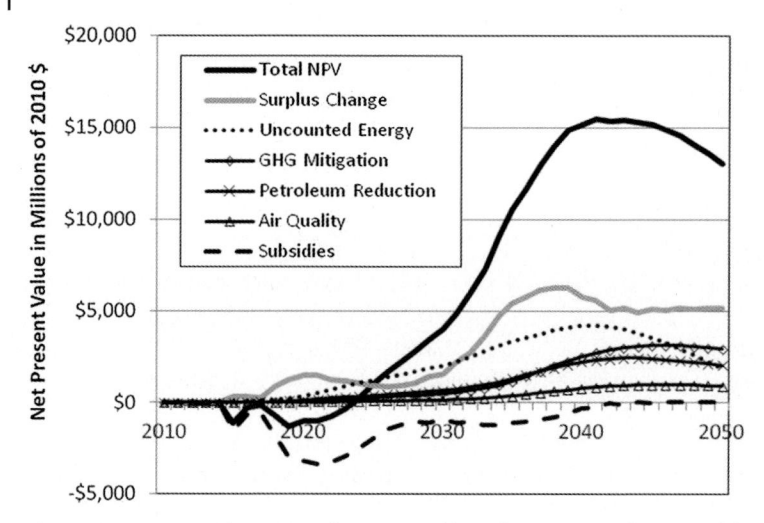

Figure 40.13 Estimated present value costs and benefits in States adopting California standards of a US transition to electric drive vehicles.

From this perspective, subsidies to the first purchasers of electric drive vehicles are more than justified by the network external benefits that they create for subsequent purchasers. Of course, this interpretation depends critically on the ultimate success of the transition such that present value benefits exceed present value costs by a substantial amount. In reality, the expected benefits of the transition would include the possibility of technological failure as well as success.

Assuming technological success, the present value of total private and social benefits of a transition to electric drive vehicles appears to exceed the excess costs[4] by approximately an order of magnitude, suggesting that even a fairly large probability of failure could justify the magnitude of investments assumed in the above analysis. The net present value of the costs illustrated in Figure 40.13 is approximately +$ 300 billion, of which total subsidies comprise –$ 40 billion. The costs and benefits shown in Figure 40.13 are from a different scenario that does not consider the effects of other countries' production of electric drive vehicles. Although such a scenario is less realistic than the one presented above, it excludes from the base case the spillover benefits to the United States from the actions of other countries and thus gives a more accurate estimate of US costs and benefits versus no transition to electric drive. Private benefits include consumers' surplus gains and energy savings not considered by car buyers at the time the vehicle is purchased. Surplus benefits derive partly from the fact that the electric drive technologies eventually become less expensive than gasoline ICE vehicles and partly from the fact that some consumers will prefer electric drive to ICE drive, all else equal. The transition enables them to own such vehicles. In the LAVE-Trans scenarios shown here, car buyers are assumed to consider only the first 3 years of fuel costs in their purchase decisions whereas the

---

4) "Excess costs" are those over and above what producers and consumers are willing to pay voluntarily. These are the explicit and implicit subsidies required to overcome the barriers to transition.

expected lifetime of a light-duty vehicle in the United States is about 14 years.[5] Other benefits are the value of reduced CO_2 emissions, reduced petroleum consumption, and improved local air quality. Of course, there is disagreement about the values of all of the social benefits; those used in Figure 40.13 are documented in [30].

Although the early deployment of infrastructure is important to PEVs and essential for HFCVs, it is not the major part of transition costs that must be explicitly or implicitly subsidized. Other assessments of the costs of a full transition to HFCVs and PHEVs have also concluded that the excess costs of the refueling and recharging infrastructure are small relative to the excess costs of the vehicles themselves (~5–20%) [6, 32–34]. The LAVE-Trans model produces similar results. The vast majority of the excess transition costs are in the vehicles.

There are too many uncertainties about market behavior and future technological progress to consider estimates by the LAVE-Trans model definitive. Rather, they suggest how the system is likely to behave and represent a particular best guess about future outcomes. It is anticipated that future research will lead to improved models and a narrowing of uncertainties about the key aspects of market behavior. The results of a sensitivity analysis of the parameters that determine market behavior are illustrated in Figure 40.14 and Figure 40.15. The diamond-shaped dots mark a centered interval containing 90% of the results.

The initial scenario leads to successful market penetration of BEVs and FCVs, as described earlier. Parameters that determine factors such as the value of fuel availability, discounting of future energy savings, importance of make and model diversity, and risk aversion of majority consumers were assigned plausible probability distributions. The transition policies that led to a successful market transition given the default parameters were held constant, that is, they were not adjusted as market

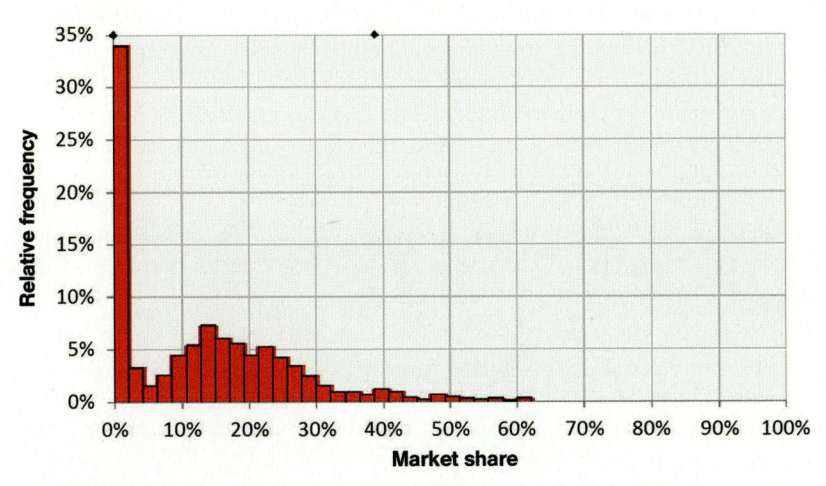

Figure 40.14 Relative frequency distribution of BEV market shares in 2050 generated by sensitivity analysis of assumptions about market behavior.

5) This assumption simulates loss-aversion behavior, as described by Greene et al. [34].

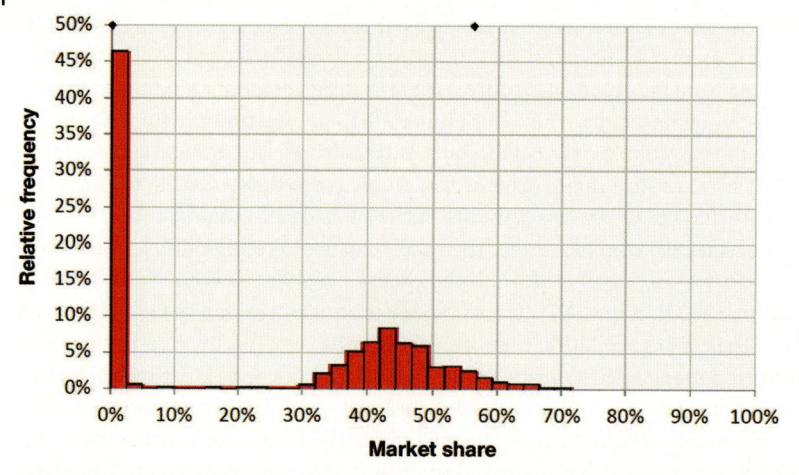

Figure 40.15 Relative frequency distribution of hydrogen fuel cell electric vehicle market shares in 2050 generated by sensitivity analysis of assumptions about market behavior.

behavior changed. The results indicate both a high degree of uncertainty about future market success and the existence of tipping points that lead to zero market penetration even in 2050. It is possible that zero market share outcomes could be "tipped" to positive market shares by making policy adjustments. Nonetheless, it is instructive to see how probable the failure mode is without an adaptive policy strategy.

The estimated market shares of both powertrain technologies are highly sensitive to assumptions about scale economies in the automotive industry, the number of innovators and early adopters and their willingness to pay for novel technology, and the value to consumers of having a diverse array of vehicle makes and models to choose from. BEVs do better if consumers are more sensitive to energy costs and less sensitive to initial price. Consumers' concerns about limited range and long recharging times are very important for the market success of BEVs. The market success FCVs is strongly dependent on the importance of fuel availability to car buyers; on the other hand, this is of much less importance for BEVs.

40.8
Conclusion

The transition to electric drive vehicles from petroleum-based ICEs poses new challenges for society. Major energy transitions of the past were driven by market forces; the transition to electric drive is motivated by an urgent need to secure public goods: energy security, environmental protection, and a sustainable energy system. It is important that the value of the transition be recognized and quantified in order to justify society's investments in vehicles and infrastructure to induce the transition.

Progress in the enabling technologies for electric drive vehicles has been not only impressive but for the most part also on schedule to achieve a transition to electric drive by 2050. The recent progress in electro-mobility technology has largely met expectations. Much remains to be done, however: both hydrogen fuel cell and plug-in electric systems need to reduce costs by the order of 50% to compete effectively with the incumbent technology.

Energy efficiency improvements to conventional vehicles, especially those that reduce the power requirements of vehicles (reduction of mass, aerodynamic drag, and rolling resistance) are key enablers of the transition to electro-mobility. Energy efficiency can reduce the amount of low-carbon energy needed by a factor of three, and load reduction is likely to favor electric powertrains over ICEs.

Current plans for the introduction of both battery electric and hydrogen fuel cell vehicles appear to be overly optimistic. Behavioral psychologists have named this very common tendency to construct plans that are very close to best-case scenarios the "planning fallacy" [35]. This does not necessarily imply that the plans will not eventually be realized, but rather that it will likely take more time and effort than initially expected. Substantial progress is still needed and several important determinants of the market's response are not well understood.

The process of replacing the "locked-in" petroleum-powered ICE vehicle technology is complex, includes powerful positive feedback mechanisms, and is fraught with uncertainty. Analysis of optimal strategies in the presence of great uncertainty has shown that adaptive strategies, strategies that respond to change in response to future developments, are more robust and can perform almost as well as the optimal strategies based on a full knowledge of future events [36].

References

1 GEA Writing Team (2012) *Global Energy Assessment – Toward a Sustainable Future*, Cambridge University Press, Cambridge, and International Institute for Applied Systems Analysis, Laxenburg.
2 International Energy Agency (IEA) (2010) *Energy Technology Perspectives 2010: Scenarios and Strategies to 2050*, OECD, Paris, p. 273.
3 International Energy Agency (IEA) (2012) *Energy Technology Perspectives 2012: Pathways to a Clean Energy System*, OECD, Paris, p. 424.
4 International Energy Agency (IEA) (2008) *World Energy Outlook 2008*, OECD, Paris, Figure 9.10.
5 Element Energy (2012) *Cost and Performance of EV Batteries. Final Report for The Committee on Climate Change*, Element Energy, Cambridge.
6 McKinsey and Company (2011) *A Portfolio of Power-trains for Europe: a Fact-based Analysis*, NOW, Berlin.
7 James, B. D., Kalinoski, J., and Baum, K. (2011) Manufacturing cost analysis of fuel cell systems, presented at the 2011 US DOE Hydrogen and Fuel Cells Program Annual Merit Review and Peer Evaluation, Washington, DC, http://www.hydrogen.energy.gov/pdfs/review11/fc018_james_2011_o.pdf (last accessed 4 January 2013).
8 James, B. D., Kalinoski, J. A., and Baum, K. N. (2010) *Mass Production Cost Estimation for Direct H_2 PEM Fuel Cell Systems for Automotive Applications: 2010 Update*, Directed Technologies, Arlington, VA.
9 Butsch, H. (2012) NOW Review, presentation to author in private meeting with Hanno Butsch, Manager International Cooperation, National Organisation Wasserstoff und Brennstoffzellentechnologie on 12 November 2012, Berlin.
10 Wipke, K., Sprik, S., Kurtz, J., Ramsden, T., Ainscough, C., and

Saur, G. (2012) *Final Results from U. S. FCEV Learning Demonstration*, NREL/CP-5600-54375, National Renewable Energy Laboratory, Golden, CO.

11 German, J. (2012) *Future Costs and Energy Efficiencies of Alternative Power Trains*, International Council on Clean Transportation, San Francisco.

12 US Department of Energy (2012) *Fuel Economy: Where the Energy Goes*, http://www.fueleconomy.gov/feg/atv.shtml (last accessed 4 January 2012).

13 Greene, D. L. and Plotkin, S. (2011) *Reducing Greenhouse Gas Emissions from U. S. Transportation*, Pew Center on Global Climate Change, Arlington, VA.

14 Fawcett, A. A., Calvin, K. V., de la Chesnaye, F. C., Reilly, J. M., and Weyant, J. P. (2009) Overview of EMF 22 U. S. transition scenarios. *Energy Econ.*, **31**, S198–S211.

15 Sperling, D. and Gordon, D. (2009) *Two Billion Cars Driving Toward Sustainability*, Oxford University Press, Oxford.

16 Yang, C., McCollum, D., and Leighty, W. (2011) Scenarios for deep reductions in greenhouse gas emissions, in *Sustainable Transportation Energy Pathways* (eds J. M. Ogden and L. Anderson), Institute of Transportation Studies, University of California at Davis, Davis, CA, Ch. 8.

17 McCollum, D. and Yang, C. (2009) Achieving deep reductions in U. S. transport greenhouse gas emissions: scenario analysis and policy implications, *Energy Policy*, **37** (12), 5580–5596.

18 Zachmann, G., Holtermann, M., Radeke, J., Tam, M., Huberty, M., Naumenko, D., and Faye, A. (2012) *The Great Transformation: Decarbonising Europe's Energy and Transport Systems*, Bruegel Blueprint Series, vol. XVI, Bruegel, Brussels.

19 Köhler, J., Grubb, M., Popp, D., and Edenhofer, O. (2006) The transition to endogenous technical change in climate-economy models: a technical overview to the Innovation Modeling Comparison Project, *Energy J.*, Special Issue (1) on Endogenous Technological Change and the Economics of Atmospheric Stabilisation, 17–56.

20 Struben, J. and Sterman, J. D. (2008) Transition challenges for alternative fuel vehicle and transportation systems, *Environ. Planning B: Planning Des.*, **35**, 1070–1097.

21 Farrell, J. and Klemperer, P. (2007) Co-ordination and lock-in: competition with switching costs and network effects, in *Handbook of Industrial Organization*, vol. 3 (eds M. Armstrong and R. Porter), North-Holland, Amsterdam, Ch. 31.

22 Nakićenović, N., Grübler, A., and McDonald, A. (eds)(1998). *Global Energy Perspectives*, Cambridge University Press, Cambridge, and International Institute for Applied Systems Analysis, Laxenberg.

23 International Energy Agency (IEA) (2012) *EV City Case Book*, OECD, Paris.

24 Electric Drive Transportation Association (2012) *Electric Drive Vehicle Sales Figures (U. S. Market) – EV Sales*, http://www.electricdrive.org/index.php?ht=d/sp/i/20952/pid/20952 (last accessed 4 January 2013).

25 Ogino, N. (2012) *Electric Vehicles in Japan*, Japan Automotive Research Institute, Fuel Cell – Electric Vehicle Research Division.

26 Fuel Cell e-Mobility (2012) *50 Hydrogen Filling Stations for Germany*, http://www.fuel-cell-e-mobility.com/article/hydrogen-filling-stations-for-germany-federal-ministry-of-transportation-and-industrial-partners-build-nationwide-network-of-filling-stations/(last accessed 4 January 2012).

27 Clean Energy Partnership (CEP) (2012) *Hydrogen Mobility Gets Going*, National Organisation for Hydrogen and Fuel Cell Technology, Berlin, http://www.google.com/url?sa=t&rct=j&q=&esrc=s&frm=1&source=web&cd=1&ved=0CC0QFjAA&url=http%3A%2F%2Fwww.now-gmbh.de%2Fde%2Fueber-die-now%2Faufgabe%2Fpublikationen-download.html%3Ftx_gopublication_piPublication%255Bdl%255D%3D1085&ei=vjK2UIzVBpK00QG4sIGICw&usg=AFQjCNGBVQEn3LlmIKyCeMCDxOo3-WTi1w (last accessed 4 January 2012).

28 California Fuel Cell Partnership (CaFCP) (2012) *A California Road Map: the Commercialization of Hydrogen Fuel Cell Vehicles*, CaFCP, West Sacramento, CA.

29 Ogden, J. M. and Nicholas, M. (2010) Analysis of a "cluster" strategy for introducing hydrogen vehicles in southern California, *Energy Policy*, **39** (4), 1923–1938.

30 National Research Council (NRC) (2013) *Transitions to Alternative Vehicles and Fuels*, National Academies Press, Washington, DC.

31 Greene, D. L., Park, S., and Liu, C.-Z. (2012) *LAVE-Trans Model Documentation*, Howard H. Baker, Jr. Center for Public Policy, University of Tennessee, Knoxville, TN.

32 National Research Council (NRC) (2008) *Transitions to Alternative Transportation Technologies: a Focus on Hydrogen*, National Academies Press, Washington, DC.

33 National Research Council (NRC) (2010) *Transitions to Alternative Vehicles and Fuels: Plug-in Hybrid Electric Vehicles*, National Academies Press, Washington, DC.

34 Greene, D. L., Leiby, P. N., James, B., Perez, J., Melendez, M., Milbrandt, A., Unnasch, S., Hooks, M., McQueen, S., and Gronich S. (2008) *Analysis of the Transition to Hydrogen Fuel Cell Vehicles and the Potential Hydrogen Energy Infrastructure Requirements*, ORNL/TM-2008/30, Oak Ridge National Laboratory, Oak Ridge, TN.

35 Kahneman, D. (2011). *Thinking Fast and Slow*, Farrar, Straus, and Giroux, New York, Ch. 23, pp. 245–254.

36 Groves, D. G. and Lempert, R. J. (2007) A New Analytic Method for Finding Policy-relevant Scenarios, *Global Environ. Change*, **17**, 73–85.

41
Nearly Zero, Net Zero, and Plus Energy Buildings – Theory, Terminology, Tools, and Examples

Karsten Voss, Eike Musall, Igor Sartori and Roberto Lollini

41.1
Introduction

In most European countries, buildings and their use account for approximately one-third of total energy consumption and associated carbon emissions. The majority of this demand is generated by living in residential buildings and the remainder by so-called nonresidential buildings, that is, for commercial uses, trade, and services. Residential buildings are the clear leader owing to their quantity. Due to the importance of the building sector and the existing concepts and technologies for major improvements, buildings are at the focus within national policies for resource and climate protection.

At present, terminology related to (nearly) zero-energy buildings (ZEBs) is important for strategy papers on energy policy in many countries. This is partly due to positive connotations of the term "zero energy." In the context of finite resources and increasing energy costs, it suggests independence, no costs, or an orientation towards the future. Also, "zero" leaves no room for discussion on quantification of suitable parameters. The interpretation of energy parameters as quantitative target definitions remains the domain of experts and offers no real basis for communication with the general public. However, at first glance, "zero" seems to doubtlessly demand the highest possible standards regardless of building type or climate, only to be superseded by the term "plus."

The term ZEB is used commercially without clear agreement on its content [1, 2]. In general, a ZEB is understood as a grid-connected, energy-efficient building that balances its total annual energy consumption by on-site generation and associated feed-in credits. To emphasize the balance concept – in contrast to an autonomous building – the term "Net" has been introduced, so that one can speak of Net ZEB and the variants nearly Net ZEB or Net plus energy building, as shown in Figure 41.1.

Since the 2010 recast of the EC Energy Performance of Buildings Directive [3], the discussion has become even more intensive. The European Association of Refrigeration, Heating and Ventilation (REHVA) published a proposal in May 2011 [4].

Transition to Renewable Energy Systems, 1st Edition. Edited by Detlef Stolten and Viktor Scherer.
© 2013 Wiley-VCH Verlag GmbH & Co. KGaA. Published 2013 by Wiley-VCH Verlag GmbH & Co. KGaA.

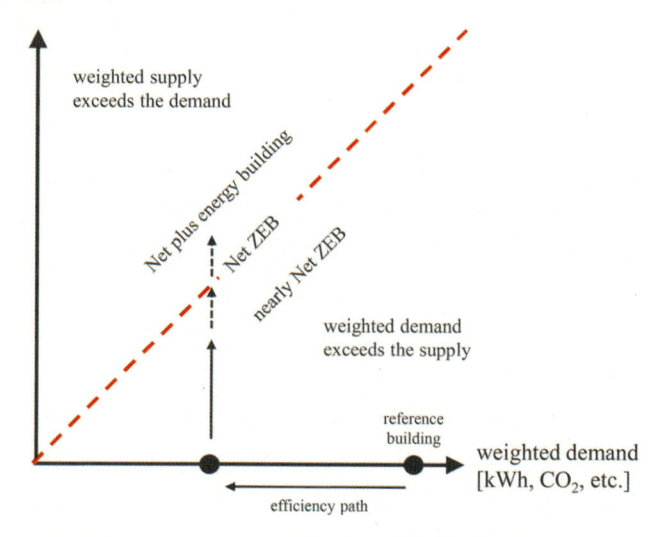

Figure 41.1 Graph representing the path towards a Net Zero Energy Building (Net ZEB), with the nearly and plus variants. Source: University of Wuppertal, EU.

The European concerted action assisting the 2010 recast of the Energy Performance of Buildings Directive (EPBD) offers a platform for member states to discuss the various national approaches to formulate relevant definitions at the building code level [5]. The ongoing International Energy Agency (IEA) activity "Towards Net Zero Energy Solar Buildings" was formed in 2008 as a scientific forum at the international level [6]. The authors of this chapter are members of the IEA subtask "Definitions and Large-Scale Implications."

Over the past few years, the IEA working group has analyzed relevant publications on the ZEB topic and has published a comprehensive review [7]. This review was followed by a recently published article addressing a consistent definition framework [8], a project database [9], and a book including a set of well-documented example buildings covering a wide range of typologies and climates [10].

41.2
Physical and Balance Boundaries

Building codes focus on a single building and the energy services that are metered. Therefore, it is possible to distinguish between a physical boundary and a balance boundary. The combination of physical and balance boundaries defines the building system boundary (see Figure 41.2).

The physical boundary identifies the building (as opposed to a buildings cluster or a neighborhood). The energy analysis addresses energy flows at the connection

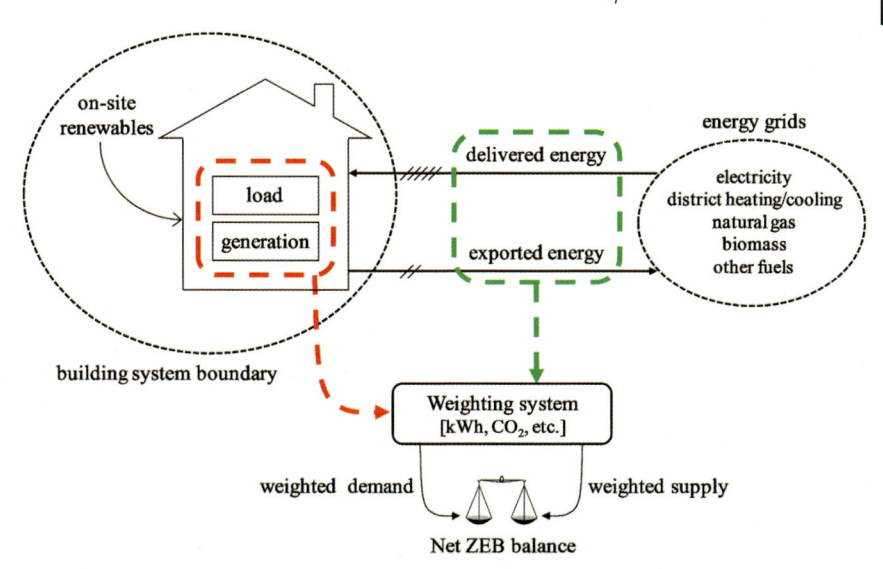

Figure 41.2 Sketch of the connection between buildings and energy grids showing the relevant terminology [8].

point to supply grids (power, heating, cooling, gas, fuel delivery chain). Consequently, the physical boundary is the interface between the building and the grids. The physical boundary therefore includes up to the meters (or delivery points). The physical boundary is also useful for identifying so-called "on-site generation" systems; if a system is within the physical boundary (within the building or building cluster distribution grid before the meter) it is considered to be on-site, otherwise it is off-site. Typical on-site generation systems are photovoltaics (PV) and micro combined heat and power (CHP), which allow energy to be exported beyond the physical boundary. The yield of solar thermal systems is typically consumed entirely on-site due to technical limitations at the connection point to district heating systems. Therefore, solar thermal systems are mostly treated as demand-reduction technology (efficiency path, *x*-axis in Figure 41.1). A typical off-site option is a share in a wind energy turbine which is financed by the building budget. This option would allow economically feasible options to balance the building energy consumption (the so-called "allowable solutions" in the UK Zero Carbon Home approach [11]), but should be considered within the primary energy factor for the imported electricity to avoid double counting. However, the EPBD addresses only energy generated on-site or nearby. Therefore, while the concept of "nearby" still needs to be better defined, off-site solutions seem to be beyond its scope.

The balance boundary identifies which energy services are considered. In the EPBD, energy balance calculations take into account the technical services for heating, cooling, ventilation, and domestic hot water (and lighting in the case of non-domestic buildings). Plug loads and central services are not included, but are typically included when metering energy use at the point of delivery. Some pilot projects also include the charging of electric vehicles on-site (before the meter [12]).

Although these loads are not related to the building performance, a holistic balance including all electric consumers on-site helps to characterize the grid interaction in more detail (see later). Electric vehicles include batteries, thereby increasing the "on-site" storage capacity.

Other forms of energy consumption that do not appear in the annual operational phase but belong to the life cycle of a building may be considered within the balance boundary, such as embodied energy/emissions related to construction materials and installations. The recently formulated definition in Switzerland and the one under development in Norway address this issue [13, 14]. The results of a recent study on the lifecycle energy balance of low-energy and Net ZEBs indicate that the embodied energy of a building increases by up to 25% when taking the step from a typical low-energy building towards a nearly or Net ZEB. This is due to the domination by structural building elements compared with energy-saving measures or generation systems [15]. The overall lifecycle energy demand is in all ZEB cases investigated much less than in low-energy buildings.

41.3
Weighting Systems

The weighting system converts the physical units of different energy forms into a common metric to facilitate the balancing process (Figure 41.2). According to the EPBD recast, the metric of the balance for a nearly ZEB is primary energy. Nevertheless, some countries prefer carbon emissions as the primary metric. Examples of weighting factors are documented in EU standards such as EN 15603, but many different factors are used in national building practice, reflecting the specific national or local power grid structure (Annex 1 in [8]). Factors develop with time and are not physical constants. Most countries typically apply factors that take only the nonrenewable component of the primary energy content into account. This is the background leading to the low conversion factors for biomass or biofuels, resulting in market stimulation for such energy supply solutions for Net ZEBs. Some countries apply politically adjusted (increased) factors in order to reflect the regionally limited availability of biomass and biofuels from sustainable forestry or agriculture (e.g., Switzerland [16]). In other countries, politically adjusted (decreased) factors are applied to electricity in order to include the expected "greening" of the power sector in accordance with national and EU road maps (e.g., Denmark [17] and Norway [18]). Such "discounting" of electricity favors all-electric solutions such as systems based on heat pumps, facilitating achieving the Net ZEB target in connection with decarbonized power grids (with a high share of renewable energy). Similarly, discounted values for the district heating/cooling grid would make the Net ZEB target more feasible in connection with thermal grids based on large shares of renewable energy, thermal cascades from industry processes, and/or waste as fuel.

Typically, symmetrical weighting factors are applied when balancing imported and exported energy; energy delivered by the grid and energy fed into the grid are given the same value. Other developments weight asymmetrically to stimulate on-site generation approaches (Germany 2012: 2.4 kWh primary energy per kilowatt hour of electricity delivered from the grid, 2.8 kWh primary energy per kilowatt hour of electricity exported to the grid [19]). Weighting factors may vary seasonally (or even at the daily or hourly level), as discussed later.

41.4
Balance Types

The Net ZEB's annual balance between weighted demand and weighted supply is often implicitly understood as the so-called import/export balance, indicated by the green line in Figure 41.2 and Figure 41.3. Weighted delivered and exported energy quantities can be used to calculate the balance when monitoring a building, provided that all consumptions are included. Separating some components of the consumption out of the balance creates the need for more sophisticated (sub-) metering.

Such quantities are known in monitoring, but in the design phase they could be calculated only if there were good estimates of "self-consumption": the share of on-site generation that is immediately consumed in the building. Self-consumption differs according to the type of generating technology, the type of building, the climate, and the user behavior because it depends on the simultaneity between generation and consumption. Currently there is insufficient knowledge about self-consumption to

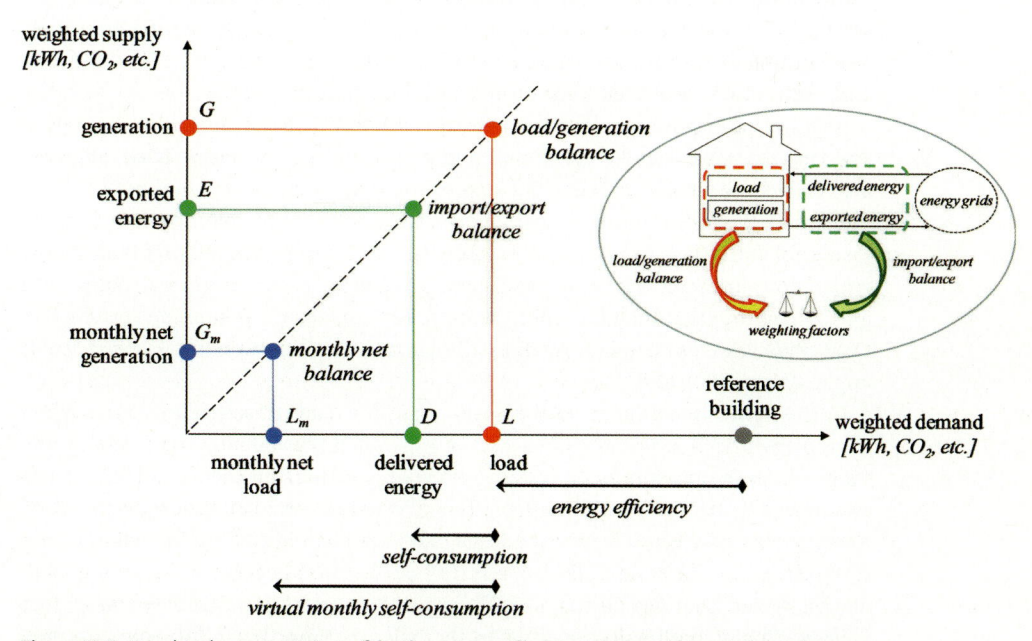

Figure 41.3 Graphical representation of the three types of balance: import/export balance between weighted exported and delivered energy, load/generation balance between weighted generation and load, and monthly net balance between weighted monthly net values of generation and load [8].

establish standardized self-consumption fractions. This is one of the points which were left open in the previous REHVA article on a Net ZEB definition [4]. In order to permit an import/export balance calculation in the design phase, planners need to have data on end uses patterns, for example, for appliances, cooking, hot water use, with sufficient time resolution. In the same way as weather data are standardized to provide reference climates for dynamic simulations, user profile data may be standardized to enable an import/export analysis under reference conditions.

As the EPBD recast mainly addresses building performance requirements in the planning phase, it focuses on the balance between weighted on-site generation and the calculated energy demand, the so-called load/generation balance (red line in Figure 41.2 and Figure 41.3). These quantities do not cross the building system boundary, so the grid interaction is disregarded. The advantage is that both quantities can be calculated independently in the absence of detailed information on time-dependent load and generation profiles with high resolution. The main difference between the two balance types is the self-consumed fraction of energy generated on-site, resulting in different numbers.

The load/generation balance in the understanding of the EPBD recast addresses generation by renewable sources only. This means that a CHP system fueled by natural gas and exporting power to the grid is not taken into account on the generation side, whereas it is typically included with its power generation on the export side of the import/export balance. Whereas solar thermal gains are counted as load reduction in the import/export balance (no heat exported), these gains are counted as on-site renewable generation within the load/generation balance and for the fraction of renewables covering the load. As the EPBD calls for a "significant share" of renewables to cover the remaining load of a Net ZEB or nearly Net ZEB, the total share of renewables needs to be clearly defined.

As most national energy codes apply calculations on a monthly basis, generation and consumption may be calculated and compared on a monthly level, allowing a so-called virtual load match to be determined. Monthly on-site generation up to the level of the monthly load is counted as virtual self-consumption (= reducing the load, efficiency path in Figure 41.1). Only the monthly residuals, that is, monthly generation surplus or remaining load, are added up to determine annual totals. Such a balancing method may be called monthly net balance (blue line in Figure 41.3). One application is in the version of the German building energy code that has applied since 2009 [20].

In the monthly net balance, the annual surplus characterizes the service taken over by the grid to overcome the seasonal mismatch between load and generation. However, in the case of multiple delivered energy forms, the annual surplus is also influenced by the substitution effect, that is, when exported electricity is also used to compensate for other forms of energy that have been imported, for example, gas or biomass. In the case of multiple forms of exported energy, for example, both electricity and heat, the annual surplus is also distorted by the different weighting factors. Finally, the result depends on the balance boundary with respect to plug loads and central building services: excluding part of the loads increases the monthly surplus (assuming constant generation). High-resolution net metering in the building operation phase typically results in a lower load/generation match and higher export than that estimated by the monthly net balance. The monthly net balance is a simplified approach for the design phase, when high-resolution profiles are not available.

41.5
Transient Characteristics

Buildings using on-site generating systems have different abilities to match the load and benefit from the availability of energy sources and the demands of the local grid infrastructure, namely the power grid and in a few cases the heating/cooling grid. Differences occur in

- the temporal match between energy generation on-site and the building load (load match)
- the temporal match between the energy transferred to a grid and the demands of a grid (grid interaction)
- the (temporal) match between the types of energy imported and exported (fuel switching).

As mentioned above, load matching and grid interaction have to be discussed with respect to the form of energy and the temporal resolution. Calculations have to be made for each form of energy separately. Simple monthly net metering is sufficient to describe and investigate the seasonal performance (Figure 41.4, example of an all-electric building), whereas high-resolution simulation or monitoring is needed to describe daily and hourly fluctuations [21, 22]. Load matching and grid interaction are almost irrelevant in the context of fuel-based energy supply but are of major importance for the electricity grid.

Increasing the load match is not an intrinsically favorable strategy for a grid-connected building. The value of the exported energy to cover loads somewhere else in the grid may be higher than losses associated with on-site storage solutions to increase matching. The value may vary depending on the season and time of day, due to the varying fraction of renewable power available in the grid. The choice between on-site storage and export will depend on such dynamic values. There is no *a priori* positive or negative implication associated with high or low load match. However, a load match calculated on monthly values (= monthly net metering) will at least give a first-order insight to characterize the service taken over by the grid to overcome the seasonal mismatch of load and generation (calculated on each single energy carrier, so without the distortions affecting the monthly net balance).

Weighting factors with seasonal/monthly variation applied within building energy code systems present a possible future method to influence the balancing results and to stimulate beneficial and sustainable developments. The factors for the power grid can address differences in the fraction of renewables. In the case of a grid with high penetration of solar power generation, large seasonal differences will be typical

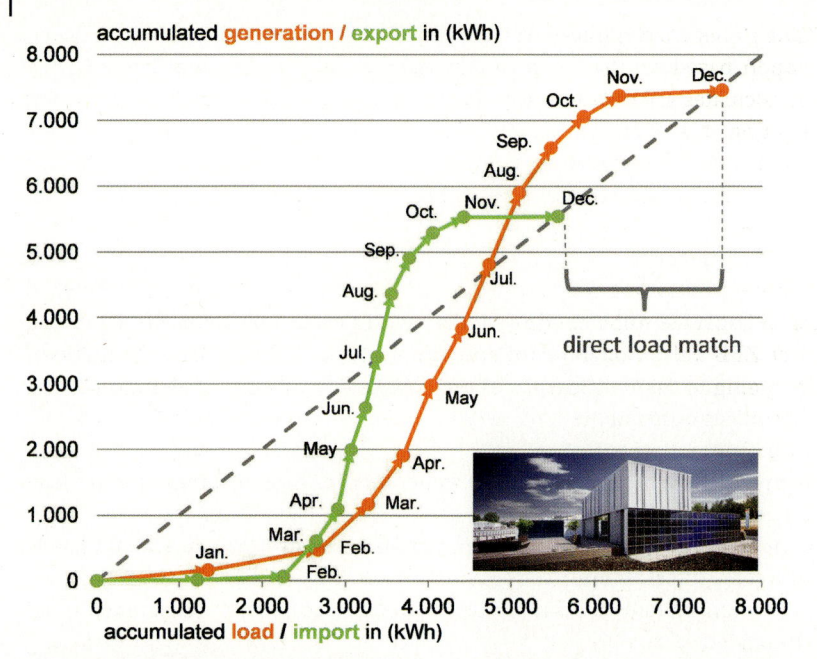

Figure 41.4 Monitoring results for a small, all-electric Net ZEB in Germany. The building is the Wuppertal University entry to the Solar Decathlon Europe 2010 in Madrid, now operated in Wuppertal [23]. The data based on 5 min resolution are expressed as a load/generation balance as well as an import/export balance including all on-site loads. Monitoring started in September 2011. 31% of the solar power is really consumed on-site. Source: University of Wuppertal, EU.

for most climates. Low weighting factors during summer as compared with higher factors during winter would stimulate building energy solutions which operate to the benefit of the grids. Time-dependent electricity tariffs are a typical measure within "smart grids" to communicate such issues at the financial level.

41.6 Tools

As aids to studying the various definition options and associated consequences, free spreadsheet tools have been developed and made publicly available by the end of 2012:

- *Net ZEB definitions evaluation tool:* After entering annual or monthly based energy demand and generation data from separate calculations, this Excel-based tool calculates and displays a set of the most relevant energy balances addressing relevant combinations of balance boundaries, balance types and weighting factors. It was made available by the IEA working group with a free download [6].

- *EnerCalC:* This Excel-based tool realizes a multi-zone, monthly energy balance calculation based on the German code DIN V 18599 in a simplified manner [24]. It calculates the energy demand for heating, ventilation, cooling, hot water, and lighting as well as the on-site power generation by PV and CHP. The 2013 edition includes a set of load/generation balances for primary energy and carbon emissions and also load match estimations. The tool is freely available [25] in German only.

A Measurement and Verification protocol for Net ZEBs [26] has been defined in terms of activities to be conducted for the assessment of the balance relevant to the Net ZEB definition set in the building project. The protocol includes 16 steps belonging to three different phases of the accomplishment of a monitoring campaign: design, implementation, and operation of the monitoring system. A standard diagram has been developed with the goal of representing buildings and possible metrics. The monitoring boundaries depend on the physical and balance boundaries set in the Net ZEB definition. Depending on the information needed (for example, net monthly balance) and the associated requirements and acceptable cost, sensors, and data acquisition system can be identified. After the planning phase, the actual monitoring can occur via spot, short, or long measurements. It is recommended to monitor for at least 2 years to correct possible malfunctioning and obtain a clearer picture of the situation.

41.7
Examples and Experiences

Over the past two decades, pilot buildings of various types an in various climates have been realized aiming towards an equalized annual energy balance [10]. About 300 have been identified and collected within a map [9]. Two school examples, new and renovated, are illustrated in Figure 41.5 and Figure 41.6.

Monitoring results of 80 international projects were collected and analyzed [27]. The major energy performance results are illustrated in Figure 41.7 as primary energy consumption compared with the credits gained by energy export from on-site energy generation. Most of the buildings consume less than 120 kWh $m^{-2}\,a^{-1}$ for their total primary energy demand as specified in the passive house concept. This results underline that equalized energy balances are reality in building practice. Energy efficiency (= reduced consumption) is given first priority in almost all projects.

As discussed before (physical and balance boundaries), plug loads and central services are not included within usual building energy balance calculations. Their importance within ZEB concepts are illustrated in Figure 41.8 based on monitoring results for those buildings delivering separated monitoring results. Owing to decreasing demands of the service technology, the importance of user related loads increases.

Figure 41.5 School at Hohen Neuendorf, Germany, 2011. The energy systems consist of 22 m^2 solar thermal collectors, 55 kW$_p$ PV and a biomass-based CHP unit. Architecture and photograph: IBUS, Berlin.

Figure 41.6 School renovation in Wolfurt, Austria, 2010. The energy system consists of 80 m^2 solar thermal collectors, 26 kW$_p$ PV and a ground coupled heat pump. Architect: G. Zweier, Wolfurt. Photograph: R. Doerler, Wolfurt.

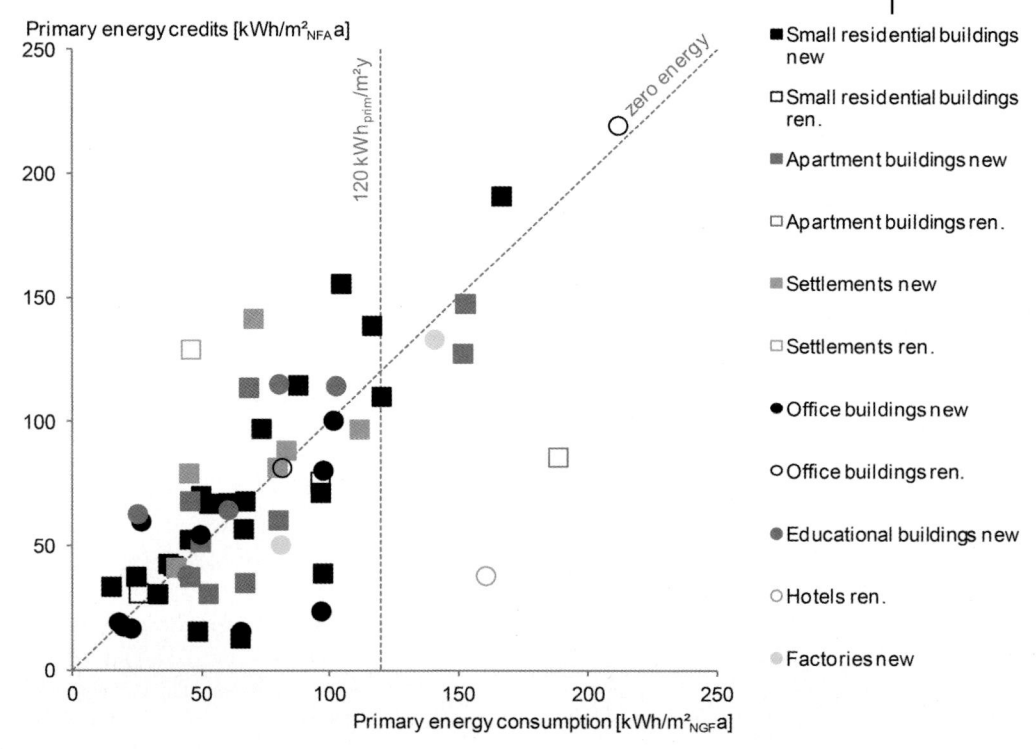

Figure 41.7 Primary energy credits for energy generation versus measured total primary energy consumption with respect to the net floor area (local primary energy factors, no climate normalization) [27]. Source: University of Wuppertal, EU.

The choice of the systems for heat generation is clearly more differentiated than for power generation. Systems range from the compact ventilation and heating units, ground or ground water coupled heat pumps to biomass boilers and co-generation plants. The use of biomass drastically reduces the primary energy needed for heating due to the low primary energy factor and thus promotes a decrease in the credits needed for energy balancing. Only few buildings generate credits for exporting heat. The majority of the zero-energy buildings apply solar thermal systems to assist hot water supply and space heating, unless other concepts, such as CHP or local heating grids, are preferred.

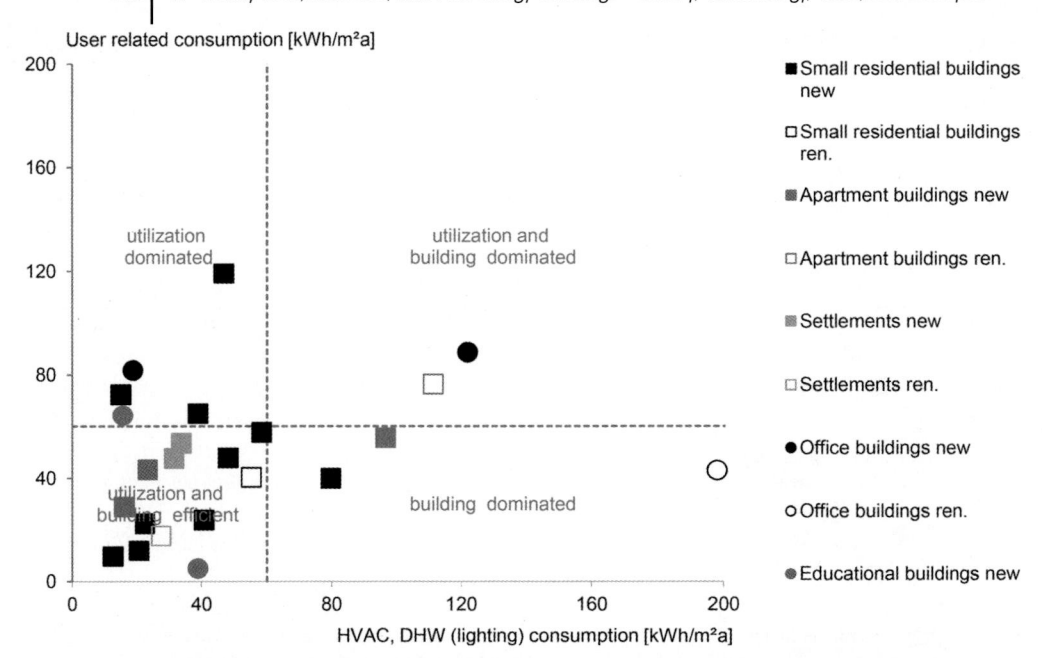

Figure 41.8 Comparison of the primary energy consumption for the technical services and the user-related consumption, such as plug loads and appliances. Only those buildings are plotted for which both types of consumption are recorded separately (local primary energy factors, no climate normalization). Optimal results are achieved for the buildings which show a high efficiency in both areas of consumption (utilization and building efficient) [27]. Source: University of Wuppertal.

Almost all investigated buildings apply solar power systems for on-site generation from renewables. For small buildings without additional power-generating capacity, an installed PV array of about 40 W peak power per square meter of net heated floor area was found to be sufficient to balance the total annual primary energy consumption (Figure 41.9). For projects with a larger energy demand (non-residential buildings or renovated buildings), this value is hardly greater, because the useful roof area decreases in relation to the related net floor area below. Especially in office buildings, further power systems such as co-generation plants or (external) wind turbines are used.

The average PV system size per unit floor area allows a rough estimation of the extra cost to balance fully the building energy demand. Taking today's typical full costs of about €2 W_p^{-1}, the PV generator increases building cost by about 4% for the example of small residential buildings (assumed construction costs €2000 m^{-2} floor area, DIN 276 cost classes 300/400).

Installed PV-capacity [W_p/m^2_{NFA}]

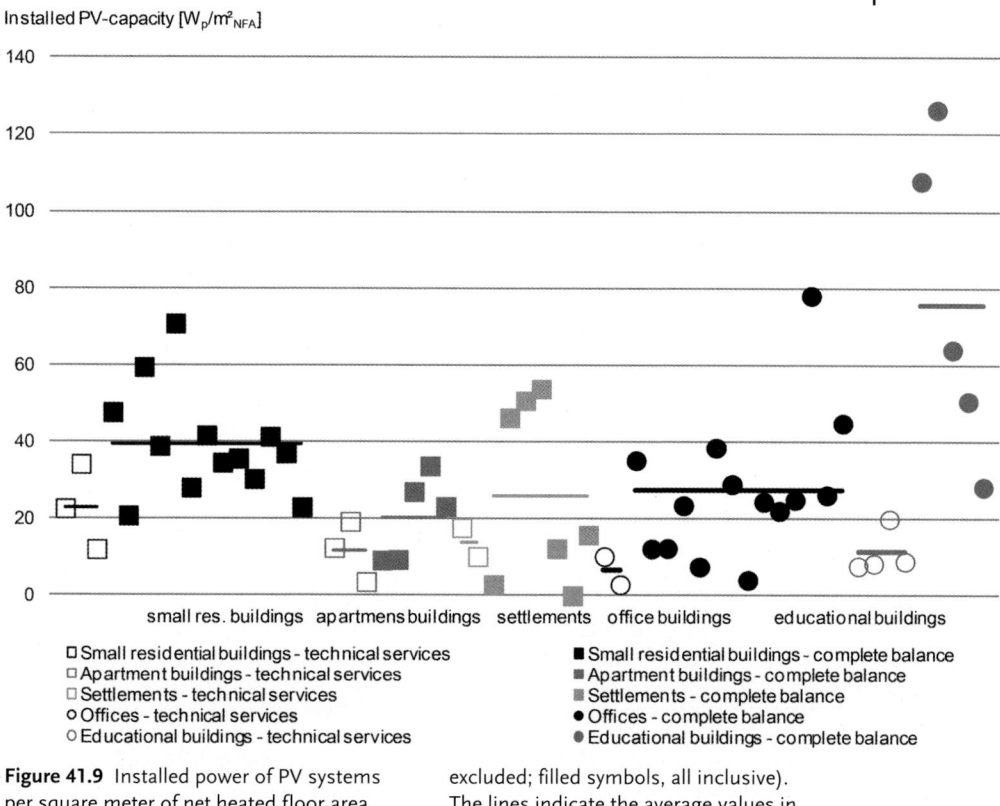

Figure 41.9 Installed power of PV systems per square meter of net heated floor area. The analysis is split by building typology and also the services to be balanced (unfilled symbols, central services only, plug loads excluded; filled symbols, all inclusive). The lines indicate the average values in each category [27]. Source: University of Wuppertal, EU.

Legend:
- □ Small residential buildings - technical services
- □ Apartment buildings - technical services
- □ Settlements - technical services
- ○ Offices - technical services
- ○ Educational buildings - technical services
- ■ Small residential buildings - complete balance
- ■ Apartment buildings - complete balance
- ■ Settlements - complete balance
- ● Offices - complete balance
- ● Educational buildings - complete balance

41.8 Conclusion

This chapter underlines the complexity of the topic and the implications of definitions and regulations for appropriate solutions. Nationally specific formulations have to clarify the balance boundaries, the balance type, and the weighting with respect to the EPBD, the already established national building energy code framework, and the strategic energy plan. As load match and grid interaction will become important characteristics in future green and smart grids, it is important that calculation procedures reflect these issues. The import/export balance including all types of on-site (before the meter) generation and loads in a harmonized way seems to be the most suitable approach in the medium to long term. However, there is a need for more knowledge on transient load patterns in the planning phase. The load/generation monthly net balance may serve as a compromise with respect to building code applications.

Asymmetric and time-dependent weighting factors for grid-based energy are important components of a future method. Such an approach would be in line with tariff systems that communicate the strategy to consumers at a financial level. However, this does not mean that a net zero or nearly net zero energy building would have net or nearly zero energy costs. This is due to the cost of using the grid and related taxation.

Acknowledgment

The work presented in this chapter was largely developed in the context of the joint IEA SHC Task40/ECBCS Annex 52: Towards Net Zero Energy Solar Buildings.

References

1 Marszal, A., Bourelle, J., Musall, E., Heiselberg, P., Gustavsen, A., and Voss, K. (2010) Net zero energy buildings – calculation methodologies versus national building codes, presented at the EuroSun Conference 2010, Graz.

2 Musall, E. and Voss, K. (2012) Nullenergiegebäude – ein Begriff mit vielen Bedeutungen. *DETAIL Green*, vol. 1, 2012, 80–85.

3 EPBD Recast (2010) Directive 2010/31/EU of the European Parliament and of the Council of 19 May 2010 on the Energy Performance of Buildings (Recast). *Off. J. Eur. Union*, 18 June.

4 Kurnitzki, J., Allard, F., Braham, D., Goeders, G., Heiselberg, P., Jagemar, L., Kosonen, R., Lebrun, J., Mazzarella, L., Railio, J., Seppänen, O., Schmidt, M., and Virta, M. 2011 How to define nearly net zero energy buildings nZEB. *REHVA J.*, vol. 23, pp. 6–12.

5 Concerted Action (2012) *Energy Performance of Buildings*, http://www.epbd-ca.eu/ (last accessed 11 January 2013); Erhorn, H. and Erhorn-Kluttig, H. (2012) The path towards 2020 – nearly zero energy buildings. *REHVA J.*, March.

6 IES Solar Heating and Cooling Programme (2012) *Current Research Projects (Tasks)*, http://www.iea-shc.org/task40/(last accessсed 11 January 2013).

7 Marszal, A., Heiselberg, P., Bourrelle, J. S., Musall, E., Voss, K., Sartori, I., and Napolitano A. (2011) Zero energy building – a review of definitions and calculation methodologies. *Energy Buildings*, **43**, 971–979.

8 Sartori, I., Napolitano, A., and Voss, K. (2011) Net zero energy buildings: a consistent definition framework. *Energy Buildings*, **48**, 220–232.

9 EnOB Research for Energy Optimized Building (2012) *International Projects on Carbon Neutral Buildings*, http://www.enob.info/en/net-zero-energy-buildings/international-projects/ (last accessed 11 January 2013).

10 Voss, K. and Musall, E. (eds) (2011) *Net Zero Energy Buildings – International Projects on Carbon Neutrality in Buildings*, ISBN 978-3-920034-80-5, DETAIL, Munich, http://shop.detail.de/eu_e/net-zero-energy-buildings.html (last accessed 11 January 2013).

11 Zero Carbon Hub (2012) *Facilitating the Mainstream Delivery of Low and Zero Carbon Homes*, www.zerocarbonhub.org (last accessed 11 January 2013).

12 Bundesministerium für Verkehr, Bau und Stadtentwicklung (2012) *Effizienzhaus Plus: Halbjahresbilanz*, http://www.bmvbs.de/DE/EffizienzhausPlus/effizienzhaus-plus_node.html (last accessed 11 January 2013).

13 Minergie (2012) *Minergie-A/A-ECO*, http://www.minergie.ch/minergie-aa-eco.html (last accessed 11 January 2013).

14 SINTEF (2012) http://www.zeb.no/ (last accessed 3 March 2013).
15 Berggren, B., and Hall, M. (2012) *LCE Analysis of Buildings – Taking the Step Towards Net Zero Energy Building*, Task Report, IEA SHCP Task 40/ECBCS Annex 52.
16 *Gebäudeenergieausweise der Kantone – Nationale Gewichtungsfaktoren*, EnDK, Bundesamt für Energie, Bern, 2009.
17 Energi Styrelsen (2012) *Analyser til Bygnigsklasse 2020*, http://www.ens.dk/da-DK/ForbrugOgBesparelser/IndsatsIBygninger/lavenergiklasser/analyser_tyvetyve/Sider/Forside.aspx (last accessed 11 January 2013).
18 Graabak, I. and Feilberg, N. (2011) CO_2 *Emissions in Different Scenarios of Electricity Generation in Europe*, Report TR A7058, SINTEF Energy Research, Trondheim.
19 Deutsches Institut für Normung (2011) *DIN V 18599, Teil 1: Allgemeine Bilanzierungsverfahren, Begriffe, Zonierung und Bewertung der Energieträger, Neufassung 12/2011*.
20 EnEV (2009) *EnEV 2009 – Energieeinsparverordnung für Gebäude*, http://www.enev-online.org/enev_2009_volltext/index.htm, Paragraph 5 (last accessed 11 January 2013).
21 Voss, K., Sartori, I., Musall, E., Napolitano, A., Geier, S., Hall, M., Karlsson, B., Heiselberg, P., Widen, J., Candanedo, J. A. and Torcellini, P. (2010), Load matching and grid interaction of net zero energy buildings, presented at the EuroSun Conference 2010, Graz.
22 Salom, J., Widén, J., Candanedo, J., Sartori, I., Voss, K., and Marszal, A. (2011) Understanding net zero energy buildings: evaluation of load matching and grid interaction indicators, presented at Building Simulation, 14–16 November, Sydney.
23 DETAIL (2011) *Solararchitektur*[4], ISBN 978-3-920034-48-5, DETAIL Green Books, Munich, http://shop.detail.de/de/solararchitektur.html (last accessed 11 January 2013).
24 Lichtmess, M. (2010) Vereinfachungen für die energetische Bewertung von Gebäuden, Dissertation, Bergische Universität Wuppertal, Fachgebiet Bauphysik und Technische Gebäudeausrüstung.
25 Federal Ministry for Economy and Technology – Energy Optimized Building, http://www.enob.info/?id=enercalc (last accessed 3 March 2013)
26 Noris, F., Napolitano, A., and Lollini, R. (2012) *Measurement and Verification Protocol for Net Zero Energy Buildings*, IEA SHCP Task 40/ECBCS Annex 52 Report, October 2012; for download see Ref. 6.
27 Musall, E., and Voss, K. (2012) The passive house concept as suitable basis towards net zero energy buildings, presented at the 16th International Passive House Conference, Hamburg.

42
China Road Map for Building Energy Conservation

Peng Chen, Yan Da, and Jiang Yi

42.1
Introduction

China has made great achievements in the field of building energy conservation with joint efforts of the community in recent years. For instance, the heating energy intensity in northern urban areas has sharply declined. Efforts aimed at building energy reduction help to offset the unavoidable increase in building energy consumption due to rapid developments and urban construction. However, the total building energy use in China is rising: the commercial energy used by the building sector per year has risen from 289 million tce (tons of standard coal equivalent; 1 tce = 29.3076 GJ) to 677 million tce from 2000 to 2010 [1]. One may pose the following questions: what level of energy consumption will Chinese buildings reach? What is the target for energy conservation? And how can it be reached it in the context of high rates of urbanization growth?

Many domestic and foreign energy research institutions are aware of the importance of these issues and they attempt to predict energy consumption in Chinese buildings through a wide range of models:

The *World Energy Outlook* issued by the IEA [2] points out that the total energy use in China will be 5.81 billion tce by 2030, of which building energy will account for 1.52 billion tce. The chief factors affecting energy use will be energy saving and emission reduction policies of the government and prices of energy resources. It is intended to limit the building energy to 1.1 billion tce in the future in order to realize the global carbon reduction target. Another report (*Energy Technology Perspectives 2010*) [3] points out that the main approach for realizing building energy conservation in China is to improve the technical level.

Research by the EIA (US Energy Information Administration) suggests that the total energy use in China will be 6.40 billion tce by 2030, of which building energy will account for 1.29 billion tce [4]; these figures are very different from the results of the IEA.

Transition to Renewable Energy Systems, 1st Edition. Edited by Detlef Stolten and Viktor Scherer.
© 2013 Wiley-VCH Verlag GmbH & Co. KGaA. Published 2013 by Wiley-VCH Verlag GmbH & Co. KGaA.

The Lawrence Berkeley National Laboratory, which has been studying building energy in China for a long time, considers the proportion of building energy in the total to be about 20% and comparatively low, and it will rise to 30%. Zhou *et al.* pointed out that building energy use in China will be 1 billion tce by 2020, and that urbanization is the chief factor for the increase in residential building energy, while the building area and the quantity of equipment lead to increases in commercial and public building energy in urban areas [5].

Some domestic organizations are also analyzing this topic: the report *The Sustainable Energy Situation in China 2020* suggests that the total energy demand in China will be around 2.32–3.1 billion tce by 2020, and the building energy demand will be between 470 and 640 million tce [6]. Actually, the total energy use in 2010 was 3.25 billion tce, and the building energy use was 677 million tce, which exceeded the limit expected. The literature also suggests that the total building area in China will be 91 billion m^2 [7], or even 118 billion m^2 [8], which is 2–3 times larger than the current amount, leading to a great increase in building energy use.

Current research is attempting to analyze the prospect of energy use in the future in China, and provide suitable policy measures and technical guidance to reduce building energy consumption successfully. However, building energy use in the future will depend on our current efforts and in the really short term. The goal is to specify the total amount of energy available for building operation in the future according to the total energy available, environmental capacity, and the demands of social and economic development for energy, rather than a prediction of the likely future. Taking into account the cap established, specific energy consumption in each sub-sector is allocated. Enhancing the building environment for rural and urban situations to support social development is analyzed, and a roadmap to reach it based on the upper limit is designed. This chapter presents our preliminary research results according to the method mentioned above.

42.2
The Upper Bound of Building Energy Use in China

The total energy use is limited by the global available energy resources and environmental capacity. The upper bound of the global carbon emission per capita and the amount of fossil energy are specified on the premise that the right to carbon emission and energy use is equal for everyone. Similarly, the upper bound of available energy for development in China can be specified while energy resources, economic and technical levels and energy possibly obtained from abroad are analyzed. With the upper bound obtained, and the energy use of social and economic development considered, the total energy available for building operation in China is specified. This section estimates the upper bound of building energy use allowed in China from the perspective of such analysis, which is the target of our building energy conservation work.

42.2.1
Limitation of the Total Amount of Carbon Emissions

Carbon emissions are mainly generated from the utilization of fossil energy. The IEA suggests that the carbon emission generated from fossil energy is about 80% of the total amount caused by human activity. Reducing the utilization of fossil energy is an important way to reduce carbon emissions.

In 2010, the total amount of carbon emission caused by energy use was 30.49 billion tons, of which China accounts for 22.3%, and the emission per capita exceeds the average level in the world [9]. The large amount of greenhouse gas (GHG) emission in China has attracted great attention worldwide, and there is an increasingly strong voice requiring China to limit its carbon emissions.

What is the target of carbon emission reduction? The IPCC (Intergovernmental Panel on Climate Change) points out that the global average temperature should not to rise by more than 2 K in order to maintain conditions suitable for human living [10,11]. We should control carbon emission progressively to reach these targets:

1. By 2020, the total amount of CO_2 emission will reach a peak value of 40 billion tons, of which 32 billion tons is caused by energy utilization. It is equal to about 15.6 billion tce fossil energy resources according to the current fossil energy structure. The global population will be 7.66 billion by 2020 according to the UN [12], and the fossil energy use will be 2 tce per capita. The fossil energy use in the United States is 9.8 tce per capita so far, five times the UN figure. In China, this number is 2.2 tce per capita, which also exceeds the UN prediction.
2. The total CO_2 emission is supposed to decrease to 48–72% of the amount in 2000 by 2050, which implies that fossil energy use should be reduced greatly unless the energy structure is adjusted and renewable or nuclear energy is used on a large scale.

The main primary energy source in China is coal, of which the carbon emission coefficient is the highest among fossil fuels, hence the total amount of fossil energy utilization must be severely restricted. On the basis of the target of global carbon emission control, the amount of fossil energy use is supposed to be limited to 2.95 billion tce if the population in China reaches 1.45 billion [13]. The common energy resources besides fossil energy include alternative energy sources such as nuclear energy and renewables, such as solar energy, wind energy, hydropower, and biomass energy. As studied by the Chinese Academy of Engineering, nuclear energy is expected to account for about 10% of primary energy, and renewable energy about 20% [8] in the future, with vigorous development of nuclear and renewable energy. Considering the contributions of these non-carbon energy resources, we calculate the upper bound of the total primary energy use to be 4.2 billion tce from the limit to carbon emission.

42.2.2
Limitation of the Total Amount of Available Energy in China

The total primary energy use in China was 3.25 billion tce in 2010, of which 68% was coal, 19% petroleum, 4.4% natural gas, and 8.6% nuclear power, hydroelectric power, and wind power [13]. The external dependence on petroleum is already over 50% [14]. It would be difficult for nuclear power, hydroelectric power, and wind power to replace fossil energy as the main source over a short period owing to constraints on resources, technology, and financial issues.

Traditional fossil energy resources are abundant in China, but the amounts per capita are small. For instance, the averages of coal, petroleum, and natural gas per capita are two-thirds, one-sixth, and one-fifteenth of those of the world, respectively [15]. Energy supply becomes a constraining factor of development in the urbanization progress. On the one hand, the annual production capacity of coal, petroleum, and natural gas is limited as it is restricted by energy storage, production safety, water resources, ecology, the environment, ground settlement, technology, and transportation; on the other hand, domestic energy production would hardly meet the rapidly growing demand, and external dependence progressively increases. However, energy import can be easily impacted owing to the restrictions of many factors such as energy-producing countries, transportation safety, and market energy price, hence it is unreasonable to satisfy domestic energy demand through the expansion of imports.

The tendency that energy use is increasing rapidly is not sustainable. It should be adjusted significantly, and the total amount of fossil energy should be controlled. The reliable energy supply capacity will be 3.93–4.09 billion tce by 2020, as concluded from studies by the Chinese Academy of Engineering. Different types of energy supply are outlined in Table 42.1 [8].

With GHG emissions and environmental restrictions considered, the energy supply will encounter a great impact. The capacity of hydropower is limited and increasing the installed capacity of nuclear energy faces many challenges. On the other hand, the other renewable technologies have not been fully developed. Therefore, energy use in China should not exceed 4 billion tce by 2020 from the viewpoint of energy supply capacity.

Table 42.1 Possible energy supply capacity in China in 2020 (×100 million tce).

	Coal	Natural gas	Petroleum	Hydraulic power	Nuclear power
Domestic production	21	2.83–3.21	3–3.29	3.27	1.63–1.86
Imported	–	1.29	4.28	–	–
	Wind power	Solar power	Solar thermal	Biomass energy	
Domestic production	0.62–0.93	0.046–0.092	0.3	1–1.45	
Imported	–	–	–	–	

42.2.3
Limitation of the Total Amount of Building Energy Use in China

Restricted by carbon emissions and available energy, the total amount of primary energy use in China in the future should be under 4 billion tce. This is the target required by long-term development instead of a temporary restriction: based on the target of global carbon emission reduction, both carbon emissions and fossil energy use in the future are supposed to decrease year by year; the energy storage is limited, and breakthrough in energy resource technology is difficult to achieve, hence the continuously increasing demand for energy can hardly be satisfied. It is imperative for us to control energy use in order to perform national obligations and ensure energy safety and sustainable development.

In the context of national total energy use restriction, the total amount of building energy should also be controlled. Building energy in China at present accounts for 20% of total energy use, the corresponding level in developed countries being 30–40% [16, 17]. Is it rational for building energy use in China to reach 30% of the total as in developed countries?

Based on the socio-economic structure in China, industry, especially manufacturing, is the major driver for development (secondary industries have accounted for 45–48% of the GDP since 2000 [13]). Manufacturing industry will remain an important economic sector supporting national development, and industrial energy will remain the major part of energy use in China, maintaining the trend of increasing year by year in the foreseeable future (the growth rate of industrial energy was 5% recently [13]). On the other hand, transportation energy use in China accounts only for about 10% (Figure 42.1) of total energy use, which is much lower than that of OECD countries both in proportion and per capita. It will certainly increase with the development of modernization.

Building energy use (non-commercial energy use in rural areas excluded) represents 20–25% of the total energy use in China [13, 18] (Figure 42.2). Assuming proper development of all sectors, it is supposed to enhance industrial energy efficiency, ensuring that industrial energy use will increase by less than 10% of the present level, and transportation energy use will increase less than twofold. Building energy use should be kept at less than 25% of social energy use.

In summary, total energy use in China should be controlled at under 4 billion tce because of restrictions on total carbon emissions and energy supply. Building energy use should be controlled at under 1 billion tce with the demands of industrial production, transportation, and living conditions considered, excluding renewable energy installed on buildings (e.g., solar thermal components, solar photovoltaic panels, wind energy).

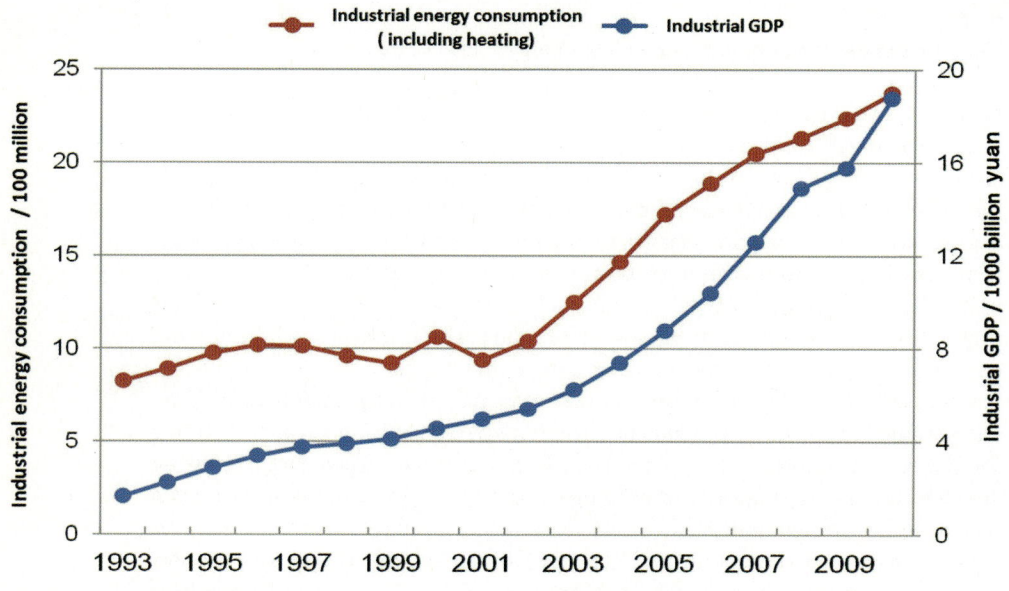

Figure 42.1 Industrial energy use and industrial GDP [13, 18].

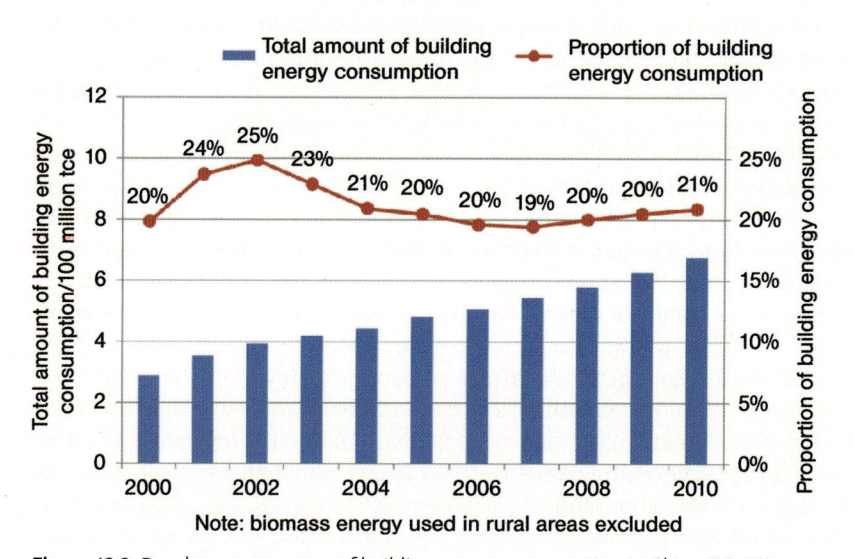

Figure 42.2 Development process of building energy consumption in China [13, 19].

42.3
The Way to Realize the Targets of Building Energy Control in China

42.3.1
Factors Affecting Building Energy Use

The question emerges of whether and how we can realize the target of building energy control in China after the upper bound of building energy use is specified.

Building energy use can be calculated as the following equation:

$$\text{total building energy use} = \text{energy use intensity} \times \text{total ownershop}$$

Energy use intensity represents the energy use per unit floor area and total ownership the total floor area. Therefore, it is necessary to study the variations of both energy use intensity and total building construction in order to study building energy use in the future. Energy use intensities in urban and rural areas differ from each other as a result of different building use patterns, environmental conditions, and so on, which makes it necessary to study variations of building energy use intensity and total building construction in the two areas.

The population in China will reach a peak of 1.47 billion by 2030–2040, when the urbanization rate will be 70% [20]. The population in urban areas will probably increase to 1 billion and that in rural areas will decrease to 0.47 billion, following the trend of social development and urbanization construction in China. Consequently, great changes will appear in total building construction in urban and rural areas.

42.3.1.1 The Total Building Floor Area
The total building floor area control is an important aspect in realizing the target of building energy conservation. The floor area of residential and commercial and public buildings in urban areas will grow further in the context of urbanization. Restricted by land and environmental resources, however, building floor area cannot increase infinitely. On the other hand, building energy use increases with the growth of floor area. There exists an upper bound of total building floor area to ensure normal operation of buildings under the constraint of total energy use.

Figure 42.3 shows the building floor space ownership per capita at present in different countries [13, 16, 17, 21–25]. It can be seen that building ownership per capita in countries and regions in Asia is very different from that in Europe and America, which is partly due to different land conditions and more probably resources obtained from overseas. Judged from the political and economic structure in the world at present, it is difficult for China to meet the demand for resources by importing. At the same time, the resources owned per capita are much less than the average of the world. In conclusion, economic development in China should be based on conservation of resources.

As the building construction is a high resource consumption industry, China should refer to the development pattern of the developed countries or regions in Asia instead of that in Europe or America, according to environmental resource

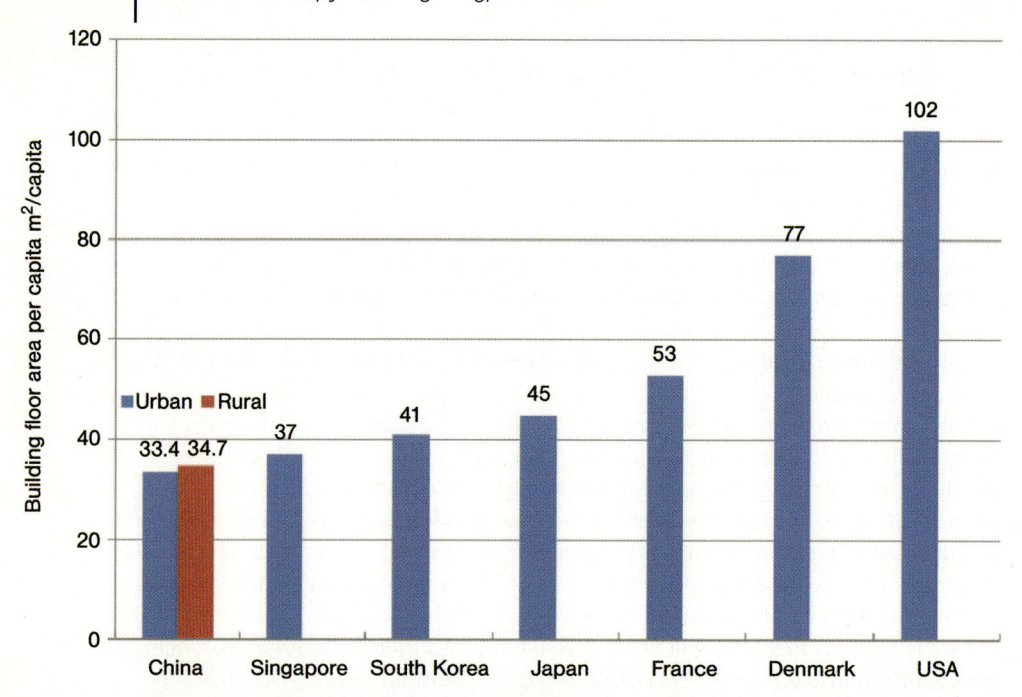

Figure 42.3 Building floor area per capita in different countries.

conditions. China should restrict the building floor area per capita to about 40 m^2, as in Japan, South Korea, and Singapore. If it is controlled to be in the range 40–45 m^2 and the population is assumed to be 1.47 billion, the calculated total building floor area is about 60 billion m^2.

The total building floor area in China is currently 45.3 billion m^2 [1], of which about 14.4 billion m^2 is in residential buildings in urban areas, about 7.9 billion m^2 in commercial and public buildings in urban areas, and about 23 billion m^2 in residential buildings in rural areas. According to the planning specifying a total floor area of 60 billion m^2, in the future the residential building floor area in urban areas should basically remain at 24 m^2 per capita as at present. The total residential building floor area will reach 24 billion m^2, thus an increase of 9–10 billion m^2 is allowed. The commercial and public buildings floor area will reach 12 m^2 per capita and 12 billion m^2 in total, thus an increase of 4 billion m^2 is allowed. In rural areas, the building floor area is expected to grow slightly from 23 billion m^2 at present to 24 billion m^2 as the population there will decrease. In this way, the total building floor area can be restricted to 60 billion m^2.

The increase in residential building floor area will be below 15 billion m^2 in this case. If the process is completed in 15–20 years, and demolition of existing buildings is not considered, the floor area of newly built constructions should be controlled between 0.8 and 1 billion m^2 per year. This is the constraint obtained from urban development, land, and resource conditions, and also the fundamental point when total building energy use is considered.

42.3.1.2 The Energy Use Intensity

The energy use intensity varies with different building energy use types. What causes the difference is the different energy use patterns and types in urban and rural areas, the different use patterns of commercial and public buildings and residential buildings in urban areas, and the different heating modes and intensities in southern and northern areas. According to the characteristics of energy use, building energy use is divided into four types: the energy use of northern urban heating (which not only includes heat used inside the building, but also includes loss in the district heating system), the energy use of urban residential buildings (excluding heating in north urban areas), the energy use of the commercial and public buildings (excluding heating in northern urban areas), and the energy use of rural residential buildings [1].

The energy use of northern urban heating refers to heating energy use in winter in the provinces, autonomous regions, and municipalities where building heating is legally required historically, including all types of district heating and decentralized heating. Classified by the sizes of heating systems and the types of energy source, it contains district heating patterns such as combined heat and power generation, district coal or gas boilers, regional coal or gas boilers, heat pump central heating of all sizes, and decentralized heating patterns such as household gas stoves, coal furnaces, air conditioner dispersed heating, and direct electric heating. Auxiliary equipment is obviously included in the corresponding end-use technology in addition to the energy source itself.

The energy use of urban residential buildings (excluding heating in northern urban areas) refers to energy use in the residential buildings in urban areas excluding heating energy use in northern urban areas. Classified by end uses, it includes the energy use of household appliances, air conditioners, lighting, cooking, domestic hot water, and heating in hot summer and cold winter zones (heating is not legally provided) in winter. The commercial energy resources mainly used are electricity, coal, natural gas, and liquefied petroleum gas.

The energy use of commercial and public buildings (excluding heating in northern urban areas) refers to energy use in commercial and public buildings caused by all activities except for heating energy use in the northern urban areas, including the energy use of air conditioners, lighting, electrical appliances, cooking, service facilities, and heating energy in winter in commercial and public buildings in urban areas in hot summer and cold winter zones. The commercial energy resources mainly used are electricity, gas, fuel oil, and coal.

The energy use of rural residential buildings refers to the energy use sustaining family life in rural areas. Classified by end uses, it includes energy used by cooking, heating, cooling, lighting, hot water, and household electricity appliances. The energy resources mainly used are electricity, coal, and biomass energy (straw and firewood). The building energy generated from the biomass energy is excluded from the calculations in this chapter since commercial energy use is the concern.

Energy-saving techniques and expectations for energy plans differ according to energy types. In the following, the current situation and energy-saving technology of different energy use types are introduced. The targets of energy conservation and total energy consumption for each of the sub-sectors mentioned above are specified; based on actual context, suitable technologies and measures are proposed.

42.3.2
The Energy Use of Northern Urban Heating

Owing to the energy intensity of district heating in urban areas of northern China, it represents a key sector in the reduction scheme. During the Eleventh Five-Year Plan period, outstanding success has been achieved through the insulation of building envelopes, the promotion of highly efficient heat sources and improvements to heating system efficiency. If electricity and coal use are allocated according to the "benefits to heat" method, heating energy use per unit area in northern China has decreased from 23.1 kgce m^{-2} (in 2000) to 16.6 kgce m^{-2} (in 2010) (kgce = kilograms coal equivalent). The total heating energy use in urban areas in northern China was 163 million tce in 2010.

With the development of urbanization, the building floor area in urban areas of northern China is predicted to increase from 9.8 billion m^2 at present to 15 billion m^2 in the future. Judging from the current situations and the effects of promoting energy-saving technology, there is some potential for energy conservation in heating energy use in urban areas of northern China:

1. *Improvement of insulation to decrease heat demand.* The current heating energy use intensity of new buildings in this century in northern China varies from 60 to 120 kWh m^{-2} in different climatic conditions, and it is capable of decreasing further compared with advanced levels of developed countries with similar climatic conditions. Heat demand could decrease to 45–90 kWh m^{-2} through improvement of exterior wall and window insulation, reduction of infiltration heat loss, introduction of quantitative ventilation windows, and highly efficient air exchange appliances with heat recovery (according to the climatic conditions, the heating demand in Shandong is 45 kWh m^{-2}, in Beijing 60 kWh m^{-2}, and in Harbin 90 kWh m^{-2}). Old buildings badly insulated in China are proportionally less in number than those in developed countries, and are easier to renovate. The renovation of these buildings according to the energy-saving standard for new buildings could obviously also decrease heating demand. A lot of renovation cases have proved that the target is feasible. For example, after envelope insulation renovation, the heating use of a certain residential building in Beijing [26] decreased from 80 to 54 kWh m^{-2}, and the room temperature is obviously higher than those without renovation. The heating use of a certain building in Shenyang [26] is below 65 kWh m^{-2}, while the room temperature is controlled around 18–20 °C.

2. *Eliminating overheating by implementing heating system reform and realizing heating regulation based on the household or the room.* It is intended to implement heating system reform, including management system reform of heating companies,

transformation from charging by area to heating quantity, and encouraging occupants to regulate actively. The terminal regulation equipment will be installed to regulate the room temperature to avoid overheating, and reducing the heating loss caused by overheating from 15–25% at present to 10% or less. For example, a certain district in Changchun [26] was reformed with terminal on–off regulation centering on room temperature control and heat allocation techniques. As a result, the heat loss caused by overheating decreased considerably. The average heating use is 85 kWh m^{-2} when only 30% of households have access to regulation, and the energy saved is 18.6% compared with the value of 105 kWh m^{-2}, which represents those buildings without renovation. By 2011, the terminal on–off regulation technique was widely applied in provinces such as Beijing, Jilin, Inner Mongolia, and Heilongjiang. The operation has resulted in positive effects after five heating periods, decreasing heating use by 10–20% compared with buildings without terminal regulation.

3. *Substantial enhancement of heat source efficiency.* There is more potential in heating sources to save energy than insulation and terminal regulation. That is mainly because (i) a power plant could supply 30–50% more heat on the condition that combined heat and power generation based on an absorption heat pump is applied when coal use and electricity generation remain the same; (ii) a gas boiler could be 10–15% more efficient when heat recovery from the exhaust is applied; (iii) exhaust waste heat exhausted from all industrial processes could represent a district heating source, which is regarded as zero energy. Most cities in northern China have developed district heating supply networks of all sizes. It is possible to explore and utilize fully the heating source mentioned if the network is taken advantage of. There already exist project cases for enhancing heating source efficiency. For example, a thermal power plant demonstration project on waste heat utilization of exhaust steam in Datong enhanced its heat supply capacity and energy efficiency remarkably with the absorption heat pump technique, recovering the waste heat of exhaust steam for heating. The heating area covered increased from 2.6 to 6.38 million m^2, while coal use did not increase and electricity generation did not decrease. In Chifeng City, industrial waste heat is applied to district heating. Waste heat exhausted by copper and cement plants that cannot be used directly is recovered. The total heat obtained from plants is 1.217 million GJ during a whole heating period, meeting a heating demand of 2.34 million m^2.

It is intended to promote the highly efficient combined heat and power generation technique further to supply 20–30 W m^{-2} of heat for 8 billion m^2, and fully utilize industrial waste heat to supply 20–30 W m^{-2} of heat for 4 billion m^2. Coal or gas boilers will serve as a backup of those systems, satisfying about 20 W m^{-2} of peak load for the 12 billion m^2 above. The area of about 3 billion m^2 where district heating is not available is heated by heat pumps, terrestrial heat, and other patterns. In conclusion, when the heating area in urban areas in northern China increases to 15 billion m^2 in the future, through enhancing the heating source efficiency, implementing heating system reform to eliminate overheating, and improving insulation to reduce heating demand, the heating intensity will possibly decrease from 16.6 kgce m^{-2} at present to 10 kgce m^{-2}, and the total energy use will decrease from 163 million tce at present to 150 million tce.

42.3.3
The Energy Use of Urban Residential Buildings (Excluding Heating in the North)

The energy use per square meter of urban residential buildings is increasing continuously and slowly. On the one hand, an increase in the number and type of appliances ownership is adding pressure in the energy demand; on the other, efficiency improvements of cooking equipment, household electrical appliances, and lighting equipment are offsetting a share of this increase. In 2010, the energy use of urban residential buildings represented 164 million tce, accounting for 24.1% of building energy use.

With the process of urbanization, over 70% of the population will live in urban areas in the future, and the total residential area will increase greatly. In the case of developing urban residential buildings reasonably, the building floor area is expected to increase from 14.4 billion m^2 at present to 24 billion m^2.

According to the climatic conditions and end-use energy types, the energy use of urban residential buildings is divided into northern China cooling, Yangtze River Basin heating and cooling, hot summer and warm winter zone cooling, household electrical appliances, cooking, domestic hot water, and lighting. From the current situations and characteristics of these parts, energy conservation of urban residential buildings can be implemented in the following respects:

1. The energy use for heating and cooling in the residential buildings in the Yangtze River Basin has been increasing sharply during recent years. Which type of heating and cooling is feasible in this area is a great controversy. Currently, the heating and cooling intensity is 10–15 kWh m^{-2} of electricity. Owing to the current really low indoor temperatures in winter, the demand for enhancing thermal comfort is increasing. Judging from the measured data, if district heating is implemented, the most efficient large-scale heat pump at present consumes 40 kWh m^{-2} of electricity during a year, and the co-generation of heating power and cooling consumes 15 kgce m^{-2}, equivalent to 45 kWh m^{-2}, while combined heat and power generation combined with individual air conditioners consumes 10 kgce m^{-2} with 10 kWh m^{-2} of electricity and 40 kWh m^{-2}. In comparison, when a dispersed air source heat pump capable of realizing the "part time and part space" use mode is implemented, it is possible to control the electricity use below 30 kWh m^{-2}.

2. With the improvement of people's living standards, the demand for cooling in summer all over will grow, and the cooling intensity will possibly increase accordingly. Recent research shows that lifestyle is the major factor affecting cooling energy use, and the type of buildings and the systems also have an influence.

3. Considering AC energy saving from two aspects, i.e. lifestyles and building system types: (i) An energy-saving lifestyle should be encouraged and maintained. The cooling mode "full-time, full-space" and "constant temperature and humidity" is discouraged, and the "part-time, part-space" and " fluctuating according to outdoors climate condition" mode is encouraged to build an indoor environment. (ii) Building types should be developed compatible with lifestyle. Residential buildings considered as energy-saving buildings equipped with high-tech equipment results in edifices with no control of opening windows and with central air conditioning. Operable windows and natural ventilation should be maintained. It is intended to develop all kinds of passive technical means for regulating the indoor environment as much as possible.

4. In this way, the cooling intensity of northern China is expected to be less than 3 kWh m^{-2} from 2 kWh m^{-2} at present, whereas in southern China it should be less than 15 kWh m^{-2} from 10 kWh m^{-2} at present.

5. For the household electrical appliances, cooking, and lighting, the following methods are adopted: (i) promoting energy-saving appliances and limiting low-efficiency appliances in the market by the market access system; (ii) promoting energy-saving lamps and prohibiting filament lamps in the market; (iii) restricting the use of high energy use appliances such as electric clothes dryers and electric dish dryers. The energy use intensity is expected to be less than 8 kWh m^{-2} for electrical appliances and 6.5 kWh m^{-2} for lighting, and 70 kgce per capita for cooking, remaining the same as present.

6. Solar energy techniques for domestic hot water are promoted, making full use of the available solar energy to meet the demand for domestic hot water. As the demand is increasing, this energy use is expected to remain at the current level 54 kgce per capita.

According to the characteristics of current energy use and analyzed from the perspective of development, the target of each part of the energy use in urban residential buildings can be achieved, as shown in Table 42.2 (the building floor area in the future is estimated from the population in each region and the urbanization level) with the premise that the techniques considered earlier are fully implemented.

In summary, under the influence of high urbanization development and people's rising living standards, both the intensity and the total amount of energy use in urban commercial and public buildings will increase. It is possible to control the total energy use below 350 million tce by guiding green and healthy ways of living.

Table 42.2 Current situations and targets of energy use of urban residential buildings.

Activity		Area or population	Energy intensity	Energy use (million tce)
Cooling in northern China	Present	6.4 billion m^2	2 kWh m^{-2}	4.1
	Future	10.0 billion m^2	3 kWh m^{-2}	9.6
Yangtze valley heating and cooling	Present	4.5 billion m^2	13 kWh m^{-2}	18.72
	Future	8.5 billion m^2	30 kWh m^{-2}	81.6
Cooling in southern China	Present	3.5 billion m^2	10 kWh m^{-2}	11.2
	Future	5.5 billion m^2	15 kWh m^{-2}	26.4
Household electrical appliances	Present	14.4 billion m^2	6.5 kWh m^{-2}	29.95
	Future	24.0 billion m^2	8 kWh m^{-2}	61.44
Cooking	Present	0.6 billion people	70 kgce per capita	42.0
	Future	1.0 billion people	70 kgce per capita	70.0
Domestic hot water	Present	0.6 billion people	54 kgce per capita	32.4
	Future	1.0 billion people	54 kgce per capita	54.0
Lighting	Present	14.4 billion m^2	5.5 kWh m^{-2}	25.34
	Future	24.0 billion m^2	6.5 kWh m^{-2}	49.92
Sum	Present	14.4 billion m^2 0.6 billion people	11.4 kgce m^{-2}	164.0
	Future	24.0 billion m^2 1.0 billion people	14.6 kgce m^{-2}	350.0

42.3.4
The Energy Use of Commercial and Public Buildings (Excluding Heating in the North)

The energy use of urban commercial and public buildings increases most rapidly among the building energy types classified. The area of these buildings increases 1.4-fold and the average energy use intensity increases 1.2-fold. The major motive for the increase is that the building energy intensity is shifting to the peak representing large-scale buildings of high energy use [27]. In 2010, the commercial and public buildings floor area accounted for 17% of total floor area, while the energy use was 174 million tce, accounting for 25.6% of total building energy use.

During the process of urbanization, the floor area of this building type will obviously increase with the improvement of public services and facilities. In buildings of this type in developed countries, the area is predicted to increase from 7.9 billion m² at present to 12 billion m². The main problem facing energy conservation is the violated understanding of the concept of "energy saving," considering "energy saving" to correspond to the application of energy-saving techniques or measures. Actual building operation data must be referred to when evaluating an energy-saving measure [28]. Based on this knowledge, it will divert "energy saving" to its reverse side where energy use increases if opinions such as "integrated with foreign countries" and "keeping advanced for several years" are stressed. The main target of energy conservation in commercial and public buildings should be to realize real energy-saving effects and sustainable development. Energy conservation in urban commercial and public buildings is achieved by the following measures:

1. For green, ecological, and low-carbon conditions as the goal of urban development, a green lifestyle is encouraged. High-energy buildings are prevented as much as possible, and the development mode of commercial buildings should be converted. The indoor temperature control method of "part time and part space" is promoted, whereas buildings with the method of "full time and full space" should be reduced. Figure 42.4 shows the energy use comparison between a certain office building and a typical building every month in Shenzhen. In 2011, the energy use per unit area of this building was 57.6 kWh m^{-2} (the electricity use was 51.8 kWh m^{-2}), while the average value was 103.7 kWh m^{-2} in comparison. The main techniques that it adopts are full usage of the natural ventilation and natural lighting, promoting the indoor temperature control method of "part time and part space."

2. It is intended to establish comprehensively the energy sub-entry measure among large-scale commercial buildings, and to convert to the energy quota management and step tariff system in order to manage energy saving based on the real energy use data.

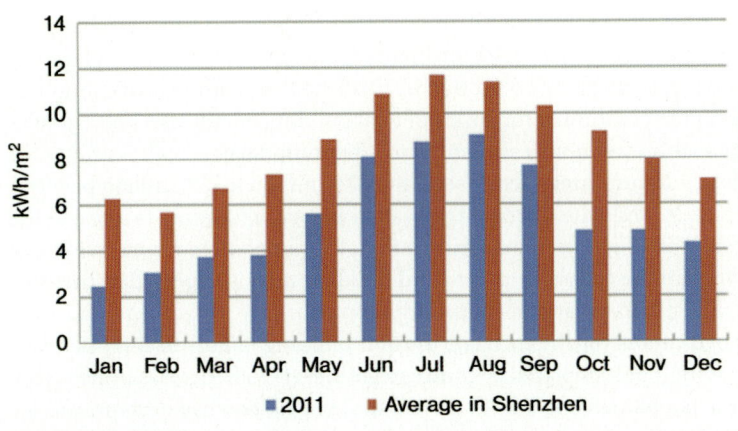

Figure 42.4 Comparison of energy use per unit area between the office building in Shenzhen and a typical building.

3. It is intended to promote the ESCO (Energy Service Company) mode, improving the current management mode of commercial building operation and encouraging energy-saving renovation.

4. It is intended to develop innovative energy-saving equipment actively and enhance system efficiency, such as the use of LED lamps, elevators with energy recovery, air conditioning systems with independent control of temperature and humidity (the energy use decreases by 30%), large-scale centrifugal chillers with direct coupling and frequency conversion, and so on.

With the techniques and measures outlined, and the current energy use situation of commercial and public buildings in urban areas referred to, what can be postulated about the energy use of this type in the future is as follows: the average energy intensity of office buildings will decrease to less than 70 kWh m^{-2} of electricity (such as the mentioned building in Shenzhen); for schools the value is less than 40 kWh m^{-2} of electricity; for large-scale shopping malls it is less than 120 kWh m^{-2} of electricity; of common shopping malls it is less than 40 kWh m^{-2} of electricity; and for hotels it is less than 80 kWh m^{-2} of electricity.

The floor area of commercial and public buildings in urban areas will increase in the future. It is possible to reduce the energy use intensity from 22.1 kgce m^{-2} at present to 20 kgce m^{-2}, and control the total energy use, which is 174 million tce at present, to less than 240 million tce through implementing the whole process management aimed at energy quotas of newly built buildings, promoting contract energy management of existing buildings, and developing advanced innovative technology.

42.3.5
The Energy Use of Rural Residential Buildings

The energy use (including biomass energy) intensity of the rural residences has already exceeded that of urban residential buildings in the same climate zone, but the service level is much lower than for urban residences. There is no distinct change in total energy use per household, while the biomass energy is likely to be replaced by commercial energy (as Figure 42.5 shows). In 2010, the rural commercial energy use was 177 million tce, accounting for 26.1% of total building energy use, and biomass energy (straw and firewood) use amounted to 139 million tce.

The population in rural areas decreased from 810 million to 670 million between 2000 and 2010 [13], while the floor area per capita increased, leading to a rise in the total floor area. With the process of urbanization, the population in rural areas will decrease further in the future, and the rural building area is expected to increase slightly, from 23 billion m^2 at present to 24 billion m^2.

There are two causes for the increase in rural building energy use: (1) biomass energy has gradually been replaced by commercial energy, with increase in electricity use; and (2) the population engaged in agricultural production live in communities as a result of the "village combining" movement, which changes the lifestyle and is negative for production and living.

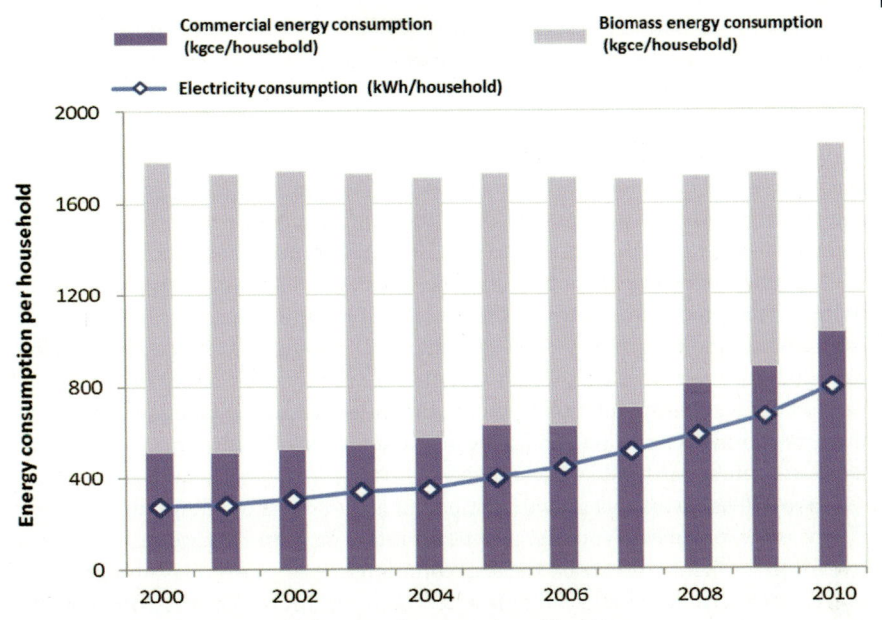

Figure 42.5 Trend of the energy intensity of rural residential buildings.

Different end energy use types considered, rural residences should make full use of biomass to meet the demand for cooking and heating in northern China, use solar energy to meet the demand for domestic hot water, and optimize natural ventilation with environmental sources to meet the demand for cooling. The energy intensity of lighting should be controlled at the same level as in urban areas when service standards are equivalent. The energy intensity of household electrical appliances will be less than 6.5 kWh m^{-2}, slightly lower than that in urban areas, although the rural residence floor area is greater than that in urban areas.

In particular, "no-coal villages" and "eco villages" should be developed in northern and southern China, respectively:

1. The techniques for no-coal villages in northern China are (i) to renovate houses and reinforce the insulation and the airtightness, thus reducing heating demand, and to use the Kang system to use fully the waste heat from cooking; (ii) to develop solar heating and domestic hot water of all types; (iii) to adopt the grain compression technique for straw and firewood to store energy in high density and combust efficiently.

2. The techniques for eco villages in southern China are (i) to renovate houses, and to make further improvements based on traditional rural residences to build comfortable indoor environments via passive methods; (ii) to employ the biomass pool to meet the demand for cooking and domestic hot water; (iii) to solve problems with combustion pollution, waste water, and so on to build a clean outdoor environment.

There have been many cases of these energy-saving techniques. For example, Shimen Village in Qinhuangdao [1] adopted many techniques such as envelope renovation, construction of biomass pool, replacement of straw gasifier for traditional stove and furnace, and reinforcement of solar energy use. The average energy use per household during a year was 2.1 tce, compared with the 3.8 tce for villages without renovation, and commercial energy (electricity, coal, and liquefied gas) decreased considerably, especially the use of coal, which was only one-tenth of that of the compared unrenovated village. The efficiency of biomass energy was enhanced, while service standards obviously also increased. For biomass energy utilization, there are many techniques and devices available at present, such as the biomass solid compression molding fuel processing technique, biomass compression molding particle burning stove, SGL gasifier and poly-generation technique, and the hypothermia methane fermentation microorganism technique. By taking advantage of the rural biomass energy resources, the demand for cooking, heating, and domestic hot water can be satisfied.

There will be no distinct growth in the floor area of rural buildings in the future. It is possible to decrease commercial energy intensity from 7.7 kgce m^{-2} at present to 4.2 kgce m^{-2}, and to decrease the total commercial energy from 177 million tce at present to 100 million tce by building a new clean energy system based on biomass energy and renewable energy, and supplemented by electricity and gas.

42.3.6
The Target of Buildings Energy Control in China in the Future

Through an analysis of current situations and energy-saving technical measures of four building energy use types, the energy use of northern urban heating, the energy use of urban residential buildings (excluding heating in the north), the energy use of urban commercial and public buildings (excluding heating in the north), and the energy use of rural residential buildings, combined with the population and the building floor area in the future, The target of the total building energy use control that will possibly be realized in China is specified with available techniques and measures.

Compared with the current building energy intensity and area, the targets of each part of energy use and floor area are shown in Table 42.3. On the basis that the building floor area increases from 45.3 to 60 billion m^2, and much specific energy-saving work is done according to characteristics of different energy use types, it is possible to control the total building commercial energy use in China to less than 840 million tce, compared with 677 million tce at present, which complies with the target of controlling building energy use to less than 1 billion tce in the future.

Table 42.3 Planning of the total energy consumption in China in the future.

Type		Floor area (billion m^2)	Intensity (kgce m^{-2})	Total amount (million tce)
Heating energy of northern urban buildings	Present	9.8	16.6	163
	Future	15.0	10	150
Urban residential building (heating excluded)	Present	14.4	11.4	164
	Future	24.0	14.6	350
Urban commercial and public building (excluding heating)	Present	7.9	22.1	174
	Future	12.0	20	240
Rural buildings	Present	23.0	7.7	177
	Future	24.0	4.2	100
Sum	Present	45.3	14.9	677
	Future	60.0	14	840

42.4 Conclusions

Building energy is related to national energy security, social stability, and sustainable development of the economy. This chapter points out the targets for building energy control in the future from top to bottom, proposes technical measures for different building types, and presents a roadmap for building energy conservation according to building energy characteristics and situations:

1. The total amount of energy use in China is restricted by both the global carbon reduction target and the energy supply capacity in China. In order to ensure national energy security, and assume responsibilities as a great power, the total energy use is intended to be controlled to less than 4 billion tce. Building energy use is intended to be controlled to less than 1 billion tce according to the energy structure based on industrial use.
2. According to the characteristics of each building energy use type, actual energy use situations. and reliable techniques or measures considered, the available targets for building energy use in China are analyzed from top to bottom. It is possible to control building energy use to less than 840 million tce in the future, including the energy use of northern urban heating, the energy use of urban residential buildings (excluding heating in the north), the energy use of urban commercial and public buildings (excluding heating in the north) and the energy use of rural residential buildings.

3. For the energy use of northern urban heating, from the three aspects of heat source, transportation and allocation, and building heating demand, it is intended to focus on enhancing heat source efficiency, implementing heating system reform to eliminate overheating, and improving insulation to reduce heating demand.
4. To guide green and healthy ways of living is the key measure to realize energy conservation in the energy use of urban residential buildings. The dispersed air source heat pump should be developed and promoted especially in the Yangtze valley, to keep heating and cooling use within 30 kWh m^{-2} during a year on the basis of further improvements to the indoor environment in winter in this region.
5. For the energy use of urban commercial and public buildings, newly built buildings should be installed with the whole process management aimed at establishing an energy quota. For existing buildings, to realize the target for energy control, contract energy management should be promoted and advanced innovative technology should be developed.
6. The energy use of rural residential buildings is the most uncertain aspect. One of the major targets of new countryside construction is to build a new clean energy system based on biomass and renewable energy, and supplemented by electricity and gas, developing "no-coal villages" in northern China and "eco villages" in southern China.

It is impossible for the building energy use in China to follow the development pattern of the developed countries in Europe and America. The way for building energy conservation in China should be as follows: beginning with the building energy characteristics in China, combined with the context of urbanization development, to implement building energy indices of all types, and beginning with real energy data, based on actual characteristics of each type, to implement and realize fully the great targets for building energy conservation from top to bottom in China.

Acknowledgments

This study was supported by the Ministry of Housing and Urban–Rural Development of the People's Republic of China, and performed under China International Technical Cooperation Project contract No. 2010DFB73870-1. The authors thank the Chinese Academy of Engineering, which organized the project "Research on China Road Map for Building Energy Reduction" to support this research.

References

1 Building Energy Research Center, Tsinghua University (2012) *Annual Report on the Development of Building Energy Saving in China 2012*, China Building Industry Press, Beijing.

2 International Energy Agency (2011) *World Energy Outlook 2011*, OECD/IEA, Paris.

3 International Energy Agency (2011) *Energy Technology Perspectives 2010*, OECD/IEA, Paris.

4 US Energy Information Administration (2011) *International Energy Outlook 2011*, EIA, Washington, DC.
5 Zhou, N., McNeil, M. A., Fridley, D., Lin, J., Price, L., de la Rue du Can, S., Sathaye, J., and Levine, M. (2007) *Energy Use in China: Sectoral Trends and Future Outlook*, Lawrence Berkeley National Laboratory, Berkeley, CA.
6 Research Group of the Sustainable Energy Situation in China 2020 (2003) *The Sustainable Energy Situation in China 2020*, China Environmental Science Press, Beijing.
7 UNDP China and Renmin University (2010) *2009–010 China National Human Development Report: Towards Low-Carbon Economy and Social Sustainable Future*, China Translation and Publishing Corporation, Beijing.
8 Research Group of China's Energy Medium and Long-Term Development Strategy Research (2011) *China's Energy Medium- and Long-Term (2030, 2050) Development Strategy Research*. Science Press, Beijing, China, 100717.
9 International Energy Agency (2011) *CO_2 Emissions from Fuel Combustion Highlights 2011*, OECD/IEA, Paris.
10 Intergovernmental Panel on Climate Change (2007) *Working Group III Fourth Assessment Report*, IPCC, Geneva.
11 Meinshausen, M., Meinshausen, N., Hare, W., Raper, S. C. B., Frieler, K., Knutti, R., Frame, D. J., and Allen, M. R. (2009) Greenhouse-gas emission targets for limiting global warming to 2 degrees C. *Nature*, **458**, 1158–1162.
12 United Nations, Department of Economic and Social Affairs (2011) *World Population Prospects, the 2010 Revision*, United Nations, New York.
13 National Bureau of Statistics of China (2011) *China Statistical Yearbook 2011*, China Statistics Press, Beijing.
14 Ministry of Land and Resources of China (2012) *Bulletin of the Chinese Land and Resources 2011*, April, http://www.mlr.gov.cn/kczygl/zhgl/201205/t20120511_1095646.htm.
15 Research Group Strategy of Sustainable Development, Chinese Academy of Sciences, *Strategy of Sustainable Development Report in China 2012*, Science Press, 2012
16 D&R International (2010) *2010 Buildings Energy Data Book*, US Department of Energy, Washington, DC.
17 European Commission (2013) *Eurostat*, http://epp.eurostat.ec.europa.eu (last accessed 11 January 2013).
18 National Bureau of Statistics of China (2012) *China Energy Statistical Yearbook 2000–2011*, China Statistics Press, Beijing.
19 Xiu, Y. (2009) Study of China's building energy efficiency based on energy data, Doctoral dissertation, Tsinghua University.
20 Research Group of Energy Research Institute (2009) *National Development and Reform Commission, China Low Carbon Development Road in 2050*, Science Press, Beijing.
21 Department of Statistics Singapore, *Yearbook of Statistics Singapore 2011*, http://www.singstat.gov.sg/publications/publications_and_papers/reference/yearbook_of_stats_2012.html.
22 Sin Chew Daily (2011) http://tech.sinchew-i.com/sc/node/228124 (last accessed 11 January 2013).
23 Korea Energy Economics Institute, Ministry of Commerce, Industry and Energy (2007) *Yearbook of Energy Statistics 2007*, KEEI, Seoul.
24 Korea National Statistical Office (2012) *Statistical Database*, http://www.kosis.kr/eng/e_kosis.jsp?listid=B&lanType=ENG (last accessed 11 January 2013).
25 Energy Data and Modeling Center, Institute of Energy Economics, Japan (2011) *EDMC Handbook of Energy and Economic Statistics in Japan, 2011*, Energy Conservation Center, Tokyo.
26 Building Energy Research Center, Tsinghua University (2011) *Annual Report on the Development of Building Energy Saving in China 2011*, China Building Industry Press, Beijing.
27 He, X. (2011) *Study on Distribution Features and Influencing Factors on Energy Use in Office Buildings by Statistical Method and Survey*, Master's dissertation, Tsinghua University.
28 Yi, J. and Da, Y. (2011) What is the real building energy saving? *Construct. Sci. Technol.*, (11), 15–23.

43
Energy Savings Potentials and Technologies in the Industrial Sector: Europe as an Example

Tobias Bossmann, Rainer Elsland, Wolfgang Eichhammer, and Harald Bradke

43.1
Introduction

In 2010, industry was responsible for 26% of European final energy use and 29% of energy-related carbon dioxide (CO_2) emissions (including indirect emissions). Energy-intensive industries, such as primary metals, nonmetallic minerals, chemicals, and pulp and paper, currently account for roughly 65% of total final industrial energy consumption. For the period up to 2050 the European Commission assumes in its EU Energy Roadmap 2050 a further increase by 19% compared with the level in 2010. No other sector features such a strong growth.

The aim of this chapter is to assess to what extent the industrial final energy demand in the European Union can be reduced within the next four decades by means of energy efficiency. The results presented here are based on a study carried out on behalf of the German Ministry of Environment [1]. It analyzes not only the contribution of specific energy efficiency technologies to energy demand and greenhouse gas (GHG) emission reductions but also the related cost savings. Although the results, in particular the cost savings, are specific for Europe, there are good indications, including from specific case studies, that worldwide similar or even higher potential exists, given the greater inefficiencies in many regions of the world. Differences may occur with respect to the cost-effectiveness of more efficient technologies, in particular in countries where energy prices are subsidized. The IEA *World Energy Outlook* [2] points to the large size of energy subsidies, which were even further increased in 2011 (the subsidies amounted to US$ 523 billion in 2011, up almost 30% on 2010). However, if subsidies are discarded from consideration, important savings on energy costs are also achievable in those countries and will benefit to the economy.

Energy savings in the European industry sector will amount to 52% or 192 Mtoe (million tons of oil equivalent) by 2050 compared with the industrial baseline demand. This equals a 43% reduction compared with today's level. Both process-specific efficiency improvements and massive energy savings by implementing efficient cross-cutting technologies (CCT) play a role. About 26% of the identified energy savings potential are based on technologies that provide steam and hot water and an additional 11% on the optimized use of efficient motor applications. About 90% of the identified energy saving potential in 2050 is cost-efficient, triggering annual net cost savings of €102 billion (€'05[1]). The sectoral primary energy demand can potentially be reduced by two-thirds. This reduction is not only based on the implementation of end-use related efficiency measures but also occurs through the shift towards a highly efficient electricity generation system that is mainly based on renewable energy sources. The related GHG emission reduction amounts to 70%.

The potential for energy efficiency improvements in industry is not equally distributed but varies across the countries of the European Union. This difference is even more significant when looking at countries outside the European Union. In many highly developed countries large energy-intensive industries already use efficient technologies that provide potential for further improvements can be realized through replacing older facilities, optimizing processes or enhanced energy management practices. In transition countries, new manufacturing facilities in energy-intensive industries are often equipped with the latest efficient technologies whereas older infrastructure is in most cases less efficient. The replacement of plants with outdated technology through new, more efficient large scale plants can produce significant energy savings. However, pure technological changes can achieve only a part of the energy savings. Additional savings are related to systems optimization and wider process changes [2].

The total final energy demand (FED) in the industry sector in the EU was 317 Mtoe in 2008, which corresponds to 27% of the total final energy demand. The bulk of this amount of energy (about 60%) is consumed by the iron and steel, chemical, nonmetallic minerals, and paper and pulp industries. All the industry branches have shown a constant energy demand over the past decades. Only for 2008 was a slight decrease in demand observed due to the economic crisis, and this was further accentuated in 2009.

In contrast to process technologies that are deployed solely in specific branches, cross-cutting technologies are spread over all industrial sectors (Figure 43.1). The cross-cutting technology most applied is the category of electric drives, accounting for 60–70% of industrial electricity consumption (Figure 43.2). Typical electric motor-driven applications include pumps, compressors, and fans, which account for 30% of the total electricity demand combined. Depending on the type of branch, the share of electric drives varies between 35 and 90% [3, 4].

---

1) All monetary values are given in real terms for the year 2005.

43.1 Introduction | 915

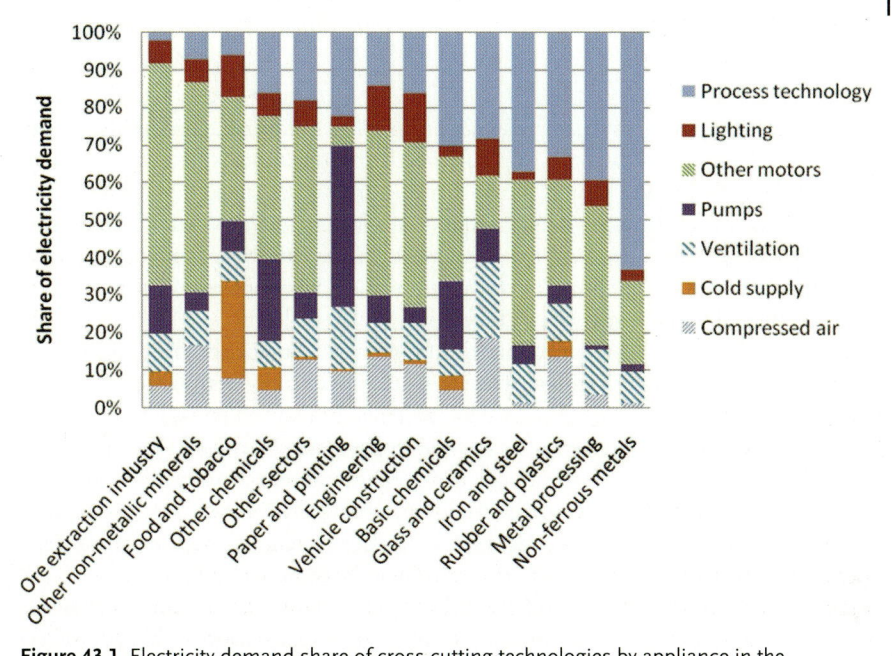

Figure 43.1 Electricity demand share of cross-cutting technologies by appliance in the European industry sector in 2008. Source: Fraunhofer ISI [5].

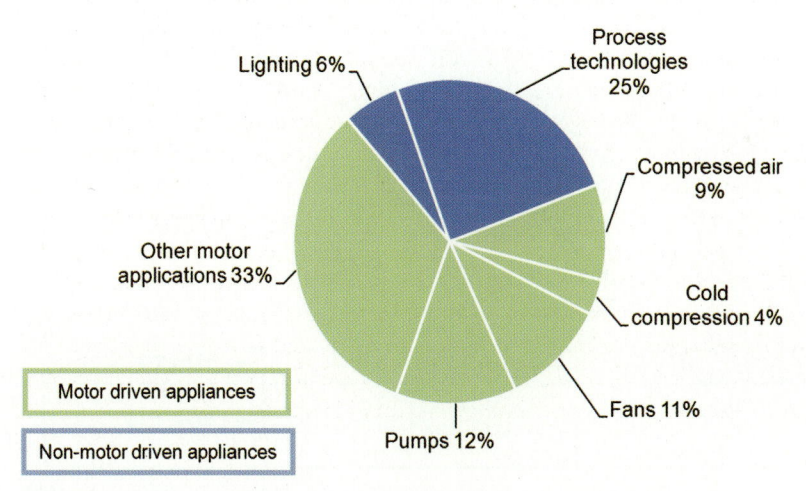

Figure 43.2 European industrial electricity demand by appliances in the industry. Source: Fraunhofer ISI 2009a.

43.2
Electric Drives

In general, electric drives convert electric energy into mechanical energy. Although a wide variety of electric drives is available, asynchronous motors are most prevalent – basically in the power range from a few hundred watts up to 5 MW. Their key advantages are that they are robust, inexpensive and very energy efficient. Therefore, about 80% of the European energy demand for electrical drives is associated with asynchronous motors [6]. Further types of electric drives are usually deployed for niche applications with special requirements.

Electric drives have been applied in the industrial sector since the mid-nineteenth century. Since then, continuous improvements in terms of energy efficiency have been accomplished, reaching a level of 90–95% [7]. Nevertheless, development towards more efficient electric drives is still in progress (Figure 43.3).

As the efficiency of the best available drive technology has already reached a level of 95%, further improvements are difficult to achieve. Nevertheless, in the light of 60–70% of industrial electricity consumption, a significant amount of overall savings can already be gained by small steps towards more efficient design.

By replacing the aluminum rotor with copper, the electrical resistance decreases. Thereby, the asynchronous motor efficiency can be increased by an additional 1.5–3.3% [8]. A simulation study [9] even computed an enhancement of efficiency

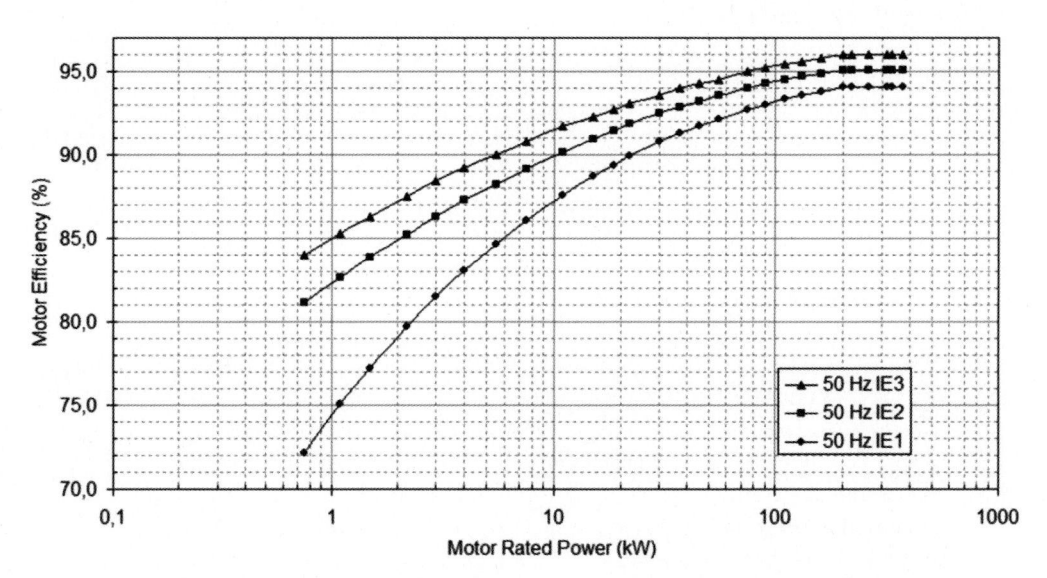

Figure 43.3 Differences in efficiency of four-pole electric motors for different efficiency standards. Note: IEC 60034 is an international standard of the International Electrotechnical Commission for rotating electrical machinery which specifies electrical efficiency classes for electric motors. IE1 equals the standard efficiency, IE3 the highest efficiency (premium efficiency). IE4 motors (super-premium efficiency) are also coming increasingly on to the market. Source: [6].

from 2.1 to 6.9%, depending on the power class of the electric drive. In lower performance categories, permanent-magnet motors can achieve an even better efficiency than the most efficient asynchronous motors. A permanent-magnet motor does not have a field winding on the stator frame, instead relying on permanent magnets to provide the magnetic field against which the rotors field interacts to produce torque [10]. Reluctance motors represent a technology of comparable efficiency that induce non-permanent magnetic poles on the ferromagnetic rotor, limiting environmental problems during the production phase.

Superconducting motors are a new type of alternating current (AC) synchronous motor that employ HTS (high-temperature superconductor) windings in place of conventional copper coils. Because HTS wire can carry significantly larger currents than copper wire, these windings are capable of generating much more powerful magnetic fields in a given volume of space. Therefore, minimum losses in conduction of electricity can be achieved [4].

Figure 43.4 depicts the energy-saving potentials in the EU that are related to an increased efficiency of electric drives. Until 2030, energy savings are concretely assigned to specific applications (e.g., pumps, fans, cold appliances). For the period up to 2050, a more general projection of energy-saving potentials has been carried out based on the results from the ADAM report [3], a large European study on climate change mitigation and adaptation. The potentials are compared with the FED of the so-called PRIMES baseline[2], which reflects a business-as-usual development. Owing to the already high efficiency of electric drives, the energy-saving potentials attributed to this technology are rather low despite their wide range of application, as can be seen in Figure 43.4.

By 2030, the overall saving potential for fans, compressed air appliances, and pumps will be almost equal, in the range 0.25–0.3 Mtoe, whereas the savings for cold appliances will be just 0.12 Mtoe. The improvement of miscellaneous electric motor driven appliances accounts for about 0.8 Mtoe. Thus, the energy demand in the European industrial sector can be reduced by 1.8 Mtoe or 0.5% by 2030. By 2050, the potential doubles to 4 Mtoe or 1% energy savings. In comparison with the energy savings from system optimization of motor-driven appliances (see the next section)s the potential saving for the motor itself is nine times lower.

Details of the methodology for the calculation of energy-saving potentials are presented by the Fraunhofer ISI [5]. The main database is represented by the preparatory studies for energy using products on Lot 11 covering electric motors. For electric motors, a stock model was used based on motor efficiency classes IE1 to IE4. The technical potential was characterized by a high diffusion rate of saving options (maximum boundary given by stock and lifetime of technologies).

The implementation of energy-saving options in cross-cutting technologies is basically very cost-effective. Thus, these options can be achieved with minimal political incentives and therefore are denoted "low-hanging fruit."

---

2) The baseline development up to 2030 is based on the energy demand projection by the European Commission in 2010 (PRIMES Reference scenario) and continued until 2050 based on the ADAM report reference scenario.

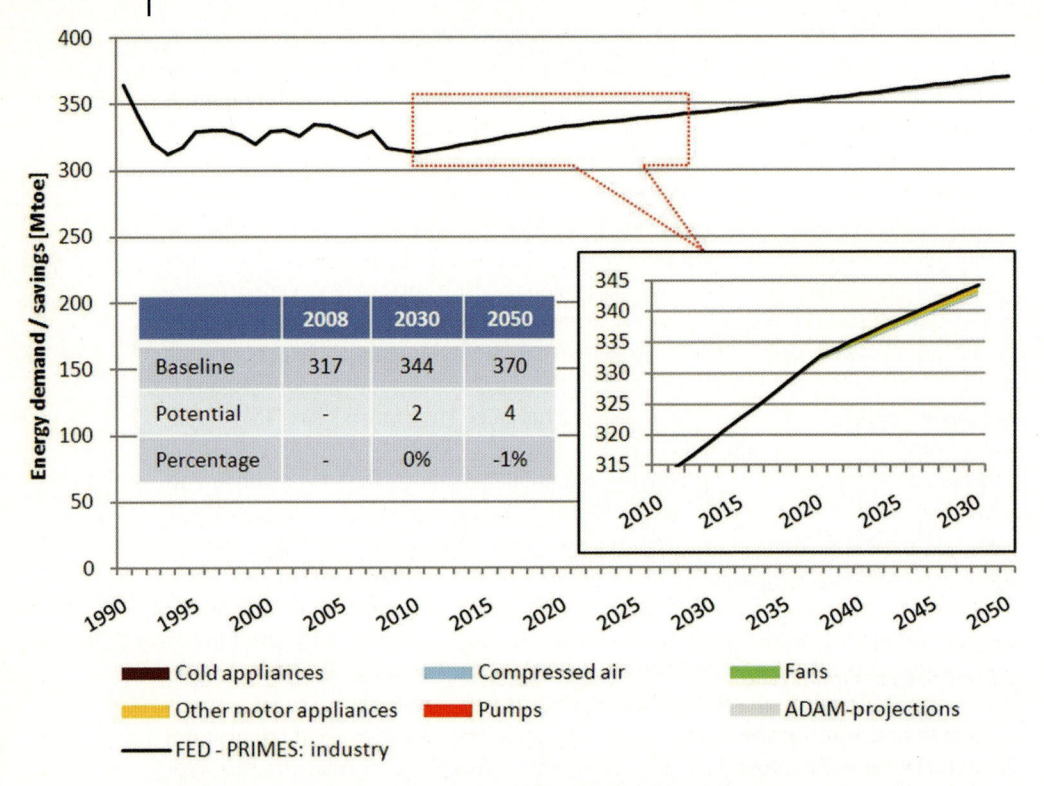

Figure 43.4 Energy-saving potentials of efficient electric drives in the European industry sector by 2050, compared with overall industrial final energy demand. Source: historical data, [7]; final energy demand projections, [11]; energy-saving potentials [3, 5].

Figure 43.5 depicts the energy-saving cost curve for more efficient electric drives for the period 2020–2050. The x-axis indicates the size of the saving potential and the y-axis represents the net costs[3] for the use of the efficiency technology. As indicated in Figure 43.5, the cost curve for electric drives is very easy to interpret. Looking at the year 2020 on the cost curve shows that the savings potential is 0.8 Mtoe in the first time interval. To attain this potential saving, the specific costs are −1056 M€'05 Mtoe^{-1}. As the efficiency of an average electric drive is nowadays about 90–95%, the specific costs change only marginally in the subsequent years in comparison with 2020. The specific costs decrease slightly down to −1156 M€'05 Mtoe^{-1} in 2050 and the savings potential from 2020 to 2050 is almost quadrupled because more efficient motors have more time to penetrate the stock of electric motors. Overall, the savings potential of electric drives is quantified as 4 Mtoe.

The quadrupling of the savings potential can be entirely translated to the equivalent evolution of the total cost savings, growing from nearly €1 billion in 2020 to more than €4 billion in 2050.

3) The net costs equal the differential costs for the investment in the efficient technology minus the savings through avoided energy use; hence negative net costs represent actual cost savings.

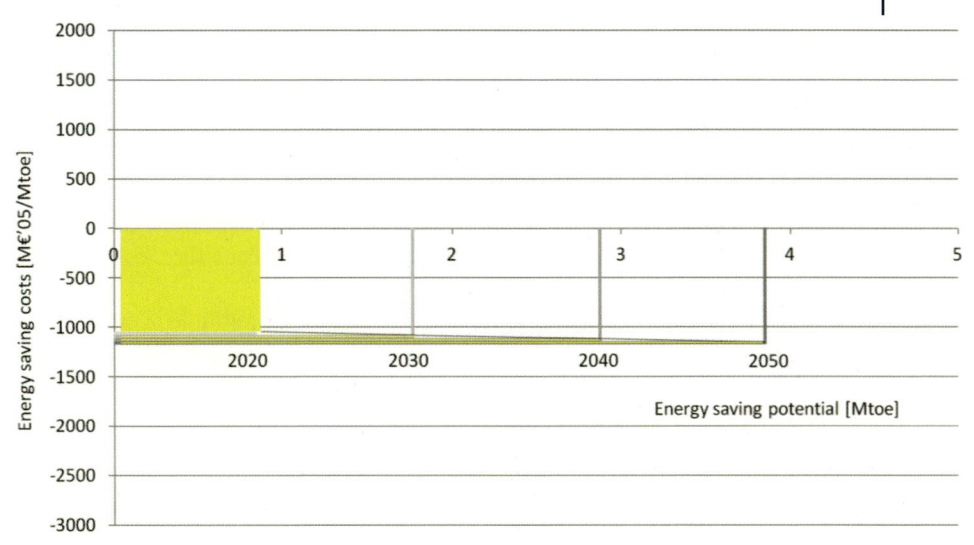

Figure 43.5 Cost curve for the implementation of highly efficient electric drives in the European industry sector. Source: Fraunhofer ISI.

43.2.1
E-Drive System Optimization

E-drive system optimization is a holistic approach that considers all elements of a technical system. Therefore, instead of solely improving the performance of physical components, the system optimization approach aims to increase the efficiency of the system as a whole by involving both technical and organizational improvements. To ensure the prevention of overlap between system optimization measures and electric drives, technical improvements of the motor itself are not considered as system optimization. Hence, no double counting of potentials takes place.

All energy efficiency measures introduced in the following could, in theory, be implemented immediately [6, 12–14].

An adjustment of speed and torque to the load requirements could be achieved by using variable-speed drives (VSDs). A VSD is a system for adjusting the rotational speed of an alternating current electric drive by controlling the frequency of the electric power supplied to the motor.

When using electric motors to drive pumps, fans, or compressed air components, the optimization of ducting leads to further improvements of energy efficiency.

The degree of efficiency in a technical system is generally determined as the product of the efficiencies of the single components. Thus, just by exclusively using high-efficiency appliances, the possibility of consuming an optimum or rather a minimum amount of electricity can be achieved.

To avoid oversizing electric drives, the motor specifications need to be matched with the requirements of the application or the whole system, otherwise the motor runs at a sub-optimal load factor, which significantly reduces the efficiency of power use.

A further controlling aspect to improve electricity efficiency is to implement a demand-related control system. This kind of system is nowadays usually designed as a closed loop control which automatically moves the system to the desired operating point and maintains it at that point thereafter by using some or all of the outputs as input parameters to optimize the system in terms of efficiency.

In addition to technical aspects, further efficiency can be achieved by proper and regular maintenance. Depending on the type of system, for example, compressed air or fans, the workload to realize this measure varies. Therefore, costs are the limiting factor to achieve the optimal potential in this case.

In addition to these measures to improve efficiency, a multitude of small/other options exist to save electricity in a motor-driven system. Measures attributed to this category are in general directly linked to specific cross-cutting technologies such as surface smoothing, and coating is related to pumps or frequent replacement of filters, which is relevant for both compressed air and fans.

The savings potentials were calculated using a scenario approach. Details of the methodology were presented by Fraunhofer ISI [5], and this section provides only an overview of the most important elements.

In order to estimate the impacts of holistic system optimization on motor systems, case studies of saving potentials in certain companies were used as the basis for our estimates. For each savings option, a technical savings potential is calculated as an average value from the case studies. This is corrected by the share of cases or companies in which the savings option is applicable and the share of companies that have already applied the option at a certain point in time. The result is the remaining technical savings potential at this point in time. The interactions between different savings options acting on the same system were taken into account by reducing the mutual savings potentials.

As illustrated in Figure 43.6, holistic improvements of motor-driven systems can lead to a fundamental decrease in electricity consumption by 2030. VSDs are estimated to have the highest savings potential at 4 Mtoe, followed by the implementation of demand-related control systems (2.2 Mtoe) and the avoidance of oversizing (2 Mtoe). In comparison with the physical improvements of electric drives, these measures can lead to a nine times higher savings potential. Putting this into perspective with respect to the baseline projection, the overall energy savings assessed for the optimization of electric motor-driven systems is 19 Mtoe or 6% by 2030 and 40 Mtoe or 11% by 2050.

The implementation of the energy-saving options in cross-cutting technologies is basically very cost-effective. Thus, these options can be achieved with minimal political incentives and therefore are denoted "low-hanging fruit."

As indicated in Figure 43.7, the energy-saving potential and the specific costs for e-drive system optimization will grow steadily until 2050. The utilization of VSDs is the most cost-effective saving measure with -912 M€'05 Mtoe^{-1} and energy savings of 3 Mtoe, which is 21% of the overall energy savings in 2020. Looking at the combination of energy-saving potential and specific costs in 2020, of the remaining options the avoidance of oversizing, demand-related control systems, and the application of high-efficiency appliances are in the second, third, and fifth places in terms of specific costs, each with an energy-saving potential of between 1.2 and 1.7 Mtoe.

43.2 Electric Drives

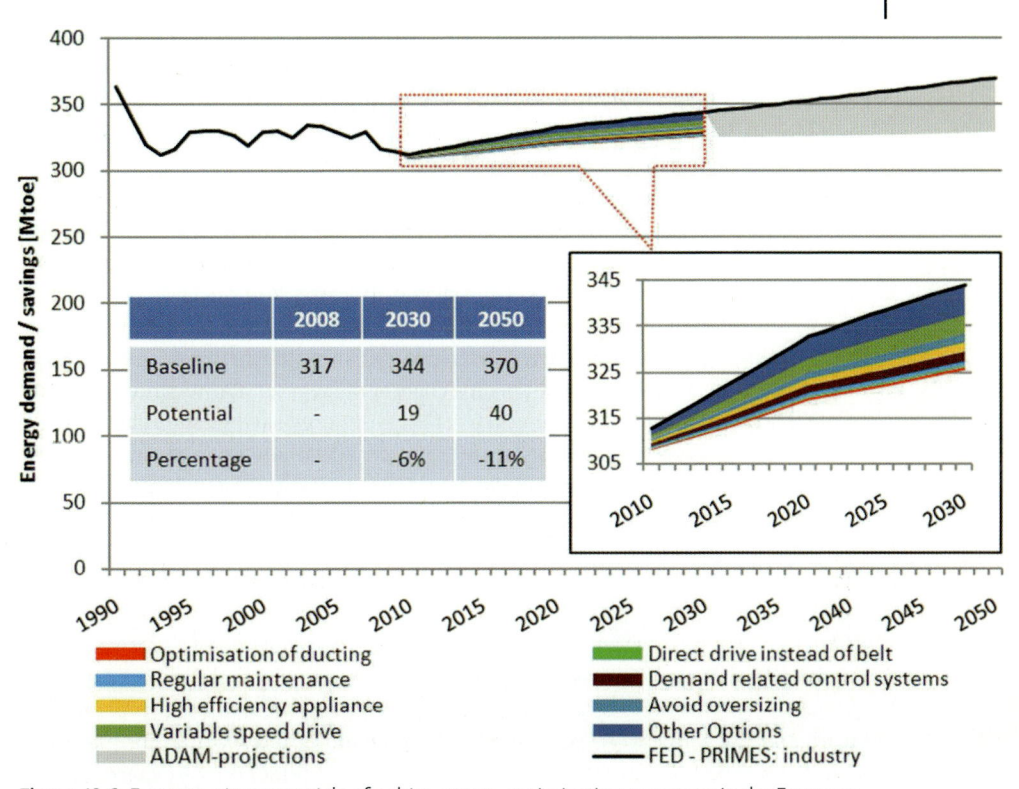

	2008	2030	2050
Baseline	317	344	370
Potential	-	19	40
Percentage	-	-6%	-11%

Figure 43.6 Energy-saving potentials of e-drive system optimization measures in the European industry sector by 2050, compared with the overall industrial final energy demand.
Source: historical data, [7]; final energy demand projections, [11]; energy-saving potentials, [5].

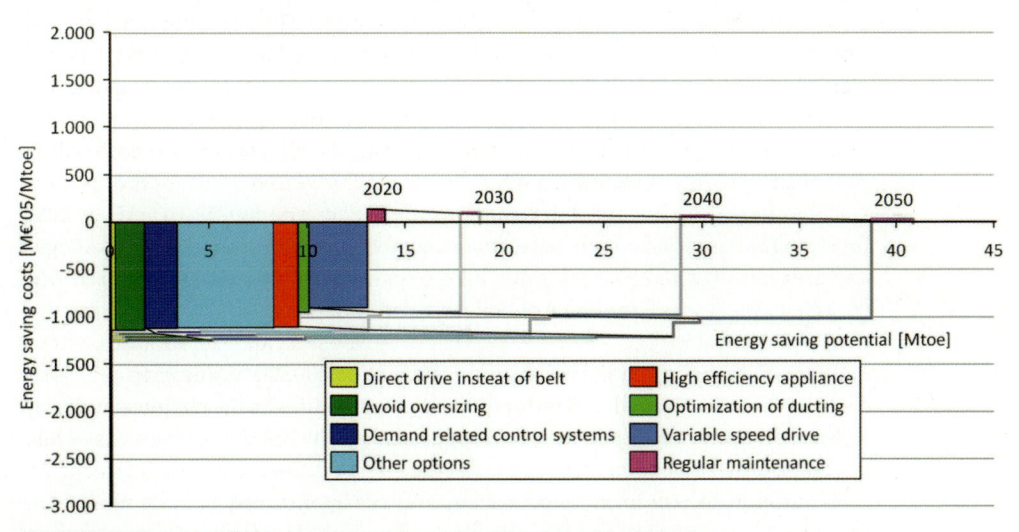

Figure 43.7 Cost curve for energy savings through e-drive system optimization.
Source: Fraunhofer ISI.

Furthermore, the utilization of direct drives instead of belts and the optimization of ducting are exclusively illustrated in the chart as cost-effective measures, but with only a minor impact regarding their energy-saving potential.

Finally, all the other energy-saving options that are not discussed in detail here also play a substantial role. These options are collectively referred to as "Other options" in Figure 43.7 with specific costs of −1113 M€'05 Mtoe^{-1} and energy savings of 4.9 Mtoe in 2020. Just the regular maintenance is not cost-effective, which results from high labor costs for maintenance specialists. These costs do not compensate for the monetary energy savings. Comparing the costs and energy-saving potentials in subsequent years with 2020, nothing surprising can be witnessed. The energy-savings potential increases for every measure, the negative specific costs increase steadily, and the positive specific costs of the regular maintenance decrease, owing to increasing electricity prices. Overall, the energy-saving potential of e-drive system optimization is quantified as 40 Mtoe in 2050. The net benefits resulting from e-drive system optimization total nearly €14 billion by 2020 (whereof less than 1% is needed to compensate for the additional costs through regular maintenance) and €45 billion by 2050

43.2
Steam and Hot Water Generation

Steam and hot water are used in industry for a wide variety of different purposes. Whereas temperatures below 100 °C tend to be used for water and space heating in the food, textile, and tobacco industries, temperatures between 100 and 500 °C are needed for many different industrial processes such as paper and poly(vinyl chloride) production (Figure 43.8). Heating at temperatures up to 1000 °C and above is very specialized and process specific, for example, in iron and steel and in glass and ceramics production [5]. Based on the predicted trend of the baseline scenario, the energy consumption of steam and hot water appliances is likely to remain more or less stable in the future. As modern appliances for steam and hot water generation already have efficiency levels of 90–95%, this technology can be described as highly developed [15].

Different types of boilers and burners are applied to generate steam and hot water for industrial use. Commonly used boilers work in the power range 100 kW–50 MW and are typically fired by oil, lignite, hard coal, electricity, natural gas (mixed with biogas), or biomass. The choice of boiler generally depends on the process requirements [3, 5, 14, 15].

When high operating temperatures in the range 200–300 °C and high pressures such as 80 bar are needed, for example, in drying processes in the chemical industry, thermal oil heaters are applied (85–89% efficiency). In contrast to water-based heat generators, thermal oil heaters use oil as the energy carrier.

In addition to technical improvements to the mentioned boilers, alternative generation concepts and greater integration of renewable energies offer substantial saving potentials [3, 5, 14, 15]:

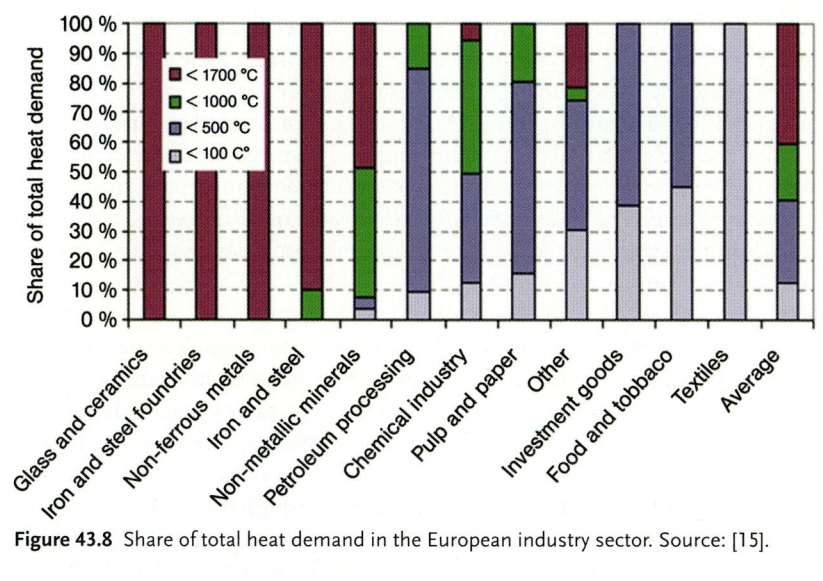

Figure 43.8 Share of total heat demand in the European industry sector. Source: [15].

Combined heat and power (CHP) generation systems can be used instead of steam boilers to provide steam for processes up to 500 °C. In CHP systems, a variety of technologies are applied such as steam back-pressure turbines, condensing turbines, gas turbines, and combined-cycle gas turbines. Their efficiency increases in that order by ~20% to an overall efficiency of > 40%.

To increase the heat produced by CHP technologies above 500 °C, one option might be to apply solid oxide fuel cells (SOFCs). Their higher operating temperatures of up to 900 °C make SOFCs suitable candidates for application with CHP.

Economizers operate in a similar way to heat exchangers, extracting residual heat from flue gases to subsequently preheat the feed water. In addition to integrated solutions, economizers can also be used to retrofit existing generation appliances.

To apply condensing heating technology, a heat exchanger is installed downstream of the economizer, which cools the flue gases below their condensation temperature. During this process, condensing heat is released, which is directly supplied to the closed heating circuit.

Depending on the age and fuel type of the burner, the operating excess air lies within the range 5–20%. Calorific energy is purged in this process. By implementing oxygen-regulation equipment, the air supply can be optimized and the energy demand minimized.

Using a continuously variable burner enables boilers to be run in a partial-load operating range, which can prevent frequent start-and-stop operation. This can reduce idling losses because the furnace no longer needs to be purged before being triggered.

The technical energy-saving potential of steam and hot water generation has been calculated using a scenario approach. Details of the methodology were presented by Fraunhofer ISI [5]; this section provides just an overview of the most important elements.

The calculation of technical energy-saving potential considers eight technology groups for the generation of heat in industry, of which only boilers represent the separate heat production (SHP), all other technologies being applied for CHP generation: steam back-pressure turbine, steam condensing turbine, gas turbine, combined-cycle, fuel cells, internal combustion engine, boilers, and other technologies.

The main input variable for the calculations is the heat demand of industry. It is derived in the first part of the model, taking into account the development of production and value added, together with certain sector-specific energy-saving options and assuming an average combustion efficiency of 85%. In the next step, the total heat demand is allocated to different temperature levels, as the possibilities and the technologies for supplying heat depend strongly on the temperature needed.

Two general groups of energy-saving options in heat generation are implemented: improved diffusion of CHP replacing separate generation of heat and electricity, and improved efficiencies in both separate and combined heat generation. We applied a methodology in accordance with Eurostat [16] that calculates the energy savings by comparing the CHP system with an alternative system that might have been in place if the CHP unit had not been built. The energy-saving potential is defined as the difference between the primary energy demands of both systems. Consequently, the choice and definition of the alternative system – the system that was replaced by the CHP plant – have a considerable influence on the results.

The technical energy-saving potential is characterized by a high diffusion rate of CHP (maximum 90% of a sector's heat consumption below 500 °C to be generated in CHP plants) and a fast EU-wide convergence of plants' mean efficiency values.

The energy-saving potential in industrial heat generation of 13% compared with the baseline, as illustrated in Figure 43.9, is due primarily to the diffusion of efficient space heating technologies, to a further diffusion of CHP technology replacing units of separate heat and electricity generation, and to efficiency improvements of separate and combined heat generation technologies. Approximately 20 Mtoe of all energy savings result from space heating, and a further 9 Mtoe result from CHP diffusion and 10 Mtoe from efficiency improvements in boiler and CHP technology. The total technical energy-saving potential will amount to 44 Mtoe by 2030 and to 95 Mtoe by 2050 compared with the baseline.

Not considered are the energy savings from the application of solar thermal energy, as this has hardly been used in industry so far. Furthermore, in the case of CHP it needs to be emphasized that the energy-saving potential technically cannot be considered as final energy because savings only arise if the comparison with a reference with separate generation of heat and electricity occurs at the level of primary energy.

As mentioned before, energy-saving options in industrial steam and hot water generation can be divided into three groups: efficient industrial space heating, further diffusion of CHP, and efficiency improvement of separate heat and power (SHP) and CHP generation.

For space heating, it is fairly easy to determine the economic potential, assuming that similar investments need to be made to those in the tertiary sector for large buildings. Thus the energy-saving potential is divided into a "low-hanging fruit" part (see the LHF share in Figure 43.10),which represents roughly one-third, and

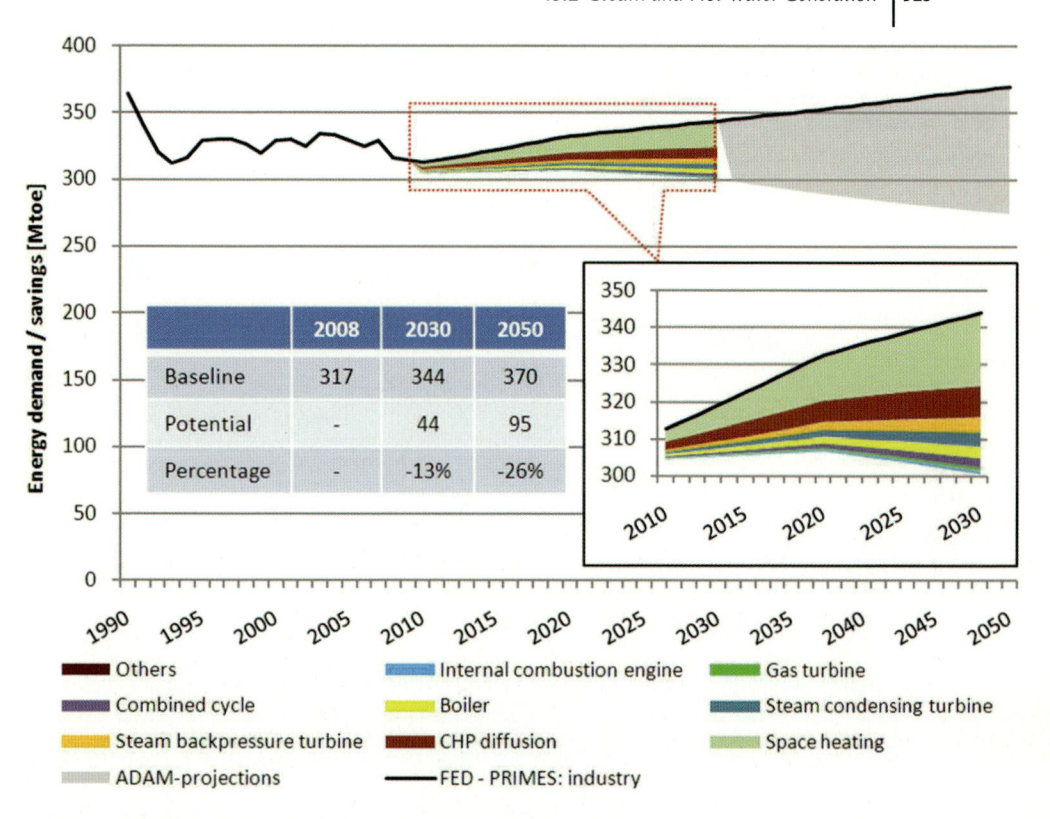

Figure 43.9 Energy-saving potential by efficient steam and hot water generation in the European industry sector until 2050 compared with the overall industrial final energy demand. Source: historical data, [7]; final energy demand projections, [11]; energy-saving potential, [5].

a "technical" part (representing measures not being economic) for the rest. While the low-hanging fruit potential further increases to 2050 from 4 to 14 Mtoe, the cost reduction involved increases from €0.4 to €10 billion. The noneconomic potential only becomes cost-efficient by 2050, if financial incentives are undertaken beforehand in order to compensate for the additional investment of the efficiency technology compared with the reference technology.

Regarding CHP, one can assume that the investment for a CHP plant is even lower than that for the construction of two separate plants that generate the same amount of heat and electricity individually. Consequently, the investment add-on for a CHP plant is equal to or even lower than zero. Hence the decisive factors for the cost-effectiveness of a new CHP plant comprise the fuel mix of the generation capacity that is displaced by the CHP plant, the price spread between the fuels used and the electricity produced, and the efficiency of the CHP and the competing SHP.

Since this is a large set of regulating tools that can be adjusted, a parameter variation was carried out in order to depict the entire range of CHP cost-effectiveness.

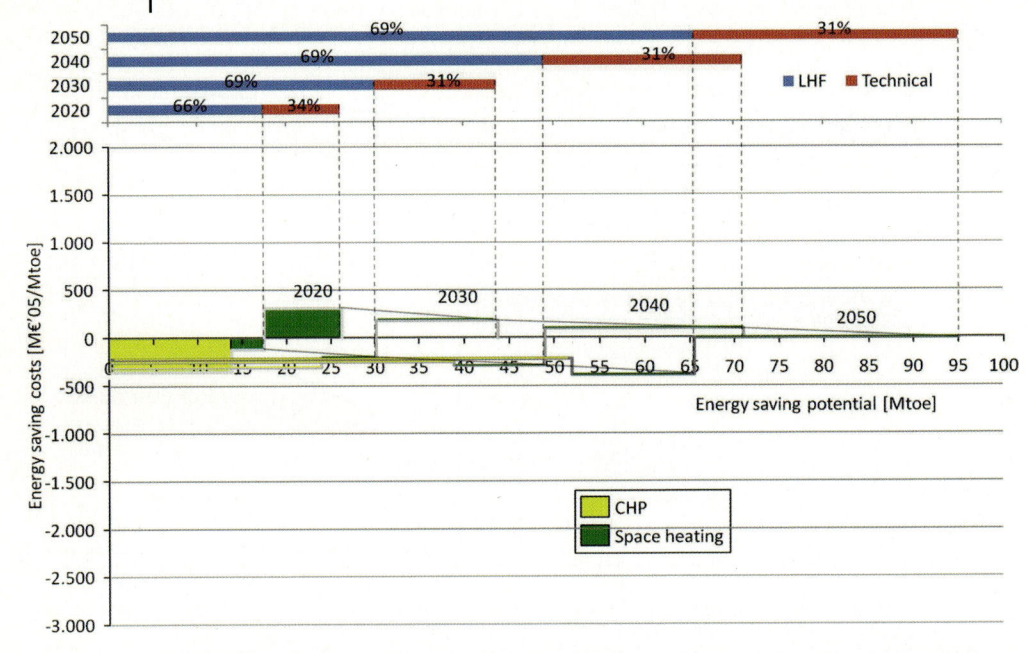

Figure 43.10 Cost curve for efficiency improvements in industrial steam and hot water generation. Source: Fraunhofer ISI.

For this present economic potential assessment, the most probable case was chosen: new SHP plants consisting of 50% hard coal-fueled (from 2030 onwards equipped with CCS technology) and 50% natural gas-fueled plants will be displaced by CHP plants with a mix of 80% biomass and 20% natural gas. For both SHP and CHP, an efficiency improvement is assumed.

Figure 43.10 shows the cost curve of the analysis and the specific cost reductions through space heating. Whereas energy-saving options for space heating experience a further decrease in specific costs, for CHP the opposite trend can be observed. This effect is driven through a decreasing fuel price spread between the fuels used in SHP and CHP and the electricity produced. Hence the cost advantage of the CHP plant is continuously compensated by relatively slower increasing fuel prices for SHP plants.

43.3
Other Industry Sectors

Apart from the energy-saving potentials identified in the different cross-cutting technologies, additional potentials are included in the process technologies of the iron and steel, nonferrous metals, chemicals, and nonmetallic minerals industries. They are briefly explained in that order in this section.

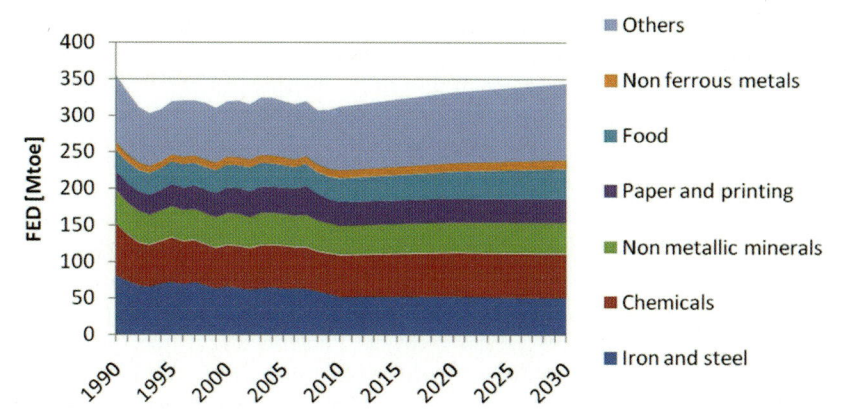

Figure 43.11 Final energy demand (FED) in the industry sector in the EU (historical and forecast). Source: 1990–2008, [7]; 2009, average value; 2010–2030, [11].

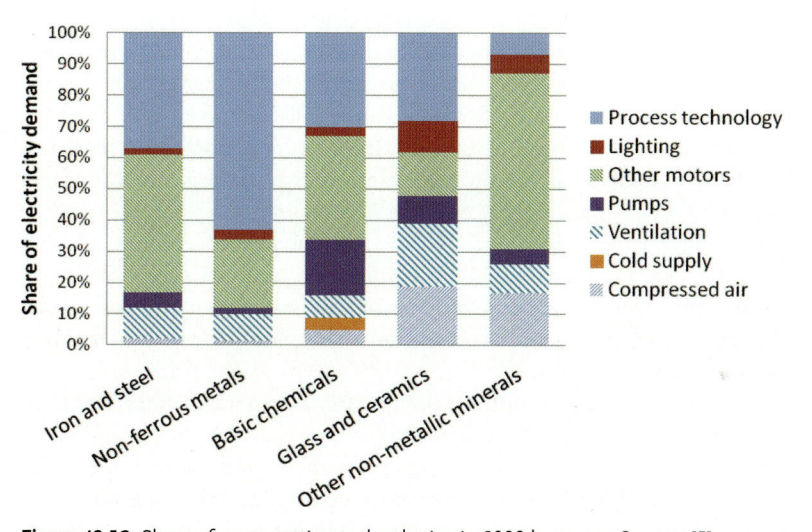

Figure 43.12 Share of cross-cutting technologies in 2008 by sector. Source: [5].

Figure 43.11 gives an overview of the individual shares of final energy demand in the different industrial sub-sectors. The iron and steel industry and the chemical industry are the main energy-consuming sub-sectors in European industry.

In order to gain an impression of the significance of the process technologies and their energy-saving potentials in the various sectors, Figure 43.12 depicts the share of cross-cutting and process technologies within the sectors. Electricity demand in the nonmetallic minerals industry (such as glass, ceramics, and cement) results mainly from cross-cutting technologies. The associated energy-saving potentials were already covered there whereas the metallic minerals industry is strongly dominated by process-specific technologies.

The iron and steel industry is the most energy-consuming industry in Europe, accounting for ~20% of the total industrial final energy demand and more than 5% of the total European energy consumption [7]. In this industry branch, two types of production processes need to be distinguished. The blast furnace route manufactures pig iron and crude steel based on the raw materials iron ore, coke, and coal. It is very energy consuming, requiring 0.29–0.36 toe t^{-1} pig iron and 0.43–0.48 toe t^{-1} crude steel [17]. Currently, about 70% of world steel is produced via the blast furnace/basic oxygen furnace route [18]. Hence blast furnaces account for a large part of energy consumption, and a particular focus is set on increasing energy efficiency. Energy saving options comprise the adoption of top pressure recovery turbines and blast furnace gas recovery. Pulverized coal injection helps reducing coke demand while combined-cycle gas turbines can be used instead of steam turbines to increase the thermal efficiency of power generation from blast furnace gas [2].

Alternatively, the electric arc furnace (EAC) uses recycled scrap, thereby avoiding the energy-intensive process of ore reduction and thus requiring only 0.07–0.12 toe t^{-1} crude steel. Hence major energy savings can be triggered through the increased proportion of scrap metal being recycled. At the same time one should note that a higher share of EAFs results in lower fuel consumption but higher overall electricity consumption. Direct current arc furnaces can significantly reduce the energy intensity if the furnace features a certain minimum production size. Gas-based direct reduced iron (DRI) is another option for less energy-intensive iron and steel making [2].

The strip casting process promises the most significant energy savings. Instead of reheating the steel for final shaping, a continuous near net shape casting is attached to the steel production process, reducing the specific energy demand by 75% to 0.002 toe t^{-1} steel. Further improvements can be achieved from heat recovery from steel rolling [23].

Among nonferrous metals, aluminum production is responsible for more than 50% of the total energy demand. Primary aluminum production involves bauxite (a type of aluminum ore) mining, production of alumina (aluminum oxide) from the bauxite, extraction of the aluminum through electrolysis, and final rolling. The production of primary aluminum requires about 1.3 toe t^{-1} of aluminum whereas the use of recycled aluminum reduces the energy demand to ~5% [20].

Energy savings can be triggered through the implementation of so-called PFPB (point feeder pre-baked) electrodes and improved operation of the furnaces and of the entire process [23]. In the long run, the integration of superconducting inductive magnet heating promises savings of up to 50% compared with conventional fuel-driven heating and melting processes [21].

The chemical industry is the second largest energy consumer of the European manufacturing industry, accounting for 57 Mtoe final energy demand in 2007. According to the baseline forecast, the chemical industry is supposed to experience a further increase within the next 20 years, thus even exceeding the iron and steel industry [11].

The chemical industry is characterized by significant heterogeneity, featuring numerous types of processes applied. Consequently, the identification of energy-saving technologies comprises a whole range of process-related measures. However, they can be traced back to a few fundamental principles, such as the application of more efficient catalysts, increased heat integration, the implementation of more energy-efficient separation units, the use of more efficient heat pumps and compressors and the adoption of advanced process automation [23]. The bulk of the energy savings in the chemical industry can be found in the sectors of refineries mainly linked to partition wall columns [5].

In this study the production of nonmetallic minerals comprises glass and cement products, accounting for nearly 14% of the industrial energy consumption.

Cement production is one of the major energy-consuming industry branches in the EU. A mixture of limestone, clay, and sand is pretreated (refining and mixing) for further processing in the furnace where the bulk of the energy is needed. The temperature increase implies chemical reactions that transform the raw material into pellets, called clinker. By adding gypsum, cement is attained. The global average energy intensity ranges between 0.07 and 0.11 toe t^{-1} cement [20]. Owing to the high energy intensity of cement production, various energy efficiency savings have already been exploited in the past (such as waste heat recovery).

A main factor for the energy intensity of cement production is the type of kiln technology employed for clinker production. Currently shaft kilns or wet/semi-dry/dry kilns are commonly used in the EU. Replacing them by dry kilns with pre-heaters and a precalciner triggers significant savings in clinker production. Additional savings result from heat recovery and efficiency improvements in the raw material production and grinding through high-efficiency classifiers and by the use of vertical roller mills [22]. The substitution of clinker through alternatives, such as fly ash, blast furnace slag, limestone, and pozzolana yields further savings [2].

There are different types of glass products, but the individual processes all include the following steps: selection of raw material (silica sand, soda ash, limestone), batch preparation (weighing and mixing of the raw materials), melting (the most energy-consuming process step) and refining, conditioning, forming, and post-processing [20]. Increased efficiency is focused mainly on the actual melting process by using oxygen as a substitute for the combustion air in the furnace and waste heat recovery from the exhaust gas, used to preheat to the combustion air.

Paper is made from pulp, which can be produced using wood or recycled paper. Pulp production can be differentiated into three alternative processes using different kinds of raw materials and producing different qualities of pulp. In the production of mechanical pulp, wood is shredded and refined to obtain a fibrous pulp. Huge amounts of waste heat are a typical by-product. Chemical pulp is also based on wood as the raw material and is produced using chemicals (sulfite or sulfate) that are used to separate the lignin content from the wood fibers in a cooking process. The lignin (around 50% of the initial wood) is then burnt in order to generate the large amounts of steam needed for this process. The third process is the production of pulp from waste paper.

Energy efficiency improvements for mechanical pulp concentrate on shredding and refining the wood and the recovery of waste heat. In the long run, large energy savings could be made by switching to water-free paper production where resin or artificial adhesive agents provide the adhesion between fibers. Long-term efficiency improvements for chemical pulp concentrate on the more efficient (energetic) use of by-products such as black liquor and the general development towards a biorefinery. The gasification of black liquor is discussed as a possible key element of such a biorefinery that would lead to significant efficiency improvements compared with the direct combustion of black liquor. However, the greater use of recovered paper has the most significant potential in several European countries.

For paper production, efforts are concentrated on the efficiency of paper drying – the process step that consumes the largest share of steam in the paper machine. Improved mechanical dewatering reduces the need for thermal drying. Although these techniques are already widespread, there is still potential for further diffusion. The shoe press technology keeps the paper inside the press for a longer period, extracting water from the paper using mechanical pressure and therefore reducing the need for thermal drying by 10–15%. Thermo-compressors increase the pressure of low-pressure waste heat, converting it into useful heat for other processes.

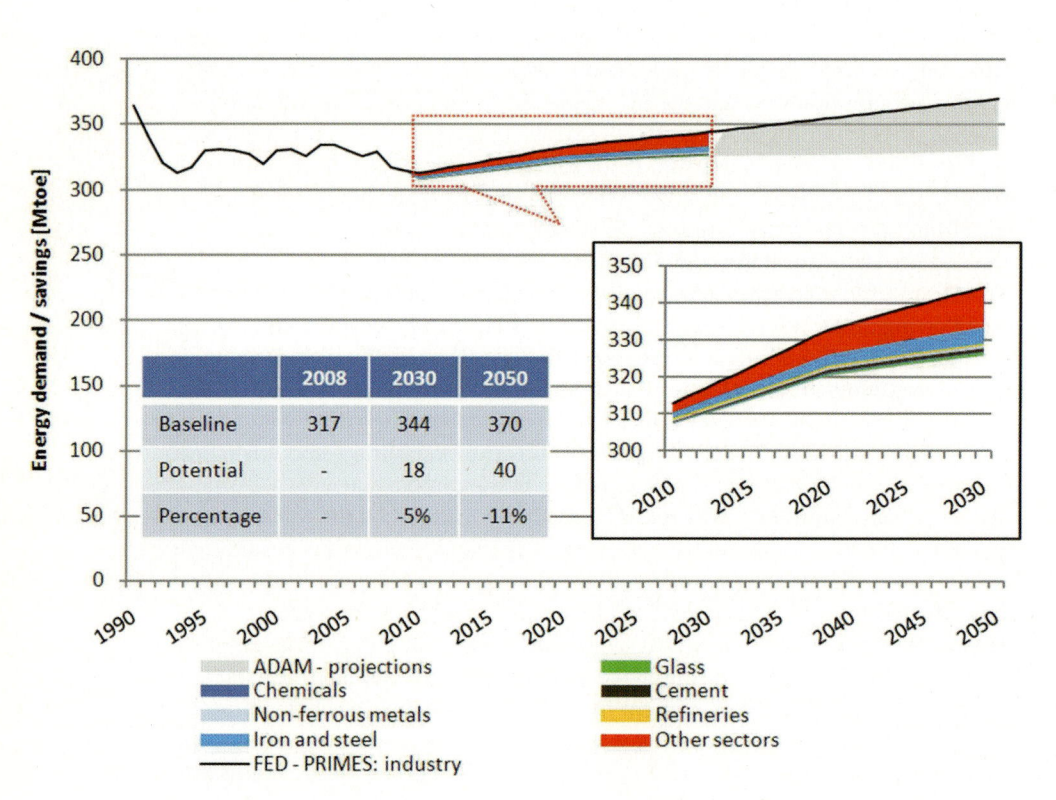

Figure 43.13 Energy savings through process technologies in the European industry sector up to 2050. Source: Fraunhofer ISI.

Better use of waste heat and heat integration means that significant steam savings of up to 20% can be realized in paper factories.

Other industry branches, such as the machinery construction, textile, food and drink, and tobacco industries feature additional energy-saving potentials that were not analyzed in detail owing to their relatively low significance. However, a rough estimate of the energy-saving potential is 12 Mtoe by 2030.

The total energy-saving potential for all industrial process technologies amounts to 18 Mtoe by 2030 and to 40 Mtoe by 2050 (Figure 43.13). This is comparable to a 5%/11% reduction relative to the baseline.

43.4
Overall Industry Sector

The total energy savings of the industry sector in Europe will be 88 Mtoe by 2030 (Figure 43.14). The entire energy savings are 26% compared with the baseline.

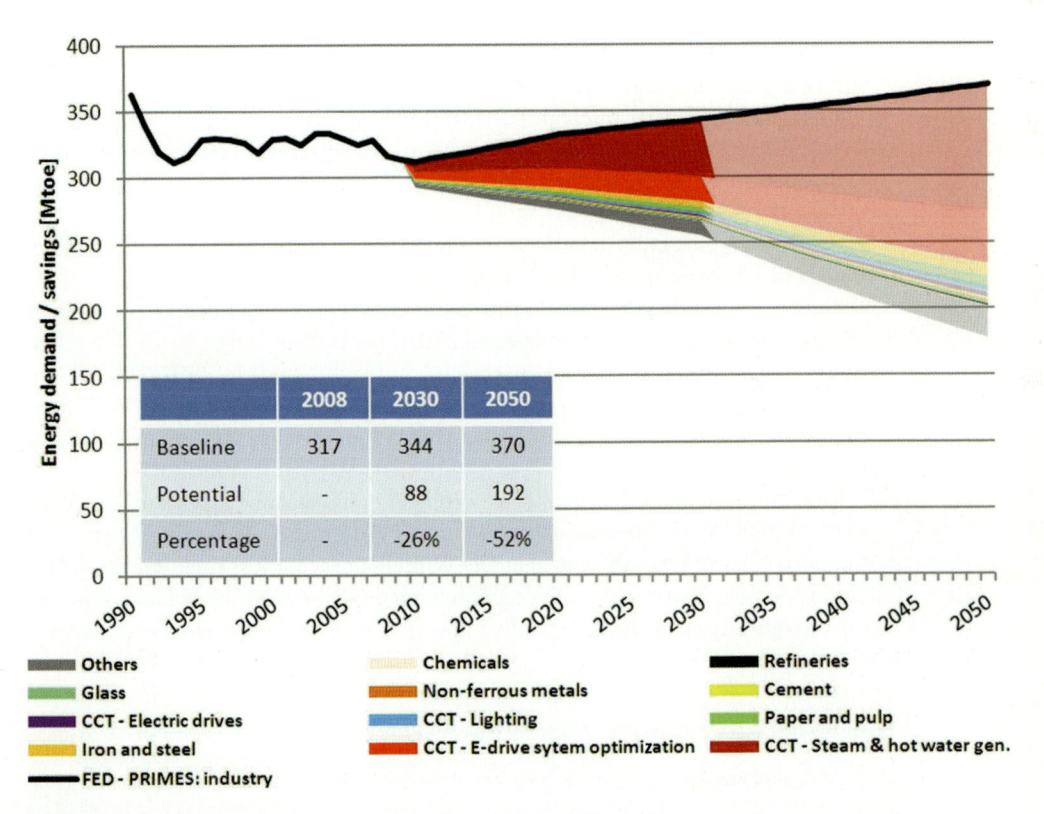

	2008	2030	2050
Baseline	317	344	370
Potential	-	88	192
Percentage	-	-26%	-52%

Figure 43.14 Total final energy-saving potential in the EU by 2050 in the industry sector. Source: Fraunhofer ISI.

Most of the short-term energy savings can be exploited by improved holistic optimization of electric motor-driven systems and energy-efficient heat generation. In the long run, further energy savings can compensate for the increasing baseline energy demand and promise even higher demand reductions. Provided that there is full implementation of the energy-saving potential by 2050, final energy demand would reach the 178 Mtoe level, representing a 52% reduction compared with the baseline.

The close relationship between materials use and energy use has not attracted much attention in the past. Looking at energy-intensive material production, however, amounting to about one-third of total industrial energy demand, it becomes obvious that the intelligent use of materials can contribute substantially to reducing per capita energy consumption. The major strategic options in this field are the following:

- Recycling and re-use of energy-intensive waste materials or used products (e.g., steel, aluminum, paper, plastics, and glass, and re-use of bottles and vehicle engines and tires).
- Substitution of highly energy-intensive materials by less energy-intensive materials or even by other technologies (e.g., steel and cement/concrete by wood, newspapers by electronic news).
- More efficient use of materials by better design and construction, improved properties of materials, oils, and solvents, and even foamed plastics and metals. This strategy is particularly important in the case of moving parts and vehicles, as lighter constructions may contribute to radically lower energy demand over the lifetime of the particular application (e.g., cars).

All three elements contribute to structural changes within industrial production, mostly in the direction of lower energy intensity of total industrial production.

Figure 43.15 shows the energy curve and Figure 43.16 shows the primary energy demand in the European industry sector. If no measures are undertaken, the baseline represents a further increase in energy demand to 592 Mtoe by 2050.

Primary energy savings in the industry sector are in two parts. By 2050, 29% of the overall baseline demand can be reduced through efficiency improvements in the power sector[4].

Even though efficient steam and hot water generation technologies (i.e., efficiency improvement of heat generation units, further CHP diffusion, and highly efficient industrial space heating) represent the bulk of the technical final energy-saving potential, their contribution to the economic savings is smaller and depends strongly on the assumptions made regarding the fuel mix of the generation capacities displaced by CHP.

---

4) In order to quantify the influence of the electricity generation mix on the primary energy savings, in the reference case, the electricity mix from the PRIMES projection of the European Commission from 2010 was used. Alternatively, a second, distinctively more ambitious electricity mix from the BMU project "EU Long-Term Scenarios 2050" was assumed, which is based to a large extent on the use of renewable energy sources (in 2050 the share of renewable energy sources for electricity generation is 92% and the median efficiency is 80%).

43.4 Overall Industry Sector

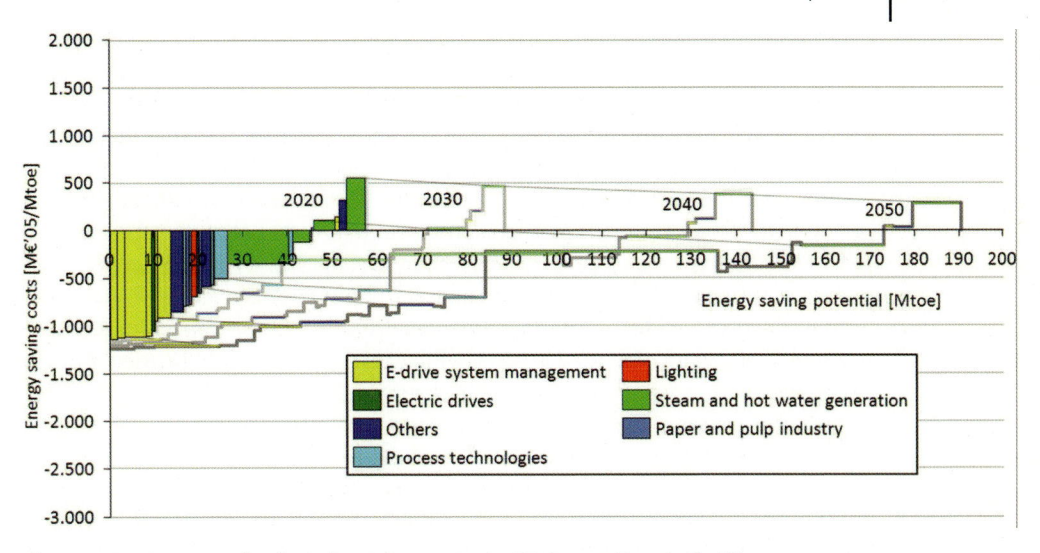

Figure 43.15 Cost curve for the industrial sector in the 27. Source: Fraunhofer ISI.

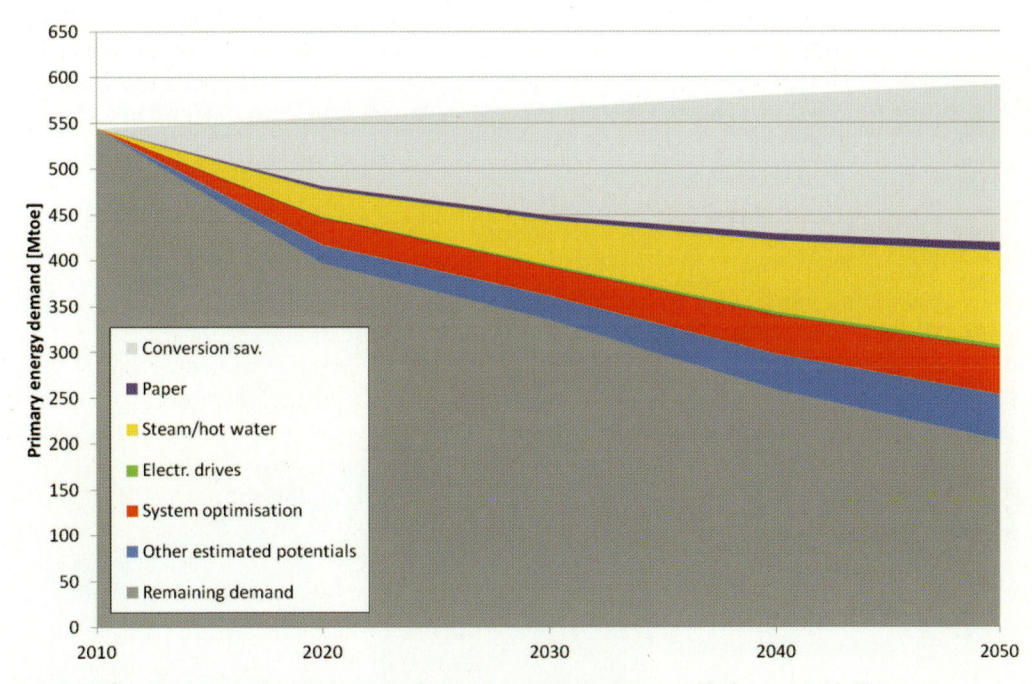

Figure 43.16 Primary energy savings in the European industry sector up to 2050 compared with the baseline energy demand. Source: Fraunhofer ISI.

In contrast, electric drive-based system optimization measures trigger an immediate cost reduction (apart from regular maintenance that causes additional labor costs), given the significant specific cost savings of more than 1000 M€'05 Mtoe^{-1} and the high energy-saving potential, as indicated in Figure 43.15. They account for roughly twice as much cost savings as benefits deriving from process technologies (nearly €14 billion versus €7 billion). Adding up all costs and benefits leads to a net cost reduction of €25 billion by 2020 and more than €100 billion by 2050. Excluding the cost benefits from CHP, which are highly sensitive regarding the price and fuel mix assumptions, reduces the net benefits to €21 and €90 billion, respectively.

Final energy-related efficiency technologies are able to deliver an additional 36% reduction compared with the baseline, which corresponds to 215 Mtoe. Although in the short term more than one-third of the savings are delivered through e-drive system optimization measures, this share declines subsequently. This is because the increasing power generation efficiency partly compensates for the significance of electricity-saving measures. Hence efficiency technologies for steam and hot water generation increase in importance, representing nearly half of the primary energy saving potential by 2050.

Figure 43.17 depicts the reduction of GHG emissions through efficiency improvements in the power sector (cf., the "conversion savings" slice) and final energy-related efficiency technologies compared with the calculated emissions from the baseline energy demand.

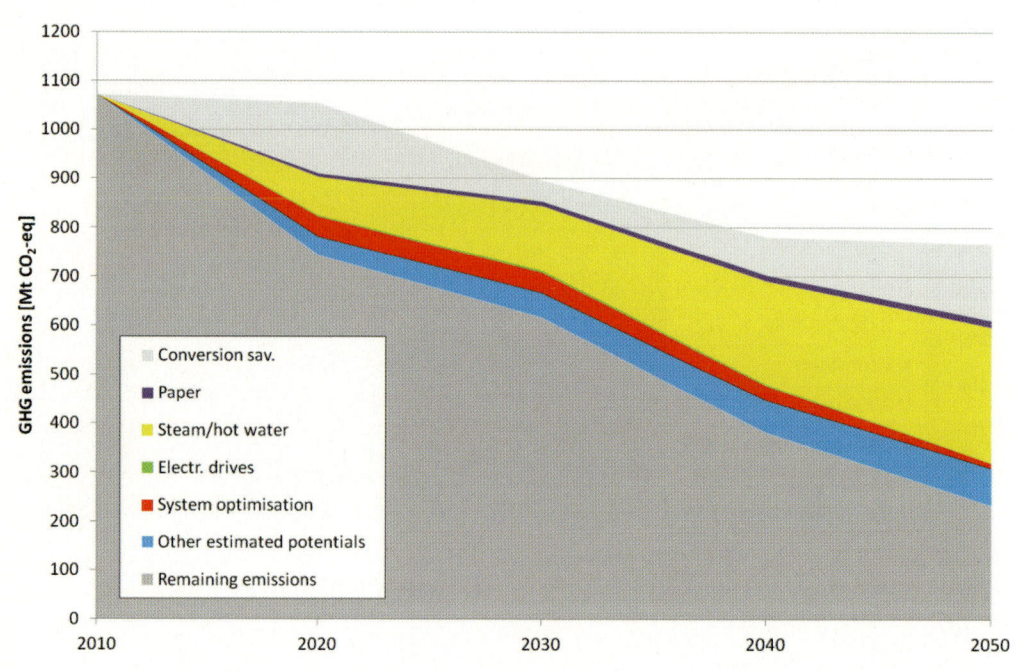

Figure 43.17 GHG emission reduction in the European industry sector up to 2050 compared with the calculated emissions from the baseline energy demand. Source: Fraunhofer ISI.

It is obvious that even in the baseline scenario GHG emission reductions will occur to a level of 767 Mt CO_2-eq. by 2050. Efficiency improvements in power generation support a decline in GHG emissions by 20% to a level of 610 Mt CO_2-eq.

The actual industry-related efficiency technologies drive a further decrease in GHG emissions by an additional 49% compared with the overall baseline, limiting the emissions to 233 Mt CO_2-eq.

An increasing share of the emission reduction potential is based on efficiency technologies in steam and hot water generation and also other process-specific efficiency technologies that trigger savings of energy carriers other than electricity. This is due to the fact that electricity savings feature a decreasing emission reduction effect because of efficiency improvements and decarbonization in the power sector.

References

1 Bossmann, T., Eichhammer, W., and Elsland, R. (2012) *Contribution of Energy Efficiency Measures to Climate Protection Within the European Union until 2050*. Policy Report and Accompanying Scientific Report, Fraunhofer ISI, Karlsruhe, http://www.isi.fraunhofer.de/isi-en/e/projekte/bmu_eu-energy-roadmap_315192_ei.php (last accessed 24 January 2013).

2 IEA (2012) *World Energy Outlook 2012*, International Energy Agency, Paris.

3 Fraunhofer ISI (2009) *ADAM Report, M1, D3: ADAM 2-Degree Scenario for Europe – Policies and Impacts*, Fraunhofer ISI, Karlsruhe.

4 Wietschel, M. et al. (2010) *Energietechnologien 2050*. Fraunhofer Verlag, Karlsruhe.

5 Fraunhofer ISI (2009) *Study on the Energy Savings Potentials in EU Member States, Candi-date Countries and EEA Countrie*, Fraunhofer ISI, Karlsruhe.

6 de Almeida, A., Ferreira, F. J. T. E., Fong, J., and Fonseca, P. (2008) *Preparatory Study for the Energy Using Products (EuP) Directive – Lot 11: Motors*. ISR – University of Coimbra, Coimbra.

7 Odyssee (2011) *Odyssee Database on Energy Efficiency Indicators*, http://odyssee.enerdata.net (last accessed 15 March 2011).

8 Deivasahayam, M., Ranganathan, G., Manoharan, S., Devarajan, N. (2009) *Energy Conservation through Efficiency Improvement in Squirrel Cage Induction Motors by using copper die cast rotors*. Karunya Journal on Research, Volume 1, Issue 1, pp. 74–85.

9 Kimmich, R., Doppelbauer, M., Kirtley, J. L., Peters, D. T. (2005) *Performance Characteristics of Driver Motors Optimized for Die-cast Copper Cages*. Energy Efficiency in Motor Driven Systems, 4th International Conference, 2005, Heidelberg, Fraunhofer Verlag, Karlsruhe.

10 Lindegger, M. (2006). Wirtschaftlichkeit, Anwendungen und Grenzen von effizienten Permanentmagnet-Motoren. Bundesamt für Energie, Bern.

11 European Commission (2010) *EU Energy Trends to 2030 – Update 2009*, European Commission, Brussels.

12 de Almeida, A., Ferreira, F. J. T. E., and Fonseca, P. (2000) *VSDs for Electric Motor Systems*, ISR – University of Coimbra, Coimbra.

13 de Almeida, A., Ferreira, F. J. T. E., and Fonseca, P. (2001) *Improving the Penetration of Energy-Efficient Motors and Drives*, ISR – University of Coimbra, Coimbra.

14 IEA (2009) *Energy Technology Transitions for Industry*, International Energy Agency, Paris.

15 Schmid, C., Brakhage, A., Radgen, P., et al. (2003) *Möglichkeiten, Potenziale, Hemmnisse und Instrumente zur Senkung des Energieverbrauchs branchen-*

übergreifender Techniken in den Bereichen Industrie und Kleinverbrauch, Fraunhofer ISI, Karlsruhe, and Forschungstelle für Energiewirtschaft (FfE), Munich.

16 European Commission (2012) *Eurostat*, http://epp.eurostat.ec.europa.eu (last accessed 24 January 2013).

17 IISI (1998) *Energy use in the steel industry*. Brussels: International Iron and Steel Institute.

18 World Steel Association (2012) *World Steel in Figures 2011*, World Steel Association, Brussels.

19 Fraunhofer ISI, Institut für Ressourceneffizienz und Energiestrategien (IRERS), and TU Berlin, Institut für Chemie (2011) *Möglichkeiten, Potenziale, Hemmnisse und Instrumente zur Senkung des Energieverbrauchs und der CO_2-Emissionen von industriellen Branchentechnologien durch Prozessoptimierung und Einführung neuer Verfahrenstechniken*, Umweltbundesamt, Berlin.

20 IEA (2007) *Tracking Industrial Energy Efficiency and CO_2 Emissions*. International Energy Agency, Paris.

21 Bührer, C., Hagemann, H., Kellers, J., et al. (2009) *Effiziente magnetische Blockerwärmung mit Gleichstrom*. Elektro Wärme International – Zeitschrift für elektrothermische Prozesse, 1/2009, pp. 19–23.

22 CSI (Cement Sustainability Initiative) and ECRA (European Cement Research Academy) (2009) *Development of State of the Art Techniques in Cement Manufacturing: Trying to Look Ahead*, CSI/ECRA Technology Paper, CSI, Geneva.

23 Fleiter, et al. (2013) *Energieverbrauch und CO_2-Emissionen industrieller Prozesstechnologien – Einsparpotenziale, Hemmnisse und Instrumente*, Stuttgart.

Subject Index

a

abandoned mining sites   644–646
absorbers
– flexible wave energy   373
– metallic foam   473–474
– point   374–375
AC voltage, three-phase   726
acceleration power   411
ACE (area control error)   733
acidic PEM electrolysis, *see* PEM electrolysis
acidification, ocean   120
active energy saving   21
active wind   754
ADAM report   917–918
adaptive management   347
adiabatic compression   662
adiabatic lapse rate   273
adsorption, pressure swing   696, 700–701
advanced AWE, electrolyzers   218
advanced batteries   579–596
aerodynamic loads   279
aerosol   120
– solar-thermal reactor   468
AGC (Automatic Generation Control)   612–613
aging power plants   409
air
– ambient air vaporizer   672
– ASHP   57, 761
– CO_2 source   826–827
– heat-transfer fluid   461–465
– lithium–air batteries   585–587
– pollution   146, 805–806
airship, fuels   681
alcohol, *see* bioethanol
algae-based biofuels   534
alkaline water electrolysis (AWE)   430–432, 440
– advanced   218

– power-to-gas   814–817
all-electric building   881–882
"allowable solutions"   877
"allowed stress range"   777
Alpha Ventus offshore wind farm   266
alternating loads   778
alternative powertrains and fuels   857
– efficiency   864
alternative process chains   838–841
alumina tubes, coated   475
ambient air vaporizer   672
ammonia
– borane   660
– conversion efficiency   701
– electrochemical conversion   691–706
– fuel cells   699
– synthesis   430
anaerobically digested gas   824
anchoring system   367
ancillary services (AS)   88
annual grid load balancing   232–236
Anthropocene age   119–121
aquifers, storage sites   642–644, 829
arable land potential   490
area control error (ACE)   733
asbestos diaphragm   432
ASHP (air source heat pumps)   57
– "smart grids"   761
Asse diapir   647
asset-heavy paradigm   769
assimilation of carbon   491–492
ASTERIX project   461, 467
astigmatism   317
Aswan/Egypt   430
asynchronous DIG   250–252
atmospheric aerosol loading   120
atmospheric air pollution   146
atmospheric boundary layer   273
atmospheric electrolyzer   430–431

Transition to Renewable Energy Systems, 1st Edition. Edited by Detlef Stolten and Viktor Scherer.
© 2013 Wiley-VCH Verlag GmbH & Co. KGaA. Published 2013 by Wiley-VCH Verlag GmbH & Co. KGaA.

atmospheric receiver   319
attenuator   376
Automatic Generation Control (AGC)   612–613
automotive fuel cell systems, *see* fuel cell vehicles
autonomy   131
availability assumptions   38–39
aviation travel   103–105

b

back contact, interdigitated   297–298
back surface field (BSF)   287
balance boundaries, ZEB   876–878
balance-of-plant (BOP) cost   222–223
balance-of-plant (BOP) system   186–187
Balance Responsible Party   502–503
balancing
– cost   83, 88–89, 260
– cross-border   733
– energy demand and supply   568–570
– large-scale   605
– real-time   743–744, 747
– ZEB   879–881
–, *see also* load balancing
balancing capacity
– hydropower   398
– marine energy   353
bank erosion   612
Barlow law   777
barriers to transition   868
base-load power, renewable   342
base metals, coated   444–445
BAT (best available technology)   98
batteries
– advanced   579–596
– EV   587
– lithium–air   585–587
– lithium–ion   185, 198, 581–582, 591–592, 853
– lithium–sulfur   584–585
– Ni–MH   23–24
– rechargeable, *see* rechargeable batteries
– redox-flow   582–583, 592–593
– smart charging   755–759
– sodium–sulfur   23, 583–584, 593–594
battery electric vehicles (BEV)   587, 853–854, 859
– market development   860
– utility index   867
BAU ("business-as-usual") reference   93
"beam-down" concept   321, 462
bell-jar solar reforming reactor   469
Bernoulli equation   786

best available technology (BAT)   98
bidirectional charging control   760
binary-cycle heat exchangers   341
biobutanol   534
biodiesel   528–529
biodiversity loss   120
bioenergy
– Scotland   57–58
– supply potential   111
bioethanol
– lignocellulosic   523
– power density   199
– production   529–530
biofuels   510–516, 625–626
– algae-based   534
– conventional   525
– cost   543–544, 544
– distribution   525
– economic feasibility   542–544, 546
– environmental impact   537, 545–546
– production   523–553, 535
– raw materials   526–527
– R&D   526–527
biogas   473, 507, 512–513, 824
– upgrading   515, 532
biohydrogen   512–513, 533
biological methanation   787–788
biological nitrogen fixation   693
biomass   485–554
– as CO_2 source   824–825
– co-combustion with fossil fuels   415–417, 507
– conversion   510, 524, 540–541
– energy efficiency   509
– environmental impact   495
– flexible power generation   499–521
– gasification   452, 531
– Germany   506
– global resources   485–497
– growth potential   494
– mechanical–thermal treatment   531
– rural China   906
– supply   525
– sustainable available potential   499
biomass to liquid (BTL) fuels   530–531, 626
biomethane   507, 512–513, 564
– production   532
biosyngas   824
bipolar plate   583
BIPV (building-integrated PV)   292–294
bird fatalities   256
Birkeland–Eyde process   694
"black start"   60
– hydropower   398

Subject Index

blades, number of   247–248
Blåsjø hydropower reservoir   392–393, 608
boil-off loss   670
boil-off rate   675
boiler
– condensing   790
– fossil-fired   180
– solar   171
BOP (balance-of-plant) cost   222–223
BOP (balance-of-plant) system   186–187
boring machines, TBM   392
Borkum West offshore wind farm   266
boron hydrides   660
Bosch, *see* Haber–Bosch process
boundaries, planetary   119–120
boundary layer, atmospheric   273
Brazil   140
BRICS   140
Bridgeman process   286–287
bridging technologies   195
brine   649–650, 830
briquetting   508
British–Irish Council   63
"brown economy"   5
Brundtland Report   121
BSF (back surface field)   287
BTL (biomass to liquids) fuels   530–531, 626
bubble formation   441
buildings   99–103
– all-electric   881–882
– BIPV   292–294
– building codes   152
– China   891–911
– commercial   904–906
– energy demand   103
– energy use   897
– floor area per capita   897–898
– public   904–906
– rural residential   906
– system boundaries   876–878
– ZEB   875–889
bulk power transport   723
– market options   732
bulk storage, near-surface   659–690
bundled concentration   162–163, 172
buses, FCV   186
"business-as-usual" (BAU) reference   93
Butler–Volmer equation   220

c

c-Si technology, *see* crystalline Si wafer-based solar cells
CAES (compressed air energy storage)   560, 567, 574–575
CAESAR experiment   469–470
California, SEGS   312
California, ZEV standard   861–862, 864–866
California Fuel Cell Partnership (CaFCP)   861–862
calorific value   774, 784, 786
"cannibalism" of VRE   78
CAP (Common Agricultural Policy)   487
capacity   558, 566
– addition   766
– mandating   148
– "on-site" storage   878
capacity factor   219
– geothermal power   344
– hydropower   382, 395
– wind turbines   249
capacity growth, offshore wind power   266
CAPEX   419–421, 544
– "smart grids"   748, 763–765
capital
– availability   261
– cost   269–270, 803
– financial   127–129
– intensity   85
caprock   829
carbon capture and sequestration (CCS)   798
carbon capture and storage (CCS)   18
– as CO_2 source   823–824
– offshore   55
– Scotland   53–56
– unnecessary with DSR   749, 764
carbon capture/cycling (CCC)   788–789
carbon dioxide, *see* CO_2
carbon emissions, China   893
carbon fiber   248
carbon formation   454
carbon hydrogenation   622
carbon intensity   29
carbon nanotubes (CNT)   586
carbon sinks, forests   491–492
Carbon Trust   357
carbonates
– carbonate–$LiAlO_2$ electrolyte   697
– PCM   467
– sodium carbonate solution   826
carbonization, hydro-thermal   508
Carnot power cycle   309
Carnot's theorem   168
Caro, *see* Frank–Caro process
cars, *see* vehicles
catalyst-coated membrane (CCM)   434, 442
catalysts, noble metal   816
catalytic activity, PEM electrolysis   433
catalytic hydrogenation   820

catalytic methanation   787
catalytic reduction, selective   698–699
catalytic systems, "structured"   456
catchment (hydrological)   146, 383
cathodic protection system (CPS)   780
caverns
– convergence   648
– hydropower   391–392
– salt   203
– storage sites   646–652, 830–831
cavity receiver   477
CCGT (combined cycle gas turbine)   731
CCS, see carbon capture and storage/ sequestration
CEDREN   605
cells
– fuel, see fuel cell
– resistance   220
– solar, see solar cells
– voltage   226, 429
cement production   929
central conversion   835
Central Europe, geological storage   636–639
central production, H_2 800
central receiver system   309
central storages   558
ceramic electrolytes, YSZ   699
ceramic foams   458, 469–474
ceramic honeycombs   458, 468–469
ceramics production   922
cereals, world production   488
cerium, see Zn–Ce redox-flow battery
certificates
– REC   6
– ROC   352
CES (constant elasticity of substitution)   33
CGE (computable general equilibrium)   31
charge/discharge time   559
charging
– EV   755–759
– uni-/bidirectional control   760
CHARM® (CH_4 airborne remote monitoring)   781–782
chemical conversion   564–565
chemical energy carriers   438
chemical energy storage   327
Chemical Engineering Plant Cost Index (CEPCI)   543
chemical fuel production   619–621
chemical hydrides   660
chemical pollution   120
chemical storage   619–628
China   140
– buildings energy conservation   891–911

– carbon emissions   893
– climate policy goals   30–31
– CO_2 emissions   29–46
– Copenhagen commitment   45
– green energy strategies   29–46
China-in-Global Energy Model (CGEM)   31–35
CHP, see combined heat and power
chrysotile diaphragm   432
CIGS   288–289, 298–300
clean technology ranking   9
climate change   120
climate-change impacts, hydropower   387–388
climate policy goals
– China   30–31
– diverging objectives   74
climate protection, Developing World   145
cluster strategy   808
CNT (carbon nanotubes)   586
CO_2
– buildings   875
– catalytic hydrogenation   820
– disproportionation   454
– methanation   204, 210–212
– methane reforming   451
– ocean sink   491–492
– recovery from air   826–827
– specific energy consumption of sources   827
– steam co-electrolysis   438
– synthetic methane production   788
– tax   495–496
– , see also carbon, greenhouse gases
co-combustion of biomass   415–417, 507
co-electrolysis   437–438
CO_2 emissions
– China   29–46
– Developing World   146–147
– gas scenario   793
– Germany   196–197
– hydropower   396–397
– marine energy   352
– reduction   40–42
– smart EV charging   760
co-evaporation process   300
co-generation   21, 195
coal
– and biomass   416
– China   893
– ex- and importers   4
– gasification   802, 834
– pulverized   508
coated alumina tubes   475

coated base metals 444–445
coating technologies, large-area 288–290
Cobb–Douglas CES 33
coefficient of performance (COP) 571
– smart grid heat pumps 761
coke-oven gas 785
cold start-up tests 189
cold storage 561–563
collector systems, linear Fresnel 309, 317–320, 457
combined cycle gas turbine (CCGT) 731
combined heat and power (CHP) 50, 58–59
– biomass 507
– demand-driven electricity commission 510–515
– distributed 557
– district heating 413–414
– fossil power plants 413–414
– geothermal 341
– industrial use 923–926, 932–934
– natural gas system 789–790
– VRE 80, 87
– ZEB 877, 883–885
combustion
– ammonia 698–699
– biomass 824–825
– co- 415–417, 507
– heating systems 835
– ICE 836–838
– turbines 835
commercial buildings 904–906
commercial floor space 101
commercialization programs, hydrogen FCV 796
Common Agricultural Policy (CAP) 487
compatibility, infrastructure 806
compensator, static VAR 715
complex metal hydrides 675
composite design simulation 225
composite materials 248
compound thin-film PV 302
comprehensive transition concept 119–136
compressed air energy storage (CAES) 560, 567, 574–575
compressed gaseous hydrogen storage 662–668
compressed natural gas (CNG) 787, 792
– as fuel 836–837
compressibility factor 773
compressors
– natural gas pipelines 778–779
– switching 754
computable general equilibrium (CGE) 31
concentrating photovoltaics (CPV) 302–303

concentrating solar high-temperature heat (CSH) 110
concentrating solar power (CSP) 95, 110, 307–338
– environmental impact 311–312
– large-scale plants 159–182, 461
– methane reforming 457
concrete, sensible heat storage 326
condenser 173–174
condensing boiler 790
conduction band 285
conductive oxides, transparent 288–289
conductor glass film, super-ionic 585
"connect and manage" approach 61
connection to shore 274–276
constant elasticity of substitution (CES) 33
consumer behavior 131–132
consumer-friendly cost structures 560
contamination
– natural gas 473
– subsurface storage 635
continuous recharge, geothermal power 345
"control hydro"/"control wind" 613
control loops, fast-reacting 410
control manifold 672
control of domestic appliances 750–755
controlled load profile 751
conventional biofuels 525
conventional petroleum 624, 852
conventional power plants 403–422
conventional vehicles 579
convergence, caverns 648
conversion
– biomass 510, 524, 540–541
– chemical 564–565
– (de)central 835
– efficiency 296, 301
– electrochemical 691–706
– hydrogen 797–798
– ortho-to-para 669–670
– "savings" 934
– solar 283, 460
– technologies 712–717
converter
– LCC technique 716
– multilevel 720
– overtopping wave energy 378
– VSC technologies 716–717, 739
cooling energy intensity 903
cooling system, CSP 169
coordination of stakeholders 807
Copenhagen commitment, China 45
cordierite honeycomb 469
core–shell structures 443

corn-based bioethanol 523
corrosion
– biomass co-combustion 417
– natural gas pipelines 778, 783
Corruption Perceptions Index 138
cost
– assumptions 38–39, 70–71
– balancing 83, 88–89, 260
– biofuel production 543–544, 544
– BOP 222–223
– capital 269–270, 803
– CEPCI 543
– competitiveness 40
– consumer-friendly structures 560
– cost of energy trajectory 357–360
– distribution 517
– drilling 346
– EGC 296–297
– electrolysis 440, 444
– energy storage systems 559, 566, 572–575
– EV 589
– excess 868
– FCV 185, 191–192, 854
– generating technologies 329
– geological storage 831
– grid-related 83, 89
– H_2 feed to natural gas grid 833
– highly efficient electric drives 919
– hydrocarbon fuel upgrading 626
– hydrogen infrastructure 210–213
– hydrogen production 223–224, 803
– hydropower 394–396
– LCOE, see levelized cost of energy
– LEC 325
– LRMC 339, 346
– marine energy 353
– maximum acceptable storage costs 573–574
– natural gas 210–212
– near-surface bulk storage 677–679
– offshore wind power 271
– power-to-gas 839
– profile 78–82, 85–88
– PV 290–291, 296, 301
– reinforcement 756–759
– risk-adjusted 259
– "smart grids" 754
– solar thermal power 328–332
– solid-state storage 677
– substation 735
– supply–cost curve 394, 396
– transition 857
– WACC 762
– water electrolysis 221–223

Cost Reduction Task Force (CRTF) 269
counter-rotating Wells turbines 377
CPS (cathodic protection system) 780
cracking
– methane 476
– natural gas 452
crashworthiness 188–189
creep tendency, salt caverns 648
Crescent Dunes solar tower system 323
crop, energy, see biomass
crop management 486
cross-border balancing 733
cross-border trading 732
cross-cutting technologies 914–915, 920, 926–927
cross-linked polyethylene (XLPE) insulation 711–712, 716
Crown Estate 355
crude oil
– ex- and importers 4
– subsurface storage 632
cryogenic liquid hydrogen storage 669–675
crystalline Si wafer-based solar cells 286–288, 297–298
– R&D 301
CSIRO 463–464
cultural perspective 131–132
culturalists 132–134
curb weight 205
current
– capacity 728
– collectors 443–445
– density 222, 225
– parasitic 219
current policy assumptions 37–38
– RE 39–40
current–voltage curve 220, 285
– alkaline electrolysis 429
– high-temperature electrolysis 437
– PEM electrolysis 434
cushion gas 828
cycling, subsurface storage 634–635
Czochralski process 286–287

d

daily winter sunshine 144
dams 385, 389
– PSH 602–603
Danish concept 243
Darcy–Weissbach equation 773–774
day–night cycles 619
DC CAPEX 748, 763–765
DC grids, meshed 718
DC supergrid concept, European 275

DCORE (dual-coil reformer)   463–464
"dead" wood   415
decarbonization, *see* CO$_2$ emissions
decentral conversion   835
decreasing market value, VRE   75–92
deforestation   491–492
degradation   743
delivery, *see* transmission, distribution
demand
- and supply   486–487, 568–570, 743–744, 747
- demand-driven electricity commission   510–515
- demand-driven power options   95
- FED   927
- flexible   709, 749
- heat   900
- peak   741, 751
- reduction   51–52
- response   86
- stoichiometric H$_2$ 622
- storage and balancing   606–607
- transmission grid changes   731–732
- weighted   879
demand-side response (DSR)   749–768
democratic question   129
dendritic growth   585
depleted oil/gas fields   642
deployment potential   109
- bioenergy   111
- geothermal power   343–345
- hydropower   397–398
- PSH   598–599
- UK marine energy   354–356
DES (distributed energy storage systems)   570–571
DESERTEC concept   181, 308
- transmission grid systems   731
deserts, solar power plants   160
deterministic scheduling   767
Developing World
- policies   147–153
- R&D   153
- RE   137
development, sustainable   121
device power up-rating   359
dewars   671
- spherical   683
DG (distributed generation)   143, 405
DHW (domestic hot water)   561, 903
diaphragm, electrolyzers   432
diapirs   646
diesel fuel
- bio-   528–529

- HHV   693
DIG/DFIG (doubly fed induction generator)   250–252, 272
dimethyl ether (DME)   533
direct ammonia fuel cells   699
direct conversion, solar radiation   283
direct drive wind generators   251–252
direct feed-in, hydrogen   212
direct-hydrogen FC systems   836
direct land use change (dLUC)   538–539
direct normal insolation (DNI)   329
direct semiconductors   288
direct steam generation
- CSP   168, 172, 316
- geothermal power   341
- heat storage   179
directly irradiated annular pressurized receiver (DIAPR)   474
directly irradiated reactors   468–476
directly irridated solar receivers (DIR)   458
discharge, relative   390
discharge times   559, 561
- CAES   785
- natural gas system   785
- PSH   785
dish solar collectors   309, 477
dish Stirling receiver   164
dishwashers   752
disproportionation, CO$_2$ 454
disruptive innovations   360
dissociation potential, water   220
dissociation pressure, metal hydrides   676
distillation residues   623
distributed CHP   557
distributed energy market   768–770
distributed energy storage systems (DES)   570–571
distributed generation (DG)   143, 405
distributed storage, hydrogen   678
distribution   709–844
- cost   517
- grid   209, 731–732
- hydrogen   796–802
- Scotland   60
district heating   88
- fossil power plants   413–415
- , *see also* residential heating
disturbance recognition and visualization tools   746
diverging objectives   74
diversification
- energy sector   17–23
- supply system   353
- water heater demand   750

DNI (direct normal insolation) 329
dogmas on sustainability 132
domestic appliances, control 750–755
domestic energy supply, China 894
domestic hot water (DHW) 561, 903
domestic industries, Developing World 149
double-circuit AC lines 736–739
doubly fed induction generator (DIG/DFIG) 250–252, 272
Douglas, Cobb–Douglas CES 33
downregulation, wind power 615
downstream migration 391
downwind wind turbines 250
drain and refill cycle, geological storage 631
drilling
– cost 346
– subsurface storage 633
– tunnel 611
drive train concept 250–252
drives
– drivetrain losses 854
– electric 853–854, 916–922
– frequency-controlled 410
– hydraulic systems 313
– variable-speed 604, 919
driving distance 579
– FCV 184
dry methane reforming 451
dryers, tumble 752
drying, thermal 930
DSR (demand-side response) 749–768
dual-coil reformer (DCORE) 463–464
duality of structure 124
ductility, salt caverns 648
ducts, steam generation inside 175–177
durability
– EV 589
– fuel cells 184, 797
– tests 190
dye-sensitized solar cells (DSSC) 298–299
dynamic positioning vessels 271
dynamic pressure 776

e

EAFs (electric arc furnaces) 928
ebb tides 363
eco-friendly vehicles 579
–, *see also* electric vehicles, zero-emission vehicles
"eco villages" 907
ecological guard rails 145–147
ecological trajectories 125
economic feasibility 68
– biofuel production 542–544, 546
– energy storage systems 572
– fossil power plants 419
– geothermal power 346
– hydrogen supply pathways 802–804
– hydropower 385–386
– near-surface bulk storage 677–679
– wind power 258–260
economic growth
– and energy transition 15
– model assumptions 35–37
economic incentive schemes 343
"economic literacy" 128
economic perspective 127–129
– "New Economics" 123
economic potential, renewable energies (RE) 141–143
economic question 129
economies, emerging 140
ECOSTAR project 331
ecosystems, marine 146
education, Developing World 152–153
EEG (Erneuerbare-Energien-Gesetz) 127, 500–502, 506–507
EEKV (energy equivalent of water) 381
EER (European Energy Regulators) 733
EESG 252
EEX (European Energy Exchange) 502
efficiency
– alternative powertrains and fuels 864
– ammonia conversion 701
– biofuel production 535
– biomass use 509
– BOP 223
– electricity 230
– electrolyzers 237
– energy, *see* energy efficiency
– Faraday 220
– four-pole electric motors 916
– Haber–Bosch process 696
– heating 901
– hydrogen FCV 795, 797
– hydropower 381
– NREL chart 303
– optical 318
– PEM electrolyzers 440
– PV conversion 296, 301
– storage 559, 566, 679–681
– theoretical total 310
– turbines 390
– vehicles 854–856
efficiency limit 284
efficiency path 876
EfW (energy from waste) 58
EGC (energy generation cost) 296–297

Subject Index

EGPS (Electricity Generation Policy Statement) 47–48
EGS (enhanced geothermal system) 339
electrodes, batteries 581
electric arc furnaces (EAFs) 928
electric drives 916–922
– technologies 853–854
electric motors 208
– four-pole 916
electric power
– capacity distribution 229
– chemical fuel production 619–621
– CO_2 emission source 196
– consumption reduction 920
– cost assumptions 70–71
– cross-cutting technologies demand 915
– demand-driven commission 510–515
– Europe 404
– gaseous biofuels 512–515
– generation in UK 48
– Germany 200–202, 501
– heat–electricity interface 87
– hydrogen production 799
– industrial use 920
– Japan 15
– LEC 325
– liquid biofuels 511–512
– low-carbon systems 743–744
– market design 73
– massive PV integration 303
– methane production 565
– necessary investments 70
– present system structure 742
– renewable 31, 501
– Scotland 47–65
– solar radiation 160
– solid biofuels 510–511
– Spain 217, 228–229
– storage 59–60, 86, 217
– sub-sea electricity grid 61
– tri-generation system 801
– wholesale and retail market 769
electric vehicles (EV) 106, 579–596
– BEVs 587, 853–854, 859–860, 867
– charging 755–759
– cost 589
– fuel cell, see fuel cell vehicles
– "on-site" storage capacity 878
– rechargeable batteries 587
– "smart grids" 755–760
electricity
– efficiency 230
– peaks 570
– storage 559–561
– trade 730
Electricity Generation Policy Statement (EGPS) 47–48
Electricity Market Reform (EMR) 352
Electricity Market Reform Expert Group (EMREG) 354–355
Electricity Networks Steering Group (ENSG) 60
electrification 112
– rate 107
electrocatalysts 235
– alkaline electrolysis 441–442
– hydrogen evolution reaction 442–443
– oxygen evolution reaction 443
electrochemical conversion, ammonia 691–706
electrochemical synthesis 697–698
electrodes, MEA 817
electrolysis 200, 203, 207, 425–450, 797, 799
– AWE 430–432, 440–442
– cost 221–223, 440, 444
– HTE 436–438, 818–819
– hydrogen pipeline system 834
– large-scale hydrogen technologies 218–226
– PEM 433–436, 442, 817–818
– physico-chemical foundations 426–430
– power-to-gas 814–820
– production simulation 224–226
– R&D 438
– specific energy consumption 428
electrolytes
– batteries 581
– carbonate–$LiAlO_2$ 697
– ion-conducting solid 583
– KOH 815
– pH 433
– YSZ ceramic 699
electrolyzers
– efficiency 237
– energy yield 222
– grid stabilization 226–236, 439
– partial load 432
– performance 219–221, 224
– pressurized 227, 441
– RE integration 819–820
electromobility 851–873
– Model Regions 861–862
EM (enriched methane) 467
embrittlement, hydrogen 674
EMEC (European Marine Energy Centre) 361
emerging economies, BRICS 140
emissions
– CO_2, see CO_2 emissions
– industrial sector 934–935

– near-zero 800–801
– NO$_x$ 495, 698
– Trading Scheme 496
– transport sector 851–852
– WTW 805–806
employment opportunities
– marine energy 354
– RE 142
EMR (Electricity Market Reform) 352
EMREG (Electricity Market Reform Expert Group) 354–355
encased receiver tubes 176
end energy carriers, Germany 621
end-member mixing analysis (EMMA) 77, 81–82
end-use technologies 834–838
endogenous growth theory 123
ENE FARMs 24
EnerCalC 883
Energiewende, *see* energy transition
energy
– active energy saving 21
– buildings 875
– capacity addition 766
– chemical 327, 438
– Chinese buildings 891–911
– co-generation 21, 195
– cost of energy trajectory 357–360
– crops, *see* biomass
– demand and supply 568–570
– distributed market 768–770
– energy–economic system 39
– FED 927
– formation energy of liquid water 426
– GEA report 851
– green, *see* green energy ...
– hydrogen, *see* hydrogen ...
– industrial savings potentials 913–936
– LCOE, *see* levelized cost of energy
– marine 351–379
– on-site generation 881
– payback ratio 400
– primary energy credits 885
– renewable, *see* renewable energies
– reserves 141
– risk-adjusted cost 259
– storage, *see* storage
– sustainable global energy system 93–117
– tidal 360–370
– TPES 244
– use in transport 854
– wave 371–378
– World Energy Outlook 824, 891, 913
– ZEB, *see* zero energy buildings

energy analysis, buildings 876–877
energy-band diagram 285
energy carriers
– ammonia 692–698
– hydrogen, *see* hydrogen
– , *see* also fuels
energy consumption 4
– Scotland 52
energy demand
– buildings 103
– global 108
– industries 100
– modeling 94
– reduction 51–52
– sector definitions 96
– transport 107
energy density
– chemical fuels 620
– cryogenic storage 669
– EV 589
– gravimetric 208
– metal hydrides 676
energy efficiency 112
– biomass 509
– Scotland 51
Energy Efficiency Action Plan 58
Energy Efficiency Directive 413
energy equivalent of water (EEKV) 381
energy from waste (EfW) 58
energy generation cost (EGC) 296–297
"energy highway" 776
energy infrastructure
– new model 143
– North Sea 51
energy-intense industries 71
energy intensity 856
– buildings 102, 897
– Chinese buildings 899–900
– cooling 903
– rural China 907
– transport 105–106
energy-intensive materials, substitution 932
energy model
– CGEM 31–35
– ISLES 61–64
energy payback ratio (EPR) 399–400
energy payback time 283
Energy Performance of Buildings Directive (EPBD) 875–880
energy recovery 698–700
Energy Roadmap 2050 913
energy saving, buildings 99
energy sector
– diversification 17–23

– transition   14
energy services, *see* power system services
energy storage systems (ESS), *see* storage
energy supply
– global   112
– primary   621
– Sweden   492
energy transition
– comprehensive concept   119–136
– German industry   67–74
– Japan   14
– supranational level   73
– , *see also* transition
energy yield, electrolyzers   222
"Engineering, Procurement and Construction" (EPC) contractor   328
engines
– gas   835
– internal combustion   836–838
– losses   854
– Stirling   170, 310
– , *see also* vehicles, transport
enhanced geothermal system (EGS)   339
enriched methane (EM)   467
ENSG (Electricity Networks Steering Group)   60
entropy   173
ENTSO-E   729–730, 735
environmental impact
– biofuel production   537, 545–546
– biomass   495
– CSP   311–312
– geothermal power   346–350
– hydrogen supply pathways   802, 805–806
– hydropower   387–388, 399
– PSH   602–604, 611–612
– "smart grids"   768
– wind power   255–257
"environmental literacy"   122
EPR (energy payback ratio)   399–400
equilibrium price   80
Erneuerbare-Energien-Gesetz (EEG)   127, 500–502, 506–507
erosion, bank   612
esters, hydrotreated   529
ethanol
– power density   199
– , *see also* bioethanol
eucalyptus   490
EURELECTRIC   419
EURO3 norm   698
Europe
– district heating   414
– geological storage   636–639

– hydropower potential   386
– industrial energy savings potentials   913–936
– natural gas infrastructure   791
– PSH capacity   599
– transmission grid   408, 729
European Association of Refrigeration, Heating and Ventilation (REHVA)   875
European DC supergrid concept   275
European Energy Exchange (EEX)   502
European Energy Regulators (EER)   733
European Integrated Hydrogen Project   663
European Marine Energy Centre (EMEC)   361
European Network of TSOs (ETSO)   733
European Union
– Common Agricultural Policy   487
– electricity market   730
– Energy Roadmap 2050   913
– Parliament Test Drive Program   193–194
– 20/20/20 targets   403
European Water Framework Directive   389
EV, *see* electric vehicles
excess cost   868
excess electricity   203
expected repair delay time   277
exploration, subsurface storage   633
externalities concept   856–857
Eyde, *see* Birkeland–Eyde process

f

FACTS (flexible AC transmission systems)   714–715, 745
Faraday efficiency   220
Faraday, Michael   725
fast-reacting control loops   410
fast response, PEM electrolysis   818
fatty acid methyl ester (FAME)   528
fatty acids, hydrotreated   529
FCCJ (Fuel Cell Commercialization Conference of Japan)   860
FCV, *see* fuel cell vehicles
FDNPP (Fukushima Dai-Ichi nuclear power plant)   13–14, 16, 67
feed-in tariffs (FIT)
– Developing World   148
– EEG   506
– Japan   21–22
– South Korea   6
– wind power   257
feedstock conversion, biomass   510
fermentation   530
– biomass   824
fertilizer production   430
filling nozzle, IR   191

Subject Index

filling stations
– hydrogen   210
– , see also refueling
final energy consumption, Scotland   52
final energy demand (FED)   927
financial capital   127–129
"financial crisis"   229
financial implementation procedure   181
financial incentives   151–152
financial security, Developing World   149
financial support, RE   407
fins   474–475
fire tests   188–189
Fischer–Tropsch process   438, 451
– biofuel production   530–531
fish-friendly power plants   391
Five-Year Plan   30, 35
"flashed" steam   341
flat mirrors   163
flat panel display (FPD)   299
"flexibility effect"   81
flexibilization capability   505
flexible AC transmission systems (FACTS)   714–715, 745
flexible blade design   248
flexible demand   709, 749
flexible EV charging   758
flexible hydropower   612–615
flexible power generation
– biomass   499–521
– fossil power plants   408–410
flexible solar cells   294–295
flexible transmission   709
flexible wave energy absorbers   373
float, spar and   375
floating point absorber   374
floating turbines   370
floating wind turbines   279–280
flood protection   387
flood tides   363
floor area   897–898
floor space, commercial/residential   101
flow   381
– two-phase   175
flow-based market coupling   86
flow variability   383–384
fluctuating RE, see variable RE
fluctuations
– RE generation   199–200
– reservoir water level   609–611
fluids
– geothermal   343
– heat-transfer, see heat-transfer fluids
– working   165–168

foams, ceramic   458, 469–474
focal point   320
food production, global   487–490
forecast errors   75, 85
forests
– carbon sinks   491–492
– energy supply   492
– , see also wood
formation energy, liquid water   426
fossil-fired boiler   180
fossil fuels
– China   29, 44, 893
– natural gas, see natural gas
– petro-mobility   851–873
– storage medium   567
fossil power plants   403–422
– economic feasibility   419
foundations, offshore wind power   267, 271, 276
four-pole electric motors   916
FPD (flat panel display)   299
Frank–Caro process   694
FreedomCAR   659
frequency-controlled drives   410
frequency regulation   600, 604
– , see also balancing
freshwater use, global   120
Fresnel mirrors   163–164
– linear   309, 317–320, 457
friction coefficient, integral   775
Fuel Cell Commercialization Conference of Japan (FCCJ)   860
fuel cell vehicles (FCV)
– cost   185, 191–192, 854
– efficiency   795, 797
– European deployment activities   192–194
– Germany   205, 207–208
– HMC   183–194
– Japan   22–23, 25
– Mercedes-Benz B-Class   208
fuel cells
– CaFCP   861–862
– direct ammonia   699
– direct-hydrogen systems   836–838
– HTE   437
– Japan   24–25
– low-temperature   218
– natural gas system   790
– passenger car powertrains   836
– PEMFC   183, 586, 700–701
– SOFC   923
– stack durability   184
fuel gas, refinery waste   455
Fuel Partnership   659

fuels
- airship 681
- alternative 857
- ammonia 692
- BTL 530–531, 626
- chemical production 619–621
- CNG 836–837
- for transport 523–524
- fossil, *see* fossil fuels
- gaseous biofuels 512–515
- hydrocarbon 619–628
- liquid biofuels 511–512, 625–626
- natural gas, *see* natural gas
- solid biofuels 510–511
- substitution of oil-based 206
- synfuel 624
- wood 493
- , *see also* transport

Fukushima Dai-Ichi nuclear power plant (FDNPP) 13–14, 16, 67
full-scale infrastructure 804
fully sustainable global energy system 93–117
furnaces, electric arc 928

g

G CAPEX 748, 763–765
"game changers" 195–196
- PV 303
gas and steam turbines 511
gas engines 835
gas fields, storage sites 640–642, 829
gas heat pump 790
gas hourly space velocity (GHSV) 822
gas insulated lines (GILs) 711–712
gas losses 635–636
gas power plants, flexibilization capability 505
gas turbine combined cycle (GTCC) plant 18
GasCam 782
gaseous biofuels 512–516
gaseous hydrocarbon fuels 620
gaseous hydrogen storage, compressed 662–668
gases
- anaerobically digested 824
- CO_2, *see* CO_2
- coke-oven 785
- cushion 828
- greenhouse, *see* greenhouse gases
- holders 664–665, 681–682
- hot gas cleaning 531
- hydrogen, *see* hydrogen
- LFG 473, 512–513, 824
- natural, *see* natural gas
- pipelines 831–834
- production 425–482
- real gas behavior 662, 773
- renewable 425
- storage 514–515, 828–831
- syngas, *see* syngas
- town gas 651, 666
- working 828

gasification 452
- biomass 531, 824–825
- coal 802, 834
- distillation residues 623

gasoline
- power density 199
- , *see also* fuels

GDP (gross domestic product)
- BRICS 140
- China 29, 36–37, 896
- Developing World 138
- South Korea 5
- sustainable global energy system 97

GEA (Global Energy Assessment) report 851
gearbox
- marine energy systems 361–363
- wind generators 250, 272

gearless-generator arrangement, gearbox 364
generation
- distributed 143, 405
- fluctuations 199–200
- "generation valleys" 784
- hot water 922–926
- mandating 148

generator
- DIG/DFIG 250–252, 272
- radial flux 251

geographic equity 149
geographical factors 806
geological storage 629–657, 828–831
- near-surface 638

geothermal power 339–350
- continuous recharge 345
- economic feasibility 346
- environmental impact 346–350
- fluids 343
- seismic risk 347
- supply potential 110

German Advisory Council on Global Change (WBGU) 121, 123
- guard rails concept 144–145

German building energy code 880
GermanHy study 206
Germany
- CO_2 emissions 196–197
- end energy carriers 621
- geological storage 636–639

- hydrogen pipelines 208–209
- natural gas 780, 832
- offshore wind power 267
- overlay grid 719
- primary energy supply 621
- RE 500–507
- resource mix 517
- seismic activity 636
- supply pattern 405
- TSOs 502–503

GHG, *see* greenhouse gases
GHSV (gas hourly space velocity) 822
Giddens, Anthony 124
GIGACELL Ni–MH batteries 24
GILs (gas insulated lines) 711–712
glass-based modules 293
glass fiber enforcement 248
glass production 922
Global Energy Assessment (GEA) report 851
global energy demand 108
global energy supply 112
global energy system, fully sustainable 93–117
global food production 487–490
global freshwater use 120
global guard rails 145–147
global hydropower potential 143
global resources, biomass 485–497
Global Trade Analysis Project (GTAP) 33
global value research 131
Global warming potential (GWP) 197
global weather model, Reanalysis 606
global ..., *see also* World ...
Goulas hydropower plant 613–615
government-driven strategies, South Korea 5–7
gradient load 406, 408–410
gradient pressure 774
grain, world production 488
graphene 586
graphite 444
gravimetric energy density 208
gravity base foundations 276, 362
"Great Transformation" 121
Green Energy Revolution (Japan) 19
green energy strategies
- China 29–46
- diverging objectives 74
- Japan 13–27, 18–22
- nuclear power technology 17
- South Korea 3–11

green growth 133
green lifestyle, China 905

green technologies, revenue potential 69
Green Technology Center Korea (GTC-K) 10
greenhouse gases (GHG)
- biofuel production 536–540
- biomass conversion 540–541
- emission reduction 195
- emissions in CSP plants 312
- GWP 197
- hydropower 396–397, 399–400
- industrial sector 934–935
- South Korea 5
- transport emissions 851–852
- well-to-wheels emissions 805–806
- , *see also* CO_2

grid
- AGC 612–613
- cost 83, 89
- distribution 731–732
- electrical/thermal 569
- European 275, 729
- extension 73
- hydrogen 209
- infrastructure 246
- IRENE-40 735, 747
- load balancing 217–240
- meshed DC 718
- micro- 732
- natural gas 832
- North Sea 63–64
- offshore 359
- overlay 709, 719
- stabilization 439
- power management 226
- services 600
- "smart", *see* "smart grids"
- stabilization 436
- sub-sea electricity 61
- "super" 709
- sustainable global energy system 112–113
- technologies 69
- transmission components 709–721
- UCTE/ENTSO-E 729–730
- vertical grid load 204
- ZEB connection 877

gross domestic product (GDP)
- BRICS 140
- China 29, 36–37, 896
- Developing World 138
- South Korea 5
- sustainable global energy system 97

gross power generation, Germany 517
ground source heat pumps (GSHP) 57
- "smart grids" 761
GROWIAN wind turbine 180

Subject Index | 951

growth potential
- biomass   494, 515
- food production   490
- hydropower   385–388

GTAP (Global Trade Analysis Project)   33
GTCC (gas turbine combined cycle) plant   18
guard rails concept, global   145–147
guide vanes   600
GWP (global warming potential)   197

h

Haber–Bosch process   694–696, 700–701
- efficiency   696

Halle mining site   644
hard coal   416
head   381
headrace   389
health impact   145
heat
- binary-cycle exchangers   341
- capture   165
- demand   900
- diversified water heater demand   750
- heat–electricity interface   87
- utilization   517
- waste   557
- waste heat utilization   58, 177

"heat grazing", rotational   348
heat of vaporization   669–670
heat pumps (HP)   57, 101
- gas   790
- "smart grids"   761–762

heat recovery steam generator (HRSG)   316
heat storage   561–563
- and utilization   517
- CSP   178
- media   165–168
- , see also thermal storage

heat-transfer fluids (HTF)   307, 314–315
- air   461–465
- molten salts, see molten salts
- sensible heat storage   326
- sodium vapor   466
- solar thermal methane reforming   461–465
- solar tower systems   321
- solid particles   467–468

heat-transfer sections   177
heated reactors   461–468
heating
- Chinese buildings   899
- combustion systems   835
- district   88, 413–415
- efficiency   901
- Northern China   900–902
- overheating   900–901
- reheating   173, 177
- REHVA   875
- residential   203, 801
- SHP   924–926
- solar thermal collectors   884–885
- solar water heater   142, 151–152

heating value, hydrogen   222
heavy petroleum   624
HEFA (hydrotreated esters and fatty acids)   529
heliostats   163, 320, 457
- fields   464

helium, liquid   671
HER (hydrogen evolution reaction)   442–443
hetero-junction cells   297
high calorific value gas (H gas)   774–775
high-enthalpy systems   349
high-flow season   388
high-performance solar modules   297–298
high-pressure tubes   663
high-temperature superconducting (HTS) cables   711–712
high-temperature superconducting (HTS) motors   917
high-temperature water electrolysis (HTE)   436–438, 818–819
high voltage AC (HVAC) cables   271
high-voltage DC (HVDC)   727
- and AC lines   736–739
- cables   274–275
- converters   715
- interconnectors   64

higher heating value (HHV)   692–693
- hydrogen   429–430

highest regulated water level (HRWL)   607–609
highly energy-intensive materials, substitution   932
hind casting   274
Hinshelwood, see Langmuir–Hinshelwood mechanism
HMC fuel cell vehicles   183–194
Hohen Neuendorf school   884
Holen PSH plant   609–611
holistic approach   228
Honda FCV   22–23
honeycombs, ceramic   458, 468–469
horizontal-axis turbine   361–363, 365, 367–370
horizontal tanks   672
HOT ELLY   436, 818
hot gas cleaning   531
hot sedimentary aquifer (HSA)   339
hot tapping   778

952 | Subject Index

hot water
- domestic   561, 903
- generation   922–926
- long-term storage   571–572

hourly grid load balancing   230–232
house-use ESS   592
HP (heat pumps)   57, 101
- gas   790
- "smart grids"   761–762

HRSG (heat recovery steam generator)   316
HRWL (highest regulated water level)   607–609
HTE (high-temperature water electrolysis)   436–438, 818–819
HTF, see heat-transfer fluids
HTS (high-temperature superconducting) cables   711–712
HTS (high-temperature superconducting) motors   917
hub height   247
human-powered travel   103
HVAC (high voltage AC) cables   271
HVDC, see high-voltage DC
hybrid cars, plug-in   106, 587, 731, 859
hybrid operation, CSP   172
hybrid power plants, geothermal power   341
hybrid tower concepts, wind power   253
hydraulic drive system   313
hydraulic yawing system   367–369
hydro-thermal carbonization (HTC)   508
hydrocarbons
- fuels   619–628
- light   798
- solar conversion   460
- sugars to   533

hydroelectric turbines   371
hydrofoil, oscillating   366
hydrogen
- and ammonia   691, 699–700
- as enabler for RE   195–216
- bio-   512–513
- chemical storage   619–628
- compressed gaseous storage   662–668
- compressors   662–663
- concentration in gases   786
- conversion   797–798, 835
- cost   223–224
- cryogenic liquid storage   669–675
- delivery   796–802
- direct-hydrogen FC systems   836
- distributed storage   678
- electrolysis production, see electrolysis
- energy carrier   438, 630
- FCV efficiency   795, 797
- feed to natural gas grid   833
- fuel upgrading   622
- gas holders   665–666
- grid load balancing   217–240
- heating value   222, 429–430
- Japan   22–23
- leakage   188
- liquid   208, 630
- "merchant"   807
- near-surface bulk storage   659–690
- near-zero emission production   800–801
- ortho-/para-   669–670
- petroleum refining   623–624
- pipeline system   208–209, 833–834
- power density   198–199
- power-to-gas   425
- production   796–802
- public acceptance   25
- RPH   838–841
- solubility in water   644
- storage   208–209, 564, 679
- storage medium   426
- strategic approach   200
- supply   211, 805–806
- synfuel   624
- synthetic methane production   788
- tank embrittlement   674
- transition from natural gas   629–657
- vehicles   796–797

hydrogen evolution reaction (HER)   442–443
hydrogen infrastructure   192, 795–811
- cost estimates   210–213

hydrogen refueling   190–191, 801
- design considerations   806–807
- filling stations   210

hydrogenation   622
- catalytic   820

hydrological cycle   382
hydropower   381–401
- cost   394–396
- economic feasibility   385–386
- efficiency   381
- environmental impact   387–388, 399
- flexible   612–615
- global potential   143
- pumped storage, see pumped storage hydropower
- supply potential   109

hydrotreated esters and fatty acids (HEFA)   529
hydrotreated vegetable oils (HVO)   529

i

I CAPEX   748, 763–765

Subject Index

IAEA (International Atomic Energy Agency) 26
IBC (interdigitated back contact) 297–298
ICCT (International Council on Clean Transportation) 862
ice formation, wind turbines 256
ICE (internal combustion engines), passenger car powertrains 836–838
ICT 515, 729
– "smart grids" 744–746, 749
ideal gas law 775
idle losses 854
IEA (International Energy Agency) 95
– geothermal power Roadmap 339, 343–344
– Technology Roadmap 494
– Wind Task 258
– World Energy Outlook 824, 891, 913
IGA (International Geothermal Association) 339, 343–344
IHA (International Hydropower Association) 398–399
incentives, financial 151–152, 343
incident angle correction 318
incident solar radiation per year (ISR) 296
inclination of roof 292
India 140
indirect land use change (iLUC) 538–539
indirectly heated reactors 461–468
indirectly irridated solar receivers 458
indoor temperature control 905
induction generator, doubly fed 250–252, 272
industrial GDP 896
industrial reformers 455
industrial round wood 493
industries
– as CO_2 source 825–826
– domestic 149
– energy-intense 71
– energy savings potentials 913–936
– energy transition 67–74
– intensity 98
– non-energy-intensive sector 914
– renewable energy 9
– secondary 895
– sustainable global energy system 97–99
industry standards 152
inertia of the system 412
information management 746
infrastructure
– compatibility 806
– full-scale 804
– grid 246
– hydrogen 192, 210–213, 795–811

– natural gas 791
– new model 143
– North Sea 51
– public acceptance 417
– scaled-up 804
– societal transition 125–127
Infrastructure Roadmap for Energy Networks in Europe (IRENE-40) 735, 747
inlet temperature, CSP 168–169
innovations, disruptive 360
in/out storage 571
inspection tools, pipelines 781
institutional change 133
institutional perspective 129–131
institutionalists 132–134
insulated gate transistors/thyristors 716
insulating duct 175–176
insulation 900
– hydrogen storage tanks 674
– XLPE 711–712, 716
– , see also zero-energy buildings
integral friction coefficient 775
integrated solar combined cycle (ISCCS) plants 315
intensity
– capital 85
– carbon 29
– energy 102, 105–106, 856, 899–900, 903, 907
inter-machine spacing 274
interconnector cross-border 733
interdigitated back contact (IBC) 297–298
interlinkages, structural 124–125
intermittent RE, see variable RE
internal combustion engines (ICE), passenger car powertrains 836–838
internal currents 219
International Atomic Energy Agency (IAEA) 26
International Council on Clean Transportation (ICCT) 862
International Energy Agency (IEA) 95
– geothermal power Roadmap 339, 343–344
– Technology Roadmap 494
– Wind Task 258
– World Energy Outlook 824, 891, 913
International Geothermal Association (IGA) 339, 343–344
International Hydropower Association (IHA) 398–399
International Solar Energy Society (IRENA) 153
internationalization 129
interstitial sites 675

intra-day markets, liquid 89
inward investment, marine energy 354
ion-conducting solid electrolytes 583
IPCC emission reduction targets 893
IPCC SRREN report
– geothermal power 339
– offshore wind power 267, 278
IR filling nozzle 191
iridium catalysts 442–443
Irish–Scottish Links on Energy Study (ISLES) 61–64
Irish Sea 62
iron production 922
– , see also steel ...
irreversibilities, thermodynamic 557
irrigation 387
ISCCS (integrated solar combined cycle) plants 315
ISES White Paper 137
isothermal compression 662
ISR (incident solar radiation per year) 296
Ivanpah solar tower system 322–323

j

jacket foundations 267, 276
Japan
– green energy strategies 13–27
– nuclear power technology 17
– offshore wind power 268
jobs from RE, *see* employment opportunities
Joule–Thomson expansion 670
Juklavatn reservoir 607–608

k

K, *see* potassium
karst 642
Kennedy Space Center (KSC) 630
kiln technology 929
Kölbel–Schulze methodology 543
Korea, South, *see* South Korea

l

L gas (low calorific value gas) 774
lamps, LED 906
land
– arable 490
– footprint 343
– use 146, 525
land use change (LUC) 120, 538–539
landfill gas (LFG) 473, 512–513, 824
Langmuir–Hinshelwood mechanism 698
large-area coating technologies 288–290
large-scale CSP plants 159–182, 461
large-scale hydrogen technologies 217–240

large-scale methanation plants 839
large-scale storage and balancing 605
latent heat storage 326–327, 562
LAVE-Trans (Light-duty Alternative Vehicle Energy Transitions) model 862–869
laws and equations
– Barlow law 777
– Bernoulli equation 786
– Butler–Volmer equation 220
– Darcy–Weissbach equation 773–774
– ideal gas law 775
– Le Chatelier principle 820
– net social value 858
– second law of thermodynamics 426
– Tafel equation 220
LCC (line-commutated converter) technique 716
leakage, hydrogen 188
leakage effects 43
LED lamps 906
lenses, optical 317
Leontief CES 33
Leuna plant, Haber–Bosch process 695
level fluctuations, reservoir water 609–611
levelized cost of energy (LCOE) 245, 357–360
– fossil power plants 420–421
– hydropower 394–395
levelized electricity cost (LEC) 325
LFG (landfill gas) 473, 512–513, 824
LHV (lower heating value) 661
LIDAR, load prediction 254–255, 274
life-supporting systems, natural 147
lifecycle assessment (LCA)
– biofuel production 537 538
– hydropower 395, 399
lifestyle, green 905
light-duty vehicles 863
– LAVE-Trans model 862–869
light hydrocarbons 798
lignite power plants 507
lignocellulosic bioethanol 523, 539
line-commutated converter (LCC) technique 716
line loading 736–739
linear concentration 162–163
linear Fresnel mirrors 309, 317–320, 457
liquefaction
– hydrogen 208
– plants 670–671
liquid biofuels 511–512, 516, 625–626
liquid-dominated systems 349
liquid helium 671
liquid hydrocarbon fuels 620
liquid hydrogen (LH_2) 630

- cryogenic storage 669–675
liquid intra-day markets 89
liquid materials, sensible heat storage 326
liquid water, formation energy of 426
liquids, organic 660
lithium–air batteries 585–587
lithium–ion batteries 185, 581–582, 853
- energy storage systems 591–592
- power density 198
lithium–sulfur batteries 584–585
lithium super-ionic conductor glass film 585
living standards, China 902
load
- aerodynamic 279
- alternating 778
- following 408–410
- leveling 591
- local network peak 757
- prediction 254–255, 274
- (un)controlled profile 751
load balancing
- grid 217–240
- hydropower 398
- ZEB 879
load factor 742
loading, line 736–739
local overexploitation 119
local storage, liquid hydrogen 671–672
locational specificity 84
long-distance trucks 106
long-run marginal cost (LRMC) 339, 346
long-term hot water storage 571–572
long-term stability, subsurface storage 635
losses 715
- biodiversity 120
- boil-off 670
- drivetrain 854
- engine 854
- gas 635–636
- idle 854
- loss-aversion behavior 869
- ohmic 445, 725, 728
- parasitic 854
- power-to-wheel 854
- thermal 319, 325
- wake 255
low calorific value gas (L gas) 774
low-carbon development 30
low-carbon systems 743–744
low-flow season 389
low-head projects 391
low-interest loans 151
low-temperature fuel cells 218
lower heating value (LHV) 661

- hydrogen 430
lowest regulated water level (LRWL) 607–609
LRMC (long-run marginal cost) 339, 346
lubrication, sea-water 369
LUC (land use change) 120, 538–539

m

magnetic flux line (MFL) 781
magnets
- PM generator 250–252, 272–273, 365
- rare earth 273
maintenance, offshore wind power 276–277
management, adaptive 347
mandating capacity/generation 148
manual reserve 411
marginal net present social value 858
marine ecosystems 146
marine energy 351–379
- cost 353
Marine Energy Array Demonstrator (MEAD) 355
Marine Energy Technology Roadmap 357
Marine Renewables Commercialization Fund (MRCF) 355
market design 73
market options, bulk power transport 732
market reform, EMR 352
markets
- constraints 406–407
- distributed 768–770
- flow-based coupling 86
- liquid intra-day 89
markup 38
massive integration, PV electricity 303
maximum acceptable storage costs 573–574
Measure–Correlate–Predict (MCP) 274
mechanical–thermal treatment, biomass 531
membranes
- CCM 434, 442
- electrolyzers 221
- membrane electrode assembly (MEA) 817
- Nafion 697, 700, 817
- proton-conducting 427
"merchant" hydrogen 807
merit order 79, 406
meshed DC grids 718
metal hydrides
- hydrogen storage 675–676
- Ni–MH batteries 23–24
metal liner 664
metal surfaces, heat capture 165
metallic foam absorbers 473–474
meteoric water 646
meteorological models 88

meteorology, offshore 274
methanation 787–788
– CO_2 204, 210
– large-scale plants 839
– power-to-gas 820
methane
– and ammonia 691, 702
– catalytic CO_2 hydrogenation 820
– CHARM® 781–782
– cracking 476
– feedstocks 453
– physical properties 775
– power-to-gas 425
– RPM 823, 838–841
– synthetic 788
methane reforming
– dry 451
– scale-up 476–477
– solar thermal 451–482
– steam 451, 453
methanol 533
– and ammonia 691, 702
MFL (magnetic flux line) 781
micro-CHPs 790, 792
micro-grids 732
microeconomic development 145
migration, fish 391
mild steel 777
mineral oil
– sensible heat storage 326
– subsurface storage 632
minerals, water-soaked 642
mining sites, abandoned 644–646
Ministry of International Trade and Industry (MITI) 15
mirror shape 163–164
mixing, vertical 603, 611
mobility
– electro-/petro- 851–873
– zero-emission 183, 217, 587
– , see also vehicles, transport
modeling
– CGEM 31–39
– German wind power potential 205–207
– hydrogen electrolysis 224–226
– improved meteorological 88
– ISLES 61–64
– mobility transition 862–870
– Reanalysis global weather model 606
– sustainable global energy system 94–95
modules, glass-based 293
Molasse Basin 640
molten salts 166–167
– as heat-transfer fluid 466–467

– nitrate 315–317
monocrystalline wafers 286
monolithic ceramic structures 458
monopile foundations 276
monorail train 24
motorized transport 851
MRCF (Marine Renewables Commercialization Fund) 355
multicatalyst systems 441
multilayer insulation 674
multilevel converter 720
multiple-junction solar cell 284
multiple rotors, marine energy 359
multitube receiver 319
"must-run" RE 406

n
N-1/N-2 769
Na super-ionic conductor (NASICON) 587
Na, see also sodium
nacelle 361–362, 368
Nafion membrane 697, 700, 817
nanodesigned catalysts 443
nanostructured carbon materials 586
NASA spherical storage tank 673
national RE 144
National Renewable Energy Action Plan (nREAP) 502
National Renewable Energy Laboratory (NREL) 184, 540, 817
natural gas
– CNG 787, 792, 836–837
– contaminated 473
– cost 210–212
– cracking 452
– (de)central conversion 835
– ex- and importers 4
– German industry 780
– grid 832
– hydrogen feed-in 212, 833
– hydrogen production 802
– infrastructure 791
– pipeline systems 668, 773–794, 831–833
– power plants 200
– R&D 784–790
– storage 631–633
– substitute 624–625
– subsurface storage 632
– synthetic 511, 532, 823
– transition to hydrogen 629–657
natural life-supporting systems 147
near-surface bulk storage
– cost 677–679
– economic feasibility 677–679

- Germany   638
- hydrogen   659–690
near-zero emission, hydrogen production   800–801
near zero-energy concept   102
nearly/net-zero energy buildings   875–889
Nesjen reservoir   607–608
net mass capacity   681–682
net present value (NPV)   858
net social value (NSV)   858
networks
- asset utilization   743
- ENSG   60
- externalities   857
- peak load   757
- transmission   723–739
"New Economics"   123
new materials development, EV   589
New Zealand, geothermal power   340
next generation turbines, offshore   278–280
Ni–MH batteries   23
- GIGACELL   24
nickel
- catalysts   454
- electrodes   441
- Raney   226
night-time operation, solar power   161–162, 178
nitrate salts   317
- sensible heat storage   326–327
nitrogen
- cycle   120
- emissions   495, 698
- fixation   693
- nitrogen-doped carbon materials   587
- nitrogen–hydrogen compounds   691
"no-coal villages"   907
"No Policy"   35
noble metal catalysts   816
noise, wind turbines   256
nominal operation cell temperature (NOCT)   291
non-cyclic changes   619
non-dispatchable RE, see variable RE
non-energy-intensive industry sector   914
non-fossil sources of primary energy   29
non-residential buildings   886
non-structured reactors   476
NorNed cable   732
North German Basin   640
North Sea
- energy infrastructure   51
- grid   63–64
- offshore wind power   267–268, 607

Northern ISLES   63
Norway
- geological storage   646
- hydropower   605
- PSH   601
"Norwegian arc process"   694
NOW study   440
nozzle, IR filling   191
NPV (net present value)   858
nREAP (National Renewable Energy Action Plan)   502
NREL (National Renewable Energy Laboratory)   184, 540, 817
- chart of cell efficiencies   303
NSV (net social value)   858
nuclear accidents   243
- , see also Fukushima
nuclear power
- China   893
- Fukushima   13–14, 16, 67
- Japan   17, 19
- plant flexibilization capability   505
- Scotland   56–57
number of blades   247–248

O

Obama, Barack   269
ocean
- acidification   120
- energy, see marine energy
- sink   491–492
OER (oxygen evolution reaction)   443
off-board storage, regenerable   659
off-grid applications   819
offshore grid, marine energy   359
offshore meteorology   274
offshore resources, RE   53–54
offshore wind power   203, 265–281
- capacity growth   266
- cost   271
- CRTF   269
- next generation turbines   278–280
- synergy with marine energy   360
ohmic losses   445, 725, 728
oil crisis, second   431
oil fields, storage sites   640–642, 829
oils
- crude   4, 632
- hydrotreated vegetable   529
- mineral   326
- palm   540
- substitution of oil-based fuels   206
- vegetable   624
on-board storage   659

on-site generation   881
on-site H$_2$ production   800, 803
on-site storage capacity   878
one-way storage   571
onshore hydroelectric turbines   371
onshore wind power   203, 243–263
open-center turbines   365
"open-loop" systems   452
operating pressure   667
– gas holders   682
operating temperature, solar tower systems   325
operational strategies
– hybrid   172, 324
– night-time operation   161–162, 178
– PSH   601–602
– turbines   390
OPEX   419–421, 544
– "smart grids"   748, 763–765
optical efficiency   318
optical lenses   317
optical path   318
optical quality, parabolic trough   314
optimization
– charging   756
– electric drives   919–922
organic liquids   660
organic photovoltaics (OPV)   298–299
organic Rankine cycles (ORC)   508
ortho-to-para conversion, hydrogen   669–670
oscillating hydrofoil   366
oscillating water column   377
oscillating wave surge converter   371–372
overexploitation, local   119
overhead lines   710
overheating   900–901
overlay grid   709
– Germany   719
overload, efficiency impact   440
overpotentials   226, 429
overtopping wave energy converter   378
ownership
– building floor space   897
– equity   149
oxygen evolution reaction (OER)   443
ozone depletion, stratospheric   120

p

packing, fuel cell stacks   184–185, 187
palm oil   540
paper, recycled   929–930
paper industry   929–930
para-hydrogen   669–670
parabolic dish   309

parabolic trough   164, 309, 312–317, 457
parasitic currents   219
parasitic losses   854
partial load, electrolyzers   432, 435
passenger car powertrains   836–838
passive house concept   883
passive wind   754
PCM, *see* phase change material
peak demand   741
– "smart grids"   751
peak electricity   230, 234
peak load, local network   757
peak shaving   591
Pearl Street system   724
pelletization   508
pellets   416
PEM (polymer electrolyte membrane) electrolysis   433–436, 442
– electrolyzers   439–440
– power-to-gas   817–818
PEMFC (polymer electrolyte membrane FC)   183
– ammonia conversion   700–701
– and Li–air batteries   586
penetration rate   81
performance
– AWE   816
– COP   571, 761
– CSP technologies   311
– electrolyzers   219–221, 224
– rechargeable batteries   590
– residential PV systems   292
– seasonal   881
– testing   188–191
perlite vacuum insulation   674
permanent magnet (PM) generator   250–252, 272–273
– tidal energy   365
permeability, rock structures   829
petro-mobility   851–873
petroleum
– refining   623–624
– (un)conventional   852
PFPB (point feeder pre-baked) electrodes   928
pH, electrolytes   433
phase change material (PCM)   165–167, 178
– carbonates   467
– energy storage systems   562–563
phase separator   174
Phasor Measurement Unit (PMU)   729
PHES (pumped hydro energy storage), *see* pumped storage hydropower
PHEV, *see* plug-in hybrids
phosphorus cycle   120

photovoltaics (PV)   110, 883–887
– building-integrated   292–294
– concentrating   302–303
– cost   290–291, 296, 301
– organic   298–299
– power density   198
– R&D   300–301
– semi-transparent modules   294
– technological design   290–300
– terawatt scale technology   283–306
– thin-film   288–290, 294–295, 298–300, 302–303
– ZEB   877
pipe storage facilities   668
pipeline systems
– hydrogen   833–834
– inspection tools   781
– natural gas   773–794, 831–833
– R&D   784–790
pistons, reciprocating   662
pitch system   254
planetary boundaries   119–120
planning permits   152
plant capacity, biofuel production   534–535
plates, bipolar   583
platinum catalysts   442–443
plug-in hybrids   106, 587, 731, 859
plus energy buildings   875–889
PM (synchronous permanent magnet) generator   250–252, 272–273
– tidal energy   365
point absorber   374–375
point concentration   162–163
point feeder pre-baked (PFPB), electrodes   928
policy
– CAP   487
– climate policy goals   30–31, 74
– current policy assumptions   37–38
– EGPS   47–48
– green energy strategies, see green energy strategies
– "No Policy"   35
– polycentric approach   130
– South Africa   154
– TEN-E   747
political participation   129
"polluter pays principle" (PPP)   495
pollution
– air   146, 805–806
– chemical   120
"poly-Si"   287
polycrystalline wafers   286
polyethylene, XLPE insulation   711–712, 716

polymer electrolyte membrane FC (PEMFC)   183
– ammonia conversion   700–701
– and Li–air batteries   586
polymer electrolyte membrane (PEM) electrolysis   433–436, 442
– electrolyzers   439–440
– power-to-gas   817–818
polysulfide dissolution   584
population growth   97
– China   897
"porcupine" concept   474–475
pore-space storage sites   639–640, 829–830
porous materials, heat capture   165
potassium hydroxide (KOH)   815
potassium nitrate   315–317
potential feed   833
"Potential Fernüberwachung"   783
poverty alleviation   142–143
power
– bulk transport   723, 732
– density   197–199
– electric, see electric power
– export   614
– geothermal, see geothermal power
– grids, see grid
– hydro-, see hydropower
– losses   715
– management   226
– nuclear, see nuclear power
– reserve balancing   260
– smoothing   24
– solar, see solar power
– thermal, see thermal power
– transformers   714
– up-rating   359
– wind, see wind power
power plants
– flexibility potential   409, 505
– fossil   403–422
power rating   247
– energy storage systems   561
– offshore   272
power system services
– CSP   155
– fossil power plants   410–413
– hydropower   398
power-to-gas   425, 571–572, 813–848
power-to-wheel losses   854
power tower, see solar tower systems
power transmission extensions, Germany   719
power/capacity ratio   325
powertrains
– alternative   857

– passenger car 836–838
"predict and provide" philosophy 744
present value annuity factor (PVANF) 572–573
pressure cycles 778
pressure gradient 774
pressure level, subsurface storage 634
pressure swing adsorption (PSA) 696, 700–701
pressure vessels, hydrogen 663–668
pressurized electrolyzers 227, 441
price pressure 44
primary energy
– China 895
– credits 885
– electric power generation 201
– Germany 621
– nonfossil sources 29
– sectoral demand 914
primary reserve/control 411
PRIMES baseline 917–918, 932
problem shifting 126
process chain alternatives 838–841
Production Tax Credit (PTC) 151, 244, 257
profile
– cost 78–82, 85–88
– (un)controlled load 751
project planning 51
"proof of value" point 358
propane 692
proton conductors
– membrane 427
– solid-state 697
PTC, see Production Tax Credit
public acceptance 68
– green energy 25
– infrastructure projects 417
public buildings 904–906
pulp 492, 929–930
pulverized coal 417, 508
pump, wave-powered 371
pumped storage hydropower (PSH) 389, 559, 567, 574–575, 597–618
– deployment 398, 598–599
– environmental impact 602–604, 611–612
– existing reservoirs 602–603
– R&D 604
PV, see photovoltaics
PVcomB 289–290
pyrolysis 454
– fast 511–512

q
Qinhuangdao, Shimen Village 908

quality, natural gas 786
quantification, VRE market value 77–83
quickly responding capacity 85
quotas 148
– wind power 257

r
radial flux generator 251
radial systems 724
radiation, solar, see solar radiation
radioactivity 17
rainforest destruction 415
ramping down 330
Raney nickel 226
Rankine cycle 172–174
ranking, clean technology 9
rapeseed 540
rapid system response 436
rare earth magnets 273
R&D, see research and development
reactive reserves 412
reactors
– bell-jar solar reforming 469
– biological methanation 788
– directly irradiated 468–476
– indirectly heated 461–468
– non-structured 476
– solar receiver/reactor concepts 456–460
– solar-thermal aerosol 468
– three-phase 821
real gas behavior 662, 773
real-time demand–supply balancing 743–744
real-world experiments 131
Reanalysis global weather model 606
rebound effect 126
receivers
– cavity 477
– development 314
– dish Stirling 164
– (in)directly irridated 458
– solar receiver/reactor concepts 456–460
– tubes 165, 175
– vacuum/atmospheric 319
– volumetric 477
reception point 256
rechargeable batteries 23, 581–587
– applications 580
– EV 587
– performance 590
– R&D 581–587
– "smart grids" 755–759
reciprocating pistons 662
recovery
– CO_2 826–827

– energy   698–700
rectification   227
recycling   98, 932
– paper   929–930
redox-flow batteries   582–583
– energy storage systems   592–593
reduction, selective catalytic   698–699
refill, *see* drain and refill cycle
refinery waste fuel gas   455
refining, petroleum   623–624
reflector materials   313
reforming
– bell-jar solar reforming reactor   469
– DCORE   463–464
– industrial reformers
– methane, *see* methane reforming
– SCORE   463–464
– steam   623
– thermodynamics   453–454
refueling, hydrogen   190–191, 210, 801, 806–807
regenerable off-board storage   659
regional factors   806
regulation, heating   900–901
reheating   173, 177
REHVA (European Association of Refrigeration, Heating and Ventilation)   875
reinforcement cost   756–759
relative discharge   390
reliability, EV   589
remote monitoring, CHARM®   781–782
Renewable Energies Act, *see* EEG
renewable energies (RE)
– base-load   342
– China   30–31
– comprehensive transition concept   119–136
– consumption   3
– cost competitiveness   40
– cost reduction   42–43
– current policy assumptions   39–40
– demand prediction   7
– deployment potentials   109
– Europe   404
– financial support   407
– generation fluctuation   199–200
– generation targets   37–38
– Germany   500–507
– hydrogen, *see* hydrogen …
– in the Developing World   137
– integration with electrolyzers   819–820
– Japan   18–22
– load following   409
– methane production   565
– national   144

– offshore resources   53–54
– photovoltaics, *see* photovoltaics
– poverty alleviation   142–143
– power density   197–199
– public acceptance   25
– rapid diffusion   21
– residual power   421–422
– Scottish routemap   52–53
– South Korean industries   9
– technical and economic potential   141–143
– transport sector   105
– variable, *see* variable RE
– wind power, *see* wind power
renewable energy certificates (REC)   6
renewable gases   425
"renewable methane", *see* biomethane
Renewable Obligation Certificates (ROC)   352
Renewable Portfolio Standard (RPS)   5, 246
renewable power hydrogen (RPH)   838–841
renewable power methane (RPM)   823, 838–841
renovated buildings   886
repair delay time, expected   277
Republic of Korea, *see* South Korea
"Reregulation Support Centre"   9
research and development (R&D)
– biofuels   526–527
– Developing World   153
– natural gas pipeline systems   784–790
– PSH   604
– PV   300–301
– rechargeable batteries   581–587
– solar thermal methane reforming   460–476
– strategies   7–9
– water electrolysis   438
reserve balancing power   260
reserves
– geothermal   344–345
– primary/secondary reserve/control   411
– strategic   631
reservoirs   383, 387
– water level fluctuations   609–611
residential floor space per capita   99
residential grid-connected PV system   290–292
residential heating   203
– tri-generation system   801
– , *see* also district heating
residential PV systems, cost   290–291
residual demand curve   79
residual generation capacity   81
residual load   80
residual power   421–422
resin-impregnated continuous filament   664

resource mix, Germany 517
retail electricity market 769
revenue, hydrogen 211
reversible cells, solid oxide 437
reversible pump turbines (RPT) 597–600
risk-adjusted cost 259
risk management 145
rivers
– flow variation 384
– protection 146
Rjukan PSH plant 609–611
robustness of technology, solar power 161, 169–170
ROC (Renewable Obligation Certificates) 352
rock caverns, hydropower 391–392
rock salt 647
rock structures, permeability 829
rocks, near-surface 638
roll-to-roll processes 295
roof inclination 292
roof installation 290–292
"rotational heat grazing" concept 348
rotors
– materials 248
– multiple 359
round wood, industrial 493
RPH (renewable power hydrogen) 838–841
RPM (renewable power methane) 823, 838–841
RPS (Renewable Portfolio Standard) 5, 246
RPT (reversible pump turbines) 597–600
run-of-river hydropower 388
runoff 385
rural residential buildings 906
Russia 140
Russian natural gas 775

S
S CAPEX 748, 763–765
SA (smart appliances) 752–753
Saccharomyces cerevisiae 529
safety
– EV 589
– safety factor 777
– subsurface storage 635
salt caverns 203
– storage sites 646–652, 830–831
salts
– molten, see molten salts
– nitrate 317, 326–327
sandstone 640
saving
– active energy 21
– conversion "savings" 934

– energy in buildings 99, 891–911
– industrial energy savings potentials 913–936
scale-up
– infrastructure 804
– PEM electrolysis 436, 442
– solar methane reforming 476–477
– solar thermal power 331
scheduling, stochastic/deterministic 767
Scholz, Roland 122
school buildings, ZEB 884
Schulze, see Kölbel–Schulze methodology
SCORE (single-coil reformer) 463–464
Scotland
– electric power 47–65
– marine energy 351
– SEA 50
Scotrenewables Tidal Turbine (SRTT) 370
sea crossings 733
SEA (Strategic Environmental Assessment) 50
sea-water lubrication 369
sealing, subsurface storage 633, 650
seasonal buffers 629
seasonal performance 881
seasonal turbines 601
second law of thermodynamics 426
second oil crisis 431
secondary batteries, see rechargeable batteries
secondary industries, China 895
"sector leakage" 42
sectoral primary energy demand 914
security
– financial 149
– supply 68, 352–353, 412–413
SEGS 312
seismic risk
– geothermal power 347
– Germany 636
selective catalytic reduction (SCR) 698–699
self-consumption 879
SEMI International Roadmap for Photovoltaics 298–299
semi-transparent solar modules 294
semiconductors
– compound 302
– direct 288
sensible heat storage 326, 561–562
sensitivity analysis 869
separate heat production (SHP) 924–926
separate turbines 599
separators 581
– plates 443–445

Subject Index

sequestration, carbon 798
"shadow price" 865
Shimen Village 908
shock waves 778
shore connection 274–276
"short-circuit effect" 412
short-term regulation 601
short-term supply curve 79
shut-down and sealing, subsurface storage 633, 650
Si hetero-junction cells 297
Si wafer-based solar cells, crystalline 286–288, 297–298, 301
SiC foams 473
significant wave heights 276
simple metal hydrides 675
simulated wind power production 607
simulation, composite design 225
single-circuit AC lines 736–739
single-coil reformer (SCORE) 463–464
single-junction solar cell 284
single-tube receiver 319
site assessment, wind power 273–274
smart appliances (SA) 752–753
smart charging, EV 755–759
"smart grids" 571, 709, 741–771
– cost 754
– energy storage 762–768
– environmental impact 768
– EV integration 755–760
– ICT 744–746, 749
– ZEB 882
smart refrigeration (SR) 753–754
smart wet appliances 752–753
SmartSim® 784
SMR (steam methane reforming) 451, 453
SNG (substitute natural gas) 624–625
SNG (synthetic natural gas) 511–513, 823
– thermochemically produced 532
social question 129
societal transition 123–132
societal willingness 858
socio-economic balance 304
socio-economic guard rails 145
socio-economic structure, China 895
socio-technical system 124
sodium carbonate solution 826
sodium nitrate 315–317
sodium reflux heat pipe solar receiver/reformer 465–466
sodium–sulfur (NaS) batteries 23, 583–584, 593–594
sodium vapor, as heat-transfer fluid 466
solar boiler 171, 176

solar cells
– crystalline Si wafer-based 286–288, 297–298, 301
– dye-sensitized 298–299
– efficiency limit 284
– fabrication 284–290
– flexible 294–295
– hetero-junction 297
– semi-transparent modules 294
solar conversion, hydrocarbons 460
Solar Decathlon Europe 2010 882
solar dish collectors 477
solar fields 164–165
solar–hybrid operation 324
solar–hydrogen stations 22–23
solar power
– CES production structure 34
– China 31
– CSH 110
– CSP, see concentrating solar power
– IRENA 153
– night-time operation 161–162, 178
– Spain 229
– supply potential 110
– , see also variable RE
solar radiation
– concentration 162–163
– direct conversion 283
– electric power 160
– incident 296
solar receivers
– receiver/reactor concepts 456–460, 465–466
– volumetric/tubular 458
solar thermal methane reforming 451–482
– R&D 460–476
solar thermal power 307–338
– aerosol reactor 468
– capacity 139
– collectors 884–885
– cost 328–332
solar tower systems 320–324
– Solar Tower Jülich (STJ) 459
– updraft 170
solar water heaters (SWH) 142, 151–152
SOLASYS reactor 471–472
solid biofuels 510–511, 516
solid electrolytes 583
solid oxide electrolyte 427, 437
solid oxide FC (SOFC) 923
solid particles
– as heat-transfer "fluid" 467–468
– non-structured reactors 476
solid-state proton conductor 697

solid-state storage   677
– sensible heat   326
SOLREF reactor   472–473
solubility, hydrogen in water   644
SOLUGAS   323
solution mining   649
sound power level   256
– wind turbines   247
sources, CO_2 emissions   196
South Africa   140
– policies   154
South Korea
– geological storage   646
– green energy strategies   3–11
– offshore wind power   268
Southern ISLES   63
soybean   540
space flight   673
Spain
– electric power   217, 228–229
– solar tower systems   322–323
– wind power   229
spar and float   375
specific energy consumption   428
– CO_2 sources   827
spherical dewar   683
spherical pressure vessels   665–666
– cost data   678
spherical solar boiler segment   176
spherical storage tank, NASA   673
Spindletop hydrogen storage site   652–653
spinning reserves   330
"split benefits"   768
sporadic buffers   629
SR (smart refrigeration)   753–754
SRTT (Scotrenewables Tidal Turbine)   370
stability, subsurface storage   635
stabilization
– grid   436, 439
– wind power   592
stack durability   184
stakeholder coordination   807
stall-regulated wind turbines   243
Standard Test Conditions (STC), solar cells   285
start–stop operation   614
static VAR compensator (SVC)   715
stationary end use   835–836
stationary storage, hydrogen   679
steady drain rate   633
steam
– CO_2 co-electrolysis   437–438
– electrolysis   428
– power plants   505

– supercritical   323
steam generation   922–926
– direct, see direct steam generation
– heat recovery   316
– inside ducts   175–177
steam reforming   623
– methane   451, 453
steel
– mild   777
– production   922
– tubular tower   253
Stirling engines   170, 310
Stirling receiver, dish   164
STJ (Solar Tower Jülich) thermal power plant   459
stochastic scheduling   767
stoichiometric H_2 demand   622
stone
– sensible heat storage   326
– , see also geological storage
storage   555–706
– capacity temperature dependency   563
– CCS, see carbon capture and storage
– central   558
– chemical   564–565, 619–628
– compressed gaseous   662–668
– cost   559, 566, 572–575
– cryogenic liquid hydrogen   669–675
– density   198–199
– distributed   678
– economic feasibility   572
– efficiency   559, 566, 679–681
– electric power   59–60, 86, 217, 559–561
– electrochemical ammonia conversion   691–706
– energy   557–577
– gases   514–515, 828–831
– geological   629–657
– heat, see heat storage
– hydrogen   208–209, 426, 564, 679
– in/out   571
– Japan   22, 24
– large-scale   605
– local   671–672
– maximum acceptable storage costs   573–574
– natural gas   631–633, 784–785
– near-surface bulk   659–690
– net mass capacity   681–682
– on-/off-board   659
– "on-site" capacity   878
– one-way   571
– period   559, 566
– pipe storage facilities   668

Subject Index

- power-to-gas 813
- pumped hydropower, *see* pumped storage hydropower
- "smart grids" 762–768
- solid-state 677
- stationary 679
- subsurface 633–636
- wind power 412

storage media
- fossil fuels 567
- heat 165–168
- hydrogen 426
- power density 197–199

strain–stress diagram 777
Strangford Lough project 363
Strategic Environmental Assessment (SEA) 50
Strategic Research Agenda for Solar Energy Technology (SRA) 300–301
strategic reserves 631
stratification, thermal 603
stratospheric ozone depletion 120
straw 415–416
structural interlinkages 124–125
structuration theory 123
structured catalytic systems 456
structured reactors 468–474
structures, monolithic ceramic 458
sub-sea electricity grid 61
sub-transmission system 724
sub-zero conditions tests 189
"substantial change" 134
substation cost 735
substitute natural gas (SNG) 624–625
substitution, highly energy-intensive materials 932
suction cassion foundations 276
sugarcane 490
sugars to hydrocarbons 533

sulfur
- lithium–sulfur batteries 584–585
- sodium–sulfur (NaS) batteries 23, 583–584, 593–594

Sun Belt 311
sun position sensor 313
sunshine, daily winter 144
"super grids" 709
super-ionic conductor glass film 585
superconductors, HTS 711–712, 917
supercritical steam cycles 323
supergrid concept, European DC 275
superheating 173, 177, 316
supply
- and demand 486–487, 568–570
- China 894
- diversification 353
- hydrogen 211
- potential 108
- primary 621
- real-time demand–supply balancing 743–744, 747
- supply–cost curve 394, 396
- supply-driven power options 95
- weather dependence 405
- weighted 879

supply security 68
- fossil power plants 412–413
- marine energy 352–353

supranational level 73
sustainability
- biomass co-combustion 415
- dogmas on 132
- energy transitions 858–859
- geothermal power 346–350
- hydrogen supply pathways 805–806
- hydropower 398–400
- transitions 130

sustainable available potential, biomass 499
sustainable development 121
sustainable global energy system 93–117
sustainable land use 146
SUVs, FCV 186
Sweden
- energy supply 492
- wheat production 488–489

SWH (solar water heaters) 142, 151–152
switchgears/substations 712–714
synchronous PM generator 250–252, 272–273
synfuel production 624
synthesis gas (syngas) 451, 798
synthetic BTL fuels 530–531
synthetic methane 788
synthetic natural gas (SNG) 511–513, 823
- thermochemically produced 532

system analysis
- biofuel production 534–544
- hydropower 398
- natural gas pipelines 791–793
- wind and flexible hydropower 612–615

system balance 536
system base price 76
system boundaries, ZEB 876–878
system inertia 412
system integration
- demand-side response 749–768
- low-carbon electricity systems 743–744
- transmission technologies 717–719

system optimization 517
– electric drives 919–922

t
T CAPEX 748, 763–765
Tafel equation 220
tanks
– hydrogen embrittlement 674
– liquid hydrogen storage 671–675
– metal hydride 676
– "table versus tank" issue 790
– temperature 682
tapping, hot 778
taxes
– carbon dioxide 495–496
– PTC, see Production Tax Credit
– relief 151
TBM (tunnel boring machines) 392
TCI (total capital investments) 542
TCO (transparent conductive oxides) 288–289
TCS (thermochemical thermal energy storage) 562
techno-economists 132–134
technology
– cross-cutting 914–915, 920, 926–927
– robustness 161, 169–170
Technology Innovation Needs Assessment (TINA) 357
technology readiness level (TRL) 533
Teesside gas storage site 652–653, 830
temperature
– and storage capacity 563
– indoor control 905
– inlet 168–169
– NOCT 291
– smart variation 754
– subsurface storage sites 636
– tanks 682
– temperature-entropy diagram 173
TEN-E (Trans-European Energy Network Policy) 747
TEPCO (Tokyo Electric Power Co., Inc.) 13–14
terawatt scale technology 283–306
tertiary reserve 411
Tesla, Nikolai 725
Texas, geological situation 653
The Geysers geothermal power plant 341
theoretical total efficiency, CSP 310
thermal drying 930
thermal grid 569
thermal losses, solar tower systems 319, 325
thermal methane reforming, solar 451–482

thermal power
– Japan 18
– Scotland 53
– solar 307–338
– VRE 80
thermal recovery time 348
thermal storage 324–328, 561–563
– commercial systems 327
– CSP 178
– , see also heat storage
thermal stratification 603
thermochemical synthesis 790
– hydrogen 797–798
thermochemical thermal energy storage (TCS) 562
thermochemically produced SNG 532
thermodynamics
– ammonia 692
– compressed gaseous storage 662–668
– cryogenic liquid storage 669–670
– irreversibilities 557
– reforming 453–454
– second law of 426
thermoneutral voltage 431
thin-film PV 288–290, 298–300
– compound semiconductor-based 302
– flexible solar cells 294–295
– R&D 302–303
Thomson, see Joule–Thomson expansion
three-phase AC voltage 726
three-phase reactor 821
tidal energy 360–370
– cost trajectory 356
timber 492
time horizons, "smart grids" 746
tip speed 247
"tipping points" 119
Tokyo Electric Power Co., Inc. (TEPCO) 13–14
top-down analysis 572–575
top-down institutional change 134
total capital investments (TCI) 542
total primary energy supply (TPES) 244
tower concepts, wind power 253–255
town gas 651
– spherical pressure vessels 666
TradeWind project 605–606
trading, cross-border 732
traffic volume 103
trajectories, ecological 125
Trans-European Energy Network Policy (TEN-E) 747
transformation paradigms 132–134
"transformative literacy" 122–123

Subject Index | 967

transformers, power 714
transient characteristics, ZEB 881–882
transition
 – barriers to 868
 – cost 857
 – hydrogen 807–808
 – in the Developing World 138–140
 – natural to hydrogen gas 629–657
 – petro-mobility to electro-mobility 856–857
 – societal 123–132
 – sustainability 858–859
 – , see also energy transition
transmission
 – development stages 724–727
 – European grid 408, 729
 – flexible 709
 – grid components 709–721
 – hydrogen 209, 796–802
 – networks 723–739
 – offshore wind power 271
 – Scotland 60
 – system tasks 727–728
 – technologies system integration 717–719
transmission system operator (TSO) 89, 410–413
 – cross-border balancing 733
 – Germany 502–503
Transparency International 138
transparent conductive oxides (TCO) 288–289
transport 103–107
 – bulk power 723, 732
 – energy use 854
 – fuels 523–524
 – GHG emissions 851–852
 – hydrogen fuels 212, 660
 – motorized, see vehicles
 – pipeline capacity 776
 – , see also fuels
Treuchtlingen mining site 644
tri-generation system 801
tripod-style foundations 267, 276
TRL (technology readiness level) 533
Tropsch, see Fischer–Tropsch process
trough, parabolic 164, 309, 312–317, 457
trucks, long-distance 106
tubular solar receivers 458
tubular steel tower, wind power 253
tumble dryers 752
tunnel boring machines (TBM) 392
tunnel drilling 611
tunneling hydropower plants 391–393
turbidity 603
turbines
 – CCGT 731
 – combustion 835
 – counter-rotating Wells 377
 – efficiency 390
 – floating 370
 – gas and steam 511
 – hydropower 390
 – offshore wind power 271
 – onshore hydroelectric 371
 – operational regime 390
 – RPT 597–600
 – "seasonal" 601
 – separate 599
 – tidal energy 361–378
 – wind, see wind turbines
two-axis tracking 308
two-phase flow 175

u

UCTE/ENTSO-E grid 729–730, 735
Ulla-Førre hydropower complex 392–393
ultra-supercritical (USC) plant 18
unbundling 502
uncontrolled load profile 751
unconventional petroleum 852
underground hydropower plants 391–393
underground lines 711–712
underground storage, see geological storage
undeveloped hydropower 385
unidirectional charging control 760
United Kingdom (UK)
 – electric power generation 48
 – marine energy 351–379
 – offshore wind power 267–268
 – smart EV charging 758–759
 – smart grid heat pumps 761
 – smart storage 762–763
 – Zero Carbon Home approach 877
up-rating, device power 359
updraft towers 170
upgrading
 – biogas 515, 532
 – fuel 622
upper energy bound 892
upstream migration 391
upwind wind turbines 250
uranium resources 17
urbanization rate, China 897, 904
Urdorf pipe storage facility 668
urea 698–699
USA
 – hydrogen transition scenario 807–808
 – , see also California, Texas
USC (ultra-supercritical) plant 18
utility index 866–867

utilization, network asset   743
"utilization effect"   81
utilization factor   217, 229–230

v

vacuum receiver   175, 319
valence band   285
value research, global   131
van der Waals radius   640
vanes, guide   600
vaporization heat   669–670
vaporizer, ambient air   672
VAR compensator   715
variability
– hydrological   383–384
– reservoir water level   609–611
variable RE (VRE)
– and biomass energy   500
– decreasing market value   75–83
– integration options   84–89, 819
– marine energy   354
– "smart grids"   743–744
variable-speed drive (VSD)   604, 919
vegetable oils   624
– hydrotreated   529
vehicle travel   103
vehicles
– CNG-fueled   837
– electric, see electric vehicles
– fuel, see fuels
– hydrogen   796–797
– light-duty   863
– plug-in hybrids   106, 587, 731, 859
– ZEV standard   861–862, 864–866
vertical grid load   204
vertical mixing   603, 611
vertical tanks   672
VGB   419
videoconferencing   103
Vindeby offshore wind farm   265–266
visualization tools   746
Volmer, see Butler–Volmer equation
voltage
– levels   741
– regulation   412, 600,
– three-phase AC   726
voltage-source converter (VSC) technologies   716–717, 739
volume production effects, solar thermal power   331
volumetric receiver   458, 477

w

wafers, mono-/polycrystalline   286

Wairakei geothermal power plant   341
wake losses   255
washing machines   752
wasp   274
waste
– recycling   932
– wood   493
waste heat   58, 177, 557
water
– CSP cooling   311
– dissociation potential   220
– diversified heater demand   750
– EEKV   381
– electrolysis, see electrolysis
– formation energy of liquid   426
– global freshwater use   120
– hot water generation   561, 903, 922–926
– level fluctuations   609–611
– meteoric   646
– oscillating water column device   377
– solar water heaters   142, 151–152
– solubility of hydrogen in   644
– supply   387
– water-soaked minerals   642
water cycle   382
water-gas shift reaction (WGS)   438, 452–454, 819
water power, see hydropower
water source heat pumps (WSHP)   57
watt-peak (unit)   283
wave energy   371–378
– cost trajectory   356
– flexible absorbers   373
– overtopping converter   378
wave heights   276
wave surge converter   371–372
WBGU (German Advisory Council on Global Change)   121, 123
– guard rails concept   144–145
weather dependence, supply pattern   405
weather model, global   606
weighted average cost of capital (WACC)   762
weighting systems, ZEB   878–879
Weissbach, see Darcy–Weissbach equation
Weizmann Institute of Science (WIS)   459, 462
well-to-tank (WTT) analysis   840–841
well-to-wheels (WTW) emissions   805–806
Wells turbines   377
wet appliances, smart   752–753
wheat prices   487
wheels, power-to-wheel losses   854
white asbestos diaphragm   432

wholesale electricity market   769
wide-area measurement   746
wind power
– capital availability   261
– CES production structure   34
– China   31
– decreasing market value   76
– design of turbines   87
– downregulation   615
– economic feasibility   258–260
– environmental impact   255–257
– fluctuations   199
– Germany   200–205
– GROWIAN   180
– linking with hydropower   612–615
– offshore, see offshore wind power
– onshore   203, 243–263
– simulated production   607
– site assessment   273–274
– Spain   229
– stabilization   592
– storage   412
– supply potential   109
– wind capacity   139
– wind farm control   255
– wind farm design   274–276
– , see also variable RE
wind speed   247
– offshore   265
wind turbines   246–255
– capacity factor   249
– floating   279–280
– offshore sites   271–273
– stall-regulated   243
wind value factors   82
WindFloat   280
"winter peak"   517
winter sunshine, daily   144

Wobbe index   773, 786–787
Wolfurt school   884
wood
– as bioenergy   491–493
– "dead"   415
– paper industry   929–930
working fluids   165–168
working gas   828
– volume   633, 640
working point, solar cells   285
working principles, photovoltaics   284–290
World, Developing, see Developing World
World Energy Outlook   824, 891, 913
world production, grain   488
WSHP (water source heat pumps)   57
WTT (well-to-tank) analysis   840–841

x
XLPE (cross-linked polyethylene) insulation   711–712, 716

y
Yangtze River Basin, residential energy consumption   902
yawing system, hydraulic   367
yeast   529
yttria-stabilized zirconia (YSZ)   437

z
Zechstein formation   646, 651
Zero Carbon Home approach   877
zero-emission mobility   183, 217, 587
Zero Emission Vehicle (ZEV) standard   861–862, 864–866
zero energy buildings (ZEB)   875–889
zero gap configuration   815
Zn–Ce redox-flow battery   593
ZSW methanation plant   821–822